SHORT PROTOCOLS IN HUMAN GENETICS

A Compendium of Methods from Current Protocols in Human Genetics

Published by John Wiley & Sons, Inc.

Published by John Wiley & Sons, Inc., Hoboken, New Jersey.
Published simultaneously in Canada.

Library of Congress Cataloging-in-Publication Data

Short protocols in human genetics : a compendium of methods from current protocols in human genetics / editorial board, Nicholas C. Dracopoli ... [et al.].
    p. ; cm.
 Includes bibliographical references and index.
 ISBN 0-471-69418-5 (cloth)
 1. Human genetics—Laboratory manuals.
 [DNLM: 1. Epistasis, Genetic. 2. Gene Therapy. 3. Genetic Techniques. QH 450 S559 2004] I. Dracopoli, Nicholas C.
 QH440.5.S54 2004
 599.93'5—dc22

                                                              2004009131

                                                              CIP

Printed in the United States of America
10 9 8 7 6 5 4 3 2 1

CONTENTS

Preface   xv
Contributors   xix

1 Genetic Mapping

1.1	**Collection of Clinical and Epidemiological Data for Linkage Studies**	**1-3**
	Establishing Evidence for Genetic Influence	1-3
	Collecting Data for Linkage Studies	1-6
1.2	**Pedigree Selection and Information Content**	**1-12**
	Strategic Approach	1-12
1.3	**Single-Sperm Typing**	**1-19**
	Basic Protocol 1: PCR Amplification of Genetic Markers from Single Sperm Cells	1-19
	Support Protocol 1: Isolation of Single Sperm Cells from Agarose Films	1-20
	Support Protocol 2: Isolation of Lysed Single Sperm Cells by Flow Cytometry	1-21
	Support Protocol 3: Multiplex Amplification from Single Sperm Cells	1-23
	Support Protocol 4: Whole-Genome Amplification of DNA from Single Cells by Primer-Extension Preamplification (PEP)	1-23
	Basic Protocol 2: Sperm-Typing Data Analysis	1-24
	Basic Protocol 3: Segregation Distortion Based on Haplotype Transmission	1-26
1.4	**Use of LINKAGE Programs for Linkage Analysis**	**1-26**
	Programs For LINKAGE Analysis	1-29
	Strategic Approach and Practical Examples	1-31
1.5	**Model-Free Tests for Genetic Linkage**	**1-55**
1.6	**Homozygosity Mapping using Pooled DNA**	**1-71**
	Basic Protocol: Amplification and Analysis of Pooled DNA	1-74
1.7	**Disease Associations and Family-Based Tests**	**1-76**
	Case-Control Design: Disease Association in Population-Based Samples	1-78
	Population Structure	1-79
	Family-Based Tests of Association and the Haplotype Relative Risk	1-80
	Family-Based Tests of LINKAGE: The TDT	1-80
	Family-Based Tests of Association: The TDT and Other Tests	1-83
	Generalizing the TDT: More Than Two Marker Alleles	1-83
	Tests of Association using Marker Haplotypes	1-86
	The SIB TDT (S-TDT)	1-86
	Pedigrees	1-90
1.8	**Analysis of Gene-Gene Interactions**	**1-90**
	Methods for Detecting Gene-Gene Interactions in Association Studies of Discrete Traits	1-92
	Methods for Detecting Gene-Gene Interactions in Association Studies of Quantitative Traits	1-96

2 Genotyping

2.1	**PCR Methods of Genotyping**	**2-1**
	Basic Protocol 1: PCR Amplification of SSLPs using End-Labeled Primers	2-2
	Alternate Protocol: PCR Amplification of SSLPs using Internal Labeling	2-4
	Basic Protocol 2: Nonradioactive Analysis of SSLPs using Silver Staining	2-4
	Basic Protocol 3: Nonradioactive Multiplex Analysis of SSLPs	2-6
	Support Protocol 1: Preparation of M13 Sequence Ladder Size Standard	2-10
	Support Protocol 2: Digoxigenin Labeling of Probes using Terminal Transferase	2-11
2.2	**Genotyping by Ligation Assays**	**2-12**
	Basic Protocol 1: Oligonucleotide Ligation Assay (OLA)	2-12

Basic Protocol 2: Ligase Chain Reaction (LCR)   2-14

Support Protocol: Preparing Modified Oligonucleotides for Ligation Assays   2-16

2.3   **Automated Fluorescent Genotyping**   **2-17**

Basic Protocol: PCR Amplification of SSLPs for Automated Fluorescent
Genotyping   2-17

Support Protocol 1: Pooling Fluorescently Labeled PCR Products for Genotype
Analysis   2-19

Support Protocol 2: Gel Electrophoresis of Pooled PCR Products   2-21

2.4   **Single-Nucleotide Polymorphism Genotyping using Microarrays**   **2-24**

Strategic Planning   2-24

Basic Protocol 1: SNP Genotyping using Allele-Specific Microarrays   2-25

Basic Protocol 2: SNP Genotyping using Generic Microarrays and Single-Base
Extension   2-30

2.5   **High-Throughput Genotyping Using the TaqMan Assay**   **2-33**

Strategic Planning   2-33

Basic Protocol: High-Throughput Genotyping using the TaqMan Assay   2-35

2.6   **High-Throughput Genotyping with Primer Extension Fluorescent
Polarization Detection**   **2-36**

Strategic Planning   2-37

Basic Protocol 1: Primer Extension Assay with Single-Plex PCR   2-37

Basic Protocol 2: Primer Extension Assay with 4-Plex PCR   2-39

3 Somatic Cell Hybrids

3.1   **Construction of Somatic Cell Hybrids**   **3-1**

Selectable Markers for Mammalian Cells   3-2

Basic Protocol: Whole-Cell Fusion of Monolayer Cells   3-4

Alternate Protocol: Whole-Cell Fusion of Suspension Donor Cells
to Monolayer Recipient Cells   3-5

Support Protocol 1: Colony Isolation Using Cloning Cylinders   3-6

Support Protocol 2: Subcloning Hybrid Cell Populations   3-7

Support Protocol 3: Cytogenetics: G-11 Staining of Metaphase
Chromosomes   3-7

Support Protocol 4: In Situ Hybridization   3-8

Support Protocol 5: Marker Analysis   3-8

4 Cytogenetics

4.1   **Chromosome Preparation from Cultured Peripheral Blood Cells**   **4-2**

Basic Protocol: Culture and Metaphase Harvest of Peripheral Blood   4-2

Support Protocol: Chromosome Slide Preparation   4-4

4.2   **Mitotic Chromosome Preparations from Mouse Cells for Karyotyping**   **4-7**

Basic Protocol 1: Culture and Metaphase Harvest of Mouse Peripheral Blood   4-7

Alternate Protocol 1: Collecting Blood by Tail Vein Method   4-10

Alternate Protocol 2: Chromosome Preparation from Fetal Liver   4-10

Support Protocol: Setting Up Timed Mouse Matings   4-12

Alternate Protocol 3: Chromosome Preparation from Bone Marrow   4-12

Alternate Protocol 4: Chromosome Preparation from Solid Adult
Tissues or Tumors   4-13

Basic Protocol 2: Giemsa Banding (G-Banding)   4-14

4.3   **Chromosome Banding Techniques**   **4-14**

Basic Protocol 1: Quinacrine Banding (Q-Banding)   4-15

Basic Protocol 2: Giemsa Banding (G-Banding) by the GTG Technique   4-16

Alternate Protocol 1: G-Banding with Wright Stain   4-17

Support Protocol 1: Aging Slides with Heat   4-17

Support Protocol 2: Aging Slides with Hydrogen Peroxide   4-18

	Basic Protocol 3: B-Pulse Replication Banding for Lymphocytes	4-18
	Alternate Protocol 2: T-Pulse Replication Banding for Lymphocytes	4-19
	Alternate Protocol 3: B-Pulse Replication Banding for Fibroblasts	4-19
	Alternate Protocol 4: T-Pulse Replication Banding for Fibroblasts	4-19
	Support Protocol 3: Visualization of BrdU Replication Banding by Fluorescent Dye	4-20
	Support Protocol 4: Visualization of BrdU Replication Banding by Light Treatment and Giemsa Staining	4-21
	Support Protocol 5: Visualization of BrdU Replication Banding by Heat Treatment and Giemsa Staining	4-21
	Basic Protocol 4: Distamycin-DAPI Staining	4-22
	Alternate Protocol 5: Hoechst 33258-Distamycin Staining	4-22
	Support Protocol 6: Destaining	4-23
4.4	**In Situ Hybridization to Metaphase Chromosomes and Interphase Nuclei**	**4-24**
	Basic Protocol: Fluorescence In Situ Hybridization to Metaphase Chromosomes	4-24
	Support Protocol 1: Amplification of Biotinylated Signals	4-28
	Support Protocol 2: Amplification of Signals from Digoxigenin-Labeled Probes	4-28
	Alternate Protocol 1: Enzymatic Detection using Horseradish Peroxidase	4-29
	Alternate Protocol 2: Enzymatic Detection using Alkaline Phosphatase	4-30
	Support Protocol 3: Biotin and Digoxigenin Labeling of Probes by Nick Translation	4-31
	Alternate Protocol 3: Ordering of Sequences in Interphase Nuclei by FISH	4-32
4.5	**High-Resolution FISH Analysis**	**4-33**
	Basic Protocol 1: Preparation of Free Chromatin with Alkaline Buffer	4-33
	Support Protocol 1: Optimization of Free Chromatin Preparation	4-34
	Alternate Protocol: Preparation of Free Chromatin from Lymphocytes by Drug Treatment	4-35
	Basic Protocol 2: High-Resolution FISH Mapping with Free Chromatin	4-36
	Basic Protocol 3: Preparation of Stretched Cellular DNA for DIRVISH Mapping	4-36
	Basic Protocol 4: High-Resolution Mapping by DIRVISH to Stretched DNA	4-37
	Support Protocol 2: Biotin and Digoxigenin Labeling of DIRVISH Probes	4-40
4.6	**Multicolor Fluorescence In Situ Hybridization (FISH) Approaches for Simultaneous Analysis of the Entire Human Genome**	**4-42**
	Production of Chromosome-Specific Paint Probes and Combinatorial Labeling	4-42
	Sky and M-FISH	4-43
	Rx-FISH	4-46
	Comparative Genomic Hybridization	4-47
4.7	**Morphology Antibody Chromosome Technique for Determining Phenotype and Genotype of the Same Cell**	**4-48**
	Basic Protocol 1: Sequential MAC Analysis Using APAAP Immunostaining and In Situ Hybridization	4-50
	Alternate Protocol 1: Immunophenotyping Using HRP-Based Detection	4-53
	Alternate Protocol 2: Immunophenotyping Using Fluorescence Detection	4-55
	Alternate Protocol 3: Genotyping by G- or C-Banding of Chromosomes	4-56
	Support Protocol 1: Preparation of Cytospin Slides	4-57
	Support Protocol 2: Preparation of In Situ Cultures	4-59
	Support Protocol 3: Preparation of Tissue Sections	4-60
	Support Protocol 4: Preparation of Blood and Bone Marrow Smears	4-61
	Basic Protocol 2: In Situ Hybridization on Previously GTG-Banded Chromosomes	4-62
4.8	**Comparative Genomic Hybridization**	**4-64**
	Basic Protocol 1: Comparative Genomic Hybridization using Directly Labeled DNA	4-64
	Alternate Protocol: Comparative Genomic Hybridization using Indirectly Labeled DNA	4-67
	Support Protocol 1: Preparation of Metaphase Chromosomes for CGH	4-67
	Support Protocol 2: Preparation of Genomic DNA for CGH	4-68

Support Protocol 3: Preparation of Labeled DNA Probes for CGH   4-68

Basic Protocol 2: Microscopy, Imaging, and Image Analysis for CGH   4-69

5 Strategies for Large-Insert Cloning and Analysis

5.1   **Pulsed-Field Gel Electrophoresis for Long-Range Restriction Mapping**   **5-2**

Basic Protocol: Constructing Long-Range Restriction Maps
of Genomic Regions   5-2

Support Protocol 1: Preparation of High-Molecular-Weight Mammalian
Genomic DNA Embedded in Agarose Blocks   5-4

Support Protocol 2: Preparation of High-Molecular-Weight Standards in
Agarose Blocks using *S. cerevisia* Chromosomes and YACS   5-5

Support Protocol 3: Preparation of BAC DNA, Restriction Digestion, and
CHEF Gel Analysis   5-7

5.2   **Screening Large-Insert Libraries by Hybridization**   **5-8**

Basic Protocol 1: Arraying Colonies and DNA at Low and High Density   5-8

Support Protocol: Preparing BAC and PAC Colony Blots for Hybridization   5-14

Basic Protocol 2: Screening BAC, PAC, and P1 Libraries by Colony
Hybridization   5-15

Alternate Protocol: Preparing and Hybridizing Colony Blots with Overgo
Oligonucleotide Probes   5-17

5.3   **Introduction of Large Insert DNA into Mammalian Cells and Embryos**   **5-19**

Basic Protocol 1: Introduction of Intact YACS into Mammalian Cells by
Spheroplast Fusion   5-19

Alternate Protocol: Introduction of Gel-Purified YAC DNA into Mammalian
Cells by Lipofection   5-23

Support Protocol 1: Introduction of a Mammalian Selectable Marker into YACS
by Homologous Recombination   5-24

Support Protocol 2: Purification of YACS by Pulsed-Field Gel Electrophoresis   5-27

Support Protocol 3: Rapid Estimation of DNA Concentration on Ethidium
Bromide/Agarose Plates   5-28

Basic Protocol 2: Introduction of Bacterial Artificial Chromosomes (BAC or
PAC) into Mammalian Cells and Mouse Embryos   5-29

Support Protocol 4: Optional Linearization and Gel Purification of BAC   5-30

5.4   **Construction of Bacterial Artificial Chromosome (BAC/PAC) Libraries**   **5-32**

Basic Protocol: Preparation of BAC/PAC Clones using pCYPAC2, pPAC4, or
pBACe3.6 Vector   5-32

Support Protocol 1: Preparation of BAC/PAC Vector for Cloning   5-36

Support Protocol 2: Preparation of High-Molecular-Weight DNA from
Lymphocytes in Agarose Blocks   5-38

Support Protocol 3: Preparation of High-Molecular-Weight DNA from Animal
Tissue Cells in Agarose Blocks   5-39

Support Protocol 4: Partial Digestion and Size Fractionation of Genomic DNA   5-40

Support Protocol 5: Modified Alkaline Lysis Miniprep for Recovery of DNA
from BAC/PAC Clones   5-44

6 Identifying Candidate Genes in Genomic DNA

6.1   **Gene Identification: Methods and Considerations**   **6-2**

Gene Identification Methods   6-3

How Well Do the Methods Work?   6-11

Strategies and Considerations   6-12

6.2   **Sequence Databases: Integrated Information Retrieval and Data Submission**   **6-14**

Introduction to *Entrez*   6-14

Data Submission: General Considerations   6-16

Submitting a Sequence to the Nucleotide Database   6-17

Submitting an Update or Correction to an Existing Genbank Entry   6-20

Submitting EST, STS, or GSS Data   6-21

Submitting High-Throughput Genome Sequences (HTGS)   6-21

6.3	**Sequence Similarity Searching Using the BLAST Family of Programs**		**6-22**
	Accessing BLAST Programs and Documentation		6-22
	Introduction to BLAST		6-23
	Examples of BLAST Searches		6-27
	Searching Strategies		6-32
	Sequence Alignment Algorithms		6-34
	Appendix A: BLAST Parameters		6-36
	Appendix B: Sequence Identifier Syntax		6-37
6.4	**Accessing the Human Genome**		**6-38**
	The Nature of the Data		6-40
	Assembling the Sequence		6-41
	Annotation		6-43
	Accessing the Data		6-46
	How to Retrieve Information: Example Questions		6-56
	Frequently Asked Questions		6-60
6.5	**Searching the NCBI Databases Using Entrez**		**6-61**
	Basic Protocol 1: Querying Entrez		6-61
	Support Protocol: Using Cubby to Save Searches and Results		6-66
	Alternate Protocol: Combine Entrez Queries		6-69
	Basic Protocol 2: Examining Structures in Entrez		6-72

7 Searching Candidate Genes for Mutations

7.1	**Amplification of Sequences from Affected Individuals**		**7-2**
	Basic Protocol 1: Amplification of RNA from Lymphocytes		7-2
	Alternate Protocol: Modified Second-Round Amplification of cDNA		7-5
	Basic Protocol 2: Amplification of Genomic DNA		7-5
7.2	**Detection of Mutations by Single-Strand Conformation Polymorphism Analysis**		**7-6**
	Basic Protocol		7-6
7.3	**Single-Strand Conformation Polymorphism Analysis Using Capillary Electrophoresis**		**7-8**
	Basic Protocol 1: Sample Preparation for Capillary Electrophoresis		7-9
	Basic Protocol 2: Automated Capillary Electrophoresis using an ABI 310 Genetic Analyzer		7-11
	Basic Protocol 3: Automated Capillary Array Electrophoresis using an ABI Prism 3100 Genetic Analyzer		7-12
7.4	**Heterozygote Detection Using Automated Fluorescence-Based Sequencing**		**7-13**
	Basic Protocol: Identifying Heterozygote Mutations from Sequence Traces		7-15
	Alternate Protocol: Large-Volume Sequencing in a 96-Well or 384-Well Plate Format for ABI 3700 DNA Analyzer		7-16
7.5	**Mutation Detection by Cycle Sequencing**		**7-19**
	Basic Protocol		7-19
7.6	**Human Mutation Databases**		**7-22**
	Types of Mutation Databases		7-22
	Submission of Data to Mutation Databases		7-23
	Accessing Data in Mutation Databases		7-25

8 Clinical Cytogenetics

8.1	**Preparation of Metaphase Spreads from Chorionic Villus Samples**		**8-2**
	Basic Protocol: Culture Method for Preparing Metaphase Spreads from Chorionic Villi		8-2
	Alternate Protocol: Direct Method for Preparing Metaphase Spreads from Chorionic Villi		8-4

8.2	**Preparation, Culture, and Analysis of Amniotic Fluid Samples**	**8-5**
	Basic Protocol: In Situ Method for the Preparation, Culture, Harvest, and Analysis of Amniotic Fluid Samples	8-5
	Support Protocol 1: Harvesting for High-Resolution Chromosome Banding using Ethidium Bromide	8-10
	Support Protocol 2: Passaging Cells from the Monolayer	8-10
	Alternate Protocol: Flask Method for the Preparation, Culture, Harvest, and Analysis of Amniotic Fluid Samples	8-11
	Support Protocol 3: Passaging Cells from Flasks	8-13
8.3	**Preparation and Culture of Products of Conception and Other Solid Tissues for Chromosome Analysis**	**8-14**
	Basic Protocol: Mechanical Disruption and Culture of Tissues for Metaphase Chromosome Analysis	8-14
	Alternate Protocol 1: Enzymatic Disruption and Culture of Tissues for Metaphase Chromosome Analysis	8-15
	Support Protocol 1: Preparation of Human Products of Conception for Culture	8-16
	Support Protocol 2: Preparation of Skin or Other Tissue Biopsies for Culture	8-17
8.4	**Analysis of Sister-Chromatid Exchanges**	**8-18**
	Basic Protocol: Analysis of Sister-Chromatid Exchanges in Mammalian Metaphase Chromosomes	8-18
8.5	**Determination of Chromosomal Aneuploidy Using Paraffin-Embedded Tissue**	**8-22**
	Basic Protocol: Preparation of Nuclear Suspensions from Paraffin-Embedded Tissue for FISH	8-22
	Alternate Protocol: Preparation of Paraffin Sections for FISH	8-25
8.6	**Preparation of Amniocytes for Interphase Fluorescence In Situ Hybridization (FISH)**	**8-26**
	Basic Protocol 1: Preparation of Uncultured Amniocytes for Interphase FISH Analysis	8-26
	Alternate Protocol: Preparation of Amniocytes Attached to a Surface for Interphase FISH Analysis	8-28
	Basic Protocol 2: Interphase FISH Analysis of Amniotic Fluid Cells	8-29
8.7	**Diagnosis of Fanconi Anemia by Diepoxybutane Analysis**	**8-31**
	Basic Protocol: Diepoxybutane Test for Postnatal Diagnosis of Fanconi Anemia	8-32
	Support Protocol 1: Working With and Disposing of Diepoxybutane	8-33
	Support Protocol 2: Giemsa Staining for Chromosome-Breakage Analysis	8-34
	Alternate Protocol 1: Diepoxybutane Test Using Fibroblast Cultures	8-35
	Alternate Protocol 2: Diepoxybutane Test for Prenatal Diagnosis of Fanconi Anemia	8-36

9 Clinical Molecular Genetics

9.1	**Multiplex PCR for Identifying Dystrophin Gene Deletions**	**9-2**
	Basic Protocol: Diagnostic Multiplex PCR to Detect Dystrophin Gene Deletions	9-2
	Alternate Protocol: Low-Cycle-Number Multiplex PCR with Radioactive Label to Detect Dosage Differences	9-7
	Support Protocol: Preparation and Storage of Stock Diagnostic PCR Mixes	9-7
9.2	**Simultaneous Detection of Multiple Point Mutations Using Allele-Specific Oligonucleotides**	**9-12**
	Basic Protocol: Screening PCR-Amplified DNA with Multiple Pooled ASOs	9-12
	Support Protocol: Stripping Old Probes and Rehybridization	9-15
9.3	**Molecular Analysis of Fragile X Syndrome**	**9-16**
	Basic Protocol 1: PCR Amplification of the Fragile X Repeat	9-16
	Basic Protocol 2: Detection of Amplification and Methylation by Direct Southern Blot Hybridization	9-19
9.4	**Analysis of Trinucleotide Repeats in Myotonic Dystrophy**	**9-21**
	Basic Protocol 1: Hybridization Analysis of PCR-Amplified DM Trinucleotide Repeats	9-21

	Support Protocol 1: Rapid Transfer of PCR Product using a Vacuum Blotter	9-23
	Support Protocol 2: Radioactive Detection of Myotonic Dystrophy (DM) Trinucleotide Repeat Expansion	9-24
	Basic Protocol 2: Hybridization Analysis of Myotonic Dystrophy (DM) Trinucleotide Repeats in Genomic DNA	9-24
9.5	**Detection of Nonrandom X Chromosome Inactivation**	**9-26**
	Basic Protocol : X Chromosome Inactivation Assay	9-26
	Support Protocol: PCR Amplification and Labeling of Digested and Undigested DNA Templates	9-27
9.6	**Molecular Analysis of Paternity**	**9-28**
	Basic Protocol 1: Analysis of VNTRs by RFLP Technology	9-28
	Basic Protocol 2: Analysis of Polymorphic Loci by PCR	9-30
	Support Protocol 1: Preparation of Genomic DNA from Whole Blood	9-33
	Support Protocol 2: Interpretation, Statistical Evaluation, and Reporting of DNA Profiles: Data from Mother and Alleged Father	9-34
	Support Protocol 3: Interpretation, Statistical Evaluation, and Reporting of DNA Profiles: Special Paternity Cases	9-35
9.7	**Amplification-Refractory Mutation System (ARMS) Analysis of Point Mutations**	**9-38**
	Basic Protocol: Analysis of Single Mutations by ARMS Tests	9-40
	Alternate Protocol: Analysis of Multiple Mutations by Multiplex ARMS	9-43
	Support Protocol: Rapid DNA Extraction from Mouthwash and Blood Samples	9-44
9.8	**Molecular Analysis of Oxidative Phosphorylation Diseases for Detection of Mitochondrial DNA Mutations**	**9-46**
	Basic Protocol 1: Screening for Mitochondrial DNA Rearrangements by Southern Blot Hybridization	9-46
	Support Protocol: Mitochondrial DNA Probe Preparation using Long-Range PCR	9-51
	Basic Protocol 2: Screening for Mitochondrial DNA Point Mutations by Restriction Analysis of PCR Products	9-52
9.9	**Single-Cell DNA and FISH Analysis for Application to Preimplantation Genetic Diagnosis**	**9-59**
	Basic Protocol 1: Analysis of Specific Gene Loci in Single Diploid Cells	9-59
	Support Protocol 1: Whole-Genome Amplification of Single Diploid Cells by Primer-Extension Preamplification (PEP)	9-64
	Support Protocol 2: Preimplantation Embryo Biopsy	9-64
	Support Protocol 3: Isolation of Blastomeres from Affected Embryos for Further Investigation	9-68
	Support Protocol 4: Isolation of Single Lymphocytes/Lymphoblastoid Cells	9-69
	Basic Protocol 2: FISH Analysis of Single Blastomeres	9-71
	Support Protocol 5: Probe Validation	9-73
9.10	**Protein Truncation Test**	**9-73**
	Basic Protocol: Nonradioactive Protein Truncation Test	9-74
	Support Protocol: Isolation and Analysis of RNA	9-79
	Alternate Protocol: Radioactive Protein Truncation Test using Coupled Transcription/Translation	9-81
9.11	**Genotyping of Apolipoprotein E (APOE): Comparative Evaluation of Different Protocols**	**9-83**
	Basic Protocol: APOE Genotyping by RFLP Analysis	9-83
	Alternate Protocol 1: APOE Genotyping by Reverse Hybridization	9-84
	Alternate Protocol 2: APOE Genotyping by Fluorescence Polarization (FP)	9-86
	Alternate Protocol 3: APOE Genotyping by SNaPshot Analysis	9-87

10 Cancer Genetics

10.1   **Metaphase Harvest and Cytogenetic Analysis of Malignant Hematological Specimens**   **10-2**

Basic Protocol: Preparation of Chromosome Spreads from Bone Marrow and Leukemic Blood Specimens   10-2

Alternate Protocol 1: Preparation of Chromosome Spreads from Chronic Lymphocyte Leukemia Bone Marrow and Peripheral Blood Specimens   10-4

Alternate Protocol 2: Preparation of Chromosome Spreads from Multiple Myeloma Bone Marrow and Peripheral Blood Specimens   10-4

Alternate Protocol 3: Preparation of Chromosome Spreads from Lymph Node Specimens   10-6

Alternate Protocol 4: Preparation of Chromosome Spreads from Spleen   10-6

10.2   **Metaphase Harvest and Cytogenetic Analysis of Solid Tumor Cultures**   **10-7**

Basic Protocol: Cytogenetic Analysis of Metaphase Cells from Solid Tumor, Lymphoma, or Effusion Samples   10-7

Support Protocol 1: Disaggregation and Culture of Solid Tumors   10-8

Support Protocol 2: Culture of Lymph Node (Lymphoma) Specimens for Cytogenetic Analysis   10-9

Support Protocol 3: Preparation of Effusion (Fluid) Specimens for Cytogenetic Analysis   10-9

10.3   **Molecular Analysis of DNA Rearrangements in Leukemias and Non-Hodgkin's Lymphomas**   **10-10**

Strategic Planning   10-10

Basic Protocol 1: Detection of Clonal Antigen-Receptor Gene Rearrangements by Southern Blot Hybridization   10-11

Alternate Protocol 1: Detection of Clonal Immunoglobulin Heavy Chain Gene Rearrangements by PCR   10-14

Alternate Protocol 2: Detection of Clonal T Cell Receptor-γ Gene Rearrangements by PCR and Denaturing Gradient Gel Electrophoresis   10-16

Basic Protocol 2: Detection of Chromosomal Translocations by Reverse Transcriptase–PCR   10-18

Support Protocol 1: RNA Isolation by the Rapid Guanidinium Method   10-20

Alternate Protocol 3: Detection of Chromosomal Translocations in DNA Samples by PCR   10-22

10.4   **Molecular Analysis of Gene Amplification in Tumors**   **10-23**

Basic Protocol: Detection of Gene Amplification by Southern Blot Hybridization Analysis   10-23

Alternate Protocol 1: Detection of Gene Amplification by Slot Blot Hybridization Analysis   10-27

Alternate Protocol 2: Detection of Gene Amplification by Differential PCR   10-28

Support Protocol: Obtaining and Processing Tumor Tissue   10-30

10.5   **Methylation-Specific PCR**   **10-30**

Basic Protocol 1: Determination of DNA Methylation Patterns by Methylation-Specific PCR   10-30

Basic Protocol 2: Determination of Methylation of CpG Sites within Methylation-Specific PCR Products   10-33

11 Transcriptional Profiling

11.1   **Oligonucleotide Arrays for Expression Monitoring**   **11-1**

Basic Protocol 1: Amplification of mRNA for Expression Monitoring and Hybridization to Oligonucleotide Array Chips   11-1

Support Protocol 1: In Vitro Transcription of Control Genes and Preparation of Transcript Pools   11-5

Alternate Protocol: Solid-Phase Reversible Immobilization Purification of cDNA and In Vitro Transcription Products   11-7

Support Protocol 2: Quantitation of cDNA   11-8

Basic Protocol 2: Data Reduction, Normalization, and Quality Assessment   11-9

11.2	**Profiling Human Gene Expression with cDNA Microarrays**	**11-12**
	Basic Protocol 1: cDNA Amplification and Printing	11-12
	Basic Protocol 2: RNA Extraction and Labeling	11-16
	Basic Protocol 3: Hybridization and Data Extraction	11-19
	Support Protocol 1: Agarose Gel Electrophoresis of ESTs	11-20
	Support Protocol 2: Fluorometric Determination of DNA Concentration	11-21
	Support Protocol 3: Coating Slides with Poly-L-Lysine	11-22
11.3	**Analysis of Expression Data: An Overview**	**11-22**
	Experimental Design	11-23
	Normalization	11-23
	Analysis	11-24
	Informatics and Databases	11-27

12 Vectors for Gene Therapy

12.1	**Biosafety in Handling Gene Transfer Vectors**	**12-2**
	Running a BL2 Lab	12-2
	Specific Vectors	12-5
12.2	**Adenoviral Vectors**	**12-5**
	Strategic Planning	12-8
	Basic Protocol: Generation of Recombinant Adenoviral Vectors using the AdEasy Method	12-12
	Alternate Protocol: Generate Recombinant Adenovirus Plasmids using AdEasier Cells	12-15
	Support Protocol 1: Preparation and Purification of High-Titer Adenoviruses	12-16
	Support Protocol 2: Adenovirus Plaque Assay	12-18
	Support Protocol 3: Preparation of Electrocompetent BJ5183 Cells	12-19
	Support Protocol 4: Preparation of Adenoviral DNA	12-20
	Support Protocol 5: Quick Agarose-Tube Dialysis	12-20
12.3	**Production of Recombinant Adeno-Associated Viral Vectors**	**12-21**
	Basic Protocol: Production of Adenovirus-Free rAAV by Transient Transfection of 293 Cells	12-21
	Alternate Protocol: rAAV Purification using Heparin Sepharose Column Purification	12-25
	Support Protocol 1: Determination of rAAV Titers by the Dot-Blot Assay	12-27
	Support Protocol 2: Infection of Cells In Vitro with rAAV and Determination of Titer by Transgene Expression	12-29
12.4	**Production of Retroviral Vectors**	**12-29**
	Basic Protocol: Production of Stable Cell Lines to Generate Vectors with Selectable Markers	12-30
	Alternate Protocol 1: Production of Stable Cell Lines to Generate Vectors without Selectable Markers	12-35
	Alternate Protocol 2: Production of Vector by Transient Transfection	12-36
	Support Protocol 1: Calcium Phosphate–Mediated Transfection of Cultured Cells	12-36
	Support Protocol 2: Assay to Titer Vectors Carrying Selectable Markers	12-37
	Support Protocol 3: Marker Rescue Assay for Helper Virus	12-38
	Support Protocol 4: Staining Cultured Cells for Alkaline Phosphatase Activity	12-39
12.5	**Production of Pseudotype-Retroviral Vectors**	**12-39**
	Basic Protocol: Pseudotype Retrovirus Production by Transient Transfection	12-40
	Alternate Protocol: Pseudotype Retrovirus Production from Stable Producer Cells	12-42
12.6	**Production of High-Titer Lentiviral Vectors**	**12-43**
	Basic Protocol: Production of High-Titer HIV-1-Based Vector Stocks by Transient Transfection of 293T Cells	12-45
	Support Protocol 1: Titration of Lentivirus GFP Vector Stocks	12-47
	Support Protocol 2: Titration of Lentivirus LacZ Vector Stocks	12-48

12.7	**Construction of Replication-Defective Herpes Simplex Virus Vectors**	**12-49**
	Basic Protocol 1: Construction of HSV-1 IE Gene-Complementing Cell Lines	12-50
	Basic Protocol 2: Construction of Replication-Defective Vectors	12-54
	Basic Protocol 3: Insertion of Foreign Gene Sequence into a Replication-Defective Genomic HSV Vector	12-57
	Support Protocol 1: Preparation of Herpes Simplex Virus Stock	12-62
	Support Protocol 2: Plaque Assay to Titer Virus	12-63
	Support Protocol 3: Isolation of Viral DNA	12-64
12.8	**Gene Delivery Using Helper Virus–Free HSV-1 Amplicon Vectors**	**12-66**
	Basic Protocol: Preparation of Helper Virus–Free Amplicon Stocks	12-66
	Support Protocol 1: Preparation of HSV-1 Cosmid DNA for Transfection	12-69
	Support Protocol 2: Titration of Amplicon Stocks	12-71
12.9	**Liposome Vectors for In Vivo Gene Delivery**	**12-74**
	Basic Protocol: Liposome Preparation by Thin-Film Hydration Followed by Extrusion	12-74

13 Delivery Systems for Gene Therapy

13.1	**Gene Transfer to Arteries**	**13-2**
	Strategic Planning	13-2
	Basic Protocol 1: Surgical Gene Delivery to Porcine Arteries	13-3
	Basic Protocol 2: Surgical Gene Delivery to Stented Porcine Iliac Arteries	13-5
	Basic Protocol 3: Surgical Gene Delivery to Atherosclerotic Rabbit Iliac Arteries	13-6
	Basic Protocol 4: Surgical Gene Delivery to Normal Murine Carotid Arteries	13-9
	Basic Protocol 5: Surgical Gene Delivery into Injured Murine Femoral Arteries	13-11
13.2	**Gene Delivery to Muscle**	**13-13**
	Basic Protocol 1: Isolation and Growth of Mouse Primary Myoblasts (without Cell Sorting)	13-13
	Alternate Protocol 1: Purification of Primary Myoblasts using Cell Sorting	13-15
	Basic Protocol 2: Infection of Primary Myoblasts with Retrovirus	13-15
	Basic Protocol 3: Implantation of Myoblasts into Skeletal Muscle	13-16
	Alternate Protocol 2: Direct Injection of Plasmid DNA into Muscle	13-18
	Support Protocol: Isolation of *lacZ*-Labeled Cells by Fluorescence-Activated Cell Sorting (FACS)	13-18
	Basic Protocol 4: DNA Vaccine Administration by Intramuscular Injection of the Quadricpeps Muscles in the Mouse	13-19
	Alternate Protocol 3: DNA Vaccine Administration by Intramuscular Injection of the Anterior Tibialis in the Mouse	13-20
13.3	**Ex Vivo and In Vivo Gene Delivery to the Brain**	**13-21**
	Basic Protocol 1: Implantation of Genetically Modified Cells into the Adult Rat Brain	13-23
	Basic Protocol 2: Implantation of Genetically Modified Cells into the Neonatal Rat Brain	13-25
	Basic Protocol 3: Implantation of Genetically Modified Cells into the Fetal Rat Brain	13-26
	Basic Protocol 4: Implantation of Collagen-Embedded Genetically Modified Fibroblasts into a Cavity in the Adult Rat Brain	13-27
	Basic Protocol 5: Direct Injection of Recombinant Viral Vectors into the Rat Brain	13-28
	Support Protocol: Preparation of Genetically Modified Cells for Grafting	13-29
13.4	**Human Hematopoietic Cell Culture, Transduction, and Analyses**	**13-30**
	Basic Protocol: Retroviral-Mediated Transduction of CD34$^+$ Cells with Stromal Support	13-30
	Alternate Protocol: Retroviral-Mediated Transduction on Fibronectin-Coated Dishes	13-35

Support Protocol 1: Maintenance of Vector-Producing Fibroblasts ... 13-35
Support Protocol 2: Collection of Cell-Free Supernatant for Transduction ... 13-36
Support Protocol 3: Enzymatic Removal of Magnetic Beads from
Immunomagnetically Selected Cells ... 13-36
Support Protocol 4: Establishing Primary Human Marrow Stromal Monolayers
from Harvested Bone Marrow ... 13-37
Support Protocol 5: Screening Media and Individual Components ... 13-38
Support Protocol 6: Plating Colony-Forming Cells from Cultures ... 13-38
Support Protocol 7: Preparing Whole-Cell Lysates of Individual CFU Colonies
for PCR ... 13-39
Support Protocol 8: Analysis of Clonal Integration in Individual Colonies ... 13-40
Support Protocol 9: Harvesting Cells for Analysis of Long-Term Bone Marrow
Cultures ... 13-43
Support Protocol 10: Freezing Hematopoietic Cells ... 13-43
Support Protocol 11: Thawing Hematopoietic Cells ... 13-44

13.5   Gene Delivery to the Airway   13-44
Strategic Planning ... 13-44
Basic Protocol 1: Isolation of Human Primary Airway Epithelial Cells ... 13-46
Basic Protocol 2: Transduction of Primary Airway Epithelial Cells ... 13-48
Basic Protocol 3: Generation of Polarized Airway Epithelial Monolayers ... 13-49
Basic Protocol 4: Gene Transfer to Polarized Airway Epithelia ... 13-51
Basic Protocol 5: Generation of Human Bronchial Xenografts ... 13-52
Basic Protocol 6: Gene Transfer to Human Bronchial Xenografts ... 13-56
Basic Protocol 7: In Vivo Gene Delivery to the Lung ... 13-57
Support Protocol 1: Harvesting of Human Bronchial Xenografts for
Morphologic Analysis to Evaluate Transgene Expression ... 13-58

13.6   Gene Delivery to the Liver   13-59
Basic Protocol 1: Recombinant Adenovirus Delivery to Mouse Liver by Tail
Vein Injection ... 13-60
Basic Protocol 2: Recombinant Adenovirus Delivery to Rabbit Liver by
Peripheral Ear Vein Injection ... 13-60
Support Protocol: Harvesting Liver Tissue for Transgene Expression Analysis ... 13-61

Appendices

A1   Reagents and Solutions   A1-1

A2   Useful Information and Data   A2-1
2A   Overview of Human Repetitive DNA Sequences ... A2-1
2B   ISCN Standard Idiograms ... A2-3
Chromosome Band Nomenclature ... A2-4
2C   Genetic Linkage Reference Maps: Access to Internet-Based Resources ... A2-22
2D   Radioisotope Data ... A2-24
2E   Centrifuges and Rotors ... A2-24

A3   Commonly Used Techniques   A3-1
3A   Isolation of Genomic DNA from Mammalian Cells ... A3-1
Basic Protocol 1: DNA Isolation from Whole Blood ... A3-1
Alternate Protocol 1: DNA Isolation from Cell Pellets ... A3-2
Alternate Protocol 2: Recovery of Genomic DNA by High-Salt
Precipitation ... A3-2
Basic Protocol 2: Isolation of DNA from Buccal Swabs ... A3-3
3B   Extraction and Precipitation of DNA ... A3-4
Basic Protocol 1: Phenol Extraction ... A3-4
Basic Protocol 2: Ethanol Precipitation of DNA ... A3-5
3C   Preparation of DNA from Fixed, Paraffin-Embedded Tissue ... A3-7
Basic Protocol: DNA Isolation using Mixed-Bed Chelating Resin ... A3-7
Alternate Protocol: DNA Isolation from Nonoptimally Fixed Tissue ... A3-7
3D   Quantitation of DNA and RNA with Absorption and Fluorescence
Spectroscopy ... A3-8

		Basic Protocol: Detection of Nucleic Acids using Absorption Spectroscopy	A3-8
		Alternate Protocol 1: DNA Detection using the DNA-Binding Fluorochrome Hoechst 33258	A3-10
		Alternate Protocol 2: DNA and RNA Detection with Ethidium Bromide Fluorescence	A3-12
	3E	Enzymatic Labeling of DNA	A3-12
		Basic Protocol 1: Uniform Labeling of DNA by Nick Translation	A3-12
		Alternate Protocol: Labeling DNA by Random Oligonucleotide-Primed Synthesis	A3-13
		Basic Protocol: 5′-End-Labeling Oligonucleotides using T4 Polynucleotide Kinase	A3-14
		Support Protocol 1: Spin-Column Procedure for Separating Radioactively Labeled DNA from Unincorporated DNTP Precursors	A3-15
		Support Protocol 2: Measuring Radioactivity in DNA and RNA by Trichloroacetic Acid (TCA) Precipitation	A3-15
	3F	Denaturing Polyacrylamide Gel Electrophoresis	A3-16
		Basic Protocol	A3-16
	3G	Analysis of DNA by Southern Blot Hybridization	A3-18
	3H	Analysis of RNA by Northern Blot Hybridization	A3-23
		Basic Protocol: Northern Hybridization of RNA Fractionated by Agarose-Formaldehyde Gel Electrophoresis	A3-23
	3I	Techniques for Mammalian Cell Tissue Culture	A3-26
		Sterile Technique	A3-26
		Culture Medium Preparation	A3-27
		Basic Protocol: Trypsinizing and Subculturing Cells from a Monolayer	A3-28
		Support Protocol 1: Freezing Human Cells Grown in Monolayer Cultures	A3-29
		Alternate Protocol: Freezing Cells Grown in Suspension Culture	A3-30
		Support Protocol 2: Thawing and Recovering Human Cells	A3-30
		Support Protocol 3: Determining Cell Number and Viability with a Hemacytometer and Trypan Blue Staining	A3-31
		Support Protocol 4: Preparing Cells for Transport	A3-33
	3J	Establishment of Permanent Cell Lines by Epstein-Barr Virus Transformation	A3-34
		Basic Protocol: Epstein-Barr Virus Transformation of Cultured Lymphocytes	A3-34
		Support Protocol 1: Preparation of Epstein-Barr Virus Stock	A3-35
		Support Protocol 2: Freezing and Reculturing of Epstein-Barr Virus-Transformed Lymphocytes	A3-36
	3K	Karyotyping	A3-37
		Basic Protocol	A3-37
A4		**Selected Suppliers of Reagents and Equipment**	**A4-1**
		References	

Index

Preface

The end of the twentieth century will likely be remembered as the beginning of the era of molecular medicine. The ability to isolate and purify genes, examine their structure and regulation, and determine the presence of pathogenic variants has placed vast power in the hands of those who study human biology and disease. Techniques of genetic analysis have therefore become important not only for the geneticist, but also for scientists and clinical investigators not formally trained in genetics.

Short Protocols in Human Genetics presents shortened versions of methods published in *Current Protocols in Human Genetics*. This compendium includes step-by-step descriptions of the principal methods covered in CPHG. As in CPHG, the scope is broad, including approaches to genetic linkage analysis, gene cloning, cytogenetics, and diagnostics. Designed for use at the lab bench, *Short Protocols* is intended for graduate students and postdoctoral fellows who are familiar with the detailed explanations found in CPHG. However, sufficient detail is provided to allow experienced investigators to use it as a stand-alone bench guide. For additional information, we recommend that the reader refer to the commentaries and detailed annotations in CPHG.

This book is not intended to substitute for a formal course of study in human genetics. Several recent texts provide a general review of human genetics, including both classical and molecular genetics (e.g., *Principles of Medical Genetics* by Thomas D. Gelehrter et al., *Thompson & Thompson Genetics in Medicine* by Robert L. Nussbaum et al., *Human Molecular Genetics* by Tom Strachen and Andrew P. Read, and *The New Genetics in Clinical Practice* by D.J.J. Weatherall). *Emery and Rimoin's Principles and Practice of Medical Genetics* by David L. Rimoin et al. is a more clinically oriented and comprehensive text of medical genetics. Finally, *Online Mendelian Inheritance in Man (OMIM)* provides an indispensable catalog of human genetic traits, and includes an up-to-date listing of gene defects that correspond with genetic disorders. Originally authored and edited for print by Victor A. McKusick and his colleagues at Johns Hopkins University and elsewhere, it has been developed for the Web by the National Center for Biotechnology Information (NCBI).

HOW TO USE THIS MANUAL

Organization

Subjects in this manual are organized by chapters, and protocols are contained in units within a chapter. Units generally describe a method and include one or more protocols with materials, steps, and references for each technique. Full references can be found in the *References* section at the end of the book. The organization of material in this manual generally follows that of CPHG. Although the order of units within chapters does not necessarily correspond to CPHG, the unit titles are the same, so that readers who own both manuals will find it easy and convenient to cross-reference CPHG when more explanatory details are required.

Many reagents and procedures are employed repeatedly throughout the manual. Instead of duplicating this information, cross-references among units are used extensively. Cross-referencing helps to ensure that lengthy and complex protocols are not overburdened with steps describing auxiliary procedures needed to prepare raw materials and analyze results. Certain units that describe commonly used techniques (e.g., gel electrophoresis) are cross-referenced in other units that describe their application. Some widely used techniques (such as quantitation of DNA) are found in *APPENDIX 3*. For most methods in molecular and cell biology, readers are referred to *Current Protocols in Molecular Biology* and *Current Protocols in Cell Biology*.

Recipes for the reagents and solutions used in each method are presented in *APPENDIX 1*. Other appendices provide useful measurements and data (*APPENDIX 2*) and the names and addresses of suppliers (*APPENDIX 4*).

Protocols

Many units in the manual contain groups of protocols, each presented with a series of steps. The *Basic Protocol* is presented first in each unit and is generally the recommended or most universally applicable approach. *Alternate Protocols* are provided where different equipment or reagents can be employed to achieve similar ends, where the starting material requires a variation in approach, or where requirements for the end product differ from those in the basic protocol. *Support Protocols* describe additional steps that are required to perform the basic or alternate protocols; these steps are separated from the core protocol because they might be applicable to other uses in the manual, or because they are performed in a time frame separate from the basic protocol steps.

Reagents and Solutions

Reagents required for a protocol are itemized in the materials list before the procedure begins. Many are common stock solutions, others are commonly used buffers or media, and others are solutions unique to a particular protocol. All recipes are provided in *APPENDIX 1*. It is important to note that the names of some of these solutions might be similar for more than one unit (e.g., lysis buffer) while the recipes differ. It is essential to prepare reagents from the proper recipes. To avoid confusion, a parenthetical listing of the unit(s) in which each recipe is used can be found next to the name of each reagent in *APPENDIX 1*, except in the case of commonly used buffers and solutions such as TE buffer.

Equipment

Special equipment and utensils are itemized in the materials list of each protocol. We have not attempted to list all items required for each procedure, but rather have noted those items that might not be readily available in the laboratory, that have specific specifications, or that require special preparation.

Standard pieces of equipment in the modern human genetics and molecular genetics laboratory are listed in the accompanying box. These items are used extensively in this manual and are not necessarily cited in the Materials list that precedes each protocol.

Commercial Suppliers

Throughout the manual, we have recommended commercial suppliers of chemicals, biological materials, and equipment. In some cases, the noted brand has been found to be of superior quality or it is the only suitable product available in the marketplace. In other cases, the experience of the author of that protocol is limited to that brand. In the latter situation, recommendations are offered as an aid to the novice experimenter in obtaining the tools of the trade. Experienced investigators are therefore encouraged to experiment with substituting their own favorite brands. Addresses, phone numbers, facsimile numbers, and Web sites of all suppliers mentioned in this manual are provided in *APPENDIX 4*.

References

Short Protocols gives only a limited number of the most fundamental references as background for each unit. These are listed at the end of the unit, with full bibliographic listings in the *References* section at the end of the book. Listings for specific references that are cited in figures and tables can also found in the *References* section. Readers who would like a more complete entry into the literature for background and application of methods are referred to the appropriate units in *Current Protocols in Human Genetics*.

SAFETY CONSIDERATIONS

Anyone carrying out these protocols may encounter the following hazardous or potentially hazardous materials: (1) radioactive substances, (2) toxic chemicals and carcinogenic or teratogenic reagents, (3) pathogenic and infectious biological agents, including samples from both healthy and sick humans, and (4) certain recombinant DNA constructs. Only limited cautionary statements are included in this manual. Users must proceed with the prudence and precaution associated with good laboratory and clinical practice. Users are responsible for understanding the dangers of

Applicator, cotton-tipped and wooden
Autoclave
Bag sealer
Balances, analytical and preparative
Beakers
Bench protectors, plastic-backed (including "blue pads")
Biohazard disposal containers and bags
Biosafety cabinet, to protect investigator from biohazardous organisms
Bottles, glass and plastic squirt
Bunsen burners
Cell harvester, for determining radioactivity uptake in 96-well microtiter plates
Clamps
CO_2 *humidified incubator*, 37°C and 5% CO_2
Computer, PC or Macintosh, and printer
Coplin jars, glass, for 75 × 25–mm slides
Cryovials, sterile (e.g., Nunc)
Cuvettes, plastic disposable, glass, and quartz
Darkroom and developing tanks
Desiccator and desiccant
Dry ice
Filtration apparatus, for collecting and washing precipitates on nitrocellulose or membrane filters
Flasks, glass (e.g., Erlenmeyer, Florence)
Forceps
Fraction collector
Freezers, 20°, 70°C, and liquid nitrogen
Fume hood
Geiger counter, or radiation survey meter
Gel dryer
Gloves, disposable plastic and asbestos
Graduated cylinders and pipets
Heating blocks, thermostatically controlled for test tubes and microcentrifuge tubes
Hemacytometer
Ice buckets
Ice maker
Incubator, 37°C
Lab coats
Laminar-flow hood, to maintain sterility for tissue culture
Light box, for viewing autoradiograms
Lint-free tissue, e.g., Kimwipes
Liquid nitrogen
Lyophilizer
Magnetic stirrer, useful with heater
Markers, including indelible markers and China-marking pencils

Microcentrifuge, Eppendorf-type, maximum speed 12,000 to 14,000 rpm
Microcentrifuge tubes, 1.5-ml
Microscopes, bright-field, noninverted
Microscope slides, glass, 75 × 25–mm, and coverslips
Mortar and pestle
Ovens, drying, hybridization, and microwave
Paper cutter, large size
Paper towels
Parafilm
Pasteur pipets and bulbs
pH meter
pH paper
Pipets, graduated and Pasteur, sterile
Pipettors, adjustable delivery, volume range 0.5- to 10-μl, 10- to 200-μl, and 200- to 1000-μl
Plastic wrap, (e.g., Saran Wrap)
Pliers, needle-nose
Polaroid camera
Policemen, rubber or plastic
Racks, test tube
Radiation shield, Lucite or Plexiglas
Radioactive ink
Radioactive waste containers, for liquid and solid waste
Refrigerator, 4°C
Ring stand and rings
Rubber stoppers
Safety glasses
Scalpels and blades
Scintillation counters, β counter and γ counter
Scissors
Shakers, orbital and platform, room temperature or 37°C
Spectrophotometer, visible and UV range
Speedvac evaporator (Savant)
Tape, masking and electrician's
Timer
Trays, plastic and glass, various sizes
UV cross-linker (e.g., Stratalinker; Strategene)
UV light sources, long- and short-wave
UV-transparent plastic wrap (e.g., Saran Wrap)
UV transilluminator
Vacuum desiccator
Vacuum oven
Water bath, 37°C
Water purification equipment
X-ray film cassettes and intensifying screens

working with hazardous materials and for strictly following the safety guidelines established by manufacturers as well as local and national regulatory agencies.

ACKNOWLEDGMENTS

The Current Protocols editorial staff at John Wiley & Sons provided us with the support and assistance needed to bring this project together. Among those who helped us are Ann Boyle, Tom Downey, Amy Fluet, Shonda Leonard, Susan Lieberman, Kathleen Morgan, Allen Ranz, Mary Keith Trawick, Joseph White, and Elizabeth Harkins. We are especially grateful to our many colleagues—in our own labs and in academic and industrial labs all over the world—who have contributed material to this manual and shared their valuable procedures and experience with the human genetics community. Finally, we appreciate and acknowledge the invaluable contribution of our colleague Dr. Donald Moir, past editor for the parent volume, *Current Protocols in Human Genetics*. His original plans for and continued development of CPHG provided a fundamental part of both *Current Protocols* and *Short Protocols in Human Genetics*.

RECOMMENDED BACKGROUND READING

Ausubel, F.M., Brent, R., Kingston, R.E., Moore, D.D., Seidman, J.G., Smith, J.A., Struhl, K. (eds.) 2004. Current Protocols in Molecular Biology. John Wiley & Sons, New York.

Bonifacino, J.S., Dasso, M., Harford, J.B., Lippincott-Schwartz, J., and Yamada, K.M. (eds.) 2004. Current Protocols in Cell Biology. John Wiley & Sons, New York.

Gelehrter, T.D., Collins, F.S., and Ginsberg, D. 1997. Principles of Medical Genetics. Lippincott, 2nd ed. Williams & Wilkins, Baltimore.

Nussbaum, R.L., McInnes, R.R., and Willard, H.F. 2004. Genetics in Medicine, 6th ed. W.B. Saunders, Philadelphia.

OMIM (Online Mendelian Inheritance in Man). *http://www.ncbi.nlm.nih.gov/entrez/query.fcgi?db=OMIM*

Rimoin, D.L., Connor, J.M., Pyeritz, R.E., Korf, B.R., and Emery, A.E.H. (eds.) 2001. Emery and Rimoin's Principles and Practice of Medical Genetics, 4th ed. Churchill Livingstone, New York.

Strachen, T. and Read, A.P. 2004. Human Molecular Genetics, 3rd ed. Garland Science, New York.

Weatherall, D.J.J. 1992. The New Genetics and Clinical Practice, 2nd ed. Oxford University Press, Oxford.

Nicholas C. Dracopoli, Jonathan L. Haines, Bruce R. Korf, Cynthia C. Morton, Anthony Rosenzweig, Christine E. Seidman, J.G. Seidman, and Douglas R. Smith

Contributors

David Adler
University of Washington
Seattle, Washington

Ellen C. Akeson
The Jackson Library
Bar Harbor, Maine

Levent M. Akyurek
National Heart, Lung, and Blood
  Institute, NIH
Bethesda, Maryland

Lisa M. Albright
Allison Park, Pennsylvania

Becky Alhadeff
New York Blood Center
New York, New York

Christopher I. Amos
University of Texas
M.D. Anderson Cancer Center
Houston, Texas

Wendy Ankener
University of Washington School
  of Medicine
Seattle, Washington

Michael Arad
Harvard Medical School
Boston, Massachusetts

Norman Arnheim
University of Southern California
Los Angeles, California

Arleen D. Auerbach
Rockefeller University
New York, New York

Charles D. Bangs
Stanford University Hospital
Stanford, California

Marie A. Barr
Jefferson Medical College
Philadelphia, Pennsylvania

Scott Barr
Harvard Medical School
Boston, Massachusetts

Andreas D. Baxevanis
National Human Genome Research
  Institute, NIH
Bethesda, Maryland

Stephen B. Baylin
Johns Hopkins Oncology Center
Baltimore, Maryland

Laurie Becker
Case Western Reserve University
  and University Hospitals
Cleveland, Ohio

Alan H. Beggs
Children's Hospital and Harvard
  Medical School
Boston, Massachusetts

Helen M. Blau
Stanford University School of
  Medicine
Stanford, California

Mark S. Boguski
National Center for Biotechnology
  Information, NIH
Bethesda, Maryland

Richard Bolin
Nexagen
Boulder, Colorado

Anne-Lise Borresen
The Norwegian Radium Hospital
Oslo, Norway

Andrew Braun
Harvard Medical School
Boston, Massachusetts

Garrett M. Brodeur
Children's Hospital of Philadelphia
Philadelphia, Pennsylvania

Terry Brown
University of Manchester Institute
  of Science and Technology
Manchester, United Kingdom

W. Ted Brown
Institute of Basic Research in
  Developmental Disabilities
Staten Island, New York

Michael C. Byrne
Genetics Institute
Cambridge, Massachusetts

Deborah E. Cabin
Johns Hopkins University School
  of Medicine
Baltimore, Maryland

Samuel S. Chong
Georgetown University Medical
  Center
Washington, D.C.

Michael Christiansen
Andersen Statens Serum Institut
Copenhagen, Denmark

Deanna Church
National Center for Biotechnology
  Information, NIH
Bethesda, Maryland

Chris D. Clark
Stanford University School of
  Medicine
Stanford, California

Richard G.H. Cotton
Mutation Research Centre
Fitzroy, Australia

Sandra L. Dabora
Brigham and Women's Hospital
Boston, Massachusetts

Paola Dal Cin
Brigham and Women's Hospital
Boston, Massachusetts

Muriel T. Davisson
The Jackson Library
Bar Harbor, Maine

Pieter J. de Jong
Roswell Park Cancer Institute
Buffalo, New York

Claire Delahunty
University of Washington School
  of Medicine
Seattle, Washington

Daniel B. Demers
Fairfax Identity Laboratories
Fairfax, Virginia

Johan T. den Dunnen
Leiden University
Leiden, The Netherlands

Sandy DeVries
University of California at San
    Francisco
San Francisco, California

Christine M. Distèche
University of Washington School
    of Medicine
Seattle, Washington

Carl Dobkin
Institute for Basic Research in
    Developmental Disabilities
Staten Island, New York

Norman A. Doggett
Los Alamos National Laboratory
Los Alamos, New Mexico

Timothy A. Donlon
Kapiolani Medical Center
Honolulu, Hawaii

Dongsheng Duan
University of Iowa
Iowa City, Iowa

John F. Engelhardt
University of Iowa
Iowa City, Iowa

Warren J. Ewens
University of Pennsylvania
Philadelphia, Pennsylvania

Lindsay Farrer
Boston University and Harvard
    Medical School
Boston, Massachusetts

David J. Fink
University of Pittsburgh School of
    Medicine and VA Medical
    Center
Pittsburgh, Pennsylvania

Maximillian T. Foulettie
Genetics Institute
Cambridge, Massachusetts

Edward A. Fox
Brigham and Women's Hospital
Boston, Massachusetts

Cornel Fraefel
Institute of Virology
University of Zurich
Zurich, Switzerland

Eirik Frengen
The Biotechnology Centre of Oslo
University of Oslo
Oslo, Norway

Scan R. Gallagher
Motorola
Tempe, Arizona

Robert M. Gemmill
Institute for Cancer Research
Denver, Colorado

James German
New York Blood Center
New York, New York

Michelle Geschwend
Stanford University School of
    Medicine
Stanford, California

Longina M. Gibas
Jefferson Medical College
Philadelphia, Pennsylvania

John Gilbert
Duke University Medical Center
Durham, North Carolina

James M. Giron
Children's Research Institute
Washington, D.C.

Joseph C. Glorioso
University of Pittsburgh School of
    Medicine
Pittsburgh, Pennsylvania

William F. Goins
University of Pittsburgh School of
    Medicine
Pittsburgh, Pennsylvania

Robert E. Gore-Langton
Georgetown University Medical
    Center
Washington, D.C.

Joe W. Gray
University of California at San
    Francisco
San Francisco, California

Anoop Grewal
Silicon Genetics
Redwood City, California

Michelle Gschwend
Whitehead Institute for Genome
    Research
Cambridge, Massachusetts

Rebecca A. Haberman
University of North Carolina
Chapel Hill, North Carolina

Jeff Hall
Sequana Therapeutics
La Jolla, California

Barbara Handelin
Integrated Genetics
Framingham, Massachusetts

Elizabeth R. Hauser
Duke University Medical Center
Durham, North Carolina

Tong-Chuan He
The University of Chicago Medical
    Center
Chicago, Illinois

Ruth A. Heim
Genzyme Genetics
Westborough, Massachusetts

Henry H.Q. Heng
Unversity of Toronto and The
    Hospital for Sick Children
Toronto, Ontario

James G. Herman
The Johns Hopkins Oncology
    Center
Baltimore, Maryland

Eric P. Hoffman
Children's Research Institute
Washington, D.C.

Ourania Horaitis
Mutation Research Center
Fitzroy, Australia

Eivind Hovig
The Norwegian Radium Hospital
Oslo, Norway

Leaf Huang
University of Pittsburgh
Pittsburgh, Pennsylvania

Rene Hubert
University of Southern California
Los Angeles, California

Thomas J. Hudson
Whitehead Institute
Cambridge, Massachusetts

Mark R. Hughes
Georgetown University Medical
    Center
Washington, D.C.

Bradley T. Hyman
Harvard Medical School/
    Massachusetts General Hospital
Charlestown, Massachusetts

Martin Ingelsson
Harvard Medical School/
    Massachusetts General Hospital
Charlestown, Massachusetts

Panayiotis A. Ioannou
The Murdoch Institute for
    Research into Birth Defects
Royal Children's Hospital
Melbourne, Australia

Michael C. Irizarry
Harvard Medical School/
    Massachusetts General Hospital
Charlestown, Massachusetts

Cynthia L. Jackson
Rhode Island Hospital & Brown
    University
Providence, Rhode Island

Laird Jackson
Jefferson Medical College
Philadelphia, Pennsylvania

John Jarcho
Brigham and Women's Hospital
    and Harvard Medical School
Boston, Massachusetts

Erica Justice-Higgins
Massachusetts General Hospital
Charlestown, Massachusetts

Norman Kaplan
National Institute of Environmental
    Health Sciences
Research Triangle Park, North
    Carolina

Charles M. Kelly
Fairfax Identity Laboratories
Fairfax, Virginia

Jae Bum Kim
Harvard Medical School
Boston, Massachusetts

Joan H.M. Knoll
Harvard Medical School
Boston, Massachusetts

Sakari Knuutila
University of Helsinki
Helsinki, Finland

Donald B. Kohn
Children's Hospital of Los Angeles
Los Angeles, California

Robert G. Korneluk
Children's Hospital of Eastern
    Ontario
Ottawa, Canada

Karen Kozarsky
SmithKline Beecham
    Pharmaceuticals
King of Prussia, Pennsylvania

David Krisky
University of Pittsburgh School of
    Medicine
Pittsburgh, Pennsylvania

Gabriele Kroner-Lux
University of North Carolina
Chapel Hill, North Carolina

Pui-Yan Kwok
Washington University
St. Louis, Missouri

Sam LaBrie
Genome Systems Inc.
St. Louis, Missouri

Bruce Lamb
Johns Hopkins University School
    of Medicine
Baltimore, Maryland

Peter Lambert
Silicon Genetics
Redwood City, California

Lars Allan Larsen
Andersen Statens Serum Institut
Copenhagen, Denmark

Charles Lee
Brigham and Women's Hospital
    and Harvard Medical School
Boston, Massachusetts

Esther P. Leeflang
University of Southern California
Los Angeles, California

Song Li
University of Pittsburgh
Pittsburgh, Pennsylvania

Peter Lichter
Deutsches
    Krebsforschungszentrum
Heidelberg, Germany

Sigbjorn Lien
Agricultural University of Norway
Aas, Norway

Stephen Little
Cellmark Diagnostics
Cheshire, United Kingdom

Janina Longtine
Brigham and Women's Hospital
Boston, Massachusetts

Karol Mackey
Molecular Research
Cincinnati, Ohio

Thomas L. Madden
National Center for Biotechnology
    Information, NIH
Bethesda, Maryland

Mani Mahadevan
Children's Hospital of Eastern
    Ontario
Ottawa, Canada

Peggy Marconi
University of Pittsburgh School of
    Medicine
Pittsburgh, Pennsylvania

Eden R. Martin
Duke University Medical Center
Durham, North Carolina

Hajime Matsukaki
Affymetrix
Santa Clara, California

Linda McAllister
Affymetrix
Santa Clara, California

John McPherson
Washington University School of
  Medicine
St. Louis, Missouri

Mark A. Micale
Case Western Reserve University
  and University Hospitals
Cleveland, Ohio

A. Dusty Miller
Fred Hutchinson Cancer Research
  Center
Seattle, Washington

Patricia Minehart Miron
Harvard Medical School and
  Brigham and Women's Hospital
Boston, Massachussetts

Jason H. Moore
Vanderbilt University Medical
  School
Nashville, Tennessee

Juliane Murphy
National Human Genome Research
  Institute, NIH
Bethesda, Maryland

Elizabeth G. Nabel
National Heart, Lung, and Blood
  Institute, NIH
Bethesda, Maryland

Gary J. Nabel
University of Michigan Medical
  Center
Ann Arbor, Michigan

Elizabeth Nanthakumar
Sequana Therapeutics
La Jolla, California

Nichole M. Napolitano
Genzyme Genetics
Westborough, Massachusetts

Deborah A. Nickerson
University of Washington School
  of Medicine
Seattle, Washington

Sarah L. Nolin
Institute for Basic Research in
  Developmental Disabilities
Staten Island, New York

Jan A. Nolta
Children's Hospital of Los Angeles
Los Angeles, California

Nassim Nouri
Affymetrix
Santa Clara, California

Kazutoyo Osoegawa
Roswell Park Cancer Institute
Buffalo, New York

B.F. Francis Ouellette
Centre for Molecular Medicine and
  Therapeutics
University of British Columbia
Vancouver, Canada

Irma Parra
Cancer Therapy and Research
  Center
San Antonio, Texas

Nila Patil
Affymetrix
Santa Clara, California

Mary C. Phelan
Thompson Children's Hospital
Chattanooga, Tennessee

Daniel Pinkel
University of California at San
  Francisco
San Francisco, California

Kim D. Pruitt
National Center for Biotechnology
  Information, NIH
Bethesda, Maryland

Ramesh Ramakrishnan
Virusys Corporation
North Berwick, Maine

Koustubh Ranade
Bristol-Myers Squibb
Princeton, New Jersey

Thomas A. Rando
Stanford University School of
  Medicine
Stanford, California

Rino Rappuoli
Chiron Corporation
Emeryville, California

Roger H. Reeves
Johns Hopkins University School
  of Medicine
Baltimore, Maryland

Willem Rens
University of Cambridge
Cambridge, United Kingdom

Carol Reynolds
Brigham and Women's Hospital
Boston, Massachusetts

Elizabeth M. Rohlfs
Genzyme Genetics
Westborough, Massachusetts

Mark T. Ross
The Sanger Centre
Cambridge, United Kingdom

Thomas Ryder
Affymetrix
Santa Clara, California

Richard Jude Samulski
University of North Carolina
Chapel Hill, North Carolina

Hong San
National Heart, Lung, and Blood
  Institute, NIH
Bethesda, Maryland

Karin Schmitt
University of Southern California
Los Angeles, California

Deborah E. Schofield
Children's Hospital and Harvard
  Medical School
Boston, Massachusetts

Rhona R. Schreck
Cedars-Sinai Medical Center
Los Angeles, California

Stuart Schwartz
Case Western Reserve University
and University Hospitals
Cleveland, Ohio

Val Sheffield
University of Iowa
Iowa City, Iowa

Yongah Shin
Harvard Medical School/
Massachusetts General Hospital
Charlestown, Massachusetts

John M. Shoffner
Scottish-Rite Children's Medical
Center
Atlanta, Georgia

Anthony P. Shuber
Integrated Genetics
Framingham, Massachusetts

Jeffrey Sklar
Brigham and Women's Hospital
Boston, Massachusetts

Paal Skytt
Andersen Statens Serum Institut
Copenhagen, Denmark

Barton E. Slatko
New England Biolabs
Beverly, Massachusetts

Birgitte Smith-Sorensen
The Norwegian Radium Hospital
Olso, Norway

Amanda C. Sozer
Fairfax Identity Laboratories
Fairfax, Virginia

Marcy C. Speer
Duke University Medical Center
Durham, North Carolina

Richard S. Spielman
University of Pennsylvania
Philadelphia, Pennsylvania

Matthew L. Springer
Stanford University School of
Medicine
Stanford, California

Jordan Stockton
Silicon Genetics
Redwood City, California

William M. Strauss
Beth Israel Hospital/Harvard
Medical School
Boston, Massachusetts

Kevin Struhl
Harvard Medical School
Boston, Massachusetts

Damir Sudar
Lawrence Berkeley National
Laboratory
Berkeley, California

Linda C. Surh
Children's Hospital of Eastern
Ontario
Ottawa, Canada

Stanley Tabor
Harvard Medical School
Boston, Massachusetts

Ludwig Thierfelder
Max-Delbruck-Centrum fur
Molekulare Medizin
Berlin, Germany

Melissa M. Thouin
Children's Research Institute
Washington, D.C.

Jeff P. Tomkins
Clemson University Genomics
Institute
Clemson, South Carolina

Didier Trono
University of Geneva
Geneva, Switzerland

Lap-Chee Tsui
University of Toronto and The
Hospital for Sick Children
Toronto, Ontario

Jeffrey B. Ulmer
Chiron Corporation
Emeryville, California

Jeffrey Vance
Duke University Medical Center
Durham, North Carolina

Rolf Vossen
Leiden University Medical Center
Leiden, The Netherlands

Jens Vuust
Andersen Statens Serum Institut
Copenhagen, Denmark

Frederic M. Waldman
University of California at San
Francisco
San Francisco, California

Douglas C. Wallace
Emory University School of
Medicine
Atlanta, Georgia

Dorothy Warburton
Columbia University
New York, New York

William Warren
Institute for Cancer Research
Surrey, United Kingdom

Jonathan C. Wasson
Washington University School of
Medicine
St. Louis, Missouri

Hugh C. Watkins
University of Oxford
Oxford, United Kingdom

Daniel E. Weeks
University of Pittsburgh
Pittsburgh, Pennsylvania

Jane M. Weisemann
National Center for Biotechnology
Information, NIH
Bethesda, Maryland

Maryann Z. Whitley
Genetics Institute
Cambridge, Massachusetts

Mark Whitmore
University of Pittsburgh
Pittsburgh, Pennsylvania

Bradford Windle
The Cancer Therapy and Research
Center
San Antonio, Texas

Rod A. Wing
Clemson University Genomics
Institute
Clemson, South Carolina

Darren Wolfe
University of Pittsburgh School of
    Medicine
Pittsburgh, Pennsylvania

Tyra G. Wolfsberg
National Center for Biotechnology
    Information, NIH
Bethesda, Maryland

Fengtang Yang
University of Cambridge
Cambridge, United Kingdom

Jiing-Kuan Yee
City of Hope National Medical
    Center
Duarte, California

Lin Zhang
University of Southern California
Los Angeles, California

Yulong Zhang
University of Iowa
Iowa City, Iowa

Romain Zufferey
University of Geneva
Geneva, Switzerland

CHAPTER 1

Genetic Mapping

The primary goal of a trait gene discovery is to identify and characterize the mutation(s), variants, or polymorphisms that give rise to specific disease phenotypes. A general strategy for trait gene discovery is diagrammed in Figure 1.0.1, which suggests a linear and temporal relationship between the steps. Although it is true that the earlier steps almost always need to precede the later steps, the earlier steps do not necessarily end when the later steps are initiated. Steps such as sample collection to increase sample size (step 5) will often continue well after genotyping (step 6) and data analysis (step 7) have begun. One example is when additional members of a family are at first difficult to contact, or are unknown to the investigator until additional information is provided by the initial family contact. A second example occurs when genetic heterogeneity is found. While the first locus is being actively pursued through fine mapping and gene identification, more families and genomic screening may be undertaken to find the second locus. A third example occurs in studying complex traits where a second data set for testing initially interesting regions may be developed while the genomic screen is underway.

After genotyping the DNA samples, it may be necessary to re-collect samples for certain individuals. This could happen if (a) DNA was mislabeled (the mistake may be detected by the apparent non-Mendelian inheritance shown by at least one marker), (b) an indication of false paternity was found that has to be confirmed, (c) DNA was lost or used up, or (d) an individual carries a very important (critical) recombination event that has to be confirmed.

KEY CONCEPTS

Mendelian Inheritance

Mendel's first law—the law of independent segregation—states that genes occur in pairs, and that during gamete formation, one of each pair is passed down to each gamete. Mendel's second law—the law of independent assortment—states that different gene pairs segregate independently of each other. Because most traits that Mendel examined were caused by single genes, human diseases caused by single genes are often referred to as Mendelian disorders. While rare, violations of Mendel's first law can occur and result in abnormal numbers of chromosomes (e.g., monosomies such as Turner syndrome or trisomies such as Down syndrome). Violations of Mendel's second law results in genetic linkage (e.g., two loci segregating together more often than expected by random chance).

Phenotype Versus Genotype

Phenotype is the description of the trait in question: e.g., clinical symptoms, quantitative measures, lab tests, or treatment outcomes. Genotype is the description of the DNA, i.e., alleles, at a particular locus.

Simple Versus Complex Traits

Any trait caused by a mutation in a single gene is said to be Mendelian. Diseases influenced by multiple genes that act in concert may be called oligogenic, if only a few genes are involved, or polygenic if many genes are involved. Multifactorial inheritance is an extension of oligogenic

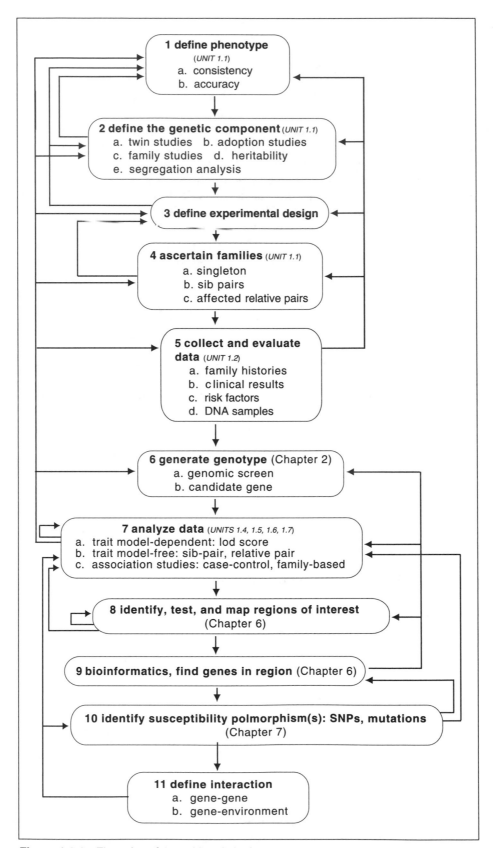

Figure 1.0.1 Flow chart for positional cloning.

or polygenic inheritance, where additional, nongenetic (environmental) factors may also be involved. Oligogenic, polygenic, and multifactorial traits are designated as complex traits.

Polymorphism

For a locus to be polymorphic, it must have at least two alleles that have population frequencies ≥ 0.01. Any allele with a frequency < 0.01 is termed a variant.

Linkage Analysis Versus Allelic Association Analysis

The first approach to testing the relationship between a specific locus and a phenotypic trait is linkage analysis to determine if two or more loci are cosegregating within a pedigree (*UNITS 1.4 & 1.5*). The second approach is allelic association analysis, which tries to match the occurrence of a trait with the occurrence of a specific allele at a specific locus (*UNIT 1.7*).

References: Haines and Pericak-Vance,1998; Ott,1999

Contributor: Jonathan L. Haines

UNIT 1.1

Collection of Clinical and Epidemiological Data for Linkage Studies

Detection of linkage between a DNA marker and a genetic locus for a disease trait is predicated on the presence of a major gene for the disease and on correct classification of subjects as affected or unaffected by a particular disease. Clinical and epidemiological data are thus necessary for: (1) determining whether the disorder has a major gene component and, if so, determining its mode of transmission; (2) selecting families appropriate for linkage studies; (3) allowing for evaluation of the presence of genetic heterogeneity by using phenotypic subsets; (4) testing for linkage or association with endophenotypes; (5) drawing correspondence between disease genotype and phenotype; and (6) deriving numerical functions that predict the probability of the phenotype given a specific genotype in at-risk persons if the phenotype is masked by reduced penetrance or variable expressivity. This unit describes approaches for establishing genetic influence and collecting linkage data.

ESTABLISHING EVIDENCE FOR GENETIC INFLUENCE

Differentiation of hereditary diseases from those that are genetic but not transmitted through the germ line (i.e., mutations in somatic tissue) is difficult and requires a large sampling of cases as well as specialized methods of analysis. Data for such analyses can sometimes be extracted from the literature; however, special attention must be given to how the case material was ascertained and diagnosed. Occasionally, the mode of inheritance of the trait can be gleaned from results of previously reported genetic analyses. In the absence of reliable published information, the following analytical approaches can be used to determine the mode of inheritance.

Twin Analysis

This method relies on the fact that monozygous (MZ) twins are genetically identical whereas dizygotic (DZ) twins share, on average, one-half of their genes. If both twins of a pair are affected by a disease, they are said to be concordant. By estimating and comparing concordance rates among MZ and DZ twins, one can get an idea of the role of genetic factors in the etiology of the disease. If the concordance rate is 100% in MZ twins and between 25% and 50% in DZ twins, then it may be concluded that the disease is strictly genetic and probably due to a single recessive (if the concordance rate is ~25%) or dominant (concordance rate ~50%) gene. For a disorder in which genetic factors are important in etiology, the concordance rate for MZ twins will be greater than for DZ twins, and the concordance rate for MZ twins reared apart will be about the same as for MZ twins reared together. Low concordance in MZ twins or equal concordance between MZ and DZ twins suggests a strong environmental influence on development of the disease. The biological limitations of twin studies include prenatal factors and the tendency for twins to share the same environment. Another important limitation is that twin studies do not provide conclusive information about the mode of inheritance.

Adoption Study

The premise of adoption studies is that if a trait has a genetic influence, the risk of illness should be higher in biological relatives (parents, siblings, or offspring) than adopted relatives living in the same household. This design controls for environmental contributions to the trait since the adoptee usually shares the same environment with adopted relatives but not with biological relatives. The basic method for an adoption study is to compare characteristics of adopted children with biological parents, adoptive parents, adoptive sibs, and biological sibs raised by other adoptive parents. There are several variations on this design. If the adoptee is selected as the proband (person with trait of interest who serves as an index case for ascertainment of study families), one can compare risk of disease in biological and adoptive relatives of affected and unaffected adoptees. Alternatively, by considering the parent as the proband, one can compare rates of illness in adopted offspring of affected and unaffected parents. Finally, in a cross-fostering approach, one compares disease risk in adoptees having well biological parents raised by affected parents with adoptees having affected biological parents raised by well adoptive parents.

The adoption study is not always optimal because: (1) obtaining an appropriate sample size can be difficult, particularly in countries where adoption records are generally unavailable; and (2) the environments of the adopted and biological relatives may not vary substantially, since adoption agencies often strive to match socioeconomic backgrounds of the adoptee and adoptive family. Moreover, results should be compared with those obtained from an adoptee control group since adoptees and their families are not representative of the general population. There is also a potential for confounding environmental influences if the adoptees are not separated from their biological parents at birth.

Recurrence Risks

Another approach for discerning a genetic basis for a disease is the computation of recurrence risks. Familial aggregation of dichotomous traits can be evaluated by examining the proportion of relatives of the probands (themselves affected with the trait) who also have the trait and comparing it with the proportion of relatives of control subjects who have the trait. If the proportion is greater among the relatives of the probands, then the trait is said to aggregate within families. It is possible that this aggregation is simply due to similar environments shared by those living together and not to a genetic etiology. To determine if a genetic etiology may be involved, additional comparisons are necessary. In particular, the pattern of familial recurrence risks is important. Letting λ_R be the ratio of the risk to relatives of type R (e.g., sibs, parents, or offspring) compared to the population risk ($\lambda_R = k_R/k$ where k_R is the risk to relatives of type R and k is the population risk), then the dropoff in ($\lambda_R - 1$) as a function of the degree of relationship R is indicative of the possible genetic modes of transmission. For example, if a single major gene (with any number of alleles) determines the trait, then ($\lambda_R - 1$) will decrease by a factor of 2 for each decreasing degree of unilineal relationship, using the parent-offspring relationship for the first-degree relative. If two major unlinked loci determine the trait, the dropoff in ($\lambda_R - 1$) will be more rapid if the effects of the loci are multiplicative, but will behave as for a single locus if the effects are additive.

Heritability

Heritability (h^2) estimation is another useful approach in determining the genetic influence of a trait. To determine heritability, first consider the total phenotypic variance of a trait (σ^2_T) which can be partitioned into genetic (σ^2_G) and nongenetic (σ^2_E) proportions. The genetic variance can be further divided into contributions from major genes and the genetic background (polygenes). Heritability is the proportion of the phenotypic variance due to the additive effects of many genes, i.e., the polygenic component (σ^2_A). The greater the value for h^2, the greater the genetic contribution to the etiology. Because the genetic component measured by heritability is polygenic, heritability estimates are only meaningful if no major genes contribute to the disorder. Heritability estimates also assume that there is no heterogeneity with regard to mode of inheritance. Heritability estimates, particularly when derived from sibs, may be inflated because of shared environmental factors. To avoid this bias, they should be derived from other relative sets including parents and offspring, and more distantly related individuals.

Segregation Analysis

A more direct approach is to compare the observed number of affected individuals in families with the number expected, assuming a particular mode of inheritance. A trait under the control of a single gene will segregate in families, and the frequencies of phenotypes should be consistent with Mendelian ratios. However, because families included in such studies are usually ascertained on the basis of having at least one affected member, it is important to apply an ascertainment correction (also see UNIT 1.2). For example, if one were evaluating the distribution of children affected with cystic fibrosis in families that were ascertained in a large clinic, the segregation ratio (i.e., proportion of affected offspring) would be much greater than the expected one-out-of-four sibs because families who (by chance) did not transmit the disease would be excluded from the sample (i.e., only families where an affected member went to the clinic would be included). In addition, it is important to examine the amount and type of deviation from the expected ratio before rejecting the genetic hypothesis. Complications such as delayed age at onset, variable expression and reduced penetrance—in addition to ascertainment errors—can produce deviations from the expected ratio.

The method to be used for calculating segregation ratios depends on the method used for ascertainment. Complex segregation analysis is done by a maximum-likelihood method. This technique permits joint consideration of several parameters of the genetic model including:

degree of dominance, penetrance, gene frequency, ascertainment probability, and measures of Mendelian transmission of the disease allele. A likelihood (i.e., numerical estimate of the fit of the pedigree data to a particular genetic transmission model) is computed and compared with likelihoods obtained from the same pedigree data under other postulated genetic transmission models. In this way, one can establish whether or not the disease has a major genetic-locus component and, if so, estimate parameters of the model that are necessary for linkage analysis: gene frequency, mode of inheritance, penetrance, and mutation rate. Several computer programs and packages for segregation analysis have been developed—e.g., PAP and Mendel.

To investigate the mode of inheritance of a disease by segregation analysis, accurate phenotypic and genealogical data on a set of families must be obtained using the approach described in the Basic Protocol. The means by which the families are ascertained must be uniform in order to correct properly for identified bias. Families that are ascertained on the basis of a positive family history or apparent pattern of transmission must be disqualified. Incidence of the disorder in the general population as well as sex and age dependencies must also be incorporated in the analyses. The number of families needed in the sample is an inverse function of the size of the pedigrees available for study. Diseases that are common in the general population or that are associated with variable expressivity or reduced penetrance will necessitate a larger sampling of families. A minimum of 25 to 50 nuclear families is usually needed to distinguish among genetic transmission models. For complex traits such as alcoholism, the actual number needed will exceed several hundred. Segregation analysis is not likely to yield meaningful results if the trait is ubiquitous (e.g., gray hair) or if family units (parents and children) are incompletely evaluated.

COLLECTING DATA FOR LINKAGE STUDIES

Strategic Approach

Accurate assessment of the phenotype (i.e., diagnosis) usually requires careful examination by a physician, especially when there is reduced penetrance (i.e., probability that a genotype will yield the predicted phenotype) and/or variable expressivity (i.e., variety in symptoms manifested and their severity). For disorders that are not associated with physical anomalies (e.g., behavioral disorders and some metabolic diseases), the diagnosis is often made after assessment of multiple pieces of information including laboratory test results. Diagnostic criteria developed by a consensus of researchers, if available, should be employed to standardize the information. Integrity of the linkage study is best maintained under double-blind conditions. Persons who establish diagnosis should not be informed of genetic marker data and DNA laboratory personnel should not be informed of the diagnosis. Steps for collecting diagnostic and epidemiological data for linkage studies are outlined below and summarized in Figure 1.1.1.

1. Formulate a list of questions that solicit all relevant diagnostic information for the trait under study, including but not limited to:

    presence or absence of key signs and symptoms
    age at examination
    age at onset of illness and/or characteristics specific to illness
    descriptive clinical data
    laboratory test results, including dates of tests
    epidemiological risk factors
    demographic data.

    Eliminate questions that cannot be answered accurately and completely for most patients.

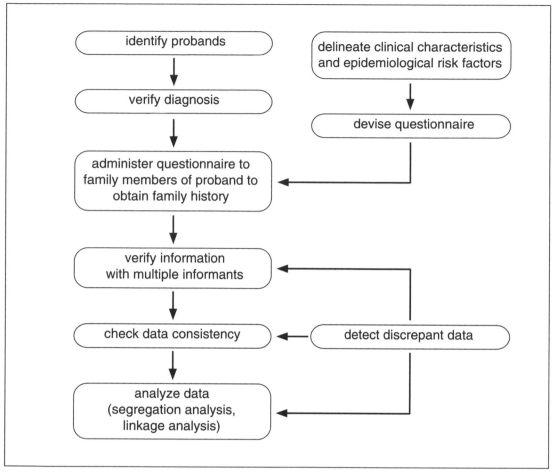

Figure 1.1.1 Flow chart for collecting diagnostic and epidemiological data for linkage studies (as described in the Strategic Approach).

2. Design the questionnaire form to be administered to individual patients. Keep it as short and specific as possible to encourage compliance. Assign unique identifiers that will enable the clinical data to be matched with corresponding identifiers used in the laboratory to track DNA marker information on individual patients and their families. To facilitate accurate and efficient data collection, use tabular formats (e.g., Fig. 1.1.2), optical scan forms (e.g., Fig. 1.1.3), or electronic forms (e-forms; see below).

3. Identify probands, e.g., from patient registries, clinic populations, referral centers, hospital records, or epidemiological surveys. Avoid ascertainment bias (*UNIT 1.2*) if the genetic transmission model is to be determined.

4. Verify diagnoses of probands using standardized criteria established by a consensus of researchers who are knowledgeable about the disorder. Proceed carefully, as inclusion of families of false-positive cases will hamper the ability to detect linkage and will introduce bias that may cause the researcher to infer falsely heterogeneity in the genetic model for the disease.

Wilson Disease Patient History

I. Patient Information

Patient name and/or ID _____

Family name and/or ID _____

Date of birth _____ Date of death if deceased _____

Cause of death _____

Age at onset if symptomatic _____ Age at diagnosis _____

II. Clinical Findings (check one)	Yes	No	Don't Know	Age at Onset
A. *Liver involvement*				
cirrhosis	____	____	____	____
jaundice	____	____	____	____
vomiting	____	____	____	____
malaise	____	____	____	____
liver histology				
B. *Neurological manifestations*	____	____	____	____
dysarthria	____	____	____	____
involuntary movements	____	____	____	____
pseudobulbar palsy	____	____	____	____
abnormal posture	____	____	____	____
intellectual decline	____	____	____	____
psychiatric disorder				
C. *Bone and joint problems*	____	____	____	____
osteoporosis	____	____	____	____
osteomalacia	____	____	____	____
joint anomalies	____	____	____	____
ligamentous laxity				
D. *Renal symptoms*	____	____	____	____
renal stones	____	____	____	____
aminoaciduria	____	____	____	____
alkaline urine	____	____	____	____
acidosis				
poor growth	____	____	____	____
E. *Kayser-Fleischer rings*				

F. *Other (list findings and age at onset)*

_____

_____

continued

Figure 1.1.2 Sample patient history form.

Wilson Disease Patient History, *continued*

III. Laboratory Findings

	Test Done (Y/N)	Result	Date	Test Done Under Treatment (Y/N)
Serum Cu level	————	————	————	————
Serum Ce level	————	————	————	————
Liver enzymes (specify)	————	————	————	————

IV. Treatment History

	Drug Given (Y/N)	Date or Age at First Treatment	Date or Age Treatment Stopped
Penicillamine	————————	————————	————————
Zinc	————————	————————	————————
Trientine	————————	————————	————————
Other: ———————	————————	————————	————————

V. Copper Exposure

	Yes	No	Don't Know
A. Is drinking water running through copper pipes at: Home	————	————	————
Work	————	————	————

B. Frequency of eating food prepared with copper utensils:

———— 3 meals per day ———— 1-6 meals per week

———— 2 meals per day ———— 1-3 meals per month

———— 1 meal per day ———— <1 meal per month

don't know

VI. Notes _____

_____

_____

_____

_____

_____

_____

_____

_____

_____

_____

_____

_____

_____

_____

_____

_____

Figure 1.1.2 (*Continued*)

Medication History

Center: ☐☐☐   Family #: ☐☐☐☐☐   Individual #: ☐☐☐☐

Please list medications taken in the past	**Please list medications currently taken and duration**
Lipitor Premarin	Pravachol - 2 years

	Have you ever taken...	Currently taking? If "Yes", duration?
Hormones:		
Oral HRT	☐ Yes ☐ No ☐ Don't Know	☐Yes ☐No Years ☐☐ Months ☐☐
Oral Contraceptives	☐ Yes ☐ No ☐ Don't Know	☐Yes ☐No Years ☐☐ Months ☐☐
Thyroids	☐ Yes ☐ No ☐ Don't Know	☐Yes ☐No Years ☐☐ Months ☐☐
Medications to lower glucose:		
Insulin	☐ Yes ☐ No ☐ Don't Know	☐Yes ☐No Years ☐☐ Months ☐☐
Oral Agents	☐ Yes ☐ No ☐ Don't Know	☐Yes ☐No Years ☐☐ Months ☐☐
Medications to lower cholesterol:		
Statins	☐ Yes ☐ No ☐ Don't Know	☐Yes ☐No Years ☐☐ Months ☐☐
Fibrate	☐ Yes ☐ No ☐ Don't Know	☐Yes ☐No Years ☐☐ Months ☐☐
Resins	☐ Yes ☐ No ☐ Don't Know	☐Yes ☐No Years ☐☐ Months ☐☐
Medications for high blood pressure or heart disease:		
Diuretics	☐ Yes ☐ No ☐ Don't Know	☐Yes ☐No Years ☐☐ Months ☐☐
Beta Blockers	☐ Yes ☐ No ☐ Don't Know	☐Yes ☐No Years ☐☐ Months ☐☐
Other	☐ Yes ☐ No ☐ Don't Know	☐Yes ☐No Years ☐☐ Months ☐☐
Medications to reduce inflammation:		
Non-Steroidal Agents (e.g. Advil, ibuprofen, Motrin, Alleve)	☐ Yes ☐ No ☐ Don't Know	☐Yes ☐No Years ☐☐ Months ☐☐
Steroidal Agents (e.g. prednisone, cortisone)	☐ Yes ☐ No ☐ Don't Know	☐Yes ☐No Years ☐☐ Months ☐☐
Weight loss drugs:	☐ Yes ☐ No ☐ Don't Know	☐Yes ☐No Years ☐☐ Months ☐☐

14351

Figure 1.1.3 Example of an optical scanning form. Alphanumeric information printed in the small boxes and responses indicated by shading boxes are read by an optical scanner. This information is translated to a data stream and can be transmitted electronically to a centralized database. Some optical scanning software can also read and store free-form block-letter text, but this requires appropriate database archival structures. The square shaded blocks in the corners of the form are reference points from which the locations of the scannable information are mapped. Each scan form is recognized by a box with a unique shading pattern, as shown in the bottom right hand corner.

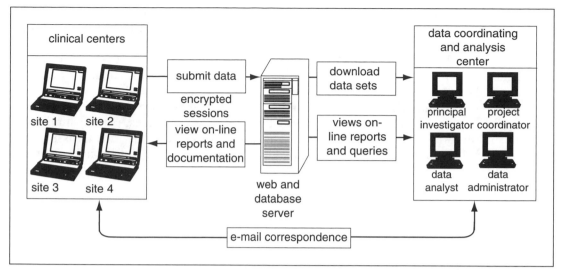

Figure 1.1.4 Sample design for collecting, transmitting, storing, and retrieving information using an electronic data management system. Questionnaire information is collected with some error checking and validation using electronic forms installed on laptop computers. Data are submitted electronically via the Internet to a database server and archived into a permanent database. These data can be accessed by personnel in the data coordinating center for further verification, generating report summaries, and analysis purposes. This scheme also permits staff involved in data collection to view individual data and reports using a Web browser.

5. Administer the questionnaire to family members and obtain a family history of the trait after the diagnosis is established for the proband. Select pedigrees that satisfy requirements for linkage study (*UNIT 1.2*).

6. Verify diagnosis of every family member participating in linkage studies (i.e., donating a blood sample) by consulting a diagnostician or skilled researcher. Do not rely on assessments from the patient or family informants, which are often inaccurate.

7. Verify information about young children, incompetent individuals, deceased persons, or other relatives unavailable for evaluation by consulting multiple reliable informants (e.g., close relatives) whenever possible.

8. Check the questionnaire information for consistency between informants and within the information provided by each individual informant. Follow up and correct discrepant observations.

Informatics

For studies involving a large number of subjects or data points, all study data should be collected through e-forms to minimize data collection errors. The quality of the data is much higher than with paper forms, since errors are caught at the source, rather than later, when the source may not be readily available. With paper forms, questions may be accidentally skipped and some answers may be illegible. Many issues regarding the tracking, shipping, and copying of paper forms are eliminated by e-forms. With paper forms, the information must be transcribed at least three times: first on the initial paper form, then in the first pass of data entry, and again in the second pass for validation.

With e-forms, data can be validated prior to submission with the use of drop-down list boxes, field entry–required checking, and numeric range checking. Automatic skip fields can be incorporated in the e-form to accelerate the interview. Data can be collected from multiple sites via a Web browser using secure socket layer (SSL) encryption. Electronic PDF forms on a laptop computer facilitate collection in the field, e.g., in the home of a study participant. The data can be submitted via the Web later. If coupled with a properly designed database management system, e-forms allow data conversion from online forms to standard formats for archiving in the database, for reporting, and for use with various statistical analysis programs. Since the data can be immediately available online in a central repository, immediate feedback can be given to research coordinators regarding their ascertainment rates, participation rates, and status of all forms for each study subject. This capability can quickly identify issues that need attention during data collection. A sample plan for collection and flow of information is shown in Figure 1.1.4.

Reference: Khoury et al., 1993

Contributor: Lindsay Farrer

UNIT 1.2

Pedigree Selection and Information Content

STRATEGIC APPROACH

Linkage analysis is a powerful tool for mapping genetic diseases and relies on detecting co-segregation of marker and disease genotypes as they are passed down through a family or through a set of families. Linkage analysis proceeds by estimating the recombination fraction between two loci, i.e., a marker locus and a disease gene. The recombination fraction is the probability that the alleles at two loci will appear in new combinations not seen in the parental generation. The further apart two loci are, the more likely they are to recombine during meiosis. If a marker and a gene are sufficiently close together on the same chromosome, then the original combination of the maternal or paternal alleles are more likely to be inherited together or co-segregate, and the loci are linked and the recombination fraction θ will be < 0.5. If loci are unlinked, then the maternal (or paternal) alleles at the two loci are inherited independently and all possible combinations of the maternal (or paternal) alleles at the two loci are seen with equal probability. In the unlinked case, $\theta = 0.5$.

There are two general approaches to linkage analysis. Nonparametric approaches, which include the affected-sib-pair method and affected-relative-pair method, use sharing of alleles among affected relatives to indicate whether two loci are linked or not linked. The idea is that two relatives affected with the same disease will share disease alleles identical by descent near the disease gene, more often than expected under the random segregation of alleles. Nonparametric methods do not require specification of a detailed genetic model, although the models may incorporate a parameter to measure genetic effect (λ). Traditional parametric approaches require full specification of the genetic model including disease allele frequency and penetrance matrix. The main implementation of this approach is the traditional lod score method (see *UNIT 1.4*), which has been used with tremendous success in the mapping of many single-gene disorders.

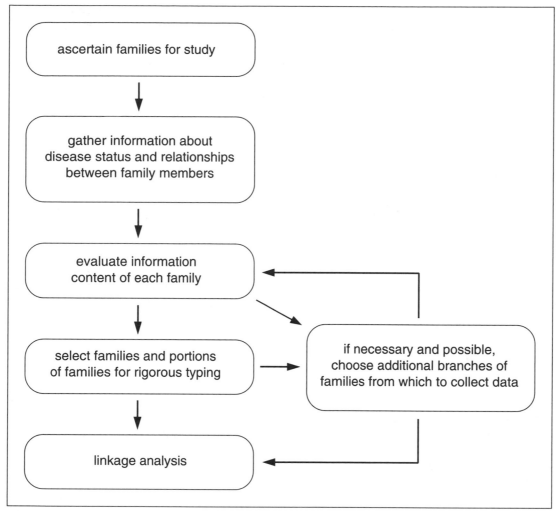

Figure 1.2.1 Flow chart showing initial steps in a linkage study.

The discussion of pedigree selection and information content begins by outlining steps in choosing families for study (Fig. 1.2.1). These steps are applicable to both nonparametric and parametric study designs.

1. Identify (i.e., ascertain) families segregating the disease gene according to the phenotype at only one locus (i.e., the disease locus) if linkage analysis is planned. Families are usually ascertained through an initial affected individual (the proband), and then data are gathered concerning disease status and the relationships of individuals to each other (see *UNIT 1.1*). A rigorously defined ascertainment scheme is required if segregation analysis is planned.

2. Compute some measure such as the expected lod score of the potential contribution of each family to the overall ability to detect linkage. It is prudent to establish that enough family data has been collected to detect linkage before extensive marker genotyping is begun. The best way to measure the potential information content of the families sampled may be via a carefully designed simulation study (see below).

3. Based on the potential information content, choose families and individuals who will be genotyped, i.e., sampled for DNA and rigorously diagnosed. To optimize the chances of detecting linkage, it may be helpful to type only a portion of each family for the marker

of interest, rather than type all members of every family that has been ascertained. Do not automatically include only large pedigrees or automatically exclude bilineal pedigrees (see below).

4. Choose branches of families from which to collect additional data, if necessary and possible. These decisions may be based on informal pedigree-expansion rules or simulation studies (see below). If segregation analysis is planned, more formal expansion rules must be followed. Simulation studies may be conducted to evaluate the effect of different expansions on the information content of the sample.

KEY CONCEPTS

Some questions to be asked in regard to collecting genetic information from families are: (1) given a single well-defined phenotype of interest, are there large families with many affected individuals that can be sampled; (2) are there mostly sibling pairs or other relative pairs; (3) are there primarily single affected members in the family; or (4) how many of each type of family can be collected? If I am only able to study a subset of the pedigrees I have identified, which ones should I study? If I wish to extend a pedigree, which branches should I study to maximize my chances of detecting linkage? As discussed below, the kind of families available for analysis dictates the kind of analyses that can be performed.

Ascertainment Bias

If testing whether or not a certain disease is recessive by determining if the proportion of affected offspring is 1/4, collecting data only from families with at least one affected child will result in a proportion that is much higher than 1/4, simply because the sampling procedure is biased toward selecting affected offspring. The collection of families is not representative of the general population, as it does not include any families without affected children. Such sampling (or ascertainment) bias can lead to incorrect conclusions if, for example, one attempts to determine the mode of inheritance, estimate penetrance, or create age-of-onset curves. If one is only interested in linkage analysis (i.e., estimating the recombination fraction), then biased ascertainment and sampling does not matter as long as the people have been ascertained solely on the basis of phenotype at the disease locus. However, if one also wishes to carry out careful and accurate segregation analysis to determine the mode of inheritance of the disease, then ascertainment and sampling must be carried out in a defined manner so that bias can be corrected.

Correctable ascertainment schemes include the following: (1) collection of nuclear families via an affected child; (2) collection of nuclear families via an affected parent; and (3) collection of nuclear families via several affected children.

Sampling bias may also be corrected for if pedigrees are collected via sequential sampling, which provides rules for extending a pedigree structure. A typical sequential sampling process is: (1) if the proband (*UNIT 1.1*) is affected, collect data on the proband and all spouses and first-degree relatives of the proband, (2) for every new affected individual sampled, sample all of that individual's spouses and first-degree relatives on whom data have not yet been collected, (3) repeat step 2 until no new affected individuals are found. Note that correct sequential sampling requires "rigid" rules by which new information is gathered only on the basis of information already gathered. According to above rules, only those portions of the pedigree connected through an affected individual may be sampled, and any indirect information heard about more distantly related affected persons must be disregarded.

General guidelines: (1) Linkage results will not be biased if people and families are ascertained and sampled according to the phenotype at only one locus. (2) Unbiased determination of the mode of inheritance (i.e., via segregation analysis) is possible only if the sampling method is precisely defined and implemented.

Information Content of Pedigrees

During a linkage study, an investigator may have to select a subset of several families for further detailed analysis. To do this effectively, the investigator must determine how much information for linkage is contained in each particular family.

Chromosomal phase

Consider an individual who has been typed at two marker loci closely linked on the same chromosome. If this individual is doubly heterozygous (1/2 at the first locus and A/B at the second locus), then that person could have a 1 and an A allele on the same chromosome or a 1 and a B allele on the same chromosome. In other words, the individual's multilocus genotype could be written as 1A/2B or 1B/2A. In linkage analysis based on recombination events, it is very important to know which of these "phases" is correct, because if a person is 1A/2B and generates a 1B gamete, that gamete is recombinant. However, if that same individual is 1B/2A and generates a 1B gamete, that gamete is nonrecombinant. Phase is considered known if it is possible to infer which chromosome carries which alleles at the two loci.

Phase information can come from two sources. First, it can be inferred from the genotypes of a person's parents. An example of this method of determining phase is presented in Figure 1.2.2, where the pedigree is segregating for a rare autosomal dominant disease with full penetrance. The disease allele is indicated by a "D" and the normal allele by a "d." The pedigree has been typed at a marker locus with 4 alleles (1, 2, 3, and 4). In this case, the phase of the mother (person 3) can be inferred to be D1/d3, because the D allele must have come from the grandfather (person 1) along with the 1 allele, and the d allele and the 3 allele must have come from the grandmother (person 2). As person 3 is a double heterozygote with known phase, it can easily be determined that person 5 must have received the d allele and the 1 allele from the mother, but person 6 must have received the D allele and the 1 allele from the mother. Therefore, person 5 received a recombinant gamete from the mother, but person 6 received a nonrecombinant gamete from the mother. Note also that person 4 is homozygous at the disease locus (d1/d4) and therefore uninformative for linkage.

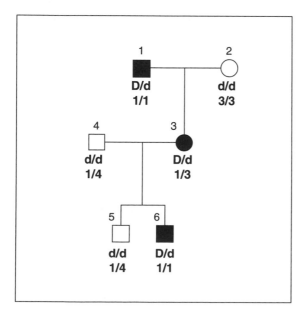

Figure 1.2.2 Three-generation family segregating for an autosomal dominant disease.

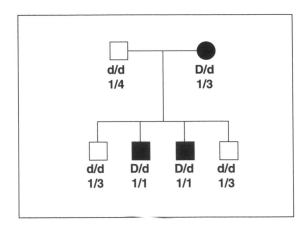

Figure 1.2.3 Nuclear family segregating for an autosomal dominant disease.

Phase can also be inferred indirectly from the genotypes of a person's children. Figure 1.2.3 illustrates another pedigree of the same loci in the figure above. If it is assumed that the marker is so closely linked to the disease that recombination between the two genes is improbable, then the most likely phase of the mother is D1/d3, because she transmits the D1 haplotype to her affected children and the d3 haplotype to her unaffected children without recombination. However, because the genotypes of the mother's parents are not known, nor is it known if the marker is as close to the disease locus as was assumed, the information about the mother's phase is indirect. In other words, knowing the children's genotypes does not eliminate any of the mother's possible phases, it only changes their probabilities.

A nuclear family consisting of two unrelated parents and one child contains no information for linkage (under the assumption of linkage equilibrium; see below) because one child is insufficient to provide phase information about the parents. For example, in the family in Figure 1.2.4, if the father is heterozygous at the disease locus, there is not enough information to determine if the father is D1/d4 or D4/d1. Therefore, this family will always give a lod score (UNIT 1.4) of zero—i.e., it contains no information about linkage. However, such a family with the addition of grandparents will contain linkage information, since phase will be known.

General guideline: More information can be obtained from those individuals for whom chromosomal phase can be inferred. Information concerning linkage comes mainly from doubly heterozygous individuals with known phase.

Information content of additional children
The information content of a certain family may be measured by the expected lod score (UNIT 1.4), and the lod score will increase for each additional child in the pedigree. Under

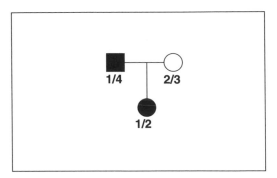

Figure 1.2.4 A nuclear family with one offspring contains no information for linkage.

optimal conditions and assuming no recombination between the disease locus and the marker locus, each additional child in a pedigree is worth ~1/10 of a significant lod score (i.e., ~0.30 lod score units) for a rare autosomal dominant disease. For a rare autosomal recessive disease, an additional affected child is worth ~0.60 lod score units and an additional unaffected child is worth ~0.13 lod score units.

Pedigree structure and information content

A three-generation pedigree in which all members are typed will usually contain more information about linkage than a comparable two-generation pedigree.

Nuclear families versus large pedigrees

A large multigenerational pedigree should contain much information for linkage, because many meioses will be segregating for the disease in a manner in which the phase is known. Additionally, large loaded pedigrees with affected individuals in several generations are more likely to be segregating for a dominant form of the disease. Thus, linkage analysis can be carried out under a dominant model, presumably without jeopardizing the results. However, large pedigrees are often ascertained in ways that preclude carrying out segregation analysis, while nuclear families can be easily ascertained to allow segregation analysis. Large pedigrees may represent a unique form of the disease, and may also be much more sensitive to misdiagnoses or genotyping errors. Furthermore, large pedigrees do not guarantee genetic homogeneity or the presence of a single major locus for the disease in question.

Pedigree extension

Parts of the pedigree in which the disease is not segregating will provide little or no additional information about linkage. Normal siblings or normal children of affected persons will typically provide some additional information, as they should have inherited the normal haplotype. However, the amount of information provided by the normal individuals will depend on several factors, including the penetrance and any age-of-onset functions of the disease in question.

Bilineal and unilineal pedigrees

A pedigree is considered bilineal if the same disease appears to be segregating in two sides of the pedigree. In other words, in a bilineal pedigree, the disease allele (or alleles, as a disease may be caused by different alleles at the same or different loci) must have entered into the pedigree at least two different times. In a unilineal pedigree, the disease appears to have entered the pedigree only once. Linkage studies for psychiatric disorders have often automatically excluded bilineal pedigrees, because of concerns about having two phenotypically similar yet genetically different diseases segregating in the same family. In addition, it has been postulated that a bilineal pedigree may be much less informative because of less information about phase and the origin of the disease allele. Simulation studies, however, have shown that the information loss because of bilineality is rather small if the bilineality is not too extensive, although there is a greater loss in phase-unknown nuclear families than in phase-known families. Bilineal pedigrees segregating two genetically distinct but phenotypically identical diseases also have little effect on the average maximum lod score and are unlikely to lead to false exclusion of linkage. Thus, bilineal pedigrees should not be automatically excluded from linkage studies. Extremely bilineal pedigrees in which both parents of the proband are affected will provide little linkage information.

Linkage Equilibrium

Consider two loci with two alleles each. If the first locus has alleles 1 and 2, and the second locus has alleles A and B, then there are four possible two-locus combinations (or haplotypes): 1A, 1B, 2A, and 2B. The two loci are in linkage equilibrium if the haplotypes occur in their expected frequencies, as predicted by the appropriate product of the individual allele frequencies. However, if, for example, haplotype 1B is much more frequent than expected, then the two loci are in linkage disequilibrium. Linkage disequilibrium can occur because of a new mutation, mixing of different populations, chance sampling in small populations, or selection of a particular haplotype. Most linkage analysis programs assume that linkage equilibrium holds, because the frequency of a particular haplotype is then simply the product of the appropriate allele frequencies.

Simulation Studies for Determining Sample Size and Power

In a simulation study, which is a way to understand the performance of a particular test of linkage, two quantities need to be estimated. The first is the size of the test, also known as the α level or the significance of the test, and is defined as the probability of obtaining a significant test statistic under the null hypothesis, i.e., the disease and marker loci are not linked. In some instances test size can be determined based on theoretical considerations. Another way to estimate the size of a test is through simulation of several thousand replicates of the data set in which there is no linkage between the disease and the marker (also called unconditional simulation). This amounts to segregating marker alleles through the pedigree without regard to the disease status of the individuals. Each replicate is analyzed using the test of linkage, and the proportion of replicates above a certain "critical" value of the test statistic is calculated. Obviously, the size of the test depends on the critical value chosen. A smaller critical value will result in a larger test size, i.e., more replicates above the critical value, and a larger critical value will result in a smaller test size, i.e., fewer replicates will be above the critical value.

The second quantity to estimate is the power of the study. The power is defined as the proportion of replicates giving a test result larger than the critical value under a specified model of linkage between disease and marker. Determination of power requires specification of a model for the relationship between a disease and a marker (also see *UNIT 1.5*). This model may be more or less completely specified depending on the type of analysis to be performed. Determination of power also requires specification of a critical value for the test of linkage, and this critical value is often chosen to provide a set test size, e.g., $\alpha = 0.05$. Again, for certain cases it may be possible to determine power theoretically. More often, power studies are performed using simulation. The estimation of power proceeds as before by simulating marker alleles through the pedigree according to the model of disease and the disease phenotypes of the family members (also known as conditional simulation, since the markers are simulated conditionally on the model of disease and the observed disease phenotypes in the family members). Power is a particularly important factor in study design because it can advise whether a given sample size is likely to result in a significant result based on a particular model for disease. If the power is high, then the sample is adequate to identify a disease locus if the model for disease is correct. If power is low for reasonable models of disease, then one must consider increasing the sample size to give the study the highest probability of reaching a conclusion.

References: Davis and Weeks, 1997; Elston et al., 1996; Martin et al., 2000; Ott, 1999

Contributors: Elizabeth R. Hauser and Daniel E. Weeks

Single-Sperm Typing

Single-sperm typing refers to the analysis of DNA sequences in individual male gametes. It allows recombination and other genetic phenomena to be examined in a single individual. One advantage of sperm typing is seen in genetic studies where a large sample size is required, as the number of sperm in a semen sample is, for all practical purposes, unlimited. Specific situations where sperm typing might be most useful include: ordering of polymorphic loci <1 to 2 cM apart (assuming reliable physical maps are not available), defining haplotypes ("setting phase") where parental information is not available, detailed examination of chromosome regions where greatly increased or decreased rates of recombination are suspected, examination of variation in recombination between individuals, measuring germ-line mutation frequencies, and analyzing segregation distortion.

NOTE: Experiments involving PCR require extremely careful technique to prevent contamination. Use of barrier pipet tips and pipets dedicated to PCR setup only is recommended.

BASIC PROTOCOL 1

PCR AMPLIFICATION OF GENETIC MARKERS FROM SINGLE SPERM CELLS

Materials (see APPENDIX 1 for items with ✓)

    Lysed sperm isolated from agarose film (Support Protocol 1) or by flow cytometry (Support Protocol 2), *or* PEP product (see Support Protocol 4)
✓ Neutralization buffer
✓ 10× amplification buffer with and without potassium
✓ 2 mM 4dNTP mix
    100 μM first-round primers (see Support Protocol 3; store at −20°C): forward and reverse
    5 U/μl AmpliTaq DNA polymerase (Perkin-Elmer, Promega)
    Light mineral oil
    100 μM second-round primers (see Support Protocol 3; store at −20°C): hemi-nested or full nested (if hemi-nesting, then either the forward or reverse primer used in the first round will also be used, depending on the orientation of the nested primer)
    Thermal cycler accommodating samples in a 96-well format (e.g., PTC-200; MJ Research) and appropriate plates or tubes

NOTE: All solutions should be prepared using ultrapure water.

1a. *For single sperm obtained from agarose films:* Prepare PCR solution for first round by adding the following reagents to each PCR tube or plate well containing sperm suspended in 5.0 μl lysis solution:

    5.0 μl neutralization buffer
    5.0 μl 10× potassium-free amplification buffer
    2.5 μl 2 mM 4dNTP mix
    2.0 μl 1 pmol/μl each primer diluted from 100 μM stock
    0.1 μl 5 U/μl AmpliTaq DNA polymerase
    Bring total volume to 50 μl with H_2O.

1b. *For lysed and neutralized single sperm obtained by flow cytometry:* Prepare reaction as described in step 1a, but replace neutralization buffer with water.

1c. *For PEP product:* Transfer a 2- to 5-μl aliquot (or other volume as determined experimentally) of each PEP reaction product to an individual plate well or tube containing PCR solution. Prepare PCR solution as in step 1a, but use 10× amplification buffer with potassium and replace neutralization buffer with water.

2. Overlay each well with light mineral oil if the thermal cycler used does not have a heated cover to minimize evaporation. Carry out first-round PCR using the following amplification cycles (optimized for multiplex PCR of as many as 16 different loci, i.e., 32 primers):

Initial step:	3 min	94°C	(denaturation)
1 cycle:	2 min	55°C	(annealing)
	1 min	72°C	(extension)
4 cycles:	15 sec	95°C	(denaturation)
	1 min	55°C	(annealing)
	1 min	72°C	(extension)
35 cycles:	15 sec	95°C	(denaturation)
	30 sec	55°C	(annealing)
	30 sec	72°C	(extension)
Final step:	indefinite	4°C	(hold).

3. Prepare PCR solution for second round:

 2.5 μl 10× potassium-containing amplification buffer
 2.0 μl 2 mM 4dNTPs mix
 1.0 μl 10.0 pmol/μl each second-round primer diluted from 100 μM stock
 0.1 μl 5 U/μl AmpliTaq DNA polymerase
 Adjust total volume with H_2O such that when the first-round PCR product (1 to 2 μl) is added, the total volume is 25 μl.

4. Add sufficient second-round PCR solution to each well of a fresh 96-well PCR plate to bring total volume to 25 μl. Transfer 1 to 2 μl of each first-round PCR product to an individual well containing PCR solution on the 96-well plate. If necessary, overlay each well with light mineral oil. Carry out second-round PCR using the following amplification cycles (annealing temperature/time may vary for different combinations of specific primer pairs):

Initial step:	3 min	94°C	(denaturation)
35 cycles:	15 sec	95°C	(denaturation)
	30 sec	55°C	(annealing)
	30 sec	72°C	(extension)
Final step:	indefinite	4°C	(hold).

If desired, store samples indefinitely at 4°C (until analysis is performed).

5. Determine the allelic state at SSLP or SNP markers using available methods (e.g., nondenaturing PAGE).

SUPPORT PROTOCOL 1

ISOLATION OF SINGLE SPERM CELLS FROM AGAROSE FILMS

Materials (see APPENDIX 1 for items with ✓)
 Semen
 0.5% (w/v) low-melting-point agarose

✓ Cell lysis solution
✓ Neutralization buffer (optional)

Minigel casting and electrophoresis apparatus (e.g., Hoefer 200 series)
Inverted phase-contrast microscope with $100\times$ objective
Paragon no. 11 scalpel blade (Maersk Medical)
PCR tube
Stereomicroscope with $4.6\times$ objective
65°C water bath or thermal cycler

NOTE: To reduce contamination, as much work as possible should be carried out in a laminar-flow hood. Standard sterile procedures should be used.

1. Mix 2 to 10 μl semen (depending on the sperm concentration) with 20 ml of 0.5% (w/v) low-melting-point agarose. Pour 2 ml onto a 10×10–cm glass plate from a minigel apparatus, and gently move the plate to produce an agarose thickness of 1 to 2 mm. If necessary, optimize sperm concentration to facilitate the efficiency of picking single sperm.

2. Let the agarose dry 20 min at 37°C or until it takes on a "sticky" quality.

3. Place the glass plate under an inverted phase-contrast microscope ($100\times$ power). Holding a Paragon no. 11 scalpel blade at an angle of 30° to 40° relative to the horizontal axis, isolate a piece of agarose containing a single sperm. Select sperm that are close to the edge of the agarose (where the agarose has not completely dried out) and cut the agarose piece to $\sim$10$\times$ the diameter of the sperm.

4. Transfer the piece of agarose to a PCR tube that contains 5 μl cell lysis solution and is fixed under a stereomicroscope ($4.6\times$ power). View under the microscope to make sure that the agarose piece is deposited into the lysis solution.

   Confirming that the piece of agarose was deposited in the well of a 96-well plate is more difficult than confirming deposition in a single PCR tube.

5. Incubate sperm in cell lysis solution 10 to 15 min at 65°C. Use the lysate immediately in the first round of PCR (see Basic Protocol 1) or add 5 μl neutralization buffer, seal tubes with tops, and store at −20°C until ready to use.

SUPPORT PROTOCOL 2

ISOLATION OF LYSED SINGLE SPERM CELLS BY FLOW CYTOMETRY

Materials (*see* APPENDIX 1 *for items with* ✓)
Semen from genotyped donor
5 μg/ml Hoechst 33342
70% (w/v) sucrose
Hoechst 33342-conjugated fluorescent beads (Flow Cytometry Standards)
10% (v/v) bleach (sodium hypochlorite; e.g., Clorox)
✓ Cell lysis solution
✓ Neutralization buffer

Biosonik III sonicator with 4-mm probe (Brownwill Scientific)
Phase-contrast and fluorescence microscopes
FACStar Plus fluorescence-activated cell sorter (Becton Dickinson)
Flexible 96-well microtiter plate (e.g., Falcon) *or* 96-tube system (e.g., GeneAmp PCR System 9600, Perkin-Elmer Cetus)
65°C water bath or thermal cycler compatible with microtiter plates or 96-tube system

NOTE: All solutions should be prepared using ultrapure water.

NOTE: To reduce contamination, as much work as possible should be carried out in a laminar-flow hood. Standard sterile procedures should be used.

1. Mix 100 μl semen from a genotyped donor with 500 μl water in a 1.5-ml microcentrifuge tube. Pellet sperm by centrifuging 1 min at 833 × *g*, room temperature. Discard supernatant and resuspend sperm in 500 μl water.

2. Sonicate suspension 1 min using a Biosonik III sonicator with 4-mm probe at minimum intensity (or other intensity determined for other sonicators) to remove sperm tails.

3. Pellet sperm by centrifuging 1 min at 833 × *g*, room temperature. Discard supernatant and resuspend sonicated sperm in 500 μl water.

4. Examine a small aliquot of sperm under a phase-contrast microscope to determine completeness of tail removal. Repeat steps 2 to 3 as necessary.

   Short tails (i.e., one-quarter the length of the head) pose no problems. Longer tails are thought to clog the nozzle of the cell sorter.

5. Centrifuge sperm 1 min at 833 × *g*, room temperature. Discard supernatant. Resuspend pellet in 100 μl of 5 μg/ml Hoechst 33342. Incubate 1 to 24 hr at 37°C. Decrease staining time when the sperm sample is small or to limit propagation of bacteria in contaminated samples. Increase staining time to improve the fluorescence signal.

6. Centrifuge sperm 1 min at 833 × *g*, room temperature. Discard supernatant and resuspend pellet in 200 μl water. Sonicate 5 sec at the same intensity as in step 2 to reduce clumping.

7. Add 200 μl of 70% (w/v) sucrose and either use immediately or store up to 1 year (or more) at −20°C until ready to proceed with sorting (step 10).

8. Align FACStar Plus fluorescence-activated cell sorter to maximize fluorescence signals using Hoechst 33342–conjugated fluorescent beads. Align fluidics settings to give one bead per sorted drop.

9. Confirm alignment of machine and fluidics settings by sorting several rows of single fluorescent beads onto a microscope slide. Under a fluorescence microscope, check that the sorting gives one bead per drop. If not, readjust fluidics settings.

10. Flush beads from machine using water or 10% (v/v) bleach followed by water until no fluorescent signal remains. Ensure that bleach is completely removed.

11. When ready to sort, dilute a portion of the stained sperm sample 100-fold with water (or adjust the flow rate so that the number of sperm that fit the sort window criteria is also ~1 cell/sec). Apply the sperm sample and reset the sort window, using side scatter on the horizontal axis and fluorescence on the vertical axis, such that the main single-sperm population is contained within the sort window and clumps of two or more sperm are excluded.

12. Sort several rows of single sperm cells onto a microscope slide. Using a fluorescence microscope, check that the sorting gives one sperm cell per sorted drop. During the sort, regularly repeat this step to confirm that the fluidics have remained steady and that single cells are still being deposited.

13. Working under a laminar flow hood, dispense 5 μl freshly mixed cell lysis solution into each well of a flexible 96-well microtiter plate or into tubes of a 96-tube system. Sort single sperm into individual solution-containing wells or tubes. Set up three wells or tubes with

20 sperm apiece and eight wells or tubes with no sperm as positive and negative controls, respectively.

If many sperm samples are to be sorted, extensive flushing of the fluidics system (as in step 10) prior to the application of each new sample is critical to reduce sample cross-contamination.

14. When finished, flush machine of all sample using 10% bleach followed by ≥5 ml water.

15. Incubate sorted sperm in the 5 μl cell lysis solution (see step 13) 10 to 15 min in a 65°C water bath or thermal cycler compatible with microtiter plates or 96-tube system. Add 5 μl neutralization buffer, seal microtiter plates with Parafilm, and store up to several months at −20°C until ready to use.

SUPPORT PROTOCOL 3

MULTIPLEX AMPLIFICATION FROM SINGLE SPERM CELLS

Sperm typing is dependent on an efficient amplification from one target molecule of DNA, and normally two rounds of PCR are needed in order to yield enough specific product for detection by ethidium bromide staining after electrophoresis. In order to achieve an overall high efficiency using the multiplex amplification approach, it is important to design primers with similar annealing temperatures and also avoid primers susceptible to primer-dimer formation with its own partner or with primers for the other loci. When large numbers of loci are to be considered, this latter task becomes quite complicated, so the authors only search for primer-dimer potential among partner-primers at a single locus. Using this approach, up to 16 loci have been successfully amplified with an amplification efficiency of more than 90% for all loci. Other factors that seem to improve results in the multiplex amplification from single cells are low primer concentrations and expanded annealing and extension steps in the first cycles of the PCR (see Basic Protocol 1 for first-round PCR). Another strategy for multiplexing involves using locus-specific primers with 5′ universal tails.

SUPPORT PROTOCOL 4

WHOLE-GENOME AMPLIFICATION OF DNA FROM SINGLE CELLS BY PRIMER-EXTENSION PREAMPLIFICATION (PEP)

Primer-extension preamplification (PEP) is a method of whole-genome amplification that can be applied to a single sperm.

Additional Materials *(also see Basic Protocol 1)*

    400 mM Poly N random 15-base primers (Operon)
    Sorted, lysed, and neutralized single sperm cells frozen in a 96-well microtiter plate (see Support Protocol 1)

NOTE: All solutions should be prepared using ultrapure water.

1. Prepare the following PEP solution (for 100 reactions):

    500 μl 10× potassium-free amplification buffer (1× final)
    500 μl 400 mM Poly N random 15-base primers (40 μM final)
    250 μl 2 mM 4dNTP mix (100 μM final each)
    100 μl 5 U/μl AmpliTaq DNA polymerase (0.1 U/ μl final)
    Adjust volume to 4 ml with H_2O.

2. Thaw microtiter plate containing lysed, neutralized single sperm cells and controls. Add 40 μl PEP reaction solution to each well containing a sperm sample or control. Mix well. Overlay each well with light mineral oil unless the thermal cycler used has a heated cover. Carry out PEP using the following amplification cycles:

Initial step:	5 min	95°C	(denaturation)
50 cycles:	2 min	37°C	(annealing)
	10 sec/°C	37° to 55°C ramp	
	4 min	55°C	(extension)
	1 min	92°C	(denaturation)
Final step:	5 min	72°C	(extension).

3. Store 96-well microtiter plate containing preamplified samples at 4°C (stable indefinitely).

BASIC PROTOCOL 2

SPERM-TYPING DATA ANALYSIS

Sample-Size Considerations

A map distance of 0.5 cM translates into a recombination fraction of 0.005, indicating that 1000 scoreable meioses would be needed to observe, on average, five recombination events. Based on the Poisson distribution (assuming no interference and independence of recombination events), in the above example there is an ~56% chance of observing at least five events and an ~88% chance of observing at least three events.

Maximum-Likelihood Estimates for Recombination Fractions and Gene Order

Two- and three-point mapping
Mathematical approaches to analysis of raw sperm-typing data in two- and three-point crosses are provided by computer programs (TWOLOC and THREELOC) that allow complete statistical analysis of the sperm-typing data.

If PCR amplification were absolutely efficient (i.e., every allele present in a sample could be detected), with no contamination by exogenous DNA, and exactly one haploid sperm per microtiter well, then only the expected parental and recombinant sperm types would be observed; however, like all experimental data, the data generated by the sperm-typing procedure are subject to error. For any locus, a small fraction of sperm cells may appear to contain none of the alleles or both of them. This can be attributed to <100% efficiency in detecting an allele present in a single sperm, contamination events, samples containing multiple or no sperm, and various combinations of these errors. The TWOLOC and THREELOC programs make use of a computer model that takes these errors into consideration. The programs provide estimates and their standard errors for four parameters: efficiency, contamination, the recombination fraction, and the fractions of samples containing zero, one, or two sperm. In addition, THREELOC provides an estimate of the coincidence coefficient specifying degree of independence between recombination events in adjacent intervals, thereby analyzing interference. Input for both programs is in the form of a table containing all observed sperm haplotypes and the number of times each haplotype was found. The programs allow analysis under a general model as well as several submodels that differ in their assumptions about the efficiency and contamination parameters. In both TWOLOC and THREELOC, the maximum likelihood for different phase/order combinations are calculated. The TWOLOC and THREELOC programs are limited in that no more than three loci can be considered at one time. They are also limited in that the data from each individual sperm donor must be analyzed separately.

All computer programs and documentation discussed here may be requested from Dr. Ken Lange, Department of Biostatistics, University of California at Los Angeles.

Multipoint mapping

Strategies for multipoint mapping of sperm typing data were developed by adapting the MENDEL program originally developed for genetic linkage analysis of human pedigree data. Like TWOLOC and THREELOC, this new version takes experimental errors into account. It estimates the same four parameters (with their standard errors) that are estimated by TWOLOC and THREELOC. In addition, it estimates marker order and male recombination fractions. The genotype of the father is taken to be the donor's actual genotype with phase and order specified. A hypothetical mother, heterozygous for all the loci being analyzed, is included to conform to the program's expectations, and each sperm cell analyzed is treated as a child. The genotype of the sperm (or child) consists of the allele combination received from the sperm donor. The genotype of the mother and the phenotypes of both parents are regarded as irrelevant. This adaptation of MENDEL makes it possible to combine multiple loci analyzed from several different individuals to construct a multipoint linkage map, even if only a subset of loci has been successfully typed for any one sperm donor. Given the computational complexity of the data when typing many loci, this program always computes likelihoods conditionally, based upon a specified phase/order combination, and does not take interference into account. It allows analysis of data combined from donors typed at different sets of loci. More complicated experimental designs, involving retesting of certain loci from PEP aliquots, are incorporated into the model. These aspects are discussed more fully in the MENDEL documentation. Input for the program is in the form of a table that includes, with each set of microtiter-well (or tube) outcomes, the donor haplotype and compatible haplotypes for an artificial mother.

Strategic approach for analysis of sperm-typing data

1. Pay careful attention to haplotype scoring, data entry, and data checking. As scoring of haplotypes can be tedious, independent scoring of data is recommended; any discrepancies should be carefully studied. In ambiguous cases, consider retyping sperm using PEP aliquots (see Support Protocol 4).

2. Choose a program appropriate to the data set, which is mainly dependent on the number of loci being typed and the experimental design.

3. Using TWOLOC, THREELOC, or MENDEL, test several different submodels derived from the general model by imposing various constraints on the contamination and efficiency parameters. The most accurate model (i.e., the general model) accounts for variable contamination rates and variable amplification efficiencies for each of the alleles at each locus; however, according to a general principle of statistical modeling, it is preferable not to "over-parameterize" a specific model as this might lead to an artificially good fit of the data to that model. Thus, it is best to have the simplest model possible that nearly completely explains the data.

It is clear from the authors' experiments and data analysis that high-resolution measurements of recombination frequency depend upon three factors: high efficiency of amplifying and detecting an allele if present in a microtiter well or tube; low frequency of contamination; and low frequency of microtiter wells or tubes with zero or more than one sperm. Loss of information from inevitable experimental errors can be offset by increasing the number of samples studied.

Direct Analysis of Linkage Data

Under circumstances where the order of a group of tightly linked markers is known, but the frequency of recombination within different intervals is to be compared, there is no need to estimate recombination fractions and associated sperm-typing errors using TWOLOC, THREE-LOC, or MENDEL. Instead, the number of recombinants within different intervals can be directly counted and compared. Consider an individual heterozygous for four tightly linked markers A, B, C, and D with a known order and phase ABCD/abcd and where recombination within the BC interval is in question. Assume that markers B and C are typed from a PEP aliquot and a recombinant sperm Bc is detected. This could be due to either a true recombination event or an error in typing. These possibilities can be distinguished if additional PEP aliquots are used to type markers A and D in the same sperm. If the sperm is found to have the genotype ABcd, then the recombination event within the BC interval is considered confirmed, while an ABcD would be nonconfirmatory. The number of confirmed recombinants within any interval can be counted and compared to the number of confirmed recombinants in other intervals. In this way, for example, a chromosome segment with a recombination hot spot can be sequentially divided into smaller and smaller intervals until the smallest interval with hot-spot activity can be defined.

BASIC PROTOCOL 3

SEGREGATION DISTORTION BASED ON HAPLOTYPE TRANSMISSION

Single sperm cell analysis is well suited for studying the paternal transmission of haplotypes, especially in cases where a preferential transmission of disease haplotypes over normal ones is suspected. The computer program Spermseg (*http://galton.uchicago.edu/~mcpeek/software/spermseg*) has been especially developed to analyze the segregation of haplotypes of single sperm cells. The core input information required in such cases is a count of all possible two-marker haplotypes and recombination rates between the markers as well as between each marker and the disease gene considered. One-marker haplotype counts may also be included, but are not necessary. Note that only markers for which a donor is heterozygous are to be included, since homozygous markers do not provide information on haplotype segregation. The program allows the marker sets used to differ between donors and even within a donor, but only two markers can be considered for a given sperm cell. The program output provides estimates of several parameters related to haplotype segregation and marker amplification efficiency. The specific models chosen, together with the confidence intervals, maximum log likelihood values corresponding to the analyzed model, and the value of the model goodness-of-fit test with corresponding simulation based *P* values are also given; however, the statistical comparison of different models is to be done outside the program, based on provided log likelihoods.

References: Arnheim and Shibata, 1997; Goradia and Lange, 1990

Contributors: Sigbjorn Lien, Esther P. Leeflang, Rene Hubert, Lin Zhang, Karin Schmitt, and Norman Arnheim

UNIT 1.4

Use of LINKAGE Programs for Linkage Analysis

This unit provides a basic introduction to human genetic linkage analysis for cases in which the genetic model (see *UNIT 1.5* for definition) is known.

KEY CONCEPTS

Lod Score

The lod (logarithm of the odds of linkage) score is also colloquially called a z score. A *two-point* lod score is calculated as the $\log_{10}$ of the ratio of the likelihood of a pedigree assuming that two loci are linked to one another at some value of θ (the recombination fraction between the two loci) to the likelihood of the pedigree assuming that the two loci are unlinked to one another (i.e., $\theta = 0.50$):

$$\text{two-point lod score} = z(\theta) = \log_{10} \frac{L(\text{Ped} \mid \theta = x)}{L(\text{Ped} \mid \theta = 0.50)}$$

where L is the likelihood and x is some value of θ ranging between 0.0 and 0.49. Two-point lod scores under a given genetic model are interpreted as follows: a lod score, $z(\theta)$, of 3.0 or greater is considered significant evidence in favor of linkage, and a lod score of -2.0 or lower is considered significant evidence in favor of nonlinkage. Any lod score that falls between -2.0 and 3.0 is inconclusive, and additional data—i.e., from more families or more individuals in the available families or increasing the informativeness of the marker locus—is required before definitive interpretation is possible.

A lod score of 3.0 is equivalent to odds of 10^3 (1000:1) in favor of linkage. However, this result is *not* equivalent to saying that the chance of an incorrect result (a false positive linkage) is 1/1000. After taking into consideration the prior probability of any two loci being linked (if the loci are "genetic" they have to lie somewhere on a chromosome and thus there is a "prior probability" that they will be close enough on the same chromosome to be linked to one another), the empiric rate of false positive evidence for linkage with a lod score of 3.0 is ~1/20 or 5%.

Two-point linkage analysis can be extended to n-point linkage analysis, allowing the simultaneous consideration of multiple loci. A multipoint lod score is analogous to a two-point lod score. The difference is that while a two-point lod score compares the recombination fraction between locus A and locus B, a multipoint lod score compares the recombination fraction between locus A and a fixed map of additional loci (B, C, ..., Z). For example, assume that the fixed map of loci includes three markers whose order is as shown in Figure 1.4.1. The investigator has a disease pedigree and wishes to determine if the disease locus falls anywhere within this map, lies on either side of locus A or locus C, or is entirely unlinked to this set of markers. If the disease locus is linked to this region, there are four possible locations numbered 0 to 3, as illustrated.

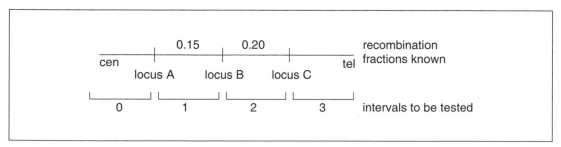

Figure 1.4.1 Genetic map showing recombination frequencies between markers A and B and between markers B and C, along with designation of each of the four intervals to be tested in a multipoint linkage analysis.

Calculation of the multipoint lod score allows assessment of these four possible locations. The multipoint lod score for disease locus positions outside the fixed map of markers is calculated as follows:

$$\text{multipoint lod score} = \log_{10} \frac{L(\text{Ped} \mid \theta_{A-B}, \theta_{B-C}, \theta_{\text{disease}=x})}{L(\text{Ped} \mid \theta_{A-B}, \theta_{B-C}, \theta_{\text{disease}=0.50})}$$

where x is the recombination fraction between the fixed map of markers and the disease locus. The denominator assumes that the disease locus is unlinked to the fixed map of markers, just as in the two-point lod score it is assumed that the two loci are unlinked to one another.

For instance, to calculate a multipoint lod score for location 0, the user might be interested in determining whether the disease locus is 5 cM proximal to locus A in the fixed map. The correct formulation of the multipoint lod score in this case is:

$$\text{multipoint lod score} = \log_{10} \frac{L(\text{Ped} \mid \theta_{\text{disease}-A=0.05}, \theta_{A-B=0.15}, \theta_{B-C=0.20})}{L(\text{Ped} \mid \theta_{\text{disease}-A=0.05}, \theta_{A-B=0.15}, \theta_{B-C=0.20})}$$

The term *location score* is often used in the literature synonymously with multipoint lod score, but there is a difference. A location score may be reported as the natural logarithm of the likelihood ratio rather than the $\log_{10}$ ratio. Natural logarithm scores (location scores) can be converted to $\log_{10}$ scores (multipoint lod scores) by dividing by 4.6 (e.g., a location score of 30.5 is equal to multipoint lod score of 6.63).

Types of Loci

The LINKAGE programs allow consideration of four types of loci for analysis. The types of loci are quantitative, codominant, binary factors, and affection status. Binary factors loci allow specification of markers that are not codominant (such as those of the ABO blood group). Quantitative trait loci describe continuous as opposed to discrete phenotypes. Examples of quantitative trait loci are creatine kinase and serum phosphorus levels. Codominant loci are loci for which each of the alleles is phenotypically discernible, such as microsatellite markers and single nucleotide polymorphisms (SNPs). Affection status loci allow specification of whether pedigree individuals are affected or unaffected with a trait, or fall at some other point in the spectrum such as possibly affected or probably affected. These variations in phenotype are described by assigning a series of liability classes.

Liability Classes

The LINKAGE programs allow for the incorporation of penetrance data for affection status loci. The penetrance values are coded in a line of three values (a liability class), with a separate penetrance indicated for each of three genotypes for the affection status. LINKAGE allows the specification of multiple liability classes for an affection status locus.

Loops

A *loop* is said to be present in a pedigree structure when an unbroken line can be drawn starting from some individual, X, in a pedigree, traversing through connecting relatives, and returning to individual X without retracing any steps. There are two types of loops: marriage loops and consanguinity loops. Consanguinity loops occur as a result of inbreeding; marriage loops occur when, for example, brothers in one pedigree have children with sisters in another pedigree. Both types of loops lead to what are termed "complex" pedigree structures. In the LINKAGE programs, loops must be broken by *doubling* individuals in a pedigree—for more information, consult the LINKAGE documentation and Terwilliger and Ott (1994). Processing of a pedigree by the LINKAGE utility program MAKEPED will detect loops and alert the user to the presence of undeclared (unbroken) loops.

Map Function

A *map function* allows conversion between genetic distances in centimorgans (cM) and recombination frequencies, under the assumption of some level of genetic interference. Positive interference is a phenomenon whereby the presence of one cross-over in an interval inhibits the formation of another cross-over in the same region; negative interference implies that the presence of one cross-over actually encourages the formation of another cross-over event within an interval. The two most frequently assumed levels of interference in human linkage analysis have been the Haldane level, which assumes no interference, and the Kosambi level, which assumes an intermediate level of interference whereby double cross-overs are rare at small genetic distance but their probability increases at larger distances.

The linkage programs MLINK and LINKMAP both require the user to input data for the recombination fraction between loci. If only map distances in cM are available, the LINKAGE utility program MAPFUN can assist the user in correctly converting between map distances and recombination frequencies.

PROGRAMS FOR LINKAGE ANALYSIS

The LINKAGE programs are most commonly used for linkage analysis. The algorithm for using these programs is illustrated in Figure 1.4.2. The programs are available free of charge from *http://linkage.rockefeller.edu*. Under the "linkage analysis programs" heading, a list of available programs can be obtained along with detailed information on downloading the programs and obtaining program documentation.

MAKEPED

MAKEPED is a batch program that will convert information on pedigree structure, affection status, and marker genotypes from a simple format, generally called a "pre-MAKEPED" or simply a "pre-" file, into a format appropriate for linkage analysis by the LINKAGE programs. MAKEPED will check the pedigree input file for various structural inconsistencies in the pedigree (e.g., two parents of the same gender), add pedigree pointers that direct the LINKAGE program's traversal of the pedigree, and add a proband field (see Example 1).

PREPLINK

PREPLINK (*http://linkage.rockefeller.edu/gui/webpreplink.html*) is a program that will prepare a parameter file for linkage analysis. PREPLINK will assist the user in ensuring the appropriate format for this parameter file by prompting the user for variables specific to the genetic model (i.e., penetrance, allele frequencies) and to the requested analysis (e.g., initial values of the recombination fraction).

LCP

The LINKAGE Control Program (LCP) prepares a batch file that will invoke the analysis programs in their appropriate order. The output from the MAKEPED and PREPLINK programs— the pedigree and parameter files, respectively—are necessary input for LCP. LCP allows the user to specify that multiple different analyses be performed.

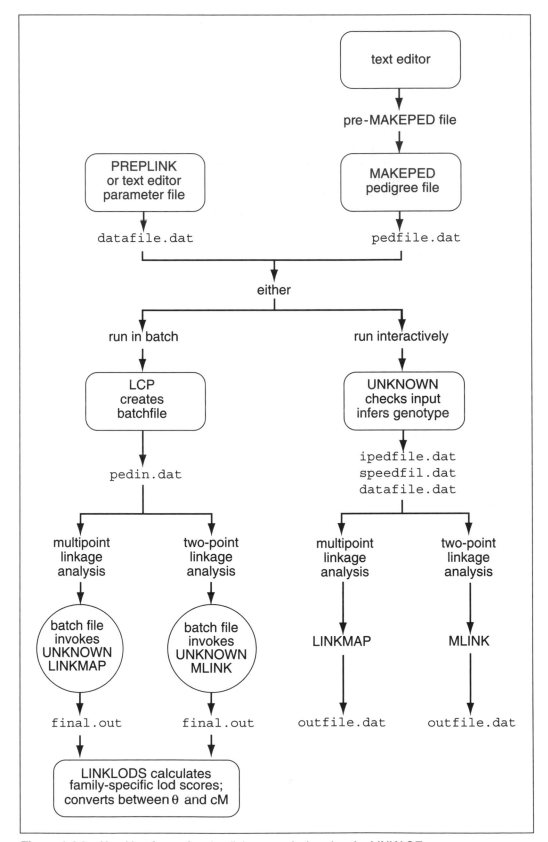

Figure 1.4.2 Algorithm for performing linkage analysis using the LINKAGE programs.

UNKNOWN

UNKNOWN must be run before invoking any of the LINKAGE programs. First, it checks the input pedigree file for data inconsistencies and alerts the user if inconsistencies are found. Second, UNKNOWN looks at the pedigree and deletes any impossible genotypes from consideration in the likelihood calculations for individuals whose disease or marker status is unknown, thereby increasing the speed of the linkage analysis.

MLINK

MLINK calculates the likelihood of the pedigree under user-specified values of θ for two loci; these likelihoods can be used to calculate lod scores. Typically, analyses performed with MLINK involve a disease or trait locus and a marker locus, although two marker loci can also be used. Analyses performed by MLINK are frequently called two-point analyses. MLINK is the only program that calculates the genetic risks for user-specified individuals.

LINKMAP

Often, an investigator will want to know where a marker or disease trait maps relative to a series of markers whose order and spacing are known. LINKMAP performs these calculations by allowing the user to fix the order and spacing of two or more markers. Likelihoods of the pedigree data are then calculated, allowing the locus whose position is unknown to be placed in all possible intervals.

LINKLODS

LINKLODS requires as input the output file from either an MLINK or LINKMAP analysis and will calculate family-specific lod scores from that file. LINKLODS will also perform the necessary subtractions to obtain a multipoint lod score from the output of an analysis performed with LINKMAP. Additionally, LINKLODS will convert θ values specified in the analysis to genetic distance, in morgans, assuming either Kosambi or Haldane levels of interference.

HOMOG

The HOMOG program package is used to investigate whether there is significant evidence in favor of genetic heterogeneity within a series of pedigrees; in other words, HOMOG will help to answer the question, "Is there more than one genetic locus responsible for the disease phenotype in these pedigrees?"

STRATEGIC APPROACH AND PRACTICAL EXAMPLES

Use of the LINKAGE programs is best illustrated through a series of four examples that guide the reader through the analysis (see Table 1.4.1).

Example 1

Two-Point Linkage Analysis of an Autosomal Dominant Disease with an Unlinked Marker

In this example, two-point linkage analysis demonstrates that a marker is unlinked to the disease locus. The disease is an autosomal dominant disease with disease allele frequency of 0.01 and

Table 1.4.1 Summary of Examples

Example	Linkage analysis	Disease	Features	Illustrates
1	Two-point	Autosomal dominant, reduced penetrance	Unlinked marker	MAKEPED for pedigree file; text editor for parameter file. UNKNOWN MLINK
2	Multipoint	Autosomal dominant, reduced penetrance	Added markers to define disease region	LINKMAP LINKLODS MAPFUN
3	Two-point	Autosomal dominant, reduced penetrance	Age-dependent penetrance	Development of age-of-onset curves. Use of liability classes.
4	Two-point	X-linked, fully penetrant		Coding of liabililty class for affection status locus

disease allele penetrance of 0.80. As part of a genomic screen for disease localization, a series of markers will be selected for genotyping in the disease pedigree. If particular candidate regions are known, these regions will usually be tested first (e.g., hypokalemic periodic paralysis and markers involved in calcium channel transport). When no candidate regions are known, markers are frequently selected on the basis of their heterozygosity and ease of genotyping (Chapter 2); the more heterozygous a marker is, the more likely it is to be informative in the pedigree and provide useful linkage information. This example utilizes the pedigree shown in Figure 1.4.3 with the data for marker 1.

Prepare the pedigree file

A simply formatted file, usually referred to as a "pre-MAKEPED file," is prepared by the user with a text editor. One line of text is prepared for every individual in a pedigree. For the example in Figure 1.4.3, the pre-MAKEPED file is illustrated in Figure 1.4.4, and is named `marker1.pre`. The first five fields are constant across any data analysis; the remaining fields contain information on affection status and/or marker data. Fields are delimited by spaces. In other words, each value must be separated by at least one space. For this example, fields 6 to 8 are for the affection status and the two alleles for marker 1. The order of the alleles in fields 7 and 8 is irrelevant; a genotype of 2 3 can also be input as 3 2.

> Field 1 = family number
> Field 2 = individual number
> Field 3 = father number
> Field 4 = mother number
> Field 5 = sex (1 = male; 2 = female)
> Field 6 = affection status (in this case, affected = 2, unaffected = 1, and unknown = 0)
> Field 7 = first allele for marker 1
> Field 8 = second allele for marker 1.

Zeros (0) must be inserted as placeholders for missing values. For example, individual 1 in Figure 1.4.3 has no marker data and is a founding parent; thus the father, mother, and allele 1 and 2 fields for marker 1 are coded as 0 in Figure 1.4.4. Although not required, it is often easier to view the data in the pre-MAKEPED file if the data are placed in right-justified columns. Although this example involves only one pedigree, multiple pedigrees may be included in the pre-MAKEPED file.

1. Invoke MAKEPED by typing the Makeped command followed by the names of the input file and the output file: `makeped marker1.pre marker1.ped`.

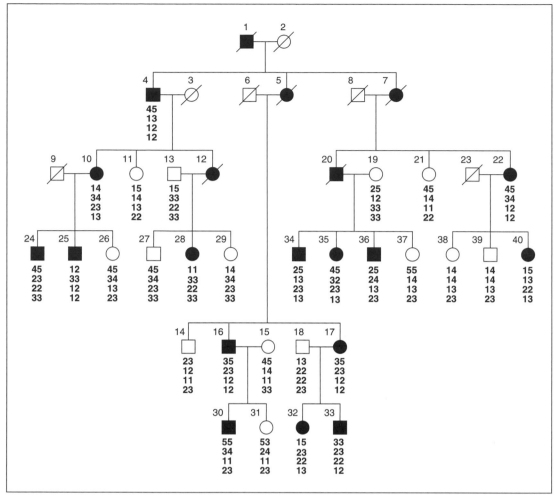

Figure 1.4.3 Pedigree of family 1113. Circles are female family members and squares are males. Shaded individuals are affected with the disease and unshaded individuals are unaffected. Diagonal lines through an individual indicate that the individual is deceased and unavailable for genotyping. Genotype results for markers 1, 2, 3, and 4, respectively, are indicated in bold beneath available individuals.

2. At the program prompt, indicate whether there are loops in the pedigree that must be identified. The pedigree in Figure 1.4.3 has no loops, so the correct response for this example is n. However, when there are loops in the pedigree, the response is y and, when prompted, it is necessary to specify whether the "doubled" individuals are to be entered interactively or read from an external file. See LINKAGE documentation for further details.

3. Respond to the program query regarding whether the proband should be selected automatically or identified by the user, again either interactively or through an external file. In a pedigree without consanguinity, all entries are 0 except for the proband, who is 1. Unless a risk calculation is being performed, the automatic selection of the proband is sufficient.

The fields of the "post-MAKEPED" file are as follows:

Field 1 = family number
Field 2 = individual number
Field 3 = father number

family number	individual number	father ID	mother ID	sex	affection status	allele 1	allele 2
1113	1	0	0	1	2	0	0
1113	2	0	0	2	1	0	0
1113	3	0	0	2	1	0	0
1113	4	1	2	1	2	4	5
1113	5	1	2	2	2	0	0
1113	6	0	0	1	1	0	0
1113	7	1	2	2	2	0	0
1113	8	0	0	1	1	0	0
1113	9	0	0	1	1	0	0
1113	10	4	3	2	2	1	4
1113	11	4	3	2	1	1	5
1113	12	4	3	2	2	0	0
1113	13	0	0	1	1	5	1
1113	14	6	5	1	1	3	2
1113	15	0	0	2	1	5	4
1113	16	6	5	1	2	3	5
1113	17	6	5	2	2	3	5
1113	18	0	0	1	1	3	1
1113	19	0	0	2	1	5	2
1113	20	8	7	1	2	0	0
1113	21	8	7	2	1	5	4
1113	22	8	7	2	2	4	5
1113	23	0	0	1	1	0	0
1113	24	9	10	1	2	4	5
1113	25	9	10	1	2	1	2
1113	26	9	10	2	1	5	4
1113	27	13	12	1	1	5	4
1113	28	13	12	2	2	1	1
1113	29	13	12	2	1	1	4
1113	30	16	15	1	2	5	5
1113	31	16	15	2	1	5	3
1113	32	18	17	2	2	1	5
1113	33	18	17	1	2	3	3
1113	34	20	19	1	2	2	5
1113	35	20	19	2	2	4	5
1113	36	20	19	1	2	2	5
1113	37	20	19	2	1	5	5
1113	38	23	22	2	1	1	4
1113	39	23	22	1	1	1	4
1113	40	23	22	2	2	1	5

Figure 1.4.4 Pre-MAKEPED file (`marker1.pre`) for Example 1 based on the pedigree for family 1113 (see Fig. 1.4.3).

family number	individual number	father ID	mother ID	pointer 1	pointer 2	pointer 3	sex	proband	affection status	allele 1	allele 2				
1113	1	0	0	4	0	0	1	1	2	0	0	Ped:	1113	Per:	1
1113	2	0	0	4	0	0	2	0	1	0	0	Ped:	1113	Per:	2
1113	3	0	0	10	0	0	2	0	1	0	0	Ped:	1113	Per:	3
1113	4	1	2	10	5	5	1	0	2	4	5	Ped:	1113	Per:	4
1113	5	1	2	14	7	7	2	0	2	0	0	Ped:	1113	Per:	5
1113	6	0	0	14	0	0	1	0	1	0	0	Ped:	1113	Per:	6
1113	7	1	2	20	0	0	2	0	2	0	0	Ped:	1113	Per:	7
1113	8	0	0	20	0	0	1	0	1	0	0	Ped:	1113	Per:	8
1113	9	0	0	24	0	0	1	0	1	0	0	Ped:	1113	Per:	9
1113	10	4	3	24	11	11	2	0	2	1	4	Ped:	1113	Per:	10
1113	11	4	3	0	12	12	2	0	1	1	5	Ped:	1113	Per:	11
1113	12	4	3	27	0	0	2	0	2	0	0	Ped:	1113	Per:	12
1113	13	0	0	27	0	0	1	0	1	5	1	Ped:	1113	Per:	13
1113	14	6	5	0	16	16	1	0	1	3	2	Ped:	1113	Per:	14
1113	15	0	0	30	0	0	2	0	1	5	4	Ped:	1113	Per:	15
1113	16	6	5	30	17	17	1	0	2	3	5	Ped:	1113	Per:	16
1113	17	6	5	32	0	0	2	0	2	3	5	Ped:	1113	Per:	17
1113	18	0	0	32	0	0	1	0	1	3	1	Ped:	1113	Per:	18
1113	19	0	0	34	0	0	2	0	1	5	2	Ped:	1113	Per:	19
1113	20	8	7	34	21	21	1	0	2	0	0	Ped:	1113	Per:	20
1113	21	8	7	0	22	22	2	0	1	5	4	Ped:	1113	Per:	21
1113	22	8	7	38	0	0	2	0	2	4	5	Ped:	1113	Per:	22
1113	23	0	0	38	0	0	1	0	1	0	0	Ped:	1113	Per:	23
1113	24	9	10	0	25	25	1	0	2	4	5	Ped:	1113	Per:	24
1113	25	9	10	0	26	26	1	0	2	1	2	Ped:	1113	Per:	25
1113	26	9	10	0	0	0	2	0	1	5	4	Ped:	1113	Per:	26
1113	27	13	12	0	28	28	1	0	1	5	4	Ped:	1113	Per:	27
1113	28	13	12	0	29	29	2	0	2	1	1	Ped:	1113	Per:	28
1113	29	13	12	0	0	0	2	0	1	1	4	Ped:	1113	Per:	29
1113	30	16	15	0	31	31	1	0	2	5	5	Ped:	1113	Per:	30
1113	31	16	15	0	0	0	2	0	1	5	3	Ped:	1113	Per:	31
1113	32	18	17	0	33	33	2	0	2	1	5	Ped:	1113	Per:	32
1113	33	18	17	0	0	0	1	0	2	3	3	Ped:	1113	Per:	33
1113	34	20	19	0	35	35	1	0	2	2	5	Ped:	1113	Per:	34
1113	35	20	19	0	36	36	2	0	2	4	5	Ped:	1113	Per:	35
1113	36	20	19	0	37	37	1	0	2	2	5	Ped:	1113	Per:	36
1113	37	20	19	0	0	0	2	0	1	5	5	Ped:	1113	Per:	37
1113	38	23	22	0	39	39	2	0	1	1	4	Ped:	1113	Per:	38
1113	39	23	22	0	40	40	1	0	1	1	4	Ped:	1113	Per:	39
1113	40	23	22	0	0	0	2	0	2	1	5	Ped:	1113	Per:	40

Figure 1.4.5 Post-MAKEPED pedigree file for Example 1 (`marker1.ped`).

Field 4 = mother number
Field 5 = identification number of first offspring (pointer 1)
Field 6 = identification number of first (or next) paternal sibling (pointer 2)
Field 7 = identification number of first (or next) maternal sibling (pointer 3)
Field 8 = sex
Field 9 = proband
Field 10 = affection status
Field 11 = first allele for marker 1
Field 12 = second allele for marker 1.

The "post-MAKEPED" file marker1.ped is depicted in Figure 1.4.5.

4. For direct input into the linkage programs, rename this post-MAKEPED pedfile.dat. For consistency, it is useful to name all pre-MAKEPED pedigree files with a .pre extension and all post-MAKEPED pedigree files with a .ped extension.

Prepare the parameter file

5. A parameter file for analysis of pedigree 1113 versus marker 1 can be prepared manually using a text editor; the file is shown in Figure 1.4.6. For this example, the parameter file consists of fourteen lines that are numbered in italics on the left to assist with this description; these line designations must not be included in a file intended for use in a linkage analysis. Text appears after the < sign in several lines; this text is optional and is primarily for the user's convenience.

Line 1:

Field 1: Number of loci in the problem to be analyzed.

Field 2: Should have a 0 unless the investigator is calculating genetic risks, in which case a 1 is inserted.

Field 3: Indicator as to whether the loci are autosomal (0) or sex-linked (1).

Field 4: Program code. Insert number associated with program to be used: 1 for CILINK; 2 for CMAP; 3 for ILINK; 4 for LINKMAP; 5 for MLINK; 6 for LODSCORE; 7 for CLODSCORE (see Table 1.4.2 for brief description of these programs).

```
line 1    2 0 0 5  << NO. OF LOCI, RISK LOCUS, SEXLINKED (IF 1) PROGRAM
line 2    0 0.0 0.0 0 << MUT LOCUS, MUT MALE, MUT FEM, HAP FREQ (IF 1)
line 3    1  2
line 4    1   2  << AFFECTION, NO. OF ALLELES
line 5    0.9900 0.010000   << GENE FREQUENCIES
line 6    1 << NO. OF LIABILITY CLASSES
line 7    0.0 0.80 0.80 << PENETRANCES
line 8    3   5  << ALLELE NUMBERS, NO. OF ALLELES
line 9    0.20 0.30 0.10 0.10 0.30
line 10   0 0  << SEX DIFFERENCE, INTERFERENCE (IF 1 OR 2)
line 11   0.0 << RECOMBINATION VALUES
line 12   1 0.001 0.001 << REC VARIED, INCREMENT, FINISHING VALUE
line 13   0.05 0.05 0.15
line 14   0.20 0.10 0.40
```

Figure 1.4.6 Parameter file for Example 1 (marker1.dat).

Table 1.4.2 Programs Available in the LINKAGE Package

Program number	Program name	Description
1	CILINK	A version of ILINK optimized to run in three-generation, CEPH-type pedigrees
2	CMAP	A version of LINKMAP optimized to run in three-generation, CEPH-type pedigrees
3	ILINK	Allows maximum likelihood estimation of various parameters used in linkage analysis—e.g., penetrance, allele frequencies, recombination fractions—for general pedigrees
4	LINKMAP	Used to calculate multipoint lod scores or location scores for a trait or marker locus vs. a fixed map of marker loci for general pedigrees
5	MLINK	Calculates two-point lod scores and performs risk evaluations for general pedigrees
6	LODSCORE	Allows calculation of maximum lod score and its associated recombination fraction in two-point linkage data for general pedigrees
7	CLODSCORE	A version of LODSCORE optimized for three-generation, CEPH-typed pedigrees

Line 2:
> Field 1: If a mutation rate is to be incorporated, enter 1; if absence of mutation is to be assumed, enter 0.
> Fields 2 and 3: If mutation is allowed, indicate male and female mutation rates, respectively; Otherwise, these values should both be 0.0.
> Field 4: If linkage disequilibrium between loci is to be incorporated in the analysis, enter 1; if the loci are to be considered in Hardy-Weinberg equilibrium, enter 0.

Line 3:
> Indicates order of loci for analysis. For a two-point analysis, indicating 1 2 on this line is the equivalent of indicating 2 1. The order of the loci is much more important for multipoint linkage analysis (see Example 2 for more detail).

Lines 4 to 9:
> Information regarding the loci. Four different types of loci can be defined for the LINKAGE analysis: continuous trait locus; affection status locus; binary factors locus; and numbered alleles locus. In this example file, six lines are required to fully describe the two loci included in the problem: lines 4 to 7 specify the affection status locus and lines 8 and 9 specify the numbered alleles locus (see LINKAGE documentation for more detail).

Line 10:
> Allows specification of options for sex-specific differences in recombination and for interference. MLINK will not support either interference or sex-specific differences in recombination.
> Field 1: 0 indicates that no differences in recombination between males and females are allowed, 1 indicates gender-specific differences in recombination that are constant across all map intervals, and 2 indicates gender-specific differences in recombination that may vary across intervals. MLINK requires a 0 in this field.

Field 2: If interference is to be allowed in the analysis, enter 1 for user-specified level of interference, 2 for Kosambi level of interference, or 0 for absence of interference.

Line 11:

Specifies the starting recombination fraction between the loci. The program expects $n - 1$ values on this line, where n is the number of loci in the problem. If there are three loci (1, 2, and 3, in order), the program expects two values on this line. For example, if line 11 reads 0.20 0.05, then the frequency of recombination between loci 1 and 2 is 20% and the frequency between loci 2 and 3 is 5%. If there are two loci, the program expects just one value, which is typically 0.0 to indicate that lod scores assuming no recombination between the loci should be calculated.

Line 12:

Field 1: Used to indicate which recombination fraction is to be varied. For the two-point analysis in this example, enter 1 because there is only one recombination fraction to vary.

Fields 2 and 3: The amount by which the starting recombination fraction is to be incremented and the stop value for calculations, respectively.

Lines 13 and 14:

These optional lines provide an opportunity to calculate lod scores at additional recombination fractions.

Field 1: The starting value of θ for calculation of lod scores.

Field 2: The value by which θ should be incremented for calculations.

Field 3: The stop value for θ.

In this example, line 13 instructs the program to calculate lod scores at $\theta = 0.05$, 0.10, and 0.15; line 14 instructs the program to calculate lod scores at $\theta = 0.20$, 0.30, and 0.40.

Perform the analysis

6. Before running the LINKAGE analysis, first copy the pedigree and parameter files into the input files expected by the programs.

```
copy marker1.ped pedfile.dat
copy marker1.dat datafile.dat
```

7. To run the linkage analysis, invoke the programs in the following order:

```
unknown
mlink
```

The output file is named outfile.dat by the LINKAGE programs. Note that each time the analysis program is run, the outfile.dat will be overwritten. Thus, it is important to copy the outfile.dat into another file so that it will not be inadvertently lost with additional analyses, as follows:

```
copy outfile.dat marker1.out
```

Interpret the results

8. The output from the two-point linkage analysis is shown in Figure 1.4.7. Notice that the lod scores are uniformly negative across all values of θ tested. Although there is no significant evidence for linkage, this result still provides useful information: the disease gene locus is excluded for the region encompassing all θ values for which the lod score is below -2.0. This interpretation assumes that the correct model is specified. For this family, these results indicate that $\sim$10 cM on either side of the marker 1 locus have been

```
Length of real variables = 8 bytes
LINKAGE (V5.20) WITH  2-POINT AUTOSOMAL DATA
ORDER OF LOCI:   1  2
------------------------------------
------------------------------------
THETAS  0.500
------------------------------------
PEDIGREE |  LN LIKE  | LOG 10 LIKE
------------------------------------
     1113   -80.910210   -35.138783
------------------------------------
TOTALS     -80.910210   -35.138783
-2 LN(LIKE) =  1.61820419737301E+0002 LOD SCORE =    0.000000
------------------------------------
------------------------------------
THETAS  0.000
------------------------------------
PEDIGREE |  LN LIKE  | LOG 10 LIKE
------------------------------------
     1113   -97.592953   -42.383991
------------------------------------
TOTALS     -97.592953   -42.383991
-2 LN(LIKE) =  1.95185905504318E+0002 LOD SCORE =   -7.245208
------------------------------------
------------------------------------
THETAS  0.001
------------------------------------
PEDIGREE |  LN LIKE  | LOG 10 LIKE
------------------------------------
     1113   -95.771287   -41.592853
------------------------------------
TOTALS     -95.771287   -41.592853
-2 LN(LIKE) =  1.91542573364822E+0002 LOD SCORE =   -6.454070
------------------------------------
------------------------------------
THETAS  0.050
------------------------------------
PEDIGREE |  LN LIKE  | LOG 10 LIKE
------------------------------------
     1113   -87.346339   -37.933952
------------------------------------
TOTALS     -87.346339   -37.933952
-2 LN(LIKE) =  1.74692677422668E+0002 LOD SCORE =   -2.795169
------------------------------------
------------------------------------
THETAS  0.100
------------------------------------
PEDIGREE |  LN LIKE  | LOG 10 LIKE
------------------------------------
     1113   -85.688952   -37.214160
------------------------------------
TOTALS     -85.688952   -37.214160
-2 LN(LIKE) =  1.71377904890767E+0002 LOD SCORE =   -2.075377
------------------------------------
------------------------------------
THETAS  0.150
------------------------------------
PEDIGREE |  LN LIKE  | LOG 10 LIKE
------------------------------------
     1113   -84.558795   -36.723340
------------------------------------
TOTALS     -84.558795   -36.723340
-2 LN(LIKE) =  1.69117589371370E+0002 LOD SCORE =   -1.584557
------------------------------------
------------------------------------
THETAS  0.200
------------------------------------
PEDIGREE |  LN LIKE  | LOG 10 LIKE
------------------------------------
     1113   -83.678711   -36.341125
------------------------------------
TOTALS     -83.678711   -36.341125
-2 LN(LIKE) =  1.67357421411557E+0002 LOD SCORE =   -1.202342
------------------------------------
------------------------------------
THETAS  0.300
------------------------------------
PEDIGREE |  LN LIKE  | LOG 10 LIKE
------------------------------------
     1113   -82.506840   -35.832189
------------------------------------
TOTALS     -82.506840   -35.832189
-2 LN(LIKE) =  1.65013679193931E+0002 LOD SCORE =   -0.693406
------------------------------------
------------------------------------
THETAS  0.400
------------------------------------
PEDIGREE |  LN LIKE  | LOG 10 LIKE
------------------------------------
     1113   -81.673555   -35.470298
------------------------------------
TOTALS     -81.673555   -35.470298
-2 LN(LIKE) =  1.63347109203385E+0002 LOD SCORE =   -0.331516
```

Figure 1.4.7 Results of two-point linkage analysis in Example 1.

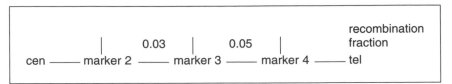

Figure 1.4.8 Genetic map showing recombination fractions for markers 2, 3, and 4.

excluded as a location for the disease gene, for a total exclusion of 20 cM (~0.6%) of the genome.

Example 2

Multipoint Linkage Analysis of an Autosomal Dominant Disease by Assessment of Additional Markers to Establish Disease Gene Location

In this example, additional markers known to be near an already-established linked marker are genotyped in the family to try to assess the location of the disease gene relative to the markers using multipoint linkage analysis. This analysis assumes the existence of a genetic map that includes a linked marker (e.g., marker 3 in Fig. 1.4.3) and two other markers that flank it (e.g., markers 2 and 4). These markers have been genotyped in the pedigree and the genotypes are shown on Figure 1.4.3. The sex-averaged recombination fractions between the markers are shown in Figure 1.4.8. Allele frequencies for these two markers are given in Table 1.4.3. The disease is an autosomal dominant disease with disease allele frequency of 0.01 and disease allele penetrance of 0.80. This example demonstrates the use of LINKMAP for multipoint linkage analysis and LINKLODS for data management. As an alternative to LINKLODS, MAPFUN can be utilized to convert between recombination fractions and morgans for multipoint linkage analysis. The pedigree file (`example2.ped`) and parameter file (`example2.dat`) for multipoint linkage analysis are shown in Figures 1.4.9 and 1.4.10, respectively. In this pedigree file, genotypes for each of the three markers are listed on the same line, in marker order 2, 3, 4 as indicated in the parameter file.

Prepare the batch file using LCP

1. On the third screen, use the arrow key to highlight LINKMAP. Press Page Down to go to the fourth screen, which allows selection of test interval options, and choose Specific Interval to specify the number of θ values in any particular interval at which likelihoods will be calculated. Press Page Down to go to the next screen. The All Intervals option can be selected if the user prefers that the same number of lod score evaluations be performed in each map interval.

2. On the fifth screen, specify whether or not there are sex differences in recombination across the map. For this example, choose No Sex Difference. Press Page Down to go to the next screen.

Table 1.4.3 Allele Frequencies for Markers 2 and 4 from Pedigree 1113 (Figure 1.4.3)

Marker	Allele			
	1	2	3	4
2	0.30	0.30	0.05	0.35
4	0.05	0.40	0.55	—

family number	individual number	father ID	mother ID	pointer 1	pointer 2	pointer 3	sex	proband	affection status	marker 2 first allele	marker 2 second allele	marker 3 first allele	marker 3 second allele	marker 4 first allele	marker 4 second allele		
1113	1	0	0	4	0	0	1	1	2	0	0	0	0	0	0	Ped: 1113	Per: 1
1113	2	0	0	4	0	0	2	0	1	0	0	0	0	0	0	Ped: 1113	Per: 2
1113	3	0	0	10	0	0	2	0	1	0	0	0	0	0	0	Ped: 1113	Per: 3
1113	4	1	2	10	5	5	1	0	2	1	3	1	2	1	2	Ped: 1113	Per: 4
1113	5	1	2	14	7	7	2	0	2	0	0	0	0	0	0	Ped: 1113	Per: 5
1113	6	0	0	14	0	0	1	0	1	0	0	0	0	0	0	Ped: 1113	Per: 6
1113	7	1	2	20	0	0	2	0	2	0	0	0	0	0	0	Ped: 1113	Per: 7
1113	8	0	0	20	0	0	1	0	1	0	0	0	0	0	0	Ped: 1113	Per: 8
1113	9	0	0	24	0	0	1	0	1	0	0	0	0	0	0	Ped: 1113	Per: 9
1113	10	4	3	24	11	11	2	0	2	3	4	2	3	1	3	Ped: 1113	Per: 10
1113	11	4	3	0	12	12	2	0	1	1	4	1	3	2	2	Ped: 1113	Per: 11
1113	12	4	3	27	0	0	2	0	2	0	0	0	0	0	0	Ped: 1113	Per: 12
1113	13	0	0	27	0	0	1	0	1	3	3	2	2	3	3	Ped: 1113	Per: 13
1113	14	6	5	0	16	16	1	0	1	1	2	1	1	2	3	Ped: 1113	Per: 14
1113	15	0	0	30	0	0	2	0	1	1	4	1	1	3	3	Ped: 1113	Per: 15
1113	16	6	5	30	17	17	1	0	2	3	2	1	2	1	2	Ped: 1113	Per: 16
1113	17	6	5	32	0	0	2	0	2	3	2	1	2	1	2	Ped: 1113	Per: 17
1113	18	0	0	32	0	0	1	0	1	2	2	2	2	2	3	Ped: 1113	Per: 18
1113	19	0	0	34	0	0	2	0	1	1	2	3	3	3	3	Ped: 1113	Per: 19
1113	20	8	7	34	21	21	1	0	2	0	0	0	0	0	0	Ped: 1113	Per: 20
1113	21	8	7	0	22	22	2	0	1	1	4	1	1	2	2	Ped: 1113	Per: 21
1113	22	8	7	38	0	0	2	0	2	3	4	1	2	1	2	Ped: 1113	Per: 22
1113	23	0	0	38	0	0	1	0	1	0	0	0	0	0	0	Ped: 1113	Per: 23
1113	24	9	10	0	25	25	1	0	2	3	2	2	2	3	3	Ped: 1113	Per: 24
1113	25	9	10	0	26	26	1	0	2	3	3	1	2	1	2	Ped: 1113	Per: 25
1113	26	9	10	0	0	0	2	0	1	3	4	1	3	2	3	Ped: 1113	Per: 26
1113	27	13	12	0	28	28	1	0	1	3	4	2	3	3	3	Ped: 1113	Per: 27
1113	28	13	12	0	29	29	2	0	2	3	3	2	2	3	3	Ped: 1113	Per: 28
1113	29	13	12	0	0	0	2	0	1	4	3	2	3	3	3	Ped: 1113	Per: 29
1113	30	16	15	0	31	31	1	0	2	3	4	1	1	2	3	Ped: 1113	Per: 30
1113	31	16	15	0	0	0	2	0	1	4	2	1	1	2	3	Ped: 1113	Per: 31
1113	32	18	17	0	33	33	2	0	2	3	2	2	2	1	3	Ped: 1113	Per: 32
1113	33	18	17	0	0	0	1	0	2	2	3	2	2	1	2	Ped: 1113	Per: 33
1113	34	20	19	0	35	35	1	0	2	1	3	2	3	1	3	Ped: 1113	Per: 34
1113	35	20	19	0	36	36	2	0	2	3	2	2	3	1	3	Ped: 1113	Per: 35
1113	36	20	19	0	37	37	1	0	2	4	2	1	3	2	3	Ped: 1113	Per: 36
1113	37	20	19	0	0	0	2	0	1	1	4	1	3	2	3	Ped: 1113	Per: 37
1113	38	23	22	0	39	39	2	0	1	1	4	1	3	2	3	Ped: 1113	Per: 38
1113	39	23	22	0	40	40	1	0	1	1	4	1	3	2	3	Ped: 1113	Per: 39
1113	40	23	22	0	0	0	2	0	2	1	3	2	2	1	3	Ped: 1113	Per: 40

Figure 1.4.9 Pedigree file for Example 2 (example2.ped) based on the pedigree for family 1113 (see Fig. 1.4.3).

```
line 1    4 0 0 5   << NO. OF LOCI, RISK LOCUS, SEXLINKED (IF 1) PROGRAM
line 2    0 0.0 0.0 0 << MUT LOCUS, MUT MALE, MUT FEM, HAP FREQ (IF 1)
line 3    1  2 3 4
line 4    1    2  << AFFECTION, NO. OF ALLELES
line 5    0.990000 0.010000   << GENE FREQUENCIES
line 6    1 << NO. OF LIABILITY CLASSES
line 7    0 0.80 0.80 << PENETRANCES
line 8    3    4  << ALLELE NUMBERS, NO. OF ALLELES
line 9    0.30 0.30 0.05 0.35
line 10   3    3
line 11   0.40 0.10 0.50
line 12   3    3
line 13   0.05 0.40 0.55
line 14   0 0   << SEX DIFFERENCE, INTERFERENCE (IF 1 OR 2)
line 15   0.5 0.03 0.05 << RECOMBINATION VALUES
line 16   1 0.0 10 << LOCUS VARIED, FINISHING VALUE, NO. OF EVALUATIONS
```

Figure 1.4.10 Parameter file for interval 0 in Example 2 (`example2.dat`).

3. On the sixth screen, the Map Specification screen, define specific properties of the analysis:

> Enter 1 at the Test Loci option to indicate that locus number 1, which codes for the disease locus, is the locus that is to be moved through various positions in a fixed map.
>
> Enter 2 3 4 at the Order of Fixed Loci option to indicate that the map order of the loci is marker 2—marker 3—marker 4.
>
> Enter 0.03 0.05 at the Recombination Fraction option to indicate that the recombination fraction between markers 2 and 3 is 0.03 and between markers 3 and 4 is 0.05. Note that these values *must* be entered in the same order as that listed in the Order of Fixed Loci option.
>
> Enter 0 at the Test Interval option to indicate that the test locus, 1, should be placed in interval 0 (proximal to marker 2) for this first analysis. See Figure 1.4.1 for interval designations.
>
> Enter 10 at the Number of Evaluations in Interval option to direct the program to take the maximum allowable amount of recombination between the test locus and marker 2 and divide it into 10 evenly spaced intervals at which likelihoods will be calculated. In interval 0, the maximum allowable recombination between the disease locus and marker 2 is 50% (that the disease is unlinked to the fixed map of loci), so likelihoods will be calculated at $\theta = 0.50, 0.45, 0.40, 0.35, \ldots, 0.0$.
>
> Press Page Down to enter this information.

This is the first of four necessary analyses; three additional analyses must done done using intervals 1, 2, and 3.

4. Without leaving the Map Specification screen, enter the additional analyses. Modify Test Interval and Number of Evaluations in Interval; other options are the same (use the Up and Down arrow keys to move between options):

To set up analysis for interval 1, change Test Interval to 1 and change Number of Evaluations in Interval to 3; press Page Down to log this analysis.

To set up analysis for interval 2, change Test Interval to 2 and change Number of Evaluations in Interval to 5; press Page Down to log this analysis.

To set up analysis for interval 3, change Test Interval to 3 and change Number of Evaluations in Interval to 10; press Page Down to log this analysis.

5. Press Control-z to exit from LCP.

If the user encounters an error message, **ERROR: Range check error,** *at least one of the program constants is too small. The problem size is larger than the executable version of LINKMAP allows, and thus the LINKMAP program must be recompiled.*

Recompile the program

6. Modify the LINKMAP program constants to suit the problem under consideration. The most frequently modified program constants are as follows:

Maxlocus is the maximum number of loci in the problem; for this example, there are four loci.

Maxall is the maximum number of alleles at any single locus; for this example, the maximum number of alleles is 4.

Maxhap is the maximum number of multilocus haplotypes. This value is obtained by multiplying together the number of alleles at each locus in the problem; for this example, Maxhap is calculated as $2 \times 4 \times 3 \times 3 = 72$.

Maxind is the maximum number of individuals in the analysis problem, which is a sum across all pedigrees; in this example, Maxind is 40.

Maxped is the maximum number of pedigrees in the problem; in this example, Maxped is 1.

Maxchild is the maximum number of full siblings in a sibship; in this example, Maxchild is 4.

Maxloop is the maximum number of loops in any single pedigree in the problem; in this example, Maxloop is 0.

Perform the analysis

7. Now that the LINKMAP program has been recompiled to accommodate the problem size, restart the analysis by typing pedin.

Calculate multipoint lod scores

The output from the analysis is found in a file called final.out and is shown in part in Figure 1.4.11. Multipoint lod scores can be obtained by hand from the information in the output file. For instance, the multipoint lod score obtained when the disease gene is proximal to marker 2 and demonstrates 40% recombination with this fixed map of markers is calculated using the numbers in bold italics in the output in Figure 1.4.11: $(-58.890286) - (-60.274212) = 1.383926$. The value of -60.274212 is the $\log_{10}$ likelihood when the disease is unlinked to the fixed map of markers and is the $\log_{10}$ of the denominator of the multipoint likelihood ratio. The value -58.890286 is the $\log_{10}$ likelihood when the disease exhibits 40% recombination with the fixed map of markers as described earlier and represents the $\log_{10}$ of the numerator of the likelihood ratio. Because these values have already been converted from their original units to $\log_{10}$ units, the $\log_{10}$ likelihood for the denominator is subtracted from the $\log_{10}$ likelihood for the numerator. These calculations can also be performed using one of the LINKAGE support programs:

8. Use LINKLODS to perform the subtraction automatically. To invoke LINKLODS, type:
linklods

9. The program will ask for the name of the input file. After an MLINK or LINKMAP analysis performed with a batch file prepared by LCP, the name of the appropriate input

```
*****************************************************************************
                    LINKMAP
        Pedigree File            : multi.TPD
        Parameter File           : multi.TDT
        Output Pedigree File     : PEDFILE.DAT
        Output Parameter File    : DATAFILE.DAT
        Log File                 : LSP.LOG
        Stream File              : LSP.STM

        Date Run                 :  5-Aug-94  23:00:06

        Sex Difference           : 0
        Test Locus               : 1
        Stop Value               : 0.00000000
        Number of Evaluations    : 10

        Locus Order              : 1 2 3 4
        Male Recomb. Fractions   : 0.50000000 0.03000000 0.05000000

*****************************************************************************

Length of real variables = 8 bytes
-----------------------------------
-----------------------------------        -----------------------------------
THETAS  0.500 0.030 0.050                   THETAS  0.200 0.030 0.050
-----------------------------------        -----------------------------------
PEDIGREE |  LN LIKE  | LOG 10 LIKE          PEDIGREE |  LN LIKE  | LOG 10 LIKE
-----------------------------------        -----------------------------------
    1113  -138.786797   -60.274212              1113  -129.315676   -56.160965
-----------------------------------        -----------------------------------
TOTALS    -138.786797   -60.274212          TOTALS    -129.315676   -56.160965
-2 LN(LIKE) = 2.77573594548582E+0002        -2 LN(LIKE) = 2.58631352404433E+0002
-----------------------------------        -----------------------------------
-----------------------------------        -----------------------------------
THETAS  0.450 0.030 0.050                   THETAS  0.150 0.030 0.050
-----------------------------------        -----------------------------------
PEDIGREE |  LN LIKE  | LOG 10 LIKE          PEDIGREE |  LN LIKE  | LOG 10 LIKE
-----------------------------------        -----------------------------------
    1113  -137.316975   -59.635878              1113  -128.058911   -55.615160
-----------------------------------        -----------------------------------
TOTALS    -137.316975   -59.635878          TOTALS    -128.058911   -55.615160
-2 LN(LIKE) = 2.74633950514413E+0002        -2 LN(LIKE) = 2.56117822981292E+0002
-----------------------------------        -----------------------------------
-----------------------------------        -----------------------------------
THETAS  0.400 0.030 0.050                   THETAS  0.100 0.030 0.050
-----------------------------------        -----------------------------------
PEDIGREE |  LN LIKE  | LOG 10 LIKE          PEDIGREE |  LN LIKE  | LOG 10 LIKE
-----------------------------------        -----------------------------------
    1113  -135.600183   -58.890286              1113  -126.992632   -55.152082
-----------------------------------        -----------------------------------
TOTALS    -135.600183   -58.890286          TOTALS    -126.992632   -55.152082
-2 LN(LIKE) = 2.71200365186978E+0002        -2 LN(LIKE) = 2.53985263822361E+0002
-----------------------------------        -----------------------------------
-----------------------------------        -----------------------------------
THETAS  0.350 0.030 0.050                   THETAS  0.050 0.030 0.050
-----------------------------------        -----------------------------------
PEDIGREE |  LN LIKE  | LOG 10 LIKE          PEDIGREE |  LN LIKE  | LOG 10 LIKE
-----------------------------------        -----------------------------------
    1113  -133.878195   -58.142437              1113  -126.271595   -54.838940
-----------------------------------        -----------------------------------
TOTALS    -133.878195   -58.142437          TOTALS    -126.271595   -54.838940
-2 LN(LIKE) = 2.67756389466871E+0002        -2 LN(LIKE) = 2.52543189780585E+0002
-----------------------------------        -----------------------------------
-----------------------------------        -----------------------------------
THETAS  0.300 0.030 0.050                   THETAS  0.000 0.030 0.050
-----------------------------------        -----------------------------------
PEDIGREE |  LN LIKE  | LOG 10 LIKE          PEDIGREE |  LN LIKE  | LOG 10 LIKE
-----------------------------------        -----------------------------------
    1113  -132.241780   -57.431753              1113  -128.172680   -55.664569
-----------------------------------        -----------------------------------
TOTALS    -132.241780   -57.431753          TOTALS    -128.172680   -55.664569
-2 LN(LIKE) = 2.64483560117904E+0002        -2 LN(LIKE) = 2.56345360523932E+0002
-----------------------------------        -----------------------------------
-----------------------------------        Additional data eliminated for brevity.
THETAS  0.250 0.030 0.050                   Data appear on screen as a single column.
-----------------------------------
PEDIGREE |  LN LIKE  | LOG 10 LIKE
-----------------------------------
    1113  -130.716504   -56.769335
-----------------------------------
TOTALS    -130.716504   -56.769335
-2 LN(LIKE) = 2.61433008316186E+0002
-----------------------------------
```

Figure 1.4.11 Partial results from analysis in Example 2.

```
Program LINKMAP

Locus Order:  1 2 3 4

NEW BASELINE LIKELIHOODS USED FOR THE FOLLOWING LOD SCORES

THETAS  0.450  0.030  0.050
Male map position:    -1.1513 (Haldane)    -0.7361 (Kosambi)
     PED        LOD
    1113        0.638
  TOTALS        0.638


THETAS  0.400  0.030  0.050
Male map position:    -0.8047 (Haldane)    -0.5493 (Kosambi)
     PED        LOD
    1113        1.384
  TOTALS        1.384


THETAS  0.350  0.030  0.050
Male map position:    -0.6020 (Haldane)    -0.4337 (Kosambi)
     PED        LOD
    1113        2.132
  TOTALS        2.132


THETAS  0.300  0.030  0.050
Male map position:    -0.4581 (Haldane)    -0.3466 (Kosambi)
     PED        LOD
    1113        2.842
  TOTALS        2.842


THETAS  0.250  0.030  0.050
Male map position:    -0.3466 (Haldane)    -0.2747 (Kosambi)
     PED        LOD
    1113        3.505
  TOTALS        3.505


THETAS  0.200  0.030  0.050
Male map position:    -0.2554 (Haldane)    -0.2118 (Kosambi)
     PED        LOD
    1113        4.113
  TOTALS        4.113


THETAS  0.150  0.030  0.050
Male map position:    -0.1783 (Haldane)    -0.1548 (Kosambi)
     PED        LOD
    1113        4.659
  TOTALS        4.659


THETAS  0.100  0.030  0.050
Male map position:    -0.1116 (Haldane)    -0.1014 (Kosambi)
     PED        LOD
    1113        5.122
  TOTALS        5.122


THETAS  0.050  0.030  0.050
Male map position:    -0.0527 (Haldane)    -0.0502 (Kosambi)
     PED        LOD
    1113        5.435
  TOTALS        5.435


THETAS  0.000  0.030  0.050
Male map position:     0.0000 (Haldane)     0.0000 (Kosambi)
     PED        LOD
    1113        4.610
  TOTALS        4.610
```

Figure 1.4.12   Output from LINKLODS for Example 2 for locus order 1 2 3 4.

```
Program LINKMAP

Locus Order:  2 1 3 4

THETAS   0.000   0.030   0.050
Male map position:      0.0000 (Haldane)      0.0000 (Kosambi)
      PED          LOD
     1113         4.610
    TOTALS        4.610


THETAS   0.010   0.020   0.050
Male map position:      0.0101 (Haldane)      0.0100 (Kosambi)
      PED          LOD
     1113         4.472
    TOTALS        4.472


THETAS   0.020   0.010   0.050
Male map position:      0.0204 (Haldane)      0.0200 (Kosambi)
      PED          LOD
     1113         4.171
    TOTALS        4.171


THETAS   0.030   0.000   0.050
Male map position:      0.0309 (Haldane)      0.0300 (Kosambi)
      PED          LOD
     1113         1.446
    TOTALS        1.446
```

Figure 1.4.13 Output from LINKLODS for Example 2 for locus order 2 1 3 4.

file is final.out. When MLINK or LINKMAP have been run directly as in Example 1, the name of the appropriate input file for LINKLODS is outfile.dat. The output from LINKLODS is in the file final.lod and is shown in Figures 1.4.12 to 1.4.15.

Graph results

10. LINKLODS also provides conversion of recombination frequencies to map distances (in morgans) assuming interference at a Haldane level (Figs. 1.4.12 to 1.4.15). To assess the support for placement of the disease gene within this known map of markers, graph the results as demonstrated in Figure 1.4.16. The y axis is the multipoint lod score and the x axis is now the map distance.

Because the LINKAGE programs require input of the recombination fractions between the loci as opposed to units in genetic distance, it is important to convert between the two units of measure accurately. This conversion is dependent upon the degree of interference the investigator chooses to assume, frequently either no interference (Haldane) or a level of interference such as provided by the Sturt mapping function, which assumes complete interference over small distances and allows the level of interference to decrease as distances increase. The MAPFUN program allows for easy conversion between recombination fraction and distance in cM.

Convert recombination fraction to map distance (an alternative approach)

11. Invoke MAPFUN at the prompt, and indicate whether map distances are to be calculated from a given θ value (respond with M) or θ values are to be calculated from a given map distance (respond with T). Values are provided under the assumption of a variety of map functions. For this example, assume a Haldane level of interference.

```
Program LINKMAP
Locus Order:  2 3 1 4
THETAS  0.030  0.000  0.050
Male map position:      0.0309 (Haldane)      0.0300 (Kosambi)
     PED         LOD
    1113        1.446
   TOTALS       1.446

THETAS  0.030  0.010  0.041
Male map position:      0.0410 (Haldane)      0.0400 (Kosambi)
     PED         LOD
    1113        1.552
   TOTALS       1.552

THETAS  0.030  0.020  0.031
Male map position:      0.0513 (Haldane)      0.0500 (Kosambi)
     PED         LOD
    1113        1.428
   TOTALS       1.428

THETAS  0.030  0.030  0.021
Male map position:      0.0619 (Haldane)      0.0601 (Kosambi)
     PED         LOD
    1113        1.093
   TOTALS       1.093

THETAS  0.030  0.040  0.011
Male map position:      0.0726 (Haldane)      0.0701 (Kosambi)
     PED         LOD
    1113        0.403
   TOTALS       0.403

THETAS  0.030  0.050  0.000
Male map position:      0.0836 (Haldane)      0.0802 (Kosambi)
     PED         LOD
    1113       -3.240
   TOTALS      -3.240
```

Figure 1.4.14   Output from LINKLODS for Example 2 for locus order 2 3 1 4.

MAPFUN converts the recombination fractions of 0.03 and 0.05 to 3.09 cM and 5.27 cM, respectively, under a Haldane level of interference. Differences between recombination fractions and genetic distances become more marked as the distances increase. Although LINKLODS will automatically convert recombination fractions to map distances, in some instances the analysis may be based on a genetic map with intermarker distances specified in morgans, or more frequently in centimorgans. The LINKAGE programs require recombination fractions as input; MAPFUN will perform the requisite conversions between morgans (or centimorgans) and recombination fractions.

Interpret the results

12. Determine a support interval about this likely placement of the disease locus. Subtract 3.0 from the highest lod score and find the associated value of θ; draw a horizontal line across the graph at a value equal to the highest lod score minus 3. All points on the map that are above this horizontal line are included within the 3-lod-unit support interval and cannot be entirely excluded from further consideration; in other words, more data are necessary to provide more confidence in the placement of the disease locus.

```
Program LINKMAP

Locus Order:  2  3  4  1

THETAS  0.030  0.050  0.000
Male map position:    0.0836 (Haldane)      0.0802 (Kosambi)
       PED        LOD
      1113      -3.240
    TOTALS      -3.240


THETAS  0.030  0.050  0.100
Male map position:    0.1952 (Haldane)      0.1816 (Kosambi)
       PED        LOD
      1113       2.253
    TOTALS       2.253


THETAS  0.030  0.050  0.200
Male map position:    0.3390 (Haldane)      0.2920 (Kosambi)
       PED        LOD
      1113       2.297
    TOTALS       2.297


THETAS  0.030  0.050  0.300
Male map position:    0.5418 (Haldane)      0.4268 (Kosambi)
       PED        LOD
      1113       1.730
    TOTALS       1.730


THETAS  0.030  0.050  0.400
Male map position:    0.8883 (Haldane)      0.6295 (Kosambi)
       PED        LOD
      1113       0.851
    TOTALS       0.851
```

Figure 1.4.15 Abbreviated output from LINKLODS for Example 2 for locus order 2 3 4 1.

13. Use the graph to interpret the location of the disease gene. For this example, the highest lod score is obtained just proximal to marker 2, where the disease locus demonstrates ~5% recombination with the fixed set of markers. The value of the maximum lod score is ~5.44. The next highest lod score, 4.54, occurs in the interval flanked by markers 2 and 3. The graph in Figure 1.4.16 is interpreted as follows: the most likely location for the disease gene locus occurs at ~5 cM proximal to marker 2. This localization is supported with odds of 7.94:1 over the next most likely position, between markers 2 and 3. The odds for support are calculated as the antilog of the difference between the maximum lod score and the highest lod score in any of the remaining intervals. For this example, this is $5.44 - 4.54 = 0.90$ and the odds are $10^{0.90}:1$ ($= 7.94$) in favor of placement of the disease gene proximal to marker 2. Also, all intervals distal to marker 3 are excluded from the 3-lod-unit support interval and thus are not likely locations for the disease gene.

14. Determine whether additional genotyping is required. For this example, the highest lod score occurs outside the map, so it is advisable to consider genotyping markers proximal to marker 2 in the family to further refine the disease gene localization.

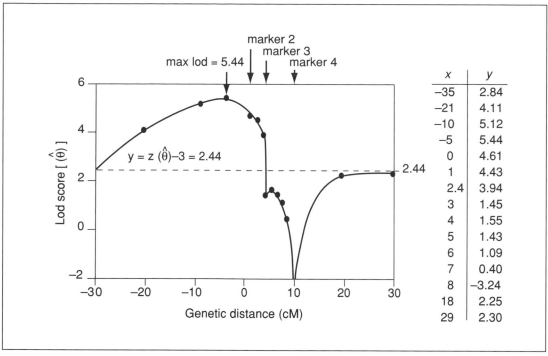

Figure 1.4.16 Graph of multipoint lod score from Example 2, with genetic distance in cM on the x axis plotted versus the lod score on the y axis. Values are taken from Figures 1.4.12–1.4.15 and a Kosambi mapping function is used. The horizontal line is drawn across the graph at $y = z(\hat{\theta}) - 3.0$ and is used in constructing a region excluded for disease gene localization with greater than 1000:1 odds. The arrows indicate multipoint lod scores when the recombination fraction, θ, between marker 1 and the marker indicated is zero.

Example 3

Two-Point Linkage Analysis of an Autosomal Dominant Disease with Age-Dependent Penetrance

This example demonstrates the incorporation of age-dependent penetrance in linkage analysis, with particular reference to the use of liability classes. If age-of-onset (AO) information on affected individuals is available, it can be used to estimate penetrance. The closer the penetrance estimates are to reality, the better; however, the true form of the distribution of the AO function is frequently unknown and thus must be approximated. Often the investigator will assume that the underlying distribution is normal or that age-of-onset increases linearly. In practice, the exact form of the penetrance function probably will have little effect on the conclusions drawn from the analysis as long as it approximates the available data. For this example, a simple approach for developing an AO curve is described and the curve is incorporated into the linkage analysis. Note that this approach is suitable only when no phenocopies (nongenetic cases of the disease) are present and when age-of-onset, as opposed to current age of affected individuals, is known. This example uses AO data for fifteen affected individuals from pedigree 1113 in Figure 1.4.3, and age-at-examination data available on ten different individuals. The range of the AO data is 17 to 55 years, spanning 38 years, and is graphed in Figure 1.4.17.

1. Determine the number of liability classes. For this example, the data are broken into six intervals based on age ranges. In practice, the number of liability classes can be established at the discretion of the user. The procedure for establishing penetrance values

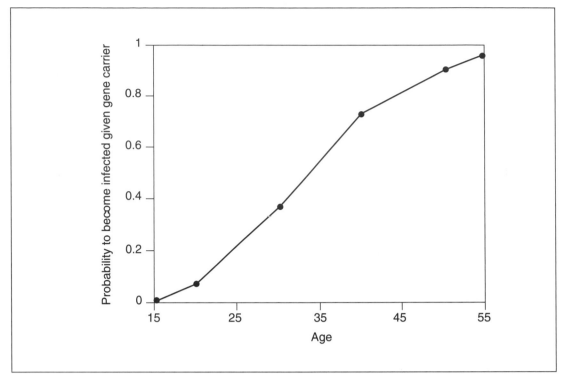

Figure 1.4.17 Cumulative distribution function for age-of-onset data for Example 3, with age (x axis) plotted against the cumulative probability of becoming affected (y axis).

for liability classes is shown in Table 1.4.4. Although no one in the current data set is younger than age 17, liability class 1 is established in case it is needed for future family members.

2. Record the number of individuals who become affected in each of the class intervals (Table 1.4.4, column 3).

3. Calculate the probability of a gene carrier becoming affected during each interval (the number of individuals who become affected during an interval divided by the total number of individuals for whom AO data is available; Table 1.4.4, column 4).

4. Calculate the cumulative probability of becoming affected during an interval (the sum of the probabilities for this and all lower classes; Table 1.4.4, column 5).

Table 1.4.4 Calculation of Penetrance Values when Penetrance is Age-Dependent

Liability class	Age range	No. of people with age-of-onset in class	Probability to become affected	Cumulative probability to become affected[a]	Input for LINKAGE liability class of susceptible genotype
1	0–14	0	0.0	0.0	0.0
2	15–24	2	$^{2}/_{15} = 0.133$	0.133	0.066
3	25–34	7	$^{7}/_{15} = 0.466$	0.599	0.366
4	35–44	4	$^{4}/_{15} = 0.266$	0.865	0.732
5	45–54	1	$^{1}/_{15} = 0.066$	0.931	0.898
6	>–54	1	$^{1}/_{15} = 0.066$	0.997	0.964

[a]Sum does not total 1.0 due to rounding error.

family number	individual number	father ID	mother ID	pointer 1	pointer 2	pointer 3	sex	proband	affection status	liability class	allele 1	allele 2		
1113	1	0	0	4	0	0	1	1	2	7	0	0	Ped: 1113	Per: 1
1113	2	0	0	4	0	0	2	0	1	7	0	0	Ped: 1113	Per: 2
1113	3	0	0	10	0	0	2	0	1	7	0	0	Ped: 1113	Per: 3
1113	4	1	2	10	5	5	1	0	2	5	1	2	Ped: 1113	Per: 4
1113	5	1	2	14	7	7	2	0	2	7	0	0	Ped: 1113	Per: 5
1113	6	0	0	14	0	0	1	0	1	7	0	0	Ped: 1113	Per: 6
1113	7	1	2	20	0	0	2	0	2	7	0	0	Ped: 1113	Per: 7
1113	8	0	0	20	0	0	1	0	1	7	0	0	Ped: 1113	Per: 8
1113	9	0	0	24	0	0	1	0	1	7	0	0	Ped: 1113	Per: 9
1113	10	4	3	24	11	11	2	0	2	4	2	3	Ped: 1113	Per: 10
1113	11	4	3	0	12	12	2	0	1	6	1	3	Ped: 1113	Per: 11
1113	12	4	3	27	0	0	2	0	2	7	0	0	Ped: 1113	Per: 12
1113	13	0	0	27	0	0	1	0	1	7	2	2	Ped: 1113	Per: 13
1113	14	6	5	0	16	16	1	0	1	4	1	1	Ped: 1113	Per: 14
1113	15	0	0	30	0	0	2	0	1	7	1	1	Ped: 1113	Per: 15
1113	16	6	5	30	17	17	1	0	2	3	1	2	Ped: 1113	Per: 16
1113	17	6	5	32	0	0	2	0	2	4	1	2	Ped: 1113	Per: 17
1113	18	0	0	32	0	0	1	0	1	7	2	2	Ped: 1113	Per: 18
1113	19	0	0	34	0	0	2	0	1	7	3	3	Ped: 1113	Per: 19
1113	20	8	7	34	21	21	1	0	2	7	0	0	Ped: 1113	Per: 20
1113	21	8	7	0	22	22	2	0	1	6	1	1	Ped: 1113	Per: 21
1113	22	8	7	38	0	0	2	0	2	6	1	2	Ped: 1113	Per: 22
1113	23	0	0	38	0	0	1	0	1	7	0	0	Ped: 1113	Per: 23
1113	24	9	10	0	25	25	1	0	2	3	2	2	Ped: 1113	Per: 24
1113	25	9	10	0	26	26	1	0	2	3	1	2	Ped: 1113	Per: 25
1113	26	9	10	0	0	0	2	0	1	3	1	3	Ped: 1113	Per: 26
1113	27	13	12	0	28	28	1	0	1	3	2	3	Ped: 1113	Per: 27
1113	28	13	12	0	29	29	2	0	2	3	2	2	Ped: 1113	Per: 28
1113	29	13	12	0	0	0	2	0	1	3	2	3	Ped: 1113	Per: 29
1113	30	16	15	0	31	31	1	0	2	3	1	1	Ped: 1113	Per: 30
1113	31	16	15	0	0	0	2	0	1	3	1	1	Ped: 1113	Per: 31
1113	32	18	17	0	33	33	2	0	2	2	2	2	Ped: 1113	Per: 32
1113	33	18	17	0	0	0	1	0	2	2	2	2	Ped: 1113	Per: 33
1113	34	20	19	0	35	35	1	0	2	3	2	3	Ped: 1113	Per: 34
1113	35	20	19	0	36	36	2	0	2	4	2	3	Ped: 1113	Per: 35
1113	36	20	19	0	37	37	1	0	2	3	1	3	Ped: 1113	Per: 36
1113	37	20	19	0	0	0	2	0	1	4	1	3	Ped: 1113	Per: 37
1113	38	23	22	0	39	39	2	0	1	2	1	3	Ped: 1113	Per: 38
1113	39	23	22	0	40	40	1	0	1	3	1	3	Ped: 1113	Per: 39
1113	40	23	22	0	0	0	2	0	2	4	2	2	Ped: 1113	Per: 40

Figure 1.4.18 Pedigree file for Example 3 (example3.ped) based on the pedigree for family 1113 (see Fig. 1.4.3).

5. Calculate the actual input for penetrance in the liability classes of the parameter file for susceptible genotypes for linkage analysis (the average height of the curve at the midpoint of the age class interval; Table 1.4.4, column 6). The penetrance value for interval i is calculated as $1/2$ [(cumulative probability to become affected during or before interval i) + (cumulative probability to become affected during or before interval $i - 1$)]. The value for the third liability class is calculated as: $1/2[0.133 + 0.599] = 0.366$.

6. Perform the analysis for this example in the c:\linkage\example3 directory. Modify the pedigree file to include the new information on liability classes. The modified example3.ped file, which includes the liability class information and uses marker 3, is shown in Figure 1.4.18.

7. Modify the parameter file to incorporate the liability classes into the analysis. Use PREPLINK to add the number of liability classes and their associated penetrance values after changing the first locus to be an affection status locus as follows:

> Choose **a** to modify the locus
> Select the locus number to modify (1)
> Select **b** to modify number of liability classes to 7
> Select **c** to modify the penetrances for the liability classes as prompted by the program
> Continue preparation of the parameter file as needed.

*The parameter file (*example3.dat*) is shown in Figure 1.4.19.*

8. Perform the two-point linkage analysis as described in the other Examples.

```
line 1    2 0 0 5  << NO. OF LOCI, RISK LOCUS, SEXLINKED (IF 1) PROGRAM
line 2    0 0.0 0.0 0 << MUT LOCUS, MUT MALE, MUT FEM, HAP FREQ (IF 1)
line 3    1  2
line 4    1    2  << AFFECTION, NO. OF ALLELES
line 5    0.990000 0.010000    << GENE FREQUENCIES
line 6    7 << NO. OF LIABILITY CLASSES
line 7    0.0 0.000 0.000   << PENETRANCES (* LIABILITY CLASS 1 *)
line 8    0.0 0.066 0.066   << PENETRANCES (* LIABILITY CLASS 2 *)
line 9    0.0 0.366 0.366   << PENETRANCES (* LIABILITY CLASS 3 *)
line 10   0.0 0.732 0.732   << PENETRANCES (* LIABILITY CLASS 4 *)
line 11   0.0 0.898 0.898   << PENETRANCES (* LIABILITY CLASS 5 *)
line 12   0.0 0.964 0.964   << PENETRANCES (* LIABILITY CLASS 6 *)
line 13   0.0 1.000 1.000   << PENETRANCES (* LIABILITY CLASS 7 *)
line 14   3    3  << ALLELE NUMBERS, NO. OF ALLELES
line 15   0.40 0.10 0.50
line 16   0 0   << SEX DIFFERENCE, INTERFERENCE (IF 1 OR 2)
line 17   0.0 << RECOMBINATION VALUES
line 18   1 0.001 0.001 << REC VARIED, INCREMENT, FINISHING VALUE
line 19   0.05 0.05 0.15
line 20   0.20 0.10 0.40
```

Figure 1.4.19 Parameter file for Example 3 (example3.dat).

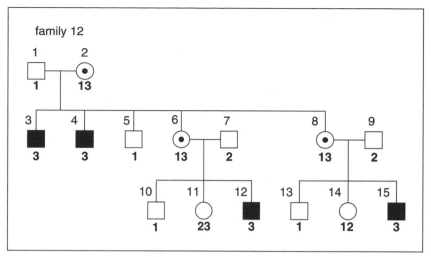

Figure 1.4.20 Pedigree for Example 4 with marker X1. A circle with a dot in the center represents an obligate carrier female.

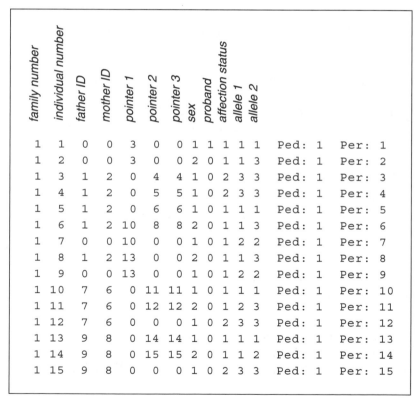

family number	individual number	father ID	mother ID	pointer 1	pointer 2	pointer 3	sex	proband	affection status	allele 1	allele 2				
1	1	0	0	3	0	0	1	1	1	1	1	Ped:	1	Per:	1
1	2	0	0	3	0	0	2	0	1	1	3	Ped:	1	Per:	2
1	3	1	2	0	4	4	1	0	2	3	3	Ped:	1	Per:	3
1	4	1	2	0	5	5	1	0	2	3	3	Ped:	1	Per:	4
1	5	1	2	0	6	6	1	0	1	1	1	Ped:	1	Per:	5
1	6	1	2	10	8	8	2	0	1	1	3	Ped:	1	Per:	6
1	7	0	0	10	0	0	1	0	1	2	2	Ped:	1	Per:	7
1	8	1	2	13	0	0	2	0	1	1	3	Ped:	1	Per:	8
1	9	0	0	13	0	0	1	0	1	2	2	Ped:	1	Per:	9
1	10	7	6	0	11	11	1	0	1	1	1	Ped:	1	Per:	10
1	11	7	6	0	12	12	2	0	1	2	3	Ped:	1	Per:	11
1	12	7	6	0	0	0	1	0	2	3	3	Ped:	1	Per:	12
1	13	9	8	0	14	14	1	0	1	1	1	Ped:	1	Per:	13
1	14	9	8	0	15	15	2	0	1	1	2	Ped:	1	Per:	14
1	15	9	8	0	0	0	1	0	2	3	3	Ped:	1	Per:	15

Figure 1.4.21 Pedigree file for Example 4 (`example4.ped`).

```
line 1   2 0 1 5  << NO. OF LOCI, RISK LOCUS, SEXLINKED (IF 1) PROGRAM
line 2   0 0.0 0.0 0 << MUT LOCUS, MUT MALE, MUT FEM, HAP FREQ (IF 1)
line 3   1  2
line 4   1   2  << AFFECTION, NO. OF ALLELES
line 5   9.9990000000E-01 1.0000000000E-04  << GENE FREQUENCIES
line 6   1 << NO. OF LIABILITY CLASSES
line 7   0 0 1.0000
line 8   0 1.0000 << PENETRANCES
line 9   3   3  << ALLELE NUMBERS, NO. OF ALLELES
line 10  0.330000 0.330000 0.340000  << GENE FREQUENCIES
line 11  0 0  << SEX DIFFERENCE, INTERFERENCE (IF 1 OR 2)
line 12  0.000 << RECOMBINATION VALUES
line 13  1 0.10000 0.45000 << REC VARIED, INCREMENT, FINISHING VALUE
```

Figure 1.4.22 Parameter file for Example 4 (example4.dat).

Example 4

Two-Point Linkage Analysis of an X-Linked Recessive Disease

In this example, MLINK is used to analyze an X-linked recessive disorder with particular reference to the difference in coding of the liability class for the affection status locus. Figure 1.4.20 shows a pedigree in which a fully penetrant X-linked recessive disorder with disease allele frequency of 0.0001 is segregating. The family has been genotyped for the three-allele marker X1. The frequency of alleles 1, 2, and 3 are 0.33, 0.33, and 0.34, respectively. Marker genotypes are shown beneath the pedigree symbols; circular pedigree symbols with dots in the center indicate a female who is an obligate carrier for this disorder. The pedigree file (example4.ped) is shown in Figure 1.4.21.

Females are coded with a 1 in the liability class because, even though they are known to be obligate carriers, they are unaffected with the disease. Look carefully at the males in the pedigree file; even though males are hemizygous at this locus and consequently have only one allele, the pedigree file has two alleles listed. For example, individual 4 has inherited the 3 allele from his mother, and in order for the linkage programs to work appropriately, his marker genotypes must be listed in the pedigree file as though he were homozygous at this locus—as a 3 3.

The parameter file (example4.dat), which was prepared by PREPLINK, is shown in Figure 1.4.22. Here, look carefully at the liability class. Although two lines are listed for penetrances in the liability class, because the disease is X-linked, the number of liability

Table 1.4.5 Two-Point Lod Scores for Family 12 versus Marker X1 (see Fig. 1.4.22)

	0.0	0.001	0.05	0.10	0.15	0.20	0.30	0.40
Two-point Lod score	2.41	2.40	2.21	2.00	1.77	1.54	1.02	0.47

classes is only one. The first line of penetrances gives penetrances for females and the second line gives penetrances for males. The female penetrance line has three values because a female can be either homozygous for the normal allele, heterozygous, or homozygous for the disease allele. The male penetrance line has only two values because a male has only two possible genotypes—either hemizygous for the normal allele or hemizygous for the disease allele. Also note that line 13 of the parameter file prompts the program to calculate lod scores only at θ values of 0.0, 0.10, 0.20, 0.30, and 0.40. This is the default specification in PREPLINK and is modified according to user specifications in LCP.

Results of the two-point linkage analysis are shown in Table 1.4.5. Because this is an X-linked disorder and thus the prior probability of linkage between the trait and marker locus is higher than for an autosomal trait, a lod score ≥ 2.0 is considered significant evidence in favor of linkage.

Reference: Terwilliger and Ott, 1994

Contributor: Marcy C. Speer

UNIT 1.5

Model-Free Tests for Genetic Linkage

Model-free procedures (see Key Concepts) for evaluating linkage are generally applied when investigating diseases and traits for which simple Mendelian inheritance patterns are not observed. In general, model-free procedures should be used to provide either a preliminary evaluation of the data for linkage or confirmatory analyses when results obtained by the usual lod score approach (*UNIT 1.4*) are unclear. The choice of model-free methods depends upon the type of data that have been gathered. GENEHUNTER PLUS (see Example 3) is an efficient tool for preliminary genomic screening of data because it can process markers from entire chromosomes jointly. However, when extended pedigrees must be divided prior to analyses with concomitant loss of power, additional follow-up studies should be implemented for any region that shows suggestive evidence for linkage by GENEHUNTER PLUS analysis (i.e., lod scores greater than perhaps 1.0 or NPL scores greater than 2). Provided that the markers are highly polymorphic, the SimIBD approach (*http://watson.hgen.pitt.edu/~davis/*) provides an easy and efficient means of testing for linkage. Affected sib pair (ASP) analyses (Example 1) can be conducted for all markers whether or not they are highly polymorphic, but they can use only data from sibships. For qualitative data (as described below), a parametric approach with the penetrance set near zero, so that only affected individuals are retained for analysis, can accommodate pedigrees that contain multiple generations of affected individuals in analyses where less polymorphic loci are available. For this type of analysis, a genetic model (dominant or recessive) is assumed, so that multiple runs must be performed and a correction for multiple tests is required. Finally, for quantitative data where the pedigrees are randomly selected, either the components-of-variance approach or the Haseman-Elston (H-E) approach (Example 2) provides excellent power as compared with parametric approaches.

When most of the data are clustered in families, ASP procedures are easily conducted and powerful (see *UNIT 1.2* for discussion of power in this context). On the other hand, if the data consist of extended families with affected individuals scattered across each family, then ASP methods will have little power and methods for pedigrees such as SimIBD or GENEHUNTER PLUS must be used. In theory, whenever appropriate data are available, multipoint analyses should be conducted. Generally available methods based on IBD cannot incorporate multipoint data. A recent adaptation of the LINKAGE program (*UNIT 1.4*) provides an approach that will allow complete use of all marker data in constructing IBD sharing for arbitrarily related individuals,

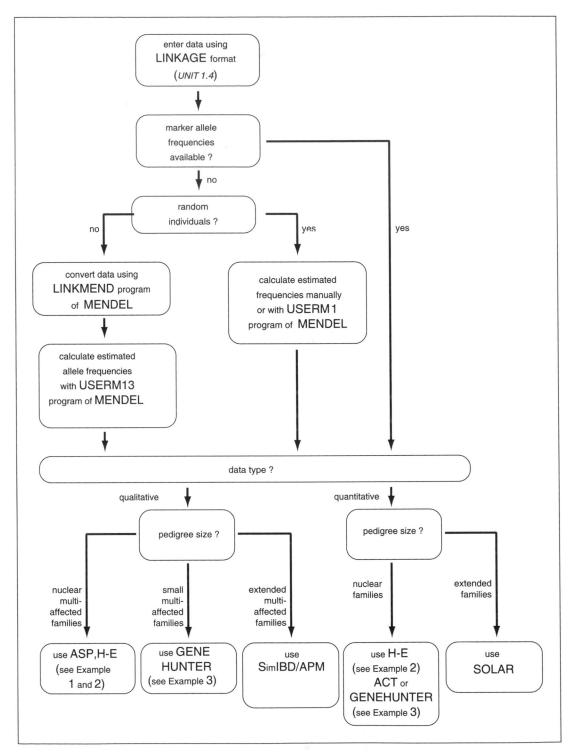

Figure 1.5.1 Decision tree for model-free linkage analysis. Abbreviations: ACT, Analysis of Complex Traits; APM, affected pedigree member method; ASP, affected sib-pair method; H-E, Haseman-Elston regression method; SOLAR, Sequential Oligogenic Linkage Analysis Routines.

but this method is computationally slow. An approximation that can be rapidly used to estimate IBD sharing from multiple linked genetic markers in nuclear families has been implemented in the SOLAR package (*http://www.sfbr.org/sfbr/public/software/solar/index.html*). Figure 1.5.1 provides a decision tree outlining the choice of methods for analysis covered in this unit.

NOTE: For all computerized analyses described in this unit, it is assumed that a Unix-based workstation will be employed. The Solaris operating system is commonly used for compiling and running most genetic software described here. In addition, GENEHUNTER PLUS requires gcc for compilation. However, SAGE has been compiled and distributed for a variety of operating systems. In general, the executable programs and source code should be kept in a directory separate from the one that is used for analysis. A softlink to the area in which the executable programs are kept can be created by typing `ln -s [executable filename]`, or the systems administrator can store the executable programs in the bin directory, where they can be used by anyone without requiring softlinks.

KEY CONCEPTS

Model-Dependent Versus Model-Free Versus Nonparametric Tests

The usual likelihood-based approaches for genetic linkage analysis require that a correct model be specified for the relationship between an individual's genotype and the corresponding chance of displaying a disease or other trait phenotype. In the commonly used lod score approach to obtaining evidence for linkage, the $\log_{10}$ of the likelihood of the data is calculated assuming a particular genetic model and a particular recombination fraction between the marker locus and disease locus. This is compared with the $\log_{10}$ of the likelihood of the data assuming the same genetic model, but with the recombination fraction between the marker and disease loci set to 50%. This approach assumes that the correct genetic model is known, i.e., that the number of alleles influencing disease susceptibility, the penetrance of each of these alleles, and the allele frequencies are correctly specified. Model-free methods, e.g., the affected-sib-pair method, do not require that the genetic model explaining disease inheritance be explicitly specified. Nonparametric (robust) procedures, additionally, avoid assumptions about the underlying statistical distribution used to determine significance.

Identity by State and Identity by Descent

Whenever a pair of individuals share the same allele at a locus, that allele is said to be identical by state (IBS). If the individuals have inherited that same allele from some common ancestor, the allele is also identical by descent (IBD). Because of Mendelian inheritance, an average of 1/4 of sib pairs will share two alleles IBD, 1/2 will share one allele IBD, and 1/4 will share no alleles IBD. To avoid confusion, in this unit the term IBD has been reserved for results obtained from analysis of a specific genomic region. The average proportion of autosomal genetic material shared between any pair of relatives is called the kinship coefficient (κ). The kinship coefficient for non-inbred pairs of individuals is $(1/2)^{R+1}$, where R is the degree of relationship between the pair, e.g., for first-degree relatives, $R = 1$ and $\kappa = 1/4$; for second-degree relatives, $R = 2$ and $\kappa = 1/8$.

IBD information is often summarized in terms of the proportion of alleles that a data set of pairs of individuals shares IBD; this proportion is indicated as π. Typically, this proportion must be estimated from the data, and the estimate of π is usually denoted by $\hat{\pi}$. This estimate is derived by calculating the sum of the probability that the pair shares two alleles IBD plus 0.5 times the probability that the pair shares one allele IBD. Figure 1.5.2 shows the pedigree of a sample family with alleles indicated. Table 1.5.1 lists the number of

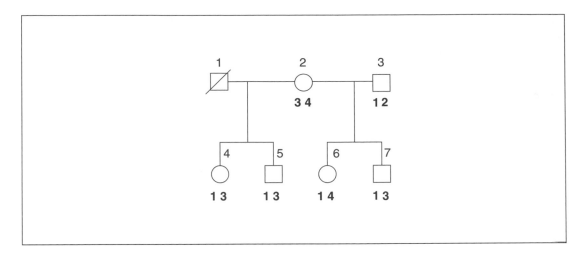

Figure 1.5.2 Sample pedigree for illustrating identity by descent and identity by state calculations. The box with the slash represents a deceased individual.

alleles shared IBS and IBD for some of the pairs of individuals in that pedigree. It is assumed that the locus under study has only four alleles.

In the sample pedigree depicted in Figure 1.5.2, it is evident that individuals 6 and 7 have inherited the same "1" allele from their father but different alleles from their mother; hence, they share one allele IBD and IBS. Although individuals 5 and 6 both have a "1" allele, this allele was not present in their common mother, individual 2; therefore, this allele "1" cannot have been inherited from the same common ancestor. Similarly, individuals 3 and 4 have a common "1" allele but have no shared ancestors. In both cases, the "1" allele is IBS but not IBD. Calculating the proportion of alleles IBD for individuals 4 and 5 is more complicated. It is obvious that a "1" allele must have been transmitted from deceased individual 1, but the entire genotype for individual 1 cannot be deduced; it could have been 1/1, 1/2, 1/3, or 1/4. Therefore, it is necessary to perform probability calculations that sum the probability of each of these possible genotypes. To calculate the proportion of alleles shared IBD for individuals 4 and 5, let p_{ij} represent the probability of genotype ij in the population from which this pedigree was sampled (with i and j each representing one of the alleles 1, 2, 3, and 4). Then, the estimate of proportion of alleles shared IBD ($\hat{\pi}$) is calculated as follows:

Table 1.5.1 Alleles Shared IBD and IBS for Some Pairs of Individuals in Figure 1.5.2

| Pair | Number of alleles shared | | Proportion of alleles shared IBD (π) | Kinship coefficient |
	IBS	IBD		
6–7	1	1	0.5	0.25
5–6	1	0	0	0.125
3 4	1	0	0	0

$$\hat{\pi} = \frac{\dfrac{3}{16}p_{11} + \dfrac{1}{16}(p_{12} + p_{13} + p_{14})}{\dfrac{1}{4}p_{11} + \dfrac{1}{16}(p_{12} + p_{13} + p_{14})}$$

The type of tedious calculation shown above is automated as a part of the SIBPAL (see Example 2) or GENEHUNTER (see Example 3) procedures. In a pedigree, individuals whose descendants have been studied but whose parents have not are called founding parents. Individuals marrying into the pedigree who are not affected with the disease being studied are one class of founding parents (often called marry-ins). In the sample pedigree shown in Figure 1.5.2, the founding parents are 1, 2, and 3.

Statistical Terms: Mean, Variance, Skewness, and Kurtosis

Four measures characterizing a distribution of values are used in this unit. Letting x_i represent the value of the ith observation with i varying from 1 to n, the *mean* ($\bar{x}$) is defined as:

$$\bar{x} = \frac{1}{n}\sum_{i=1}^{n} x_i$$

The usual estimator of the *variance* ($\hat{\sigma}^2$) is defined as:

$$\hat{\sigma}^2 = \frac{1}{n-1}\sum_{i=1}^{n}(x_i{}^*)^2$$

where $x_i{}^*$ represents $x_i - \bar{x}$. Variance is a measure of the spread of values about the common mean. The standard deviation is the square root of the variance.

The *skewness* is calculated by an equation analogous to that for variance, except that $x_i{}^*$ is cubed rather than squared, and then divided by the cube of the standard deviation. For a normal distribution, the skewness is 0, indicating that the distribution is symmetrical about its mean. Positive skewness suggests that most values are clustered at the low end of the distribution but the remaining values tend to be relatively high.

The *kurtosis* is also calculated by an equation analogous to the one for variance, but $x_i{}^*$ is raised to the fourth power and divided by the fourth power of the standard deviation, and the value 3 is then subtracted. Kurtosis reflects the flatness or peakedness of the curve of distribution relative to the normal distribution, which has zero kurtosis.

Hypothesis Testing

Linkage analysis generally compares two competing hypotheses—that of linkage and that of no linkage. This is done by means of hypothesis tests, which contrast two statements about a set of data, e.g., a *null hypothesis* (of no linkage) and an *alternate hypothesis* (of linkage). The null hypothesis is constructed by generating a statistic in which some of the parameters that are used to characterize the data are set to particular values (i.e., constrained). For genetic studies, the alternate hypothesis is usually constructed using a broader set of constraints on the parameters that characterize the data than those used to construct the null hypothesis, so that the alternate hypothesis typically includes the null hypothesis as a possibility. A simple alternate hypothesis can be postulated in which the parameters that were constrained to a particular value under the null hypothesis are constrained to an alternative value under the alternate hypothesis. For example, it is possible to test a null hypothesis in which the mean of a distribution is 0 against a simple alternate hypothesis that the mean is -1.0, or against a one-sided alternate hypothesis in which the mean of the distribution is ≤ 0, or against the two-sided alternate hypothesis that the mean is any value. The *significance* of a hypothesis test is the probability of rejecting the null hypothesis given that it is actually true, i.e., the probability of making a false judgment against the null hypothesis. The *power* of a hypothesis test is the probability of correctly rejecting the null hypothesis given that a simple alternate hypothesis is true. Hypothesis tests are constructed by developing a test statistic for evaluating a specific null hypothesis, then comparing this statistic to a reference distribution that would be generated from similar data based on the assumption that the null hypothesis is true. The significance of the test statistic is assessed by comparing the observed value to the values provided by the reference distribution, and represents the probability of having obtained the observed test statistic (or one that is more extreme) if the null hypothesis were true. In many cases, a suitable reference distribution for assessing significance can be obtained from theoretical calculations. For the cases discussed in this unit, the reference distributions are either the normal or the χ^2 distribution, for which any standard statistics text or book of mathematical tables can be consulted. Standard statistical software packages, e.g, Statistical Analysis System (SAS; available from SAS Institute), can also be used to assess significance.

Example 1

Simple Methods for Linkage Testing in Affected Sib Pairs

Three tests using IBD information are widely used for genetic linkage testing with affected pairs of sibs. These can be carried out without relying on a computer program. The *means test* evaluates whether there is an excess proportion of estimated alleles shared IBD ($\hat{\pi}$) among affected individuals. The *proportions test* compares, among sib pairs, the proportion sharing both alleles IBD to 1/4, which is the proportion expected under the null hypothesis of no linkage. The *goodness-of-fit test* compares the proportions of sibs sharing zero, one, or two alleles IBD to 1/4, 1/2, and 1/4, respectively (again, the proportions expected under the null hypothesis). The power (UNIT 1.2) of each of these tests depends upon the underlying genetic mechanism involved in the disease etiology. Constructing the optimal test for linkage requires knowledge of the genetic mechanism, and this is not generally feasible a priori. However, for most alternate hypotheses, the means test is more powerful than the other tests.

For some highly polymorphic loci, e.g., the human histocompatability leukocyte antigen (HLA) system and microsatellite markers, IBD sharing can be immediately identified. As an example, consider the data presented in Table 1.5.2, in which genetic linkage between the HLA-DR locus and insulin-dependent diabetes mellitus (IDDM) is documented. Among the 137 affected sib pairs, 59% were observed to share two alleles IBD, 34% were observed to share one allele IBD, and 7% were observed to share no alleles IBD. Let n represent the number of affected

Table 1.5.2 IBD Sharing for the HLA System Among 137 Sib Pairs Affected by Insulin-Dependent Diabetes Mellitus[a]

	f_2	f_1	f_0	total
n	81	46	10	137
Expected ratios	1/4	1/2	1/4	
Expected numbers	34	69	34	

[a]Definitions: n = number of affected sib pairs; f_2 = sib pairs sharing 2 alleles IBD; f_1 = sib pairs sharing 1 allele IBD; f_0 = sib pairs sharing 0 alleles IBD.

sib pairs, f_2 the proportion of sib pairs sharing two alleles IBD, f_1 the proportion of sib pairs sharing one allele IBD, and f_0 the proportion of sib pairs sharing no alleles IBD. The means tests compares the mean proportion of alleles shared IBD among affected sib pairs to 0.5, i.e., it tests whether $f_2 + 1/2f_1 > 0.5$. Using data from Table 1.5.2, the test statistic (t_2) is calculated as follows:

$$t_2 = (2f_2 + f_1 - 1)\sqrt{2n}$$
$$t_2 = [2(0.59) + 0.34 - 1]\sqrt{2(137)} = 8.61$$

The proportions test compares f_2 to 1/4. Using data from Table 1.5.2, the test is calculated as follows:

$$t_1 = \left(f_2 - \frac{1}{4}\right)^2 4\sqrt{\frac{n}{3}}$$
$$t_1 = (0.59 - 0.25)4\sqrt{\frac{137}{3}} = 9.19$$

These two tests can be compared with a normal distribution provided that $n > 60$. Finally, a goodness-of-fit test can be constructed by comparing observed to expected numbers of individuals in each of the allele-sharing categories. Using the data from Table 1.5.2, the goodness-of-fit test is constructed as follows:

$$\chi^2 = 2n\left[2\left(f_2 - \frac{1}{4}\right)^2 + \left(f_1 - \frac{1}{2}\right)^2 + 2\left(f_0 - \frac{1}{4}\right)^2\right]$$
$$\chi^2 = 2(137)[2(0.59 - 0.25)^2 + (0.34 - 0.5)^2 + 2(0.07 - 0.25)^2] = 88.12$$

The test statistic obtained from this equation can then be compared to a χ^2 distribution having two degrees of freedom.

The three tests above (using the data from Table 1.5.2) all yield p values <0.0001. The slightly more significant test statistic derived from the proportions test (9.19) as compared to that obtained from the means test (8.61) suggests that the effects of a locus linked to the HLA-DR locus are recessive-acting. Although mathematically simple, the goodness-of-fit test is generally less powerful than the means test and is not often used for sib-pair analyses.

Example 2

Using SIBPAL Software to Perform Affected-Sib-Pair Analyses

The tests described in Example 1 can be easily used when IBD is readily apparent, e.g., when the markers are highly polymorphic and parental genotypes are available. However, in many situations these conditions are not fulfilled and IBD can only be estimated. This is a computationally complex task requiring computer programs. Moreover, when IBD sharing is calculated from data, the formulae given by the equations above are conservative and will have decreased power compared with tests that estimate the variance of IBD sharing from the observed data.

SIBPAL
SIBPAL, one of the modules of the SAGE package of programs, can perform several model-free tests for genetic linkage, two of which are described here. The SIBPAL procedure provides an

efficient algorithm for estimating by using all data available in nuclear families. Tests available in SIBPAL include the means tests (Example 1) and the Haseman-Elston (H-E) test.

The Haseman-Elston method

The H-E method was originally developed for quantitative traits, but has been adapted for use with qualitative traits; the latter application is the focus of the description here. In the original formulation of H-E, Y_{ij} may be allowed to represent the phenotype (usually a quantitative value) for the jth individual ($j = 1,2$) in the ith pair of relatives from a pedigree. The H-E test then uses the transformation $Z_i = (Y_{i1} - Y_{i2})^2$. Subsequently, the values of Z_i are regressed on $\hat{\pi}_i$, where i indexes the sib pair. The regression coefficient for any set of pairs of relatives depends only upon the recombination fraction and the genetic variance attributable to any putative genetic locus that is linked to the marker being considered. Linkage is tested by determining whether this regression coefficient is negative. A more detailed discussion is available in the SIBPAL documentation.

Performing an analysis using SIBPAL

Prior to performing an analysis, data for use by SIBPAL must first be processed using the Family Structure Program (FSP), another module of SAGE. To obtain a copy of SIBPAL and FSP or the entire SAGE (v. 3.1) package, write to Attention: SAGE, Robert Elston, Ph.D., Case Western Reserve University School of Medicine Department of Epidemiology and Biostatistics, Metro Health Sciences Center, 2500 Metro Health Drive, Cleveland, Ohio 44109. The FSP and SIBPAL modules are extremely well documented, so their use should be straightforward once they are compiled. They are distributed with detailed directions concerning installation, which will have to be performed by a programmer.

The SAGE package requires specific column-related formats for input. Prior to running SIB-PAL it is necessary to preprocess the data using FSP. FSP will create pointers that are needed to divide the data into nuclear pedigrees. In addition, FSP will check for inconsistencies in the pedigree structure, and provide summary statistics. It can also create output for use by model-dependent linkage and segregation analysis programs, which are also included in the SAGE package.

Here the steps used in running SIBPAL are described, starting with a data file provided in LINKAGE format containing data based on the pedigree illustrated in Figures 1.4.3 and 1.4.9.

1. Modify the data to fit the FSP format. FSP can use data that are in standard LINKAGE format once the parents of founding individuals are set to blank (instead of 0s as required by MAKEPED; UNIT 1.4). For example, in Figure 1.4.9, the father and mother IDs (which are 0s) in lines 1, 2, 3, 6, 8, 9, 13, 15, 18, 19, and 23 must be replaced with blanks as shown in Figure 1.5.3. It is also necessary to modify the marker data. The simplest coding scheme for the marker genotypes requires that the allele numbers follow a nine-character format in which the first four characters are occupied by the first allele number (left justified) followed by a slash (/) and then the next allele number (left justified). If this coding scheme is used, the allele numbers must be ordered on each line, e.g., 2 . . ./3 . . . , where each dot represents a blank. For instance, the recoded genotype 3 . . ./2 . . . , which is indicated for individual 16 in Figure 1.4.9, would be read by SIBPAL as missing. Missing marker data should be indicated by blanks (but 0s will also be read as missing). The reorganized data needed to run FSP and SIBPAL are illustrated in Figure 1.5.3.

2. Create the parameter file for FSP, illustrated in Figure 1.5.4. This parameter file can be created de novo by completing queries on the SAGE World Wide Web server at *http://darwin.cwru.edu/sagegui/main-menu.html*. The first line of the FSP parameter file is a title card, which can have up to 80 characters. To construct nuclear families for SIBPAL, the second line should have a 1 in column 5 and 0s in columns 10, 15, and 20. Column 25 indicates the number of records per individual in the data file (which is 1 in this example). Columns 26 and 27 indicate the symbols for males and females

family number	individual number	father ID	mother ID	sex	affection status	marker 1		marker 2		marker 3	
1113	1	0	0	1	2						
1113	2	0	0	2	1						
1113	3	0	0	2	1						
1113	4	1	2	1	2	1	/3	1	/2	1	/2
1113	5	1	2	2	2						
1113	6	0	0	1	1						
1113	7	1	2	2	2						
1113	8	0	0	1	1						
1113	9	0	0	1	1						
1113	10	4	3	2	2	3	/4	2	/3	1	/3
1113	11	4	3	2	1	1	/4	1	/3	2	/2
1113	12	4	3	2	2						
1113	13	0	0	1	1	3	/3	2	/2	3	/3
1113	14	6	5	1	1	1	/2	1	/1	2	/3
1113	15	0	0	2	1	1	/4	1	/1	3	/3
1113	16	6	5	1	2	2	/3	1	/2	1	/2
1113	17	6	5	2	2	2	/3	1	/2	1	/2
1113	18	0	0	1	1	2	/2	2	/2	2	/3
1113	19	0	0	2	1	1	/2	3	/3	3	/3
1113	20	8	7	1	2						
1113	21	8	7	2	1	1	/4	1	/1	2	/2
1113	22	8	7	2	2	3	/4	1	/2	1	/2
1113	23	0	0	1	1						
1113	24	9	10	1	2	2	/3	2	/2	3	/3
1113	25	9	10	1	2	3	/3	1	/2	1	/2
1113	26	9	10	2	1	3	/4	1	/3	2	/3
1113	27	13	12	1	1	3	/4	2	/3	3	/3
1113	28	13	12	2	2	3	/3	2	/2	3	/3
1113	29	13	12	2	1	3	/4	2	/3	3	/3
1113	30	16	15	1	2	3	/4	1	/1	2	/3
1113	31	16	15	2	1	2	/4	1	/1	2	/3
1113	32	18	17	2	2	2	/3	2	/2	1	/3
1113	33	18	17	1	2	2	/3	2	/2	1	/2
1113	34	20	19	1	2	1	/3	2	/3	1	/3
1113	35	20	19	2	2	2	/3	2	/3	1	/3
1113	36	20	19	1	2	2	/4	1	/3	2	/3
1113	37	20	19	2	1	1	/4	1	/3	2	/3
1113	38	23	22	2	1	1	/4	1	/3	2	/3
1113	39	23	22	1	1	1	/4	1	/3	2	/3
1113	40	23	22	2	2	1	/3	2	/2	1	/3

Figure 1.5.3   Data file representing pedigree shown in Figure 1.4.3, modified for use with SIBPAL.

```
                    FSP RUN, EXAMPLE3

                    1   0   0   0 112

         (T5,A1,T1,I4,3X,A2,4X,A2,4X,A2,4X,A1)
```

Figure 1.5.4 Parameter file for an FSP run, corresponding to data in Figure 1.5.3.

(1 and 2 in this example). The last line must have the FORTRAN format for reading in the data, starting and ending with parentheses. The data that must be read include a study ID in character format (up to five characters are allowed), the family ID in integer format (up to five integers allowed), the individual's and parents' IDs in character format (up to eight characters each), and the sex (one character). The study ID may be blank. To read in the data that is used in Example 2 of *UNIT 1.4*, for instance, the format (T5,A1,T1,I4,3X,A2,4X,A2,4X,A2,4X,A1) could be used (see Fig. 1.5.4). In this format, T indicates transferring to a particular column, X indicates that one skips columns, A indicates character format, and I indicates integer format. Because there is no study ID, a blank field is read in column 5.

3. Invoke the SAGE program by typing sage31 at the prompt (or whatever alias the programmer has created for invoking SAGE). Once SAGE is invoked, a menu will appear to run the SAGE modules. At the first prompt, type fsp to invoke the FSP program. At the next prompts, type fsp.par to indicate the name of the parameter file to run FSP, and example3.pre to indicate the data file to analyze. Although FSP contains many data checks, some errors that cannot be detected may cause the program to crash. Common problems that can cause an error without the display of a clear error message include failing to provide a value where needed in the parameter file (fsp.par) or providing an incorrect format statement in that file. If FSP aborts without providing a display showing that it has processed any families, recheck the parameter file. If FSP or SIBPAL abort without completing their runs, check for and remove any core dumps, as these will consume disk space unless deleted.

4. At the completion of an FSP run, two files are generated. The first, called fsp.inf, provides detailed statistics concerning the pedigree structure. The second, fsp.lnk, includes pointers describing the nuclear family structure of the data.

5. Create the parameter file (sibpal.par) for use by SIBPAL, illustrated in Figure 1.5.5. The requirements for this parameter file are clearly described in the documentation. In line 1, a title (up to 80 characters) is needed. Line 2 can be blank: it allows the user to ask for optional output and/or to reflect the data that have been read in. Line 3 should have:

> a 2 in column 5 to indicate a disease outcome,
> a 3 in column 10 to indicate three markers,
> a 1 in column 15 to indicate a single disease outcome,
> a 1 in column 20 to indicate univariable regressions only,
> a 0 in column 25 to indicate that no weights will be used,
> a 0 in column 30 to suppress plotting of the data, and
> a 0 in column 35 to indicate the absence of covariates.

The weighting scheme provides theoretically more powerful tests for linkage, provided ≥300 sib pairs are available for analysis. The format statement needed for line 3 is (T5,A1,T1,I4,3X,A2,T31,F1.0,T36,A9,4X,A9, 4X,A9). In this statement, a blank field is read for the study ID (T5,A1), the family ID is in column 1

```
                ANALYSIS FOR A SIMPLE MENDELIAN DISEASE
            1   1   1   1
            2   3   1   1   0   0   0
            dis   0   0   2   1

            (T5,A1,T1,I4,3X,A2,T31,F1.0,T36,A9,4X,A9,4X,A9)
```

Figure 1.5.5   SIBPAL parameter file for an affected-sib-pair analysis.

(T1,I4), the individual ID is read as A2, and the disease phenotype is read in real format as F1.0. The three markers have been read using character format (A9).

6. Create a locus description file, illustrated in Figure 1.5.6. Because the marker is an autosomal codominant system (see *UNIT 1.4*), a simple parameter file describing the allele frequencies can be used. For the first marker, the first line is the marker name (e.g., M2) and the second through fifth lines are $1 = 0.3$, $2 = 0.3$, $3 = 0.05$, and $4 = 0.35$ (to indicate the allele frequencies). The next two lines have semicolons to indicate that all of the allele numbers are given and that the genotypes should be automatically generated. For marker systems in which one allele is dominant with respect to another, a more complex data structure is required in which the second genotype is replaced by lines such as $1 = \{1/1, 1/2\}$, which would indicate that phenotype 1 results from the 1/1 genotype or the 1/2 genotype.

7. SIBPAL is now ready to run. Type sage31 at the prompt, then sibpal once the SAGE menu appears. The user will be asked to provide names for the parameter file, locus description file, data file, and fsp-pointer file (previously named fsp.lnk by the FSP program). During the run, messages are provided indicating that the IBD status of

```
                    M2
                    1 = 0.3
                    2 = 0.3
                    3 = 0.05
                    4 = 0.35
                    ;
                    ;
                    M3
                    1 = 0.4
                    2 = 0.1
                    3 = 0.5
                    ;
                    ;
                    M4
                    1 = 0.05
                    2 = 0.4
                    3 = 0.55
                    ;
                    ;
```

Figure 1.5.6   Locus description file for SIBPAL analysis.

Table 1.5.3 Results from SIBPAL Means Test Analysis[a]

Locus	No. sibs affected	Pairs	Mean	Std. Dev.	Std. Err.	T values	p values[b]
M2	0	2	0.796	0.065	0.046	6.429	0.0117*
M2	1	14	0.317	0.164	0.044	4.174	0.0005**
M2	2	6	0.519	0.298	0.122	0.156	0.4406
M3	0	2	0.875	0.177	0.125	3.000	0.0477*
M3	1	14	0.310	0.237	0.063	2.999	0.0048**
M3	2	6	0.547	0.252	0.103	0.459	0.3314
M4	0	2	0.668	0.177	0.125	1.340	0.1561
M4	1	14	0.398	0.209	0.056	1.832	0.0442*
M4	2	6	0.433	0.326	0.133	−0.506	1.0000

[a]Definitions: Std. Dev., standard deviation; Std. Err., standard error.

[b]Single asterisk indicates $p < 0.05$. Double asterisk indicates $p < 0.01$.

the markers has been derived and then other messages are displayed to show progress in completing the analyses using each of the phenotypes (in this case, only one was specified). If the program does not provide these messages, there may be a mistake in the parameter file (`sibpal.par`), commonly resulting from an incorrect format statement. Also, once the job has been run, output files (`sibpal.sum`, `sibpal.out`, and optionally `sibpal.opt`) have to be deleted or moved before SIBPAL can be run again.

Interpretation of results

The primary results are written to unit 22. The salient results returned by SIBPAL are shown in Table 1.5.3 for the means test and in Table 1.5.4 for H-E linear regression analysis. Some results have been rounded off to three significant digits.

In Table 1.5.3, the first row provides an analysis for marker M2 that includes all sib pairs in which neither sib was affected (there were only two such pairs). The mean proportion of alleles at this marker locus was 0.796 and the standard deviation was 0.065 with the standard error of the mean being 0.046. The T value is calculated as the mean minus its expected value (0.5) divided by the standard error of the mean; here $T = (0.796 − 0.5)/0.046 = 6.429$. Comparing this value with a t distribution that has one degree of freedom provides a p value of 0.0117. The fourth and seventh rows provide similar means-test results for markers M3 and M4, including all sib pairs where neither sib was affected, and comparing the proportion of alleles among such unaffected sib pairs to 0.5 in the column labeled T values. The second, fifth, and eighth rows provide the same test with respect to markers M2, M3, and M4 for discordant

Table 1.5.4 Results from SIBPAL Haseman-Elston Linear Regression Analysis

Trait	Locus	Effective D.F.[a]	Full sibs π mean	Regress Y on π			
				T values	p values[b]	Intercept	Slope
dis	M2	13	0.416	−2.805	0.007**	1.083	−1.074
dis	M3	13	0.426	−2.830	0.007**	1.030	−0.925
dis	M4	13	0.432	−0.841	0.208	0.800	−0.380

[a]D.F., degrees of freedom.

[b]Double asterisk indicates $p < 0.01$.

(affected-unaffected) sib pairs. The third, sixth, and ninth rows do the same with respect to the three markers for sib pairs in which both sibs were affected.

Under the hypothesis of genetic linkage, the concordant pairs are expected to share a larger fraction of alleles IBD than 0.5, and the discordant pairs to share a smaller fraction of alleles IBD than 0.5. The column labeled p values reports results from one-sided hypothesis tests to evaluate that the mean number of alleles IBD among concordant pairs is >0.5 and that the mean number of alleles IBD among discordant pairs is <0.5. For genetic effects with low penetrance, only the affected sib pairs would provide appreciable information for detecting genetic linkage. However, because the disease gene under study in this example is fully penetrant, both the discordant and concordant affected sib pairs provide evidence for linkage. There are more discordant pairs than concordant pairs, so the power to detect linkage from this pairing is greater than for the concordant affected pairs. The results from Table 1.5.3 suggest that markers M2 and M3 are linked to the disease gene. For diseases with low penetrance, the affected sib pairs contain virtually all linkage information, and results from studies of discordant sib pairs or concordant unaffected sib pairs are not meaningful.

Table 1.5.4 provides results from the Haseman-Elston regression analysis. The first column indicates that a disease phenotype is under consideration, the second indicates the marker locus being considered, the third provides an estimate of the degrees of freedom available for constructing a hypothesis test, the fourth indicates the IBD sharing among all sib pairs (including both concordant and discordant pairs), the fifth shows results of the t test for the regression of concordance for disease on IBD sharing at the marker locus, and the sixth provides the p value from this test by comparing the T value to a t distribution that has thirteen degrees of freedom. Finally, the seventh and eighth columns provide estimates of the regression coefficients. These results provide evidence that markers M2 and M3 are linked to the disease locus (indicated by the asterisks in the sixth column). There is no significant evidence that marker M4 is linked to the disease locus.

Example 3

Analysis using GENEHUNTER and GENEHUNTER PLUS

Model-free genetic linkage analysis can also be performed using exact methods to calculate the IBD sharing among relatives from extended families. This section describes the use of GENE-HUNTER and GENEHUNTER PLUS for this purpose, as this software is relatively easy to use, and easily performs analyses that include many markers. GENEHUNTER software uses a modified version of the Green-Lander algorithm, where computational time required to complete an analysis is linear in the number of markers, so that, in general, entire chromosomes can be analyzed in a single run. This facilitates genome scanning because the output from a run of a single chromosome can be easily managed. However, the computational requirements (both time and memory) increase exponentially with the number of pedigree members, limiting the application of this algorithm to families of ~ 25 subjects or less. Larger families must be divided into smaller ones in order to use GENEHUNTER, and this division usually leads to some loss of information. When using model-free methods that only weight affected individuals, the first step in pedigree size reduction is to eliminate unaffected individuals. However, because unaffected individuals often contain some information about potential marker genotypes for untyped individuals, this type of reduction may also lead to some loss of information. Therefore, unaffected siblings of affected individuals should not be deleted if the parents have not been genotyped and the unaffected siblings have different genotypes from any other retained sibling in the sibship.

GENEHUNTER PLUS is a modification of GENEHUNTER that allows for incomplete marker information in constructing a test for linkage. Simulation studies and applications to data have shown that this modification leads to a more powerful test for linkage than the NPL score test that is given by GENEHUNTER. Online documentation and a PDF version of the documentation for GENEHUNTER 2.0 can be obtained from *http://linkage.rockefeller.edu/soft/gh/*.

Performing an analysis using GENEHUNTER and GENEHUNTER PLUS
These programs can be obtained using any Web browser from *http://www.fhcrc.org/labs/kruglyak/Downloads/index.html* and clicking the appropriate link to download ghp-1.2.tar.gz (if that is the most current version of the software). Once the files have been downloaded, the files must be uncompressed and detarred by typing gunzip ghp-1.2.tar.gz and tar xvpf ghp-1.2.tar. They may have to be recompiled using the Make utilities that are provided.

1. Organize the data. The pedigree structure provided by Figure 1.4.9 is too large to be analyzed directly by GENEHUNTER PLUS. It must first be cut into smaller families. One way to cut the family is to ignore common parentage of individuals 4, 5, and 7. Then, a revised version of the pedigree file given by Figure 1.4.4 will have individuals 1 and 2 deleted, and the parents of individuals 4, 5, and 7 will be indicated as missing. The family identification numbers will also need to be renumbered so that the descendants of 4, 5, and 7 are in independent families as shown in Figure 1.5.7.

2. Run GENEHUNTER. GENEHUNTER uses a C-shell script language in order to call subroutines. To enter the scripting language, type ghp. Online help can provide assistance in the selection of commands. However, a set of commands (here called script.in)

Family ID	Individual ID	Father ID	Mother ID	Sex	Affection	Marker1 allele1	Marker1 allele2	Marker2 allele1	Marker2 allele2	Marker3 allele1	Marker3 allele2
1	3	0	0	2	1	0	0	0	0	0	0
1	4	0	0	1	2	1	3	1	2	1	2
1	9	0	0	1	1	0	0	0	0	0	0
1	10	4	3	2	2	3	4	2	3	1	3
1	11	4	3	2	1	1	4	1	3	2	2
1	12	4	3	2	2	0	0	0	0	0	0
1	13	0	0	1	1	3	3	2	2	3	3
1	23	0	0	1	1	0	0	0	0	0	0
1	24	9	10	1	2	3	2	2	2	3	3
1	25	9	10	1	2	3	3	1	2	1	2
1	26	9	10	2	1	3	4	1	3	2	3
1	27	13	12	1	1	3	4	2	3	3	3
1	28	13	12	2	2	3	3	2	2	3	3
1	29	13	12	2	1	4	3	2	3	3	3
2	5	0	0	2	2	0	0	0	0	0	0
2	6	0	0	1	1	0	0	0	0	0	0
2	14	6	5	1	1	1	2	1	1	2	3
2	15	0	0	2	1	1	4	1	1	3	3
2	16	6	5	1	2	3	2	1	2	1	2
2	17	6	5	2	2	3	2	1	2	1	2
2	18	0	0	1	1	2	2	2	2	2	3
2	30	16	15	1	2	3	4	1	1	2	3
2	31	16	15	2	1	4	2	1	1	2	3
2	32	18	17	2	2	3	2	2	2	1	3
2	33	18	17	1	2	2	3	2	2	1	2
3	7	0	0	2	2	0	0	0	0	0	0
3	8	0	0	1	1	0	0	0	0	0	0
3	19	0	0	2	1	1	2	3	3	3	3
3	20	8	7	1	2	0	0	0	0	0	0
3	21	8	7	2	1	1	4	1	1	2	2
3	22	8	7	2	2	3	4	1	2	1	2
3	23	0	0	1	1	0	0	0	0	0	0
3	34	20	19	1	2	1	3	2	3	1	3
3	35	20	19	2	2	3	2	2	3	1	3
3	36	20	19	1	2	4	2	1	3	2	3
3	37	20	19	2	1	1	4	1	3	2	3
3	38	23	22	2	1	1	4	1	3	2	3
3	39	23	22	1	1	1	4	1	3	2	3
3	40	23	22	2	2	1	3	2	2	1	3

Figure 1.5.7 Pedigree data needed for GENEHUNTER PLUS analysis.

can also be run in batch mode by attaching a file when ghp is invoked, by typing ghp <
script.in. The following is an example of a set of commands to run GENEHUNTER.
Note that number lines should not be supplied to GENEHUNTER and are only used in
this text for clarity.

```
 1. photo ghp_example3.out
 2. max bits 20
 3. skip large off
 4. load example3.par
 5. map function kosambi (default)
 6. increment step 5 (default)
 7. off end 0.10
 8. ps on
 9. haplotype off
10. score all (default)
11. scan example3.gpre
12. total stat het
13. npl_ghp_example3.ps
14. lod_ghp_example3.ps
15. inf_ghp_example3.ps
16. quit
```

In order, these commands tell GENEHUNTER to (1) store output from the run in a file
ghp_example3.out; (2) increase the maximum memory size to 20 bits from the
default of 16; (3) turn off the option to skip large pedigrees; (4) load the linkage locus
description file example3.par, which is the same as that given in Figure 1.4.11; (5) use
a Kosambi map function; (6) use five steps between markers; (7) compute lod scores 10
recombination units past the end of each set of markers; (8) provide PostScript files; (9) *not*
provide most probable haplotypes of all individuals; (10) use the all-test statistic (default);
(11) analyze the data in example3.gpre shown in Figure 1.5.7; (12) give results over
all pedigrees analyzed including heterogeneity tests; (13-15) name of PostScript files for
NPL scores, lod scores, and marker informativity maps, respectively; and (16) quit the
program. Note, that the default map function for GENEHUNTER is Kosambi, which is
appropriate for mice. A Kosambi map function is considered acceptable for humans. Note
that GENEHUNTER skips the first distance in the file (on line 15) example3.par
shown in Figure 1.4.11 (if the first number in line 2 is a 1, indicating the trait locus), as
it assumes that this is a distance to be estimated between the disease locus and the set of
marker loci, each of which have fixed locations.

Results from this analysis are given in Table 1.5.5. The first column gives positions on
the chromosome in centimorgans (cM), with the first marker set to 0. The second column
gives the lod score under the assumed model given by the example3.par file. These
lod scores are less than those given by Figures 1.4.13 and 1.4.17, reflecting the loss of
information that occurred when the pedigrees were divided into smaller families. The
next column gives the lod score under the parametric model given by example3.par,
with an added parameter to model possible genetic heterogeneity. In the third column, the
parameter alpha reflects the evidence for heterogeneity in the data; an alpha of 1.0 reflects
no hetereogeneity, while one near 0 indicates a great deal of heterogeneity, with very
few families showing evidence for linkage. The fourth column gives the nonparametric
linkage score (NPL). High NPL scores support evidence for linkage. The NPL score is
normally distributed for completely informative marker loci. Thus, an NPL score is not
equivalent to a lod score. An NPL score can be converted to the same units as a lod score
by squaring it and dividing by 4.6. For less than fully informative loci, the NPL score is
generally less powerful than the ASM test available from GENEHUNTER PLUS. The

Table 1.5.5 Results from GENEHUNTER Analysis

Position (cM)	Lod score	(alpha,hlod)	NPL score	p value	Information
−10.14	3.475	(1.000,3.475)	2.013	0.0593	0.3204
−8.11	3.594	(1.000,3.594)	2.235	0.0398	0.3784
−6.08	3.685	(1.000,3.685)	2.480	0.0269	0.4463
−4.05	3.726	(1.000,3.726)	2.746	0.0164	0.5267
−2.03	3.653	(1.000,3.653)	3.033	0.0106	0.6247
0.00	2.829	(0.878,2.852)	3.343	0.0083	0.7632
0.62	2.759	(0.895,2.774)	3.151	0.0095	0.7345
1.24	2.644	(0.901,2.657)	2.956	0.0116	0.7245
1.86	2.462	(0.890,2.477)	2.760	0.0162	0.7249
2.48	2.139	(0.855,2.168)	2.567	0.0231	0.7357
3.09	−0.282	(0.375,1.321)	2.380	0.0312	0.7646
4.15	−0.189	(0.385,1.134)	1.779	0.0818	0.7164
5.20	−0.334	(0.384,0.887)	1.275	0.1302	0.6963
6.25	−0.695	(0.361,0.543)	0.884	0.1612	0.6930
7.31	−1.415	(0.200,0.063)	0.611	0.1862	0.7069
8.36	−4.992	(0.000,−0.000)	0.453	0.2150	0.7503
10.39	−1.334	(0.007,−0.001)	0.429	0.2206	0.6151
12.42	−0.379	(0.323,0.091)	0.405	0.2272	0.5188
14.44	0.119	(0.550,0.250)	0.376	0.2342	0.4397
16.47	0.421	(0.840,0.427)	0.348	0.2404	0.3729
18.50	0.614	(0.997,0.613)	0.320	0.2492	0.3158

fifth column provides an estimate of the p value for the NPL score, assuming that it is normally distributed. The last column gives the information content for the set of markers that have been studied; an information content of 1.0 indicates complete information, while an information of 0.0 would be completely noninformative.

In this example, there is relatively high informativity within a few cM of the three markers that were typed, so the p value that was derived from the NPL score should be accurate for this region. However, for more distant parts of the chromosome, where marker informativity was low (i.e., less than ~0.6), the p values are likely to be quite conservative. To provide more accurate p values, one can apply the ASM test from GENEHUNTER PLUS.

3. Run the ASM program. Assuming that GENEHUNTER PLUS has been downloaded and run with the ghp command, all of the files that are needed to run this program should be available. The options for ASM include linear versus exponential models for the effect that increased allele sharing has upon risk for disease development. A linear model is reasonable for additive or dominant models, while an exponential model would be more powerful for a recessive mode of inheritance. The program searches at each location for a value of a parameter δ that reflects the increased allele sharing among pairs of relatives, due to linkage of the disease and marker loci. One can select ranges for a grid search, which would be appropriate when population prevalences restrict the range of the δ parameter. One is not required to specify ranges for the search. For the current analysis, if one types asm lin, the results in Table 1.5.6 are obtained.

Table 1.5.6 Output from GENEHUNTER PLUS Analysis

Position (cM)	NPL score	KC score	KC lod score	δ value
−1.01E+01	2.01E+00	2.12E+00	9.73E−01	1.06E+00
−8.11E+00	2.23E+00	2.19E+00	1.04E+00	1.06E+00
−6.08E+00	2.48E+00	2.27E+00	1.11E+00	1.06E+00
−4.05E+00	2.75E+00	2.34E+00	1.19E+00	1.06E+00
−2.03E+00	3.03E+00	2.41E+00	1.26E+00	1.06E+00
0.00E+00	3.34E+00	2.48E+00	1.33E+00	1.06E+00
6.19E−01	3.15E+00	2.43E+00	1.28E+00	1.06E+00
1.24E+00	2.96E+00	2.37E+00	1.22E+00	1.06E+00
1.86E+00	2.76E+00	2.30E+00	1.15E+00	1.06E+00
2.48E+00	2.57E+00	2.21E+00	1.06E+00	1.06E+00
3.09E+00	2.38E+00	2.10E+00	9.58E−01	1.06E+00
4.15E+00	1.78E+00	1.96E+00	8.36E−01	1.06E+00
5.20E+00	1.28E+00	1.79E+00	6.99E−01	1.06E+00
6.25E+00	8.84E−01	1.59E+00	5.52E−01	1.06E+00
7.31E+00	6.11E−01	1.37E+00	4.10E−01	1.06E+00
8.36E+00	4.53E−01	1.18E+00	3.01E−01	1.06E+00
1.04E+01	4.29E−01	1.15E+00	2.87E−01	1.06E+00
1.24E+01	4.03E−01	1.12E+00	2.72E−01	1.06E+00
1.44E+01	3.76E−01	1.08E+00	2.55E−01	1.06E+00
1.65E+01	3.48E−01	1.05E+00	2.39E−01	1.06E+00
1.85E+01	3.20E−01	1.01E+00	2.21E−01	1.06E+00

References: Almasy and Blangero, 1998; Blackwelder and Elston, 1985; Haseman and Elston, 1972; Kruglyak and Lander, 1995; Risch, 1990

Contributor: Christopher I. Amos

UNIT 1.6

Homozygosity Mapping Using Pooled DNA

Genetic linkage mapping is the identification of regions of the genome that are shared in common by related individuals affected with a genetic disorder at a level of significance greater than can be explained by chance. The DNA pooling method described below is particularly suited to mapping autosomal recessive disorders in cases where affected individuals would be expected to share a homozygous genomic region by descent from a common founder (Fig. 1.6.1). Such cases include extended inbred kindreds where pedigree information indicates that the affected individuals share a common ancestor. The method is also applicable to cases where it is likely that affected individuals inherit the disorder from a common founder because they have the same narrowly defined phenotype and are from the same genetically isolated population.

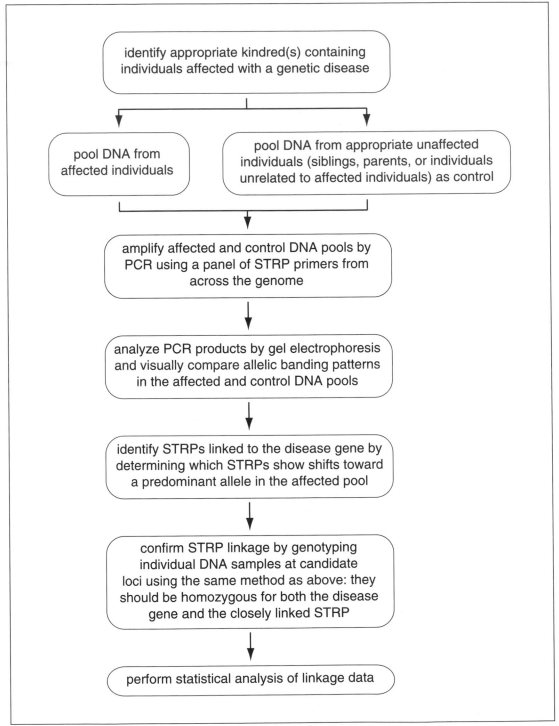

Figure 1.6.1 Flow chart for homozygosity mapping of recessive disease genes with short tandem repeat polymorphic markers (STRPs) using pooled DNA.

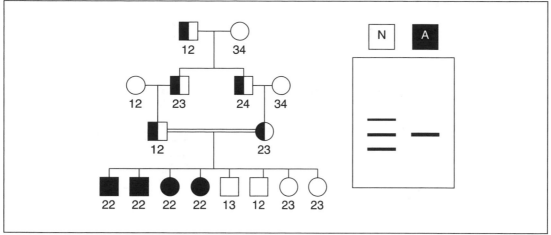

Figure 1.6.2 Example of an inbred pedigree and expected gel pattern (schematic drawing) for an STRP linked to an autosomal recessive disorder.

KEY CONCEPTS

DNA Pooling

One method to simplify the search for genomic regions that have been inherited from a common founder (identity by descent; see *UNIT 1.5*) is to genotype physically pooled DNA samples from related affected individuals. Identity by descent at a disease locus can be observed by pooling equal molar amounts of DNA from affected individuals and analyzing the pooled DNA sample with short tandem repeat polymorphic markers (STRPs; also known as simple sequence length polymorphisms, SSLPs) distributed across the genome. The frequency of a given allele in the pooled sample will be indicated by the intensity of the allelic band on the electrophoretic gel. The number and relative frequency of alleles at each STRP can be compared to those observed for a control DNA pool consisting of DNA from unaffected individuals. Because DNA samples are assigned to pools based on the phenotype of the individual (i.e., affected versus unaffected), STRPs linked to the disease locus will have a different allele frequency in the two pools (Fig. 1.6.2), as estimated from the number and intensity of bands on the gel. In contrast, unlinked markers will show similar allele frequencies in the two DNA pools (Fig. 1.6.3). Use of DNA pooling greatly reduces the amount of genotyping that is required to identify a disease gene locus.

Homozygosity Mapping

The basis for homozygosity mapping is that in offspring of consanguineous matings a portion of the genome is homozygous due to identity by descent from the shared ancestor. Homozygous regions are random between different offspring of these matings, except at a common disease locus shared by the affected individuals. Affected offspring from several consanguineous relationships can be used to identify a genomic interval that is homozygous in all affected individuals and thus likely to harbor the disease-causing gene. Homozygosity mapping can be most effectively applied to extended inbred kindreds and/or isolated populations where all affected individuals would be expected to inherit two copies of the gene from a distant ancestor or founder. In such cases all affected individuals are expected to be not merely homozygous, but homozygous for the same shared allele for markers near the responsible gene.

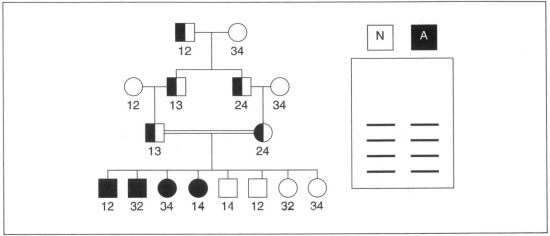

Figure 1.6.3 Example of an inbred pedigree and expected gel pattern (schematic drawing) for an STRP not linked to an autosomal recessive disorder.

BASIC PROTOCOL

AMPLIFICATION AND ANALYSIS OF POOLED DNA

Materials (*see* APPENDIX 1 *for items with* ✓)

Individual DNA samples from affected and unaffected (control) individuals
✓ TE buffer, pH 7.4
Primers for short tandem repeat polymorphic markers (STRPs; Research Genetics)
✓ 10× stock amplification buffer
✓ 2.0 mM 4dNTP mixture
5 U/l *Taq* DNA polymerase
Light mineral oil
✓ Formamide loading buffer
Rain-X (UNELKO) or Sigmacote (Sigma)
✓ Binding solution

96-well microtiter plate for use in thermal cycler *or* 0.5-ml microcentrifuge tubes
Thermal cycler, preferably accommodating 96-well microtiter plates
Equipment for denaturing polyacrylamide gel electrophoresis: e.g., Life Sciences Model S2 sequencing-gel apparatus
Water bath or heat block, 95°C

1. Dilute individual DNA samples to 100 ng/μl with TE buffer (pH 7.4) based on initial spectrophotometric readings (A_{260}; APPENDIX 3D). Take a second A_{260} reading to confirm that all samples are 100 ng/μl ±10 ng/μl.

2. PCR amplify each DNA sample with one or two STRPs (UNIT 2.1) to test for equal amplification of all samples.

3. Combine equal amounts of DNA from each affected individual into a single tube. Use 2 μg in 20 μl per individual (usually enough to complete a 10-cM-density genome screen), giving a final DNA concentration of 100 ng/μl (total DNA, not per individual). Dilute the sample to 20 ng/μl by adding 4 vol water.

4. Create a control DNA pool in the same manner by combining DNA from unaffected individuals. Dilute the control pool to a final DNA concentration of 20 ng/μl.

5. Add 2 μl of pooled DNA (40 ng) to individual microtiter plate wells or 0.5-ml microcentrifuge tubes, with alternate wells or tubes containing affected and control pooled DNA. If unfamiliar with the allele pattern of the STRP markers to be used, include a third well containing 20 ng DNA from a single individual to facilitate interpretation of the pooled sample by providing a reference for allele patterns.

6. Prepare a primer mixture containing 1.25 pmol/μl of each primer (primers for two or three STRPs can be added to each well provided they amplify markers with nonoverlapping allele size ranges). Add 2 μl of the mixture (2.5 pmol of each) to the relevant affected and control wells.

7. Prepare the PCR solution (for 96 reactions):

    100 μl 10× stock amplification buffer
    125 μl 2 mM 4dNTP mix
    375 μl H$_2$O
    5 μl 5 U/μl *Taq* DNA polymerase.

8. Add 6 μl of PCR solution to each well of the 96-well microtiter plate containing the template DNA and the primers (total volume, 10 μl). Overlay each well with light mineral oil if the cycler does not have a heated cover. Perform PCR amplification using following conditions:

Initial step:	3 min	94°C
35 cycles:	30 sec	94°C
	30 sec	55°C
	30 sec	72°C.

    Add 5 μl formamide loading buffer to each of the PCR-amplified wells.

9. Prepare glass plates for pouring of polyacrylamide gel by treating the long gel plate (42 × 34–cm for Life Sciences Model S2) with Rain-X or Sigmacote. Treat short glass plate (40 × 34–cm) with binding solution. In each case, treat the side of the plate that will be in contact with the gel.

10. Prepare denaturing 6% polyacrylamide gel (19:1 acrylamide/bisacrylamide; APPENDIX 3F), using a sharkstooth comb to create wells. Allow gel to polymerize. Prerun gel for 10 min at 60 W.

11. Immediately prior to loading, heat samples 3 min at 95°C using a water bath or heat block and place on ice. Load 4 μl of each sample into separate wells of the gel, alternating control and affected samples to facilitate comparison of allele patterns. Electrophorese at 60 W constant power for 1 hr, 45 min.

12. Silver stain (UNIT 2.1) and score gel by visually comparing the DNA pool from affected samples with the control pool. Keep the following in mind to achieve accurate identification of STRPs linked to the disease.

    a. Consider mode of inheritance. DNA pooling works best for autosomal recessive disorders in which all affected individuals are likely to be homozygous at the disease locus by descent from a common founder. In such instances, with STRPs tightly linked to the disease locus, one should expect to see a single predominant allele in the affected DNA pool.

    b. Consider pedigree structure. A recombination event between the marker and the disease locus will result in different allele frequencies in the affected DNA pool depending upon where in the pedigree the recombination event occurred. For example, in a pedigree that consists of a single large nuclear family whose parents are first cousins, a recombination event between the parent and one offspring will result in the observation of a single predominant allele with a second minor allele (found in a single individual). However, a single recombinant event in the preceding generation can result in two equally predominant alleles in the affected DNA pool. Therefore, it is important to consider the pattern of alleles one might expect to observe for a given pedigree.

    c. Consider data from neighboring markers. In instances where a marker that is truly linked to a disease phenotype is identified, neighboring STRPs should provide supporting data. A cluster of markers that each give an allele pattern suggesting linkage is far stronger evidence for linkage than a single marker.

    d. Recognize uninformative markers. A marker that shows a single major band in the control pool is uninformative. Should a first-pass screen of the genome fail to identify a linked region, additional STRPs near uninformative markers should be utilized.

References: Hastbacka et al., 1992; Houwen et al., 1994; Lander and Botstein, 1987

Contributor: Val C. Sheffield

UNIT 1.7

Disease Associations and Family-Based Tests

When patients with a particular disease are studied, they may also exhibit other characteristics that differentiate them from unaffected controls. Such *disease associations* have been studied intensively in the hope that they will provide a better understanding of various aspects of the disease. In particular, associations with alleles at known loci (often so-called "marker" loci) can be used to locate genes contributing to disease susceptibility. A disease association does not necessarily imply linkage with the disease locus. Family-based tests, such as the TDT and S-TDT provide valid tests of linkage between a disease and a marker. When the disease is complex, these tests may sometimes be more successful than conventional linkage methods in detecting or localizing genes that predispose an individual to the disease.

KEY CONCEPTS

Marker Locus and Allele

A *marker locus* is a locus whose location on a chromosome is known, and for which there are different identifiable genotypes defined by the *alleles* at that locus. Marker loci are usually highly polymorphic and the allelic variation is not functionally significant. A marker locus is denoted by M, and the possible alleles at that locus are denoted numerically (e.g., M_1, M_2, M_3). In current studies, two types of markers are often used: microsatellite markers and single nucleotide polymorphisms. Microsatellite markers are usually regions of DNA containing a variable number of tandem repeats (e.g., dinucleotides or tetranucleotides) that allow genotyping by PCR. A second highly abundant class of DNA variants consists of single nucleotide polymorphisms, or SNPs. Although SNPs usually have only two alleles, they are so common in the genome ($\sim$1 per 1000 nucleotides) that they have become very valuable tools for studies of linkage and association.

Association

A *population association* between two traits or phenotypes (e.g., hair color and eye color) exists if the population frequency of persons with both black hair and blue eyes is not equal to the product of the population frequency of persons with black hair and the population frequency of persons with blue eyes. As an algebraic consequence of such an association, the frequency of blue eyes differs between those persons with black hair and those with other hair color. Association between alleles A_1 and A_2 (at some genetic locus A) and alleles B_1 and B_2 (at some locus B) exists when the population frequency of the gamete A_1B_1 is not equal to the product of the separate population frequencies of the A_1 allele and the B_1 allele. This *allelic association* is a statistical concept that implies no specific cause or physical genetic linkage. The statistical measure of association, often called the coefficient of gametic disequilibrium (δ), is defined by:

$$\delta = A_1B_1 \text{ gamete frequency} \\ - (A_1 \text{ frequency}) \times (B_1 \text{ frequency})$$

In the particular case where A is a disease locus and B is a marker locus, the existence of an allelic association between the two loci implies that the frequency of the allele B_1 among persons with the disease (cases) differs from the frequency among persons without the disease (controls). In testing for an association between disease and marker loci, it is therefore often more convenient to test for a difference between these two frequencies than to attempt to test δ directly.

Linkage

In contrast to association, *linkage* between two loci A and B is a genetic concept: two loci are linked if they are so close on the same chromosome that the segregation of genes at one locus through meiosis is not independent of segregation of genes at the second locus. Of particular interest is tight linkage, implying that the two loci are very close to each other on the chromosome.

Linkage Disequilibrium

An allelic association between two linked loci A and B might occur if the A_1 allele arose recently by mutation on a B_1-bearing chromosome, and insufficient time has passed since the original mutation for the association between A_1 and B_1 to decay through recombination. In this case, the loci are said to be in *linkage disequilibrium*, and the level of disequilibrium, or association, is influenced by the tightness of the linkage between them. However, since association can arise for reasons other than this, the existence of association does not necessarily imply linkage.

Case-Control Versus Family-Based Studies

A *case-control* study investigates the association between a disease and some other characteristic, either phenotypic or genetic, using data from unrelated patients and unrelated controls. Conclusions are drawn from a comparison of marker allele or genotype frequencies in patients (cases) and controls. This procedure rests on a key assumption about matching: it is assumed that the two groups do not differ in marker frequencies for any reason that is *not* a consequence or correlate of disease. It is essential that controls be matched to affected individuals with respect to various features that might give rise to artifactual differences. Since ethnic groups often differ in marker allele frequencies, matching is clearly needed with respect to ethnic background. It may also be needed for other features, such as age, demographic features, or

Table 1.7.1   Numbers of Marker Alleles M_1 and M_2 Among n Affected Patients and n Controls in Random Samples of Unrelated Individuals

	M_1	M_2	Total
Affected	x_1	$2n - x_1$	$2n$
Control	x_2	$2n - x_2$	$2n$
Total	$x_1 + x_2$	$4n - x_1 - x_2$	$4n$

sex distribution, if these features are correlated with marker frequency differences. A *family-based* study uses data from within a family. Limiting the analysis to family members helps to avoid the matching problems experienced in case-control studies, which we now discuss.

CASE-CONTROL DESIGN: DISEASE ASSOCIATION IN POPULATION-BASED SAMPLES

Measures of disease associations with alleles at a marker locus are typically applied to data such as those in Table 1.7.1. Because these data refer to unrelated individuals, the analysis is referred to as a case-control study, and the results demonstrate an example of population association. Although there is no requirement that the number of patients and controls be equal, this assumption is made for algebraic simplicity both in Table 1.7.1 and in the comparisons that follow. If there are multiple alleles at the M locus and only one allele (M_1) is tested for association, all other alleles of the marker locus are combined as M_2.

It is customary to present disease associations, especially those involving human leukoocyte antigen (HLA), as relative risks. Using the data in Table 1.7.1, an allelic relative risk, for allelic differences associated with the disease, can be defined as the *cross-product ratio*:

$$\text{allelic relative risk} = \frac{x_1(2n - x_2)}{x_2(2n - x_1)}$$

Strictly speaking, this equation defines an odds ratio rather than a relative risk if controls represent unaffected individuals rather than a true random sample from the population. However, the two are essentially identical for rare diseases. If the frequencies of alleles are equal among patients and controls, there is no association between disease and marker, and the expected value of the allelic relative risk is 1.

As an example, suppose that the allele M_1 occurs 80 times and the allele M_2 occurs 20 times among 50 affected individuals, while M_1 occurs 55 times and M_2 occurs 45 times among 50 control individuals. The allelic relative risk $= (80 \times 45)/(20 \times 55) = 3.27$. There is thus a substantial allelic association between the disease and the allele M_1.

It is essential to test the statistical significance of any observed value of the allelic relative risk in order to interpret its value. This may be done either by calculating an approximate 95% confidence interval for the relative risk, or, more exactly, by forming the standard two-by-two contingency table chi-square statistic, calculated with the data as in Table 1.7.1. This test statistic, which is called the population-based contingency statistic (PBCS), reduces to:

$$\text{PBCS} = \frac{4n(x_1 - x_2)^2}{\left[(x_1 + x_2)(4n - x_1 - x_2)\right]}$$

Use of the population-based contingency statistic to infer linkage from an observed association rests on the assumption that the patient and control groups have been suitably matched. However, apart from imperfect matching, there is a second, more subtle, way that observed disease associations may arise "spuriously" (that is, in the absence of linkage): differences may arise as a result of *population structure*.

POPULATION STRUCTURE

Population structure is a general description for admixture, stratification, or heterogeneity, reflecting departure from random mating within a population. In order to illustrate the effects of population structure on various statistics for assessing linkage and association, it is necessary to consider a model that contains all the relevant features of both the genes and the populations. In the above discussion of classical population association studies, the possibility of a gene locus for disease was only implicit. To analyze the role of population structure, the relationship between the marker locus M and the disease D is now made explicit. For simplicity, it is assumed that a single disease locus has alleles D and d, that only the DD individuals become affected (i.e., the disease is recessive), and that all of the DD individuals do so (i.e., the disease is fully penetrant). These assumptions are not necessary, but they simplify the mathematical analysis. Additionally, the marker locus is assumed to have alleles M_1 and M_2, and the recombination fraction between M and D is denoted by θ.

In generation 0, individuals are assumed to live in a collection of s subpopulations, with relative sizes $\alpha_1, \alpha_2, \ldots, \alpha_s$. In subpopulation i, the frequency of D is denoted p_i, the frequency of M_1 is q_1, and the coefficient of gametic disequilibrium (see above) in subpopulation i is denoted by δ_i.

The offspring of individuals from these subpopulations move to a common area, where they mate at random without regard to their mates' subpopulation of origin. Of course, admixture in real human populations is unlikely to be as extreme and abrupt as this. The point of this model is to illustrate how the effects of the population structure in generation 0, which might not be easily detected in human populations, can influence conclusions drawn from association studies where the data are taken from individuals in generations 1 and 2.

The effects of population structure can best be exhibited by considering the mean value of $x_1 - x_2$, whose square appears in the numerator of the population association test statistic given (PBCS). If data are taken from affected children in generation 2 in the model described above, it can be shown for a recessive disease that:

$$
\text{mean}\,(x_1 - x_2) \\
= 2n\left(\frac{\left[\sum \alpha_i p_i^2 q_i\right] - q\left[\sum \alpha_i p_i^2\right]}{\sum \alpha_i p_i^2}\right) \\
+ 2n\left(\frac{(1 - \theta)\sum \alpha_i p_i \delta_i}{\sum \alpha_i p_i^2}\right)
$$

where summations are over all subpopulations. In this expression, the first term does not contain δ_i, so that this term represents association due solely to population structure.

Three strategies have recently been proposed for dealing with the problem of population structure when testing for association. The first, called genomic control (GC), takes advantage of the observation that if structure exists, then test statistics at the candidate marker, as well as unlinked markers, will be elevated. Thus, by typing multiple unlinked markers in the same cases and controls, the increase in the test statistic due to population structure can be estimated and taken into account. A second approach, structure association (SA), assumes that the population

Table 1.7.2   Marker Alleles M_1 and M_2 Among $2n$ Transmitted and $2n$ Nontransmitted Alleles in Parents of n Affected Children[a]

	M_1	M_2	Total
Transmitted	w	$2n - w$	$2n$
Nontransmitted	y	$2n - y$	$2n$
Total	$w + y$	$4n - w - y$	$4n$

[a]Reprinted with permission from Spielman et al. (1993).

is composed of an unknown number of homogeneous subpopulations. A third approach is to use family data. These methods are based on the transmission of alleles from parent to offspring, so population structure is no longer an issue.

FAMILY-BASED TESTS OF ASSOCIATION AND THE HAPLOTYPE RELATIVE RISK

For family data, the direct analog of the population-based relative risk defined allele relative risk (above) is the haplotype relative risk. In each nuclear family, genotypes at the marker locus are obtained for parents and affected children. Each parent of an affected child has one transmitted allele and one nontransmitted allele. The data used are the total number of M_1 and M_2 alleles in the transmitted and the nontransmitted categories, as shown in Table 1.7.2, where w is the number of M_1 alleles among parental alleles transmitted to affected offspring and y is the number of M_1 alleles among nontransmitted parental alleles.

The haplotype relative risk is defined as:

$$\text{haplotype relative risk} = \frac{w(2n - y)}{y(2n - w)}$$

Similarly, the natural statistic to test for association using data from this table is found by analogy to the population-based contingency statistic defined for population-based contingency statistic (PBCS). This statistic compares w (the number of M_1 alleles transmitted to affected offspring) with y (the number of M_1 alleles not transmitted) and is called the AFBAC (affected family-based controls) statistic. It is defined as:

$$\text{AFBAC} = \frac{4n(w - y)^2}{[(w + y)(4n - w - y)]}$$

The analogy between the population-based relative risk and contingency test statistic on the one hand, and the family-based haplotype relative risk and AFBAC statistic on the other, is now complete.

FAMILY-BASED TESTS OF LINKAGE: THE TDT

Because the goal in current research is frequently to localize disease genes, it is desirable to develop a test that is formulated specifically for linkage and that is free of spurious effects resulting from population structure. It is possible to do this by making use of the same kind of data that were provided in Table 1.7.2 for the analysis of association. For a test of linkage, however, it is convenient to rewrite the entries of Table 1.7.2 as in Table 1.7.3. The data in Table 1.7.3 are presented in four transmission categories, where $a = $ the number of times an

Table 1.7.3 Combinations of Transmitted and Nontransmitted Marker Alleles M_1 and M_2 Among $2n$ Parents of n Affected Children[a]

Transmitted allele	Nontransmitted allele		
	M_1	M_2	Total
M_1	a	b	$a + b$
M_2	c	d	$c + d$
Total	$a + c$	$b + d$	$2n$

[a]Reprinted with permission from Spielman et al. (1993).

M_1M_1 parent transmits M_1 to affected offspring, $b = $ the number of times an M_1M_2 parent transmits M_1 to affected offspring, $c = $ the number of times an M_1M_2 parent transmits M_2 to affected offspring, and $d = $ the number of times an M_2M_2 parent transmits M_2 to affected offspring.

The entries in Tables 1.7.2 and 1.7.3 are related by the equations $w = a + b$ and $y = a + c$. This enables us to rewrite the AFBAC statistic in terms of a, b, c, and d as the following:

$$\text{AFBAC} = \frac{4n(b-c)^2}{[(2a+b+c)(b+c+2d)]}$$

This statistic is not designed to test for linkage between the marker and the disease. A statistic that is designed to test for linkage is the transmission/disequilibrium test statistic (TDT). This statistic is also calculated from the entries in Table 1.7.3, as:

$$\text{TDT} = \frac{(b-c)^2}{(b+c)}$$

The TDT statistic is distributed approximately as chi-square with 1 degree of freedom (df) when disease and marker are unlinked, so that linkage is tested by referring the value of this statistic to chi-square tables with 1 df. It uses only the data entries b and c in Table 1.7.3. This is expected, because these quantities derive from heterozygous M_1M_2 parents, and only heterozygous parents can give information about linkage.

Validity: Comparison with the Population Relative Risk Test

The TDT is valid as a test of linkage regardless of population structure. Further, the TDT is valid as a test of linkage whether the data come from families that are simplex (one affected offspring) or multiplex (two or more affected offspring), a mixture of the two, or multigenerational. This can be seen by comparing it with the population relative risk test statistic. The equation on page 1-79 shows the mean value of the quantity $x_1 - x_2$ appearing in the numerator of the population association test statistic (PBCS), when the disease of interest is recessive and the population sampled has structure caused by admixture. The mean of the analogous quantity $b - c$, appearing in the numerator of the TDT statistic, is:

$$\text{mean } (b - c) = 2n \frac{(1 - 2\theta)[\sum \alpha_i p_i \delta_i]}{\sum \alpha_i p_i^2}$$

This equation reveals two important properties of the TDT. First, the mean value of the numerator of the TDT statistic, unlike the mean value of the population relative risk test statistic (page 1-79), has no term representing association due solely to admixture (the first term in the population relative risk statistic). This confirms that the TDT procedure is valid in structured populations. Second, the expression in this equation, unlike the second term in the population relative risk statistic, is zero when $\theta = 1/2$ (i.e., when disease and marker are unlinked). This confirms that the TDT is a valid test of linkage and that the population-based contingency test is not. The expression in this equation resembles the *second* term of the population relative risk statistic, and when $\theta = 0$ (i.e., there is no recombination between marker and disease loci) the two are identical. This suggests that the total population relative risk test statistic partitions into a component measuring the spurious association due to population structure, and another component that is exactly equal, when recombination is small enough to be ignored, to the TDT statistic.

Power: Comparison with Other Tests for Linkage

The existence of association between disease and marker alleles is essential for the use of the TDT as a test of linkage. *If no such association exists, the TDT has no power to detect linkage.* This makes it important to compare the TDT with tests for linkage that do not rely on association, in particular those tests that depend on sharing of marker alleles (identical by descent) in affected sib pairs. For simple diseases, sharing methods appear best and have revealed important linkages. However, for complex diseases, the TDT might be preferable.

Other Considerations

Mode of inheritance
Although the TDT was originally derived under the assumption of a recessive disease, and the expression in TDT statistic is calculated under this assumption, the test is valid whatever the mode of inheritance.

Segregation distortion
If the TDT detects an excess of allele M_1 in transmissions from heterozygous parents to affected offspring, the model given above suggests that there is evidence for linkage. In principle, however, the finding could result instead from some preferential transmission in the meiotic process itself. This phenomenon, known as segregation distortion, will produce excess transmissions of the M_1 allele to the unaffected as well as to the affected offspring. This possibility can be tested by examining transmissions from heterozygous parents to unaffected offspring, ideally the sibs of the affected individuals in the TDT. If there is a deficiency of transmissions of M_1 to unaffected sibs, there is no evidence for segregation distortion and no new test is necessary. If there is an excess of transmissions of M_1 to unaffected sibs, a two-by-two contingency chi-square may be used to test for differential transmission of M_1 to affected and unaffected sibs.

Incomplete genotype data
Consider the case where there are only two alleles at M, and where the available parent, if informative, is heterozygous (M_1M_2). If the offspring is M_1M_1, it is clear that this parent transmitted M_1. If the offspring is M_2M_2, it is clear that this parent transmitted M_2. Thus, it might initially appear that data from incomplete families can be used in these cases. However, if the offspring is heterozygous, it is not possible to determine which allele was transmitted from the available parent, so the data must be discarded. Including informative and discarding uninformative families in this way leads to bias in the TDT. Under the null hypothesis (no linkage), M_1 and M_2 are equally likely to be transmitted by the available parent. However, as the frequency of M_1 increases in the population, the likelihood increases that M_1 is the allele transmitted by the unavailable parent. When this happens, data on transmission from the

available M_1M_2 parent will be used if that parent transmits M_1, but will be discarded if that parent transmits M_2. The net result is that the higher the frequency of M_1 in the population, the greater the *apparent* transmission of M_1 from heterozygotes. The families that give rise to the problem are those where the child is homozygous or heterozygous for the alleles in the available parent. However, if the offspring has an allele (e.g., M_3) that is not present in the available parent, no combination will be discarded and no bias will occur if the data are used.

Reconstruction of parental genotypes

In cases where parental genotypes are not available, one might be tempted to try to reconstruct them from the offspring genotypes and then, for families where this reconstruction is possible, proceed with the TDT as though the parental information were available. This procedure is, however, not usually valid. There is an ascertainment bias in using data from families where reconstruction of parental genotypes is possible, since only certain offspring genotypes allow reconstruction.

FAMILY-BASED TESTS OF ASSOCIATION: THE TDT AND OTHER TESTS

The TDT may also be used with appropriate data as a valid test of association, even in subdivided populations. Specifically, while data from multiplex families are allowed when the TDT is used as a test of linkage, data from only one affected offspring per family are allowed when the TDT is used as a test of association. The TDT procedure assumes independent transmissions of marker alleles from a parent to two or more affected offspring, a condition that holds under the null hypothesis when the the TDT is used as a test of linkage. This condition does not necessarily hold, however, when the TDT is used as a test of association. In the latter case, even under the null hypothesis of no population association, the transmissions to two affected offspring will sometimes *not* be independent, particularly if disease and marker are linked.

The AFBAC statistic (page 1-81) has also been proposed as a test of association. As with the TDT, and for the same reasons, the statistic may be used only with data from simplex families. Even when simplex family data are used, the statistic has a chi-square distribution under the hypothesis of no association only if the parental population being sampled is in Hardy-Weinberg equilibrium. In a structured population, this is unlikely. Thus, the test will sometimes fail to reject the null hypothesis when the TDT would do so.

GENERALIZING THE TDT: MORE THAN TWO MARKER ALLELES

In the above discussion it has been assumed that there are only two marker alleles, M_1 and M_2. In practice, this is generally not the case, and the procedures described above can be extended to a case with an arbitrary number k of marker alleles ($M_1, M_2, \ldots, M_k$). The mathematical complexity of these procedures is considerably greater than that for the two-allele case.

The data to be analyzed will be of the form shown in Table 1.7.4, which is the natural k-allele generalization of the two-allele Table 1.7.3, and which describes the number of parents of affected children in each of the various transmission/nontransmission categories indicated. Note that, because of the arbitrary number of marker alleles, a more complex notation than that used in Table 1.7.3 is needed. Also, in order to conform with the notation of Table 1.7.3, the total number of parents is denoted by $2n$ as in Table 1.7.3.

Table 1.7.4 Combinations of Transmitted and Nontransmitted Marker Alleles $M_1, M_2, \ldots, M_k$ Among $2n$ Parents of n Affected Children

	Nontransmitted allele				
Transmitted allele	M_1	M_2	$\ldots$	M_k	Total
M_1	n_{11}	n_{12}	$\ldots$	n_{1k}	$n_{1.}$
M_2	n_{21}	n_{22}	$\ldots$	n_{2k}	$n_{2.}$
$\ldots$	$\ldots$	$\ldots$	$\ldots$	$\ldots$	$\ldots$
M_k	n_{k1}	n_{k2}	$\ldots$	n_{kk}	$n_{k.}$
Total	$n_{.1}$	$n_{.2}$	$\ldots$	$n_{.k}$	$2n$

Multiallelic Tests of Linkage

The null hypothesis that disease and marker loci are unlinked implies that for the data in Table 1.7.4, the mean value of n_{ji} is the same as that of n_{ij}, for all (i, j) combinations. Thus, one possible test for linkage between disease and marker loci is to test for symmetry in the data matrix of Table 1.7.4. However, for the case of k marker alleles, this test has $k(k-1)/2$ degrees of freedom (df) and the use of such a test runs the risk of a "swamping" effect: one or a few markers with a strong effect might not be detectable in a global test that includes many markers with no effect or only a small effect. It is therefore agreed, on the whole, that such tests should not be used.

At the next level of simplification, tests of linkage focus on the data provided by the row totals $(n_{1.}, \ldots, n_{k.})$ and the column totals $(n_{.1}, \ldots, n_{.k})$ in Table 1.7.4. In a population of heterozygous parents whose genotypes each contain the allele M_i, these tests compare the number $(n_{.i})$ of transmissions of the allele M_i $(n_{i.})$ with the number of transmissions of an allele other than M_i. The basis of these tests is the fact that when the null hypothesis is true, the mean value of $n_{i.}$ is the same as that of $n_{.i}$ for all i, and the test is in effect a test of the equality of these two means.

One test exploiting this fact is the generalized TDT (GTDT) statistic. Using this test, one first calculates the values of d_i (for $i = 1, 2, \ldots, k$), defined by $d_i = n_{i.} - n_{.i}$. The sum of these values is necessarily zero, so that without loss of information one ignores one arbitrarily chosen marker allele (say allele k) and forms a vector $\mathbf{d}'$ defined by:

$$\boldsymbol{d}' = (d_1, d_2, \ldots, d_{k-1})$$

If the null hypothesis that disease and marker loci are unlinked is true, the estimate of the variance of d_i is $n_{i.} + n_{.i} - 2n_{ii}$ and the estimate of the covariance between d_i and d_j is $-(n_{ij} + n_{ji})$. These variance and covariance estimates are formed into a matrix V, and the GTDT test statistic is then defined as:

$$\mathrm{GTDT} = \boldsymbol{d}' V^{-1} \boldsymbol{d}$$

Under the null hypothesis, the GTDT statistic has asymptotically a chi-square distribution with $k-1$ df, and thus can be used as a test of linkage by referring the observed value of the statistic to tables of significance points of this distribution.

Although the GTDT statistic is the most natural generalization of the TDT statistic, its calculation requires the inversion of a large and possibly sparse matrix. A test statistic that is very similar to GTDT but that does not involve inversion of a matrix, denoted T_{mhet}, is defined by:

$$T_{mhet} = [(k-1)/k]\Sigma_i[\frac{(n_{i.} - n_{.i})^2}{(n_{i.} + n_{.i} - 2n_{ii})}]$$

In a nonstratified population, this statistic has a distribution close to chi-square with $k - 1$ df under the null hypothesis, and, like the GTDT statistic, also reduces to the two-allele TDT statistic when $k = 2$.

A rather different test, denoted maxTDT, focuses on a "most significant" marker allele instead of on all alleles jointly, and is computed as follows. For each i ($i = 1, 2, \ldots, k$), all alleles other than allele i are lumped as "non-i," so that a "two-allele" TDT statistic can be calculated, giving a total of k different TDT statistics. The largest of these statistics (denoted maxTDT) is used as the test statistic. In terms of the entries in Table 1.7.4, the maxTDT statistic is the largest (as i takes successively the values $1, 2, \ldots, k$) of the various possible "two-allele" TDT statistics:

$$\frac{(n_{i.} - n_{.i})^2}{(n_{i.} + n_{.i} - 2n_{ii})}$$

As expected, this test statistic reduces to the TDT statistic when $k = 2$. The maxTDT statistic has 1 df, and thus largely avoids the swamping effect. However, one may not use chi-square tables to test for its significance, because such a deliberately chosen largest TDT statistic does not have a chi-square distribution. The significance points of the maxTDT statistic have been found by simulation, and significance values are given in Table 1.7.5.

Multiallelic Test of Association

As discussed above, the two-allele TDT statistic may be used in testing for association. The same is true for the GDTDT statistic, the T_{mhet} statistic, and the maxTDT statistic,

Table 1.7.5 Significance Points for the Statistic maxTDT as Determined by Simulation

k	Type I error			
	5%	1%	0.1%	0.01%
2	3.84	6.64	10.83	15.13
3	5.49	8.46	12.70	17.39
4	6.10	9.10	13.49	18.03
5	6.51	9.51	13.78	18.26
6	6.88	9.87	14.21	18.55
7	7.15	10.15	14.48	18.68
8	7.41	10.38	14.74	19.02
9	7.61	10.60	14.87	19.33
10	7.82	10.81	15.10	19.51
11	7.99	11.01	15.39	19.86
12	8.15	11.17	15.51	20.14

provided that (as with the TDT) only one affected sib in each family is used in the analysis.

TESTS OF ASSOCIATION USING MARKER HAPLOTYPES

Traditionally, association tests have been single-locus tests, testing each marker locus independently. Several tests have been proposed that test for association with marker haplotypes composed of alleles at multiple markers. These tests look for association between disease status and a particular combination or combinations of alleles at different loci, i.e., marker haplotypes. Intuitively, one might expect marker haplotypes to provide more information than the markers individually. One often imagines that association arises due to the presence of a susceptibility allele that occurs initially on one, or few, ancestral haplotypes. The amount of association between the susceptibility allele and a particular marker allele or marker haplotype depends strongly on the frequency of the associated allele or haplotype. The strongest associations are found when the positively associated allele or haplotype has low frequency in the population. A haplotype necessarily has lower frequency than any of frequencies for the individual alleles in the haplotype. Thus, there is a potential for greater association when examining haplotypes. A difficulty to be overcome in testing for haplotype associations is that haplotypes are not generally observed directly. Instead, one typically has information on an individual's genotypes. In some cases, the haplotypes can be inferred with certainty from the genotype data, but when the individual is heterozygous at more than one locus, the haplotype assignment is ambiguous. Genotype information from family members can often aid in haplotype inference. An additional difficulty with haplotype association tests is that there can be many haplotypes to consider with multiple markers and/or multiple alleles, and often there is no a priori knowledge of which haplotype might be associated. Haplotypes with low frequency will contribute small numbers of observations to the analysis, and caution must thus be used in applying and interpreting the results of haplotype tests. Often, haplotypes with small numbers can be pooled to provide valid asymptotic tests, but this pooling could reduce power if negatively and positively associated haplotypes are pooled. Typically global as well as individual haplotype tests are produced. Multiple testing should be considered in interpreting many tests for individual haplotypes.

THE SIB TDT (S-TDT)

The TDT as described above uses data from families where marker genotypes are available for father, mother, and affected offspring. When diseases with onset in adulthood or old age are studied, it may be impossible to obtain genotypes for markers in the parents of the affected offspring. This difficulty has limited the applicability of the TDT. The "sib TDT" (or S-TDT) method overcomes this problem by using marker data from unaffected sibs instead of parents, thus allowing application of the principle of the TDT to sibships without parental data. This extension of the TDT is potentially valuable for studying diseases of late onset, such as type 2 (non-insulin-dependent) diabetes, cardiovascular diseases, Alzheimer disease, and other diseases associated with aging. The data appropriate for the S-TDT consist of marker genotypes for sibships that meet two requirements: first, there must be at least one affected and one unaffected member of the sibship, and second, the members of the sibship must not all have the same genotype. A sibship with one affected and one unaffected, of different marker genotypes—the "minimal" sibship, discussed later—meets these requirements. The units of observation are the marker genotypes of the offspring, affected and unaffected, for each family.

In essence, the S-TDT determines whether the marker allele frequencies among affected offspring differ significantly from the frequencies in their unaffected sibs. Because the comparison between affected and unaffected is carried out within families, disease association without linkage will not result in such differences; the marker allele frequencies will, apart from random sampling effects, be the same in the affected and unaffected sibs unless linkage is also present (including the case in which the marker itself is responsible). Thus, the null hypothesis tested by the S-TDT is that disease and marker are unlinked.

Consider a family with a affected and u unaffected sibs, with $t = a + u$. Suppose that in this sibship the number of sibs who are of genotype M_1M_1 is r and the number of sibs of genotype M_1M_2 is s, so that the number of sibs of genotype M_2M_2 is $t - r - s$. For the moment, we assume that any marker genotype is either M_1M_1, M_1M_2, or M_2M_2. Our aim is to test whether one particular marker, e.g., M_1, occurs significantly often in affected sibs, conditional on the totals a, u, r, and s within this family. If the null hypothesis that disease and marker are unlinked is true, then, conditional on these totals, the number of affected sibs who are M_1M_1, and the number who are M_1M_2, both have hypergeometric distributions. From this we can calculate the mean:

$$(2r + s)a/t$$

and the variance:

$$au\{4r(t - r - s) + s(t - s)\} / \{t^2 (t - 1)\}$$

for the number of M_1 alleles among affected sibs within this family. Taking into account every family in the data, we compute a total null hypothesis mean A and variance V of the number of M_1 alleles in affected sibs in the entire data by simple summation:

$$A = \Sigma(2r + s)a / t$$

$$V = \Sigma au\{\frac{4r(t - r - s) + s(t - s)}{t^2(t - 1)}\}$$

where in both cases, summation is over all families in the sample.

The test of significance is then carried out by using A and V, together with the observed number, Y, of M_1 alleles observed among affected sibs, from which a z score, defined by:

$$z = \frac{Y - A}{\sqrt{V}}$$

is computed. From the z score an approximate p value is calculated, using a normal distribution approximation. It is customary to make a continuity correction, and the p value is calculated from:

$$z' = \frac{|Y - A| - \frac{1}{2}}{\sqrt{V}}$$

Because the calculations are all carried out by first finding "within-family" means and variances, which are then summed over all families (A and V above) to produce an overall mean and an overall variance, potential problems resulting from population structure are eliminated, as is true of the original TDT.

As an example of the calculations involved, consider a family with the data shown in Table 1.7.6. Here $r = 4$, $s = 2$, $a = 5$ and $t = 9$. The contribution of this family to the overall mean A is $(8 + 2) \times 5/9 = 5.56$ and to the overall variance V is $5 \times 4 \times (48 + 14)/648 = 1.91$.

Table 1.7.6 Sample Family Data for S-TDT

| | \multicolumn{4}{c}{No. of sibs with genotype:} | | | |
	M_1M_1	M_1M_2	M_2M_2	Total
Affected	3	0	2	5
Not affected	1	2	1	4
Total	4	2	3	9

Combination of the TDT and the S-TDT into an Overall Test

In a single collection of families there might be some that can be analyzed only by the TDT and others suitable for analysis by the S-TDT. We now show, for such cases, how data from all families in this collection may be used jointly in one overall procedure that tests for linkage in the presence of association.

Consider first the case of only two marker alleles, M_1 and M_2. For any family to be used in the test, the genotype of at least one affected offspring in the family must be available. Further, the data for other family members must define one or other of the following three groups.

1. Genotypes available for both parents, but not for unaffected sibs.
2. Genotypes available for at least one unaffected sib, but not for both parents,
3. Genotypes available for both parents and at least one unaffected sib.

Group 3 families meet the requirements for both the TDT and the S-TDT. Recent research suggests that in sufficiently large samples, the TDT is at least as powerful as the S-TDT, in cases where either could be used. We therefore combine families in group 3 with those in group 1 and ignore the unaffected sibs in families in group 3. "Group 1" henceforth refers to this combined group.

If data in Group 1 only were analyzed, the relevant (chi-square) TDT statistic is given. However, for our present purposes it is more convenient (and equivalent) to use as test statistic the number X of transmissions of the M_1 allele among the $b + c$ alleles transmitted by heterozygous M_1M_2 parents in this group to their affected children. When marker and disease are unlinked, X has a binomial distribution with mean $(b + c)/2$ and variance $(b + c)/4$.

The S-TDT procedure is appropriate for families in group 2. As described above, the test statistic is the number, Y, of M_1 alleles among affected sibs. When marker and disease are unlinked, Y has a distribution with mean A and variance V, respectively.

When data are available from both group 1 and group 2, a natural test statistic is W, the sum of X and Y. Under the null hypothesis that disease and marker are unlinked, W has a mean A_{comb} and variance V_{comb}, given respectively by:

$$A_{comb} = \frac{b+c}{2} + A$$

$$V_{comb} = \frac{b+c+V}{4}$$

The test of significance is now carried out with a z' statistic (including a correction for continuity), calculated from the formula:

$$z' = \frac{|W - A_{\text{comb}}| - \frac{1}{2}}{\sqrt{V_{\text{comb}}}}$$

The null hypothesis that disease and marker are unlinked is rejected if z' departs significantly from zero, as judged by standard z tables.

As an example of the use of the combined procedure, suppose that, for families in Group 1, the total number of transmissions $b + c$ from heterozygous M1M2 parents is 300, and that of these, 166 are transmissions of M1. Further, suppose that for families in Group 2, there are 215 M1 alleles among the affected sibs and that A and V, calculated from Equations 1.7.16. and 1.7.17 respectively, are 197.4 and 188.6. Thus W = 166 + 215 = 381.

Under the null hypothesis that disease and marker are unlinked, the mean number of transmissions of M_1 from M_1M_2 parents is 150 and the variance is 75. Then $A_{\text{comb}} = 150 + 197.4 = 347.4$, and $V_{\text{comb}} = 75 + 188.6 = 263.6$. The z' value, calculated using Equation 1.7.21, is thus:

$$\frac{|381 - 347.4| - \frac{1}{2}}{\sqrt{263.6}}$$

which is significant as a two-sided test at the 5% significance level. Further examples of the use of this combined test are given in Spielman and Ewens (1998).

A computer program for the TDT, the S-TDT, and the combined test procedures is available at *http://genomics.med.upenn.edu/spielman/TDT.htm*.

Validity of the S-TDT as a Test of Association

The S-TDT is proposed above as a test of linkage between marker and disease. However, the necessary and sufficient condition that the TDT be valid also as a test of association is that the data are entirely from simplex families (one affect offspring; one or both parents heterozygous for the marker). The analogous necessary and sufficient requirement for the S-TDT to be valid as a test of association is that the data consist of the "minimal S-TDT configuration" of exactly one affected and one unaffected sib in each family, the two sibs having different marker genotypes.

Multiple Alleles

The maxTDT and the GTDT statistics generalize the TDT when k alleles exist at the marker locus, none being of special interest. There are analogous generalizations of the S-TDT for families where affected and unaffected sib data form the core of the test. For such data, the analog of the maxTDT is as follows. For allele M_1, all other alleles are grouped as "non-M_1," the number of M_1 alleles among affected sibs is determined, and then equations analogous to equations for A and V are used compute a z score. An analogous procedure is then performed for each of the k marker alleles in turn. Each z score is then squared, and the largest of these squared scores is referred to Table 1.7.5 for significance. This might be called the maxS-TDT method. A parallel procedure may be used for the combined test. Here we calculate k different z scores from equations analogous to equations for A_{comb} and V_{comb}. Each z score is then squared, and the significance of the largest squared z score is determined by reference to Table 1.7.5. There are two drawbacks to this latter approach. First, as with the GTDT, a significant effect for one (or several) marker alleles might be obscured by the presence of many other alleles with little or no association. Second, it is not apparent how to combine an S-TDT procedure with the GTDT. For these reasons, the maxS-TDT approach is recommended.

PEDIGREES

A second property common to the TDT and the S-TDT concerns data from pedigrees that contain several sibships—e.g., sets of cousins in two sibships, or a set of sibs and their aunts/uncles. In such pedigrees, the TDT is valid as a test of linkage, but not as a test of association, when based on data from all heterozygous parents and their offspring. The corresponding result for the S-TDT is that data from separate sibships can also be combined to give a test that is valid for linkage, but not for association.

The difficulty is that contributions of multiple nuclear families or sibships to the statistic are not independent when there is linkage, even if the null hypothesis of no allelic association is true. Thus, tests such as the TDT and S-TDT, which treat nuclear families or sibships, respectively, as independent, are not valid—i.e., they will not have the correct significance level.

For fine-mapping and candidate gene studies, it is desirable to establish linkage disequilibrium; therefore, tests have been developed to test for both linkage and association in extended pedigrees. One such test is the pedigree disequilibrium test (PDT; software available at *http://wwwchg.mc.duke.edu/software*). The tests properly account for correlations within pedigrees that arise from linkage between disease and marker loci. The test statistics all take a similar form:

$$T = \frac{\left(\sum_{i=1}^{N} X_i \right)^2}{\sum_{i=1}^{N} X_i^2}$$

where X_i is a measure of association for the ith pedigree, $i = 1, \ldots, N$, with mean 0 under the null hypothesis. The tests differ in the way that association is measured with pedigrees.

References: Abecasis et al., 2000; Nelson et al., 2001; Pritchard et al., 2000; Rabinowitz and Laird, 2000; Ritchie et al., 2001

Contributor: Warren J. Ewens, Richard S. Spielman, Norman L. Kaplan, and Eden R. Martin

UNIT 1.8

Analysis of Gene-Gene Interactions

The goal of this unit is to introduce gene-gene interactions or epistasis as a significant complicating factor in the search for disease susceptibility genes. A commonly used textbook definition of epistasis is one gene masking the effects of another gene. A classic example of biological epistasis (i.e., where the gene-gene interaction has a biological basis) comes from studies of the shape of seed capsules from crosses of a plant called the shepard's purse (Fig. 1.8.1). Crosses from doubly heterozygous plants yielded Mendelian ratios of 15 triangular capsules to one oval capsule. It is generally believed that there are two pathways with dominant loci that lead to the triangular shape. It is only when both pathways are blocked by recessive alleles that the oval-shaped seed capsule is produced. This is an example of a recessive-by-recessive interaction, since having two recessive genotypes leads to a different phenotype compared to the result of only one from either locus.

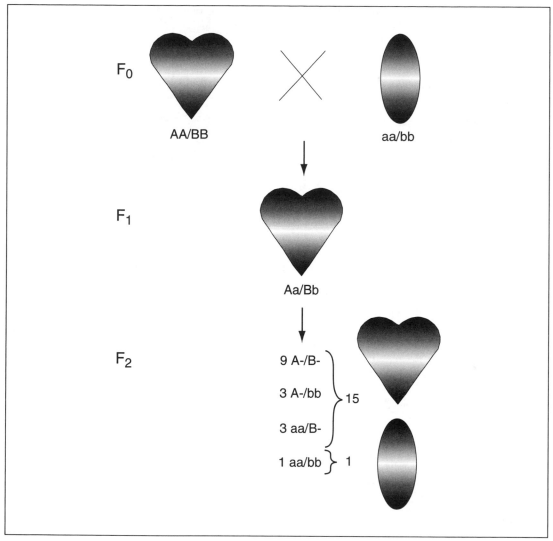

Figure 1.8.1 An F1 cross of the shepard's purse plant that leads to a 15:1 phenotypic ratio of triangular to oval shaped seeds. Depicted are two genes, A and B, each with two alleles (A or a, B or b). Only the plants homozygous for the a and b alleles have oval-shaped seeds. This is an example of recessive-by-recessive epistasis.

A simple example of statistical epistasis (or epistacy) in the form of penetrance functions is presented in Table 1.8.1. A penetrance function is simply the probability (P) that an individual will have the endpoint (D) given a particular genotype or combination of genotypes (G) from multiple loci (i.e., $P[D|G]$). Table 1.8.1 illustrates a penetrance function that relates two single nucleotide polymorphisms (SNPs), each with two alleles and three genotypes, to risk of disease. In this example, the alleles each have a biological population frequency of $p = q$ $= 0.5$ with genotype frequencies of p^2 for AA and BB, $2pq$ for Aa and Bb, and q^2 for aa and bb, consistent with Hardy-Weinberg equilibrium. Thus, assuming the frequency of the AA genotype is 0.25, the frequency of Aa is 0.5, and the frequency of aa is 0.25, then the marginal penetrance of BB (i.e., the effect of just the BB genotype on disease risk) can be calculated as $(0.25 \times 0) + (0.5 \times 0) + (0.25 \times 1) = 0.25$. This means that the probability of disease given the BB genotype is 0.25, regardless of the genotype at the other genetic variation. Similarly, the marginal penetrance of Bb can be calculated as $(0.25 \times 0) + (0.5 \times 0.5) + (0.25 \times 0)$ $= 0.25$. Note that for this model, all of the marginal penetrance values (i.e., the probability

Table 1.8.1 Penetrance Values for Combinations of Genotypes from Two SNPs Exhibiting Interactions in the Absence of Independent Main Effects

	Table penetrance			Margin penetrance
	AA (0.25)	Aa (0.50)	aa (0.25)	
BB (0.25)	0	0	1	0.25
Bb (0.50)	0	0.50	0	0.25
bb (0.25)	1	0	0	0.25
Margin penetrance	0.25	0.25	0.25	

of disease given a single genotype, independent of the others) are equal, which indicates the absence of main effects (i.e., the genetic variations do not independently affect disease risk). This is true despite the table penetrance values not being equal. Here, risk of disease is greatly increased by inheriting exactly two high-risk alleles (e.g., a and b are defined as high risk).

Epistasis is difficult to detect and characterize using traditional parametric statistical methods, such as linear and logistic regression, because of the sparseness of the data in high dimensions. That is, when interactions among multiple polymorphisms are considered, there are many multilocus genotype combinations that have very few or no data points. For example, with two SNPs that each have three genotypes, there are nine two-locus genotype combinations (e.g., Table 1.8.1). In the case of three SNPs, there are 27 three-locus genotype combinations. Thus, as each additional SNP is considered, the number of multilocus genotype combinations goes up exponentially. The result of this added dimensionality is that exponentially larger sample sizes are needed to have enough data to estimate the interaction effects.

METHODS FOR DETECTING GENE-GENE INTERACTIONS IN ASSOCIATION STUDIES OF DISCRETE TRAITS

Logistic Regression

Logistic regression models the probability of disease (p) as a linear function of independent variables. A logit transformation of p, $\ln[p/(1-p)]$, is used to prevent p from taking on values of <0 or >1. By expressing the linear function in terms of exponentials, p can be modeled as $p = (e^{\alpha+\beta X})/(1 + e^{\alpha+\beta X})$, where α and β are regression coefficients (i.e., parameters) and X is an independent variable. For a discrete independent variable such as a polymorphism, an odds ratio relating genotypes to probability of disease can be estimated from e^β. The independent main effects of two polymorphisms, A and B, can be modeled as $p = (e^{\alpha+\beta_1 A + \beta_2 B})/(1 + e^{\alpha+\beta_1 A + \beta_2 B})$. The interaction between A and B can be modeled by adding a product term of the form $\beta_3 AB$ to the equation. A test of the null hypothesis of no interaction can be carried out by testing whether $\beta_3 = 0$. Rejection of this null hypothesis provides evidence for an interaction on a multiplicative scale. The advantage of logistic regression is that interactions can be modeled relatively easily, the statistical theory is very well characterized, and the approach can be implemented on a standard desktop computer using a variety of freely and commercially available statistical packages. As mentioned above, an important disadvantage is that very large sample sizes are needed to accurately estimate the parameters in the model when there are many independent variables.

Several free computer programs help to evaluate the power of a planned gene-gene interaction study using logistic regression. The Power program (*http://dceg.cancer.gov/POWER/*) allows estimation of sample size and power for two-locus interactions in both cohort and case-control studies. Another program, called Quanto, is available from *http://hydra.usc.edu/gxe/* for the estimation of sample size and power in matched case-control, case-sibling, case-parent, and case-only designs.

Multifactor Dimensionality Reduction

The multifactor dimensionality reduction (MDR) method was developed specifically to improve the power to detect gene-gene interactions in epidemiological study designs over that provided by logistic regression. The MDR approach is nonparametric in that no parameters are estimated and is free of any assumed genetic model. Figure 1.8.2 illustrates the general procedure to implement the MDR method. In step 1, the data are divided into a training set

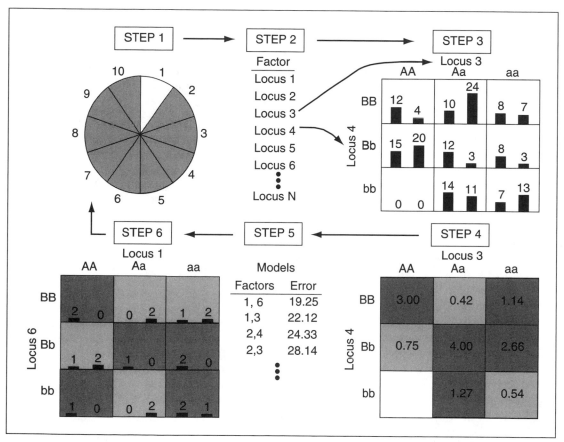

Figure 1.8.2 Summary of the general steps involved in implementing the MDR method. In step 1, the data are divided into a training set (e.g., 9/10 of the data) and an independent testing set (e.g., 1/10 of the data) as part of cross-validation. In step 2, a set of *N* genetic and/or discrete environmental factors is then selected from the pool of all factors. In step 3, the *N* factors and their possible multifactor classes or cells are represented in *N*-dimensional space. In step 4, each multifactor cell in the *N*-dimensional space is labeled as high-risk if the ratio of affected individuals to unaffected individuals (the number in the cell) exceeds some threshold *T* (e.g., *T* = 1.0) and low-risk if the threshold is not exceeded. In steps 5 and 6, the model with the best misclassification error is selected and the prediction error of the model is estimated using the independent test data. Steps 1 through 6 are repeated for each possible cross-validation interval. Bars represent hypothetical distributions of cases (left) and controls (right) with each multifactor combination. Dark-shaded cells represent high-risk genotype combinations while light-shaded cells represent low-risk genotype combinations. No shading or white cells represent genotype combinations for which no data were observed.

(e.g., 9/10 of the data) and an independent testing set (e.g., 1/10 of the data) as part of cross-validation. Second, a set of n genetic and/or environmental factors are selected. The n factors and their possible multifactor classes are represented in n-dimensional space; for example, for two loci with three genotypes each, there are nine possible two-locus-genotype combinations. The ratio for the number of cases to the number of controls is then calculated within each multifactor class. Each multifactor class in n-dimensional space is then labeled as "high-risk" if the cases to controls ratio meets or exceeds some threshold (e.g., ≥ 1), or as "low-risk" if that threshold is not exceeded; thus reducing the n-dimensional space to one dimension with two levels ("low-risk" and "high-risk"). The collection of these multifactor classes composes the MDR model for the particular combination of factors. Among all of the two-factor combinations, a single MDR model that has the fewest misclassified individuals is selected. This two-locus model will have the minimum classification error among the two locus models. In order to evaluate the predictive ability of the model, prediction error is estimated using ten-fold cross-validation. This entire procedure is performed ten times, using different random number seeds, to reduce the chance of observing spurious results due to chance divisions of the data.

For studies with more than two factors, the steps of the MDR method are repeated for each possible model size (i.e., each number of loci and/or environmental factors), if computationally feasible. The result is a set of models, one for each model size considered. From this set, the model with the combination of loci and/or discrete environmental factors that minimizes the prediction error is selected. Prediction error is a measure of how well the MDR model predicts risk status in the independent test sets. The prediction error is calculated as the average of the prediction errors across each of the ten cross validation subsets. Hypothesis testing of the best model(s) can then be performed by evaluating the magnitude of the cross-validation consistency and prediction error estimates using permutation testing. Here, the disease labels are randomized and the entire MDR analysis repeated. An MDR software package is available free from *http://phg.mc.vanderbilt.edu/Software/MDR/* for the Unix and Linux operating systems.

As an example, first consider a simple nonlinear gene-gene interaction model in the form of a penetrance function (Table 1.8.2). Here, disease is associated with being heterozygous at one locus, or the other, but not both. As with the model in Table 1.8.1, disease susceptibility for each individual genotype is the same. Thus, there is a strong interaction effect in the absence of any main effects. Table 1.8.3 illustrates an example case-control dataset simulated from this interaction model. Table 1.8.4 comprises contingency tables for single-locus tests of association. Note that neither locus A nor locus B has statistically significant single-locus effects as evaluated by using Fisher's exact test. The genotype distributions look very much the same between cases and controls as expected given the genetic model. Figure 1.8.3A illustrates the distribution of cases (left bars) and controls (right bars) for each two-locus genotype combination. Since the number of cases and controls is balanced, the numbers can be directly compared. With MDR, a new single genotype variable that is a combination of the

Table 1.8.2 Penetrance Values for Combinations of Genotypes from Two SNPs Exhibiting Interactions in the Absence of Independent Main Effects

	Table penetrance			Margin penetrance
	AA (0.25)	Aa (0.50)	aa (0.25)	
BB (0.25)	0	1	0	0.5
Bb (0.50)	1	0	1	0.5
bb (0.25)	0	1	0	0.5
Margin penetrance	0.5	0.5	0.5	

Table 1.8.3 Simulated Two-Locus Genotypes for Ten Cases and Ten Controls

Two-Locus Genotypes	
Cases	Controls
AABb	AaBb
AaBB	AABB
Aabb	AaBb
aaBb	Aabb
aaBb	aaBB
Aabb	AaBb
AaBB	aabb
AABb	AAbb
aaBb	AaBb
AABb	aaBB

two is generated. Genotype combinations that have more cases than controls are pooled into one group and those with more controls than cases are pooled into another (Fig. 1.8.3B). In doing so, a new variable with two levels, group 1 and group 2 (Fig. 1.8.3C), is created. An assessment can then be done on whether this new variable is associated with cases-control status. As described above, MDR uses cross-validation and permutation testing to evaluate each grouping. However, for illustration purposes, think about this in terms of a contingency table as in the single-locus case (Table 1.8.5). Here, everyone in group 1 is affected and everyone in group 2 is unaffected. If a χ^2 test of association with one degree of freedom were carried out, $\chi^2 = 20$ ($P < 0.001$) would be obtained. Thus, the new MDR variable captures the information about the interaction.

Table 1.8.4 Contingency Table for Locus A and Locus B[a]

Genotype	Observed (expected) counts		Total
	Cases	Controls	
AA	3 (3)	3 (3)	6
Aa	4 (4)	4 (4)	8
aa	3 (3)	3 (3)	6
Total	10	10	20
BB	2 (2.5)	3 (2.5)	5
Bb	6 (5)	4 (5)	10
bb	2 (2.5)	3 (2.5)	5
Total	10	10	20

[a]P > 0.05, Fisher's exact test.

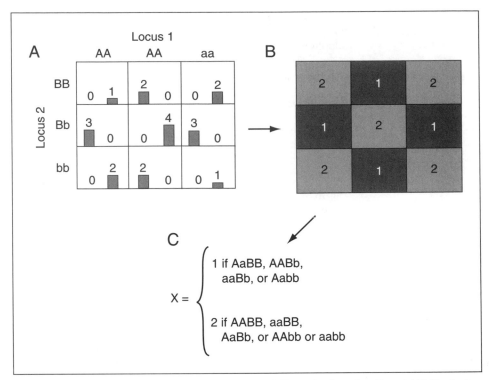

Figure 1.8.3 A simple illustration of how the MDR method works. Panel (**A**) illustrates the distribution of cases (left bars) and controls (right bars) for each two-locus genotype combination from the simulated data in Table 1.8.3. Those genotype combinations with more affected than unaffected subjects are pooled into group 1 and shaded light gray while those with more unaffected than affected subjects are pooled into group 2 and shaded dark gray (**B**). In (**C**), a new variable is formed with two levels that correspond to genotypes in groups 1 and 2. This new variable can then be evaluated using cross-validation and permutation testing.

METHODS FOR DETECTING GENE-GENE INTERACTIONS IN ASSOCIATION STUDIES OF QUANTITATIVE TRAITS

Linear Regression

Linear regression is a parametric statistical approach for modeling a continuous outcome variable (Y) as a linear function of discrete and/or continuous predictor variables (X_1, X_2, etc.). The linear model relating X to Y looks something like $Y = \beta_0 + \beta_1 X_1 + \beta_2 X_2 + \varepsilon$, where

Table 1.8.5 Contingency Table for the MDR Variable[a]

Genotype combination	Observed (expected) counts		Total
	Cases	Controls	
Group 1	10 (5)	0 (5)	10
Group 2	0 (5)	10 (5)	10
Total	10	10	20

[a]$P < 0.001$, χ^2 test (1 df).

β_0 is the intercept, β_1 and β_2 are the regression coefficients, and ε is the unexplained error in the model. In this model, the slope or regression of Y on X_1 is constant across the range of values for X_2. This means that the relationship between Y and X_1 is independent of X_2. Thus, the effects of the two predictor variables are purely additive. Deviations from additivity (i.e., interaction) can be measured by including a product term in the model as done above for logistic regression. Here, the term $\beta_3 X_1 X_2$ would be added to account for any interaction. The presence of an interaction term in the linear model allows there to be a different regression relationship between Y and X_1 for each value of X_2. Thus, the null hypothesis of no interaction is equivalent to $\beta_3 = 0$. For genetic studies, it is customary to encode polymorphisms as dummy variables that specify certain types of genetic effects. Each polymorphism with N genotypes should be encoded by $N-1$ dummy variables.

As with logistic regression, advantages of linear regression are that interactions can be modeled relatively easily, the statistical theory is very well characterized, and the approach can be implemented on a standard desktop computer using a variety of commercially available or free statistical packages. However, an important disadvantage is that very large sample sizes are needed to accurately estimate the parameters in the model when there are many independent variables.

Combinatorial Partitioning Method

The combinatorial partitioning method (CPM) is one of the few alternatives to linear regression that has been developed. The CPM simultaneously considers multiple polymorphic loci to identify combinations of genotypes that are most strongly associated with variation in the quantitative trait. Genotypes from multiple loci are pooled into a smaller number of classes, thereby avoiding the increased dimensionality associated with modeling of interactions. First, all possible multilocus genotypes are identified and this multilocus genotype space is divided into partitions or groups. This partitioning serves to collapse multiple dummy variables needed to encode multiple polymorphisms and their interactions into a single variable with two or more levels. This new independent variable can then be evaluated using simple linear regression as described above. The benefit of this data reduction is that only two parameters need to be estimated regardless of the number of polymorphisms. Second, all possible ways of partitioning the multilocus genotypic state space are evaluated, and the partitions that explain the most variation are identified. While CPM may provide a powerful alternative to linear regression, there are several important limitations. First, the approach is quite computationally intensive since it combinatorially sifts through many genotype partitions. Second, there is no software package available for this method. Finally, the power and type I error rate for CPM has not been fully evaluated.

References: Gauderman, 2002; Hirschhorn et al., 2002

Contributor: Jason H. Moore

CHAPTER 2

Genotyping

Simple sequence length polymorphisms (SSLPs) are highly informative markers that can be typed by the polymerase chain reaction (PCR) and which are widely distributed throughout eukaryotic genomes. Each SSLP is based on the variable numbers of di-, tri-, or tetranucleotide repeats at a particular location and can easily be characterized (genotyped) using PCR primers that anneal to single-copy DNA flanking the repetitive element. This technology is easily automated and permits the very-high-throughput typing necessary for high-resolution mapping of eukaryotic genomes. High-resolution SSLP-based maps have been developed for the human genome (Genethon: *http://www.genethon.fr/php/index_us.php*; Genome Database: *http://www.gdb.org*) and mouse genome (Whitehead/MIT Genome Center: *http://www.genome.wi.mit.edu*).

UNITS 2.1 & *2.3* describe basic methods for genotyping using SSLPs. These methods are suitable for the analysis of all di-, tri-, and tetranucleotide repeats. The basic protocols in *UNIT 2.1* are designed for low- or moderate-throughput genotyping. *UNIT 2.3* describes high-throughput genotyping using automated fluorescent genotyping. Additional high-throughput genotyping methods are presented in *UNITS 2.4–2.6*.

A second class of polymorphism widely used in human and animal genetic mapping is based on single-nucleotide substitutions and small deletions. *UNIT 2.2* describes two methods, the oligonucleotide ligation assay (OLA) and ligase chain reaction (LCR), that can be used to type these polymorphisms in all cases where the sequence flanking the nucleotide substitution or deletion is known. These are highly efficient techniques that utilize the potential of DNA amplification by PCR to permit automated high-throughput analysis of polymorphisms without the need to screen DNA fragments on gels.

Contributor: Nicholas C. Dracopoli

UNIT 2.1

PCR Methods of Genotyping

Simple sequence length polymorphisms (SSLPs) are derived from DNA elements containing very short simple-sequence repeats such as $(CA)_n$, which are interspersed throughout eukaryotic DNA. Their usefulness as genetic markers stems from their polymorphism in length—that is, the variation between alleles in the number of repeats. The single most important step in developing a robust assay for each marker by these methods is the choice of PCR primers flanking the repeat; the programs PRIMER (available at no charge from Mark Daly, Whitehead Institute, Cambridge, Mass.; *mjdaly@genome.wi.mit.edu.*), OLIGO (National Biosciences), and MacVector (Eastman Laboratory Research Products) are useful for this purpose.

NOTE: Experiments involving PCR require extremely careful technique to prevent contamination.

PCR AMPLIFICATION OF SSLPs USING END-LABELED PRIMERS

Materials *(see APPENDIX 1 for items with ✓)*

  20 µM forward and reverse primers (store at −20°C)
  ✓ 10× T4 polynucleotide kinase buffer
  10 mCi/ml [γ-^{32}P]ATP (3000 Ci/mmol)
  50 U/µl T4 polynucleotide kinase
  ✓ Template DNA: 5 to 20 ng/µl genomic DNA in TE buffer (see recipe for buffer) or H$_2$O
  ✓ 10× PCR amplification buffer containing 15 mM MgCl$_2$
  ✓ 1.25 mM 4dNTP mix
  5 U/µl *Taq* DNA polymerase
  Light mineral oil
  ✓ 2× formamide loading buffer (store in 10-ml aliquots ≤6 months at −20°C)
  Labeled DNA size marker: e.g., γ-^{32}P-labeled *Msp*I digest of pBR322 (with 12 bands in 100- to 250-bp range) and/or 1-bp-resolution γ-^{32}P-labeled M13 sequence ladder

  65°C water bath
  96-well microtiter plate suitable for use in a thermal cycler
  UV-transparent plastic wrap (e.g., Saran Wrap)
  Centrifuge and rotor with microplate carrier (e.g., Beckman or Sorvall)
  Thermal cycler accommodating 96-well microtiter plate
  Used X-ray film *or* Whatman 3MM filter paper

1. To end-label enough primer for amplification of 100 samples, pipet the following into a clean 1.5-ml microcentrifuge tube (also see *APPENDIX 3E*):

     10 µl 20 µM forward *or* reverse primer
     2.5 µl 10× T4 polynucleotide kinase buffer
     10 µl 10 mCi/ml [γ-^{32}P]ATP (3000 Ci/mmol)
     1 µl 50 U/µl T4 polynucleotide kinase
     1.5 µl H$_2$O.

   Incubate 30 min at 37°C.

   If one primer contains repeat elements (e.g., Alu), use the other primer in the labeling reaction to provide a cleaner banding pattern.

2. Inactivate enzyme by heating reaction mixture 10 min at 65°C. If desired, store labeled primer at −20°C up to 1 week.

3. Pipet 5 µl template DNA into each well of a 96-well microtiter plate appropriate for the thermal cycler to be used. Cover plate with lid or UV-transparent plastic wrap and store on ice to prevent evaporation.

4. Prepare sufficient PCR mix for the number of DNA samples to be amplified by mixing the following (per 100 samples, plus 10% excess to allow for wastage during pipetting):

     220 µl 10× PCR amplification buffer
     352 µl 1.25 mM 4dNTP mix
     44 µl 20 µM forward primer
     55 µl 20 µM reverse primer
     27.5 µl labeling reaction (from step 2)
     11 µl 5 U/µl *Taq* polymerase
     940.5 µl H$_2$O.

Table 2.1.1 Thermal Cycling Parameters for Amplification of SSLPs

Process	Temperature	Perkin-Elmer 9600 cycler	Generic thermal cycler
Initial denaturation	94°C	3 min	3 min
Cycling (30×)	94°C	15 sec	30 sec
	55°C	30 sec	2 min
	72°C	30 sec	2 min
Final extension	72°C	5 min	7 min
Hold	4°C	—	—

5. Add 15 μl PCR mix to each well of the microtiter plate. Keep the PCR plate on ice or on a refrigerating block prior to starting the thermal cycling, to reduce the chance of primer misannealing.

6. Overlay each well with 60 μl light mineral oil to minimize evaporation. Centrifuge plate briefly at 1000 to 2000 rpm in a Beckman or Sorvall plate carrier, 4°C.

7. Program thermal cycling parameters in accordance with the primers selected and the thermal cycling device used; standard PCR conditions for a reaction containing primers with a T_m of 60°C are shown in Table 2.1.1.

   Usually, an annealing temperature 5°C below the minimum T_m of the PCR primers should be tried. This can be gradually adjusted upwards if nonspecific bands are identified.

8. Start cycle without microtiter plate. When block temperature reaches 94°C, add microtiter plate to thermal cycler and complete cycle. When cycle is completed, add 20 μl of 2× formamide loading buffer to each well. If necessary, store plates ≤2 weeks at −20°C in a Plexiglas box before or after addition of loading buffer (preferable to use immediately).

9. Prepare a 6% denaturing polyacrylamide gel (*APPENDIX 3F*). Prerun gel 1 hr at 40 W (constant power). Adjust power during run to maintain the gel temperature at ~55°C.

10. Shortly before loading the gel, heat-denature samples 5 min at 94°C and quickly cool on ice. Rinse gel wells to remove urea by pipetting buffer in and out of wells. Load 2 μl sample per well. Include a labeled DNA size marker (e.g., γ-^{32}P-labeled *Msp*I digest of pBR322) on each gel. To compare PCR products for a large set of DNAs run on different gels, load either the PCR products of one or two reference DNAs or a 1-bp-resolution γ-^{32}P-labeled M13 sequence ladder on each gel.

11. Run gel 2 to 3 hr at constant power, maintaining a temperature of 55°C, until the xylene cyanol is close to the bottom of the gel.

12. Remove gel sandwich from electrophoresis apparatus and lay flat, with the silanized plate on top. Slowly pry apart plates using a long metal spatula (the gel should stick to the bottom plate). Lay a piece of used X-ray film or Whatman 3MM paper on the gel, taking care to prevent air bubbles from forming between them. Slowly peel the gel from the glass plate by lifting the support. Wrap gel and support with UV-transparent plastic so that gel is completely covered. Carry out autoradiography for 2 to 24 hr at −70°C using one intensifying screen.

PCR AMPLIFICATION OF SSLPs USING INTERNAL LABELING

When testing many primers on a limited number of DNA samples, this protocol minimizes manipulations and use of radioactivity compared to the end-labeling protocol (Basic Protocol 1).

Additional Materials (*also see Basic Protocol 1; see* APPENDIX 1 *for items with* ✓)

DNA template: 5 to 20 ng/μl genomic DNA in 10 mM Tris·Cl/0.2 mM EDTA, pH 7.3 (store at 4°C)

✓ 10× cold nucleotide mix

Primer mix: 100 ng/μl each forward and reverse primers (store at −80°C)

10 μCi/μl [α-^{32}P]dCTP (800 Ci/mmol)

1. Pipet 1.5 μl DNA template into each well of a 96-well microtiter plate. Cover plate with lid or plastic wrap (e.g., Saran wrap) and store on ice.

2. Prepare sufficient PCR mix for the number of DNA samples to be amplified by mixing the following (per 100 samples, plus 10% excess):

> 110 μl 10× PCR amplification buffer
> 110 μl 10× cold nucleotide mix
> 11 μl primer mix
> 7.7 μl [α-^{32}P]dCTP
> 6.6 μl 5 U/μl *Taq* polymerase
> 689.7 μl H$_2$O.

> *[α-^{35}S]dATP can be used, but this generates a weaker signal and requires an extra step to dry the gel before autoradiography.*

3. Add 8.5 μl PCR mix to each well. Overlay each well with 30 μl light mineral oil. Carry out thermal cycling, electrophoresis, and autoradiography (see Basic Protocol 1, steps 6 to 12).

BASIC PROTOCOL 2

NONRADIOACTIVE ANALYSIS OF SSLPs USING SILVER STAINING

Use sterile, deionized, double-distilled water in all recipes and steps. Use ~200 ml of each solution in the following steps.

Materials (*see* APPENDIX 1 *for items with* ✓)

Denaturing polyacrylamide gel of PCR-amplified SSLPs (see Basic Protocol 1, steps 3 to 11)

10% (v/v) ethanol

1% (v/v) nitric acid

✓ Silver nitrate staining solution

✓ Developing solution

10% (v/v) acetic acid

Staining apparatus: e.g., see Figure 2.1.1
Whatman filter paper
80°C vacuum oven

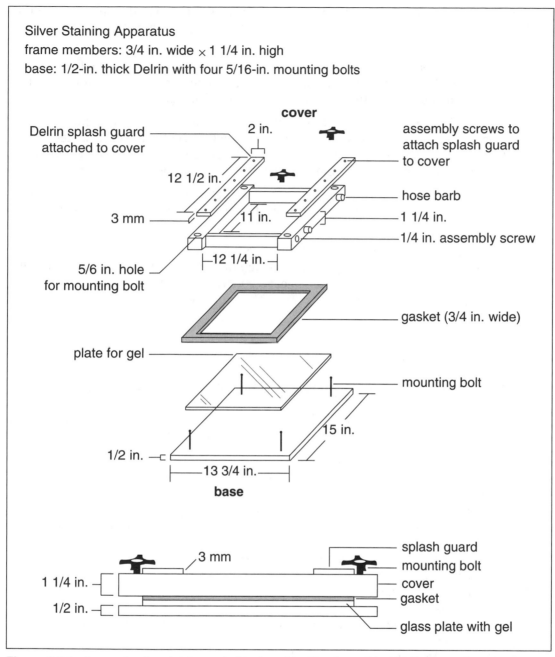

Silver Staining Apparatus
frame members: 3/4 in. wide × 1 1/4 in. high
base: 1/2-in. thick Delrin with four 5/16-in. mounting bolts

Figure 2.1.1 Apparatus for staining gels (designed by Bender et al., 1994). The splash guards are attached to the cover frames by glue and assembly screws to provide a leak-proof seal.

1. After electrophoresis, carefully disassemble the apparatus and separate the glass plates. Place the plate with the gel on the base of the staining apparatus. Carefully place the upper frame and gasket over the edges of the gel and secure the mounting bolts.

   If there are small pinholes in the gel that will allow reagents to get trapped between the gel and the glass plate, they can be filled by adding (dropwise) a little Sequagel-6 (National Diagnostics) containing a high concentration of ammonium persulfate. The gel solution should polymerize rapidly.

2. Wash the gel 5 min in 10% ethanol, then decant the solution. Oxidize the gel 3 min in 1% nitric acid, then decant the solution. Rinse the gel 5 sec in water, then decant the water.

3. Add silver nitrate staining solution to the gel and cover the apparatus to avoid exposure to light. Incubate 20 min, then decant the silver nitrate solution and dispose of it according to safety regulations. Rinse the gel 5 min in water and decant.

4. Add developing solution to reduce the gel (it initially turns brownish yellow). Gently rock the apparatus so the solution washes over the gel. Decant and add fresh developing solution. Continue changing until the solution no longer changes color and remains clear (after solution remains clear ~2 min, bands will begin to develop).

5. Decant developing solution after image appears but before gel turns brown (avoid overdeveloping). Stop developing process by incubating gel 1 min in 10% acetic acid. Decant solution. Rinse gel 5 sec with water.

6. Transfer gel to Whatman filter paper by placing the paper directly on the developed gel and gently lifting the gel off the plate. Vacuum dry the gel 1 hr at 80°C. Record the results. Trim the excess gel and store as a permanent record (gels have been stored >2 years with no apparent loss of information).

BASIC PROTOCOL 3

NONRADIOACTIVE MULTIPLEX ANALYSIS OF SSLPs

The protocol is summarized in the flow chart in Figure 2.1.2. This method offers the best data interpretation because each lane of the gel includes internal size standards.

Materials (see APPENDIX 1 for items with ✓)

Template DNA: 10 ng/μl genomic DNA in H_2O (not TE buffer) from each individual to be genotyped

Primer mixes: for each of 24 SSLPs to be amplified, mix equal volumes of 6.6 μM forward primer and 6.6 μM reverse primer (final 3.3 μM each); store indefinitely at −20°C

✓ 10× PCR amplification buffer containing 15 mM $MgCl_2$
✓ 2.5 mM 4dNTP mix
5 U/μl *Taq* DNA polymerase
Light mineral oil (optional)
Qiaex Gel Extraction Kit (Qiagen) containing solutions QX1 and QX3 and Qiaex glass bead suspension
✓ 2× formamide loading buffer
M13 sequence ladder size standard (see Support Protocol 1)
✓ TBE buffer
✓ Prehybridization solution
Digoxigenin-labeled probes (made from primers used to amplify SSLPs and from M13 anti-universal primer; see Support Protocol 2)
✓ Oligo wash buffer
✓ Blocking solution
Alkaline phosphatase–conjugated anti-digoxigenin Fab fragments (Boehringer Mannheim)
✓ Detection buffer
✓ Substrate buffer

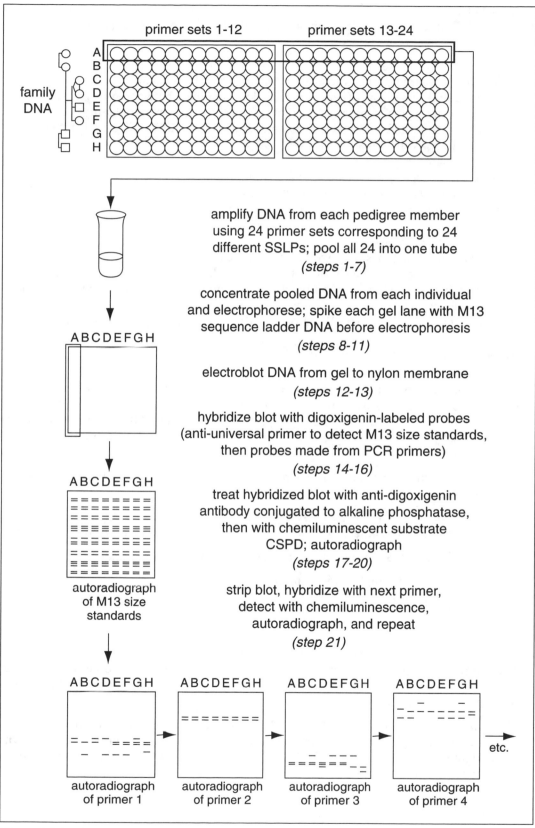

Figure 2.1.2 Flow chart for nonradioactive multiplex analysis of SSLPs.

✓ 0.2 mM CSPD

✓ 2 mM EDTA/0.2% (w/v) SDS, preheated to 70°C (made from sterile stock solutions, see recipes; store at room temperature)

96-well microtiter plates

0.2-ml thin-walled PCR tubes arrayed in 96-well (8 × 12) format

MicroAmp Full Plate Cover (Perkin-Elmer; decontaminate between uses by washing with soap and water, soaking in 10% bleach overnight, then autoclaving)

Thermal cycler with heated cover, accommodating 0.2-ml thin-walled PCR tubes arrayed in 96-well format, preheated to 94°C

42°C water bath

Blotting filter paper (Owl Scientific Plastics preferred)

TE 90 GeneSweep Sequencing Gel Transfer Unit (Hoefer)

Hybond-N membrane (critical choice; Amersham)

80°C oven

Stratalinker (Stratagene)

38 × 300–mm glass tubes, silanized (e.g., *CPMB* APPENDIX 3)

Hybridization oven (e.g., Model 400 Hybridization Incubator, Robbins Scientific)

Rolling shaker (e.g., Model V5250 Incubator, Robbins Scientific, or Model 400 set at room temperature)

Transparent plastic film (Saran Wrap preferred) or two thin polypropylene sheets

Film cassette

Autoradiography film (e.g., Hyperfilm-MP, Amersham)

70°C incubator

1. For a 24-plex configuration (8 template DNA samples with 24 primer sets), set up 0.2-ml thin-walled PCR tubes arrayed in 96-well format as shown in Figure 2.1.2 and place on ice. Pipet 5 μl of the first family member's template DNA into each of the 24 tubes comprising row A. Pipet the second family member's DNA into row B, and continue until all eight rows are filled.

2. Pipet 15 μl of the first twelve primer mixes into separate wells of a 96-well microtiter plate on ice.

   Mixes are prepared in sets of twelve because most thermal cyclers can only accommodate one plate at a time and the master mix (step 3) should be prepared immediately before amplification.

3. Mix the following in a 1.5-ml microcentrifuge tube on ice (490 μl final):

   140 μl 10× PCR amplification buffer
   112 μl 2.5 mM 4dNTP mix
   231 μl H_2O
   7 μl 5 U/μl *Taq* DNA polymerase.

4. Add 35 μl of this master mix to each well of the microtiter plate from step 2. Then pipet 5 μl from the first primer mix into each of the eight tubes comprising column 1 from step 1. Repeat for columns 2 through 12, until each of the twelve primer mixes has been added to each of the eight template DNAs.

5. Cover tubes with MicroAmp Full Plate Cover and quickly load into thermal cycler preheated to 94°C. If thermal cycler does not have a heated cover, overlay each well with 30 μl light mineral oil. Carry out PCR using the following amplification cycles (also see Basic Protocol 1):

Initial step:	5 min	94°C	(denaturation)
30 cycles:	10 sec	94°C	(denaturation)
	30 sec	55°C	(annealing)
	30 sec	72°C	(extension)
Final extension:	5 min	72°C	(extension)
Final step:	indefinitely	4°C	(hold).

These parameters are ideal for primers with a T_m of 60°C, although they have been successfully used for primers with T_m values ranging from 55°C to 70°C. If the PCR fails, the annealing temperature can be optimized for a particular T_m.

6. Repeat steps 2 to 5 to amplify the DNA in columns 13 to 24 with the remaining twelve primer mixes.

7. Combine 10 μl from each of the 24 different PCRs for a single individual's DNA (i.e., from each of the 24 tubes comprising a horizontal row) into one microcentrifuge tube (240 μl total).

8. Add 3 vol (720 μl) solution QX1 and 10 μl glass bead suspension, vortex, and incubate 5 min at room temperature. Microcentrifuge ~30 sec at maximum speed. Aspirate and discard supernatant. Add 0.5 ml solution QX3, vortex, then microcentrifuge ~30 sec at maximum speed. Aspirate and discard supernatant. Microcentrifuge again at maximum speed and aspirate all traces of supernatant. Resuspend pellet in 20 μl water and incubate 5 min at 42°C to elute DNA from glass beads. If necessary, store eluted DNA indefinitely at −20°C.

9. Prepare 5% denaturing polyacrylamide gel (*APPENDIX 3F*) using 0.4-mm spacers. Prerun 20 to 30 min at 70 W, constant power.

10. Microcentrifuge pooled DNA samples 1 min at maximum speed (to pellet beads) and then mix:

    0.5 to 2.0 μl pooled DNA sample
    1 vol 2× formamide loading buffer
    0.2 μl M13 sequence ladder size standard.

11. Shortly before loading gel, heat-denature pooled DNA samples 2 to 3 min at 94°C and quickly cool on ice. Rinse gel wells to remove urea by pipetting TBE buffer in and out of wells. Load 1 to 3 μl of sample mixture per lane. Run gel at 70 W for ~1.5 to 2 hr in TBE buffer. To visualize an SSLP with a reference allele of ~100 bp, run gel until xylene cyanol FF is ≥15 cm from bottom of gel. For SSLPs larger than ~200 bp, run gel until xylene cyanol FF is ≥5.5 cm from bottom of gel.

12. Remove gel sandwich from electrophoresis apparatus and lay flat. Remove top silanized plate. Wet a piece of blotting filter paper with TBE buffer. Drain off excess buffer and lay paper on gel in a smooth wave, avoiding bubbles. Slowly peel gel from glass plate by lifting paper. If gel fails to stick to wet paper, blot paper with a dry sheet to remove excess buffer. Place paper and gel on TE 90 GeneSweep Sequencing Gel Transfer Unit, gel-side-up. With gloved hand, wet a piece of Hybond-N membrane with TBE buffer. Carefully lay membrane onto gel surface in one smooth motion, avoiding bubbles. Wet another piece of blotting filter paper with TBE buffer and lay on top of membrane, creating a sandwich of paper, gel, membrane, and paper.

13. Follow manufacturer's instructions for electroblotting using GeneSweep apparatus. After transfer is complete (~5 min), remove top layer of paper and peel membrane from gel. Remove any traces of gel residue stuck to sides of the membrane with a gloved hand. Bake

membrane at 80°C for 20 min, then cross-link with UV, DNA side up, at 120 mJ/cm^2 in Stratalinker set to automatic cross-link.

Transfer can also be performed by Southern blotting using capillary action (APPENDIX 3G).

14. Roll membrane to form a tube. Insert into silanized 38 × 300–mm glass tube and add 25 ml prehybridization solution (for 800-cm^2 blot; adjust for specific blot size). Incubate in hybridization oven 15 min at 37°C, then discard solution by turning glass tube upside down.

15. Add 10 pmol of digoxigenin-labeled probe(s) to 10 ml prehybridization solution and mix (final concentration should be ~1 pmol/100 cm^2 membrane). Add solution to tube containing blot and incubate ≥1 hr in 37°C hybridization oven.

Use probe made from M13 anti-universal primer for first hybridization in order to assess overall quality of gel and transfer.

For overnight hybridizations, amount of probe can be reduced 2-fold. If two SSLP markers are >50-bp apart in size (i.e., their pattern of bands does not overlap), they can be detected simultaneously by combining probes (each at 1 pmol/100 cm^2 membrane).

16. Wash blot six times with 100 ml oligo wash buffer at room temperature, shaking tube vigorously end-to-end several times before discarding each wash. Immediately add 50 ml freshly prepared blocking solution and agitate on rolling shaker 30 min at room temperature. Discard solution.

17. Mix 5 μl alkaline phosphatase–conjugated anti-digoxigenin Fab fragments with 50 ml detection buffer (1:10,000 dilution). Add to blot and agitate on rolling shaker 30 min at room temperature. Discard antibody solution.

18. Rinse blot quickly with 50 ml detection buffer, then wash three times at room temperature with 80 ml detection buffer, agitating 5 min on rolling shaker each time.

19. Rinse blot quickly with 25 ml substrate buffer and discard buffer. Add 10 ml of 0.2 mM CSPD and agitate 10 min at 37°C.

20. Remove blot from tube, leaving it damp but not dripping, and wrap in transparent plastic film or lay between two thin polypropylene sheets. Roll out bubbles using a glass pipet as a rolling pin. Within 6 to 8 hr after addition of CSPD, place wrapped membrane, DNA side up, in film cassette and expose to autoradiography film 30 min (typically) at 37°C.

21. To strip the blot, transfer to a 38 × 300–mm silanized glass tube, add 100 ml of 2 mM EDTA/0.2% SDS (preheated to 70°C), and incubate 10 min at 70°C with agitation. Discard solution and repeat. Store damp at room temperature, sealed in the 38 × 300–mm tube or wrapped in Saran Wrap, until they are probed again. Repeat steps 14 to 21 using other digoxigenin-labeled probes.

SUPPORT PROTOCOL 1

PREPARATION OF M13 SEQUENCE LADDER SIZE STANDARD

The "A" lane of an M13 sequence ladder is used in Basic Protocol 2 as an internal standard in each lane of the gel that is used to separate the SSLP-PCR products.

M13 DNA, DTT, Sequenase reaction buffer, Sequenase T7 DNA polymerase, and termination mix are included in the Sequenase Version 2.0 DNA Sequencing Kit (U.S. Biochemical).

Materials (*see* APPENDIX 1 *for items with* ✓)

M13mp18 single-stranded DNA (U.S. Biochemical) in TE buffer
0.5 pmol/µl M13 universal sequencing primer (5'-GTAAAACGACGGCCAGT-3'; e.g., U.S. Biochemical)
✓ 5× Sequenase reaction buffer (U.S. Biochemical)
✓ TE buffer
✓ 0.1 M DTT
✓ 7.5 µM 4dNTP mix
13 U/µl Sequenase Version 2.0 T7 DNA polymerase (U.S. Biochemical)
Termination mix: 80 µM each dGTP, dATP, dCTP, dTTP, plus 8 µM ddATP in 50 mM NaCl (prepared mix available from U.S. Biochemical)
✓ 2× formamide loading buffer

1. Combine in microcentrifuge tube:

    38 µg M13mp18 DNA
    24 µl 0.5 pmol/µl M13 universal sequencing primer
    40 µl 5× Sequenase reaction buffer
    TE buffer to 200 µl.

    Anneal by heating 2 min at 65°C, then cool slowly to <35°C. Microcentrifuge briefly at maximum speed.

2. Add to cooled, annealed mixture:

    20 µl 0.1 M DTT
    8 µl 7.5 µM 4dNTP mix
    32 µl H$_2$O
    45 µl TE buffer
    5 µl Sequenase T7 DNA polymerase.

    Leave at room temperature 7 min.

3. Preheat a 1.5-cm microcentrifuge tube containing 200 µl termination mix to 37°C. Add mixture from step 2 and incubate 5 min at 37°C. Stop reaction by adding 320 µl of 2× formamide loading buffer. Store indefinitely at −20°C.

SUPPORT PROTOCOL 2

DIGOXIGENIN LABELING OF PROBES USING TERMINAL TRANSFERASE

The M13 anti-universal primer and either the forward or reverse primer used to amplify each SSLP in Basic Protocol 2 are tailed using digoxigenin-11-dUTP. It is important to use only one primer for a given SSLP as a hybridization probe in order to detect only one strand of the PCR product; if both the CA strand and GT strand are detected simultaneously, their different migration patterns will complicate genotype interpretation.

Materials (*see* APPENDIX 1 *for items with* ✓)
✓ 5× terminal transferase reaction buffer, pH 6.6
25 mM CoCl$_2$
6.6 µM forward or reverse primer used to amplify each SSLP (see Basic Protocol 2)
6.6 µM M13 anti-universal primer (5'-ACTGGCCGTCGTTTTAC-3')
1 mM digoxigenin-11-dUTP (Boehringer Mannheim)
10 mM dATP

50 U/μl terminal transferase (Boehringer Mannheim)

✓ 0.2 M EDTA, pH 8.0

1. For each primer (including M13 anti-universal primer) prepare reaction mix:

>   6 μl 5× terminal transferase reaction buffer
>   6 μl 25 mM CoCl$_2$
>   15 μl 6.6 μM primer
>   1 μl 1 mM digoxigenin-11-dUTP
>   1 μl 10 mM dATP
>   1 μl 50 U/μl terminal transferase.

2. Incubate 20 min at 37°C, then stop reaction by adding 2 μl of 0.2 M EDTA. Store labeled primers indefinitely at −20°C.

>   *Labeling in this manner results in an average tail length of 50 bp (range 10 to 100 bp) with ~5 digoxigenin-11-dUTP nucleotides per tail. The labeling reaction can be checked by dot blotting.*

References: Dib et al.,1996; Weber and May,1989; Weissenbach et al.,1992

Contributors: Thomas J. Hudson, Chris D. Clark, Michele Gschwend, and Erica Justice-Higgins

---

UNIT 2.2

Genotyping by Ligation Assays

Ligation assays determine whether two adjacent 15- to 20-base oligonucleotide primers can be joined when they are hybridized to a target template such as a PCR product or genomic DNA sample. Primer-joining by DNA ligase requires complementary base-pairing at the junction of the hybridized oligonucleotides. Even a single-base-pair mismatch at this junction will prevent the ligation of two adjacent primers (see Fig. 2.2.1). Thus, assaying for the success of ligation will establish the existance of a single nucleotide sequence change. Ligation assays are also useful for genotyping known insertions or deletions in a target, because two primers must also lie directly next to one another to be ligated.

NOTE: Experiments involving PCR require extremely careful technique to prevent contamination.

BASIC PROTOCOL 1

OLIGONUCLEOTIDE LIGATION ASSAY (OLA)

The OLA uses three oligonucleotide primers to genotype single-nucleotide variations—two 5′ biotinylated primers (one for each of the alternative alleles) and one adjacent 3′ digoxigenin-labeled primer (see Fig. 2.2.1).

Materials (see APPENDIX 1 *for items with* ✓ *)*

✓ Streptavidin-coated microtiter plates

✓ Blocking buffer: 0.5% (w/v) BSA in PBS (see recipe for PBS)

Modified oligonucleotide primers (see Support Protocol): 5 μM stocks of two allele-specific primers, labeled at the 5′ end with biotin, and of a common reporter primer, labeled at the 3′ end with digoxigenin and phosphorylated at the 5′ end

✓ OLA ligation mix

Amplified target DNA (UNIT 7.1), 80 to 1500 bp long

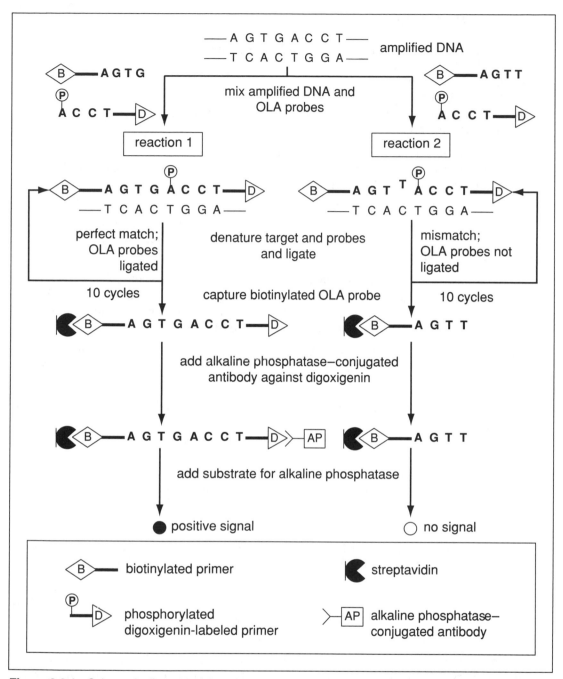

Figure 2.2.1 Schematic diagram of the oligonucleotide ligation assay (OLA). To genotype a nucleotide substitution (e.g., G to T, shown here), two OLA reactions are assembled with aliquots of the amplified DNA product. One detects the G allele (reaction 1) and the other detects the T allele (reaction 2). In the presence of complementary targets, DNA ligase covalently joins the biotinylated probes to the digoxigenin-labeled probes (left side of figure). If there is a probe/target mismatch, the probes are not joined (right side of figure). To determine the outcome of the ligation, the biotinylated probes are captured on a support, the support is washed, and an enzyme-linked immunoassay for digoxigenin is performed. The presence or absence of a color signal indicates whether the probes were ligated on the amplified target. In this example, the target was homozygous for the G substitution and signal was obtained from only that reaction. Homozygous DNA targets yield a positive signal in only one of the two OLA reactions (either reaction 1 or 2) and heterozygous targets produce positive signals in both reactions 1 and 2.

0.1% (v/v) Triton X-100
Mineral oil
5 U/μl thermostable DNA ligase (e.g., Epicentre Technologies)
100 mM EDTA/0.1% (v/v) Triton X-100
✓ Tris/NaCl wash buffer
0.01 M NaOH/0.05% (v/v) Tween 20
Alkaline phosphatase–labeled anti-digoxigenin antibodies (Boehringer Mannheim)
✓ Colorimetric substrate

96-well microtiter plates, flexible, round-bottom (Falcon)
Thermal cycler accommodating 96-well microtiter plate *or* heating block modified to accommodate a 96-well microtiter plate
Microtiter plate spectrophotometer

1. Remove the streptavidin solution from streptavidin-coated microtiter plates by flicking into the sink and tap the plates dry (well-side-down) on a paper towel. Add 200 μl blocking buffer to each well and incubate ≥20 min at room temperature before use in step 5.

2. Prepare a separate ligation mix for each allele to be genotyped by adding 2 μl of the appropriate biotinylated oligonucleotide and 2 μl of the digoxigenin-labeled reporter oligonucleotide to 600 μl of OLA ligation mix containing 2 μl of 5 U/μl thermostable DNA ligase (sufficient to type 48 DNA samples in one microtiter plate).

3. Add 10 μl of the ligation mix for allele 1 to every well in the odd-numbered columns of a flexible round-bottom 96-well microtiter plate. Add the mix for allele 2 to even-numbered columns.

4. Dilute 10 μl of the amplified target DNA with 90 μl of 0.1% Triton X-100. Add 10 μl to each well and cover reactions with 60 μl mineral oil (not required if thermal cycler has heated lid). Perform ligation in a 96-well thermocycler using the following program:

$$\text{10 cycles:} \quad \begin{array}{lll} 30\ \text{sec} & 93°\text{C} & \text{(denaturation)} \\ 2\ \text{min} & 58°\text{C} & \text{(annealing/ligation).} \end{array}$$

Add 10 μl of 100 mM EDTA/0.1% Triton X-100 to each reaction to stop the ligation.

5. Remove the blocking solution from the streptavidin-coated microtiter plate by flicking into the sink and tapping it dry. Wash plate once with Tris/NaCl wash buffer and tap dry.

6. Transfer ligation reactions to the streptavidin-coated plate and incubate ≥30 min at room temperature to capture the biotinylated oligonucleotides. Wash plate twice with 0.01 M NaOH/0.05% Tween 20 (to remove unligated reporter primers) and twice with Tris/NaCl wash buffer.

7. Prepare a 1:1000 dilution of alkaline phosphatase–labeled anti-digoxigenin antibody in blocking buffer. Add 30 μl to each well and incubate 30 min at room temperature. Wash plate five times with Tris/NaCl wash buffer and add 25 to 50 μl colorimetric substrate to each well. Incubate 10 to 15 min at room temperature and read the plate in a microtiter plate spectrophotometer.

BASIC PROTOCOL 2

LIGASE CHAIN REACTION (LCR)

By performing ligation reactions on each of the complementary strands of a DNA target, it is possible to exponentially amplify the products of a ligation assay (see Fig. 2.2.2). LCR can be

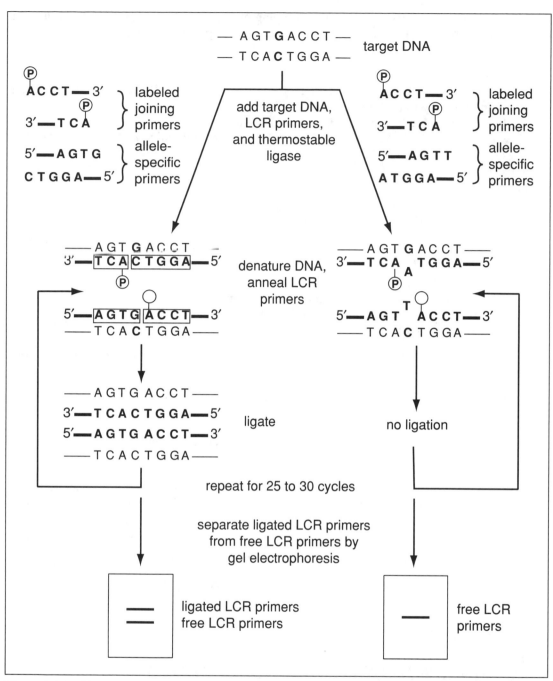

Figure 2.2.2 Flow chart of the ligase chain reaction (LCR) protocol. Phosphate groups in this procedure are radioactively tagged. The four primers added to the reaction should be designed such that a single-base 3′ overhang is formed at the junction (as illustrated by the boxed primers). This configuration is believed to minimize the occurence of nonspecific ligation. A 2-bp overhang can be added to the 5′ end of the allele-specific primer to minimize target-independent ligation of the primers.

used to increase the sensitivity of DNA genotyping by ligation. As few as 200 molecules of a specific DNA target can be detected when radioisotopic reporters are employed.

Materials *(see APPENDIX 1 for items with ✓)*

✓ LCR mix
✓ Oligonucleotide primers for LCR: allele-specific primers and ^{32}P-labeled joining primers complementary to both target strands (Fig. 2.2.2)
   5 U/μl thermostable ligase (e.g., Epicentre Technologies)
   Target DNA: 10 μl of PCR product (*UNIT 7.1*) or ~10 ng genomic DNA sample (*APPENDIX 3A*)
   Mineral oil
   Formamide
   DNA molecular size markers
   Automated thermal cycler or 96-well heating block

1. Prepare separate ligation reactions for each allele (each contains two allele-specific primers and two labeled joining primers):

   100 μl LCR mix
   40 fmol of allele-specific primers for each strand
   40 fmol of ^{32}P-labeled joining primers for each strand
   0.5 U thermostable ligase.

2. Mix 10 μl (~10 to 50 ng) target DNA with each allele-specific ligation mix. Cover reactions with 30 μl mineral oil.

3. Carry out LCR in an automated thermal cycler or 96-well heating block using the following amplification cycles:

   25 to 30 cycles: 30 sec   94°C   (denaturation)
                       2 min   65°C   (annealing/ligation).

   Beyond 30 cycles, background ligation events begin to amplify exponentially. To decrease non-specific ligation at the joining site, primers with single-base-pair 3′ overhangs should be used.

4. Mix 4 μl of each reaction with 4 μl formamide. Denature by heating 2 min at 100°C and load onto a 10% denaturing polyacrylamide gel (*APPENDIX 3F*). Include appropriately labeled molecular size markers on the gel. Electrophorese 2 hr at 60 W.

5. Dry the gel, expose to film, develop, and analyze results (see Fig. 2.2.2).

SUPPORT PROTOCOL

PREPARING MODIFIED OLIGONUCLEOTIDES FOR LIGATION ASSAYS

Modified oligonucleotide primers are commercially available or can be prepared in laboratories with an automated DNA synthesizer and prior expertise in DNA synthesis.

Biotinylated Primers

Incorporation of 5′ biotin during synthesis. A 5′ biotin phosphoramidite is incorporated into the primer during synthesis. Detailed instructions provided by the manufacturer of the DNA synthesizer should be followed.

Addition of biotin after synthesis. A 5′ aminohexylphosphate linker is incorporated during oligonucleotide synthesis (follow synthesizer instructions). The biotin is added after the primer is deprotected by preparing a reaction mixture containing 100 μl (200 μM) oligonucleotide, 50 μl of 50 mM biotin-*N*-hydroxysuccinimide ester in dimethylformamide, and 50 μl of 1 M sodium bicarbonate/sodium carbonate buffer, pH 9.0 (see *APPENDIX 1*), and incubating 1 hr at room temperature. The biotinylated primer is purified by reversed-phase HPLC on a C18 column using a 30-min linear gradient of 90:10 to 60:40 triethylamine acetate/acetonitrile and a flow rate of 1 ml/min.

Phosphorylated and Digoxigenin-Labeled Primers

Addition of digoxigenin after synthesis. The oligonucleotide primer is synthesized on a 3′-amine-on column, and a 5′-phosphate-on phosphoramidite is added during the completion of the synthesis. Once deprotected, this primer contains a 3′ amine group which is reacted with 40 mM digoxigenin-3-*O*-methylcarbonyl-ε-aminocaproic acid *N*-hydroxysuccinimide ester as described above for biotin, except that the reaction is incubated overnight. The modified primer is then purified by reversed-phase HPLC.

3′-Tailing of primers with digoxigenin-11-UTP. The primer is assembled with a 5′-phosphate-on phosphoramidite added during the synthesis. Following deprotection, the primer is evaporated to dryness and resuspended in water. An aliquot is labeled at the 3′ end by mixing 500 pmol phosphorylated primer, 4 μl of 5× terminal transferase stock buffer, 2 μl digoxigenin-11-UTP, 1 μl of 40 μM dATP, and water to 19 μl. One microliter (35 to 70 U) terminal transferase is added and the reaction is incubated from 1 hr to overnight at 37°C. The terminal transferase must be heat-inactivated by incubating 15 min at 65°C. Because multiple digoxigenin molecules are added to the primer, this method may yield a stronger signal than that obtained by direct chemical modification (above). Digoxigenin labeling reagents and oligo labeling kits can be purchased from Boehringer Mannheim.

References: Nickerson et al., 1990; Wiedmann et al., 1994

Contributors: Deborah A. Nickerson, Wendy Ankener, Claire Delahunty, and Pui-Yan Kwok

UNIT 2.3

Automated Fluorescent Genotyping

Genotyping on the scale described in this unit is required when processing a large number of samples that need to be typed with multiple microsatellite markers. For example, a genome-wide search in a disease linkage study may involve testing 300 markers on a population of 500 individuals, which would require the generation of 150,000 genotypes. The technology described in this unit uses the Perkin-Elmer automated DNA sequencing systems, but can be modified to other automated fluorescent sequencers.

BASIC PROTOCOL

PCR AMPLIFICATION OF SSLPs FOR AUTOMATED FLUORESCENT GENOTYPING

There are two primary concerns in choosing primers. First is the ability to combine primers labeled with the same fluorochrome such that the products that each locus generates will not overlap in the acrylamide gel analysis. In each multiplexing scheme the allele size ranges for each locus of a single color must not overlap. The authors allow a 20-bp gap between marker systems using dinucleotide repeats and as much as a 40-bp gap between marker systems using tetranucleotide repeats. Second, it is necessary to determine that the primer's amplification product will be of sufficient quality to allow for rapid and unambiguous peak calling and genotyping. Several companies—e.g., Perkin-Elmer and Research Genetics—offer human genomic mapping panels that are optimized for marker density and performance.

NOTE: Experiments involving PCR require extremely careful technique to prevent contamination.

Materials (see APPENDIX 1 *for items with* ✓)

✓ 4 ng/µl patient DNA samples in TE buffer (see recipe for buffer)
  10× PCR buffer (Perkin-Elmer)
✓ 10 mM 4dNTP mix (see recipe or purchase from Perkin-Elmer)
  25 mM MgCl$_2$
  20 µM fluorochrome-labeled forward primer for each microsatellite marker
  20 µM reverse primer for each microsatellite marker
  5 U/µl AmpliTaq Gold DNA polymerase (Perkin-Elmer)

  96-well plates (Robbins)
  Drying oven (optional)
  Multichannel pipettor and sterile reagent trays
  96-well strip caps (Robbins) or 96-well plastic plate-sealing film (Costar)
  5-ml and 50-ml tubes
  Thermal cycler(s) equipped to handle 96-well plates

1. Arrange 4 ng/µl patient DNA samples on a master 96-well plate in the desired order (e.g., group family members together so that all will appear on the same gel). Arrange samples down the column of wells from A1 to H1 and from row 1 to row 12 to facilitate gel loading.

2. Pipet 5 µl of each patient DNA (20 ng) into the appropriate well of as many replica plates as are required for the number of markers to be amplified.

   For short-term storage, cap the plates or use plate-sealing film and refrigerate for up to a few days. For long-term storage, dry the plates in a 70°C oven for 15 to 30 min and store at room temperature. Rehydrate by adding a 10-min incubation at 95°C before PCR cycling.

3. Label a series of 5-ml tubes for each primer-specific reaction mix and one 50-ml tube for the master reaction mix minus primers. Include MgCl$_2$ concentration and primer name on all tubes.

4. In the 50-ml tube, prepare the master reaction mix minus primers for all of the plates. Keep the tube and all ingredients on ice. For each plate add (based on 110 reactions/plate to allow for losses):

   220 µl 10× PCR buffer
   220 µl 10 mM 4dNTP mix
   220 µl 25 mM MgCl$_2$
   220 µl 5 U/µl *Taq* DNA polymerase
   715 µl H$_2$O (or 1265 µl if plates were dried down)
   Total volume = 1595 µl/plate (or 14.5 µl/reaction).

   This reaction mix assumes that all primers need 2.5 mM MgCl$_2$, but MgCl$_2$ may need to be optimized. If the MgCl$_2$ requirements for the primers differ, individual master reaction mixes must be made. Any change in MgCl$_2$ volume must be compensated for by adjusting the amount of water.

5. Dispense 1595 µl master reaction mix into each of the 5-ml tubes. To each tube, add 27.5 µl of the appropriate fluorochrome-labeled forward primer and 27.5 µl of the appropriate reverse primer to make the primer-specific PCR mix for each plate.

   Keep primers out of the freezer only as long as they are needed. Keep in an ice bucket with a lid over them to protect the fluorescent label from fluorescent overhead lights.

6. Pour primer-specific PCR mix into a multipipettor reagent tray. Label a 96-well plate with the marker locus name of the primer pair, e.g., D2S126. Aliquot 15 µl (20 µl if plates were dried down) of each primer-specific mix into the appropriately labeled 96-well plate containing DNAs for the population to be genotyped.

7. After all plates have been filled, cap the wells securely using strip caps or plate-sealing film and put each plate into a thermal cycler. Enter the correct program for each primer pair into each machine and begin the program.

Activation step:	12 min	95°C
16 cycles:	30 sec	94°C
	30 sec	66°C, decreasing by 1°C/cycle
	30 sec	72°C
22 cycles:	30 sec	94°C
	30 sec	50°C
	30 sec	72°C
Final step	indefinitely	4°C.

   The annealing temperature decreases from 66° to 50°C over the first 16 cycles to provide a selective amplification of the most stringent products, and is then held constant to allow amplification to a level that is detectable on a gel. PCR conditions, particularly amplification temperature, may need to be optimized for each primer pair, DNA template, and thermal cycler.

8. Keep the PCR product plates on ice and prepare PCR product pools (see Support Protocol 1), or place them in the appropriate refrigerator (used only to store PCR products) until needed. Run pooled products on polyacrylamide gels (see Support Protocol 2).

SUPPORT PROTOCOL 1

POOLING FLUORESCENTLY LABELED PCR PRODUCTS FOR GENOTYPE ANALYSIS

The amount of each marker to be included in a multiplexed set is determined by spot checking the PCR products for each locus on an agarose gel. A minimum of two samples are taken from each plate for analysis.

Materials

Fluorochrome-labeled PCR products (see Basic Protocol)
4× BB loading dye: 0.167% (w/v) bromphenol blue/30% (v/v) glycerol
50 µg/ml DNA size standard (e.g., DNA Mass Ladder, Life Technologies)
Genescan 500 Tamra size standard (Perkin-Elmer)

Beckman CS-6 centrifuge and PTS-2000 rotor (or equivalent) with 96-well plate adaptors
96-well plates
Multichannel pipettor
96-well plastic plate-sealing film (Costar)
90°C oven

1. If the fluorochrome-labeled PCR product plates have been refrigerated and have condensation on the caps or sealing film, centrifuge in a Beckman CS-6 centrifuge with

96-well plate adapters for a few seconds at $800 \times g$, at room temperature, to bring down condensate.

2. Cut a piece of Parafilm $\sim 4 \times 20$–cm and dot 2 μl of 4× BB loading dye on the Parafilm for each sample to be loaded. Add 8 μl of PCR product from the negative and positive control wells to the corresponding 2-μl 4× BB dots. Mix each sample prior to loading on agarose. For each row of agarose gel used, load one lane with 5 μl of 50 μg/ml DNA size standard.

3. Load the 10 μl of each mix onto a 2% agarose gel containing 0.2 mg/ml ethidium bromide (*APPENDIX 3G*). After loading all samples, run the gel at 90 mV for ~ 20 min or until the bands migrate about halfway through the gel.

4. Photograph the gel to determine the band intensity (the brighter the band, the stronger the PCR reaction and the more product contained in the corresponding well). Label the photograph with the primer name corresponding to the plate sampled.

5. Look up the fluorescent dye label for each marker in the set. Determine pooling amounts using the starting amounts shown in Table 2.3.1 for each dye, adjusting the product amount to best reflect the agarose gel results. Arrange PCR product plates into groups based on the pooling volume from lowest to highest. If PCR product plates were stored in a refrigerator between loading and pooling, centrifuge briefly at $800 \times g$ to collect any condensate.

The dye labels are specific to each primer of a microsatellite marker. The primer will be labeled with one of three dyes: Fam (6-carboxyfluorescein), Hex (4,7,2',4',5',7'-hexachloro-6-carboxyfluorescein), or Tet (4,7,2',7'-tetrachloro-6-carboxyfluorescein). The dye labels vary in their intensity, with Hex approximately half as bright as either Fam or Tet. Determining pooling amounts will require some experience with the process, and may vary depending on the instrument used. Pooling too much product will result in overloaded gels, and pooling too little will result in low signal. The fluorescent range of the PCR products should be between 200 and 2000 fluorescent units on the ABI instrument's scale.

6. Label pooling plate and final gel-loading plate. When all the PCR product plates are laid out on the work area, carefully remove their caps or sealing films so that no droplets come out of the individual wells. Before proceeding, check that all plates are properly oriented

Table 2.3.1 Suggested Pooling Volumes

Dye[a]	Band intensity	Suggested pooling amount (μl)
Fam	No band	0.1
	Light	0.04
	Good	0.02
	Bright	0.01
Tet	No band	0.1
	Light	0.04
	Good	0.02
	Bright	0.01
Hex	No band	0.2
	Light	0.08
	Good	0.04
	Bright	0.02

[a]Abbreviations: Fam, 6-carboxyfluorescein; Hex, 4, 7, 2', 4', 5', 7'-hexachloro-6-carboxyfluorescein; Tet, 4, 7, 2', 7'-tetrachloro-6-carboxyfluorescein.

Table 2.3.2 Example Illustrating Preparation of Pooling Set

Marker	Final amount needed for acrylamide gel (μl)	Amount taken from PCR product plate for the pooling plate (μl)
1	0.01	1
2	0.02	2
3	0.08	8
4	0.1	10
5	0.02	2
6	0.08	8
7	0.1	10
8	0.04	4
9	0.02	2
Total	0.47	47 (+ 153 μl H_2O)

and that the pooling and gel-loading plates have the same positioning as all of the PCR product plates that are going to be pooled.

7. To the pooling plate, add the amount of water that will be required to give a final volume of 200 μl per well. Using a multichannel pipettor, transfer the appropriate amounts from each of the PCR product plates (each marker) into the pooling plate. When finished transferring all of the markers, mix the pooling plate contents well.

> *See Table 2.3.2 for an example for pooling a hypothetical nine-marker set. In this example, the total PCR product for all markers loaded on the gel is 0.47 μl. This value must not exceed 2 μl, as excess PCR salts will interfere with electrophoresis.*

8. Transfer 2 μl from the pooling plate into the gel-loading plate. If the gel is to be loaded at a later time, seal the gel-loading plate with plate-sealing film and refrigerate it. If the gel is to be loaded on the same day, set up the gels (see Support Protocol 2) before proceeding to step 9. Store the individual marker PCR product plates and the pooling plates covered with caps or plate-sealing film in a PCR product refrigerator until the genotype data has been successfully collected from the gel, in case a rerun is required.

9. Add 1 μl Genescan 500 Tamra size standard into each well of the gel-loading plate. Seal the plate with sealing film. Centrifuge the plate for a few seconds at 800 × *g*, room temperature, to remove air bubbles from the wells.

10. Heat plate 5 min in a 90°C oven (longer heating may cause plate to melt). After heating the plate, *immediately* put it on ice and pack the ice down to assure even cooling of the samples. Proceed with gel electrophoresis (see Support Protocol 2).

SUPPORT PROTOCOL 2

GEL ELECTROPHORESIS OF POOLED PCR PRODUCTS

Materials *(see APPENDIX 1 for items with ✓)*

Gel-loading plate (see Support Protocol 1)
✓ 6% (w/v), 24-cm well-to-read acrylamide gel
✓ 1× TBE electrophoresis buffer

Automated fluorescent sequencer (e.g., ABI 373 or 377 for four-color detection, Perkin-Elmer; or Pharmacia, LICOR for single-color detection) with computer and mouse

Genescan 672 Collection, Analysis, and Genotyper software (Perkin-Elmer)

60-ml syringe with needles

Lane-indicator strip, plastic (Perkin-Elmer; optional)

Micropipettor

Gel-loading micropipettor tips

1. Turn on the sequencer and restart the computer.

2. Remove the clips from a 6%, 24-cm polymerized well-to-read acrylamide gel and wash with water to clean excess acrylamide and dust from the glass plates. Slowly remove the comb and rinse away loose acrylamide with water. Prop plates up to dry or dry with a clean Kimwipe.

   For one 96-well gel-loading plate, the procedure is described below for three 32-well gels. Two 48-well gels can also be used.

3. Open the chamber door on the front of the automated fluorescent sequencer. Place the lower buffer reservoir on the bottom of the chamber. Set a clean gel in the chamber so it sits flat against the laser housing with the bottom of the plates held in position by the lower reservoir. Make sure that the gel is in the correct position with the eared plate on the inside.

4. Look at the scanning region of the gel for any debris that may be clinging to the plates and wipe clean with a damp Kimwipe, if necessary. Lock the gel in place with the black brace and shut the chamber door.

5. Adjust the photomultiplier tube (PMT) voltage. Using the 373A keypad, select the Main Menu option, choose Calibration, and then choose Configure to verify the run parameters. Check the default parameters and change, if necessary, to read:

   > Run time: 5 hr 30 min at 1070 V
   > Will not stop at datacom errors
   > Filter wheel B
   > Laser power at 40 mA/30 mW.

6. Using the mouse on the keyboard drawer, drag down the Windows menu in the Genescan 672 Collection software and select Scan and Map. From the 373A Main Menu, use the keypad to select Start Pre-run then Plate Check. Begin the plate check scan by selecting Full Scan.

   If the read region of the gel is clean, the Scan will appear as flat lines and the map will be entirely gray across the width of the gel. If there are sharp peaks seen on the Scan and colored peaks on the Map, there is debris or dust on the glass plates in the laser's read region. The electrophoresis chamber must be opened and both sides of the glass plates wiped clean. This is most effectively done with a single horizontal swipe across each side with a damp Kimwipe. Recheck the gel for a clean scan as described above and repeat cleaning if necessary until the read region is clean.

   A scan that shows a peak containing all four colors is caused by an irregularity in the gel, so it cannot be wiped away. However, the problem will usually correct itself as the buffer runs through the gel. In the case of these types of spurious peaks, note the x-axis reading of the peak in the scan window. This value is equivalent to the channel number and can be used to determine what lane the peak artifact will be seen in. Determine the lane number for a 36-well comb corresponding to the channel value where the gel irregularity occurs and skip that lane when loading.

7. Using the mouse, bring the tip of the arrow on the monitor to the lowest colored line seen in the scan. In the lower left corner of the scan are the x and y coordinates of the arrow's position. Check the PMT value as the y coordinate of the lowest scan line.

   The value of y should be within 20 units of 800. The y value is adjusted through the PMT value in the Configure menu on the 373A keypad. The Configure menu is accessed through Calibration in the Main Menu. Raising the PMT raises the y value.

8. Repeat steps 2 to 7 until each gel is set up in electrophoresis chambers with clean scan regions and correct PMT values.

9. Attach the upper buffer chamber to the secured glass plates by turning the wing clamps until the gasket flattens and the clamps are hand-tight. Add $1 \times$ TBE electrophoresis buffer to the upper buffer chamber, being careful not to drip any buffer down into the read region, or that will need to be cleaned and rescanned. Fill the upper buffer chamber to ~ 0.5 cm above the top of the well line and ~ 1 cm from the top of the chamber. Fill the lower buffer reservoir to ~ 1 cm from the top with $1 \times$ TBE buffer. Make sure that the wire in the lower chamber is covered. Check that the gasket in the upper buffer chamber is creating a good seal and is not leaking.

   About 1.5 liters total $1 \times$ TBE buffer are needed per machine.

10. Using a 60-ml syringe with a needle, flush the wells with $1 \times$ TBE. Position the needle so that the tip is just above the point where the two glass plates come together in the upper buffer chamber.

11. Place a plastic lane indicator strip, if available, on the outside glass plate so that the numbers correspond to the wells in the gel. Inspect the gel for broken well arms, bubbles, or misshapen wells. Note any wells (lanes) that cannot be loaded due to these problems.

12. Put the gel-loading plate in an ice bucket and pull the plastic seal back from the first four columns of the gel loading plate (A1 to H4). Set the micropipettor at 3 μl. Attach a gel-loading micropipet tip to the micropipettor and take up the sample in well A1. Fit the flat surface of the tip between the glass plates and about halfway inside the first well. Slowly release the sample into the well, pulling the tip out as the well fills. Be careful not to overfill the well, allowing sample to spill over into the next well. If a neighboring well becomes contaminated, skip it and make a note of the lane number.

13. Continue loading samples from the gel-loading plate, moving down the column to well H1, and then moving to the next column beginning at well A2 and ending at well H2. Load the first four columns of the plate onto the first of the three gels. If it appears that a gel is being misloaded, take exact notes on which wells are being loaded in which lanes.

14. When the gel has been loaded, take the plastic lane indicator strip off the glass plate and plug in the electrical fittings for the upper and lower buffer chambers. Close the door to the electrophoresis chamber. On the 373A Sequencer keypad, press the Choose Run button. Start electrophoresis by selecting Genescan Run and immediately begin the collection software. Using the mouse, place the arrow on the green Collect button shown on the monitor and click the mouse button to begin the Genescan Collection software.

   The Collect button on the monitor will begin flashing to verify that data collection has started. If the green button does not start flashing, press the Stop Run button on the 373A keypad and get help from someone experienced in using the equipment.

15. Load the remaining two gels by repeating steps 9 through 14 for wells A5 through H8 and A9 through H12. When all three machines have been loaded and the Genescan collection started, take a final look to see that the collection buttons are flashing. When the run is complete, check for run errors.

The Genescan run should not have any errors over the collection period, and the 373 keypad display should show an elapsed run time of 5.5 hr.

16. Shut off the 373A sequencer and open the door to the electrophoresis chamber. Hold the gel, with the upper buffer chamber attached, against the laser housing. Release the black brace by lifting the latches on both sides and pull it away from the gel. Lift the gel and upper buffer chamber out of the electrophoresis chamber. Carry them to the sink and pour out the buffer. Separate the upper buffer chamber from the gel.

17. Gently lift the lower buffer reservoir out of the electrophoresis chamber and pour out the used buffer. Rinse both chambers and place under the sequencer. Wipe away any buffer that may have spilled in the electrophoresis chamber using a damp paper towel. Close the door to the electrophoresis chamber and repeat for the remaining gels.

18. Analyze the gels and assign alleles using the Genescan Analysis and Genotyper software as detailed in the manufacturer's manuals.

References: Litt and Luty, 1989; Weber and May, 1989

Contributors: Jeff Hall and Elizabeth Nanthakumar

UNIT 2.4

Single-Nucleotide Polymorphism Genotyping Using Microarrays

NOTE: Use only molecular biology grade water (i.e., nuclease-free; BioWhittaker) for these applications, including as a component of any solution.

STRATEGIC PLANNING

In order to avoid contamination, great pains must be taken to prevent items from migrating from one area to another. The following is a list of equipment and reagents that should be assembled in designated areas before beginning the experiment.

Pre-PCR Clean Room

Microcentrifuge
1.5-ml microcentrifuge tubes
Micropipettors and appropriate barrier tips: 0.2 to 2, 1 to 10, 2 to 20, 50 to 200, and 100 to 1000 μl
Microseal "A" film (MJ Research)
1 to 10-μl multichannel pipettors and appropriate barrier tips
96-well polypropylene V-bottom microplates (MJ Research)
Seal and Sample aluminum foil lids (Beckman)
1-ml deep-well titer plates (Beckman)
0.2-ml thin-wall tubes and caps, strip of 8 (MJ Research)
Tube rack
Vortex mixer
All PCR reagents and components except the target DNA

PCR Staging Room

Micropipettors and appropriate barrier tips: 0.2 to 2, 1 to 10, 2 to 20, 20 to 200, and 100 to 1000 µl

5 to 50-µl multichannel pipettor and appropriate barrier tips

Swinging-bucket centrifuge (Sorvall)

Tube rack

Vortex mixer

Target DNA, AmpliTaq Gold, and T3 and T7 primers

Main Laboratory

Low-copy area

1 to 10-µl micro- and multichannel pipettor and appropriate barrier tips

Seal and Sample aluminum foil lids (Beckman)

Vortex mixer

Medium-copy area

Micropipettor and appropriate barrier tips: 0.2 to 2 µl

1 to 10-µl multichannel pipettor and appropriate barrier tips

Vortex mixer

High-copy area (near cycler)

100-bp DNA ladder (Life Technologies)

Gel apparatus and 4% NuSieve 3:1 Plus agarose gel, 24-well format (FMC Bioproducts)

Loading dye

1.5-ml microcentrifuge tubes (USA Scientific)

Microcon-10 microconcentrators (Millipore)

Micropipettors and appropriate barrier tips: 0.2 to 2, 1 to 10, 2 to 20, 50 to 200, and 100 to 1000 µl

Microcentrifuge

$20\times$ TBE stock solution (BioWhittaker)

Thermal cycler (e.g., MJ Research DNA Engine or Tetrad; Perkin Elmer 2400, 9600, or 9700)

Vortex mixer

All PCR products and hybridization, washing, and staining reagents

BASIC PROTOCOL 1

SNP GENOTYPING USING ALLELE-SPECIFIC MICROARRAYS

The GeneChip HuSNP probe array contains oligonucleotide probes representing the two alleles for each of 1494 SNP markers. A schematic of the assay is shown in Figure 2.4.1. Consult the HuSNP Mapping Assay User Manual, available with the HuSNP array and reagent kit (Affymetrix), for additional information.

Materials

GeneChip HuSNP reagent kit (Affymetrix):

Multiplex Primer Pools 1 to 24

Labeling Primer, Biotin-T7

Labeling Primer, Biotin-T3

Control oligonucleotide B1

HuSNP Reference DNA

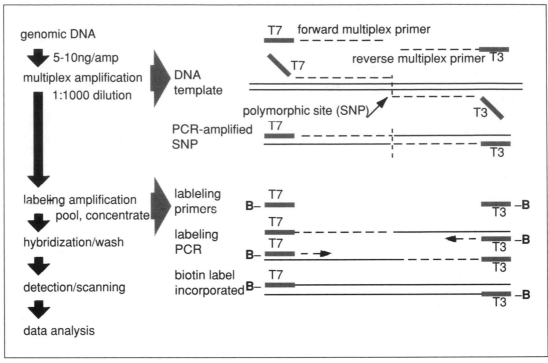

Figure 2.4.1 The HuSNP Assay. The multiplex amplification assay contains marker-specific primers designed for amplification of each SNP from genomic DNA; SNP-specific primers also contain T7 and T3 sequences at their ends; consequently, all of the products from the multiplex PCR have the same T7 and T3 sequences in common. The labeling amplification assay is performed with biotinylated T7 and T3 primers which target the common T7 and T3 sequences at the ends of the multiplex PCR products; consequently, all of the products from the labeling PCR have a biotin molecule on their 5′ ends.

PCR master mix I (Table 2.4.1):

> 10× Buffer II, 25 mM MgCl$_2$, and AmpliTaq Gold DNA polymerase (Perkin Elmer; supplied together)
> 2.5 mM dNTPs: 25 µl each 100 mM dNTP stock solution (Pharmacia Biotech)/900 µl H$_2$O; store at −20°C

4 ng/µl genomic DNA
PCR master mix II (Table 2.4.2)
4% NuSieve agarose gel (optional; FMC Bioproducts)

Table 2.4.1 PCR Master Mix I[a]

Stock solution	For 1 reaction (µl)	For 28 reactions (µl)	Final concentration[b]
10× buffer II	1.25	35	1×
25 mM MgCl$_2$	2.5	70	5 mM
2.5 mM dNTPs	2.5	70	0.5 mM
5 U/µl AmpliTaq Gold	0.25	7	0.1 U/µl
H$_2$O	1.75	49	
Final volume	8.25	231	

[a]This recipe is for a single DNA sample.

[b]The final concentration is that of the multiplex-PCR amplification reaction.

Table 2.4.2 Labeling PCR Master Mix II[a]

Stock solution	For 1 reaction (μl)	For 28 reactions (μl)	Final concentration[b]
10× buffer II	2.5	70	1×
25 mM MgCl$_2$	4	112	4 mM
2.5 mM dNTPs	4	112	0.4 mM
10 μM biotin-T7 primer	2	56	0.8 μM
10 μM biotin-T3 primer	2	56	0.8 μM
5 U/μl AmpliTaq Gold	0.5	14	0.1 U/μl
H$_2$O	7.5	210	
Final volume	22.5	630	

[a]This recipe is for a single DNA sample.

[b]The final concentration is that of the labeling PCR amplification reaction.

Hybridization mix (Table 2.4.3):
  5 M tetramethylammonium chloride (TMAC; Sigma)
  1 M Tris·Cl, pH 7.8 (*APPENDIX 1*)
  1% Tween 20 (Pierce): mix thoroughly with H$_2$O and filter through a 0.2-μm filter; store at room temperature
  0.5 M EDTA, pH 8.0 (*APPENDIX 1*)
  10 mg/ml herring sperm DNA (*APPENDIX 1*)
  50× Denhardt's Solution (*APPENDIX 1*)

Amplification reaction tubes
Low-speed swinging-bucket centrifuge with titer plate carrier
Thermal cycler (e.g., MJ Research DNA Engine or Tetrad; Perkin Elmer 2400, 9600, or 9700)
96 × 1–ml deep-well titer plate and aluminum foil lid (Beckman)
Microcon-10 microconcentrators (Millipore)
10-μl non-barrier pipet tip (i.e., yellow tip)

Table 2.4.3 Hybridization Mix

Stock solution	Volume[a] (ml)	Final concentration
5 M TMAC	81.00	3 M
Control oligonucleotide B1[b]	1.35	2 nM
1 M TrisCl, pH 7.8	1.35	10 mM
1% Tween 20	1.35	0.01%
0.5 M EDTA, pH 8.0	1.35	5 mM
10 mg/ml herring sperm DNA	1.35	100 μg/ml
50× Denhardt solution	13.5	5×
H$_2$O	3.75	
Final volume	105.00	

[a]The volumes specified are sufficient for one sample.

[b]For information on this hybridization control, refer to the Affymetrix HuSNP Mapping Assay User Manual.

GeneChip HuSNP probe arrays (Affymetrix)
GeneChip Hybridization Oven 320 or 640, 44°C (Affymetrix)

NOTE: Absolutely no template DNA or PCR product should be allowed into the designated pre-PCR clean room; gloves and gowns are recommended to minimize the risk of PCR carryover.

1. For each genomic DNA sample, label 24 amplification reaction tubes, one for each of the 24 multiplex primer pools, using either a 96-well plate or an 8-tube strip format. Aliquot 3 µl of each Multiplex Primer Pool into the appropriately labeled amplification reaction tube.

2. In the pre-PCR clean room, prepare enough PCR master mix I for 28 reactions (Table 2.4.1).

3. In the PCR staging room, for each genomic DNA sample, add 33.75 µl of 4 ng/µl genomic DNA (enough for 27 reactions) to a 222.75-µl aliquot of PCR master mix I. Vortex thoroughly to mix. Add 9.5 µl to each of the 24 wells or tubes containing 3 µl Multiplex Primer Pools (step 1). Seal tubes or plate, vortex gently, and spin down (<2000 rpm) in a swinging-bucket centrifuge.

   Each 12.5-µl reaction contains 5 ng genomic DNA and 50 nM SNP specific primers.

4. In the high-copy area, start the multiplex PCR amplification reaction using the following conditions for MJ thermal cyclers (for Perkin-Elmer instruments, use 50 sec annealing cycles with and without ramp):

Initial step:	5 min	95°C	(denaturation)
30 cycles:	30 sec	95°C	(extension)
	55 sec	52°C + 0.2°C/cycle	(annealing with ramp)
	30 sec	72°C	(extension)
5 cycles:	30 sec	95°C	(extension)
	55 sec	58°C	(annealing)
	30 sec	72°C	(extension)
1 cycle:	7 min	72°C	(final extension)
Final step:	indefinitely	4°C	(hold).

5. In the pre-PCR clean room, for each genomic DNA sample, set up and label a second set of 24 amplification reaction tubes and prepare enough labeling PCR master mix II *without* AmpliTaq Gold for 28 reactions (Table 2.4.2). Vortex thoroughly to mix.

6. Add AmpliTaq Gold polymerase immediately prior to setting up the labeling PCR in the PCR staging room. Dispense 22.5 µl of completed master mix II to the fresh tubes/plates set up for labeling PCR (see step 5) and move the tubes to the low-copy area of the main laboratory (away from the thermal cyclers).

7. Fill 24 wells of a 96 × 1–ml deep-well titer plate with 1.0 ml molecular grade (i.e., nuclease-free) water.

   To minimize the risk of PCR carryover, a stock of deep-well titer plates can be filled with water in the pre-PCR clean room, covered with aluminum foil, and kept out in the main laboratory.

8. Upon completion of the multiplex PCR reaction (step 4), centrifuge the multiplex amplification reactions at low speed (i.e., ~1000 × g) in a swinging-bucket centrifuge with a plate carrier.

9. In the medium-copy area of the main laboratory, carefully remove the caps from the tubes. Using a 1- to 10-µl multichannel pipet fitted with barrier tips, transfer 1.0 µl of all

24 multiplex amplification reactions to the plate containing water. Seal the wells tightly with an aluminum foil lid, vortex gently, and microcentrifuge.

10. In the low-copy area of the main laboratory, use a 1- to 10-μl micropipettor (preferably multichannel) fitted with barrier tips to dispense 2.5 μl of the 1:1000 dilutions (step 9) into the labeling amplification tubes/plates containing PCR master mix II (step 6). Seal tubes with caps and microcentrifuge.

11. Start the labeling PCR amplification reaction in the thermal cycler using the following profile (optimized for MJ and Perkin-Elmer thermal cyclers):

Initial step:	8 min	95°C	(denaturation)
40 cycles:	30 sec	95°C	(extension)
	90 sec	55°C	(annealing)
	30 sec	72°C	(extension)
1 cycle:	7 min	72°C	(final extension)
Final step:	indefinite	4°C	(hold).

12. *Optional:* To confirm the PCR amplifications, in the high-copy area of the main laboratory, analyze 1.5 μl from each reaction on a 4% NuSieve agarose gel (APPENDIX 3G).

   A closely clustered family of bands running at ~100 bp should be visible in all 24 lanes.

13. In the high-copy area of the main laboratory, remove 23 μl of each PCR product and pool together in two Microcon-10 microconcentrators. Microcentrifuge at 13,000 × g for 20 min at room temperature, or until the volume is reduced at least 10-fold (<60 μl total). Recover the sample by reversing the filters and microcentrifuging 3 min at 3,000 × g. Combine the concentrates and adjust the volume by adding water to 60 μl. Aliquot 30 μl for hybridization and store remaining sample at −20°C.

14. Prepare the hybridization mix according to the volumes given in Table 2.4.3. If running multiple samples, prepare enough hybridization mix for one additional sample to compensate for pipetting inaccuracies.

15. In a microcentrifuge tube, combine 30 μl of the concentrated labeled DNA (from step 13) with 105 μl hybridization mix. Denature samples at 95°C for 5 to 10 min. Place tubes immediately in an ice-water bath. Incubate for 2 min on ice. Microcentrifuge briefly.

16. Insert a 10 to 200-μl non-barrier pipet tip into the upper septum on a HuSNP probe array as a vent. Using a 250-μl non-barrier pipet tip, mix the hybridization sample by pipetting up and down a few times in the sample tube (i.e., resuspend any material that dropped out of solution) and inject the sample into the bottom septum of the probe array.

17. Hybridize the probe array in a GeneChip Hybridization Oven 320 or 640 at 44°C overnight (~16 hr), rotating at 40 to 50 rpm on the rotisserie. Wash, stain, and scan the array as described in the HuSNP Mapping Assay User Manual.

18. Begin GeneChip data analysis by assigning an experiment name to a probe array and creating an .exp file. Scan the probe array to create a .dat file or image file.

   From this .dat file, the software automatically generates a .cel file by demarcating individual cells. A probe cell is the area on the surface of the array containing a unique oligonucleotide sequence. The software averages the pixel intensities within each probe cell, producing a .cel file.

19. Produce the data output file, the .chp file, by running an analysis on a .cel file, using the appropriate GeneChip algorithm for that probe array.

The automated analysis provides genotype calls (AA, AB, or BB) for those markers that pass the criteria. The final results of the analysis are listed in the .chp *file and can be saved as a* .txt *file to import into project databases for further analysis.*

BASIC PROTOCOL 2

SNP GENOTYPING USING GENERIC MICROARRAYS AND SINGLE-BASE EXTENSION

A schematic of the assay described here is shown in Figure 2.4.2. Further information regarding the design of the Affymetrix GenFlex Tag array can be obtained from the Affymetrix web site (*http://www.Affymetrix.com*).

Materials *(see* APPENDIX 1 *for items with* ✓ *)*

    20 ng/µl genomic DNA
    PCR mix (Table 2.4.4):

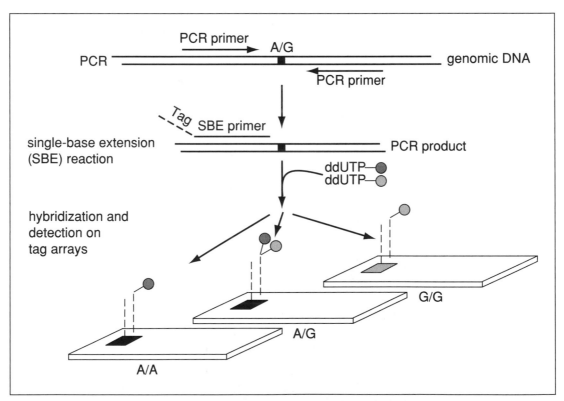

Figure 2.4.2 TAG-SBE genotyping assay. Marker-specific primers are designed for amplification of each SNP from genomic DNA; all SNPs with the same pair of variant bases (e.g., A/G SNPs) are pooled. The double-stranded PCR products serve as templates for the SBE reaction. Each SBE primer is chimeric with a 5′-end complementary to a unique tag synthesized on the array and a 3′-end complementary to the genomic sequence and terminating one base before a polymorphic SNP site. Thus, each SBE primer is uniquely associated with a specific tag (location) on the array. SBE primers corresponding to multiple markers are added to a single reaction tube and extended in the presence of pairs of ddNTPs labeled with different fluorophores; for example, an A/G biallelic marker is extended in the presence of biotin-labeled ddUTP and fluorescein-labeled ddCTP. The labeled multiplex SBE reaction products are pooled and hybridized to the tag array. Three hybridization patterns are shown, corresponding to three genotypes AA, AG, and GG.

Table 2.4.4 PCR Mix for Multiplex Locus-Specific Amplification

Stock solution	Volume (μl)	Final concentration
10× buffer II	2.5	1×
25 mM MgCl$_2$	5	5 mM
25 mM dNTPs	1	1 mM
5 μM (each) primer mix	2.5	0.5 μM each
20 ng/μl genomic DNA	2.5 (total)	2 ng/μl[a]
5 U/μl AmpliTaq Gold	0.4	0.08 U/μl[b]
H$_2$O	11.1	
Final volume	25	

[a] 50 ng genomic DNA total.

[b] 2 U AmpliTaq Gold total.

10× Buffer II, 25 mM MgCl$_2$, and 5 U/μl AmpliTaq Gold (Perkin Elmer; supplied together)

25 mM dNTPs: 25 μl each 100 mM dNTP stock solution (Pharmacia Biotech)/900 μl H$_2$O; store at −20°C

5 μM each locus-specific primer for each SNP marker

10 U/μl exonuclease I (Amersham Life Sciences)

1 U/μl shrimp alkaline phosphatase (Amersham Life Sciences)

Extension mix (Table 2.4.5):

SBE primer mix: 20 nM hybrid SBE primers with TAG/SNP for each marker

5× Thermo Sequenase buffer (Amersham Biosciences): 260 mM Tris·Cl/65 mM MgCl$_2$, pH 9.5

1 nmol/μl fluorescein-N6-ddNTPs (New England Nuclear)

Biotin-N6-ddUTP, biotin-11-acyclo-ddCTP, or biotin-N6-ddATP (NEN)

1 nmol/μl remaining two ddNTPs

6.4 U/μl Thermo Sequenase Gold (Amersham Biosciences)

Table 2.4.5 Extension Mix

Stock solution	Volume (μl)	Final concentration
Template	6	
SBE primer mix (20 nM each primer)	2.5	1.5 nM each
5× Thermo Sequenase buffer[a]	6.6	1×
1 nmol/μl fluorescein-ddNTP	0.8	25 pmol/μl
x nmol/μl biotin-ddNTP[b]	0.5	7.6 or 3.8 pmol/μl[b]
1 nmol/μl other two cold-ddNTPs	0.3 each	10 pmol/μl each
6.4 U/μl Thermo Sequenase Gold[c]	0.4	0.08 U/μl
H$_2$O	15.6	
Final volume	33	

[a] 260 mM Tris-Cl, pH 9.5/65 mM MgCl$_2$.

[b] $x = 0.5$ nmol/μl for biotin-ddUTP or biotin-11-acyclo-dCTP (7.6 pmol/μl final); $x = 0.25$ nmol/μl for biotin-ddATP (3.8 pmol/μl final).

[c] 2.56 U Thermo Sequenase total.

Table 2.4.6 Hybridization Mix

Stock solution	Volume (μl)	Final concentration
5 M TMAC	72.0	3 M
12× MES buffer, pH 6.7	5.0	50 mM
1% Triton X-100	1.2	0.01%
10 mg/ml herring sperm DNA	1.2	100 μg/ml
5 nM fluorescein-c213 control oligonucleotide	1.2	50 pM
20 mg/ml BSA	3.0	500 μg/ml
Prepared sample	29.4	
Final volume	120.0	

100 μg/ml glycogen (Boehringer Mannheim)
8 M LiCl
Absolute ethanol, −20°C
✓ 6× SSPE-T (see recipe) containing 0.5 mg/ml acetylated BSA
Hybridization mix (Table 2.4.6):
      5 M tetramethylammonium chloride (TMAC; Sigma)
      12× MES buffer, pH 6.7 (*APPENDIX 1*)
      1% Triton X-100
      10 mg/ml herring sperm DNA (*APPENDIX 1*)
      5 nM fluorescein-c213 control oligonucleotide
      20 mg/ml acetylated BSA
✓ 1× SSPE-T
✓ Staining solution (fresh)

S-300 column (Pharmacia Biotech)
GenFlex Tag arrays (Affymetrix)
FS400 fluidic station (Affymetrix)
Agilent scanner (Affymetrix)
GeneChip software (Affymetrix)

1. Plan pools of SNPs with same base composition at the polymorphic site (e.g., A/G, T/C). Prepare a primer mix containing 1 μM of each forward (F) and reverse (R) locus-specific primer for each SNP marker to be amplified in the pool (up to 30 markers).

2. Prepare the locus-specific multiplex PCR mix to a final volume of 25 μl according to Table 2.4.4. Perform PCR using the following cycling conditions:

Initial step:	10 min	96°C	(denaturing)
40 cycles:	30 sec	94°C	(extension)
	40 sec	57°C	(annealing)
	90 sec	72°C	(extension)
1 cycle:	10 min	72°C	(final extension)
Final step:	indefinitely	4°C	(hold).

3. To degrade and dephosphorylate unused primers and dNTPs, add 1 μl of 10 U/μl exonuclease I and 1 μl of 1 U/μl shrimp alkaline phosphatase to 25 μl PCR products and incubate mixture at 37°C for 1 hour. Inactivate the enzymes by incubating at 100°C for 15 min. Apply the sample to an S-300 column according to manufacturer's instructions to replace buffer with water.

4. Prepare the single-base extension (SBE) reaction as outlined in Table 2.4.5. Run the SBE reaction as follows:

Initial step:	3 min	96°C	(initial denaturation)
45 cycles:	20 sec	94°C	(denaturation)
	11 sec	58°C	(annealing)
Final step:	indefinite	4°C	(hold).

5. Combine all extension reactions (25 μl each) and mix with 30 μl of 100 μg/ml glycogen. Precipitate by adding 18.75 μl of 8 M LiCl and 1125 μl of absolute ethanol, prechilled to −20°C; mix well. Microcentrifuge at maximum speed for 15 min at room temperature. Decant the supernatant and dry the samples at 40°C for 40 min. Resuspend the samples in 33 μl water.

6. Denature the prepared sample at 100°C for 10 min. Snap cool on ice for 2 to 5 min.

7. Prehybridize the Tag array with 6× SSPE-T containing 0.5 mg/ml of acetylated BSA, at 42°C for 15 min. Hybridize with 120 μl hybridization mix (Table 2.4.6) at 42°C for 2 hr on a rotisserie at ~40 rpm.

   For each DNA sample, all extension mixes (A/G, A/T, A/C, G/T, G/C, and T/C) can be hybridized to a single array.

8. Rinse the array with 1× SSPE-T twice for 10 sec each, then wash with 1× SSPE-T for 15 to 20 min at 40°C on a rotisserie at ~40 rpm. Wash on a FS400 fluidic station 10 times with 6× SSPE-T at 22°C. Stain the array at room temperature with 120 μl staining solution on a rotisserie for 15 min at ~40 rpm. After staining, wash the probe array 10 times with 6× SSPE-T on the FS400 fluidic station at 22°C.

9. Scan the arrays on an Agilent scanner to capture emissions for wavelengths of 530 and 570 nm. Use GeneChip Software to convert the image files into digitized files. Analyze the image files further using GeneChip Software to give a list of intensity values for every probe on the array.

References: Fan et al., 2000; Lindblad-Toh et al., 2000; Wang et al., 1998

Contributors: Nila Patil, Nassim Nouri, Linda McAllister, Hajime Matsukaki, and Thomas Ryder

UNIT 2.5

High-Throughput Genotyping Using the TaqMan Assay

This unit describes the use of the 5′-nuclease allelic discrimination assay, or TaqMan assay, for genotyping with single nucleotide polymorphisms (SNPs; Fig. 2.5.1).

STRATEGIC PLANNING

Probes and primers are designed using the PrimerExpress software (Applied Biosystems) that is supplied with the ABI PRISM 7700 sequence detection machine, using the following rules:

1. Probes for both alleles have a T_m of 67° to 70°C. The difference in the T_m values of the two probes should be <1°C.

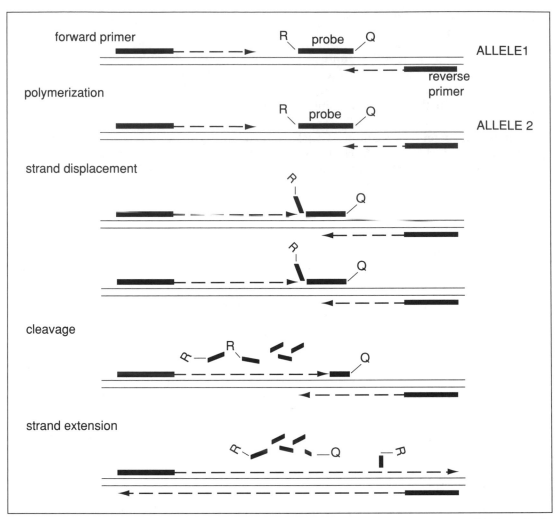

Figure 2.5.1 Schematic of the TaqMan PCR. The case for a heterozygote is shown. Each probe is specific for one allele and is marked by a different fluorescent dye, called the reporter (R), at the 5′ end. When probes are in solution or are annealed at their specific targets, they do not fluoresce because of the presence of a common quencher (Q) at their 3′ ends. However, degradation of the annealed probes by polymerase releases the reporter from the quencher, resulting in a net increase in fluorescence, which is measured at the end of the PCR.

2. Probes are designed to anneal to the same strand, i.e., they should not be complementary to each other.

3. Probes do not have a G at the 5′ end because this nucleotide interferes with fluorescence from the reporter dye that will be attached to this end.

4. Probes are designed using the strand that has fewer Gs than Cs. As far as possible, a run of ≥3 Gs is avoided.

5. The SNP is located approximately in the middle of the probe.

6. Probes can be between 19 and 40 nt in length, although this will be determined by the T_m requirement above. In the author's experience, shorter probes show better discrimination than longer ones. Probes >35 nt do not seem to work well. For probes that need to be long

to meet the T_m requirement, modified probes should be used instead. These have recently become available from Applied Biosystems.

7. Probes are synthesized with FAM or VIC fluorescent reporter dyes at the 5′ end and a TAMRA quencher dye at the 3′ end.

8. Primers should have a T_m values that are ∼7°C lower than those of the corresponding probes. The 3′ end of the primers should be as close to the probe as possible without overlapping. The entire amplicon should be <150 nt in length.

BASIC PROTOCOL

HIGH-THROUGHPUT GENOTYPING USING THE TAQMAN ASSAY

Materials (see APPENDIX 1 for items with ✓)

✓ 15 ng/µl genomic DNA in 10 mM Tris·Cl, pH 7.5 (see recipe)/0.2 mM EDTA (see recipe)
PCR mix (see Table 2.5.1)
96-well optical plate and caps (Applied Biosystems)
Plate centrifuge
ABI PRISM 7700 sequence detection system (Applied Biosystems)
Thermal cycler
Statistical software package (e.g., SPSS; *http://www.spss.com*)

1. In each well of rows B through H of a 96-well optical plate, dispense 2 µl (30 ng) of genomic DNA. Reserve row A for controls—in four wells, pipet 2 µl of buffer alone for "no DNA" controls, and in remaining eight wells, pipet 2 µl of genomic DNA for positive controls of the two types of homozygotes (four wells each). If the allele-calling software of the ABI PRISM 7700 will be used, increase positive controls to eight wells each.

2. Dispense 23 µl of PCR mix per well. Cover the plate using the optical caps. Centrifuge briefly in a plate centrifuge. Obtain a "pre-PCR read" of fluorescence using the ABI PRISM 7700 sequence detection machine. Use the "Plate Read" format, making sure that the Use Spectral Compensation for Endpoint box (under Advanced Options in the Instrument and Diagnostics menu) is checked.

Table 2.5.1 PCR Mix for Use in TaqMan Genotyping

Stock solution	For 1 reaction (µl)	Final concentration
20 µM primers	1.125	900 nM
20 µM probes	0.125	100 nM
Master-mix[a]	12.5	
Water	9.25	
Total volume	23	

[a]Master-mix containing PCR buffer, nucleotides, AmpliTaq Gold DNA polymerase, uracil-*N*-glycosylase, and the reference dye, ROX, can be purchased from Applied Biosystems. If genotypes will be called by pooling all the samples from different plates, a single lot of master-mix should be used because lot variability in the master-mix leads to reproducible differences in the final fluorescence values.

3. Start PCR on a thermal cycler using the following conditions:

1 cycle:	2 min	50°C	
1 cycle:	10 min	95°C	(denaturation)
40 cycles:	15 sec	94°C	(denaturation)
	1 min	62°C	(annealing/extension)
Final step:	indefinite	4°C	(hold).

4. Read fluorescence again to obtain the "post-PCR" read. If using the allele-calling software supplied with the ABI PRISM 7700, then read fluorescence using the "Allelic Discrimination" format, in which genotype calls can be made automatically using their algorithm or by visual inspection of the data.

If large-scale genotyping (e.g., >500 samples) has been performed, assigning genotypes after pooling data from a number of plates appears to be more accurate. In this case, follow the remaining steps of this protocol.

5. Export the raw data (both pre- and post-PCR) from the plate read format into a statistical package. Calculate normalized fluorescence values as: (reporter fluorescence – background)/(reference fluorescence – background); the reporter fluorescence will be in the FAM or VIC channels and the reference fluorescence will be in the ROX channel.

6. For each plate, compare the mean pre-PCR normalized fluorescence for each reporter dye across plates using a nonparametric test (e.g., Kruskal-Wallis test). If the pre-PCR fluorescence for a plate (or a set of plates) is found to be significantly different from the other plates, adjust post-PCR fluorescence values accordingly. For example, if the mean fluorescence for the FAM reporter dye for a particular plate is 1.5-fold greater than that for the other plates, divide the post-PCR FAM fluorescence for each well in this plate by this factor.

7. Use K-means clustering to automatically classify the corrected post-PCR data into four groups—three genotypes and one "no DNA" control. Most statistical packages provide clustering algorithms and these can be directly used.

In general, clusters will be obvious and no further processing is necessary. It is important to examine a scatter-plot of the data to make sure that the genotype assignment makes sense. Outliers, for example, can defeat the clustering algorithm and result in classification that is obviously incorrect. Excluding such outliers generally results in correct classification.

References: Landegren et al., 1998; Ranade et al., 2001

Contributor: Koustubh Ranade

UNIT 2.6

High-Throughput Genotyping with Primer Extension Fluorescent Polarization Detection

This assay is based on a DNA sequencing reaction that determines the nature of the one base immediately 3' to the sequencing primer (called the SNP primer in the protocol). The design of the SNP primer is such that it anneals immediately upstream of the polymorphic site on the DNA being genotyped. When the target DNA is incubated with the appropriate dye-labeled terminators and DNA polymerase, the SNP primer is extended by the complementary base as

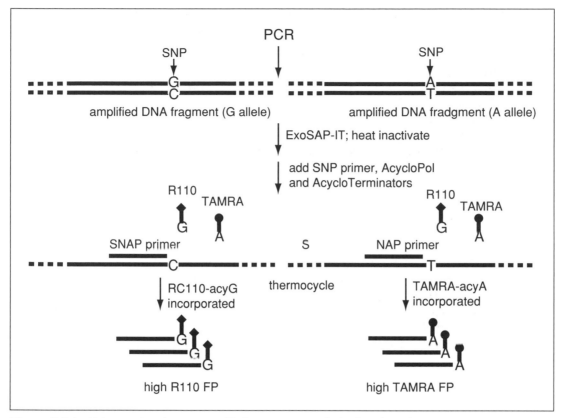

Figure 2.6.1 Primer extension assay with fluorescence polarization detection.

dictated by the allele(s) present at the polymorphic site. By determining which terminator is incorporated, the allele present in the target DNA can be inferred (Fig. 2.6.1).

STRATEGIC PLANNING

Care must be taken to design robust PCR assays and avoid SNP primers that contain repetitive or duplicated sequences that may be present in the PCR products. In order to achieve high-throughput genotyping, PCR primers for all assays should be designed with similar annealing temperatures so that a universal set of reaction conditions can be used for all assays. SNP primers should also be designed with similar (but lower) annealing temperatures. The PCR primers can be selected with the Primer3 program (*http://www.genome.wi.mit.edu/ genome_software/other/primer3.html*) using parameters optimized for robust amplification. All of the SNPs in public databases have been put through this design process, and the PCR and SNP primers for SNPs with viable assay designs are freely available at the Washington University (*http://snp.wustl.edu*) and the Perkin-Elmer Web sites (*http://las.perkinelmer.com/content/ forms/SNPDatabase/welcome.asp*).

BASIC PROTOCOL 1

PRIMER EXTENSION ASSAY WITH SINGLE-PLEX PCR

Although the single-plex approach is more costly than the 4-plex approach (see Basic Protocol 2), it is more flexible and easier to optimize. It is the protocol of choice when one needs to obtain data from a particular SNP on short notice or if an SNP fails in the multiplex approach.

Materials *(see* APPENDIX 1 *for items with* ✓ *)*

   0.8 ng/µl genomic DNA of interest

✓ 10× PCR buffer (may be purchased from Invitrogen): 200 mM Tris·Cl, pH 8.4 (see recipe)/500 mM KCl

✓ 2.5 mM 4dNTP mix

   5 U/µl Platinum *Taq* DNA polymerase (Invitrogen)

   PCR primer mix: 0.2 µM each forward and reverse PCR primers (may be synthesized by Integrated DNA Technologies)

   AcycloPrime FP SNP Detection Kit (Perkin-Elmer) containing:

      Exo-SAP-IT kit

      10× Exo-SAP-IT PCR clean-up reagent

      AcycloPol DNA polymerase

      AcycloTerminator mix: six combinations, each containing two labeled terminators (R110-G/TAMRA-A; R110-G/TAMRA-C; R110-G/TAMRA-T; R110-C/TAMRA-A; R110-C/TAMRA-T; R110-A/TAMRA-T) plus the other two unlabeled terminators

      10× reaction buffer

   1 µM SNP primer: 20- to 25-mer complementary to either the sense or antisense strand of the target DNA and designed to anneal with its 3′ end immediately adjacent to the polymorphic site (synthesized, e.g., by Integrated DNA Technologies)

   384-well black PCR plates

   Multichannel pipettor or liquid-handling workstation

   Sealing mats for 384-well PCR plates

   Thermal cycler with heated cover

   FP plate reader: Victor^2 or EnVision (Perkin-Elmer); LJL Analyst (Molecular Devices); or Ultra (Tecan)

   Software to create an *x-y* scatter plot: e.g., Microsoft Excel or SNPScorer (Perkin-Elmer)

1. Dispense 3 µl of 0.8 ng/µl genomic DNA per well of a 384-well black PCR plate using a multichannel pipettor or a liquid-handling workstation. Centrifuge plates briefly at 500 × g, room temperature, in a benchtop centrifuge, then air dry overnight at room temperature. Stack and wrap plates with plastic film and keep in vacuum desiccator until use.

2. For each 384-well plate, assemble a master mix containing:

      225 µl 10× PCR buffer
      315 µl 25 mM MgCl$_2$
      90 µl 2.5 mM 4dNTP mix
      9 µl 5 U/µl Platinum *Taq* DNA polymerase
      711 µl H$_2$O.

   Add 3 µl master mix to each well containing the dried genomic DNA.

3. Add 3 µl PCR primer mix to the PCR reaction mixture. Seal the plate with a sealing mat and centrifuge briefly at 500 × g, room temperature.

4. Perform hot-start PCR with the following thermal cycling conditions:

1 cycle:	2 min	95°C	(denaturation)
35 cycles:	10 sec	92°C	(denaturation)
	20 sec	56°C	(annealing)
	30 sec	68°C	(extension)
1 cycle:	10 min	68°C	(final extension)
Final step:	indefinite	4°C	(hold).

5. Dilute 90 μl of 10× Exo-SAP-IT PCR clean-up reagent with 810 μl water and add 2 μl of this 1× reagent to the PCR product mixture. Seal the plate with a sealing mat and centrifuge briefly at 500 × g, room temperature. Incubate 1 hr at 37°C. Heat inactivate the enzymes at 80°C for 15 min. Cool and hold the reaction mixture at 4°C until ready for the next step.

6. Prepare a master mix of the following reagents:

   900 μl 10× reaction buffer
   11.25 μl AcycloPol DNA polymerase
   225 μl AcycloTerminator mix
   1113.75 μl H_2O.

   Add 5 μl to each PCR product.

7. Add 8 μl of 1 μM SNP primer to each well. Seal the plate with a sealing mat and centrifuge the plate briefly at 500 × g, room temperature. Perform the primer extension reaction with the following thermal cycling conditions:

1 cycle:	2 min	95°C	(denaturation)
20 cycles:	15 sec	95°C	(denaturation)
	30 sec	55°C	(extension)
Final step:	indefinite	4°C	(hold; keep in dark).

8. Determine the FP values of the reaction mixtures by reading the plates in a Victor[2], Analyst, Ultra, or EnVision FP plate reader according to the manufacturer's instructions. Export the FP values for the two dyes to a spreadsheet (e.g., Microsoft Excel) or SNPScorer and create an x-y scatter plot. Assign the genotype of each sample by noting its position in the plot.

BASIC PROTOCOL 2

PRIMER EXTENSION ASSAY WITH 4-PLEX PCR

All of the PCR primers should be designed using the same parameters so that they work under the same conditions. The 4-plex PCR protocol is most suitable for a large-scale project where many markers are being typed.

Materials (see APPENDIX 1 for items with ✓)

   0.8 ng/μl genomic DNA of interest
✓ 10× PCR buffer (may be purchased from Invitrogen): 200 mM Tris·Cl, pH 8.4 (see recipe)/500 mM KCl
✓ 2.5 mM 4dNTP mix
   5 U/μl Platinum *Taq* DNA polymerase (Invitrogen)

PCR primer mix: 0.22 μM each of the eight PCR primers for the four SNPs in the set (may be synthesized by Integrated DNA Technologies)

AcycloPrime FP SNP Detection Kit (Perkin-Elmer) containing:
    Exo-SAP-IT kit
    10× Exo-SAP-IT PCR clean-up reagent
    AcycloPol DNA polymerase
    AcycloTerminator mix: six combinations, each containing two labeled terminators (R110-G/TAMRA-A; R110-G/TAMRA-C; R110-G/TAMRA-T; R110-C/TAMRA-A; R110-C/TAMRA-T; R110-A/TAMRA-T) plus the other two unlabeled terminators
    10× reaction buffer

0.2 μM SNP primer: 20- to 25-mer complementary to either the sense or antisense strand of the target DNA and designed to anneal with its 3′ end immediately adjacent to the polymorphic site (synthesized, e.g., by Integrated DNA Technologies)

384-well clear PCR plates and black PCR plates
Multichannel pipettor or liquid handling workstation
Sealing mats for 384-well PCR plates
Thermal cycler with heated cover
FP plate reader: Victor^2 or EnVision (Perkin-Elmer); Analyst (Molecular Devices); or Ultra (Tecan)
Software to create an *x-y* scatter plot: e.g., Microsoft Excel or SNPScorer (Perkin-Elmer)

1. Dispense 8 μl of 0.8 ng/μl genomic DNA per well of a 384-well clear PCR plate using a multichannel pipettor or a liquid handling workstation. Centrifuge plates briefly at 500 × g, room temperature, in a benchtop centrifuge, then air dry overnight at room temperature. Stack and wrap plates with plastic film and keep in vacuum desiccator until use.

2. For each 384-well plate, assemble a master mix containing:

    450 μl 10× PCR buffer
    630 μl 25 mM MgCl$_2$
    180 μl 2.5 mM 4dNTP mix
    18 μl 5 U/μl Platinum *Taq* DNA polymerase
    522 μl H$_2$O.

Add 4 μl master mix to each well containing the dried genomic DNA.

3. Add 7 μl PCR primer mix to the PCR reaction mixture. Seal the plate with a sealing mat and centrifuge the plate briefly at 500 × *g*, room temperature.

4. Perform hot-start PCR with the following thermal cycling conditions:

1 cycle:	2 min	95°C	(denaturation)
35 cycles:	10 sec	92°C	(denaturation)
	20 sec	58°C	(annealing)
	30 sec	68°C	(extension)
1 cycle:	10 min	68°C	(final extension)
Final step:	indefinite	4°C	(hold).

5. Dilute 200 μl of 10× Exo-SAP-IT PCR clean-up reagent with 1800 μl water and add 5 μl of this 1× reagent to the PCR product mixture. Seal the plate with a sealing mat and centrifuge briefly at 500 × *g*, room temperature. Incubate 1 hr at 37°C. Heat inactivate the enzymes at 80°C for 15 min. Cool and hold the reaction mixture at 4°C until ready for the next step.

6. Add 10 μl of 0.2 μM SNP primer to each well of a 384-well black PCR plate. Transfer 3 μl of the processed PCR product (from step 5) to each well of this plate.

7. Prepare a master mix of the following reagents:

> 900 μl 10× reaction buffer
> 11.25 μl AcycloPol DNA polymerase
> 76.5 μl AcycloTerminator mix
> 2162.25 μl H$_2$O.

Add 7 μl to each well in the plate prepared in step 6. Seal the plate with a sealing mat and centrifuge the plate briefly at 500 × g, room temperature.

8. Perform the primer extension reaction with the following thermal cycling conditions:

1 cycle:	2 min	95°C	(denaturation)
20 cycles:	15 sec	95°C	(denaturation)
	30 sec	55°C	(extension)
Final step:	indefinite	4°C	(hold; keep in dark).

9. Determine the FP values of the reaction mixtures by reading the plates in a Victor^2, Analyst, Ultra, or EnVision FP plate reader according to the manufacturer's instructions. Export the FP values for the two dyes to a spreadsheet (e.g., Microsoft Excel) or SNPScorer and create an x-y scatter plot. Assign the genotype of each sample by noting its position in the plot.

References: Chen et al., 1999; Hsu and Kwok, 2003; Hsu et al., 2001; Kwok, 2002

Contributor: Pui-Yan Kwok

CHAPTER 3

Somatic Cell Hybrids

Since the mapping of the thymidine kinase gene using somatic cell genetics over 25 years ago, this powerful technology has played a primary role in gene mapping. The methodology has experienced a near-constant state of evolution since its inception, with a variety of modifications occurring as molecular genetic techniques have emerged. This trend continues today in conjunction with the huge mapping effort underway as a part of the Human Genome Project. Undoubtedly somatic cell genetics will continue to be a fundamental tool for human genetic analysis.

Probably the most significant advance in the use of somatic cell hybrids has occurred with development of molecular genetic methods. DNAs can now be prepared from the cell lines in the hybrid panel and a Southern blot of the panel DNAs prepared. A probe for a particular gene can be hybridized to a filter and the concordance and discordance of a hybridizing fragment with the chromosomes present in a cell line can be determined. Similar to the requirement that rodent and human proteins be able to be differentiated, it is necessary in these experiments to be able to identify species-specific restriction fragments in instances where the human probe cross-hybridizes with a rodent sequence. Numerous genes have been mapped successfully using this technology. Of particular value, this approach permits the tracking of a certain-sized restriction fragment corresponding to the probe in instances where either a gene family or pseudogene are also detected by the probe.

Other developments in molecular technologies have impacted somatic cell genetics. The polymerase chain reaction (PCR) has brought about a new way to screen panels that requires significantly smaller amounts of precious hybrid DNAs. PCR of hybrid DNAs using human-specific repetitive elements (e.g., *Alu* sequences) as primers demonstrated that a PCR karyotype of a cell line could be generated for determining chromosomal content in hybrids with a restricted number of chromosomes. Over the many years of hybrid mapping, a few errors and confusions in assignments resulted from portions of human chromosomes retained in hybrid cells and not detected by conventional cytogenetic analysis. These chromosome fragments were often derived from breakage during the fusion process and may have become stably incorporated into rodent chromosomes. The implementation of chromosomal in situ hybridization using either C_0t_1 DNA or total human genomic DNA as probe has permitted characterization of hybrid panels at a level not previously possible.

The unit provided in this chapter presents the most fundamental method for generating somatic cell hybrids by whole-cell fusion using adherent recipient cells with either adherent or suspended donor cells. It describes a number of selectable markers used in the generation of hybrids, as well as three methods commonly used to characterize the resulting cell lines.

Contributor: Cynthia C. Morton

UNIT 3.1

Construction of Somatic Cell Hybrids

NOTE: All incubations are performed in a humidified 37°C, 5% CO_2 incubator using pre-warmed medium, unless otherwise indicated.

SELECTABLE MARKERS FOR MAMMALIAN CELLS

Cell lines with a number of auxotrophic, temperature-sensitive, and other mutations that allow direct selection for human chromosomes (Table 3.1.1) may be used as recipient cell lines with donor cells whose chromosomes carry complementary markers. If such endogenous markers are unavailable, a dominant-acting bacterial selectable marker may be introduced exogenously into donor cells. Choice of donor cell lines depends on their ultimate use. Typical cell types include primary fibroblasts, lymphoblastoid cell lines, and lymphocytes. There is no formal repository for donor cell lines containing marked chromosomes, but they can be requested from individual investigators (see Athwal et al., 1985; Warburton et al., 1990; and Kurdi-Haidar et al., 1993).

When selection is applied following fusion, the donor chromosome carrying the marker gene required for growth is retained, often with additional donor chromosomes. The best way to obtain a monochromosomal hybrid is to use a marker that selects for the chromosome of interest. This type of selection is considered to be a positive (forward) selection for the donor chromosome. For certain chromosomes this can be done by complementation of a mutation in a rodent chromosome by a gene on the human chromosome of interest. For example, HAT medium can be used to select hybrid cells containing the human X chromosome in an HPRT$^-$ recipient cell line, because it blocks normal DNA synthesis and forces cells to rely on HPRT, the gene for which resides on the X chromosome. In some experimental systems, it may be desirable to perform a subsequent negative (or reverse) selection; an example of this negative selection would be to eliminate an X chromosome by propagating an HPRT$^+$ cell line in medium containing 6-thioguanine. A number of selectable markers available for use in making somatic cell hybrids are described below.

Ouabain

Ouabain acts by inhibiting the Na$^+$/K$^+$ ATPase. Rodent cells can grow in a concentration of ouabain that will kill human cells; therefore, selection by ouabain sensitivity is an effective way to remove human donor cells. Use at a final concentration of 1 to 2 μM in complete medium.

Thymidine Kinase (TK)

TK is a salvage enzyme that allows synthesis of dTTP from thymine. Most mammalian cell lines express TK, so this marker requires a special TK$^-$ cell line as the recipient. For forward selection (TK$^-$ to TK$^+$), supplement complete medium with 100 μM hypoxanthine, 0.4 μM aminopterin, 16 μM thymidine, and 3 μM glycine (HAT medium). This blocks normal formation of dTDP from dCDP, forcing cells to use the TK pathway. For reverse selection (TK$^+$ to TK$^-$), supplement complete medium with 30 μg/ml 5-bromodeoxyuridine (BrdU), which kills TK$^+$ cells.

Hypoxanthine-Guanine Phosphoribosyltransferase (HPRT)

HPRT is a salvage enzyme that catalyzes conversion of hypoxanthine to inosine monophosphate (IMP), which in turn is converted to dGTP or dATP. For forward selection (HPRT$^-$ to HPRT$^+$), supplement complete medium with 100 μM hypoxanthine, 0.4 μM aminopterin, 16 μM thymidine, and 3 μM glycine (HAT medium). The aminopterin inhibits de novo purine synthesis. When cells are removed from HAT medium, they should be grown in medium containing hypoxanthine and guanine before being transferred to nonselective medium. For reverse selection (HPRT$^+$ to HPRT$^-$), supplement complete medium with 40 μg/ml 6-thioguanine.

Adenine Phosphoribosyltransferase (APRT)

APRT is a salvage enzyme that catalyzes conversion of adenine to AMP using adenine and 5-phosphoribosyl-1-pyrophosphate as substrates. For forward selection (APRT$^-$ to APRT$^+$), supplement complete medium with 100 μM adenine, 0.8 μM aminopterin, and 16 μM thymidine (AAT medium). As in forward selection with HPRT, aminopterin serves to inhibit de novo purine synthesis. For reverse selection (APRT$^+$ to APRT$^-$), supplement complete medium with 30 μg/ml 2,6-diaminopurine.

Xanthine-Guanine Phosphoribosyltransferase (XGPRT, gpt)

XGPRT is a bacterial enzyme that does not have a mammalian homolog, allowing it to function as a dominant-acting selectable marker in mammalian cells. For selection, prepare complete medium containing dialyzed FBS, 250 μg/ml xanthine, 15 μg/ml hypoxanthine, 10 μg/ml thymidine, 2 μg/ml aminopterin, 25 μg/ml mycophenolic acid, and 150 μg/ml L-glutamine. Aminopterin and mycophenolic acid both block de novo GMP synthesis. Selection uses medium with xanthine but not guanine. The amount of mycophenolic acid varies with cell type and can be determined by titration in the absence and presence of guanine.

Aminoglycoside Phosphotransferase (neo, G418, APH)

Expression of the bacterial APH gene in mammalian cells results in detoxification of G418. To use, prepare a stock of G418 in a highly buffered solution (e.g., 100 mM HEPES, pH 7.3). For selection, prepare 100 to 800 μg/ml G418 in complete medium. Note that cells will divide once or twice in the presence of lethal doses of G418, so the effects of the drug take several days to become apparent. G418 is the most widely used dominant-acting selection system.

Hygromycin-B-Phosphotransferase (HPH)

The HPH gene (isolated from *E. coli* plasmid pJR225; Gritz and Davies, 1983) detoxifies hygromycin-B, a protein synthesis inhibitor, by phosphorylation. For selection, supplement complete medium with 10 to 400 μg/ml hygromycin-B (many cell lines require 200 μg/ml). Note that vectors that efficiently express the HPH gene are not as widely available as those that express G418 resistance; therefore, this marker is often used in applications where two different selectable markers are needed.

Histidinol Dehydrogenase (hisD)

In medium lacking histidine and containing histidinol, only cells possessing the *hisD* gene product are able to grow, as mammalian cells cannot synthesize histidine, but *hisD* allows oxidation of histidinol to histidine. Because histidinol is also a potent inhibitor of histidyl-tRNA synthetase, it is possible to take advantage of this selection scheme even when the medium contains histidine, provided that the histidinol concentration is high enough. For selection, supplement complete medium containing histidine with 4 to 5 mM histidinol (Sigma) *or* medium lacking histidine with $\leq$0.5 mM histidinol. Note that histidinol will prevent growth of most but not all recipient cells lines in medium containing histidine; therefore, recipient cells should be tested before fusion is attempted.

BASIC PROTOCOL

WHOLE-CELL FUSION OF MONOLAYER CELLS

This protocol describes PEG-induced fusion for production of hybrid cells.

Materials

Confluent recipient cells (Table 3.1.1)
Medium appropriate for recipient cells, with and without serum (*APPENDIX 3I*), 37°C
Appropriate selective agent (see Selectable Markers for Mammalian Cells)
Donor cells containing chromosomes of interest and appropriate genetic markers
✓ 50% (w/v) PEG 1000, 37°C

100-mm tissue culture plates
25-cm^2 tissue culture flasks
Inverted phase-contrast microscope

1. If unknown, determine selection conditions for the recipient cell line by splitting a confluent 25-cm^2 tissue culture flask of recipient cells 1/10 into medium containing serum and several different levels of selective agent. Incubate 10 days, changing medium and checking flasks for viable cells twice weekly. Identify the lowest concentration of selective agent that prevents cell growth.

2. In the afternoon of the day before fusion, prepare cells by plating equal amounts (usually 5×10^5 to 1×10^6) of donor and recipient cell lines into one 100-mm tissue culture plate containing 10 ml medium with serum. Also prepare control plates containing donor and recipient cells for exposure to PEG with and without selective medium. Incubate overnight.

   Donor and recipient controls should each die under selection. Cells with PEG alone should recover and become confluent. Perform the latter control when a new cell line or a new batch of PEG is used.

3. Check cells under an inverted phase-contrast microscope. They should be ∼70% confluent, with good cell-to-cell contact but no crowding. Prepare new plates with increased numbers of cells if density is insufficient.

4. Aspirate medium from plate and rinse twice with serum-free medium, aspirating completely each time. Tilt plate a short time and aspirate last of medium.

5. Add 2 ml of 50% (w/v) PEG solution, 37°C. Tilt plate back and forth to distribute solution quickly and evenly over cells. Let stand 2 min (time is critical). If necessary, optimize

Table 3.1.1 Common Recipient Cell Lines Used in Fusion

Cell line	Type	Marker(s)[a]	Source
A9	Mouse (L cell)	HPRT$^-$, APRT$^-$	ATCC #CCL 1.4
L-M TK$^-$	Mouse (L cell)	TK$^-$	ATCC #CCL 1.3
RAG	Mouse	HPRT$^-$	ATCC #CCL 142
E 36	Hamster	HPRT$^-$	Gillin et al., 1972
CHOK1	Chinese hamster ovary[b]	Pro$^-$	ATCC #CCL 61

[a]Descriptions of markers, with selection conditions and references, can be found in this section. Abbreviations: HPRT, hypoxanthine-guanine phosphoribosyltransferase; APRT, adenine phosphoribosyltransferase; TK, thymidine kinase; Pro, proline.

[b]Numerous auxotrophic and other mutant derivatives of Chinese hamster ovary and other Chinese hamster cell lines are available from individual investigators (e.g., Kao et al., 1976).

fusion by performing pilot experiments in which PEG concentrations are varied from 44% to 50% and incubation times from 2 to 3 min for monolayers and 2 to 4 min for suspensions.

6. Carefully add 10 ml serum-free medium to one side of the plate. Gently swirl to dilute PEG. Aspirate medium and discard. Repeat twice more with 10 ml serum-free medium and once with 10 ml serum-containing medium. Perform these washes carefully, as cell membranes are delicate at this point.

7. Feed with 10 ml medium with serum and incubate 36 to 48 hr (i.e., allow to double twice).

8. Trypsinize cells and subdivide among five to ten 100-mm plates. To ensure selection, plate cells at a density low enough that they can divide several times before confluence. To each plate, add 10 ml medium containing serum and appropriate level of selective agents (step 1). Incubate at 37°C.

9. Replace medium every 5 days until colonies appear. Check plates for colonies ~10 to 14 days (up to 3 weeks) after transferring to selective medium. Avoid disturbing the cells or allowing them to grow too large, as dislodged cells may form sibling colonies.

10. Isolate colonies using cloning cylinders (see Support Protocol 1).

11. Expand, characterize (see Support Protocols 3 to 5), and freeze (APPENDIX 3I) clones as soon as possible to minimize labor involved in maintaining and expanding undesired cell lines. Keep passages to a minimum to avoid loss of chromosomes and rearrangements (three passages should provide cells for one cryovial, or one 6-well plate should provide two cryovials). Use multiple characterization methods to optimize monitoring of cell lines, and repeat characterization each time cells are thawed and expanded to check for rearrangements and segregation. If necessary, subclone hybrids to isolate lines that have segregated additional chromosomes (see Support Protocol 2).

12. Upon rapid thawing, transfer cells to 25-cm^2 tissue culture flasks. If cells recover poorly after freezing, increase the amount of serum in the freezing medium.

ALTERNATE PROTOCOL

WHOLE-CELL FUSION OF SUSPENSION DONOR CELLS TO MONOLAYER RECIPIENT CELLS

This procedure is used for fusion of two cell lines where a suspension culture or lymphocytes in suspension are used as the donor cell line.

Additional Materials (also see Basic Protocol)

> Donor cell line growing in suspension *or* fresh lymphocytes from whole blood
> (APPENDIX 3B)
> 15-ml snap-cap polypropylene tubes

1. Trypsinize adherent recipient cells (APPENDIX 3I) and resuspend in medium. Count donor and recipient cells using a hemacytometer (APPENDIX 3I) and mix 2×10^6 cells from each in a 15-ml snap-cap polypropylene tube. Maintain one flask of recipient cells to verify that selection is working.

2. Centrifuge 5 min at $400 \times g$, room temperature. Add 10 ml serum-free medium appropriate for recipient cells, centrifuge again, and aspirate *all* medium carefully, tilting tube to remove the last drops.

3. Loosen pellet by flicking tube gently. Pipet 0.3 ml of 50% PEG solution, 37°C, down the side of the tube while mixing gently by flicking or swirling. Incubate 3 min at room temperature. If necessary, optimize fusion conditions (see Basic Protocol, step 5).

4. Add 8 ml serum-free medium and, as quickly as possible, centrifuge 5 min at 400 × g.

5. Carefully aspirate medium and gently resuspend cells in 10 ml serum-free medium. Centrifuge again, aspirate medium, and resuspend pellet in 10 ml medium with serum, room temperature. Perform these washes carefully, since cell membranes are delicate at this point.

6. Add 3 ml cell suspension to each of three 100-mm tissue culture plates. Add 1 ml cell suspension to another 100-mm plate as a control. Incubate all plates 36 to 48 hr.

7. Feed cells with medium containing serum and selective agent (or without selective agent for control).

8. Prepare colonies and characterize as described (see Basic Protocol, steps 9 to 12).

SUPPORT PROTOCOL 1

COLONY ISOLATION USING CLONING CYLINDERS

Additional Materials (also see Basic Protocol)

    Cells on plates
✓ PBS, 37°C
    Sterile vacuum grease (place in glass petri dish and sterilize by autoclaving)
    5% (w/v) trypsin/2% (w/v) EDTA, 37°C (store in aliquots at −20°C)

✓ Cloning cylinders
    6- or 24-well tissue culture plates

1. Examine undersides of plates and circle colonies with a laboratory marker.

2. Evaluate colonies using an inverted phase-contrast microscope and select healthy colonies of 100 to 500 cells that are well separated from each other on the same plate. Pick colonies from several different plates. To avoid analyzing sibling colonies, assign each colony a number reflecting the fusion, plate, and colony of origin.

3. Remove medium from plate and rinse cells with 5 ml PBS, 37°C.

4. Dip one end of a sterile cloning cylinder into a petri plate containing sterile vacuum grease to coat with just enough grease to attach the cloning cylinder to the plate without contaminating the colony. Using sterile forceps, place cylinder over a selected colony using marks on bottom of dish as guides.

5. Holding the cloning cylinder in place with forceps, add 2 to 3 drops 5% (w/v) trypsin/2% (w/v) EDTA, 37°C, to the colony using a sterile plugged Pasteur pipet. Incubate 5 min at 37°C. If the cylinder leaks, push down on top with side of forceps and add more trypsin/EDTA.

6. Examine plate under inverted phase-contrast microscope to see if colonies are sufficiently trypsinized (i.e., cells have round morphology and some are floating).

7. Pipet up and down gently several times with a sterile plugged Pasteur pipet to disperse cells.

8. Transfer very rapidly growing lines such as A9 to 6-well plates. Transfer more slowly growing lines such as some CHO derivatives to 24-well plates and expand from there. Keep track of the number of passages, as approximately five are required to grow cells from a single colony to multiple confluent 100-mm plates ready for large-scale DNA isolation.

SUPPORT PROTOCOL 2

SUBCLONING HYBRID CELL POPULATIONS

Using this procedure, hybrids can be cloned again to isolate lines that have segregated additional chromosomes.

Materials

Cloned hybrid cells (see Basic Protocol or Alternate Protocol)
Serum-containing medium appropriate for hybrid cell line (APPENDIX 3I)
96-well microtiter plate

1. Count cloned hybrid cells (APPENDIX 3I) and dilute to 5 cells/ml with serum-containing medium. Pipet vigorously to ensure a single-cell suspension.

2. Transfer diluted cells into a 96-well plate at 0.2 ml/well (an average of 1.0 cell/well). Incubate, changing medium twice weekly, until colonies are apparent when held up to the light.

3. Discard any well containing more than one colony. Expand wells containing a single colony as needed.

SUPPORT PROTOCOL 3

CYTOGENETICS: G-11 STAINING OF METAPHASE CHROMOSOMES

Cytogenetics provides the most direct means of identifying the chromosome of interest. An assortment of cytogenetic methods for preparing and staining chromosomes is described in UNITS 4.1 & 4.3. An alternative to these techniques is to use the differential stain G-11. This protocol (provided by H.F.L. Mark) is a simple means to distinguish human chromosomes (which will stain blue with red paracentromeric regions) from rodent chromosomes (which will stain magenta with blue centromeres). Neither this nor the staining and banding methods discussed in UNITS 4.1 & 4.3 can identify small fragments of human chromosomes.

Materials

Cells of human-rodent hybrid lines (see Basic Protocol and Alternate Protocol)
Cells of human donor and rodent recipient lines used to construct hybrids
✓ Giemsa stain
Hydrion buffer: one Hydrion buffer capsule, pH 11.00 (Micro Essential Laboratory), in 100 ml H_2O

Coplin jars
60°C water bath
100-ml staining tray
Bright-field microscope equipped with 10× lens for scanning and 100× oil-immersion lens for analysis

NOTE: Double-distilled water should be used throughout this procedure.

1. Prepare metaphase slides from human-rodent hybrid cell lines to be characterized and from human donor and rodent recipient lines (*UNIT 4.1*). Optimize chromosome density to obtain unambiguous color differention between human and rodent chromosomes without sacrificing banding quality, which is inversely proportional to density.

2. Place slides in a Coplin jar, cover with water, and presoak 1 to 3 hr at 60°C.

3. Prepare staining solution by mixing 5 ml Giemsa stain and 95 ml Hydrion buffer in a Coplin jar. Prewarm to 37°C in a 100-ml staining tray. If the color change is too rapid, the amount of Giemsa can be reduced to 2% (2 ml in 98 ml Hydrion buffer).

4. Place eight air-dried metaphase slides from each batch of cells in staining solution. Remove the first slide after 5 min and other slides at 5-min intervals up to 40 min. Rinse each slide immediately in a Coplin jar of water and allow to air dry.

5. Examine slides under a bright-field microscope.

SUPPORT PROTOCOL 4

IN SITU HYBRIDIZATION

In situ hybridization (*UNIT 4.4*) requires relatively few cells, can be utilized to detect small fragments of human chromosomes not evident by conventional cytogenetics, and does not require extensive cytogenetic experience. Total human genomic DNA, typically labeled with biotin or digoxigenin, is hybridized to metaphase chromosome spreads of the hybrid cell line. Whole human chromosomes as well as additional human chromosome fragments are detected using enzymatic or fluorescent detection. Commericial chromosome-specific painting probes are available for identification of many human chromosomes and can be used to determine the chromosome content of hybrid cell lines. This requires a larger number of metaphase spreads, but the simultaneous detection of differentially labeled probes with different fluorophores requires less time and fewer materials than a separate hybridization for each probe.

SUPPORT PROTOCOL 5

MARKER ANALYSIS

Hybrids can be tested for the presence of particular human chromosomes by assaying for markers known to reside on those chromosomes. Generally, two markers (one per chromosome arm) are tested. This can be done most efficiently by PCR, using primer pairs obtained from the literature. Alternatively, Southern blot analysis (*APPENDIX 3G*) may be performed if necessary. At the back of this unit, Table 3.1.2 lists a set of PCR primers that can be used to analyze somatic cell hybrids; another set has been published by Abbott and Povey (1991). One well of a 24-well plate can yield enough hybrid-cell DNA for five to ten PCR reactions. By testing for multiple markers (which give PCR products of different sizes) in a single tube, one row of a 24-well plate yields more than enough DNA to analyze a hybrid. To obtain DNA:

1. Remove medium and wash plates with PBS.

2. Lyse cells by adding 100 μl PCR buffer with nonionic detergents (see recipe in *APPENDIX 1*) to each well of a 24-well plate.

3. Transfer the solution (which will be viscous) to a microcentrifuge tube and incubate 1 hr at 55°C and 10 min at 95°C.

4. Divide the DNA solution into aliquots in microcentrifuge tubes and store frozen at −20°C. Typically, use 10 to 20 µl per PCR reaction.

PCR with primers to interspersed repetitive elements (IRSs) can be used to produce a "PCR karyotype" for each chromosome, namely a specific set of amplified products visualized by agarose gel electrophoresis. This pattern is compared to the pattern from a hybrid panel of known composition, such as the panel of reduced hybrids available from the NIGMS Human Genetic Mutant Cell Repository. In reality, direct analysis is still useful to determine the chromosome content for hybrids that contain one to two chromosomes. However, the products of this IRS-PCR method can be used as probes for in situ hybridization to normal human metaphase spreads. This provides an alternative method to detect small fragments not seen using conventional cytogenetics, and may pick up rearrangements that would not be picked up using human DNA as a probe to metaphase spreads of the hybrid.

Although it is very rapid, marker analysis has the disadvantage that the presence of the chromosome is inferred from presence of the marker. It will therefore not determine whether the chromosome is intact. If the karyotype of the hybrid has not yet stabilized, further segregation may occur subsequent to analysis.

References: Abbot and Povey, 1991; Athwal et al.,1985; Gritz and Davies,1983; Kurdi-Haidar et al.,1993; Theune et al.,1991; Warburton et al.,1990

Contributors: Cynthia L. Jackson

Table 3.1.2 PCR Primers for Characterization of Cell Hybrids[a]

Gene symbol	Chromosome region	PCR primer	Primer sequence	Amplicon (bp)[b]	Conditions (°C/mM)[c]	Gene region[b,d]
NGFB	1p22.1	NGFB1 NGFB2	5'-GATGCCAGATTAGGGATCTGCTGG-3' 5'-ACTCCTGCTTCTGGCAGCTGCAG-3'	497	65/1.5	A
REN	1q32 or q42	REN1 REN2	5'-CCTGCCAAGAAACCAGTCATGAAG-3' 5'-ATCTTGTGTCCAGTGACGCTAGCA-3'	285	60/1.5	B
POMC	2p23	POMC1 POMC2	5'-GACCCAAGAGTCTCTTGACTTGAG-3' 5'-CTCTGCACTAAGCCTCCAAAACTG-3'	593	60/1.5	B
TGFA	2p13	TGFA1 TGFA2	5'-GATCTGAGCCCTGCATCTTTCCT-3' 5'-GATCTCCAGGAGAACAGGGGATAC-3'	180	60/1.5	A
IL1A	2q12-q21	IL1A1 IL1A2	5'-AACTTAGCCACTGGTTCTGGCTGA-3' 5'-GAATCTTGGCAGCTCCTGGCAACT-3'	396	60/1.5	C
ALPP	2q37	ALPP1 ALPP2	5'-GACCATGGTCATCATGAAAGCAGG-3' 5'-GTGCCAGCATTACTCCCATTTGAC-3'	279	60/1.5	A/B
GLB1	3p21-pter	GLB1 GLB2	5'-TGATGAAAGCCTGTGCTTTGAG-3' 5'-AAAGCTTCCATTCCAGCCCTG-3'	245	60/2.0	D
GLUT2	3q26	GLUT2A GLUT2B	5'-AAACAATAAGGGAACCGTCTGT-3' 5'-CTCTATAATCCATTCCACATGAA-3'	550	58/2.5	D
QDPR	4p15.3	QDPR1 QDPR2	5'-GTCACTAACCTGTCTCAGTGTGG-3' 5'-GTAGTCAAGATGACAGCCACTGTC-3'	297	65/1.5	D
AFP	4q11-q13	AFP1 AFP2	5'-GATGCACCTGACCCACTTTATAAAG-3' 5'-GAGATTGTCTGACCGATTCAGACTC-3'	321	60/1.5	C
C9	5p14-p12	C9A C9B	5'-AGCTGTTGGCTTCTCTGAGCTCCA-3' 5'-TTTCCGTGGATAAGCAGTTCTGGC-3'	309	60/1.5	D
HMGCR	5q13.3-q14	HMGCR1 HMGCR2	5'-GCCCGACAGTTCTGAACTGGAACA-3' 5'-GAACCTGAGACCTCTCTGAAAGAG-3'	160	65/1.5	D

Gene	Location	Primer	Sequence	Size	Conditions	Code
TNFA	6p21.3	TNFA1 TNFA2	5'-CAGGGTCCTACACACAAATCAGTCA-3' 5'-AAGAGAACCTGCCTGGCAGCTTGT-3'	417	65/1.5	C
MAS1	6q24-q27	MAS1 MAS2	5'-CGGTCACAGTTGAGACTGTCGTCT-3' 5'-TTAGTATCTCATGCATATGGGATGAG-3'	157	60/1.5	D
EGFR	7p13-p12	EGFR1 EGFR2	5'-CCAGTCATGAGCGTTAGACTGACT-3' 5'-TAGATGACTCAAGGCAGAGACACT-3'	495	60/1.5	D
CFTR	7q31-q32	C16B C16D	5'-GTTTTCCTGGATTATGCCTGGCAC-3' 5'-GTTGGCATGCTTTGATGACGCTTC-3'	97	60/2.0	B
NEFL	8p21	NEFL1 NEFL2	5'-GCTCCAGGACCTCAATGACCGCTT-3' 5'-AAAGGTCGGGCTTGGTCACGTCCA-3'	490	65/1.5	B
DEF	8p23-p22	DEF3A DEF3B	5'-CTTGCAGAAAAGAAAAATGAGCTC-3' 5''GTAACAAGGCATTTATTTGAGATGAG-3'	NR	60/1.5	NR
LHRH	8p21-11.1	LHRH1 LHRH2	5'-GGGCCAGAAGGAATGACCATTAC-3' 5'-CATTCACACACAGCACTTTATTATGG-3'	NR	60/2.0	NR
TG	8q24	TG1 TG2	5'-GAGCCTGTTCCCTCCAAAGATACA-3' 5'-CTTACCGAAGATATTGGCCGACAC-3'	392	65/1.5	C
RLN1	9pter-q12	RLN1A RLN1B	5'-CAGAGCTACAGCAGTATGTACCTG-3' 5'-TCAAACAGTGCCACGTAGGGTCGT-3'	178	60/1.5	B
IFNA	9p22-p13	IFNA1 IFNA2	5'-CTCATTGACTAATGCATCATCTCACAC-3' 5'-TCAGGTTATACTGTCAGGCTTGGCAT-3'	348	65/1.5	D
ALDOB	9q21.3-q22.2	ALDOB1 ALDOB2	5'-AGCCTAGCTCCAGTGCTTCTAGTA-3' 5'-CTTTGGATGAGGAGCCGATATTG-3'	189	65/1.5	D
FNBR	10p11.2	FNBR1 FNBR2	5'-CTGAAAGACAAGTATGTTGAGAGTTGC-3' 5'-GAATGTGACTAGTGTGAAACAAGATGGG-3'	581	65/1.5	D
PLAU	10q24-qter	PLAU1 PLAU2	5'-ATCTGATGCTCTTCAGCTGGGCCT-3' 5'-CTGGAGGACAAACAGAGGGATGTCTT-3'	289	65/1.5	A/B
CAT	11p13	CAT1 CAT2	5'-ATATCACGTTGCTGCCCATGAGGT-3' 5'-GCATTTGCACATCTAGCACACAGGA-3'	219	65/1.5	A/B
CLG	11q21 q22	CLG1 CLG2	5'-AGTACAGGAGCCGAACAGCCATC-3' 5'-GGTGACTCCTAGCAGATTATTTTGG-3'	302	65/2.0	C

(continued)

Table 3.1.2 (Continued)

Gene symbol	Chromosome region	PCR primer	Primer sequence	Amplicon (bp)[b]	Conditions (°C/mM)[c]	Gene region[b,d]
KRAS2	12P12.1	KRAS1 / KRAS2	5'-GATTCCTACAGGAAGCAAGTAGTAA-3' / 5'-CTATAATGGTGAATATCTTCAAATGATTT-3'	179	65/3.0	B
F8VWF	12pter-p12	F8VWF1 / F8VWF2	5'-GACACTAACGGAGGATACCGCTGAG-3' / 5'-CACAAAGTCTTCTCACACAGGGCC-3'	NR	60/1.5	NR
PAH	12q22-q24.2	PAH1 / PAH2	5'-GGACCTGCTTCATTCAAGCTTCATA-3' / 5'-GGATTAGGTGCAGAGTTTTATTACCT-3'	334	65/2.0	D
RB1	13q14.2	RB1 / RB2	5'-GAGGAAACAATCTGCTACAACT-3' / 5'-CCAGCTTCTACTCGAACA-3'	334	50/1.5	A/B
NP	14q11.2	NP1 / NP2	5'-AGCAGAGCGAGTAACTCACAGTAG-3' / 5'-GCTACACTGAAATGCATGACATACAT-3'	344	65/1.5	A
PI	14q32.1	PI1 / PI2	5'-GAAGCTCTCCAAGGCCGTGCATAA-3' / 5'-GTTGAGGAGGCGAGAGGCAGTTATT-3'	222	65/1.5	B
B2M	15q21 q22.2	B2M1 / B2M2	5'-CACCCAGTCTAGTGCATGCCTTCT-3' / 5'-TGAGAAGGAAGTCACGGAGCGAGA-3'	357	65/1.5	A
HBA	16p13.3	HBA1 / HBA2	5'-ATCAACTGTCAGGAAGACGGTGTC-3' / 5'-TACAGGCATGAGTCATCACACCTG-3'	350	65/1.5	D
HPR	16q22.1	HPR1 / HPR2	5'-CTGACCATCTGAAGTATGTCATGCT-3' / 5'-GCATCGCCATAGCAGGTGTCTTC-3'	186	65/1.5	B
LCAT	16q22.1	LCAT1 / LCAT2	5'-TCATTGAGTAAGCTGACACTGAGCA-3' / 5'-TTCAGCTTGATGCTGGACATGATG-3'	222	65/1.5	A/B
TP53	17p13.1	TP53A / TP53B	5'-TTCCTCTTCCTGCAGTACTC-3' / 5'-AGACCTCAGGCGGCTCATAG-3'	397	55/1.5	A/B
MPO	17q21-q23	MPO1 / MPO2	5'-CACTTCCTGCATTGAACCTGGCTT-3' / 5'-CTCAAGGTCACATAGCTAGCAAGC-3'	417	65/1.5	B

Gene	Location	Primer	Sequence	Size	Annealing/Mg	Type
GAS	17q	GAS1 GAS2	5'-ATGCTAGTCGGTGTAGAGCCATG-3' 5'-TTGTACCTCATAGGGCTGCGTGA-3'	297	65/1.5	C
TS	18pter-q13 or q21.3-qter	TS1 TS2	5'-TTGCCACTGGCAAATGTAACTGTGC-3' 5'-ACACTCTACATCATGATCGATGGTG-3'	289	65/2.0	D
LDLR	19p13.2-p13.1	LDLR1 LDLR2	5'-AGCTGGATCACTTGAGTTCAGGAGT-3' 5'-CTATCTGTACAGGGACGCATTTACG-3'	497	65/1.5	D
APOC2	19q13.2	APOC2A APOC2B	5'-AGGGCAAAGATCGATAAAGCAGGAAT-3' 5'-CCACCCTAACTCTAAGCAGAAGCT-3'	297	65/1.5	A
PRIP	20pter-p12	PRIP1 PRIP2	5'-GRCACAACACTGAACCTCTGGCTA-3' 5'-CTATGAACTTGACCTAATTCTGGT-3'	412	65/3.0	D
ADA	20q13.11 or 20q13.2-qter	ADA1 ADA2	5'-GTAAGAAGTACCAGCAGCAGACCG-3' 5'-GGATTCCAGTTCCAAGCCTTAAGA-3'	251	65/1.5	A
APP	21q21.2	APP1 APP2	5'-GCTGTATCAAACTAGTGCATGAATAG-3' 5'-GCAGAAGCAATCTGTACAGTAA-3'	295	65/1.5	D
CD18	21q22.3	CD18A CD18B	5'-CACAGCTCTTGAGGATGTCACCAA-3' 5'-TCAGACTGATGTCCTGACTTGCAC-3'	208	65/1.5	D
IGLC2	22q11.1-q11.2	IGL1 IGL2	5'-CTGAGGAGCTTCAAGCCAACAAGG-3' 5'-TTCATGCGTGACCTGGCAGCTGTA-3'	229	65/1.5	B
STS	Xp22.32	STS1 STS2	5'-CACAAGGTCAGTAATGCTGCAGG-3' 5'-CCACATTGTTGAATTGAGTCACGATAG-3'	308	65/1.5	D
PGK1	Xq113	PGKA1A PGK1B	5'-CTTAGCATTTTCTGCATCTCCACTTG-3' 5'-CATGCTGAGTAGTGAAACAGTGACA-3'	288	65/1.5	D
G6PD	Xq28	G6PD1 G6PD2	5'-CTACTTGGTTATCTAGTAGCCTTCTC-3' 5'-GTTAATTGCTCAGTGTGATCAGATCGG-3'	275	65/1.5	C

[a] Data from Theune et al., 1991, and Dubois and Naylor, 1993.

[b] NR, not reported.

[c] Figures represent annealing temperature (°C) and optimal Mg^{2+} concentration in PCR reaction cocktail.

[d] A indicates primer sequence is from an intron, B indicates primer sequence is from an exon, C indicates primer sequence is from 5' flanking region, D indicates sequence is from 3' untranslated region of message.

CHAPTER 4

Cytogenetics

The goal of this chapter on cytogenetics (i.e., the study of chromosomes) is to present both classical and new technologies. Protocols in *UNIT 4.1* are essential for the initiation of clinical and research applications, as peripheral blood is the most easily obtainable human tissue from which chromosomes are prepared. A detailed procedure for slide making is provided, as therein lies perhaps the most critical and variable aspect of any cytogenetic method.

UNIT 4.2 concerns cytogenetics of the mouse. As the mouse is increasingly becoming the model organism for further study of human developmental and neoplastic disorders, the need to assess the karyotype has become pronounced. Although mouse cytogenetics is not a skill that is common, it seems likely that its value will dictate its development in many laboratories.

UNIT 4.3 includes a variety of the most commonly used banding techniques and discussions of their applications. Several protocols for nonisotopic in situ hybridization are provided in *UNIT 4.4*, including detection of molecular probes by fluorescence (FISH) and enzymatic methods. Application of FISH for the ordering of probes in interphase cells is also detailed.

Various modifications on FISH protocols are used for extended mapping of probes on decondensed chromatin, searching for differences in the relative abundance of DNA sequences between two different genomes, combining immunohistochemistry with probe detection, and determining the orientation of DNA sequences on a chromosome. In *UNIT 4.5*, high-resolution FISH analysis is presented for ordering of genes in the same chromosomal region; chromatin fibers are released from the nucleus, providing a method for mapping probes with increased precision compared to that possible with intact interphase nuclei. A commentary unit on multicolor FISH methods is provided as *UNIT 4.6*. It discusses, from a practical standpoint, the design of probe sets for individual identification of each human chromosome in a karyotype, as well as different systems for analysis.

The widespread adoption of nonradioactive methods such as FISH has made possible the assessment of phenotype and genotype on a single interphase or mitotic cell. *UNIT 4.7* provides a variety of protocols for the technique known as MAC (Morphology Antibody Chromosome), whereby immunocytochemical or cytochemical methods for phenotypic analysis of cells is combined with genotyping by FISH or chromosome banding. In addition, a protocol describing FISH to preparations that have been previously analyzed using G-banding is presented. This technique will be of great value in retrospective analyses of both clinical and research materials.

Protocols pertaining to comparative genomic hybridization (CGH) are provided in *UNIT 4.8*. CGH is a powerful technique for assessing amplifications and deletions of genes, and has been rapidly adopted in many laboratories with a focus on the complexities of genomic rearrangements in tumors.

Undoubtedly, molecular cytogenetic techniques will continue to have an expanding role in clinical cytogenetics and an integral role in the molecular dissection of the human genome. It can be anticipated that these methods will provide valuable insight into the very nature of genetic disorders.

Contributor: Cynthia C. Morton

Chromosome Preparation from Cultured Peripheral Blood Cells

Although chromosomes may be obtained from other cells, human peripheral blood leukocytes are most amenable to synchronization, and thus to high-resolution analysis of their chromosomes. A troubleshooting guide to obtaining good chromosome preparations is found in Table 4.1.1 at the end of this unit.

BASIC PROTOCOL

CULTURE AND METAPHASE HARVEST OF PERIPHERAL BLOOD

Materials (*see* APPENDIX 1 *for items with* ✓)

Heparinized whole blood obtained via Vacutainer (Becton Dickinson) or syringe with preservative-free sodium heparin (25 U/ml)

✓ Complete RPMI/10% FBS medium containing 50 µg/ml gentamycin sulfate in place of penicillin and streptomycin

100× phytohemagglutinin-M (PHA; Life Technologies), reconstituted in sterile deionized water (store at 4°C)

✓ 10 µM methotrexate (optional)

✓ 1 mM thymidine (optional)

10 µg/ml Colcemid (Life Technologies)

75 mM KCl (store ≤2 weeks at room temperature)

Fixative: 3:1 (v/v) HPLC-grade absolute methanol/glacial acetic acid (prepare fresh)

15-ml sterile disposable conical polypropylene centrifuge tubes (do not use polystyrene)

TB syringe with 21-G needle (VWR Scientific; do not use preattached 25-G needle)

NOTE: All incubations are performed in a humidified 37°C, 5% CO_2 incubator unless otherwise specified. All reagents and equipment coming into contact with live cells must be sterile.

1. Collect peripheral blood by venipuncture into a sodium heparin Vacutainer or a syringe with 25 U preservative-free sodium heparin per milliliter blood (do not use other anticoagulants). Initiate cultures as soon as possible (preferred) or store up to ≤4 days at 4°C. Ship at room temperature.

2. Using a TB syringe with a 21-G needle, inoculate 0.25 ml whole blood into a sterile 15-ml polypropylene centrifuge tube containing 5 ml complete RPMI with 10% FBS and gentamycin. Use 0.2 ml whole blood for newborns ≤3 weeks old, and do not exceed 0.5 ml for any one tube. Add 0.05 ml of reconstituted 100× PHA.

   A single culture typically yields three to five full-slide preparations (or more if only part of the slide is used).

3a. *For standard samples:* Incubate 2 to 4 days (3 days optimal) with tubes tilted at 45° (angle increases gas exchange and prevents dense packing).

3b. *For newborns:* Harvest as described either directly or following a 1- to 2-day culture (2 days optimal).

3c. *For older patients*: Harvest at 3 or 4 days as leukocytes from this age group do not seem to respond as quickly to PHA.

8. Remove all but 0.5 ml of the supernatant and resuspend the brown clumpy pellet in remaining supernatant gently but thoroughly by drawing it up and down with a Pasteur pipet. Avoid drawing in too great a volume as cells will stick permanently to the glass. Do not press the pipet tip against the bottom. Add 1 ml fixative and immediately mix gently. Adjust volume to 5 ml with fixative and mix thoroughly. Centrifuge as in step 5.

9. Aspirate supernatant, resuspend pellet in 5 ml fixative, and centrifuge as in step 5.

10. Remove supernatant and resuspend pellet in a volume of fixative sufficient to produce a light milky suspension. Allow to stand 30 min at room temperature or overnight at 4°C (longer fixations often improve chromosome spreading).

11. Prepare slides and analyze chromosome spreads (see Support Protocol).

SUPPORT PROTOCOL

CHROMOSOME SLIDE PREPARATION

This protocol works for a wide range of cell cultures: peripheral blood, bone marrow (UNIT 10.1), ascites and pleural effusions, amniotic fluid (UNIT 8.2) and tissue flask harvests (UNITS 8.1, 8.3 & 10.2), somatic cell hybrids (UNIT 3.1) or radiation hybrids, lymphoblastoid cell lines, and nonhuman and hybridoma cultures—in short, any culture harvest that results in a fixed suspension of mitotic cells.

Materials

Fixed cultures (see Basic Protocol)
Fixative: 3:1 (v/v) absolute methanol/glacial acetic acid (both AR grade, J.T. Baker)

Microscope slides (one end frosted) stored in 100% methanol (absolute AR grade, J.T. Baker) in Coplin jars
Zeiss Standard phase-contrast microscope with 16× Ph2 objective and condenser ring (or equivalent)

1. Remove slide from methanol and polish with lint-free tissue (a clean slide is essential). Dip slide once in methanol and then several times in deionized water until the methanol is gone and a thin uniform film of water covers the slide.

2. Holding the frosted end between the thumb and finger, position the slide with the one long edge parallel to the bench top, and blot the lower long edge on a paper towel to draw off excess water. Keeping the lower long edge in contact with the paper towel, lower the opposite edge until the slide forms a 30° angle with the bench top, with the film of water facing up (Fig. 4.1.2).

3. From a Pasteur pipet held in a horizontal position 1 to 2 in. above the slide, place three evenly spaced drops of cell suspension onto the slide, moving successively toward the *frosted* end. Apply so that drops strike the tilted slide one-third of its width from the elevated long edge (Fig. 4.1.2) and burst on the water film, spreading out evenly as they strike. If discrete areas of cells are observed at the drop sites, surrounded by areas with few cells, hold the slide at a lower angle (i.e., <30°) when the drops are applied.

4. Blot off excess fixative and tilt the slide at a 30° angle as in step 2. Flood with fresh fixative, dropwise, using a Pasteur pipet. Start at the elevated corner of the nonfrosted end and move toward the frosted end, placing drops on the upper edge of the slide, and creating a uniform front of fixative.

5. Blot the lower long edge again and wipe off the back of the slide. Turn so that the nonfrosted end is elevated 30° with respect to the frosted end, with the cell side facing up, and air dry

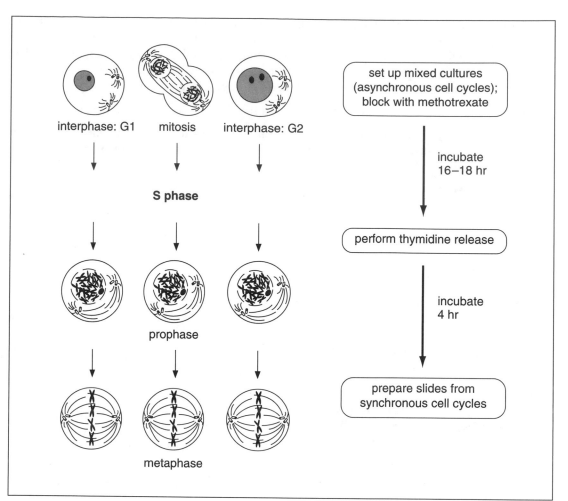

Figure 4.1.1 Cell culture synchronization. Two- or three-day asynchronous lymphocyte cultures are blocked by addition of methotrexate, which inhibits thymidine synthesis. Depletion of the thymidine pool prevents the cells from completing replication, and cells accumulate in the S phase of the cell cycle. Subsequent addition of thymidine releases a synchronized wave of cells to complete replication and proceed through G2 and into mitosis.

4. *Optional:* For longer chromosomes and more mitotic cells, synchronize cells in the following manner (Fig. 4.1.1). The day before harvest, add 0.05 ml of 10 μM methotrexate (10^{-7} M final) to block DNA replication. Incubate 16 to 18 hr (but not longer than 18 hr). On the following day, add 0.05 ml of 1 mM thymidine (10^{-5} M final) to release the methotrexate block. Incubate ~4 hr (duration is critical).

5. Within 3 to 4 days of culture (step 3) or immediately after synchronization (step 4), initiate harvest by adding 25 μl of 10 μg/ml Colcemid (0.05 μg/ml final). Incubate 30 min. Centrifuge 8 min at $180 \times g$, room temperature. Discard supernatant.

6. Add 6 ml of 75 mM KCl at room temperature and resuspend cells gently. Let stand 15 min at room temperature. To optimize conditions, adjusting the volume of KCl relative to the pellet volume will have more of an effect than increasing time.

7. Add 10 to 12 drops fixative with a Pasteur pipet and mix well. Centrifuge as in step 5.

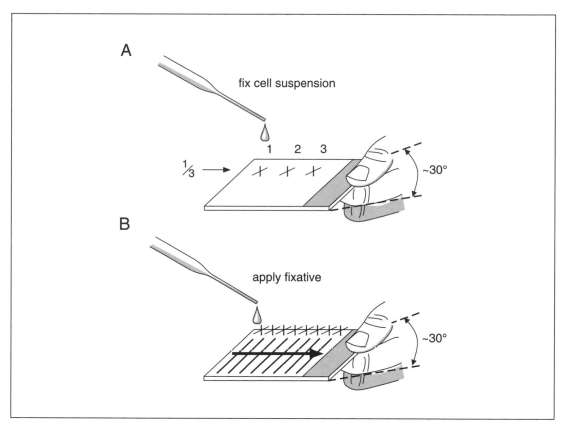

Figure 4.1.2 Chromosome slide preparation. (**A**) After blotting the long edge of the slide to obtain a thin uniform layer of water, the slide is tilted to ~30° and three separate drops of fixed cell suspension are applied starting away from and proceeding toward the frosted end. (**B**) After application of the cell suspension, the slide is flooded with fixative across the top edge, again proceeding toward the frosted end. It is important to avoid pooling of excess fluid on the surface of the slide and to obtain a thin even film of fixative to ensure uniform drying.

to induce spreading. Refer to Table 4.1.1 if ambient conditions are not conducive to good spreading (i.e., 20° to 22°C with a relative humidity of 50%).

6. Examine slides for good chromosome spreading and morphology by phase-contrast microscopy. Select slides that have a moderate and uniform density of chromosomes that are evenly dispersed and nonoverlapping. Chromosomes should be flat with sharp margins and uniform dark, crisp contrast in order to achieve good banding or hybridization signals.

7. Store slides in a clean dry container in the dark at room temperature (short term) or frozen at −70°C (long term). Use slides for FISH within several weeks and for autoradiographic in situ hybridization up to 1 year.

Reference: Barch et al., 1997

Contributors: Charles D. Bangs and Timothy A. Donlon

Table 4.1.1 Troubleshooting Guide to Chromosome Preparation

Problem	Possible cause	Solution
General		
No or few mitoses; cell debris	Bad specimen	Obtain fresh specimen
Hemolysis	Insufficient medium	Prepare fresh medium; start new cultures; reduce amount of blood per culture
	Cultures contaminated	Prepare fresh medium; obtain fresh specimen; improve sterile technique
Few mitoses	Poor gas/nutrient exchange	Tilt culture tubes slightly; mix cultures daily
No mitoses; many polynucleated cells; few T cell blasts	Omission or failure of mitogen	Add PHA at culture setup; use fresh PHA
No mitoses; few polynucleated cells; many T cell blasts	Omission or failure of mitotic arrest	Add Colcemid before harvest; use fresh Colcemid
	Hypotonic solution too hypoosmotic; mitoses lysed	Prepare hypotonic solution correctly
	Treatment too rough during harvest; mitoses ruptured	Resuspend and mix more gently
	Culture synchronization not released	Release cultures with thymidine on harvest day; make fresh thymidine; omit synchronization
No mitoses; cell debris present	Fixative reagents are tainted	Use fresh, high-grade methanol and acetic acid
Slide-making		
Poor spreading; excessive contrast; cytoplasm encapsulation	Slide-drying too fast	Slow drying; increase humidity by breathing gently on slides; chill slides; dry on wet paper towel
Poor spreading; cells too dense	Cell suspension too concentrated	Dilute cell suspension with additional fixative
Poor spreading	Hypotonic solution too saline; cells not swelling	Make hypotonic solution correctly
	Incomplete fixation	Refix or fix overnight at 4°C
	Fixative reagents tainted	Use new, high-grade methanol and acetic acid
Poor spreading; chromosomes "sticky"	Excessive ethidium bromide exposure	Make fresh ethidium bromide; reduce concentration; reduce exposure time; omit ethidium bromide
Over-spreading; broken metaphases; low contrast	Slide-drying too slow	Warm slide on arm or slide warmer; wave slide more gently
	Hypotonic solution too hypoosmotic; cells lysing	Make fresh hypotonic solution

Mitotic Chromosome Preparations from Mouse Cells for Karyotyping

NOTE: In all protocols, use aseptic technique until chromosomes are harvested.

BASIC PROTOCOL 1

CULTURE AND METAPHASE HARVEST OF MOUSE PERIPHERAL BLOOD

Lymphocytes from mouse peripheral blood can be stimulated to proliferate with any of the following mitogens: the T cell stimulators phytohemagglutinin (PHA) and concanavalin A (Con-A), and the B and T cell stimulators lipopolysaccharide (LPS) and pokeweed mitogen.

Materials *(see APPENDIX 1 for items with ✓)*
✓ Complete RPMI
✓ PHA solution
✓ 750 μg/ml LPS solution
✓ FBS
  Female (preferred) mice (Table 4.2.1), 7 to 10 weeks old (optimal; or 5 weeks to 2 years if needed)
✓ 500 USP U/ml sodium heparin solution
✓ 50 μg/ml colchicine solution
  0.075 M (0.56% w/v) KCl solution, fresh, 37°C
  Fixative solution: 3:1 (v/v) methanol/glacial acetic acid, fresh

  Sterile 16 × 125–mm polystyrene tissue culture tubes with screw cap (Fisher)
  Heparinized 75-μl micro-hematocrit capillary tubes (Fisher)
  Sterile 12 × 75–mm snap-cap tubes (Fisher)
  5-ml conical glass centrifuge tubes (e.g., Kimax, available from Fisher)
  Precleaned microscope slides (e.g., Fisher Colorfrost)
  Phase microscope with 10× or 20× objective for scanning slides; 40× if typing unstained chromosomes

1. For each mouse sampled, set up one 16 × 125–mm tissue culture tube containing:

   > 0.95 ml complete RPMI
   > 0.1 ml PHA solution
   > 0.1 ml of 750 μg/ml LPS solution
   > 0.15 ml FBS.

   Refrigerate tubes until inoculation.

2. Collect two heparinized 75-μl micro-hematocrit capillary tubes of blood per mouse from the retro-orbital sinus (Fig. 4.2.1; up to 300 μl blood can be withdrawn from an ∼30-g mouse). Inoculate into a 12 × 75–mm snap-cap tube containing 0.1 ml of 500 USP U/mL sodium heparin solution.

   NOTE: *If unskilled in this procedure, collect blood by the tail vein (see Alternate Protocol 1).*

3. Cap tubes tightly and swirl gently to mix blood and heparin. Keep at room temperature until inoculation in culture medium (not more than 2 hr).

Table 4.2.1 Strain Response to PHA Stimulation of Lymphocyte Proliferation[a]

Poor	Intermediate	Good
AU/SsJ	A/HeJ	129/J
BDP/J	A/J	AKR/J
C3H/HeJ	CBA/J	BALB/cJ
CBA/CaJ	LP/J	BUB/BnJ
CBA/H-T6J		C57BL/6J
CE/J		C57BL/10J
DBA/1J		C57BL/10Sn
DBA/2J		C57BL/KS/J
I/LnJ		C57BR/cdJ
LG/J		C57L/J
LT/ChRe		C58/J
MA/J		CE/J
NZB/BlNJ		HRS/J
P/J		PL/J
RF/J		PRO/Re
SEA/GnJ		RIII/J
SEC/ReJ		WB/Re
SJL/J		WC/RE
SM/J		WH/Re
ST/bJ		
SWR/J		

[a]Adapted from Heiniger et al. (1975). Printed from Akeson and Davisson (2000) with permission from Oxford University Press.

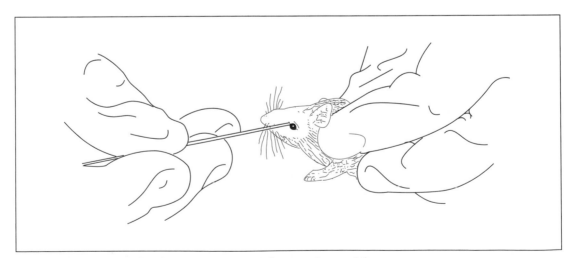

Figure 4.2.1 Blood collection from the orbital sinus or plexus of the mouse.

4. Inoculate 0.2 ml blood/heparin mixture into each prepared culture tube. Cap tubes tightly and swirl gently to mix cells and medium.

5. Incubate culture tubes 41 to 43 hr at an ~45° angle in a shaking water bath at 37°C (the angle maximizes gas exchange and prevents dense packing). Set the shaking speed so that the basket moves back and forth ~32 to 35 times per minute. Hand shake the culture tubes at least three times daily to resuspend the cells.

   If a large number of premitotic blast cells (large pale-staining nuclei) are seen in the chromosome slides, cells were not cultured long enough. If mostly late metaphase chromosomes are seen, the culture time was too long.

6. Add 0.15 ml of 50 μg/ml colchicine solution to each culture tube and incubate 15 to 20 min at 37°C. Transfer each culture to a 5-ml conical glass centrifuge tube and centrifuge 10 min at $400 \times g$, room temperature, in a clinical benchtop centrifuge.

7. Remove supernatant and gently add 2 to 3 ml of prewarmed 0.075 M KCl. Gently suspend the cells by pipetting with a Pasteur pipet. Incubate 15 min at 37°C. Centrifuge 10 min at $500 \times g$, room temperature. Remove supernatant without disturbing the pellet, being careful not to remove the buffy coat (white layer) on top of the pellet.

8. Gently add 3 to 4 ml fixative solution down the side of the tube. Pipet gently but rapidly to mix thoroughly. Cap tubes and allow to sit ≥30 min at room temperature. If necessary, tubes can be stored up to 2 hr at 4°C.

9. Centrifuge cells 10 min at $400 \times g$, room temperature, remove the fixative, and resuspend the cells in 2 to 3 ml fresh fixative. Repeat twice.

10. Centrifuge cells 8 min at $400 \times g$, room temperature, and resuspend in ~1/3 to 1/2 ml fresh fixative. If slides are not made immediately, do not perform this last wash until just prior to making slides. If slides are to be made the next day, wash cells at least twice before leaving overnight.

11. Immerse precleaned slides in fixative for ≥15 min prior to use. Wipe slides dry with a Kimwipe or other lint-free tissue.

12. Apply a small drop of cell suspension onto the surface of a slide with a Pasteur pipet and allow it to spread. If bubbles appear at the edges, use less sample. As soon as the drop begins to contract and Newton's rings are visible (rainbow colors around the edge of the drop), blow or otherwise pass air across the slide surface to achieve rapid drying, which is critical to obtaining well-spread metaphase chromosomes.

   Optional: To increase the spreading of the chromosomes, hold the slide in a flat position and drop a very small drop of fixative onto the cell suspension once it has started to dry (i.e., the "bomb" method).

13. Repeat step 12 until all of the sample is on the slide. (More drops from a dilute suspension are preferred over less drops of a concentrated suspension.) Monitor cell concentration and chromosome spreading and contrast by phase microscopy. Evaluate slide by cell concentration rather than number of metaphases.

14. Age slides 7 to 10 days at room temperature prior to G-banding (see Basic Protocol 2). Age slides 2 to 4 weeks at room temperature (authors use a desiccator) to prevent overdenaturation during FISH. For long-term storage (e.g., 4 weeks to 1 year), freeze slides at −20°C.

COLLECTING BLOOD BY TAIL VEIN METHOD

Materials (see APPENDIX 1 for items with ✓)

    Female mouse, 7 to 10 weeks old
    70% (v/v) ethanol
✓ 500 USP U/ml sodium heparin solution

    Jar
    Desk lamp
    Restraining device
    Razor blade
    Sterile 12 × 75–mm snap-cap tube (Fisher)

1. Prewarm a mouse in a large jar placed beneath a desk lamp for 1 to 2 min (i.e., until it rubs its nose and shows excessive activity). Do not overheat as the mouse can go into shock and die.

2. Place mouse in a restraining device so that the tail is free (Fig. 4.2.2). Wash the tail with 70% ethanol and wipe it twice with a clean Kimwipe. Gently cut across the vein on the side of the tail about one inch from the base of the tail with a razor blade wiped with 70% ethanol. Allow one drop of blood to drop off.

3. Collect 0.15 to 0.3 ml blood (5 to 10 drops) into a 12 × 75–mm snap-cap tube containing 0.1 ml of 500 USP U/ml sodium heparin solution, letting the blood run down the side of the tube into the heparin solution. Be careful not to touch the tail to the mouth of the tube.

4. Culture heparinized blood as described (see Basic Protocol 1).

ALTERNATE PROTOCOL 2

CHROMOSOME PREPARATION FROM FETAL LIVER

Chromosome spreads from fetal liver are often of better quality than from peripheral lympho-cytes, with longer chromosomes that band as well or better. The mitotic index in fetal liver is

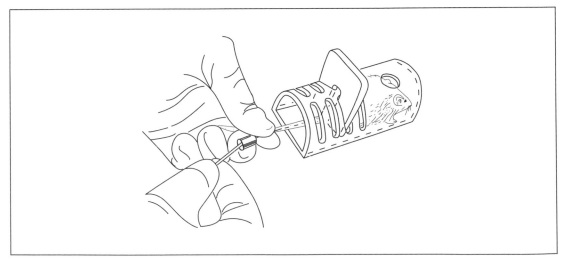

Figure 4.2.2   Blood collection from the tail vein of the mouse.

highest at day 14 and decreases until birth. After birth, mitotic activity resumes and suitable preparations can be made from livers of mice up to 5 days of age. For embryos younger than 14 days gestation, almost any tissue will give sufficient metaphases for karyotyping using this procedure.

Additional Materials (*also see Basic Protocol 1; see* APPENDIX 1 *for items with* ✓)
    Female mouse, pregnant with 14-day-old embryos (see Support Protocol)
✓ 0.5% (w/v) colchicine solution
✓ EDTA buffer with 0.025% (w/v) colchicine (optional)
    Culture medium, any kind, 37°C
    6:1 (v/v) methanol/glacial acetic acid (optional)

    Surgical tools for removing embryonic livers
    60-mm petri dish
    Small glass dish (~35-mm diameter)
    Dewecker iris scissors
    15-ml conical centrifuge tubes (e.g., Corning)

1a. *For injection:* Inject a pregnant female mouse with 0.1 ml of 0.5% colchicine solution intraperitoneally to disrupt spindle formation. Wait 10 to 15 min and euthanize by CO_2 asphyxiation or cervical dislocation. Remove embryos and euthanize by decapitation.

1b. *For noninjection:* To avoid injecting the pregnant female, collect cells as described in steps 2 to 4, and incubate the supernatant containing suspended cells in 1 ml EDTA buffer with 0.025% colchicine for 10 min at 37°C. Proceed to step 5.

2. Remove embryonic livers by placing forceps on both sides of the liver and pinching so that the liver pops up through the skin. Clean away any extraneous tissue and place the liver in a 60-mm petri dish with prewarmed culture medium.

3. Wash once to remove any blood and debris. Place cleaned liver in a small (~35-mm-diameter) glass dish with 1 pipetful medium from a 9-in. Pasteur pipet.

4. Mince tissue with Dewecker iris scissors. Pipet several times against the bottom of the dish with a broken-tip Pasteur pipet to further break up the tissue. Try not to create any bubbles, as cells may become trapped in them. Tilt the dish to allow chunks of tissue to settle, and transfer supernatant with suspended cells (leaving tissue pieces) to a 15-ml conical centrifuge tube.

5. Centrifuge 5 min at $400 \times g$, room temperature, in a clinical benchtop centrifuge. Remove supernatant and discard. Resuspend pellet by flicking bottom of tube with finger, add 1 pipetful prewarmed 0.075 M KCl solution, mix gently, and let stand 12 to 15 min at room temperature (longer times rupture cells).

6. Centrifuge as before and remove supernatant. Rinse twice with $^1/_2$ pipetful of fixative solution without disturbing the pellet. Add $^1/_2$ pipetful of fresh fixative and let stand 5 to 10 min or until the pellet is completely white.

7. Remove supernatant. Resuspend pellet by flicking bottom of tube with finger. Add $^1/_2$ pipetful of fresh fixative and pipet gently. Centrifuge as before. Repeat for a total of three washes.

8. Remove supernatant and add ~3 to 5 ml fixative. Let larger clumps settle and do not mix. If desired, use 6:1 methanol/acetic acid to improve spreading.

9. Prepare slides as described (see Basic Protocol 1, steps 11 to 15). More than one slide can be made from this suspension. Determine optimal amount of suspension per slide using phase-contrast microscopy to evaluate cell concentration (not number of metaphases).

SETTING UP TIMED MOUSE MATINGS

To set up timed matings, place a 6- to 8-week-old male mouse into a cage containing one to two female mice in the late afternoon. Identify females in estrus by the appearance of the vagina, which should be gaping and pink with pronounced striations. Early the next morning (by 9 a.m.) check the females for the presence of a copulation plug in the vagina (vaginal plug), which looks like a grain of rice. If needed, use a blunt metal probe to spread the vaginal lips apart to see the plug. Remove the male and remate in the late afternoon if no plug is evident. Designate the day the plug is found as day 0. Count days to determine the date the female should be sacrificed to provide 14-day-old embryos. By palpating the mother's abdomen, the embryos can be felt as small bumps by ∼12 to 14 days after impregnation.

ALTERNATE PROTOCOL 3

CHROMOSOME PREPARATION FROM BONE MARROW

This method does not require cell culturing, is easier for the novice or for the investigator who analyzes chromosomes infrequently, and works for mice of any age and gender. However, preparations do not G-band as consistently well as those from peripheral lymphocytes or fetal liver.

Additional Materials (also see Basic Protocol 1; see APPENDIX 1 *for items with* ✓ *)*
  ✓ 0.5% (w/v) colchicine solution
    Surgical tools
    1-ml disposable syringes with 23-G needles
    15-ml conical centrifuge tubes (Corning)

1. Inject mouse with 0.1 ml of 0.5% colchicine solution intraperitoneally. Wait 30 min (20 min in young mice, e.g., 3 weeks old, but never longer than 1 hr).

2. Sacrifice mouse by CO_2 asphyxiation or cervical dislocation, and remove femur(s) and tibia(s).

3. Cut off just enough of the bone heads to insert a 23-G needle into the marrow cavity.

4. Flush out cells into a 15-ml conical centrifuge tube using a 1-ml syringe filled with 0.075 M KC1 solution.

5. Incubate 15 min at room temperature (37°C if room is cool).

6. Centrifuge 5 min at 400 × g, room temperature, in a clinical benchtop centrifuge. Remove supernatant and add 0.5 ml fixative solution without disturbing the pellet. Remove fixative after 3 to 4 sec and add 2 ml fresh fixative without disturbing the pellet.

7. Allow tubes to sit 30 min at room temperature. Refrigerate if cells are to be left in fixative >30 min.

8. Centrifuge 5 min at 400 × g, room temperature. Remove fixative and resuspend in fresh fixative. Repeat for a total of three washes, using just enough fixative in the last addition to make a thin cell suspension.

9. Prepare and evaluate slides as described (see Basic Protocol 1, steps 11 to 15). More than one slide can be made from this suspension. Determine optimal amount of suspension per slide using phase-contrast microscopy to evaluate cell concentration (not number of metaphases).

CHROMOSOME PREPARATION FROM SOLID ADULT TISSUES OR TUMORS

Metaphase chromosomes can also be obtained from any adult mouse tissue that is mitotically active (e.g., spleen or tumors). Metaphase yields are usually lower than with bone marrow or fetal liver, and banding is not as consistently good as with peripheral lymphocyte cultures or fetal liver.

Additional Materials (also see Basic Protocol 1)

> Culture medium, any kind, 37°C
> Surgical tools for removing tumor or tissue
> 47-mm petri dish
> Small glass dish (~35-mm diameter)
> Dewecker iris scissors
> 15-ml conical centrifuge tubes (e.g., Corning)

1. Sacrifice mouse by CO_2 asphyxiation or cervical dislocation, and remove solid tumor(s) or tissue of interest.

2. Clean away any extraneous tissue and place the tumor in a 47-mm petri dish with pre-warmed culture medium.

3. Wash once to remove any blood and debris. Place cleaned tumor into a small (~35-mm-diameter) glass dish with 1 pipetful of medium from a 9-in. Pasteur pipet.

4. Mince tissue with Dewecker iris scissors. Pipet several times against the bottom of the dish with a broken-tip Pasteur pipet to break up the tissue more. Try not to create any bubbles as the cells may become trapped in them.

5. Tilt the dish to allow the chunks of tissue to settle. Transfer supernatant with suspended cells (leaving tissue pieces) to a 15-ml conical centrifuge tube.

6. Centrifuge 5 min at $400 \times g$, room temperature, in a clinical benchtop centrifuge.

7. Remove supernatant and resuspend cells in 2 ml medium.

8. Add 0.2 ml of 50 μg/ml colchicine solution and incubate 30 to 60 min at 37°C.

9. Centrifuge as before and remove supernatant. Add 2 to 3 ml prewarmed 0.075 M KCl solution, gently resuspend, and incubate 15 to 20 min at 37°C.

10. Centrifuge as before and remove supernatant without disturbing the pellet. Gently add 3 to 4 ml fixative solution down the side of the tube and pipet gently but rapidly to prevent cell clumping.

11. Cap the tubes and allow to sit ≥30 min at room temperature. Refrigerate if cells will be held ≥30 min.

12. Centrifuge cells 5 min at $400 \times g$, room temperature, remove the fixative, and resuspend the cells in fresh fixative. Repeat for a total of three washes.

13. Remove supernatant and add fixative to dilute the cell suspension for making slides. Let larger clumps settle before making slides. Do not remix cell suspension.

14. Prepare and evaluate slides as described (see Basic Protocol 1, steps 11 to 15). More than one slide can be made from this suspension. Determine optimal amount of suspension per slide using phase-contrast microscopy to evaluate cell concentration (not number of metaphases).

GIEMSA BANDING (G-BANDING)

Materials *(see APPENDIX 1 for items with ✓)*

  Slides of metaphase chromosomes (see Basic Protocol 1 or Alternate Protocols 1 to 4)
✓ 2× SSC, 60° to 62°C
  0.9% (w/v) NaCl
✓ Trypsin/Giemsa solution
  Phosphate buffer: 1:1 (v/v) Gurr's phosphate buffer, pH 6.8 (Bio/medical Specialties), in deionized water

  Coplin jars
  Bright-field microscope with 20× objective, 100× oil-immersion objective, and a green filter

1. Age metaphase chromosome slides 7 to 10 days at room temperature.

2. Incubate slides 1.5 hr in Coplin jars (no more than seven per jar) containing 2× SSC, 60° to 62°C.

3. Transfer all slides to 0.9% NaCl at room temperature. Rinse each slide thoroughly in fresh NaCl and drain.

4. Stain slides 5 to 7 min in trypsin/Giemsa solution. Remove the metallic film that forms on the stain surface with a cotton ball before placing the slides in the Coplin jar, or float the film off with running water before removing the slides.

5. Transfer all slides to fresh phosphate buffer.

6. Rinse slides individually and thoroughly in two changes of phosphate buffer. Shake off excess liquid and blow dry with an air jet.

7. View banded chromosomes with a bright-field microscope and 20× objective. Apply immersion oil directly to slide and perform chromosome analysis with a 100× oil-immersion objective. Use a green filter for better band definition. To keep slide, wash off immersion oil with clean acetone.

   The characteristic banding patterns (landmarks) at different stages of chromosome condensation are presented elsewhere (Cowell, 1984; Sawyer et al., 1987; http://www.informatics.jax.org).

References: Cowell, 1984; Evans, 1996; Hogan et al., 1994; Lee et al., 1990

Contributors: Ellen C. Akeson and Muriel T. Davisson

UNIT 4.3

Chromosome Banding Techniques

All of the techniques presented here can be applied to both metaphase and prometaphase (extended, high-resolution) chromosome preparations from any tissue source (*UNIT 4.1* and Chapter 8). Choice of staining technique will vary with the application and available equipment (bright-field versus fluorescence microscopy; Table 4.3.1). Additional techniques for chromosome identification include G-11 banding (*UNIT 3.1*) and fluorescent in situ hybridization with chromosome-specific probes (*UNIT 4.4*). Karyotype construction and methods of analyzing chromosomes are presented in *APPENDIX 3K* and in units dealing specifically with the cytogenetic analysis of different tissues (e.g., *UNIT 8.2*).

Table 4.3.1 Chromosome Banding Techniques and their Applications[a]

Banding type	Stain (technique)	Microscope used	Uses and advantages
Q-banding	Quinacrine (QTQ)	F	ID of all chromosomes and bands; reveals polymorphisms on chromosomes 3, 4, 13, 14, 15, 21, 22, and Y; easily destained for sequential staining
G-banding	Giemsa (GTG)	B	ID of all chromosomes and bands; permanent stain; simple photography
	Wright	B	ID of all chromosomes and bands; permanent stain; simple photography
R-banding	Giemsa (RHG)	B	ID of all chromosomes and bands; visualization of ends of chromosomes and small positive R-bands
	CH3/DA	F	ID of all chromosomes and bands; visualization of ends of chromosomes and small positive R-bands
Replication banding[b]	Hoechst	F	ID of all chromosomes and bands, and of inactive, late-replicating X chromosome
	Hoechst and Giemsa	B	ID of all chromosomes and bands, and of inactive, late-replicating X chromosome
DA−DAPI staining	Distamycin A/DAPI and distamycin A/Hoechst	F	ID of centromeric heterochromatin regions of chromosomes 1, 9, 15, 16, and Y; useful in evaluation of chromosome 15–derived markers

[a] Abbreviations: B, bright-field; CH3/DA, chromomycin A/distamycin A; DA, distamycin A; DAPI, $4'$,6-diamidino-2 phenylindole; F, fluorescent; ID, identification.

[b] Depending on timing of BrdU incorporation, a G- or Q-type banding pattern can be obtained with highlighting of late-replicating X chromosome.

CAUTION: The following hazardous agents are used in this unit: BrdU, DAPI, distamycin A, H_2O_2, human lymphocytes and fibroblasts, Hoechst 33258, quinacrine, and xylene.

NOTE: All incubations are performed in a humidified 37°C, 5% CO_2 incubator unless otherwise specified.

BASIC PROTOCOL 1

QUINACRINE BANDING (Q-BANDING)

Materials (*see* APPENDIX 1 *for items with* ✓)

Air-dried slides of metaphase chromosomes (UNIT 4.1)
✓ Quinacrine staining solution
✓ McIlvaine buffer, pH 5.6 (pH between 5.4 and 5.6 is critical)
Immersion oil, low fluorescence

Coplin jars
Coverslips, no. 0 or 1
Fluorescence microscope with 430- to 460-nm excitation filters, <510-nm barrier filters, and camera with film

1. Place air-dried slide of metaphase chromosomes in a Coplin jar containing quinacrine staining solution, 5 min at room temperature.

2. Rinse slide by dipping several times into a Coplin jar filled with water. Repeat with fresh water. Air dry slide. If desired, store dry up to weeks protected from light.

3. Mount slide using McIlvaine buffer, pH 5.6. Add a no. 0 or 1 coverslip and gently squeeze excess buffer from under coverslip by blotting gently with a paper towel.

4. View and photograph using a fluorescence microscope with appropriate filters. Use an iris diaphragm or funnel stop to reduce fluorescence glare if available. If using Kodak Technical Pan film at ASA 200, typically use exposures in the range of 15 to 30 sec with a 100× objective. If the camera has an automatic exposure system, set it for spot metering, place a chromosome under the spot and adjust the ASA setting as needed. Periodically run a test film (e.g., when lamp is changed).

5. After viewing, remove coverslip and store the slide dry, protected from light. Remount before examining again.

BASIC PROTOCOL 2

GIEMSA BANDING (G-BANDING) BY THE GTG TECHNIQUE

G-banding can be followed by destaining (see Support Protocol 6) and Q-banding (see Basic Protocol 1).

Materials (see APPENDIX 1 for items with ✓)
✓ HBSS
✓ Trypsin solution
  70% and 90% (v/v) ethanol
  2% (v/v) Giemsa staining solution (Bio/medical Specialties or Fisher; dilute in water immediately before use)
  Aged slides of metaphase chromosomes (see Support Protocols 1 and 2)
  Xylene or Hemo-De

  Coplin jars
  Bright-field microscope and green interference filter
  Fine-grained film (e.g., Kodak Technical Pan)

NOTE: Because the stain is difficult to remove from skin, it is advisable to wear gloves when working with Giemsa.

1. Prepare a series of Coplin jars containing the following solutions at room temperature:

   HBSS
   trypsin solution
   HBSS
   70% ethanol
   90% ethanol
   2% Giemsa staining solution
   water.

2. Place aged slide of metaphase chromosomes in HBSS ~10 sec.

3. Transfer slide to trypsin solution and incubate 30 sec for bone marrow, 60 sec for lymphocytes or amniocytes, and 90 sec for other cells from long-term tissue culture. If needed, determine optimal trypsinization time using three to five identical slides. Avoid insufficient

trypsinization (evenly stained slides with no bands) and over-trypsinization (pale "puffy" chromosomes with staining around the outside of the chromosome).

4. Rinse jars successively in HBSS, 70% ethanol, and 90% ethanol, dipping three to four times in each jar. Air dry. If necessary store up to several hours.

5. Place slide in Giemsa staining solution for 4 min or as empirically determined for optimal staining.

6. Place slide in water ~30 sec. Air dry. Store months or years.

7. Apply immersion oil directly to the slide and view with a bright-field microscope. Use a green interference filter for black-and-white photography. Use a fine-grained film with ASA setting of 50 to 100 and short exposure time. To store slides, wash off oil with fresh xylene or Hemo-De.

ALTERNATE PROTOCOL 1

G-BANDING WITH WRIGHT STAIN

This type of G-banding produces a sharp, crisp G-banding pattern and can be used if the Giemsa method proves inadequate. It can also be followed by destaining (see Support Protocol 6) and Q-banding (see Basic Protocol 1).

Additional Materials (*also see Basic Protocol 2; see* APPENDIX 1 *for items with* ✓)
✓ Bacto Trypsin solution
✓ Sorensen phosphate buffer, pH 6.8
✓ Wright stain (prepare fresh)

1. Place aged slide of metaphase chromosomes in Bacto Trypsin solution 20 to 30 sec, then rinse by agitating briefly in Sorensen phosphate buffer, pH 6.8.

2. Stain slide in horizontal position 40 to 60 sec with freshly prepared Wright stain (solution will feel slightly warm from exothermic reaction between water and ethanol).

3. Rinse by agitating briefly in water and air dry. Store up to years.

4. View and photograph as described (see Basic Protocol 2, step 7).

SUPPORT PROTOCOL 1

AGING SLIDES WITH HEAT

Time, heat, and drying cause an alteration in chromosomal material (probably protein denaturation) that affects banding. Underaged slides result in fuzzy banding. Overaged slides do not band. Techniques for manipulating the aging of chromosome slides vary widely. Optimal aging conditions may vary with cell type or tissue source.

Incubate air-dried slide of metaphase chromosomes 2 days at 55°C or 20 min at 90° to 95°C (using dry oven or slide warmer). If it is necessary to reduce time of incubation, increase temperature. If incubation time will be longer than 2 days (e.g., over a weekend), decrease temperature. Optimal times and temperatures must be established empirically in each laboratory.

AGING SLIDES WITH HYDROGEN PEROXIDE

When immediate banding of slides is required, the effects of aging can be obtained with hydrogen peroxide treatment.

Materials

Air-dried slides of metaphase chromosomes (prepare fresh; *UNIT 4.1*)
15% (v/v) H_2O_2 (dilute 30% H_2O_2 1:1 with water immediately before use)

Coplin jars
50°C hot plate or slide warmer

1. Flood freshly prepared slide with 15% H_2O_2. Leave peroxide in contact with slide 7 min.

2. Place slide in Coplin jar filled with water and rinse under running tap water 2 min.

3. Place slide on 50°C hot plate or slide warmer 1 hr to overnight. Cool to room temperature and proceed with banding.

BASIC PROTOCOL 3

B-PULSE REPLICATION BANDING FOR LYMPHOCYTES

This protocol will produce an R-type banding pattern in lymphocyte chromosomes. The reduced staining of the inactive X chromosome allows differentiation of the two X chromosomes in females. Cells are exposed to BrdU during late S phase.

Materials

Phytohemagglutinin (PHA)-treated lymphocyte cultures (*UNIT 4.1*)
10 mM BrdU (30.7 mg in 10 ml H_2O; store in 1-ml aliquots $\leq$6 months at -20°C)
10 mM 2′-deoxycytidine (dC; 28 mg in 10 ml H_2O; store in 1-ml aliquots $\leq$6 months at -20°C)
10 μg/ml Colcemid (e.g., Life Technologies; store at 4°C)
Tissue culture vessel of desired size

NOTE: All live cells treated with BrdU should be protected from 313-nm light, which is emitted by fluorescent light sources. All reagents and equipment coming into contact with live cells must be sterile.

1. Set up PHA-treated lymphocyte cultures in a tissue culture vessel of desired size and incubate 2 days.

2. At 9 a.m. on the third day, add 10 mM BrdU and 10 mM dC to each tissue culture vessel to 100 μM final (each). Incubate ~6 hr.

3. Add 10 μg/ml Colcemid to 0.05 μg/ml and incubate 30 min at 37°C.

4. Harvest metaphase cells and prepare chromosome slides.

5. Proceed with visualization of BrdU replication banding (see Support Protocols 3 to 5).

T-PULSE REPLICATION BANDING FOR LYMPHOCYTES

This method produces G- or Q-type banding. The prominent staining of the inactive X chromosome allows differentiation of the two X chromosomes in females. Cells are exposed to BrdU during early S phase.

Additional Materials (*also see Basic Protocol 3*)

Tissue culture medium with and without 10 μM thymidine (from 1 mM thymidine stock; APPENDIX 1), 37°C

1. Set up PHA-treated lymphocyte cultures and incubate.

2. Between 8:00 and 10:00 p.m. on the second day of incubation, add 10 mM BrdU and 10 mM dC to each vessel to 100 μM final (each). Incubate 10 to 12 hr.

3. At 8:00 a.m. on the third day of incubation, transfer cells and medium to a sterile centrifuge tube. Centrifuge 8 min at 180 × g, room temperature. Discard supernatant.

4. Add 2 ml fresh tissue culture medium without BrdU and dC, 37°C. Resuspend pellet by flicking tube.

5. Centrifuge 8 min at 180 × g. Remove and discard supernatant.

6. Add 5 ml fresh tissue culture medium containing 10 μM thymidine, 37°C. Incubate 5 hr to allow late-replicating DNA to incorporate thymidine.

7. Add 10 μg/ml Colcemid to 0.05 μg/ml and incubate 30 min.

8. Harvest metaphase cells and prepare chromosome slides.

9. Proceed with visualization of BrdU replication banding (see Support Protocols 3 to 5).

B-PULSE REPLICATION BANDING FOR FIBROBLASTS

1. Add 10 mM BrdU and 10 mM dC to each vessel of actively growing fibroblasts (~24 hr after passage) to 100 μM final (each). Incubate 10 to 12 hr. If expected banding patterns are not achieved, determine optimal incubation time with a range of pulse times (e.g., 6 to 13 hr at 1 hr increments).

2. Add 10 μg/ml Colcemid to 0.05 μg/ml and incubate 30 min.

3. Harvest metaphase cells using techniques for cultures of attached cells, and prepare chromosome slides.

4. Proceed with visualization of BrdU replication banding (see Support Protocols 3 to 5).

T-PULSE REPLICATION BANDING FOR FIBROBLASTS

Additional Materials (*also see Basic Protocol 3*)

Tissue culture medium with and without 10 μM thymidine (from 1 mM thymidine stock; APPENDIX 1), 37°C

4

1. At ~5:00 p.m on starting day of procedure, add 10 mM BrdU and 10 mM dC to a flask containing actively growing fibroblasts (~8 to 10 hr after passage) to 100 μM final (each). Incubate ~16 hr. If recommended incubation times with BrdU and dC do not produce expected banding patterns, determine optimal incubation time empirically (see Alternate Protocol 3, step 1).

2. Remove BrdU/dC-containing medium and rinse cells with fresh 37°C tissue culture medium without BrdU and dC. Add fresh medium containing 10 μM thymidine, 37°C. Incubate ~6 hr.

3. Add 10 μg/ml Colcemid to 0.05 μg/ml final and incubate 30 min.

4. Harvest metaphase cells and prepare chromosome slides.

5. Proceed with visualization of BrdU replication banding (see Support Protocols 3 to 5).

SUPPORT PROTOCOL 3

VISUALIZATION OF BrdU REPLICATION BANDING BY FLUORESCENT DYE

Materials (*see APPENDIX 1 for items with* ✓)
    95%, 70%, and 30% (v/v) ethanol
✓ PBS
✓ Hoechst 33258 staining solution A
    Metaphase slides of BrdU-substituted chromosomes (see Basic Protocol 3 or Alternate Protocols 2 to 4)
✓ McIlvaine buffer, pH 7.5
    Fluorescence microscope with appropriate filters, 100× Zeiss Neofluar or Olympus S-planapo objective, and camera with film (e.g., Kodak Technical Pan)

1. Prepare a series of Coplin jars containing the following solutions and then incubate each metaphase slide of BrdU-substituted chromosomes as shown:

> 2 min each in 95%, 70%, and 30% ethanol
> 2 min in water
> 5 min in PBS
> 10 min in Hoechst 33258 staining solution A
> 2 min in PBS
> two rapid rinses in water (separate jars).

Keep stained slide protected from light at all times. Dry the slide if it will not be used immediately.

2. Mount in McIlvaine buffer, pH 7.5, add a coverslip, and view with a fluorescence microscope equipped with appropriate filter set (Hoechst absorbs at 356 nm and emits at 465 nm).

3. Photograph immediately using exposures in the range of 15 to 30 sec with a 100× objective and Kodak Technical Pan film at ASA 200. If necessary, adjust the ASA or use a faster film (e.g., Kodak Tri-X). If the camera has an automatic exposure system, set it for spot metering and place a chromosome in the spot.

4. Store the slide protected from light for future reexamination by removing the coverslip and air drying the slide. Use McIlvaine buffer to mount the slide and add a coverslip before examining again.

VISUALIZATION OF BrdU REPLICATION BANDING BY LIGHT TREATMENT AND GIEMSA STAINING

Materials *(see APPENDIX 1 for items with ✓)*

   Metaphase slides of BrdU-substituted chromosomes (see Basic Protocol 3 or Alternate
      Protocols 2 to 4)
✓ PBS
✓ Hoechst 33258 staining solutions A and B
✓ 2× SSC, 60° or 65°C
   4% (v/v) Giemsa staining solution (Bio/medical Specialties or Fisher; dilute in water
      immediately before use)

   100 × 100 × 15–mm square plastic dishes with covers
   24 × 60–mm coverslips
   20-W fluorescent light
   Bright-field microscope with green interference filter and camera with fine-grained film
      (e.g., Kodak Technical Pan)

1. Presoak metaphase slide of BrdU-substituted chromosomes 5 min in PBS and stain
   10 min in Hoechst 33258 staining solution A.

2. Drain slide and place in a 100 × 100 × 15–mm square dish with two slides or glass rods
   at the bottom as supports. Cover slide with 3 to 4 drops Hoechst staining solution B. Place
   a 24 × 60–mm coverslip on the slide.

3. Cover dish and expose to a 20-W fluorescent light 5 to 7.5 cm from the surface of the
   dish a minimum of 6 hr (or overnight). Put thin layer of PBS at bottom of dish to prevent
   drying.

4. Rinse coverslip off slide under warm running tap water (do not try to pry it off).

5. Place slide in prewarmed 2× SSC for 15 min. Rinse well in running water.

6. Stain 7 min in 4% Giemsa staining solution.

7. Rinse in water and air dry.

8. Examine under bright-field microscope. Photograph using a green interference filter for
   black-and-white photography. Use a fine-grained film with an ASA setting of 50 to 100
   and a short exposure time.

VISUALIZATION OF BrdU REPLICATION BANDING BY HEAT TREATMENT AND GIEMSA STAINING

This technique is simpler than visualization by light treatment and Giemsa staining, but does
not work quite as consistently.

Materials *(see APPENDIX 1 for items with ✓)*

   Metaphase slides of BrdU-substituted chromosomes (see Basic Protocol 3 or Alternate
      Protocols 2 to 4)
✓ 1.0 M sodium phosphate buffer, pH 8.0
   4% (v/v) Giemsa staining solution (Bio/medical Specialties or Fisher; dilute in water
      immediately before use)

87° to 90°C water bath
Bright-field microscope with green interference filter and camera with fine-grained film
  (e.g., Kodak Technical Pan)

1. Heat slide 10 min at 87° to 92°C in 1.0 M sodium phosphate buffer, pH 8.0. Rinse briefly by agitating slide in water.

2. Stain 5 to 7 min in 4% Giemsa staining solution.

3. Rinse in water and air dry.

4. Examine under a bright-field microscope. Photograph using a green interference filter for black-and-white photography. Use a fine-grained film with an ASA setting of 50 to 100 and a short exposure time.

BASIC PROTOCOL 4

DISTAMYCIN-DAPI STAINING

DAPI-banding can be followed by destaining (see Support Protocol 6) and Q-banding (see Basic Protocol 1).

Materials (*see* APPENDIX 1 *for items with* ✓)
  Air-dried slides of metaphase chromosomes (*UNIT 4.1*)
✓ McIlvaine buffer, pH 7.0 and 7.5
✓ Distamycin A staining solution
✓ DAPI staining solution
  Immersion oil, low fluorescence

  Humidified chamber (e.g., petri dish with moist paper towel)
  Coverslips, no. 0 or no. 1
  Fluorescence microscope with appropriate filters and objectives (see Support Protocol 3)

1. Immerse an air-dried slide of metaphase chromosomes 5 min in McIlvaine buffer, pH 7.0, room temperature. Drain.

2. Place slide in a humidified chamber. Add 3 to 6 drops (100 to 200 μl) of distamycin A staining solution and spread by applying coverslip. Let stand 10 min at room temperature. If necessary, reduce distamycin A incubation time to enhance visualization of DAPI banding.

3. Rinse briefly (~10 sec) in McIlvaine buffer, pH 7.0. Expose to DAPI staining solution 10 min at room temperature, then rinse again in McIlvaine buffer, pH 7.0.

4. Mount slides in McIlvaine buffer, pH 7.5, using no. 0 or no. 1 coverslip.

5. View with a fluorescence microscope using low-fluorescence immersion oil and Hoechst filters (DAPI absorbs at 355 nm and emits at ~450 nm). Photograph as in Support Protocol 3 (step 3). Note that using a faster film may cause some loss of definition in the banding.

ALTERNATE PROTOCOL 5

HOECHST 33258-DISTAMYCIN STAINING

Additional Materials (*also see Basic Protocol 4; see* APPENDIX 1 *for items with* ✓)
✓ PBS
✓ Hoechst 33258 staining solution A
  Mounting buffer: 1:1 (v/v) glycerol/McIlvaine buffer, pH 7.5

Coplin jars
Spray bottle (optional)

1. Immerse air-dried slide 5 min in PBS at room temperature.

2. Immerse slide 12 min at room temperature in a Coplin jar containing Hoechst 33258 staining solution A.

3. Rinse slide 4 min in PBS, then 1 min with a stream of water. Air dry. If necessary, store in a covered box.

4. Put 3 drops (100 μl) distamycin A staining solution on slide, add coverslip, and let stand 5 min. Rinse in McIlvaine buffer, pH 7.5, using a spray bottle if desired.

5. Mount in mounting buffer and let sit 2 min. View and photograph as described (see Support Protocol 3, step 2 and 3). Note using a faster film may cause some loss of definition in the banding.

SUPPORT PROTOCOL 6

DESTAINING

Materials

Stained slides, unmounted or nonpermanently mounted
95% (v/v) ethanol
Xylene or Hemo-De (Fisher)
1:1 (v/v) xylene/ethanol (if xylene is used)
Fixative: 3:1 (v/v) methanol/acetic acid
Methanol, reagent grade

1a. *If slide is mounted in aqueous buffer:* Carefully lift and remove coverslip (use running water if buffer has dried out). Proceed to step 2a or 2b if immersion oil has been used, or to step 3 if it has not.

1b. *If slide is mounted in glycerol:* Soak 2 hr in water to remove coverslip, then soak 2 min in ethanol and allow to dry. Proceed to step 2a or 2b if immersion oil has been used, or to step 3 if it has not.

1c. *If slide is unmounted:* Proceed to step 2a or 2b if covered with immersion oil or to step 3 if not.

2a. *If xylene is used:* Remove oil by soaking slide 10 min in xylene followed by 1 min in 1:1 xylene/ethanol and 1 min in 95% ethanol.

2b. *If Hemo-De is used:* Soak slide 10 min in Hemo-De.

 Hemo-De is less hazardous than xylene.

3. Soak slide 10 min in fixative to remove stain.

4. Dip slide in methanol 1 min. Air dry.

References: Latt, 1976; Sahar and Latt, 1980; Schweitzer, 1981; Sumner, 1990; Sumner and Evans, 1973

Contributors: Rhona R. Schreck and Christine M. Distèche

In Situ Hybridization to Metaphase Chromosomes and Interphase Nuclei

Slides of metaphase chromosomes can come from cell suspensions (e.g., *UNIT 4.1*) or from adherent cells such as fibroblasts (*UNIT 8.3*) and amniocytes (*UNIT 8.2*). Nonisotopic probes are generally detected by fluorescence or enzymatic methods (Figs. 4.4.1 and 4.4.2).

NOTE: Many of the steps in these procedures must be performed with minimal exposure to ambient light.

BASIC PROTOCOL

FLUORESCENCE IN SITU HYBRIDIZATION TO METAPHASE CHROMOSOMES

Materials (*see APPENDIX 1 for items with ✓*)

Slide or coverslip containing metaphase human chromosomes (e.g., *UNIT 4.1*)
Denaturation solution: 70% (v/v) deionized formamide/2× SSC (pH 7.0), 70°C (store up to 2 weeks at 4°C if reusing with repetitive-element probes, otherwise prepare fresh)
70% (ice-cold), 80%, 95%, and 100% (v/v) ethanol (prepare 70% and 80% from 95% ethanol)
Cot-1 DNA (Life Technologies)
Nonisotopically labeled DNA probe: single-copy, chromosome-paint, *or* repetitive-element probe (Oncor or Imagenetics; see Support Protocol 3)
✓ Deionized formamide (also American Bioanalytical)
✓ Master hybridization mix

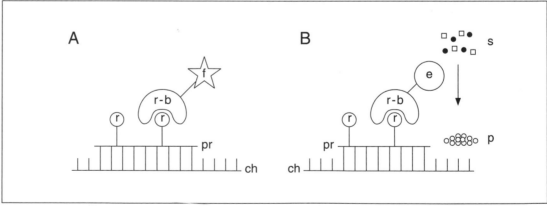

Figure 4.4.1 Detection of hybridized probe. (**A**) Fluorescence in situ hybridization (FISH). (**B**) Enzymatic detection. Abbreviations: ch, chromosome; e, enzyme (e.g., AP or HRP); f, fluorochrome (e.g., fluorescein, rhodamine, Texas Red); p, colored precipitate product; pr, probe labeled with reporter molecule; r, reporter molecule (e.g., biotin, digoxigenin); r-b, reporter-binding molecule (e.g., avidin, streptavidin, digoxigenin antibody); s, soluble substrate. Arrow indicates reaction catalyzed by enzyme.

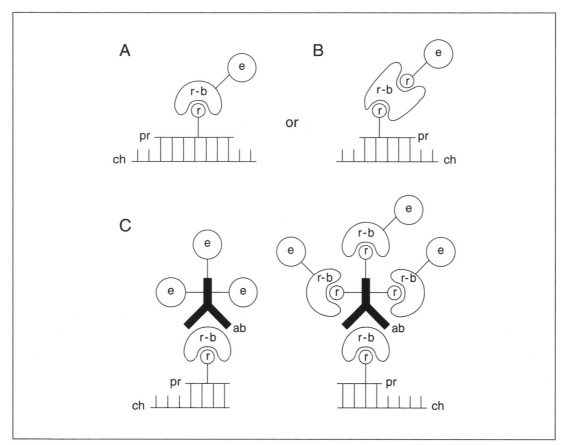

Figure 4.4.2 Enzyme-mediated detection of reporter molecules. (**A**) For direct detection, enzyme is conjugated to reporter-binding molecule. (**B**) For a two-step indirect procedure, reporter-binding molecule is applied first, followed by incubation with reporter-conjugated enzyme. (**C**) For signal amplification, incubation with reporter-binding molecule is followed by incubation with an antibody to reporter-binding molecule. The antibody may be conjugated with enzyme or reporter molecule. In the latter case, there is an incubation with enzyme-conjugated reporter-binding molecule before addition of substrate. Fluorochrome or isotope can be substituted for enzyme molecule. Abbreviations: ab, antibody to reporter-binding molecule conjugated with enzyme or reporter molecule; ch, chromosome; e, enzyme (e.g., AP or HRP); pr, probe labeled with reporter molecule; r, reporter molecule (e.g., biotin, digoxigenin); r-b, reporter-binding molecule (e.g., avidin, streptavidin, anti-digoxigenin).

50% (v/v) formamide (not deionized)/2× SSC
✓ 2×, 1×, and 4× SSC, pH 7.0
✓ Biotin detection solution, digoxigenin detection solution, *or* biotin/digoxigenin detection solution
0.1% (v/v) Triton X-100/4× SSC
✓ DAPI *or* propidium iodide staining solution
✓ Appropriate antifade mounting medium

Coplin jars
72°, 39°, and 42°C water baths
Phase-contrast microscope
22-mm^2 coverslips
Rubber cement
Moist chamber (Fig. 4.4.3)

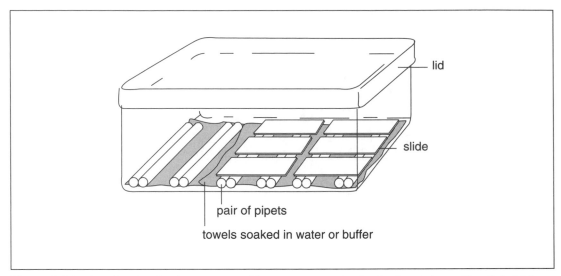

lid

slide

pair of pipets

towels soaked in water or buffer

Figure 4.4.3 A moist chamber is constructed from any conveniently sized container with a tight lid (e.g., Tupperware or equivalent). Slide supports are made by attaching pairs of 5- or 10-ml pipets to the bottom, spaced so they support the slide at both ends, or by using a stainless steel cake rack. The bottom is lined with absorbent paper soaked with water or (for longer incubations) a solution of identical osmolarity and formamide concentration as the hybridization solution. For use with light-sensitive reagents, the moist chamber is wrapped in aluminum foil, painted black, constructed from a black container, or otherwise shielded from ambient light.

Slide box with desiccant (Baxter Scientific)

Nail polish

Fluorescence microscope with epi-illumination and filter set(s) appropriate for fluorochrome(s) used, including dual-band-pass filter (fluorescein/Texas Red) or triple-band-pass filter (fluorescein/Texas Red/DAPI; Omega Optical or Chroma Technology)

Ektar-1000, Ektachrome-400, or Technical Pan 2415 film (Kodak), or CCD camera

1. Soak slide or coverslip containing metaphase chromosomes 2 min in 70°C denaturation solution (check with thermometer) in a Coplin jar in a 72°C water bath. Do not leave the solution in the water bath longer than 30 min.

2. Rinse slide 2 min in a Coplin jar containing ice-cold 70% ethanol to stop denaturation. Continue dehydration by incubating slide 2 min (each) in room temperature 80%, 95%, and 100% ethanol. Allow slide to air dry.

3. Using a phase-contrast microscope, check the denatured chromosome preparation, which should have evidence of internal structure, with no hollow or transparent chromosomes.

4. Lyophilize 20 to 150 ng nonisotopically labeled DNA probe (20 ng chromosome-specific repetitive-sequence probes, 50 to 100 ng cDNA and whole-chromosome paint probes, or 100 to 150 ng single-copy genomic probes, including YACs). If probe contains repetitive elements interspersed with unique sequences (i.e., single-copy genomic clones in various vectors, probes with interspersed repetitive and unique sequences, and whole-chromosome paint probes), suppress the repetitive elements by adding 10 μg human Cot-1 DNA prior to lyophilization.

   If total yeast DNA containing a YAC is utilized, determine the proportion of human-insert DNA to yeast DNA and increase the total amount of probe DNA to achieve the appropriate concentration of human DNA during hybridization.

5. Resuspend pellet in 10 μl deionized formamide. Heat 10 min at 72°C to denature DNA. If probe does not require suppression, place on ice. If probe requires suppression, place at 37°C and preanneal 30 min to several hours (use longer times for larger probes and those with a large amount of repetitive sequence).

6. Add 10 μl master hybridization mix to denatured probe. Mix well, spin briefly at high speed in a microcentrifuge, and place on denatured chromosome preparation. Cover with 22-mm^2 coverslip, remove any large air bubbles with gentle pressure, seal with rubber cement, and incubate overnight (~14 to 18 hr) in a moist chamber at 37°C. For centromeric and whole-chromosome paint probes, decrease time to as little as 4 hr.

7. During last 30 min of hybridization, warm 50 ml of 50% formamide/2× SSC wash solution and 50 ml of 2× SSC in Coplin jars in a 39°C water bath.

8. Remove slide from moist chamber. Peel off rubber cement, carefully remove coverslip, and wash hybridized slide 15 min each in 39°C 50% formamide/2× SSC, 39°C 2× SSC, and room temperature 1× SSC. For centromeric and whole-chromosome paint probes, increase temperature to 42°C and decrease salt concentration to 0.4× to 1× SSC.

9. Allow slide to equilibrate 5 min in 4× SSC at room temperature. Remove slide and drain excess buffer. Do not allow slide to dry at any point in the procedure.

10. Add 50 μl biotin detection solution, digoxigenin detection solution, or biotin/digoxigenin detection solution and cover with a 22-mm^2 square of Parafilm. Incubate 45 min in an aluminum foil–wrapped moist chamber (detection reagents and DAPI are light sensitive) at 37°C.

11. Soak slide sequentially, 10 min each, in aluminum foil–wrapped Coplin jars containing room temperature 4× SSC, 0.1% Triton X-100/4× SSC, and 4× SSC.

12. Add 50 μl DAPI or propidium iodide staining solution to counterstain the chromosomes, cover with 22-mm^2 square of Parafilm, and allow to stain 10 min at room temperature. Rinse briefly in 1× SSC in a Coplin jar to remove excess stain. Blot slide, but do not dry.

> *DAPI generates a G-band-like pattern (UNIT 4.4) and can be viewed simultaneously with fluorescein and either rhodamine or Texas Red. Propidium iodide stains more uniformly and can be viewed simultaneously with fluorescein, but not with rhodamine and Texas Red.*

13. Add 7 μl appropriate antifade mounting medium—e.g., DABCO for green fluorescence, *p*-phenylenediamine or Vectashield for red fluorescence (will also work for fluorescein). Coverslip, squeeze out excess antifade medium, and seal with nail polish. Store at −20°C in slide box with desiccant. Examine immediately (preferred) or store several weeks at 4° or −20°C.

14. Examine slide using a fluorescence microscope with epi-illumination and filter set appropriate for the fluorochrome used.

15. Photograph using either Ektar-1000 (for prints) or Ektachrome-400 (for slides) color film (DAPI exposure: ~2 sec; dual-band-pass filter: 30 to 90 sec; triple-band-pass filter: 3 to 8 sec). For bright signals (e.g., satellite sequences, some unique probes), use black-and-white film can be used at ASA 200. Alternatively, prepare digitized images with a CCD camera or confocal laser microscope.

16a. *For DAPI-banded chromosomes:* Determine map position on ≥10 hybridized chromosomes that show signals on both chromatids. Compare position of hybridization signal (viewed with dual- or triple-band-pass filter) to DAPI-banded chromosome image to determine exact chromosome band location.

16b. *For PI-stained chromosomes:* Utilizing photo enlargements, projections from 35-mm slides, or digitized images of ten or more hybridized chromosomes, obtain FLPter value by comparing the fractional length of the total chromosome relative to the end of the short arm, utilizing ISCN chromosome ideograms (*APPENDIX 2B*) as standards (Lichter et al., 1990).

SUPPORT PROTOCOL 1

AMPLIFICATION OF BIOTINYLATED SIGNALS

Amplification should be used only when the probe yields weak fluorescence with very little background.

Additional Materials (*also see Basic Protocol*)

> 1 to 3 μg/ml biotinylated anti-avidin antibodies (Vector Laboratorics) in 4× SSC/1% (w/v) BSA (fraction V)
> Biotin amplification solution: 2 to 5 μg/ml fluorescein-avidin DCS (Vector Laboratories) in 4× SSC/1% (w/v) BSA (fraction V)

1. Hybridize slide with biotinylated probe, wash, and perform first round of signal detection as described (see Basic Protocol, steps 1 to 11).

2. If slide has been mounted and sealed, remove coverslip by breaking the seal with a needle or scalpel, lifting off the nail polish, and soaking slide 15 min in 0.1% Triton X-100/4× SSC in an aluminum foil-wrapped Coplin jar with gentle agitation. If coverslip is loose, carefully lift it off. Repeat wash twice.

3. Drain excess wash solution from slide and add 50 μl of 1 to 3 μg/ml biotinylated anti-avidin antibody. Cover with 22-mm^2 square of Parafilm, place in moist chamber, wrap moist chamber in aluminum foil, and incubate 30 min at 37°C.

4. Remove Parafilm and wash 15 min in 0.1% Triton X-100/4× SSC with gentle agitation.

5. Drain excess solution from slide and add 50 μl biotin amplification solution. Cover with Parafilm and incubate 30 min at 37°C in a foil-wrapped moist chamber.

6. Remove Parafilm and wash 15 min in 0.1% Triton X-100/4× SSC, then 15 min in 4× SSC, both with gentle agitation. If desired, repeat steps 3 to 6 to increase signal (background will also increase significantly).

7. Counterstain, mount, and examine slide as described (see Basic Protocol, steps 12 to 16).

SUPPORT PROTOCOL 2

AMPLIFICATION OF SIGNALS FROM DIGOXIGENIN-LABELED PROBES

Additional Materials (*also see Basic Protocol*)

> 10 μg/ml Fab fragment of sheep anti-digoxigenin (Boehringer Mannheim) in 4× SSC/1% (w/v) BSA (fraction V)
> Digoxigenin amplification solution: 3.5 to 7.0 μg/ml fluorescein-conjugated rabbit anti-sheep IgG (Sigma) in 4× SSC/1% (w/v) BSA (fraction V)

1. Hybridize slide with digoxigenin-labeled probe and wash (see Basic Protocol, steps 1 to 8).

2. Add 50 μl of 10 μg/ml Fab fragment of sheep anti-digoxigenin to slide and cover with 22-mm² square of Parafilm. Incubate 30 min at 37°C.

3. Remove Parafilm. Wash in 4× SSC, then 0.1% Triton X-100/4× SSC, then 4× SSC, 15 min each.

4. Drain excess 4× SSC and add 50 μl digoxigenin amplification solution to slide. Cover with 22-mm² square of Parafilm, place in moist chamber, wrap chamber in aluminum foil, and incubate 30 min at 37°C.

5. Repeat washes as in step 3.

6. Counterstain, mount, and examine slide as described (see Basic Protocol, steps 12 to 16).

ALTERNATE PROTOCOL 1

ENZYMATIC DETECTION USING HORSERADISH PEROXIDASE

In situ hybridization using an enzymatic system produces a stable colored precipitate, thus providing a permanent signal. Indirect methods (Fig. 4.4.2) are preferred to direct enzyme-coupled probes because enzyme coupling can interfere with probe penetration. Table 4.4.1 summarizes the most commonly used enzyme-substrate pairs used for ISH. This method describes detection of biotinylated probes with HRP. Digoxigenin-labeled probes can also be visualized with an HRP-conjugated antibody.

4

Table 4.4.1 Commonly Used Enzyme-Substrate Combinations for Detection of In Situ Hybridized Probes

Enzyme	Substrate[a]	Color[b]
Alkaline phosphatase	5-Bromo-4-chloro-3-indolyl phosphate (BCIP) and nitro-blue tetrazolium (NBT)	Bluish-purple precipitate
	Naphtol-AS-MX-phosphate and fast red TR	Red precipitate and red fluorescence
	Naphtol-AS-MX-phosphate and fast blue BN (Boehringer)	Blue precipitate
	Naphtol-AS-MX-phosphate and fast green BN (Boehringer)	Green precipitate
	Vector Red	Red precipitate
	Vector Black	Black precipitate
	Vector Blue	Blue precipitate
Horseradish peroxidase	3,3-Diaminobenzidine tetrahydrochloride (DAB)	Brown precipitate[c]

[a]The main suppliers for these reagents are Boehringer Mannheim, Life Technologies, Promega, Sigma, and Vector Laboratories, but other suppliers deliver equivalent reagents.

[b]The detection systems listed result in hybridization signals that are generally analyzed by conventional bright-field microscopy. A fluorescence microscope is needed when utilizing the fluorescent signals produced by fast red. Optimization of the fluorescence detection procedure has been reported (Speel et al., 1992).

[c]Can be intensified by silver deposition.

Additional Materials (*also see Basic Protocol; see* APPENDIX 1 *for items with* ✓)

Biotinylated hybridization probe

Blocking solution: 1% (w/v) BSA in PBS

✓ Streptavidin solution

0.1% (v/v) Tween 20/PBS, 42°C

✓ Biotinylated horseradish peroxidase (HRP) solution

3% H_2O_2

DAB substrate solution: 500 μg/ml 3,3′-diaminobenzidine tetrahydrochloride (DAB) in PBS, prepare fresh

✓ PBS

✓ 90% (v/v) glycerol *or* appropriate antifade mounting medium

24 × 60–mm coverslips

42°C shaking water bath

1. Hybridize slide of chromosome preparation with biotinylated probe and wash as described (see Basic Protocol, steps 1 to 8).

2. Remove slide from 1× SSC. Drain as much buffer as possible, but do not allow slide to dry at any time in the procedure. Add 200 μl blocking solution and place 24 × 60–mm coverslip on top of applied solution. Place slide in an aluminum foil-wrapped moist chamber and incubate 30 min at 37°C.

3. Remove slide from chamber, tilt it to let coverslip slide off, and drain off as much blocking solution as possible. Add 200 μl streptavidin solution, coverslip, and incubate 30 min at 37°C in the moist chamber.

4. Remove slide from chamber, tilt it to let the coverslip slide off, and place slide in Coplin jar containing 42°C 0.1% Tween 20/PBS. Agitate 5 min at 42°C in shaking water bath. Repeat wash twice.

5. Drain thoroughly. Add 200 μl biotinylated HRP solution, coverslip, and incubate 30 min at 37°C in the moist chamber.

   Steps 3 to 5 can be accomplished in a single step by incubating the slide with 3 μg/ml streptavidin-conjugated HRP.

6. Repeat wash as in step 4.

7. Drain thoroughly. Add $^1/_{200}$ vol of 3% H_2O_2 (0.015% final) to DAB substrate solution. Immediately add 200 μl DAB substrate solution, coverslip, and incubate 10 to 20 min in the dark at room temperature.

8. When the colored precipitate becomes visible by eye, wash slide 5 min in room temperature PBS to stop the reaction.

9. Apply a fluorescent counterstain to identify chromosomes (see Basic Protocol, step 12) if desired.

10. Mount slide in 90% glycerol or appropriate antifade mounting medium (see Basic Protocol, step 13). View and photograph with a phase-contrast microscope.

ALTERNATE PROTOCOL 2

ENZYMATIC DETECTION USING ALKALINE PHOSPHATASE

Additional Materials (*also see Basic Protocol; see* APPENDIX 1 *for items with* ✓)

✓ Biotinylated alkaline phosphatase (AP) solution

3 μg/ml streptavidin-conjugated AP (Life Technologies)

✓ Alkaline phosphatase buffer, pH 9.5, 42°C
✓ NBT/BCIP substrate solution
✓ PBS
✓ 90% glycerol *or* appropriate antifade mounting medium

24 × 60–mm coverslips
42°C shaking water bath

1a. *For multistep detection:* Hybridize slide of chromosome preparation with biotinylated probe, wash, block, and incubate in streptavidin solution as described (see Alternate Protocol 1, steps 1 to 4). Remove slide from 0.1% Tween 20/PBS and drain without allowing to dry. Add 200 μl biotinylated AP solution, coverslip, and incubate 30 min at 37°C in a moist chamber.

1b. *For single-step detection:* Hybridize, wash, and block as above, and then incubate with 3 μg/ml streptavidin-conjugated AP.

2. Remove slide from chamber, tilt it to let the coverslip slide off, and place in a Coplin jar containing 42°C 0.1% Tween 20/PBS. Agitate 5 min at 42°C in shaking water bath. Repeat wash twice.

3. Transfer slide to a Coplin jar containing 42°C alkaline phosphatase buffer, pH 9.5. Agitate 5 min at 42°C. Repeat once.

4. Place slide in an aluminum-foil-wrapped Coplin jar containing 50 ml freshly prepared NBT/BCIP substrate solution. Incubate in the dark, usually 15 to 60 min, at 37°C or at room temperature (to slow down reaction) until color development is suitable.

5. Wash slide in PBS 5 min at room temperature to stop reaction.

6. Apply a fluorescent counterstain to identify chromosomes (see Basic Protocol, step 12) if desired.

7. Mount in 90% glycerol or appropriate antifade mounting medium (see Basic Protocol, step 13). View and photograph with a phase-contrast microscope.

SUPPORT PROTOCOL 3

BIOTIN AND DIGOXIGENIN LABELING OF PROBES BY NICK TRANSLATION

Using nick translation, probes do not need to be isolated or linearized.

Materials (see APPENDIX 1 for items with ✓)

Nick translation kit (e.g., Boehringer Mannheim) *or* equivalent materials:
✓ 10× nick translation buffer
  1 μg probe DNA (e.g., cloned inserts in plasmids, phage, cosmids, P1, or single YACs)
  1 mM biotin-16-dUTP *or* 1 mM digoxigenin-11-dUTP (Boehringer Mannheim)
✓ 1 and 4 mg/ml DNase I (also from Worthington)
  5 U/ml DNA polymerase I, endonuclease-free (Boehringer Mannheim)
✓ 0.5 M EDTA
✓ 10% (w/v) SDS
✓ 10 mg/ml sonicated salmon sperm DNA
✓ 3 M sodium acetate, pH 5.2
  95% and 70% (v/v) ethanol, ice-cold

15° and 65°C water baths
65° heating block

1. Add the following, in the order listed, to a 1.5-ml microcentrifuge tube on ice:

   10 μl 10× nick translation buffer (1× final)
   1 μg probe DNA
   5 μl 1 mM biotin-16-dUTP *or* 1 mM digoxigenin-11-dUTP (0.05 mM final)
   Sterile H_2O to give a final total volume of 100 μl
   10 μl 1 μg/ml DNase I
   4 μl 5 U/ml DNA polymerase I.

   Mix and microcentrifuge briefly to pool ingredients, then incubate 2 to $2^1/_2$ hr at 15°C.

2. Boil a 5-μl aliquot of reaction product 3 min and let stand 2 min on ice (set aside remaining 95 μl at −20°C). Electrophorese in an agarose minigel to check probe size (should be 100 to 500 bp with majority at 250 to 300 bp). If the majority of probe DNA is >800 bp, add 5 μl of 4 μg/ml DNase I to reaction mix, incubate 30 min at 15°C, and check size again.

3. When correct molecular-size distribution is achieved, inactivate enzymes in reaction mix by adding 3 μl of 0.5 M EDTA and 1 μl of 10% SDS, and heating 10 min at 65°C.

4. Ethanol precipitate labeled probe by adding 1 μl of 10 mg/ml sonicated salmon sperm DNA, $^1/_{10}$ vol 3 M sodium acetate, and $2^1/_2$ vol 95% ethanol to reaction mix. Incubate ≥1 hr at −20°C or 10 min in dry ice.

5. Microcentrifuge 30 min at maximum speed, 4°C. Remove supernatant and rinse DNA pellet with ice-cold 70% ethanol. Air dry and resuspend pellet in 200 μl sterile water (5 ng/μl final). Store at 4°C (stable >1 year). If desired (e.g., new lots of DNase I or DNA polymerase I are used, poor hybridization is obtained), check incorporation (e.g., see *CPMB UNIT 3.18*).

ALTERNATE PROTOCOL 3

ORDERING OF SEQUENCES IN INTERPHASE NUCLEI BY FISH

FISH interphase mapping is useful for ordering sequences that localize to the same chromosome band. Interphase chromatin is less condensed than metaphase chromatin and permits ordering of probes that are not resolvable on metaphase chromosomes. At least three DNA probe sequences are necessary to determine relative order, which is linear in the range of 50 kb to ~1 Mb on interphase chromatin. Hybridization conditions are the same as those described (see Basic Protocol).

FISH interphase mapping is performed on G1 or G2 cells, usually present in adequate numbers in standard cytological preparations harvested for metaphase chromosomes. Hybridization with a single probe results in one signal for each chromosome in G1 cells and two signals for each chromosome (both chromatids) in G2 cells. Slides enriched for G1 cells can be prepared by collecting the buffy coat from peripheral blood specimens, treating cells from the buffy coat with hypotonic solution, fixing the cells, and preparing slides (*UNIT 4.1*).

Denaturation, hybridization, and detection are performed as described except that three probe sequences are hybridized to the same slide—either two biotin-labeled and one digoxigenin-labeled probe, two digoxigenin-labeled and one biotin-labeled probe, or one biotin-labeled probe, one digoxigenin-labeled probe, and one dually labeled probe. Signals are visualized with complementary fluorochromes.

Analysis of 25 to 50 interphase cells generally gives an interpretable order. If adjacent probes are the same color (e.g., red-red-green), then an additional hybridization, with the label on one of the adjacent probes of the same color changed, is necessary to determine probe order. The color sequence then becomes red-green-green or green-red-green. If the third probe is co-labeled with both biotin and digoxigenin, it produces yellow fluorescence and the order of red, green, and yellow signals can be determined directly. If the color order is not evident, it is possible that the probes are too far apart for interphase mapping. As the distance between two probes increases, especially in the 1.5- to 3-Mb range, there is a greater probability that the localization of the signals will be influenced by packing of chromatin in the nucleus.

References: Lichter et al., 1988, 1990; Pinkel et al., 1988

Contributors: Joan H.M. Knoll and Peter Lichter

UNIT 4.5

High-Resolution FISH Analysis

Using free chromatin for FISH analysis allows mapping of genes that are ≤ 1 Mb apart (the resolution limit for FISH with metaphase chromosomes; *UNIT 4.4*).

NOTE: All incubations are performed in a humidified 37°C, 5% CO_2 incubator unless otherwise specified. All reagents and equipment coming into contact with live cells must be sterile.

BASIC PROTOCOL 1

PREPARATION OF FREE CHROMATIN WITH ALKALINE BUFFER

Materials (*see* APPENDIX 1 *for items with* ✓)

Fibroblast culture *or* 10 ml fresh human peripheral or cord blood
✓ Fibroblast or lymphocyte culture medium
Trypsin/EDTA solution (Life Technologies)
✓ Alkaline buffer
Fixative: 3:1 (v/v) methanol/glacial acetic acid (prepare fresh)

60-mm tissue culture plates
25-cm² tissue culture flasks
15-ml screw-cap polystyrene tubes
Microscope slides, chilled 5 to 10 min on ice
Phase-contrast microscope
Slide box

1. Determine optimal time of alkaline treatment for the particular cell type to be used (see Support Protocol 1).

    Only a fraction of cells should give rise to free chromatin in order to prevent aggregation of chromatin fibers. Slides with aggregated preparations should not be used for FISH mapping. Besides treatment time, the cell number can also be adjusted if necessary (steps 4a and 3b).

For fibroblasts (adherent cultures)

2a. Incubate fibroblast cultures in a 60-mm tissue culture plate containing 4 ml fibroblast medium with serum until cells are confluent. If desired, culture undisturbed 2 to 4 days post-confluency to enrich for G1 cells, which exhibit a less complex hybridization pattern than G2 cells.

3a. Rinse culture with serum-free fibroblast medium. Add 1 ml trypsin/EDTA solution to each 60-mm plate. Incubate 30 to 60 sec at 37°C. Stop trypsinization by adding 1 ml fibroblast medium with serum as soon as cells begin to detach from the plate. Avoid prolonged trypsinization.

4a. Quickly divide into five to ten 0.2- to 0.4-ml aliquots ($\sim 10^4$ cells) in 15-ml culture tubes.

For lymphocytes (suspension cultures)

2b. Isolate lymphocytes from peripheral or cord blood by low-speed centrifugation (i.e., 5 min at $10 \times g$, room temperature) or unit-gravity sedimentation. Add 0.5 to 0.8 ml lymphocytes (10^5 to 10^6 cells) to a 25-cm² tissue culture flask containing 20 ml lymphocyte culture medium with serum. Incubate 48 hr. If desired, add thymidine to a final concentration of 0.3 mg/ml and incubate an additional 16 to 18 hr to synchronize cells in G1. Wash thymidine-treated cells in lymphocyte medium with serum and culture an additional 10 hr.

3b. Divide cell suspension into 4-ml aliquots ($\sim 10^4$ cells) in 15-ml screw-cap polystyrene tubes. Centrifuge 7 min at $180 \times g$, room temperature.

4b. Resuspend each pellet in 0.3 ml lymphocyte medium with serum.

5. Add 2 ml alkaline buffer to each aliquot, mix contents gently by tapping the tube, and incubate at room temperature 2 to 10 min (see step 1).

6. Add 3 ml fixative to each tube to stop alkaline treatment. Mix gently but thoroughly. Centrifuge 7 min at $180 \times g$ to collect free chromatin.

7. Resuspend pellet in 4 ml fresh fixative and incubate 10 min at room temperature. Centrifuge again.

8. Resuspend pellet in 0.2 ml fixative and mix by gently tapping the tube. Drop free chromatin suspension on prechilled slide. Air dry quickly by waving slide back and forth.

9. Examine slide with a phase-contrast microscope to check quality of preparation. Determine if a good preparation (two to ten free chromatin structures per 100× field) has been obtained. Adjust cell number and treatment time if necessary.

10. Prepare additional slides according to the optimized procedure. For easy comparison, prepare free chromatin and metaphase chromosome spreads (*UNIT 4.1*) on the same slide. Dry good-quality slides 1 day at room temperature. Store up to several months at −20°C in a slide box sealed with Parafilm (avoid dehydration which can irreversibly damage free chromatin structures).

SUPPORT PROTOCOL 1

OPTIMIZATION OF FREE CHROMATIN PREPARATION

For each cell type, a time course is performed to determine the optimal duration of alkaline treatment. The optimal pH (10 to 11) and KCl concentration (usually 0.4% to 1% w/v) can also be determined by holding two of the three factors constant. It is normally unnecessary to determine optimal KCl concentration and pH exhaustively because it is easy to control duration.

Additional Materials (*also see Basic Protocol 1*)

 4% Giemsa stain: 2 ml Giemsa stain (Fisher)/48 ml Gurr, pH 6.8 (BDH), prepare fresh

1. Using $\sim 5 \times 10^4$ fibroblasts or lymphocytes that will be analyzed by FISH, perform steps 2a to 4a or 2b to 4b of Basic Protocol 1.

2. Prepare five 15-ml tubes each containing 2 ml alkaline buffer. Transfer 0.2 to 0.4 ml fresh cell suspension ($\sim 10^4$ cells) dropwise to each tube and mix contents gently by tapping tube.

3. After 2, 4, 6, 8, and 10 min, add 3 ml fixative to one of the five tubes and mix immediately to stop treatment.

4. Centrifuge 7 min at $180 \times g$, room temperature. Resuspend in 3 ml fresh fixative and incubate 5 min at room temperature. Centrifuge again, resuspend in 0.3 ml fixative, and mix thoroughly.

5. Place a drop of each suspension on a prechilled slide. Air dry slide quickly by waving it back and forth. Stain 5 min at room temperature with 4% Giemsa stain.

6. Examine each slide with a microscope to determine quality of free chromatin structures, which should have an elongated spindle or rope-like structure with smooth edges, and should have minimal aggregation (contact). Identify optimal duration of alkaline buffer treatment. If there is little free chromatin, increase incubation time. If there is aggregation, decrease incubation time.

ALTERNATE PROTOCOL

PREPARATION OF FREE CHROMATIN FROM LYMPHOCYTES BY DRUG TREATMENT

Even though free chromatin can be easily prepared by alkaline buffer, drug-treatment seems to produce more consistent results with lymphocytes, and the morphology of free chromatin produced appears to be more homogeneous.

Additional Materials (*also see Basic Protocol 1*)

 10 ml fresh human peripheral or cord blood
 ✓ 5 mg/ml m-AMSA, 10 mg/ml ethidium bromide, *or* BrdU and 2′-deoxycytidine
 0.4% (w/v) KCl
 4% Giemsa stain: 2 ml Giemsa stain (Fisher)/48 ml Gurr, pH 6.8 (BDH), prepare fresh

CAUTION: m-AMSA and ethidium bromide are mutagens and should be handled with care.

1. Isolate lymphocytes from 10 ml fresh human peripheral or cord blood by low-speed centrifugation (5 min at $10 \times g$, room temperature) or unit-gravity sedimentation.

2. Collect lymphocytes by aspiration and transfer 0.5 to 0.8 ml isolated cells to 20 ml lymphocyte medium with serum in a 25-cm^2 tissue culture flask. Incubate 50 hr.

3. Add 40 µl of 5 mg/ml m-AMSA (final 10 µg/ml), 20 µl of 10 mg/ml ethidium bromide (final 10 µg/ml), or 100 µM BrdU $\times$ 100 µM 2′-deoxycytidine (final concentrations). Incubate 2 hr.

 If necessary, test concentrations of the drug (e.g., 5, 10, and 20 µg/ml m-AMSA) with small portions of the culture at ~ 46 hr incubation (step 2). After 2 hr incubation with the different drug concentrations, carry out steps 4 to 9 to identify the drug concentration yielding optimal free chromatin. Use the optimal drug concentration to treat the remaining culture at 50 hr incubation (step 3). Optimization is not typically required.

4. Transfer at most 4 ml culture ($\sim10^4$ cells) to a 15-ml screw-cap polypropylene tube. Collect drug-treated cells by centrifuging 7 min at $180 \times g$, room temperature.

5. Resuspend cell pellet in 0.3 ml lymphocyte culture medium with serum. Add 5 ml of 0.4% KCl solution and mix well. Incubate at 37°C for 10 min (up to 20 to 30 min if yield of free chromatin is low).

6. Add 0.1 to 0.2 ml freshly prepared fixative. Mix gently by inverting tube. Centrifuge again.

7. Discard supernatant. Add a few drops of fresh fixative and tap bottom of tube gently to loosen pellet. Add 5 ml fixative, resuspend cells, and incubate 20 min at room temperature.

8. Centrifuge again, discard supernatant, and resuspend pellet in 0.5 ml fresh fixative.

9. Place two drops of free chromatin suspension on a prechilled slide. Air dry slide quickly by waving it back and forth. Examine with a phase-contrast microscope. Alternatively, stain with 4% Giemsa for 5 min at room temperature and examine with a bright-field microscope. Determine if a good preparation has been obtained (two to ten free chromatin structures per 100× field). If the concentration of free chromatin is not optimal, increase or reduce fixative in subsequent preparations.

10. Prepare additional slides using optimal conditions (also see Basic Protocol 1, step 10).

BASIC PROTOCOL 2

HIGH-RESOLUTION FISH MAPPING WITH FREE CHROMATIN

Materials

Slides of free chromatin preparations (see Basic Protocol 1 or Alternate Protocol)
Denaturing solution: 70% deionized formamide (*APPENDIX 1*)/2× SSC (pH 7.0;
  APPENDIX 1), 70°C
55°C oven
70°C water bath
Ektachrome P1600 film (Kodak; high speed film essential)

1. Bake slides of free chromatin preparations 2 hr at 55°C.

2. Incubate slides of free chromatin preparations 3 to 4 min in 70°C denaturing solution.

3. Denature probe(s) and hybridize to free chromatin as described for FISH to metaphase chromosomes (*UNIT 4.4*), using signal amplification for cosmid or phage probes (amplification is optional for YAC and repetitive-sequence probes; unnecessary if a CCD camera is used).

4. Examine slides with a microscope equipped with epifluorescence optics and a dual- (FITC/Texas Red) or triple-band-pass (FITC/Texas Red/DAPI) filter. Photograph slides using Ektachrome P1600 film and ASA 3200 exposure setting. Determine probe order or distance between probes.

BASIC PROTOCOL 3

PREPARATION OF STRETCHED CELLULAR DNA FOR DIRVISH MAPPING

Stretched DNA is deleted of chromatin proteins and lacks nuclear structure. Use of stretched DNA for direct visual hybridization (DIRVISH) mapping improves resolution and results in virtually straight lines of signal.

Materials *(see APPENDIX 1 for items with ✓)*

    Cultured cells (or other suitable cells)
✓ PBS
✓ Lysis buffer
    Fixative: 3:1 (v/v) methanol/glacial acetic acid (prepare fresh)
    N_2 gas

    $25 \times 75 \times 1$–mm glass microscope slides
    Moist chamber (Fig. 4.4.3)
    Heat-sealable bags
    Drierite, 8-mesh (Fisher)

1. Harvest cultured cells and resuspend in PBS to give a concentration of 100 to 5000 cells in 2 μl (5×10^4 to 2.5×10^6 cells/ml).

2. Place 2 μl cell suspension at one end of a $25 \times 75 \times 1$–mm glass slide and allow to air dry.

3. Apply 5 to 8 μl lysis buffer to dried cells on slide and incubate 5 min in a moist chamber at room temperature.

4. Gently tilt slide to vertical position with DNA at the upper end. Allow DNA to stream toward other end of slide, then air dry almost completely.

5. Cover slide with 400 μl fixative at, or shortly before, the time that the DNA stream dries. Wait 1 min.

6. Tilt slide to let excess fixative drain off. Let slide air dry. Mark the area around the DNA stream with an etching pen to help identify the area of interest.

7. Store slides under N_2 gas in a heat-sealable bag at $-20°C$, with a small amount of Drierite to absorb excess moisture (stable >1 year).

BASIC PROTOCOL 4

HIGH-RESOLUTION MAPPING BY DIRVISH TO STRETCHED DNA

Direct visual hybridization (DIRVISH) is used to map the relative positions of two or more probes to a resolution of 1 kb. Some prior knowledge of the relative positions of the probes is required for this protocol—i.e., FISH analysis of metaphase chromosomes or interphase nuclei (*UNIT 4.4*) should show that the probes are tightly linked. The size of one of the probes should be known so it can be used as an internal reference, unless the order of probes is the only thing being determined or there is a size estimate for one of the probes and estimates of gap and overlap distances are the only information desired.

The following combination gives the best DIRVISH results: biotin detected with fluorescein-conjugated avidin (green), digoxigenin detected with rhodamine-conjugated anti-digoxigenin (red; also Texas Red), an equal mix of biotin and digoxigenin (yellow). Other colors may be generated by adjusting the balance of biotin-dUTP and digoxigenin-dUTP in the labeling reaction (e.g., orange can be obtained by using 3 parts digoxigenin and 1 part biotin). As the number of different colors increases, an imaging system may be required to objectively analyze the colors.

Materials (see APPENDIX 1 for items with ✓)

    100 μg/ml RNase: 2 mg/ml DNase-free RNase (*APPENDIX 1*) in 2× SSC; prepare fresh
    Glass slides containing streams of stretched DNA (see Basic Protocol 3)
✓ 4× and 2× SSC
    70% (v/v) deionized formamide (*APPENDIX 1*)/2× SSC, 70°C
    70% (v/v) ethanol, ice-cold
    100% and 90% (v/v) ethanol, room temperature
    Labeled probes (see Support Protocol 2)
    Sonicated genomic DNA (100 to 1000 bp; *APPENDIX 1*) *or* C_0t_1 DNA (e.g., Life
       Technologies)
✓ Hybridization cocktail
    Rubber cement
    50% (v/v) nondeionized formamide/2× SSC
✓ Preavidin block solution
    5 μg/ml fluorescein-avidin DN (Vector) in 4× SSC/1% (w/v) BSA
    4× SSC/0.1% (v/v) Triton X-100
✓ PN buffer
    NGS/PN solution: 4% (v/v) normal goat serum (NGS, Vector) in PN buffer (store 1-ml
       aliquots at 4°C)
    5 μg/ml biotinylated anti-avidin D antibody (Vector) in NGS/PN solution
    10 μg/ml mouse anti-digoxigenin antibody (Boehringer Mannheim) in NGS/PN solution
    10 μg/ml digoxigenin-labeled polyvalent anti-mouse Ig F(ab′)$_2$ fragment (Boehringer
       Mannheim) in NGS/PN solution
    25 μg/ml rhodamine-conjugated anti-digoxigenin Fab fragment (Boehringer Mannheim)
       in NGS/PN solution
    Vectashield antifade mounting medium (Vector)
    Nuclear stains (e.g., 1 μg/ml DAPI; optional)

    22 × 40–mm coverslips, no. 1
    Polyethylene Coplin jars
    Moist chamber (Fig. 4.4.3)
    45° and 70°C water baths
    Rubber cement
    Epifluorescence microscope with triple-band-pass filter (DAPI, fluorescein, and Texas
       Red or rhodamine) and 100× oil objective, N.A. 1.4 (e.g., PlanApo, Nikon)
    Ektachrome ASA 400 slide film (Kodak) or digital imaging system (e.g., Oncor VI 50)
       and appropriate software
    Stage micrometer with 10-μm division scale

1. Apply 50 μl of 100 μg/ml RNase to an area on a glass slide containing streams of stretched DNA. Cover with a 22 × 40–mm no. 1 coverslip. Incubate 1 hr at 37°C in a moist chamber.

2. Set up five polyethylene Coplin jars containing 2× SSC, 70% formamide/2× SSC, and 70%, 90%, and 100% ethanol, respectively. Heat jars with SSC and formamide/SSC to 70°C; chill jar with 70% ethanol on ice or in ethanol from a −20°C freezer. Leave jars of 90% and 100% ethanol at room temperature.

3. Gently slide coverslip to side of slide until edge hangs over and can be grasped. Gently lift coverslip off slide. Immerse slide momentarily in 2× SSC, 70°C.

4. Immerse slide 2 min in 70% formamide/2× SSC, 70°C. Agitate slide slightly with forceps.

5. Quickly transfer slide to ice-cold 70% ethanol and dehydrate 1 min, agitating slightly. Continue dehydration sequentially with room-temperature 90% and 100% ethanol. Air dry slide.

6. Thoroughly mix each labeled probe (20 ng for cosmid or smaller probes; 200 ng for YACs or other large probes) with 10 μg sonicated genomic DNA or C_0t_1 DNA in a 1.5-ml microcentrifuge tube. Place multiple probes in the same tube. Dry in a Speedvac evaporator.

7. Resuspend dried pellet in 2 μl deionized water and mix well. Add 8 μl hybridization cocktail and mix well. Denature probe 5 min at 70°C, then place on ice.

8. Add to an area of slide containing stretched DNA and cover with a 22 $\times$ 40–mm coverslip. Gently press on coverslip with forceps to force out air bubbles, seal edges with rubber cement, and incubate overnight (~18 hr) at 37°C in a moist chamber.

9. Set up two Coplin jars containing 40 ml of 50% formamide/2$\times$ SSC and two Coplin jars containing 40 ml of 2$\times$ SSC and heat to 45°C.

10. Remove slide from moist chamber. Peel off rubber cement and carefully lift off coverslip with forceps. Immerse slide 3 min in each Coplin jar of formamide/2$\times$ SSC and 2 min in each jar of 2$\times$ SSC. If necessary, slide can be stored at least 3 hr in 4$\times$ SSC, room temperature.

11. Remove slide from buffer and drain off as much buffer as possible. Do not allow slide to dry out at any point during the detection procedure.

12. Apply 50 μl preavidin block solution. Cover with a 22 $\times$ 40–mm coverslip, avoiding air bubbles. Incubate 10 min at room temperature in a moist chamber.

13. Carefully remove coverslip. Apply 25 μl of 5 μg/ml fluorescein-avidin DN and cover with a coverslip. Incubate 20 min at room temperature in a moist chamber.

14. Carefully remove coverslip. Immerse slide sequentially for 2 min each in Coplin jars containing 4$\times$ SSC, 4$\times$ SSC/0.1% Triton X-100, 4$\times$ SSC, and PN buffer at room temperature.

15. To amplify the signal, apply 50 μl NGS/PN solution to slide and cover with coverslip. Incubate 10 min at room temperature in a moist chamber.

16. Carefully remove coverslip. Apply 25 μl of 5 μg/ml biotinylated anti-avidin D antibody to area containing hybridized DNA and cover with coverslip. Incubate 20 min at room temperature in a moist chamber.

17. Wash as in step 14 and then repeat fluorescein detection of biotin as in steps 12 to 15.

18. Carefully remove coverslip. Apply 25 μl of 10 μg/ml mouse anti-digoxigenin antibody to slide and cover with coverslip. Incubate 20 min at room temperature in a moist chamber.

19. Repeat steps 14 and 15.

20. Carefully remove coverslip. Apply 25 μl of 10 μg/ml digoxigenin-labeled anti-mouse Ig F(ab′)$_2$ fragment to slide and cover with a coverslip. Incubate 20 min at room temperature in a moist chamber.

21. Repeat steps 14 and 15.

22. Carefully remove coverslip. Apply 25 μl of 25 μg/ml rhodamine-conjugated anti-digoxigenin Fab fragment to slide and cover with coverslip. Incubate 20 min at room temperature in moist chamber.

23. Repeat step 14.

24. Place 10 μl Vectashield antifade mounting medium on slide and cover with a 22 $\times$ 40–mm coverslip. Press on coverslip using forceps and wipe away excess mounting medium. If desired, add a nuclear stain such as 1 μg/ml DAPI to visualize DNA streams.

25. View slide under an epifluorescence microscope using an appropriate filter set (e.g., a triple-band-pass filter for simultaneous detection of DAPI, fluorescein, and rhodamine) and a 100× oil objective with N.A. 1.4. Locate proper focal plane and scan fields for strings of signal beginning at the site of cell lysis and progressing to the other end of the slide. Store up to many days in the dark at 4°C.

26. Photograph slide using Ektachrome ASA 400 slide film and appropriate exposure times (e.g., 15 to 60 sec with a triple-band-pass filter). Alternatively, use a digital imaging system with appropriate software.

   The Oncor system includes software for determining contour length for each signal and producing a map.

27. Choose images with signals for all probes to create an accurate map of the relative probe positions (images lacking one or more signals cannot be interpreted as accurate data). If necessary (e.g., analysis of rearrangements), statistically analyze data.

28. Measure the distance between the start of a colored probe-specific signal and the end of the signal to provide a reasonably accurate length measurement of virtually straight signals. Measure gaps from the end of one signal to the start of a neighboring signal (e.g., in a digitized image or using a ruler to measure lengths on a slide projection of the image).

29. Identify regions of overlap between two probes—i.e., regions with a uniform though sometimes mottled blend of two probe colors. For three colors, this may require repeating the experiment with two colors at a time.

30. Normalize all distances to the reference probe distance.

   The reference probe serves as an internal control for the extent of DNA stretching (e.g., if the signal for a 40-kb cosmid probe is 10 μm, then a 5-μm gap between this probe and the next would be 20 kb). This assumes that stretching is uniform, which is less likely over longer distances.

31. Average the normalized distance information from at least ten images to create a map of the probes. Photograph a stage micrometer with a 10-μm division scale to calibrate distance measurements for absolute distances when determining the amount of DNA extension.

SUPPORT PROTOCOL 2

BIOTIN AND DIGOXIGENIN LABELING OF DIRVISH PROBES

Probes labeled by random oligonucleotide priming are not used for DIRVISH because of the difficulty in controlling labeled fragment and template DNA size, and because template DNA remains completely unlabeled.

Materials (see APPENDIX 1 for items with ✓)
   ✓ Nick translation buffer
   DNA to be labeled for use as DIRVISH probes
   1 mM biotin-16-dUTP or digoxigenin-11-dUTP (Boehringer Mannheim)
   ✓ DNase I working solution
   10 Ci/ml [α-^{32}P]dCTP (3000 Ci/mmol; Du Pont NEN)
   5 U/ml DNA polymerase I (Promega)
   10× stop NT buffer: 0.5% (w/v) SDS/0.25 M EDTA, pH 8.0
   ✓ Sephadex G-50
   0.25% (w/v) bromphenol blue (store at room temperature)
   ✓ Bio-Gel P-60
   2% (w/v) agarose minigel, ≤0.5-mm thick
   100- to 1000-bp DNA molecular size markers

14° and 100°C water baths
1-cc tuberculin syringe
Glass wool
9-in. (22-cm) Pasteur pipet
Heat-sealable bags

1. Add the following to a 1.5-ml microcentrifuge tube (total 100 µl):

   10 µl nick translation buffer
   2 µg DNA to be labeled
   3 µl 1 mM biotin-16-dUTP or 1 mM digoxigenin-11-dUTP (or 1.5 µl of each)
   H_2O to yield 93.8 µl
   4 µl DNase I working solution
   0.2 µl 10 mCi/ml [α-^{32}P]dCTP
   2 µl 5 U/ml DNA polymerase I.

   Tap tube with finger to mix ingredients thoroughly. Incubate 2 to 4 hr at 14°C.

2. Add 10 µl of 10× stop NT buffer and incubate 5 min at 37°C to stop reaction. If necessary, store samples at −20°C.

3. Remove 2 µl sample ($^1/_{50}$ vol) and place in a microcentrifuge tube as a "before" sample to determine percent incorporation (step 8).

For biotin-labeled probes

4a. Plug narrow end of 1-cc tuberculin syringe with glass wool. Fill syringe to top with Sephadex G-50 and place inside a 15-ml centrifuge tube.

5a. Centrifuge 10 sec at ~300 × g, room temperature. Transfer column to a new 15-ml tube.

6a. Add 1 µl of 0.25% bromphenol blue (migrates at approximately same position as unincorporated nucleotides) to labeled DNA and transfer to top of column. Centrifuge again.

7a. Check effluent with Geiger counter to make sure that it contains ^{32}P (i.e., labeled DNA). If the DNA does not elute, wash Sephadex with 100 µl water and centrifuge again.

For digoxigenin- and double-labeled probes

4b. Plug narrow end of a 9-in. Pasteur pipet with glass wool. Fill pipet with Bio-Gel P-60 and allow gel to settle to ~1 in. from top. Position column over a 15-ml centrifuge tube. Do not use a Sephadex spin column for digoxigenin-labeled probes.

5b. Add 1 µl of 0.25% bromphenol blue to labeled DNA and transfer to top of column when water level has just reached the top of the gel. Allow sample to enter gel, then add ~100 µl water to start sample through column.

6b. Continue to add water to column to elute probe, monitoring radioactivity in effluent using a Geiger counter. When the first peak (incorporated nucleotides) is about to emerge (i.e., when counts begin to increase), carefully move column to a 1.5-ml microcentrifuge tube and collect sample until peak has left column (i.e., when counts have dropped and leveled off).

7b. Because the probe is collected in a large volume of water, dry fraction containing probe in a Speedvac evaporator and resuspend in water to obtain desired probe concentration.

8. Remove $^1/_{50}$ vol. of probe solution as "after" sample and place in a microcentrifuge tube. Count before and after samples in a scintillation counter to determine percent incorporation (must be ~10% to 20% to achieve a contiguous string of signal).

9. Remove a small aliquot of probe solution and heat 3 min at 100°C to denature. Electrophorese on a 2% (w/v) agarose minigel with 100- to 1000-bp size marker. Stain in ethidium bromide to visualize and photograph size markers next to ruler.

10. Briefly blot the gel dry, seal in a heat-sealable bag, and place in a cassette with X-ray film. Autoradiograph 6 hr or as long as necessary. If desired, decrease exposure time by using more ^{32}P tracer in the initial reaction. When performing autoradiography on a gel that still retains water, use a gel that is ≤0.5 cm thick so that it will fit in the film cassette.

11. Use probes that are 100 to 1000 bp with an average between 500 and 1000 bp and a maximum of 1500 bp (standard FISH probes in the range of 100 to 500 bp are also suitable). If probe size is not optimal, adjust DNase I in subsequent reactions to make smaller or larger fragments. Store at −20°C (stable at least 1 year).

References: Heng and Tsui, 1994; Parra and Windle, 1993

Contributors: Henry H.Q. Heng, Lap-Chee Tsui, Bradford Windle, and Irma Parra

UNIT 4.6

Multicolor Fluorescence In Situ Hybridization (FISH) Approaches for Simultaneous Analysis of the Entire Human Genome

PRODUCTION OF CHROMOSOME-SPECIFIC PAINT PROBES AND COMBINATORIAL LABELING

Fundamental to most multicolor FISH approaches is their employment of chromosome-specific paints—DNA probe sets that specifically bind to sites all along a given chromosome. Although these paint probes can be derived from sources such as somatic cell hybrid DNAs and chromosome microdissections, chromosome flow sorting has now become the method of choice for obtaining consistent and accurate chromosomal template DNA. In bivariate chromosome flow sorting, metaphase chromosome preparations are stained with chromomycin A3 (stains GC-rich regions) and Hoechst 33258 (stains AT-rich regions). The chromosome preparations are then passed through a fluorescence-activated cell sorter equipped with two 5-W argon ion lasers. One laser is set to excite Hoechst (300 mW at 351 to 364 nm) while the other excites chromomycin (300 mW at 458 nm). Relative fluorescence measurements are collected for a large number of chromosomes and graphically plotted by an integrated computer system to produce a characteristic flow karyotype (Fig. 4.6.1). The chromosome-containing fluid stream is broken into a series of minute drops such that some droplets contain a single chromosome. An electric charge is applied to droplets containing a chromosome of interest. When these droplets pass between two high-voltage plates, they are deflected into one of two clean microcentrifuge tubes. Approximately 300 to 500 chromosomes are collected and subjected to two rounds of degenerate oligonucleotide primed (DOP) PCR. The first round amplifies the chromosomal DNA with unlabeled dNTPs, while the second incorporates a haptenized or fluorochrome-conjugated nucleotide (e.g., biotin-14-dATP or fluorogreen-11-dUTP). Usually, unlabeled human Cot-1 DNA is also added to the painting probe mixture to suppress the hybridization of highly repetitive DNAs, although in certain circumstances, a self-preannealing of the probe has been shown to suffice.

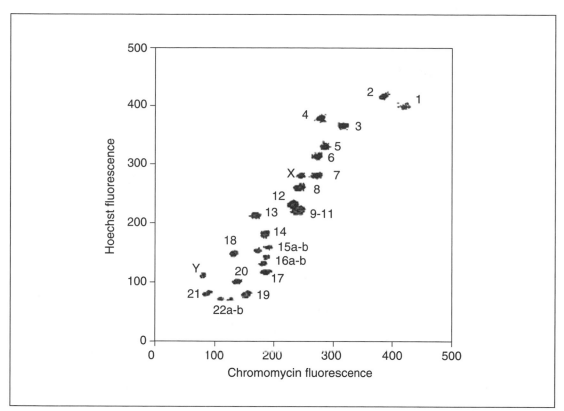

Figure 4.6.1 Bivariate human flow karyotype produced on a fluorescence-activated cell sorter using a chromosome suspension from a normal male after staining with Hoechst 33258 and chromomycin A3. Here, chromosomes 9 to 11 are unresolved by size and base pair composition. Chromosomes 15, 16, and 22 are resolved into their separate homologs. Courtesy of P.C.M. O'Brien.

To analyze the entire genome simultaneously with human chromosome paint probes, one must be able to discriminate each human chromosome by color. As there are not enough spectrally distinct fluorochromes to individually label all 24 human chromosomes, a labeling strategy must be employed to color-differentiate each human chromosome with the limited number available. One labeling strategy discerns targets by combining fluorochromes in different ratios (i.e., two fluors used at 75:25, 50:50, and 25:75); however, this strategy is not used routinely for large-scale commercial probe production primarily because of the difficulty in consistently recreating the fluorochrome ratios. The more widely employed labeling method is the combinatorial scheme, where different combinations of fluorochromes are used, often in equimolar quantities. The total number of combinatorial permutations for a combinatorial labeling scheme is defined by the equation $2^n - 1$, where n represents the number of fluorochromes used. In this manner, all 24 human chromosomes can be differentially labeled using only five fluorochromes. In fact, as many as 31 chromosomes could be labeled and individually discriminated with the same five fluorochromes.

SKY AND M-FISH

When all human chromosomes are differentially labeled in a combinatorial fashion and mixed together, the result is a 24-color probe set. Hybridization of this probe set to a given metaphase spread allows detection of each human chromosome in a distinct hue. As the emission spectra for the commonly used haptens are clustered and partially overlap between 480 and

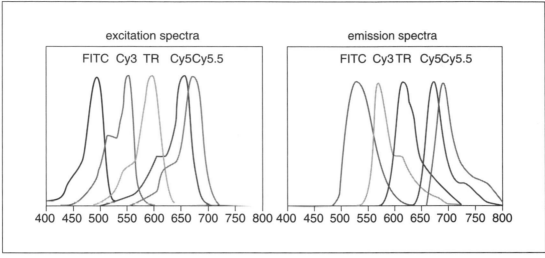

Figure 4.6.2 The absorption and emission spectra of five commonly used fluorochromes in multicolor FISH experiments. The *x* axis represents wavelength (nm) and the *y* axis indicates fluorescence intensity (in arbitrary units). Fluorescein isothiocyanate (FITC), Cy2, and Spectrum Green share similar spectra. Cy3, Spectrum Orange, and rhodamine share similar spectra. Texas Red (TR), Cy3.5, and Spectrum Red share similar spectra. Cy5 and Spectrum Far Red share similar spectra.

700 nm (Fig. 4.6.2), modified fluorescence microscopy and specialized software are needed to recognize the color combinations for each chromosome. Spectral karyotyping (SKY; Schröck et al., 1996) and multifluor/multiplex FISH (M-FISH; Speicher et al., 1996) are two systems designed for this function.

With the SKY system, fluorescent light passes through a triple-band-pass dichroic filter, which allows for the simultaneous excitation and detection of all fluorochromes used. Fluorescence emissions from the sample are split into two separate beams by the interferometer. The beams are then recombined with an optical path difference that forces them out of register and causes them to interfere with each other. Fourier transformation then translates the information from the interferometer and determines the complete spectrum projected onto each pixel of an integrated CCD camera. Each pixel's spectrum is then compared to the spectra of all five fluorochromes used in the experiment to determine the fluorochrome combination producing that particular spectrum. Once the fluorochrome combination has been determined, it is compared to a known probe-labeling scheme (Fig. 4.6.3) to identify the corresponding chromosome material. The spectral image of the metaphase spread can be karyotyped using specific pseudocolors for maximal visual discrimination of the chromosomes.

The M-FISH system uses five single-band-pass excitation/emission filter sets where each filter set is specific for one of the five fluorochromes used in the combinatorial labeling process. Five separate exposures are taken (one for each filter set) and combined to create a multicolor image. Most M-FISH systems now have motorized filter set selectors, which minimize any image registration shifts that may occur between exposures. For each pixel, the presence/absence of signal from each of the five fluorochromes is assessed and then matched to the probe-coding scheme to determine which chromosome material is present. As with SKY, an image of the metaphase spread is converted into a karyotype and specific pseudocolors are used to maximize visual discrimination of the chromosomes.

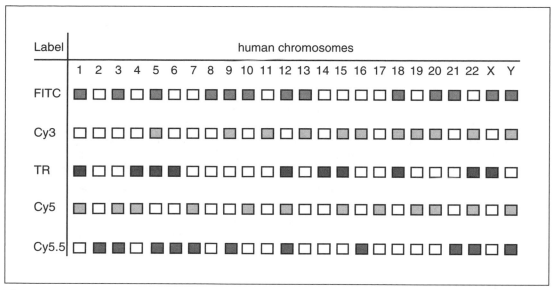

Figure 4.6.3 A combinatorial labeling scheme for the production of a 24-color FISH experiment. Each chromosome painting probe has a distinct spectral signature based on its fluorochrome composition.

SKY systems are available from Applied Spectral Imaging. M-FISH modules are available for cytogenetic image analyses platforms of Applied Imaging, Leica Microsystems, Metasystems, Perceptive Scientific Instruments, and Vysis (distributed by Applied Imaging). SKY uses FITC, rhodamine (comparable to Cy3), Texas Red, Cy5, and Cy5.5. For M-FISH, companies often use unique fluorochromes and fluorochrome combinations. Because SKY and M-FISH differ only in the way they capture and analyze multicolor images, the same probe set used for SKY could also theoretically be used in M-FISH experiments.

SKY and M-FISH systems permit rapid identification of interchromosomal rearrangements (i.e., chromosomal translocations between nonhomologous chromosomes) without prior knowledge of the specific chromosome aberrations involved. Both systems are also capable of identifying the chromosomal origins of extra chromosomal material. Chromosome regions containing heterochromatin-associated moderately/highly repetitive DNA will either be unlabeled or labeled with a mixture of fluorochromes, yielding uninformative data. In cases with double minutes, determining their chromosomal origin by SKY or M-FISH will also indicate the chromosomal origin of the respective DNA amplifications. However, these systems were not specifically designed to detect intrachromosomal amplifications, deletions, or inversions.

The SKY system reports a 98% pixel classification accuracy with a chromosomal translocation detection sensitivity of 0.5 to 1.5 Mb (Schröck et al., 1996). Speicher et al. (1996) and Eils et al. (1998) claim a similar specificity and sensitivity for M-FISH systems. Of course, with either system, the sensitivity is highly dependent on the quality of the metaphase spreads obtained for testing. Next to chromosome integrity after pretreatments and DNA denaturation, chromosome length appears to be the most important factor affecting system resolution. Hence, short chromosomes obtained from samples such as tumors will naturally have decreased sensitivity. Further complications can also occur during the analysis of overlapping chromosomes or with the translocation of relatively small amounts of chromosomal material. In these situations, overlapping fluorochrome signals may result in the erroneous assignment of a chromosome region. It may therefore be desirable to confirm many SKY/M-FISH results with more conventional single- or dual-color FISH experiments.

Rx-FISH

Although gibbon and human DNA share 98% homology, human chromosomes are extensively rearranged with respect to gibbon chromosomes, so that a probe derived from a gibbon chromosome will hybridize to a specific set of human subchromosomal regions (Fig. 4.6.4). When all gibbon chromosomes are combinatorially labeled and simultaneously hybridized to human metaphase chromosomes, a characteristic banding pattern emerges. The Rx-FISH probe set (Applied Imaging) is generated from the chromosomes of a gibbon species known to have the most extensively rearranged karyotype compared to humans. Chromosome-specific paint probes from all 27 gibbon chromosomes are combinatorially labeled with three fluorochromes: Cy3, Cy5, and FITC. When hybridized onto normal human chromosomes, the Rx-FISH probe set produces a pattern of 90 bands in eight distinct colors for a human haploid genome. Seven colors are generated from the combinatorial labeling process, and regions of no hybridization (usually blocks of centromeric heterochromatin) are black. Human chromosomes 15, 18, 21, 22, X, and Y do not exhibit banding patterns and are each painted in a single color. One factor

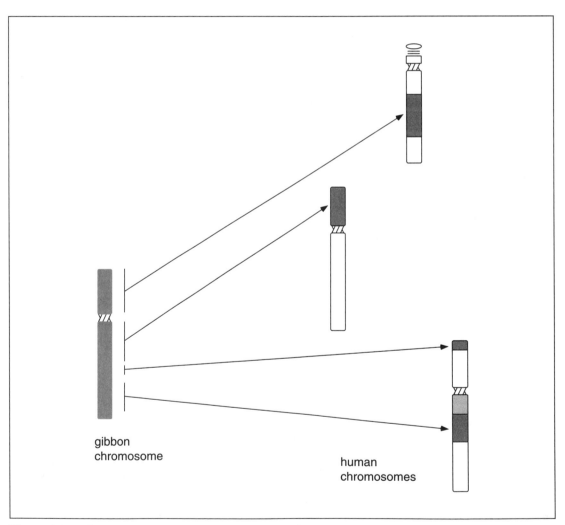

Figure 4.6.4 Schematic representation of subregional homologies between a gibbon chromosome and several human chromosomes. In this example, a chromosome-specific gibbon painting probe hybridizes to four distinct homologous segments on three different human chromosomes.

limiting the resolution of the Rx-FISH probe set is the number of chromosomal rearrangements that naturally separate the karyotype of this gibbon species from humans. By incorporating probes from other gibbon species, the banding resolution can be increased from 90 to 98 bands. Additional bands are produced on human chromosomes 2, 5q, 10p, 11q, 14q, and 16p. Adding two more fluorochromes in the combinatorial labeling scheme could also increase the resolution of the Rx-FISH probe set; however, the increased benefits to the user may not warrant the rise in probe cost and need for new hardware and software for processing FISH data. Rx-FISH images can be captured with any appropriate three-filter system and classified manually. Alternatively, the Rx-FISH system provided by Applied Imaging enables automatic image analyses.

Although Rx-FISH has the potential of identifying interchromosomal rearrangements, translocation identifications must involve chromosome regions of different colors. Translocations involving bands of the same color will be missed by this system. Rx-FISH also has the potential of identifying the origin of extrachromosomal material. However, when such chromosomes are painted in one color, they may indicate several possible chromosomal origins. The major advantage of Rx-FISH appears to be its ability to detect certain chromosome inversions. Paracentric inversions that occur within the confines of a single color band, however, will go undetected. Wherever feasible, Rx-FISH should be combined with traditional banding and other molecular cytogenetic techniques to establish the most accurate karyotypic interpretation.

It has been argued that the color banding patterns produced by Rx-FISH are not as sensitive to chromosome contraction state as traditional banding techniques. If so, this could prove to be useful in tumor cytogenetics, where a low mitotic index, highly condensed chromosomes, complex chromosomal rearrangements, and the multiclonal origin of tumor cells often limit the information gathered by standard G-banding techniques

COMPARATIVE GENOMIC HYBRIDIZATION

Comparative genomic hybridization (CGH; *UNIT 4.8*) allows the detection of intrachromosomal amplifications or deletions. CGH scans the entire human genome for differences in DNA sequence copy number in a single hybridization experiment. This technique has been especially useful for screening potential sites of proto-oncogenes (with amplifications) or tumor suppressor genes (with deletions) in various tumor samples. CGH can also detect gains or losses due to unbalanced chromosome translocations. In a typical CGH experiment, DNA from a test tissue sample is labeled with one hapten and combined in equimolar ratios with normal reference genomic DNA, which is labeled with a second hapten. The probe mixture is concurrently hybridized to normal "target" metaphase chromosomes. If the DNA from the unknown sample is directly labeled/indirectly detected with a green fluorochrome, and the normal reference DNA is directly labeled/indirectly detected with a red fluorochrome, most chromosome regions will appear in yellow fluorescence. DNA amplifications in the unknown sample will increase the green-to-red molar ratio at the amplified locus. Conversely, DNA deletions in the unknown sample will decrease the green-to-red ratio. CGH analysis programs determine the relative red and green fluorescence intensities along the whole chromosome length, subtract background signals, and graph the remaining fluorescence profiles. These programs are available from Applied Imaging, Leica Microsystems, Perspective Scientific Instruments, and Vysis (distributed through Applied Imaging).

Unlike the previously mentioned FISH techniques, CGH utilizes genomic DNA rather than metaphase chromosomes. This means that the specimen to be tested does not have to be fresh or even have intact nuclei. Cell-culturing artifacts are no longer a consideration, and very high-molecular-weight DNA is not required. Moreover, when a given tumor sample is rare and only available in limited amounts of archived paraffin-embedded material, DOP-PCR can be used reliably to amplify and label as little as a few nanograms of the isolated template DNA.

The detection sensitivity of CGH depends on the size of the affected chromosome region, the degree of amplification or deletion, the proportion of cells in the sample carrying the chromosome gain/loss, and the contraction of the "target" metaphase chromosomes. When optimal metaphase chromosomes are prepared or purchased and the tested sample is assumed to have a relatively homogenous genetic composition (not usually the case for most tumor specimens), detection sensitivity becomes primarily determined by the product of the copy number variation and size of the aberrant chromosome region. In such ideal cases, a 50% copy number increase (i.e., an increase from two to three copies of a nonrepetitive DNA sequence in diploid cells) could theoretically be detected if the duplicated chromosome region is at least 2 Mb. Practically, DNA amplifications are usually detected when the size of the aberrant chromosome region exceeds 5 Mb. FISH experiments on chromosome spreads from the specimen being tested with a panel of probes from the region of interest may serve to narrow down the region of involvement indicated by CGH.

References: Cremer et al., 1988; Kallioniemi et al., 1994; Lichter et al., 1988; Pinkel et al., 1988; Schröck et al., 1996; Speicher et al., 1996

Contributors: Charles Lee, Willem Rens, and Fengtang Yang

UNIT 4.7

Morphology Antibody Chromosome Technique for Determining Phenotype and Genotype of the Same Cell

In planning a MAC experiment, a number of choices must be made (see outline in Table 4.7.1)—i.e., the method of specimen preparation, the method of phenotyping cells (also see Table 4.7.2), the method of genotyping cells (also see Table 4.7.3), and whether the phenotype and genotype will be analyzed simultaneously or sequentially. Phenotyping by immunoanalysis with APAAP and Giemsa counterstaining is suitable for all kinds of cells and preparations, and is one of the most sensitive techniques for detecting antigens. Alternatively, immunoreagents can be conjugated with horseradish peroxidase or a fluorochrome. Genotyping by in situ hybridization (ISH) is suitable for both interphase and metaphase cells, whereas chromosome banding can only be used for metaphase cells. It is also possible to use ISH and chromosome banding in combination. Generally, simultaneous detection of phenotype and genotype is faster than sequential detection, but less accurate. Simultaneous detection of APAAP immunophenotype and enzymatic ISH signals is recommended when only antigenic properties are studied in a large number of interphase cells using repeat-sequence probes. It is difficult to detect simultaneous signals when metaphase chromosomes are genotyped by FISH with whole chromosome paint probes and when interphase cells are genotyped using a single-copy probe after APAAP and immunoperoxidase immunophenotyping, as the APAAP and immunoperoxidase reactions prevent detection of small enzymatic or fluorescent signals. However, after immunofluorescence phenotyping, it is possible to simultaneously detect even single-copy probes by fluorescence. Simultaneous fluorescence detection of phenotype and ISH signals is recommended when only antigenic properties are studied in a large number of interphase cells using unique-sequence hybridization probes (e.g., cosmids, contigs, YACs). Sequential detection of phenotype and genotype is recommended when metaphase cells from cytospin or in situ culture preparations are genotyped using FISH and whole chromosome paint probes, when precise cellular morphology or histology is required on sections and smears, or when metaphase cytospin or in situ culture preparations are studied using G-banding.

Table 4.7.1   Combinations of Preparative and Analytical Techniques for MAC Analysis[a]

Function	Method			
Specimen preparation	Cytospin (Support Protocol 1)			
	In situ culture (Support Protocol 2)			
	Tissue sections (Support Protocol 3)			
	Blood and bone marrow smears (Support Protocol 4)			
Phenotyping	APAAP (Basic Protocol 1)			
	Immunoperoxidase (Alternate Protocol 1)			
	Immunofluorescence (Alternate Protocol 2)			
Genotyping	System	Probe	Detection	Cell type
	ISH	Repeat-sequence	Enzymatic	I, M
	ISH	Repeat-sequence	Fluorescent	I, M
	ISH	Single-copy	Fluorescent	I, M
	ISH	Whole chromosome paint	Fluorescent	M
	G-banding	NA	Giemsa	M
	Q-banding	NA	Fluorescent	M
	C-banding	NA	Giemsa	M

[a]Abbreviations: APAAP, alkaline phosphatase anti–alkaline phosphatase; I, interphase; ISH, in situ hybridization; M, metaphase; NA, not applicable.

4

Table 4.7.2   Techniques Used for MAC Phenotyping

Method	Protocol	Cell type	Comments[a]
APAAP immunocyto-chemistry	Basic 1	All kinds	A: Sensitive in detection of weak antigens; all kinds of cells can be used; permanent preparations; fair cellular morphology D: Small in situ hybridization signals are not detectable in simultaneous analysis; time-consuming
Immunoperoxidase	Alternate 1	Other than myeloid cells	A: Faster than APAAP R: Endogenous peroxidase activity cannot be fully eliminated D: Small in situ hybridization signals are not detectable in simultaneous analysis
Immunofluorescence	Alternate 2	All kinds	A: All kinds of cells can be used; fast; small hybridization signals can be detected simultaneously R: Fluorescence microscope required D: Fluorescence fades over time; antibodies of weak antigens do not work

[a]A, advantage(s); D, disadvantage(s); R, restriction(s).

Table 4.7.3 Techniques Used for MAC Genotyping

Method	Protocol	Comments[a]
In situ hybridization	Basic 1	A: Either interphase or metaphase cells can be studied; success rate higher than in chromosome banding R: Specific probes needed
Chromosome banding	Basic 1 and Alternate 3	A: Gives an overview of chromosomal changes R: Metaphase cells needed D: Success rate low; small number of cells available for analysis
Chromosome banding followed by in situ hybridization	Basic 2	A: Gives an overview of chromosomal changes and allows targeted analysis of selected areas in the genome R: Metaphase cells and specific probes needed D: Low success rate

[a] A, advantage(s); D, disadvantage(s); R, restriction(s).

NOTE: All incubations are performed in a humidified $37°C$, 5% CO_2 incubator unless otherwise specified.

NOTE: Table 4.7.4 presents a MAC troubleshooting guide.

BASIC PROTOCOL 1

SEQUENTIAL MAC ANALYSIS USING APAAP IMMUNOSTAINING AND IN SITU HYBRIDIZATION

See Fig. 4.7.1 for an outline of this procedure and Fig. 4.7.2 for an illustration of the APAAP detection scheme. Other substrates for alkaline phosphatase and other chromogens are also suitable (see Table 4.4.1).

Materials (see APPENDIX 1 for items with ✓)
   Air-dried slide containing cells for analysis (see Support Protocols 1 to 4)
✓ Formaldehyde-buffered acetone fixative, 4°C
   0.05% (v/v) Triton X-100 (BDH) in PBS
✓ PBS
✓ TBS/FBS
   Mouse anti-human primary antibody diluted 1:10 to 1:50 in TBS/FBS
✓ Rabbit anti-mouse Ig antiserum solution
   APAAP mouse immunocomplex (Dakopatts) diluted 1:25 to 1:50 in TBS/FBS
✓ APAAP substrate solution (prepare immediately before use)
✓ 5% Giemsa stain
   95% ethanol
   3:1 (v/v) methanol/acetic acid fixative
   0.01 N HCl, 37°C
   0.1 to 1 mg/ml pepsin in 0.01 N HCl (add pepsin immediately before use)

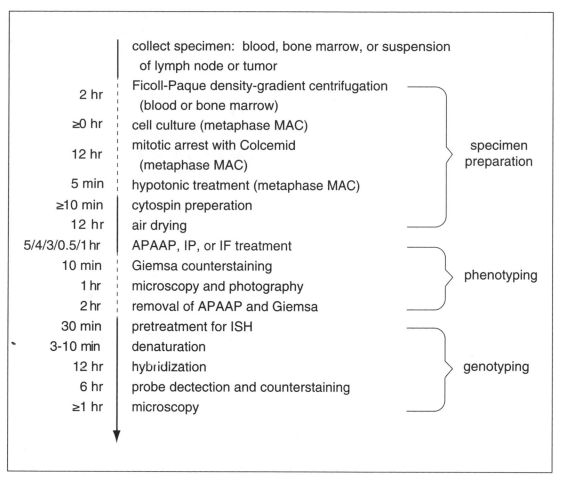

	collect specimen: blood, bone marrow, or suspension of lymph node or tumor
2 hr	Ficoll-Paque density-gradient centrifugation (blood or bone marrow)
≥0 hr	cell culture (metaphase MAC)
12 hr	mitotic arrest with Colcemid (metaphase MAC)
5 min	hypotonic treatment (metaphase MAC)
≥10 min	cytospin preperation
12 hr	air drying
5/4/3/0.5/1 hr	APAAP, IP, or IF treatment
10 min	Giemsa counterstaining
1 hr	microscopy and photography
2 hr	removal of APAAP and Giemsa
30 min	pretreatment for ISH
3-10 min	denaturation
12 hr	hybridization
6 hr	probe dectection and counterstaining
≥1 hr	microscopy

specimen preparation

phenotyping

genotyping

Figure 4.7.1 Outline and time requirements for sequential MAC method with ISH. Solid line indicates steps that are performed for all protocols; dotted line indicates optional steps.

Biotin- or digoxigenin-labeled repeat-sequence probe, single-copy probe, or whole chromosome paint probe (*UNIT 4.4*)
Harris hematoxylin stain (Papanicolaou solution, Merck) diluted 1:1 in H_2O
0.0025% (v/v) ammonia solution
Mounting medium: Entellan (Merck) or other standard cytogenic mounting medium for enzymatic detection; antifade mounting medium for fluorescence detection (*APPENDIX 1*)

Coplin jars
Cytospin filters (Shandon/Lipshaw)
Moist chamber (Fig. 4.4.3)

0.45-μm filter membrane (Millipore)
Bright-field microscope equipped with camera, stage micrometer, and 100× dry objective
Microscope equipped with camera, stage micrometer, fluorescence epiillumination, appropriate filter sets, and 100× objective (optional, for fluorescence detection)

1. Fix an air-dried slide containing cells for analysis 1 min in a Coplin jar containing 4°C formaldehyde-buffered acetone fixative (cytoplasm and cell membrane structure must be intact).

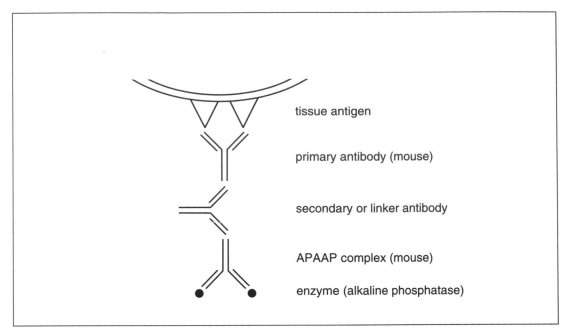

tissue antigen

primary antibody (mouse)

secondary or linker antibody

APAAP complex (mouse)

enzyme (alkaline phosphatase)

Figure 4.7.2 Schematic illustration of alkaline phosphatase anti-alkaline phosphatase (APAAP) immunostaining.

2. Wash 2 min under cold running tap water.

3. *Optional:* If an antigen other than a membrane antigen is being studied, permeabilize the cell membrane by adding 0.05% (w/v) Triton X-100 in PBS to the cells and incubating 10 min at room temperature. Wash the slide 10 min in two changes of PBS.

4. Rinse briefly in a Coplin jar containing TBS/FBS, then wash 10 min in another Coplin jar containing TBS/FBS.

5. Use a cytospin filter to blot excess liquid from the slide.

6. Add 15 μl diluted primary mouse anti-human antibody to cells on the slide and incubate 30 min at room temperature in a moist chamber. Repeat steps 4 and 5.

7. Add 15 μl diluted rabbit anti-mouse Ig antibody (secondary or linker antibody; see Fig. 4.7.2) to the cells. Incubate the slide 30 min at room temperature in a moist chamber. Repeat steps 4 and 5.

8. Add 15 μl diluted APAAP mouse immunocomplex to the cells. Incubate the slide 30 min at room temperature in a moist chamber. Repeat steps 4 and 5. If the APAAP reaction is weak, intensify the signal by repeating steps 7 and 8.

9. Filter the APAAP substrate solution through an 0.45-μm filter directly onto the slide. Incubate the slide 20 min at room temperature in a moist chamber. Wash 2 min under cold running tap water. Air dry overnight at room temperature.

   Slides can be counterstained with Giemsa (steps 10 to 12) to visualize chromosomal morphology. If this is not necessary, proceed to step 13.

10. Counterstain slide 10 min in a Coplin jar containing 5% Giemsa stain. Wash briefly in cold running tap water.

11. Examine slide (without mounting) using a bright-field microscope equipped with a dry 100× objective. Photograph the cells using the stage micrometer to note the exact coordinates of positive cells. Use Ektachrome 64T for color slides.

12. Soak the slide 10 min in a Coplin jar containing 95% ethanol. Incubate in a Coplin jar containing 3:1 methanol/acetic acid fixative 1 to 12 hr at 20°C.

13. Pretreat the slide as follows:

> Incubate 10 min at 37°C in 0.01 N HCl
> Incubate 4 to 10 min at 37°C in 0.1 to 1 mg/ml pepsin in 0.01 N HCl
> Rinse briefly in each of three Coplin jars containing distilled water at room temperature.

Optimize probe penetration by varying the pepsin centration and treatment time. Also see conditions in Support Protocols 2 to 4.

14. Air dry the slide 10 min at room temperature.

15. Hybridize a biotin- or digoxigenin-labeled repeat sequence probe (for simultaneous or sequential detection) to the slide using the procedure described in steps 1 to 8 of the Basic Protocol in *UNIT 4.4*. Alternatively, use single-copy or whole chromosome paint probes (requires sequential detection).

For enzymatic detection

16a. Follow the procedure described in Alternate Protocol 1 of *UNIT 4.4* to enzymatically detect the hybridized biotinylated probe (other methods can also be used).

17a. Incubate the slide 30 sec at room temperature in Harris hematoxylin stain. Rinse briefly in 0.0025% (v/v) ammonia at room temperature. Wash briefly in water at room temperature.

18a. Mount the slide using Entellan or any standard cytogenic mounting medium. Do not use Entellan for simultaneous analysis because it may remove the APAAP reaction product.

19a. Use a 100× oil immersion objective (or dry objective if slide is not mounted) to examine the slide and photograph the same fields that were photographed in step 11.

For fluorescence detection

16b. Follow the procedure described in steps 8 to 11 of the Basic Protocol in *UNIT 4.4* and the amplification described in Support Protocols 1 or 2 of *UNIT 4.4* to detect the hybridized biotinylated probe by fluorescence.

17b. Counterstain fluorescently labeled samples with DAPI or propidium iodide (steps 12 to 13 of the Basic Protocol in *UNIT 4.4*).

18b. Mount the slide using antifade mounting medium.

19b. Examine the slide with a 100× objective and, for sequential analysis, study the same fields that were photographed in step 11.

ALTERNATE PROTOCOL 1

IMMUNOPHENOTYPING USING HRP-BASED DETECTION

Bound primary antibody can be detected using horseradish peroxidase (HRP) instead of APAAP, which is more time consuming, to detect bound primary antibody (see Fig. 4.7.3). Some antibodies (e.g., anti–B cell markers) are actually detected better. Additionally, immunoperoxidase detection is more sensitive than indirect immunofluorescence. Immunoperoxidase phenotyping and ISH signals can be studied simultaneously or sequentially with either

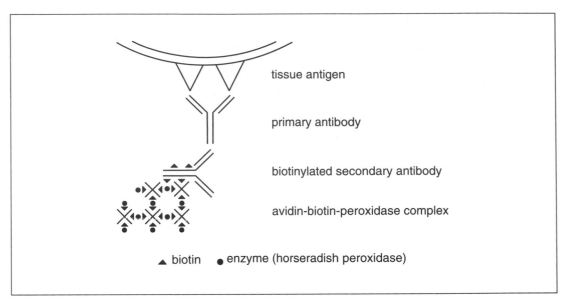

Figure 4.7.3 Schematic illustration of immunoperoxidase staining.

enzymatic or fluorescent detection of hybridized probes. Repeat sequence probes can be used for simultaneous and sequential detection; single-copy probes and paint probes can be used for sequential detection. If banding is used to genotype cells with endogenous peroxidase activity, the cells should be phenotyped using APAAP or immunofluorescence.

Additional Materials (*also see Basic Protocol 1; see* APPENDIX 1 *for items with* ✓)

1% (v/v) H_2O_2 in methanol
✓ Acetone/formaldehyde fixative
PBS/FBS: 0.08%, 0.2%, and 5% (v/v) filter-sterilized (1.2-μm filter) FBS in PBS
Normal horse serum diluted 1:50 in PBS/5% (v/v) FBS
Mouse anti-human primary antibody diluted 1:10 to 1:50 in PBS/5% FBS
Biotinylated anti-mouse IgG secondary antibody diluted 1:250 in PBS/5% FBS
ABC kit (Vectastain) containing avidin DH and biotinylated horseradish peroxidase H (HRP)
✓ Immunoperoxidase substrate solution

1. If necessary, block endogenous peroxidase activity by incubating the slide 30 min at room temperature in 1% (v/v) H_2O_2 in methanol.

2. Fix an air-dried slide containing cells for analysis 1 min in a Coplin jar containing 4°C acetone/formaldehyde fixative. Rinse the slide 2 to 3 min under cold running tap water.

3. *Optional:* If an antigen other than a membrane antigen is being studied, permeabilize cell membrane by adding 0.05% (w/v) Triton X-100 in PBS and incubating 10 min at room temperature. Wash the slide 10 min in two changes of PBS.

4. Wash slide briefly in a Coplin jar containing PBS/0.08% FBS and blot with a cytospin filter. Keep the cells wet until step 10.

5. Add 25 μl diluted normal horse serum to the cells. Incubate 30 min at room temperature in a moist chamber. Drain any excess serum and blot the slide briefly.

6. Add 20 μl mouse anti-human primary antibody to the cells and incubate 1 hr at room temperature in a moist chamber. Rinse briefly in PBS/0.2% FBS, then wash 10 min in fresh PBS/0.2% FBS. Blot the slide.

7. Add 20 μl diluted biotinylated anti-mouse IgG secondary antibody. Incubate 30 min at room temperature in a moist chamber. Rinse and blot as in step 6.

8. Mix equal volumes of avidin DH and biotinylated HRP to form avidin-biotin complex. Dilute the complex 1:160 in PBS/5% FBS.

9. Add 20 μl diluted avidin-biotin complex to the cells. Incubate the slide 30 min at room temperature in a moist chamber. Wash briefly in a Coplin jar containing PBS/0.2% FBS and blot.

10. Incubate slide 20 min at room temperature in a Coplin jar containing immunoperoxidase substrate solution. Wash briefly in a Coplin jar containing PBS and again briefly in a Coplin jar containing distilled water. Air dry overnight at room temperature.

11. Counterstain the slide with Giemsa (optional; see Basic Protocol 1, steps 10 to 12) and perform ISH (see Basic Protocol 1, step 13 onward). Use a standard cytological mounting medium and view with a dry 100× objective, as Entellan and immersion oil both remove the brown reaction product.

ALTERNATE PROTOCOL 2

IMMUNOPHENOTYPING USING FLUORESCENCE DETECTION

Immunofluorescence phenotyping can be used for either sequential or simultaneous analysis with enzymatic or fluorescence detection of in situ hybridized probes. Additionally, although it is less sensitive than APAAP or HRP, it is faster and, with careful selection of fluorochromes, allows simultaneous detection of fluorescently labeled single-copy ISH probes with immunofluorescence phenotyping results. Single-copy or repeat-sequence probes are suitable for ISH with interphase nuclei, and whole chromosome paint probes are suitable for ISH with metaphase chromosomes.

Additional Materials (also see Basic Protocol 1; see APPENDIX 1 for items with ✓)
 ✓ Acetone/formaldehyde fixative
   PBS/FBS: 0.2% (v/v) filter-sterilized FBS in PBS
   Mouse anti-human primary antibody diluted 1:10 to 1:30 in PBS/0.2% FBS
   Fluorochrome-conjugated goat anti-mouse IgG secondary antibody (Cappel) diluted
     1:200 in PBS/0.2% FBS
 ✓ 0.5% (w/v) quinacrine dihydrochloride

1. Fix air-dried slide containing cells for analysis 1 min in a Coplin jar containing 4°C acetone/formaldehyde fixative and wash 2 to 3 min under cold running tap water.

2. *Optional:* If an antigen other than a membrane antigen is being studied, permeabilize the cell membrane by adding 0.05% Triton X-100 in PBS to the cells and incubating 10 min at room temperature. Wash the slide 10 min in two changes of PBS.

3. Wash the slide briefly in a Coplin jar containing PBS/0.2% FBS. Blot the slide with a cytospin filter. Keep the cells wet until the completion of the protocol.

4. Add 20 μl diluted mouse anti-human primary antibody to the cells. Incubate 30 min at room temperature in a moist chamber. Wash twice briefly in a Coplin jar containing PBS/0.2% FBS, then blot.

5. Add 20 μl diluted fluorochrome-conjugated goat anti-mouse IgG secondary antibody to the cells. Incubate 30 min at room temperature in a moist chamber. Wash briefly in a Coplin jar containing PBS/0.2% FBS.

6. Counterstain 10 min in a Coplin jar containing 0.5% quinacrine dihydrochloride. Evaluate Q-bands before proceeding with FISH.

7a. *For sequential analysis:* Examine the slide using a fluorescence microscope equipped with an appropriate filter, then remove the fluorescence (see Basic Protocol 1, step 12) and proceed to ISH (see Basic Protocol 1, step 13 onward).

7b. *For simultaneous analysis:* Proceed directly to ISH (see Basic Protocol 1, step 13 onward). For fluorescent probes, be sure to use different fluorochromes than were used for immunophenotyping.

ALTERNATE PROTOCOL 3

GENOTYPING BY G- OR C-BANDING OF CHROMOSOMES

Cytospin preparations and in situ cultures (see Support Protocols 1 and 2) can be used to generate metaphase chromosomes that can be genotyped using any of the chromosome banding techniques described in *UNIT 4.3* (Fig. 4.7.4). Any of the banding methods can be used with any of the phenotyping methods. If immunofluorescence phenotyping is performed, Q-banding may be performed without any refixation after immunostaining. However, G- and C-banding, as well as sister chromatid differential staining (*UNIT 8.4*), require acidic refixation after phenotyping. G- and C-banding are viewed with a bright-field microscope, and can be combined with APAAP or HRP immunostaining. If desired, ISH can performed after banding (see Basic Protocol 2).

Additional Materials *(also see Basic Protocol 1)*

Cells from cytospin or in situ culture preparation (see Support Protocols 1 and 2)
60°C oven

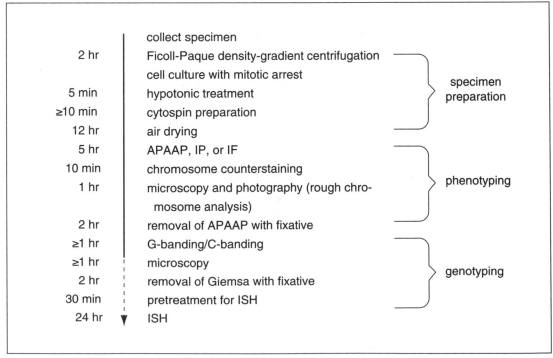

Figure 4.7.4 Outline and time requirements for MAC method with chromosome banding. Solid line indicates steps that are performed for all protocols; dotted line indicates optional steps.

1. Phenotype cells of cytospin or in situ culture preparations using APAAP (see Basic Protocol 1, steps 1 to 10), HRP (see Alternate Protocol 1), or immunofluorescence detection (see Alternate Protocol 2).

2. Incubate the slide 10 min in a Coplin jar containing 95% ethanol.

3. Incubate 1 to 12 hr at 20°C in a Coplin jar containing 3:1 methanol/acetic acid fixative.

4. Air dry 18 to 24 hr in a 60°C oven.

5. Perform G- or C-banding (*UNIT 4.3*). For G-banding, increase trypsinization time to 10 min.

6. *Optional:* Carry out in situ hybridization (see Basic Protocol 2).

SUPPORT PROTOCOL 1

PREPARATION OF CYTOSPIN SLIDES

Mitotically active cells are treated with mitogen to increase the number of metaphase chromosomes (Fig. 4.7.5); if the cells are not mitotically active (e.g., pure granulocytes) or if interphase cells are being analyzed, mitogen stimulation is eliminated. Cytospin preparations of suspension cells are superior to dropped or air-dried cells (*UNIT 4.1*) and smear preparations.

Materials *(see APPENDIX 1 for items with ✓)*

Specimen (e.g., heparinized blood, bone marrow specimen, suspended tumor cells) or cultured cells (suspension culture or trypsinized monolayer cells)
Immunomagnetic beads coated with desired antibodies (e.g., Dynal Dynabeads M-450, Miltenyi Biotec MiniMACSO; optional)
✓ Complete RPMI/20% FBS
Mitogen: 12-*O*-tetradecanoylphorbol-13-acetate (TPA, Sigma; for B lymphocytes), phytohemagglutinin (PHA, Sigma; for T lymphocytes), or other mitogen or growth factors (optional)
Colcemid (optional)
✓ MAC hypotonic solution, 20°C

50-ml tissue culture flask (Nunc)
10-ml disposable conical polypropylene centrifuge tube
Cytospin 2 (Shandon/Lipshaw)
Ethanol-cleaned glass slides
Cytospin filters (Shandon/Lipshaw)

1. If necessary, remove erythrocytes from the specimen by Ficoll-Hypaque density-gradient centrifugation (*UNIT 10.3*) or dextran sedimentation. If desired, further treat isolated mononuclear leukocytes with immunomagnetic beads coated with appropriate monoclonal antibodies to isolate specific cell subsets.

2. Resuspend mononucleated cells at 10^6 cells/ml in complete RPMI/20% FBS. Add 5 to 10 ml cell suspension into a 50-ml tissue culture flask. If desired, add mitogen to the cultures to increase the number of metaphases (e.g., 0.5 μg/ml PHA for T lymphocytes, 2 μg/ml TPA for B lymphocytes in chronic lymphocytic leukemia).

3. Incubate 12 hr to 4 days in a humidified 37°C, 5% CO_2 incubator.

4. For metaphase cells, add Colcemid to a final concentration of 0.4 μg/ml as soon as culture is started (for bone marrow cells) or during exponential growth phase (for other cell types). Continue incubating an additional 12 hr at 37°C. Optimize conditions, if necessary, using longer incubation times and higher Colcemid concentrations to achieve a maximum

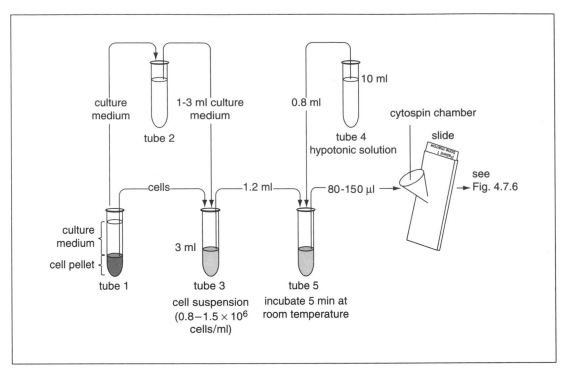

Figure 4.7.5 Diagram of hypotonic treatment and cytospin slide preparation.

number of mitoses, and shorter times to avoid adverse affects on chromosome morphology, length, and banding.

5. Transfer the cells and medium into a disposable 10-ml conical polypropylene centrifuge tube (tube 1 in Fig. 4.7.5) and centrifuge 10 min at 180 × g, room temperature. Pour all supernatant (medium) into another tube (tube 2 in Fig. 4.7.5) and save it.

6. Resuspend the cell pellet and count the cells using a hemacytometer (APPENDIX 3I). Add enough supernatant to make a suspension at 0.8–1.5 × 10⁶ cells/ml (tube 3 in Fig. 4.7.5).

7. Transfer 1.2 ml cell suspension (enough for 12 slides) and 0.8 ml hypotonic solution into a disposable 10-ml conical centrifuge tube (tube 5 in Fig. 4.7.5) and mix (4–7.5 × 10⁵ cells/ml final). Incubate 5 min at room temperature. Use the suspension in step 6 or a similar suspension of granulocytes, cultured suspension cells, or trypsinized monolayer cells that are not mitogen-stimulated for preparation of interphase cells.

8. Assemble a cytospin slide holder using an ethanol-cleaned slide and cytospin filter. Transfer 80 to 150 µl hypotonic cell suspension into the cytospin chamber and centrifuge 5 to 10 min at 400 × g, room temperature.

9. For additional slides, repeat steps 7 and 8 for a new aliquot of cell suspension. Prepare only one aliquot at a time. For a typical experiment, prepare 10 to 20 slides.

10. Air dry cytospin-prepared slides (Fig. 4.7.6) 12 hr at room temperature. Store up to 3 weeks at room temperature, or −20° or −70°C, for MAC phenotyping and genotyping (slides stored 2 to 5 days at room temperature are optimal). If frozen, dry slides 12 to 24 hr at room temperature prior to immunocytochemical procedures and in situ hybridization.

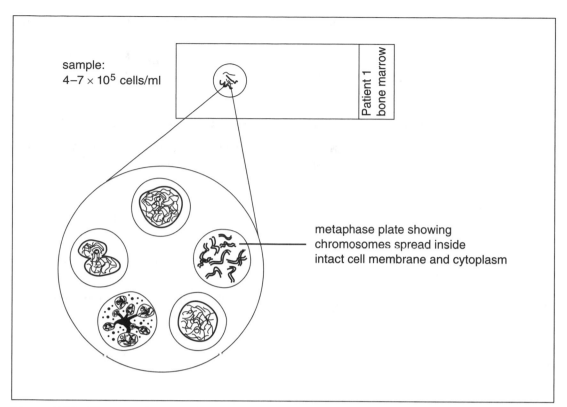

sample:
$4–7 \times 10^5$ cells/ml

Patient 1
bone marrow

metaphase plate showing
chromosomes spread inside
intact cell membrane and cytoplasm

Figure 4.7.6 Diagram of a cytospin preparation.

SUPPORT PROTOCOL 2

PREPARATION OF IN SITU CULTURES

MAC hypotonic solution can be applied directly to a monolayer of cells. Because mitotic cells are not lost during enzymatic digestion and washing, and trypsin-sensitive antigens are not affected, this procedure is suitable if the number of monolayer tumor cells is low (10^3 to 10^4 cells) and the mitotic index is also low.

Materials *(see* APPENDIX 1 *for items with* ✓ *)*
    Adherent cells (e.g., tumor biopsy or cells, fibroblasts)
✓ Complete RPMI/20% FBS
    Proteolytic enzymes (for biopsy)
    Colcemid
✓ MAC hypotonic solution
    0.01 N HCl, 37°C
    Pepsin: 0.1 to 1 mg/ml in 0.01 N HCl, 37°C
    15-ml flaskette chamber slide (Nunc)

1. Prepare a suspension of cells at a concentration of $\sim3.3 \times 10^5$ cells/ml in complete RPMI/20% FBS medium (see Support Protocol 1, steps 1 and 2, and APPENDIX 3I). For a tumor biopsy, first mince the sample in complete RPMI/20% FBS supplemented with proteolytic enzymes to facilitate tissue disaggregation.

2. Add 3 ml cell suspension to a 15-ml flaskette chamber slide and culture the cells in a humidified 37°C, 5% CO_2 incubator until the cells form a non-overlapping monolayer and/or mitotic figures are observed.

3. Add Colcemid to a final concentration of 0.4 μg/ml and incubate an additional 12 hr (see Support Protocol 1, step 4 for optimization).

4. Pour off the culture medium and mix 2 ml culture medium with 2 ml MAC hypotonic solution. Add mixture back to the chamber and incubate 5 min at 20°C

5. Pour off the hypotonic solution and remove the slide from the chamber.

6. Air dry slide 2 days at room temperature. Store as in Support Protocol 1, step 10.

7. To genotype by ISH after immunochemical or cytochemical phenotyping, remove the reaction product (see Basic Protocol 1, step 12) and pretreat slides (individually until conditions are optimized) as follows:

   Incubate 10 to 30 min at 37°C in 0.01 N HCl
   Incubate 10 to 30 min at 37°C in 0.1 to 1 mg pepsin/ml 0.01 N HCl
   Wash in three changes of water.

   Perform on one slide at a time to determine optimal acid treatment time and pepsin concentration and treatment time.

8. Perform ISH using repeat sequence, single-copy, or whole chromosome paint probes with either enzymatic or fluorescence detection (see Basic Protocol 1, steps 14 onward).

SUPPORT PROTOCOL 3

PREPARATION OF TISSUE SECTIONS

In principle, all immunostaining methods are suitable for paraffin and cryostat sections. Both simultaneous and sequential analysis of phenotype and ISH signals are appropriate. Sections allow histology and morphology to be used for cellular characterization. Because two or three consecutive sections represent the same area in the tissue or tumor, phenotyping can be done on one section and genotyping on the next, if analysis is not needed at the single-cell level. ISH yields better results on a section without prior phenotyping. Because only interphase cells are studied, repeat sequence or single-copy probes are used for hybridization.

Material (see APPENDIX 1 for items with ✓)
   5-μm paraffin-embedded or cryostat tissue sections
   Xylenes
   Graded alcohol series: 70%, 95%, and 100% ethanol
   1% (v/v) H_2O_2 in methanol (optional)
   100% methanol (optional)
   1 M sodium thiocyanate, 72°C
   0.01 N to 0.1 N HCl, 37°C
   1 to 5 mg/ml pepsin in HCl, 37°C
✓ Poly-L-lysine-coated microscope slides
   60°C and 72°C ovens

1. Mount 5-μm paraffin-embedded or cryostat tissue sections on poly-L-lysine-coated microscope slides. If using cryostat sections, proceed to step 4.

2. Incubate the slide overnight in a 60°C oven.

3. Remove paraffin by incubating slide in three changes of xylenes, 10 min each.

4. Incubate 5 min each in 100%, 95%, and 70% ethanol. Air dry 12 to 24 hr at room temperature.

5. *Optional:* If phenotypic analysis uses HRP, block endogenous peroxidase (if necessary) by incubating the slide 30 min at room temperature in 1% H_2O_2 in methanol. Then wash 5 min in methanol and air dry 12 to 24 hr at room temperature.

6a. *To genotype paraffin-embedded sections:* Remove phenotypic staining (see Basic Protocol 1, step 12; skip for simultaneous analysis) and pretreat as follows:

> Incubate 10 min at 72°C in 1 M sodium thiocyanate
> Incubate 10 to 30 min at 37°C in 0.01 to 0.1 N HCl
> Incubate 10 to 30 min at 37°C in 1 to 5 mg/ml pepsin in 0.01 to 0.1 N HCl
> Wash in three changes of water.

Perform on one slide at a time to optimize concentrations and treatment times for both HCl and pepsin. Use HCl of the same concentration for washes with and without pepsin.

6b. *To genotype cryostat sections:* Remove phenotypic staining (see Basic Protocol 1, step 12) and pretreat as follows:

> Incubate 5 min at 72°C in 1 M sodium thiocyanate
> Incubate 10 min at 37°C in 0.01 N HCl
> Incubate 10 to 20 min at 37°C in 0.1 mg pepsin/ml 0.01 N HCl
> Wash in three changes water.

Perform on one slide at a time to optimize pepsin treatment time.

7. Perform ISH (see Basic Protocol 1, step 14 onward) using repeat sequence or single-copy probes.

SUPPORT PROTOCOL 4

PREPARATION OF BLOOD AND BONE MARROW SMEARS

Making good smears (Fig. 4.7.7) is fast but requires experienced hands. Blood or bone marrow smears can be used for sequential immunophenotyping and genotyping in the same way as cytospin preparations; however, because smears do not contain mitotic cells, they are not equivalent to mitogen-stimulated cytospin preparations.

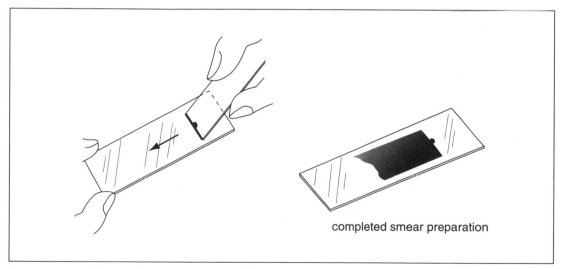

completed smear preparation

Figure 4.7.7 Diagram showing the preparation of a smear.

Materials

> Heparinized blood or bone marrow aspirate
> 0.01 N HCl, 37°C
> 0.1 to 1 mg/ml pepsin in 0.01 N HCl, 37°C
> Ethanol-cleaned microscope slides

1. Place a drop of heparinized blood or bone marrow aspirate near the right edge of an ethanol-cleaned slide.

2. Take a second slide (the spreader) and place it on the first slide to the left of the drop.

3. Hold the spreader slide at an angle of ~40° and move it to the right until it just touches the drop and the drop spreads along the edge of the spreader.

4. Push the spreader slide sideways to the left to spread the drop over the surface of the first slide.

5. Air dry the slide 1 to 12 hr at room temperature.

6. To genotype smears by in situ hybridization, remove the phenotypic staining (see Basic Protocol 1, step 12) and pretreat slides as follows:

> Incubate 10 min at 37°C in 0.01 N HCl
> Incubate 10 to 30 min at 37°C in 0.1 to 1 mg/ml pepsin in 0.01 N HCl
> Wash in three changes of water.

Perform on one slide at a time until pepsin concentration and treatment time are optimized.

7. Perform ISH (see Basic Protocol 1, step 14 onward) using repeat sequence or single-copy probes.

BASIC PROTOCOL 2

IN SITU HYBRIDIZATION ON PREVIOUSLY GTG-BANDED CHROMOSOMES

A GTG-banded MAC preparation or a standard cytogenetic preparation can be used for ISH. Whole chromosome paint probes are generally the probes of choice for tumor cytogenetics or prenatal diagnosis when marker chromosomes and complicated chromosomal aberrations may be present. Cosmid and α-satellite (centromere) repeats are not as useful for these cases, and in the authors' hands, they have not worked well after GTG banding. Only repeat-sequence probes are used for enzymatic detection of hybridization signals; single-copy, repeat-sequence, and whole chromosome paint probes can be used for fluorescence detection. This protocol can be used with unmounted slides ≤1 week old or Entellan-mounted slides that have been stored ≥10 years at room temperature.

Materials (see APPENDIX 1 for items with ✓)

> Unmounted or Entellan-mounted slide with GTG-banded metaphase cells (Alternate Protocol 3 or UNIT 4.3)
> Xylene
> 70%, 80%, 90%, and 100% (v/v) ethanol
> ✓ Formaldehyde-buffered acetone fixative, 4°C
> 2× SSC (APPENDIX 1)/0.2% (v/v) Tween, 42°C
> Whole chromosome paint probe (UNIT 4.4)
> Coplin jars

Table 4.7.4 Troubleshooting Guide for MAC

Problem	Possible cause	Solution
Poor spreading of chromosomes inside cell membrane	Hypotonic treatment not strong enough	Decrease proportion of culture medium in the hypotonic solution; increase force × g in cytocentrifugation
Cell membrane broken	Hypotonic treatment too strong	Increase proportion of culture medium in the hypotonic solution; decrease × g in cytocentrifugation
Fuzzy chromosome morphology/but good immunostaining	Poor chromosome fixation	Increase formaldehyde concentration in fixation
Poor immunostaining/but good chromosome morphology	Inadequate antigen fixation	Decrease formaldehyde concentration in fixation
	Weak antigen expression	Increase antibody concentration; in APAAP, repeat procedures for antibody detection; minimize washing condition after antibody coupling; if possible, use two different antibodies simultaneously (e.g. anti-CD19 and anti-CD22, both of which recognize normal B cells)
No chromosome G-banding (GTG)	Poor chromosome fixation or preparation too old	Perform banding soon after immunostaining
No hybridization signal	Poor penetration of the probe	Increase pepsin (all preparations) and/or NaSCN (except cytospin preparations) concentrations and/or duration of treatment
	Poor DNA denaturation	Decrease formaldehyde concentration
	Poor hybridization	Change denaturation conditions (see *UNIT 4.4*)
		Change hybridization conditions
	Inadequate probe detection	Change hybridization conditions (see *UNIT 4.4*) test the conditions with a standard cytogenetic preparation
Signal background	Too high probe concentration	Decrease probe concentration
	Inadequate washing	Change washing method (see *UNIT 4.4*)
	Other causes (see *UNIT 4.4*)	

1. Soak Entellan-mounted slidesin xylene at room temperature. If necessary (e.g., the slide has been stored several weeks or more), soak the slide 1 to 4 weeks to remove the coverslip.

2. Incubate 5 min each in 100%, 90%, 80%, and 70% ethanol at room temperature. Air dry 10 min at room temperature.

3. Fix 1 min in a Coplin jar containing 4°C formaldehyde-buffered acetone fixative.

4. Wash extensively in cold tap water. Air dry 12 hr at room temperature.

5. Wash three times, 10 min each, in 2× SSC/0.2% Tween, 42°C. Wash 5 min in warm tap water (37° to 40°C). Air dry 10 min at room temperature.

6. Incubate 5 min each in 70%, 80%, 90%, and 100% ethanol. Air dry 10 min at room temperature.

7. Perform ISH using whole chromosome paint probe (see Basic Protocol 1, step 13 onward).

References: Fletcher et al., 1991; Knuutila et al., 1994

Contributor: Sakari Knuutila

UNIT 4.8

Comparative Genomic Hybridization

BASIC PROTOCOL 1

COMPARATIVE GENOMIC HYBRIDIZATION USING DIRECTLY LABELED DNA

This procedure is outlined in Figure 4.8.1. Directly labeled probes give less background than indirectly labeled probes.

Materials (see APPENDIX 1 for items with ✓)

    20 μg/ml fluorescein-labeled test probe DNA (e.g., whole-genomic tumor DNA; see Support Protocol 3)

    20 μg/ml Texas Red-labeled reference probe DNA (normal whole-genomic DNA; see Support Protocol 3)

    1 μg/μl human Cot-1 DNA (Life Technologies)

✓ 3 M sodium acetate, pH 5.2

    70% and 85% (v/v) ethanol (room temperature) and 100% ethanol (ice-cold and room temperature)

✓ Master hybridization solution

    Slides containing normal human metaphase chromosomes (*UNIT 4.1*; also see Support Protocol 1)

✓ Denaturation solution

✓ Proteinase K solution

✓ Hybridization wash buffer, 45°C and room temperature

✓ 2× SSC, 45°C and room temperature

✓ PN buffer

✓ DAPI in antifade medium

    Diamond pen
    45°, 70°, and 75°C water baths
    37°C slide warmer
    18- and 22-mm² coverslips
    Rubber cement
    Moist chamber: empty pipet-tip holder containing wet Kimwipe or equivalent (Fig. 4.4.3)

NOTE: As the quality of the metaphases is a critical parameter for the success of this experiment, it is often best to make large batches of slides and empirically optimize all parameters, including storage conditions, for each batch.

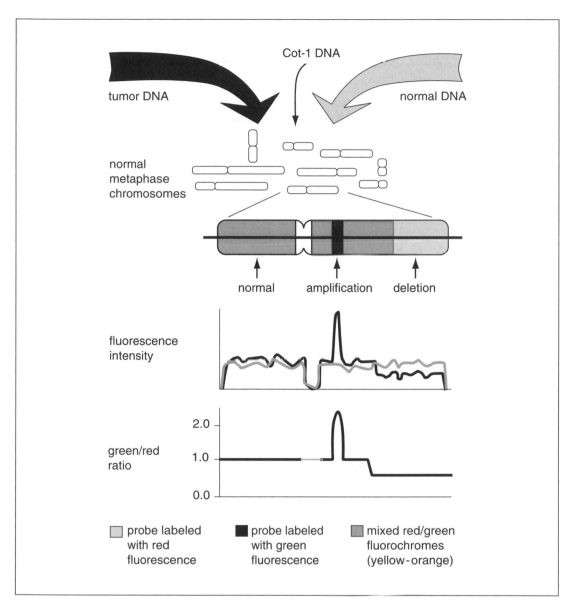

Figure 4.8.1 Comparative genomic hybridization (CGH). Tumor DNA (labeled with green fluorochrome) and normal reference DNA (labeled with red fluorochrome) are hybridized to normal human metaphase chromosomes in the presence of Cot-1 DNA (to suppress binding of labeled DNA to repetitive elements). Regions of DNA gain are seen as an increased green fluorescence intensity on the target chromosomes; regions of DNA loss are seen as increased red fluorescence intensity. Regions on the chromosome that are stained equally for both green and red indicate equal copy number for tumor and reference DNA. A ratio of 1.0 indicates no copy-number change. A high green-to-redo (>1.0) indicates DNA gain, and a low green-to-red ratio (<1.0) indicates DNA deletion (Kallioniemi et al., 1993; reprinted with permission of Academic Press).

1. Combine the following in a 1.5-ml microcentrifuge tube:

   10 µl of 20 µg/ml fluorescein-labeled test probe DNA
   10 µl of 20 µg/ml Texas red-labeled reference probe DNA
   20 µl of 1 µg/µl human Cot-1 DNA.

2. Add $^1/_{10}$ vol of 3 M sodium acetate, pH 5.2, mix, then precipitate DNA by adding 2.5 vol of 100% ethanol and mixing again. Microcentrifuge 30 min at maximum speed, 4°C. Increase the amounts of test and reference probe DNA if the resulting hybridization signals appear too dim.

3. Decant the supernatant and blot excess liquid from the inside of the tube, taking care to avoid blotting the pellet. Allow pellet to air dry 10 min.

4. Add 10 µl master hybridization solution and resuspend the pellet gently with a pipet tip. Vortex, then microcentrifuge 10 sec at maximum speed to bring liquid to bottom of tube. Keep dissolved pellet at room temperature while performing steps 5 to 9.

5. Outline area on back of slide containing metaphase chromosomes using a diamond pen. If necessary, breathe on the slide to aid visualization of the area containing metaphase.

6. Prewarm denaturation solution to 73°C in a Coplin jar placed in a 75°C water bath (check temperature inside the jar). Prewarm slides <2 min on a 37°C slide warmer and place them (no more than three to four at a time) in the denaturation solution. Incubate 2.5 to 10 min at 73°C. Determine optimal time for each batch of slides.

7. Dehydrate slides by immersing them 2 min successively in Coplin jars containing 70%, 85%, and 100% ethanol at room temperature. Remove from ethanol and air dry in a vertical position to minimize spotting.

8. *Optional:* Immerse slides in proteinase K solution for 5 to 7 min at room temperature. Repeat step 7. Use higher times to reduce background and increase probe penetration. Use lower times (or skip completely) to preserve chromosome morphology and banding.

9. Place slides on a 37°C slide warmer. Denature probe mix from step 4 by heating 5 min at 70°C, then (with slides still on warmer) immediately add 10 µl probe mix to the area of each slide containing the metaphases. Cover with 18-mm^2 coverslip, removing any air bubbles by gently lowering the coverslip onto the slide at an angle using fine-tipped forceps. Seal coverslip edges with rubber cement.

10. Incubate in a moist chamber at 37°C for 2 to 3 days (preferably >36 hr; can be >3 days as long as the hybridization mixture does not dry out).

11. Peel off rubber cement and gently remove coverslips. Wash slides by immersing successively for 10 min each in three changes of hybridization wash buffer, 45°C. Alternatively, peel away the rubber cement and allow the coverslip to float off during the first wash.

12. Wash slides by immersing 10 min in 45°C 2× SSC, then 10 min in room temperature 2× SSC.

13. Wash slides by immersing successively for 5 min each in two changes of room temperature PN buffer and two changes of room temperature distilled water. Air dry.

14. Apply 8 µl DAPI in antifade medium and cover with a 22-mm^2 coverslip. Store up to 2 to 3 weeks at 4°C.

15. Examine and analyze slides (see Basic Protocol 2).

ALTERNATE PROTOCOL

COMPARATIVE GENOMIC HYBRIDIZATION USING INDIRECTLY LABELED DNA

CGH hybridization with indirectly labeled probes generally shows more variation in fluorescence intensity than CGH with directly labeled probes, and additional steps and reagents for secondary detection are required. However, these disadvantages are offset by the fact that directly conjugated nucleotides are much more costly.

Additional Materials (*also see Basic Protocol 1; see* APPENDIX 1 *for items with* ✓)

    20 μg/ml biotin-labeled test probe DNA (e.g., whole-genomic tumor DNA; see Support Protocol 3)
    20 μg/ml digoxigenin-labeled reference probe DNA (normal whole-genomic DNA; see Support Protocol 3)
    Blocking solution: 4× SSC/1% (w/v) BSA
✓ Detection solution
✓ 4× SSC
    4× SSC/0.1% (v/v) Triton X-100
    24 × 50–mm coverslips

1. Prepare probe mix (see Basic Protocol 1, step 1) substituting biotin-labeled test probe DNA for fluorescein-labeled test probe DNA and digoxigenin-labeled reference probe DNA for Texas Red–labeled reference probe DNA.

2. Prepare probes and metaphases, then carry out hybridization (see Basic Protocol 1, steps 2 to 12).

3. Incubate slides 5 min in blocking solution at room temperature.

4. Blot excess liquid from slides but do not allow to dry out. Add 100 μl detection solution to area containing metaphases on each slide. Place 24 × 50–mm coverslip loosely on slide so that it floats on the drop of solution, then incubate 60 min at room temperature.

5. Wash slides by immersing successively for 10 min each at room temperature in 4× SSC, in 4× SSC/0.1% Triton X-100, and again in 4× SSC. Finally, immerse for 5 min each in two changes of distilled water, then air dry.

6. Counterstain and examine slides (see Basic Protocol 1, steps 14 to 15).

SUPPORT PROTOCOL 1

PREPARATION OF METAPHASE CHROMOSOMES FOR CGH

It is best to use metaphases from a male donor, regardless of the sex of the tumor or reference DNA. Slides that work well for FISH with centromeric or unique-sequence probes may not work optimally for CGH. If CGH does not work well using the normal slide preparation procedure (see UNIT 4.1), try the following modifications.

1. In the PHA treatment (see UNIT 4.1, Basic Protocol, step 3a), set the incubation time at 72 hr.

2. For synchronization (see UNIT 4.1, Basic Protocol, step 4), use a 17-hr incubation time after addition of methotrexate. Remove the methotrexate (before adding thymidine) by centrifuging 8 min at 180 × g, aspirating the medium, and adding 5 ml HBSS. Repeat once and aspirate the HBSS. After thymidine addition, incubate 5 hr.

3. In the Colcemid treatment (see *UNIT 4.1*, Basic Protocol, step 5), add Colcemid to a final concentration of 0.1 µg/ml and set the incubation time to 10 min.

4. In the hypotonic treatment (see *UNIT 4.1*, Basic Protocol, step 6), make the 75 mM KCl fresh before use and set the incubation time at 20 min.

5. In the first fixative treatment (see *UNIT 4.1*, Basic Protocol, step 7), allow the pellet to incubate 30 min in fixative at 4°C before continuing.

6. In the fixation step (see *UNIT 4.1*, Basic Protocol, step 10), repeat the fixation and centrifugation four to six times.

7. Before preparing a large batch of slides, check the quality of the preparation. Drop an aliquot onto fresh ethanol-cleaned slides that have been wiped once with a lint-free Kimwipe and placed flat on the laboratory benchtop. Examine under a phase-contrast microscope. A satisfactory preparation has medium to low cell density, many well-spread metaphases, and little cytoplasm or debris.

8. When slide-making (see *UNIT 4.1*, Support Protocol) has been completed, store the slides at −20°C in nitrogen-filled sealed bags for at least 2 to 3 weeks before testing them in the CGH procedure.

9. Test slides by running standard CGH reaction with and without proteinase K treatment and then examining slides. Determine optimal times for denaturation and proteinase K treatment.

SUPPORT PROTOCOL 2

PREPARATION OF GENOMIC DNA FOR CGH

High-molecular-weight whole-genomic DNA (>4 kb) from a healthy normal donor (reference DNA) and from the tumor tissue being studied (test DNA) is required for successful CGH. The test and reference DNA must be matched with respect to the sex of the donor, though they need not be from the same patient. Normal reference DNA is prepared from blood (e.g., *APPENDIX 3A, UNIT 10.3*) or mammalian tissue (*CPMB UNIT 2.2*). The test (tumor) DNA is prepared from clinical samples (*UNIT 10.4*) or cell lines (*APPENDIX 3A*).

Purity of the DNA is not as critical for CGH as for other techniques, but degraded DNA should be avoided because it will yield probes that are too small after nick translation (see Support Protocol 3).

When evaluating tumor samples for CGH studies, it is critical that the tumor specimen contain ≥50% tumor cells. Samples may be enriched for the tumor cell population by examining hematoxyline and eosin-stained thin sections of the frozen block to identify the areas containing tumor, then trimming the normal tissue from the block.

SUPPORT PROTOCOL 3

PREPARATION OF LABELED DNA PROBES FOR CGH

Materials

✓ 10× nucleotide mix

Labeled dUTP for normal reference DNA: 1 µmol/ml Texas-Red-5-dUTP (direct label; Du Pont NEN) *or* digoxigenin-11-dUTP (indirect label; Boehringer Mannheim)

Labeled dUTP for test (tumor) DNA: 1 µmol/ml fluorescein-12-dUTP (direct label; Du Pont NEN)

Labeled dATP for test (tumor) DNA: biotin-14-dATP (indirect label) in BioNick 10×
   dNTP mix (Life Technologies)
Test (tumor) and normal (reference) genomic DNA (see Support Protocol 2)
5 to 10 U/μl DNA polymerase I (Life Technologies)
BioNick 10× enzyme mix (Life Technologies)
1% agarose minigel
15° and 70°C water baths

1a. *To label with conjugated dUTPs:* Add nick translation reagents in the following order to
a 0.5-ml microcentrifuge tube on ice:

> 5 μl 10× nucleotide mix (20 μM each dATP, dCTP, and dGTP, final)
> 1 μl labeled dUTP for test or reference DNA (1 nmol final)
> 1 μg test or reference DNA
> H$_2$O to give final volume of 50 μl
> 1 μl 5 to 10 U/μl DNA polymerase I (0.1 to 0.2 U/μl final)
> 3 μl BioNick 10× enzyme mix (adjust depending on DNA quality and resulting
>   probe size).

1b. *To label with biotin-14-dATP:* Prepare the reaction in step 1a, except substitute 5 μl
BioNick 10× dNTP mix containing biotin-14-dATP for the nucleotide mix, and leave out
the labeled dUTP.

2. Mix well, microcentrifuge briefly, and incubate at 15°C for 45 to 90 min (optimize based
on DNA quality; generally 60 min).

3. Stop reaction by heating 10 min at 70°C. Electrophorese 7 μl of the reaction mix on a
nondenaturing 1% agarose minigel along with bacteriophage size markers, stain gel with
ethidium bromide, view with UV illumination, and estimate size of labeled probes (should
be 300 bp to 3 kb for best results).

BASIC PROTOCOL 2

MICROSCOPY, IMAGING, AND IMAGE ANALYSIS FOR CGH

Direct visual inspection may be used for initial assessment of the CGH hybridization to provide
a preliminary interpretation of chromosomal changes. View metaphase spreads hybridized by
CGH using a fluorescence microscope with epiillumination, filter sets appropriate for the
fluorochromes used, and a 63× or 100× oil-immersion lens. Changes in relative copy number
of DNA sequences are determined from changes in the relative intensities of red and green
fluorochromes along individual chromosomes. Differences in the relative intensities of the
two fluorochromes can be visualized by using a dual-band-pass filter that transmits red and
green fluorescence simultaneously, along with a filter for the DAPI counterstain. Large regional
deletions along a chromosome will appear red, and large regional gains will appear green. Gene
amplifications >5- to 7-fold can be detected as small regions of bright green fluorescence.
Chromosomal regions with normal relative copy number will appear yellow-orange. Small
regions of gain or deletion (<10 to 20 Mb) may not be visible to the human eye. These
can be detected more reliably by measuring relative fluorescence intensity ratios along the
chromosomes by digital image analysis. High-quality CGH shows good DAPI chromosomal
banding, evenly hybridized green and red colors along the length of the chromosomes, and
minimal background staining.

High-resolution quantitative information can only be derived from digital imaging. Because
CGH analysis is a relatively new procedure, most successful CGH analysis systems have been
developed in research laboratories. Only a few companies (Applied Imaging, MetaSystems,

Oncor Image Systems, and Vysis) produce systems for the commercial market that are claimed to have CGH analysis capability. These systems vary greatly in sophistication, are undergoing rapid evolution, and must therefore be evaluated at the time of purchase to ensure that they actually have the required capability.

For each experiment, select three to five metaphase spreads with (1) optimal spreading with few sharply bent, touching, or overlapping chromosomes, (2) minimal background staining, (3) a relatively smooth, nongranular hybridization pattern, and (4) minimal variation of chromosomal condensation across each metaphase.

CGH specificity is dependent on confirmation of all changes by inverse hybridizations (i.e., repeat hybridizations with probes that have the tags reversed). When the green-to-red ratios are compared between the standard and inverse hybridizations, copy-number changes should be in opposite directions.

A normal-versus-normal hybridization should be run with each data set as a control. Centromeric and heterochromatic regions, as well as the p arms of acrocentric chromosomes, should be excluded from interpretation. Caution should be exercised in the interpretation of the GC-rich regions at 1pter, 19, and 22. These three chromosome regions, as well as centromeric and telomeric regions of all chromosomes, can give false ratio changes if the probe sizes or hybridization conditions are not optimal. Data from at least four copies of each chromosome should be analyzed.

For an extensive discussion of various aspects of image analysis and data interpretation, see *CPHG UNIT 4.6*.

References: DuManoir et al., 1995; Kallioniemi et al., 1994

Contributors: Sandy DeVries, Joe W. Gray, Daniel Pinkel, Frederic M. Waldman, and Damir Sudar

CHAPTER 5

Strategies for Large-Insert Cloning and Analysis

This chapter provides tested protocols for a variety of large-insert cloning and analysis techniques. Together, these methods provide effective means to isolate genomic regions in ordered sets of overlapping large-insert clones (YAC, BAC, PAC) and to manipulate those clones in a variety of ways. For the purposes of this chapter, large-insert cloning is considered to include any host-vector system capable of faithfully carrying human DNA inserts in excess of 40 kb.

Long-range restriction mapping (*UNIT 5.1*) permits partial characterization of the DNA between anchors without cloning it. Although this method is no longer preferred for physical mapping of human chromosomal regions, it may be useful for analyzing other genomes lacking abundant clone resources. Long-range restriction mapping is used to determine the distances between genomic landmarks when the DNA between them has not been isolated in overlapping clones; e.g., the locations of CpG islands, which are likely landmarks for expressed genes, may be determined.

A variety of total-genome large-insert clone libraries have been constructed, and several companies offer screening services, DNA pools, high-density array filters, clone libraries, and individual clones. Screening may also be performed as a research collaboration with a group having access to DNA pools and large-insert clones. Many investigators may wish to construct large-insert clone libraries from their organism of interest. Proven protocols to accomplish this are presented in *UNIT 5.4* for BAC libraries.

Large-insert BACs and YACs are typically analyzed in a variety of ways. First, the clones or portions of the clones are screened, probed, fingerprinted, fragmented, or amplified by PCR to determine marker content. Second, the clones or PCR-amplified portions are used as probes for chromosomal walking, in FISH assays to determine the chromosomal location of an insert, or to identify cDNAs encoded by the region. In all cases, a DNA preparation is needed. While YACs do not provide an abundant source of DNA, many methods provide an adequate amount of DNA for PCR amplification. BAC DNA may be separated from the endogenous *E. coli* chromosomal DNA by standard alkaline lysis procedures similar to those used for plasmid and cosmid DNA preparation (*UNIT 5.4*). These procedures can be readily adapted to produce sufficient DNA for exhaustive restriction mapping or shotgun sequencing experiments. YAC DNA, on the other hand, is more difficult to obtain in pure form. YACs are carried primarily as single-copy, linear chromosomes in a yeast host, and yeast cell densities in overnight cultures are significantly lower than *E. coli* cell densities. Intact YACs are required for restriction mapping (*UNIT 5.1*). Where high-molecular-weight DNA is required, agarose-block DNA preparations are recommended; procedures for preparing these are described in *UNIT 5.1*.

One of the basic methods currently in use for screening large-insert libraries uses filter hybridization (*UNIT 5.2*) for detection of clones containing the marker in question. Preparation of YAC (or BAC) colony filters is not difficult but requires considerable investment in time and equipment if it is to be carried out in duplicate on multi-genome-equivalent libraries containing up to 400,000 clones. Such filters are readily available from resource centers and commercial suppliers.

Retrofitted YACs may be introduced into mammalian cells (*UNIT 5.3*). By recombining several YACs and by retrofitting the resulting YAC with a mammalian selectable marker, it is possible to introduce large genomic regions—either intact or containing known mutations, insertions, or deletions—into mammalian cells and to study the function of genes in their normal long-range genomic context.

Contributors: Donald T. Moir and Douglas R. Smith

UNIT 5.1

Pulsed-Field Gel Electrophoresis for Long-Range Restriction Mapping

This unit describes procedures for generating long-range restriction maps of genomic DNA (>500 kb to >5 Mb) and for analysis of large insert clones.

NOTE: All reagents and equipment coming into contact with live cells must be sterile.

BASIC PROTOCOL

CONSTRUCTING LONG-RANGE RESTRICTION MAPS OF GENOMIC REGIONS

Materials (*see* APPENDIX 1 *for items with* ✓)

    Agarose-embedded DNA samples (see Support Protocol 1)
    Rare-cutting restriction endonuclease (Table 5.1.1) and 10× buffer
    1 mg/ml BSA, acetylated and nuclease-free
    100 mM spermidine trihydrochloride
✓ TE buffer
    Agarose (ultrapure; e.g., Life Technologies)
✓ Electrophoresis buffer (e.g., 0.5× TBE), 10°C
✓ 6× gel loading buffer
    DNA size standards (see Support Protocol 2)
✓ 0.5 µg/ml ethidium bromide

    100-mm petri plates
    Water bath at appropriate temperature for restriction digestion (Table 5.1.1)
    Pulsed-field gel electrophoresis apparatus (with optional cooling unit; e.g., Bio-Rad systems)
    Constant-voltage power supply
    Pulse generator
    Positively charged nylon membranes

1. Place DNA sample embedded in agarose block in a petri plate and cut into eight ∼25-µl (1 × 4 × 6–mm) pieces containing ∼3 µg of DNA each, handling the sample carefully to avoid damage that may translate into distorted bands.

2. In a microcentrifuge tube on ice, mix the following (for final volume of 100 µl, including the agarose block):

        10 µl 10 × restriction endonuclease buffer
        10 µl 1 mg/ml acetylated BSA
        4 µl 100 mM spermidine trihydrochloride
        50 µl sterile H_2O.

Add one agarose block to reaction mix and incubate 5 min on ice.

3. Add 15 U of a rare-cutting restriction endonuclease, mix, and incubate 60 min on ice to allow the enzyme to diffuse into the agarose matrix.

4. Incubate tube 5 hr in a water bath at the appropriate temperature (see Table 5.1.1 or supplier's recommendations). For reactions incubated at 50°C (e.g., *Bss*HII and *Sfi*I digestions), cover the mixture with 2 to 3 drops of mineral oil to prevent evaporation.

5. Stop digestion by diluting with 1 ml TE buffer. If the sample is not to be used immediately, refrigerate at 4°C (samples can be stored for several months).

6. *Optional:* For double digestion, allow first restriction buffer to diffuse into TE buffer ≥30 min at room temperature. Remove TE/restriction buffer and repeat steps 2 to 5 with a second restriction enzyme under conditions suitable for the new enzyme.

7. Prepare an agarose gel suitable for the PFGE device being used (see manufacturer's instructions). Cast gel on a flat, level surface to avoid band distortion.

Table 5.1.1 Cleavage Sites and Reaction Temperatures for Restriction Endonucleases with Rare Cleavage Specificities

Endonuclease	Cleavage site[a]	Incubation temperature (°C)
*Asc*I	GG↓CGCGCC	37
*Bss*HII	GC↓GCGC	50
*Cla*I	AT↓CGAT	37
*Eag*I	C↓GGCCG	37
*Mlu*I	A↓CGCGT	37
*Nae*I	GCC↓GGC	37
*Nar*I	GG↓CGCC	37
*Not*I	GC↓GGCCGC	37
*Nru*I	TCG↓CGA	37
*Pac*I	TTAAT↓TAA	37
*Pae*R7I	C↓TCGAG	37
*Pme*I	GTTT↓AAAC	37
*Pvu*I	CGAT↓CG	37
*Rsr*II	CG↓G(A/T)CCG	37
*Sac*II	CCGC↓GG	37
*Sal*I	G↓TCGAC	37
*Sfi*I	GGCCN₄↓NGGCC	50
*Sgr*AI	CPu↓CCGGpyG	37
*Sma*I	CCC↓GGG	25
*Srf*I	GCCC↓GGGC	37
*Sse*8387I	CCTGCA↓GG	37
*Swa*I	ATTTAAAT	25
*Xho*I	C↓TCGAG	37
*Xma*III	C↓GGCCG	37

[a] Abbreviations: N, any nucleotide (G, A, T, C); Pu, purine (G or A); Py, pyrimidine (C or T); A/T, A or T.

High agarose concentrations, e.g., 1.0% to 1.5%, slow migration of molecules but result in sharper banding. Typically, 1% agarose is useful for DNA molecules up to 3 Mb, but 1.2% to 1.5% may be used for improved band tightness. For extremely large DNA molecules, i.e., >2 Mb, 0.5% to 0.9% is typical. A chromosomal-grade agarose is highly recommended to provide optimal separations.

8. Add 20 μl of 6× gel loading buffer to the sample tube. Plan sample distribution on the gel such that samples that will be compared directly are loaded in adjacent lanes (PFGE separations nearly always involve some distortion in the migration path) and size standards are distributed over the gel to assist in size determination.

9. Carefully remove the comb. Fill wells with 0.5× TBE buffer and load DNA sample into the well using two spatulas, one to position the sample and one to open the well to permit the sample to slide in (make sure that no air bubbles are trapped under the blocks as these can cause serious distortion in migration patterns).

10. Place gel in chamber of PFGE apparatus and add 2.5 liters of 0.5× TBE buffer. Allow buffer to precool to 10°C, e.g., by recirculating the buffer with a cooling unit or refrigerated water bath, or by placing the apparatus in a refrigerator.

11. Close lid on gel box (for Bio-Rad systems), which automatically connects leads from power supply. For typical separations and systems, use an electric field strength of 5 V/cm, a pulse time of 90 sec, and a running time of 48 hr (also see manufacturer's instructions).

12. When the separation run is complete, turn off power supply, drain away buffer, and stain gel with 0.5 μg/ml ethidium bromide for 2 to 24 hr. Photograph the gel.

13. Transfer PFGE-separated DNA to a positively charged nylon membrane. For optimal transfer of large DNAs from pulse-field gels, use an acid depurination step, a denaturing transfer solution, or UV nicking.

14. Hybridize filter with radiolabeled probe and autoradiograph. Use a probe that is free of repeat elements (or has been preannealed with unlabeled genomic DNA to block their hybridization) and vector sequences.

SUPPORT PROTOCOL 1

PREPARATION OF HIGH-MOLECULAR-WEIGHT MAMMALIAN GENOMIC DNA EMBEDDED IN AGAROSE BLOCKS

This protocol is suitable for living cells from most mammalian cell lines that can be made to single-cell suspensions, as well as disaggregated tissues in which there is no cell death. Frozen cell pellets should not be used unless they are the only source of DNA available, because the freeze-thaw process will usually render samples unsatisfactory.

Materials (*see* APPENDIX 1 *for items with* ✓)
   1.0% (w/v) low gelling/melting temperature agarose (e.g., InCert; FMC Bioproducts)
      prepared in HBSS
   70% (v/v) ethanol
   Mammalian cells
✓ HBSS, 37°C
✓ Cell lysis buffer
✓ TE buffer
✓ PMSF/TE wash buffer (make fresh)

50° and 42°C water baths

Plexiglas block mold (e.g., Bio-Rad 10-well sample-plug mold; supplied with most PFGE systems)

Beckman GPR centrifuge with GH-3.7 swinging-bucket rotor (or equivalent)

1. Melt an appropriate quantity of 1.0% low gelling/melting temperature agarose by placing tube containing agarose in a beaker half-filled with water and heating just to boiling point, taking care not to lose water volume from the agarose. Quickly place in a 50°C water bath to cool.

2. Clean block mold with 70% ethanol, assemble, and place on crushed ice to chill (also see manufacturer's instructions).

> *The Bio-Rad mold produces 225-μl agarose blocks that are cut into 1 × 4 × 6–mm pieces as described in the Basic Protocol. Each gel piece will contain ~3 μg DNA. For tetraploid or polyploid cells, the protocol must be adjusted to achieve this amount of DNA and avoid overloading.*

3. Harvest cells by centrifuging 8 min at $150 \times g$ (800 rpm in Beckman GH-3.7 rotor), room temperature. Remove and discard supernatant. Resuspend cell pellet in 10 ml of 37°C HBSS and mix to produce a single-cell suspension. Count the cells using a hemacytometer (*APPENDIX 3I*).

4. Wash cells twice more by centrifugation with HBSS. Calculate the final volume of suspension required to yield a cell concentration of 4×10^7 cells/ml. Resuspend washed cells in a volume of HBSS slightly less than the amount calculated, measure the actual volume of suspension by carefully pipetting into a separate tube, and add HBSS until the calculated volume is reached.

5. Transfer melted agarose from 50°C water bath to a 42°C water bath (equilibration is not critical). Place the tube of cells in the 42°C water bath, add an equal volume of 42°C agarose, and mix. Keep the cell/agarose suspension at 42°C.

6. Pipet 225 μl of cell/agarose suspension into each slot of the chilled block mold. Allow blocks to solidify 5 min on ice. Using two spatulas, transfer the blocks to a tube containing 20 ml cell lysis buffer (sufficient for ten 225-μl blocks). Incubate 24 hr at 50°C.

7. Decant lysis buffer and replace with 20 ml fresh lysis buffer. Incubate another 24 hr at 50°C.

8. Dialyze blocks twice, each time for 1 hr at room temperature against 40 ml TE buffer, inverting the tube gently several times. Dialyze a final time in the same fashion at 50°C.

9. Replace TE buffer with freshly prepared PMSF/TE wash buffer and incubate 30 min at 50°C. Repeat once.

10. Replace PMSF/TE with fresh TE buffer and store blocks at 10°C.

SUPPORT PROTOCOL 2

PREPARATION OF HIGH-MOLECULAR-WEIGHT STANDARDS IN AGAROSE BLOCKS USING *S. CEREVISIAE* CHROMOSOMES AND YACS

S. cerevisiae strains usually contain 16 chromosomes ranging in size from ~200 to >2000 kb, which provide a highly convenient range of size standards for PFGE.

Materials *(see APPENDIX 1 for items with ✓)*

    S. cerevisiae cells of desired strain

  ✓ YPD plates and medium *or* AHC plates and medium

✓ 50 mM EDTA, pH 7.5
   70% (v/v) ethanol
✓ SCE buffer
   Zymolyase solution: 100 U/ml Zymolase 100T (dry powder; ICN Biomedicals)/100 mM
      mercaptoethanol (2-ME) in SCE buffer
   1.4% (w/v) low gelling/melting temperature agarose prepared in SCE buffer
   2-ME solution: 70 mM 2-ME in SCE buffer
✓ Cell lysis buffer
✓ TE buffer

   30°C incubator
   30°C shaking water bath
   250-ml centrifuge bottles
   Sorvall centrifuge with GSA rotor (or equivalent)
   50-ml conical centrifuge tube, screw-cap
   Plexiglas block mold (Bio-Rad)
   Beckman GPR centrifuge with GH-3.7 rotor (or equivalent)
   42° and 50°C water baths

1. Streak *S. cerevisiae* strain of choice to prepare single colonies on a YPD plate. Incubate 1 to 2 days at 30°C. If preparing chromosomes from a YAC-bearing yeast strain, substitute a yeast growth medium that selects for at least one YAC arm—e.g., AHC plates and medium for URA3- and TRP1-bearing YACs.

2. Pick a single colony and inoculate into 5 ml liquid medium in a sterile test tube. Incubate overnight at 30°C with aeration.

3. Inoculate 200 ml medium in a 500-ml sterile flask with 2 ml of overnight culture. Incubate overnight at 30°C with aeration.

4. Transfer culture to 250-ml centrifuge bottle and centrifuge 10 min at $5800 \times g$ (6000 rpm in Sorvall GSA rotor), room temperature, to harvest cells.

5. Resuspend cell pellet in 15 ml of 50 mM EDTA, pH 7.5, and transfer to a 50-ml conical screw-cap centrifuge tube. Determine cell concentration by making a 10-, 100-, or 1000-fold dilution in 50 mM EDTA and counting cells using a hemacytometer (*APPENDIX 3I*).

6. Clean block mold with 70% ethanol, assemble, and chill in crushed ice.

7. Pellet cells by centrifuging 10 min at $900 \times g$ (2000 rpm in Beckman GH-3.7 rotor), room temperature. Resuspend cells in zymolyase solution at a concentration of 4×10^9 cells/ml.

8. Melt 1.4% low gelling/melting temperature agarose and cool in a 50°C water bath.

9. Place 1 ml cell suspension (4×10^9 cells) in a 15-ml tube and place tube in a 42°C water bath. Add 1 ml melted agarose, mix, and return to water bath.

10. Distribute cell/agarose mixture into block molds on ice and allow to solidify 5 min. Transfer blocks from the mold into 20 ml of 2-ME solution and incubate overnight at 37°C.

    A 225-μl block will be cut into into $1 \times 4 \times 6$–mm pieces as for the samples. At 2×10^9 cells/ml, the final DNA concentration is ~33 ng/μl.

11. Pour off solution and replace it with 30 ml cell lysis buffer. Incubate overnight at 50°C.

12. Check to see that the white, opaque blocks have become translucent (indicating that most cells have lysed). Replace solution with fresh cell lysis buffer and incubate an additional 24 hr at 50°C.

13. After lysis is completed, dialyze blocks against TE buffer by filling tube with TE and changing every 30 min to 1 hr, for four changes total. Fill tube with fresh TE buffer and store blocks at 10°C.

14. Analyze blocks by PFGE (see Basic Protocol) using 5 V/cm electric field strength, 90-sec pulse time, and 36- to 48-hr running time.

SUPPORT PROTOCOL 3

PREPARATION OF BAC DNA, RESTRICTION DIGESTION, AND CHEF GEL ANALYSIS

Materials (see APPENDIX 1 for items with ✓)

✓ LB medium with antibiotic (same concentration as solid medium on which bacteria are growing)

E. coli strain DH10B streaked cultures of desired BAC clones grown on solid LB/antibiotic/IPTG (25 µg/ml)/X-gal (50 µg/ml); typical antibiotic concentration for most common BAC vectors is 12.5 µg/ml chloramphenicol or 25 µg/ml kanamycin

✓ GTE solution, ice cold

✓ NaOH/SDS solution

✓ Potassium acetate solution

Isopropanol

70% (v/v) ethanol

✓ TE buffer, pH 8.0

10 U/µl *Not*I restriction enzyme and buffer (NEB)

100× BSA (NEB)

✓ 6× gel loading buffer

1% (w/v) electrophoresis-grade agarose gel prepared in 0.5× TBE (APPENDIX 1)

DNA size standards (see Support Protocol 2)

✓ 0.5 µg/ml ethidium bromide

15-ml test tubes

37°C incubator with shaker

Table-top centrifuge with swinging bucket rotor

37°C incubator or water bath

Bio-Rad CHEF unit (DR-II, DR-III, or Mapper) with chiller and pump

UV light box

1. Inoculate 3 ml of LB medium with antibiotic in 15-ml test tube with a single colony from a streaked *E. coli* strain DH10B plate and grow 16 to 18 hr at 37°C with shaking (250 rpm).

2. Pellet bacteria 10 min at 1770 × g (3000 rpm in a swinging bucket rotor) and remove supernatant. Add 200 µl ice-cold GTE solution, resuspend bacteria, and transfer to a 1.5-ml microcentrifuge tube. Add 400 µl of NaOH/SDS solution and mix by inverting tube. Let tubes sit for 7 min on ice.

3. Add 300 µl potassium acetate solution, mix by inverting tube, and let sit for 7 min on ice. Microcentrifuge 20 min at maximum speed. Transfer supernatant to new 1.5-ml microcentrifuge tube and repeat microcentrifugation if any debris is visible.

4. Add 600 µl isopropanol, mix by inverting tube, and let sit for 20 min on ice. Pellet the DNA by microcentrifuging 25 min at maximum speed. Remove supernatant and add 1 ml of 70% ethanol. Microcentrifuge 15 min at maximum speed. Completely remove 70%

ethanol and air dry the tubes for 1 hr in a laminar flow hood or several hours on the benchtop. Resuspend BAC DNA in 35 μl TE buffer, pH 8.0.

5. Use the following typical recipe for a restriction digestion (e.g., *Not*I) on BAC DNA. Increase the DNA concentration if needed. Increase the concentration of enzyme to 0.75 to 1.0 μl, if desired, to enhance digestion.

> 1.5 μl *Not*I buffer
> 0.2 μl 100× BSA
> 0.5 μl 10 U/μl *Not*I
> 7.8 μl sterile H_2O
> 5.0 BAC DNA
> Total volume = 15 μl.

Incubate ≥5 hr at 37°C. After incubation, stop reactions by adding 0.2 μl of 6× gel loading buffer and storing on ice until loading CHEF gel.

6. Prepare a 1% agarose gel with 0.5× TBE and let cool to ~50°C before pouring. After gel is solidified, add size standards prepared in agarose (Support Protocol 2) and seal wells with molten agarose.

7. Add 2.5 liters of prechilled (14°C) 0.5× TBE buffer to fill the electrophoresis chamber. After placing the gel within the chamber, add the restricted BAC samples containing loading dye with a pipet as if loading a normal horizontal submarine gel. Run the gel using the following conditions: 5 to 15 sec switch time, linear ramp, 15 hr run time, and 6 V/cm (200 V).

8. Following electrophoresis, stain gel with 0.5 μg/ml ethidium bromide as described in the Basic Protocol. View gel and document on a UV light box.

> *When observing the CHEF gel, insert bands should be clear and in most cases located above the vector band, which is 7.5 kb for pBeloBAC11 and its derivatives. The vector band should be clearly visible in all lanes. In the case of humans, there should generally be one insert band above 100 kb since most human BAC libraries have an average insert size greater than 100 kb.*

9. If desired, transfer the BAC DNA from the gel by Southern blotting and hybridize to a probe (see Basic Protocol, steps 13 and 14).

References: Birren and Lai, 1993; Gemmill, 1990; Schwartz and Cantor, 1984; Smith, 1990; Vollrath and Davis, 1987

Contributors: Robert M. Gemmill, Richard Bolin, Jeff P. Tomkins, and Rod A. Wing

UNIT 5.2

Screening Large-Insert Libraries by Hybridization

BASIC PROTOCOL 1

ARRAYING COLONIES AND DNA AT LOW AND HIGH DENSITY

Random plating is not recommended for filter preparation and storage of YAC, BAC, and PAC libraries, because limited numbers of filters can be produced from the primary transformants, and any form of amplification can introduce bias into the library due to differences in clonal propagation rates. Instead, the primary transformants should be arrayed into microtiter plates for long-term storage at −80°C. This facilitates preparation of library copies for distribution and allows easy communication of screening data between groups in the form of positive plate addresses. In addition, the microtiter plate format allows for manual and automated preparation of low- and high-density hybridization filters using devices with multiple metal or plastic prongs. This basic protocol describes three methods for filter preparation.

Replica Plating YAC Transformants

The procedure most commonly used to transform *S. cerevisiae* cells with high-molecular-weight genomic DNA ligated to YAC vector arms requires enzymatic removal of the yeast cell wall. Consequently, transformants are fragile and must be grown initially within a supporting layer of agar. This precludes lifting primary YAC clones directly onto a membrane for hybridization screening. One way to overcome this problem is to use a metal colony replicator (or metal "velvet"), in which thousands of pins are used to transfer cells protruding from the top agar of a primary transformation plate onto the surface of a selective agar plate ("recovery plate"). A 40,000-pin metal colony replicator is available from ICRT. The agar overlay is kept to a minimal thickness and allowed to dry following colony growth so colonies can be easily contacted by the replicator. The replicator is sterilized by immersion in 99% industrial-grade alcohol or 95% ethanol, then flamed to remove the alcohol. Alternatively, sterile autoclaved fabric velvets may be used, but colonies will spread out more upon transfer.

When colonies have grown, they can be transferred to a nylon membrane by placing the dry membrane in contact with the colonies for 15 min. Colonies are then lysed in situ to release DNA for hybridization. Several agar recovery-plate copies can be made from a single impression of a primary transformation plate. A filter lift can be made from each recovery plate and positive clones can be isolated from one of the agar copies. Agar copies can be stored several weeks at 4°C. To freeze an agar copy, it can be placed colony-side down on a Whatman 3MM filter paper soaked in freezing medium (YPD medium containing 30% glycerol). The filter sandwich is placed between polycarbonate plates and frozen at −70°C. Positive clones are isolated by cutting out a small area of both membranes and placing them separately on agar to produce new colonies.

Unfortunately, there are limits to both the number of filters that can be produced and the number of clones per filter, and neither agar plates nor frozen filters offer ideal storage conditions for clones. Such filters are therefore not suitable for large-scale genomic mapping experiments that require many hybridizations to the same clone set and a robust and easily accessed storage facility for clones.

To generate unlimited numbers of filters for hybridization screening of YAC libraries, it is necessary to pick primary YAC clones from transformation plates and place them into microtiter dishes. Clones can be stored in these dishes indefinitely at −70°C in the presence of 20% (v/v) glycerol. Recovery of clones following freezing may be improved by freezing in nonselective medium (e.g., YPD medium).

Manual and Machine-Assisted Gridding of Low- to Medium-Density Colony and DNA Filters

Low-density filter arrays of YAC, BAC, and PAC colonies can be produced manually using metal or plastic devices with 96 or 384 solid or hollow pins (Fig. 5.2.1). These devices can also be used for filter arrays of YAC DNA samples—e.g., total yeast DNA from YAC clones, or *Alu*-PCR products from YACs or pools of YACs. Gridding tools are available from a number of commercial suppliers (V&P Scientific, Sigma, Nunc, Dynatech, Genetix, Techne).

5

Figure 5.2.1 A 96-pin metal replicating device, which can be used to inoculate microtiter plates or to grid large colonies at low density.

The requirements placed upon the arraying device are somewhat different depending on whether colonies or DNAs are to be arrayed. Because even a few cells will grow into a colony, the delivery of cells can be variable without serious consequences for the final filter. However, differences in the volume of DNA solution added to each spot will be reflected directly in the intensity of signal from hybridization.

Key issues to consider when choosing an arraying device for DNA solutions are the volume of solution delivered, area occupied by the resulting spot, and variability in transfer volume from pin to pin. Hollow pins and plastic pins deliver more volume than do solid metal pins. These are helpful for inoculating microtiter dishes with YAC clones or delivering dilute solutions of total yeast DNA from YAC clones. However, for delivery of relatively concentrated IRS-PCR products, the smaller delivery volume of solid metal pins is more useful because it results in a stronger signal achieved with modest total volume (thus increasing the number of filters that can be produced), and it permits a higher density of spots. The pin-to-pin variability is also lower with metal pins. If more DNA is needed to provide adequate signals, it is preferable to spot three, four, five, or more times with pins that deliver a small volume so that the area of the spots and volume of sample consumed are kept to a minimum. This can be achieved using a machine that provides a template for the arraying device or with a robotic device (see below).

Prior to use, a metal 96- or 384-pin replicator is sterilized by immersing the pins in 95% ethanol. The ethanol should be just deeper than the depth of culture in the microtiter plate. The replicator is then passed briefly and with great caution through a flame to remove the excess ethanol. Some plastic replicators may be sterilized by autoclaving.

The multi-pin replicator is used to copy cells or DNA from a microtiter dish onto a nylon membrane. Uncharged (e.g., Hybond N; Amersham) or positively charged membranes (e.g Hybond N+; Amersham) can be used for YAC colonies, but positively charged membranes should be used for BAC and PAC colonies and for DNA. For YAC colony gridding, it is necessary to disturb the layer of cells at the bottom of the well with the pins of the replicator. For colony grids, the nylon filter can be placed onto a nutrient agar plate of the appropriate type prior to gridding. Care should be taken to ensure that all pins make contact with the filter surface and that the replicator does not move along the filter surface once the pins are in place. After gridding, the agar plate and filter are incubated for 27 hr at 30°C (YACs) or 12 to 16 hr at 37°C (BACs, PACs) in an inverted position. Colonies are then lysed by an appropriate method. Low-density grids prepared in this manner are suitable for hybridization analysis of relatively small genomes, for preparative purposes, or for identification by hybridization of a clone tracked to a single microtiter dish by PCR screening.

Robotic Gridding of High-Density Colony and DNA Filters

High-density colony and DNA arrays can be prepared using custom-built and commercially available robotic systems. Gridding robots are available commercially from Genetix, BioRobotics, Genomic Solutions, and Beckman. These systems have additional capabilities, such as colony picking, as either standard features or options. The characteristics of these gridding systems are summarized in Table 5.2.1. Figure 5.2.2 shows an example of a gridding robot comprising an x-y-z platform built on a linear induction drive platform with an accuracy of 5 μm. The robot bed is designed to accept the source microtiter plates in their −80°C storage racks, and the recipient filters on agar plates (8 × 12 cm) in racks of twelve.

The procedures involved in robotic colony gridding, and their variations, are as follows. In preparation for each gridding run, the sterilization bath is set up with fresh 95% ethanol. The ink and sonication baths are optional but recommended. The ink [e.g., autoclaved 1% (v/v) Higgins Black Magic ink] is used to create a grid precisely matching the final colony pattern,

Table 5.2.1 Comparison of Commercially Available Gridding Systems[a]

Supplier	Genetix	BioRobotics	Genomic Solutions	Beckman[b]
Robot name	Q Bot	Bio Grid	Flexsys	Biomek2000
96- & 384-well gridding	Both	Both	Both	Both
Filter size (number)	22 × 22 (15)	22 × 22 (4) 8 × 12 (24)	22 × 22 (3) 8 × 12 (8)	8 × 12 (11)[c]
Automatic filter loading	No	No	Optional	Optional
Automatic plate loading (number)	Yes (72)	Yes (24)	Optional (112)[d]	Optional
Tool drying	No	Hot air	No	Optional fan
Air filtration	Optional	Optional	Optional	No
Barcode reader	Optional	Optional	No	No
Additional capabilites:				
colony picking	Yes	No	Optional	No
replicating	Yes	Yes	Yes	Yes
rearraying	Yes	No	Optional	Yes
liquid handling	No	No	No	Yes
microarraying on glass	No	No[e]	Optional	No

[a] For further details of these systems, consult the WWW sites given in *APPENDIX 4*.

[b] The Beckman Biomek2000 robot is primarily a liquid-handling system with optional gridding capability.

[c] The Biomek2000 basic platform has twelve available positions for either microtiter plates or 8 × 12–cm filter plates, so the maximum number of filters that can be gridded is eleven. This capacity can be increased using the optional side loader.

[d] The Flexsys autoloader holds fifteen trays, each of which can accommodate either eight source microtiter plates, eight 8 × 12–cm filter plates, or one 22 × 22–cm filter plate.

[e] An alternative system is marketed by BioRobotics for this purpose.

Figure 5.2.2 Colony gridding robot. The robot is shown in the process of gridding 16 384-well plates in a 4 × 4 density onto 36 8 × 12–cm nylon membranes on agar plates. At the rear of the robot are stations for filter inking, pin sterilization, and sonication.

as an aid to scoring hybridization data. The sonication bath, containing 0.1% (v/v) Decon Neutracon (Merck), is used to clean the pins after ink gridding and at the end of each gridding run. If a sonication bath is unavailable, the gridding pins should be brushed gently with the Decon solution. The source plates for gridding are thawed in a single layer on the benchtop at room temperature and then placed onto the robot bed. Some systems (see Table 5.2.1) permit automated plate selection and lid removal, while in others lids are removed manually just prior to commencement of the run. The recipient nylon filters (see above for membrane types) are also placed upon the robot bed, either on nutrient agar plates or on Whatman 3MM filter paper moistened with nutrient broth. The robot is equipped with the appropriate gridding tool, which should ideally have floating rather than fixed pins (Fig. 5.2.3). This ensures good contact between filter and pins at all positions while minimizing the risk of filter damage, and also ensures that the pins contact the bases of the microtiter-plate wells. The latter is important for YAC libraries since yeast cells settle rapidly in liquid medium. When gridding many copies of a YAC plate, the robot should be programmed such that the gridding pins follow a spiral pattern around the well (only practical for 96-well plates). Alternatively, the plates should be agitated on a commercial plate shaker just before gridding. Finally, the gridding area is closed off both for user safety and to minimize contamination. Some designs include an air-filtration device to minimize contamination. For others, the robot interior is irradiated with UV light prior to gridding.

The robot is initialized and a program is selected to carry out the following steps:

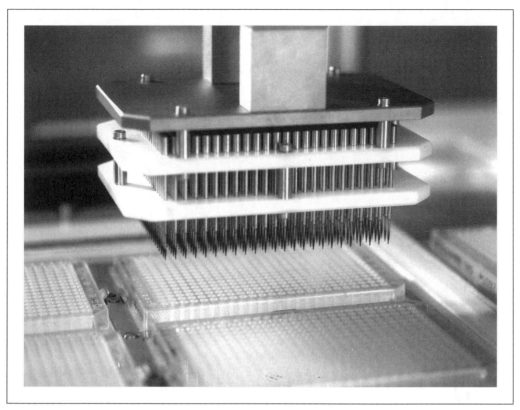

Figure 5.2.3 Tool with 384 floating pins for high-density robotic colony gridding. Stainless steel pins have conical tips tapering to a diameter of 0.4 mm. They are suitable for gridding to a density of at least 5×5 (i.e., 25 384-well plates represented on an 8×12–cm area).

1. immerse pins in the sterilization bath
2. evaporate ethanol
3. immerse pins into the first microtiter plate
4. grid onto membrane 1
5. repeat steps 3 and 4 for all subsequent filters
6. repeat steps 1 to 5 for all subsequent plates

Some robots are equipped with a drying apparatus (forced hot air or halogen lamp) for ethanol evaporation, while others use air drying. In both cases, the precise sterilization and drying times should be established empirically. The spotting position of one microtiter plate is slightly offset from that of the next, allowing large numbers of clones to be accommodated in a relatively small area. The appropriate density of colonies depends on the total number of clones to be gridded, the number of copies of each clone on a filter, the diameter of the pin tips, and the availability of automated image analysis systems for scoring. In practice, the presence of an ink grid allows accurate manual scoring of 4×4-density 384-well grids (i.e., 16 384-well plates on an 8×12–cm filter, or 6×16 384-well plates on a 22×22-cm filter). The use of higher-density grids requires the use of an automated system for scoring positive signals or of a multiple spotting scheme for each clone (see below).

The microtiter-plate addresses of positive clones or DNA pools are identified from their coordinates on the grid (Fig. 5.2.4). At very high densities, coordinate assignment can be problematic, but can be assisted in various ways, including the use of an ink grid. Another approach favored by some researchers is to grid each plate twice, using a gridding order designed to indicate the positive plate number by the relative positions of the two hybridization signals within the

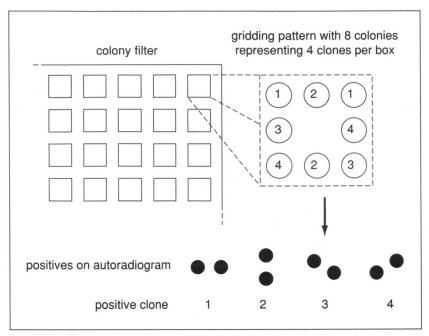

Figure 5.2.4 Identification of positive colonies aided by colony duplication and a complex spotting pattern. Each clone is spotted as shown in the expanded box, and the positive clone is identified by the relative positions of the hybridization signals. The empty central position can be spotted with ink as an additional aid to scoring.

square (Fig. 5.2.4). The obvious drawback with this approach is that it halves the number of clones represented on a filter.

SUPPORT PROTOCOL

PREPARING BAC AND PAC COLONY BLOTS FOR HYBRIDIZATION

Materials (*see* APPENDIX 1 *for items with* ✓)

PAC or BAC colonies arrayed on nylon filters (see Basic Protocol 1)
✓ LB agar plates containing 30 μg/ml kanamycin (PACs) or 20 μg/ml chloramphenicol (BACs) made in 8×12–cm (Genomic Solutions) or 22×22–cm plates (Nunc)
✓ 10% sodium dodecyl sulfate (SDS)
Denaturing solution: 0.5 N NaOH/1.5 M NaCl
✓ 1 M Tris·Cl, pH 7.4
1× neutralizing solution: 0.5 M Tris·Cl, pH 7.4/1.5 M NaCl
✓ 2× SSC
2× SSC/0.1% SDS
Whatman 3MM filter paper
Plastic trays
Orbital shaker

1. Incubate filters carrying gridded BAC or PAC clones at 37°C on LB agar plates containing antibiotic for 12 to 16 hr, until colonies are ∼0.5 to 1 mm in diameter.

2. Soak two pieces of Whatman 3MM filter paper in 10% SDS (until moist, but not so wet that liquid will wash onto surface of filter) and place in a plastic tray. Transfer the

nylon filters onto this, colony side up, avoiding bubbles, and leave for 4 min at room temperature.

3. Transfer the filters onto two pieces of Whatman 3MM filter paper soaked in denaturing solution, and leave for 10 min at room temperature. Air dry the filters 10 min on a fresh, dry piece of Whatman 3MM filter paper.

4. Transfer several filters one-by-one to a container of neutralizing solution and ensure that all are immersed completely. Soak for 5 min with gentle agitation on an orbital shaker. Repeat once. Replace with fresh neutralizing solution if pH rises above 8.0.

5. Treat the filters in the same manner using the following solutions:

    $0.1 \times$ neutralizing solution for 5 min
    $2 \times$ SSC/0.1% SDS for 5 min
    $2 \times$ SSC for 5 min
    50 mM Tris·Cl, pH 7.4, twice for 5 min each.

6. Air dry the filters at least 1 hr, colony side up, on fresh, dry Whatman 3MM filter paper. Store filters at room temperature between pieces of 3MM paper.

7. Before use, cross-link the DNA for 2 min, colony side down, with a 254-nm UV light source (120,000 mJ/cm^2). Continue with hybridization to colony blots (see Basic Protocol 2 or Alternate Protocol).

BASIC PROTOCOL 2

SCREENING BAC, PAC, AND P1 LIBRARIES BY COLONY HYBRIDIZATION

Materials (*see APPENDIX 1 for items with* ✓)

High-density filters for a BAC, PAC, or P1 library (commercially available, or prepared using Basic Protocol 1 and the Support Protocol)
50 ng DNA for radiolabeling reaction (free of any vector sequences)
10 to 40 μCi [α-^{32}P]dNTP (3000 Ci/mmol)
✓ Prehybridization solution
✓ $5 \times$ SSC (*APPENDIX 1*) containing 2.5 mg/ml sonicated human genomic DNA (prepare from 10 mg/ml stock; optional)
✓ Hybridization solution
✓ Wash solution: 1 mM Tris·Cl, pH 8.0/1% *N*-lauroylsarcosine (see both recipes, prepare fresh daily)
✓ 1 mM Tris·Cl, pH 8.0

Paper towels
Backing film (old autoradiographs)
Autoradiography film
Static Guard (anti-static spray)
Saran Wrap or other UV-transparent plastic wrap
Hybridization oven with tubes and bottles
100°C water bath or incubator
Autoradiography equipment

1a. Label 40 to 50 ng of probe DNA with 30 to 40 μCi [α-^{32}P]dNTP by random oligonucleotide-primed synthesis (*APPENDIX 3E*). Purify probe away from unincorporated [α-^{32}P]dNTP and check radioactivity (successful probes should have $5-7 \times 10^6$ cpm).

1b. Alternatively, as a hybridization probe, radioactively label one or more PCR products, such as STSs, during the polymerase chain reaction.

2. Roll filters together in a tube and place in a hybridization bottle. Add 15 ml prehybridization solution and seal bottle. Place bottle in a 55°C hybridization oven and rotate at 20 to 30 rpm for at least 1 hour. Add balance tubes as necessary, depending on the instrument. The rolled filters should expand to the limits of the hybridization tube. For more than six filters per tube, increase solution volume by 2 ml per filter. The use of a nylon mesh will also aid in liquid flow.

3. Heat the probe to 100°C for 5 min. For hybridization of probes containing repetitve elements, reassociate the probe 30 to 60 min at 65°C in 5× SSC containing 2.5 mg/ml sonicated human genomic DNA.

4. Add probe to 15 ml hybridization solution. Remove the hybridization tube from the oven and replace the prehybridization solution with the probe/hybridization solution. Seal the tube, return it to oven, and incubate 12 to 16 hr at 55°C.

5. Remove tube from oven and set oven to 25°C. Replace hybridization solution with 40 ml wash solution and return tube to the 25°C oven to wash for 15 min. Follow with four washes using 40 ml of 1 mM Tris·Cl, pH 8.0. If background is too high, raise wash temperature up to 42°C.

6. Remove the filters from the tube and unroll. Place the individual filters on paper towels, blotting dry most of the liquid. Put each filter, DNA side up, on an old piece of film. Spray briefly with Static Guard and cover with Saran Wrap. Expose the filters to autoradiography film for ~12 hr. Develop film. Between uses, carefully wrap filters in Saran Wrap and store at −20°C. For troubleshooting, see Table 5.2.2.

7. Determine plate and well addresses of positive clones. If using commercially available filters, obtain the positive clones from the supplier and streak each for single-colony isolation. If using filters prepared by the Support Protocol, retrieve positive clones from stored multiwell plates and streak those.

8. Prepare a "patch plate" with patches for each positive and a negative control. Perform a standard colony lift, lysis, and hybridization experiment to confirm that each positive clone contains the sequence of interest (see Support Protocol).

Table 5.2.2 Troubleshooting Hybridization to Colony Blots

Problem	Possible cause	Solution
No positives	Probe insufficiently radioactive	Test labeling method
		Test integrity of source DNA for labeling
	Filter quality poor	Repeat experiment on new filters
High background	Unincorporated label not removed	Test probe purification method
	Washing not sufficient	Repeat washes; increase stringency
	Filter quality poor	Repeat experiment on new filters
Thousands of positives	Repetitive sequence in probe	Locate repetitive element in probe DNA and avoid
		Include blocking DNA (e.g., salmon sperm, human placental, or C_0t_1 DNA) in a preannealing reaction with probe or in hybridization
	Vector sequence in probe	Locate vector sequence in probe DNA and avoid in future probe designs

PREPARING AND HYBRIDIZING COLONY BLOTS WITH OVERGO OLIGONUCLEOTIDE PROBES

The short length of overgo probes (Fig. 5.2.5) is advantageous when there is limited sequence from which to design a probe. These probes can be used for identifying specific clones, including YAC, BAC, PAC, and Pl clones on high-density filters, or for Southern blot analysis. When designing overgos, the sequence that is used must first be screened for presence of repetitive elements and low-complexity DNA sequences. Useful resources are Repeatmasker (*http://ftp.genome.washington.edu/cgi-bin/RepeatMasker*) and the Genome Sequencing Center (*http://genome.wustl.edu/gsc*).

Materials (see APPENDIX 1 for items with ✓)

    Two appropriate overlapping 24-mer oligonucleotides for generating each overgo probe
✓ 2 mg/ml bovine serum albumin (BSA)
✓ Overgo labeling buffer
    10 μCi/μl [α-^{32}P]dATP (~3000 Ci/mmol)
    10 μCi/μl [α-^{32}P]dCTP (~3000 Ci/mmol)
    Klenow fragment of *E. coli* DNA polymerase I
    10% (w/v) trichloroacetic acid (TCA)
    95% ethanol
✓ Church's hybridization buffer, 60°C
✓ 20× SSC
✓ 10% SDS
    Target DNA immobilized on nylon colony blot filters (see Basic Protocol 1 and Support Protocol)

    Glass microfiber filter disk (A/E fiber filter; Gelman)
    Vacuum filter holder
    Sephadex G-50 microspin column (e.g., NICK Spin Columns; Amersham Pharmacia Biotech)

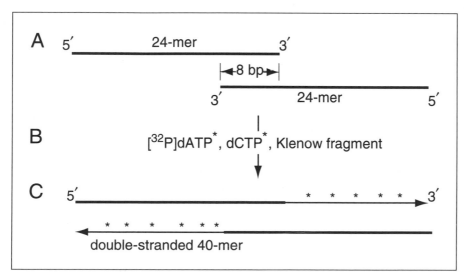

Figure 5.2.5 Two 24-mer oligonucleotides with an 8-bp overlap are annealed to create two 16-bp overhangs (**A**). Klenow fragment is used to fill in the overhang, concurrently introducing radionucleotides (**B**). The final product (**C**) is a double-stranded 40-mer of high specific activity.

Hybridization oven
Autoradiography equipment and supplies

1. For each overgo probe to be generated, transfer 10 pmol of each overlapping oligonu-cleotide in a total combined volume of 4 μl water to either a tube or a microtiter-plate well.

2. Cap each tube or well and heat the paired oligonucleotides 5 min at 80°C to denature the oligonucleotides, then 10 min at 37°C to allow overhangs to form. Store the annealed oligonucleotides on ice until they are labeled. If labeling is not done within 1 hr, repeat these two incubations before proceeding.

3. Prepare a master mix containing the following for each overgo to be labeled (plus 10% more to compensate for losses in transfer):

   0.5 μl 2 mg/ml BSA
   2.0 μl overgo labeling buffer
   0.5 μl 10 μCi/μl (∼3000 Ci/mmmol) $[\alpha\text{-}^{32}P]dATP$
   0.5 μl 10 μCi/μl (∼3000 Ci/mmmol) $[\alpha\text{-}^{32}P]dCTP$
   1.5 μl H_2O
   1 μl 2 U/μl Klenow fragment.

4. Pipet 6 μl master mix into each tube containing annealed oligonucleotide pairs and incubate 1 hr at room temperature.

5. To quickly check labeling, spot 1 μl onto a glass microfiber disk and use a Geiger counter to measure the cpm. Wet the top of the vacuum filter holder with 10% TCA and then place the disk onto the filter holder such that the filter becomes wet. Apply a gentle vacuum and wash the filter with ∼1 ml of 10% TCA followed by 1 to 2 ml of 95% ethanol. Allow to dry for 30 sec. Turn off the vacuum, remove the filter, and measure cpm again.

   Approximately 40% to 60% of the labeled nucleotides should remain bound to the glass filter. It is likely that probes retaining only 10% of the labeled nucleotides will work for most applications; however, if probes are to be pooled together, best results are obtained when all probes are labeled approximately equally.

6. For best results, remove unincorporated nucleotides using a Sephadex G-50 microspin column (following the manufacturer's protocol) before proceeding with hybridization.

7. Prehybridize the nylon filters by wetting with 60°C Church's hybridization buffer and rolling the filters into a hybridization bottle containing 20 ml of the same prewarmed buffer. Incubate the filter ≥ 1 hr at 60°C.

8. Denature the labeled probes by heating to 90°C for 10 min and then placing on ice for 10 min. Remove 5 ml hybridization buffer from the bottle containing filters and place in a 15-ml tube. Add the probe(s) and transfer back to the hybridization bottle. Incubate overnight at 60°C.

9. Drain the buffer from the bottle and add 200 ml of 2× SSC/0.1% SDS. Return the hybridization bottle to the incubation oven for 10 min. Drain the SSC solution from the bottle.

10. Transfer the filters to 2 liters 60°C 1.5× SSC/0.1% SDS and wash for 30 min with gentle shaking. Repeat with 2 liters 60°C 0.75× SSC/0.1% SDS.

11. Wrap filters in plastic wrap and autoradiograph (usually overnight) at −70°C with an intensifying screen. Determine plate and well addresses of positive clones.

12. When required, strip probes from the filters by washing in 2 liters 85°C 0.2× SSC/0.1% SDS for 30 min. Check by autoradiography.

References: Bentley et al., 1992; Nelson et al., 1989

Contributors: Mark T. Ross, Sam LaBrie, John McPherson

UNIT 5.3

Introduction of Large Insert DNA into Mammalian Cells and Embryos

An overview of the procedure is provided in Figure 5.3.1.

NOTE: All reagents and equipment coming in contact with live cells must be sterile. All tissue culture incubations should be performed in a humidified 37°C, 5% CO_2 incubator unless otherwise specified.

BASIC PROTOCOL 1

INTRODUCTION OF INTACT YACS INTO MAMMALIAN CELLS BY SPHEROPLAST FUSION

Materials (see APPENDIX 1 for items with ✓)

    Target cells: adherent mammalian cells in tissue culture (e.g., mouse ES cells)
✓ SD dropout medium and plates: SD –Ura –Trp –His
    Yeast strain carrying YAC tagged with *neo* (G418-resistance) selectable marker (see Support Protocol 1)
    1 M sorbitol (store at room temperature)
✓ SCE solution
    20 mg/ml Zymolyase 20T (ICN Biomedicals) in SCE solution; titrate new stocks to determine amount of enzyme that gives 90% of spheroplasts in 45 to 60 min
✓ 10% (w/v) SDS
    ST solution: 1 M sorbitol/10 mM Tris·Cl, pH 7.5 (APPENDIX 1)
✓ PBS
    0.05% trypsin/0.53 mM EDTA (Life Technologies)
    Complete culture medium with serum appropriate to mammalian cells (APPENDIX 3I) with and without 200 to 800 μg/ml G418 (Life Technologies), room temperature
    Culture medium appropriate to mammalian cells without additives or serum, room temperature
    50% (w/v) polyethylene glycol (PEG) 1500 (fusion-tested; Boehringer Mannheim)
    Appropriate PCR primers (Table 5.3.1)
    Appropriate radiolabeled probes (Table 5.3.2; see APPENDIX 3E for radiolabeling)
    Mouse Cot-1 DNA (Life Technologies)
✓ 20× SSC

    100-mm-diameter and 24-well tissue culture plates
    15-ml glass culture tubes with caps
    30°C incubator with roller drum and shaker
    500-ml Erlenmeyer culture flask with cap
    Beckman TJ-6 centrifuge with TH-4 rotor and buckets (or equivalent)
    Phase-contrast microscope
    Cloning cylinders (UNIT 3.1)
    65°C water bath

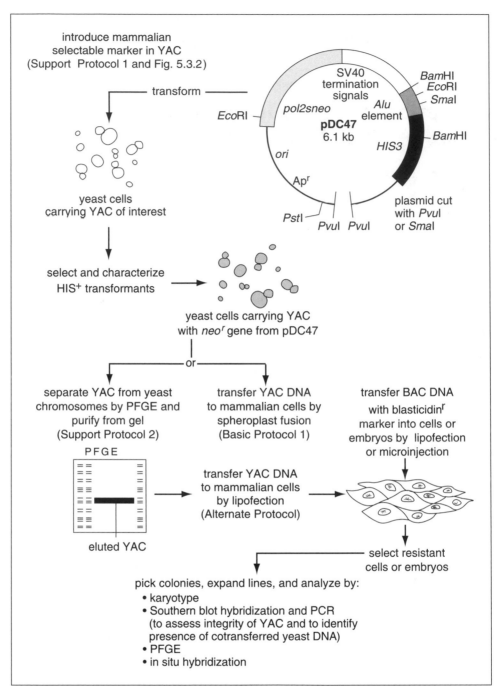

Figure 5.3.1 Marker tagging of YAC and introduction into mammalian cells.

1. Begin expanding target mammalian cells in 100-mm tissue culture plates to provide 10^7 cells at 90% confluence in ~3 days (sufficient for five fusions at 2×10^6 cells per fusion).

2. Two days before mammalian cells will reach 90% confluence, inoculate 5 ml of SD –Ura –Trp –His medium in a 15-ml glass culture tube with a single colony of yeast carrying a YAC tagged with *neo*. Incubate overnight at 30°C in a roller drum.

3. The next day, inoculate 100 ml SD –Ura –Trp –His medium in a 500-ml Erlenmeyer flask with the yeast culture from step 2 and grow overnight at 30°C with vigorous shaking.

Table 5.3.1 PCR Primers Used to Assess Integrity of YACs Before and After Transfer into Mammalian Cells

Primers[a]	Sequence	Product size (bp)
NEO1	5′-GATTGCACGCAGGTTCTCCG-3′	654
NEO2	5′-CCAACGCTATGTCCTGATAG-3′	
URA3A	5′-AATGCACACGGTGTGGTG-3′	277
URA3B	5′-CGTCTCCCTTGTCATCTAAACC-3′	
XIST1	5′-GGGACCTAACTGTTGGCTTTATCAG-3′	203
XIST2	5′-GAAGTGAATTGAAGTTTTGGTCTAG-3′	

[a]PCR conditions—30 cycles: 95°C, 1 min; 60°C, 1 min; 72°C; 2 min.

4. On morning of fusion, add 25 ml fresh SD –Ura –Trp –His medium to yeast culture and incubate at 30°C with vigorous shaking for 1.5 hr or until OD_{600} is between 1.2 and 1.5.

5. Divide yeast culture among four 50-ml conical centrifuge tubes and centrifuge 5 min at $500 \times g$ (2000 rpm in Beckman TH-4 rotor), room temperature. Resuspend each pellet in 5 ml of 1 M sorbitol and pool cells into one 50-ml conical centrifuge tube.

6. Centrifuge again 5 min at $500 \times g$ and remove supernatant. Add 20 ml of 1 M sorbitol, centrifuge 5 min at $500 \times g$, and remove supernatant. Resuspend cells in 20 ml SCE solution and add 100 μl of 20 mg/ml Zymolyase 20T. Incubate 30 min at 30°C.

7. Place 10 μl of 10% SDS on a microscope slide and add 10 μl of cell suspension. Check spheroplast formation with a phase-contrast microscope. Lysed spheroplasts will appear as dark gray ghosts; cells that are not lysed (i.e., were not spheroplasted) are highly refractile.

8. Continue 30°C incubation, checking spheroplast formation every 10 to 15 min until 85% to 95% of cells are spheroplasted.

> *From this point on, extreme care must be taken to avoid lysing the delicate spheroplasts. Very slow, gentle resuspensions are necessary, requiring as much as 1 hr of gentle rocking.*

9. Centrifuge spheroplasts 7 min at $250 \times g$ (1000 rpm in Beckman TH-4 rotor), 4°C. Decant supernatant, add 20 ml ST solution, then rock very gently to resuspend pellet. Repeat wash and gently resuspend final pellet in 10 ml ST solution.

Table 5.3.2 Probes Used to Assess Integrity of YACs Before and After Transfer into Mammalian Cells

Probe targets	Plasmid	Restriction endonucleases	Size (kb)	Reference
TRP1	pYAC4	*Pst*I/*Xba*I	0.6	Burke et al., 1987
URA3	YEP24	*Hin*dIII	1.2	Botstein et al., 1979
Acentric YAC arm	pBR322	*Pvu*II/*Sal*I	1.4	Burke et al., 1987
Centric YAC arm	pBR322	*Pvu*II/*Eco*RI	2.3	Burke et al., 1987
neo	pDC47	*Xba*I/*Eco*RI	0.8	D.E.C., unpub. observ.
Alu element	pDC47	*Bam*HI	0.33	D.E.C., unpub. observ.
LINE element	pBP90	*Cla*I/*Nco*I	0.4	Pavan et al., 1990
B1 element	pB1con	*Pst*I/*Sst*I	0.3	J. McKee-Johnson, unpub. observ.
Ty1	pJEF1271	*Xho*I	6	Eichinger and Boeke, 1988

10. Dilute 10 µl spheroplast suspension in 990 µl ST solution and count on a hemacytometer (*APPENDIX 3I*). Adjust concentration to 0.8–1.4×10^8 spheroplasts/ml with ST, if necessary. Keep spheroplasts on ice, rocking gently every 5 to 10 min to keep in suspension.

11. Wash target mammalian cells twice with 5 ml PBS per 100-mm dish. Add 1 ml trypsin/EDTA and incubate at 37°C until cells detach from dish. Resuspend $\sim 10^7$ cells in 12 to 14 ml complete culture medium with serum and transfer to a 15-ml conical centrifuge tube. Determine concentration using a hemacytometer (*APPENDIX 3I*).

12. Centrifuge target cell 3 min at $250 \times g$, 4°C, and resuspend at 2×10^6 cells/ml in culture medium without serum. For each fusion to be performed (at least four to create a time course for PEG treatment), transfer a 1-ml aliquot to a 15-ml conical centrifuge tube, and centrifuge 3 min at $250 \times g$. Do not remove supernatant.

13. Add 1×10^8 spheroplasts in ST solution on top of supernatant. Centrifuge 3 min at $250 \times g$ and carefully remove supernatant by aspiration, leaving 50 to 100 µl medium. Gently triturate pellet in remaining liquid to thoroughly resuspend cells.

14. Add 2 ml of 50% PEG 1500 and mix by gently pulling the solution up and down 3 times with a 5-ml pipet. Immediately after mixing, begin timing fusion.

    Optimal exposure time to PEG may be cell-line-dependent and must be determined empirically; an exposure time course should be used in each set of fusions. Individual fusion reactions should stand for 30, 60, 90, or 120 sec.

15. At end of time period, add 5 ml culture medium without serum, invert gently three times, and centrifuge 3 min at $250 \times g$. Resuspend pellet gently in complete culture medium with serum and plate each fusion in four to ten 100-mm tissue culture plates at 2–5×10^5 cells/plate.

    The number of plates and number of cells plated depends on the growth rate and the time required for G418 killing of sensitive cells. Density should be low enough that the plates do not overgrow before G418 killing.

16. Between 12 and 18 hr after fusion, feed cells with complete medium containing serum and 200 to 800 µg/ml G418 (varies with cell type). Continue incubation in complete medium with serum and G418 for 8 to 14 days, until most cells have died and colonies of G418-resistant (i.e., neo[r]) cells are visible.

17. Use cloning cylinders to transfer individual colonies into separate wells of 24-well tissue culture plates containing complete medium with serum and G418. For ES-cell fusions, select several dozen clones to allow for the high percentage that will not survive subcloning and expansion without undergoing differentiation. Expand cells for freezing and DNA analysis.

18. Prepare DNA from approximately 10^6 cells from a single G418-resistant clone (*APPENDIX 3A*).

19. Analyze cells for the presence of *URA3* and *neo* marker sequences by PCR amplification using the primers listed in Table 5.3.1, along with DNA from the parental yeast strain with and without the YAC, as controls. Use the following PCR cycle parameters:

30 cycles:	1 min	95°C
	1 min	60°C
	2 min	72°C

20. Determine presence or absence of any known genes or markers on the YAC by PCR or by agarose gel electrophoresis and Southern blot hybridization (*APPENDIX 3G*) using appropriate radiolabeled probe DNA.

21. Prepare "*Alu* profiles" (to determine integrity of the YAC) by digesting DNA prepared in step 18 with several restriction endonucleases, making Southern blots, and probing with a human *Alu* probe to sample segments spanning the YAC. The *Alu* probe must be preannealed with mouse repetitive DNA. Mix:

> 50 ng probe DNA
> 100 µg mouse Cot-1 DNA
> 25 µl 20× SSC
> H_2O to 100 µl.

Boil 10 min and cool 1 min on ice. Incubate 10 min at 65°C. Add preannealed probe to hybridization buffer and proceed with Southern blot hybridization.

22. Determine presence or absence of yeast DNA by hybridizing a Southern blot with yeast Ty repetitive element probe (the number of Ty-positive restriction fragments provides an indicator of the amount of yeast information present in the cell).

23. Prepare metaphase spreads from cell line (*UNIT 8.2*), band chromosomes (*UNIT 4.3*), and analyze karyotype (*APPENDIX 3K*).

ALTERNATE PROTOCOL

INTRODUCTION OF GEL-PURIFIED YAC DNA INTO MAMMALIAN CELLS BY LIPOFECTION

Additional Materials (*also see Basic Protocol 1; see APPENDIX 1 for items with ✓*)

PFGE gel slice (see Basic Protocol 2) containing intact YAC DNA with *neo* selectable marker (see Support Protocol 1)
✓ Dialysis buffer I
1 U/µl β-agarase (New England BioLabs)
Lipofectin (Life Technologies) or other cationic lipid for DNA transfection (e.g., Transfectam, Promega)
Control DNA for optimizing lipofection (any neo^r plasmid expressed by the target cell, e.g., pDC47)

40° and 65°C water baths
35-mm-diameter tissue culture plates
5-ml polystrene tubes

1. Transfer PFGE gel slice containing YAC to a 50-ml conical tube and dialyze at 4°C against three changes of dialysis buffer I (30 to 40 ml final) over 10 to 12 hr.

2. Remove all dialysis buffer, divide gel slice into 0.5- to 2.0-ml segments, and place in microcentrifuge tubes. Incubate tubes 15 to 20 min at 65°C until agarose is melted completely, then equilibrate 5 min at 40°C.

3. Add 10 U β-agarase and mix by stirring gently with the pipet tip. Incubate 60 to 120 min at 40°C until agarose is completely digested. Test by placing tube 5 min on ice and examining for solid agarose. If solid agarose remains, remelt at 65°C, cool to 40°C, add an additional 10 U β-agarase, and continue incubation at 40°C until agarose is fully digested. If YAC DNA is to be stored >24 hr, heat-inactivate enzyme 10 min at 55°C and store at 4°C.

4. Test integrity of purified YAC DNA by running an aliquot on a pulsed-field gel and performing Southern blot hybridization.

5

5. To optimize conditions, 18 to 24 hr before lipofection, plate $2–4 \times 10^5$ target mammalian cells in culture medium onto each of six 35-mm tissue culture plates. If necessary (depending on cell type), adjust cell number to achieve 80% to 90% confluence.

6. In 5-ml polystyrene tubes, prepare six solutions containing 0, 5, 10, 20, 40, or 80 μg cationic lipid plus 100 ng control DNA in 200 μl of the appropriate culture medium *without* serum. Incubate 45 min at room temperature to allow stable lipid/DNA complexes to form. After this incubation, add 0.8 ml culture medium without serum or antibiotics.

7. Aspirate medium from target cells. Add 1 ml lipid/DNA complex to each plate and culture 4 to 18 hr (depending on cell type).

8. Refeed cells with appropriate medium containing serum and antibiotics. Incubate 24 to 48 hr. Trypsinize cells and replate each original plate into one 100-mm tissue culture plate in medium containing 200 to 800 μg/ml G418. After 8 to 14 days, count G418-resistant colonies and select the lipofection conditions that result in the highest yield.

9. For final transformation, plate target mammalian cells in medium with serum and antibiotics. Start the cultures 24 to 48 hr before transfection to achieve ~80% confluence.

10. Mix optimal amount of Lipofectin (determined in step 8) with ~100 ng purified YAC DNA (from step 3). Incubate 45 min at room temperature to allow lipid/DNA complexes to form.

   Concentration of gel-purified YAC DNA can be estimated by titration on ethidium agarose plates (see Support Protocol 3). If concentration of purified YAC DNA is not known, use a range of volumes (e.g., 100, 200, 400, and 1000 μl).

11. Add appropriate serum-free culture medium to a volume of 1 ml. Rinse cells once with serum-free medium and add lipid/DNA complexes to cultured cells. Incubate 4 to 18 hr (depending on cell type).

12. Refeed cells with appropriate complete medium containing serum. Incubate until 24 to 48 hr after lipofection is completed. Trypsinize cells and replate in several 35-mm tissue culture plates (typically two to four plates per original well).

   The number of wells depends on the growth rate and the time required for G418 killing of sensitive cells. Density should be low enough that the plates do not overgrow before G418 killing.

13. Feed cultures with appropriate complete medium containing serum and 200 to 800 μg/ml G418. Select, clone, expand, and analyze transfectants (see Basic Protocol 1).

SUPPORT PROTOCOL 1

INTRODUCTION OF A MAMMALIAN SELECTABLE MARKER INTO YACS BY HOMOLOGOUS RECOMBINATION

Materials (*see* APPENDIX 1 *for items with* ✓)

pDC47 (ATCC #87028; Fig. 5.3.2) or other integrating plasmid with *HIS3* selectable marker

Appropriate restriction endonucleases and buffers for linearizing integrating plasmid (Fig. 5.3.2)

✓ TE buffer, pH 7.5

YAC-bearing yeast strain containing *HIS3* mutation

✓ SD dropout plates and medium: SD –Ura –Trp and SD –Ura –Trp –His

✓ YPD medium

✓ Lithium acetate/Tris/EDTA solution

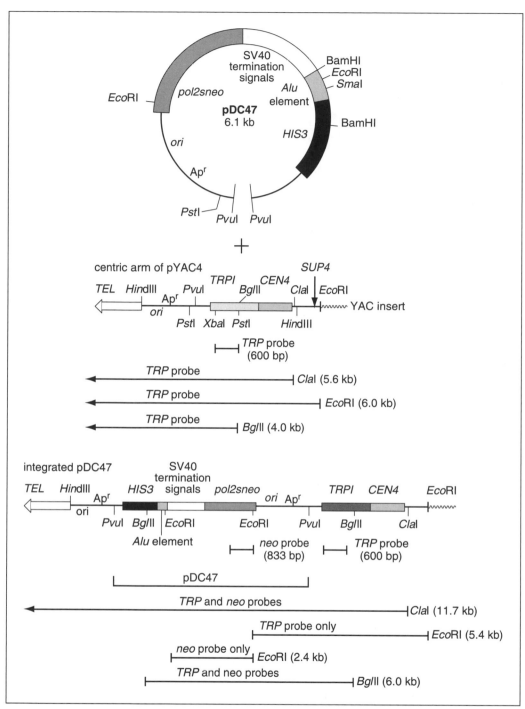

Figure 5.3.2 Integration of the pDC47 plasmid into a YAC. pDC47 contains a neo^r gene under the control of the *pol2s* promoter, which has been shown to be highly expressed in ES cells. The centric arm of the pYAC4 cloning vector contains plasmid sequences that can serve as a target for integration when pDC47 is linearized with *Pvu*I. Sizes of *Eco*RI, *Cla*I, and *Bgl*II fragments in the unmodified YAC vector are shown. Integration of pDC47 in pYAC4 alters sizes of several fragments identified with probes for *TRP1* and *neo*. Following proper integration, digestion with *Cla*I, *Eco*RI, and *Bgl*II followed by probing with *TRP1* will produce 11.7-, 5.4-, and 6-kb bands, respectively. These *Cla*I and *Bgl*II fragments are also recognized by the *neo* probe, which detects a 2.4-kb *Eco*RI fragment after appropriate targeting. Restriction fragments containing the telomere may be slightly larger than the drawing indicates, due to the expansion of telomere repeats in yeast.

✓ 40% PEG solution

Radiolabeled probes for *neo*, *URA3*, *TRP1*, and repetitive elements (Table 5.3.2; see APPENDIX 3E for radiolabeling)

✓ 0.5 μg/ml ethidium bromide

*Cla*I, *Eco*RI, and other restriction endonucleases appropriate for analysis of transformation products (see Fig. 5.3.2)

30°C incubator and shaker
Centrifuge with Beckman TH4 rotor
42°C water bath
Nitrocellulose or nylon membrane
X-ray film

1. Linearize 20 μg pDC47 with appropriate restriction endonucleases (e.g., *Pvu*I to target integration in YAC vector arm or *Sma*I to target *Alu* sequences). Concentrate DNA by ethanol precipitation and resuspend in 20 μl TE buffer.

2. Streak YAC-bearing yeast strain containing *HIS3* mutation onto an SD –Ura –Trp plate and incubate at 30°C for 3 days or until colonies appear.

3. Pick a yeast colony and inoculate into 5 ml YPD medium in a 15-ml culture tube. Grow overnight to saturation in a shaking incubator at 30°C. Inoculate 50 ml SD –Ura –Trp medium in a 150-ml flask with 0.5 ml saturated culture and continue growing to an OD_{600} of 1 to 2 (~6 to 10 hr, depending upon the strain).

4. Centrifuge 5 min at 500 × *g* (2000 rpm in Beckman TH4 rotor), room temperature. Discard supernatant, add 10 ml freshly prepared lithium acetate/Tris/EDTA solution, and resuspend pellet by vortexing vigorously. Repeat centrifugation and resuspension.

5. Incubate cells 30 min without shaking at 30°C and then overnight at 4°C.

6. Centrifuge 5 min at 500 × *g*, room temperature. Add 1 ml lithium acetate/Tris/EDTA solution and resuspend by vortexing vigorously. Aliquot 100 μl of cells per microcentrifuge tube for transformations.

7. Add 3 to 10 μg linearized pDC47 DNA in ≤10 μl TE buffer (step 1) to 100 μl competent yeast cells. Vortex vigorously for 2 sec and incubate 10 min at 30°C.

8. Add 0.5 ml 40% PEG solution to each tube and briefly vortex at medium speed. Incubate 60 min at 30°C.

9. Heat-shock cells by heating tubes 5 min at 42°C.

10. Add sufficient sterile water to fill tube, vortex, and microcentrifuge 5 sec at high speed, room temperature. Decant supernatant and resuspend yeast cells by vortexing at medium speed in 1 ml sterile water. Microcentrifuge 5 sec at high speed, room temperature.

11. Resuspend cells in 50 μl sterile water and spread on an SD –Ura –Trp –His agar plate. Incubate 3 to 4 days at 30°C.

12. Streak each colony on a one-eighth segment of an SD –Ura –Trp –His plate. Incubate 2 to 3 days at 30°C until colonies form.

13. Select individual colonies from strains that are His[+] Trp[+] Ura[+] and expand in SD –Ura –Trp –His medium.

14. Prepare high-molecular-weight DNA in agarose blocks (*UNIT 5.1*) from yeast cells grown from individual colonies. Run on a pulsed-field gel under conditions that will separate chromosomes in the size range of the YAC. Include DNA from the parental yeast strain without a YAC and from the nontransformed YAC-bearing strain as controls.

15. Stain gel with ethidium bromide, photograph, and blot onto nitrocellulose or nylon membrane.

16. Hybridize filter with radiolabeled probe for *neo*. Strip and hybridize sequentially with probes for *URA*, *TRP*, and/or repetitive elements from the appropriate species. After each stripping, expose filter to X-ray film for 12 to 24 hr before reprobing to assure that the previous probe has been completely removed.

17. Digest DNA from individual transformants with several restriction endonucleases including *Cla*I and *Eco*RI. Carry out agarose gel electrophoresis and Southern blotting. Hybridize, autoradiograph, and strip sequentially using radiolabeled probes for *neo*, *TRP1*, and species-specific repetitive DNA elements.

18. Perform repetitive-element "profiles" (see Basic Protocol 1, step 21). Include the same controls used for PFGE (see step 14 of this protocol).

19. Verify that all bands identified with the *TRP1* and *neo* probes are of the predicted size (see Fig. 5.3.2) and the repetitive-sequence probe gives the same profile of bands in parental and modified YACs.

SUPPORT PROTOCOL 2

PURIFICATION OF YACS BY PULSED-FIELD GEL ELECTROPHORESIS

Materials (see APPENDIX 1 for items with ✓)

    YAC-bearing yeast strain
✓ 0.5× TBE buffer
    Low gelling/melting temperature agarose (e.g., SeaPlaque GTG grade, FMC Bioproducts)
    High-molecular-weight size standards in agarose (*UNIT 5.1*)
    0.1 mg/ml ethidium bromide in 0.5× TBE buffer
✓ Dialysis buffer II
    10 mg/ml poly-L-lysine (Sigma; optional)
    Agarase: e.g., β-agarase (New England Biolabs) *or* GELase (Epicentre Technologies)

    Dialysis membrane
    Customized gel casting plate (optional; Fig. 5.3.3)
    Pulsed-field gel electrophoresis (PFGE) apparatus, prechilled to 4°C
    Short- or medium-wave UV transilluminator with camera attachment
    Rulers (preferably fluorescent)
    65° and 40°C heating blocks or water baths

1. Prepare YAC DNA in agarose blocks and dialyze in 0.5× TBE buffer (see *UNIT 5.1*, Support Protocol 2; use 0.5× TBE in place of TE for dialysis).

2. In a 4°C cold room, pour a 0.5% to 0.8% low gelling/melting temperature agarose gel in 0.5× TBE buffer, using a gel comb with a single large tooth (or teeth taped together) to make a gel trough and small teeth at either end to make wells for molecular size standards.

    A gel of this type takes ~1 hr to solidify, and will be delicate and not adhere well to the glass casting plate. One approach is to use a glass casting plate with Velcro strips attached near its edges (Fig. 5.3.3). The gel sets into the Velcro so that the casting plate grips and supports the gel during handling. An alternative is to use a frosted glass casting plate.

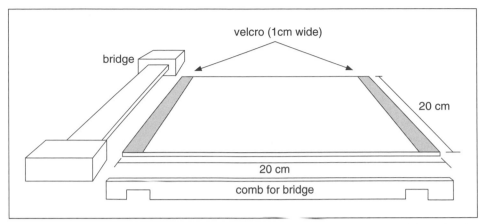

Figure 5.3.3 Gel casting plate with Velcro strips for PFGE. Parallel strips of Velcro hooks are glued with silicone adhesive to the edges of a glass plate. Agarose is then poured over the Velcro-covered surface and the comb is clipped to the bridge and positioned over the plate. When it sets, the Velcro hooks secure the gel to the plate.

3. Equilibrate agarose blocks against 0.5× TBE using three changes (1000 vol total) over 30 min. Load YAC DNA blocks into gel trough lengthwise, placing them as close together as possible without significant compression. Seal plugs in trough with a little melted agarose (the same mixture as used to pour gel). Load molecular size standards in outer wells.

4. Place glass plate in a PFGE apparatus containing prechilled 0.5× TBE and let gel equilibrate for 10 to 15 min. Run gel using appropriate parameters (switching time, voltage, and run time) to allow maximum resolution in the size range of the DNA fragments of interest.

5. After electrophoresis is complete, slice off the left and right edges of the gel, including size standards and a small portion of sample DNA. Carefully slide these gel slices into a glass or plastic dish containing 0.1 mg/ml ethidium bromide in 0.5× TBE and allow slices to stain 30 to 60 min. Replace buffer with 0.5× TBE buffer and destain 30 to 60 min.

6. Place slices side-by-side with a ruler atop a UV transilluminator and photograph. Place two rulers on either side of the remaining unstained gel. Line up a third ruler along the width of the gel using the photograph as a reference. With a clean scalpel, slice out a strip of the section of the gel containing the YAC bands, as large as 1 × 16 cm.

7. Transfer slice to a 50-ml conical centrifuge tube and dialyze at 4°C against at least three changes of dialysis buffer II (30 to 40 ml total) over the course of 12 hr. Use DNA as soon as possible (within several days to a week). If the DNA is to be used for transfection, add 10 µl of 10 mg/ml poly-L-lysine.

8. Divide gel slice into 0.5- to 2.0-ml sections and transfer into small conical tubes. Melt 10 min at 65°C and equilibrate 5 min at 40°C. Add 10 U agarase, mix gently, and allow to digest 60 to 120 min at 40°C. For long-term storage, heat-inactivate agarase 10 min at 55°C.

SUPPORT PROTOCOL 3

RAPID ESTIMATION OF DNA CONCENTRATION ON ETHIDIUM BROMIDE/AGAROSE PLATES

Materials (*see* APPENDIX 1 *for items with* ✓)

    0.8% (w/v) agarose in H$_2$O

✓ 10 mg/ml ethidium bromide

1 μg/μl DNA stock solution (store frozen; can be thawed repeatedly if not contaminated)
Unknown DNA sample
60-mm-diameter petri plates

1. Prepare 100 ml of 0.8% agarose in water. Cool to 50°C and add 10 μl of 10 mg/ml ethidium bromide in water. Mix by swirling and pour into eight 60-mm petri plates. Allow to harden and dry slightly 2 to 3 hr at room temperature. Wrap in Parafilm and store ≤1 month at 4°C.

2. Starting with 1 μg/μl stock solution, prepare seven 1:1 serial dilutions to provide eight standard solutions ranging in concentration from 1 μg/μl to 7.8 ng/μl. Place 0.5 μl of each standard solution along with 0.5 μl unknown DNA in separate spots on an ethidium bromide/agarose plate. Allow to stand 15 min.

3. Photograph plate using UV transillumination. Match fluorescence intensity of unknown to that of standard dilutions to estimate concentration.

BASIC PROTOCOL 2

INTRODUCTION OF BACTERIAL ARTIFICIAL CHROMOSOMES (BAC OR PAC) INTO MAMMALIAN CELLS AND MOUSE EMBRYOS

Materials (*see* APPENDIX 1 *for items with* ✓)

Selective plate
✓ LB medium with appropriate antibiotic (selective medium)
RNase A (Sigma)
Qiagen Midi-prep kit (Qiagen) including the following:
    P1 buffer
    P2 buffer (before use, be sure SDS is thoroughly resuspended; if it has
        precipitated, warm to 37°C)
    P3 buffer, prechilled
    Midi-tip 100 columns
    QBT buffer
    QC buffer
    QF buffer, prewarmed to 65°C
Isopropanol
✓ Pronuclear injection buffer (PIB)
Purified BAC DNA *or* PFGE gel slice containing linearized BAC DNA (Support Protocol 2)
✓ Dialysis buffer I
70% ethanol
Mouse embryos (and transgenic core facility)
Mammalian cells
Blasticidin (ICN Biomedical) dissolved in complete medium at 1 to 4 μg/μl to make a 1000× solution

37°C incubator for bacterial culture
High-speed centrifuge (e.g., Sorvall RC-series with GSA rotor or equivalent)
50-ml high-speed centrifuge tubes
15-ml Corex glass centrifuge tubes

1. Prepare a starter culture by picking a single colony from a selective plate and inoculating 5 ml of LB medium containing the appropriate antibiotic (selective medium). Grow for several hours at 37°C with shaking until turbid.

2. Inoculate 100 ml selective medium with 0.5 ml starter culture. Grow 14 hr at 37°C with vigorous shaking (200 rpm).

3. Divide the culture into two 50-ml high-speed centrifuge tubes. Centrifuge cells 20 min at 4500 × g (Sorvall GSA rotor 5500 rpm), 4°C. Decant culture medium supernatant and save pellet.

4. Dissolve 0.2 g RNase A in 20 ml P1 buffer. Resuspend each pellet in 10 ml P1 plus RNase buffer. Add 10 ml P2 buffer to each tube. Gently invert several times to mix thoroughly. Leave 5 min at room temperature.

5. Add 10 ml chilled P3 buffer and immediately invert gently several times to mix thoroughly. Incubate 15 min on ice. Centrifuge 15 min at 20,000 × g (13,000 rpm), 4°C. Immediately remove supernatants to clean tubes and centrifuge again. Pool the supernatants from the two tubes.

6. Equilibrate a Qiagen midi-tip 100 column with 4 ml QBT buffer following the directions provided by the manufacturer. Apply the supernatant solution to the column and allow it to enter the resin by gravity flow. Wash the column with 10 ml QC buffer two times.

7. Elute DNA into a 15-ml Corex glass centrifuge tube by applying 1 ml QF buffer pre-warmed to 65°C. Repeat four times (5 ml total). Add 3.5 ml isopropanol and centrifuge 30 min at 15,000 × g (11,500 rpm), 4°C. Discard supernatant and save pellet.

8. Gently add 2 ml of 70% ethanol (room temperature) and centrifuge 10 min at 15,000 × g (11,500 rpm), 4°C. Discard supernatant and allow pellet to air dry 10 min. Dissolve DNA in 200 μl PIB. Determine DNA concentration spectrophotometrically (APPENDIX 3D).

9. If desired, linearize and gel purify the BAC DNA according to Support Protocol 4.

10a. *To introduce BAC DNA into mouse embryos:* Resuspend DNA at 100 μg/ml in PIB. Carry out microinjection in a qualified transgenic core facility.

10b. *To transfect mammalian cultured cells:* Using BACs made with the pPAC4 or pBACe3.6 vectors, carry out the lipofection method described in the Alternate Protocol, steps 5 to 8, substituting 1 to 4 μg/ml blasticidin (added from 1000× stock) for G418 in step 8 (most sensitive cells are killed in 1 week).

11. Prepare genomic DNA from blasticidin-resistant cells or from transgenic mice for PCR and Southern blotting.

12. Determine the presence or absence of any known genes or markers on the BAC by PCR or Southern blotting to detect polymorphisms between the marker on the BAC and the corresponding allele in the recipient cell line. Estimate copy number using a probe molecule adjacent to the vector:insert junction of the BAC.

SUPPORT PROTOCOL 4

OPTIONAL LINEARIZATION AND GEL PURIFICATION OF BAC

Materials (see APPENDIX 1 for items with ✓)
   BAC DNA (from Basic Protocol 2, step 8)
   *Not*I or *Asc*I restriction enzyme and appropriate buffer
✓ Dialysis buffer I
   β-agarase
   25:24:1 (v/v/v) phenol/chloroform/isoamyl alcohol

24:1 (v/v) chloroform/isoamyl alcohol

✓ 3 M sodium acetate

100% ethanol

✓ Pronuclear injection buffer (PIB)

50-ml conical tubes

1.5-ml microcentrifuge tubes

40°, 55°, and 65°C water baths

1. To determine whether the genomic insert contains *Not*I or *Asc*I restriction sites, digest 5 μg of BAC DNA with 10 U of restriction enzyme in the appropriate buffer under conditions recommended by the manufacturer of the enzyme.

2. Load a pulsed-field gel and run under appropriate conditions to resolve fragments from 5 to 200 kb. Stain the pulsed-field gel with ethidium bromide, photograph, and determine fragment sizes (procedures all described in *UNIT 5.1*).

> *For Not*I *in most BAC vectors, if the insert contains no Not*I *sites, two bands will be visible, representing the vector (typically 11 to 20 kb) and insert (typically 80 to 200 kb). For Asc*I *using the pPAC4 vector, if the insert contains no Asc*I *sites, two bands will be visible, representing a 2.4-kb internal vector band and the insert plus 14.3 kb vector. For Asc*I *using the pBACe3.6 vector, if the insert contains no Asc*I *sites, there will be a single band the size of the insert plus 11.5 kb vector (Asc*I *cuts the vector once). In all cases, the presence of additional bands indicates that the insert has been cut by the enzyme and the BAC cannot be linearized using that enzyme. If the insert is cut by both enzymes, circular BACs must be used in the subsequent procedures and the resulting mice/cell lines analyzed as discussed below.*

3. Digest 20 μg of BAC DNA with the appropriate restriction enzyme. When digestion is complete, add gel loading buffer.

4. Prepare a low-melting-point gel for pulsed-field gel electrophoresis (see Basic Protocol 2). After placing the gel in the electrophoresis apparatus, load the BAC DNA as for a conventional gel. Proceed through step 6 of Support Protocol 2.

5. Transfer gel slice containing the BAC to a 50-ml conical tube and dialyze at 4°C against three changes of dialysis buffer I (30 to 40 ml total) over 10 to 12 hr.

6. Remove all dialysis buffer and divide the gel slice into small segments (∼0.5 ml) in 1.5-ml microcentrifuge tubes. Heat tubes for 15 to 20 min at 65°C until melted completely, then equilibrate to 40°C for 5 min.

7. Add 10 U β-agarase and mix by stirring gently with the pipet tip. Incubate 60 to 120 min until no solid agarose is visible in the tube. If solid agarose remains after 120 min, remelt at 65°C, cool to 40°C, add 10 U of agarase, and continue the incubation.

8. Heat-inactivate the β-agarase 10 min at 55°C. Extract the DNA one time with 25:24:1 phenol/chloroform/isoamyl alcohol and two times with 24:1 chloroform/isoamyl alcohol (*APPENDIX 3B*). Add 0.1 volume of 3 M sodium acetate and precipitate DNA by adding 2 volumes of 100% ethanol and incubating 30 min at –20°C.

9. Recover DNA by microcentrifuging at maximum speed. Decant ethanol completely and air dry pellet for 30 min. Resuspend in 100 μl of PIB. Test integrity of purified BAC DNA by Southern blot analysis of a pulsed-field gel containing an aliquot of the DNA.

References: Antoch et al., 1997; Cabin et al., 1995; Huxley et al., 1991; Pavan et al., 1990; Sikorski and Hieter, 1989

Contributors: Roger H. Reeves, Deborah E. Cabin, Bruce Lamb, and William M. Strauss

Construction of Bacterial Artificial Chromosome (BAC/PAC) Libraries

Vectors used in this unit are illustrated in Figure 5.4.1.

CAUTION: To prevent shearing, use sterile wide-bore pipet tips for all steps involving the handling of genomic DNA.

BASIC PROTOCOL

PREPARATION OF BAC/PAC CLONES USING pCYPAC2, pPAC4, OR pBACe3.6 VECTOR

Materials *(see APPENDIX 1 for items with ✓)*

≥2 to 10 ng/µl size-fractionated *Mbo*I- or *Eco*RI-digested genomic DNA (see Support Protocol 4)

10 to 50 ng/µl pBACe3.6, pCYPAC2, or pPAC4 vector DNA prepared for cloning (see Support Protocol 1)

✓ 1 Weiss U/µl T4 DNA ligase and 5× buffer (Life Technologies; see APPENDIX 1 for buffer)

✓ 0.5 M EDTA, pH 8.0

10 mg/ml proteinase K

✓ 100 mM PMSF solution

✓ TE buffer, pH 8.0

TE/PEG solution: 0.5× TE buffer, pH 8.0, containing 30% (w/v) polyethylene glycol 8000 (PEG 8000)

Electrocompetent bacterial cells (ElectroMAX DH10B; Life Technologies)

✓ SOC medium (Life Technologies; also see APPENDIX 1, but reduce yeast extract to 0.5%)

✓ LB plates containing 5% (w/v) sucrose and either 25 µg/ml kanamycin (for PAC clones) or 20 µg/ml chloramphenicol (for BAC clones):

100 × 15–mm petri dishes for test transformation

22 × 22–cm trays for picking colonies

✓ LB medium containing 25 µg/ml kanamycin (PAC clones) or 20 µg/ml chloramphenicol (BAC clones)

*Not*I restriction endonuclease and buffer (New England Biolabs)

1% (w/v) agarose solution (ultra pure; Life Technologies)

✓ 0.5× TBE buffer

Low-range PFG markers in agarose containing a mixture of lambda *Hin*dIII fragments and lambda concatemers (New England Biolabs)

✓ 0.5 µg/ml ethidium bromide in 0.5× TBE buffer

80% (v/v) glycerol, sterile

Dry ice/ethanol bath

16° and 37°C water baths

0.025-µm-pore-size microdialysis filters (Millipore), 25- and 47-mm diameter

Wide-bore pipet tips, sterile

Disposable microelectroporation cuvettes with a 0.15-cm gap (Life Technologies or equivalent)

Electroporator (Cell Porator equipped with a voltage booster, Life Technologies, or equivalent)

15-ml snap-cap polypropylene tubes, sterile

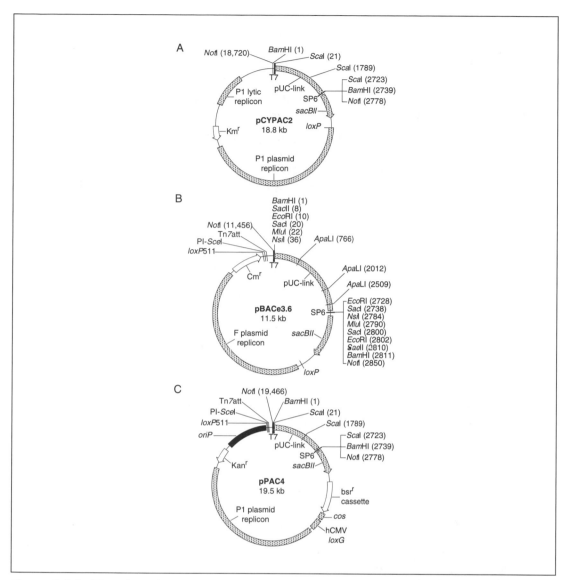

Figure 5.4.1 The pCYPAC2 (**A**), pBACe3.6 (**B**), and pPAC4 (**C**) vectors contain common features for positive selection of cloned inserts. The insert-containing *E. coli* cells can be grown in preference over cells containing plasmids without an insert by including sucrose in the medium. Sucrose is converted into toxic metabolites by levansucrase encoded by the *sacBII* gene present in all three vectors. The pUC-link segment contains a functional high-copy-number plasmid that includes the ampicillin resistance gene (Ap^r; not shown). The pUC-link sequence is removed in the cloning procedure and replaced by the digested genomic DNA fragment in the PAC and BAC clones. The *loxP* recombination sites are not used in cloning. (A) The pCYPAC2 vector contains the P1 plasmid replicon, which maintains recombinant clones at ~1 copy/cell. Alternatively, the P1 lytic replicon can be induced to provide higher copy numbers by adding the *lac* inducer IPTG into the medium. The kanamycin resistance gene (Km^r) is also present in this vector. (B) The pBACe3.6 vector contains the chloramphenicol resistance gene (Cm^r) and the F plasmid replicon, maintaining recombinant clones at 1 copy/cell. The multiple cloning sites flanking the pUC-link segment allow positive selection for cloned inserts using any of the restriction enzymes *Bam*HI, *Sac*II, *Eco*RI, *Sac*I, *Mlu*I, or *Nsi*I. The Tn*7*att sequence permits specific Tn*7*-based retrofitting of BAC clones. (C) The pPAC4 vector contains four elements from pCYPAC2: the P1 plasmid replicon, the Km^r gene, the *sacBII* gene, and the pUC-link. The P1 plasmid replicon ensures low-copy-number maintenance of recombinant clones; the P1 lytic replicon is removed. In addition, the *oriP* and bsr^r cassette have been included to facilitate the use of PAC clones for functional analysis of genes carried on the cloned inserts. The Tn*7*att sequence is also present, enabling specific retrofitting of PACs.

Strategies for Large-Insert Cloning and Analysis

Short Protocols in Human Genetics

Unit 5.4

Page 5-33

Orbital shaker, 37°C

Automated plasmid isolation system (AutoGen 740, Integrated Separation Systems, optional)

Flexible plastic 96-well plate (Falcon or equivalent)

Contour-clamped homogeneous electrical field (CHEF) apparatus (Bio-Rad) or field-inversion gel electrophoresis (FIGE) apparatus (Bio-Rad or equivalent)

Digital imager (Alpha Innotech IS1000 or equivalent)

1. Mix ~50 ng of each 120- to 300-kb size fraction of *Mbo*I- *or Eco*RI-digested genomic DNA (Support Protocol 4) with 25 ng of *Bam*HI- or *Eco*RI-digested and dephosphorylated pBACe3.6 vector DNA *or* 50 ng of *Bam*HI-digested and dephosphorylated pCYPAC2 or pPAC4 vector DNA prepared for cloning (Support Protocol 1). Include ligation controls (reactions with vector only or without ligase).

    Steps 1 to 14 present the small-scale pilot experiment.

    These quantities yield a 10:1 molar ratio of vector to insert, with a low final genomic DNA concentration (1 ng/µl) to favor circle formation over concatemerization.

2. To each tube, add 10 µl of 5× T4 DNA ligase buffer and sterile water to bring the total volume to 50 µl. Mix very gently. Add 1 Weiss unit T4 DNA ligase and incubate at 16°C for 4 hr for *Eco*RI-*Eco*RI cloning or 8 hr for *Bam*HI-*Mbo*I cloning.

3. To stop the reaction, add 1 µl of 0.5 M EDTA, pH 8.0, and 1 µl of 10 mg/ml proteinase K and incubate at 37°C for 1 hr. To inactivate proteinase K, add 1 µl of 100 mM PMSF solution and incubate at room temperature for 1 hr.

4. Spot the ligation mixture on the middle of a 25-mm-diameter, 0.025-µm-pore-size micro-dialysis filter floating on sterile distilled deionized water in a disposable petri dish. Dialyze for ≥2 hr at room temperature.

5. Carefully recover the dialyzed ligation mixture using a sterile wide-bore pipet tip and transfer to a microcentrifuge tube. Discard the water from the petri dish and pour ~15 ml TE/PEG solution in the dish. Transfer the empty membrane onto this solution, return the ligation mixture to the membrane, and continue dialysis for ~5 hr at room temperature (or overnight at 4°C) until it is equilibrated.

6. Transfer the membrane onto a petri dish cover, recover the ligation mixture (typically ~8 µl after concentration) from the membrane using a sterile wide-bore pipet tip, and transfer it to a microcentrifuge tube. Keep on ice and transform the solution into *E. coli* as soon as possible.

7. Precool disposable microelectroporation cuvettes (0.15-cm gap). Thaw (on ice) the required amount of electrocompetent bacterial cells: $(n + 1) \times 20$ µl cells, where n is the number of electroporations. Perform test transformations in duplicate or triplicate. Transfer 20-µl aliquots into precooled microcentrifuge tubes. Keep on ice.

8. Using a sterile wide-bore pipet tip, transfer 2 µl ligation mixture to each aliquot and gently mix with the pipet tip. Transfer the solution into a precooled electroporation cuvette, placing the droplet carefully between the electrodes and avoiding the formation of any air bubbles. Place the cuvette into the electroporation chamber and deliver a pulse according to the following conditions:

    voltage booster settings:
    resistance on voltage booster, 4000 Ω

    Cell-Porator settings:
    voltage gradient, 13 kV/cm
    capacitance, 330 µF

impedance, low Ω
charge rate, fast

9. Collect the droplet of cells and dilute immediately into 500 μl SOC medium in a 15-ml snap-cap polypropylene tube. Incubate 1 hr at 37°C in an orbital shaker at 200 rpm.

10. Spread the entire aliquot of cells on a 100 × 15–mm LB plates containing 5% sucrose and either 20 μg/ml chloramphenicol (for BAC clones) or 25 μg/ml kanamycin (for PAC clones). Use plates that have been dried under a hood to speed absorption. Grow overnight at 37°C, count colonies, and estimate the titer per transformation.

11. Pick 40 clones with a sterile tooth pick and grow the cells overnight in 1.5 ml LB medium containing 20 μg/ml chloramphenicol (BAC clones) or 25 μg/ml kanamycin (PAC clones). Extract DNA using an automated plasmid isolation system (AutoGen 740) or the modified alkaline lysis procedure (see Support Protocol 5).

12. Dissolve DNA (0.5 to 1 μg) in 100 μl TE buffer. Digest 5 to 10 μl DNA with 0.1 U NotI in a 20-μl volume to separate the vector and insert DNA fragments. Perform the reaction in a flexible 96-well plate at 37°C for 2 hr.

13. Analyze digested DNA using a CHEF or a FIGE apparatus. Use a 1% agarose gel, 0.5× TBE buffer, low-range PFG markers, and the following conditions:

    For CHEF: 14°C, 6 V/cm, 16 hr, 0.1 to 40 sec pulse time, 120° angle for CHEF

    For FIGE: room temperature, 180 V forward voltage, 120 V reverse voltage, 16 hr with 0.1 to 14 sec switch time linear shape.

14. Stain the gel with 0.5 μg/ml ethidium bromide solution and calculate the molecular weight of inserts using a digital imager.

    Most of the clones will give a vector band (8.7 kb for pBACe3.6, 15.5 kb for pCYPAC2, and 16.8 kb for pPAC4) and the insert band. If the insert contains internal NotI sites, additional bands will be seen. Incomplete digestion usually leads to a characteristic doublet, where the two bands differ by the size of the vector.

    Steps 15 to 20 present the scaled-up procedure for creating the library.

15. Repeat the ligation procedure using a size fraction of genomic DNA that gives the desired average insert size and cloning efficiency. Scale up the ligation reaction to a total of 500 to 1000 μl using the entire remaining electroeluted insert DNA, but perform in 250-μl aliquots in two to four tubes.

16. Dialyze and concentrate the ligation mixture as described, but use a 47-mm-diameter microdialysis filter for each 250-μl aliquot (yielding ~40 μl from each filter after concentration).

17. Add 12 μl concentrated DNA to tubes containing ~110 μl cells. Thaw enough electrocompetent cells to complete all transformations using the entire recovered ligation mixture.

18. Perform identical transformations at 20 μl per electroporation and collect the transformed cells from 10 electroporations into a 50-ml disposable centrifuge tube containing 5 ml SOC medium at room temperature. Incubate the cells on an oribital shaker at 200 rpm at 37°C for 1 hr.

19. Add 800 μl of 80% sterile glycerol solution and mix very well. Spread 200 μl on two individual 100 × 15–mm LB plates containing 5% sucrose and the appropriate antibiotic to examine the titer from each tube.

20. Freeze the remaining cells in a dry ice/ethanol bath and keep at −80°C until colony picking can be scheduled. To pick colonies, spread culture at 1600 clones/plate on 22 × 22–cm

5

LB plates containing 5% sucrose and the appropriate antibiotic. Grow overnight at 37°C. Store frozen cells up to 1 year.

SUPPORT PROTOCOL 1

PREPARATION OF BAC/PAC VECTOR FOR CLONING

Additional Materials (also see Basic Protocol; see APPENDIX 1 for items with ✓)

pBACe3.6, pCYPAC2, or pPAC4 stock in *E. coli* DH10B cells (P. de Jong)

✓ LB plates containing:
> 25 μg/ml kanamycin (for PAC) or 20 μg/ml chloramphenicol (for BAC)
> 5% (w/v) sucrose and either kanamycin or chloramphenicol
> 5% (w/v) sucrose, 100 μg/ml ampicillin, and either kanamycin or chloramphenicol

*Bam*HI and *Eco*RI restriction endonucleases and 10× buffers (New England Biolabs or equivalent)

0.7% (w/v) agarose gels (for standard electrophoresis)

Calf intestine alkaline phosphatase (AP; Boehringer Mannheim)

10 mg/ml proteinase K (Boehringer Mannheim)

95% (v/v) ethanol

1.0% (w/v) agarose solution (ultrapure, LifeTechnologies; for CHEF system)

✓ 6× gel loading buffer, not containing xylene cyanol FF

1-kb ladder or lambda *Hin*dIII markers

T4 polynucleotide kinase (New England Biolabs)

30% (w/v) polyethylene glycol (PEG) 8000

1.5-mm-thick electrophoresis comb for CHEF apparatus

Dialysis tubing of 3/4-in. diameter, MWCO 12,000 to 14,000 daltons (Life Technologies or equivalent)

Dialysis clip

1. To test the vector DNA, streak recombinant DH10B cells harboring BAC or PAC vectors onto an LB plate containing chloramphenicol or kanamycin, respectively. Incubate overnight at 37°C.

2. Isolate five single colonies and inoculate into separate 15-ml snap-cap polypropylene tubes containing 3 ml LB medium with the appropriate antibiotic. Grow cultures overnight at 37°C.

3. Use 1.5 ml of each culture to prepare DNA using an automated plasmid isolation system (AutoGen 740) or the modified alkaline lysis protocol (see Support Protocol 5). Store the remainder of each culture at 4°C.

4. Resuspend DNA in 100 μl TE buffer. Digest 5 μl DNA from each preparation with *Not*I, 5 μl with *Bam*HI, and 5 μl with *Eco*RI, using the manufacturers' recommended conditions. Analyze in a 0.7% agarose gel. Discard clones that contain any rearrangements of the vector.

   PAC vectors do not contain EcoRI sites. Otherwise all enzymes should liberate pUC19-link from the vectors, producing fragments of ~2.7 kb (the pUC19 stuffer fragment) and ~8.7 kb (pBACe3.6), 15.5 kb (pCYPAC2), or 16.8 kb (pPAC4).

5. Select a culture and dilute into 1 liter LB medium containing the appropriate antibiotic. Grow to saturation at 37°C. Using standard methods (e.g., *CPMB UNITS 1.7 & 1.8*), prepare a crude lysate by the alkaline lysis or cleared lysate method, and purify vector plasmid by CsCl/ethidium bromide equilibrium centrifugation.

6. To establish the minimal amount of restriction enzyme required for complete digestion of the vector, first digest 50 ng vector DNA with 0.1, 0.2, 0.5, and 1 U of either *Bam*HI (BAC or PAC vectors) or *Eco*RI (pBACe3.6) in separate 10-μl reactions for 1 hr at 37°C. Perform conventional agarose gel electrophoresis and ethidium bromide staining to view the digestion.

7. To establish the maximal amount of DNA in the reaction (and the minimum scale-up volume), digest 100, 200, and 400 ng vector DNA with 2, 4, and 8 times the minimal amount of enzyme in separate 10-μl reactions (total 9 reactions) for 1 hr at 37°C. Perform conventional agarose gel electrophoresis and ethidium bromide staining to view the digestion.

8. Incubate ~30 μg vector DNA with the defined amount of the appropriate enzyme at 37°C for 15 min. Add 1 U calf intestine AP and continue incubating at 37°C for 1 hr. To stop the reaction, add 0.5 M EDTA to a final concentration of 15 mM, add 10 mg/ml proteinase K to 200 μg/ml final, and incubate at 37°C for 1 hr.

9. To inactivate proteinase K, add 100 mM PMSF solution to 2 mM final and incubate at room temperature for 1 hr. During the incubation, clean a 1.5-mm-thick electrophoresis comb with 95% ethanol and cover teeth with autoclave tape to create a large preparative slot with sufficient space to load the reaction mixture. Leave enough empty wells for a size marker on both sides and at least one empty well between the markers and the preparative slot.

10. Prepare a 1% agarose gel in a CHEF mold using the preparative comb. While the gel is solidifying, add 2 liters of 0.5× TBE buffer to the CHEF apparatus tank and equilibrate the unit at 14°C.

11. Place the gel in the precooled unit. Add 6× gel loading buffer to the samples (final 1×) and load samples in the large preparative well. Load 1-kb ladder or lambda *Hin*dIII markers in both side wells. Perform electrophoresis at 14°C and 6 V/cm for 16 hr with 0.1 to 40 sec pulse time at 120° angle.

12. Cut away two flanking gel slices that contain the markers plus 1 to 2 mm of the central vector lane. Wrap the remaining preparative lane in plastic wrap and store at 4°C. Stain the flanking slices with ethidium bromide to detect the major vector fragment.

13. Cut linear vector DNA lacking the pUC-stuffer fragment from the central lane and recover by electroelution (see Support Protocol 4). Load sample into 3/4-in.-diameter dialysis tubing (MWCO 12,000 to 14,000 Da) and dialyze once against 1 liter TE buffer at 4°C for >2 hr. Recover solution from the dialysis tubing and precipitate DNA with sodium acetate and either ethanol or isopropanol (APPENDIX 3B). Dissolve DNA in 100 μl TE buffer.

14. To assess dephosphorylation, place four microcentrifuge tubes on ice and number them 1 to 4. Add 4 μl of 5× T4 DNA ligase buffer and 50 ng digested and AP-treated vector DNA to the tubes and adjust the volume to 20 μl with sterile water.

15. Add 1 U T4 polynucleotide kinase to tubes 3 and 4 and incubate all four tubes at 37°C for 1 hr. Heat at 65°C for 20 min, then place on ice. Add 1 U T4 DNA ligase to tubes 2 and 4 and incubate all four tubes at 16°C for >4 hr.

16. Run samples in a 0.7% agarose gel, stain the gel with ethidium bromide, and view.

Closed circular DNA should not be seen in samples 1, 2, and 3. Sample 4 should have closed circular DNA but no linear DNA.

17. Perform a ligation using all the remaining vector DNA solution in a 200-μl reaction volume with 4 U T4 DNA ligase. Purify the linear monomeric vector DNA away from the background of ligated vector by PFGE in a CHEF apparatus (see steps 10 through 13).

18. Perform test ligations using vector DNA alone (25 ng pBACe3.6, 50 ng PAC vectors) and vector plus 50 ng control DNA (see Basic Protocol, steps 1 to 3). Dialyze against water and 30% PEG 8000, and perform transformation as described (see Basic Protocol, steps 4 to 9).

19. Spread transformed cells on LB plates containing 5% sucrose and either kanamycin or chloramphenicol, as appropriate. Grow overnight at 37°C.

These plates are used to determine the level of recombinant clones and nonrecombinant clones. If the number of colonies obtained from vector DNA alone is high (>100), vector must be prepared again, beginning with enzyme titration (step 6). If the plasmid is isolated from these background clones and analyzed by PFGE after NotI digestion, smaller vector sizes will be observed. Presumably, star activity causes nonspecific digestion in the sacB gene (either promoter or coding sequence), thus inactivating the gene, and the nonspecifically digested, nondephosphorylated DNA is circularized by self-ligation. Thus, these colonies can grow on sucrose plates without insert DNA.

20. To examine the complete removal of stuffer fragment from the vector, spread the transformed cells with self-ligated vector on LB plates containing 5% sucrose, 100 μg/ml ampicillin, and either kanamycin or chloramphenicol, as appropriate. Grow overnight at 37°C.

These plates are used to determine the remaining level of undigested vector containing the pUC-link. Ideally, there should be zero Amp^R colonies.

SUPPORT PROTOCOL 2

PREPARATION OF HIGH-MOLECULAR-WEIGHT DNA FROM LYMPHOCYTES IN AGAROSE BLOCKS

Materials (see APPENDIX 1 for items with ✓)

Healthy human volunteer
✓ PBS, ice cold
✓ 1× RBC lysis solution
InCert agarose (FMC Bioproducts)
✓ Proteinase K lysis solution
✓ TE$_{50}$ buffer: 10 mM Tris·Cl, pH 8.0/50 mM EDTA (see individual recipes)
✓ 0.1 mM PMSF solution: 100 mM PMSF solution diluted 1/1000 in TE$_{50}$ buffer (see individual recipes) immediately before use
✓ 0.5 M EDTA, pH 8.0

Blood-drawing equipment
Blood collection tubes containing EDTA
Automated hematology counter
50-ml conical screw-cap polypropylene tubes, sterile
Refrigerated centrifuge with rotor/adapters for 50-ml tubes (e.g., Sorvall RI6000D centrifuge with H-1000B swinging-bucket rotor)
Roller mixer (Robbins Scientific or equivalent)
50°C water bath
10 × 5 × 1.5–mm disposable DNA plug mold (Bio-Rad)

1. Use blood-drawing equipment to obtain 45 ml venous blood from a healthy human volunteer in blood collection tubes containing EDTA. Mix well to avoid clot formation. Count the total number of lymphocytes using an automated hematology counter (should be 2.25–3.3 × 10^8 in 45 ml).

2. Transfer blood to two 50-ml conical screw-cap polypropylene tubes, add 10 ml ice-cold PBS to each, and mix gently. Centrifuge at 1876 × g for 5 min at 4°C. Discard the

supernatant using a 10-ml disposable pipet, being careful not to remove any lymphocytes from the fuzzy coat layer. Repeat wash with ice-cold PBS ten times.

3. Discard the final supernatant, mix the cell suspension well, and divide into four 50-ml tubes. Add 25 ml of $1 \times$ RBC lysis solution to each tube, mix gently on a roller mixer, and incubate on mixer ~20 min at room temperature.

4. Visually monitor the progress of RBC lysis (cloudiness will disappear as RBCs lyse). When it appears complete (usually <30 min), centrifuge 10 min at $208 \times g$, 4°C. Discard the supernatant by gentle inversion, taking care not to disturb the lymphocyte pellet. Rinse the inside of the tubes with 2 ml ice-cold PBS, being careful not to disturb the pellet. Discard the supernatant with a micropipet tip.

5. Resuspend lymphocytes in 10 ml ice-cold PBS and combine in one 50-ml tube. Centrifuge 5 min at $208 \times g$, 4°C, and discard the supernatant carefully to remove most of the remaining lysate debris. Repeat wash until most of the red color is removed. Resuspend the pellet in ice-cold PBS at 1×10^8 cells/ml (~600 μg DNA/ml).

6. Dissolve 0.1 g InCert agarose in 10 ml PBS (1%) in a microwave oven and keep at 50°C in a water bath. Mix the cell suspension well and transfer 400 μl to a microcentrifuge tube. Warm the tube by hand for 3 min. Add 400 μl of 1% molten InCert agarose, mix gently, and transfer as quickly as possible to a $10 \times 5 \times 1.5$–mm disposable DNA plug mold using a micropipet tip (45 ml blood should yield ~45 plugs). Avoid making any bubbles.

7. Place the molds on ice for 30 to 60 min to solidify the agarose. Extrude the plugs from the molds directly into a 50-ml tube (<50 plugs/tube) containing 50 ml proteinase K lysis solution.

8. Stand the tube in a 50°C water bath, mix periodically, and incubate for 48 hr. Replace solution with fresh proteinase K lysis solution after the first 24 hr. Rinse plugs several times with sterile distilled, deionized water.

9. Add 50 ml TE_{50} buffer and mix on a roller mixer at 4°C for 24 hr, replacing the solution with fresh TE_{50} buffer at least twice. To inactivate proteinase K, wash plugs twice for 2 hr each with 50 ml of 0.1 mM PMSF solution on the roller mixer at 4°C. Repeat the 24-hr TE_{50} wash.

10. Store DNA plugs in 0.5 M EDTA at 4°C (stable at least 1 year).

SUPPORT PROTOCOL 3

PREPARATION OF HIGH-MOLECULAR-WEIGHT DNA FROM ANIMAL TISSUE CELLS IN AGAROSE BLOCKS

Although DNA from cultured cell lines (see Support Protocol 2) has been extensively used in constructing cosmid libraries, passage in tissue culture may result in chromosomal rearrangements. It is also often difficult to obtain enough lymphocytes from small model organisms (e.g., mice, rats). Here, an alternate method is described for preparation of DNA plugs from animal tissue cells (e.g., spleen, kidney, and brain are all good sources of high-molecular-weight DNA).

Additional Materials (also see Support Protocol 2)
>   Healthy animal (e.g., ~5-week-old mice, rats)
>   Sterile dissecting tools
>   Sterile Dounce homogenizer
>   15-ml conical screw-cap polypropylene tubes, sterile (Corning or equivalent)
>   Counting chambers (VWR)

1. Euthanize mice or rats in a plastic bag or desiccator containing CO_2 gas. Using sharp sterile scissors, immediately remove desired tissue(s) (e.g., spleen, kidney, brain) and transfer them to a petri dish on ice. Rinse tissues with ice-cold PBS and remove fat and other associated tissues with sterile forceps.

2. Transfer 1 to 2 spleens, kidneys, or brains to a precooled sterile Dounce homogenizer on ice (homogenize different tissue types separately). Add ~2.5 ml ice-cold PBS and grind ~5 times on ice. Transfer the supernatant to a chilled 50-ml conical screw-cap polypropylene tube, taking care to leave behind as much tissue as possible. Add more PBS to the homogenizer and repeat homogenization four more times, adding the PBS supernatant to the chilled conical tube until most tissue is disrupted.

3. Remove debris from the homogenizer with forceps and repeat step 2 for all spleens, kidneys, and brains.

4. Fill all tubes to 50 ml with ice-cold PBS. Stand tubes on ice for 2 to 3 min and carefully transfer supernatants to new 50-ml tubes. Centrifuge 10 min at $208 \times g$ (e.g., 1000 rpm in a Sorvall H1000B rotor), 4°C. Discard supernatant by gentle inversion, taking care not to disturb the cell pellet.

5. Add ~1 ml ice-cold PBS to pellet, resuspend cells by pipetting gently, and remove large debris that cannot be suspended with a micropipet. Fill tubes to 50 ml with ice-cold PBS, mix gently, and repeat centrifugation (step 4). Decant the supernatant and resuspend the cells in the residual PBS on ice.

6. Add 1 ml ice-cold PBS and transfer to a 15-ml conical screw-cap polypropylene tube. Rinse the 50-ml tubes with 1 ml ice-cold PBS and add to the 15-ml tubes. Prepare 20 µl of a 20-fold dilution of the cell suspension in a 1.5-ml microcentrifuge tube, and count the number of cells using a counting chamber.

7. Assuming that 40% of cells counted are RBCs (lacking chromosomal DNA), subtract the estimated number of RBCs and then adjust the concentration of cells that contain chromosomal DNA to 1×10^8 cells/ml by adding ice-cold PBS (a single cell carries 6 to 10 pg DNA).

8. Embed cells in agarose and extract high-molecular-weight DNA (see Support Protocol 2, steps 6 to 10).

SUPPORT PROTOCOL 4

PARTIAL DIGESTION AND SIZE FRACTIONATION OF GENOMIC DNA

The PAC vectors utilize *Bam*HI restriction sites for cloning. However, it is not advisable to use *Bam*HI for partial digestion of genomic DNA, because there may be regions of DNA relatively devoid of the *Bam*HI site. Instead, it is preferable to use the four-base cutter *Mbo*I, which produces four-base overhangs that are compatible with the *Bam*HI overhangs on the vector. Since pBACe3.6 has *Bam*HI and *Eco*RI cutting sites for cloning, it possible to use either *Mbo*I or *Eco*RI for partial digesion. Using this protocol, ~300 µl of a 2 to 20 ng/µl solution of size-fractionated DNA is obtained.

Materials *(see APPENDIX 1 for items with ✓)*

  Agarose plugs with embedded high-molecular-weight genomic DNA stored in 0.5 M
    EDTA (see Support Protocols 2 and 3)
✓ 0.5× TBE buffer, sterile
  95% (v/v) ethanol

1% (w/v) agarose gel (ultra pure; Life Technologies) in 0.5× TBE buffer
Low-range PFG markers in agarose containing a mixture of lambda HindIII fragments and lambda concatemers (New England Biolabs)
✓ 0.5 M EDTA, pH 8.0
✓ 0.5 µg/ml ethidium bromide solution in 0.5× TBE buffer (see individual recipes)
✓ 1× MboI buffer
10 U/µl MboI restriction endonuclease (Life Technologies)
1 M MgCl$_2$
10 mg/ml proteinase K (Boehringer Mannheim)
✓ 10% (w/v) N-lauroylsarcosine
✓ TE$_{50}$ buffer: 10 mM Tris·Cl, pH 8.0/50 mM EDTA (see individual recipes)
✓ 100 mM PMSF solution
10 mg/ml BSA (New England Biolabs)
✓ 10× EcoRI endonuclease/methylase buffer
0.1 M spermidine
20 U/µl EcoRI endonuclease (New England Biolabs)
40 U/µl EcoRI methylase (New England Biolabs)
✓ TE buffer, pH 8.0
✓ 1× TAE buffer (optional)
Lambda DNA

15- and 50-ml conical screw-cap polypropylene tubes, sterile
Contour-clamped homogeneous electrical field (CHEF) apparatus (Bio-Rad or equivalent) with 1.5-mm-thick, 20-well comb
Digital imager (Alpha Innotech IS1000 or equivalent)
Disposable γ-ray-sterilized inoculating loops
Dialysis tubing, 3/4-in. diameter, MWCO 12,000 to 14,000 Da (Life Technologies or equivalent)
Dialysis clips
Submarine gel electrophoresis apparatus (Bio-Rad Sub-Cell GT DNA Electrophoresis Cell, 31-cm length × 16-cm width, or equivalent)
Wide-bore pipet tips

1. Transfer six DNA-containing agarose plugs to a 50-ml conical screw-cap polypropylene tube containing 50 ml sterile 0.5× TBE buffer and dialyze at 4°C for ≥3 hr.

2. Clean a 1.5-mm-thick comb with 95% ethanol and cover teeth with autoclave tape to create a large preparative slot with sufficient space to place the DNA plugs. Leave enough empty wells for PFG markers on both sides and at least one empty well between the markers and the preparative slot.

3. Place the preparative comb along the short axis of a CHEF mold and prepare a 1% agarose gel of sufficient thickness to cover the sample gel blocks when laid along their length. While the gel is solidifying, add 2 liters of 0.5× TBE buffer to the CHEF apparatus tank and equilibrate the unit at 14°C. Remove the comb gently and load the plugs in the large preparative slot. Load low-range PFG markers in the outer lanes on each side of the preparative lane. Do not seal the wells with 1% agarose.

4. Place the gel in the precooled unit and run the gel along the long axis at 14°C and 4 V/cm for 10 hr with a 5-sec pulse time. Remove the plugs from the preparative slot and store at 4°C in 50 ml TE buffer (up to 2 weeks) or in 0.5 M EDTA, pH 8.0 (up to 1 year).

5. Stain the gel in 0.5 µg/ml ethidium bromide solution and examine on a digital imager to check for sheared DNA (embedded chromosomal DNA cannot migrate out of the plug).

To digest with MboI

6a. Transfer pre-electrophoresed DNA plugs into 50 ml 1× *Mbo*I buffer and dialyze at 4°C overnight.

7a. Cut a DNA plug into four pieces and transfer to separate microcentrifuge tubes with a disposable γ-ray-sterilized inoculating loop. Add 400 μl of 1× *Mbo*I buffer and 2.5 U *Mbo*I to each tube. Keep the plugs on ice for 1 hr.

8a. Add 5 μl of 1 M $MgCl_2$ to each tube and leave 15 min on ice followed by 20 to 40 min at 37°C. Test for optimal partial digestion by using various incubation times or amounts of enzyme.

9a. Immediately place the tubes on ice and add 150 μl of 0.5 M EDTA, 30 μl of 10 mg/ml proteinase K, and 75 μl of 10% *N*-lauroylsarcosine. Mix well and incubate at 37°C for 1 hr.

10a. Pour the solution and plugs to a petri dish and remove the solution with a micropipettor. Rinse plugs with TE buffer. Transfer plugs to a 15-ml conical screw-cap polypropylene tube using a disposable γ-ray-sterilized inoculating loop.

11a. Add 15 ml TE_{50} buffer and 15 μl of 100 mM PMSF solution and dialyze three times for 20 min each at 4°C. Dialyze twice for 30 min each with TE_{50} buffer at 4°C. Proceed to step 12.

To digest with EcoRI and EcoRI methylase

6b. Transfer pre-electrophoresed DNA plugs into separate microcentrifuge tubes using a disposable γ-ray-sterilized inoculating loop.

7b. Add 25 μl of 10 mg/ml BSA, 50 μl of 10× *Eco*RI endonuclease/methylase buffer, and 13 μl of 0.1 M spermidine, and mix well.

8b. Test for optimal partial digestion by varying 1 to 2 U *Eco*RI endonuclease and 0 to 200 U *Eco*RI methylase. Adjust the volume to 500 μl with sterile water, and put the tubes on ice for 1 hr to allow the enzymes to penetrate into the plugs.

9b. Incubate the tubes at 37°C for 2.5 hr and then put on ice.

10b. Add 150 μl of 0.5 M EDTA, 30 μl of 10 mg/ml proteinase K, and 75 μl of 10% *N*-lauroylsarcosine, and mix well. Incubate samples at 37°C for 1 hr.

11b. Wash plugs as described (steps 10a to 11a) and then proceed to step 12.

12. Determine the optimal partial digestion condition using a CHEF apparatus. Use the same conditions as in the Basic Protocol, step 13.

13. Take two new entire plugs, cut each into four pieces, and carry out eight identical partial digestions using the optimized conditions. Store plugs in TE_{50} buffer at 4°C until use (up to a month).

14. Clean a 1.5-mm-thick, 20-well comb with 95% ethanol and cover four to six teeth in the middle of the comb with autoclave tape to create a wide tooth for preparative use. In a small CHEF gel mold, pour a 1% agarose gel of sufficient thickness to cover the sample gel blocks when laid in along their length. Allow the gel to solidify ~1 hr at room temperature. While the gel is solidifying, pour 2 liters of 0.5× TBE buffer in the CHEF tank and equilibrate the unit at 14°C.

15. Gently remove the comb and array the partially digested DNA (eight small agarose pieces) in the large preparative slot. Load low-range PFG markers on each side of the preparative lanes, and fill the remaining space in the preparative and marker slots with molten 1% agarose.

5

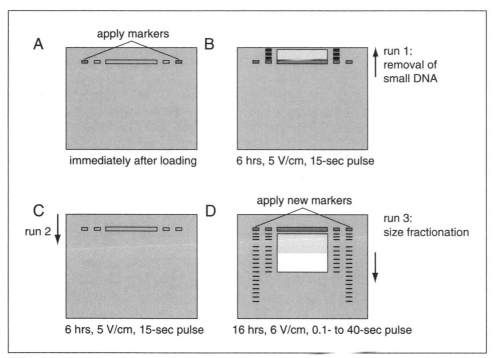

Figure 5.4.2 Isolation of insert DNA from agarose blocks by PFGE. (**A**) Agarose blocks containing partially digested DNA are applied to the center wells of a 1% agarose gel, and low-range PFG markers are applied to flanking wells. The DNA separation occurs in three CHEF stages, all with pulse directions at a 120° angle. (**B**) The initial direction of the field allows the DNA to migrate from the wells toward the nearest gel edge (1 cm away from the well). Fragments <120 kb are run out of the gel. (**C**) The same conditions are then used after turning the direction of the gel, bringing all fragments remaining in the gel back to the original starting wells. (**D**) After the second run, new marker DNA is applied to additional flanking wells that were not previously used. High-molecular-weight fragments are then resolved in the third run. After this procedure (not shown), the flanking marker lanes are removed from the gel and stained with ethidium bromide to indicate the location of the size ranges. The first set of markers are used to check the electrophoresis condition for the first run by comparing to the second set. The second set of markers are used to identify the size range after the final electrophoresis. Gel slices are cut from the genomic DNA lanes by horizontal cuts at 0.5-cm intervals to obtain gel slices in the range of 150 to 500 kb.

16. Size fractionate the DNA by three sequential stages of gel electrophoresis with one exchange of polarity, as outlined in Figure 5.4.2. In stage 1, orient the gel so that the field forces the DNA to migrate from the wells toward the nearest gel edge. Perform gel electrophoresis at 14°C for 6 hr using 5.0 V/cm, with a 15-sec pulse time at 120° angle in 0.5× TBE buffer. Optimize voltage, if necessary, between 4.5 to 5.5 V/cm so that fragments <120 kb run out of the gel.

17. Change the electrophoresis buffer, rotate the gel 180° in the tank, and run stage 2 using the same electrophoresis conditions to bring all fragments remaining in the gel back to the original starting wells.

18. Apply new marker DNA to flanking wells that were not previously used. In stage 3, resolve high-molecular-weight fragments at 6 V/cm for 16 hr with 0.1- to 40-sec pulse time.

19. Cut the outer lanes containing markers plus 1 to 2 mm from each side of the preparative lane to assess the success of the partial digestion. Wrap the remaining portions of the preparative lane in plastic wrap and store at 4°C.

20. Stain the outer portions of the gel with ethidium bromide and examine with a fluorescent ruler on a digital imager to ascertain the size ranges.

    The stained portion of the preparative lane should contain a broad smear extending from ~120 kb to >1 Mb.

21. Slice genomic DNA lane by cutting horizontally at 0.5-cm intervals to obtain size-fraction blocks in the range of 150 to 500 kb. Cut an ~1-mm-wide slice from each size-fraction block and store size-fractionated DNA blocks in 0.5 M EDTA at 4°C until use.

    If DNA will be eluted within a few days, it is possible to store blocks in sterile 0.5× TBE at 4°C and omit the dialysis in step 24.

22. Load 1-mm slices directly into separate wells in 1% agarose gel and perform electrophoresis with 0.5× TBE buffer, using the following conditions in a CHEF apparatus: 120° angle, 6 V/cm, for 16 hr with 0.1- to 40-sec pulse time, at 14°C.

23. Stain the remaining gel pieces in 0.5 μg/ml ethidium bromide solution, and reassemble them on a digital imager. Take a picture with a fluorescent ruler to ascertain the size fractionation and cutoff point.

24. Stain the analytical gel from step 22 in 0.5 μg/ml ethidium bromide solution and take a picture on the imager to assess the quality and size distribution of the DNA fragments in each block. Select one or two size-fraction blocks in the 150- to 200-kb size range and dialyze against 15 ml sterile 0.5× TBE buffer for 3 hr.

25. Cut an ~10-cm-long piece of dialysis tubing (one per fraction block) and rinse with sterile water. Close one end with a dialysis clip and remove the residual water completely.

26. Place the fraction block and 300 μl sterile 0.5× TBE buffer in the dialysis tubing. Completely remove air bubbles and seal the other end of the tubing with a clip. Orient the long axis of the gel parallel to the tubing.

27. Prepare 0.5× TBE buffer in a submarine gel electrophoresis tank and immerse the bag in a shallow layer of the 0.5× TBE. Place a plastic cover on the top of the clip to keep it down (additional dialysis bags can be electroeluted simultaneously). Pass electric current through the dialysis bag(s) with 3 V/cm (100 V for the Sub-Cell GT DNA Electrophoresis Cell) for 3 hr.

28. Reverse the polarity of the current for 30 sec to release the DNA from the wall of the bag. Transfer the bag to 1 liter TE buffer and dialyze at 4°C for ≥2 hr to completely remove borate ions, which may inhibit the ligation reaction.

29. Open the bag and carefully transfer all of the solution to a fresh microcentrifuge tube using a wide-bore pipet tip. Load 3 to 5 μl DNA into a 0.7% to 1.0% agarose gel in 0.5× TBE or 1× TAE buffer and perform electrophoresis with various amounts (5 to 50 ng) of lambda DNA. Estimate the DNA concentration using lambda DNA as a standard. Keep the eluted DNA at 4°C up to 10 days (do not freeze).

SUPPORT PROTOCOL 5

MODIFIED ALKALINE LYSIS MINIPREP FOR RECOVERY OF DNA FROM BAC/PAC CLONES

Materials *(see APPENDIX 1 for items with ✓)*

    BAC or PAC clones (see Basic Protocol)

✓ LB medium or terrific broth containing 25 μg/ml kanamycin (for PAC clones) or 20 μg/ml chloramphenicol (for BAC clones)

✓ Resuspension solution
  Alkaline lysis solution: 0.2 N NaOH/1% (w/v) SDS (prepare before use)
  Precipitation solution: 3 M potassium acetate, pH 5.5 (autoclave and store indefinitely at 4°C)
  Isopropanol
  70% (v/v) ethanol
✓ TE buffer, pH 8.0
  *Not*I restriction endonuclease and buffer (New England Biolabs)

  Toothpicks, sterile
  12- to 15-ml snap-cap polypropylene tubes
  Forceps, sterile
  Orbital shaker, 37°C
  1.5-ml microcentrifuge tubes or 2-ml screw-cap tubes

1. Using a sterile toothpick, inoculate a single isolated BAC or PAC clone into 2 ml LB medium or terrific broth containing either 25 μg/ml kanamycin (PAC) or 20 μg/ml chloramphenicol (BAC) in a 12- to 15-ml snap-cap polypropylene tube. Remove toothpick using sterile forceps. Grow overnight ($\leq$16 hr) at 37°C in an orbital shaker at 200 rpm.

2. Remove snap cap from the tube and centrifuge 5 min at $1600 \times g$. Discard supernatant and resuspend (vortex) pellet in 0.3 ml resuspension solution. Add 0.3 ml alkaline lysis solution and gently shake tube to mix contents. Let stand at room temperature ~5 min (suspension should change from very turbid to almost translucent).

3. While gently shaking, slowly add 0.3 ml precipitation solution. Stand tube on ice $\geq$5 min. Centrifuge 15 min at $16,000 \times g$, 4°C, to pellet the white precipitate and place tube on ice.

4. Using a micropipettor or disposable pipet, transfer supernatant to a 1.5-ml microcentrifuge tube containing 0.8 ml isopropanol. Mix by inverting a few times. Stand on ice $\geq$5 min. Avoid transferring any of the white precipitate. If desired, store supernatant overnight at −20°C.

5. Centrifuge 15 min at maximum speed in a microcentrifuge at room temperature. Aspirate as much of the supernatant as possible. Add 0.5 ml of 70% ethanol and invert tube several times to wash the DNA pellet. Centrifuge 5 min at maximum speed, 4°C. Aspirate as much of the supernatant as possible using a gel-loading pipet tip. Air dry pellet at room temperature until the pellet becomes translucent.

6. Resuspend pellet in 100 μl TE buffer by allowing the solution to sit, occasionally tapping the bottom of the tube gently to mix the contents (may take >1 hr). Do not mix with a pipet.

7. Use 5 to 10 μl DNA for digestion with 0.1 U *Not*I, or 18 μl DNA for more frequent-cutting enzymes such as *Bam*HI or *Eco*RI. Analyze insert size by pulsed-field gel electrophoresis.

References: Albertson et al. ,1990; Ioannou et al. ,1994; Osoegawa et al. ,1998; Shizuya et al. ,1992; Sternberg ,1990

Contributors: Kazutoyo Osoegawa, Pieter J. de Jong, Eirik Frengen, and Panayiotis A. Ioannou

CHAPTER 6

Identifying Candidate Genes in Genomic DNA

A variety of databases provide essential tools in the search for disease genes. These databases, including those for genomic DNA and expressed sequences, have grown dramatically over the past several years. Databases containing protein sequence and three-dimensional protein structure are also growing rapidly. Directions to these databases and the types of information that can be obtained from them is provided in *UNIT 6.2*. These databases should be consulted prior to the initiation of any other approach toward the identification of disease genes. After consultation with this collection of databases, a list of candidate genes can be generated that provides a handle for searching for a disease gene; the investigator searches each candidate gene, using the methods discussed in Chapter 7, for disease-causing mutations. However, this screening of candidate genes can be a long and frustrating process because one is frequently searching for a mutation as small as a single nucleotide in more than a million base pairs of genomic DNA. Added to the difficulties associated with searching for a mutation in a large segment of DNA is the problem that the candidate gene list is likely to be imperfect. One issue is that the current version of the human genome is not complete. Although most of the gaps in the sequence are "small," they are sometimes large enough to include part of or even an entire gene. Another perhaps more serious problem is that not all genes have been identified in genomic DNA. That is, in any stretch of human DNA a set of genes has been annotated; these annotated genes provide the candidate gene list. While using the candidate gene list, we must recognize these gene annotations are fallible and most entries should be verified independently. Where is the source of error? How do these genes become annotated? This chapter describes protocols that have been used to define genes in genomic DNA. Although the gene annotation indicates how the gene was identified, the investigator will sometimes find, using the methods outlined below, that the annotations are incomplete or inaccurate. By applying the methods presented in this chapter, one should be able to identify all of the genes encoded by a DNA sequence.

Genes are identified in genomic DNA by several different methods. The most accurate, and perhaps easiest methods, involve demonstrating that the DNA sequence encodes a cDNA. Such gene identification can be done in silico, using the methods outlined in *UNITS 6.1, 6.2, & 6.3*.

New software packages are rapidly being added to the list of gene finding programs. *UNIT 6.1* describes the problems associated with identifying genes in a large DNA sequence as well as some of the advantages and disadvantages of current algorithms being used for gene prediction. An overview on accessing the human genome can be found in *UNIT 6.4*. For a description on the retrieval of information from NCBI databases using Entrez, see *UNIT 6.5*.

In summary, definition of genes within a segment of human DNA is an important step in the positional cloning process. Assessment of the relevance of newly identified genes to a pathologic process will be assisted by knowledge of gene structure and encoded protein.

Contributors: J. G. Seidman and Christine Seidman

6

Gene Identification: Methods and Considerations

With the announcement of the completion of a "working draft" of the sequence of the human genome in June 2000 and the Human Genome Project targeting the completion of sequencing in 2002, investigators are faced with the challenge of developing a strategy to deal with the oncoming flood of both unfinished and finished data. These data undergo what can best be described as a "maturation process," starting as single reads off a sequencing machine, passing through a phase in which they become part of an assembled (yet incomplete) sequence contig, and finally ending up as part of a finished, completely assembled sequence with an error frequency of <1 in 10,000 bases. Even before sequencing data reaches this highly polished state, investigators can begin to ask whether or not given stretches of sequence represent coding or noncoding regions. The ability to make such determinations is of great relevance in the context of systematic sequencing efforts, since all of the data being generated by these projects are, in essence, "anonymous" in nature—nothing is known about the coding potential of particular stretches of DNA as they are being sequenced. Consequently, automated methods will become increasingly important in annotating the human and other genomes in order to increase the intrinsic value of these data as they are deposited into the public databases.

In considering the problem of gene identification, it is important to briefly review the basic biology underlying what will become, in essence, a mathematical problem. At the DNA level, upstream of a given eukaryotic gene there are promoters and other regulatory elements that control the transcription of that gene. The gene itself is discontinuous, being comprised of both introns and exons. Once this stretch of DNA is transcribed into an RNA molecule, both ends of the RNA are modified, with the 5' end being capped and a polyA signal being placed at the 3' end. The RNA molecule reaches maturity when the introns are spliced out, based on short consensus sequences found both at the intron-exon boundaries and within the introns themselves. Once splicing has occurred and the start and stop codons have been established, the mature mRNA is transported through a nuclear pore into the cytoplasm, at which point translation can take place.

While the process of moving from DNA to protein is obviously more complex in eukaryotes than in prokaryotes, the mere fact that it can be described in its entirety in eukaryotes would lead one to believe that predictions can confidently be made as to the exact positions of introns and exons. Unfortunately, the signals that control the process of moving from the DNA level to the protein level are not very well defined, precluding their use as foolproof indicators of gene structure. For example, upwards of 70% of promoter regions contain a TATA box, but because the remainder do not, the presence (or absence) of a TATA box in and of itself cannot be used to assess whether a region is a promoter. Similarly, during end modification, the polyA tail may be present or absent, or may not contain the canonical AATAAA. Adding to these complications is the fact that an open reading frame is required *but is not sufficient* to identify a region as an exon. Given these and other considerations, there is at present no straightforward method that will allow 100% confidence in the prediction of intron or exon status. Despite this, a combinatorial approach, relying on a number of methods, can be used to increase the confidence with which gene structure is predicted.

Briefly, gene-finding strategies can be grouped into three major categories. *Content-based methods* rely on the overall, bulk properties of a sequence in making their determinations. Characteristics considered here include the frequency at which particular codons are used, the periodicity of repeats, and the compositional complexity of the sequence. Because different organisms use synonymous codons with different frequency, such clues can provide insight to help determine which regions are more likely to be exons. In *site-based methods*, by contrast, the focus is on the presence or absence of a specific sequence, pattern, or consensus. These methods are used to detect features such as donor and acceptor splice sites, binding sites for transcription factors, polyA tracts, and start and stop codons. Finally, *comparative methods* make determinations based on sequence homology. Here, translated sequences are subjected to database searches against protein sequences (e.g., BLASTX; UNIT 6.3) in order to see whether a previously characterized coding region corresponds to a region in the query sequence. While this is conceptually the most straightforward of the three approaches, it is restrictive in that most newly discovered genes do not have gene products that match anything in the protein databases. Also, the modular nature of proteins and the fact that there are only a limited number of protein motifs makes it difficult to predict anything more than just exonic regions in this way. The reader is referred to a number of excellent reviews detailing the theoretical underpinnings of these various classes of methods (Claverie, 1997a,b, 1998; Guigo, 1997; Snyder and Stormo, 1997; Rogic et al., 2001). While many of the gene prediction methods belong strictly to one of these three classes of methods, most of those that will be discussed here combine the strength of different classes of methods in order to optimize their predictions.

Given the complexity of the problem at hand and the range of approaches for tackling it, it is important for investigators to appreciate when and how each particular method should be applied. A recurring theme in this unit will be the fact that each method will perform differently *depending on the nature of the data*. Put another way, while one method may be best for human finished sequence, another may be better for unfinished sequence, or for sequence from another organism.

GENE IDENTIFICATION METHODS

GRAIL

GRAIL—Gene Recognition and Analysis Internet Link (*http://compbio.ornl.gov/tools/index. shtml*)—is the elder statesman of gene prediction techniques, since it is among the first one developed in this area and enjoys widespread usage. As more and more has become known about gene structure in general and better Internet tools have become more widespread, GRAIL has continuously evolved in order to keep in step with the current state of the field.

Two basic versions of GRAIL will be discussed here. GRAIL 1 makes use of a neural network method to recognize coding potential in fixed-length (100-base) windows considering the sequence itself, without looking for additional features such as splice junctions or start and stop codons. An improved version of GRAIL 1, called GRAIL 1a, expands on this method by considering regions immediately adjacent to regions deemed to have coding potential, resulting in better performance in both finding true exons and eliminating false positives. Either GRAIL 1 or GRAIL 1a is appropriate for searching for single exons. A further refinement led to the second version, called GRAIL 2, in which variable-length windows are used and contextual information (e.g., about splice junctions, start and stop codons, polyA signals) is considered. Since GRAIL 2 makes its prediction by taking genomic context into account, it is appropriate for determining model gene structures.

In what follows, the output of each method discussed will be shown using the same set of input data as the query so as to highlight the methods' strengths and weaknesses. The sequence that will be considered is that of a human BAC clone RG364P16 from 7q31, which was established as part of the systematic sequencing of chromosome 7 (GenBank AC002467). For the purposes of this example, a client-server application called XGRAIL has been used. This software, which runs on the UNIX platform, allows for graphical output of GRAIL 1/1a/2 results, as shown in Figure 6.1.1. Because the DNA sequence in question is rather large and

6

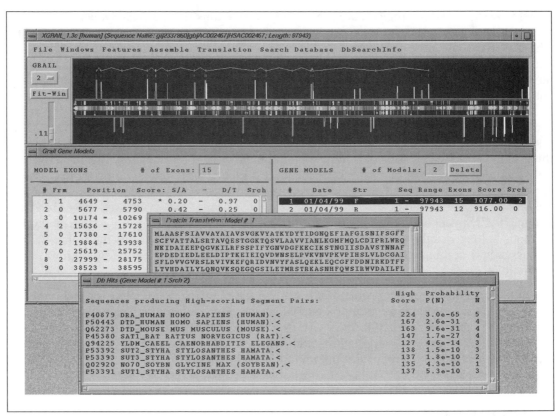

Figure 6.1.1 XGRAIL output obtained using the human BAC clone RG364P16 from 7q31 as the query. The upper window shows the results of the prediction, with the histogram representing the probability that a given stretch of DNA is an exon. The various colored bars in the center represent features of the DNA (e.g., arrows represent repetitive DNA, and vertical bars represent repeat sequences). Exon and gene models, protein translations, and the results of a genQuest search using the protein translation are shown.

is likely to contain at least one gene, GRAIL 2 was selected as the method. The large upper window presents an overview of the ~98 kb making up this clone, and the user can selectively turn on or off particular markings that identify features within the sequence (described in the figure legend). Of most importance in this view is the prediction of exons at the very top of the window, with the histogram representing the probability that a given region represents an exon. Information on each of these predicted exons is shown in the Model Exons window, and the model exons can be assembled and shown both as Model Genes and as a Protein Translation. Only putative exons with acceptable probability values (as defined in the GRAIL algorithm) are included in the gene models. The protein translation can, in turn, be searched against the public databases to find sequence homologs using a program called genQuest (integrated into XGRAIL), and these are shown in the Db Hits window. In this case, the fifteen exons in the first gene model (from the forward strand) are translated into a protein that shows significant sequence homology to a group of proteins thought to be involved in sulfate transport.

Most recently, the authors of GRAIL have released GRAIL-EXP (*http://grail.lsd.ornl.gov/grailexp/*), which is based on GRAIL but uses additional information in making the predictions, including a database search of known complete and partial gene messages. The inclusion of this database search in deducing gene models has greatly improved the performance of the original GRAIL algorithm.

FGENEH/FGENES

FGENEH, developed by Victor Solovyev and colleagues, is a method that predicts internal exons by looking for structural features such as donor and acceptor splice sites, potential coding regions, and intronic regions both 5′ and 3′ to the putative exon. The method makes use of *linear discriminant analysis*, a mathematical technique that allows data from multiple experiments to be combined. Once the data are combined, a linear function is used to discriminate between two classes of events—here, whether a given stretch of DNA is or is not an exon. In FGENEH, results of the linear discriminant approach are then passed to a dynamic programming algorithm that determines how to best combine these predicted exons into a coherent gene model. An extension of FGENEH, called FGENES, can be used in cases where multiple genes are expected in a given stretch of DNA.

The Sanger Centre Web server provides a very simple front end for performing FGENES searches. The query sequence (again, the BAC clone from 7q31 in this discussion) is pasted into the query box, an identifier is entered, and the search can then be performed. The results are returned in a tabular format, as shown in Figure 6.1.2. The total number of predicted genes and exons (2 and 33, respectively) are shown at the top of the output. The information for each gene (G) then follows. For each predicted exon, the strand (Str) is given, with $+$ indicating the forward strand and $-$ indicating the reverse. The Feature list in this particular case includes initial exons (CDSf), internal exons (CDSi), terminal exons (CDSl), and polyA regions (PolA). The nucleotide region for the predicted feature is then given as a range. In the current example, the features of the second predicted gene are shown in reverse order, since the prediction is based on the reverse strand. Predicted proteins based on the information in the table are given at the bottom of the output in FASTA format. The definition line for each predicted protein gives the range of nucleotide residues involved as well as the total length of the protein and the direction ($+$ or $-$) of the predicted gene.

MZEF

MZEF stands for "Michael Zhang's Exon Finder," named after its developer at the Cold Spring Harbor Laboratory. The predictions rely on a technique called *quadratic discriminant analysis* (Zhang, 1997). Imagine a case where the results of two types of predictions are plotted against each other on a simple *x-y* graph (for instance, splice site scores versus exon length). If the relationship between these two sets of data is nonlinear or multivariate, the resulting graph will look like a swarm of points. Points lying in only a small part of this swarm will represent a "correct" prediction, so a quadratic function is used to separate the correctly-predicted points from the incorrectly predicted points (hence the name of the technique). In the case of MZEF, the measured variables include exon length, intron-exon and exon-intron transitions, branch sites, 3′ and 5′ splice sites, and exon, strand, and frame scores. MZEF is intended to predict internal coding exons and does not give any other information about gene structure.

There are two implementations of MZEF currently available. The program can be downloaded from the CSHL FTP site for UNIX command-line use or accessed through a Web front end. The input is a single sequence, read in only one direction (either the forward or the reverse strand); to perform an MZEF analysis on both strands, the program must be run twice. Returning to the example of the chromosome 7 BAC clone, MZEF predicts a total of 27 exons in the forward strand (Figure 6.1.3). Focusing in on the first two columns of the table, the region of the prediction is given as a range, followed by the probability that the prediction is correct (P). Predictions with $P > 0.5$ are considered correct and are included in the table. The differences in predictions between methods are immediately evident. MZEF is again geared towards finding single exons, so the exons are not shown in the context of a putative gene, as they are in GRAIL 2 or FGENES. Moreover, as will be discussed later, the exons predicted by these methods are not the same.

```
Number of predicted genes:     2 In +chain:    1 In -chain:    1
Number of predicted exons:    33 In +chain:   23 In -chain:   10
Positions of predicted genes and exons:
G Str Feature    Start         End    Weight   ORF-start  ORF-end

 1 +   1 CDSf    3413  -     3594     2.50      3413  -     3592
 1 +   2 CDSi    4606  -     4753     1.73      4607  -     4753
 1 +   3 CDSi    5677  -     5790     1.91      5677  -     5790
 1 +   4 CDSi    9956  -    10033     2.55      9956  -    10033
 1 +   5 CDSi   10174  -    10269     1.86     10174  -    10269
 1 +   6 CDSi   11486  -    11592     1.81     11486  -    11590
 1 +   7 CDSi   13595  -    13664     3.39     13596  -    13664
 1 +   8 CDSi   15636  -    15728     2.38     15636  -    15728
 1 +   9 CDSi   17380  -    17610     1.97     17380  -    17610
 1 +  10 CDSi   19884  -    19938     2.72     19884  -    19937
 1 +  11 CDSi   25607  -    25752     3.18     25609  -    25752
 1 +  12 CDSi   28092  -    28175     3.04     28092  -    28175
 1 +  13 CDSi   40915  -    40981     1.00     40915  -    40980
 1 |  14 CDSi   41081  -    41262     1.42     41083  -    41262
 1 +  15 CDSi   51053  -    51131     1.31     51053  -    51130
 1 +  16 CDSi   55392  -    55442     0.95     55394  -    55441
 1 +  17 CDSi   60609  -    60692     1.52     60611  -    60691
 1 +  18 CDSi   64433  -    64600     3.71     64435  -    64599
 1 +  19 CDSi   68964  -    69064     3.15     68966  -    69064
 1 +  20 CDSi   69448  -    69531     3.48     69448  -    69531
 1 +  21 CDSi   70971  -    71044     3.04     70971  -    71042
 1 +  22 CDSi   73696  -    74083     2.25     73697  -    74083
 1 +  23 CDSl   74150  -    74731     2.94     74150  -    74728
 1 +     PolA   75218                 4.18

 2 -     PolA   82006                 4.57
 2 -   1 CDSl   82727  -    82738     1.32     82730  -    82738
 2 -   2 CDSi   83132  -    83197     2.58     83132  -    83197
 2 -   3 CDSi   83319  -    83461     2.79     83319  -    83459
 2 -   4 CDSi   87607  -    87661     3.62     87608  -    87661
 2 -   5 CDSi   89473  -    89706     2.93     89473  -    89706
 2 -   6 CDSi   90330  -    90425     1.75     90330  -    90425
 2 -   7 CDSi   92005  -    92097     1.79     92005  -    92097
 2 -   8 CDSi   92190  -    92259     1.39     92190  -    92258
 2 -   9 CDSi   93728  -    93834     2.05     93730  -    93834
 2 -  10 CDSi   95221  -    95316     2.27     95221  -    95316

Predicted proteins:
>FGENES 1.5 AC002467       1 Multiexon gene    3413  -   74731    1087 a Ch+
MLSRPTVGSGFPTSCLSTDGVHSTVSLWGRMGYKEKRSLKINLTGRESKATRAENQTDLV
RFLPPELPPVSLFSEMLAASFSIAVVAYAIAVSVGKVYATKYDYTIDGNQEFIAFGISNI
FSGFFSCFVATTALSRTAVQESTGGKTQVAGIISAAIVMIAILALGKLLEPLQKSVLAAV
<remainder of output truncated>
```

Figure 6.1.2 FGENES output obtained using the human BAC clone RG364P16 from 7q31 as the query. The columns, going from left to right, represent the gene number (G), strand (Str), feature (described in the main text), start and end points for the predicted exon, a scoring weight, and start and end points for corresponding open reading frames (ORF-start and ORF-end). Each predicted gene is shown as a separate block. Following the tables are protein translations of any predicted gene products.

GENSCAN

GENSCAN, developed by Chris Burge and Sam Karlin, is designed to predict complete gene structures. As such, like a number of the other gene identification algorithms GENSCAN can identify introns, exons, promoter sites, and polyA signals. Like FGENES, GENSCAN does not assume the input sequence will represent one and only one gene, or one and only one exon: it can accurately make predictions for sequences representing either partial genes or multiple genes separated by intergenic DNA. The ability to make these predictions accurately when a sequence is in a variety of contexts makes GENSCAN particularly useful for gene identification.

GENSCAN relies on what the authors term a "probabilistic model" of genomic sequence composition and gene structure. By looking for gene structure descriptions that match or are consistent with the query sequence, the algorithm can assign a probability to the chance that

```
          Internal coding exons predicted by MZEF
          Sequence_length: 97943 G+C_content: 0.391

          Coordinates    P     Fr1    Fr2    Fr3   Orf  3ss    Cds    5ss
          4606-4753    0.548  0.475  0.614  0.444  212  0.531  0.547  0.538
          5469-5543    0.557  0.588  0.461  0.600  121  0.499  0.594  0.622
          7353-7630    0.826  0.584  0.520  0.549  122  0.498  0.585  0.632
          10174-10269  0.546  0.605  0.443  0.442  122  0.517  0.552  0.515
          13595-13664  0.998  0.552  0.463  0.608  121  0.564  0.570  0.736
          15636-15728  0.534  0.444  0.432  0.544  221  0.488  0.500  0.636
          16654-16749  0.904  0.541  0.398  0.458  122  0.534  0.531  0.615
          17380-17610  0.940  0.614  0.470  0.442  122  0.518  0.569  0.594
          18736-18797  0.597  0.417  0.550  0.603  221  0.536  0.618  0.619
          19884-19938  0.866  0.434  0.406  0.537  221  0.550  0.504  0.657
          24126-24225  0.969  0.655  0.543  0.539  122  0.532  0.622  0.559
          25607-25752  0.977  0.551  0.452  0.466  122  0.530  0.542  0.647
          28107-28175  0.966  0.438  0.412  0.662  221  0.492  0.579  0.562
          37600-37687  0.605  0.328  0.610  0.434  212  0.515  0.549  0.586
          38297-38434  0.946  0.558  0.511  0.441  122  0.528  0.540  0.559
          50415-50823  0.632  0.557  0.451  0.470  122  0.543  0.533  0.519
          55133-55173  0.873  0.375  0.489  0.530  221  0.531  0.524  0.702
          57112-57175  0.518  0.562  0.424  0.469  122  0.514  0.530  0.618
          61089-61182  0.602  0.438  0.552  0.456  212  0.556  0.549  0.700
          64433-64600  0.980  0.614  0.552  0.505  122  0.517  0.599  0.606
          68964-69064  0.941  0.316  0.579  0.564  211  0.513  0.534  0.558
          69448-69531  0.997  0.565  0.444  0.364  122  0.536  0.523  0.705
          70971-71044  0.948  0.448  0.300  0.507  121  0.575  0.462  0.656
          73696-74083  0.968  0.487  0.594  0.498  212  0.552  0.574  0.536
          77911 77972  0.596  0.467  0.593  0.434  212  0.480  0.549  0.602
          80338-80413  0.944  0.467  0.464  0.590  221  0.507  0.555  0.662
          97197-97358  0.738  0.597  0.497  0.523  122  0.521  0.586  0.545
```

Figure 6.1.3 MZEF output obtained using the human BAC clone RG364P16 from 7q31 as the query. The columns, going from left to right, give the location of the prediction as a range of included bases (Coordinates), the probability value (P), frame preference scores (Fr$_i$), an ORF indicator showing which reading frames are open, and scores for the 3′ splice site, coding regions, and 5′ splice site.

a given stretch of sequence represents an exon, promoter, and so forth. The "optimal exons" are the ones with the highest probability and represent the parts of the query sequence having the best chance of actually being exons. The method will also predict "sub-optimal exons," stretches of sequence having a probability value that is acceptable but not as good as those of the optimal exons. GENSCAN's developers encourage users to examine both sets of predictions in order not to miss alternatively spliced regions of genes or other nonstandard gene structures.

Using the human BAC clone from 7q31 again, the query can be issued directly from the GENSCAN Web site, using *Vertebrate* as the organism, the default suboptimal cutoff, and *Predicted Peptides Only* as the print option. The results for this query are shown in Figure 6.1.4. The output indicates that there are three genes in this region, with the first gene having eleven exons, the second gene thirteen, and the third gene ten. The most important columns in the table are those labeled Type and P. The Type column indicates whether the prediction is for an initial exon (Init), an internal exon (Intr), a terminal exon (Term), a single-exon gene (Sngl), a promoter region (Prom), or a polyA signal (PlyA). The P column gives the probability that this prediction is actually correct. GENSCAN exons having a very high probability value (P > 0.99) are 97.7% accurate where the prediction matches a true, annotated exon. These high-probability predictions can be used in the rational design of PCR primers for cDNA amplification or for other purposes where extremely high confidence is necessary. GENSCAN exons that have probabilities in the range of 0.50 to 0.99 are deemed to be correct most of the time; the best-case accuracies for P values >0.90 is on the order of 88%. Any predictions <0.50

```
Gn.Ex Type S  .Begin  ...End  .Len Fr Ph I/Ac Do/T CodRg P....  Tscr..
----- ---- - ------  ------  ---- -- -- ---- ---- ----- -----  ------

1.01 Init +   4697   4801  105  1  0   64   80   103 0.651   7.58
1.02 Intr +   5725   5838  114  0  0   48   91   116 0.993   7.62
1.03 Intr +  10004  10081   78  1  0   61   70    78 0.809   2.13
1.04 Intr +  10222  10317   96  0  0   94   87   117 0.999  11.49
1.05 Intr +  11534  11640  107  1  2  118   62    31 0.953   1.59
1.06 Intr +  13643  13712   70  2  1   88  111    32 0.950   3.77
1.07 Intr +  15684  15776   93  2  0   45   98    59 0.782   1.84
1.08 Intr +  16702  16797   96  0  0   70  100    26 0.709   1.29
1.09 Intr +  17428  17658  231  0  0   69   79   233 0.911  17.55
1.10 Intr +  19932  19986   55  2  1   90   94    29 0.805   1.33
1.11 Term +  25128  25375  248  1  2   48   48   167 0.867   3.67
1.12 PlyA +  25382  25387    6                               1.05

2.00 Prom +  26739  26778   40                              -7.05
2.01 Init +  27929  28093  165  1  0   77   94    65 0.948   5.68
2.02 Intr +  28140  28223   84  2  0   69   64   142 0.901   9.00
2.03 Intr +  29931  30071  141  2  0  126   38    55 0.262   3.93
2.04 Intr +  52002  52164  163  2  1   99   17   149 0.194   7.53
2.05 Intr +  53036  53243  208  0  1   48   -2   191 0.028   3.31
2.06 Intr +  58789  58968  180  1  0   82   35   127 0.411   4.86
2.07 Intr +  59932  60222  291  1  0   69   20   255 0.369  12.13
2.08 Intr +  63258  63277   20  0  2  102   86   -16 0.527  -5.06
2.09 Intr +  64481  64648  168  0  0   47   86   162 0.939  10.90
2.10 Intr +  69012  69112  101  1  2   56   75   115 0.967   5.91
2.11 Intr +  69496  69579   84  0  0   25  115    57 0.615   1.20
2.12 Intr +  71019  71092   74  2  2  105   90   -21 0.950  -2.91
2.13 Term +  73744  74779 1036  1  1   85   44   805 0.960  66.40
2.14 PlyA +  75266  75271    6                               1.05

3.11 PlyA -  75947  75942    6                               1.05
3.10 Term -  83049  82945  105  0  0   77   38    68 0.831  -1.87
3.09 Intr -  83245  83180   66  1  0  113   94    43 0.948   5.58
3.08 Intr -  83509  83367  143  2  2  108   69    88 0.995   8.05
3.07 Intr -  87709  87655   55  1  1   50  115    63 0.988   2.83
3.06 Intr -  89754  89539  216  0  0  110   42   182 0.727  13.58
3.05 Intr -  90488  90378  111  2  0   25  100   169 0.499  11.46
3.04 Intr -  92145  92053   93  0  0  109   59    52 0.893   3.64
3.03 Intr -  92307  92238   70  2  1  101   67    38 0.955   1.27
3.02 Intr -  93882  93776  107  0  2   70   68    84 0.640   2.69
3.01 Intr -  95364  95269   96  0  0   68   75   106 0.661   6.59
Predicted peptide sequence(s):


>AC002467.seq|GENSCAN_predicted_peptide_1|430_aa
MLAASFSIAVVAYAIAVSVGKVYATKYDYTIDGNQEFIAFGISNIFSGFFSCFVATTALS
RTAVQESTGGKTQVAGIISAAIVMIAILALGKLLEPLQKSVLAAVVIANLKGMFMQLCDI
PRLWRQNKIDAVIWVFTCIVSIILGLDLGLLAGLIFGLLTVVLRVQFPSWNGLGSIPSTD
<remainer of output truncated>
```

Figure 6.1.4 GENSCAN output obtained using the human BAC clone RG364P16 from 7q31 as the query. The columns, going from left to right, represent the gene and exon number (Gn.Ex), the type of prediction (Type), the strand on which the prediction was made (S, with + indicating the forward strand and − as the reverse), the beginning and endpoint for the prediction (Begin and End), the length of the prediction (Len), the reading frame of the prediction (Fr), several scoring columns, and the probability value (P). Each predicted gene is shown as a separate block; notice that the third gene has its exons listed in reverse order, reflecting the fact that the prediction is on the reverse strand. Following the tables are the protein translations for each of the three predicted genes.

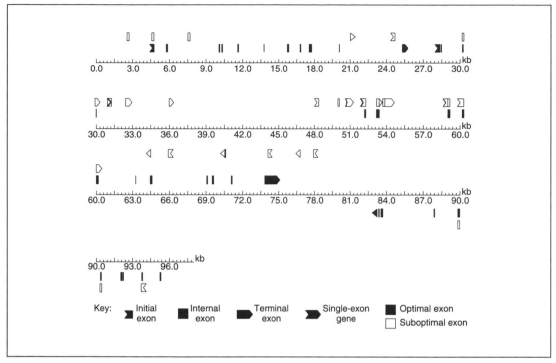

Figure 6.1.5 GENSCAN output in graphical form, obtained using the human BAC clone RG364P16 from 7q31 as the query. Optimal and suboptimal exons are indicated, and the initial and terminal exons show the direction in which the prediction is being made ($5' \rightarrow 3'$ or $3' \rightarrow 5'$).

should be discarded as unreliable, and those data are not given in the table. An alternative view of the data is shown in Figure 6.1.5. Here, both the optimal and suboptimal exons are shown, with the initial and terminal exons showing the direction in which the prediction is being made ($5' \rightarrow 3'$ or $3' \rightarrow 5'$). This view is particularly useful for large stretches of DNA, as the tables become harder to interpret as more and more exons are predicted.

In the near future, a new program named GenomeScan will be available from the Burge laboratory at MIT. GenomeScan assigns a higher score to putative exons that overlap BLASTX hits than to comparable exons for which similarity evidence is lacking. Regions of higher similarity (e.g., according to BLASTX E-value) are accorded more confidence than regions of lower similarity, since weak similarities sometimes do not represent homology. Thus, the predictions of GenomeScan tend to be consistent with all or almost all of the regions of high similarity that have been detected, but may sometimes ignore a region of weak similarity that either has weak intrinsic properties (e.g., poor splice signals) or is inconsistent with other extrinsic information. The accuracy of GenomeScan tends to be significantly higher than that of GENSCAN when a moderate or closely related protein sequence is available.

PROCRUSTES

Living up to its namesake, PROCRUSTES (*http://www-hto.usc.edu/software/procrustes*) takes genomic DNA sequences and "forces" them to fit into a pattern as defined by a related target protein. Unlike the other gene prediction methods that have been discussed, the algorithm does not use a DNA sequence on its own to look for content- or site-based signals. Instead, the algorithm requires that the user identify putative gene products *before* the prediction is made, so that the prediction will represent the best fit of the sequence to its putative transcription product. The method uses a spliced alignment algorithm to sequentially explore all possible exon assemblies, looking for the best fit of predicted gene structure to candidate protein. If the candidate protein is known to arise from the query DNA sequence, correct gene structures can be predicted with an accuracy of $\geq 99\%$. By making use of candidate proteins in the course of the prediction, PROCRUSTES can take advantage of information available about this protein, or related proteins, in the public databases to better determine the location of the introns and the exons in this gene. PROCRUSTES can handle cases where there are either partial or multiple genes in the query DNA sequence.

The input to PROCRUSTES is through a Web interface and is quite simple. The user needs to supply the nucleotide sequence along with as many protein sequences as are relevant. PROCRUSTES will treat these protein sequences as similar, though not necessarily identical, to that encoded by the DNA sequence. Typical output (not shown here) includes an aligned map of the predicted intron-exon structure for all target proteins, probability values, a list of exons with their starting and ending nucleotide positions, translations of the gene model (which may not be the same as the submitted protein sequences), and a "spliced alignment" showing any differences between the predicted protein and the target protein. The nature of the results makes PROCRUSTES a valuable method for refining results obtained by other methods, particularly in the context of positional candidate efforts.

GeneID

The current version of GeneID (*http://www1.imim.es/geneid.html*) finds exons based on measures of coding potential. The original version of this program was amongst the fastest in that it used a rule-based system to examine the putative exons and assemble them into the "most likely gene" for that sequence. GeneID uses position-weight matrices to assess whether or not a given stretch of sequence represents a splice site or a start or stop codon. Once this assessment is made, models of putative exons are built. Based on the sets of predicted exons that GeneID develops, a final refinement round is performed, yielding the most probable gene structure based on the input sequence.

The interface to GeneID is through a simple Web front-end, where the user pastes in the DNA sequence and specifies whether the organism is either human or *Drosophila*. The user can specify whether predictions should be made only on the forward or reverse strand, and available output options include lists of putative acceptor sites, donor sites, and start and stop codons. Users can also limit output to only first exons, internal exons, terminal exons, or single genes, for specialized analyses. It is recommended that the user simply select All Exons to assure that all relevant information is returned.

GeneParser

GeneParser uses a slightly different approach in identifying putative introns and exons. Instead of predetermining candidate regions of interest, GeneParser computes scores on all "subintervals" in a submitted sequence. Once each subinterval is scored, a neural network approach is used to determine whether each subinterval contains a first exon, internal exon, final exon, or intron. The individual predictions are then analyzed to find the combination that represents the most likely gene. There is no Web front end for this program, but the program itself is freely available for use on Sun, DEC, and SGI-based systems.

HMMgene

HMMgene predicts whole genes in any given DNA sequence using a hidden Markov model (HMM) method geared towards maximizing the probability of an accurate prediction. The use of HMMs in this method helps to assess the confidence in any one prediction, enabling HMMgene to not only report the "best" prediction for the input sequence, but alternative predictions on the same sequence as well. One of the strengths of this method is that, by returning multiple predictions on the same region, the user may be able to gain insight onto possible alternative splicings that may occur in a region containing a single gene.

The front-end for HMMgene requires an input sequence, with the organismal options being either human or *C. elegans*. An interesting addition is that the user can include known annotations, that could be from one of the public databases or based on experimental data that the investigator is privy to. Multiple sequences in FASTA format can be submitted as a single job to the server. Examples of sequence input format and resulting output are given in the documentation file at the HMMgene Web site.

HOW WELL DO THE METHODS WORK?

As is already evident from the preceding discussion, different methods produce different types of results—in some cases, lists of putative exons are returned but these exons are not in a genomic context; in other cases, complete gene structures are predicted, but possibly at a cost of less reliable individual exon predictions. Looking at the absolute results for the 7q31 BAC clone, anywhere between one and three genes are predicted for the region, and those one to three genes have anywhere between 27 and 34 exons. In the cases of similar exons, the boundaries of the exons are not always consistent. Which method is the "winner" in this particular case is not important; what is important is the variance in the results.

Since different methods will perform better or worse depending on the system being examined, it is important to be able to quantify the performance of each of these algorithms. Several studies have systematically examined the rigor of these methods using a variety of test data sets (Burset and Guigo, 1996; Claverie, 1997a; Snyder and Stormo, 1997; Rogic et al., 2001). Before discussing the results of these studies, some definition of terms is in order.

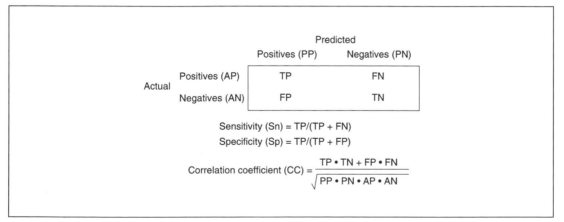

Figure 6.1.6 The matrix in this figure shows how both sensitivity and specificity are determined from four possible outcomes, giving a tangible measure of the effectiveness of any gene prediction method. FN, FP, TN, and TP are false and true negatives and positives. (Figure adapted from Burset and Guigo, 1996, and Snyder and Stormo, 1997.)

For any given prediction, there are four possible outcomes: detection of a true positive, a true negative, a false positive, or a false negative (Figure 6.1.6). Two measures of accuracy can be calculated based on the ratios of these occurrences: a *sensitivity* value, reflecting the fraction of actual coding regions that are correctly predicted as being coding regions, and a *specificity* value, reflecting the overall fraction of the prediction that is correct. In the best-case scenario, the methods will optimize the balance between sensitivity and specificity, in order to be able to find all true exons without becoming so sensitive as to pick up an inordinate number of false positives. An easier-to-understand measure that combines the sensitivity and specificity values is called the *correlation coefficient*. Like all correlation coefficients, its value can range from –1, meaning that the prediction is always wrong, through zero, to +1, meaning that the prediction is always right.

As a result of a Cold Spring Harbor Laboratory meeting on gene prediction ("Finding Genes: Computational Analysis of DNA Sequences," March 1997), a Web site called the "Banbury Cross" (*http://igs-server.cnrs-mrs.fr/igs/banbury*) was created. The intent behind its creation was two-fold: to allow groups actively involved in program development to post their methods for public use, and to allow researchers actively deriving fully characterized, finished genomic sequence to submit such data for use as "benchmark" sequences. In this way, the meeting participants created an active forum for dissemination of the most recent findings in the field of gene identification. Using these and other published studies, Jean-Michel Claverie at CNRS in Marseilles compared the sensitivity and specificity of fourteen different gene identification programs (Claverie, 1997a, and references therein), including all of those discussed here except PROCRUSTES. PROCRUSTES was not considered, because its method is substantially different from those of other gene prediction programs. In examining data from these disparate sources, either the best performance found in an independent study *or* the worst performance reported by the developers of the method themselves were used in making the comparisons. Based on these comparisons, the best overall individual exon finder was deemed to be MZEF and the best gene structure prediction program was deemed to be GENSCAN. (Back-calculating as best as possible from the numbers reported in the Claverie paper, these two methods gave the highest correlation coefficients within their class, with $CC_{MZEF} \sim 0.79$ and $CC_{GENSCAN} \sim 0.86$.)

Since these gene-finding programs are undergoing constant evolution, adding new features and incorporating new biological information, the idea of a comparative analysis of a number of representative algorithms was recently revisited. Using an independent data set containing 195 sequences from GenBank in which intron-exon boundaries have been annotated, GENSCAN and HMMgene appeared to perform the best, both having a correlation coefficient of 0.91.

STRATEGIES AND CONSIDERATIONS

Given these statistics, it can be concluded that both MZEF and GENSCAN are particularly suited for differentiating introns from exons at different stages in the maturation of sequence data. However, this should *not* be interpreted as a blanket recommendation to use only these two programs in gene identification. Remember that these results represent a compilation of findings from different sources, so keep in mind that the reported results may not have been derived from the same data set. It has already been stated numerous times that any given program can behave better or worse depending on the input sequences. It has also been demonstrated that the actual performance of these methods is highly sensitive to G+C content, with no pattern emerging across all of the methods as to whether a method's predictive powers improve or degrade as G+C content is raised. For example, Snyder and Stormo (1997) reported that GeneParser (Snyder and Stormo, 1993) and GRAIL2 (with assembly) performed best on test sets having high G+C content (as assessed by their respective CC values), while GeneID (Guigo et al., 1992) performed best on test sets having low G+C content.

Most gene identification programs share several major drawbacks of which users need to be keenly aware. Since most of these methods are "trained" on test data, they will work best in finding genes most similar to those in the training sets (that is, they will work best on things similar to what they have "seen" before). Often methods have an absolute requirement to predict both a discrete beginning and an end to a gene, meaning that these methods may miscall a region that consists of either a partial gene or multiple genes. The importance given to each individual factor in deciding whether a stretch of sequence is an intron or an exon can also influence outcomes, as the weighing of each criterion may be either biased or incorrect. Finally, there is the unusual case of genes that are transcribed but not translated—so-called "noncoding RNA genes."

It is becoming evident that no one program provides the foolproof key to computational gene identification. The correct choice of program will depend on the nature of the data and where in the pathway of data maturation that data lies. Based on the studies described above, some starting points can be recommended. In the case of incompletely assembled sequence contigs (prefinished genome survey sequence), MZEF provides the best jumping-off point, because for sequences of this length one would expect no more than one exon. In the case of finished or nearly finished sequences, where much larger contigs provide a good deal of contextual information, GENSCAN or HMMgene would be an appropriate choice. In either case, users should supplement these predictions with results from *at least* one other predictive method, as consistency amongst methods can be used as a qualitative measure of the robustness of the results. Furthermore, use of comparative search methods, such as BLAST or FASTA, should be considered an absolute requirement, with users targeting both dbEST and the protein databases for homology-based clues. PROCRUSTES should be used when some information regarding the putative gene product is known, particularly when the cloning effort is part of a positional candidate strategy.

A good example of the combinatorial approach is illustrated in the case of the gene for cerebral cavernous malformation (CCM1) located at 7q21 to 7q22; here, a combination of MZEF, GENSCAN, XGRAIL, and PowerBLAST (Zhang and Madden, 1997) was used in an integrated fashion in the prediction of gene structure (Kuehl et al., 1999).

A combinatorial method developed at the National Human Genome Research Institute links most of the methods described in this unit into a single tool. This tool, named GeneMachine, allows users to query multiple exon and gene prediction programs in an automated fashion. A suite of Perl modules are used to run MZEF, GENSCAN, GRAIL2, FGENES, and BLAST. RepeatMasker and Sputnik are used to find repeats within the query sequence. Once GeneMachine is run, a file is written that can subsequently be opened using NCBI Sequin, in essence using Sequin as a workbench and graphical viewer. Using Sequin also has the advantage of presenting the results to the user in a familiar format—basically the same format that is used in Entrez for graphical views. The most noteworthy feature of GeneMachine is that the process is fully automated; the user is only required to launch GeneMachine and then open the resulting file with NCBI Sequin. GeneMachine also does not require users to install local copies of the prediction programs, enabling users to pass off to Web interfaces instead and reducing the overhead of maintaining the program—albeit with the tradeoff of slower performance. Annotations can be made to GeneMachine results prior to submission to GenBank, thereby increasing the intrinsic value of the data.

6

References: Rogic et al., 2001; Zhang and Madden,1997

Contributor: Andreas D. Baxevanis

UNIT 6.2

Sequence Databases: Integrated Information Retrieval and Data Submission

First it is important to appreciate not only the variety of available data but also how rapidly it is expanding. The biomedical literature, as represented by MEDLINE, is growing at a rate of 400,000 articles per year. MEDLINE, produced by the U.S. National Library of Medicine (NLM), currently contains bibliographic information and journal abstracts from more than 11,000,000 publications indexed from ~4,300 biomedical journals.

GenBank is a comprehensive repository of sequence data and associated annotation, built and distributed by the National Center for Biotechnology Information (NCBI) at the NLM in Bethesda, Maryland. GenBank is part of an international collaboration with the DNA Database of Japan (DDBJ) in Mishima and the European Molecular Biology Laboratory (EMBL) Data Library at the European Bioinformatics Institute near Cambridge, United Kingdom. Because these partners exchange data daily, what we refer to here as simply "GenBank" actually contains sequences from GenBank, EMBL, and DDBJ.

In the 1990s, new types of sequence data became a major part of the database entries. For example, "expressed sequence tags" or ESTs are partial sequences that are derived from automated, "single-pass" sequencing on clones that are randomly selected from cDNA libraries, usually with the intent of surveying expressed genes. These data present certain analytic challenges (Boguski et al., 1993) that are dealt with in *UNIT 6.3*, and they also require special procedures for batch submission to sequence databases [see discussion of submitting expressed sequence tags (EST), sequence tag site (STS), or genome survey sequence (GSS) data]. Short genomic sequences, including those derived from "exon trapping" or "exon amplification" and genomic survey sequences present analytical challenges similar to those for ESTs. There are special procedures for submission of these "GSS" entries, similar to the procedures for EST submission. Large-scale sequencing of complete genomes has required additional methods for submission and has spurred development of new ways to access and display the data.

Database homology searching is described in detail in *UNIT 6.3*. The present unit focuses on the wide variety of sequence, structure, and bibliographic databases—as well as how they have been integrated (Figure 6.2.1) into a single, easy-to-use retrieval environment called *Entrez*, which provides users with "one stop shopping" over the Internet.

INTRODUCTION TO *ENTREZ*

Entrez (Figure 6.2.1) ties together a diverse set of information resources: it accesses nucleotide and protein sequences from a number of databases as well as genome data from the NCBI Genomes division, three-dimensional structures from MMDB (Molecular Modeling Database of NCBI), human disease data from OMIM (Online Mendelian Inheritance in Man), genetic locus data from LocusLink, and bibliographic citations from PubMed (discussed below). *Entrez* is available through the World Wide Web (*http://www.ncbi.nlm.nih.gov/Entrez/*) and in a client/server version. In the case of the client/server application, Network *Entrez* (*NetEntrez*), the client runs on the user's local machine and interacts with the server at NCBI. Versions of the client program are available for Macintosh, Microsoft Windows, Unix/X-windows and some

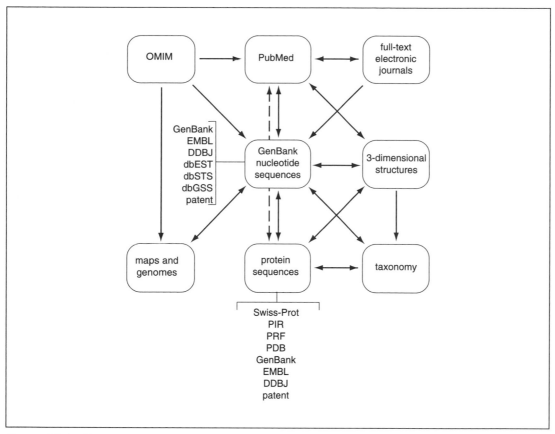

Figure 6.2.1 Data sources and links that constitute the *Entrez* system.

other systems. *Entrez* for the World Wide Web (*WebEntrez*) works through standard WWW browsers. *NetEntrez* and *WebEntrez* have similar functionality, but there are differences in the way that they are used. *WebEntrez* has the advantage of additional links—e.g., to WebBLAST outputs and to full-text articles for some journals. The *NetEntrez* interface is more intuitive and is often easier to use for complex queries. *Entrez* can be used in many ways and some common scenarios are presented below.

Entrez substantially lowers the activation energy of a search by combining all activities in an easy-to-use system with a mouse-driven, graphical user interface. *Entrez* establishes all of the requisite links between DNA sequences, the proteins they encode, and the literature that describes their biology. The literature component of *Entrez* comes from the PubMed database, which includes abstracts. While an abstract is not a substitute for the full text of a paper, one can frequently go an amazingly long way in the analysis and interpretation of sequence homology results by the information contained in an abstract. For an increasing list of journals, the Web version of *Entrez* also has links to full-text articles. For many human disease genes, there are also links from PubMed records to the description of the disease in the Online Mendelian Inheritance in Man (OMIM) database. The PubMed database of bibliographic information contains citations from MEDLINE and PreMEDLINE (preliminary citations that eventually become MEDLINE records).

Entrez not only contains explicit links between different data sources, but also implicit or computed links within the data itself. Daily, the entire set of DNA and protein sequence databases are compared among one another and all significant homologies are computed and stored in the system. Thus, it is possible to instantly retrieve the homologs of any sequence in

Entrez with the click of a mouse, without having to manually perform a new database search. Thus *Entrez* contains within it thousands of answers to questions that have not yet been asked.

In addition to precomputed sequence homologies, *Entrez* performs a similar operation on PubMed records. It is possible to compute a statistical relationship between two articles based on their frequency of use of significant terms. To indicate what constitutes a significant term, consider two negative examples: i.e., the terms "novel" and "important" are not significant terms because they appear in most scientific papers and thus are not of use in distinguishing one paper from another in terms of specific content. The set of "related articles" (sometimes referred to as "neighbors") is recomputed daily. This constitutes a very powerful information resource. One can enter *Entrez* via an author name or keyword and retrieve not only a specific paper but an entire bibliography of related material and then branch out (or "neighbor out") to other areas of the information space and effortlessly go back and forth between the literature and the sequences. Further examples can also be found on the NCBI Coffee Break web site (*http://www.ncbi.nlm.nih.gov/Coffeebreak/*). These pages describe recent biological discoveries and include tutorials demonstrating the linkage between various components of *Entrez* and other bioinformatics tools.

Finding information on sequence homologs using the Web version of *Entrez* is made even easier by the connections between *Entrez* and Web-based BLAST output. For example, it is possible to do a BLAST search using WebBLAST, and the hit list will have hypertext links to protein or nucleotide entries in *Entrez*.

Other features of *Entrez* include taxonomy-based sequence retrieval; from the Web page for the NCBI Taxonomy Browser that enables one to scan through organism names and phylogenetic trees and select the full set of nucleotide or protein sequences for any species. There are now links in *Entrez* from sequences and the literature to protein three-dimensional structures in NCBI's Molecular Modeling Database (MMDB). A three-dimensional structure viewer called Cn3D has also been added to the system. Cn3D can be run as a client/server program or as a "Helper" application integrated into a standard Web browser.

Another special viewer is available for the *Entrez* Genomes division. The genome-level views presented by *Entrez* are built from sequences of complete chromosomes, from composites of sequence fragments, and from integrated genetic and physical maps. *Entrez* allows the user to visualize the sequence information at varying levels of detail, either graphically or as text. The chromosome views are tightly linked to the other *Entrez* component databases, allowing the user to jump effortlessly between maps, sequences, and bibliographic components.

DATA SUBMISSION: GENERAL CONSIDERATIONS

Currently, journals prefer not to publish sequence data because it takes up a great deal of space and is much more usable in electronic form. Most journals, however, require proof in the form of a database accession number that the authors of a publication have sent their data to GenBank, EMBL, or DDBJ. Accession numbers are unique identifiers (consisting of one letter followed by five digits or two letters followed by six digits) that serve as confirmation that the data has been submitted and allow the scientific community to retrieve it. Making sequences available by depositing them in the public databases is recognized as an important responsibility of researchers to the community.

GenBank, the EMBL Data Library (*http://www.ebi.ac.uk/Submissions/index.html*), and the DDBJ (*http://sakura.ddbj.nig.ac.jp/*) are partners in the international database collaboration; authors may submit their sequences to whichever database is most convenient without regard to where papers describing those sequences may be published. Sequences should be submitted to a single site; data submitted to one site are exchanged with the others on a daily basis.

Sequences by themselves have little meaning; hence, thorough, accurate biological annotation is very important. One important source of annotation is the published article referring to the sequence; indeed, NCBI (*http://www.ncbi.nlm.nih.gov/Genbank/index.html*) has made every effort to link citations to the sequences via MEDLINE and *Entrez*. However, electronic annotation obtained from authors at the time of submission is becoming even more critical as the amount of data grows, because the only realistic way of searching through this much data is by computer.

Sequences submitted to any of the sequence databases go through a series of validation checks conducted by the database staff. Accession numbers are returned to the authors, usually within 24 hr, or less, if there are no problems with the submission(s). The accession number should be included in any manuscript that deals with the sequence, preferably in a footnote on the first page of the article (or in the manner prescribed by the individual journal). The full text of the provisional record is passed back to the authors for their review prior to being released. Authors may request that the data be kept confidential until publication.

SUBMITTING A SEQUENCE TO THE NUCLEOTIDE DATABASE

Submission Methods

As mentioned above, a sequence can be submitted to any of the three sequence databases (see discussion of Data Submission: General Considerations). A submission to GenBank is described in this section, but readers should keep in mind that the example could apply to DDBJ or EMBL submissions. Researchers should submit their sequences to whichever database is most convenient. The important point is to only submit to one database.

If the researcher has access to the World Wide Web, the easiest way to submit data to GenBank is to use the BankIt form on the NCBI home page (or for DDBJ and EMBL submissions, use Sakura or WebIn, respectively). BankIt (*http://www.ncbi.nlm.nih.gov/BankIt/*) allows one to simply copy and paste most of the submission information and data directly out of one's word processor or sequence-analysis package into an electronic form. BankIt was designed to handle the most common types of GenBank submissions, such as mRNAs or short genomic records.

NCBI also has developed a platform-independent submission program called Sequin, which runs stand-alone or over the network. Sequin is suitable for a wide range of sequence lengths and complexities, including traditional (gene-sized) nucleotide sequences, segmented entries (e.g., genomic sequences of a spliced gene where not all of the intronic sequences have been determined), long (genome-sized) sequences with many annotated features, and sets of related sequences (i.e., population, phylogenetic, or mutation studies of a particular gene, region, or viral genome). It also has a number of built-in validation functions for enhanced quality assurance. The validator checks such things as missing organism information, incorrect coding-region lengths (compared to the attached protein sequence), internal stop codons in coding regions, mismatched amino acids, or nonconsensus splice sites. Sequin can present various views of the data, including GenBank-, FASTA-, and EMBL-formatted files, and a graphical view of the sequence including features and alignments. A sequence editor is built into Sequin; this editor automatically adjusts feature intervals as the sequence is edited. Additional capabilities include sequence-analysis functions—e.g., a coding region translator, an ORF finder, and a function that will replace a sequence with the sequence of the complementary strand and simultaneously adjust the positions of the features.

Sequin is available from NCBI by anonymous FTP; see *http://www.ncbi.nlm.nih.gov/ Sequin/index.html* for further details on Sequin, including full detailed documentation.

BankIt is best suited for submitting one or a small number of sequences. Sequin provides more of a data "workbench" environment that can accommodate large as well as small contigs with detailed annotation that can be added over time and which can finally be used to submit data to GenBank. In addition to these methods, NCBI provides streamlined submission procedures for large data sets—e.g., expressed sequence tags (ESTs), sequence-tagged sites (STSs), genome survey sequences (GSSs), and high-throughput genome sequences (HTGSs). NCBI provides a means for interacting with the laboratory information-management systems of genome-sequencing centers to ensure the efficient, timely, and accurate submission of their data. These options are discussed in more detail in subsequent sections (see discussion of Submitting EST, STS, or GSS Data, and Submitting High-Throughput Genome Sequences).

Instructions and Tips for Preparing Sequence Submissions

The most important item is, of course, the new DNA sequence itself. It should be free of sequencing errors, cloning artifacts, etc.; standard things to check for in this regard are described below. In the special case of "single-pass" sequences such as ESTs, the data often contain ambiguities and sequencing errors; in these cases it is important to note it as "single-pass" data.

What does one need to know about a sequence to submit it to GenBank? The sequence record contains references to the article in which the sequence is published, information on the submitters, and information on the source organism. The record also should describe the sequence as fully as possible. Significant regions of the sequence, such as coding region, promoter region, and transcription boundaries, are labeled with defined feature names. These features are detailed in the GenBank Feature Table document, available by anonymous FTP from *ftp://ncbi.nlm.nih.gov* in the directory */genbank/docs*.

The DEFINITION line is a short description of the biological sequence in the record and the organism from which it was derived. The LOCUS field contains the length of the sequence, the type of sequence, and the date that it was released to the public (or re-released following any updates, corrections, or modifications). The ORGANISM field in the SOURCE section contains a complete phylogenetic classification. The REFERENCE section contains one or more citations, the name and address of the submitter, and the date that the sequence was originally submitted to the database. The COMMENT field may be used to include annotations that cannot be accommodated in the FEATURES table. REFERENCEs and COMMENTs that apply to the complete record are referred to as "descriptors." Whenever possible, annotations should be expressed as "legal" FEATURES, using the proper syntax, because this is the only manner in which they are computer-readable. A more complex example, illustrating detailed annotation of a human breakpoint cluster region, is the GenBank record with the accession number U07000, which can be obtained from *Entrez* (see Introduction to *Entrez*, above) or the e-mail server.

Further details on submission to the sequence databases can be found on the database Web sites and in Kans and Ouellette (1998). The following are seven important questions that should be asked when preparing a sequence for submission.

1. What organism is the source of this DNA?

Taxonomic classification of the organism from which the sequence was isolated is very important when using GenBank to ensure that the correct genetic code is used when translating the DNA to generate the protein product from a coding sequence (CDS) feature (see item 4). If the wrong taxa are used, or if a mitochondrial origin is not indicated, then an erroneous translation will be produced and may go undetected. Correct taxonomic classifications are also important for retrieval purposes via *Entrez*.

A sequence submission in most cases need only include the genus and species, which the database staff use to obtain the complete classification from NCBI's Taxonomy Database. If precise information is not available, the researcher should send in the best conjecture as to the genus, species, and/or common name, and the database staff will attempt to determine the proper classification.

Because there is great biodiversity among microbial organisms used for research, it is often important to include the strain name or number with the DNA sequence for microbial submissions. These will appear as a qualifier, for example, /strain="S288c," and are quite useful when doing sequence comparisons or trying to explain certain sequence conflicts. Strain information is also important in the context of genetic mapping data from the mouse (e.g., /strain="BALB/c") and other organisms.

2. Is this a genomic or mRNA (cDNA) sequence?

The sequenced molecule is most often DNA, but the annotation should indicate the biological identity of the starting material. For example, cDNA (i.e., an mRNA that was made into a DNA molecule using reverse transcriptase so that it could be cloned) should be annotated in the database as an mRNA. HIV is an interesting case. HIV sequences, as well as those from other retroviruses, may be classified in different ways: as cDNA corresponding to genomic RNA (therefore as "RNA"); as nonintegrated, nonproviral DNA (therefore as single-stranded DNA or "ssDNA"); or as integrated, proviral DNA, (therefore as double-stranded DNA or "dsDNA").

3. If this is a genomic sequence, are there introns for which the sequence has not been determined?

In many cases only parts of a gene have been sequenced—e.g., the exons with only a few nucleotides of flanking intron. In this case it is necessary to create what is known as a "segmented entry." What this means is that the researcher will create individual submissions for each discrete sequence and indicate the relationships between them—i.e., their linear order. Additionally, if the approximate distances (in base pairs or kilobase pairs) between the segments of a set are known, this information should be included. This will permit the complete gene to be reassembled automatically by software programs. It should be noted that this is an example of a type of entry for which Sequin is the best tool.

4. Is there a coding sequence (CDS) or other RNA product (rRNA or tRNA) on this DNA? If so, what are their positions in the DNA sequence?

If this is an mRNA, or if it is a DNA molecule encoding a protein product, it is important to indicate that a coding region is present and to supply its start and stop codons. RNA products also have boundaries or coordinates that should be indicated. An important check that is done with all submissions is to verify that the coordinates submitted correspond with the indicated feature(s). If there are any discrepancies, these are resolved by the database annotation staff in consultation with the submitter.

As described above under question 1, a common problem occurs when the taxonomic information is incorrect or incomplete, leading to the use of an improper genetic code. Fifteen different genetic codes are presently in use. Any codon usage that is not represented in this list must use the translation exception qualifier (/transl_except) to indicate the discrepancy.

Sequin has a built-in tool for analyzing open reading frames. The user can set the minimum size of the open reading frames and the genetic code and the "ORF Finder" will graphically display all open reading frames in a sequence. There is also a separate Web version of the ORF Finder (*http://www.ncbi.nlm.nih.gov/gorf/gorf.html*).

5. If there is a CDS, what is gene name and the name of the encoded protein? If the gene product is an enzyme, what is the EC number?

Putting a protein or gene product name on a CDS is the best way of labeling the protein encoded by the DNA. It is possible to add more than one qualifier, but a */product*, */gene*, and */EC_number* (if the product is an enzyme) are quite sufficient. This is the label that will be carried over to the protein databases—e.g., GenPept or Swiss-Prot. Enzyme Commission (EC) numbers can be obtained or verified using the World Wide Web at *http://www.expasy.ch/enzyme/enzyme-search-ec.html*.

6. Is there information about your sequence for which there does not seem to be an appropriate place on the forms?

There are sometimes problems finding the appropriate place to put a given feature in the BankIt or Sequin forms. In such cases, just attach a note to the submission and the database staff will find the correct place or contact you for more information.

7. Is this sequence ready for submission? Has it been tested for the presence of vector contamination, mitochondrial DNA, ribosomal and tRNAs, and repetitive (e.g., Alu) elements?

Potential problems can be detected by using the BLAST programs to compare a new sequence against special data sets that represent common "contaminants." The network-aware version of Sequin will allow you to run BLASTN against vector and mitochondrial databases. The VecScreen system has recently been developed for quickly finding vector contamination in sequences. VecScreen is automatically run on BankIt sequences and it can be run from within Sequin. It can also be used independently from its own web page (*http://www.ncbi.nlm.nih.gov/VecScreen/VecScreen.html*). Other validation routines are built into both the Sequin and BankIt submission tools. These routines include a comparison of the conceptual translation of a coding region with the submitted amino acid sequence, a check that actual data length matches the given length, a check for correct use of qualifiers on features, and a check for illegal characters. Although the database staff applies a number of validation tests to each new sequence, it is more expeditious for authors themselves to catch any problems prior to submission.

SUBMITTING AN UPDATE OR CORRECTION TO AN EXISTING GENBANK ENTRY

Authors are encouraged to submit updates or corrections to their sequence records. In fact anyone who notices a problem, error, or omission in a database record should bring this to the attention of the database staff. It is not uncommon for the authors to forget to notify the databases that a previously confidential sequence may now be released. If an accession number is seen in a published article and the sequence cannot be retrieved, this is almost certainly the case. The databases will release the data if the complete journal citation is supplied, including the full title of the paper, or if a copy of the title page plus the page showing the accession number is faxed to the databases.

Update information consisting of a simple revision, such as a citation change or release of information, can be submitted by sending an e-mail message in paragraph form explaining the change. The message must include the accession number of the sequence to be updated along with all update, correction, or publication information.

Updates can also be submitted using BankIt or Sequin. See the BankIt or Sequin Web pages for details (see Submitting a Sequence to the Nucleotide Database, Submission Methods).

SUBMITTING EST, STS, OR GSS DATA

Expressed-sequence tags (ESTs), sequence-tagged sites (STSs), and genome survey sequences (GSSs) are usually submitted to GenBank and the specialized databases dbEST, dbSTS, or dbGSS as batches of dozens to thousands of entries, with a great deal of redundancy in the citation, submitter, and library information. For these data, there are special procedures designed to improve the efficiency of the submission process. A special "tagged flat file" input format is used. Documents describing the format can be obtained by sending a request to *info@ncbi.nlm.nih.gov*. They are also available from the dbEST, dbSTS, or dbGSS Web pages at: *http://www.ncbi.nlm.nih.gov/dbEST/*, *http://www.ncbi.nlm.nih.gov/dbSTS/*, and *http://www.ncbi.nlm.nih.gov/dbGSS/*, respectively. Once the data are ready to submit, they should be e-mailed to *batch-sub@ncbi.nlm.nih.gov*. For large data sets, an account can be set up into which the data can be deposited by FTP. For more information write to *info@ncbi.nlm.nih.gov*.

SUBMITTING HIGH-THROUGHPUT GENOME SEQUENCES (HTGS)

Some groups may find that their needs are not conveniently met by the standard means for data submission—e.g., a genome center that has its own internal information system or a group assembling a very large contig. The staff of NCBI, EBI, or DDBJ are happy to discuss these issues and make special arrangements with such groups so that their data can be conveniently incorporated into the databases. This includes setting up automatic exchange of data, creation of special FTP accounts, and the generation of tools to ensure data exchange in the most useful format. At NCBI, for example, FTP accounts have been set up for all submitting genome sequencing centers and a variety of tools have been created for accelerating the submission of high-throughput genome sequences (HTGS).

For large sequencing operations that wish to develop an integrated direct submission system, or for those researchers who have a data submission problem that does not seem to be addressed by the standard tools, e-mail can be sent to *info@ncbi.nlm.nih.gov*. The situation should be explained in as much detail as possible and a phone number and postal address should be provided so that the GenBank staff can contact the investigators to discuss their situation.

References: Baxevanis et al., 1997; Boguski et al., 1993

Contributors: Jane M. Weisemann, Mark S. Boguski, and B.F. Francis Ouellette

6

Sequence Similarity Searching Using the BLAST Family of Programs

Over the years, a number of algorithms have been implemented that allow searching of sequence databases. The most useful of these tools should share the following characteristics: (1) Speed. Because today's databases are so large, the programs must be fast in order to process megabases of sequence in seconds. (2) Sensitivity. The programs must report all potentially interesting similarities. (3) Rigorous statistics. The programs must provide a way to evaluate the significance of the results. (4) Ease of use. Scientists with no formal training in sequence-analysis algorithms should understand how to use the programs and interpret the results. Advanced users should have the option to tailor the programs to their needs. (5) Access to up-to-date databases. The doubling time of GenBank is currently ~16 months. It is important to search the most recent version of the database.

The BLAST (Basic Local Alignment Search Tool) family of sequence similarity search programs satisfies the above criteria. In short, users input either a nucleotide or amino acid query sequence, and search a nucleotide or amino acid sequence database. The program returns a list of the sequence "hits," alignments to the query sequence, and statistical values. This unit describes how to choose an appropriate BLAST program and database, perform the search, and interpret the results.

ACCESSING BLAST PROGRAMS AND DOCUMENTATION

The National Center for Biotechnology Information (NCBI) supports several versions of BLAST at no charge to the public. BLAST 2.0, or gapped BLAST, is the standard version of BLAST that allows a user to search a sequence database with a nucleotide or protein sequence of interest. BLAST 2.0 places gaps into the query and target sequences so that separate areas of similarity between the two sequences can be returned as one alignment. Position-specific iterated BLAST (PSI-BLAST) is an iterative BLAST search, optimized for finding distantly related sequences. Pattern-hit initiated BLAST (PHI-BLAST) searches for conserved sequence patterns or motifs in proteins (Zhang et al., 1998). "BLAST 2 sequences" produces an alignment of two sequences entered by the user. On the "Specialized BLAST" pages, researchers can search sequences that are not in GenBank. At present, these databases include finished and unfinished microbial genomes, *P. falciparum* (the human malaria parasite), human genome sequences, and immunoglobulin sequences (IgBLAST). VecScreen identifies potential vector contamination by performing BLAST searches against a database of vector sequences.

The easiest and most popular way to access the BLAST suite of programs is through the NCBI World Wide Web site, at *http://www.ncbi.nlm.nih.gov/BLAST/*. All versions of BLAST are accessible from this site, and can be used to query all sequence databases available at the NCBI. Documentation, which includes an overview of BLAST, frequently asked questions (FAQs), a "What's New" page, the BLAST manual, and a list of references, is also available here.

For users who want to run BLAST against private local databases or downloaded copies of NCBI databases, the NCBI offers a stand-alone version of the BLAST program. BLAST binaries and documentation are provided for the latest versions of IRIX, Solaris, DEC OSF1, and Win32 systems.

BLAST can also be run as a client-server program, in which the user installs client software on a local machine that communicates across the network with a server at NCBI. This setup is useful for researchers who run large numbers of searches on NCBI databases, because they can automate the process to run on their local computer.

The NCBI BLAST e-mail server is the best option for people without convenient access to the Web. A similarity search can be performed by sending a properly formatted e-mail message containing the nucleotide or protein query sequence to *blast@ncbi.nlm.nih.gov*. The query sequence is compared against a specified database and the results are returned in an e-mail message. For more information on formulating e-mail BLAST searches, please send a message consisting of the word HELP to the same address, *blast@ncbi.nlm.nih.gov*.

INTRODUCTION TO BLAST

Basic Versus Advanced BLAST Searches

BLAST 2.0 is available in a basic or advanced version. In both versions, the user can select the type of BLAST program and the database to be searched, and choose whether to filter the query sequence to mask low-complexity regions (see below). The advanced version allows the user to change parameters as well. For most researchers, the Basic version, which uses the default parameters, is adequate. For a discussion of BLAST parameters, see Appendix A at the end of this unit.

BLAST Programs

Five types of the BLAST program have been developed to support sequence similarity searching using a variety of nucleotide and protein sequence queries and databases. These programs are listed and described in Table 6.3.1. The type of BLAST search to be carried out is dependent on the type of information that is desired.

Table 6.3.1 BLAST Search Programs

Program	Query sequence	Database sequence	Comments
BLASTP	Protein	Protein	Can be run in standard mode or in a more sensitive iterative mode (PSI-BLAST), which uses the previous search results to build a profile for subsequent rounds of similarity searching.
BLASTN	Nucleotide (both strands)	Nucleotide	Parameters optimized for speed, not sensitivity; not intended for finding distantly related coding sequences. Automatically checks complementary strand of query.
BLASTX	Nucleotide (six-frame translation)	Protein	Very useful for preliminary data containing potential frameshift errors (ESTs, HTGs, and other "single-pass" sequences).
TBLASTN	Protein	Nucleotide (six-frame translation)	Essential for searching protein queries against EST database. Often useful for finding undocumented open reading frames or frameshift errors in database sequences.
TBLASTX	Nucleotide (six-frame translation)	Nucleotide (six-frame translation)	Should be used only if BLASTN and BLASTX produce no results. Restricted for search against EST, STS, HTGS, GSS, and Alu databases.

6

NCBI Databases

One frequent mistake in sequence similarity searching is failure to search an up-to-date database. The NCBI produces GenBank and updates it daily. It also shares data on a daily basis with the DNA Data Bank of Japan (DDBJ) and the European Molecular Biology Laboratory (EMBL). A search of the NCBI databases using the BLAST Web page, client, or e-mail server guarantees access to the most recent database. The NCBI supports a number of databases for sequence similarity searching. These databases are subject to change, and a current list and description are available on the NCBI Web site at *http://www.ncbi.nlm.nih.gov/BLAST/ blast_databases.html.*

Peptide sequence databases for BLASTP and BLASTX

nr. The nr (nonredundant) database is the most comprehensive. GenBank, DDBJ, and EMBL are nucleotide sequence databases. If any nucleotide sequence in any of the databases is annotated with a coding sequence (CDS), this CDS appears in nr. The nr database also contains protein sequences obtained from PDB (sequences associated with 3-D structures in the Brookhaven Protein Data Bank), Swiss-Prot (a curated database of protein sequences), PIR (Protein Identification Resource, a comprehensive collection of protein sequences), and PRF (Protein Research Foundation). Although nr may contain multiple copies of similar sequences, identical sequences are merged into one entry. To be merged, two sequences must have identical lengths and every residue at every position must be the same. There are nr databases for both peptide and nucleotide sequences. The peptide database is automatically selected for BLASTP and BLASTX searches.

month. The month database receives its sequences from the same sources as nr, but contains only those sequences released within the last 30 days. All sequences in month are also present in nr.

Drosophila genome. This database contains *Drosophila melanogaster* genome proteins provided by Celera and the Berkeley Drosophila Genome Project (BDGP). *Drosophila melanogaster* sequences submitted by other researchers are not in this database.

Yeast. The yeast database provides access to the NCBI's *Saccharomyces cerevisiae* reference sequences. The sequences and annotations of the sixteen yeast chromosomes and one mitochondrion are provided by the Saccharomyces Genome Database (SGD; *http://genome-www.stanford.edu/Saccharomyces/*).

E. coli. The E. coli database is derived from the complete sequence of *Escherichia coli* K-12.

pdb. pdb contains the sequences that are derived from the Brookhaven Protein Data Bank, a database of 3-D structures.

Alu. The *Alu* database contains six frame translations of representative *Alu* repeats from all *Alu* subfamilies (APPENDIX 2A). If a query sequence containing an *Alu* repeat is used in a BLAST search of nr, month, or HTGS, many of the resulting high scoring hits will also contain *Alu* sequences. If an unexpected number of high scoring hits are returned with a human query, it may be useful to search the *Alu* database with this query to identify the location of any *Alu* repeats that might be responsible for these potentially misleading alignments.

Nucleotide sequence databases for BLASTN, TBLASTN, and TBLASTX

nr. The nr database contains all nucleotide sequences present in GenBank, EMBL, and DDBJ. It also contains nucleotide sequences obtained from PDB (sequences associated with 3-D structures in the Brookhaven Protein Data Bank). nr comprises only sequences that are normally well annotated, so it does not contain EST (expressed sequence tag), STS (sequence tagged site), GSS (genome survey sequence), or HTG (high throughput genomic) sequences. nr is no longer nonredundant, and may contain multiple copies of identical sequences.

month. The month database contains all nucleotide sequences present in GenBank, EMBL, DDBJ, and PDB that were released within the last 30 days. Unlike nr, it also contains EST, STS, GSS, and HTG sequences released within the last month.

EST. EST accesses a nonredundant copy of all ESTs present in GenBank, EMBL, and DDBJ. ESTs are short sequences, a few hundred nucleotides in length, which are derived by partial, single-pass sequencing of inserts of randomly selected cDNA clones. Since the number of ESTs is increasing rapidly, it is an important database to search for novel cDNAs.

STS. STS contains a nonredundant copy of all STSs present in GenBank, EMBL, and DDBJ. An STS is a short unique genomic sequence that is used as a sequence landmark for genomic mapping efforts.

HTGS. HTGS contains "unfinished" DNA sequences generated by the high-throughput sequencing centers. A typical HTG record might consist of all the first-pass sequence data generated from a single cosmid, BAC, YAC, or P1 clone. The record is composed of two or more sequence fragments that have a total length of ≥ 2 kb and contain one or more gaps. The sequences are normally updated by the sequencing centers as more data become available. A single accession number is assigned to this collection of sequences. The accession number does not change as the record is updated, and only the most recent version of the record remains in GenBank. Phase 1 HTG sequences are unordered, unoriented contigs with gaps. Phase 2 HTG sequences are ordered, oriented contigs with or without gaps. All HTG records contain a prominent warning that the sequence data is unfinished and may contain errors. When a record is considered finished, it becomes a Phase 3 HTG and is moved to the nr database with the same accession number. HTGS is a valuable source of new genomic sequences not yet in nr.

GSS. GSS includes short, single-pass genomic data identified by various means. Many of the sequences have been mapped.

Drosophila genome, yeast, E. Coli, and pdb. The contents of these databases are analogous to the peptide databases described elsewhere in this unit (see discussion of Peptide Sequence Databases for BLASTP and BLASTX).

Alu. The *Alu* database contains representative *Alu* repeats from all *Alu* subfamilies (APPENDIX 2A). If a query sequence containing an *Alu* repeat is used in a BLAST search of the above nucleotide databases, many of the resulting high-scoring hits will also contain *Alu* sequences. It may be useful, especially with a genomic sequence query, to search the *Alu* database with this query to identify the location of any *Alu* repeats that might be responsible for these high scoring and potentially misleading hits. An easier alternative is to filter the sequence for human repeats, an option available on the Advanced BLAST search.

Vector. The Vector database contains nucleotide sequences of a number of standard cloning vectors. New sequences should be screened against the Vector database to assure that they do not contain any vector contamination.

Mito. The Mito database contains complete genomic sequences from >125 organisms. Nuclear-derived sequences may be screened against the Mito database to assure that they do not contain any mitochondrial contamination.

Formatting the Query Sequence

Users of the BLAST Web page can initiate a search either by entering the sequence itself, or, if the sequence is already in the sequence database, by entering the accession number or gi (see Appendix B at the end of this unit). The preferred format for entering new sequences is the so-called FASTA format; however, if the sequence is not in FASTA format or the sequence is interspersed with numbers and spaces, BLAST will still accept the query. A sequence in FASTA format begins with a single-line description, followed by lines of sequence data. The description line is distinguished from the sequence data by a ">" symbol (greater than) in the first column. An example sequence in FASTA format is:

```
>aaseq Human choroideremia protein
MADNLPTEFDVVIIGTGLPESILAAACSRSGQRVLHIDSRSYYGGNWASFSFSGLLSWLKE
YQQNNDIGEESTVVWQDLIIIETEEAITIRKKDETIQHTEAFPYASQDMEDNVEEIGALQKN
PSLGVSNTFTEVLDSALPEESQLSYFNSDEMPAKHTQKSDTEISLEVTDVEESVEKEKYCG
DKTCMHTVSDKDGDKDESKSTVEDKADEPIRNRITYSQIVKEGRRFNIDLVSKLLYSQGLL
IDLLIKSDVSRYVEFKNVTRILAFREGKVEQVPCSRADVFNSKELTMVEKRMLMKFLTFCL
EYEQHPDEYQAFRQCSFSEYLKTKKLTPNLQHFVLHSIAMTSESSCTTIDGLNATKNFLQC
LGRFGNTPFLFPLYGQGEIPQGFCRMCAVFGGIYCLRHKVQCFVVDKESGRCKAIIDHFGQ
RINAKYFIVEDSYLSEETCSNVQYKQISRAVLITDQSILKTDLDQQTSILIVPPAEPGACA
VRVTELCSSTMTCMKDTYLVHLTCSSSKTAREDLESVVKKLFTPYTETEINEEELTKPRLL
WALYFNMRDSSGISRSSYNGLPSNVYVCSGPDCGLGNEHAVKQAETLFQEIFPTEEFCPPP
PNPEDIIFDGDDKQPEAPGTNNVVMAKLESSEESKNLESPEKHLQN
```

Alternatively, the user can enter the database accession number or gi. The above sequence is in fact already in the database with a gi identifier of 116365 and a Swiss-Prot accession number of P26374. Either of these sequence identifiers can be entered on the BLAST page. The type of identifier (for nucleotide or protein sequence) must match the type of query sequence used in that search. Sequences in manuscripts are often referred to by their GenBank accession number; however, this accession number refers to a nucleotide (not protein) sequence, and cannot be used to initiate a BLASTP or TBLASTN search, both of which require a protein sequence query. The identifier of the protein sequence encoded by a nucleotide accession number can be obtained by searching the NCBI's Entrez nucleotide database, at *http://www.ncbi.nlm. nih.gov/Entrez.*

Filtering Sequences

Both nucleotide and protein sequences may contain regions of low complexity, i.e., regions with homopolymeric tracts, short-period repeats, or segments enriched in one or only a few residues. Such low-complexity regions commonly give spuriously high BLAST scores that reflect compositional bias rather than significant position-by-position alignment. For example, two protein sequences that contain low-complexity regions rich in the same amino acids may produce high-scoring alignments in those regions even though other parts of the proteins are entirely dissimilar.

Filtering the query sequence (that is, replacing the repeated sequence with strings of n for nucleotide sequence or X for protein sequence) can eliminate potentially confounding matches, such as hits to low-complexity proline-rich regions, poly(A) tails, or Alu elements present in the database. By default, all searches performed through the NCBI BLAST Web page, BLAST clients, e-mail server, and stand-alone programs automatically filter low-complexity sequence in the query. This filtering can be turned "off" at the Basic and Advanced BLAST Web pages. BLASTN queries are filtered with DUST (R.L. Tatusov and D.J. Lipman, pers. comm.). Other BLAST queries use SEG. Filtering for human repeats is "off" by default, and can be turned "on" with a checkbox on the Advanced BLAST page. By default, filtered sequence is represented in the final BLAST report as a string of n or X (e.g., nnnnnnnnnn or XXXXXXXXX). An option to mask low-complexity or human repeat sequence only for the lookup table is also available on the Advanced BLAST page. When this box is checked, the sequences are masked only during the initial phase of the BLAST search, and are not replaced by n or X in the BLAST report.

Receiving the BLAST Results

By default, BLAST results are available from a Web browser window. The document will have hypertext links that make it easier to analyze the results. In the middle of the afternoon, the BLAST server may be busy. Thus, it is sometimes more efficient to receive the BLAST results by e-mail. E-mail results are sent either as plain text or in HTML format. The HTML formatted results must be opened in a Web browser, as the resulting document contains hypertext links.

In general, BLAST queries are processed in the order they are received; however, the NCBI WWW BLAST server is a shared resource and it would be unfair for a few users to monopolize it. To prevent this, the server keeps track of how many queries are in the queue for each user and penalizes those users with many queries in the queue.

EXAMPLES OF BLAST SEARCHES

BLASTP

The BLASTP program compares a protein query to a protein database. The Swiss-Prot database has been selected and the program has been changed to BLASTP. For an example, a query in FASTA format has been entered in the input box of the human choroideremia protein, implicated in hereditary blindness. The top of the results page begins with some header information about the type of program (BLASTP), the version (2.0.5), and a release date. The version and release date will change as the program is updated. Also listed are a reference, the query definition line, and a summary of the database used. Figure 6.3.1 presents a graphical overview of the results. The database hits are shown aligned to the query, which is the numbered bar at the top. The next three bars show high-scoring database matches that align to the query sequence throughout its length. The next nine bars show lower-scoring matches that align to two regions of the query—one from residue 3 or 4 of the query up to about residue 60, and the other from about residues 220 to 500. For these nine bars, the solid segments (on the left and right) indicate aligned regions; the cross-hatched segment indicates that the two alignments (i.e., the solid portions) are from the same database sequence, but that there is no alignment where the bar is cross-hatched. If there is no cross-hatching between two or more bars in the same row (i.e., the thirteenth row), this denotes different alignments with unrelated database sequences. These bars are presented on the same line to conserve space in the graphical view. Moving the cursor over a bar in the graphic ("mousing over") causes the identifier and definition line of the database match to be shown in the window. If the results of a BLAST search are sent by e-mail, this graphical view is not included.

The hit list produced by the BLASTP search is shown in Figure 6.3.2. Each line of the hit list is composed of four fields. The first field contains the database designation, accession number, and locus name for the matched sequence, separated by vertical bars (see Appendix B at the end of this unit for more information). The second field contains a brief textual description of the sequence, the definition line. The content of the definition line varies between and within databases, but usually includes information on the organism from which the sequence was derived, the type of sequence (e.g., mRNA or DNA), and some information about function or

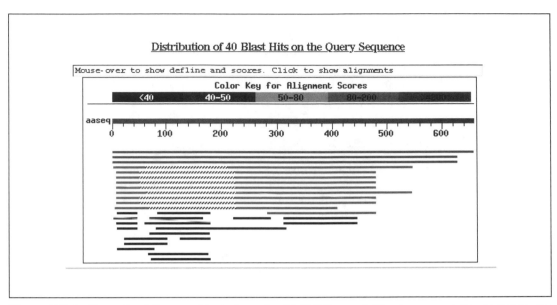

Figure 6.3.1 Example of the graphical view of a BLASTP report. The bars are color coded by the strength of the database match. The strongest matches (those with a bit score >200) are red, followed by pink (bit score 80 to 200), green (50 to 80), blue (40 to 50), and black (<40).

Sequences producing significant alignments:		Score (bits)	E Value
sp|P26374|RAE2_HUMAN	RAB PROTEINS GERANYLGERANYLTRANSFERASE COM...	1223	0.0
sp|P24386|RAE1_HUMAN	RAB PROTEINS GERANYLGERANYLTRANSFERASE COM...	881	0.0
sp|P37727|RAE1_RAT	RAB PROTEINS GERANYLGERANYLTRANSFERASE COMPO...	856	0.0
sp|P39958|GDI1_YEAST	SECRETORY PATHWAY GDP DISSOCIATION INHIBITOR	127	6e-29
sp|P50397|GDIB_MOUSE	RAB GDP DISSOCIATION INHIBITOR BETA (RAB G...	124	5e-28
sp|P21856|GDIA_BOVIN	RAB GDP DISSOCIATION INHIBITOR ALPHA (RAB ...	122	1e-27
sp|P50398|GDIA_RAT	RAB GDP DISSOCIATION INHIBITOR ALPHA (RAB GD...	122	1e-27
sp|P31150|GDIA_HUMAN	RAB GDP DISSOCIATION INHIBITOR ALPHA (RAB ...	122	2e-27
sp|P50395|GDIB_HUMAN	RAB GDP DISSOCIATION INHIBITOR BETA (RAB G...	121	4e-27
sp|Q10305|YD4C_SCHPO	PUTATIVE SECRETORY PATHWAY GDP DISSOCIATIO...	121	4e-27
sp|P50399|GDIB_RAT	RAB GDP DISSOCIATION INHIBITOR BETA (RAB GDI...	120	7e-27
sp|P32864|RAEP_YEAST	RAB PROTEINS GERANYLGERANYLTRANSFERASE COM...	98	5e-20
sp|P50396|GDIA_MOUSE	RAB GDP DISSOCIATION INHIBITOR ALPHA (RAB ...	80	9e-15
sp|Q49398|GLF_MYCGE	UDP-GALACTOPYRANOSE MUTASE	35	0.35
sp|P24588|AK79_HUMAN	A-KINASE ANCHOR PROTEIN 79 (AKAP 79) (CAMP...	35	0.46
sp|P36225|MAP4_BOVIN	MICROTUBULE-ASSOCIATED PROTEIN 4 (MICROTUB...	34	0.79
sp|Q46337|SOXA_CORSP	SARCOSINE OXIDASE ALPHA SUBUNIT	34	0.79
sp|P30599|CHS2_USTMA	CHITIN SYNTHASE 2 (CHITIN-UDP ACETYL-GLUCO...	33	1.4
sp|P53911|YNN6_YEAST	HYPOTHETICAL 49.4 KD PROTEIN IN NAM9-FPR1 ...	32	2.3
sp|P75499|GLF_MYCPN	UDP-GALACTOPYRANOSE MUTASE	32	2.3
sp|P37747|GLF_ECOLI	UDP-GALACTOPYRANOSE MUTASE	32	3.1
sp|P40142|TKT_MOUSE	TRANSKETOLASE (TK) (P68)	32	4.0
sp|P10587|MYSG_CHICK	MYOSIN HEAVY CHAIN, GIZZARD SMOOTH MUSCLE	32	4.0
sp|P50137|TKT_RAT	TRANSKETOLASE (TK)	32	4.0
sp|Q02455|MLP1_YEAST	MYOSIN-LIKE PROTEIN MLP1	31	5.2
sp|P52538|DNBI_HSV6Z	MAJOR DNA-BINDING PROTEIN (MDBP)	31	6.9
sp|P52338|DNBI_HSV6U	MAJOR DNA-BINDING PROTEIN (MDBP)	31	6.9
sp|P37637|YHIV_ECOLI	HYPOTHETICAL 111.5 KD PROTEIN IN HDED-GADA...	31	9.0
sp|Q02469|FRDA_SHEPU	FUMARATE REDUCTASE FLAVOPROTEIN SUBUNIT PR...	31	9.0
sp|Q01550|TANA_XENLA	TANABIN	31	9.0
sp|P49731|MIS5_SCHPO	MIS5 PROTEIN	31	9.0

Figure 6.3.2 Example of the hit list from a BLASTP report.

phenotype. The definition line is often truncated in the hit list to keep the display compact, but is fully displayed in association with the local alignments (see below). The third field contains the alignment score in bits. Higher scoring hits are found at the top of the list. In very general terms, this score is calculated from a formula that takes into account the alignment of similar or identical residues, as well as any gaps that must be introduced in order to align the sequences. A key element in this calculation is the "substitution matrix," which assigns a score for aligning any possible pair of residues. The BLOSUM-62 matrix is the default for BLAST, and it works well for most searches. The fourth field contains the Expect (E) value, which provides an estimate of statistical significance. E values reflect how many times one expects to see such a score occur by chance. A statistician would consider an E value <0.05 to be significant; however, an inspection of the alignments (see below) is required to determine biological significance. The maximum number of hits displayed in the hit list is set at a default value of 500. This number can be changed on the advanced BLAST page with the *descriptions* option.

For the first hit in the list, the database designation is sp (for Swiss-Prot), the accession number is P26374, the locus name is RAE2_HUMAN, the definition line begins RAB PROTEINS, the score is 1223, and the E value is 0.0.

The higher the score and the lower the E value, the more statistically significant is the hit. Note that the first thirteen hits all have very low E values ($<10^{-14}$) and are either RAB proteins or GDP dissociation inhibitors. The next eighteen database matches have much higher E values (≥ 0.35), meaning that roughly one match with that score would be expected by chance. Furthermore, the definition lines no longer show the same consistency. Given the combination of inconsistent definition lines and high E values, caution should be used in assuming an evolutionary relationship between the query and the 18 lower-scoring database hits. It would be necessary to examine the alignments in order to make a final determination.

The pairwise alignments between the query and the database hits are displayed below the hit list. Figure 6.3.3 shows one sequence alignment between the query sequence and a database hit; other alignments are omitted for brevity. The alignment is preceded by the sequence identifier, the full definition line, and the length, in amino acids, of the database sequence. Next comes the score of the match in bits (the raw score is in parentheses), the Expect value (E value) of the hit, the number of identical residues in the alignment (identities), the number of conservative substitutions (positives) according to the scoring system (e.g., BLOSUM-62), and, if applicable, the number of gaps in the alignment. Finally the actual alignment is shown, with the query on top and the database match labeled as subject (Sbjct). Since both the query and the match are amino acid sequences, the numbers at left and right refer to the position in the amino acid sequence. The center line between the two sequences indicates the similarities between the sequences. If the residue is identical between the query and the subject, the residue itself is shown; conservative substitutions, as judged by the substitution matrix, are indicated with a + sign. One or more dashes (−) within a sequence indicates insertions or deletions. Query residues masked for low complexity are replaced by Xs (see fourth and eleventh blocks, Fig. 6.3.3). By default, a maximum of 500 pairwise alignments are displayed. This number can be changed on the advanced BLAST page with the *alignments* option.

Results generated on the NCBI BLAST Web page contain text hyperlinks that assist the user in navigating around the page. Clicking on a colored bar in the graphical view (Fig. 6.8.3) moves the user to the sequence alignment (Fig. 6.3.3). The database sequence identifiers, visible both in the one-line descriptions of the database matches (Fig. 6.3.2) and in the sequence alignment (Fig. 6.3.3), link to the Entrez record that describes the sequence. Entrez may provide more information about the sequence, including links to relevant publication abstracts in PubMed.

6

```
sp|P37727|RAE1_RAT RAB PROTEINS GERANYLGERANYLTRANSFERASE COMPONENT A 1 (RAB ESCORT
                   PROTEIN 1) (REP-1)
                   Length = 650

  Score =  856 bits (2188), Expect = 0.0
  Identities = 433/632 (68%), Positives = 491/632 (77%), Gaps = 13/632 (2%)

Query: 1    MADNLPTEFDVVIIGTGLPESILAAACSRSGQRVLHIDSRSYYGGNWASFSFSGLLSWLK  60
            MADNLP++FDV++IGTGLPESI+AAACSRSGQRVLH+DSRSYYGGNWASFSFSGLLSWLK
Sbjct: 1    MADNLPSDFDVIVIGTGLPESIIAAACSRSGQRVLHVDSRSYYGGNWASFSFSGLLSWLK  60

Query: 61   EYQQNNDIGEESTVVWQDLIHETEEAITLRKKDETIQHTEAFPYASQDMEDNVEEIGALQ  120
            EYQ+NND+ E+++ WQ+ I E EEAI L  KD+TIQH E F YASQD+ +VEE GALQ
Sbjct: 61   EYQENNDVVTENSM-WQEQILENEEAIPLSSKDKTIQHVEVFCYASQDLHKDVEEAGALQ  119

Query: 121  KNPSLGVSNTFTEVLDSA----LPEESQLSYFNSDEMPAKHTQKSDTEISLEVTDVEESV  176
            KN +   S   E ++A   LP  +   S E+PA+ +Q   E S EV D E +
Sbjct: 120  KNHASVTSAQSAEAAEAAETSCLPTAVEPLSMGSCEIPAEQSQCPGPESSPEVNDAEATG  179

Query: 177  EKEKYCGDKTCMHTVXXXXXXXXXXXXTVEDKADEPIRNRITYSQIVKEGRRFNIDLVSK  236
            +KE        V+D   + P +NRITYSQI+KEGRRFNIDLVS+
Sbjct: 180  KKEN--------SDAKSSTEEPSENVPKVQDNTETPKKNRTTYSQIIKEGRRFNIDLVSQ  231

Query: 237  LLYSQGLLIDLLIKSDVSRYVEFKNVTRILAFREGKVEQVPCSRADVFNSKELTMVEKRM  296
            LLYS+GLLIDLLIKS+VSRY EFKN+TRILAFREG VEQVPCSRADVFNSK+LTMVEKRM
Sbjct: 232  LLYSRGLLIDLLIKSNVSRYAEFKNITRILAFREGTVEQVPCSRADVFNSKQLTMVEKRM  291

Query: 297  LMKFLTFCLEYEQHPDEYQAFRQCSFSEYLKTKKLTPNLQHFVLHSIAMTSESSCTTIDG  356
            LMKFLTFC+EYE+HPDEY+A+     FSEYLKT+KLTPNLQ+FVLHSIAMTSE++  T+DG
Sbjct: 292  LMKFLTFCVEYEEHPDEYRAYEGTTFSEYLKTQKLTPNLQYFVLHSIAMTSETTSCTVDG  351

Query: 357  LNATKNFLQCLGRFGNTPFLFPLYGQGEIPQGFCRMCAVFGGIYCLRHKVQCFVVDKESG  416
            L ATK FLQCLGR+GNTPFLFPLYGQGE+PQ FCRMCAVFGGIYCLRH VQC VVDKES
Sbjct: 352  LKATKKFLQCLGRYGNTPFLFPLYGQGELPQCFCRMCAVFGGIYCLRHSVQCLVVDKESR  411

Query: 417  RCKAIIDHFGQRINAKYFIVEDSYLSEETCSNVQYKQISRAVLITDQSILKTDLDQQTSI  476
            +CKA+ID FGQRI +K+FI+EDSYLSE TCS VQY+QISRAVLITD S+LKTD DQQ SI
Sbjct: 412  KCKAVIDQFGQRIISKHFIIEDSYLSENTCSRVQYRQISRAVLITDGSVLKTDADQQVSI  471

Query: 477  LIVPPAEPGACAVRVTELCSSTMTCMKDTYLVHLTCSSSKTAREDLESVVKKLFTPYTET  536
            L VP  EPG+  VRV ELCSSTMTCMK TYLVHLTC SSKTAREDLE VV+KLFTPYTE
Sbjct: 472  LAVPAEEPGSFGVRVIELCSSTMTCMKGTYLVHLTCMSSKTAREDLERVVQKLFTPYTEI  531

Query: 537  EINEEELTKPRLLWALYFNMRDSSGISRSSYNGLPSNVYVCSGPDCGLGNEHAVKQAETL  596
            E   E++ KPRLLWALYFNMRDSS ISR  YN LPSNVYVCSGPD GLGN++AVKQAETL
Sbjct: 532  EAENEQVEKPRLLWALYFNMRDSSDISRDCYNDLPSNVYVCSGPDSGLGNDNAVKQAETL  591

Query: 597  FQXXXXXXXXXXXXXXXXXXXXXDGDDKQPEAP  628
            FQ                   DGD  Q E P
Sbjct: 592  FQQICPNEDFCPAPPNPEDIVLDGDSSQQEVP  623
```

Figure 6.3.3 Example of a BLASTP alignment.

The score (to the right of the one-line description; Fig. 6.3.2) provides a link within the page to the alignment between that hit and the query.

BLASTX

BLASTX searches involve a six-frame translation of a nucleotide sequence queried against a protein database. Researchers most often use a BLASTX search if they have sequenced a DNA molecule and do not know the reading frame of the coding sequence, the beginning or end of the open reading frame, or the function of the encoded protein. It is also useful for analyzing preliminary sequence data containing potential sequencing or frameshift errors, such as ESTs, HTGs, or GSSs. A BLASTX search of the nr protein database can quickly highlight any similarities between all the open reading frames in a nucleotide sequence and any characterized proteins.

When a query is submitted against the nr protein database, the output is similar to that from BLASTP. At the top is information about the search, including the query, database, and type of BLAST program used. Next is the graphical overview representing the alignment of the database hits to the query sequence, followed by a list of the hits, including their definitions, scores, and E values. Pairwise alignments between the query and hit sequences follow.

TBLASTN

TBLASTN searches involve a protein sequence queried against the six-frame translation of a nucleotide sequence database. TBLASTN searches are especially useful for finding novel similarities to a protein sequence in the open reading frames from uncharacterized, sometimes lower-quality nucleotide sequences, such as those found in EST, STS, HTGS, and GSS databases. The translations of these sequences are not present in nr because the sequences are not annotated. The most common use of TBLASTN is to search for ESTs that are similar to a protein of interest. These ESTs might represent orthologs of the protein from a different species, or just additional members of a gene family.

The following example shows a TBLASTN search of HTGS, a database that contains unfinished and uncharacterized genomic sequences generated by genome sequencing centers. Since HTGs are genomic and often contain introns, the results can be more difficult to interpret than those from an EST search; however, HTGs can also provide insight into the chromosomal organization of genes.

In this case, the six-frame translation of HTGS is searched with the amino acid sequence of the disintegrin-like domain of mouse ADAM 1. TBLASTN output is similar to that of BLASTP and BLASTX. It displays the graphical overview of the alignment, the list of database hits, and selected pairwise alignments between the query sequence and the hits. For TBLASTN searches, the top line (Query) of the alignment contains protein sequence, and the location refers to amino acids. The bottom line (Sbjct) of the alignment contains a translated nucleotide sequence, and the location refers to the nucleotide location.

BLASTN

An example here is a BLASTN search of accession number U93237 against the human EST database. U93237 is the 9.2-kb genomic sequence of the human menin (MEN1) gene for multiple endocrine neoplasia type I. This example demonstrates the power of a BLAST search to confirm exons in a genomic sequence, as well as the danger of accepting results without further investigation. The matches of base pairs 4000 to 9000 confirm the locations of exons 2 to 10 of the MEN1 gene. The large number of matches of base pairs 500 to 1000 and 3000 to 3600 appear at first glance to be interesting matches, but are actually caused by the presence of *Alu* elements in the query and EST sequences. One can determine this by checking the definition lines of the matches in this region, as many of these ESTs are annotated as being similar to *Alu* elements. It is important to remember, though, that the definition line summaries contain truncated definition lines, and that it is necessary to look at the full definition line (provided in the alignment section) to see the *Alu* annotation. One way to avoid the hits due to *Alu* elements is to filter the sequence for human repeats. This option is available only from the Advanced BLAST page.

Since both the query and the subject are nucleotide sequences, the numbers refer to nucleotide sequence locations. Nucleotides in the query sequence that have been masked for low complexity are replaced by the letter n (e.g., nnnnnnn in the third block). BLASTN checks the plus and minus strand of the query against the plus strand of the database sequence. The alignment presents the plus strand of the query to the minus strand of the database sequence, which is equivalent to showing the minus strand of the query against the plus strand of the database sequence.

BLASTN is optimized for speed, not sensitivity. The algorithm uses a simple system to assess the alignments; matches are given a positive score, mismatches a negative score. BLASTN should only be used when a direct comparison of nucleotide sequences is desired. BLASTN results should not be used to make any predictions about the functions of the encoded proteins. Generally, searches that involve protein comparisons are more sensitive, owing to the larger

6

alphabet (twenty residues versus four nucleotides) and the degeneracy of the genetic code (two codons, with different third bases often coding for the same protein). The more sophisticated scoring system used with protein comparisons also takes into account similarities between different residues.

PSI-BLAST

A major new feature of the BLAST version 2.0 is the iterative search. Position-specific iterated BLAST (PSI-BLAST) first performs a normal BLASTP run, but then internally computes a profile using the most significant hits. A profile may be understood as a table that lists the probability of finding a residue at each position in a conserved protein domain. Conserved regions between the query and the most-significant matches are used to compute the profile. These regions are presumably important to the function of the protein. The profile is then used to perform another search (or iteration), at which point the new hits are also added into the profile. The cycle may then be repeated, with a new profile computed from the results of the last iteration. Each PSI-BLAST iteration takes into account the most significant sequences found in the previous round (i.e., those below the threshold for inclusion in the profile). The algorithm is more sensitive than BLASTP, and more likely to lead to the discovery of distantly related sequences. At present, such iterative searching is only available with the BLASTP program.

The PSI-BLAST output is different from the other programs. The PSI-BLAST report shows one-line summaries from the first PSI-BLAST iteration (i.e., the first search using the profile). The one-line summaries are divided into two groups, those with an E value better (lower) than that used to build the profile (the default value is 0.001), and those with a worse E value. Matches better than the threshold, which will be included in the profile, are checked. The user may check (or uncheck) matches, causing them to be included (or excluded) from the profile used in the next iteration. Matches found on a previous iteration are marked with a green ball. New sequences are marked with a yellow "new". The page also alerts the user when convergence is achieved (i.e., no new sequences below the threshold for inclusion in the profile were found in that iteration).

If one wants to report the E value for a hit found with a PSI-BLAST search, one should use the E value calculated during the first iteration that the match is below the threshold for inclusion in the profile (i.e., when the hit is marked with a "new" flag); however, if the hit is found in the first BLASTP run, then the reported E value should be from this run. In either case, in subsequent iterations, the hit itself is used to compute the profile, and the reported E value is not representative of the true value.

SEARCHING STRATEGIES

Filtering for Low-Complexity Regions

Low-complexity regions tend to result in misleading database search results. The residue frequencies within such regions differ dramatically from the database as a whole, disrupting the statistics used by BLAST. This can produce alignments that are based purely on compositional bias rather than a significant position-by-position alignment. By default, low-complexity sequences are filtered out of query sequences and replaced by strings of n (nucleotide) or X (protein). A BLASTP search of a proline-rich, DNA-directed RNA polymerase (Swiss-Prot accession P11414) against the nr database illustrates the differences in output resulting from a filtered or unfiltered query sequence. Twenty-four database matches with an E value better than 10 are found if filtering is enabled; 4873 are found if filtering is disabled.

Filtering for Human Repeats

LINES, SINES, retroviruses, and other repeated elements in the query sequence also result in misleading database search results. The Advanced BLAST page contains a checkbox to filter human repeats. This selection is "off" by default. If human repeat filtering is turned "on," the hits on the left side of the graphical view will disappear.

Reporting BLAST Results

The most reliable indicator of the importance of a BLAST alignment is the Expect (E) value (the number of chance database matches one expects to see at the same score). This number takes into account the length of the query and size of the database as well as the scoring system. Neither the bit nor the raw score is a reliable indicator, as their significance cannot be judged independently of the information used to calculate the E value. Normal intuition fails when one is faced with databases of the current size (e.g., the EST database contains ~700 million base pairs at this time), as chance alignments are likely and it is impossible for a user to know at what score matches become significant. Statistical significance also varies from database to database (e.g., a hit with a certain score may be statistically significant in a search of the relatively small month database, but not in a search of nr), so that reporting a certain score without the context of the database size can be misleading. The percentage identity is also a poor indicator of statistical significance for the same reasons; additionally, there is really no way to estimate an E value or even a score from a percentage identity, making it impossible to even guess whether a hit could be due to chance or not.

Analyzing mRNA Sequences

1. Identify the open reading frame to obtain the protein sequence. Perform a BLASTX search of the cDNA sequence against the nr protein database. Any resulting hits may identify the correct reading frame for the protein sequence and provide information about its function.

2. Determine whether the protein sequence is similar to that of any previously identified proteins. Perform a BLASTP and a PSI-BLAST (iterative) search against the nr protein database using the protein sequence as a query.

3. Determine whether the protein sequence is similar to the translation products of any uncharacterized DNA sequences whose translations are not in nr. Perform a TBLASTN search of the EST, STS, GSS, and HTGS databases.

4. After the initial search, perform a monthly search of the month nucleotide and protein sequence databases to provide information about newly added sequences. A BLASTP search of the month protein database will reveal newly characterized protein sequences. A TBLASTN search of the month nucleotide database will reveal new open reading frames in nr, EST, HTGS, STS, and GSS. PSI-BLAST searches of nr should also be conducted occasionally, as new sequences may be added to the profile that will, in turn, help identify more distantly related sequences.

5. If the protein sequence contains multiple functional domains, it may be useful to perform the searches with each of these domains individually.

Analyzing Genomic DNA Sequences

1. Identify a potential transcribed segment. Perform a BLASTN search of the nucleotide sequence against the DNA sequences in the EST database to determine if any mRNAs are derived from the genomic sequence. EST hits will highlight the location of exons. A

Table 6.3.2 Substitution Matrices for Short Sequences

Query length (amino acids)	Matrix
<35	PAM-30
35–50	PAM-70
50–85	BLOSUM-80
>85	BLOSUM-62

BLASTN search against nr may identify transcripts or previously characterized genomic segments.

2. Identify potential open reading frames. Perform a BLASTX search against the nr protein database. Any resulting hits may identify the correct reading frame of a protein sequence and provide information about its function. If no results are found with the previous methods, a TBLASTX search (translated nucleotide query versus a translated nucleotide database) against EST, GSS, HTGS, and STS may identify other open reading frames.

3. After the initial search, a monthly search of the month nucleotide (with BLASTN) and protein (with BLASTX) sequence databases will provide information about newly added sequences.

4. Any potential transcribed sequences can be analyzed as per the above instructions for mRNA sequences.

Searching Short Sequences

Short sequences can only produce short alignments. Such short alignments need to be relatively strong (i.e., have a high percentage of matching residues) to rise above the background noise. Short, but strong, alignments are more easily detected using a matrix with a higher relative entropy than that of the default BLOSUM-62. Relative entropy is basically the average information available per position to distinguish the alignment from chance. The BLOSUM series of matrices does not include any that are suitable for the shortest queries, so it is recommended to use the PAM matrices instead. These matrices may be chosen from a menu on the Advanced BLAST page. Suggested matrices for different query lengths are shown in Table 6.3.2 (S.F. Altschul, pers. comm.).

SEQUENCE ALIGNMENT ALGORITHMS

There are two types of pairwise sequence alignments, global and local. Global strategies attempt to align along the entire length of both sequences, while local alignments focus on sequence similarity over shorter regions.

Heuristic algorithms implement certain shortcuts to allow a comparison against an entire database within a reasonable time. They are not exhaustive searches in the sense that they do not explore possibilities that seem unlikely to be interesting. Some programs allow the user to adjust parameters that increase the sensitivity of the search (i.e., explore more possibilities), but this must be balanced against the slower speed of the search. Studies have shown that heuristic algorithms, with default settings, approach the sensitivity of exhaustive searches. BLAST and FASTA are two popular heuristic programs. The rest of this discussion will focus on the newest version (2.0) of BLAST.

BLAST calculates a statistical significance for every alignment it produces. Significance is expressed in terms of the E value, which is the number of matches with a given score (the calculation of the score is described below) that one would expect by chance. The E value depends on the size of the database and the length of the query, as increasing either of these parameters increases the number of chance matches. Calculation of the E value is performed with Karlin-Altschul statistics (Karlin and Altschul, 1990, 1993).

The raw score of an alignment is calculated with a scoring system, which is a table of residue substitution scores and penalties for the existence and extension of a gap. For normal protein/protein comparisons, a matrix (i.e., table of residue substitution scores) provides a score for each possible alignment of two amino acids. For example, L aligned with I gets a positive score, while L aligned with E gets a negative score. BLOSUM-62 is the default matrix, and experimentation has shown that it is among the best for detecting weak protein similarities. PSI-BLAST uses a table that lists the frequency of finding a certain residue at each position in a conserved protein domain. This table is calculated "on the fly" for each iteration from the best matches to the database. BLASTN uses a simple scheme where matches have a positive score and mismatches have a negative score. Gaps are opened or extended when one sequence is longer than another over a certain region. BLAST charges the score $-a$ for the existence of a gap and $-b$ for each letter in a gap, so that a gap of length k letters is penalized $-(a + bk)$. Gap costs of this form are known as affine gap costs. These gap costs may be changed by the user.

The scoring system determines the raw score of an alignment. A bit score can be calculated from the raw score using the following formula:

$$S_{\text{bit}} = [\lambda \times S_{\text{raw}} - \ln(K)]/\ln(2)$$

where λ and K are Karlin-Altschul parameters that depend on the scoring system. The statistical significance corresponding to the resulting bit score is independent of the scoring system used. This allows one to compare the bit scores obtained from two different BLAST runs, each performed using different matrices or gap extension and existence values. Nevertheless, it is more meaningful to compare the results of different BLAST runs using their E values rather than their bit scores, as the E value takes into account the size of the database.

A word-based approach is used to find matches in BLAST. An ungapped search, with BLASTN, may be used to demonstrate the concept of a word. First a list is made of all the subsequences in the nucleic acid query sequence that are of a certain length (called the word size), and the positions are recorded. The database is then scanned, looking for words that are in the list. If a match is found between a query and a database word, the alignment is extended (without allowing gaps) until the corresponding score drops a certain amount below the maximum found, at which point that maximum score (and the corresponding alignment) is used. The aligned region is a local alignment. With BLASTN, only exact matches are extended and the default word size is 11 bp.

The procedure is more complicated for the other programs (BLASTP, BLASTX, TBLASTN, and TBLASTX). Exact word matches are not required and the default word size is three residues. A match between a query word and a database word is based on the scoring matrix, which specifies whether two residues result in a positive, negative, or neutral match. If three residues in a query word are compared with three residues in a database sequence, and the score is above a certain threshold, this word warrants further consideration. The default behavior in BLAST version 2.0 is that two words (on the query) must match with two words (on the database sequence) within a window of 40 residues before the match is extended, as described for BLASTN. The effect of looking for two word matches is to eliminate random matches, increasing the speed of BLAST by a factor of three.

Ungapped alignments are used as a starting point for the gapped alignments that BLAST produces. The eleven highest-scoring letters are determined and the center letter is used as the start of the gapped alignment. The alignment is then extended, with gaps allowed. This extension is not exhaustive; if the score drops by more than a certain amount from the maximum found so far, that maximum (and the corresponding alignment) is used.

The default BLAST parameters are tuned for a balance of speed and sensitivity. It is possible that significant matches may be missed, especially if small databases are searched. In that case, it may be advisable to use a shorter word size for BLASTN or a lower-threshold word for the other programs.

It is important to remember that BLAST calculates similarity. Homology is an evolutionary relationship that must be proved by further analysis.

References: Zhang and Madden, 1997; Zhang et al., 1998

Contributors: Tyra G. Wolfsberg and Thomas L. Madden

APPENDIX A: BLAST PARAMETERS

The Advanced BLAST page offers the possibility to change a number of BLAST parameters. The most important options supported are listed below.

ORGANISM (pull-down menu) limits the search by organism. A pull-down menu provides a list of the most common organisms. Others may be entered in a text-box.

EXPECT (pull-down menu) also known as the *E*-value, the statistical significance threshold for reporting matches; the default value is 10, such that 10 alignments are expected to be found merely by chance, according to the stochastic model of Karlin and Altschul (1990). Lower EXPECT thresholds are more stringent, leading to fewer chance matches being reported. Fractional values are acceptable.

FILTERING (checkbox) masks off segments of the query sequence that have low compositional complexity and/or that contain human repeats. Compositional complexity is determined by the SEG program or, for BLASTN, by the DUST program (R.L. Tatusov and D.J. Lipman, pers. comm.). Filtering is only applied to the query sequence (or its translation products), not to database sequences. By default, a filtered sequence is represented in the final BLAST report as a string of n for nucleotide queries or X for protein (e.g., nnnnnnnnnn or XXXXXXXXX). An option to mask low-complexity or human repeat sequence only for the lookup table is also available on the Advanced BLAST page. When this box is checked, the sequences are masked only during the initial phase of the BLAST search, and are not replaced by n or X in the BLAST report.

GENETIC CODE (pull-down menu) selects the genetic code to be used for a BLASTX translation of the query. Default value is the "universal" code. The choice of code is determined by the source of the query sequence (e.g., species specific, mitochondrial).

MATRIX (list) allows one to change the matrix from the default of BLOSUM-62. This is useful for searches against short sequences (see Searching Short Sequences). It is also possible to change the gap existence and extension penalties, but it is recommended to use the defaults for a given matrix.

GRAPHICAL OVERVIEW (checkbox) shows an overview of the database sequences aligned to the query sequence. The score of an alignment is indicated by one of five different colors, which divides the range of scores into five groups. If more than one alignment is displayed for a database sequence, the two alignments are connected by a cross-hatched bar. Placing the cursor over a match causes the definition and score to be shown in the window at the top; clicking on a bar takes the user to the associated alignments. The overview is "on" by default.

DESCRIPTIONS (pull-down menu) restricts the number of one-line descriptions to the number specified. The default is 100.

ALIGNMENTS (pull-down menu) restricts the number of database hits for which alignments are shown. Several alignments may be associated with one database sequence, so the number of alignments may actually be larger than this number. The default is 50.

APPENDIX B: SEQUENCE IDENTIFIER SYNTAX

The syntax of sequence header lines used by the NCBI BLAST server depends on the database from which each sequence was obtained. Table 6.3.3 lists the identifiers for the databases from which the sequences were derived. For example, an identifier might be *gb|M73307|AGMA13GT*, where the gb tag indicates that the identifier refers to a GenBank sequence, M73307 is its GenBank accession number, and AGMA13GT is its GenBank locus.

NCBI assigns gi identifiers for all sequences contained within NCBI's sequence databases. The gi identifier provides a uniform and stable naming convention whereby a specific sequence is assigned its unique gi identifier. If a nucleotide or protein sequence changes, a new gi identifier is assigned, even if the accession number of the record remains unchanged. Thus, gi identifiers provide a mechanism for identifying the exact sequence that was used or retrieved in a given search.

For searches of the nr protein database where the sequences are derived from conceptual translations of sequences from the nucleotide databases the gi syntax is *gi|gi_identifier*. An example would be *gi|451623 (U04987) env [Simian immunodeficiency*..., where 451623 is the gi identifier and U04987 is the accession number of the nucleotide sequence from which it was derived.

Users may select the *-gi* option for BLAST output, which will produce a header line with the gi identifier concatenated with the database identifier of the database from which it was derived. For example, *gi|176485|gb|M73307|AGMA13GT* would be used for a match from a nucleotide database, and *gi|129295|sp|P01013|OVAX_CHICK* for a protein database.

Table 6.3.3 Identifier Syntax for Sequence Databases

Database	Identifier Syntax[a]
GenBank	gb | accession | locus
EMBL Data Library	emb | accession | locus
DDBJ (DNA Database of Japan)	dbj | accession | locus
NBRF PIR	pir || entry
Protein Research Foundation	prf || name
SWISS-PROT	sp | accession | entry name
Brookhaven Protein Data Bank	pdb | entry | chain
Patents	pat | country | number
GenInfo Backbone Id	bbs | number
General database identifier	gnl | database | identifier
NCBI Reference Sequence	ref | accession | locus

[a]The bar ("|") separates different fields; in some cases, a field is left empty, even though the original specification called for including this field. To make these identifiers backwards-compatible for older parsers, the empty filed is denoted by an additional bar ("||").

6

The gnl (general) identifier allows databases not listed in Table 6.3.3 to be identified with the same syntax. An example here is the PID identifier *gnl|PID|e1632*. PID stands for Protein-ID, and the e in e1632 indicates that this ID was issued by EMBL. As mentioned above, use of the *-gi* option produces the NCBI gi (in addition to the PID), which users can also use to retrieve sequences of interest.

Accessing the Human Genome

The majority of the sequence for the human genome is now available. Regardless of the researcher's area of interest, it is quite likely that they will want to use some aspect of this data; however, there are several challenges to doing this effectively:

1. There are >4 Gb raw sequence data.

2. The finished and draft sequence data are of varying quality.

3. There are frequent updates of both sequence and map data.

4. Multiple groups provide views of the data, each generated semi-independently.

5. There are diverse map coordinate systems.

6. There are diverse data types, including gene function and phenotype information.

Data of this complexity present challenges to groups that distribute the data and resources for public use, as well as to the users of these resources. The providers must define adequate processes to track, analyze, and integrate information that are easy to use by scientists with a wide variety of research interests. The research community that uses this information needs to be familiar with the complexities and possible limitations of the data in order to effectively use the different Web sites presenting data.

Due to the dynamic nature of Web sites, some of the information presented here may change rapidly. Thus, while the authors provide a general overview of many processes and sites, more current information may be available at individual Web sites (Table 6.4.1). To assist in a general understanding of the terminology presented here, please consult the following glossary:

BAC (bacterial artificial chromosome) commonly used clone type for the human genome project. BAC vectors can hold large inserts, typically 80 to 200 kb, and propagate in *E. coli* as a single-copy episome.

BLAST (basic local alignment search tool) a method for performing sequence comparisons. Either protein sequences or nucleotide sequences can be used.

BLAT hashing algorithm to allow rapid searching of large amounts of genome sequence. Also allows for accurate alignment of transcribed sequences by looking at splice site information.

e-PCR (electronic PCR) program that searches a given sequence for the presence of primer pairs. These primers must be in the proper orientation and a specified distance apart to define a match.

ExoFish a technique that took whole genome shotgun (WGS) reads from the puffer fish, *Tetraodan nigroviridis*, and aligned them to the human genome in an effort to annotate genes based on sequence conservation.

Fingerprint the pattern of bands produced by a clone when restricted by a particular enzyme, such as *Hind*III.

Table 6.4.1 Useful Web Addresses and FTP sites

Website	URL
UCSC Genome Browser	*http://genome.ucsc.edu*
Sequencing Information	
Celera Genomics	*http://www.celera.com*
Analysis Tools	
BLAT	*http://genome.ucsc.edu/cgi-bin/hgBlat?command=start*
BLAST	*http://www.ncbi.nlm.nih.gov/BLAST*
E-PCR	*http://www.ncbi.nlm.nih.gov/genome/sts/epcr.cgi*
RepeatMasker	*http://repeatmasker.genome.washington.edu*
Sim4 (mRNA to genomic alignment tool)	*http://globin.cse.psu.edu*
Spidey (mRNA to genomic alignment tool)	*http://www.ncbi.nlm.nih.gov/spidey*
SSAHA	*http://www.sanger.ac.uk/Software/analysis/SSAHA*
Maps	
BAC FingerPrint Map	*http://genome.wustl.edu*
Other Annotation Sources and Viewers	
Celera Genomics	*http://www.celera.com*
DAS	*http://www.biodas.org*
The Genome Channel	*http://compbio.ornl.gov/channel*
Incyte Genomics	*http://www.incyte.com*
FTP sites	
Ensembl	*ftp://ftp.ensembl.org/pub/current_human/data*
NCBI	*ftp://ftp.ncbi.nih.gov/genomes/H_sapiens*

FISH (fluorescent in situ hybridization) in this case, BAC clones are fluorescently labeled and hybridized to chromosome spreads. In this way a clone can be mapped to a discrete cytogenetic band.

Gi an integer assigned by GenBank/EMBL/DDBJ to uniquely identify a particular instance of a sequence in the repository.

LocusID unique identifier given to all of the transcripts, proteins, and models associated with a given locus.

RH (radiation hybrid) map a physical mapping method that estimates linkage and distance relative to radiation-induced chromosome breakage. This is analogous to meiotic mapping.

Sequence searching algorithms also known as hashing algorithms (e.g., BLAT, SSAHA). This means the programs divide the genome into defined "words," usually of 10 to 12 letters. These words are then indexed based on their location and stored in memory. These algorithms allow for a very rapid search of the genome; however, the result of this speed is a loss of sensitivity. Thus, hashing algorithms are very useful for performing DNA searches of highly related DNA. When performing searches with more highly divergent sequences, it is better to use the slower, but more sensitive BLAST search.

SNP (single nucleotide polymorphism) when comparing the same sequence from two individuals, there can often be single base-pair changes. These can be useful genetic markers.

SSAHA a hashing algorithm developed for rapid searching of large amounts of genome sequence.

SSLP simple sequence length polymorphisms.

STS (sequence tag site) short sequences (200 to 500 bp) are produced in general. Oligonucleotide primers are generated such that this sequence can be amplified using PCR to produce a discrete band when analyzed by electrophoresis.

WGS (whole genome shotgun) a sequencing method by which an entire genome is cut into chunks of discrete sizes (usually, 2, 10, 50, and 150 kb) and cloned into an appropriate vector. The ends of these clones are sequenced. The two ends from the same clone are referred to as mate pairs. The distance between two mate pairs can be inferred if the library size is known and has a narrow window of deviation.

THE NATURE OF THE DATA

The Input

Two methods have been used to sequence the human genome. A commercial company, Celera Genomics (*http://www.celera.com*), employed whole genome shotgun (WGS) sequencing. This is a nonhierarchical approach that involves sequencing both ends of clones that have been obtained from libraries with inserts of different defined sizes. This sequence data and assembly is available only by subscription and is not reviewed here. The publicly funded Human Genome Sequencing Consortium used a hierarchical approach based on sequencing individual clones and then assembling the sequence based on sequence overlap and map position of the clones.

Publicly funded sequencing centers are required to deposit data from clone-based sequencing projects into public databases (GenBank/EMBL/DDBJ) as soon as contigs are ≥ 2 kb (*http://www.hugo-international.org/hugo/bermuda.htm*). The sequences submitted by large-scale sequencing centers include keywords (Table 6.4.2) that describe the status of the sequence. These keywords support retrieval of specific types of records (e.g., finished, different draft stages) using term-based searches. Sequencing centers also submit the quality scores associated with each base, information that can be quite useful when trying to assess the validity of sequence alignments.

Table 6.4.2   High Throughput Genome Sequencing (HTGS) Keywords

Phase 0	Light pass coverage of a clone. Generally only $1\times$ coverage.
Phase 1	$4 \times -10 \times$ coverage of a BAC clone. Order and orientation of the fragments unknown.
Phase 2	$4 \times -10\times$ coverage of a BAC clone. Order and orientation of the fragments is known.
Phase 3	Completely finished sequence
HTGS_draft	A draft project is either phase 1 or 2 that has exceeded a specified quality standard. Generally, this translates to 3- to 4-fold sequence coverage of the BAC clone in high quality bases.
HTGS_fulltop	Added to a record when the center responsible for finishing the clone has added sufficient new shotgun coverage for their finishing process to begin
HTGS_activefin	Added when the center responsible for finishing actually begins the process of finishing the sequence
HTGS_cancelled	Added to clones that will never be finished. Clones may not be finished for various reasons.

Submitted sequences are assigned an accession number, version number, and gi number. Both the version and gi change when an updated sequence is submitted (the version increments and a new unique gi number is provided); however, the accession number is a stable identifier for the sequence record and will always retrieve the most current record. The combination of accession and version numbers, or the gi number by itself, provides unambiguous identification of a sequence, as it exists at a specific point in time.

Draft sequence is first submitted as unordered unoriented fragments separated by gaps of unknown length. These sequence records may be updated many times before reaching a finished state. Sequence updates vary considerably and may include significant changes, including merged fragments, reordered or reoriented fragments, new sequence, deletion of contaminating or duplicated sequence, or split fragments. In addition, accession numbers can be completely withdrawn by the sequencing centers if a significant problem is found (e.g., rearranged clones). Contaminants (e.g., DNA from *E. coli*, phage, and other organisms being sequenced) can inadvertently be included in the submitted sequence of a particular clone either through laboratory error (incorrectly labeling a plate or tube), sequencing error (lane tracking), or associating a sequence with the wrong identifier in a database. As these errors are detected, they are removed in subsequent updates to the sequence. Thus, when using sequence data, it is critical to note the accession *and* version (or gi) numbers of the sequence entry. If the data change, previous versions can then be retrieved and used to identify the differences.

ASSEMBLING THE SEQUENCE

Ideally, the sequence of the human genome will be available as finished chromosomes; however, this is not possible until there is a complete finished sequence of the human genome. Indeed, even chromosomes 20, 21, and 22, all of which have been represented as "finished," still contain sequence gaps due to current limitations in cloning and sequencing technology. Over 95% of the human genome is already on hand as a combination of finished and draft sequence records and thus a considerable amount of information is available for use today; however, use of these data in their original submitted form is hampered by several factors:

1. The volume of data.

2. The draft sequence includes fragments in unknown order.

3. There exists redundancy in submission.

4. There exists only inconsistent or partial annotation.

The availability of assembled sequence has several benefits. Assembly reduces redundancy and can help eliminate many clone-level errors, such as contamination or improper annotation. Most notably, an assembly increases the likelihood of identifying gene structures as the overall length of contiguous sequence is significantly increased.

Presently, NCBI generates the publicly available genome assembly that is displayed in the NCBI, University of California-Santa Cruz (UCSC), and Ensembl resources described below; however, for over 1 year there were two assemblies, one generated by UCSC and used in their browser and the Ensembl browser, and one assembly generated by NCBI and displayed in the NCBI Map Viewer. Figure 6.4.1 illustrates the NCBI general assembly and annotation process flow.

When assessing the publicly available genome assembly, the user should be aware that there are several complicating factors to be dealt with in a clone-by-clone sequence assembly, including:

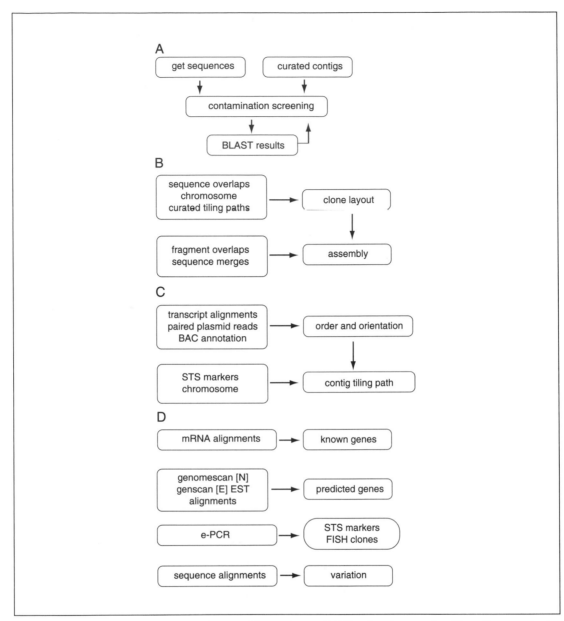

Figure 6.4.1 Genome assembly and annotation process. (**A**) The data freeze algorithm. On a particular date, a list of gi numbers is generated that represents a snapshot of HTG sequence on that day. In addition, hand-curated contigs representing finished sequences are also "frozen." These sequences are screened for contaminants. The resulting sequences are used to generate a set of "hit" files representing BLAST results using the HTG sequence. (**B**) The contig assembly algorithm. Using overlap and map information, a clone layout is assembled. The sequence from these clones is merged where there are sequence overlaps to generate assembled contigs. Linking information can be used to organize contigs into scaffolds. (**C**) The Contig placement step. Alignments are used to order and orient the assembled pieces. STS content information is used to place and orient contigs along the chromosome. (**D**) The annotation step. Features (right column) are annotated on the assembled sequence using alignment and computational methods (left column).

1. Clone mapping errors.

2. Most fragments in the available draft sequence are unordered and unoriented.

3. The many small, interspersed repetitive elements.

4. Numerous large-scale segmental duplications.

5. Large- and small-scale polymorphism.

Assembly of these clones into a nonredundant reference sequence is confounded by the problems above. One way to facilitate sequence comparisons is to segregate clones into defined bins. It is clear the early attempts to assemble the genome gave rise to different products; however, refinement of heuristics, the addition of more finished sequence, and the incorporation of clone tiling path information caused the two assemblies to converge to the current single assembly. The clone tiling paths are provided by chromosome editors and are based on a variety of experimental mapping efforts. Quality control checks, manual review, and regular communication among NCBI, UCSC, Ensembl, and the chromosome editors further improve subsequent assemblies.

Differences may be observed between independently generated assemblies, or assemblies generated by the same group over time. These differences arise because assemblies are generated at different times (called freezes) and thus start with different versions of the underlying sequence which are updated daily by genome sequencing centers. In order to perform a meaningful comparison of any assembly, the set of specific sequences used (i.e., represented as gi or accession.version) must be known. In general, a freeze is associated with a particular calendar date. Even if the sequencing center updates the sequence record and that sequence record is assigned a new version and gi, the data in the freeze will remain unaffected.

ANNOTATION

Annotation represents the location of sequence landmarks and includes information ranging from the location of repeats, CpG islands, genes, variation, other markers, and binding sites to regions encoding protein domains or structural motifs. A great deal can be gained by automating the annotation of the human genome to place most of the commonly used features onto the genome.

There are numerous annotation efforts underway by both public resources as well as commercial vendors. Public annotation efforts include those available through the Distributed Annotation System (DAS), the Genome Channel, Ensembl, the UCSC Genome Browser, and the NCBI Map Viewer.

Features of Annotations

Ensembl, the UCSC Genome Browser, and the NCBI Map Viewer use automated systems that place various features into nucleotide coordinates. Both Ensembl and the NCBI have evolving internal annotation pipelines. The UCSC site incorporates data from numerous individuals on a collaborative basis and is amenable to displaying differing types of annotations. In addition, UCSC and Ensembl exchange many data sets.

The algorithms behind the generation of many of the features are consistent among the three sites. The program RepeatMasker is used to identify common repetitive elements scattered throughout the genome (A.F. Smit and P. Green, unpub. observ.). Sequence Tagged Sites (STSs) are placed on the sequence using the program e-PCR. In addition, UCSC and Ensembl use BLAT to place the sequences from which many of the STS markers are derived on the genome. Placement of STSs on the genome is critical for the integration of sequence data

6

with other nonsequence-based maps, such as genetic and radiation hybrid (RH) maps. Many features, such as single nucleotide polymorphisms (SNPs), expressed sequence tags (ESTs), mRNAs, mouse WGS reads, and genomic clones can be placed on the genome assembly by using standard DNA sequence alignment methods such as BLAST, BLAT, or SSAHA (see glossary at beginning of unit, Sequence Searching algorithms). The integration of cytogenetic data is achieved through alignment of sequence from clones that have been FISH mapped. The integration of nonsequence-based maps with the sequence provides a powerful mechanism to access the relevant sequence, based on marker or cytogenetic data.

The identification of known genes within the genome assembly provides critical landmarks and functional context to the sequence data, which in turn makes it easier to traverse to other rich sources of gene and protein information, including publications, OMIM, RefSeq, SWISS-PROT, and LocusLink. The promise inherent in the human genomic sequence is to identify new genes. The quality of gene annotation must be evaluated in terms of how well the method addresses difficulties such as unfavorable signal-to-noise ratio (coding sequences are minute in the landscape of genomic DNA), repetitive regions around splice sites, pseudogenes, gene families, alternative splicing, sequencing errors/gaps in the sequence, genome assembly errors, and mRNA sequence errors.

These problems make identifying known genes in the genome nontrivial and the accurate identification of novel genes extremely difficult.

The process in which these annotation pipelines differ the most lies in the generation of gene models. The three groups use different approaches, as well as different alignment and prediction programs, to carry out gene annotation. In general, these different approaches make use of mRNA alignments, protein alignments, and prediction programs at different points in the pipeline. The resulting annotated genes are often called "models" as they are derived in bulk and represent an interpretation of the genome sequence data. Gene models may also have corresponding transcript and protein models; sometimes multiple transcript and protein models, representing alternate splicing, are provided for a single gene.

UCSC Gene Models

UCSC is not generating gene models, assigning identifiers, or tracking genes from one build to the next, although UCSC does provide the user with much of the underlying support to do this. Presently, all available mRNAs (RefSeqs and GenBank/EMBL/DDBJ) and ESTs are aligned to the genome. ESTs with evidence of splicing are discriminated from ESTs with no such evidence. In addition, the output of several gene prediction programs (e.g., GenScan, Fgenesh++, Ensembl, Acembly) as well as additional features of the genome are contributed by collaborators and are available in the UCSC Genome Browser.

Ensembl Gene Models

Ensembl uses a hierarchical approach to identifying gene models that utilizes the most reliable information first, and then works down to models that are purely predicted. These models are based on protein alignments rather than nucleotide alignments. Initially, Ensembl tries to identify known genes by identifying the best protein alignment in the genome. These protein alignments are then refined using a program known as GeneWise. This program is a collection of algorithms that uses mRNA alignment information and attempts to define splice sites and UTRs to generate a gene model. The Ensembl pipeline then looks at regions of the genome for which there is no known gene but for which there are strong protein hits. GeneWise is then used on these regions to produce gene models.

Table 6.4.3 Accession Formats

Accession no.	Sequence type	FTP site
NCBI Accession Format		
NC_123456	Chromosome	*ftp://ncbi.nlm.nih.gov/genomes*
NT_123456	Genomic Contig	*ftp://ncbi.nlm.nih.gov/genomes/H_sapiens*
XM_123456	mRNA, model	*ftp://ncbi.nlm.nih.gov/genomes/H_sapiens/RNA*
XP_123456	Protein, model	*ftp://ncbi.nlm.nih.gov/genomes/H_sapiens/protein*
XR_123456	Untranslated RNA, model	*ftp://ncbi.nlm.nih.gov/genomes/H_sapiens/RNA*
NG_123456	Genomic Region, curated	*ftp://ncbi.nlm.nih.gov/refseq*
NM_123456	mRNA, Known Gene	*ftp://ncbi.nlm.nih.gov/refseq*
NP_123456	Protein, Known Gene	*ftp://ncbi.nlm.nih.gov/refseq*
Ensembl Accession Format		
ENSG0000XXXX	Genes	*ftp://ftp.ensembl.org/pub/current/data/fasta/cdna*
ENST0000XXXX	Gene transcripts	*ftp://ftp.ensembl.org/pub/current_human/data/fasta/cdna*

GenScan is used to produce gene models, in regions for which there are no protein-derived annotations (Burge and Karlin, 1997). After all of these gene models are produced, the protein sequences are compared to each other to produce a nonredundant set of protein models for the human genome. Ensembl is also using human curation to refine gene models. Models are given unique identifiers that facilitate tracking and database searching (Table 6.4.3).

NCBI Gene Models

NCBI also uses a hierarchical approach to produce gene models, but relies primarily on nucleotide alignments rather than protein alignments. The goal is to provide gene models based on varying levels of evidence, starting with models that represent known genes, to models that are purely predicted. One effort involves leveraging information by propagating features from manually curated genomic RefSeq records or by using a nucleotide alignment approach. Regions of finished sequence that are difficult to annotate using the automatic alignment approach, typically due to closely related gene family members, are treated manually resulting in a genomic RefSeq record (Table 6.4.3), which provides annotation to the automatic annotation pipeline. The majority of known genes, and their corresponding mRNAs and proteins, are defined by aligning the reference sequence (RefSeq) and GenBank mRNA records to the assembled genomic sequence. Alternate mRNA models derived from the available transcript sequence data are grouped under the same gene when they share one or more exons or introns on the same strand. If mRNA sequences align to more than one location on the genome, the best alignment is selected and annotated on the contig. If the alignments cannot be distinguished, gene models are produced at both locations and each model is given a different identifier.

Gene models are predicted by aligning human ESTs and mRNAs to the genome. These alignments are "chained" together when sufficient evidence suggests these transcripts are derived from the same gene. Additional predictions are derived from GenomeScan, a gene prediction program that uses a combination of gene prediction and homology assessments. The genomic sequence is segmented by putative gene boundaries, based on the alignment of RefSeq mRNAs and "chained" gene models. GenomeScan predictions that don't overlap chained models are assigned an accession number and tracked between builds.

Contig annotation points to an existing RefSeq record (NM_ accession prefix) when there are minimal discrepancies between the genomic and transcript sequences (i.e., <4 mismatches, alignment >90% of the RefSeq transcript length, and <3 small gaps). However, since it is difficult to automatically resolve sequence differences between an aligning mRNA and the genome, NCBI provides separate records with a distinct accession format (Table 6.4.3) representing the mRNA and protein sequence that are predicted on the genome. These sequences correspond to the genomic sequence, the only exception being where a small sequence difference changes the reading frame relative to the supporting mRNA and EST data; in this case, the model protein sequence is adjusted to provide the protein product that corresponds to the aligning mRNA data. Thus, a modeled mRNA or protein sequence might reflect sequence differences present in the genomic sequence, or may be the first instantiation of a previously unidentified paralogous gene. Models are tracked between builds by their associations with known genes via connections to LocusLink. When a model unambiguously corresponds to an existing known gene, it is associated with the preexisting LocusID. Novel interim LocusLink identifiers are assigned when there are ambiguities due to conflicting alignments in gene families, pseudogenes, or regions with more significant sequence differences. NCBI's RefSeq curation staff is working to resolve many of these ambiguous results.

ACCESSING THE DATA

While there are similarities between the three sites, both in terms of available content and services offered (Table 6.4.4), they do adopt distinct styles and users must learn where to click in order to carry out a query or modify a display. Thus, although all three sites provide query support, configurable displays, and data download options, there are large differences in the details of these features at each site.

Understanding the data available and how it is presented is key to accessing genome data at these three sites. Both Ensembl and UCSC focus on displaying features that have been placed in nucleotide coordinates and have adopted a convention of displaying maps horizontally, with the various features displayed as "tracks" relative to the sequence. NCBI is displaying feature information in both sequence- and nonsequence-based coordinates and provides maps in a vertical format. For example, STS features can be viewed on the sequence map (nucleotide coordinates), radiation hybrid maps (cRay coordinates), and genetic maps (centimorgan coordinates). The ability to view all coordinate systems in the Map Viewer allows users to visually inspect the maps simultaneously.

Constructing integration sequence- and nonsequence-based maps is a difficult task as the scale of these maps is quite different; however, this integration is quite useful as some features, such as phenotypes, cannot always be placed directly on the genome sequence. It is imperative to remember that at the UCSC and Ensemble sites, features will be available by accessing a particular track and at NCBI, features are accessible by viewing a particular map. Table 6.4.5 shows various features that have been placed on the genome and the identifiers under which the information can be obtained at the various sites. For the NCBI site, the coordinate system of the particular map is shown in parentheses. Also, map types for which there are multiple reference types are shown only under their generic name (such as Radiation Hybrid map).

The three Web sites discussed here have many common characteristics. Table 6.4.5 shows a breakdown of the various features that have been indexed for simple searching.

UCSC

The Genome Browser is based upon the idea of simplifying access to the assembled human genome. All information available at this site is displayed in terms of the genome.

Table 6.4.4 Annotations Provided by the Various Browsers[a]

Feature	UCSC Genome Browser	Ensembl	NCBI
STS	STS Markers	Markers	STS (Mb), Radiation Hybrid (cRay), Genetic (cM), Whitehead-YAC Map (ordinal)
Genes	Ensembl genes, Fgenesh++ genes, full mRNA, and known genes	GenScan models, mRNA alignments, and transcripts	Genes_Cytogenetic (cM), Genes_Sequence (Mb), GScan (Mb; prediction), and RNA (Mb)
Cytogenetic	Cytogenetic band	CytoView	Ideogram (cytogenetic band), FISH Clone (cytogenetic bands), and Mitelman Breakpoint (cytogenetic bands)
ESTs	ESTs and spliced ESTs	EST Alignments	UniGene (Mb)
SNPs	Overlap SNPs and random SNPs	SNP	Variation (Mb)
BAC clones	Coverage, FPC Contigs, and Tiling Pathm	Tiling Path	GenBank Map (Mb)
Phenotype	NA	NA	Morbid (cytogenetic bands)
GC content	GC percent	GC content	NA
Repeats	RepeatMasker and simple Repeats	Repeats	NA
Mouse WGS reads	MouseBLAT and Exonerate	mouse traces	NA
RNA genes	RNA genes	tRNAs	NA
CpG Islands	NA	CpG Islands	NA
Sequence Assembly	Assembly and gaps	Contigs	Contigs (Mb)
Mouse Conserved Synteny	Mouse conserved synteny	NA	Separate Web site[b]
Tetraodon WGS	ExoFish	NA	NA
User Defined Annotation	Add user-defined tracks	DAS	Third party annotation

[a] Abbreviations: cM, centimorgan; Mb, megabase; NA, not applicable.

[b] See *http://www.ncbi.nlm.nih.gov/Homology*.

A unique aspect of the Genome Browser is that previous versions of the genome assembly are readily accessible. In addition, a program is provided that will convert the coordinates between two different assemblies. This is quite useful due to the way the UCSC browser receives its genome annotation. Since much of the annotation is provided from outside sources, all tracks are not always available on the most recent assembly. In addition, this feature allows users to directly see how the genome assembly has changed between versions.

Text-based queries

Queries in the Genome Browser are straightforward and intuitive. Queries are performed on the Genome Browser home page by entering a request in the text box labeled "genome position." Text-based queries can be performed using a number of different types of terms, such as

Table 6.4.5 Services Available at Each Web Site

Service	UCSC	Ensembl	NCBI
Query			
By position in genome	+	+	+
Nucleotide query (by alignment)	+	+	+
Protein query (by alignment)	+	+	+
Text	+	+	+
Text, advanced	−	+	+
Display Data			
Annotated feature sequence	−	+	+
Assembled sequence	+	+	+
Graphical	+	+	+
Tabular	−	+	+
Download			
Other Map data for region	−	−	+
Sequence region	+	+	+
Upload Custom Map	+	+	−
Change Display Configuration			
Add/Remove tracks/maps	+	+	+
Change order of track/map	−	−	+
Jump to different chromosome	+	+	+
Scroll along chromosome	+	+	+
Scalebar (ruler)	+	+	+
Specify coordinates to view	+	+	+
Zoom	+	+	+
FTP			
Assembled sequence	+	+	+
Map location, nonsequence based	−	−	+
Map location, sequence based	+	+	+
Model mRNA sequence	−	+	+
Model protein sequence	−	+	+
Links			
FAQ	+	+	+
Help documentation	−	+	+
Jump to other browser site	to Ensembl	to UCSC	−[a]
Statistics	+	+	+
View Previous builds	+	−	−

[a]Links to UCSC and Ensembl browsers are provided through LocusLink.

gene name, accession number, author name, or gene family name. Each word in the query is considered separately. For a single word, any item in the database that includes the word as a substring is considered a match. When a query contains multiple words, only items that match all words are returned; however, the browser does not allow location- and text-based queries to be mixed. Thus, while one can ask for all of the zinc fingers in the genome, one cannot specifically ask for all of the zinc fingers on chromosome 19.

Results. Many queries will return more than one result. Result formats in the Genome Browser depend on whether a single entity or multiple entities are returned in response to a query. Queries that return a single answer go straight to a graphical display in the genome browser. When these queries return multiple answers, a list is returned. Items in this list are hot-linked to the graphical display. The query term is not highlighted in any way when the graphical display is reached. For example, if a user queries a specific accession number, the accession list will have to be manually reviewed in order to identify the query term used.

Location-based queries

Using the same query, users can also jump to a specific location in the genome. Location queries can be based on base pair coordinates, cytogenetic coordinates, or on specific marker names. Users can jump to a specific base pair range by typing the chromosome number, a colon, and then the base pair range of interest (e.g., chr9:2954936-2989310). A single cytogenetic band can be specified to perform a cytogenetic search; however, queries for cytogenetic ranges (e.g., 5p15 or 5p15.1) are not currently supported. Cytogenetic locations are converted to base pair positions and the user is sent to this range in the browser. In order to perform a marker-based query, a single marker name is used, and the browser centers the view on this marker. Thus, range queries cannot be performed for either cytogenetic or marker-based location queries.

Results. By their nature, location-based queries should have only one result. In the Genome Browser, location-based queries automatically take the user to the graphical display of the genome.

Sequence-based queries

Sequence comparisons can be performed starting with either nucleotide or protein queries, using a program called BLAT (see glossary at beginning of unit, Sequence Searching algorithms). Archival genome assemblies are available for searching. The query options are DNA, protein, translated RNA, and translated DNA. These can either be specified, or BLAT will guess based on the sequence composition.

Results. Sequence-based queries with multiple alignments also return a list of identified matches. Clicking on the Details link brings up the alignment display where mismatches and gaps are highlighted. In addition, the genomic context of the alignment can be accessed by clicking on the browser link; this returns a graphical display of the region of the genome identified in the sequence search. The BLAT results are displayed in a track labeled Your Sequence from BLAT Search. It is possible to include multiple sequences in a single BLAT search.

Customizing the display

The Genome Browser displays the chromosome horizontally with all features displayed as "tracks" below the chromosome, regardless of strand. Navigation is very straightforward and intuitive. Buttons along the top of the window allow the user to move right or left along the chromosome, as well as zoom in or out. Clicking on the chromosome base position labels will also zoom the display in 3× and center the map at that position. Navigation control below the graphical display window allows a user to quickly jump to another chromosome or chromosomal base-pair position. In addition, buttons on either side of the display allow the user to expand or contract the display on either side. Features are displayed in tracks, which are controlled by pull-down menus located at the bottom of the page. The menu area allows the user to turn tracks on or off, or to change the display between full and dense modes.

If the track is turned on, the display can be toggled between a full or dense display by simply clicking on the track label in the graphical view. The dense display is more compact as less information is displayed. For example, a full display of the EST map includes all of the ESTs that align to a particular region. In contrast, the dense display shows an EST aligned to a region, with no indication of the number of alignments.

The UCSC genome browser allows users to add their own annotations to the display. This does not incorporate the new data into the underlying database, but does allow the user to visualize their own annotations in the context of the UCSC annotation during a particular session. User defined annotations can be provided in a variety of formats as described on the UCSC Web site (*http://genome.ucsc.edu/goldenPath/help/customTrack.html*). Users can make their annotations public by placing a file in one of the supported formats on a Web site. They can then construct links that will open up the browser with their custom tracks in place.

Downloading sequence

The sequence of the region being displayed in the browser can be easily downloaded. The blue navigation bar at the top of the graphical display page contains a link labeled DNA. The download page displays the bases that the user is requesting. The user has the option of downloaded sequence that is masked or unmasked for repeats, or which has the case or color follow any built-in track.

Data formats

In addition to the default graphical view, users can also access a tabular format for data from a particular region. The blue navigation bar at the top of the graphical display page contains a link labeled Tables. This link sends the user to a page that allows the user to formulate simple queries to the underlying database. In this form, the user can specify the genomic region of interest, the table of interest, and the particular fields in that table.

Viewing gene model evidence

The UCSC site is not generating gene models; however, it should be noted that this site is displaying alignments of sequences to the genome. Similar alignments are the basis for gene models at the other two sites.

FTP

All of the data displayed in the Genome Browser are available via FTP. In addition to downloading the assembled sequence, all of the underlying database tables that support the browser are available. These tables include all of the raw data that underlie the features that are displayed in the browser (see Table 6.4.1 for FTP information).

Ensembl

Ensembl is also largely focused on displaying features on the assembled genome sequence. This information is displayed in a variety of contexts, known as views (Table 6.4.6).

Ensembl has been developed from open-source software. Not only are the data freely available, but the entire software package can be downloaded and executed on any UNIX-based machine.

Text-based queries

At the top of the Ensembl front page a simple form labeled Search is available for text-based querying. A pull-down menu allows the user to restrict the query to a particular type of data (e.g., gene, clone, marker, disease), or to search against all available data. The search engine for Ensembl is based on AltaVista. Thus, wildcard and Boolean searches (i.e., using AND/OR) can be performed. For example, "clath* AND vesicles" is a valid search option.

Text-based queries can be limited to a single chromosome by using the Export Data option. From the Ensembl front page, the export data link will go to a query form. From this page, complex queries can be performed that allow the user to restrict the query to a particular chromosome or region of the chromosome. In addition, other restrictions can be applied to the query, such as requiring the result to be associated with a disease or a particular domain.

Results. Text-based queries are returned as lists of results. If the full data set was searched, then results are grouped by resource. The list provides a brief description of the result, with

Table 6.4.6 Ensembl Views

View	Description
AnchorView	Allows the user to select two features on the same chromosome and then displays the region of the genome between these two features
ContigView	A graphical view of the assembled genome displaying various annotated features. This view is equivalent to the UCSC Genome Browser and the NCBI MapViewer.
CytoView	A clone-oriented view of the assembled genome. The display is limited to cytogenetic information, contigs, clones, genes, and repeats.
DiseaseView	An association of phenotypes (based on OMIM), HUGO gene nomenclature, and Ensembl genes
DomainView	Index of proteins that contain similar protein domains as based on InterPro
FamilyView	Index of genes that have been classified as being in the same gene family based on an unpublished algorithm by Enright and co-workers
GeneView	A collection of all available information for each known and predicted gene. This information includes links to external resources, views of alignment evidence, and links to other Ensembl views.
MapView	There is a MapView page for each chromosome. This provides textual summary information, graphical histograms of chromosome features such as GC content and gene density, as well as quick links into other views for the chromosome.
SageView	Collection of gene expression information based on mapping of SAGE tags
TransView	View alignment evidence supporting gene models

links to the appropriate display (e.g., a hit in the Disease Index is linked to DiseaseView, while a match to the Gene Index is linked to GeneView).

Location-based queries

There are multiple ways to access the genome using a location-based query. At the top of the Ensembl front page is a pull-down menu labeled Browse Chromosome. This allows a user to specify a chromosome and a base-pair range to begin browsing. In addition, AnchorView can be accessed using the site map. The AnchorView page is a form that allows the user to specify a chromosome. Once the chromosome is specified, two features (such as bands, contigs, markers, or genes) can be specified to define a region of interest. These features need not be of the same type, but they must be on the same chromosome.

A location-based query can also be performed using the MapView pages. From the Ensembl front page, click on a chromosome. This will bring the user to a MapView page. From this page, the user has multiple options by which to define a location. Clicking on any of the images (the histograms or the ideogram) will take the user to the ContigView for that region. Alternatively, the user can define two markers. After the two markers are selected, the user can hit the Lookup button, and this will send the user to the ContigView page for the region bounded by these markers.

Results. All location-based queries are restricted to a chromosome and will take you to the ContigView of the requested region. If the region requested is large (>1 Mb), then an overview box provides granular view showing only genes, markers, and DNA contigs. Both the contig overview and the contig detailed view are shown for regions <1 Mb.

6

Sequence-based queries

Ensembl provides two different methods for performing sequence-based searches of the genome, SSAHA and BLAST (see glossary at beginning of unit).

On the Ensembl front page, there is a link to the SSAHA page. SSAHA can only be used for nucleotide queries. There is a text box in which a user can cut and paste a FASTA sequence. Alternatively, a user can upload a file from a local computer. Currently, SSAHA only searches the current assembly. There are many SSAHA options that can be adjusted.

Results. SSAHA queries return a list of alignments showing sequence identifiers, alignment positions on the subject and query, alignment orientation, alignment length, and the percent identity of the alignment. The subject sequence identifier is hot linked to ContigView; however, the alignment positions are not represented within the ContigView graphic.

A BLAST page is also accessible from the Ensembl front page. From this page, DNA or protein comparisons can be performed. Multiple databases are available for querying, including the assembled genome, confirmed or predicted cDNAs, and confirmed or predicted peptides. The usual BLAST options are available.

BLAST output is displayed in a graphical and tabular manner. Moving the mouse over the alignments in the graphical view produces a pop-up menu, from which the user gets some information about the alignment, and has the option of viewing the actual alignment, or jumping to the ContigView for the alignment. In this case, going to the ContigView produces a view that shows a graphical representation of the BLAST alignment to the genome. Viewing the alignment shows a typical, text-based view of the pair-wise alignment between the subject and the query.

Customizing the display

This category refers only to the detailed view in ContigView. Above the detailed view window are a series of buttons. The center button gives the user zoom control, which ranges from a minimum of 50 bp to a maximum of 1 Mb. Buttons are also available that allow the user to scroll either right or left a defined number of bases: a Window unit (defined by the zoom control), or 1 or 2 Mb.

The user can also control what annotations are displayed in the window. Ensembl has divided the annotations into different Categories, Features, Decorations, and DAS sources. Features include things such as mRNAs, CpG Islands, GenScan models and SNPs (to name a few). Decorations consist of calculable entities on the sequence, such as length and GC content, although the contigs and tiling path are in this category as well. DAS sources are a set of external annotations and can include some which are user-defined. For more information on DAS go to the Web site (Table 6.4.1).

Users can use these pull-down menus to turn tracks on or off. At the bottom of the features and decorations menus is an advanced users option. These bring the user to a new Web page that allows the user to define the tracks to be viewed, their color, and the extent of their display.

Moving the mouse over a displayed feature in ContigView produces a pop-up menu that will contain more information about the feature, and provide a list of relevant links. The ruler produces a pop-up menu that controls zoom navigation or allows the user to bring the display back to center.

The Jump To menu at the top of the detailed view allows the user to see other views of this region in either the CytoView or at the UCSC browser. A link to MapView is also provided in this menu, but this just returns the user to the chromosome-specific MapView page.

6

Downloading the sequence

The Export menu found at the top of the detailed view allows the user to download sequence from a region of interest. Clicking an option from this menu takes the user to a new page (ExportView) were the region can be defined and downloaded as either a FASTA file, or in EMBL Flat File format.

Display formats

The Export page described above also allows the user to export annotation information in tabular formats as lists. The user can restrict the list by using the check boxes provided in the form. For example, a user can produce a list of the known disease genes between bases 34488143 and 35488143 on chromosome 5. In addition graphical representations of the data can be exported in four different graphical formats (PNG, PS, SVG, and WMF).

Viewing gene-model evidence

Two different levels of gene-model evidence can be viewed. In GeneView, the alignment and surrounding splice site information is available. TransView provides protein similarity search data based on these models. Gene models that identify proteins from extant species are given a higher score than gene models that appear to be novel.

FTP

Ensembl software is available for download, as well as the assembled genome, cDNAs, and peptides. The data are available in both FASTA format as well as GenBank and EMBL format. In addition, all of the annotation data are available as dumps of SQL from the MySQL database that stores the Ensembl data. These tables should be easily imported into most databases (see Table 6.4.1 for FTP information).

NCBI

NCBI provides several complimentary resources that facilitate access of the human genome data starting from diverse questions. These resources represent a significant data set and are extensively cross-linked so it is easy to navigate across this information space. Together, the key related resources listed in Table 6.4.7 make it possible to retrieve information about gene function, disease associations, clone availability, expression, variation, gene structure, genomic context, and the genomic transcript and protein sequences.

The Human Genome Resources page (HGR) is the primary point of access to NCBI's suite of resources that treat human sequence data. The MapViewer resource is the primary point of access for the assembled annotated human genome data. This resource can be accessed from HGR, LocusLink, and Entrez based on alignment of accessions to the assembled sequence. Table 6.4.5 provides an overview of the services provided by the MapViewer resource including support of simple and advanced queries, sequence-based queries, and general browsing. Information is presented in two main views: query results are provided as a genome overview, and detailed information is provided on chromosome-specific map views. Both sequence and map data can be downloaded for a defined region; the full genome data set is also available on the FTP site.

Text-based query

Information is retrieved by searching with an identifier (e.g., accession or MIM number), a marker name (e.g., D7S2742), or a text term or phrase (e.g., gene symbol, gene name, disease name). Complex text queries are supported including use of multiple query terms, wildcards (e.g., *), Boolean operators (i.e., AND, OR, NOT), field restrictions, and predefined properties. Thus, it is possible to query for a gene family, to restrict the query to family members on one or more chromosomes, or to those with a known disease association. Details about the basic and advanced query options are available in the resource help documentation.

6

Table 6.4.7 Human Genome Resources Available at NCBI

Resource	Description	URL
Clone Registry	Records clone sequence status and distribution information	*http://www.ncbi.nlm.nih.gov/genome/clone*
dbSNP	Database of polymorphisms, small-scale insertions/deletions, and polymorphic repetitive elements	*http://www.ncbi.nlm.nih.gov/SNP*
Human Genome Guide	Overview of available human genome data with links to related resources and tutorials	*http://www.ncbi.nlm.nih.gov/genome/guide/human*
LocusLink	Descriptive information about genetic loci for select organisms with extensive links to related resources and sequence data	*http://www.ncbi.nlm.nih.gov/LocusLink*
Map Viewer	Integrated views of chromosome maps including the annotated genome assembly. Includes maps based on several coordinate systems.	*http://www.ncbi.nlm.nih.gov/cgi-bin/Entrez/map_search?*
OMIM	Catalog of human genes and disorders	*http://www.ncbi.nlm.nih.gov/entrez/query.fcgi?db=OMIM*
RefSeq	NCBI nonredundant reference sequence database. An ongoing curation effort and reagent for genome annotation.	*http://www.ncbi.nlm.nih.gov/LocusLink RefSeq.html*
UniGene	Clusters of related transcript sequences, with tissue expression information, and links to related resources	*http://www.ncbi.nlm.nih.gov/UniGene/index.html*
UniSTS	A unified nonredundant database of sequence tagged sites (STSs).	*http://www.ncbi.nlm.nih.gov/genome/sts.index.htm*

Results. Initial query results are shown on the Genome View page as both a graphical representation of the relative locations of the query results in the genome, and as a tabular summary of the chromosomal location, the identified match, and the specific map or maps that contain the query match. Several options are provided to view the query results in the Map Viewer:

1. Click on a chromosome of interest in the graphical overview to view all results for that chromosome.

2. Click on the Map Element link in the table to view that locus on all maps listed (e.g., a gene query may return both the cytogenetic and sequence map).

3. Click on a map name in the table to see that locus on the selected map.

Clicking on any of these elements takes the user to the Map Viewer graphical display. Once in this display, the query term will be highlighted. Users might feel the transition to the Map Viewer graphical display is a bit slow. This is caused by the many calculations that are being performed in order to display maps on different coordinate systems. Common elements on different maps are also being calculated.

Location-based query

Location queries can only be carried out from individual chromosome views; the genome overview page does not support this query. Enter locations into the Region Shown boxes located in the side column of the display page, or alternatively, this can be carried out using the Maps&Options form (see Customizing the Display below). Position queries can be done in sequence or nonsequence coordinates, including centimorgan, century, cytogenetic bands. This interface also supports entry of marker names, including gene symbols.

Results. The Map Viewer graphical display is redrawn such that the coordinates or terms provided delimit the display. In essence, this carries out a precisely defined zoom with the view centered on the region of interest. The position or terms used to define the view are not highlighted.

Sequence-based query

Users can query the genome starting with a nucleotide or protein sequence by using the human genome BLAST page, which can be accessed from the Map Viewer, the BLAST home page, or the HGR page. Several blast databases are provided to facilitate identifying matches to the genome, transcripts, or proteins. BLAST result pages have been customized to facilitate navigation to the Map Viewer or LocusLink.

Results. Following a query against the genome database, the BLAST result page includes a link to view hits in the context of the genome. Clicking on the Genome View button brings up the genome overview page with the location of the BLAST result highlighted on the chromosome overview, as well as summarized in the tabular display. The tabular summary includes a link to the Map Viewer as well as back to the BLAST alignment page (Score column). The Map Element link brings up a Map Viewer display that is zoomed in to the region containing the hit. The locations of the BLAST hits are highlighted, and a text summary indicates the percent identity and corresponding region of the query sequence. Links to view each aligned region are also provided. This provides a powerful mechanism to determine if genomic sequence is available for a newly identified cDNA, or to use sequence similarity to identify other related genomic regions, transcripts, or proteins.

Customizing the display

Chromosome-oriented graphical views are displayed in a vertical orientation. A thumbnail map, located in the left-side column, provides a rough indication of the region displayed and is used for coarse scrolling up and down the chromosome (click on a region to view). Brief labels are provided on all maps to provide landmarks at lower resolution views. Additional text information can be displayed by mousing over an object of interest. More detailed text information is provided for the right-most map (termed the Master Map). Links are provided to the sequence data and related NCBI resources to facilitate navigation to expanded information sources.

The Map Viewer provides several zoom, navigation, and simple map display controls directly on the display page, with more advanced controls provided on a separate page (e.g., click on the Maps&Options link to bring up the Maps & Options box). The level of resolution can be altered using the side column zoom control box or alternatively, by providing a numerical range or bounding markers such as marker names, gene symbols, or cytogenetic bands in the Region Shown text boxes. In addition, both zoom and navigational changes can be accomplished by simply clicking over a map at a point of interest. This brings up a menu with predefined zoom levels, the option to center the view again, or the option to open another browser window displaying the sequence. This provides a convenient approach to quickly drill down to a high-resolution view of a region of interest.

Click on the Maps&Options link to customize the Map Viewer display. This brings up a new window where the number and order of maps to display (up to a total of ten) can be

6

defined. Maps are organized by the coordinate system used to define the map (see Table 6.4.4 for available maps). For display purposes, the features annotated on the assembled sequence are treated as different map views, similar to the Ensembl and UCSC tracks. The sequence coordinates are directly comparable because all of the sequence maps are generated using the same genome assembly. The Maps&Options toolbox also provides choices to add scale rulers, view verbose or condensed text descriptions, modify the page length, or define the region shown. There is also an option to display connections between maps.

Downloading the sequence

Clicking on the link to Download/View Sequence/Evidence opens a new window where the user can redefine the size of the region and choose to download or view the assembled genomic sequence in FASTA or standard Flat File format. The Seq link on the gene display also provides access to this download capability. In addition, a link to the download page is provided in the Data as Table View (see below).

Display formats

Follow the left column link for Data as Table View to access a tabular report that corresponds to the defined graphical display. This page provides a list of all objects found in the region being displayed and includes the option to save all, or map specific subsets of the data to a local file.

Viewing gene model evidence

Two resources are provided to highlight the evidence supporting any given gene model. The ev link, provided on sequence maps (when in the Master Map position), calls up the Evidence Viewer page where more detailed information is provided on the transcript alignments that were used to define the annotated genes. This page provides a graphical representation of aligning transcripts, the sequence alignments, predicted translation, and highlights sequence differences both graphically and within the aligned sequences.

An additional display, called Model Maker (follow the mm link from the gene sequence map), shows a graphical representation of all transcript alignments (mRNA and EST), GenomeScan model, and the gene model annotated in that location on the contig. Model Maker offers a unique service in that it allows users to select individual exon pieces, as defined by the alignments, to reconstruct a model. Thus, should it be decided that a given exon should or should not be included in the proposed gene model, Model Maker can be used to reconstruct the gene to include or exclude any predicted exon.

FTP

All data available at the NCBI is in the public domain and available on the FTP site. This includes the data presented in the Map Viewer and includes sequence data, chromosome scaffold data, and map data (see Table 6.4.1 for FTP information).

HOW TO RETRIEVE INFORMATION: EXAMPLE QUESTIONS

Sample questions illustrating how to retrieve information at the Ensembl, NCBI, and UCSC Web sites are included here. It should be noted that both the Ensembl and NCBI Web sites were designed so that information can be retrieved in many ways. The answers to some of these questions may be obtained in ways other than those specified here.

How Can I Retrieve Information Using an Official Gene Symbol or Name?

UCSC
Action. From the Genome Browser home page, type the gene name into the query bar.

Result. Query results are listed with links to the genome browser. There will be no designation in the Genome Browser of which gene in the region corresponds to the query.

Ensembl

Action. From almost any page at the Ensembl site, type the gene name in the query box. The choices are to either search all of the indices, or restrict the query to a particular feature type, in this case, known genes. The query terms will not be highlighted in ContigView.

Result. A list of results itemized by resource will be retrieved. All results will be linked to the view associated with that resource (i.e. GeneView or DiseaseView).

NCBI

Action. Query LocusLink to search for expanded descriptive information associated with known genes. This can be done from the LocusLink home page, the HGR homepage, or the NCBI home page. Alternatively, query the Map Viewer to go directly to the genome location of a known gene. The query can be restricted by chromosome location or other properties. Simple queries can be carried out from many pages including the HGR homepage and LocusLink. The advanced query page is available directly from the Map Viewer.

Result. With LocusLink, a list of all records associated with the gene name is returned. The LocusID is linked to expanded gene report pages. With MapViewer, the query results are represented in a graphical genome overview and in a tabular report. Click on a chromosome or individual query result to access the detailed chromosome-specific display where query matches are highlighted.

I Have Mapped a Phenotype Between Two Markers On Chromosome 4. How Can I Determine The Physical Size of This Region and Get All of The Candidate Genes Within This Region?

UCSC

Action. Query for each marker individually and note the base-pair position of each marker. Use the base-pair position of these markers to perform a base-pair-range query (e.g., 4:567897-577898).

Result. Completion of the query will bring the user to the Genome Browser graphical display if valid base-pair coordinates were specified. To obtain a list of candidate genes in this region, use the Tables link in the blue navigation bar above the graphical display window. With this feature, the user can obtain a list of genes, or predicted genes and their locations within the candidate region.

Ensembl

Action. From the Ensembl home page, choose AnchorView. Select the marker type and type the marker name in the text boxes.

Result. If both markers have been placed on the genomic sequence, the result of this query will go to ContigView. If the size of the region is >1 Mb, only the overview will be shown. If the region is 1 Mb or less, both the Overview and Detailed View will be displayed. To obtain the entire list of candidate genes from this region, click on the Export option at the top of the detailed view menu, and select Gene List.

NCBI

Action. Query for the two markers on the Map Viewer query page using the Boolean OR (e.g., D4S2931 OR D4S819).

Result. After carrying out a query, click on chromosome 4 in the genome overview display to bring up the graphical display of all maps for which a match was found. The query term (or alias name) is highlighted on all displayed maps on which it is found. The equivalent physical location of the region in the genome assembly can be found by comparison to the STS sequence map and the Genes_seq map. If the query markers have been placed directly on the sequence map by alignment, then there is a direct correspondence between the mapped markers and the sequence coordinates. If the markers have not been directly placed on the assembled genome, then positions can be inferred by looking at the placement of neighboring markers that are placed on both an RH map and the sequence map. Candidate genes are identified by direct comparison between the Genes_seq and STS maps once the region on the STS sequence map has been determined.

How Do I Find All of The Pax Family Genes That Are Associated With a Disease And at Least One SNP?

UCSC
Action. This is a difficult task in the UCSC browser. Because wild cards cannot be used in the search engine, the user must already have a list of the Pax gene family members. From the Genome Browser front page, query for each gene.

Result. If each gene has been mapped to a single location in the genome, the query will result in a graphical view of the region surrounding the gene. To identify the SNPs in the gene, make sure the two SNP tracks (Random SNPs and Overlap SNPs) are turned on.

Ensembl
Action. From the Ensembl front page, select Export Data. On the resulting query form, select Gene List. In the Region section of the form, select Entire Genome. In the Restrict Selection section, select Disease Genes only. Scroll to the Include Disease data section, and select either the disease description or OMIM number. Scroll down the page to the Include SNP data section. In the text boxes at the top of this section, specify the distance of the flanking sequence to be searched, and check the required SNP Ids. At the bottom of the form, select the desired format.

Result. This query will result in a list of all Ensembl gene identifiers and SNP Ids. There will be one row for each SNPid returned. If HTML format is selected, the gene identifiers will be hot-linked to GeneView. If OMIM numbers were selected, they will be hot-linked to DiseaseView. The SNPids are not hot-linked, but can be cut and pasted into other SNP databases.

NCBI
Action. Use the Map Viewer Advanced Search page. Supply the wild card query PAX* and select the options to search only records With a Disease Known and With Variation Known.

Result. Ten genes are returned with this query. Click on the Map element link to bring up the zoomed chromosome display. Use the Display Settings toolbox to add the variation map.

I Have a Protein Sequence from *C. elegans*. How Do I Determine If There Is a Homolog in the Human Genome?

UCSC
Action. From the UCSC front page, select the BLAT search option. Paste the protein into to the text box, and select Protein from the Query Type pull-down menu.

Result. Protein BLAT was designed to identify sequences of 80% similarity over at least 20 amino acids. So, in general, a *C. elegans* protein may be too diverged to identify a homolog by BLAT. However, if the protein is highly conserved, a list of matches and their genome positions will be returned.

Ensembl

Action. From the Ensembl front page, select BLAST. Either a search against peptide databases (divided into confirmed or predicted peptides) or a translating search of genome can be performed.

Result. This query will return a graphic of the genome location of the BLAST hits, as well as a tabular list of hits. Placing the mouse over hits in the graphic will produce a pop-up menu that provides more information about the blast hit, as well as links to ContigView and to view the alignment. Clicking on a link in the tabular list will result in a pairwise view of the sequence alignment.

NCBI

Action. Carry out a BLAST query against the human genome. You can choose to query directly against the model protein, or to carry out a translating BLAST search against the genome assembly.

Result. If you query against the model protein database (XP_ accessions), then your query results page will include links to LocusLink; you can navigate to view the location of that locus on the genome by following the yellow Maps button link located near the top of the expanded gene report page. Queries against the genome database return results pages that provide direct links to the Map Viewer to view the results in the context of the genome.

How Do I Identify a BAC Clone That Contains My Favorite Gene?

UCSC

Action. Perform a text query using the gene name or symbol.

Result. This query will likely return a list of features that are associated with this genome (i.e. Known Genes, GenBank mRNAs). Click on the link under the Known Genes header. This will lead to a graphical view of the gene in the browser. Using the pull-down menus below this display, turn on the Assembly track and click on the refresh button. This will provide the accession number that was actually used in constructing that part of the genome. Currently, there is no way to get the actual clone name from this browser; however, the accession number can be copied and used to query the NCBI clone registry.

Ensembl

Action. From the Ensembl front page, search the gene index for the gene name or symbol.

Result. Go to the resulting GeneView page. In the genome location section, click on the link to ContigView. From the Jump To pull-down menu, select CytoView. This will provide a view of the clone set in the region containing the gene of interest.

NCBI

Action. Carry out a text query using the gene symbol or name. Alternatively, use the sequence to carry out a BLAST query against the assembled genome sequence.

Result. Once you have retrieved a chromosome display of your query result, add the GenBank map, as the Master Map, to the display using the Display Settings toolbox. When the Genes_sequence and GenBank map are aligned, with the query term highlighted, it is easy to determine which accession or accessions contributed toward the assembly in that region of the genome. The descriptive text provided for the GenBank map includes links to the Clone Registry, where distribution information is available.

I Just Cloned a Novel Human cDNA. How Do I Determine Its Location in The Genome?

UCSC
Action. Use the BLAT search option to query the genome assembly.

Result. A list of matches to the genome and their locations. This list will contain links to a view of the alignment and to the graphical view in the browser. The location of the BLAT hits will not be displayed in the browser.

Ensembl
Action. Using SSAHA, perform a sequence search against the assembled genome. Alternatively, a BLAST search can also be performed.

Result. When using SSAHA, if the cDNA is contained in the assembled genome, a list of matches to the genome and the genome location will be returned. The genome locations will be linked to ContigView, but the SSAHA hits will not be displayed. When using BLAST, a graphical view of the genome with the locations of the hits represented by arrows will be displayed, with a tabular list of alignments shown below. The links to ContigView from this page will display the BLAST hit locations as another feature in the ContigView.

NCBI
Action. Submit the cDNA sequence as a BLAST query against the genome sequencing.

Result. The BLAST result page includes a link to the Genome View where the distribution of the hit against the genome is graphically displayed. The display includes an indication of percent identity and the corresponding query coordinates. Thus, it is possible to visualize a general intron/exon structure and it is readily apparent if a section of the query sequence did not align to the genome.

FREQUENTLY ASKED QUESTIONS

Why Has My Gene Changed Chromosomes?

The location of a contig, and its annotated features, may change over time. The nature of this change varies:

1. The contig location changes slightly to a nearby position.

2. The contig moves to different arm of the chromosome.

3. The contig is relocated to a different chromosome.

4. A previously unplaced contig is localized on a chromosome.

Small order changes arise because the genome is regularly reassembled to incorporate new sequence data. Contig locations may shift as additional finished sequence becomes available and assembly gaps are filled. Larger changes may occur if the chromosomal location of the underlying GenBank sequences changes. This is infrequent but can happen for several reasons:

1. The GenBank record was updated to change chromosome location annotation.

2. A human genome project chromosome committee (or other personal communication) provided a new chromosome location.

3. The assembly process was modified to use input information differently or to incorporate new sources of chromosome location information.

Why Can I Find My Gene At One Web Site And Not Another?

Genome assemblies are based on "freezes" of the data at different points in times. All of the assemblies at NCBI are given a build number that is associated with a list of gi's from a particular date. Ensembl, UCSC, and NCBI run independent annotation pipelines, so each site could be displaying information annotation on different builds. For example, a gene may be annotated on one Web site and not the other because of a later freeze date that captured more finished sequence for that gene, or that captured a newly available defining mRNA sequence. Assembly differences across a repeat or draft sequence region can impact annotation in that region if the mRNA that defines the gene can no longer align across the region. And, small differences in each sites annotation pipeline, such as use of different alignment parameters and cut-offs, can result in some genes appearing on only one Web site.

What Does "N" Represent in the Contig Sequences?

NCBI contigs may include fragments, derived from draft sequence submissions, which can be ordered relative to other available sequence regions, but which do not yet provide fully contiguous sequence through the region. A series of 100 N's is added to the record to represent the gap position. Gaps between contigs are arbitrarily set at 10,000 nucleotides unless true gap size information is available for finished chromosomes.

Why Can't I Find My Gene In The Genome?

Failure to retrieve information based on either text or sequence based queries can occur if either the text-query term is not indexed for retrieval or if the genome sequence is simply not yet available for the gene of interest. If using a text query, consider trying alternate names as an approach to find the gene. If no significant hits are returned following a sequence-based query, then the gene of interest may be located in a region for which there is still very limited genomic sequence available. The genome sequence is not finished and thus it is possible that some cDNA BLAST queries will fail to identify the gene in the assembled genome.

References: Kent, 2002; McPherson et al., 2001

Contributors: Deanna Church and Kim D. Pruitt

UNIT 6.5

Searching the NCBI Databases Using Entrez

BASIC PROTOCOL 1

QUERYING ENTREZ

The Entrez Web interface is located at *http://www.ncbi.nlm.nih.gov/Entrez* (Fig. 6.5.1). Most of the Web pages at the NCBI Web site provide a direct link to Entrez, either in a blue bar running across the top of the page or in the left-hand sidebar. Note that the example presented

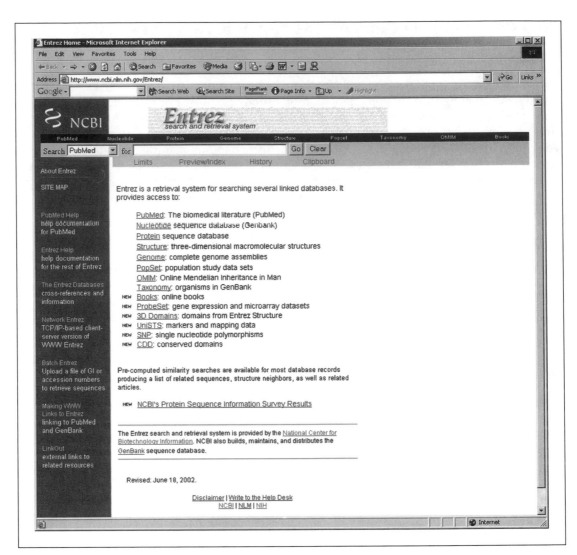

Figure 6.5.1 The Entrez Web site.

in Basic Protocol 1 searches the PubMed database, however, a similar search procedure works for any Entrez database.

Note that there is an alternative implementation to the Web-based version of Entrez, called Network Entrez. This is the fastest of the Entrez programs in that it makes a direct connection to an NCBI "dispatcher." The graphical user interface features a series of windows, and each time a new piece of information is requested, a new window appears on the user's screen. Since the client software resides on the user's machine, it is up to the user to obtain, install, and maintain the software, downloading periodic updates as new features are introduced. The installation process itself is fairly trivial. Network Entrez comes bundled with Cn3D, the graphical three-dimensional viewer that will be described later in this unit (see Basic Protocol 2). The Network Entrez implementation will not be discussed specifically here, but the logic in issuing queries and navigating through the NCBI information space is essentially identical.

Materials

An up-to-date Web browser, such as Netscape Communicator or Internet Explorer

Select and search an Entrez database

1. Begin at the Entrez home page (*http://www.ncbi.nlm.nih.gov/Entrez;* Fig. 6.5.1). For this example, select PubMed from the Search drop-down list.

2. In the For text box, enter the following:
   `atherosclerosis [MH] AND aspirin [NM].`

   Using Boolean operators such as AND, OR, *and* NOT *is the simplest way to query the Entrez system. Please note that all Boolean operators must be capitalized for the query to return the expected results.*

   [MH] *and* [NM] *in the search query are qualifying terms used to search for the subject and substance name, respectively. A list of available search qualifiers is given in Table 6.5.1. Note that when using qualifiers, the brackets ("[]") are required.*

3. Select Go. Running the query in April 2002 returned 443 papers (Fig. 6.5.2).

   The user can insert additional terms and qualifiers (Table 6.5.1) into the query to generate a more specific search.

4. Search for a specific author within the set of returned papers via one of the following methods:

   a. Select the individual page number links (Select page: 1 2 3 4 . . .) and inspect the list by eye.

   b. Select Author from the Sort drop-down list, then select Display and scroll through the alphabetical listings.

   c. Enter the Author's name in the For text box, in addition to the original search criteria (AND Smith), then select Go.

   For this step, use option "c" above. Enter AND Cayatte *in addition to the primary query. The For text box should now read:* atherosclerosis [MH] AND aspirin [NM] AND Cayatte. *Select Go.*

View an individual database record

5. Select the author hyperlink to display the Abstract view of the selected paper. The Abstract view presents the name of the paper, the list of authors, their institutional affiliation, and the abstract itself, in standard format. See Figure 6.5.3 for the Abstract view of Cayatte et al.

6. To change the display, select the drop-down list next to the Display button. Select Citation and click Display. Switching to this format produces a similar looking entry; however, the cataloging information, such as the MeSH terms and indexed substances relating to the entry, is now displayed below the abstract.

   Select MEDLINE from the drop-down list and click Display. This selection produces the MEDLINE/MEDLARS layout, with two-letter codes corresponding to the contents of each field going down the left-hand side of the entry (e.g., the author field is denoted by the code AU). Entries in this format can be saved and easily imported into third-party bibliography management programs, such as EndNote and Reference Manager.

7. Select Abstract from the drop-down list and click Display to return to the Abstract view.

8. To view the full text of an article, select the Full-Text online hyperlink located under the name of the publisher. With the proper individual or institutional privileges, the user can view the entire text of the paper, including all figures and tables.

6

Table 6.5.1 Entrez Boolean Search Statements

General syntax	
`search term `**`[tag]`**` Boolean operator`[a] `search term `**`[tag]`**`...`	
where	
[tag] =	
`[AD]`	Affiliation
`[ALL]`	All fields
`[AU]`	Author name
	`O'Brien J [AU]` *yields all of* O'Brien JA, O'Brien JB, etc.
	`` ``O'Brien J" [AU]`` *yields only* O'Brien J
`[RN]`	Enzyme Commission or Chemical Abstract Service numbers
`[EDAT]`	Entrez date
	`YYYY/MM/DD, YYYY/MM, or YYYY`
`[IP]`	Issue of journal
`[TA]`	Journal title, official abbreviation, or ISSN number
	`Journal of Biological Chemistry`
	`J Biol Chem`
	`0021-9258`
`[LA]`	Language
`[MAJR]`	MeSH Major Topic
	One of the *major* topics discussed in the article
`[MH]`	MeSH Terms
	Controlled vocabulary of biomedical terms (*subject*)
`[PS]`	Personal name as subject
	Use when name is subject of article, e.g., `Varmus H [PS]`
`[DP]`	Publication date
	`YYYY/MM/DD, YYYY/MM, or YYYY`
`[PT]`	Publication type
	`Review`
	`Clinical Trial`
	`Lectures`
	`Letter`
	`Technical Publication`
`[SH]`	Subheading
	Used to modify MeSH Terms
	hypertension [MH] AND toxicity [SH]
`[NM]`	Substance name
	Name of chemical discussed in article
`[SI]`	Secondary source ID
	Names of secondary source databanks and/or accession numbers of sequences discussed in article
`[TW]`	Text words
	All words and numbers in the title and abstract, MeSH terms, subheadings, chemical substance names, personal name as subject, and MEDLINE secondary sources
`[UID]`	Unique Identifiers (PMID/MEDLINE numbers)
`[VI]`	Volume of journal

[a]Boolean operator = AND, OR, or NOT

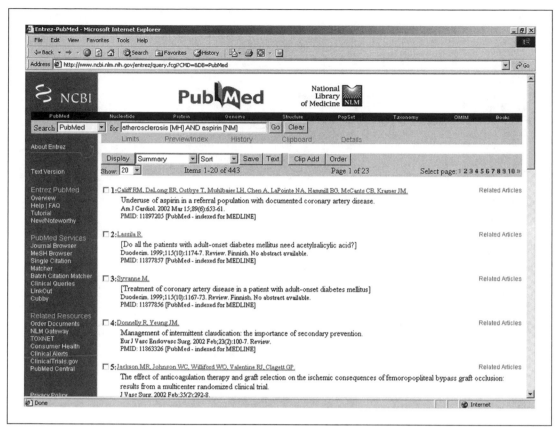

Figure 6.5.2 A text-based Entrez query using Boolean operators against PubMed. The initial query is shown in the Search text box near the top of the window. Each entry gives names of authors, title of the paper, and citation information. Clicking on the author name hyperlink can retrieve the actual record. See text for details.

Find related material

9. Select the Related Articles link on the upper right-hand corner of the abstract display (Fig. 6.5.3).

   In April 2002, Entrez indicated that there were 162 papers of similar subject matter associated with the original Cayatte reference. The list returned appears much like the one in Figure 6.5.2. The first paper in the list is the same Cayatte paper because, by definition, it was most related to itself (the "parent"). The order of the following entries is based on statistical similarity. Thus, the entry closest to the parent is deemed to be the closest in subject matter to the parent.

10. Click the Back button on the browser to return to the Abstract view.

11. Select Books. This link will take the user to a heavily hyperlinked version of the original citation.

12. Select the hyperlink for atherosclerosis. Five book thumbnails are displayed that can take the user to full-text books that are available through NCBI.

13. Select the two items hyperlink for the book Molecular Biology of the Cell. The two items that display are *Cells Import Cholesterol by Receptor-mediated Endocytosis* and *Normal and Mutant LDL Receptors*. Select the hyperlink for *Cells Import Cholesterol by Receptor-mediated Endocytosis*.

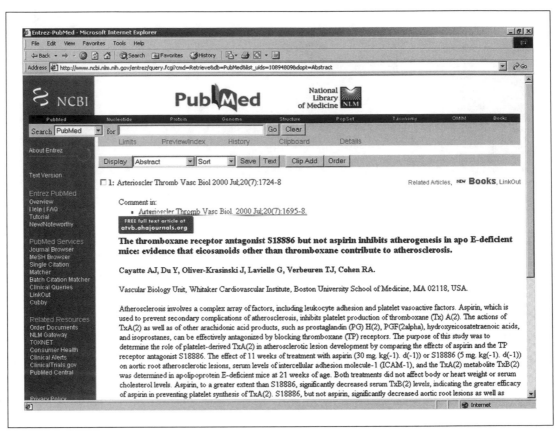

Figure 6.5.3 An example of a PubMed record in Abstract format as returned through Entrez. This Abstract view is for the selected paper by Cayatte et al., 2000. This view provides links to Related Articles, Books, LinkOut, and the actual full-text journal paper. See text for details.

14. Select the Back button three times to return to the heavily hyperlinked version of the original citation.

15. Select the final link in the series, LinkOut, located in the upper right of the display (Fig. 6.5.3).

16. Save the search results in Cubby (see Support Protocol).

SUPPORT PROTOCOL

USING CUBBY TO SAVE SEARCHES AND RESULTS

When using Entrez, Cubby is currently available in the blue sidebar when searching the following databases: PubMed, Protein, Nucleotide, PopSet, and Books.

Important Cubby Notes

1. Links to Related Articles cannot be stored as a Cubby Stored Search.

2. History numbers, often used to combine searches (e.g., #1 AND #2; see Alternate Protocol), cannot be stored in the Cubby.

3. Dates and date ranges are not recommended for stored searches.

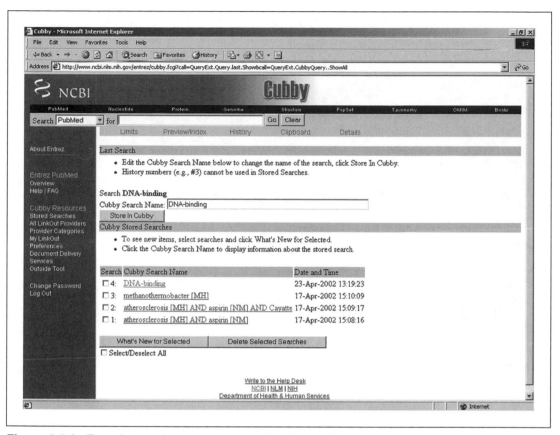

Figure 6.5.4 Entrez's search storage area, called Cubby. Entrez provides this search storage area to save queries. The saved queries can be recalled and updated without having to re-enter the query, therefore, providing an efficient way to update search results. See text for details.

4. Stored searches are numbered and listed in descending order by the date and time they were originally stored.

5. The maximum number of stored searches is 100 per user.

6. Stored searches cannot be edited.

Materials

An up-to-date Web browser, such as Netscape Communicator or Internet Explorer

Register and log in

1. After executing a search in PubMed, Protein, Nucleotide, PopSet, or Books, click Cubby from the sidebar (Figure 6.5.2).

2a. If not registered with Cubby, click Register from the I Want to Register for Cubby command.
Enter the following:

  a. **User Name** (3 to 10 characters)

  b. **Password** (6 to 8 characters)

  c. **Mother's Maiden Name** or **Pet's Name** in the event the password is forgotten.
Click Register. Once registered, you will automatically be logged into Cubby.

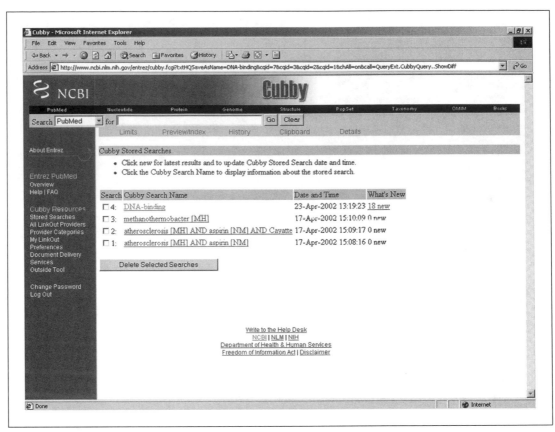

Figure 6.5.5 Cubby updates stored searches and indicates the number of new items since the last query. See text for details.

2b. If registered, but not already logged in, select Cubby from the sidebar, enter your **User Name** and **Password**, and click Login. The login will remain active for 12 hr.

Store a Cubby search

3. The most recent search displays in the Last Search section of the screen. To change the name of the search to something more manageable, edit the query in the Cubby Search Name text box.

    If the last search is not displayed, your system may not be configured to accept Cookies. Contact the System Administrator if this problem continues.

4. Select the Store in Cubby button. See Figure 6.5.4.

Retrieve and update a Cubby search

5. Select a stored search or searches by placing a check in the Search check box preceding the Cubby Search Name (see Fig. 6.5.4). For this example, check all of the Search check boxes.

6. Select the What's New for Selected button. The Cubby displays the list of selected searches, indicating the number of new items entered since the last query. If no additions have been made to the query, 0 new will display in the What's New column. See Figure 6.5.5.

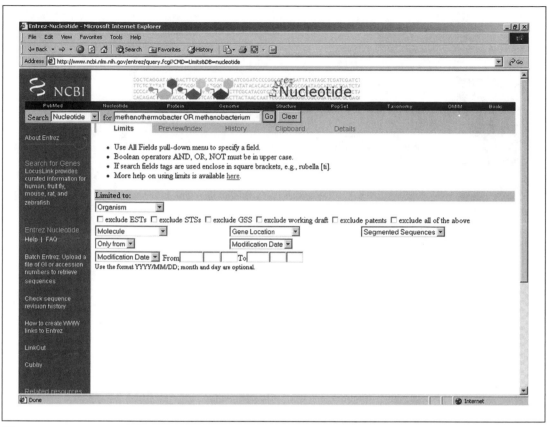

Figure 6.5.6 Using the Limits feature of Entrez to limit a search to a particular organism. See text for details.

7. To view the new information, select the number new link (in this case, for the cubby search named "DNA-binding," 18 new). The date and time of the query is now updated to reflect the current date and time. If this link is not selected, the date and time of the query will not be updated.

ALTERNATE PROTOCOL

COMBINE ENTREZ QUERIES

There is another way to perform an Entrez query, involving some built-in features of the system. Consider an example in which the user is attempting to find all genes coding for DNA-binding proteins in methanobacteria. Although this example is for the nucleotide database, the general strategy works equally well for other Entrez databases.

Materials

An up-to-date Web browser, such as Netscape Communicator or Internet Explorer

Execute multiple queries

1. Open a Web browser and go to the Entrez Web page (*http://www.ncbi.nlm.nih.gov/Entrez;* Fig. 6.5.1). Select Nucleotide from the Search drop-down list and enter the term DNA-binding in the For text box. Select Go.

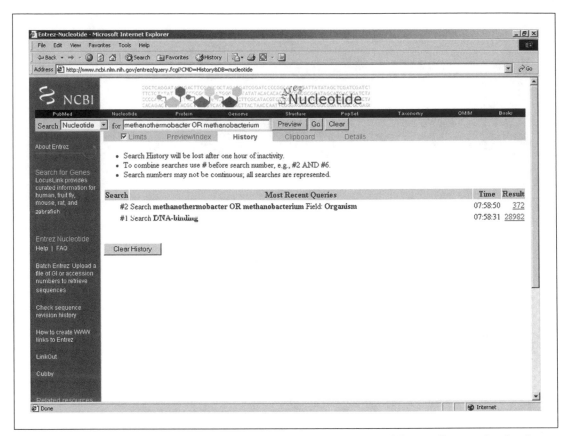

Figure 6.5.7 Combining individual queries using the History feature of Entrez. See text for details.

2. To narrow the query, select the Limits hyperlink, which is located directly below the For text box.

3. To limit the search by organism, select Organism from the Limited to drop-down list (Fig. 6.5.6).

4. Enter `methanothermobacter OR methanobacterium` in the For text box (Fig. 6.5.6).

5. Select Go.

Combine selected queries

6. Click the History hyperlink, located below the For text box. The History page displays the user's most recent queries (Fig. 6.5.7).

   The list shows the individual queries, whether those queries were field-limited, the time at which the query was performed, and how many entries that individual query returned.

7. To combine the two queries into one query, use their query numbers. `Enter #1 AND #2` in the For text box. Click Preview to regenerate a table, showing the new, combined query as `#3`, containing three entries. Click Go to show the three entries in the nucleotide format (Fig. 6.5.8). As in Basic Protocol 1, there are a series of hyperlinks to the upper right of each entry, four are shown for the first entry, *Methanobacterium thermoautotrophicum* tfx gene.

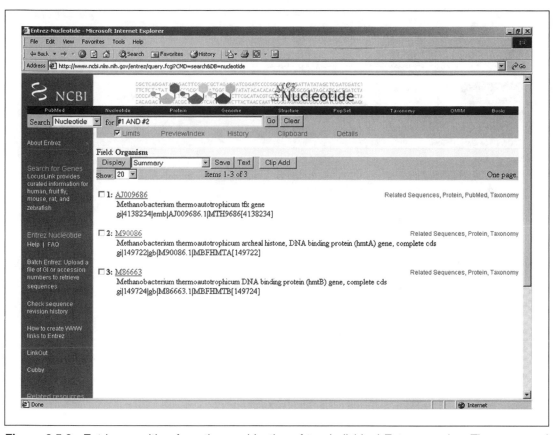

Figure 6.5.8 Entries resulting from the combination of two individual Entrez queries. The command producing the results is shown in the text box near the top of the window. The information on the individual queries that were combined is given in Figure 6.5.6. See text for details.

Explore material related to search results

8. Click the Related Sequences link to display all sequences similar to that of the *Methanobacterium thermoautotrophicum* tfx gene at the nucleotide level, in essence, showing the results of a precomputed BLAST search.

9. Click the Back button on the browser and select the Protein link for the *Methanobacterium thermoautotrophicum* tfx gene. Clicking this link displays CAA08778, the tfx protein. Select GenPept from the drop-down list and click Display. The GenPept entry that corresponds to *Methanobacterium thermoautotrophicum* tfx gene's conceptual translation displays (Fig. 6.5.9).

   Notice that, within the entry itself, the scientific name of the organism is represented by hypertext. Clicking on the Methanothermobacter thermautotrophicus link displays the NCBI Taxonomy database, which provides information on this organism's lineage.

10. Return to the GenPept display (Fig. 6.5.9) and select Graphics from the drop-down list and click Display. The Graphics view (Fig. 6.5.10) is one of the most useful views at this level. This view attempts to show graphically all of the features described within the entry's feature table, providing a very useful overview, particularly when the feature table is very long.

11. Click the Back button on the browser until the results for the #1 AND #2 query are displayed (Fig. 6.5.8). Select the PubMed link for the *Methanobacterium*

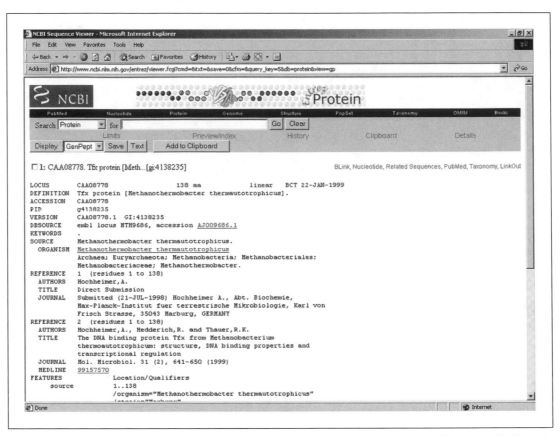

Figure 6.5.9 The protein neighbor for the *M. thermoautotrophicum* tfx gene. Clicking on the Protein hyperlink next to the first entry in Figure 6.5.8 leads the user to this GenPept entry. See text for details.

thermoautotrophicum tfx gene. Selecting the PubMed link takes the user back to the bibliographic entry corresponding to this GenBank entry.

12. Click the Back button on the browser once to view the combined search results (Fig. 6.5.8) and select the Taxonomy link for the *Methanobacterium thermoautotrophicum* tfx gene. Select the *Methanothermobacter thermautotrophicus* hypertext to display the taxonomy for the gene.

BASIC PROTOCOL 2

EXAMINING STRUCTURES IN ENTREZ

For the example below, assume that the user is trying to find information regarding the structure of HMG-box B from rat, whose PDB accession number is 1HMF.

Materials

An up-to-date Web browser, such as Netscape Communicator or Internet Explorer

1. Go to the Entrez Web page (*http://www.ncbi.nlm.nih.gov/Entrez;* Fig. 6.5.1) and select Structure from the Search drop-down list. Enter 1HMF in the For text box. Click Go.

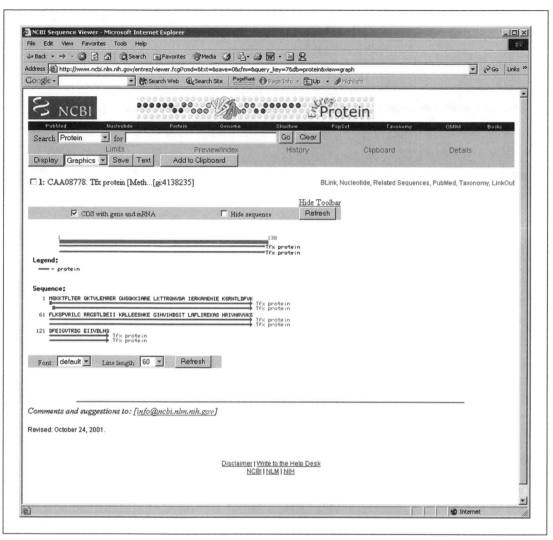

Figure 6.5.10   The Graphics view of the *Methoanobacterium thermoautotrophicum* tfx gene.

2. Click the 1HMF hypertext. The structure summary page displays, and the user will immediately note the decidedly different format than any of the pages displayed so far (Fig. 6.5.11).

> *This page shows the definition line from the source Molecular Modeling Database (MMDB) document (which is derived from PDB), as well as links to PubMed and to the taxonomy of the source organism. The graphic below the header schematically illustrates the protein as a bar of length 77 (meaning 77 amino acids), below which is a bar showing the position of a defined domain within the protein (here, the HMG box, a DNA-binding domain).*

3. Click on the upper bar corresponding to the full-length protein. This displays a table of four neighbors, as assessed by the Vector Alignment Search Tool (VAST).

4. To glean initial impressions about the shape of the protein, download Cn3D by clicking Get Cn3D 4.0. The application will walk the user through the installation program.

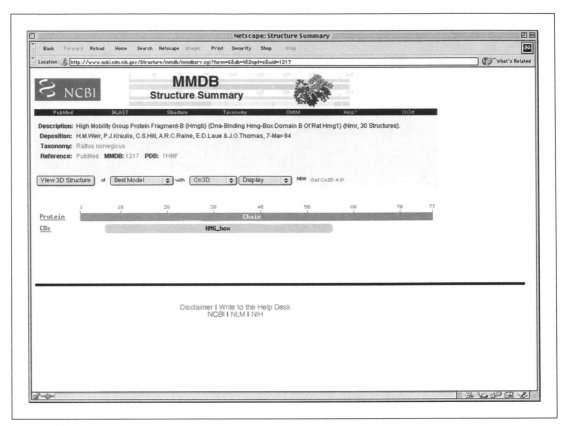

Figure 6.5.11 The structure summary for 1HMF, resulting from a direct query of the structures accessible through the Entrez system. The entry shows header information from the corresponding MMDB entry, links to PubMed and to the taxonomy of the source organism.

5. Once installed, use the Web browser's Back button to return to the 1HMF structure summary page. Click on View 3D Structure. This will launch the Cn3D viewer once the three-dimensional coordinates of 1HMF have been downloaded from the NCBI server.

6. Cn3D will produce two windows, one showing the structure of 1HMF, the other showing the sequence. The user can highlight any part of the sequence shown in the sequence window, and the corresponding part of the structure will appear in yellow. The user can also adjust the display of the structure by selecting options in the Style > Rendering Shortcuts and Style > Coloring Shortcuts sub-menus.

7. Rotate the structure by moving the mouse while holding down the mouse button. To zoom in or out, hold down the Apple (Mac) or Command key (PC) while dragging the mouse.

References: Altschul et al., 1990; Madej et al., 1995

Contributors: Juliane Murphy and Andreas D. Baxevanis

CHAPTER 7

Searching Candidate Genes for Mutations

Identification of a candidate gene that may be responsible for an inherited disease initiates a search for disease-causing mutations. As the human gene map improves and as new methods become available for positional cloning, the number of genes that might be screened for disease-causing mutations is rapidly increasing. Autosomal dominant disease-causing mutations are distinguished from polymorphisms by their functional consequences, their cosegregation with disease in a family, and their absence in unaffected and unrelated individuals. Precise definition of the functional consequences of a mutation may require extensive biochemical and human or animal model studies, but substantial support for the hypothesis that a mutation is etiologic for a disease can be obtained from genetic analyses of family members and unaffected individuals.

Disease-causing mutations can be classified into two broad groups: those that cause a significant change in gene structure (large deletions, insertions, inversions, or duplications) and those that produce only a minimal change in DNA structure (small deletions, insertions, inversions, duplications, and missense mutations). Dominant, recessive, and X-linked disease-causing mutations have been found in both classes, and a priori one cannot guess whether a candidate gene is more likely to have a large deletion or insertion or a single nucleotide change. Fortunately, methods for identifying mutations that grossly alter gene structure are well established. Standard and specialized cytogenetic techniques (Chapter 4) can readily identify mutations that grossly perturb chromosome structure. Techniques that assess genomic organization, including pulsed-field gel electrophoresis (UNIT 5.1), Southern blot hybridization (APPENDIX 3G), and restriction mapping, can be used to identify large deletions, insertions, inversions, or duplications. Protocols for these techniques are also described in *Current Protocols in Molecular Biology* (CPMB: Ausubel et al., 2004). It is particularly worthwhile to apply these techniques when cloned cDNA or cloned gene segments are available prior to or concomitant with the commencement of analyses for point mutations within a candidate gene.

Quite different strategies are required to identify mutations that produce only a minimal change in genomic structure. Identifying point mutations in candidate genes can be laborious. Candidate genes range in size from a few thousand to hundreds of thousands of base pairs in length, and the coding sequence may be segmented into many exons. Searching for disease-causing mutations in a large gene may require identification of a single nucleotide change among thousands of nucleotides. A further complexity for autosomal dominant conditions is that affected individuals are usually heterozygous for dominant diseases or compound heterozygotes for recessive diseases, so that both a normal and a mutant sequence are found at any particular point of the gene.

All of the methods described in this chapter depend upon the amplification of candidate-gene or cDNA segments by PCR to produce nonradioactive amplified DNA. Procedures for producing and characterizing this amplified DNA are described in UNIT 7.1.

One method, single-strand conformational polymorphism (SSCP) analysis (UNIT 7.2, UNIT 7.3), detects mutations by the conformational changes they produce in DNA molecules, which cause altered electrophoretic mobility of mutated DNA molecules compared to normal sequences. Some mutations alter the conformation of the DNA molecule and result in a dramatic mobility shift, while others produce little change. Change in conformation is determined in part by the nucleotide sequence flanking the mutation. Augment differences in mobility, failure to detect a mutation needs to be critically assessed if a gene appears to be a particularly good candidate for bearing disease-causing mutations.

Mutations can also be detected by direct sequencing of PCR-amplified DNA. In this method, each exon of a candidate gene from affected individuals is PCR amplified and sequenced (UNIT 7.4). Because the techniques for DNA sequence analysis have improved dramatically in the past few years, direct sequencing of PCR-amplified DNA from affected persons is now a useful approach for mutation screening. The primary advantage of this method is that it detects all or nearly all mutations.

Most of the protocols for identifying missense mutations (RNase protection, chemical cleavage, FAMA) in this chapter permit analyses of as many as 30,000 to 60,000 base pairs of DNA on a single gel. DHPLC can be used to screen this number of base pairs of DNA in a 12- to 18-hr period. These methods are appropriate for rapid screening of large candidate genes. Ultimately each protocol leads to DNA sequencing of a gene region shown to contain a mutation. Cycle sequencing (UNIT 7.5) provides a method for determining DNA sequence of shorter DNA segments. The ability to sequence PCR-amplified DNA segments allows one to characterize mutations identified by other procedures. Cycle sequencing can also be used to directly screen shorter segments for mutations.

In summary, this chapter outlines different methods for detecting mutations in affected individuals. The power of these methods is demonstrated by the large number of disease causing mutations that have been identified. A database containing descriptions of many of these mutations has been created (see UNIT 7.6).

Contributors: J.G. Seidman and Christine Seidman

UNIT 7.1

Amplification of Sequences from Affected Individuals

Target sequences that may be present at very low copy number in patient samples are amplified by the polymerase chain reaction (PCR; *CPMB* Chapter 15; Fig. 7.1.1) using nested primers for two successive rounds of PCR. In many instances, it is desirable to confine analysis to the coding sequence and the splice-donor and splice-acceptor sites in the immediate flanking intron sequences, where disease-causing mutations are most likely to be found. Primers should be positioned so as to provide sufficient overlap for detection of mutations near the end of each segment, because point mutations within primer sequences are typically not detectable after PCR. It may be desirable for primers to be situated with their 3′ ends close to the exon/intron boundary to minimize inclusion of intronic sequences, which are more likely to contain noncoding polymorphisms.

BASIC PROTOCOL 1

AMPLIFICATION OF RNA FROM LYMPHOCYTES

Materials (see APPENDIX 1 for items with ✓)

Transformed B cells from saturated Epstein-Barr virus (EBV) cultures (APPENDIX 3J) or nontransformed lymphocytes isolated by Ficoll-Hypaque gradient centrifugation (UNIT 10.3)

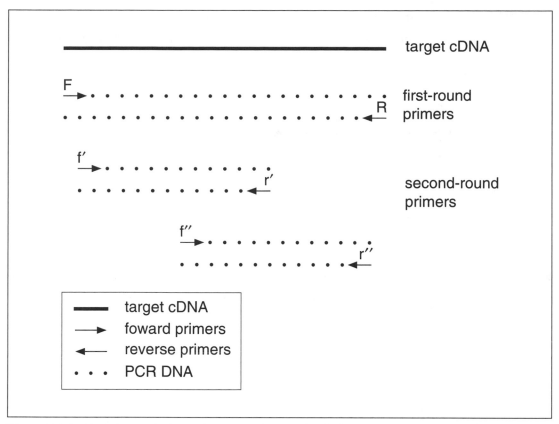

Figure 7.1.1 Use of nested primers for successive rounds of PCR. (**A**) Primer R may be used for reverse transcription of cDNA from an mRNA template; it is also used with primer F to generate the first-round PCR product. (**B**) Arrangement of nested inner primers for amplification of sections containing possible mutations.

✓ PBS, ice cold
✓ Diethylpyrocarbonate (DEPC)-treated H_2O
✓ 1.25 mM 4dNTP mix
✓ 10× PCR amplification buffer containing 15 mM $MgCl_2$
    25-mer oligonucleotide primers (Fig. 7.1.1): outer (reverse and forward) and inner
      (reverse and forward)
    RNasin (Promega)
    Reverse transcriptase [e.g., Moloney murine leukemia virus (MoMuLV)-RT]
    Taq DNA polymerase (Perkin-Elmer Cetus)
    Mineral oil
    Sieving agarose (e.g., Nusieve, FMC Bioproducts)
    DNA molecular size markers

    1.5- and 0.5-ml polypropylene microcentrifuge tubes, clean and RNase-free
    Thermal cycler
    42°C water bath
    Beckman JS-4.2 rotor or equivalent

1. Collect transformed B cells from 30 ml of saturated EBV culture or nontransformed lymphocytes extracted by Ficoll-Hypaque gradient centrifugation from 20 to 30 ml fresh whole blood. (The yield from these amounts is $\sim 10^9$ cells.)

2. Wash cells once with ice-cold PBS. Centrifuge 5 min at $300 \times g$, $4°C$.

3. Extract total RNA from the cells by the single-step guanidinium thiocyanate method (*UNIT 10.3*).

4. Resuspend RNA in 100 μl DEPC-treated water in microcentrifuge tubes. Store in two 50-μl aliquots at $-70°C$ (stable for ≥ 1 year). Expect the RNA concentration to be ~ 1 mg/ml (100 μg total).

5. Set up the following 20-μl reaction in a 0.5-ml microcentrifuge tube:

   2 μl 1.25 mM 4dNTP mix
   2 μl 10× PCR amplification buffer
   500 ng reverse (antisense) outer primer (60 pmol)
   20 U RNasin
   2 μg (~ 2 μl) RNA (from step 4)
   50 U reverse transcriptase
   DEPC-treated H_2O to 20 μl.

   Incubate 45 min at $42°C$. Heat-inactivate the reverse transcriptase 10 min at $90°C$. After cooling, microcentrifuge briefly at top speed. Intersperse control reactions (identical except for absence of RNA) with test samples and process in parallel. Store cDNA product overnight at $4°C$ until amplified, or indefinitely at $-20°C$.

6. Set up the following 100-μl reaction in a 0.5-ml microcentrifuge tube:

   16 μl 1.25 mM 4dNTP mix
   8 μl 10× PCR amplification buffer
   500 ng forward (sense) outer primer (60 pmol)
   20 μl cDNA (step 5)
   DEPC-treated H_2O to 100 μl.

   Add 2.5 U *Taq* DNA polymerase. Mix, microcentrifuge briefly at top speed, and overlay with 2 drops of mineral oil if appropriate for thermal cycler model.

7. Carry out PCR using the following amplification cycles:

40 cycles:	30 sec	94°C	(denaturation)
	1 min	55°C	(annealing)
	2 min	72°C	(extension)
Final step:	indefinitely	4°C	(hold).

8. Dilute 10 μl first-round PCR product in 1 ml water. Use 5 μl of this 1:100 dilution as the template for the second PCR amplification. Save both the first PCR product and the dilution and store them indefinitely at $4°C$.

9. For each sample, set up the following 50-μl reaction in a 0.5-ml microcentrifuge tube:

   8 μl 1.25 mM 4dNTP mix
   5 μl 10× PCR amplification buffer
   250 ng each forward and reverse inner primers (30 pmol)
   5 μl 1:100 dilution of first-round PCR product (from step 8)
   H_2O to 50 μl.

   Add 1.25 U *Taq* DNA polymerase. Mix, microcentrifuge briefly at top speed, and overlay with mineral oil if appropriate for thermal cycler model. Intersperse further control reactions (identical except for absence of first amplification product) with test samples and process in parallel.

10. Carry out PCR using the following amplification cycles:

30–40 cycles:	30 sec	94°C	(denaturation)
	1 min	55°C	(annealing)
	1 min	72°C	(extension)
Final step:	indefinitely	4°C	(hold).

11. Electrophorese 10 μl from each second-round reaction on a sieving agarose gel stained with ethidium bromide (*APPENDIX 3G*). Include a lane that contains a known quantity of molecular size markers. Make a photograph of the gel.

ALTERNATE PROTOCOL

MODIFIED SECOND-ROUND AMPLIFICATION OF cDNA

Applications such as heteroduplex analysis or chemical cleavage may require $5'$-^{32}P-labeled product. For these applications, the second-round PCR is performed as described, but using an end-labeled primer. The $5'$ ends of either or both primers are labeled prior to use in the PCR reaction by the action of T4 polynucleotide kinase in the presence of $[\gamma$-^{32}P]ATP (*APPENDIX 3E*).

Alternatively, the second PCR can be performed using primers containing restriction enzyme sites to facilitate subcloning of the final product, e.g., for sequencing. The enzyme sites selected must not be present in the target DNA sequence and should generate cohesive ends that will facilitate cloning into the polylinker of the chosen vector. Two measures are necessary to ensure efficient restriction of the PCR product. First, the oligonucleotide primers should have a "cap" of three nucleotide residues $5'$ to the enzyme site sequence (which is itself $5'$ to the target DNA sequence). Second, because of the tendency of the PCR reaction to produce product with incomplete ends, the second-round PCR product should be filled in by treatment with Klenow fragment to ensure that the complete enzyme site is generated.

BASIC PROTOCOL 2

AMPLIFICATION OF GENOMIC DNA

Materials (see APPENDIX 1 for items with ✓)
    Genomic DNA (*APPENDIX 3A*)
✓ 1.25 mM 4dNTP mix
✓ 10× PCR amplification buffer containing 15 mM MgCl$_2$
    Oligonucleotide primers: reverse and forward
    Taq DNA polymerase (Perkin-Elmer Cetus)
    Mineral oil
    Sieving agarose (e.g., Nusieve, FMC Bioproducts)
    DNA molecular size markers

    1.5- and 0.5-ml polypropylene microcentrifuge tubes
    Thermal cycler

1. Dilute each genomic DNA sample to be analyzed to 100 ng/μl and heat-denature 5 min at 95°C. Store DNA samples indefinitely at 4°C.

2. For each sample, set up the following 50-μl reaction in 0.5-ml microcentrifuge tubes:

> 8 μl 1.25 mM 4dNTP mix
> 5 μl 10× PCR amplification buffer
> 250 ng each reverse and forward primers
> 200 ng genomic DNA (2 μl from step 1)
> Sterile H_2O to 50 μl.

Add 1.25 U *Taq* DNA polymerase. Mix, microcentrifuge briefly at top speed, and overlay with mineral oil if appropriate for thermal cycler model. Intersperse control reactions (identical to reactions except for absence of DNA) and process in parallel.

3. Carry out PCR using the following amplification cycles:

30 cycles:	30 sec	94°C	(denaturation)
	1 min	55°C	(annealing)
	1 min	72°C	(extension)
Final step:	indefinitely	4°C	(hold).

4. Electrophorese 10 μl from each second-round reaction on a sieving agarose gel stained with ethidium bromide (*APPENDIX 3G*). Include a lane that contains a known quantity of molecular size standard.

Contributor: Hugh C. Watkins

UNIT 7.2

Detection of Mutations by Single-Strand Conformation Polymorphism Analysis

BASIC PROTOCOL

Materials (*see* APPENDIX 1 *for items with* ✓)

> PCR primers A and B (1 OD$_{260}$/ml; 1 pmol/μl): forward and reverse primers designed to amplify <220 bp of DNA region of interest
> 5 U/μl T4 polynucleotide kinase and 10× buffer (*APPENDIX 3E*)
> 10 mCi/ml [γ-^{32}P]ATP (3000 Ci/mmol; Amersham)
> ✓ 2 mM 4dNTP mix
> 5 U/μl *Taq* DNA polymerase
> ✓ 10× PCR amplification buffer
> 25 to 250 μg/ml human genomic DNA from affected and unaffected individuals (see
>   UNIT 7.1)
> Mineral oil
> Low gelling/melting temperature agarose (e.g., NuSieve GTG agarose, FMC
>   Bioproducts)
> 2% dimethyldichlorosilane (BDH Diagnostics; store at room temperature)
> ✓ 50% (w/v) acrylamide stock solution
> ✓ 10× TBE buffer
> 10% (w/v) ammonium persulfate (APS; prepare immediately before use)
> TEMED
> 0.1% (w/v) SDS/10 mM EDTA (pH 8.0)
> ✓ 2× formamide loading buffer

65°C water bath

Thermal cycler

DNA sequencing gel apparatus with 31 × 38.5–cm glass plates, 0.4-mm spacers, and sharkstooth comb

Waterproof tape

90°C heating block

Whatman 3MM filter paper

UV-transparent plastic wrap (e.g., Saran Wrap)

1. Label PCR primers A and B by mixing the following (enough for 20 PCR tubes):

> 5 μl 1 OD_{260}/ml PCR primer A
> 5 μl 1 OD_{260}/ml PCR primer B
> 5 μl 10× T4 polynucleotide kinase buffer
> 5 to 10 μl [γ-^{32}P]ATP (50 to 100 μCi)
> 2 μl 5 U/μl T4 polynucleotide kinase
> H_2O to 50 μl.

Incubate 30 to 60 min at 37°C. Heat 10 min at 65°C to inactivate the kinase (*APPENDIX 3E*).

PCR products can be labeled directly by adding 0.1 μCi [α-^{32}P]dCTP to each PCR reaction and reducing unlabeled dCTP to one-tenth of that normally used. Gels may also be silver-stained.

2. Prepare a PCR mix for 20 reactions (960 μl total):

> 100 μl PCR primer A
> 100 μl PCR primer B
> 100 μl 2 mM 4dNTP mix
> 100 μl 10× PCR amplification buffer
> 4 μl 5 U/μl *Taq* DNA polymerase
> 506 μl H_2O
> 50 μl labeled primer mix (step 1).

Mix well and aliquot 48 μl to each of 20 tubes. Add 2 μl sample DNA (50 to 500 ng) to each tube. Add 3 drops of mineral oil and amplify using appropriate thermal cycling conditions. Include control reactions with DNA from a known normal individual and, if possible, from an individual known to carry a mutation in the target sequence (*UNIT 7.1*).

PCR products can be labeled directly by substituting 2 μl of 10 mCi/ml [α-^{32}P]dCTP and 44 μl water for 50 μl of labeled primer mix.

3. Confirm amplification of the target sequence by analyzing 5 to 10 μl of the PCR products on a 2% to 3% minigel made with low gelling/melting temperature agarose (*APPENDIX 3G*).

4. Prepare glass plates by cleaning thoroughly with detergent, water, and ethanol. Treat one plate with 2% dimethyldichlorosilane: carefully apply a thin film with a Kimwipe moistened with dimethyldichlorosilane, allow the film to dry, and wipe with ethanol. Insert spacers and seal the sides and bottom with waterproof tape.

5. Prepare gel solution. For a 4.5% (0.4-mm thick) nondenaturing polyacrylamide gel, mix:

> 6.3 ml 50% (or higher) polyacrylamide stock solution
> 3.5 ml 10× TBE buffer
> 59 ml H_2O.

Degas, then add 1 ml of 10% APS and mix well. Add 50 μl TEMED and mix again. If necessary, add glycerol, 5% or 10% (v/v), to the gel mix to improve resolution. Do not use a denaturant such as urea.

6. Pour gel immediately. Remove any bubbles by tapping the plates. Insert comb, clamp well, and lay flat at least 45 min to polymerize. For a sharkstooth comb, insert the comb with the teeth pointing upward to form a single well the width of the gel.

   It is important that the gel be of uniform thickness. Support the plates at the sides and in the middle to help ensure uniformity.

7. Chill gel, 0.5× TBE buffer, and tank to 4°C.

8. Remove the clamps, the tape from the bottom of the gel, and the comb. Attach the plates to the sequencing apparatus. For a sharkstooth comb, replace the comb with the teeth pointing downward and just in contact with the gel surface. Add 0.5× TBE to top and bottom tanks. Flush out the wells and the bubbles from the bottom of the gel using a syringe with a needle.

9. Dilute PCR products 1/10 in 0.1% SDS/10 mM EDTA (pH 8.0). Add 10 μl diluted sample to an equal volume of 2× formamide loading buffer. Heat 3 min at 90°C, then place sample immediately on ice to prevent the DNA strands from reannealing. Alternatively, before heating, add 2 vol formamide loading buffer directly to the completed PCR mixture or load PCR samples directly from the 90°C heating block.

10. Immediately load 2 to 5 μl of each PCR sample. Include lanes with PCR-amplified positive and negative control DNAs and undenatured PCR product (to indicate the position of double-stranded DNA on the gel). Electrophorese until bromphenol blue has just run off the bottom of the gel—typically 2 to 4 hr at 30 W, 4°C, or overnight at room temperature at low power (e.g., 6 W).

    Gels containing glycerol generally run more slowly. A 6% acrylamide gel containing 10% glycerol is best run overnight at 6 to 8 W.

11. After electrophoresis, disconnect power and remove plates from the apparatus. Remove tape from the sides and lay plates flat. Pry plates apart (the gel should stick to one plate). Transfer gel to a sheet of Whatman 3MM paper and cover with plastic wrap. Dry 45 min at 80°C in a gel dryer.

12. Autoradiograph 4 hr to overnight at room temperature without intensifying screens.

Reference: Spinardi et al., 1991

Contributors: William Warren, Eivind Hovig, Birgitte Smith-Sorensen, and Anne-Lise Borresen

UNIT 7.3

Single-Strand Conformation Polymorphism Analysis Using Capillary Electrophoresis

Single-strand conformation polymorphism (SSCP) is one of the most frequently used mutation detection methods (see also *UNIT 7.2*). It is possible to distinguish between a wild-type and a mutant DNA fragment by allowing the DNA to be separated in a nondenaturing polymer by capillary electrophoresis; and by labeling each strand with a different label, it is possible to distinguish the two strands (Fig. 7.3.1). The conformation is temperature dependent and, therefore, the gel electrophoresis must be performed at different temperatures to achieve a high sensitivity. A number of problems and possible solutions have been listed in Table 7.3.1.

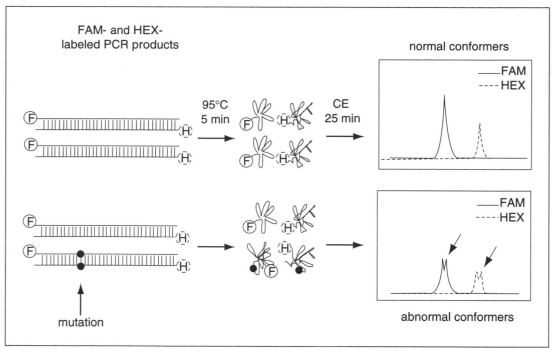

Figure 7.3.1 Schematic representation of the assay principle. DNA fragments are amplified by PCR using fluorescent primers. The labeled fragments are denatured and cooled to 4°C. The single-stranded DNA fragments are separated using nondenaturing capillary electrophoresis. The DNA fragments fold in sequence-specific secondary or tertiary structures when entering the nondenaturing environment. Mutant fragments display a different peak pattern in the electropherogram due to different migration of the abnormal conformers (arrows).

BASIC PROTOCOL 1

SAMPLE PREPARATION FOR CAPILLARY ELECTROPHORESIS

Materials (*see* APPENDIX 1 *for items with* ✓)

    5′-FAM-labeled forward primer (e.g., MWG-Biotech)
    5′-HEX-labeled reverse primer (e.g., MWG-Biotech)
✓ dNTPs (10 mM each)
    10× PCR buffer (Roche Diagnostics)
    Pwo DNA polymerase (Roche Diagnostics)
    20 to 60 ng/μl genomic DNA
    Milli-Q water
    25 mM MgSO$_4$
    2% (w/v) agarose gel
    7% (v/v) GeneScan polymer (Applied Biosystems)
    10× genetic analyzer buffer with EDTA (Applied Biosystems) or TBE (APPENDIX 1)
    87% (v/v) glycerol solution (Merck)
    Deionized formamide
    0.3 N NaOH
    DNA size standard (e.g., GeneScan 500 ROX, Applied Biosystems)

Table 7.3.1 Troubleshooting

Problem	Possible cause	Solution
Several DNA fragments in PCR product	Unspecific PCR	Optimize the PCR conditions by changing the annealing temperature and/or the $MgSO_4$ concentration, or design a new primer pair
No signal detected	Current is blocked by air bubble or clotted capillary	Check the block for air bubbles Change the capillary
Broadening of peaks in internal standard	Low-resolution	Rerun in a new capillary
Sample signal is too low	Weak PCR product	Re-inject with a longer injection time Apply more sample
Sample signal is off scale	Strong PCR product	Re-inject with a shorter injection time Dilute the sample
Peaks resulting from late injections are much weaker than peaks resulting from earlier injections	Renaturing of the samples (this has been observed for long PCR fragments—1.2 kbp)	Repeat denaturing and cooling

Software for T_m calculation (e.g., AnnHyb, *http://annhyb.free.fr/*; or Oligo, Molecular Biology Insights; optional)
0.2-ml PCR tubes
Thermal cycler (e.g., PTC200 DNA engine, MJ Research)
50-ml polypropylene tubes
ABI PRISM genetic analyzer (Applied Biosystems)

1. Calculate the theoretical T_m ($T_{m(theoretical)}$) for the PCR primers. Use software for the calculation, or use any commonly used standard calculation. Use the following annealing temperature: $T_{m(theoretical)} - 5°C$.

2. Prepare a PCR master mix for five 25-μl reactions. Mix the following (multiplied by 5) on a water/ice bath:

   > 1.25 μl 20 μM 5′-FAM-labeled forward primer (final 1 μM)
   > 1.25 μl 20 μM 5′-HEX-labeled reverse primer (final 1 μM)
   > 0.5 μl (10 mM each) dNTPs (final 200 μM)
   > 2.5 μl 10× PCR buffer (final 1×)
   > 0.1 μl 5 U/μl *Pwo* DNA polymerase (final 0.5 U/25 μl)

3. Prepare five 0.2-ml PCR tubes on water/ice and add the following to each respectively labeled tube:
   a. 1 μl 20 to 60 ng/μl genomic DNA, 17.9 μl milli-Q water, 0.5 μl 25 mM $MgSO_4$
   b. 1 μl 20 to 60 ng/μl genomic DNA, 17.4 μl milli-Q water, 1.0 μl 25 mM $MgSO_4$
   c. 1 μl 20 to 60 ng/μl genomic DNA, 16.9 μl milli-Q water, 1.5 μl 25 mM $MgSO_4$
   d. 1 μl 20 to 60 ng/μl genomic DNA, 16.4 μl milli-Q water, 2.0 μl 25 mM $MgSO_4$
   e. 1 μl 20 to 60 ng/μl genomic DNA, 15.9 μl milli-Q water, 2.5 μl 25 mM $MgSO_4$

4. Add 5.6 μl of the PCR master mix to each 0.2-ml PCR tube. Vortex. Keep on water/ice.

5. Amplify DNA fragments using a thermal cycler programmed with the following:

1 cycle:	4 min	94°C	(denaturation)
34 cycles:	20 sec	94°C	(denaturation)
	20 sec	x°C	($T_{m(theoretical)} - 5$°C; annealing)
	40 sec	72°C	(extension)
1 cycle:	7 min	72°C	(extension)

Start the thermal cycler with the lid open. Transfer the tubes directly from the ice/water bath to the heating block as soon as the temperature is >85°C. Close the lid.

6. Analyze 10 μl PCR products by agarose gel electrophoresis in a 2% agarose gel (*APPENDIX 3G*). Choose the $MgSO_4$ concentration that results in clear sharp bands of the correct size.

7. Prepare PCR master mix for 48 25-μl reactions. Mix on water/ice.

    1.25 μl 20 μM 5′-FAM-labeled forward primer (final 1 μM)
    1.25 μl 20 μM 5′-HEX-labeled reverse primer (final 1 μM)
    0.5 μl 10 mM each dNTPs (final 200 μM)
    2.5 μl 10× PCR buffer (final 1×)
    0.5 to 2.5 μl 25 mM $MgSO_4$ (final 0.5 to 2.5 mM)
    15.9 to 17.9 μl milli-Q water
    0.1 μl 5 U/μl *Pwo* DNA polymerase (0.5 U/25 μl)

Mix well, and dispense 24 μl to 0.2-ml PCR tubes. Add 1 μl of 20 to 100 ng genomic DNA. Keep on ice.

8. Amplify DNA fragments using a thermal cycler using the temperature profile in step 5. Start the thermal cycler with the lid open. Transfer the tubes directly from the ice/water bath to the heating block as soon as the temperature is >85°C. Close the lid.

9. Analyze 10 μl PCR product by agarose gel electrophoresis in a 2% agarose gel (*APPENDIX 3G*).

10. Add the following to a 50-ml polypropylene tube:

    3.56 g 7% GeneScan polymer
    500 μl 10× genetic analyzer buffer with EDTA or TBE
    287 μl 87% glycerol.

Add milli-Q purified water up to 5 ml. Mix by slow stirring using a magnetic stirrer. Avoid bubbles. Load the polymer onto the ABI PRISM genetic analyzer according to the instructions from the manufacturer.

11. Mix 1 μl PCR product (0.1 to 0.5 μg/ul DNA) with 10 μl deionized formamide, 0.5 μl of 0.3 N NaOH, and 0.5 μl DNA size marker.

12. Denature the PCR products for 5 min at 95°C, and cool the mixture to 4°C (alternatively, transfer to an ice/water bath after denaturation).

BASIC PROTOCOL 2

AUTOMATED CAPILLARY ELECTROPHORESIS USING AN ABI 310 GENETIC ANALYZER

For low-throughput SSCP analysis, a single capillary electrophoresis instrument such as the ABI 310 genetic analyzer (Applied Biosystems) or the P/ACE 5510 (Beckman Coulter) may often be adequate. Analysis must be carried out at different temperatures, since appearance of aberrant conformers is dependent upon the temperature in a sequence-specific way. In fact, some mutations will display aberrant conformers only at subambient temperatures. To this end, a simple cooling aggregate may be used (Larsen et al., 1999).

Materials

PCR products (see Basic Protocol 1)
5% nondenaturing polymer, with 5% glycerol added (see Basic Protocol 1, step 10)
10× genetic analyzer buffer with EDTA (Applied Biosystems)
Milli-Q purified water

ABI PRISM 310 genetic analyzer (Applied Biosystems)
Genetic analyzer capillaries, 50-μm i.d. (Applied Biosystems)
GeneScan analysis software (Applied Biosystems)
Genotyper software (optional, Applied Biosystems)

1. Place the PCR tubes in the sample tray. Prepare the ABI PRISM genetic analyzer for a normal run according to the manufacturer's instructions. Install a 50-μm-i.d. genetic analyzer capillary with a length to detector of 36 cm. Load the syringe with 5% nondenaturing polymer, and load the buffer chambers with 1× genetic analyzer buffer. Prepare the sample sheet.

   If TBE buffer is used in the 5% nondenaturing polymer, the genetic analyzer buffer must be substituted with 1× TBE.

2. Run the capillary electrophoresis at 15°C, 25°C, and 35°C, using the following common parameters:

injection time:	12 sec
injection voltage:	15 kV
run voltage:	13 kV
run time:	20 min

   Use filter set D for detection of the FAM, HEX, and ROX labels.

   If a cooling plate is installed, room cooling is essential in order to avoid condensed water, which may cause short-circuiting and may damage the instrument. Caution must be taken if working with the instrument with open doors due to the high voltage applied during electrophoresis.

3. Analyze the results using GeneScan analysis software according to the manufacturer's instructions. Define a size standard for each of the electrophoresis temperatures. The sizes of the peaks in the standard are defined as the scan points divided by 10 (e.g., a peak at scan point 3550 is defined as having a size of 355). Use the internal DNA size standard for alignment of the peaks. Compare the unknown samples with a normal sample with a known DNA sequence. Alternatively, perform automated data analysis using Genotyper software according to the manufacturer's instructions.

4. Re-amplify all samples with split peaks or aberrant migration using unlabeled primers, and detect the mutation by DNA sequencing (*UNIT 7.5*).

BASIC PROTOCOL 3

AUTOMATED CAPILLARY ARRAY ELECTROPHORESIS USING AN ABI PRISM 3100 GENETIC ANALYZER

For medium- to high-throughput SSCP analysis, a multicapillary electrophoresis instrument such as the ABI 3100 instrument is an excellent choice. To attain a high detection rate (~90%), it is suggested to run the SSCP analysis at three temperatures.

7

Materials

> PCR products (see Basic Protocol 1)
> 5% nondenaturing polymer, with 5% glycerol added (see Basic Protocol 1, step 10)
> 10× genetic analyzer buffer with EDTA (Applied Biosystems)
> Milli-Q purified water
>
> ABI PRISM 3100 genetic analyzer with 16 capillaries (Applied Biosystems)
> 36-cm GeneScan capillary array (50-μm i.d.; Applied Biosystems)
> GeneScan NT version 3.7 (Applied Biosystems)
> Genotyper NT version 3.7 (optional, Applied Biosystems)

1. Install a 16-capillary array (50-μm i.d.) and a length to detector of 36 cm. Prepare the instrument for a normal run according to the manufacturer's instructions. Load the syringe with 5% nondenaturing polymer, and load the buffer chambers with 1× genetic analyzer buffer. Prepare the sample sheet. Place the tubes in the sample tray.

2. Run the capillary electrophoresis using three run-modules with electrophoresis temperatures of 18°C, 25°C, and 35°C, and the following common parameters:

injection time:	10 sec
injection voltage:	3 kV
run voltage:	15 kV
run time:	25 min.

   Use filter set D for detection of the labels FAM, HEX, and ROX.

3. Analyze the results using GeneScan analysis software according to the manufacturer's instructions. Define a size standard for each of the electrophoresis temperatures. The sizes of the peaks in the standard are defined as the scan points divided by 10 (e.g., a peak at scan point 3550 is defined as having a size of 355). Use the internal DNA size standard for alignment of the peaks. Compare the unknown samples with a normal sample with a known DNA sequence. Alternatively, perform automated data analysis using Genotyper software according to the manufacturer's instructions.

4. Re-amplify all samples with split peaks or aberrant migration patterns using unlabeled primers, and detect the mutation by DNA sequencing (*UNIT 7.5*).

References: Larsen et al., 1999; Walz et al., 2000

Contributors: Lars Allan Larsen, Michael Christiansen, Jens Vuust, and Paal Skytt Andersen

UNIT 7.4

Heterozygote Detection Using Automated Fluorescence-Based Sequencing

Identifying mutations at the sequence level is the critical final step in identifying pathologic mutations in autosomal dominant disorders as well as in identifying carriers of autosomal recessive and X-linked disorders. In addition, identification of the rapidly expanding collection of single-nucleotide polymorphisms (SNPs), which are now being used in the study of complex genetic traits in large populations, also requires a method for identifying heterozygote sequence variation. Various considerations for troubleshooting are given in Table 7.4.1.

Table 7.4.1 Troubleshooting Guide to Fluorescence-Based Sequencing

Problem	Possible cause	Solution
Basic sequencing procedure		
Poor-quality sequence	Faint or fuzzy PCR amplicon	Design better PCR primers, optimize PCR conditions
	Primer problem	Design internal primers for sequencing
	Concentration of template or primer incorrect	Adjust concentration
	Contamination problem	Try gel purification
	Unable to identify	Repeat sequencing (it may work the second or third time)
Unable to align sequence with consensus in GAP4	Consensus sequence contains gaps	Edit ends; edit prior files; inspect for insertion or deletion
Unable to align sequences using Sequencher	Noisy ends or insertion/deletion	Use "align interactively" command; inspect for insertion or deletion; trim ends
Mutation seen in only one direction	Sequencing chemistry results in variable peak heights	Try alternative method of verification (restriction digestion, allele-specific hybridization, HD, DGGE, SSCP, or DHPLC), repeat sequence analysis with alternative chemistry
	Artifact/noise	As above, plus repeat with controls
Mutation seen with nonsequencing method is not detected with sequencing	Mosaicism	Isolate mutant band, reamplify, and sequence
	Variation located close to primer	Interpretation of sequence traces near the primer is difficult and may be improved by spin-column purification of sequencing reactions prior to ABI run
Large-volume sequencing procedure		
High background	Poor PCR product	Optimize primers/PCR conditions
	Non-specific priming during sequencing	Try other primer or design a sequencing primer
	Insufficient amount of template	Load more PCR product
Chromatogram trace amplitudes too high with early termination	Excess template in cycle sequencing reaction	Adjust PCR template volume
Trace suddenly becomes uninterpretable	Mispriming	Check PCR product for nonspecific bands
	Repetitive sequence	Try other direction
	Deletion/insertion	Try other direction
		Check by other methods
Smearing and overlapping of waves	Insufficient amount of template	Load more PCR product
	Capillary artifact	Repeat the reaction or try other direction
All 4 colors combine at one position	Air bubble capillary artifact	Repeat the reaction or try sequencing from the other direction

IDENTIFYING HETEROZYGOTE MUTATIONS FROM SEQUENCE TRACES

Materials (see APPENDIX 1 for items with ✓)

  2.5 to 25 ng/µl genomic DNA (isolated from blood) in TE buffer, pH 7.4
✓ TE buffer, pH 7.4
  Oligonucleotide primers (Research Genetics):
      250 µM stocks in TE buffer, pH 7.4
      10 µM stocks in water (deionized, distilled, and autoclaved)
✓ 10× PCR amplification buffer with 1.5 mM $MgCl_2$ (Perkin-Elmer; use 0.01% (w/v)
      gelatin)
  AmpliTaq Gold DNA polymerase (Perkin-Elmer)
  10 mM 4dNTP mix (New England Biolabs)
  BigDye terminator cycle sequencing reaction kit (Perkin-Elmer), including ready
      reaction mix
  Linear acrylamide (Ambion)
  Loading buffer: 5:1 (w/v) deionized formamide/25 mM EDTA, pH 8.0, with 50 mg/ml
      blue dextran

  PCR tubes (strip tubes or 96-well plates)
  Thermal cycler (e.g., MJ Research)
  PCR purification spin columns (Qiagen or Wizard PCR Prep; Promega)
  ABI 377 DNA analysis system (Perkin-Elmer)
  Software (select one): GAP4, PolyPhred, SeqMan (DNAStar), Sequencher (Gene Codes)
  Computer: Unix machine for GAP4 or PolyPhred, Macintosh or PC for Sequencher or
      SeqMan

1. Design oligonucleotide primers for amplification. Select primers with an annealing temperature near 57°C and a length of 18 to 22 bp. Position the primers so that at least 6 bp of flanking intronic sequence are amplified with each exon. Perform BLAST searches (see *UNIT 6.3*) for each primer to eliminate those encoding repetitive elements such as *Alu* or long interspersed repetitive element (LINE) sequences. Use the same primers for both PCR amplification and sequencing.

   Primers must be designed for the PCR amplification and sequencing of exons or other genomic regions of interest. Although there are many computer programs to assist with primer design, the Wisconsin Package primer design program (Genetics Computer Group, http://www.gcg.com) and the WI primer 3 program (http://www-genome.wi.mit.edu/cgi-bin/primer/primer3_www.cgi) are both user-friendly and generate good primers.

   The above parameters have been used to design primers that amplify exons ranging in size from 174 to 421 base pairs.

2. Set up 100-µl amplification reactions in PCR tubes using:

      50 to 500 ng genomic DNA
      1 µM primer (use 10 µM stocks)
      150 µM 10 mM 4dNTP mix
      1 µl AmpliTaq Gold DNA polymerase
      10 µl 10× PCR amplification buffer with 1.5 mM $MgCl_2$.

3. Perform PCR using the following amplification cycle:

Initial step:	12 min	95°C	(denaturation)
35 cycles:	30 sec	94°C	(denaturation)
	30 sec	57°C	(annealing)
	45 sec	72°C	(extension)
Final step:	4 min	72°C	(extension).

4. Purify PCR product (with or without gel purification) using PCR purification columns (e.g., Qiagen). Assess the quality and concentration of the purified amplicon by running it on an agarose gel (*APPENDIX 3G*) with standards of known concentration. Alternatively, measure the OD_{260} (*APPENDIX 3D*) to quantitate the DNA prior to sequencing.

5. Prepare a working primer stock by diluting the 250 μM stock to 2.5 to 3.2 μM with water. Set up the cycle sequencing reaction for the amplicon of interest according to the manufacturer's BigDye sequencing protocol, with these minor modifications:

> 2.4 μl BigDye terminator ready reaction mix
> 1 to 1.3 pmol oligonucleotide primer from working stock
> 3 to 100 ng purified amplicon as the template (>20 ng is better)
> water to a total volume of 7.4 μl.

6. Place tubes in a thermal cycler and cycle as follows:

25 cycles:	10 sec	96°C	(denaturation)
	5 sec	50°C	(annealing)
	4 min	60°C	(extension).

Store samples at 4°C until ready for purification.

7. To purify the sequencing extension products, perform two rounds of ethanol precipitation (*APPENDIX 3B*) or spin-column purification (*APPENDIX 3E*). Use linear acrylamide as the carrier during ethanol precipitation to make the DNA pellets visible. Dry the pellets and resuspend them in 1.2 to 2 μl loading buffer.

8. Load the samples onto an ABI 377 sequencing gel. Sequence from both directions with inclusion of controls.

9. Transfer ABI chromatogram data to a computer for analysis with the selected software tools and analyze the data to detect mutations.

> *Because GAP4 and PolyPhred run only on Unix platform machines, a file transfer program such as Fetch is required for importing sequence data from a Macintosh computer to the appropriate Unix host. Fetch can be obtained free for noncommercial uses via e-mail request (Fetch@dartmouth.edu). For Sequencher or SeqMan, ABI chromatogram files can be imported directly into those programs on a Macintosh computer. SeqMan and Sequencher also run on a PC/Windows platform.*

ALTERNATE PROTOCOL

LARGE-VOLUME SEQUENCING IN A 96-WELL OR 384-WELL PLATE FORMAT FOR ABI 3700 DNA ANALYZER

Sequencing is the most sensitive method to define mutations in autosomal dominant and other genetic diseases. DNAs from a large cohort of individuals are often screened by direct sequencing of the major candidate genes comprising numerous exons.

Short Protocols in Human Genetics · Searching Candidate Genes for Mutations

Page 7-16 · **Unit 7.4**

Materials *(see APPENDIX 1 for items with ✓)*

 $10\times$ *Taq* buffer, without magnesium chloride (Roche)
✓ 25 mM $MgCl_2$
✓ 1.25 mM 4dNTPs mixture
 8 pmol/µl amplification oligonucleotide primers
 Taq DNA polymerase
 Pfu polymerase (Stratagene)
 Tth XL 3.3× buffer (Applied Biosystems, Roche)
 25 mM magnesium acetate
 Tth XL DNA polymerase (Roche)
 Genomic DNA
 DNA molecular weight marker
 1 U/µl shrimp alkaline phosphatase (USB)
 10 U/µl exonuclease I (USB)
 Dilution buffer (Exo/ShrAP): 50 mM Tris·Cl, pH 8.0
 5× BigDye diluent (prepare with Milli-Q dH_2O, then filter sterilize; Applied
   Biosystems): 400 mM Tris·Cl in Milli-Q H_2O, and 10 mM $MgCl_2$, pH 9
 BigDye Terminator sequencing kit (Applied Biosystems)

 96- or 384-well PCR plates
 Thermal controller/cycler
 PERFORMA DTR 96 Well Plates (Edge BioSystems)
 ABI 3700 automated DNA capillary sequencer

1. Assemble a PCR mixture for a 25-µl reaction on ice. Use one of two reaction cocktails. Both utilize a high-fidelity DNA polymerase mixture.

   PCR reaction cocktail A:

     2.5 µl 10× *Taq* buffer, without magnesium chloride
     1.5 µl 25 mM $MgCl_2$ (1.5 mM)
     4 µl 1.25 mM 4dNTPs (200 µM each dNTP)
     2.5 µl each 8 pmol/µl primer (final 0.8 µM)
     0.5-0.7 U *Taq* DNA polymerase
     0.05-0.07 U *Pfu* polymerase
     ddH_2O to 25 µl

   PCR reaction cocktail B:

     7.5 µl *Tth* XL 3.3× buffer
     1 µl 25 mM magnesium acetate
     4 µl 1.25 mM 4dNTPs (200 µM each dNTP)
     2.5 µl of each 8 pmol/µl primer (final 0.8 µM)
     0.25-0.5 µl *Tth* XL DNA polymerase (Roche)
     ddH_2O to 25 µl

   The best results are usually achieved with amplicons of 200 to 800 bp in size.

   Position the primers 50 to 100 bases from the exon-intron boundaries. The oligonucleotide primers' size is usually 18 to 25 bases. It is advisable to keep the difference in melting temperatures within the primer-pair <2°C. The calculated annealing temperature is 5°C below the oligonucleotide T_m but should be adjusted on an experimental basis.

2. Pipet 23 μl reaction cocktail into wells of a 96- or 384-well plate containing 50 to 100 ng genomic DNA in 2 μl/well (total volume 24 μl/well). Keep all reagents and samples on ice.

3. Centrifuge briefly. Move the plate from ice into the thermal cycler preheated to the denaturation temperature. Perform PCR using the following amplification cycle:

Initial step:	2.5 min	95°C	(initial denaturation)
35 cycles:	20 sec	95°C	(denaturation)
	30 sec	annealing temperature	(primer specific)
	1 min	72°C	(extension)
1 cycle:	5 min	72°C	(final extension).

Hold and store at 4°C.

4. Run a small sample of the PCR product on an agarose gel to assess quality and yield. Use an appropriate DNA molecular weight marker to verify product size and estimate its concentration. Assuming 10 to 15 ng PCR product/100 bp are needed for each sequencing reaction, transfer 2 to 5 μl of PCR product from the PCR plate into a new 96- or 384-well plate for purification and cycle sequencing.

5. Prepare the following enzyme cocktail to get rid of the oligonucleotides and dNTPs in the PCR mixture. Prepare excess master mix for the number of reactions required (typically, prepare 240 μl of cocktail for a 96-well reaction). Pipet 2 μl of enzyme cocktail into each well of the PCR plate:

   0.1 μl 1 U/μl shrimp alkaline phosphatase
   0.1 μl 10 U/μl exonuclease I
   1.8 μl dilution buffer.

6. Briefly pulse-centrifuge the products to bring down any residual fluid along the sides of the wells/tubes. Place the PCR tubes/plates into a thermal controller/cycler using the following parameters:

1 cycle:	15 min	37°C	(enzyme incubation)
	15 min	80°C	(heat inactivate enzymes)
		4°C	(hold).

Store samples temporarily at 4°C or frozen at –20°C until subsequent sequencing.

7. To the enzymatically cleaned PCR product, directly add cycle sequencing reagents individually or as a ready-made mixture/cocktail (no change of PCR plates is necessary at this point). Assemble reagents and scale up for the number of wells/reactions. Reaction per well:

   4-7 μl enzymatically cleaned PCR product
   1-2 μl sequencing primer (4 to 8 pmol, 4 pmol recommended)
   2 μl BigDye Terminator
   6 μl 1× BigDye diluent
   ddH$_2$O up to 21 μl total volume.

Avoid excessive exposure of BigDye reagent to light, as it may result in fading of fluorescence signal.

Choose primers that are nested sequencing primers; these should be designed whenever PCR primers fail to provide the expected results. Nested primers are of particular advantage when the PCR product is very long or of poor quality (i.e., contains extra bands).

8. Briefly centrifuge plate to bring down the contents. Place in thermal cycler and run the thermal cycler using the following parameters:

25 cycles:	10 sec	96°C	(rapid thermal ramp)
	5 sec	50°C	(rapid thermal ramp)
	4 min	60°C	(rapid thermal ramp).

Rapid thermal ramp and hold at 4°C until ready to purify.

9. When ready to purify, briefly centrifuge the plate to spin down well contents. Purify cycle sequencing products using PERFORMA DTR 96 Well Plates per manufacturer's instructions.

10. Dry at 80°C for ~1 hr and store at −20°C protected from light until ready to be analyzed (up to 3 weeks). Resuspend in 10 μl water/well immediately before loading into an ABI 3700 automated DNA capillary sequencer.

11. Transfer ABI chromatogram data and analyze (see Basic Protocol, step 9).

Reference: Nickerson et al., 1997

Contributors: Sandra L. Dabora, Michael Arad, Scott Barr, and Jae Bum Kim

UNIT 7.5

Mutation Detection by Cycle Sequencing

BASIC PROTOCOL

Candidate genes are screened for mutations by a DNA sequencing procedure known as cycle sequencing followed by fractionation on a denaturing polyacrylamide gel and visualization by autoradiography. DNA segments on the order of 200 bp from 10 to 30 individuals can be screened on each gel. Common problems and their possible causes and remedies are listed in Table 7.5.1.

Materials *(see APPENDIX 1 for items with* ✓ *)*

    Blood or other DNA source from individual(s) to be tested
    5 M sodium acetate, pH 4.8
    Isopropanol (pure)
✓ TE buffer, pH 7.4
    Molecular weight markers: e.g., φX-174 plasmid digested with *Hae*III (Hoefer Pharmacia Biotech or Life Technologies)
    12.5 pmol oligonucleotide sequencing primer
✓ 10× T4 polynucleotide kinase reaction buffer
    10 μCi/μl [γ-^{32}P]ATP (3000 Ci/mmol), 10 μCi/μl [γ-^{33}P]ATP (3000 Ci/mmol), *or* 10 μCi/μl [γ-^{35}S]ATP (1000 Ci/mmol)
    5 U/μl T4 polynucleotide kinase
    Dideoxynucleotides: 100 μM ddGTP (*or* 7-deaza-dGTP for GC-rich templates), 600 μM ddATP, 1000 μM ddTTP, and 600 μM ddCTP
    2 U/μl *Taq* DNA polymerase
✓ 10× cycle sequencing reaction buffer
    Mineral oil (if needed, depending on thermal cycler)
✓ Stop solution
    Thermal cycler and thermal cycler tubes

7

Table 7.5.1 Guide to Troubleshooting Cycle Sequencing

Problem	Possible cause	Solution
Bands are seen at same position in all 4 lanes, especially near primer	Template impurity	See if problem exists with control DNA; if not, repurify DNA
	Incorrect primer annealing temperature	Reduce primer annealing temperature; use longer (more stable) primer; increase annealing and extension time
	Impure or old reagents, ddNTP amount too high, or dNTP amount too low	Prepare fresh reagents and readjust dNTP/ddNTP ratios
Bands below a certain site on gel are very dark, and bands above that are very faint	Secondary structure of template impeding extensions	Use higher reaction temperatures
A dark band is present across all 4 lanes of gel; bands below and above it are equally dark	Secondary structure of reaction products causing anomalous electrophoresis position of products in gel	Use higher gel temperatures or formamide gel; use base analogs or TdT in reaction; check purification procedure for PCR products
Intensity of bands at bottom of gel is too low	Proportion of reaction products terminating near bottom of gel too low	Increase ddNTP/dNTP ratio in sequencing mixes
One lane of reaction failed or is weak or smeary	Possible cycler fault in one reaction slot	Check cycler performance
	Pipeting error	Review procedure
All lanes are smeary or show high backgrounds	Template contaminated with other DNA molecules	Run control (template); repeat DNA preparation
	Nonspecific priming, possibly of contaminating DNA molecules	Gel-purify DNA fragment or use a different oligonucleotide (e.g., an internal primer)
	Primer impurity	Check primer purity
	Cycler problem	Check cycler function
	Reagent problem	Review procedure; make fresh reagents
	Primer/template ratio too high or low	Optimize ratio
	Incorrect priming temperature	Calculate T_m as guideline for reaction
Entire sequence is light	Insufficient template or primer in reaction	Double DNA and primer quantities in reaction
	Mismatches between primer and template	Recheck primer and template
	Cycler error; incorrect or inadequate cycling conditions	Check cycler performance; increase time or number of cycles; optimize temperature
	Oligonucleotide inadequately labeled	Check for problems with labeling reaction (e.g., inactive kinase) by measuring label incorporation into TCA-precipitatable material (*APPENDIX 3E*), which should be ~40%–50%
	Template contaminated with excess salt or other reaction-inhibiting material	Ethanol precipitate and resuspend template DNA
	Taq DNA polymerase inactive	Purchase new enzyme
	Reaction buffer defective	Prepare fresh buffer
Oil covering reaction makes it difficult to load gel	Excess oil present	Remove oil or use hot-top apparatus; microcentrifuge reaction before loading gel

1. Isolate and purify human genomic DNA using standard methods (e.g., APPENDIX 3A and UNITS 9.8 & 10.4), resuspending the DNA in water or TE buffer (pH 7.4) at the end of the procedure.

   Only 10 to 20 fmol of DNA template is needed per cycle sequencing reaction (see APPENDIX 3D for DNA quantification methods). It is convenient to calculate the equivalent amount of double-stranded DNA (i.e, how much will constitute 10 fmol) as follows: ng DNA template = 10 fmol × N × 6.6 × 10^{-3}, where N is the number of base pairs of sequencing template.

2. To prepare a DNA template for sequencing, PCR amplify segment of interest from genomic DNA as described in UNIT 7.1.

3. Add 50 μl of 5 M sodium acetate, pH 4.8, and 100 μl pure isopropanol to the 50-μl PCR reaction. Incubate 10 min at room temperature. Microcentrifuge 10 min at maximum speed (∼16,000 × g), room temperature. Discard supernatant and air dry.

4. Resuspend the pellet in 50 μl water or TE buffer (pH 7.4). Quantify the amount of resuspended DNA by analyzing a 5-μl aliquot on an agarose gel and comparing the resulting band to a known amount of a size marker (e.g., 1 μg of φX-174 plasmid digested with *Hae*III).

5. Prepare a 25-μl primer labeling reaction mixture containing:

   12.5 pmol oligonucleotide sequencing primer
   2.5 μl 10× T4 polynucleotide kinase reaction buffer
   1.0 μl 5 U/μl T4 polynucleotide kinase (5 U total)
   3.0 μl 10 μCi/μl [γ-^{32}P]ATP or [γ-^{33}P]ATP (30 μCi total) *or* 2.0 μl 10 μCi/μl [γ-^{35}S]ATP (20 μCi total)
   H$_2$O to 25 μl total volume.

   Incubate 30 min at 37°C (12.5 pmol of labeled primer is sufficient for 100 sequencing reactions). Store the labeled primer up to 1 week at −20°C.

   The sequencing primer is derived from sequence immediately 5′ to the sequence being screened for mutations. A PCR primer (step 2) or a nested primer can be used. Because only a single primer is used, amplification is linear, not exponential.

6. For each DNA template, prepare four thermal cycler tubes and add 3 μl of *one* of the following to each tube: 100 μM ddGTP, 600 μM ddATP, 1000 μM ddTTP, or 600 μM ddCTP. Store tubes on ice.

7. Prepare sequencing cocktail by placing in a fresh tube:

   10 to 20 fmol sequencing template (DNA solution from step 5)
   1 μl 5′-end-labeled sequencing primer (from step 6)
   4 μl 10× cycle sequencing reaction buffer
   1 μl 2 U/μl *Taq* DNA polymerase (2 U total)
   H$_2$O to 30 μl total volume.

   Place 7-μl aliquots of cocktail into each of the four cycle sequencing reaction tubes prepared in step 6. For GC-rich templates, dGTP should be replaced with 7-deaza-dGTP to minimize sequence compression.

8. Add mineral oil (if necessary), close caps, and start thermal cycling. Cycling conditions will vary for different templates, sequencing primers, thermal cycler designs, etc. Begin with standard PCR conditions and optimize the sequencing reaction by varying reaction temperatures, cycling times, and/or number of cycling rounds (see CPMB UNIT 15.1).

Standard PCR conditions to start with are, e.g., 30 cycles of 94°C, 55°C, 72°C with cycling times of, e.g., 1 min, 1 min, 2 min, in Perkin-Elmer Cetus 4800 thermal cycler or 15 sec, 15 sec, 30 sec in Perkin-Elmer Cetus 9600 or MJ Research PTC-200.

9. After completion of thermal cycling, add 5 μl stop solution to each reaction, heat denature 5 min at 95°C, and cool rapidly to <4°C.

10. Analyze sequencing reaction products on a 6% or 8% denaturing polyacrylamide sequencing gel (*APPENDIX 3F*) by loading 5 μl of the denatured reaction and running the gel as far as required. When screening the DNA of multiple individuals simultaneously, load the G reactions together on the gel (followed by A, T, and C reactions) to facilitate identification of mutations.

11. Dry gel and autoradiograph overnight.

Reference: Murray, 1989

Contributor: Ludwig Thierfelder

UNIT 7.6

Human Mutation Databases

TYPES OF MUTATION DATABASES

General Mutation Databases

General mutation databases contain pooled information on variation across the whole genome. Some collect only mutations, others only single-nucleotide polymorphisms (SNPs; see Criteria for Data Submission, below, for the distinction between mutations and polymorphisms). These databases have developed tools for analyzing existing data collections while providing consistent user interfaces to all genes. None of the general databases are sufficient in themselves for the needs in medical genetics, as they are not run by experts on every gene but rather by bioinformatics experts who can present the data collected in a multitude of ways. Due to the effort involved in acquiring variation data on every gene described, only general information about published mutations can be included in these databases. Data that are unpublished, however, remain unreported in general databases. The distinguishing features of these databases can be seen in Table 7.6.1. For examples of general mutation databases, see Accessing Data in Mutation Databases, below.

Table 7.6.1 Current Characteristics of Locus-Specific Databases and Central Databases

Characteristics	Locus-specific databases	Central databases
Mutations collected	Published/unpublished	Published
Polymorphisms	Usually collected	Usually not collected
Patient phenotype data	Usually collected	Usually not collected
Assay data	Usually collected	Usually not collected
Multiple occurrences of mutation	Often collected	Never collected
Ethnicity/geographical distribution	Often collected	Never collected

Locus-Specific Mutation Databases (LSDBs)

LSDBs concentrate on variation within a single gene and are usually run by a consortium of collaborating researchers with scientific expertise in a particular gene or phenotype. Some curators are responsible for a number of LSDBs at a single Web site that may be specific for a particular disease or group of related genes. Due to the expertise of the curator, LSDBs often contain far more information on each mutation than general mutation databases (see discussion above), and some contain such extensive information that they have become known as knowledgebases; the phenylalanine hydroxylase database is an example of such a knowledgebase. LSDBs generally contain a greater number of mutations in each gene than general databases because unpublished mutations are included as they are curated by the expert(s) presenting the database. As unpublished mutations can often equal published mutations, this is a huge amount of information to be made available. The typical characteristics of these databases can be seen in Table 7.6.1.

There are now over 270 LSDBs available on the World Wide Web, and there are likely to be many more. Due to the different curators' expertise in informatics or ability/inability to spend time or money, LSDBs vary enormously in content and presentation (Claustres et al., 2002). Some LSDBs are nothing more than a collection of mutations and their references, while others are fully interactive databases with a wealth of knowledge. Details of a review of LSDBs can be seen in the paper by Claustres et al. (2002). The primary list of LSDBs and links is available on the HGVS Web site. This review also gives some detailed recommendations for database content.

SUBMISSION OF DATA TO MUTATION DATABASES

Methods for Submission

Mutations are entered into databases by several means. One of these is publication searching—most general mutation databases search the literature on a weekly or monthly basis by a computerized search procedure to acquire new data. Mutations can also be entered into databases by direct submission.

Investigators generally submit mutation data to LSDBs either by directly contacting the curators themselves or by filling in an electronic form available on the database. This is usually done before publication, or often when no publication is intended. Direct submission is also invited by the SNP databases. In order to collect the maximum number of mutations, the authors of this unit urge the reader to submit any unpublished mutations to the appropriate databases. The HUGO MDI has developed an entry form that helps with quality control of data as a guide to the submission of mutations in databases (Fig. 7.6.1). The review by Claustres et al. (2002) showed that 16% of all LSDBs use this form for collection of mutations in their databases.

Criteria for Data Submission

Before material can be entered in a database, it must be ensured that the data are of the highest quality. Nomenclature must conform to standards and quality control checks must be done.

Nomenclature
The issue of nomenclature has been discussed at two levels: (1) the term "mutation" and (2) how to designate a mutation. "Mutation," throughout biology, means a change in the DNA sequence; however, in clinical medicine, which is the field to which this unit pertains, "mutation" refers to a base change that causes disease, whereas "polymorphism" refers to a harmless change. This problem has been discussed in the article by Cotton and Scriver (1998) and a recommendation from Victor McKusick is that all base changes be referred to as allelic variants.

SAMPLE ALLELE VARIANT ENTRY FORM

Enter allelic variants/mutations and polymorphisms (SNP's) in the fields below; information will be added to the database after the entry has been checked by the database curators. The date of each submission will be recorded, and you will receive confirmation by email that submission has been received and approved.
Please add as much descriptive information as possible to the comment field; in particular, please try to give as complete a description as possible of the result of the mutation in terms of the RNA and protein products.
Email us with any problems.

Submitted by-
First name:
Last name:
Email:
Laboratory:
General information
Gene/Locus name:
Gene symbol:
Category: **mutation causing disease** ☐   **polymorphism not causing disease** ☐   **don't know** ☐
Mutation/Polymorphism name (to follow nomenclature guide):
Aliases/Published name:
Mutation type   **AA substitution** ☐   **stop codon** ☐   **RNA splicing** ☐   **deletion** ☐ **insertion** ☐
                   duplication ☐   **no change** ☐   **unknown** ☐   **Other list_____**
DNA
Location (exon/intron):
Nucleotide (or starting nt):
 # bases inserted or deleted:
Original base(s):
New base(s):
Protein
Mutation name (to follow nomenclature guide):
Old aa:
New aa:
Residue #:
Quality Assurance Checklist for Mutation/Polymorphism
(Yes-Y/No-N/Not checked-NC)
Mutation/polymorphism found on repeat PCR sample (not an artifact) **Yes** ☐   **No** ☐   **Not Checked** ☐
Mutation segregates with trait) **Yes** ☐   **No** ☐   **Not Checked** ☐
Parents, grandparents, siblings, or other family members checked for carrier status/result)) **Yes** ☐ **No** ☐ **Not Checked** ☐
No other mutation found on same allele (in cis)) **Yes** ☐   **No** ☐   **Not Checked** ☐
Mutation affects conserved residue) **Yes** ☐   **No** ☐   **Not Checked** ☐
Information content for splicing mutations) **Yes** ☐   **No** ☐   **Not Checked** ☐
Expression analysis confirms phenotypic effect) **Yes** ☐   **No** ☐   **Not Checked** ☐
Number of normal individuals tested/frequency) **Yes** ☐   **No** ☐   **Not Checked** ☐
Other databases
Mutation in HGMD   **Yes** ☐   **No** ☐
Mutation in LSDB   **Yes** ☐   **No** ☐
Detection
cDNA analyzed for variant) **Yes** ☐   **No** ☐
Genomic DNA analyzed for variant) **Yes** ☐   **No** ☐
Detection method:
Detection conditions: (sequence of primers, PCR and electrophoresis conditions,...):
Extent of DNA analyzed:
Diagnosis method developed: (ASO, etc.):
Ability to performs tests for others   **Yes** ☐   **No** ☐
Comments:
Haplotype association:
Population association:
Geographic association:
Other:
References:

Figure 7.6.1   Proposed HUGO Mutation Database Initiative allele variant entry form for mutation databases. From *http://www.hgvs.org/entry.html.*

7

Table 7.6.2 Quality Control of Data Submitted to Databases

Features	Comments
PCR errors	Has the change been found in a second PCR sample amplified from the patient's DNA?
Incomplete scanning	Even if one mutation is found, scanning should not be stopped because a second mutation that may be disease-causing might be there. If using a method that detects only ≤80% of the mutations, some might be missed.
Phenotype	Is the change always present when the phenotype is present?
Present in 100 normals?	Ideally, to prove that a mutation is disease-causing, the change should be shown not to be present in 100 normal chromosomes.
Expression analysis	Ideally this should be done to prove the substitution affects activity.
Conserved residue	Is the base change in a conserved residue? If so, it will more likely be disease-causing.
A polymorphism?	Is the frequency of the base change reported in >1% of the normal population?
Reporting errors	Typographical errors
	Misreading a gel
	Incorrect assignment of base position
	Incorrect deduction of amino acid change

Regarding designation, the HUGO MDI Nomenclature Working Group has finalized and published recommendations based on the systematic name for the DNA change. These have now become widely accepted. For example, the systematic name c.2674C→A means that the C at the 2674th base of the cDNA is changed to A. The corresponding trivial name T265Y uses the single-letter code for amino acids and means threonine 265 is changed to tyrosine. If this same mutation were described in genomic DNA the mutation would be designated g.2674C→A. If on the other hand it was a deletion of C at position 2674 it would be designated g.2674delC. More details and examples may be found on the Recommendations page of the HGVS Web site.

Quality control

Base changes in genes need to be reliably reported in databases. Therefore, a few points should be checked before submitting a mutation, to ensure the highest quality of data in the database. The quality of published mutations that enter mutation databases directly should be high, although the authors have shown that this is not necessarily the case. The most important features to look for in quality control are shown in Table 7.6.2.

ACCESSING DATA IN MUTATION DATABASES

Mutation data can be accessed in many ways: through the general mutation databases of OMIM, HGMD, and others, or through the more than 270 LSDBs. Searching for a mutation in either of these is a relatively simple and straightforward process. One should familiarize oneself with OMIM and HGMD, as well as the LSDB of the gene of interest, and see the differences in the information contained therein. Some examples are shown in the following paragraphs.

General Mutation Databases

Online Mendelian Inheritance in Man (OMIM)

OMIM documents inherited disorders and includes all known genes; however, only selected published mutations are included. These include the first mutation described in a locus, the most common (dozen or so) mutations at a locus, mutations of historical interest, mutations showing an unusual mechanism of mutagenesis or unusual pathogenesis, or any other mutation of interest. The information is shown as text—especially clinical information with reference information, links to MEDLINE articles, and sequence information. As of Dec. 3, 2002, OMIM contained 14053 entries in total. To access mutation data in OMIM, one searches for the disease name, gene name, or OMIM accession number. The result is a gene or genes, along with the assigned accession number(s). By clicking each gene, a table of contents for that gene is obtained. Note the section Allelic Variants/View of Allelic Variants; this is where any mutation information available in OMIM will be found.

Human gene mutation database (HGMD)

HGMD became publicly available online in 1996. It is a collection or index of all the published germline mutations in nuclear genes along with their references and summaries of mutations in each gene. Maps and reference sequences are also included. There is, however, a considerable lag from publication time to appearance on the public version of HGMD due to commercial contracts. As of September 2, 2002, HGMD contained 30641 mutations in 1245 genes and provided 1100 reference cDNA sequences.

The database may be searched either by disease, gene name, gene symbol, or GDB or OMIM accession numbers. HGMD tabulates the information and categorizes it by mutation type (e.g., small deletions or small insertions, among others) or phenotype (with links to OMIM matching these phenotypes). Choosing one of the mutation types, e.g., nucleotide substitutions, leads to another table listing all the nucleotide substitutions with their relative location, the substitution, the phenotype, and the associated reference and link to PubMed. Distribution of mutations can be assessed at a glance by consulting the mutation maps provided; these are graphical displays of mutations in the coding regions. Information such as frequency of different types of mutations and links to any LSDB available for that gene are also provided.

Human genome variation database (HGVbase)

This database was previously known as HGBASE. HGVbase attempts to summarize all known sequence variations (i.e., mutations and SNPs, indels, simple tandem repeats, and other sequence alternatives) in the human genome and to facilitate research into how genotypes affect common diseases, drug responses, and other complex phenotypes. All sequence variations are presented with details of how they are physically and functionally related to the closest neighboring gene. At present the majority of data are of the SNP type. Data are harvested (with permission), submitted from all major public genome databases, or extracted from literature published over the previous decade or so. Submissions received from research groups are also welcomed. Extensive annotation and manual review of every record is undertaken to address potential errors and deficiencies.

dbSNP

The most common genetic variations are single nucleotide polymorphisms (SNPs), which occur approximately once every 100 to 300 bases. The term SNP is loosely used by the dbSNP database and implies "minor genetic variation"; thus dbSNP includes disease-causing clinical mutations as well as neutral polymorphisms, although it mainly contains polymorphisms. There has recently been great interest in SNP discovery, and detection of SNPs is expected to facilitate large-scale association genetics studies. Established September, 1998, by the National Center for Biotechnology Information (NCBI) to serve as a central repository for SNPs, microsatellite repeats, and small insertion/deletion polymorphisms, this database now contains over 4 million human SNPs. Although not part of GenBank, the sequences of dbSNP records are expected to be contained within the sequences of one or more GenBank records. Submissions are invited from all laboratories; the major contributors are laboratories associated with the grants program of the National Human Genome Research Institute (NHGRI) of the United States.

National and Ethnic Mutation Databases

Databases showing lists of mutations in particular countries or specific ethnic groups ensure complete collection of mutations and are useful in patient management and diagnostics, acting as a diagnostic tool for clinicians, geneticists, counselors, and health care planners. One excellent example is the Arab Genetic Disease database, which contains over 1000 unique, annotated entries of disorders that occur in Arab populations. Although there are not many such databases available online, there are a few in development—e.g., the Finnish, Korean, and Chinese disease databases. The HGVS encourages creation of such databases.

Locus-Specific Databases (LSDBs)

An LSDB may have a vast amount of information or very little, depending on the gene. Since there is no standard, the presentation, user interfaces, and content of LSDBs vary almost as much as the genetic loci covered by the more than 270 LSDBs available. LSDBs vary from simple flat-file databases consisting of a mutation list plus published references to interactive knowledgebases with information such as ethnic or geographical origin of patient, detection method, restriction enzyme sites, clinical information, and more (Claustres et al., 2002). The phenylalanine hydroxylase knowledgebase (PAHdb) is a model LSDB. The database is run by a consortium of over 80 specialists in PAH. Mutation data have been collated from both published articles and personal communications from the PAH mutation analysis consortium. The database aims to provide users with access to up-to-date information about mutations at this locus. It may be searched by mutation name, polymorphic haplotype, population, geographic location, gene region, codon number, mutation type, substitution, phenotype, and author's name. Data may be submitted to the database via an electronic form. Information about the disease phenylketonuria is available for patients and their families as well as links to other sources of information. Two equally excellent but completely different LSDBs are the Syllabus of Human Hemoglobin Variants and the Blood Group Antigens databases. The authors of this unit hope that other LSDBs follow these examples.

A complete listing of LSDBs (and other mutation databases) may be found on the HGVS Web site—a database of databases that has also been published. Databases are listed in alphabetical order by approved HUGO gene symbol; a link to the OMIM entry for that gene is also provided. Other such lists are available on the HGMD, OMIM, and other Web sites, although these are not as comprehensive.

Steps in creating a locus-specific database
Many curators have been encouraged by the HUGO MDI (now the HGVS) to create and curate a mutation database for their gene(s) of interest if there is a need for such a database. In order to ensure smooth parsing between databases, various criteria need to be addressed, and the HUGO MDI working groups have made recommendations for nomenclature (see Criteria for Data Submission, above), content of databases, and quality control. These recommendations can all be viewed online by following the Recommendations link on the HGVS Web site. Steps to follow for creating an LSDB are, for the time being:

1. Contact Ourania Horaitis at the HGVS office (*horaitis@medstv.unimelb.edu.au*) to determine if someone else is already working on a new database for the gene.

2. Form a consortium of interested researchers and clinicians to spread the load, and ensure longevity of the database by forming a group to find a new curator, if necessary; the original

Table 7.6.3   Core Information (Essential Content) as Suggested by Scriver et al. (1999, 2000)

Item no.	Information to include
1	A unique identifier must be issued by the database for each allele
2	The source/report of the data
3	The context of the allele: reference and sequence accession no.—DDBJ/EMBL/GenBank
4	The name of the allele using the guidelines in Antonarakis et al. (1998) and den Dunnen and Antonarakis (2000): e.g., 1425→A
5	Description of the change as fully as possible
6	A document showing the "biography" of the database and a map of its details
7	A minimum set of links to other databases should be included for mutual compatibility. e.g., OMIM
8	The date when the database was last updated

investigator(s) may not be involved with the gene in a few years. Recently the PAH database designated subcurators for specific areas of the database, further strengthening the database and spreading the load.

3. Collect all mutations described by all investigators involved with the gene, including polymorphisms, and name them according to the recommendations on the HGVS Web site. One might wish to publish this data as a review for a specialized journal such as Human Mutation (which features reviews of mutations), or another appropriate journal. This is a good incentive to collect the relevant information, since a useful publication will be produced.

4. Include "Core" information (Table 7.6.3). To make the database more meaningful and useful, one should include all the elements of the Core information. Of course other data particular to the disease or gene of interest may be included, and it is encouraged that the following be included: HUGO-approved gene symbol; chromosome region; gene information; location exon/intron; phenotype (if any); RNA effect; protein effect; detection method; clinical information about disease for clinicians and patients; geographic origin; ethnic origin; expression studies; list of contributors to database; and list of associations and organizations. The review by Claustres et al. (2002) also makes a list of suggestions for the ideal LSDB.

5. Determine the software to be used. The HGVS hopes to be able to provide ideal "kit" software that addresses all the criteria for an LSDB in future; however, for the time being there are three LSDB-specific software applications currently made available by HGVS members, which are described individually under Software for Creating and Curating LSDBs, below. It is necessary to contact the curators/creators directly in order to use any of them. Look at the databases running under particular software applications to see which, if any, are appropriate for the purposes at hand. Contact information is available on the respective Web sites. Alternatively it is possible to create a simple flat-file database using HTML. Claustres et al. (2002) showed that 72% of all LSDBs use this format even though it is not ideal.

6. Once created, register the LSDB for inclusion in the HGVS online list and the HGVS newsletter, so that it is visible to the community. See the HGVS Web site for information on how this is done.

Software for creating and curating LSDBs

MuStaR (mutation storage and retrieval): This was developed by Alastair Brown in 1999 (Brown et al., 2000) for use with the *PAX6* and later the *PAX2* genes. It has more recently been used at other sites, e.g., in setting up the SHOX Mutation Database. The curation software is based on Microsoft Access, and Web sites can be created for the database for both Unix and Windows platforms; the curation software itself is restricted to Windows PCs. This software is freely available to the academic community.

Universal mutation database (UMD): This was developed by Christophe Beroud in 1994 (Beroud et al., 2000) and used for a large number of genes including *p53*, *WT1*, *BRCA1*, and *BRCA2*. The software is based on 4th Dimension and uses the 4D language. UMD is currently freely available. UMD's software, while mutation- and patient-oriented, is hosted by a central server and therefore is not amenable to use on LSDB curators' local systems.

MUTbase: This was developed by Mauno Vihinen and co-workers in 1999. Currently the MUTbase system is used for over 30 immunodeficiency genes including *ADA*, *AIRE*, *BLM*, *BTK*, *C2*, *CYBA*, *CYBB*, *IKBKG*, *IL12RB1*, *ITGB2*, *JAK3*, *NCF1*, *NCF2*, *PNP*, *RAG1*, *RAG2*, *SH2D1A*, and *ZAP70*. MUTbase is patient-based and includes clinical information for each subject. Further, the software runs only on its central server, not on individual curators' local systems.

References: Claustres et al., 2002; Cotton and Scriver, 1998

Contributors: Ourania Horaitis and Richard G.H. Cotton

7

CHAPTER 8

Clinical Cytogenetics

Cytogenetic analysis provides a rapid screen for large-scale genetic rearrangements associated with syndromes of abnormal development. Discovery of the mitogenic effects of phytohemagglutinin and development of a simple and rapid means of growing lymphocytes led to the introduction of chromosomal analysis into the repertoire of the hospital clinical laboratory. Fluorescence in situ hybridization (FISH) has become a routine clinical test used to detect many microdeletion syndromes and to characterize rearranged chromosomes.

Chapter 4 provides protocols for obtaining preparations of human chromosomes, staining, and in situ hybridization, while this chapter outlines clinical application of these techniques.

Techniques for culturing peripheral blood lymphocytes and staining of chromosomes to elicit banding patterns are provided in UNITS 4.1 & 4.3. Adaptation of these techniques for prenatal cytogenetic analysis with chorionic villus tissue is presented in UNIT 8.1 or with amniotic fluid cells in UNIT 8.2. Chorionic villus sampling allows for diagnosis in the first trimester, but is limited by a slightly higher rate of complications. Both chorionic villus sampling and amniotic fluid analysis are subject to misleading results of pseudomosaicism. A protocol for culture and cytogenetic analysis of fibroblast cells is provided in UNIT 8.3. This is useful in the study of products of conception, stillborn infants, and for analysis of a second tissue in the study of possible chromosomal mosaicism.

Two special protocols are provided for the study of disorders characterized by an increased rate of chromosome breakage. Bloom's syndrome is an autosomal recessive disorder characterized by short stature, a photosensitive skin rash, and increased incidence of malignancies. It is associated with a high frequency of sister chromatid exchange as described in UNIT 8.4. Fanconi anemia is an autosomal recessive disorder characterized by aplastic anemia and congenital malformations. The most sensitive test for affected individuals and carriers, presented in UNIT 8.7, is the detection of increased rates of chromosome breakage when cells are grown in the presence of diepoxybutane.

Techniques of molecular analysis are being increasingly incorporated into routine cytogenetic studies. Fluorescence in situ hybridization bridges the gap between microscopic analysis of chromosome structure and detection of chromosome changes at the DNA level. While the basic approach to FISH is provided in UNIT 4.4, a means of detecting aneuploidy in interphase cells using FISH (as well as a description of probes for chromosome-specific repeated sequences) is outlined in UNIT 8.5. This is useful when dividing cells are not available for cytogenetic study—for example in archived pathological material. It can also be adapted for rapid prenatal detection of major aneuploidies (UNIT 8.6), although this does not substitute for more thorough analysis of chorionic villus or amniotic fluid cells by full chromosomal analysis.

Forty years ago it was believed that the diploid chromosome number in humans was 48. Now we are well on our way towards a complete map of the human genome. Cytogenetic analysis has always been at the leading edge of the clinical application of genetic knowledge. Although it can be expected that there will be an increasing melding of cytogenetic and molecular genetic techniques, it is likely that the overview of the genome provided by study of the chromosome will remain important for many years.

Contributor: Bruce Korf

Preparation of Metaphase Spreads from Chorionic Villus Samples

BASIC PROTOCOL

CULTURE METHOD FOR PREPARING METAPHASE SPREADS FROM CHORIONIC VILLI

The mesenchymal cores of chorionic villi contain a variety of cellular elements—fibroblasts, endothelial cells, and macrophages—and an abundant intercellular matrix. Fibroblasts of the villous core are capable of rapid proliferation in vitro and are suitable material for cytogenetic diagnostic purposes.

NOTE: All incubations are performed in a humidified 37°C, 5% CO_2 incubator unless otherwise specified.

Materials *(see* APPENDIX 1 *for items with* ✓ *)*

   15 to 20 mg chorionic villus samples
✓ HBSS
✓ Trypsin/EDTA/HBSS solution
✓ Collagenase solution
   Complete medium: Chang in situ medium (Irvine Scientific) or Amniomax (Life Technologies), 37°C
   10 μg/ml Colcemid (e.g., Life Technologies)
   1% (w/v) sodium citrate, 37°C
   Fixative: 3:1 (v/v) methanol/acetic acid, 4°C (prepare fresh)

   Watchmaker's fine forceps, sterile
   4.5-ml cryotubes (Nunc)
   Rocking table tube mixer (blood mixer)
   Sorvall GLC-2B centrifuge equipped with HL-4 rotor and six 15-ml swinging buckets (or equivalent)
   Glass coverslips, flame sterilized
   35-mm plastic petri plates
   Inverted microscope *or* dissecting stereomicroscope
   Hot-steam humidifier
   Air pump

1. Carefully separate villi from maternal decidua under an inverted phase-contrast microscope or dissecting stereomicroscope using sterile watchmaker's fine forceps. Wash villi by transferring them through changes of HBSS until the villi are clean. Divide the sample into two parts—one part for culturing (step 2) and the other for direct preparation (see Alternate Protocol, step 1).

2. Transfer 4 to 10 mg dissected and washed villi to a 4.5-ml sterile cryotube that contains 4 ml trypsin/EDTA/HBSS solution. Incubate 60 min on a rocking table tube mixer at 37°C. Centrifuge 5 min at 800 × *g*, room temperature.

3. Gently aspirate and discard trypsin solution. Add 1.5 to 2 ml collagenase solution and incubate 30 min on the rocking table tube mixer at 37°C.

4. Centrifuge 5 min at 800 × g. Gently aspirate and discard all but ~0.5 ml of the collagenase solution.

5. With a sterile Pasteur pipet, gently pipet villi and remaining collagenase solution up and down ten times to break villi into a single-cell suspension. Add 3 ml prewarmed Chang in situ medium to the suspension.

6. Place a 0.5-ml aliquot of the single-cell suspension on the surface of a flame-sterilized coverslip in two to six labeled 35-mm plastic petri dishes. If the cell suspension on the coverslip is too dense (as observed with an inverted microscope), dilute with Chang in situ medium. Incubate 24 hr.

7. Add 2 ml complete Chang in situ medium to each dish and incubate until cells attach to coverslip (2 to 3 days).

8. After cells attach to coverslip, aspirate medium, add 2 ml fresh complete Chang in situ medium, and incubate.

   The best results are obtained when only the mesenchymal cores are cultured. If there is a lot of debris or maternal cell contamination (floating cells), the coverslip should be carefully transferred to another dish with 2 ml Chang in situ medium.

   After 1 to 2 days, nest-like colonies of fibroblastic cells should be present. Coverslips should be ready for harvesting 5 to 7 days after plating, depending upon cell growth and mitotic activity (i.e., number and size of colonies and number of dividing cells).

9. Add 10 μl Colcemid to each dish and incubate 1 hr. Remove and discard medium from the dish and add 3 ml of prewarmed 1% sodium citrate. Incubate 20 min.

10. Gently add 2 ml of 4°C fresh fixative to the dish and let stand 2 min at room temperature.

11. Aspirate sodium citrate/fixative solution and replace with 2 ml of 4°C fresh fixative. Let stand 20 min. Repeat two additional times, incubating the cells in fixative for 20 min the first time and 10 min the second time.

12. Aspirate fixative, remove excess liquid, and dry coverslips one at a time. Expose surface of coverslip to hot steam ~1 to 3 sec. Immediately start the drying process by directing a very gentle flow of air from the air pump onto the surface of the coverslip. Observe the surface of the coverslip for uniform drying and adjust the airflow if necessary

    Correct drying of the coverslips is critical to success of the experiment.

13. Check the coverslip with an inverted microscope to assess the degree of chromosome spreading. Adjust the drying procedure if necessary to obtain good metaphase spreads.

14. Stain the chromosomes using the trypsin-Giemsa banding procedure (*UNIT 4.3*).

15. Photograph and analyze 7 to 10 metaphase spreads. Photograph and karyotype (*APPENDIX 3K*) two spreads.

8

DIRECT METHOD FOR PREPARING METAPHASE SPREADS FROM CHORIONIC VILLI

Metaphase spreads are obtained directly from spontaneously mitotic trophoblastic cells (Langerhans cells of the cytotrophoblast) or from trophoblasts after a limited period (i.e., only several hours) of incubation.

Additional Materials (see APPENDIX 1 for items with ✓)

    Chorionic villus samples, dissected and washed (see Basic Protocol)
✓ Complete RPMI/20% FBS containing 1% gentamycin, 37°C
✓ 10 uM fluorodeoxyuridine (FUdR)
✓ 1 mM thymidine
    60% (v/v) acetic acid in H_2O (prepare fresh)

✓ Slide-making supplies
    Slide warmer
    65° or 90°C oven

1. Place remainder of villi not used in the culturing method into a 35-mm petri plate containing 3 ml complete RPMI/20% FCS prewarmed to 37°C. Incubate villi ≥ 3 hr (never <3 hr).

2. Add 0.1 ml of 10 mM FUdR (3.3×10^{-7} M final concentration) to synchronize the culture. Incubate 15 to 17 hr.

3. Add 0.1 ml of 1 mM thymidine (3.3×10^{-5} M final concentration). Incubate 5 hr (incubation time in thymidine can vary from 3 to 8 hr without adversely affecting metaphase yield).

4. Add Colcemid to a final concentration of 0.5 μg/ml and incubate 2 hr. Remove dish from incubator and gently aspirate medium with a Pasteur pipet.

5. Slowly add 3 ml of 37°C 1% sodium citrate. Incubate 20 min. Gently add 0.5 ml of 4°C fixative to stop the hypotonic action of sodium citrate.

6. Aspirate sodium citrate/fixative solution and add 2 ml of fresh 4°C fixative. Rinse villi by rocking the dish, then aspirate fixative. Add 2 ml fixative and refrigerate overnight (or for 20 min if the result is needed within 24 hr).

7. Remove fixative and let plate stand 1 to 2 min at room temperature to allow remaining fixative to evaporate.

8. Add 100 to 200 μl of 60% acetic acid. Observe release of cells from villi under an inverted phase-contrast microscope. Agitate gently. When release of cells is apparent (by single cells coming off of the tissue), proceed to slide-making. Check the number of single floating cells.

Preparing slides using bent-pipet rake

9a. Place a labeled slide on surface of slide warmer set to ~45°C. Place 100 μl fixed cell suspension on slide and gently touch bent-pipet rake to cell suspension to spread it across slide by surface tension.

10a. Without touching slide surface with rake, move slide slowly with the other hand so that the rake spreads the fixed cell suspension along the length of the slide to cover the central 50% to 60% of the slide. Make three to five back-and-forth trips, taking 30 to 60 sec for each trip, to spread and dry the suspension (should take 3 to 5 min).

11a. Remove slide from slide warmer and blot any remaining fluid by tapping edge of slide. Use an inverted microscope to check for mitotic spreads. Incubate slides overnight in a 65°C oven or 10 min at 90°C before staining (step 12).

Preparing slides using mechanical slide-making apparatus

9b. Set apparatus temperature to 45°C and place labeled slides in slide holders. Add 100 μl of fixed cell suspension to each slide and lower the combs. (The apparatus can prepare six slides at one time.)

10b. Run apparatus 4 to 5 min. Remove slides and blot any remaining fluid by tapping edge of slide. Use a phase-contrast microscope to check for mitotic spreads. Incubate slides overnight in a 65°C oven or 10 min at 90°C before staining. Proceed to step 12.

12. Stain the chromosomes using trypsin-Giemsa banding procedure (*UNIT 4.3*). Analyze the results from 7 to 10 metaphase spreads.

Reference: Brambati et al., 1987

Contributors: Laird Jackson, Longina M. Gibas, and Marie A. Barr

UNIT 8.2

Preparation, Culture, and Analysis of Amniotic Fluid Samples

NOTE: To ensure successful culturing for each case, it is advisable to culture each sample using two different lots of medium, each with a separate lot of fetal bovine serum (FBS). Preferably, the two setups should be placed in separate incubators supplied by independent sources of CO_2.

NOTE: All incubations are performed in a humidified 37°C, 5% CO_2 incubator unless otherwise specified.

BASIC PROTOCOL

IN SITU METHOD FOR THE PREPARATION, CULTURE, HARVEST, AND ANALYSIS OF AMNIOTIC FLUID SAMPLES

Materials (see APPENDIX 1 for items with ✓)

    10 to 20 ml amniotic fluid
    95% ethanol
✓ 20% Chang medium, room temperature and 37°C
    10 mg/ml Colcemid (e.g., Life Technologies)
    Hypotonic solution: 0.075 M KCl *or* 0.7% (w/v) sodium citrate
    6:1 and 3:1 (v/v) methanol/glacial acetic acid fixative (ACS grade; prepare fresh daily)
    Permount

    20-ml syringes or tubes approved for cell culture
    15-ml centrifuge tubes
    Centrifuge (e.g., Fisher Marathon 21K with four-place rotor)
    35-mm tissue culture plates
    22-mm^2 glass coverslips (no. 1; Becton Dickinson Clay Adams Gold Seal or Corning), soaked in 95% ethanol overnight

8

9-in. (23-cm) Pasteur pipet, autoclaved
Inverted microscope
Fine forceps
Blue pads *or* paper towels
60°C slide warmer
Red china pencil
3-in. × 1-in. × 1-mm microscope slides (Curtin Matheson)

1. Obtain amniotic fluid in 20-ml syringes or tubes that have been approved for cell culture. Maintain samples at room temperature until they are processed. Do not refrigerate samples or expose them to heat, as temperature extremes may be deleterious to cell growth.

    The volume should be ~20 ml for samples obtained from amniocentesis performed later than 14 gestational weeks and 10 ml for samples obtained from amniocentesis performed earlier than 14 weeks. A general rule is that ~1 ml of fluid per gestational week is obtained. Process samples on same day of arrival.

2. Gently shake original vessel containing amniotic fluid sample to resuspend cells.

3. For samples >15 ml, transfer equal amounts of fluid to two 15-ml centrifuge tubes; transfer samples <15 ml to one centrifuge tube. Centrifuge 10 min at 109 × *g*, room temperature.

4. While tubes are centrifuging, label 35-mm culture plates on tops and bottoms. Label five plates for samples obtained from routine amniocentesis (14.0 to 22.0 gestational weeks) and four plates for samples obtained from early or late amniocentesis (earlier than 14.0 gestational weeks or later than 22.0 gestational weeks, respectively).

5. Flame-dry sterile 22-mm² coverslips (ensure that they are completely dry) and place one in all but the last culture plate.

6. Remove tubes from centrifuge, taking care not to disturb cell pellets. Draw off supernatant from each tube, leaving ~1 ml fluid above pellet. Transfer supernatant to another tube and save for any biochemical testing at 4°C.

For samples from routine amniocentesis with >15-ml volume

7a. Gently resuspend the pellet in one of the two tubes with 0.7 ml of room temperature 20% Chang medium. Tap gently on side of tube to resuspend pellet. Distribute cell suspension equally to the center of the coverslips in plates 1 and 2. Be very careful to confine the suspension to the coverslip.

8a. Resuspend the pellet in the second tube with 1.2 ml of room temperature 20% Chang medium. Distribute this cell suspension equally to the three remaining plates, again being careful to keep the suspension on the coverslips in plates 3 and 4.

9a. Add 2.0 ml of room temperature 20% Chang medium to plate 5 (the monolayer backup culture). Incubate cultures 24 to 72 hr.

For all samples from early amniocentesis and for any sample with <15-ml volume

7b. Add 1.6 ml of room temperature 20% Chang medium to the cell pellet and gently resuspend as in step 7a.

8b. Distribute cell suspension equally to four culture plates labeled 1 through 4 (see step 5). Be careful to keep cell suspension contained on the glass coverslips in the first three plates.

9b. Add 2 ml of room temperature 20% Chang medium to the fourth culture plate (the monolayer backup culture). Incubate cultures 24 to 72 hr.

For samples from late amniocentesis

7c. Add 0.7 ml of room temperature 20% Chang medium to each tube and gently resuspend each pellet as in step 7a.

8c. Distribute the cell suspension from one tube to plates labeled 1 and 2, each containing a glass coverslip.

9c. Distribute the cell suspension from the second tube equally between plates 3 and 4, one that holds a coverslip and one that does not (the monolayer). Add 2 ml of room temperature 20% Chang medium to the monolayer. Incubate cultures 24 to 72 hr.

10. Gently add 2 ml of prewarmed 20% Chang medium to the culture plates containing the glass coverslips. Incubate cultures 2 days.

11. Aspirate medium with an autoclaved Pasteur pipet into a vacuum flask set up in the hood. Be sure to move the pipet around the periphery of the plates so the cells are not disturbed.

12. Add 2 ml of prewarmed 20% Chang medium. Incubate cultures. After this first feeding, feed cell cultures on a weekly basis. Do not feed cells on coverslips that will be harvested the same day, so as not to disturb actively dividing cells that are rounded up and less attached. Optimally, feed cells 24 to 48 hr before harvesting to promote optimal growth. If the culture is particularly "junky" or bloody, gently agitate the plate before aspirating medium.

13. On the fifth day in culture, check primary cultures periodically for colony formation and cell growth using an inverted microscope. Check passed cultures starting 24 hr after passage.

   The day of the initial setup is the first day of culture. Most primary cultures are ready for harvest after 5 to 9 days in culture. If after 5 days of culture, colonies are forming and cells are mitotically active, growth is considered good. Under an inverted microscope, check for appearance of refractile doublets. Harvest the cells (see Support Protocol 1) when a significant number of mitotic doublets are present on the periphery of the cell colonies and there is still enough space between them to distinguish separate colonies.

14. Add 50 μl of 10 mg/ml Colcemid to each 2-ml coverslip culture. Gently rotate plate to mix. Incubate 20 min.

15. Remove plates from incubator and gently add ∼1 ml hypotonic solution around inner edge of plate. Let stand 10 to 12 min.

16. Carefully aspirate the hypotonic solution/medium mix from around edge of plate using a Pasteur pipet. Add 2 ml hypotonic solution as in step 15 and let stand 12 min.

17. Gently add ∼1 ml of fresh 6:1 fixative around inner edge of plate and let stand 10 min. Then, carefully aspirate fixative/hypotonic mix from around edge of plate.

18. Add 2 ml of fresh 3:1 fixative. Let stand 12 min, then aspirate. Repeat this procedure with fresh 3:1 fixative two or three times for 10 min each time. Do not remove final fixative.

19. Lift coverslip out of plate with fine forceps and place edge on a paper towel to drain excess fixative. Blow gently on cell side of coverslip. Lean corner of coverslip (cell-side-up) against the cover of the 35-mm plate placed on top of a wet blue pad or wet paper towels to create a humid environment while the drying process begins (Fig. 8.2.1A). Let sit 2 to 3 min on wet pad or towels.

20. Lean the coverslip (cell-side-up) against the 35-mm plate cover on a 60°C slide warmer until dry (∼10 min). When coverslips are completely dry, hold between thumb and first finger and gently label back with a red china pencil.

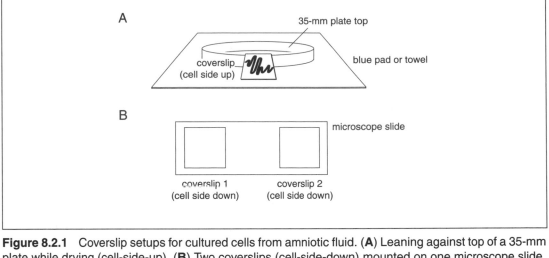

Figure 8.2.1 Coverslip setups for cultured cells from amniotic fluid. (**A**) Leaning against top of a 35-mm plate while drying (cell-side-up). (**B**) Two coverslips (cell-side-down) mounted on one microscope slide.

21. Leave coverslips cell-side-up for at least 4 hr, but for no more than 24 hr, on the 60°C slide warmer. GTG-band coverslips immediately or store them ≤2 weeks at room temperature before banding.

22. Stain coverslips in Coplin jars for coverslips using the standard recipes and times for staining slides (*UNIT 4.3*). As a guideline for first-time GTG-banding, try 1.5 to 2 min in 0.025% (w/v) trypsin and 5 min in 4% (w/v) Giemsa staining solution.

23. After coverslips are stained and completely dry, mount them cell-side-down on labeled 3-in. × 1-in. × 1-mm glass microscope slides using 1 or 2 drops Permount per coverslip. Place two coverslips on one slide (Fig. 8.2.1B).

24. Begin analysis of first coverslip. Select a cell for counting in which chromosomes are spread well and appear to be complete (e.g., 46 chromosomes in most cases). For an overview of the analytical procedure, see Figure 8.2.2.

25. Record colony number and vernier indices for each metaphase counted or analyzed. Select, at most, one metaphase spread per colony for microscopic examination.

If no abnormal cells are found

26a. Count a total of 12 colonies and visually analyze five metaphases selected from at least two coverslips if no abnormality is found.

> *Counting 12 normal colonies on primary cultures rules out cell mosaicism of >25% at a 95% confidence limit. If 12 colonies are not found on the coverslips made from the primary culture, colonies from passaged material may be used to complete the count. In this laboratory, the count of two cells from passaged material is equated to one colony from primary material. Thus, for a normal case, 12 colony equivalents—i.e., 12 colonies from primary material, 24 colonies from passaged material, or some combination of colonies from primary and passaged material equal to 12 colony equivalents—should be counted.*

27a. Photograph the five metaphases and karyotype one metaphase.

> *Preferably, one of the coverslips used for a metaphase analysis should be from an ethidium bromide harvest (see Support Protocol 1). In a karyotype, each band of every chromosome should be clearly visible, without overlaps with other chromosomes. If one metaphase spread cannot meet these requirements, additional chromosome pairs from other metaphases that have "clean" banding in the problem spots of the first karyotype should be included. These additional*

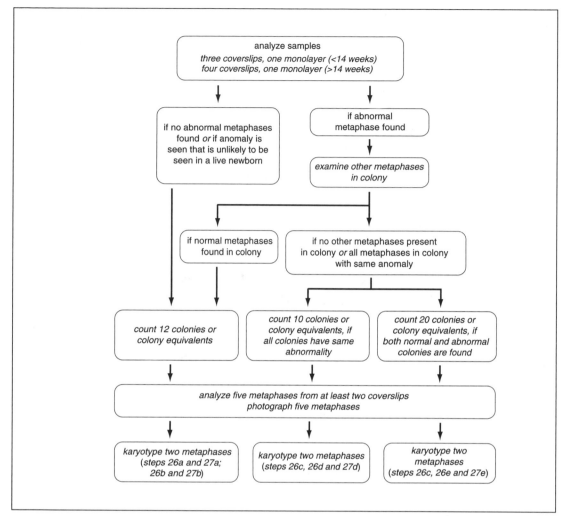

Figure 8.2.2 Flow chart of analysis of amniotic fluid samples. In this context, abnormal metaphases are those with (a) trisomies of chromosomes 8, 9, 13, 18, or 21; (b) sex chromosome aneuploidy; (c) marker chromosomes; and (d) unbalanced structural rearrangements. Adapted with permission from Frederick R. Bieber, Brigham & Women's Hospital, Boston.

pairs of chromosomes should be positioned alongside the corresponding chromosome pair of the first karyotype and the metaphase spread from which they came should be indicated.

If an abnormal cell is found

26b. Examine other metaphases in the same colony. If the other metaphases are normal, count a total of 12 colonies and visually analyze five metaphases selected from at least two coverslips.

27b. Photograph the five metaphases and karyotype two metaphases.

If an abnormal cell is found and there are no other metaphases in the same colony or if all metaphases in the colony have the same anomaly (nonmosaic)

26c. Check other colonies for the anomaly.

If all colonies contain cells with the same anomaly (nonmosaic culture)

26d. Count 10 colonies (or colony equivalents) and visually analyze five metaphases from at least two coverslips.

8

27d. Photograph the five visually analyzed metaphases and karyotype two metaphases.

If only some colonies contain cells with anomalies (mosaic culture)

26e. Count at least 20 colonies or colony equivalents and visually analyze five metaphases (two metaphases from one cell line and three metaphases from the other cell line) from at least two coverslips.

27e. Photograph the five visually analyzed metaphases and karyotype two metaphases (one representing each cell line).

> *A case is diagnosed as truly mosaic if the same abnormality is found in two or more tissue culture plates (e.g., tissue culture plates for the in situ method or flasks for the flask method, see Alternate Protocol). A case is diagnosed with pseudomosaicism if an abnormality is found in only one plate, even if the abnormality is found in different colonies using the in situ method, as mitotic cells may become unattached and migrate within the plate.*

SUPPORT PROTOCOL 1

HARVESTING FOR HIGH-RESOLUTION CHROMOSOME BANDING USING ETHIDIUM BROMIDE

This protocol is used to obtain metaphase chromosomes that are longer than those usually found in the standard harvesting protocol (~800 bands compared to ~500 bands). Longer banding is useful for identifying small chromosome abnormalities, but long chromosomes are harder to spread and tend to have more overlapping areas with other chromosomes. For complete analysis, prepare chromosomes using both harvesting techniques.

Additional Materials (see APPENDIX 1 for items with ✓)

    Cell cultures ready to harvest (see Basic Protocol)
✓ 0.5 mg/ml ethidium bromide

1. Add 25 μl of 0.5 mg/ml ethidium bromide to a culture that is ready for harvesting (i.e., from Basic Protocol, step 13). Incubate 70 min.

2. Add 50 μl of 10 mg/ml Colcemid and incubate 20 min.

3. Proceed with harvest, begin with Basic Protocol, step 15.

SUPPORT PROTOCOL 2

PASSAGING CELLS FROM THE MONOLAYER

If harvest of coverslips does not yield an adequate number of analyzable metaphase chromosome preparations, additional mitoses can be obtained by passaging cells from the monolayer to new coverslips.

Additional Materials (see APPENDIX 1 for items with ✓)

    Cell monolayer cultures ready to be passaged (see Basic Protocol)
✓ HBSS *without* calcium and magnesium, 37°C
    Trypsin/EDTA solution: 0.05% (w/v) trypsin/0.53 mM EDTA·4Na (Life Technologies), 37°C

1. Scan monolayer to be passaged (i.e., from Basic Protocol, steps 10a, 10b, and 10c) using an inverted microscope. Determine number of coverslips to which monolayer will be passaged by calculating that, at 40× magnification, one field of cells is sufficient to cover one glass coverslip.

Do not passage a monolayer with only one colony unless it is appropriate in size (i.e., fills one microscope field at 40× magnification) to passage to one slip.

2. Place sterile (completely dry) coverslips in labeled 35-mm plates.

3. Aspirate medium from the monolayer culture. Wash cells with 2 ml prewarmed HBSS to thoroughly remove serum.

4. Aspirate HBSS from monolayer culture. Add 0.5 ml of prewarmed trypsin/EDTA solution to monolayer. Incubate 3 to 4 min.

5. Check for release of cells under the inverted microscope—these cells should be rounded up and floating. If all cells are not released, tap on bottom of plate. If cells still do not release, incubate monolayer for an additional 1 min.

6. Add enough prewarmed 20% Chang medium to set up each plate with 0.5 ml fluid. Pipet suspension up and down to break up cell clumps.

7. Distribute 0.5 ml of the cell suspension to each coverslip. Add 2 ml of prewarmed 20% Chang medium to the original monolayer.

8. Incubate all plates at least 2 hr. If coverslips are set up late in the day, incubate overnight.

9. Add 2 ml of prewarmed 20% Chang medium to the plates containing the glass coverslips.

If passaging cells to 25-cm² flasks, a total of 5 ml cell suspension should be used to initiate secondary cultures. One confluent monolayer may be passaged to one 25-cm² flask (see Support Protocol 3). When culturing cells in flasks or flaskettes, cells may be suspended and plated in the final volume of growth medium, because they should use the entire bottom of the vessel as a surface for growth.

ALTERNATE PROTOCOL

FLASK METHOD FOR THE PREPARATION, CULTURE, HARVEST, AND ANALYSIS OF AMNIOTIC FLUID SAMPLES

This method generally takes longer than the in situ method, but because the support medium is larger, more metaphases can be obtained.

***Additional Materials** (also see Basic Protocol)*

Trypsin/EDTA solution: 0.05% (w/v) trypsin/0.53 mM EDTA·4Na (Life Technologies), 37°C

25-cm² tissue culture flasks (Falcon; Fisher)

1. Obtain amniotic fluid sample, transfer to 15-ml tubes, and centrifuge as in Basic Protocol, steps 1 to 3.

2. While tubes are centrifuging, label 25-cm² flasks. Prepare three flasks for samples obtained from routine amniocentesis (14.0 to 22.0 weeks) and two flasks for samples obtained from early or late amniocentesis (<14.0 weeks or >22.0 weeks).

3. Take care not to disturb pellets when removing tubes from the centrifuge. Draw off the supernatant from each tube, leaving ~1 ml fluid above pellet. Transfer supernatant to another tube (keep at 4°C prior to AFP testing) and save for any biochemical testing.

4. If the fluid was divided into two centrifuge tubes, resuspend one of the pellets with 3 ml of prewarmed 20% Chang medium, tap gently on side of tube to resuspend pellet, then transfer this first cell suspension to the second tube and resuspend second pellet. If

8

the fluid was contained in one centrifuge tube only, gently resuspend pellet in 3 ml of prewarmed 20% Chang medium. Tap gently on side of tube to resuspend pellet.

5. Equally distribute cell suspension to flasks. Add prewarmed 20% Chang medium to bring the volume to 5 ml in each flask. Incubate flasks 4 to 6 days.

6. Using an autoclaved Pasteur pipet, aspirate medium into a vacuum flask set up in the hood. Tilt flask so medium can be removed by placing pipet in one corner to avoid disturbing cells.

7. Add 5 ml of prewarmed 20% Chang medium to the flask and incubate. Handle flasks as infrequently as possible, but change medium (when yellow) at least once weekly.

8. Under an inverted microscope, check flasks periodically after 5 or 6 days in culture. Harvest cells when a significant number of mitotic doublets are present on the periphery of cell colonies and there is still enough space between them to distinguish separate colonies (usually at 6 to 12 days). If flasks become overgrown, passage cells from one flask to multiple flasks (see Support Protocol 3).

9. Feed cultures 12 to 22 hr before harvesting. When cells are ready for harvest, add 100 μl of 10 mg/ml Colcemid to each flask and incubate 20 min. Do not harvest all flasks from a case. Always save one flask as a backup that can be passaged if more material is required.

10. Aspirate medium from flask. Rinse flask with 2 ml prewarmed trypsin/EDTA solution to remove serum and put this wash into a 15-ml centrifuge tube for use in step 14.

11. Add another 2 ml prewarmed trypsin/EDTA solution to flask and incubate 5 min.

12. Check for release of cells under an inverted microscope—these cells should be rounded up and floating. If all cells are not released, tap on bottom of plate. If cells still do not release, incubate flask an additional 5 min.

13. Add 2 ml of prewarmed 20% Chang medium to help rinse cells from flask surface. Pipet suspension up and down to break up cell clumps.

14. Transfer suspension to centrifuge tube containing the 2 ml trypsin/EDTA solution used to rinse the flask in step 10. Centrifuge 10 min at $109 \times g$, room temperature.

15. After centrifuging cells, decant supernatant (or aspirate very carefully), leaving a small amount of fluid (<0.5 ml) in which to resuspend the pellet.

16. Add 1 ml prewarmed hypotonic solution drop by drop, letting the liquid slide slowly down side of tube. Tap gently to mix. Then add an additional 6 ml prewarmed hypotonic solution, also letting the solution slide down side of tube before coming in contact with cell mixture. Cap tube and incubate 10 min.

17. Add 5 drops of 3:1 freshly made fixative and mix gently by inverting the tube. Centrifuge 10 min at $70 \times g$, room temperature.

18. Remove supernatant by decanting or careful aspiration and add 1 ml of 3:1 fixative drop by drop, letting it run slowly down side of tube. Gently resuspend pellet, taking care not to splash cells all over the tube.

19. Add 6 ml of 3:1 fixative, letting it run slowly down side of tube. Cap tube and chill ≥30 min at −20°C.

20. Centrifuge 10 min at $70 \times g$. Remove the supernatant and add 7 ml of 3:1 fixative. Add the first milliliter drop by drop and gently resuspend before adding the remaining 6 ml.

8

21. Repeat step 20 until the fixative has been changed three or four times (for cultures that have visible debris, or if cytoplasm around metaphase spreads has clouded stained chromosomes on previous preparations).

22. After the last fixative change, resuspend the pellet in a small volume of 3:1 fixative (usually ~0.5 ml) so the desired concentration for dropping suspension onto slides is obtained (*UNIT 4.1*).

23. Prepare three to five slides, each containing 10 to 30 metaphases, from each flask and stain by GTG-banding (*UNITS 4.1 & 4.3*). As a guideline for first-time GTG-banding, try 1.5 to 2.0 min in 0.025% (w/v) trypsin and 5 min in 4% (w/v) Giemsa stain.

24. Begin analysis of first slide. Select a cell for counting in which chromosomes are spread well and appear to be complete (e.g., 46 chromosomes in most cases).

If no abnormal cells are seen

25a. Count 20 cells, ideally 10 from each of two flasks. Analyze five metaphases under the microscope.

> *If the volume of amniotic fluid sample was ≥16.0 ml, this rules out mosaicism of >20% at a confidence level of 95%.*

26a. Photograph the five visually analyzed metaphases and karyotype one of these metaphases. See Basic Protocol, steps 26 and 27, for discussion on karyotyping.

If an abnormality is seen in every cell

25b. Count 10 cells with at least one count from the second flask. Analyze five metaphases.

26b. Photograph the five metaphases and karyotype one.

If a single numerical abnormality is found

25c. Harvest a third flask and count 20 metaphases from the two flasks in which the original abnormality was not found. Visually analyze five metaphases (two metaphases from one cell line and three metaphases from the other cell line).

> *If the abnormality is not found in either of the two other flasks, interpret the result as pseudomosaicism or culture artifact. If the abnormality is found in one of the two additional flasks, interpret the result as true mosaicism.*

26c. Photograph the analyzed metaphases and karyotype two, with one karyotype representing each cell line.

If a single structural abnormality is found

25d. Score for that abnormality in 20 cells from the second flask. Visually analyze five metaphases (two metaphases from one cell line and three from the other).

> *If the abnormality is not found in the second flask, interpret the result as pseudomosaicism. If the abnormality is found in the second, interpret the result as true mosaicism.*

26d. Photograph the five metaphases from each cell line and karyotype one from each of the two cell lines.

SUPPORT PROTOCOL 3

PASSAGING CELLS FROM FLASKS

1. After a culture has grown enough to be passaged, a confluent or close to confluent flask (see Alternate Protocol, step 8), aspirate medium from flask.

2. Rinse flask with 2 ml prewarmed trypsin/EDTA to remove serum that would inhibit trypsin and save this wash in a 15-ml centrifuge tube.

3. Add 2 ml prewarmed trypsin/EDTA solution to the flask and incubate 5 min.

4. Check for release of cells under an inverted microscope—these cells should be rounded up and floating. If all cells are not released, tap on bottom of plate. If cells still do not release, incubate flask an additional 5 min.

5. Add 2 ml of prewarmed 20% Chang medium to help rinse cells from the flask surface. Pipet suspension up and down to break up cell clumps.

6. Transfer suspension to centrifuge tube containing the 2 ml of trypsin/EDTA solution first used to rinse the flask. Centrifuge 10 min at 109 × g, room temperature.

7. Decant supernatant (or aspirate very carefully) and resuspend cells in prewarmed 20% Chang medium. Determine volume according to number of flasks which will be set up.

8. Incubate flasks until harvested and processed as in Alternate Protocol, steps 8 to 26.

Reference: Hsu et al., 1992

Contributor: Patricia Minehart Miron

UNIT 8.3

Preparation and Culture of Products of Conception and Other Solid Tissues for Chromosome Analysis

In clinical settings, chromosome studies are usually performed on solid tissue other than solid tumors for one of two reasons: the tissue biopsy is the only tissue available from the patient, or tissues other than the standard peripheral blood lymphocytes must be examined because of suspected mosaicism.

NOTE: All reagents and equipment coming into contact with live cells must be sterile. All incubations are performed in a humidified 37°C, 5% CO_2 incubator unless otherwise specified.

BASIC PROTOCOL

MECHANICAL DISRUPTION AND CULTURE OF TISSUES FOR METAPHASE CHROMOSOME ANALYSIS

Materials (*see* APPENDIX 1 *for items with* ✓)

    Tissue sample: 2- to 3-mm^3 fetal tissue, 2-mm-deep × 2-mm-diameter skin biopsy, or 3-mm^3 tissue biopsy, fresh or stored at 4°C (see Support Protocols 1 and 2)

    ✓ HBSS containing standard antibiotic solution—e.g., Life Technologies PSN antibiotic mixture—at 5× usual concentration (store at –55°C or 4°C when opened)

    Complete culture medium/15% to 20% FBS (APPENDIX 1; preferably supplemented Ham F-10 or a-MEM from Life Technologies)

    35-mm petri plate
    Dissection instruments: needles, scissors, fine forceps, and scalpel
    25-cm^2 plastic tissue culture flasks

8

1. Using sterile dissecting instruments, place tissue sample in 35-mm petri plate containing ~5 ml HBSS with 5× antibiotics. Leave ≥30 min at room temperature.

2. Transfer tissue sample to a fresh sterile 35-mm petri plate containing ~2 ml complete culture medium/FBS (supplemented media such as Chang medium, UNIT 8.2 or Irvine Scientific, will speed up growth). Mince tissue into fine fragments using either fine scissors and forceps or a scalpel.

3. Transfer small pieces of tissue and a minimal amount of complete culture medium/FBS (~1 ml per flask) into two or three 25-cm^2 plastic tissue culture flasks using a sterile Pasteur pipet, distributing pieces evenly over bottom (widest surface) of flask.

4. Carefully stand flask on end (bottom surface vertical and cap pointed up), close cap tightly, and incubate several hours to overnight.

5. Add 4 ml complete culture medium/FBS by letting it flow down side of flask opposite the tissue, being careful not to dislodge attached pieces of tissue. Cap flask tightly (to avoid contamination) and incubate 2 days with flask lying flat (surface to which tissue is attached horizontal).

6. Examine explant for contamination as evidenced by cloudy appearance, acidity of culture medium, or obvious fungal growth. If none is seen, loosen caps and continue incubation.

7. After 5 days (and up to 3 weeks), check for growth from explant—i.e., a small number of viable cells, usually fibroblasts, growing out from the tissue fragment.

   Cells that grow in cultures of amnion and some other tissues may be epithelial, rather than fibroblastic, in nature. These cells do not survive trypsinization as well as fibroblast cultures and should be harvested as soon as possible for chromosome preparations.

8. When growth begins, replace medium twice per week. Wash away large explants of tissue if possible. When several colonies are well-expanded and in log phase of growth (i.e., rounded up mitotic cells and empty space between colonies), harvest cultures directly from initial culture and prepare metaphase chromosome spreads by the flask method as for amniotic fluid cultures (UNIT 8.2). Perform chromosome banding (UNIT 4.3) and analyze karyotype (APPENDIX 3K).

ALTERNATE PROTOCOL 1

ENZYMATIC DISRUPTION AND CULTURE OF TISSUES FOR METAPHASE CHROMOSOME ANALYSIS

Additional Materials (see APPENDIX 1 for items with ✓)
- ✓ Collagenase solution
- ✓ Trypsin/EDTA solution

  15-ml centrifuge tube
  20-G needle or fine Pasteur pipet
  IEC Clinical centrifuge or equivalent

1. Place selected pieces of tissue in individual 35-mm petri plates (one for each type of tissue). Add ~1 ml collagenase solution and incubate 1 hr.

2. Using a pipet or syringe, transfer collagenase solution containing loose cells to a sterile 15-ml centrifuge tube. Cap and leave at room temperature. Add 1 ml trypsin/EDTA solution to tissue remaining in plate and incubate 1 hr.

8

3. Break up remaining clumps of cells by aspirating the solution up and down using a syringe fitted with 20-G needle or fine Pasteur pipet. Add to centrifuge tube containing cell suspension from step 2.

4. Add complete culture medium/FBS to ~12 ml total. Centrifuge 8 min at $70 \times g$, room temperature. Discard supernatant and resuspend pellet in 1 to 2 ml complete culture medium/FBS.

5. Set up cell suspension on four coverslips or in two 25-cm^2 flasks as for amniotic fluid cultures, checking for contamination after 1 day. Prepare metaphase chromosome spreads by the in situ or flask method starting from feeding of cells before harvesting (*UNIT 8.2*). Perform chromosome banding and analyze karyotype.

SUPPORT PROTOCOL 1

PREPARATION OF HUMAN PRODUCTS OF CONCEPTION FOR CULTURE

Materials (see *APPENDIX 1* for items with ✓)

Products of conception
✓ HBSS or Ringers solution

Dissecting instruments: needles, scissors, fine forceps, and scalpel
Dissecting microscope
Plastic petri plate or stainless steel container

1. Place products of conception in sterile or clean closed containers for transport.

2. If the specimen is to be processed quickly, do not add fluid because enough fluid is usually present in the specimen itself to keep it moist. Use specimen within 24 hr, but if the specimen cannot be used immediately, store <5 days at 4°C. Do not use any additional solution and do not freeze.

3a. For larger or complete specimens, if fetal parts are easily recognizable, as with a whole or disrupted large fetus, remove them carefully from the container and place in a suitably sized plastic petri plate or stainless steel container. Also search for pieces of placenta or fetal membranes.

   IMPORTANT NOTE: *Although most samples will not be sterile, so that real sterile technique is impossible, all dishes and instruments should be clean and alcohol sterilized.*

3b. For very small or partial specimens, place entire specimen in a large petri plate. Teasing specimen carefully with dissecting needles or forceps, search for small embryos or embryonic parts and for pieces of placenta, fetal membranes, and cord.

   Much of the specimen may be maternal decidua, which will not give the fetal genotype and should not be used. Decidual tissue may look membranous and resemble the fetal membranes (i.e., amnion and chorion), but is much more fragile. If necessary, check identity of small fragments of questionable placenta or fetal tissues using a dissecting microscope. The presence of branching chorionic villi distinguishes placental from decidual tissue.

   In some complete early specimens, the embryonic sac may be completely surrounded by chorionic villi and/or decidua so that the specimen must be cut open to find the embryo.

   In some incomplete specimens, no identifiable fetal tissue may be found.

4. If specimen is a fresh, unmacerated fetus, dissect ~2 to 3 mm^3 of muscle from a limb for culture. If tissues from other organs are desired because of possible mosaicism, take

skin, kidney, lung, and/or liver, all of which grow readily in culture. If the specimen is a fresh small embryo, dissect a 2- to 3-mm³ portion once examination or photography of the specimen is complete. If the embryo or fetus is macerated, dissect a 2- to 3-mm³ piece of fetal membrane or chorionic villi.

In many cases of spontaneous abortion and some cases of induced abortion, embryonic or fetal death may occur several days before expulsion, with the result that fetal material will be discolored, fragile, and soft (macerated) and will usually not contain enough viable cells for culture. In such cases the placental tissue—fetal membrane and chorionic villi—is usually more viable.

SUPPORT PROTOCOL 2

PREPARATION OF SKIN OR OTHER TISSUE BIOPSIES FOR CULTURE

Materials *(see APPENDIX 1 for items with ✓)*

Tissue of interest
✓ HBSS
Dissecting instruments: sterile scissors and forceps or scalpel

Collect tissue biopsy using sterile technique. Carefully trim fat using sterile forceps and fine scissors or scalpel. Place biopsy in sterile centrifuge tubes containing sufficient HBSS or unsupplemented culture medium to cover and store at 4°C if necessary. Do not freeze.

Skin biopsies, the usual choice when fibroblast cultures are needed from a living patient, should be ~2 mm deep and 2 mm in diameter. Other tissues may become available during surgery (fascia or other connective tissues are the easiest to culture); in some cases it may be necessary to study specific tissues (e.g., gonadal biopsies in cases of sex-chromosome mosaicism). For these other tissues the biopsy should be a small piece of tissue (~3 mm³). At autopsy, as in stillbirth or neonatal death, a larger selection of tissues is available for culture. If no other special factors need be considered, the best choice is a piece of deep muscle. This is less likely to become contaminated than skin and can be viable for up to 5 to 7 days after death.

The samples are now ready to be cultured to prepare metaphase chromosome spreads. The explants should be checked for outgrowth after 3 to 4 days in culture.

Direct Preparation of Chorionic Villi

Fresh chorionic villi from either a spontaneous or induced abortion can be used directly to prepare metaphase chromosome spreads. Prepare chorionic villi from chorionic villus biopsy and prepare metaphase chromosome spreads directly from the chorionic villi as described in *UNIT 8.1*. Band and analyze as described in *UNIT 4.3 & APPENDIX 3K*.

Such preparations will generally not yield large numbers of metaphase figures and will not be of the highest quality. However, compared to the methods described above, this requires less medium and culture time and is also less susceptible to misdiagnosis due to inadvertent overgrowth of maternal cells. Direct preparations are also suitable for screening for particular aneuploidies by in situ hybridization on interphase nuclei (*UNIT 4.4*).

References: Lee, 1991; Priest, 1991; Rooney and Czepalkowski, 1992

Contributor: Dorothy Warburton

8

Analysis of Sister-Chromatid Exchanges

NOTE: Use disposable latex gloves when handling BrdU, Hoechst 33258 dye, and Giemsa stain. BrdU and Hoechst 33258 dye are light-sensitive materials. Wrap all containers, including tissue culture flasks, centrifuge tubes, and Coplin jars, in foil. Use indirect lighting (e.g., light from an adjacent room) or a shielded low-wattage light bulb when working with these reagents.

NOTE: All incubations are performed in a 37°C, 5% CO_2 humidified incubator unless otherwise specified.

BASIC PROTOCOL

ANALYSIS OF SISTER-CHROMATID EXCHANGES IN MAMMALIAN METAPHASE CHROMOSOMES

Two requirements for the cytogenetic analysis of sister-chromatid exchanges (SCEs) in somatic cells are (1) a population of actively proliferating cells that will provide an adequate number of metaphases and (2) sister chromatids that in some way are differentially labeled or stained in the metaphases. The principle of this method is shown in Figure 8.4.1, and a troubleshooting guide is given in Table 8.4.1.

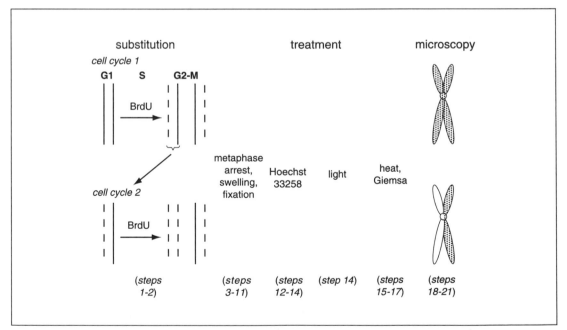

Figure 8.4.1 Diagram depicting the principle by which sister chromatids are stained differentially. BrdU substitution in the newly replicated DNA occurs throughout two cell division cycles. If metaphase chromosomes of cells that have been substituted with BrdU are dyed with the fluorochrome Hoechst 33258 and exposed to light, Giemsa will stain the bifilarly substituted chromatid distinctly lighter purple than its unifilarly substituted sister. Chromosomes with unifilar substitution in both sister chromatids, or with no substitution, i.e., in metaphases of cells that had passed through only one or no cell division cycle in the presence of BrdU, will fail to show sister-chromatid differentiation. Solid line, unsubstituted strand of DNA; broken line, BrdU-substituted strand of DNA.

Materials *(see APPENDIX 1 for items with ✓)*

    Sterile heparinized blood

    Mitogen-containing culture medium: complete RPMI/15% FBS (*APPENDIX 1*)
        supplemented with 2% (v/v) phytohemagglutinin A (PHA; Life Technologies)

✓ 2 mM 5-bromo-2′-deoxyuridine (BrdU) stock solution

    10 μg/ml Colcemid (e.g., Life Technologies; store at 4°C)

    Hypotonic solution: 0.075 M KCl, 37°C (store ≤2 weeks at 4°C)

    Fixative: 3:1 (v/v) absolute methanol/glacial acetic acid (freshly prepared)

✓ 250 μg/ml Hoechst 33258 stock solution

    Giemsa stain (Harleco Modified Azure Blend Type, Baxter Scientific; Fisher; Gurr
        Improved R66, Bio/medical Specialties)

    Gurr buffer, pH 6.8, from tablets (Bio/medical Specialties)

    25-cm^2 tissue culture flasks

    15-ml conical glass or polypropylene centrifuge tubes

    Standard tabletop centrifuge

    Polyethylene transfer pipets

    Precleaned microscope slides, wiped with soft tissue paper and dipped into
        room-temperature water just before use

    Black plastic microscope slide box (VWR Scientific)

    24 × 50–mm coverslips, wiped with soft tissue paper just before use

    Coplin jars

    Lamp equipped with a 60-W incandescent bulb

    37° and 60°C water baths

    Rubber cement

    Scalpel blades

    Mounting medium (e.g., Permount)

    Kodak TMax 100 Professional Film

1. Add 0.3 ml sterile heparinized whole blood to a 25-cm^2 tissue culture flask containing 10 ml mitogen-containing medium. Incubate 24 hr.

2. Add 50 μl of freshly thawed, sterile 2 mM BrdU solution to the culture to give a final BrdU concentration of 10 μM (3 μg/ml). Incubate culture for an additional 48 hr.

    Cells must undergo two rounds of DNA replication in BrdU-containing medium to make possible the desired staining pattern, so a rough approximation of the duration of the cell cycle for the cell type under investigation is needed. Many mammalian cells in culture have cell cycles of ~24 hr; the timing indicated here will often prove successful.

3. Add 150 μl of 10 μg/ml Colcemid (0.15 μg/ml final concentration) to the culture and incubate 30 min.

4. Transfer culture to a 15-ml conical tube. Centrifuge 10 min at ~180 × g. Remove all but 0.2 ml of the supernatant and resuspend cell pellet by gently tapping the tip of the tube.

    Cultures of attached cells will have to be removed from the flask by trypsinization (APPENDIX 3I).

5. Slowly add 6 ml of 37°C hypotonic solution while gently tapping the tip of the tube. Incubate 12 min in a 37°C water bath.

6. Add 4 drops of freshly prepared fixative to cell suspension while gently tapping the tube. Invert tube once for thorough mixing. Centrifuge 10 min at ~180 × g.

7. Remove all but ~0.2 ml of supernatant fluid and resuspend pellet by tapping the tip of the tube. Add 5 ml fixative, at first drop by drop and then more rapidly, always keeping cells in suspension by tapping the tube. Incubate 20 min at 4°C.

8. Centrifuge 10 min at $\sim$180 $\times$ g. Remove supernatant fluid and add 5 ml fixative, mixing gently. Centrifuge 10 min at $\sim$180 $\times$ g. Repeat this step two to three times, until the cell pellet is colorless.

9. Remove supernatant fluid and resuspend cell pellet in a small volume of fixative, sufficient to produce a slightly cloudy suspension.

10. Shake excess water from a microscope slide. Draw a few drops of cell suspension into a polyethylene transfer pipet. Holding the tip of the pipet 2.5 cm (1 in.) above the slide, place 3 drops of suspension onto the slide—each in a different location along the slide's surface such that each drop can spread independently. Stand the slide at a slant to dry.

11. Evaluate quantity and quality of metaphases using phase-contrast microscopy (e.g., at 16$\times$ magnification). Age slides 2 to 3 days at room temperature in the dark (e.g., in a black plastic microscope slide box).

    Excessive chromosome spreading can be reduced by decreasing the temperature of the water in which the slides are standing. Adjust the volume of fixative, if necessary, to give an even distribution of well-spread metaphases. Dense cell suspensions that crowd cells on the slide seem to interfere with differential staining. See UNIT 4.1 for additional details on slide preparation.

    Preparation of four slides is adequate to permit varying the staining conditions. The remaining cell suspension can be stored in 5 to 10 ml fixative at −20°C. When preparing slides at a later date, the suspension should be brought to room temperature and centrifuged to change to fresh fixative.

12. Add 1 ml of 250 μg/ml 33258 Hoechst stock solution to 9 ml water (store up to 10 to 14 days at 4°C). Add 1 ml diluted Hoechst 33258 solution to 49 ml water in a Coplin jar, for a final dye concentration of 0.5 μg/ml.

13. Place several aged slides in the Coplin jar containing diluted Hoechst 33258 solution and incubate 12 min at room temperature. Rinse slides well in water.

14. Place a 24 $\times$ 50–mm coverslip onto each wet slide. Seal edges of coverslip with rubber cement. Expose slides to light overnight by placing 38 cm (15 in.) below a lamp equipped with a 60-W incandescent bulb.

15. Peel rubber cement from slides. Lift a corner of each coverslip and flip (rather than slide) it off using a scalpel blade. Incubate slides by adding at staggered intervals of 10 to 12 min in a prewarmed Coplin jar of water exactly 2 hr at 60°C.

16. Add 1 ml Giemsa stain to 49 ml Gurr buffer, pH 6.8, in a Coplin jar at room temperature. Fill two Coplin jars with room-temperature water for rinsing.

17. Remove a slide from hot water, wave it in the air to cool and dry, and place it in the buffered Giemsa solution for 3.5 min. Remove excess stain by swirling the slide in each of the two Coplin jars of water (change water when it becomes colored), and wave it in the air to dry.

18. Examine cells with a high/dry objective (e.g., 63$\times$ or 80$\times$) for the quality of staining. Determine specifically whether the sister chromatids in at least some of the metaphases differ in their staining intensity.

    Conditions can be modified slightly to improve the results and to adjust for changes in reagents (e.g., aging of stock solutions or variations in commercial lots of a reagent; see Table 8.4.1).

19. Mount the best chromosome preparations under a coverslip using an appropriate mounting medium.

20. Scan the slide for well-spread, intact metaphases using a low-power objective (e.g., 16$\times$). Switch to high-power objective (e.g., 100$\times$ oil-immersion) when such a metaphase is

8

Table 8.4.1 Troubleshooting Guide for Analysis of Sister-Chromatid Exchange

Problem	Possible cause	Solution
No differentially stained test chromosomes while control slide is acceptable	Defective BrdU or not enough incorporation during DNA replication	Prepare fresh solution or use new lot of BrdU
	Arrested metaphase cells did not undergo two rounds of replication in BrdU-containing medium because of an unusually long cell cycle	Determine length of cell division cycle for particular cell type, and adjust time in BrdU appropriately
Too many third-division cells	Cell cycle shorter than 24 hr	Determine length of cell division cycle for particular cell type, and adjust time in BrdU appropriately
Both control and test cells exhibit no differentiated metaphases	33258 Hoechst treatment	Use freshly prepared stock solution or new lot of dye; obtain a lot that has been proven to work
	Inadequate light exposure	Vary duration of light exposure and/or wattage of light source
	Lot-to-lot variation of Giemsa stain	Try a different lot of Giemsa stain, or try a different source
Differentially stained chromatids, but both too purple	Preparation is over-stained	Reduce staining time
Differentially stained chromatids, but both too faintly stained	Non-optimal Giemsa stain concentration and/or staining time	Increase Giemsa concentration, and decrease staining time; dip slide in water and return to Giemsa for additional 2 min
Insufficient contrast between light and dark sister chromatids	Non-optimal Giemsa stain concentration and/or staining time	Increase Giemsa concentration or staining time
Pale, "washed out" Giemsa-stained mitotic and interphase cells	Overexposure to light	Reduce duration of light exposure and/or wattage of light source

located. Determine whether the sister chromatids are stained differentially, and whether the contrast in the staining intensity is adequate to permit recognition of SCEs. Scan each chromosome in a metaphase with sister-chromatid differential staining to make certain that all chromatid segments are differentially stained (i.e., that the cell has not passed more than two rounds of replication in BrdU; see Fig. 8.4.1).

21. Record the location on the slide (i.e., the coordinates of the microscope stage) of informatively stained second-division cells. Count and record the total number of chromosomes and SCEs. Continue the study until 15 informative cells have been located and scored. Determine the mean and the range of SCEs/metaphase. Photograph one or two of the best cells for further analysis and filing using a high-quality fine-grained film such as Kodak TMax 100 Professional Film.

For metaphases with very high numbers of SCEs, photographs (e.g., 5 × 7–in. or 13 × 18–cm prints) facilitate accurate scoring. With a photographic print in hand, the metaphase should be re-examined through the microscope for comparisons and decisions as to sites of exchanges, and

8

the SCEs should be marked on the print with a wax pencil. The combination of direct microscopy and photography helps make decisions concerning SCEs near/in centromeres and telomeres, and helps detect exchanges that are very close together, such as those characteristic of Bloom's syndrome (BS).

References: Latt et al., 1984; Perry and Wolff, 1974

Contributors: James German and Becky Alhadeff

UNIT 8.5

Determination of Chromosomal Aneuploidy Using Paraffin-Embedded Tissue

Tissue processed in pathology departments is generally fixed in formalin and embedded in paraffin blocks. These blocks are usually stored for a minimum of ~20 years, providing a source of potentially useful information. In general, the techniques described here work well on tissue fixed in formalin or paraformaldehyde. Some other fixatives (e.g., alcohol) may be acceptable whereas some fixatives (e.g., Zenker's) appear suboptimal. However, experience with these fixatives is minimal, and it is worth analyzing tissues prepared in them if the need arises.

BASIC PROTOCOL

PREPARATION OF NUCLEAR SUSPENSIONS FROM PARAFFIN-EMBEDDED TISSUE FOR FISH

The number of hybridization signals per nucleus can be determined by standard fluorescence in situ hybridization (FISH; *UNIT 4.4*) using DNA probes for centromeric DNA.

Materials (see APPENDIX 1 for items with ✓)
  Paraffin-embedded tissue of interest (*CPMB UNIT 14.1*)
  Xylene (e.g., ChemPure, Curtin Matheson)
  100%, 95%, 80%, 70%, and 50% (v/v) ethanol
✓ Pepsin solution
✓ PBS
✓ 2× SSC
  Acetone
  10 to 20 ng biotin- or dioxigenin-labeled centromeric DNA probe (e.g., Oncor) for chromosome of interest (*UNIT 4.4*)
  Hybridization solution: 65% (v/v) formamide (e.g., Oncor) in 2× SSC (e.g., Hybrisol VI, Oncor)
  Wash solution: 50% (v/v) formamide (e.g., Oncor) in 2× SSC, pH 7.0 (prepare just prior to use and prewarm to 39°C)
  0.1× SSC, 39°C
  Phosphate-buffered detergent (PBD; e.g., Oncor), room temperature

  Microtome with knives
  15-ml glass conical centrifuge tubes
  Fisher Marathon 21K centrifuge with swinging-bucket rotor (or equivalent)
  37° to 43°C water bath (with agitation if possible)
  5-ml syringes with 18- and 21-G needles

Monofilament nylon mesh (105-μm; Small Parts)
Coplin jars
22 × 22–mm glass coverslips
Rubber cement
90°C oven
Humidified slide-incubation chamber (*UNIT 4.4*)
Fluorescence microscope with appropriate filters and beam splitter (e.g., Zeiss Axiphot microscope with BP450-490 excitation filter, FT510 beam splitter, and SP520 barrier filter)

NOTE: The probe and hybridization solution should be kept on ice to prevent deterioration of signal intensity over multiple uses. If there will be a significant delay between preparation of the probe mixture and applying it to the target DNA, then the probe mixture should also be stored on ice.

1. Using a microtome, cut one to three 50-μm-thick sections (for most types of tissue, two 1 × 1–cm sections should yield enough nuclei to perform hybridization with ≥10 probes) of paraffin-embedded tissue of interest and place in a 15-ml glass conical centrifuge tube.

2. Add 5 ml xylene and let stand 5 min with occasional manual agitation. Centrifuge 5 min at 200 × *g*, room temperature. Aspirate and discard supernatant.

3. Repeat step 2 using the following wash solutions:

   Xylene
   100% ethanol
   95% ethanol
   70% ethanol
   50% ethanol
   Distilled water (two changes).

   If the procedure was successful, no firm, white, waxy flecks of paraffin should remain in the tissue. If only a few fragments of paraffin remain, mechanically remove them. If it is not possible to remove paraffin fragments, specimens must be dehydrated again by performing washes in the reverse order (starting with water and ending with the first xylene wash), then repeating deparaffinization and rehydration.

4. Add 1 ml pepsin solution and incubate 30 min in a 37°C water bath with agitation.

5. Using a 5-ml syringe equipped with an 18-G needle, vigorously draw the specimen up and down several times to disperse the clumps (some clumps will always remain).

6. Filter specimen through a monofilament nylon mesh into a 15-ml conical centrifuge tube.

7. Rinse the digestion tube and mesh by adding 4 ml PBS to the tube. Pour this PBS through the mesh and into the tube from step 6, then pour an additional 4 ml PBS through the mesh and into the same tube. Discard mesh.

8. Centrifuge 10 min at 200 × *g*, room temperature. Aspirate and discard supernatant. Resuspend pellet in 2 ml PBS. Vigorously draw the specimen up and down repeatedly into a 5-ml syringe equipped with a 21-G needle.

9. Centrifuge 10 min at 200 × *g*, room temperature. Aspirate and discard supernatant. Using a Pasteur pipet, resuspend the pellet containing the isolated nuclei in PBS and pipet onto glass microscope slides.

8

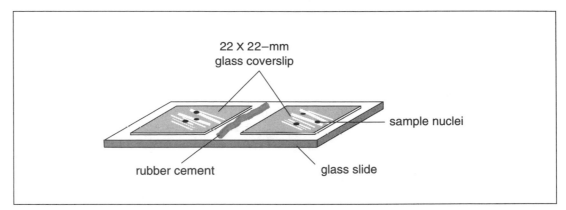

Figure 8.5.1 Placement of nuclei on a glass slide to perform two separate hybridizations simultaneously.

Two different probes can be hybridized on the same slide. If this is desired, pipet nuclei onto both ends of the slide (Fig. 8.5.1).

10. Air dry slides completely (≥24 hr). Store ≤1 month open to the air at room temperature. If long-term storage is unavoidable, store slides in a desiccator at 4°C.

11. Dehydrate slides by immersing for 2 min in each of the following room-temperature solutions, set up in 50-ml Coplin jars:

    2× SSC
    70% ethanol
    80% ethanol
    95% ethanol
    100% ethanol
    Acetone.

    Air dry the slides. Store for several days at room temperature if necessary.

12. Add 1 to 1.25 µl biotin- or digoxigenin-labeled centromeric DNA probe (*UNIT 4.4*) of chromosome of interest (10 ng/µl final) to 15 µl hybridization solution (for a 22 × 22–mm glass coverslip).

13. Pipet probe mixture onto slide and cover with glass coverslip, minimizing bubbles. Apply rubber cement to edges of coverslip to seal.

    If using two different probes on the same slide, apply one probe, coverslip (usually a 22 × 22–mm coverslip works best), and sealant before applying the second probe to prevent the probes from mixing (see Fig. 8.5.1).

14. Incubate slides 5 to 10 min in 90°C oven to denature probe and target DNA.

    If the nuclei are fragile due to some unknown factor of tissue fixation or storage, denaturation times can be shortened to ≤5 min or temperatures decreased as low as 75°C with minimal loss of final hybridization signal intensity.

15. Incubate slides overnight (12 to 18 hr) in a humidified slide-incubation chamber set up in a 37°C oven to allow hybridization of probe with target DNA.

16. Wash slides two times, 15 min each, in 50-ml Coplin jars containing 39°C wash solution (*UNIT 4.4*). Frequently agitate manually.

17. Transfer slides to a Coplin jar containing 39°C 0.1× SSC and incubate 30 min in a 39°C water bath with agitation.

8

18. Transfer slides to a Coplin jar containing room-temperature PBD and incubate 2 min at room temperature with gentle agitation. Store slides in PBD ≤2 weeks at 4°C, but do not allow to air dry from this point on.

19. Perform probe detection and chromosome counterstaining as described in UNIT 4.4.

20. View slides with fluorescence microscope and tabulate the number of primary hybridization signals (greatest intensity) per nucleus; ignore signals of less intensity.

ALTERNATE PROTOCOL

PREPARATION OF PARAFFIN SECTIONS FOR FISH

Additional Materials (see APPENDIX 1 for items with ✓)
 10%, 20%, or 30% (w/v) sodium bisulfite (Sigma)
 25 mg/ml proteinase K (Sigma; store at −20°C)
✓ Silanized glass microscope slides (e.g., Sigma)
 65°C oven

1. Using a microtome, cut 2- to 10-μm-thick sections of paraffin-embedded tissue and apply to silanized glass microscope slides by floating section in water.

   The optimal thickness of the section depends upon the overall cellularity of the tissue and the amount of intercellular substance (e.g., connective tissue), and usually needs to be individualized for each tissue type (and occasionally from case to case). Ideally, overlap of nuclei should be minimized while cutting as thick a section as possible to increase the amount of each nucleus being examined (average nuclear diameter is ~5 to 15 μm). In addition, large amounts of intercellular substance are more conspicuous on thicker tissue sections and may partially obscure the hybridization signals.

2. Air dry slides and bake overnight at 65°C to ensure tissue adherence to slides.

3. Deparaffinize the tissue sections by placing slides in 50-ml Coplin jars containing xylene for 10 min, followed by two changes of 100% ethanol, 5 min each. Agitate occasionally. Let slides air dry. If sodium bisulfite treatment is required, proceed to step 4; if not, proceed to proteinase K treatment in step 5.

4. *For sodium bisulfite treatment:* Transfer slides to a Coplin jar containing prewarmed 10%, 20%, or 30% sodium bisulfite and incubate 10 to 30 min in a 43°C water bath. Rinse and dehydrate slides in Coplin jars at room temperature using the following regimen:

    2× SSC—three changes, 1 min each
    70% ethanol—2 min
    80% ethanol—2 min
    90% ethanol—2 min
    100% ethanol—2 min.

   Air dry the slides.

   Sodium bisulfite theoretically enhances proteinase K activity by increasing access of the enzymes to the tissues. Those tissues most resistant to digestion require longer times and higher concentrations of pretreatment. The necessity of performing this step is determined largely by trial and error—if hybridization signals are weak to absent, this step should be done.

5. Add 400 μl of 25 μg/ml proteinase K to 40 ml of 2× SSC, place in a Coplin jar, and prewarm to 37°C in water bath.

6. Place slides in the proteinase K/SSC solution and incubate 10 to 40 min in a 37°C water bath.

8

Clinical Cytogenetics   *Short Protocols in Human Genetics*

Unit 8.5   Page 8-25

7. Rinse slides in three changes of 2× SSC, 1 min each, at room temperature.

8. Dehydrate slides by immersing in the following wash fluids for 2 min each, at room temperature:

> 70% ethanol
> 80% ethanol
> 95% ethanol
> 100% ethanol
> Acetone.

9. Air dry the slides, then proceed with hybridization as described in Basic Protocol, steps 12 to 20.

Reference: Schofield and Fletcher, 1992

Contributor: Deborah E. Schofield

UNIT 8.6

Preparation of Amniocytes for Interphase Fluorescence In Situ Hybridization (FISH)

Whereas classical cytogenetics yields a diagnosis from amniotic fluid cells in 7 to 12 days, interphase analysis by FISH permits detection of specific aneuploidies in ≤48 hr. The FISH protocol describes indirect detection but can be used with directly labeled DNA probes as well (see *UNIT 4.4* for a more detailed discussion of FISH).

NOTE: All incubations are performed in a humidified 37°C, 5% CO_2 incubator unless otherwise specified.

BASIC PROTOCOL 1

PREPARATION OF UNCULTURED AMNIOCYTES FOR INTERPHASE FISH ANALYSIS

Materials (see *APPENDIX 1* for items with ✓)

Amniotic fluid, stored at room temperature in the dark, if neccesary
Minimum essential medium (MEM; Life Technologies) with L-glutamine, ribonucleosides, and deoxyribonucleosides, without serum or antibiotics (room temperature)
✓ Hypotonic solution
0.25% (v/v) Triton X-100 (store indefinitely at room temperature)
✓ PBS
100% ethanol, −20°C
Acetone, ice cold
Fixative: 3:1 (v/v) methanol/glacial acetic acid
15-ml screw-cap conical centrifuge tubes
Centrifuge (e.g., IEC Centra-HN)
Cytocentrifuge (e.g., Cytospin III; Shandon/Lipshaw)

8

1. Place 1 to 5 ml of fresh amniotic fluid in a 15-ml screw-cap centrifuge tube.

2. Gently shake tube to resuspend cells. Wash cells by adding 1 to 2 vol MEM without serum, then centrifuge 5 min at $200 \times g$, room temperature.

3. Remove tube from centrifuge, taking care not to disturb cell pellet. Aspirate all but 1 ml of supernatant, then resuspend pellet in remaining liquid.

4. Add 10 ml hypotonic solution. Mix well and incubate 30 min at room temperature. Centrifuge 5 min at $200 \times g$, then repeat step 3.

For laboratories with access to a Cytospin cytocentrifuge

5a. Dilute samples with sufficient hypotonic solution to obtain optimal concentration of nuclei.

> *The final volume for these dilutions should be 0.5 ml/slide if a single funnel is used and 1.0 ml/slide if a double funnel is used (see step 6a). As a general rule 2 to 3 ml of amniotic fluid from a 15- to 17-gestational-week pregnancy can be used to make up to five slides with 75 to 100 scorable nuclei on each slide.*

> *This step is critical, but optimal cell concentration must be determined through trial and error. If the cell suspension is too dense, there will be excessive overlapping nuclei and cellular cytoplasm, leading to decreased hybridization. Suspensions that are too dilute will result in slides with insufficient nuclei, leading to a more tedious analysis.*

6a. Label slides with patient's laboratory identifier (e.g., name and accession number) and place in Cytospin holder. Place funnel on holder and close holder. Place holder in Cytospin.

> *Both single and double Cytospin funnels can be used. Use of the double funnel has been found to be more efficient and cost-effective as it allows placement of cells in two distinct areas on the slide. This permits hybridization with two different probes when appropriate.*

7a. With slide and holder in Cytospin, slowly and carefully pipet sample into funnel, allowing any air in funnel to be displaced. Ensure that there are no bubbles at slide-funnel interface.

8a. Set Cytospin to 1500 rpm and centrifuge 4 min. When instrument has stopped spinning, remove slides from holder and air dry. Proceed to step 9.

For laboratories without access to a Cytospin cytocentrifuge

5b. Add 1 ml room temperature fixative drop by drop, letting it run slowly down side of tube, and gently mix. Add 10 ml room temperature fixative, letting it run slowly down side of tube. Cap tube and incubate 15 min at room temperature.

6b. Centrifuge 10 min at $200 \times g$, room temperature.

7b. Remove tube from centrifuge, taking care not to disturb cell pellet. Aspirate all but 1 ml of supernatant, then resuspend pellet in remaining liquid.

8b. Repeat steps 5b to 7b two times, omitting the 15-min incubation in fixative. Resuspend pellet in a small volume of fixative (usually 0.5 ml) to obtain desired cell concentration for slide preparation. Prepare 1 to 3 chromosome slides (*UNIT 4.1*) per sample. Proceed to step 9.

9. Mix 100 μl of 0.25% Triton X-100 and 40 ml PBS in a Coplin jar. Place slide in jar and soak 5 min at room temperature.

10. Transfer slide to a Coplin jar containing 100% ethanol prechilled to −20°C, then place jar and slide in −20°C freezer for 30 min (to fix slides). (It is not necessary to rinse slide or allow it to dry between the Triton and ethanol treatments.)

11. Immerse slide for a few seconds in ice-cold acetone in a Coplin jar, then let slides air dry (to permeabilize cell membranes). Use immediately for FISH analysis (see Basic Protocol 2 and *UNIT 4.4*), or store at room temperature until the next day (or for longer periods at 4° or −20°C).

ALTERNATE PROTOCOL

PREPARATION OF AMNIOCYTES ATTACHED TO A SURFACE FOR INTERPHASE FISH ANALYSIS

Because some time may be required for attachment of cells to the coverslip and, in some cases, for initial growth, turnaround time may be longer with this protocol than with Basic Protocol 1.

NOTE: Depending on the protocol to be used in subsequent FISH analysis, cells can be initially attached either to slides or coverslips. For the sake of brevity, only coverslips are mentioned in the protocol steps.

Additional Materials (also see Basic Protocol 1)
    Chang in situ medium (Irvine Scientific) supplemented with 1% (v/v)
        penicillin/streptomycin (10,000 U/ml penicillin/10,000 µg/ml streptomycin;
        BioWhittaker) or Amniomax complete medium (Life Technologies)

1. Prepare amniotic fluid for culturing by performing *UNIT 8.2*, Basic Protocol, steps 1 to 8, using either Chang in situ medium or Amniomax medium at step where plates/coverslips are set up.

2. Place tissue culture plates with coverslips in incubator. Leave undisturbed for 24 hr to allow cells to attach firmly to the coverslips.

3. After 24 hr of incubation, gently add an additional 1.5 ml of the same medium (Chang or Amniomax) that is already in the culture. Continue incubation.

4. After an additional 24 to 48 hr of incubation, carefully remove plates from incubator. Gently remove medium from plate without tipping and add 2 ml room temperature hypotonic solution around edge of plate. Let stand 20 min at room temperature.

5. Gently add 2 ml of room temperature fixative to the hypotonic solution, adding dropwise around edge of plate. Let stand 5 min at room temperature, then gently aspirate and discard solution.

6. Gently add ∼2 ml of fresh room temperature fixative as in step 5. Let stand 20 min at room temperature, then gently remove and discard solution. Repeat this procedure two more times.

7. Gently add ∼2 ml of fresh room temperature fixative to the first plate and put a lid on it. Repeat for each plate.

8. Dry coverslips as in *UNIT 8.2* Basic Protocol, steps 19 to 21. Use immediately for FISH analysis (see *UNIT 4.4*) or store at room temperature until the next day (or longer 4° or −20°C).

BASIC PROTOCOL 2

INTERPHASE FISH ANALYSIS OF AMNIOTIC FLUID CELLS

Materials (see APPENDIX 1 for items with ✓)

    Slide or coverslip containing amniotic fluid cells (see Basic Protocol 1 or Alternate Protocol)

    2× SSC, 37°C

    70%, 80%, and 100% (v/v) ethanol, ice cold

✓ Denaturation solution, 70°C

    Biotin- or digoxigenin-labeled DNA probe (see also *UNIT 4.4*): 10 ng/μl α-satellite probe (e.g., D18Z1 for chromosome 18; Oncor or Vysis) *or* 10 to 100 ng/μl prepackaged unique-sequence probe (premixed with blocking DNA and hybridization solution; Oncor or Cambio)

✓ Hybridization solution

✓ 65% formamide wash solution

✓ 50% formamide wash solution

✓ PN buffer

✓ Fluorescein-labeled avidin

✓ 5 μg/ml biotinylated anti-avidin antibody working solution

✓ DAPI staining solution (2 μg/ml working solution)

    2 μg/ml propidium iodide

✓ Phenylenediamine dihydrochloride antifade mounting medium (see recipe for antifade mounting medium)

    37°C slide warmer

    43°, 70°, and 72°C water baths

    24 × 50–mm plastic and 24 × 60–mm glass coverslips

    Rubber cement

    Moist chamber (*UNIT 4.4*)

    Fluorescence microscope with epi-illumination and filter set appropriate for fluorochrome used (*UNIT 4.4*)

1. Soak slides or coverslips containing amniotic fluid cells 1 hr in 37°C 2× SSC solution in a Coplin jar kept in a 37°C water bath.

2. Rinse slides 2 min in ice-cold 70% ethanol in a Coplin jar. Continue dehydration by rinsing slides 2 min each in 80% and 100% cold ethanol. Allow slides to air dry at room temperature.

3. Soak slides exactly 2 min in exactly 70°C denaturation solution in a Coplin jar in a 72°C water bath (raise the temperature of the water bath 1°C for every additional slide placed in the Coplin jar).

    This step is critical to success.

4. Repeat step 2. Prewarm denatured slides to 37°C on a slide warmer until ready for hybridization.

For α-satellite repetitive-DNA probes

5a. Prewarm the 10 ng/μl α-satellite DNA probe and hybridization solution in separate 1.5-ml microcentrifuge tubes for 5 min in a 37°C water bath.

    This step is for repetitive DNA probes obtained from commercial sources. Refer to the product data sheet for any modifications of this procedure that are specfic to a particular probe. See

UNIT 4.4 for details about working with different probes and protocols for labeling probes in the laboratory.

6a. For each slide to be hybridized, mix 1.5 μl prewarmed probe and 30 μl prewarmed hybridization solution in a fresh 1.5-ml microcentrifuge tube.

7a. Incubate probe/hybridization solution mixture 5 min in a 70°C water bath to denature probe. Quickly chill to 4°C on ice. Vortex briefly, then microcentrifuge 2 to 3 sec at maximum speed.

8a. Apply ~31.5 μl of probe/hybridization solution mix (15 ng DNA) to denatured chromosome preparation on each slide from step 4. Cover mixture with a 24 × 50–mm plastic coverslip, removing all air bubbles with gentle pressure. Seal with rubber cement. Incubate overnight (~14 to 18 hr) at 37°C in a moist chamber (*UNIT 4.4*).

   Although the procedure calls for overnight incubation, the time of hybridization can be as brief as 2 to 4 hr when α-satellite DNA probes are utilized.

9a. Wash slides 15 min with intermittent agitation in 65% formamide wash solution in a Coplin jar kept in a 43°C water bath. Proceed to step 10.

For prepackaged unique-sequence probes

5b. For each slide to be hybridized, transfer 10 μl of 10 to 100 ng/μl prepackaged probe into a 1.5-ml microcentrifuge tube.

   Commercial unique-sequence probes are packaged premixed with both blocking (Cot-1) DNA and 50% formamide hybridization solution. Many commercial unique-sequence probes (e.g., Oncor) do not require denaturation.

6b. Incubate prepackaged probe 5 min in a 37°C water bath.

   Refer to the product data sheet for any modifications of this procedure that are specific to a particular probe. See UNIT 4.4 for details about working with different probes and protocols for labeling probes in the laboratory.

7b. Apply ~10 μl prepackaged probe (100 to 1000 ng DNA) to denatured chromosome preparation on each slide from step 4.

8b. Cover mixture with a 24 × 50–mm plastic coverslip, removing all air bubbles with gentle pressure. Seal with rubber cement. Incubate overnight (~14 to 18 hr; see annotation in step 8a) at 37°C in a moist chamber.

9b. Wash slides for 15 min with intermittent agitation in 50% formamide wash solution in a Coplin jar kept in a 43°C water bath. Proceed to step 10.

10. Wash slides for 8 min with intermittent agitation in 2× SSC in a Coplin jar kept in a 37°C water bath.

11. Rinse slides briefly in PN buffer in a Coplin jar at room temperature. Remove slides from jar and drain excess fluid but do not allow slides to dry.

12a. *For biotin-labeled probes:* Apply 60 μl fluorescein-labeled avidin to hybridized chromosome preparation on each slide. Place a 24 × 50–mm plastic coverslip over solution, removing all air bubbles with gentle pressure. Incubate 20 min at 37°C in a moist chamber.

12b. *For digoxigenin-labeled probes:* Apply 60 μl fluorescein-conjugated Fab fragment of sheep anti-digoxigenin to hybridized chromosome preparation on each slide. Place a 24 × 50–mm plastic coverslip over solution, removing all air bubbles with gentle pressure. Incubate 20 min at 37°C in a moist chamber.

8

13. Wash slides for 2 min with intermittent agitation successively in each of three separate Coplin jars containing PN buffer.

14a. *To amplify signal (if needed) for biotin-labeled probes:* Apply 60 μl of 5 μg/ml biotinylated anti-avidin antibody to hybridized chromosome preparation on each slide. Place a 24 × 50–mm plastic coverslip over preparation, removing all air bubbles with gentle pressure. Incubate 20 min at 37°C in a moist chamber. Repeat steps 13, 12a, and 13 again.

14b. *To amplify signal (if needed) for digoxigenin-labeled probe:* Apply 60 μl of 5 μg/ml rabbit anti-sheep antibody followed by fluorescein-labeled anti-rabbit antibody, washing as needed between steps.

> *In some situations where weak signals are obtained, a second round of amplification may be required, in which case step 14 should be repeated.*

15. Apply 21 μl of 2 μg/ml DAPI staining solution and 21 μl of 2 μg/ml propidium iodide to hybridized chromosome preparation on each slide. Incubate at room temperature for 5 min in darkness.

> *DAPI allows for the best cellular visualization and propidium iodide provides a uniformly red background on the interphase chromosomes against which to view the yellow-green fluorescein signals.*

16. Rinse slides 3 min in PN buffer in a Coplin jar to remove excess stain.

17. Apply 21 μl phenylenediamine dihydrochloride antifade mounting medium to hybridized chromosome preparation on each slide. Cover with a 24 × 60–mm glass coverslip.

18. Examine slides with fluorescence microscope and photograph cells that demonstrate informative results. Analyze interphase cells by counting number of hybridization signals in each of 50 cells for each probe utilized.

References: ACMG, 1993; Klinger et al., 1992

Contributors: Stuart Schwartz, Mark A. Micale, and Laurie Becker

---

UNIT 8.7

Diagnosis of Fanconi Anemia by Diepoxybutane Analysis

Fanconi anemia (FA) is an autosomal recessive syndrome characterized clinically by progressive bone marrow failure and a high risk of malignancies, particularly acute myelogenous leukemia (AML) and squamous cell carcinoma.

Although the pathophysiology of FA remains unknown, hypersensitivity to the clastogenic (chromosome-breaking) effect of DNA cross-linking agents provides a unique marker for the diagnosis of FA. This cellular characteristic can be utilized as a diagnostic test to identify the preanemic patient as well as the patient with aplastic anemia or leukemia who may or may not have the classic physical stigmata associated with FA. A variety of chemical agents can be used to test for DNA cross-link sensitivity. Diepoxybutane (DEB; 1,3-butadiene diepoxide) analysis is the preferred test for FA diagnosis, as it has the highest sensitivity and specificity, while other agents have higher rates of false-positive and false-negative test results.

CAUTION: DEB is a potential carcinogen and precautions should be taken when handling this compound (see Support Protocol 1).

NOTE: All reagents and equipment coming into contact with live cells must be sterile. All incubations are performed in a humidified 37°C, 5% CO_2 incubator unless otherwise specified.

BASIC PROTOCOL

DIEPOXYBUTANE TEST FOR POSTNATAL DIAGNOSIS OF FANCONI ANEMIA

The preferred tissue for laboratory diagnosis of FA by DEB testing is peripheral blood. At a concentration of 0.1 µg/ml, DEB induces multiple chromosomal breaks and exchanges in FA cells, but has little clastogenic effect on cells from non-FA individuals.

Materials (*see* APPENDIX 1 *for items with* ✓)

    Peripheral blood: collect in a preservative-free sodium heparin Vacutainer tube (e.g., Fisher)

    Complete RPMI/15% FBS medium (APPENDIX 1)/1% (w/v) phytohemagglutinin (PHA; Burroughs Wellcome)

✓ PBS

    Diepoxybutane (1,3-butadiene diepoxide, DEB; Aldrich; store at 4°C)

    1 µg/ml Colcemid (e.g., Life Technologies; store at 4°C)

    0.075 M KCl, 37°C

    Fixative: 3:1 (v/v) methanol/glacial acetic acid, fresh

    10-ml syringe equipped with 18 1/2-G needle

    25-cm^2 tissue culture flasks

    15-ml sterile conical-bottom centrifuge tubes with caps

    IEC clinical centrifuge or equivalent

    Microscope slides (store in 70% ethanol)

    Inverted microscope

1. Drop peripheral blood (18 to 25 drops; 0.4 to 0.5 ml) from a 10-ml syringe equipped with an 18 1/2-G needle into 25-cm^2 tissue culture flasks containing 10 ml complete RPMI/15% FBS/1% PHA. Set up duplicate cultures for DEB studies and duplicate cultures to serve as untreated controls (a total of four flasks). Incubate cultures 24 hr (APPENDIX 3I).

2. Aliquot PBS to three 15-ml sterile tubes: 10 ml to tube 1, 6 ml to tube 2, and 4.5 ml to tube 3.

3. Just prior to addition to culture, add 10 µl stock DEB to tube 1, cap, and mix well. Transfer 4 ml from tube 1 to tube 2, cap, and mix well. Transfer 0.5 ml (500 µl) from tube 2 to tube 3, cap, and mix well.

4. Add 25 µl of the solution from tube 3 to two of the peripheral blood cultures that have been incubating 24 hr (from step 1; final concentration of DEB in the flask is 0.1 µg/ml.). Incubate cultures an additional 48 to 72 hr.

5. Add 2 ml of 1 µg/ml Colcemid to each 10-ml culture and incubate 20 min.

6. Transfer contents of each culture flask to a 15-ml conical-bottom centrifuge tube, cap, and centrifuge 10 min at ~150 × g.

7. Remove most of the supernatant. Resuspend each pellet in the residual supernatant by flicking the tube with finger, and carefully add 5 ml prewarmed 0.075 M KCl to swell cells without breakage. Incubate 10 min in a 37°C water bath. Centrifuge 10 min at 150 × g, room temperature.

8. Remove the supernatant and gently add 1 ml fresh fixative without disturbing the pellet. Remove fixative immediately and repeat this step.

9. Immediately break up the pellet by striking the tube several times with an index finger. Quickly add fixative to resuspend the cells without getting clumps. Bring volume to 8 to 10 ml with fixative. Break up any clumps by pipetting with a Pasteur pipet. Let stand 30 min at room temperature.

10. Centrifuge 10 min at 150 × *g*, room temperature. Remove supernatant and resuspend pellet in 3 to 5 ml fixative. Let stand 10 min at room temperature. Repeat this step one or two more times either immediately, or after letting cells stand in fixative overnight at room temperature.

11. Centrifuge 10 min at 150 × *g*, room temperature. Remove the supernatant, break the pellet, and add enough fixative (~0.25 to 0.5 ml) to give the desired cell concentration for slide-making.

    An optimal cell concentration provides a maximum number of cells in a microscope field without crowding metaphase spreads.

12. Drop a few drops of the cell suspension onto a clean microscope slide from a height sufficient to make well-spread metaphases (*UNIT 4.1*). Make at least six slides per culture (it is recommended to practice on cells not used for diagnosis).

    This step is critical to success.

13. Examine slides on an inverted microscope to determine quantity and quality of metaphases. Stain slides with Giemsa (see Support Protocol 2).

14. Analyze 50 to 100 Giemsa-stained metaphases from each DEB-treated preparation (25 metaphases for cases with very high breakage; *APPENDIX 3K*). Score for chromosome number and for the numbers and types of structural abnormalities. If chromosomal breakage is increased above the normal range, analyze an untreated preparation. Photograph abnormal cells for later analysis.

15. Interpret results according to published values for baseline and DEB-induced chromosomal breakage in PHA-stimulated peripheral blood lymphocytes.

    Table 8.7.1 presents data for baseline and treated cultures from normal and affected individuals.

SUPPORT PROTOCOL 1

WORKING WITH AND DISPOSING OF DIEPOXYBUTANE

DEB (1,3-butadiene diepoxide) is a potential carcinogen and precautions should be taken when handling this compound. This section describes how to work with DEB and the proper disposal of wastes.

Work space. All work with DEB must be done using a chemical fume hood or a class II Biological Safety Cabinet. The investigator should wear a laboratory coat and gloves when working with DEB. A bottle of 6 M HCl (~50 ml) should be kept at hand to use to rapidly inactivate DEB in case of spills.

Cell cultures. Cell culture work involving DEB should be done in a class II Biological Safety Cabinet. Once the cells are fixed they may be processed outside the safety cabinet. Spent medium and wash solutions should be disposed of by adding 6 M HCl and discarding as biohazardous waste. Used disposable pipets and culture flasks should be rinsed with 6 M HCl

8

Table 8.7.1 Range of Chromosomal Breakage Observed in Peripheral Blood Lymphocytes and Cultured Fetal Cells

Tissue	Diagnosis	Mean breaks/cell		No individuals tested
		Baseline	DEB-treated[a]	
Peripheral blood	FA	0.02–0.85[b]	1.10–23.9	98
	Non-FA	0.00–0.12	0.00–0.36	124
Fibroblasts	FA	0.20–0.36	0.68–1.10	3
	Non-FA	0.00–0.08	0.00–0.07	4
Amniocytes	FA	0.18–0.45	0.69–0.96	10[c]
	Non-FA	0.00–0.12	0.00–0.14	48[c]
Chorionic villus cells	FA	0.30–0.46	1.00–1.40	9[c]
	Non-FA	0.00–0.10	0.00–0.14	39[c]

[a]DEB concentration is 0.1 µg/ml in peripheral blood cultures, and 0.01 µg/ml in fibroblast, amniotic fluid, and chorionic villus cell cultures.

[b]Data show the range for mean breaks/cell in each category.

[c]Number of prenatal diagnoses performed in pregnancies of couples who had a previously affected child (A.D.A., unpub. observ.).

and discarded in autoclavable biohazard bags. Micropipet tips should be discarded into a small bottle of 6 M HCl.

Waste disposal. The stock solution of DEB should be disposed of as hazardous chemical waste without dilution or chemical inactivation. Disposable plasticware should be rinsed with 6 M HCl and discarded in autoclavable biohazard bags. Medium and wash solutions should be treated with 6 M HCl and disposed of as biohazardous waste.

SUPPORT PROTOCOL 2

GIEMSA STAINING FOR CHROMOSOME-BREAKAGE ANALYSIS

Giemsa stain is a histological stain with a particular affinity for nuclei and chromosomes. The procedure is simple and results in unbanded reddish-blue to purple nuclei and chromosomes.

Materials

    Gurrs buffer, pH 6.8, tablets (Bio/medical Specialties)
    Giemsa stain (e.g., Life Technologies)
    Metaphase chromosome slide (see Basic Protocol)
    Permount histological mounting medium
    Coplin jars

1. Dissolve one Gurrs buffer, pH 6.8, tablet in 1 liter water.

2. Add 4 ml fresh Giemsa stain to 46 ml Gurr's buffer in a staining jar. Use stain for ~1 hr or to stain two groups of ten slides; then prepare fresh stain.

3. Stain metaphase chromosome slide 5 min. Rinse by dipping several times in a Coplin jar containing Gurrs buffer (step 1).

4. Rinse slide by dipping several times in a Coplin jar containing water. Air dry the slide and check slide for stain intensity.

   Stain intensity should be sufficient so that gaps and breaks in the chromosomes appear as achromatic (nonstained) regions.

5. Put a coverslip on the slide using Permount histological mounting medium.

ALTERNATE PROTOCOL 1

DIEPOXYBUTANE TEST USING FIBROBLAST CULTURES

If peripheral blood is not available for testing, fibroblast cultures can be initiated from skin biopsies. Small pieces of lung or skin taken from abortuses or stillborns can also be used for testing. DEB is used at a ten-fold lower concentration for fibroblast cultures because the higher concentration is too toxic for FA fibroblasts and makes it difficult to collect sufficient metaphases for study.

Additional Materials (also see Basic Protocol; see APPENDIX 1 for items with ✓)

    Tissue biopsies (e.g., full-thickness skin sample taken by a 3-mm punch)
✓ Complete DMEM/20% FBS
    0.051 M (0.38% w/v) KCl

    60-mm tissue culture plates
    Disposable scalpels

1. Place the tissue biopsy in 0.5 ml complete DMEM/20% FBS in a 60-mm tissue culture plate and cut it into 1-mm pieces with two sterile disposable scalpels.

2. Score the bottoms of at least three additional 60-mm plastic tissue culture plates by making horizontal and vertical lines with the point of a sterile scalpel. Add a few drops of complete DMEM/20% FBS to the plate with the biopsy pieces. Use a sterile Pasteur pipet to transfer 5 to 6 biopsy pieces into each 60-mm plate. Place the tissue pieces at the intersections of the scored lines.

3. Incubate the plates, without any additional tissue culture medium, 15 to 30 min to allow the pieces of tissue to adhere to the plates. Add 5 ml complete DMEM/20% FBS to each plate and incubate the cultures.

4. Change the medium every 3 to 4 days without disturbing the tissue pieces. Use an inverted microscope to inspect the cultures for evidence of cell growth (it may take from 1 week to 1 month to observe good cell growth).

5. When a large patch of fibroblasts can be seen surrounding the tissue pieces in the plate, passage the cells by treating them with $1\times$ trypsin/EDTA solution (*APPENDIX 3I*), taking care not to dislodge the biopsy pieces as they will continue to grow in culture. Transfer cells to a sterile 25-cm^2 tissue culture flask in complete DMEM/15% FBS.

6. Trypsinize cells again when the first passage culture reaches confluency (*APPENDIX 3I*). Subculture at a 1:3 split ratio in complete DMEM/15% FBS. Proceed with DEB-testing for FA when at least two second-passage cultures reach confluency.

7. Plate cells in 5 ml complete DMEM/15% FBS in 25-cm^2 flasks at a density of 3×10^5 cells per flask (two treated and two untreated control flasks are required for this analysis). Incubate 24 hr.

8. Make DEB dilutions as described (see Basic Protocol, steps 2 and 3), except add a fourth tube with 4.5 ml PBS. Make the final dilution of DEB for fibroblast cultures by taking 0.5 ml (500 μl) from tube 3 and adding it to tube 4.

9. Add 25 μl of the solution from tube 4 to 10 ml fresh complete DMEM/15% FBS. Replace the growth medium in two cultures with this DEB-containing medium (5 ml in each flask; 0.01 μg/ml final). Add 25 μl solution from tube 3 to 10 ml fresh complete DMEM/15% FBS. Replace the growth medium in two extra cultures, if available, with this DEB-containing medium (5 ml in each flask; 0.1 μg/ml final concentration). Replace growth medium in two untreated control cultures with fresh medium.

10. Incubate cells until the cultures are nearly confluent ($\sim$72 hr).

11. Trypsinize cells (APPENDIX 3I) and subculture at a 1:2 or 1:3 split ratio into fresh complete DMEM/15% FBS without DEB to collect sufficient mitotic cells for cytogenetic studies. Re-feed reserve cultures with medium containing freshly diluted DEB.

12. Harvest cultures at the first mitotic wave after subculture ($\sim$24 to 48 hr; when metaphase cells appear rounded). At 3 hr prior to harvest, add 1 ml of 1 μg/ml Colcemid to the DEB-treated cultures and the untreated controls (in 5 ml medium). Incubate 3 hr.

13. Trypsinize cells (APPENDIX 3I). Transfer the contents of the culture flask to a sterile 15-ml centrifuge tube, cap, and centrifuge 10 min at $150 \times g$.

14. Remove most of the supernatant. Resuspend the pellet in the remaining supernatant and slowly add 5 ml of 0.051 M KCl. Incubate 10 min in a 37°C water bath. Centrifuge 10 min.

15. Fix cells and prepare slides as described (see Basic Protocol steps 9 to 12; in step 9, slowly add 3 to 5 ml of fixative).

16. Analyze 100 Giemsa-stained metaphases from untreated and from DEB-treated cultures for chromosomal breakage as described (see Basic Protocol, step 13, and Support Protocol 2). Interpret results according to published values for baseline and DEB-induced chromosomal breakage in FA fibroblasts (Table 8.7.1).

ALTERNATE PROTOCOL 2

DIEPOXYBUTANE TEST FOR PRENATAL DIAGNOSIS OF FANCONI ANEMIA

FA can be diagnosed in pregnancies at risk by studying cultured trophoblast cells obtained by chorionic villus sampling (CVS) at 9 to 12 weeks of gestation, or amniotic fluid cells obtained by amniocentesis at 15 to 17 weeks. FA can also be diagnosed prenatally from a fetal blood sample using methods similar to those described previously for peripheral blood. Results with fetal blood can be obtained within 3 days from the time the sample is collected so this method is of value in pregnancies where a prenatal diagnosis could not otherwise be completed before the last part of the second trimester.

Study Fetal Cells Obtained by CVS or Amniocentesis

Additional Materials

Fetal cells, enough to establish cultures in two 25-cm^2 flasks

Growth medium: 1:1 mixture of Chang medium and complete RPMI/15% FBS (see individual recipes in APPENDIX 1)

1. Establish fetal cell cultures in growth medium in 25-cm² flasks and incubate (*UNITS 8.1 & 8.2*).

2. Trypsinize and subculture primary cultures when flasks contain several large, rapidly growing colonies (*APPENDIX 3I*). Do not allow colonies to become overgrown. Use a split ratio of 1:2 or 1:3. Incubate 24 hr.

3. Replace the growth medium in at least two flasks (from two different primary cultures) with fresh growth medium containing 0.01 μg/ml of DEB, and 0.1 μg/ml of DEB in two extra flasks, if available, freshly diluted as described (see Alternate Protocol 1, steps 8 and 9).

   The remaining flasks are untreated controls for baseline chromosome breakage studies.

4. Incubate the cultures, harvest and fix the cells, and prepare metaphase spreads as described (see Alternate Protocol 1, beginning of step 10).

5. Analyze a total of 100 metaphases from each group (baseline and DEB-treated; 50 metaphases from each of two primary flasks in each group). Interpret results according to published values for baseline and DEB-induced chromosomal breakage in FA fetal cells obtained by CVS or amniocentesis (Table 8.7.1).

Study Cells from Fetal Blood

Additional Materials

Heparinized fetal blood
✓ Complete RPMI/15% FBS

1. Add ~0.25 ml heparinized fetal blood (collected in a preservative-free sodium heparin Vacutainer tube; e.g., Fisher) to a 25-cm² flask containing 10 ml complete RPMI/15% FBS. Set up two to four flasks for baseline and DEB-treated cultures, depending on the amount of fetal blood collected.

2. Incubate, treat, harvest, and analyze fetal blood cultures as described (see Basic Protocol, steps 2 to 15).

References: Auerbach, A.D. 1993; Auerbach et al., 1989

Contributor: Arleen D. Auerbach

8

CHAPTER 9

Clinical Molecular Genetics

Research in human genetics is aimed largely at the identification of genes responsible for human disease and characterization of mutations. This can provide insight into the pathophysiology of genetic disorders as well as new means of diagnosis. Molecular diagnosis involves the detection of pathogenic mutations in DNA or RNA samples. It can offer very precise diagnosis, sometimes predicting clinical course in a disorder with variable expression. It also obviates the need to biopsy clinically affected tissue, because any source of DNA, such as peripheral blood or buccal lining cells, can be used. In addition, prenatal diagnosis can be offered based on DNA analysis from chorionic villus biopsy tissue or amniotic fluid cells.

Methods of direct mutation analysis differ with specific disorders. The intention in this chapter is not to describe all tests in current use, but rather to present examples of specific modes of testing, which can often be applied to other diseases, using appropriate DNA probes or PCR primers. For example, detection of deletions in the dystrophin gene in males with Duchenne/Becker muscular dystrophy (UNIT 9.1), which is based on multiplex PCR, can be generalized to other X-linked disorders in which deletion is a common mechanism of mutation. Another multiplex PCR approach, UNIT 9.2, has been used to identify point mutations responsible for cystic fibrosis, but it is applicable to any disorder in which point mutations commonly occur. UNIT 9.7 presents another protocol for detection of point mutations, in this case based on use of PCR primers that specifically amplify mutant or wild type alleles. It, too, should be widely applicable to the task of efficient analysis of clinical specimens for known mutations. Along similar lines, UNIT 9.11 presents protocols for the analysis of single nucleotide polymorphisms of the apolipoprotein E gene (ApoE).

Several disorders have recently been identified in which variation in the size of a triplet repeat region is the basis for gene mutation. Detection of one such mutation, involved in fragile X syndrome, is described in UNIT 9.3. UNIT 9.4 describes the approach to another triplet repeat disorder, myotonic dystrophy.

In recent years, molecular analysis has also been extended to the mitochondrial genome, where mutations lead to disorders of energy metabolism. Protocols for analysis of mitochondrial mutations are provided in UNIT 9.8.

PCR technology and advances in in vitro fertilization have enabled the detection of mutations in some genes in single cells isolated from early embryos at the blastocyst stage (UNIT 9.9). UNIT 9.10 describes the protein truncation assay, an approach to the detection of mutations that lead to premature termination of translation. This has provided a general approach to finding mutations and is particularly useful for large genes with a wide diversity of mutations, a high proportion of which cause protein truncation.

Molecular genetic testing is also increasingly being used for identity testing. This is addressed in this chapter by a unit describing paternity testing (UNIT 9.6). On a molecular level, it is now also possible to determine which of the two X chromosomes is inactivated in females (UNIT 9.5).

The field of molecular diagnosis is evolving rapidly, as knowledge of the human genome is increasing. New diagnostic tests are in constant development, creating a great challenge both to the diagnostic laboratory and to the practicing physician in keeping current. Computer databases available on the internet are playing an increasingly vital role in providing point of care access to information in genetic disorders and genetic testing.

Contributor: Bruce R. Korf

UNIT 9.1

Multiplex PCR for Identifying Dystrophin Gene Deletions

This unit presents a multiplex PCR approach for assaying deletions in the dystrophin gene (Fig. 9.1.1). The endpoints of many of these deletions can be identified, allowing prediction of disease severity (e.g., DMD versus BMD) based on effects on the translational reading frame (Fig. 9.1.2).

NOTE: Experiments involving PCR require extremely careful technique to prevent contamination.

BASIC PROTOCOL

DIAGNOSTIC MULTIPLEX PCR TO DETECT DYSTROPHIN GENE DELETIONS

Materials (see APPENDIX 1 for items with ✓)

    Reaction mixes A, B, and C (see Support Protocol)
    5 U/μl *Taq* DNA polymerase (e.g., Perkin-Elmer Cetus AmpliTaq or Native Taq)
    50 ng/μl template DNA in TE buffer (pH 8.0; *APPENDIX 1*), prepared from cell lines, whole blood, preserved tissues (*APPENDIX 3A*), chorionic villus biopsy, amniotic fluid, or other clinical specimens
    Mineral oil
    1.5% (w/v) agarose minigel containing 0.5 μg/ml ethidium bromide
✓ Electrophoresis buffer (TAE or TBE) containing 0.5 μg/ml ethidium bromide
✓ 10× gel loading buffer

    Dedicated micropipets for setting up reactions
    Two thermal cyclers, preheated to 94°C
    General-use micropipets for analyzing reaction products

1. Thaw each of the three reaction mixes (A, B, and C) on ice and calculate the amount of each that will be needed to amplify the number of samples to be analyzed (each 50-μl reaction requires 44.5 μl reaction mix). Calculate enough for 2.5 extra reactions to accommodate positive (from male with intact dystrophin gene) and negative (no template) controls, with 0.5 extra reaction volumes to account for sample loss while pipetting.

2. On ice, aliquot the required amount of each mix to separate 1.5-ml microcentrifuge tubes labeled A, B, and C. Add 0.5 μl per reaction of 5 U/μl *Taq* DNA polymerase to each tube. Mix gently by pipetting up and down and microcentrifuge 5 to 10 sec at high speed to collect the fluid at the bottom of the tubes.

3. Label the lids of appropriate reaction tubes for thermal cycling. Place tubes on ice and aliquot 45 μl prepared reaction mixes to each tube.

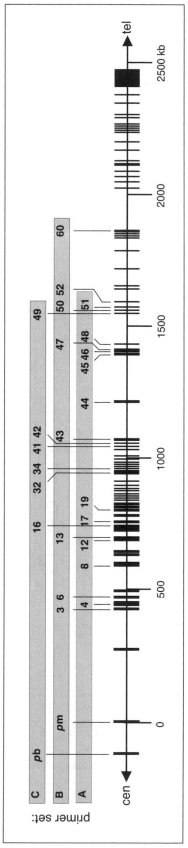

Figure 9.1.1 Schematic diagram of the dystrophin gene showing relative locations of exons amplified by each of the three standard diagnostic assays A, B, and C (enclosed by gray bars). Vertical lines indicate approximate locations of exons as determined by Coffey et al. (1992); thin lines indicate location is approximate. Abbreviations: *pb*, brain-specific promoter (Boyce et al., 1991), *pm*, muscle-specific promoter; cen, centromeric side of gene; tel, telomeric side of gene.

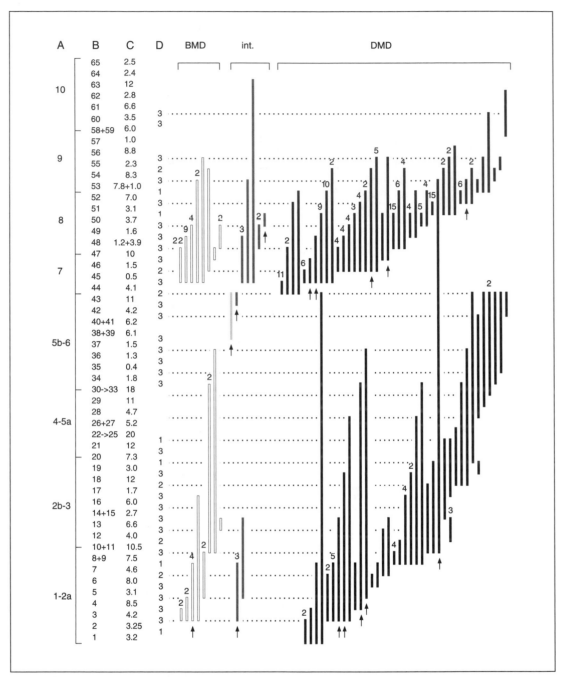

Figure 9.1.2 Correspondence of dystrophin gene deletions (and their effects on the translational reading frame) with clinical phenotypes of patients with DMD and BMD. Vertical bars indicate extent of deletions in 273 independent cases of DMD/BMD. (**A**) dystrophin cDNA probes used for detection of deletions (probe 11-14 is not shown because no deletions were detected with this probe); (**B**) exon numbers; (**C**) exon-containing *Hind*III fragments (sizes in kb); (**D**) exon border type. Bars indicate extent of deletions with associated phenotypes indicated above. "int" describes both young patients with clinical conditions clearly milder than typical DMD and intermediate patients who became wheelchair-bound between 13 and 15 years of age. Numbers above the bars indicate the number (if >1) of independent cases with similar deletions. Arrows indicate deletions that do not follow the reading-frame rule relating phenotype to genotype. Modified from Koenig et al. (1989) with permission by University of Chicago Press and American Journal of Human Genetics. © 1989 by the American Society of Human Genetics. All rights reserved.

4. Add 5 µl of 50 ng/µl template DNA to each tube. Add 5 µl sterile water to the last tube for a negative control.

5. If the thermal cycler being used doesn't have a heated lid, add one drop (~25 to 50 µl) mineral oil to each tube and cap tightly. Avoid splashing the contents of one tube into another as the drops are added. Microcentrifuge the tubes 5 to 10 sec at high speed to collect all of the aqueous reaction components under the layer of oil, ensuring that the oil completely covers the surface of the mixture.

6. Preheat a thermal cycler to 94°C, add the tubes containing reaction mix A, and amplify under the following conditions:

Initial step:	6 min	94°C	(denaturation)
25 cycles:	30 sec	94°C	(denaturation)
	30 sec	53°C	(annealing)
	4 min	65°C	(elongation)
Final step:	7 min	65°C	(elongation)

Store at 4°C until analysis.

7. Preheat another thermal cycler to 94°C and amplify all the reactions with mixes B and C using the following conditions:

Initial step:	6 min	94°C	(denaturation)
25 cycles:	30 sec	94°C	(denaturation)
	4 min	65°C	(annealing and elongation)
Final step:	7 min	65°C	(elongation)

Store at 4°C until analysis.

8. Prepare 1.5% agarose minigels (gel volume ~25 ml) containing 0.5 µg/ml ethidium bromide and electrophoresis buffer containing an equal concentration of ethidium bromide.

9. Run all reactions from each primer set on the same gel to allow for direct comparison between positive and negative controls and patient sample. To load gels, first place 2-µl spots of 10× loading buffer on a sheet of Parafilm (spot only enough loading buffer for one gel at a time to avoid evaporation loss). Remove 15 µl of PCR products by inserting a pipet tip through the oil layer into the aqueous phase below while expelling a bubble of air as the tip passes through the oil. Wipe off excess oil from the tip with a tissue before mixing the reaction product with one of the spots of loading buffer on the Parafilm. Immediately load this mixture onto the gel.

10. Run minigels 15 min at 2 V/cm, then ~90 min at 5 V/cm until bromphenol blue is ~1 cm from the bottom of the gel. Photograph and record the results. Be careful to inspect each gel visually as it is photographed to ensure that all bands are reproduced accurately.

11. Store remaining reaction products at 4°C until after the gels are photographed and interpreted. If there are problems with gel analysis, run the samples again within several weeks after amplification.

12. Record and interpret the PCR results on a standard form such as the one illustrated in Figure 9.1.3.

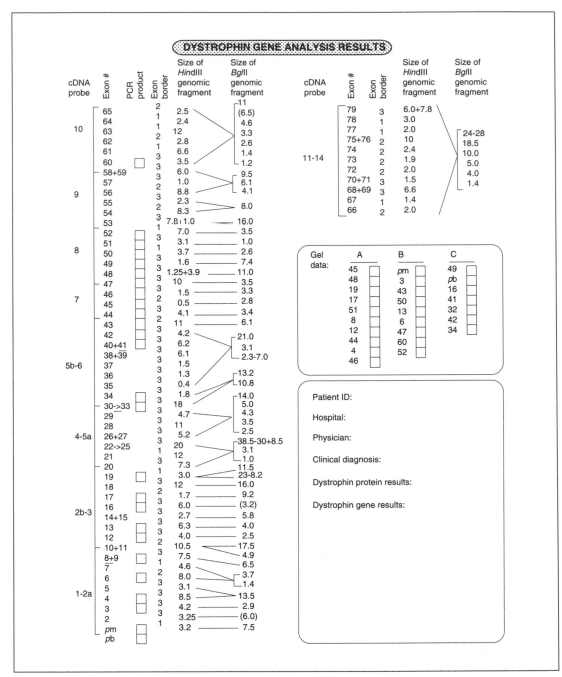

Figure 9.1.3 Standard form for recording results of dystrophin gene analysis. Indicated at left are cDNA probes 1-2a through 11-14 and the exons they contain, boxes for each expected PCR product (to be checked + or − depending on assay results), and the known exon borders and sizes of *Hind*III and *Bgl*I restriction fragments containing each exon. Brackets indicate uncertainty in order of restriction fragments and correspondence to particular exons. The gel data box can be used to directly check off presence or absence of PCR products by comparison to photographs of gels. The patient information box can be modified to include any relevant information for a particular site. For deletions with unambiguous endpoints, the effects on the translational reading frame can be predicted by comparing exon border types at the two ends of the deletions. Deletions that bring together two exons with similar borders are predicted to maintain the reading frame and are consistent with a diagnosis of Becker muscular dystrophy (BMD); those that join dissimilar borders create a frameshift and are generally associated with Duchenne muscular dystrophy (DMD).

LOW-CYCLE-NUMBER MULTIPLEX PCR WITH RADIOACTIVE LABEL TO DETECT DOSAGE DIFFERENCES

The following protocol addresses problems associated with detecting dosage diffences in males with dystrophin gene duplications or in female carriers of deletions through the addition of a radioactive label.

Additional Materials

10 mCi/ml [α-^{32}P]dCTP (3000 mCi/mmol)
Thin 2% agarose gel: pour hot (80°C) and, if necessary, prewarm molds

Centricon 100 concentrator (Amicon)
Centrifuge with fixed-angle rotor (23° to 45°): e.g., Beckman JA-18.1
Whatman 3MM filter paper
Densitometer or PhosphorImager

1. Set up PCR reactions exactly as described (see Basic Protocol, steps 1 to 5), except add 5 µCi [α-^{32}P]dCTP (0.5 µl of 10 mCi/ml) per reaction in step 2. Amplify 18 cycles in a thermal cycler as described (see Basic Protocol, step 6).

2. *Optional:* Remove the aqueous PCR product from beneath the mineral oil and dilute in 2 ml water in the barrel of a Centricon 100 concentrator assembled according to the manufacturer's recommendations. Remove unincorporated nucleotides by centrifuging 30 min at 1000 × g in a fixed-angle rotor. Resuspend sample with 2 ml water and centrifuge again. Invert concentrator and centrifuge concentrate into retentate cup.

3. Separate one-quarter of the purified PCR product (typically ~10 µl, depending on the angle of the rotor used for the Centricon column) on a thin 2% agarose midi- or maxigel with uniform thickness.

4. Rinse the gel briefly in water to wash away superficial radioactivity. Blot dry on several sheets of Whatman 3MM filter paper.

5. Place the gel on a sheet of Whatman 3MM filter paper on a preheated gel dryer and cover with a sheet of plastic wrap. Dry the gel thoroughly 30 min to 1 hr at 80°C.

6. Autoradiograph 2 to 24 hr. Make several different exposures to ensure that one is in the linear range of the film. Analyze the relative intensities of the bands using a densitometer, comparing to the normal controls. Alternatively, use a PhosphorImager.

PREPARATION AND STORAGE OF STOCK DIAGNOSTIC PCR MIXES

If primers are of high quality, they can often be used without any special purification; however, purification on polyacrylamide gels (*CPMB UNIT 2.12*) is recommended to ensure maximum efficiency and reproducibility of amplification. If primer purification involves a final ethanol precipitation step, extra care should be taken to wash away as much salt as possible, because multiplexing many primers together may result in significantly elevated levels of contaminating salts. Detection of dosage in males with duplications or in female heterozygotes with deletions is particularly sensitive and may require purified primers to ensure accuracy.

Materials (*see* APPENDIX 1 *for items with* ✓)

    Oligonucleotide PCR primers (Tables 9.1.1, 9.1.2, 9.1.3, and 9.1.4)
    Double-distilled, sterile H_2O
✓ PCR buffer: $10\times$ PCR amplification buffer containing 15 mM $MgCl_2$
✓ 2.5 mM 4dNTP mix

1. Dilute primers to 400 ng/μl in double-distilled, sterile water. Store unused primers in tightly closed 1.5-ml microcentrifuge tubes at $-20°C$. Store some primers separately to allow mixing new combinations for special uses. For any primer pair employed alone or in combination with a limited number of other primers, use 0.5 μl each primer (~0.5 μM final assuming 26 bases per primer average) in a 50-μl PCR reaction.

2. In 1.5-ml microcentrifuge tubes labeled "pri-A," "pri-B," and "pri-C," make primer mixes (assume 1.0 μl per pair in a 50-μl reaction) for each of the three diagnostic assays. Combine equal amounts of each primer listed in Tables 9.1.1 (for "pri-A"), 9.1.2 (for "pri-B"), and 9.1.3 (for "pri-C") in the tubes.

3. In 1.5-ml microcentrifuge tubes, prepare final-reaction master mixes in batches of 25 reactions/tube as described below.

Mix A	Mix B	Mix C	Component
637.5 μl	662.5 μl	712.5 μl	Double distilled H_2O, sterile
125 μl	125 μl	125 μl	PCR buffer
100 μl	100 μl	100 μl	2.5 mM 4dNTP mix
250 μl (pri-A)	225 μl (pri-B)	175 μl (pri-C)	Mixed primers

    Prepare multiple reaction mixes at the same time and store tightly sealed (e.g., with Parafilm) ≤3 months at $-20°C$ or for longer periods of time at $-70°C$. Adjust size of each aliquot according to the caseload of the laboratory, but avoid more than three freeze-thaw cycles.

References: Beggs et al., 1990; Chamberlain et al., 1988

Contributor: Alan H. Beggs

Table 9.1.1 PCR Primers for Diagnostic Set A[a]

Exon[b]	Forward primer ($5'\rightarrow 3'$)	Reverse primer ($5'\rightarrow 3'$)	Primer location[c]	Product size (bp)
45	AAACATGGAACATCCTTGTGGGGAC	CATTCCTATTAGATCTGTCGCCCTAC	I/I	547
48	TTGAATACATTGGTTAAATCCCAACATG	CCTGAATAAAGTCTTCCTTACCACAC	I/I	506
19	TTCTACCACATCCCATTTTCTTCCA	GATGGCAAAAGTGTTGAGAAAAAGTC	I/I	459
17	GACTTTCGATGTTGAGATTACTTTCCC	AAGCTTGAGATGCTCTCACCTTTTCC	I/I	416
51	GAAATTGGCTCTTTAGCTTGTGTTTC	GGAGAGTAAAGTGATTGGTGGAAAATC	I/I	388
8	GTCCTTTACACACTTTACCTGTTGAG	GGCCTCATTCTCATGTTCTAATTAG	I/I	360
12	GATAGTGGGCTTTACTTACATCCTTC	GAAAGCACGCAACATAAGATACACCT	I/I	331
44	CTTGATCCATATGCTTTTACCTGCA	TCCATCACCCTTCAGAACCTGATCT	I/I	268
4	TTGTCGGTCTCCTGCTGGTCAGTG	CAAAGCCCTCACTCAAACATGAAGC	I/I	196
46	GCTAGAAGAACAAAAGAATATCTTGTC	CTTGACTTGCTCAAGCTTTTCTTTTAG	E/E	148

[a] All primers are from Chamberlain et al. (1990) except those for exon 46 which are from Beggs et al. (1991).

[b] Exons numbered according to Koenig et al. (1989). Forward and reverse are relative to the coding sequence.

[c] I = primer sequences inside or overlapping the neighboring intron and splice signals; E = primers that bind entirely within the exon.

Table 9.1.2 PCR Primers for Diagnostic Set B[a]

Exon[b]	Forward primer (5′→3′)	Reverse primer (5′→3′)	Primer location[c]	Product size (bp)
pm[d]	gaagatcTAGACAGTGGATACATAACAAATGCATG	TTCTCCGAAGGTAATTGCCTCCCAGATCTGAGTCC	I/E	535
3	TCATCCATCATCTTCGGCAGATTAA	CAGGCGGTAGAGTATGCCAAATGAAAATCA	I/I	410
43	GAACATGTCAAAGTCACTGGACTTCATGG	ATATATGTGTTACCTACCCTTGTCGGTCC	I/I	357
50	CACCAAATGGATTAAGATGTTCATGAAT	TCTCTCTCACCCAGTCATCACTTCATAG	I/I	271
13	AATAGGAGTACCTGAGATGTAGCAGAAAT	CTGACCTTAAGTTGTTCTTCCAAAGCAG	I/I	238
6	CCACATGTAGGTCAAAAATGTAATGAA	GTCTCAGTAATCTTCTTACCTATGACTATGG	I/I	202
47	CGTTGTTGCATTTGTCTGTTTCAGTTAC	GTCTAACCTTTATCCACTGGAGATTTG	I/I	181
60	AGGAGAAATTGCGCCTCTGAAAGAGAACG	CTGCAGAAGCTTCCATCTGGTGTTCAGG	E/E	139
52	AATGCAGGATTTGGAACAGAGGCGTCC	TTCGATCCGTAATGATTGTTCTAGCCTC	E/E	113

[a] All primers are from Beggs et al. (1990).

[b] Exons numbered according to Koenig et al. (1989). Forward and reverse are relative to the coding sequence.

[c] I = primer sequences inside or overlapping the neighboring intron and splice signals; E = primers that bind entirely within the exon.

[d] pm = muscle-specific promoter and first exon. Lowercase nucleotides create an artificial *Bgl*II site.

Table 9.1.3 PCR Primers for Diagnostic Set C[a]

Exon[b]	Forward primer (5′→3′)	Reverse primer (5′→3′)	Primer location[c]	Product size (bp)
49	GTGCCCTTATGTACCAGGCAGAAATTG	GCAATGACTCGTTAATAGCCTTAAGATC	I/I	439
pb	TCTGGCTCATGTGTTTGCTCCGAGGTATAG	CTTCCATGCCAGCTGTTTTTCCTGTCACTC	I/E	332
16	TCTATGCAAATGAGCAAATACACGC	GGTATCACTAACCTGTGCTGTACTC	I/I	290
41	GTTAGCTAACTGCCCTGGGCCCTGTATTG	TAGAGTAGTAGTTGCAAACACATACGTGG	I/I	270
32	GACCAGTTATTGTTTGAAAGGCAAA	TTGCCACCAGAAATACATACCACACAATG	I/I	253
42	CACACTGTCCGTGAAGAAACGATGATGG	CTTCAGAGACTCCTCTTGCTTAAAGAGAT	E/E	195
34	GTAACAGAAAGAAAGCAACAGTTGGAGAA	CTTTCCCCAGGCAACTTCAGAATCCAAA	E/E	171

[a] All primers are from Kunkel et al. (1991) except those for exon 49 which are from Beggs et al. (1990).

[b] Exons numbered according to Koenig et al. (1989). pb = brain-specific promoter described by Boyce et al. (1991). Forward and reverse are relative to the coding sequence.

[c] I = primer sequences inside or overlapping the neighboring intron and splice signals; E = primers that bind entirely within the exon.

Table 9.1.4 PCR Primers to Amplify Miscellaneous Dystrophin Gene Exons[a]

Exon[b]	Forward primer (5′–3′)	Reverse primer (5′–3′)	Primer location[c]	Reaction conditions[d]	Product size (bp)	Ref.[e]
pb-2	CTTTCAGGAAGATGACAGAATC	CCTGTCACTCCATCATGCC	E/E	60/72	95	1
pm-2	TCACTTTCCCCCTACAGGACTC	ACAGTCCTCTACTTCTTCCCA	E/E	55/72	229	2
pm-3	CTTTCCCCCTACAGGACTCAG	GTCCTCTACTTCTTCCCACC	E/E	60/72	223	1
2-1	AGATGAAAGAGAAGATGTTCAAAAG	AATGACACTATGAGAGAAATAAAACGG	E/I	55/63	174	3
2-2	GAAAGAGAAGATGTTCAAAAG	CTTAGAAAATTGTGCATTTAC	E/E	60/72	60	1
3-2	GGCAAGCAGCATATTGAGAAC	CCCTGTCAGGCCTTCGAGGAG	E/E	60/72	81	1
5	GTTGATTTAGTGAATATTGGAAGTAC	CTGCCAGTGGAGGATTATATTCCAAA	E/E	55/63	93	3
7-1	TATTTGACTGGAATAGTGTGGTTTGC	CTTCAGGATCGAGTAGTTTCTCTA	E/E	55/63	111	3
7-2	CTATTTGACTGGAATAGTGTG	CAGGATCGAGTAGTTTCTC	E/E	60/72	109	1
7-3	AGACCTATTTGACTGGAATAGT	CAGGATCGAGTAGTTTCTCTATGCC	E/E	55/68	113	4
8-2	ATGTTGATACCACCTATCCAG	TCTTTAGTCACTTTAGGTGGCC	E/E	50/63	140	5
8-3	CCTATCCAGATAAGAAGTCC	CTTTAGGTGGCCTTGGCAAC	E/E	60/72	117	1
9	ATCACGGTCAGTCTAGCACA	TGAAGGAAATGGGCTCCGTGTA	E/E	55/68	126	4
11-1	GTACATGATGGATTTGACAGC	CATGCTAGCTACCCTGAGGC	E/E	60/72	166	1
11-2	GGATTTGACAGCCCATC	GCTTTGTTTTTCCATGC	E/E	55/68	169	4
11-3	CAAAATAAAACTCAAAAACCACACC	CTTCCAAAACTTGTTAGTCTTC	I/I	53/65	303	6
14	GTATTGGGAGATC	CTGTTCTTCAGTAAGACGTTGCC	E/E	55/68	102	4
20-1	GTGTTAATGCAGATAGCATCAAAC	ACAAATTTTTAACTGACTTTTAATTG	E/E	51/65	239	7
20-2	GCAGATAGCATCAAACA	CTGACTTTTAATTGCTG	E/E	55/68	216	4

continued

Table 9.1.4 PCR Primers to Amplify Miscellaneous Dystrophin Gene Exons[a], continued

Exon[b]	Forward primer (5′–3′)	Reverse primer (5′–3′)	Primer location[c]	Reaction conditions[d]	Product size (bp)	Ref.[e]
20-3	TGGCTTTCAGATCATTTCTTTC	AAATACCTATTGATTATGCTCC	I/I	53/65	357	6
21-1	GAAGTCAACCGGCTATCAGGTCCT	TCTGTAGCTCTTTCTCTCTGGCCT	E/E	65/65	175	8
21-2	GTTCCTGGATGCAGACTTTGTG	TTGGAAAATGTCAAGTTAGCC	E/E	60/72	130	1
21-3	GATGAAGTCAACCGGCTATCA	CACAAAGTCTGCATCCAGGAA	E/E	55/68	114	4
21-4	GCAAAATGTAATGTATGCAAAG	ATGTTAGTACCTTCTGGATTTC	I/I	53/65	319	6
22	TTGACACTTTGCCACCAATGCGCTATC	CAATTCCCCGAGTCTCTGCTCCATG	E/E	51/65	140	7
23-1	GCTTTACAAAGTTCTC	CTGAATTTTTCGGAGT	E/E	55/68	213	4
23-2	GTCATAACTGATAGAAGATCATC	TTTACAGTTTACAGTGTATCGTTAG	I/I	53/65	423	6
24	AATCACATACAAACCC	TCTGCACTGTTTCAGC	E/E	55/68	114	4
25	CAATTCAGCCCAGTCTAAAC	CTGAGTGTTAAGTTCTTTGAG	E/E	60/72	113	1
26	GTCTATGCCAGAAAGGAG	CTTCATCTCTTCAACTGC	E/E	55/68	171	4
27-1	GCTAAAGAAGAGGCCCAAC	GGCCTCTTGTGCTACAGGTGG	E/E	60/72	113	1
27-2	AGAGCTAAAGAAGAGG	AAAGGCTTGCATTTCC	E/E	55/68	183	4
28	GAAGTTTGGACATGTTGG	AGCACCTCAGAGATTTCC	E/E	55/68	135	4
29-1	CATTCAGAGGATAACCCAAATCAGATT	GTTCCCTCCAACGAGAATTAAATGT	E/E	65/65	117	8
29-2	AATTTGATGCGACATTC	CTCTTCATGTAGTTCC	E/E	55/68	150	4
29-3	CCAATGTATTTAGAAAAAAAAGGAG	GCAAATTAGATTAAAGAGATTTTCAC	I/I	53/65	242	6
30	GCTGTAAGGCAAAAG	CTGGGCTTCCTGAGGC	E/E	55/68	159	4
32	AAATTACAAGATGTCTCC	CACACAATGATTTAGCTG	E/E	55/68	174	4
33-1	GTCTGAGTGAAGTGAAGTCTG	CAAAGCTGTTACTCTTTCATC	E/E	60/72	133	1
33-2	CTTGTATAAAAGTCTGAG	ACCTTTGCTCCCAGC	E/E	55/68	155	4
34-1	AGAAAGAAAGCAACAG	CCCCAGGCAACTTCAG	E/E	55/68	164	4
34-2	CAGAAATATAAAAGTTCCAAATAAGTG	CATGTTAATACTTCCTTACAAAATC	I/I	53/65	338	6
35-1	AGAGATTGAGAAACAG	AACAAAAGATTTAACC	E/E	55/68	175	4
35-2	CCGTTTCATAAGCATTAAATC	AGCTTCTAGCCTTTTCTC	I/I	53/65	271	6
37	CACTGCAGGAAATTAGTAGAGC	AATGGAGGCCTTTCCAGTCTT	E/E	55/68	172	4
39-1	CAACTTACAACAAAGAATCACAG	CTTGAGAGCATTATGTTTTGTC	E/E	60/72	91	1
39-2	AATGAAGACAATGAGG	CTTGAGAGCATTATG	E/E	55/68	138	4
40-1	GGTATCAGTACAAGAGGCAG	CCTTTCATCTCTGGGCTCAG	E/E	60/72	95	1
40-2	GAGGTCTCAAAGAAG	TTCATCTCTGGGCTC	E/E	55/68	153	4
41-1	ATTGATCGGGAATTGC	CAAAGTTGAGTCTTCG	E/E	55/68	183	4
41-2	TGTGGTTAGCTAACTGCCCT	GAGTAGTAGTTGCAAACACATACG	I/I	53/65	276	6
42-2	CACACTGTCCGTGAAGAAACGATGATG	TTAGCACAGAGGTCAGGAGCATTGAG	E/E	60/72	155	9
42-3	GAGCCAACTCAGATCCAGCTCA	AATTTGTGCAAAGTTGAG	E/E	55/68	195	4
44-2	CGATTTGACAGATCTGTTGAG	GCATGTTCCCAATTCTCAGG	E/E	60/72	125	1
45-2	GCGGCAAACTGTTGTCAGAACA	GGCATCTGTTTTGAGGATTGC	E/E	50/63	73	5
45-3	CTGGAGCTAACCGAGAGGTGC	CATTCCTATTAGATCTGTCG	I/I	60/60	428	10
45-4	CTCCAGGATGGCAATGGCAG	CTGTCTGACAGCTGTTTGCAG	E/E	60/72	164	1
46-2	ATTTGTTTTATGGTTGGAGG	GCTTTTCTTTTAGTTGCTGC	E/E	60/72	83	1
46-3	CTAGAAGAACAAAAGAATATC	TTGCTGCTCTTTTCCAGGTTC	E/E	55/68	121	4
50-2	ATTAAGCATGTTGCTGAGAGGGAACTG	GCCTCTGCTGCAGACAAATCACATTTC	I/I	55/63	712	11
51-2	CCCAAAATATTTTAGCTCCTACTCAG	CTTGATTATACTTAGGCTGAATAGTG	I/I	55/63	409	12
51-3	CTGCTCTGGCAGATTTCAAAC	GTCACCCACCATCACCCTCTG	E/E	60/72	98	1
52-2	AATGCAGGATTTGGAACAGAGGCGTCC	TTCGATCCGTAATGATTGTTCTAGCCTC	E/E	60/72	113	1
53-1	TTGAAAGAATTCAGAATCAGTGGGATG	CTTGGTTTCTGTGATTTTCTTTTGGATTG	E/E	60/72	212	9
53-2	GAAAGAATTCAGAATCAGTGGGATG	CTTCCATGCCTCAAGCTTGGCTCT	E/E	55/68	212	4
54-1	AGGATTCAGAAGCTGTTTACGAAGT	AATCCTCATGGTCCATCCAGTTTCA	I/I	55/72	329	13
54-2	GACCTCCGCCAGTGGCAGAC	GAATGCTTCTCCAAGAGGC	E/E	60/72	136	1
55	ATGAGTTCACTAGGTGCACCATTCT	TGTTCAATTGGATCCACAAGAGTGC	I/I	55/72	303	13
56-1	GGTGAAATTGAAGCTCACAC	GTAACAGGACTGCATCATCG	E/E	60/72	100	1
56-2	CTCCAAGGTGAAATTGA	GAGACTTTTTCCGAAG	E/E	55/68	173	4
57	GTCCGATTTGGAAGC	GTACATCGTTCTGCT	E/E	55/68	157	4
59	GAGGCCACGGATGAGCTGG	GGTGATCTTGGAGAGAGTC	E/E	60/72	103	1
60	TAAATATTCTCATCTTCCAATTTGC	TTACTGTAACAAAGGACAACAATG	I/I	58/72	231	14
61-1	AATGAGAGAACATAATTTCTCTCC	AATCAAGATGCAATAAAGTTAAGTG	I/I	58/72	163	14
61-2	ACATAATTTCTCTCCTTTTCC	CAAGATGCAATAAAGTTAAGTG	I/I	53/65	150	6
62	TAATGTTGTCTTTCCTGTTTGCG	ATACAGGTTAGTCACAATAAATGC	I/I	58/72	185	14

continued

Exon[b]	Forward primer (5′–3′)	Reverse primer (5′–3′)	Primer location[c]	Reaction conditions[d]	Product size (bp)	Ref.[e]
63	TACTCATGGTAAATGCTAAAGTC	TAGCAAAGTAACTTTCACACTGC	I/I	58/72	194	14
64-1	TTTCTGATGGAATAACAAATGCTC	ATCAAGATCTTCAAATACTGGCC	I/I	58/72	151	14
64-2	TTCTGATGGAATAACAAATGCT	CCTACTTTTTATTCTAAGCAAAGA	I/I	53/65	180	6
65	TATGAGAGAGTCCTAGCTAGG	TAAGCCTCCTGTGACAGAGC	I/I	58/72	347	14
66-1	CAGGGAGGATCCGTGTCCTG	GTCTTCCAAATGTGCTTTAC	E/E	60/72	68	1
66-2	TCTGCTTTGATTCTTCATAATAGG	ATCTAGAACTAGGGTAATTAGCC	I/I	58/72	173	14
66-3	GTCTAGTAATTGTTTTCTGCTTTG	ATAAGAACAGTCTGTCATTTCCC	I/I	53/65	211	6
67	AATTGCTACTGGAATTGAGTTGG	AAGAATAAATATGTTACCTAGAAGG	I/I	58/72	286	14
68	TAATCGAACTGATATACACCTCC	ACTAACAGCAACTGGCACAGG	I/I	58/72	352	14
69-1	TCAAATTAGAACGTGGTAGAAGG	GAACTAACTCTCACGTCAGGC	I/I	58/72	242	14
69-2	GAACGTGGTAGAAGGTTTATTAAA	CTAACTCTCACGTCAGGCTG	I/I	53/65	231	6
70-1	TCCTAAATCTGATCTCACCATG	ATCAAACAAGAGTGTGTTCTGC	I/I	58/72	210	14
70-2	TGGTCATTAGTTTTGAAATCATC	CATCAAACAAGAGTGTGTTCT	I/I	53/65	237	6
71	AAAGCGGTGTGTCTCCTTCACC	ATGTGTTGGTGGTAGCAGCAC	I/I	58/72	125	14
72	GATGGTATCTGTGACTAATCAC	ATTTCAATCAATATTTGCCTGGC	I/I	58/72	144	14
73	ACGTCACATAAGTTTTAATGAGC	ATGCTAATTCCTATATCCTGTGC	I/I	58/72	202	14
74	ACCAAAACCTTTGATTTTATTTTCC	TTTCTATGTGTGCAAGTGTATGC	I/I	58/72	248	14
75	TCTTTTTTACTTTTTTGATGC	AGTGCTCTCTGAGGTTTAG	I/I	58/72	344	14
76-1	AAGTAATTCTGTTTTCTTTTGGATG	CTTCAGACAACAAAATCTGAGAG	I/I	58/72	196	14
76-2	GTAATTCTGTTTTCTTTTGGATG	CTACCTTTCTTCAGACAACAAAAT	I/I	53/65	198	6
77	TAATCATGGCCCTTTAATATCTG	TAATCATGGCCCTTTAATATCTG	I/I	58/72	270	14
78	TTCTGATATCTCTGCCTCTTCC	AATGAGCTGCAAGTGGAGAGG	I/I	58/72	220	14
79-1	AGAGTGATGCTATCTATCTGCAC	TGCATAGACGTGTAAAACCTGCC	I/E	58/72	349	14
79-2[e]	GAAAGATTGTAAACTAAAGTGTGC	GGATGCAAAACAATGCGCTGCCTC	E/E	65/65	123-137	15
79-3[e]	atgATCAGAGTGAGTAATCGGTTGG	atatcgatCTAGCAGCAGGAAGCTGAATG	E/E	65/65	78/82	16

[a]This table lists PCR primers for miscellaneous exons. Some use different primers to amplify the same exons included in the standard diagnostic mix, and therefore can be used to confirm single-exon deletions. Others may be useful in special cases for determining exact deletion endpoints and effects on the translational reading frame. These primers should be used in combination with another control (e.g., undeleted) pair following standard PCR protocols (see CPMB UNIT 15.1).

[b]Exons numbered according to Koenig et al. (1989) and Roberts et al. (1992b). Alternative primers to amplify exons detected in one of the diagnostic sets are indicated by "−1", "−2", "−3" or "−4".

[c]I, primer sequences inside or overlapping the neighboring intron and splice signals; E, primers that bind entirely within the exon.

[d]Reaction conditions indicated are annealing temperature (1 min)/elongation temperature (3 min) from original reference. Note that many of these primers will work under different conditions and may potentially be multiplexed in various combinations for particular needs.

[e]References: (1) Coffey et al., 1992; (2) Sakuraba et al., 1991; (3) Beggs et al., 1991; (4) Prior et al., 1995; (5) Hentemann et al., 1990; (6) den Dunnen, pers. comm.; (7) Covone et al., 1992; (8) Meng et al., 1991; (9) Abbs et al., 1991; (10) Blonden et al., 1989; (11) Kilimann et al., 1992; (12) Speer et al., 1989; (13) K. Friedman, pers comm; (14) Lenk et al., 199?; (15) Beggs and Kunkel, 1990; (16) Roberts et al., 1989.

[f]These primers amplify length polymorphisms in the 3′ untranslated region and effectively delimit the 3′ portion of the gene. Lowercase nucleotides create an artificial restriction site and are not derived from dystrophin sequence.

Simultaneous Detection of Multiple Point Mutations Using Allele-Specific Oligonucleotides

BASIC PROTOCOL

SCREENING PCR-AMPLIFIED DNA WITH MULTIPLE POOLED ASOs

In the following protocol, pools of radiolabeled allele-specific oligonucleotide (ASO) probes are hybridized to dot blots containing polymerase chain reaction (PCR)-amplified DNA products generated from one or more loci (Fig. 9.2.1). This approach is particularly powerful when used to screen for rare alleles and deletions/insertions, but it is not appropriate for screening commonly occurring alleles.

A significant advantage in using this procedure is that designing ASO probes only involves the necessity to observe two parameters: (1) all ASO probes in a single pool should be the same length (this protocol is optimized for a 17-base ASO probe) and (2) the position of the discriminating single-base mismatch between the oligonucleotide probe and target sequence should not be located at the end of the ASO (this protocol is based on the single-base mismatch being located 5 to 7 bases from the 5′ end of the ASO probe).

Table 9.2.1 presents a troubleshooting guide for this procedure.

Materials *(see APPENDIX 1 for items with ✓)*
   ✓ 10× T4 polynucleotide kinase buffer (or buffer supplied by enzyme manufacturer), stored at −20°C
   2 μM allele-specific oligonucleotide (ASO) working solutions in 1× TE buffer (APPENDIX 1), for all normal and mutant sequences (prepare from 100 μM stocks or pooled lyophilized ASOs, stored at −80°C)
   10 mCi/ml [γ-^{32}P]dATP (∼3000 Ci/mmol), stored at −20°C
   10 U/<μl T4 polynucleotide kinase, stored at −20°C
   ✓ 2× SSC
   ✓ Denaturing solution, freshly prepared
   PCR-amplified DNA from genes of interest (UNITS 7.1 & 9.1 and CPMB UNIT 15.1)
   Positive control: PCR-amplified DNA from a known homo- or heterozygote (preferred), or a cell line or clone (UNIT 7.1 and CPMB UNIT 15.1)
   Negative control: PCR-amplified DNA from an individual without the mutation (preferred), or a cell line or clone (UNIT 7.1 and CPMB UNIT 15.1)
   ✓ TMAC hybridization solution, freshly prepared
   ✓ TMAC wash solution, freshly prepared, at room temperature and 52°C
   37°C heat block
   Dot-blot apparatus (volumes provided are for 6-mm dots)
   Nylon membrane: e.g., Biotrans+ (ICN Biomedicals) or Biodyne (Pall)
   Whatman 3MM filter paper
   80°C vacuum oven
   Sealable bag, Tupperware container, or hybridization tube
   52°C shaking water bath or hybridization oven
   Kodak Biomax MS film
   X-ray film cassettes and intensifying screens

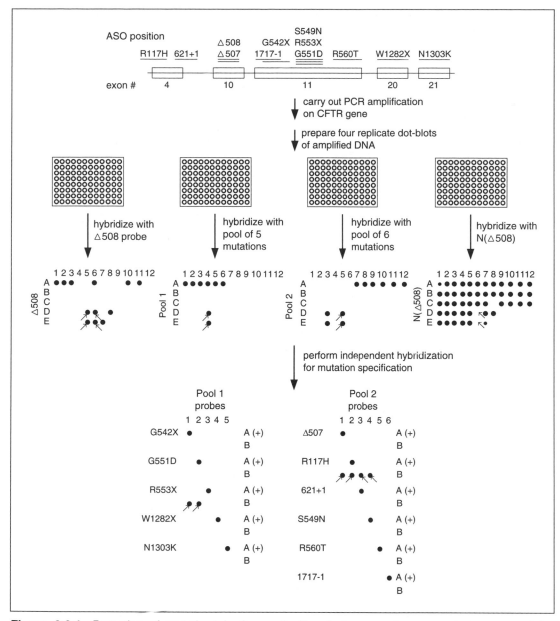

Figure 9.2.1 Detection of mutations in the cystic fibrosis transmembrane conductance regulator (CFTR) gene using allele-specific oligonucleotides.

To label individual ASOs with ^{32}P

1a. For each ASO to be labeled, prepare the following labeling reaction:

> 0.75 μl H₂O
> 1.00 μl 10× T4 polynucleotide kinase buffer
> 5.00 μl 2 μM ASO working solution
> 3.00 μl 10 mCi/ml [γ ^{32}P]ATP
> 0.25 μl 10 U/μl T4 polynucleotide kinase.

Incubate 1 hr at 37°C in a heat block.

2a. Add 90 µl water, mix, microcentrifuge briefly to collect solution at bottom of tube, and use immediately or store at −20°C.

To label ASO pools with ^{32}P

1b. For each ASO pool to be labeled, determine the total number of picomoles in the pool. Note that the ratio of total picomoles ASO to the volume of 10 mCi/ml γ-^{32}P]ATP should be 3.3. Adjust total reaction and reagent component volumes in step 1a to accommodate this ratio. Incubate 1 hr at 37°C in a heat block.

2b. Add 4 vol. water, mix, microcentrifuge briefly to collect solution at bottom of tube, and use immediately or store at −20°C.

3. Prepare a dot-blot apparatus according to the manufacturer's instructions, cleaning if necessary.

4. Cut a nylon membrane to the appropriate size to fit the dot-blot apparatus. Mark the dry membrane in asymmetric corners so that it can be reoriented after hybridization.

5. Wet the membrane by floating it on water (it should wet immediately), then submerge until use. Partially fill a container with 2× SSC.

6. For each sample to be dotted onto the membrane, add 92 µl freshly prepared denaturing solution to 8 µl PCR-amplified DNA from the gene of interest (up to 11 products) or multiplexed PCR products (use 200 ng each product). Mix by vortexing. Also prepare positive, negative, and PCR-contamination controls (i.e., no DNA).

7. Place the membrane on the dot-blot apparatus gasket and assemble the manifold. Do not apply vacuum until just before loading the samples.

8. When all samples have been mixed with denaturing solution, apply the vacuum to the dot-blot apparatus. Add the entire volume of each sample to a well, avoiding bubbles on the membrane. Be careful not to touch or puncture the filter.

9. With the vacuum still on, remove the upper block of the apparatus, quickly remove the membrane, and blot excess liquid from the membrane onto Whatman 3MM filter paper.

10. Place the membrane into the container of 2× SSC for 5 min and then blot dry on Whatman 3MM filter paper.

11. Place a piece of plastic wrap over the gasket while the vacuum is still on to remove any excess liquid trapped in the apparatus. Blot any other visible liquid with a paper towel.

12. Fix the DNA onto the membrane by incubating 1 hr in an 80°C oven.

13. Wet the membrane in water for at least 5 min, then place in a sealable bag, Tupperware container, or hybridization tube.

14. Add an appropriate amount (e.g., 10 ml) TMAC hybridization solution containing the ^{32}P-labeled ASO probes (determine probe concentration empirically, but should be between 0.03 to 0.15 pmol/ml) to each membrane and seal the container. If necessary, reduce background hybridization by adding ≥20-fold excess unlabeled ASO corresponding to the "normal" allele. Incubate 4 hr to overnight, with shaking, in a 52°C water bath or hybridization oven.

15. Remove the membrane from the bag and wash with vigorous agitation as follows: 20 min in 200 ml (1 liter for multiple membranes) TMAC wash solution, room temperature, followed by 20 min in 300 ml (1 liter for multiple membranes) prewarmed TMAC wash solution, 52°C. If background is high, add an initial 5-min room temperature wash step.

16. Blot membranes on Whatman 3MM filter paper to dry and expose to BioMax MS film 15 min to 1 hr at room temperature in an X-ray film cassette.

STRIPPING OLD PROBES AND REHYBRIDIZATION

Membranes can be stripped of previously hybridized probe and rehybridized several times, depending on the amount of DNA (control and experimental) on the membranes. If the signal-to-noise ratio allows reading of the appropriate hybridization signals from positive- and negative-control DNA, and the amount of experimental and control DNA on the membrane is equivalent, then the experimental results can be interpreted.

Table 9.2.1 Troubleshooting Guide for Mutation Analysis Using Allele-Specific Oligonucleotides Probes

Problem	Possible cause	Solution
Hybridization signal from sample is low or absent	Specific activity of labeled ASO probe is low	Repeat the labeling procedure with fresh ASO and label
	Probe is partially hydrolyzed or degraded	Prepare new probe
	No denaturing solution added to PCR-amplified DNA sample	Denature, spot, and hybridize a second aliquot from amplified sample
	Probe not added to the hybridization cocktail	Add probe to hybridization solution and proceed as usual
	No DNA spotted onto membrane	Repeat denaturation, spotting, and hybridizations with new aliquot of amplified sample
Background signal is high	Hybridization and wash temperature are not high enough	Do not allow membrane to completely dry out, return damp membrane to a new wash solution and repeat washes at given temperature, omit room temperature wash
	Labeled probe remains on membrane after the washes have been done	Perform an additional 5-10 min wash at the required wash temperature
	Agitation of washes is too slow	Repeat washes with increased agitation, check to be sure that membrane has not become stuck to side of wash container
	Inadequate volume of wash solution	Add additional wash solution and repeat
Signal in individual wells not even; signal present between wells	Insufficient vacuum during blotting resulted in leakage or inconsistent blotting	Repeat blotting step and proceed as usual
Row-to-row inconsistencies in background signal	Inadequate denaturation or incorrect amount of denaturant added during blotting	Repeat denaturation, blotting, and hybridization steps with new aliquot of amplified sample and freshly prepared solutions
Weak positive signal from a sample with acceptable amplification	Possible contamination during DNA extraction, amplification, or blotting	Re-extract sample and repeat the process with the original and the re-extracted sample
	Cross-hybridization with a sequence variant	If weak signal is reproducible, analyze sequence with alternate technology

Additional Materials (also see Basic Protocol; see APPENDIX 1 for items with ✓)

Hybridized membrane (see Basic Protocol)

✓ TMAC wash solution, freshly prepared and prewarmed to 15°C above previous hybridization temperature

Agitating water bath, 15°C above previous hybridization temperature

1. Add previously hybridized membrane to 300 ml prewarmed TMAC wash solution in a washing dish. Wash 1 hr with vigorous agitation in a water bath at 15°C above previous hybridization temperature.

2. Remove membrane from the wash and blot dry on Whatman 3MM filter paper.

3. Autoradiograph to confirm that the previously hybridized probe has been removed.

4. Hybridize membranes as described (see Basic Protocol, steps 13 to 16) or store dry at room temperature. Store indefinitely in resealable bags.

References: Shuber et al., 1992, 1997

Contributors: Nichole M. Napolitano, Elizabeth M. Rohlfs, and Ruth A. Heim; Barbara Handelin and Anthony P. Shuber (membrane stripping and rehybridization)

UNIT 9.3

Molecular Analysis of Fragile X Syndrome

BASIC PROTOCOL 1

PCR AMPLIFICATION OF THE FRAGILE X REPEAT

PCR analysis of the CGG repeat region is faster than Southern analysis (see Basic Protocol 2) and accurately determines the size of normal (6 to 45 repeats), intermediate (~45 to 55 repeats), and premutation alleles (~55 to 200 repeats; Fig. 9.3.1). It also detects small changes in repeat number from one generation to another. Substitution of 7-deaza-2′-dGTP for dGTP in the PCR reaction allows amplification of full mutations (>200 repeats), which is unsuccessful with standard PCR protocols.

NOTE: Experiments involving PCR require extremely careful technique to prevent contamination.

Materials (see APPENDIX 1 for items with ✓)

10× PCR amplification buffer: 100 mM Tris·Cl, pH 8.3 (*APPENDIX 1*)/500 mM KCl

✓ 25 mM $MgCl_2$

✓ 10 mM dNTP mix of dATP, dCTP, dTTP and 7-deaza-2′-deoxyguanosine-5′-triphosphate (7-deaza-2′-dGTP; Roche Diagnostics)

10 μM oligonucleotide primers 1 and 3 in 10 mM Tris·Cl, pH 7.5 (*APPENDIX 1*)/1 mM EDTA:

Primer 1: 5′-GACGGAGGCGCCGCTGCCAGG-3′
Primer 3: 5′-GTGGGCTGCGGGCGCTCGAGG-3′

DMSO (Sigma)

5 U/μl *Taq* DNA polymerase (Applied Biosystems)

0.3 to 1 μg/μl purified sample DNA (*APPENDIX 3B*) in TE buffer, pH 7.5 (*APPENDIX 1*)

SequaGel-6 (National Diagnostics; optional)

✓ 0.67× and 2.8× TBE buffer

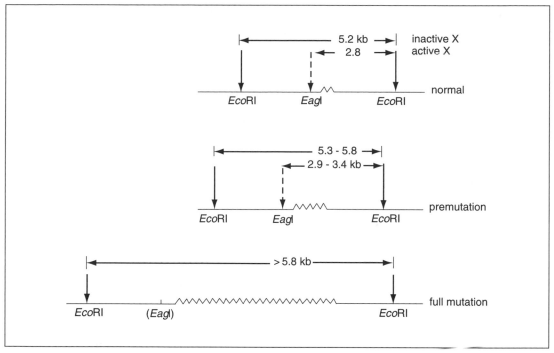

Figure 9.3.1 The fragile X region with normal, premutation and full mutation CGG repeats. The repeat region is represented by a jagged line. Restriction sites are indicated by solid arrows for *Eco*RI and a dashed arrow for methylation sensitive *Eag*I.

   ✓ Loading buffer
      ^{32}P-labeled DNA size markers (e.g., pBR322 digested with *Msp*I, λ digested with *Bst*EII, or mixture of DNA sample from several heterozygous women with known allele sizes amplified by PCR)
      0.4 N NaOH
   ✓ 2× SSC
      Quick Light Hybridization Kit (Orchid Biosciences) containing:
            Quick Light buffer concentrate
            Wash I and II solutions
            Hybridization solution
      Quick Light Genome Mapping Probe Kit (Orchid Biosciences) containing:
            Alkaline phosphatase-labeled $(CGC)_n$ oligonucleotide probe
            Lumi-Phos spray
      Thermal cycler
      100° and 55°C water baths
      Vertical gel apparatus (e.g., BRL V15.17, Life Technologies)
      Powder-free gloves
      Positively charged nylon membrane (Biotrans+, ICN Biomedicals), 12 × 16 cm
      Whatman 3MM filter paper
      Blotting paper (Owl Separation Systems), 12 × 16 cm
      Electroblot apparatus (Integrated Separation Systems)
      80°C vacuum oven

1. Prepare PCR reaction mix (enough for $n + 1$ for <20 sample, for $n + 2$ for ≥20 samples) in a 1.5-ml microcentrifuge tube by combining the following for each reaction:

      1 μl 10× PCR amplification buffer (1× final)
      0.3 μl 25 mM MgCl$_2$ (0.75 mM final)

0.8 µl 10 mM dNTP mix (200 µM final each dNTP)
5 µl of 10 µM primer 1 (0.5 µM final)
5 µl of 10 µM primer 3 (0.5 µM final)
1 µl DMSO (10% final)
0.05 µl 5 U/ml *Taq* DNA polymerase (0.25 U)
4.85 µl H_2O.

2. Transfer 9 µl PCR reaction mix to the bottom of a labeled 0.2-ml microcentrifuge tube.

3. Dilute sample DNA to ∼30 to 100 ng/µl, vortex, and add 1 µl DNA (50 to 100 ng works well) to each PCR reaction.

4. Place sample in thermal cycler and perform the following amplification cycles:

Initial step:	4 min	94°C	(heating)
30 cycles	1 min	94°C	(denaturation)
	1 min	63°C	(annealing)
	2 min	72°C	(elongation)

5. Prepare a 6% denaturing polyacrylamide gel containing 8.3 M urea in a small vertical apparatus (e.g., BRL V15.17; see APPENDIX 3F for details). Alternatively, prepare using SequaGel-6. Prerun the gel in 0.67 × TBE buffer ∼30 min at a constant ∼500 V.

6. Combine 2 to 4 µl of each PCR product with 2 vol loading buffer and mix by vortexing. Place samples in a 100°C water bath for 1 min, then on ice (at 4°C). Load samples onto gel, including ∼0.4 nCi of ^{32}P-labeled DNA size markers in two wells.

7. Electrophorese gel ∼90 min at 500 V in 0.67× TBE buffer. When electrophoresis is complete, trim gel to 12 × 16 cm to fit on the precut membrane.

8. Wearing powder-free gloves, place a piece of plastic wrap onto the gel. Lift one edge of the gel with a spatula so it adheres to the plastic wrap and transfer the gel from the plate to the plastic wrap. Lay the gel, plastic-wrap-side-down, on the bench with the gel facing up. Place the positively charged precut nylon membrane (12 × 16 cm) onto the gel, avoiding bubbles. Wet three sheets of precut blotting paper (12 × 16 cm) with 2.8× TBE buffer and place on top of the gel.

9. Place the lower portion of the electroblotter onto the paper and invert the stack so that the gel is facing up. Remove the plastic wrap from the gel. Place three sheets of filter paper soaked in 2.8× TBE on top of the gel to complete the sandwich. Roll a pipet across the gel sandwich to remove any air bubbles (critical). Place the top of the electroblot apparatus onto the sandwich. Electroblot 30 min at a constant current setting of 3 mA/cm^2 according to manufacturer's instructions.

10. Separate the nylon membrane from the sandwich, place it DNA-side-up on Whatman 3MM filter paper moistened with 0.4 N NaOH, and let stand ∼10 min.

11. Wash membrane 5 min in 100 ml of 2× SSC. Blot dry with Whatmann 3MM and paper and place membrane 15 min in a vacuum oven at 80°C.

12. Hybridize and probe using the Quick-Lite Hybridization and Genome Mapping kits as follows (alternatively follow Basic Protocol 1 of UNIT 9.4):

    a. Warm all wash solutions to 55°C before starting.

    b. Prewet the membrane in 35 ml Wash I solution at 55°C for 5 min.

    c. Decant the solution and add 30 ml hybridization solution with 4 µl (CGC)$_n$ probe labeled with alkaline phosphatase. Hybridize at 55°C for 20 min.

    d. Wash membrane in 50 ml Wash I solution at 55°C for 20 min. Pour off solution.

e. Wash membrane in 50 ml Wash II solution at 55°C for 20 min.

f. Rinse the membrane twice in 50 ml 1× Quick-Light buffer for a total of 5 min at ambient temperature.

g. Place the membrane in plastic envelope sealed on one side and spray the membrane six to eight times with Lumi-Phos.

13. Remove any air bubbles from the plastic bag, seal it, and clean excess Lumi-Phos from the outside of the sealed plastic bag. Expose to X-ray film at 37°C for an appropriate amount of time (related to the amount of probe used).

BASIC PROTOCOL 2

DETECTION OF AMPLIFICATION AND METHYLATION BY DIRECT SOUTHERN BLOT HYBRIDIZATION

Both degree of CGG repeat amplification and extent of methylation can be detected by genomic Southern blot hybridization.

NOTE: Since restriction fragments that contain the full mutation may be >10 kb and heterogeneous in size, the original genomic DNA must be of high molecular weight to insure detection of the full mutation.

Materials (*see* APPENDIX 1 *for items with* ✓)

    0.5 to 1 μg/ml genomic DNA (APPENDIX 3A)
✓ 5× restriction buffer
    0.1 M DTT (Sigma)
    3 U/ml RNase A (Sigma)
    100 U/μl *Eco*RI (New England Biolabs)
    50 U/μl *Eag*I (New England Biolabs)
✓ 10× Ficoll loading buffer
✓ 1× TAE buffer, pH 8.5
✓ Denaturing buffer
✓ Transfer buffer
✓ Neutralizing buffer
✓ Hybridization buffer
    30 to 50 μCi/ml [α-^{32}P] StB12.3 probe (3000 mCi/mmol, Amersham; or unlabeled from Dr. J.L. Mandel, Institut de Chimie Biologique; *mandeljl@igbmc.u-strasbg.fr*)
✓ 2× SSPE/0.1% (w/v) SDS (prepare fresh)
✓ 1× SSPE/0.1% (w/v) SDS (prepare fresh)
✓ 0.1× SSPE/1.0% (w/v) SDS, 60°C (prepare fresh)
✓ 20× SSPE

    Whatman 3MM filter paper
    Nytran SuPercharge Turboblotter Rapid Downward Transfer System System and membrane (Schleicher & Schuell)
    80°C oven
    60°C hybridization oven
    UV-transparent plastic wrap

1. Add ~10 μg genomic DNA to a labeled 1.5-ml microcentrifuge tube. Bring total volume to 20 μl with sterile H$_2$O (the genomic DNA should be viscous, which makes accurate volume measurement difficult). Bring total volume to 29 μl with sterile water. Stir with pipet tip; do not vortex.

2. Add (in the order listed):

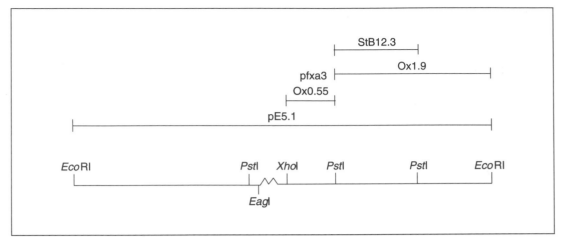

Figure 9.3.2 Restriction map of the fragile X region. Probes used in Southern analysis of region are shown.

> 8 μl 5× restriction buffer
> 0.4 μl 0.1 M DTT
> 0.4 μl 3 U/μl RNase A.

3. Add 1 μl (100 U) *Eco*RI and 1 μl (50 U) *Eag*I to each tube (see Fig. 9.3.2 for restriction map) and microcentrifuge 2 sec at maximum speed to ensure mixing. Incubate overnight (16 to 20 hr) at 37°C.

4. Add 4 μl of 10× Ficoll loading buffer, vortex, and spin briefly.

5. Load samples (~40 μl total volume) onto a 0.8 × 10–cm well of a 14 × 11 × 0.6–cm 0.8% (w/v) agarose gel in 1× TAE buffer containing 0.25 μg/ml ethidium bromide.

6. Electrophorese agarose gel in 1× TAE buffer at ~1.2 V/cm until the dye front reaches the end of the gel, overnight (usually ~18 hr).

7. Incubate the agarose gel in denaturing buffer twice for 30 min each. Wash the gel in transfer buffer for 15 min. Transfer DNA to Nytran membrane using Turboblotter Rapid Downward Transfer System according to manufacturer's directions. Wash membrane 5 min in 1× neutralizing buffer. Bake membrane 30 min at 80°C. Alternatively, use a standard Southern blotting protocol (*APPENDIX 3G* or *CPMB UNIT 2.9A*).

8. Prehybridize blot 6 to 18 hr in 10 to 20 ml hybridization buffer without probe at 60°C.

9. Prepare an α-^{32}P-labeled StB12.3 by incorporating 3000 Ci/mmol [α-^{32}P]dCTP to ~10^6 cpm/ng using an oligo-labeling kit. Denature ~50 ng of the labeled probe by boiling 2 to 5 min, then place on ice. Dilute into 10 ml hybridization buffer to ~5×10^6 cpm/ml and use to replace prehybridization buffer. Hybridize 18 to 20 hr at 60°C.

10. Wash the membrane 5 min with 100 ml of 2× SSPE/0.1% SDS, room temperature. Replace with 500 ml of 1× SSPE/0.1% SDS and incubate 5 min at room temperature. Replace with 125 ml prewarmed 0.1% SSC/1.0% SDS and incubate 60 min at 30°C. Rinse membrane in 2× SSPE at room temperature.

11. Blot the membrane dry, wrap in UV-transparent plastic wrap, and autoradiograph 1 to 10 days at −70°C.

References: Maddalena et al., 2001; Rousseau et al., 1991

Contributors: Sarah L. Nolin, Carl Dobkin, and W. Ted Brown

Analysis of Trinucleotide Repeats in Myotonic Dystrophy

BASIC PROTOCOL 1

HYBRIDIZATION ANALYSIS OF PCR-AMPLIFIED DM TRINUCLEOTIDE REPEATS

In this protocol, PCR amplification is used to detect DM protomutations and expansions <3 kb.

NOTE: Experiments involving PCR require extremely careful technique to prevent contaminations.

Materials *(see APPENDIX 1 for items with ✓)*

✓ 10 mM dNTP mix
✓ 10× PCR amplification buffer with 15 mM MgCl$_2$ (store up to 6 months at 4° or −20°C)
   10 ng/μl PCR primer 1a: (5′-TGGAGGATGGAACACGGACGG-3′) or primer 1b
     (5′-CAGAGCAGGGCGTCATGCACA-3′; less sensitive to Mg and more consistent)
   10 ng/μl PCR primer 2: (5′-GAAGGGTCCTTGTAGCCGGGAA-3′)
   5 U/μl *Taq* DNA polymerase
   DNA sample: 50 to 500 ng purified genomic DNA, Chelex-boiled lysates (Welsh et al.,
     1991) from chorionic villus sample or amniocytes, or NaOH-boiled cell lysates
     (Kogan et al., 1987)
   Mineral oil
   1.8% (w/v) agarose gel
✓ 1× TBE buffer
   10 mg/ml ethidium bromide
✓ 10× gel loading buffer
   Molecular size markers
✓ Prehybridization/hybridization buffer, 50°C
   1 μM alkaline phosphatase-conjugated (CTG)$_{10}$ oligonucleotide
   Wash buffer I: 1× SSC *(APPENDIX 1)*/0.5% (w/v) SDS, warmed to 50°C
   Wash buffer II: 0.25× SSC *(APPENDIX 1)*/0.5% (w/v) SDS, warmed to 50°C
   Wash buffer III: 1× SSC *(APPENDIX 1)*, room temperature
   Lumi-Phos 530 substrate solution (Boehringer Mannheim, Lumigen)

   0.5-ml PCR tubes
   15 ×10–cm casting tray with 0.75-mm-thick comb
   Nylon membrane, positively charged (e.g., Biotrans+, ICN Biomedicals)
   Thermal cycler (e.g., Perkin Elmer 480)
   Spray applicator
   Kodak X-AR autoradiographic film

1. Label 0.5-ml PCR tubes, one for each DNA sample plus positive (DNA with known large DM expansion) and negative (water in place of DNA) controls.

2. For each sample, prepare the following PCR reaction mix or prepare a master mix to be kept on ice for multiple samples (prepare enough for *n* +1 samples):

4 μl 10 mM dNTP mix
2.5 μl 10× PCR amplification buffer with 15 mM MgCl$_2$
5 μl 10 ng/μl PCR primer 1 (product from 1b is 64 kb larger than from 1a)
5 μl 10 ng/μl PCR primer 2
7.25 μl autoclaved H$_2$O
0.25 μl 5 U/μl *Taq* DNA polymerase.

3. Aliquot ≤1 μg (1 μl) DNA sample into the tube using a separate DNA pipettor. If necessary, cover reactions with mineral oil.

4. Carry out PCR using the following amplification cycles (optimized for Perkin Elmer model 480 thermocycler):

Initial step:	3 min	96°C (temperature critical)
30 cycles:	1 min	96°C
	1 min	60°C
	1.5 min	72°C
Final step	10 min	72°C

5. Prepare 73 ml of a 1.8% agarose gel (use <1% for vacuum-blot transfer; see Support Protocol 1) in 1× TBE buffer. Add 10 mg/ml ethidium bromide to 0.5 μg/ml (final). Pour the gel in a 15 × 10–cm tray with a 0.8-mm-thick comb.

6. Take 5 μl of each PCR product, add 10× gel loading dye to achieve a 1× final concentration, and apply to the gel. Include one lane of appropriate molecular size markers. Electrophorese the gel at 80 V, 90 mA, in 1× TBE buffer that does not contain ethidium bromide.

7. Photograph the gel when the bromophenol blue dye front has traveled ∼3.5 cm from the wells.

8. Stop electrophoresis when the bromophenol blue dye front has traveled ∼5.5 cm from the wells and the xylene cyanol dye front has traveled ∼2 cm.

9. Transfer CTG-PCR product overnight by capillary transfer (provides more complete transfer and optimizes smear analysis of large DM expansion) or for a few hours by vacuum blotting (see Support Protocol 1) to a positively charged nylon membrane. Fix the DNA to the membrane via the preferred method (i.e., with baking or UV cross-linking). Handle the membrane with blunt-end forceps at the edges; do not leave fingerprints (even gloved).

10. Prehybridize membrane in prehybridization/hybridization buffer 1.5 hr to overnight at 50°C.

11. Discard prehybridization/hybridization buffer. Add 20 μl alkaline phosphatase–conjugated (CTG)$_{10}$ oligonucleotide to 20 ml fresh prehybridization/hybridization buffer (∼1 nM final probe concentration) and hybridize to membrane 1.5 to 2 hr at 50°C. Prewarm wash buffers I and II to 50°C.

12. Remove membrane from prehybridization/hybridization buffer and drain off excess liquid. Wash membrane using the following regimen:

Wash buffer I—two times, 5 min each, 50°C
Wash buffer II—two times, 5 min each, 50°C
Wash buffer III—one time, 5 min, room temperature.

13. Place membrane on clean surface. Do not let it dry out. Spray Lumi-Phos 530 substrate solution sparingly from a height of ∼12 to 18 inches in a fume hood. Cover the membrane completely and evenly.

14. Wrap membrane in plastic wrap or acetate protective sheets. Seal edges well to prevent excess Lumi-Phos 530 from leaching out and producing artifacts.

15. Expose blot to Kodak X-AR autoradiographic film 3 hr to overnight at 37°C.

SUPPORT PROTOCOL 1

RAPID TRANSFER OF PCR PRODUCT USING A VACUUM BLOTTER

NOTE: The following protocol must be followed exactly to ensure even transfer.

Additional Materials *(also see Basic Protocol; see* APPENDIX 1 *for items with* ✓ *)*

Agarose gel of PCR product (see Basic Protocol; ≤1% agarose works best)
0.5 M NaOH/1.5 M NaCl
✓ 20× and 2× SSC

Filter paper (Whatman 3MM)
Vacuum transfer apparatus
Masking tape
80°C oven *or* UV cross-linker

1. Following step 8 of Basic Protocol 1, denature agarose gel of PCR product 10 to 15 min in 500 ml of 0.5 M NaOH/1.5 M NaCl.

2. Handling the membrane with blunt forceps only at edges, cut to 15 × 20 cm. Cut filter paper to 25 × 20 cm, with rounded corners.

3. Separate vacuum apparatus. Place porous vacuum plate with rough side down. Place filter paper on the porous plate. Wet it with water and remove air bubbles. Place dry nylon membrane on top of filter paper.

4. Place gasket on top of membrane. Be sure that gasket covers the O ring of the vacuum stage. Check the pressure by putting some water on the filter and setting the pressure gauge to 40 cm H_2O.

5. Reduce pressure to zero and place gel onto gasket window. Cover wells in gel with masking tape. Remove air bubbles between gel and membrane by rolling a glass rod or 10-ml pipet across surface of gel.

6. Cover top of gel with a small amount of 20× SSC and increase pressure to 40 cm H_2O.

7. Fill tank with fresh 20× SSC to completely immerse gel, at ~0.5- to 1-cm buffer depth over the gel. Leave pressure at 40 cm H_2O for 1 hr. Be sure gel is submerged in buffer throughout this time.

8. At end of transfer, leave pressure on and pour off buffer. Turn off pressure and remove gel. Remove nylon membrane. Do not rinse.

9. Fix DNA by baking membrane 15 min at 80°C or by UV cross-linking. Rinse fixed nylon membrane 3 to 5 min in 2× SSC. Hybridize membrane (see Basic Protocol 1, steps 10 to 15).

SUPPORT PROTOCOL 2

RADIOACTIVE DETECTION OF MYOTONIC DYSTROPHY (DM) TRINUCLEOTIDE REPEAT EXPANSION

Hybridizations can also be performed using a $(CTG)_{10}$ oligonucleotide that is ^{32}P-end-labeled using T4 polynucleotide kinase.

Additional Materials (*also see Basic Protocol 1*)

PCR product blotted onto positively charged nylon membrane (see Basic Protocol 1 or Support Protocol 1)

10 ng/ml column-purified [γ-^{32}P]dATP end-labeled $(CTG)_{10}$ oligonucleotide (specific activity 10^8 to 10^9 cpm/μg; *APPENDIX 3E*) diluted to 10 ng/ml in prehybridization/hybridization buffer (*APPENDIX 1*)

1. Prehybridize membrane in prehybridization/hybridization buffer 1.5 hr at 50°C.

2. Hybridize membrane with 10 ng/ml column-purified [γ-^{32}P]dATP end-labeled $(CTG)_{10}$ oligonucleotide in prehybridization/hybridization buffer 2 hr at 50°C.

3. Wash membrane using the following regimen (radioactive label is more tolerant of high stringency than enzymatic analysis):

    Wash buffer I—two times, 5 min each, 55°C
    Wash buffer II—two times, 5 min each, 55°C
    Wash buffer III—one time, 3 to 5 min, 55°C.

4. Expose membrane to X-AR autoradiographic film 0.5 to 3 hr at room temperature.

BASIC PROTOCOL 2

HYBRIDIZATION ANALYSIS OF MYOTONIC DYSTROPHY (DM) TRINUCLEOTIDE REPEATS IN GENOMIC DNA

Genomic DNA hybridization is specifically useful for identifying DM expansions >250 bp. It is used for ruling out E3 to E4 expansions (Table 9.4.1), where only one allele is visible on CTG-PCR and when there is no detectable protomutation or expansion after hybridization of the CTG-PCR product with the $(CTG)_{10}$ oligonucleotide (see Basic Protocol 1).

NOTE: Expansions such as E3 and E4 will not transfer well by vacuum blotting.

Table 9.4.1 DM Classification System

Category	Expansion characteristics
E0	Presence of expansion detected by PCR hybridization but not detectable by genomic DNA hybridization
E1	<0–1.5 kb
E2	1.5–3.0 kb
E3	3.0–4.5 kb
E4	4.5–6.0 kb

Materials *(see APPENDIX 1 for items with ✓)*

    Purified genomic DNA

    50 U/µl *Eco*RI restriction enzyme and 10× restriction endonuclease buffer

    0.6% agarose gel

✓ 1× TBE buffer

✓ 10× gel loading buffer

    Molecular-size markers (e.g., λ-*Hin*dIII digested DNA)

    1.0 µg/ml ethidium bromide in 1× TBE buffer

    0.4 M NaOH

    Neutralization solution: 0.2 M Tris·Cl, pH 7.5 (*APPENDIX 1*)/2× SSC (*APPENDIX 1*)

✓ Prehybridization/hybridization buffer

    10^8 to 10^9 dpm/µg column-purified [α-^{32}P]dCTP-pGB2.2 *or* -p750 probe (available from R. Korneluk; *APPENDIX 3E*), labeled by random-hexamer priming and column-purified

    Wash buffer IV: 2× SSC (*APPENDIX 1*)/0.1% (w/v) SDS, room temperature

    Wash buffer V: 0.1× SSC (*APPENDIX 1*)/1.0% (w/v) SDS

    20 ×20–cm casting tray with 1.5-mm-thick comb

    Nylon membrane, positively charged (e.g., Biotrans+, ICN Biomedicals)

    Filter paper (Whatman 3MM)

1. Digest purified genomic DNA (2 to 5 µg per lane) with 50 U (possibly higher and with 4 mM spermidine for genomic DNA prepared with an automated DNA extractor) *Eco*RI 3 hr at 37°C.

2. While the DNA is being digested, prepare a 0.6% agarose gel in 1× TBE buffer without ethidium bromide. Pour gel in a 20 × 20–cm casting tray with a 1.5-mm-thick comb.

3. Add 10× gel loading buffer to the digested DNA samples (1× final concentration) and load samples onto the gel. Include one lane with appropriate molecular size markers.

4. Electrophorese the gel in 1× TBE buffer without ethidium bromide for 20 to 22 hr at 35 V, 40 mA, then increase to a maximum of 45 V, 50 mA for an additional 5 to 8 hr until xylene cyanol dye front reaches bottom of gel.

5. Stain the gel 20 to 30 min in 500 ml of 1× TBE buffer containing 1.0 µg/ml ethidium bromide. Photograph the stained gel.

6. Denature the gel 30 min in 500 ml of 0.4 M NaOH (use other protocols for other membranes).

7. Transfer DNA to positively charged nylon membrane by capillary transfer in ≥500 ml of 0.4 M NaOH. Transfer 6 hr to overnight.

8. After transfer, remove nylon membrane from gel and incubate 15 min in neutralization solution with gentle shaking.

9. Blot membrane lightly on filter paper. Fix DNA by baking the membrane 30 min at 80°C or by UV cross-linking.

10. Prehybridize membrane in prehybridization/hybridization buffer ≥1 hr at 65°C.

11. Hybridize membrane overnight at 65°C using column-purified [α-^{32}P]dCTP-labeled pGB2.2 at 10^6 cpm/ml in prehybridization/hybridization buffer.

12. After hybridization, wash the membrane with increasing stringency using the following regimen:

    Wash buffer IV—two times, 10 min each, room temperature
    Wash buffer V—two to three times, 10 min each, 65° to 70°C.

13. Blot membrane dry with filter paper and wrap in plastic.

14. Expose hybridized membrane to X-AR autoradiographic film with intensifying screen overnight at −80°C.

Reference: Mahadeyan et al., 1992

Contributors: Linda C. Surh, Mani Mahadevan, and Robert G. Korneluk

UNIT 9.5

Detection of Nonrandom X Chromosome Inactivation

BASIC PROTOCOL

X CHROMOSOME INACTIVATION ASSAY

NOTE: High-quality, high-molecular weight genomic DNA must be utilized for this assay.

Materials
NEBuffer 1 (New England Biolabs)
10 U/μl *Hpa*II restriction endonuclease
100 ng/μl genomic DNA sample (*APPENDIX 3A*) isolated, e.g., from peripheral blood or oral mucosa
100 ng/μl DNA sample that has previously shown a pattern of highly skewed X chromosome inactivation (contact authors at *ehoffman@childrens-research.org*)

RNase/DNase-free microcentrifuge tubes
Thermal cycler or water bath

1. Prepare the following two reaction mixtures in labeled, sterile RNase/DNase-free microcentrifuge tubes, combining the reagents in the order indicated (total volume: 12.5 μl). Pipet up and down vigorously after each addition to ensure mixing.

   Reaction 1: HpaII predigestion reaction:
   5.25 μl deionized H_2O
   1.25 μl NEBuffer 1
   1.0 μl 10 U/μl *Hpa*II
   5 μl 100 ng/μl isolated DNA product (~500 ng DNA)

   Reaction 2: negative control (no HpaII):
   6.25 μl deionized H_2O
   1.25 μl NEBuffer 1
   5 μl 100 ng/μl isolated DNA product (~500 ng DNA).

2. As a positive control, prepare the above two reaction mixes using a DNA sample which has previously shown a pattern of highly skewed X-chromosome inactivation for enzyme digestion.

3. In a thermal cycler or water bath, incubate reactions 120 min at 37°C.

4. Finish with a 30 min incubation at 65° to deactivate the *Hpa*II. Store up to 48 hr at 4°C, or at −20°C for extended periods of time.

5. Proceed with PCR amplification, dye labeling, and gel electrophoresis of the restriction fragments (see Support Protocol).

SUPPORT PROTOCOL

PCR AMPLIFICATION AND LABELING OF DIGESTED AND UNDIGESTED DNA TEMPLATES

The authors and others have found that the most accurate method for detection is the use of primers tagged covalently with either fluorescent molecules or infrared-absorbing dyes. Depending on the strength of the dye and coupling efficiency, one must empirically identify the correct dilution of labeled primer with unlabeled primer, and/or the correct PCR cycle number, so that the amount of labeled PCR product is in the linear range of the detection system used. Here, an example is provided of the forward primer synthesized with the LI-COR IR labeled dye (1 pmol/µl), with 20 cycles of PCR, and detection of products on an IR sequencer. Other labeling and detection methods can be substituted as mentioned below.

NOTE: Labeled primers are light-sensitive and should be kept covered or in the dark whenever possible.

Materials *(see APPENDIX 1 for items with* ✓ *)*

    Red *Taq* DNA polymerase and 10× PCR reaction buffer (Sigma)
✓ 1.25 mM dNTP mix (1.25 mM each dNTP)
✓ 1 pmol/µl labeled Met-F forward primer (see recipe for primers in APPENDIX 1)
✓ 5 pmol/µl unlabeled Met-R1 reverse primer (see recipe for primers in APPENDIX 1)
    50 ng/µl *Hpa*II-predigested and positive and negative-control DNA (see Basic Protocol)
    IR2 stop solution/loading dye (LI-COR)
    100-bp DNA ladder (Life Technologies)

    96-well PCR plates (e.g., Perkin-Elmer)
    Thermal cycler
    LI-COR 4200S DNA Analyzer

1. Prepare the following master mix for each reaction, for single-day use (note that there are two reactions per sample, and also positive controls; see Basic Protocol):

        6.0 µl deionized H$_2$O
        1.25 µl 10× PCR reaction buffer
        1.0 µl 1.25 mM dNTP mix
        1.0 µl 1 pmol/µl labeled Met-F forward primer
        0.25 µl 5 pmol/µl unlabeled Met-R1 reverse primer
        1.0 µl 1 U/µl Red *Taq* DNA polymerase.

2. Place a 10.5-µl aliquot of the master mix for each sample into the appropriate well of a prelabeled PCR plate.

3. Add 2 µl of 50 ng/µl *Hpa*II-digested and positive and negative control DNA samples prepared in the Basic Protocol to the correspondingly labeled wells of the PCR plate containing the master mix.

4. Place the PCR plate in a thermocycler and perform PCR under the following conditions (store up to 1 week at −20°C):

1 cycle:	3 min	94°C	(denaturation)
19 cycles:	30 sec	94°C	(denaturation)
	30 sec	60°C	(annealing)
	45 sec	72°C	(extension)
Final step:	indefinitely	4°C	(hold)

5. Thaw the PCR products if they have been in storage and add 3 μl of IR2 stop solution/loading dye to each of the 12.5-μl samples, pipetting vigorously up and down to mix.

6. Denature 3 min at 93°C in the thermal cycler and immediately place on ice.

7. Run cut PCR products beside uncut products on the LI-COR 4200S DNA Analyzer with molecular weight markers on the outside lanes. Follow instrument operations manual to load and electrophorese the samples on an acrylamide denaturing gel (products ~250 to 300 bp).

8. Manually assign lanes from the completed image files. Quantitate the relative activation/inactivation ratios of androgen-receptor alleles for each sample by first identifying the two peaks corresponding to the two X chromosome alleles in the undigested sample, then normalizing the peak densities to compensate for any unequal amplification of alleles. Multiply this normalization factor by the digested sample peak ratio.

Reference: Allen et al., 1992

Contributors: Melissa M. Thouin, James M. Giron, and Eric P. Hoffman

UNIT 9.6

Molecular Analysis of Paternity

BASIC PROTOCOL 1

ANALYSIS OF VNTRs BY RFLP TECHNOLOGY

Materials (*see* APPENDIX 1 *for items with* ✓)
> DNA from individuals to be tested (e.g., mother, child, alleged father, and human control; see Support Protocol 1), generally isolated from whole blood
> Restriction enzyme and appropriate buffer
> 0.6% (w/v) agarose minigel
> DNA molecular size markers
> 20-cm, 0.7% to 1.0% analytical agarose gel
> ✓ 1× TBE buffer (Life Technologies)
> ✓ 10 mg/ml ethidium bromide (store in the dark up to 1 year at room temperature)
> Radiolabeled or nonisotopically labeled DNA probe (Table 9.6.1) for VNTR sequence identification
> 0.2 N NaOH
> ✓ 0.2 M Tris·Cl, pH 7.5/2× SSC
> ✓ 2× SSC
>
> Nylon membrane (e.g., PALL Biodyne)
> Whatman 3MM filter paper
> X-ray film (e.g., X-Omat film; Kodak)
> 50°C incubator or water bath

Table 9.6.1 Commonly Used Probes for RFLP-Based Paternity Testing

Probe	Locus	Chromosomal location	Compatible restriction enzymes
3′HVR	D16S85	—	*Hae*III, *Pvu*II, *Pst*I
CMM101	D14S13	14q	*Hae*III, *Pst*I
EFD52	D17S26	17q	*Hae*III
g3	D7S22	7q36-qter	*Hinf*I
MS31	D7S21	7p22-qter	*Hinf*I, *Pst*I
MS43A	D12S11	12q24.3-qter	*Pst*I, *Hinf*I
pAC415	D7S467	7q	*Hae*III, *Pst*I
pAC424	D6S132	6q27	*Hae*III, *Hinf*I, *Pst*I
pAC425	D1S339	1p36	*Hae*III, *Hinf*I, *Pst*I
pH30	D4S139	4q35	*Hae*III
pL336	D1S47	—	*Pst*I, *Hae*III
pL427-4	D21S112	—	*Pst*I
TBQ7	D10S28	10pter-p13	*Hae*III, *Pvu*II
V1	D17S79	—	*Hae*III, *Pst*I
YNH24	D2S44	2p	*Hae*III, *Pst*I, *Hinf*I

1. Digest ~1 µg each DNA sample to completion with a selected restriction enzyme. Follow manufacturer-specified reaction conditions for the enzyme (beware of star activity).

2. Load 1 µl each digested DNA as well as digested and undigested controls onto a 0.6% (w/v) agarose minigel. Electrophorese and photograph the gel as described (see Support Protocol 1, steps 9 and 10). Add more enzyme and extend incubation if digestion is incomplete.

3. Load DNA molecular size markers, mother sample, child sample, alleged father sample, and human control sample onto a 20-cm, 0.7% to 1.0% analytical agarose gel immersed in 1× TBE buffer. Load the DNA molecular size markers in the first and last wells (or next to the first and last samples) and in the center of the gel if analyzing more than one case.

4. Electrophorese at ~40 V overnight or until the bromphenol blue tracking dye migrates to the edge of the gel, recirculating the buffer if desired.

5. Stain gel with ethidium bromide and photograph (minimize UV exposure) to document DNA digestion, migration distances, and comparisons of DNA concentrations (see Support Protocol 1, steps 9 and 10).

6. Transfer DNA from the gel to a nylon membrane by capillary transfer, electroblotting, or vacuum blotting, using standard Southern blotting techniques (*APPENDIX 3G*).

7. Hybridize probe to the membrane following standard protocols. Use ~2 × 10^5 cpm probe per milliliter hybridization solution.

8. Wash the blots under high-stringency conditions depending upon the probe (also see *APPENDIX 3G*).

9. Pour off final wash solution and blot excess liquid from membrane with Whatman 3MM filter paper. Wrap membrane in plastic wrap and expose to X-ray film (e.g., sandwich style) 12 to 48 hr for radioactive probes or less for nonradioactive systems.

10. Once the autoradiogram has been generated, strip the probe from the membrane by washing the membrane 30 min in 350 ml of 0.2 N NaOH at 50°C with constant gentle agitation. Pour off NaOH and repeat wash. Rinse blot with 350 ml of 0.2 M Tris·Cl, pH 7.5/2× SSC for ~30 min at room temperature with constant gentle agitation. Rinse briefly with ~300 ml of 2× SSC. Rehybridize with another single-locus probe or dry, wrap in plastic, and store up to 5 yr at room temperature.

11. Inspect developed autoradiograms for quality and adequate band intensity before labeling the film with sample names.

12. Determine band sizes (e.g., automatically using a Numonics digitizer in conjunction with DNAVIEW). If two independent determinations of the band sizes are performed, compare the independent size measurements for accuracy and average the values. Verify the sizes of the human control bands each time.

BASIC PROTOCOL 2

ANALYSIS OF POLYMORPHIC LOCI BY PCR

While Southern-based RFLP testing is a powerful state-of-the-art means of establishing biological relationships, the use of the polymerase chain reaction (PCR) has gained popularity in routine paternity testing. Commercially available kits for amplification of polymorphic regions contain protocols using the vendor's premade solutions of enzyme, buffer, and primers, as well as their recommendations for DNA concentrations, time, and temperature cycles for the amplification itself; however, it is important to note that different loci can require refinements of the general amplification conditions, which should be determined by the laboratory performing the work. All protocols for analysis by PCR start with genomic DNA. Hence, if RFLP analysis is to be done together with PCR, the initial biological sample must be divided before the restriction digest (see Basic Protocol 1, step 1), and the portion to be used for PCR kept intact in order to set up the amplification.

Table 9.6.2 presents a comparison of paternity testing by RFLP (see Basic Protocol 1) versus PCR.

Table 9.6.2 Comparison of Paternity Testing by RFLP Versus PCR Methodologies

RFLP	PCR
Routinely used in paternity testing since 1990.	Newer technique, used by a small number of paternity labs.
Small sample required (DNA equivalent of 0.1–0.5 ml whole blood).	Minimum sample required (DNA equivalent of <0.1 ml whole blood). Beneficial for low-abundance samples.
Requires high molecular weight DNA.	Degraded samples can be analyzed.
Actual hands-on testing time can be as little as one week.	Requires even less hands-on testing time than RFLP (e.g., as little as 2–3 days).
DNA characteristics isolated, sized, and reported.	DNA isolated and many copies of characteristics made before analysis.
Relatively long-established standards and guidelines available for its use in paternity testing. General court acceptance.	Recently established standards and guidelines for its use in paternity testing. Less court acceptance.
High degree of discrimination; large number of falsely accused men excluded or high probability of paternity values obtained.	Characteristics less polymorphic, less helpful in determining relationships, requiring more loci be tested to reach standard criterion.

Table 9.6.3 Powers of Exclusion for Allele-Specific
Oligonucleotide Probes Commonly Used for Paternity
Testing by PCR

System	Chromosomal location	Average power of exclusion[a]
LDLR	19p13.1–13.3	0.087–0.239
GYPA	4q28–31	0.120–0.190
HBGG	11p15.5	0.181–0.274
D7S8	7q22–31.1	0.161–0.237
Gc	4q11–13	0.132–0.298
DQA1 (DQα)	6p21.3	0.476–0.575

[a]The indicated range is over three major racial groups typically recognized in North
America as Black, Caucasian, and Hispanic.

Allele-Specific Oligonucleotide (ASO) Probes

The most widely used ASO-based system is the commercially available AmpliType PM +
DQA1 PCR Amplification and Typing kit (Perkin-Elmer), which contains complete instructions and the reagents necessary to amplify and type six independent loci (Table 9.6.3).

Sequence-specific probe strips are not routinely used in paternity testing due to the relatively
low power of exclusion of the combined systems and the relative cost of purchasing the kit.
They are, however, useful in forensic identity testing, where entire genetic patterns rather
than single alleles are matched and where different statistical data result. Polymarker results
for three individuals, along with negative and positive controls for the analysis, are shown in
Table 9.6.4.

Amplified Fragment Length Polymorphisms (AmpFLPs)

AmpFLPs can be divided into two categories based on the size of the repetitive unit which
is amplified upon hybridization of primers to undigested genomic DNA from each tested
individual. Long tandem repeats (LTRs) and short tandem repeats (STRs) are loci that have
variable numbers of tandem repeats and that are analyzed by PCR.

Long tandem repeats
LTRs have repeat units that range from 10 to as many as 80 base pairs, with allele sizes between
300 and 1000 bp, and can have heterozygosity values of ~95%.

Table 9.6.4 Multilocus Typing for Three Individuals

Individual	LDLR	GYPA	HBGG	D7S8	Gc
A	B	A, B	A, B	A, B	A, B
B	B	A	B	A	A, C
C	B	A	B	A, B	A, C
Neg. control	—	—	—	—	—
Pos. control	B	A, B	A	A	B

Table 9.6.5 Short Tandem Repeat (STR) Loci Commonly Used for Paternity Testing by PCR

STR	Chromosomal location	Gene identification	K562 allelic pattern[a]
CSF1PO	5q33.3–34	CSF-1 receptor protooncogene	10, 9
TH01	11p15.5	Tyrosine hydroxylase	9.3, —
TPOX	2p23–2pter	Thyroid peroxidase	9, 8
vWA	12p12pter	von Willebrand antigen	16, 16
LPL	8p22	Lipoprotein lipase	12, 10
F13A01	6p24–25	Coagulation factor XIII subunit a	5, 4
F13B	1q31–q32.1	Coagulation factor XIII subunit b	10, —
D16S539	16q24qter	NA[b]	12, 11
D7S820	7q	NA[b]	11, 9
D13S317	13q22–q31	NA[b]	8, —
D5S818	5q21–q31	NA[b]	12, 11
FES/FPS	15q25qter	c-fes/fps proto-oncogene	12, 10

[a]Number of repeats at each locus. Dashes indicate the conservative convention of indicating that only a single allele is detectable but that the genotype may or may not represent a homozygous condition.

[b]NA = not available.

Short tandem repeats

Most STRs used in identity testing have repeats that are four base pairs in length with allele sizes between 100 and 350 bp, and heterozygosities generally ranging from 65% to 85%. Table 9.6.5 gives a list of some of the commonly used STRs. Additional information for many of these loci can be found at *http://www.promega.com/tbs/tmd008/tmd008.html*.

Electrophoresis and Detection

Manual systems

With manual systems, an aliquot of the PCR product is most commonly electrophoresed through a polyacrylamide gel, although there are some high-resolution agaroses available, such as MetaPhor and NuSieve (FMC BioProducts). The exact conditions will depend on the size of the products to be separated and the desired resolution (e.g., *APPENDIX 3F*, *CPMB UNIT 2.5A*, or Sambrook et al., 1989).

Semiautomated systems

Semiautomated systems for the separation, detection, and analysis of LTR and STR PCR products are also available and include gel-based systems that combine electrophoresis, detection, and analysis into one unit (PCR fragments must be fluorescently labeled), and fluorescence imaging systems.

Automated systems

At the time of this writing, the only completely automated system for LTR and STR analysis is by capillary electrophoresis (CE). As the name implies, electrophoresis is carried out in a microcapillary tube rather than between glass plates. The disadvantage of CE is that only single capillary systems are currently available, so analysis proceeds one sample at a time. However, because CE can be carried out at much higher voltage, separation of fragments is completed in ~20 min which helps to offset this reduced throughput.

PREPARATION OF GENOMIC DNA FROM WHOLE BLOOD

Materials (*see* APPENDIX 1 *for items with* ✓)

   Whole blood samples from individuals to be tested (e.g., mother, child, alleged father, and human control)
   ✓ Cell shocking solution
   ✓ Nuclear lysis buffer
   ✓ 10% (w/v) sodium dodecyl sulfate (SDS)
   20 mg/ml proteinase K (Boehringer Mannheim)
   Saturated NaCl solution
   100% and 70% ethanol, room temperature
   ✓ TE buffer, pH 7.5
   0.6% (w/v) agarose minigel
   DNA concentration standards
   10 mg/ml ethidium bromide (store dark up to 1 year at room temperature)

   EDTA Vacutainer tube (purple top)
   56°C incubator or water bath

1. Collect blood samples (250 μl minimum) from each individual to be tested in an EDTA Vacutainer tube (purple top) using standard phlebotomy techniques. Label tubes with appropriate identification information, including each individual's name and the draw date as well as other information (e.g., transfusion and bone marrow transplant history).

2. Add 250 to 500 μl of each blood sample to a labeled microcentrifuge tube. Freeze samples at least 1 hr at −70°C to lyse red blood cells. If necessary, store microcentrifuge tubes up to years at −70°C. Store remaining blood in Vacutainer up to several months at 4°C.

3. Thaw each sample to room temperature and microcentrifuge 1 min at ∼10,000 × g, room temperature. Remove supernatant, leaving 100 μl containing nucleated cells, and add 900 μl cell shocking solution. Mix by inversion 7 to 10 min until cell lysis is complete. Microcentrifuge 1 min at ∼10,000 × g, room temperature. Discard supernatant.

4. Add 300 μl nuclear lysis buffer to pellet and mix by gentle pipetting. Add 20 μl of 10% SDS and 10 μl of 20 mg/ml proteinase K. Mix by gentle inversion and incubate 2 hr to overnight at 56°C.

5. Add 100 μl saturated NaCl solution. Shake vigorously 15 sec. Allow tube to sit on the bench 5 min and microcentrifuge 4 min at ∼10,000 × g, room temperature. Transfer ∼400 μl DNA-containing supernatant to a clean, labeled microcentrifuge tube.

6. Add 2 vol room-temperature 100% ethanol. Mix by inversion until the DNA precipitates. Microcentrifuge 1 min at ∼5000 × g, room temperature. Discard supernatant.

7. Rinse DNA pellet with 70% ethanol to remove residual salt. Gently pipet 500 μl of 70% ethanol over the pellet and gently invert the tube. Microcentrifuge 1 min as above, starting at 1% power and gradually increasing speed to 100% power (∼10,000 × g). Remove ethanol with a pipettor. Repeat the wash. Allow the pellet to air dry ∼5 min or until the edges of the pellet become clear.

8. Resuspend each DNA pellet in 20 to 80 μl TE buffer, pH 7.5, depending on the pellet size. Incubate 2 hr at 56°C with periodic gentle mixing.

9. Load 1 μl of each DNA sample onto a 0.6% (w/v) agarose minigel. Include DNA concentration standards. Electrophorese at ∼7 V/cm until the bromphenol blue tracking dye

has migrated 3/4 of the way down the gel. Stain gel by gently agitating 10 to 30 min in ~50 ml H_2O containing 10 μl of 10 mg/ml ethidium bromide, and photograph. If desired, rinse stained gel two times in water for 15 min each to reduce background.

10. Use the gel to assess the quantity and condition of the DNA. Compare the concentration of the test samples to known concentration standards (3 to 10 μg DNA is usually obtained from 500 μl whole blood).

SUPPORT PROTOCOL 2

INTERPRETATION, STATISTICAL EVALUATION, AND REPORTING OF DNA PROFILES: DATA FROM MOTHER AND ALLEGED FATHER

This procedure outlines the interpretation and evaluation of single-locus DNA profiles for paternity determination when samples from both the mother and the alleged father are available.

1. Compare the sizes of the child's DNA bands with those from the mother and alleged father. If the bands belonging to the parent do not fall within the resolution threshold (delta; expressed as percent molecular weight and determined independently in each laboratory, e.g., ~2.5% of band size), do not consider them a match.

2. For each DNA probe tested, determine whether the alleged father can be excluded at that particular locus. For a given probe, if the child's paternal band (i.e., band not matching the mother's) matches a band from the alleged father, do not exclude the alleged father as the biological father at that locus (for a detailed list of possible nonexcluding patterns, see Morris et al., 1989). If the child's paternal band does not match any bands from the alleged father, exclude the alleged father at that particular locus.

3. After reviewing data from all probes tested, determine whether the alleged father can be included or excluded as the biological father according to the following guidelines:

   a. An exclusion with a minimum of two probes constitutes an exclusion of paternity. Prepare paternity report as described in step 7.

   b. If the alleged father cannot be excluded by a minimum of two probes, perform a statistical evaluation of paternity data as outlined in steps 4 to 6. An inclusion of paternity should require a probability of paternity of at least 99%.

4. Calculate the paternity index (PI) for each probe (i.e., chance of the alleged father producing a child of the given phenotype compared to a random unrelated individual of similar race producing the child) for which the alleged father (AF) cannot be excluded using the equation:

$$PI = X/Y$$

where X is the probability that the child's phenotype could result from a mating of the mother and alleged father (i.e., the product of the frequencies of fragments obtained from the child being contributed by mother and alleged father), and Y is the probability that the child's phenotype could result from a mating of the mother and an unrelated randomly selected individual from a similar racial background as the alleged father (i.e., product of the frequencies of fragments from the child being contributed by mother and a random individual from the relevant racial population).

> *DNAVIEW incorporates the population frequency of the child's paternal allele into the Y value using a refined coelectrophoresis model (Brenner and Morris, 1989). This calculation incorporates a sigma value which estimates the amount of measurement imprecision as a percentage of*

molecular weight. This value must be determined by each laboratory. Typical values range from 0.4% to 1.2%. See CPHG UNIT 14.4 for alternative methods of obtaining allele frequencies.

5. Calculate the combined paternity index ($PI_{combined}$), which is the product of the individual PI values calculated for each probe.

6. Calculate the probability of paternity value (*W*), based on the combined PI value calculated in step 5.

$$W = \frac{p \times PI_{combined}}{p \times PI_{combined} + 1 - p}$$

where *p* is the prior probability which is often given the neutral value of 0.5 (see Support Protocol 3 for a discussion of the prior probability).

7. Issue a written report on laboratory letterhead. Include the following information on every standard paternity report:

> Names of the individuals tested
> Relationship of the individuals tested
> Race of the mother and alleged father
> Draw date
> Laboratory case number
> *For RFLP:* Name of the locus tested, probe name, and restriction enzyme used
> *For PCR:* Name of each DNA locus amplified, and name or source of the primer pair used
> Resulting band sizes for each individual tested
> *For exclusion:* If the alleged father is excluded as the biological father, the inconsistencies should be indicated
> *For nonexclusion:* If the alleged father cannot be excluded as the biological father, the individual PI, combined PI, prior probability used, and the probability of paternity should be reported
> Signature of a laboratory director.

SUPPORT PROTOCOL 3

INTERPRETATION, STATISTICAL EVALUATION, AND REPORTING OF DNA PROFILES: SPECIAL PATERNITY CASES

While the vast majority of paternity cases involve a mother, child, and alleged father, there are circumstances where one or both of the child's parents are unavailable for testing (i.e., family studies or kinship studies). Additionally, other questions regarding biological relationship (e.g., twin zygosity, sibling relatedness, likelihood of incestuous relationships) occur. These questions can be answered using VNTR DNA typing.

Computer programs are commercially available for analyzing the genetic information from family studies (e.g., DNAVIEW, Traver Paternity Software). Upon input of the relationship hypothesis and an alternate hypothesis (i.e., monozygotic versus dizygotic twins), and the genotypes from each individual, the program will generate an algebraic formula describing the likelihood (chance) of the relationship hypothesis versus the likelihood of the alternate hypothesis.

Motherless Cases

When the tested man matches with the child and the mother has not been tested, the PI is typically reduced by a factor of two. When the alleged father is heterozygous and is of the same race as the untested mother, the paternity index is calculated as PI $= 1/4q$, where the term q is defined as the probability of finding the shared allele (Q), in a random sampling of the man's racial population. Typically, this is taken as the mathematical frequency of that allele in the population. If the alleged father is homozygous, PI $= 1/2q$. If the alleged father and child are both homozygous, PI $= 1/q$.

If the race of the untested mother differs from that of the alleged father, the equations cannot be simplified as above and must incorporate the frequencies of both of the child's alleles from the two racial populations (Traver, 1996).

When no mother is tested, one must be careful to establish that the alleged father is not a close biological relative of the mother. If the mother is not tested and the alleged father is her relative, the alleged father may be falsely included as the biological father when the characteristics he shares with the child were in fact contributed to the child by the mother.

Fatherless Cases

If the alleged father is unavailable for testing, various known biological relatives of the alleged father can be tested in his place to determine a likelihood ratio or relationship index (RI); however, it must be kept in mind that the relationship index describes the possible relationship between the child and the tested individuals and not directly with the untested man.

When selecting which individuals to test, care must be taken to ensure that the tested individuals are in fact biologically related to the untested alleged father. It may also be prudent to test close male relatives of the alleged father who could be accused of paternity in order to exclude them as potential alleged fathers.

Testing both alleged paternal grandparents. The RI incorporates the number of obligate paternal alleles (OPAs) that the grandparents share with the child and can be simplified as RI $=$ (no. OPAs in grandparents)$/4q$.

Testing only one alleged paternal grandparent. This type of analysis can be performed as above by counting the number of possible obligate paternal alleles in the alleged grandparent (0, 1, or 2), and adding $2q$. This will take into account the probability of the untested grandparent having the OPA. The numerical result is then divided by four. If the tested grandparent is heterozygous and contains the obligate paternal allele, the equation is RI $= 1/2 + 1/2q$. If the tested alleged grandparent does not contain the obligate paternal allele, the equation is RI $= 0.5$.

Testing the alleged father's full siblings. By testing a number of the alleged father's known full siblings one can attempt to reconstruct the genotypes of the alleged grandparents. When one sibling is tested, only two of the four grandparental alleles are identified. If the sibling is homozygous and shares an allele with the child, the relationship equation is RI $= 1/2 + 1/2q$. If the alleged father's sibling is heterozygous and shares the obligate paternal allele with the child, RI $= 1/2 + 1/4q$. If the child does not share a characteristic with the sibling of the alleged father, the relationship index is 0.5.

Twintyping

According to Brenner (1997), the equation used to evaluate the genetic evidence favoring dizygosity when the twins are heterozygous, is RI $= (1 + p + q + 2pq)/4$, where p and q are the frequencies of the two bands in the profile. If the identical profiles are homozygous with a band frequency of q, the equation becomes $RI = (1 + q)^2/4$.

Paternity Establishment with Related Alleged Fathers

If the alleged father cannot be excluded as the biological father and there is a substantive allegation that a close relative of the alleged father could also be the father, the relative should also be tested. If this is not possible, the following equations can be used to estimate the likelihood of paternity of the tested man to that of his untested relative.

If the child has a single obligate paternal allele that it shares with the heterozygous alleged father, the genetic odds in favor of the tested man being the father over that of his brother, father, or son, would be $RI = 2/(1 + 2q)$. If the alleged father is homozygous for the shared allele, the relationship index is $RI = 2/(1 + q)$.

Single Exclusions

Because of the low but not insignificant mutation rate in the VNTR probe systems, an exclusion from paternity cannot be based on an exclusion in a single DNA system. An exception is the case involving closely related alternative fathers where one is excluded in only one system and the other is excluded in none.

If only a single exclusion is found after extended testing, one of three conclusions can be drawn:

1. The alleged father is not biologically related to the child but demonstrates the obligatory paternal fragment in the systems tested purely by chance. Such an event is rare, with a frequency directly related to the frequency of the shared alleles in the population.

2. The alleged father is not the biological father but is biologically related to the child. If the mother had relations with a close relative of the alleged father (i.e., brother, father, or son) the laboratory should, if at all possible, test that individual for paternity.

3. The alleged father is the true biological father and is excluded at a single location due to a mutational event that occurred either in the alleged father's germ cells or in the early development of the child.

If no relative of the tested man is alleged to be the biological father of the child, then a nonexclusionary paternity report should be issued incorporating the suspected mutational event.

Although mutation rates do not appear to differ among racial groups, the paternal and maternal mutation rates at a given locus may differ significantly (Endean, 1995; Henke et al., 1993; Kelley, unpub. observ.). To calculate a mutation PI for a single discrepant system, a mutation rate (μ) is used. The numerator in the mutation PI is μ. The denominator is the average rate of exclusion for nonfathers (ARE), which is a function of the heterozygosity (h) of the test system. This value can be determined experimentally or it can be calculated as $ARE = h^2(1 - 2hH^2)$, where H is the homozygosity $(1 - h)$ of the system (Garber and Morris, 1983; Brenner and Morris, 1989; Endean, 1989). The average mutation PI for a discrepant system (Gjertson and Endean, 1996) is given by $PI = \mu/ARE$. Table 9.6.6 lists the paternal mutation rate, ARE, and mutation PI for several commonly used probes for human DNA digested with the enzyme HaeIII.

The mutation PI can be combined with the remaining nonexclusionary systems to obtain an overall result as described in Support Protocol 2. Additional testing may be needed in light of a downward adjustment in the cumulative PI. The extra testing can also allow opportunity

Table 9.6.6 Paternal Mutation Rates, AREs, and Mutation PIs for Commonly Used Single-Locus Probes[a]

Probe/locus	Rate (μ)	ARE[b]	Mutation PI
pYNH24/D2S44	0.0015	0.909	0.00165
pCMM101/D14S13	0.0018	0.928	0.00194
TBQ7/D10S28	0.0009	0.929	0.00097
EFD52/D17S26	0.0036	0.894	0.00403
pH30/D4S139	0.0071	0.867	0.00819
pAC415/D7S467	0.0008	0.872	0.00092
pAC424/D6S132	0.0002	0.785	0.00025
pAC425/D1S339	0.0050	0.785	0.00637

[a] Values obtained with human DNA digested with the enzyme *Hae*III.

[b] The AREs shown were calculated using the average heterozygosity values for Black, Caucasian, Hispanic, and Asian databases.

to uncover a second exclusion, should it exist, when the tested man is in fact a relative of the biological father.

Use of the Prior Probability

Prior probability is an assigned numerical value for the nongenetic evidence and its assigned value is for the "trier of fact" to determine based on known, or knowable, nongentic circumstances surround the conceptive event. It ranges from nearly zero (impossibility) to ~100% (near certainty). The laboratory, which has no knowledge of the social evidence surrounding a paternity case, will most frequently report a neutral prior probability of 50%. Some laboratories choose to report a prior probability in the neighborhood of 75%, representing the laboratory's nonexclusion rate or the percentage of times that mothers truly identify the biological father of the child (Morris, 1993). Upon request, the laboratory can recompute the probability of paternity for any particular prior probability the court may request. As the strength of the genetic evidence increases, the nongenetic evidence has less effect on the overall probability of paternity.

References: AABB, 1997; Parentage Testing Accreditation Requirements Manual, 1995; Allen et al., 1990

Contributors: Amanda C. Sozer, Charles M. Kelly, and Daniel B. Demers

UNIT 9.7

Amplification-Refractory Mutation System (ARMS) Analysis of Point Mutations

The ARMS technique is based upon the observation that oligonucleotides that are complementary to a given DNA sequence, except for a mismatched 3′ terminus, will not function as PCR primers under appropriate conditions. For a robust and reliable ARMS test, it is necessary to determine a combination of primer sequence and reaction conditions that will generate a detectable ARMS product from the target allele while minimizing false priming at the nontarget allele.

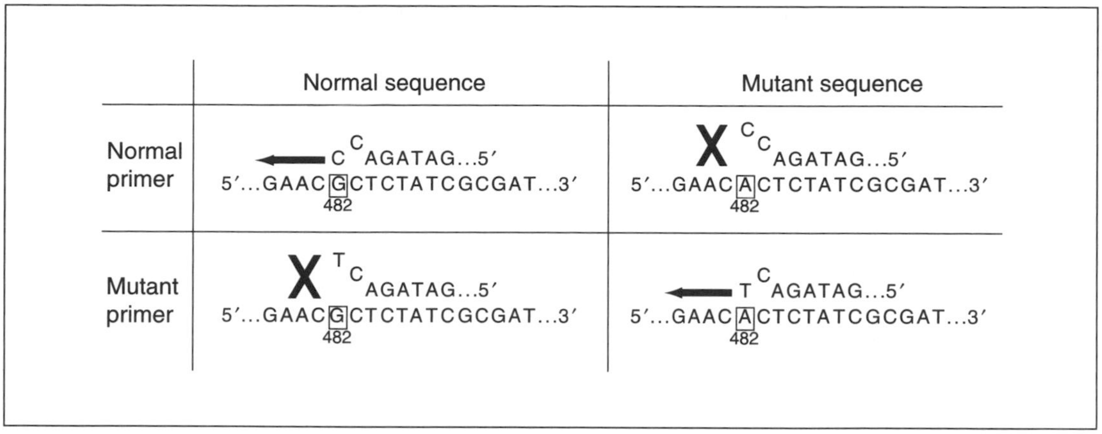

Figure 9.7.1 ARMS primer sequences for a single ARMS test. Sequences of the ARMS primers and target DNA sequences around the *R117H* mutation of the *CFTR* gene (G→A at position 482; Dean et al., 1990). The base that is altered is indicated in the normal and mutant DNA sequences by a box. The presence of an arrow indicates that primer/target combinations can be extended by *Taq* DNA polymerase; an "X" indicates extension does not occur. Bases in the ARMS primers that are not complementary to the target are shown displaced from the target sequence. A single mismatch (in this case a C/C) at the penultimate base is not sufficient to prevent extension whereas a primer with two adjacent mismatches, at the terminal and the penultimate base, is not extended.

The following guidelines are used to generate ARMS primers that will detect point mutations when used in combination with the reaction conditions outlined below (see Basic Protocol). An example is given in Figure 9.7.1.

1. ARMS primers should be oligonucleotides of ~30 or more bases. Primers <28 bases long should be avoided. Longer primers (up to 60-mers) may be used.

2. For the mutant-specific primer (M), the 3′ terminal base of the ARMS primer should be complementary to the mutation; for the normal-specific primer (N), the 3′ terminal base should be complementary to the corresponding normal sequence.

3. Additional deliberate mismatches should normally be introduced at the penultimate base of the ARMS primer to increase the specificity of the ARMS reaction. Because different mismatches have been found to have different destabilizing effects, it is necessary to consider both terminal and penultimate mismatches together. If the mutation-induced terminal mismatch is strong, a weak additional mismatch should be selected, and vice versa, as indicated in Table 9.7.1.

4. The remainder of the ARMS primer should be complementary to the target sequence.

5. The design of the common primer is straightforward. The primer should be 30 bases long; selected to have ~50% G + C content, no 3′ complementarity with the ARMS primer or the internal control primers, and no repeated or unusual sequences (e.g., runs of a single base, palindromes); and, for use with the control PCR reactions suggested in this unit, should give a 150- to 250-bp PCR product.

NOTE: High-quality water (e.g., tissue culture grade) should be used in all solutions. Sigma double-processed tissue culture water has been shown to work well.

Table 9.7.1 ARMS Primer Additional Mismatch Selection[a]

Terminal mismatch	Coding strand nucleotide corresponding to penultimate nucleotide in the primer			
	A	G	C	T
AA	A	G	A	G
AG	C	T	A	G
AC	G	A	C	T
TT	C	T	A	G
TG	G	A	T	C or T
TC	C	T	A	G
CC	C	T	A	G
GG	A	G	A	G

[a]The table indicates which base to include at the penultimate position of the ARMS primer. The left-hand column lists the 3′ terminal mismatches specified (i.e., the mutation in the coding strand with the normal base in the anti-coding strand) and the top row indicates the target base in the coding strand corresponding to the penultimate base in the primer.

BASIC PROTOCOL

ANALYSIS OF SINGLE MUTATIONS BY ARMS TESTS

Materials *(see APPENDIX 1 for items with ✓)*

    50 μM normal ARMS primer
    50 μM mutant ARMS primer
    50 μM common primer
✓ 50 μM each control primer A and B (or other appropriate pair)
✓ 1 mM 4dNTP
✓ 10× PCR amplification buffer containing 12 mM $MgCl_2$
    Control DNA samples of known genotype (10 to 50 ng/μl in H_2O)
    Light mineral oil
    5 U/μl *Taq* DNA polymerase (Perkin-Elmer Cetus AmpliTaq or equivalent enzyme
        without 3′ to 5′ proofreading action)
    Nusieve 3:1 agarose (FMC Bioproducts) or equivalent
✓ 1× TBE buffer containing 0.5 μg/ml ethidium bromide
✓ Loading buffer
    DNA molecular size markers
    Test DNA samples (10 to 50 ng/μl in H_2O; see Support Protocol)
    Perkin-Elmer Cetus thermal cycler (480 or TC) and suitable reaction tubes

1. Follow the ARMS primer guidelines outlined in the introduction to design normal, mutant, and common ARMS primers suitable for the mutation of interest. Select an appropriate control primer pair. Obtain all primers and prepare 50 μM solutions of each.

2. Prepare normal (N) and mutant (M) ARMS reaction premixes according to the guidelines presented in Table 9.7.2. Dispense 40-μl aliquots of the premix into reaction tubes suitable for use in a thermal cycler. If necessary, store up to several days at room temperature or several months at −20°C.

Table 9.7.2 ARMS Reaction Premixes[a]

Reagent	Normal ARMS reaction (N)	Mutant ARMS reaction (M)
50 μM common primer	10 μl	10 μl
50 μM normal ARMS primer	10 μl	—
50 μM mutant ARMS primer	—	10 μl
50 μM control primer A	10 μl	10 μl
50 μM control primer B	10 μl	10 μl
1 mM 4dNTP mix	50 μl	50 μl
10× PCR amplification buffer	50 μl	50 μl
H$_2$O	260 μl	260 μl

[a]Volumes are for ten ARMS reactions.

3. Add 5 μl of 10 to 50 ng/ml control DNA to each tube of an N and M premix pair at room temperature. If possible, set up separate reactions containing DNA from normal, heterozygotic, and homozygotic individuals. Include a negative control test in which no DNA is added to the N and M tubes.

4. Add one drop mineral oil to each reaction tube and cap firmly. Microcentrifuge 10 sec at high speed.

5. Dilute *Taq* DNA polymerase by adding 12.5 μl of 10× PCR amplification buffer and 5 μl *Taq* DNA polymerase (1 U/5 μl final) to 107.5 μl sterile water (sufficient for ten tests). Mix by pipetting.

6. Place tubes in a thermal cycler block. Incubate 5 min at 94°C. Minimizing the time spent away from the block, remove one tube and add 5 μl *Taq* DNA polymerase dilution to the lower (aqueous) phase through the mineral oil layer. Return tube to 94°C block and repeat process until enzyme has been added to all tubes.

7. Stop the 94°C hold program and immediately run the following amplification program:

35 cycles:	1 min	94°C	(denaturation)
	1 min	60°C	(annealing)
	1 min	72°C	(extension)
1 cycle:	10 min	72°C	(extension)

Store up to 1 week at room temperature.

If a rapid thermal cycler such as the Perkin-Elmer Cetus 9600 is used, follow the manufacturer's recommendations to alter the cycling times to suit the characteristics of this type of machine.

At the end of the amplification program, it is advisable to leave the reaction tubes until the block temperature has reached 50°C to prevent the formation of artifact products.

8. Prepare a 3% agarose gel using 4.5 g Nusieve 3:1 agarose and 150 ml of 1× TBE buffer with 0.5 μg/ml ethidium bromide (150 ml sufficient for at least ten tests).

9. Label a 0.5-ml microcentrifuge tube for each reaction. Add 10 μl loading buffer to each tube.

10. Transfer 25 μl ARMS reaction from beneath the oil layer into the corresponding labeled tube, preferably uncapping the tubes in a fume hood or separate laboratory to avoid contamination with ARMS-product aerosols. Mix by pipetting.

11. Load 20 μl each reaction per lane. Include a DNA molecular size marker in one lane of the gel. Electrophorese 1 to 2 hr at 120 V and photograph under UV transillumination.

12. Analyze each lane for the presence of the expected-size control and ARMS fragments (control fragments should be 360, 813, or 825 bp for primer pairs A and B, C and D, or E and F, respectively). For both the normal and the mutant ARMS reactions, there are three potential outcomes:

    Expected outcome. ARMS product when target allele is present in sample; no visible product when target allele is absent.

    Nonspecific outcome. ARMS product when target allele is present in sample; ARMS product still visible when target allele absent. (This is testable only if a homozygous sample for the nontarget allele is available.)

    Insensitive outcome. ARMS product absent or faint even when target is present; no visible product when target allele is absent.

    a. If the expected result is obtained for both the normal and mutant ARMS primers, then the optimization is complete and the ARMS test is ready for use—omit steps 13 and 14 and proceed to step 15.

    b. If one or both of the ARMS primers is nonspecific, proceed to step 13a.

    c. If one or both of the ARMS primers is insensitive, proceed to step 13b.

For nonspecific ARMS primer(s)

13a. Increase the reaction specificity by repeating the procedure starting at step 2 with two sets of reactions. Reduce the amount of ARMS primer in the two reaction mixes by two- and five-fold, respectively.

14a. If nonspecific bands are still observed, increase the strength of the additional mismatch at the penultimate base (see Table 9.7.3 for guidelines). Repeat the procedure starting at step 1.

For insensitive ARMS primer(s)

13b. Increase sensitivity by repeating the procedure starting at step 2 with two sets of reactions. Increase the amount of ARMS primer or reduce the concentration of the control primers in the two reaction mixes by two- and four-fold respectively.

Table 9.7.3 ARMS Primer Mismatch Strength Table[a]

Destabilization strength	Mismatch pairing(s)
Maximum	GA, CT, TT
Strong	CC
Medium	AA, GG
Weak	CA, GT
None	AT, GC

[a]Mismatches are grouped according to destabilization strength. At least two mismatch strengths are available for each base.

14b. If bands for the expected ARMS products are still weak or absent, decrease the strength of the additional mismatch at the penultimate base. Repeat the procedure starting at step 1. Decrease the additional mismatch strength by following the guidelines shown in Table 9.7.3.

> *Changing the additional mismatch in an ARMS primer often causes a large alteration in the specificity and sensitivity of an ARMS reaction. Altering the primer concentration has a smaller effect and can be used to fine-tune the reaction.*

> *Following optimization of the ARMS test using DNA samples of known genotype, it is recommended that a large batch of ARMS reactions be made and divided into aliquots. The individual ARMS reactions are stable at least 6 months at $-20°C$.*

15. Using the information obtained from the pilot experiments to optimize reaction conditions, analyze test DNA samples by following steps 2 through 12 but replacing the control DNA with the test samples to be analyzed. If possible, include both negative (without DNA) and positive (known genotype) samples. If problems are encountered, refer to Table 9.7.4.

ALTERNATE PROTOCOL

ANALYSIS OF MULTIPLE MUTATIONS BY MULTIPLEX ARMS

It is often necessary to analyze a particular gene for the presence or absence of several different point mutations. One option is to carry out a series of single ARMS tests; the alternative approach—described in this protocol—is to develop a multiplex ARMS test that will allow simultaneous analysis of several mutations.

The approach is the same as that used for the single ARMS tests in that most of the variables are standardized and only primer sequences and concentrations are altered to achieve the desired result. The principle difference is that there are several primer combinations to be optimized simultaneously, which increases the complexity of the procedure. In view of this increased complexity it is not possible to outline a standard protocol for multiplex ARMS test development. The basic steps in the protocol are the same as for the single test development with the following differences:

1. In general, use less destabilizing mismatches in an ARMS test that is part of a multiplex than if it had been a single ARMS test. Use Table 9.7.1 (mismatch selection) followed by Table 9.7.3 (mismatch strength) to select less destabilizing mismatches.

2. The different ARMS products in the reaction are distinguished on the basis of size. It is best to plan the ARMS reactions so there is a ≥50-bp difference in the ARMS products' sizes. In some cases it will be necessary to use a control PCR amplification that gives a larger product; two alternatives are listed among the oligonucleotide primers (see control primer recipe in APPENDIX 1).

3. Combine several ARMS reactions into two multiplex reactions. Split the normal and mutant reactions so both multiplex reactions contain normal ARMS primers for some reactions and mutant ARMS primers for others.

4. In testing for an autosomal recessive inherited disease gene, the PCR control reaction can be omitted if at least four mutations are analyzed simultaneously and both reaction tubes contain at least two normal specific ARMS reactions.

5. If the multiplex test is for several mutations in close proximity in the same exon, it is possible to use a single common primer in combination with a number of ARMS primers.

6. The multiplex protocol is essentially the same as for the single ARMS tests, except that it may be necessary to repeat the full procedure (steps 1 to 15) multiple times because there are several amplification reactions to optimize simultaneously.

SUPPORT PROTOCOL

RAPID DNA EXTRACTION FROM MOUTHWASH AND BLOOD SAMPLES

The following protocol describes a rapid DNA extraction method for blood and mouthwash samples that provides DNA compatible with the ARMS reaction conditions described in this unit.

Materials *(see APPENDIX 1 for items with ✓)*

    170 mM ammonium chloride (prepare fresh)
    Blood, fresh or frozen, with or without anticoagulants *or* mouthwash sample
    10 mM NaCl/10 mM EDTA
    4% (w/v) sucrose in H_2O
    50 mM sodium hydroxide
✓ 1 M Tris·Cl, pH 7.5

    1.5-ml screw-cap microcentrifuge tube
    Rotator
    25-ml plastic sample tube, sterile
    Low-speed centrifuge
    Boiling water bath

To extract DNA from blood

1a. Add 800 µl freshly prepared 170 mM ammonium chloride to 200 µl blood in a 1.5-ml screw-cap microcentrifuge tube. Mix 20 min on a rotator. Microcentrifuge 2 min at high speed to obtain a white cell pellet. Discard supernatant.

2a. Wash cell pellet with 300 µl of 10 mM NaCl/10 mM EDTA. Microcentrifuge 15 sec at high speed to pellet cells. Repeat wash three times to remove all visible hemoglobin.

To extract DNA from mouthwash samples

1b. Vigorously agitate 10 ml of 4% (w/v) sucrose in the mouth for 20 sec to produce a suspension of buccal epithelial cells. Collect suspension in a sterile 25-ml plastic sample tube. Centrifuge 10 min at $1200 \times g$, room temperature, to collect cells. Discard supernatant.

> *To avoid any potential contamination problems, it is recommended that food and drink be avoided in the 30 min prior to sampling.*

2b. Resuspend cells in 500 µl of 10 mM NaCl/10 mM EDTA and transfer to screw-cap microcentrifuge tubes. Microcentrifuge 15 sec at high speed and discard supernatant.

3. Resuspend pellet in 500 µl of 50 mM sodium hydroxide and vortex 10 sec to resuspend white cell pellet. Incubate 20 min in boiling water bath.

4. Add 100 µl of 1 M Tris·Cl to neutralize and vortex 5 sec.

5. Microcentrifuge 15 sec at high speed to remove cell debris. Store supernatant (5 µl contains ~100 ng DNA, sufficient for one ARMS reaction) at −20°C until use (up to −20°C).

References: Ferrie et al., 1992; Newton et al., 1989

Contributor: Stephen Little

Table 9.7.4 Troubleshooting Guide

Problem	Possible cause	Solution
ARMS and control bands faint and/or absent in all reactions	Inactive enzyme	Use fresh batch of *Taq* DNA polymerase
	Component omitted from reaction	Repeat ARMS from reaction preparation
ARMS bands faint and/or absent in all reactions; control bands normal	Failure to denature high-G+C ARMS product	Increase denaturation temperature to 96°C
		Add 10% (v/v) glycerol to mix
	Insensitive ARMS test	Increase primer concentration, decrease mismatch strength
ARMS and control bands faint and/or absent in some reaction pairs	Sample-borne PCR inhibition	Dilute sample 1:100
	Insufficient DNA in affected samples	Add 10× more sample
ARMS bands absent in some reactions; control bands normal	Control reaction too efficient	Reduce concentration of control primers
Faint additional bands same size as ARMS products	Contamination	Check negative controls
	Excess enzyme	Check enzyme dilution
	Nonspecific ARMS test	Reduce primer concentration, increase mismatch strength
Faint additional bands, different size from ARMS products	Excess enzyme	Check enzyme dilution
	Cooling artifacts	Allow reaction tubes to cool to room temperature in PCR block
	Spurious priming elsewhere in genome	Identify specific primer(s) causing problem and redesign
Streakiness in some lanes	Excess DNA	Add less DNA
PCR products ~50–60 bp	Primer dimers	Ensure hot start is used, check 3′ ends of primers for complementarity
First lanes on gel give extra bands; later lanes give faint bands	Insufficient mixing of enzyme dilution	Increase mixing

Molecular Analysis of Oxidative Phosphorylation Diseases for Detection of Mitochondrial DNA Mutations

Oxidative phosphorylation (OXPHOS) diseases are caused by inherited or spontaneously occurring mutations in the mitochondrial (mtDNA) or nuclear DNA. To date, 17 mtDNA mutations with confirmed pathogenicity and 33 mutations with provisional pathogenicity are known (Wallace et al., 1995; Table 9.8.1). The complexities associated with the diagnosis of OXPHOS diseases, and the selection of mtDNA mutations and of specific tissues for analysis, are beyond the scope of this unit. Refer to Shoffner and Wallace (1995) for a detailed discussion.

BASIC PROTOCOL 1

SCREENING FOR MITOCHONDRIAL DNA REARRANGEMENTS BY SOUTHERN BLOT HYBRIDIZATION

The detection of mtDNA rearrangements is important in the diagnosis of various classes of OXPHOS diseases.

NOTE: Banding patterns produced by supercoiled mtDNA, polymorphisms, and the wide array of possible breakpoints can confound test interpretation; see Wallace et al. (1995) for a catalog of mtDNA rearrangements and mtDNA polymorphisms. Confirming the identity of abnormally sized molecules is essential to correct interpretation of the Southern blot analysis.

Materials (see APPENDIX 1 for items with ✓)

    Genomic DNA from skeletal muscle (*UNIT 10.4*) or leukocytes (*APPENDIX 3A*)
    10× restriction buffer (supplied with restriction enzyme)
    *Bam*HI and *Eco*RV restriction endonucleases (8 to 12 U/µl; Life Technologies)
    5.0 M sodium chloride (NaCl)
    100% and 70% ethanol
✓ 6× gel loading buffer with Ficoll
    0.8% or 1% agarose gel
    DNA molecular weight markers: e.g., kilobase ladder and λ *Hin*dIII digest (Life Technologies; size ranges 75 to 12,216 bp and 564 to 23,130 bp, respectively)
✓ 1× TBE buffer
✓ Ethidium bromide solution
✓ Prehybridization/hybridization buffer
✓ 2× SSC (prepare fresh)
    Salmon sperm DNA (e.g., Stratagene)
    >1 ×10^8 cpm/µg [α-^{32}P]dCTP mtDNA probe (see Support Protocol)
✓ 0.5×, 1×, and 2× SSC/0.1% (w/v) SDS (prepare fresh), 65°C
    Centrifuge (~20,000 × *g*)
    55°C heat block
    14 × 11 × 0.6–cm and 25 × 20 × 0.6–cm gel electrophoresis apparatuses (e.g., Life Technologies)

Table 9.8.1   mtDNA Mutation Summary[a]

HGM mutation designation[b,c]	Nucleotide position	Gene/AA	Homo-plasmy[d]	Hetero-plasmy[d]	Status[e]
MTRNR1∗ADPD956–965 insertion	5-bp insertion between 956 and 965	12S rRNA	+	−	Provisional
MTRNR1∗DEAF1555G	A to G: 1555	12S rRNA	+	−	Confirmed
MELAS:1642A	G to A: 1642	tRNA^{Val}	−	+	Provisional
MTRNR2∗ADPD3196A	G to A: 3196	16S rRNA	+	+	Provisional
MTTL1∗MELAS3243G	A to G: 3243	tRNA$^{Leu(UUR)}$	−	+	Confirmed
MTTL1∗MM3250C	T to C: 3250	tRNA$^{Leu(UUR)}$	−	+	Provisional
MTTL1∗MM3251G	A to G: 3251	tRNA$^{Leu(UUR)}$	−	+	Confirmed
MTTL1∗MELAS3252G	A to G: 3252	tRNA$^{Leu(UUR)}$	−	+	Provisional
MTTL1∗MELAS3256T	C to T: 3256	tRNA$^{Leu(UUR)}$	−	+	Confirmed
MTTL1∗MMC3260G	A to G: 3260	tRNA$^{Leu(UUR)}$	−	+	Confirmed
MTTL1∗MELAS3271C	T to C: 3271	tRNA$^{Leu(UUR)}$	−	+	Provisional
MTTL1∗PEM3271Δ	Single-bp deletion: 3271	tRNA$^{Leu(UUR)}$	−	+	Provisional
MTTL1∗MELAS3291C	T to C: 3291	tRNA$^{Leu(UUR)}$	−	+	Provisional
MTTL1∗MM3302G	A to G: 3302	tRNA$^{Leu(UUR)}$	−	+	Confirmed
MTTL1∗MMC3303T	C to T: 3303	tRNA$^{Leu(UUR)}$	+	+	Provisional
MTND1∗NIDDM3316A	G to A: 3316	ND1: Ala to Thr	+	−	Provisional
MTND1∗LHON3394C	T to C: 3394	ND1: Tyr to His	+	−	Provisional
MTND1∗ADPD3397G	A to G: 3397	ND1: Met to Val	+	−	Provisional
MTND1∗LHON3460A	G to A: 3460	ND1: Ala to Thr	+	+	Confirmed
MTND1∗LHON4136G	A to G: 4136	ND1: Tyr to Cys	+	−	Provisional
MTND1∗LHON4160C	T to C: 4160	ND1: Leu to Pro	+	−	Provisional
MTND1∗LHON4216C	T to C: 4216	ND1: Tyr to His	+	−	Confirmed
MTTI∗FICP4269G	A to G: 4269	tRNA^{Ile}	−	+	Provisional
CPEO:4285C	T to C: 4285	tRNA^{Ile}	−	+	Provisional
MTTI∗MICM4300G	A to G: 4300	tRNA^{Ile}	−	+	Provisional
MTTI∗FICP4317G	A to G: 4317	tRNA^{Ile}	?	?	Provisional
CME:4320	C to T: 4320	tRNA^{Ile}	−	+	Provisional
MTTQ∗ADPD4336C	T to C: 4336	tRNA^{Gln}	+	−	Provisional
MTND2∗LHON4917G	A to G: 4917	ND2: Asp to Asn	+	−	Provisional
MTND2∗LHON5244A	G to A: 5244	ND2: Gly to Ser	−	+	Provisional
MTTW∗DEMCH5549A	G to A: 5549	tRNA^{Trp}	−	+	Provisional
MTTN∗CPEO5692G	A to G: 5692	tRNA^{Asn}	−	+	Provisional
MTTN∗CPEO5703G	G to A: 5703	tRNA^{Asn}	−	+	Provisional
MTCOI∗LHON7444A	G to A: 7444	COI: termination to Lys	+	−	Provisional
MTCOI∗DEAF7445G	A to G: 7445	COI or tRNA$^{Ser(UCN)}$	+	+	Confirmed
MTTS1∗AMDF7472+C	single bp insertion (C): 7472	tRNA$^{Ser(UCN)}$	+	+	Provisional
MTTS1∗MERME7512C	T to C: 7512	tRNA$^{Ser(UCN)}$	+	+	Provisional
EM: 8272−8289 insertion (9 bp)	9 bp (C)$_5$ TCTA insertion between 8272 and 8289	Non-coding region between COII and tRNA^{Lys}	?	+	Provisional
MTTK∗MERRF8344G	A to G: 8344	tRNA^{Lys}	−	+	Confirmed
MTTK∗MERRF8356C	T to C: 8356	tRNA^{Lys}	−	+	Confirmed
SN: 8851C	T to C: 8851	ATP6: Trp to Arg	−	+	Provisional

continued

Table 9.8.1 mtDNA Mutation Summary[a], continued

HGM mutation designation[b,c]	Nucleotide position	Gene/AA	Homo-plasmy[d]	Hetero-plasmy[d]	Status[e]
MTATP6∗NARP8993C	T to C: 8993	ATP6: Leu to Pro	−	+	Confirmed
MTATP6∗NARP8993G	T to G: 8993	ATP6: Leu to Arg	−	+	Confirmed
MTATP6∗LHON9101C	T to C: 9101	ATP6: Ile to Thr	+	−	Provisional
MTATP6∗FBSN9176C	T to C: 9176	ATP6: Leu to Pro	+	+	Provisional
MTCO3∗LHON9438A	G to A: 9438	COIII: Gly to Ser	+	−	Provisional
CrMy: 9487–9501 deletion	15-bp deletion: 9487–9501	COIII	−	+	Provisional
MTCO3∗LHON9738T	G to T: 9738	CoIII: Ala to Thr	+	−	Provisional
MTCO3∗LHON9804A	G to A: 9804	COIII: Ala to Thr	+	−	Provisional
MTCO3∗PEM9957C	T to C: 9957	COIII: Phe to Leu	−	+	Provisional
MTTG∗MHCM9997C	T to C: 9997	tRNA^{Gly}	?	+	Provisional
MTTG∗CIPO10006G	A to G: 10,006	tRNA^{Gly}	?	?	Provisional
SD: 10044G	A to G: 10,044	tRNA^{Gly}	−	+	Provisional
LHON+SD: 11696G	A to G: 11,696	ND4: Val to Ile	+	+	Provisional
MTND4∗LHON11778A	G to A: 11,778	ND4: Arg to His	+	+	Confirmed
MTTS1∗CIPO12246G	C to G: 12,246	tRNA$^{Ser(AGY)}$	?	?	Provisional
MTTL2∗CPEO12311C	T to C: 12,311	tRNA$^{Leu(CUN)}$	+	+	Provisional
CPEO: 12315A	G to A: 12,315	tRNA$^{Leu(CUN)}$	−	+	Provisional
MTND5∗LHON13708A	G to A: 13,708	ND5: Ala to Thr	+	−	Confirmed
MTND5∗LHON13730A	G to A: 13,730	ND5: Gly to Glu	−	+	Provisional
MTND6∗LDYT14459A	G to A: 14,459	ND6: Ala to Val	+	+	Confirmed
MTND6∗LHON14484C	T to C: 14,484	ND6: Met to Val	+	+	Confirmed
MTTE∗MDM14709G	A to G: 14,709	tRNA^{Glu}	−	+	Provisional
MTCYB∗LHON15257A	G to A: 15,257	Cytochrome *b:* Asp to Asn	+	−	Confirmed
MTCYB∗LHON15812A	G to A: 15,812	Cytochrome b: Val to Met	+	−	Confirmed
MTTT∗LIMM15923G	A to G: 15,923	tRNA^{Thr}	?	+	Confirmed
MTTP∗MM15990T	C to T: 15,990	tRNA^{Pro}	−	+	Provisional

[a] Refer to Wallace et al. (1995) for primary references for each of these mtDNA mutations.

[b] At the time of submission, HGM designations (MITOMAP, *http://Infinity.gen.emory.edu/mitomap.html*) were not assigned for the following: MELAS:1642A (Taylor et al., 1996), CPEO:4285C (Silvestri et al., 1996), CME:4320C (Santorelli et al., 1995), EM:8272-8289 insertion (9 bp) (Fabrizi et al., 1995), CrMy:9487-9501 deletion (Keightly et al., 1996), SD:10044G (Santorelli et al., 1996), LHON+SD:11696G (De Vries et al., 1996), CPEO:12315A (Fu et al., 1996).

[c] Human Gene Map designations are given for each mutation. In order of appearance in the designation, the first descriptor is the chromosome (MT), the second is the mtDNA gene in which the mutation is located, the third is the phenotype designation (see below), the fourth is the mutation location, and the fifth is the nucleotide change. For example, the designation for the mutation that causes myoclonic epilepsy and ragged-red fiber disease is MTTK∗MERRF8344G. The five components of this descriptor are placed in parentheses and indicated by superscripts: (MT)[1](TK)[2]∗(MERRF)[3](8344)[4](G)[5]. HGM, Human Gene Map; AA, amino acid change in the polypeptide. Phenotype designations: AD, Alzheimer's disease; ADPD, Alzheimer's disease plus Parkinson's disease; CIPO, chronic intestinal pseudoobstruction with myopathy and ophthalmoplegia; CME, cardiomyopathy and encephalopathy; CPEO, chronic progressive external ophthalmoplegia; CrMy, cramps and myalgias; DEAF, maternally inherited deafness or aminoglycoside-induced deafness; DMDF, diabetes mellitus and deafness; EM, encephalomyopathy; FICP, fatal infantile cardiomyopathy plus a MELAS-associated cardiomyopathy; LDYT, Leber's hereditary optic neuropathy plus dystonia; LHON, Leber's hereditary optic neuropathy; LHON+SD, Leber hereditary optic neuropathy and spastic dystonia; LIMM, lethal infantile mitochondrial myopathy; MELAS, mitochondrial encephalomyopathy, lactic acidosis, and stroke-like episodes; MERRF, myoclonic epilepsy and ragged-red fiber disease; MM, mitochondrial myopathy; MMC, maternally inherited myopathy and cardiomyopathy; NARP, neurogenic muscle weakness, ataxia, and retinitis pigmentosa; PEM, progressive encephalomyopathy; SD, sudden death.

[d] To assist in interpreting the results of mtDNA mutation analysis, the status of heteroplasmy or homoplasmy is designated by + when present, − when cases have not been identified, and ? when the study did not determine whether the mutation was heteroplasmic or homoplasmic.

[e] Provisional, a single pedigree has been identified that harbors the mutation; confirmed, at least two independent pedigrees are known that harbor the mtDNA mutation.

MP4 camera (Polaroid) and type 55 film or Eagle Eye II still video imaging system (Stratagene)

Fluorescent ruler

Zeta-probe blotting membrane (Bio-Rad)

Positive-pressure DNA transfer apparatus (Stratagene)

UV Stratalinker (Stratagene)

Blotting paper (Schleicher & Schuell)

60° to 80°C hybridization oven with rotisserie (Hybaid) and appropriate hybridization bottles

PhosphorImager (Molecular Dynamics) *or* Kodak X-Omat AR film (for conventional autoradiography)

1. Add ~2.0 to 2.5 μg skeletal muscle genomic DNA to a labeled 1.5-ml microcentrifuge tube. If only leukocyte genomic DNA is available, use 4.0 to 4.5 μg.

2. Select either *Bam*HI or *Eco*RV for the analysis of each sample (see Table 9.8.2 for restriction sites, fragment sizes, and commonly encountered polymorphisms). Add 40 μl of the appropriate 10× restriction buffer. Bring volume to 400 μl with sterile deionized, distilled water.

3. Add 20 U (~2 μl) of the selected restriction enzyme and incubate overnight (16 to 20 hr) at 37°C. The next morning, add an additional 20 U enzyme and continue incubation an additional 2 hr at 37°C.

4. Precipitate DNA by adding 10 μl of 5.0 M NaCl and 800 μl of 100% ethanol. Mix contents of microcentrifuge tubes by inversion and incubate ≥1 hr at 4°C.

5. Centrifuge samples 1 hr at ~20,000 × g, 4°C. Decant supernatant and wash pellet with 1.0 ml of 70% ethanol. Centrifuge 30 min at 20,000 × g, 4°C. Decant supernatant and dry samples using a vacuum desiccator.

6. Resuspend DNA in 30 μl sterile deionized, distilled water. If necessary, heat 30 to 90 min in a 55°C heat block to hasten resuspension.

7. Add 6 μl of 6× Ficoll loading buffer, vortex, and spin briefly.

8. Load samples (~36 μl total volume) onto a 0.8% (w/v) agarose gel (25 × 20 × 0.6–cm) for samples digested with *Bam*HI or a 1.0% (w/v) agarose gel (14 × 11 × 0.6–cm) for samples digested with *Eco*RV. Include DNA molecular weight markers in separate lanes. Use 1× TBE without ethidium bromide as the electrophoresis buffer.

9. To maximize detection of small rearrangements, electrophorese *Bam*HI digestions at ~0.35 V/cm until the 1 2-kb marker has migrated 12 to 14 cm (~36 hr). Electrophorese the *Eco*RV digestions at ~0.35 V/cm until the 1.0-kb marker is ~1 cm from the end of the gel (~18 hr).

10. When electrophoresis is complete, record image of size marker migration photographically (Polaroid MP4 camera with type 55 film) or digitally (Stratagene Eagle Eye II). Align a fluorescent ruler with size standards to record migration distances and to compare the migration of the size standards with the mtDNA migration on the autoradiogram.

11. Depurinate, denature, and neutralize the gel using standard methods.

12. Transfer the DNA to a Zeta-probe membrane using a positive-pressure blotting apparatus according to the manufacturer's instructions. Confirm that transfer of DNA from the gel to the membrane is successful by staining with ethidium bromide and inspecting for fluorescence by transillumination with UV light.

13. Cross-link DNA to the membrane using a UV light source (Stratalinker).

14. Heat the hybridization oven and the prehybridization solution to 65°C. Wet membrane in 2× SSC, apply mesh supplied with hybridization oven to the membrane to ensure access of buffer, and insert membrane into a glass hybridization bottle. Refer to manufacturer's instructions for complete details of this procedure.

15. Denature sonicated salmon sperm DNA 10 min at 98°C and add to prewarmed prehybridization solution (20 ml per membrane is generally sufficient) at a final concentration of 10 μg/ml. Add the prehybridization solution to the glass bottle containing the membrane and rotate 2 to 4 hr at 65°C.

16. Denature [α-^{32}P]dCTP-labeled mtDNA probe (~5 × 10^5 to 5 × 10^6 cpm/ml hybridization buffer) and the sonicated salmon sperm DNA (final concentration 10 μg/ml) by heating 5 min at 95°C.

17. Warm 11 ml (Hybaid small hybridization bottle) or 25 ml (Hybaid large hybridization bottle) of hybridization solution to 65°C, then add the denatured, sonicated salmon sperm and the mtDNA probe. Replace the prehybridization solution with the hybridization solution in the glass bottle and rotate 18 to 20 hr at 65°C.

18. Wash membrane three times with 2× SSC/0.1% SDS for 30 to 60 min each time. Wash with 1× SSC/0.1% SDS for 30 to 60 min, then wash with 0.5× SSC/0.1% SDS for 30 to 60 min. Perform all washes at 65°C.

19. Rinse the membrane in distilled, deionized water and seal in a plastic pouch. Perform autoradiography using Kodak X-Omat AR film (1 to 3 days at −20°C) or PhosphorImager analysis (few hours following manufacturer's recommendations).

Table 9.8.2 mtDNA Polymorphisms Encountered with *Eco*RV and *Bam*HI

Common fragments produced	Polymorphisms
EcoRV (recognition site GATATC)	
Positions 12,871 to 3179: 6877 bp	16,274 (G→A): site gain The 6877-bp fragment is digested into two fragments or 3403 bp (positions 16,274 to 12,871) and 3474 bp (positions 16,274 to 3179).
	13,404 (T→C): site gain The 6877-bp fragment is digested into two fragments of 6344 bp (positions 13,404 to 3179) and 533 bp (positions 12,871 to 13,404).
Positions 12,871 to 6734: 6137 bp	6734 (G→A): site loss The restriction site producing the 6137- and 3555-bp fragments is lost, resulting in a 9692-bp fragment (positions 3179 to 12,871).
Positions 6734 to 3179: 3555 bp	—
BamHI (recognition site: GGATCC)	
Position 14,258, producing a linearized mtDNA molecule: 16,569 bp	16,491 (T→G): site gain or 16,505 (T→A): site gain
	16,389 (G→A): site gain When the 14,258 restriction site is present, two fragments are produced of 2141 bp (positions 16,389 to 14,258) and 14,428 bp (positions 14,258 to 16,389); this is a commonly encountered polymorphism in Caucasian populations.
	14,258: site loss
	13,368 (G→A): site gain

MITOCHONDRIAL DNA PROBE PREPARATION USING LONG-RANGE PCR

The mtDNA primers (Cheng et al., 1994) used for this amplification are located within the cytochrome *b* gene. Only 300 bp of this gene are excluded from the amplification.

Additional Materials (*also see Basic Protocol 1; see* APPENDIX 1 *for items with* ✓)

✓ 10 mM 4dNTP mixture
  Primer 1: 5′-TGAGGCCAAATATCATTCTGAGGGGC-3′
  Primer 2: 5′-TTTCATCATGCGGAGATGTTGGATGG-3′
  Ampliwax PCR Gem 50 (Perkin-Elmer)
  Expand Long Template PCR System (Boehringer Mannheim) containing 1.75 U/μl *Taq* DNA polymerase/*Pwo* DNA polymerase enzyme mixture and PCR buffers
  Total genomic DNA (APPENDIX 3A, UNIT 10.4) from leukocytes or skeletal muscle of normal individual (~1 μl)
  DNA molecular weight markers: e.g., λ *Hind*III digest (Life Technologies; size range 564 to 23,130 bp)
✓ 1× TBE buffer containing 0.5 μg/ml ethidium bromide
  Qiaex II gel extraction kit (Qiagen)
  [α-^{32}P]dCTP (3000 mCi/mmol; Amersham)
  High Prime DNA labeling kit (Boehringer Mannheim)

1. Prepare the PCR reaction mixture by combining the following reagents (total volume 30 μl per reaction) in a 0.5-ml microcentrifuge tube:

    1.5 μl 10 mM 4dNTP mixture
    0.5 μl 20 μM primer 1
    0.5 μl 20 μM primer 2
    27.5 μl H$_2$O.

    For <20 samples, prepare sufficient reaction mixture (minus Taq DNA polymerase/Pwo DNA polymerase mixture) for n + 1 samples, then divide into separate tubes.

2. Add one Ampliwax PCR Gem 50 to each tube. Place tube in thermal cycler and heat to 80°C for 10 min to melt wax. Allow tube to cool until wax hardens. Add the following reagents on top of the wax (20 μl total volume):

    5 μl 10× PCR amplification buffer
    50 to 250 ng total genomic DNA from normal individual (~1 μl)
    0.5 μl 1.75 U/μl *Taq* DNA polymerase/*Pwo* DNA polymerase mixture
    13.5 μl H$_2$O.

3. Place the tube in the thermal cycler and amplify the mtDNA using the following parameters:

Initial step:	120 sec	94°C	(heating)
34 cycles:	10 sec	94°C	(denaturation)
	10 min	68°C	(annealing and elongation)
Cycle 35:	10 sec	94°C	(denaturation)
	4 min	72°C	(annealing and elongation)

4. Add 5 μl of 6× Ficoll loading buffer to 25 μl PCR reaction. Load sample (30 μl final volume) onto a 0.8% (w/v) agarose gel (14 × 11 × 0.6–cm) and electrophorese at

~0.35 V/cm. Include DNA molecular weight markers in separate lanes. Use 1× TBE containing ~0.5 μg/ml ethidium bromide as the electrophoresis buffer.

λ HindIII is a useful marker for identification of the 16.3-kb mtDNA fragment.

5. Identify the 16.3-kb mtDNA fragment by ultraviolet fluorescence of the ethidium bromide-stained bands. Remove the region of agarose containing the 16.3-kb mtDNA band and purify the fragment from the gel using the Qiaex II Gel Extraction Kit.

6. To prepare radiolabeled mtDNA probe, label with [α-^{32}P]dCTP (to a final specific activity >1 × 10^8 cpm/μg) using the High Prime DNA Labeling Kit (Boehringer Mannheim) according to manufacturer's instructions or by random oligonucleotide-primed synthesis (see *APPENDIX 3E*).

BASIC PROTOCOL 2

SCREENING FOR MITOCHONDRIAL DNA POINT MUTATIONS BY RESTRICTION ANALYSIS OF PCR PRODUCTS

mtDNA point mutations often produce (site gain, SG) or abolish (site loss, SL) a naturally occurring restriction endonuclease site, making mutation detection straightforward. When the mutation does not alter a naturally occurring restriction site, an oligonucleotide primer that contains a mismatch with the template sequence is used to introduce an informative restriction site into the PCR-amplified mtDNA fragment (Fig. 9.8.1).

NOTE: Preventing contamination of the PCR is essential to accurate detection of mtDNA mutations, particularly when present at low copy numbers in patient tissues. Use appropriate precautions, including washing of pipets with 10% Clorox and UV irradiating all tubes and pipets. In addition, mtDNA is highly polymorphic, and a detailed understanding of mtDNA polymorphisms is often necessary for test interpretation.

Materials (*see APPENDIX 1 for items with ✓*)

    10× PCR amplification buffer: 100 mM Tris·Cl, pH 8.3 (*APPENDIX 1*)/50 mM KCl/25 mM MgCl$_2$
✓ 1.25 mM 4dNTP mixture
    10 pmol/μl forward and reverse oligonucleotide primers (see Table 9.8.3)
    5 U/μl *Taq* DNA polymerase
    Genomic DNA from leukocytes (*APPENDIX 3A*) or skeletal muscle (*UNIT 10.4*), or mitochondrial DNA from platelets
    Mineral oil (nuclease free)
    Agarose gel
    10× restriction buffer (supplied with restriction enzymes)
    BSA
    Spermidine
    Restriction endonucleases for mutation detection (see Table 9.8.3)
✓ 6× gel loading buffer with Ficoll
    DNA molecular weight markers: e.g., DNA marker V (Boehringer Mannheim; size range 11 to 587 bp)
✓ 1× TBE buffer
✓ Ethidium bromide solution

    UV Stratalinker (Stratagene)
    Thermal cycler (Perkin-Elmer)

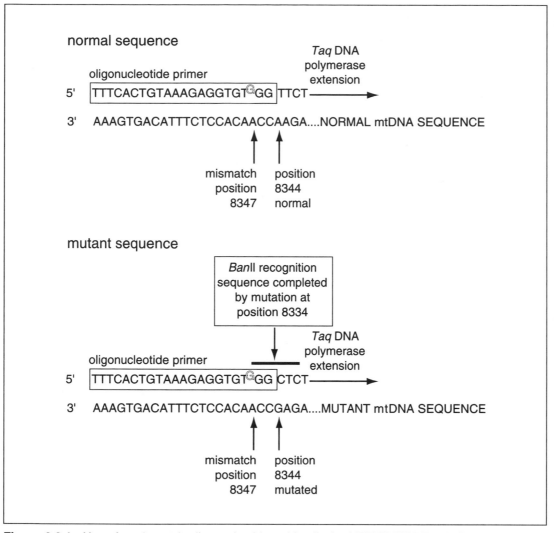

Figure 9.8.1 Use of a mismatch oligonucleotide to identify the MTTK*MERRF8344G mutation. Illustrated is the initial amplification cycle, in which an oligonucleotide is used to introduce a base change at position 8347 of the mtDNA. In normal mtDNA molecules from the patient, introduction of the base change does not introduce a *Ban*II restriction site into the amplified fragment. However, when the MTTK*MERRF8344G mutation is present at position 8344, a *Ban*II site (GPGCQC) is produced, thus allowing recognition of the mutation in the patient sample.

1. Prepare the PCR reaction mixture by combining the following reagents (total volume 48.75 μl per reaction):

   5 μl 10× PCR amplification buffer
   8 μl 1.25 mM 4dNTP mixture
   1.5 μl 10 pmol/μl forward primer
   1.5 μl 10 pmol/μl reverse primer
   32.75 μl H₂O.

   Irradiate the reaction mixture with UV light (254 nm) for 5 min.

   For <20 samples, prepare sufficient reaction mixture (minus Taq DNA polymerase) for n + 1 samples. For >20 samples, prepare reaction mixture for n + 2 samples.

Table 9.8.3 Testing Parameters for mtDNA Mutations[a]

HGM mutation designation	Primer 1 (forward) (5′→3′)	Primer 2 (reverse) (5′→3′)	Annealing temperature (°C)	Restriction endonuclease[b]	Gels for mtDNA mutation detection
12S 956–965 insertion	GCCTTTCTCTATTAGCTCTTAG (np 664–683)	AAGAGGTGGTGAGGTTGATC (np 1227–1246)	51	AluI (5-bp insertion)	8% polyacrylamide
MTRNR1*DEAF1555G	AGACGTTAGGTCAAGGTG (np 1319–1336)	GTTTAGCTCAGAGCGGTC (np 1677–1660)	50	BsmAI (SL)	2% SeaKem agarose
MELAS:1642A	CCCACTCCCATACTACT (np 461–477)	TTGCGGTACTATATCTAT (np 1773–1790)	50	MboII (SG)	4% polyacrylamide
MTRNR2*ADPD3196A	CCCGATGGTGCAGCCGC (np 3007–3023)	GGCTACTGCTCGCAGTG (np 3701–3717)	55	SspI (SG)	2% SeaKem agarose
MTTL1*MELAS3243G	CCCGATGGTGCAGCCGC (np 3007–3023)	GCATTAGGAATGCCATTGCG (np 3351–3370)	55	HaeIII (SG)	8% polyacrylamide
MTTL1*MM3250C	GGTTTGTTAAGATGGCAGAG[G]CCGG (np 3225–3249)	CACGTTGGGGCCTTTGCGTA (np 3404–3423)	45	NaeI (SG)	8% polyacrylamide
MTTL1*MM3251G	CCCGATGGTGCAGCCGC (np 3007–3023)	GGCTACTGCTCGCAGTG (np 3701–3717)	51	MboI (SG)	2% SeaKem agarose
MTTL1*MELAS3252G	ACATGCTAAGACTTCACCAG (np 2851–2870)	TGTAAAGTTTTAAGTTTTATG[T]GA (np 3253–3276)	51	MaeIII (SG)	8% polyacrylamide
MTTL1*MELAS3256T	TGGCAGAGCCGGTAA[A]CG (np 3237–3255)	CTAGGGTGACTTCATATGAG (np 3726–3745)	51	MaeII (SG)	8% polyacrylamide
MTTL1*MMC3260G	GAGCCCGGTAAATCGC[T]TA (np 3242–3259)	GGCTACTGCTCGCAGTG (np 3701–3717)	55	DdeI (SG)	8% polyacrylamide
MTTL1*MELAS3271C[c]	CCCGATGGTGCAGCCGC (np 3007–3023)	GAATTGAACCTCTGACT[C]TAA (np 3272–3292)	55	DdeI (SG)	8% polyacrylamide
MTTL1*PEM3271Δ[c]	CCCGATGGTGCAGCCGC (np 3007–3023)	GAATTGAACCTCTGACT[C]TAA (np 3272–3292)	55	DdeI (SG)	8% polyacrylamide
MTTL1*MELAS3291C	AAAACTTAAAAACTTTACAGTCAG-AGGTT[GG]AT (np 3259–3290)	GGGGTTCATAGTAGAAGAGCGATGG (np 3545–3569)	67	BamHI (SG)	8% polyacrylamide
MTTL1*MM3302G	TTCAAATTCCTCCCTGTACG (np 3108–3127)	GGCTACTGCTCGCAGTG (np 3701–3717)	55	DdeI (SG)	8% polyacrylamide
MTTL1*MMC3303T	TCAGAGGTTCAATTCCTCTTT[A]TTA (np 3278–3301)	GGCTAGAGGTGGCTAGAAT[T]AA (np 3615–3637)	51	AseI (SG)	8% polyacrylamide

MTND1*NIDDM3316A[d]	AAGGTTCGTTTGTTCAACGA (np 3029–3048)	AGCGAAGGGTTGTGTAGTAGCC (np 3437–3456)	55	HaeIII (SL)	8% polyacrylamide
MTND1*LHON3394C	GAGCCCGGTAATGCGTTA (np 3242–3259)	ccacaagcttGGCTACTGCTCGCAGTG (np 3701–3717 plus 10-bp tail)	51	HaeIII (SG)	8% polyacrylamide
MTND1*ADPD3397G	CCCGATGGTGCAGCCGC (np 3007–3023)	ccacaagcttGGCTACTGCTCGCAGTG (np 3701–3717 plus 10-bp tail)	51	RsaI (SG)	2% SeaKem agarose
MTND1*LHON3460A	TTCAAATTCTCCCTGTACG (np 3108–3127)	ccacaagcttGGCTACTGCTCGCAGTG (np 3701–3717 plus 10-bp tail)	51	BsaHI (SL)	2% SeaKem agarose
MTND1*LHON4136G	CGCCGCAGGCCCCTTCGCCC (np 3951–3970)	TTTCTTAGGTTTGAGGGGG[C]ATGC (np 4244–4267)	52	SphI (SG)	8% polyacrylamide
MTND1*LHON4160C[e]	CGCCGCAGGCCCCTTCGCCC (np 3951–3970)	TTTTTCATAGGAGGT[CC]ATG (np 4161–4180)	42	StyI (SG)	8% polyacrylamide
MTND1*LHON4216C	CTCAACCTAGGCCTCCTA (np 3598–3615)	ATTTTGGATTCTCAGGGATG (np 4351–4370)	51	NlaIII (SG)	2% SeaKem agarose
MTTI*FICP4269G	CGCCGCAGGCCCCTTCGCCC (np 3951–3970)	ACTCTATCAAAGTAACTCTTTTATCAGA[A]A (np 4270–4299)	55	SspI (SL)	8% polyacrylamide
CPEO:4285C	CCTACCACTCACCCTAGCATTAC (np 4185–4207)	GCGAGCTTAGCGCTGTGATGAGTG (np 4519–4542)	55	AluI (SG)	12% polyacrylamide
MTTI*MICM4300G	AGAGTTACTTTGATAG[G]G (np 4281–4298)	GTGCGAGCTTAGCGCTGTGATG (np 4523–4544)	50	HphI (SG)	8% polyacrylamide
MTTI*FICP4317G	GCATTACTTATATG.atatg (np 4201–4219)	GATGGTAGAGTAGATGACGG (np 4489–4508)	55	AflII (SG)	2% SeaKem agarose
CME: 4320T	CCTACCACTCACCCTAGCATTAC (np 4185–4207)	GTGCGAGCTTAGCGCTGTGATG (np 4523–4544)	60	MnlI(SG)	12% polyacrylamide
MTTQ*ADPD4336C	CGCCGCAGGCCCCTTCGCCC (np 3951–3970)	AAAGATGGTAGAGTAGATGACGG (np 4489–4511)	55	AvaII (SG)	2% SeaKem agarose
MTND2*LHON4917G	GCACCCCTCTGACATCC (np 4831–4847)	GTGTTAGTCATGTTAGCTTG (np 5174–5193)	51	MaeI (SG)	8% polyacrylamide
MTND2*LHON5244A	GCACCCCTCTGACATCC (np 4831–4847)	CGGTCGGCGAACATCAGTGG (np 5898–5917)	51	HpaII (SL)	2% SeaKem agarose
MTTW*DEMCH5549A	CTCATCGCCCTTACCACGCT (np 5454–5473)	GTATTGCAACTTACTGA[AAA]G (np 5550–5569)	50	HindIII (SL)	8% polyacrylamide
MTTN*CPEO5692G	AAATGACAGTTTGAACATAC (np 5409–5428)	GTTGATTAGGGTGCTTA[A]CT (np 5697–5722)	47	AluI (SL)	8% polyacrylamide
MTTN*CPEO5703G	GACTGCAAAACCCCACTCTG (np 5591–5610)	TGATTTTCATATTGAATTGC (np 5797–5816)	51	DdeI (SL)	8% polyacrylamide

continued

Table 9.8.3 Testing Parameters for mtDNA Mutations[a], continued

HGM mutation designation	Primer 1 (forward) (5'→3')	Primer 2 (reverse) (5'→3')	Annealing temperature (°C)	Restriction endonuclease[b]	Gels for mtDNA mutation detection
MTCOI*LHON7444A	CATACACCACATGAAACATCC (np 7240–7260)	GGGAAGTAGCGTCTTGTAG (np 7610–7628)	55	XbaI (SL)	2% SeaKem agarose
MTCOI*DEAF7445G	CTTCCCACAACACTTTCTCGG (np 7178–7198)	TGTAAAGGATGCGTAGGGATG (np 7821–7840)	55	XbaI (SL)	2% SeaKem agarose
MTTS1*AMDF7472+C	ACATAAAATCTAGACAAAAAGG AAGGAATCGAACC[A]CCC (np 7432–7471)	CTTCTATGATAGGGGAAGTAGCGTC (np 7616–7640)	60	XcmI (SG)	8% polyacrylamide
MTTS1*MERME7512C	ACCTGGAGTGACTATATGGATG (np 7375–7396)	GACAAAGTTATGAAATGGTTTTTC TAATACCTT[C]T[C]GA (np 7513–7550)	60	XhoI (SG)	8% polyacrylamide
EM: 8272–8289 insertion (9 bp)	Same as MTTK*MERRF8344G	Same as MTTK*MERRF8344G	50	BanII (insertion)	8% polyacrylamide
MTTK*MERRF8344G	GGTATACTACGGTCAATGCTCT (np 8155–8176)	TTTCACTGTAAAGAGAGGTGT[G]GG (np 8345–8366)	50	BanII (SG)	8% polyacrylamide
MTTK*MERRF8356C	GGTATACTACGGTCAATGCTCT (np 8155–8176)	ATTTAGTTGGGGCATTTCACT[C]TA (np 8357–8380)	55	XbaI (SG)	8% polyacrylamide
SN:8851	TAAACCTAGCCATGGCCATCCCC [G]TA (np 8825–8850)	GTGTTGTCGTGCAGGTAGAGGCTT [C]CT (np 9177–9203)	55	BsiWI (SG)	8% polyacrylamide
MTATP6*NARP8993G[f]	CCTAGCCATGGCCATCC (np 8829–8845)	CAGATAGTGAGGAAAGTTGA (np 9840–9859)	55	HpaII (SG)	2% SeaKem agarose
MTATP6*NARP8993C[f]	CCTAGCCATGGCCATCC (np 8829–8845)	CAGATAGTGAGGAAAGTTGA (np 9840–9859)	55	HpaII (SG)	2% SeaKem agarose
MTCO1*LHON9101C	Same as MTATP6*NAR8993G	Same as MTATP6*NAR8993G	55	HphI (SG)	2% SeaKem agarose
MTATP6*FBSN9176C	GCCCTAGCCCACTTCTTAC (np 8896–8914)	GTGTTGTCGTGCAGGTAGAGGCTT [C]CT (np 9177–9203)	60	ScrfI (SG)	12% polyacrylamide
MTCO3*LHON9438A	ATCCAAGCCTACGTTTTC (np 9151–9168)	CCGTAGTGGGGCGATTTA (np 9564–9581)	50	StuI (SL)	2% SeaKem agarose
MTCO3*LHON9738T	ACTCCTAAACACATCCG[G]ATT (np 9596–9616)	GGAGACTCGAAGTACTCTT[CC]GG (np 9739–9760)	55	BspEI (SG)	2% SeaKem agarose
MTCO3*LHON9804A	TCTCCCTTCACCATTTCC (np 9756–9773)	GAGTGATAGACGAAGTAG (np 9849–9866)	48	MaeIII (SG)	8% polyacrylamide
MTCO3*PEM9957C	CATTTTGTAGATGTGGTTTGA[G]TA (np 9933–9956) or GCTCAGCATAGTCTAATAGAAAA (np 9645–9667)	GGGCAATTTCTAGATCAAATAA TAAGAAGGT (np 10,239–10,269) or ATCAATAGATGGAGACATAC [CT]AA (np 9958–9981)	55	RsaI (SG) or DdeI (SG)	12% polyacrylamide

9

MTTG*MHCM9997C	AGCCGCCGCCTGATACTGGCAT (np 9914–9935)	GAATGTTGTCAAAACTAGTTAATTGG (np 10,022–10,047)	55	BfaI (SG)	8% polyacrylamide
MTTG*CIPO10006G	GGGTCTTACTCTTTTAGT[T]TAA (np 9984–10,005)	GTAGTAAGGCTAGGAGGGTG (np 10,088–10,107)	53	DraI (SG)	8% polyacrylamide
SD: 10044G	CTTTGGCTTCGAAGCGCGCGCCT (np 9902–9924)	GGGCAATTTCTAGATCAAATAATAA (np 10,245–10,269)	60	MaeII (SG)	12% polyacrylamide
MTND4*LHON11778A	CCCACCTTGGCTATCATC (np 11,141–11,158)	GGTAAGGCGAGGTTAGCG (np 11,851–11,868)	51	MaeIII (SG)	2% SeaKem agarose
MTTS1*CIPO12246G	GAACTGCTAACTCATGCCCCI[GA]GT (np 12,221–12,245)	CTTTTATTTGGAGTTGCACCA[G]AAT (np 12,309–12,334)	60	EcoRI (SG)	8% polyacrylamide
MTTL2*CPEO12311C	GTTCATACACCTATCCCCCA (np 12,070–12,089)	TACTTTTATTTGGAGTTGCAC[TT]AA (np 12,312–12,336)	51	AflII (SG)	8% polyacrylamide
CPEO: 12315	ACTCATGCCCCATGTCTAA (np 12,230–12,249)	TACTTTTATTTGGAGT[G]GCA (np 12,336–12,317)	50	BanI (SL)	8% polyacrylamide
MTND5*LHON13708A	CGCAGCAGTCTGCGCCC (np 13,197–13,213)	GGGGATTGTGCGGGTGTGTG (np 13,932–13,950)	55	BstNI (SL)	2% SeaKem agarose
MTND5*LHON13730A	TCATCGAAACGCAAACATA (np 13,520–13,539)	TGCTAGGGTAGAATCCGAGT (np 13,911–13,930)	51	Tsp5091 (SG)	2% SeaKem agarose
MTND6*LDYT14459A	ATGCCTCAGGATACTCCTCAATAGCC[G]TC (np 14,430–14,458)	CTTCTAAGCCTTCTCCTATT (np 14,595–14,615)	51	MaeIII (SG)	8% polyacrylamide
MTND6*LHON14484C	AAACAATGGTCAACCAG[G]AAC (np 14,191–14,210)	TTTTTTTAATTTATTTAGGGGG[CC]TG (np 14,485–14,509)	45	MvaI (SG)	2% SeaKem agarose
MTTE*MDM14709G	CACGGACTACAACCACGACCAAT [C]ATA (np 14682–14708)	GGAGGTCGATGAATGAGTGGT (np 14,790–14,810)	60	NdeI (SG)	12% polyacrylamide
MTCYB*LHON15257A	GAATCTGAGGAGGCTACTCA[T]TA (np 15,234–15,256)	GATCCGTTTCGTGCAAG (np 15,343–15,360)	51	MseI (SG)	8% polyacrylamide
MTCYB*LHON15812A	CTAAGCCAATCACTTTATTG (np 15,704–15,723)	TTCAATTAGGGAGATAGTTGG (np 15,845–15,865)	42	RsaI (SL)	8% polyacrylamide
MTTT*LIMM15923G	GCGGTTGTTGATGGGTGAGT (np 14947–14966)	CCACATCACTCGAGACGTAA (np 14,947–14,966)	60	AccI (SG)	2% SeaKem agarose
MTTP*MM15990T	GGCGACCCAGACAATTA (np 15,497–15,513)	TTTAAATTAGAATCTTA[T]CT[C]TG (np 15,991–16,013)	45	DdeI (SG)	8% polyacrylamide

[a] Refer to Wallace et al. (1995) for primary references for each of these mtDNA mutations. np, nucleotide position.

[b] SL, site loss; SG, site gain.

[c] DdeI analysis does not differentiate between these two mutations. The single nucleotide deletion mutation (MTTL1*PEM3271Δ) can be detected by mtDNA sequencing.

[d] The oligonucleotides used for the MTTL1*MELAS3243G reaction can be used. Digestion of the mtDNA fragment with HaeIII will detect both the MTTL1*MELAS3243G and MTND1*NIDDM3316A mutations.

[e] Add dimethyl sulfoxide to the PCR buffer to a final concentration of 1% in the standard buffer when testing for this mutation.

[f] When HpaII is used for mutation detection, a site gain is produced if either the TG or TC mutation is present at position 8993 of the mtDNA. To differentiate between these two mutations, digest the PCR fragment with AvaI. This restriction endonuclease produces a site gain only when the TG mutation is present.

2. Add *Taq* DNA polymerase (0.25 μl per reaction) to the reaction mixture, mix thoroughly by vortexing the solution, and pipet 49 μl mixture to the bottom of labeled 0.5-ml microcentrifuge tubes. Place tubes with reaction mixture in the Stratalinker and irradiate again with 240,000 J (a few seconds) of UV light (254 nm).

3. Add ~25 to 50 ng genomic DNA from leukocytes or skeletal muscle or 5 to 10 ng mtDNA from platelets to each tube. Also include a control without DNA.

   When testing blood samples, analysis of both leukocyte and platelet mtDNA from the same individual is useful to confirm the results as well as to screen for laboratory errors. Test results should be concordant in the leukocyte and platelet samples.

4. Overlay the PCR reaction mixture with 2 drops nuclease-free mineral oil and cap the microcentrifuge tube tightly.

5. Place the sample in a thermal cycler and perform the following amplification cycles:

Initial step:	120 sec	94°C	(heating)
30 cycles:	60 sec	94°C	(denaturation)
	60 sec	annealing temperature (see Table 9.8.3)	
	60 sec	72°C	(elongation)

6. Separate 10 to 15 μl of reaction product on an agarose or agarose/NuSieve (FMC Bioproducts) gel to confirm that the PCR generated an appropriately sized fragment, that the PCR control had no reaction product, and that the amplification did not generate additional bands of a size that would interfere with interpretation of the restriction endonuclease results.

7. Add ~10.0 to 15.0 μl amplified mtDNA fragment to a labeled 0.5-ml microcentrifuge tube.

8. Add the following reagents to the tube (for <20 samples, prepare sufficient mixture for $n+1$ samples; for >20 samples, prepare sufficient mixture for $n+2$ samples):

   2 μl appropriate 10× restriction buffer (Table 9.8.3)
   BSA and spermidine according to manufacturer's recommendations
   Sterile, deionized, distilled H_2O to 20 μl
   10 U selected restriction enzyme (Table 9.8.3).

   Incubate overnight (16 to 20 hr) at the temperature appropriate for the restriction enzyme. The next morning, add an additional 10 U enzyme and continue incubation for an additional 2 hr.

9. Add 4 μl of 6× Ficoll loading buffer, vortex, and centrifuge briefly.

10. Load sample (~30 μl total volume) onto an appropriate gel for electrophoresis (Table 9.8.3). Include size markers and a positive control on the gel during electrophoresis to confirm fragment sizes produced by mtDNA digestion with the restriction endonuclease. Use 1× TBE containing ~0.05 μg/ml ethidium bromide as the electrophoresis buffer.

References: Hammans et al., 1991; Wallace, et al., 1995

Contributor: John M. Shoffner

Single-Cell DNA and FISH Analysis for Application to Preimplantation Genetic Diagnosis

All analyses of specific gene loci described in this unit may employ PEP (see Support Protocol 1) as the initial amplification step prior to primary PCR. However, some of the analyses do not require PEP and these may begin with primary PCR. Only those for the factor VIII gene (hemophilia) and DMD gene include PEP as an integral part of the analyses.

Additional information on interphase FISH can be found in *UNITS 4.4 & 8.6.*

NOTE: The procedures for molecular analysis described in Basic Protocol 1 and its accompanying support protocols require extreme precautions to prevent contamination with foreign DNA. Technologists should wear disposable outer clothing, including caps, masks, shoe covers, and powder-free gloves to prevent contamination of the embryology and DNA amplification work areas. Aerosol-barrier pipet tips (e.g., Continental Lab Products) should be used in all pipetting steps.

BASIC PROTOCOL 1

ANALYSIS OF SPECIFIC GENE LOCI IN SINGLE DIPLOID CELLS

NOTE: Use HPLC-grade water (e.g., Fisher) to prepare all solutions and in all protocol steps as required.

Materials (see APPENDIX 1 for items with ✓)

Single cells or blastomeres in 0.5-ml tubes, lysed and neutralized with oil or wax overlay (see Support Protocols 2, 3, and 4) and corresponding control tubes
✓ 10× potassium-free amplification buffer
✓ 25 mM 4dNTP mix (Boehringer Mannheim)
20 μM forward and reverse primers (Table 9.9.1)
5 U/μl *Taq* DNA polymerase (AmpliTaq from Perkin-Elmer)
✓ 10× PCR amplification buffer containing 15 mM $MgCl_2$
Light mineral oil (Perkin-Elmer)
20 μM forward and reverse inner (nested) primers for secondary PCR (Table 9.9.2)
DNA fragments from known homozygous *CFTR* ΔF508 mutant genomic DNA *or* *HEXA* exon 11 +TATC mutant genomic DNA *and* from homozygous normal DNA, PCR-amplified (30 cycles starting with 20 to 50 ng genomic template) using the inner primers for the appropriate locus (see Table 9.9.2)
✓ 10× TBE buffer
✓ DNA gel loading buffer
Size marker: φX-174 plasmid digested with *Hae*III
✓ 0.1 μg/ml ethidium bromide
✓ 5× *Dde*I digestion buffer
Restriction enzymes: *Dde*I, *Bcl*I, and *Hae*III
10× One-Phor-All Buffer Plus (Pharmacia Biotech)
0.5-ml clarified polypropylene PCR reaction tubes (Out Patient Services)

Table 9.9.1 Primer Sequences and Concentrations for Primary PCR

Locus	Primer	Sequence (5′ to 3′)[a]	Final concentration[b] (μM)	Volume[b] for 10 reactions (μl)	Product size (bp)
CFTRΔF508	1F	GACTTCACTTCTAATGATGAT	0.8	20	193/190
	1R	CTCTTCTAGTTGGCATGC	0.8	20	
DMD exon 4	1F	TTGTCGGTCTCCTGCTGGTCAGTG	0.25	6.25	196
	1R	CAAAGCCCTCACTCAAACATGAAGC	0.25	6.25	
DMD exon 8	1F	GTCCTTTACACACTTTACCTGTTGAG	0.25	6.25	360
	1R	GGCCTCATTCTCATGTTCTAATTAG	0.25	6.25	
DMD exon 12	1F	GATAGTGGGCTTTACTTACATCCTTC	0.25	6.25	331
	1R	GAAAGCACGCAACATAAGATACACCT	0.25	6.25	
DMD exon 45	1F	AAACATGGAACATCCTTGTGGGGAC	0.25	6.25	547
	1R	CATTCCTATTAGATCTGTCGCCCTAC	0.25	6.25	
DMD exon 48	1F	TTGAATACATTGGTTAAATCCCAACATG	0.25	6.25	506
	1R	CCTGAATAAAGTCTTCCTTACCACAC	0.25	6.25	
FVIII intron 18	HA1F	CTACCTGGCTTAGTAATGGCTC	0.8	20	429
	HA1R	AAAGGAATAAATTCCTTTTCCC	0.8	20	
HEXA exons 11 & 12	TS-F	GGTGTGGCGAGAGGATATTCCA	0.8	20	580/584
	TS-R	TCTCTCAGGCCTGAAAGAAGGG	0.8	20	
ZFX & ZFY	Z868F	ACCRCTGTACTGACTGTGATTACAC	0.4	10	495
	Z1338R	GCACYTCTTTGGTATCYGAGAAAGT	0.4	10	

[a]R = A or G; Y = C or T.

[b]Also see Basic Protocol steps 2a and 2b.

Table 9.9.2 Primer Sequences for Secondary (Nested or Hemi-Nested) PCR

Locus	Primer	Sequence (5′ to 3′)[a]	Final concentration (μM)[b]	Volume for 10 reactions (μl)[b]	Product size (bp)
CFTRΔF508	2F	TGGGAGAACTGGAGCCTT	0.8	20	154/151[c]
	2R	GCTTTGATGACGCTTCTGTAT	0.8	20	
DMD exon 4	1F	TTGTCGGTCTCCTGCTGGTCAGTG	0.25	6.25	168
	2R	CTGTGTCACAGCATCCAGACCTTGT	0.25	6.25	
DMD exon 8	2F	TCATGGACAATTCACTGTTCATTAA	0.25	6.25	321
	1R	GGCCTCATTCTCATGTTCTAATTAG	0.25	6.25	
DMD exon 12	1F	GATAGTGGGCTTTACTTACATCCTTC	0.25	6.25	301
	2R	TATGTTGTTGTACTTGGCGTTTTAG	0.25	6.25	
DMD exon 45	2F	GCTCTTGAAAAGGTTTCCAACTAAT	0.25	6.25	506
	1R	CATTCCTATTAGATCTGTCGCCCTAC	0.25	6.25	
DMD exon 48	1F	TTGAATACATTGGTTAAATCCCAACATG	0.25	6.25	444
	2R	AATGAGAAAATTCAGTGATATTGCC	0.25	6.25	
FVIII intron 18	HA2F	ATCAAAGGATTCGATGGTATCT	0.8	20	339
	HA2R	TTTCCTTTTTAGCAATTTTTCT	0.8	20	
HEXA exon 11	TS11F	AACTGGTCACCAAGGCCGGCTT	0.8	20	118/122[d]
	TS11R	CCTTCAAATGCCAGGGGTTCCA	0.8	20	
HEXA exon 12	TS12F	CAGGTACCCCTGAGCAGAAGGC	0.8	20	176
	TS12R	GGTGGCTAGATGGGATTGGGTC	0.8	20	
ZFX & ZFY	Z916F	AYAACCACCTGGAGAGCCACAAGCT	0.4	10	344
	Z1233R	TGCAGACCTATATTCRCAGTACTGGCA	0.4	10	

[a]R = A or G; Y = C or T.

[b]Also see Basic Protocol, step 5.

[c]Fragment with ΔF508 deletion mutation is 3 bp shorter.

[d]Fragment with +TATC insertion mutation is 4 bp longer.

Thermal cycler (PTC-100 or -200 from MJ Research; DNA Thermal Cycler or DNA Thermal Cycler 480 from Perkin-Elmer)
PAGE minigel casting and electrophoresis apparatus (MiniProtean II, Bio-Rad)

1. If appropriate, perform PEP (*UNIT 1.3*) on lysed and neutralized single cells or blastomeres (see Support Protocol 1).

   When deletions within the DMD gene are to be detected by comparative analysis of different exons, it is necessary to prepare HaeIII-digested PCR products of the ZFX and ZFY genes in parallel with the PCR products of the relevant DMD exon PCR products. Therefore, steps 2 to 7 and 8d must be carried out using the materials and conditions for the ZFX and ZFY genes while the corresponding steps for the DMD gene are being executed.

To perform primary PCR where PEP has not been performed (direct single-locus PCR)

2a. Prepare the following ten-reaction master mix A for every nine cells or blastomeres to be analyzed:

   50 μl 10× potassium-free amplification buffer (1× final)
   5 μl 25 mM 4dNTP mix (0.25 mM each dNTP final)
   6.25 to 20 μl 20 μM forward primer (0.25 to 0.8 μM final; see Table 9.9.1)
   6.25 to 20 μl 20 μM reverse primer (0.25 to 0.8 μM final; see Table 9.9.1)
   2 μl 5 U/μl *Taq* DNA polymerase (1 U per reaction final)
   H_2O to 400 μl.

3a. Add 40 μl master mix A to each tube containing lysed and neutralized single diploid cells and to control tubes, expelling the solution under the oil layer.

To perform primary PCR from PEP reaction product

2b. Prepare the following 10-reaction master mix B:

   50 μl 10× PCR amplification buffer containing 15 mM $MgCl_2$ (1× final)
   5 μl 25 mM 4dNTP mix (0.25 mM each dNTP final)
   6.25 to 20 μl 20 μM forward primer (0.25 to 0.8 μM final; see Table 9.9.1)
   6.25 to 20 μl 20 μM reverse primer (0.25 to 0.8 μM final; see Table 9.9.1)
   2 μl 5 U/μl *Taq* DNA polymerase (1 U per reaction final)
   H_2O to 450 μl.

3b. Add 45 μl master mix B to each 5-μl aliquot of single cell/blastomere PEP reaction product in a fresh 0.5-ml PCR reaction tube. Overlay reaction solutions with 50 μl light mineral oil.

4. Perform primary PCR using the thermal cycling conditions for each locus as indicated in Table 9.9.3.

5. Prepare a 10-reaction master mix that will be sufficient for nine secondary PCRs:

   50 μl 10× PCR amplification buffer containing 15 mM $MgCl_2$ (1× final)
   5 μl 25 mM 4dNTP mix (0.25 mM each dNTP final)
   5 to 20 μl 20 μM forward inner primer (0.2 to 0.8 μM final; see Table 9.9.2)
   5 to 20 μl 20 μM reverse inner primer (0.2 to 0.8 μM final; see Table 9.9.2)
   2 μl 5 U/μl *Taq* DNA polymerase (1 U per reaction final)
   H_2O to 480 μl.

   Alternatively, use one primer from the primary PCR in combination with the opposing internally nested primer (i.e., hemi-nested PCR)

6. Add 48 μl master mix from step 5 to a fresh reaction tube containing 2 μl primary PCR product. Overlay reaction solutions with 50 μl light mineral oil.

Table 9.9.3 Conditions for Primary PCR

Locus	Initial denaturation Temp. (°C)	Initial denaturation Time (sec)	No. of cycles	Denaturation Temp. (°C)	Denaturation Time (sec)	Annealing Temp. (°C)	Annealing Time (sec)	Extension Temp. (°C)	Extension Time (sec)	Final extension Temp. (°C)	Final extension Time (sec)	Soak Temp. (°C)
CFTRΔF508	93	180	20	92	45	40	45	72	90	72	300	4
DMD (all exons)	94	120	20	94	30	52	60	65	60	—[a]	—[a]	4
FVIII intron 18	92	180	20	92	30	49	30	72	30	72	180	4
HEXA exons 11 & 12	93	180	20	92	40	54	60	72	90	72	300	4
ZFX/ZFY	97	120	20	94	45	49	60	72	60	—[a]	—[a]	4

[a]No final extension step.

7. Perform secondary PCR using the thermal cycling conditions for each locus as indicated in Table 9.9.4.

To analyze for the CFTR ΔF508 or the HEXA exon 11+TATC mutations by heteroduplex formation

8a. Mix a 5-µl aliquot of each amplified test sample and control with an equal volume of previously amplified DNA fragments known to be either homozygous normal or homozygous *CFTR* ΔF508 or *HEXA* exon 11+TATC mutant (using both normal and mutant recommended).

9a. Heat mixed samples 10 min at 93°C to denature all DNA strands, then cool to 65°C and maintain that temperature 10 min to allow reannealing of single strands.

10a. Using a MiniProtean II PAGE minigel casting apparatus, prepare an 8% to 10% nondenaturing polyacrylamide minigel in 1× TBE buffer (also see *UNIT 7.2*).

11a. Add 1 µl of 10× DNA gel loading buffer to each sample and load on minigel. Include a size marker in one lane.

12a. Electrophorese samples 30 min at 200 V, then stain gel 10 min in 0.1 µg/ml ethidium bromide. Interpret the gel to detect the mutation, which differ from normal by 3 and 4 bp, respectively

Table 9.9.4 Conditions for Secondary PCR

Locus	Initial denaturation Temp. (°C)	Initial denaturation Time (sec)	No. of cycles	Denaturation Temp. (°C)	Denaturation Time (sec)	Annealing Temp. (°C)	Annealing Time (sec)	Extension Temp. (°C)	Extension Time (sec)	Final extension Temp. (°C)	Final extension Time (sec)	Soak Temp. (°C)
CFTRLΔF508	93	180	30	92	45	40	45	72	90	72	300	4
DMD (all exons)	94	120	30	94	30	52	30	65	60	—[a]	—[a]	4
FVIII intron 18	92	180	30	92	30	46	30	72	30	72	180	4
HEXA												
exon 11	93	180	30	92	40	57	60	72	90	72	300	4
exon 12	93	180	30	92	40	60	60	72	90	72	300	4
ZFX/ZFY	97	60	30	94	45	58	45	72	60	—[a]	—[a]	4

[a]No final extension step.

To analyze for the HEXA exon/intron 12 point mutation by DdeI digestion

8b. Add 4 μl of 5× *Dde*I digestion buffer to 15 μl of each amplified test sample and control, followed by 20 to 25 U *Dde*I. Incubate 90 min at 37°C.

9b. Using a MiniProtean II PAGE minigel casting apparatus, prepare a 10% nondenaturing polyacrylamide minigel in 1× TBE (also see *UNIT 7.2*).

10b. Add 2 μl of 10× DNA gel loading buffer to each sample and load on minigel. Include a size marker in one lane.

11b. Electrophorese samples 30 min at 200 V, then stain gel 10 min in 0.1 μg/ml ethidium bromide. Interpret the gel to detect the mutation, which is indicated by the addition of an 85-bp band to the digestion products.

To detect restriction-site polymorphism within intron 18 of the FVIII gene by BclI digestion

8c. Add 30 U of *Bcl*I to each amplified test sample and control, directly under oil. Incubate 60 min at 50°C.

9c. Using a MiniProtean II PAGE minigel casting apparatus, prepare an 8% nondenaturing polyacrylamide minigel in 1× TBE (also see *UNIT 7.2*).

10c. Mix 10 μl of each digestion product with 1 μl 10× DNA gel loading buffer and load on a minigel. Include a size marker in one lane.

11c. Electrophorese samples 30 min at 200 V (also see *UNIT 7.2*), then stain gel 10 min in 0.1 μg/ml ethidium bromide. Interpret the gel to detect the mutation, which is indicated by the presence of a 339-bp band as well as 222- and 117-bp bands (heterozygote), or just the 339-bp band (homozygous affected).

To discriminate between the ZFX and ZFY genes by HaeIII digestion

8d. Add 5 μl of each amplified test sample and control to 5 μl of 2× One-Phor-All Buffer Plus containing 5 U HaeIII. Incubate 90 min at 37°C.

9d. Using a MiniProtean II PAGE minigel casting apparatus, prepare a 5% or 8% nondenaturing polyacrylamide minigel in 1× TBE (also see *UNIT 7.2*).

10d. Add 1 μl of 10× DNA gel loading buffer to each digestion product and load onto minigel. Include a size marker in one lane.

11d. Electrophorese samples 20 or 30 min at 200 V, then stain gel 10 min in 0.1 μg/ml ethidium bromide. Interpret the gel to detect the mutation; a female cell yields fragments of 300 and 44 bp and a male cell yields fragments of 300, 216, 84, and 44 bp.

To detect deletions within the DMD gene by comparative analysis of different exons

8e. Prepare a 1.5% or 3% agarose gel in 0.5% TBE.

9e. Add 1 μl of 10× DNA gel loading buffer to each amplified test sample and control, then load onto the agarose gel.

10e. Load *ZFX/ZFY* products digested with *Hae*III on the same gel. Include a size marker in one lane.

11e. Electrophorese samples 20 to 30 min at 10 V/cm. Stain gel 20 min in 0.1 μg/ml ethidium bromide. Interpret the gel to detect the mutation (see product sizes in Table 9.9.2).

WHOLE-GENOME AMPLIFICATION OF SINGLE DIPLOID CELLS BY PRIMER-EXTENSION PREAMPLIFICATION (PEP)

Set up the PEP reaction as in *UNIT 1.3*, Support Protocol 2, using HPLC-grade water (e.g., Fisher; see *SUPPLIERS APPENDIX*) to prepare all solutions and in all protocol steps. In step 1, prepare PEP solution for 10 reactions instead of 100. In step 2, replace the sperm cells in microtiter wells with single diploid cells or blastomeres in 0.5-ml tubes. In step 4, carry out PEP with the amplification cycling steps switched as follows:

Initial step:	5 min	95°C	(denaturation)
50 cycles:	1 min	92°C	
	2 min	37°C	
	10 sec/°C ramp	37°C to 55°C	
	4 min	55°C	
Final step:	5 min	72°C	(extension)

SUPPORT PROTOCOL 2

PREIMPLANTATION EMBRYO BIOPSY

NOTE: All solutions and equipment coming into contact with cells must be sterile, and proper aseptic technique should be used accordingly. Media and solutions are filtered through 0.22-μm pore-size Millex-GV filters (Millipore); light paraffin oil is filtered through 0.45-μm pore-size tissue culture filters (Nalgene). In both cases, a small volume of the initial filtrate is discarded before collecting the final filtered product. Glass microcapillaries are dry-sterilized at 120°C for 3 hr. Finished micropipets and PCR tubes are irradiated 15 min in a UV cross-linker at maximum energy.

NOTE: Use Milli-Q purified water (meeting College of American Pathologists Type I reagent-grade water standards) or other embryo-tested ultrapure water in all reagents and medium that come in contact with embryos.

NOTE: Pipets can be made in advance of the experiment.

Materials *(see APPENDIX 1 for items with ✓)*
- ✓ Acid Tyrodes solution
- ✓ Embryo culture and wash medium (see recipe for culture and micromanipulation media)
  Light paraffin oil for cultures (EM Science), filter sterilized, then washed and equilibrated with culture medium
- ✓ Biopsy medium (see recipe for culture and micromanipulation media)
- ✓ Blastomere wash medium (see recipe for culture and micromanipulation media)
  Cell lysis buffer: 200 mM KOH/50 mM DTT (prepare fresh)
  Human embryos
  Light mineral oil for PCR (Perkin-Elmer) or Chill-Out 14 liquid wax (MJ Research)
- ✓ Neutralization buffer

  Micropipet puller (Sutter Instruments P-97 or Narishige PB-7)
  Microforge (Narishige MF-9 or Research Instruments MF42)

Glass microcapillaries (0.7-mm-i.d. × 0.87-mm-o.d. Pyrex glass from Vitrocom or 0.75-mm-i.d. × 1.0-mm-o.d., borosilicate glass from Sutter Instruments), dry-sterilized 3 hr at 120°C

UV cross-linker

Tissue culture plates (Falcon): 60 × 15, 50 × 9, and 35 × 10 mm

Humidified 37°C, 5% CO_2 incubator (e.g., Forma)

37°C warm plate

0.5-ml clarified polypropylene PCR reaction tubes (Out Patient Services), irradiated 15 min in UV cross-linker at maximum energy

Olympus IX70 inverted microscope with Hoffman modulation contrast condenser (40-mm WD) and objectives (HMC EF10×, HMC 20× LWD), mechanical X-Y stage and Brook stage warmer (video camera, monitor, and videocassette recorder are optional)

Diamond pen

Antivibration table (Newport)

Two dual instrument holders (Narishige)

Drilling and biopsy micropipet controllers consisting of:
> Micrometer drive
> Gastight syringe (Hamilton)
> Support base (base, micrometer drive and syringe are available preassembled from Stoelting)
> Luer adapters
> Polyethylene tubing (PE90, Intramedic)

Holding micropipet controller consisting of:
> Gastight threaded-plunger syringe (Hamilton)
> Luer adapters
> Polyethylene tubing (PE90, Intramedic)
> Stand clamp

Two upright joystick-type micromanipulators (Narishige; electronic coarse control and hydraulic fine control)

Two stereoscopic zoom microscopes with transmitted-light base and (preferably) wide-field oculars

Modeling clay (optional)

Micropipet controller consisting of mouthpiece (Fisher) connected through a 0.22-μm Millex-GV filter (Millipore) or equivalent to one end of a 60-cm length of Tygon tubing with the other end fitted to either a microinstrument holder (Leica) or microcapillary pipet holder (Microcap, Drummond)

Laminar-flow hood

Heating block

1. Using a micropipet puller, a microforge, and glass microcapillaries, prepare the following micropipets (all with straight shanks) and sterilize by irradiating 15 min in a UV cross-linker at maximum energy (o.d.'s are for 0.7 × 0.87–mm microcapillaries and may have to be adjusted slightly upward for 0.75 × 1.0–mm microcapillaries):

    Holding: 65 to 80 μm o.d. (larger will work), 25 to 30 μm i.d., heat-polished blunt end
    Drilling: 5 to 10 μm o.d.
    Biopsy: 40 to 50 μm o.d. (for <8-cell embryos) or 30 to 40 μm o.d. (for >8-cell embryos), heat-polished blunt end
    Cell-transfer: 50 to 60 μm o.d.

Make sufficient holding, drilling, and biopsy micropipets to allow at least several spares of each type or size. One set of micropipets may be reused for all embryos in one case, provided they are not broken, remain free of debris, and continue to work effectively.

2. Make embryo-handling pipets, consisting of hand-pulled Pasteur pipets (~120 mm i.d.) or equivalent machine-pulled capillaries (same as those used in IVF for stripping cumulus cells for oocytes and handling embryos).

3. Thaw an aliquot of acid Tyrodes solution and keep at room temperature.

4. For each embryo prepare the following.

   a. One 60 × 15–mm tissue culture plate (for culture of embryos selected for biopsy) with a 20-μl drop of culture medium and an overlay of paraffin oil that has been washed and equilibrated with culture medium. Place in a humidified 37°C 5% CO_2 incubator at least 1 hr before use.

   b. One inverted lid of a 50 × 9–mm tissue culture plate (for biopsy) containing a central 20-μl drop of biopsy medium with a 3-ml overlay of equilibrated paraffin oil sufficient to completely cover the drop. Spread the drop with a disposable pipet tip prior to adding oil. Place on a 37°C warm plate in a laminar flow hood or other sterile environment (e.g., under the lid from one 150-mm culture plate) until use.

   c. One 60 × 15–mm tissue culture plate (for culture after biopsy) with a 20-μl drop of culture medium and an overlay of equilibrated paraffin oil. Place in a 37°C humidified 5% CO_2 incubator at least 1 hr before use.

   d. One 35 × 10–mm tissue culture plate (the "embryo wash" plate) containing 2 ml culture medium. Place in a 37°C 5% CO_2 incubator at least 1 hr before use to prewarm and gas-equilibrate the medium. (Alternatively, add four drops of culture medium in the embryo culture plate—step 4a—and pass the embryo through three drops to wash and add to the fourth for culture.)

   e. One 35 × 10–mm tissue culture plate (the "blastomere wash" plate) containing 2 ml blastomere wash medium. Place this next to the stereoscopic microscopes used in blastomere isolation and keep at room temperature.

   f. Three 0.5-ml clarified polypropylene PCR reaction tubes (two for blastomeres, one for blank) containing 5 μl cell lysis buffer. Prepare two additional blanks for each complete diagnostic procedure.

5. Examine embryos on an inverted microscope with Hoffman modulation contrast at 200× to 400× magnification and assess stage of development and morphological grade.

6. Select only those embryos suitable for micromanipulation (6 to 10 cells with morphological grade appropriate for embryo transfer) and place each into a separate drop of medium under oil in a 60 × 15–mm culture plate (prepared in advance in step 5a). Number the embryos consecutively by marking the bottom and lid of the plate with a diamond pen. Return embryos to the incubator.

   This step is for convenience only and may be omitted if suitable embryos are transferred directly from the original drop to the biopsy drop (step 13).

7. Working on an antivibration table, mount the holding micropipet (usually on the left side for right-handed persons) and drilling and biopsy micropipets (together, usually on the right side) in dual instrument holders.

8. Connect holding, drilling, and biopsy micropipets to their respective microsyringe controllers via polyethylene tubing.

9. Adjust instrument holders to a shallow angle with the micropipets pointing toward the illuminated field of the inverted microscope.

10. By adjusting the microsyringe controllers, push oil into the micropipets to the ends of the tips, leaving no air gap or bubbles. Check for leaks.

11. Using coarse and fine micromanipulators, lower micropipets into view for inspection under the inverted microscope and individually adjust positions of drilling and biopsy micropipets to be in focus and to project the same distance from the dual instrument holder.

12. Place the drilling micropipette in a drop of acid Tyrodes solution and fill by suction to produce a liquid column of ~5 mm along the micropipet shank.

13. Preheat the stage of the inverted microscope to 37.5°C (or other predetermined offset temperature needed to keep the medium at 37°). Transfer the first embryo into the drop of biopsy medium on the inverted tissue culture plate lid that was prepared in step 5b. Place the lid on the heated stage.

14. With the embryo in focus at low power (100×), secure the embryo to the holding pipet by applying suction, then orient embryo as necessary so that the blastomeres to be removed are at the 3 o'clock position.

15. Increase magnification to 200× to 400× for zona drilling, according to personal preference. With the drilling pipet against or near the zona pellucida, slowly expel acid Tyrodes solution, allowing adequate time for each adjustment of the microsyringe controller to result in fluid flow from the micropipet. Just before the zona is dissolved through, stop the flow of acid Tyrodes and push the micropipet through into the subzonal space. Ensure that the size of the dissolved opening is just sufficient for entry of the biopsy micropipet. Move the drilling micropipet out of the drop.

16. At lower magnification (100×), bring the biopsy micropipet into focus near the embryo. Select a blastomere for biopsy that contains an identifiable nucleus. Perform the biopsy by placing the biopsy micropipet through the zona opening and applying suction to the selected blastomere, while at the same time tugging on the cell, first to loosen its adhesion to neighboring blastomeres and then to maneuver it out through the hole and into the bore of the micropipet.

17. Expel the cell from the biopsy micropipet and move the expelled cell out of the working field using the mechanical stage.

18. For embryos at developmental stages of ≥8 cells, isolate a second blastomere in the same manner as the first.

19. Release the embryo from the holding micropipet, raise the micropipets out of the drop, and move the plate containing the biopsied embryo and blastomeres to a pair of stereoscopic microscopes (which together with the cell transfer pipet are used to recover each blastomere in turn for washing and transfer to a separate reaction tube). In advance, optimize magnification and focus settings for blastomere recovery and washing on one microscope and for cell transfer on the second; do not readjust settings.

20. Consecutively wash each blastomere briefly in 2 ml blastomere wash medium, using the blastomere wash plate prepared in step 4e. Avoid repeated pipetting.

21. Uncap PCR tube and place horizontally on a stereoscopic microscope. Immobilize the tube by fixing it to a piece of modeling clay or holding it down with one finger. Adjust to clearly visualize the air-liquid interface near the upper tube wall.

22. Using a cell transfer micropipet with mouthpiece controller, transfer each single blastomere in a minimum volume of solution into a 0.5-ml PCR tube containing 5 µl cell lysis buffer (prepared in step 4f) by holding the micropipet holder in one hand and maneuvering the micropipet containing the single cell into the reaction tube until the tip is near the liquid

surface (guiding the manipulations by observing the tube from the side). While observing through the oculars, extend the tip into the lysis buffer and bring into sharp focus near the upper wall of the tube. Expel fluid slowly under gentle pressure until the cell is seen leaving the micropipet as a projectile. Before or during the transfer, visually confirm that each blastomere biopsied contains a nucleus and that the blastomere has been expelled into the lysis buffer. Immediately number tubes with a permanent marker according to the embryo number on the culture plate. Rinse the transfer pipet with blastomere wash medium before removing a second blastomere from the biopsy drop still containing the embryo.

23. To serve as a blank reaction for each embryo biopsied, take up a small volume of medium from the blastomere wash plate in the same transfer pipet and expel into a fresh PCR tube containing lysis buffer.

24. Immediately give the dish with the remaining embryo to the embryologist, who will wash the embryo three times and then transfer it into a new drop of embryo culture and wash medium.

25. Repeat steps 14 to 24 for subsequent embryos.

26. When all blastomeres and blank samples are in reaction tubes, open tubes in a laminar-flow hood one at a time and add 50 µl PCR oil or liquid wax to completely cover reagents. Microcentrifuge a few seconds to collect wax or oil above the lysis buffer.

27. Incubate tubes in a heating block at 65°C for 15 to 30 min, cool to room temperature, then add 15 µl neutralization buffer below the oil or wax interface.

SUPPORT PROTOCOL 3

ISOLATION OF BLASTOMERES FROM AFFECTED EMBRYOS FOR FURTHER INVESTIGATION

NOTE: Permission to perform these analyses must be explicitly obtained from patients.

NOTE: All solutions and equipment coming into contact with cells must be sterile, and proper aseptic technique should be used accordingly. Media and solutions are filtered through 0.22-µm pore-size Millex-GV filters (Millipore); light paraffin oil is filtered through 0.45-µm pore-size tissue culture filters (Nalgene). In both cases, a small volume of the initial filtrate is discarded before collecting the final filtered product. Glass microcapillaries are dry-sterilized at 120°C for 3 hr. Finished micropipets and PCR tubes are irradiated 15 min in a UV cross-linker at maximum energy.

NOTE: Use Milli-Q purified water (meeting College of American Pathologists Type I reagent-grade water standards) or other embryo-tested ultrapure water in all reagents and medium that comes in contact with embryos.

Materials *(see* APPENDIX 1 *for items with* ✓ *)*
> Human embryos (affected with a genetic disease of interest; see Basic Protocol 1) in
>   drops of culture medium (see Support Protocol 2)
> ✓ Acid Tyrodes solution
> ✓ Embryo culture and wash medium (see recipe for culture and micromanipulation media)
> ✓ Blastomere wash medium (see recipe for culture and micromanipulation media)
>   Cell lysis buffer: 200 mM KOH/50 mM DTT (prepare fresh)
>   Light mineral oil for PCR (Perkins-Elmer) or Chill-Out 14 liquid wax (MJ Research)

Embryo handling pipets (see Support Protocol 2, step 2)

Two stereoscopic zoom microscopes with transmitted-light base and (preferably) wide-field oculars

35 × 10–mm tissue culture plates (Falcon)

Cell transfer pipets (50 to 60 μm o.d.; see Support Protocol 2, step 1)

Micropipet controller consisting of mouthpiece (Fisher) connected through a 0.22-μm Millex-GV filter (Millipore) or equivalent to one end of a 60-cm length of Tygon tubing with the other end fitted to either a microinstrument holder (Leica) or microcapillary pipet holder (Microcap, Drummond)

0.5-ml clarified polypropylene PCR reaction tubes (Out Patient Services), irradiated 15 min in UV cross-linker at maximum energy

1. Using an embryo handling pipet, transfer affected embryos one at a time from the culture drops in minimal volume of medium into 5-μl drops of acid Tyrodes solution on the inverted lid of a 35 × 10–mm tissue culture plate.

2. Observe embryos continuously and transfer into fresh embryo culture and wash medium as soon as the zona pellucida has fully dissolved (a few seconds to a minute). If the zona does not dissolve in the first drop of acid Tyrodes, transfer to a second drop.

3. Using an embryo handling pipet, immediately transfer zona-free embryos to 5-μl blastomere wash medium on the inverted lid of a 35 × 10–mm culture plate. Transfer to a second drop before dispersing blastomeres (e.g., by pipetting in and out of a close fitting pipet and/or picking at the embryo with a cell transfer pipet).

4. Using a cell transfer pipet with mouthpiece controller, transfer single embryonic cells into 0.5-ml PCR tubes containing cell lysis buffer. Cover with PCR oil or liquid wax (see Support Protocol 2, steps 22 to 26) and store at −70°C until required.

SUPPORT PROTOCOL 4

ISOLATION OF SINGLE LYMPHOCYTES/LYMPHOBLASTOID CELLS

This protocol describes manual methods used for the isolation of single cells from lymphoblastoid cell lines or freshly prepared lymphocytes as models for single-cell DNA amplification.

CAUTION: Universal precautions should be observed in handling all biological materials of human origin. Caution should be exercised when pipetting cells potentially contaminated with Epstein-Barr virus (EBV) or other pathogenic viruses.

NOTE: All solutions and equipment coming into contact with cells must be sterile, and proper aseptic technique should be used accordingly. Media and solutions are filtered through 0.22-μm pore-size Millex-GV filters (Millipore); light paraffin oil is filtered through 0.45-μm pore-size tissue culture filters (Nalgene). In both cases, a small volume of the initial filtrate is discarded before collecting the final filtered product. Glass microcapillaries are dry-sterilized 3 hr at 120°C. Finished micropipets and PCR tubes are irradiated 15 min in a UV cross-linker at maximum energy.

NOTE: Use HPLC-grade water (e.g., Fisher) to prepare all solutions and in all protocol steps where water is called for.

Materials *(see APPENDIX 1 for items with ✓)*

Primary lymphocytes or EBV-transformed lymphoblastoid cell lines prepared from male and female gamete donors of each case under investigation

RPMI 1640 medium with HEPES (Sigma), without serum or protein supplement

Cell lysis buffer: 200 mM KOH/50 mM DTT (prepare fresh)
Modeling clay (optional)
Light mineral oil for PCR (Perkin-Elmer) or Chill-Out 14 liquid wax (MJ Research)
✓ Neutralization buffer

Micropipet puller (Sutter Instruments P-97 or Narishige PB-7)
Microforge (Narishige MF-9 or Research Instruments MF42)
Glass microcapillaries (0.7-mm i.d. × 0.87-mm o.d., Pyrex glass from Vitrocom or
    0.75-mm i.d. × 1.0-mm o.d., borosilicate glass from Sutter Instruments).
IEC Clinical centrifuge or equivalent
0.5-ml clarified polypropylene PCR reaction tubes (Out Patient Services), irradiated 15
    min in UV cross-linker at maximum energy
35 × 10–mm tissue culture plates (Falcon)
Two stereoscopic zoom microscopes with transmitted-light base and (preferably)
    wide-field oculars (dark-field illumination may also be used)
Micropipet controller consisting of mouthpiece (Fisher) connected through a 0.22-μm
    Millex-GV filter (Millipore) or equivalent to one end of a 60-cm length of Tygon
    tubing, with the other end fitted to either a microinstrument holder (Leica) or a
    microcapillary pipet holder (Microcap, Drummond)
Laminar-flow hood
Heating block

1. Using a micropipet puller, a microforge, and glass microcapillaries, prepare a 25-μm-o.d.
   micropipet in advance and sterilize by irradiating 15 min in a UV cross-linker at maximum
   energy.

2. Wash lymphocytes or transformed lymphoblastoid cells three times by centrifuging 5
   min at ~1200 × g in an IEC Clinical centrifuge at room temperature, removing the
   supernatant, and resuspending the cell pellet in HEPES-buffered RPMI medium without
   protein supplement. Determine cell viability by trypan blue dye exclusion (APPENDIX 31)
   and use only preparations with high viability.

3. Add 5 μl cell lysis buffer to 0.5-ml PCR tubes (thin-walled or 0.2-ml PCR tubes may
   require modification of the method) under sterile conditions and seal with cap.

4. Place three (or more) 5-μl drops of HEPES-buffered RPMI medium without protein
   supplement and one 5-μl drop of appropriately diluted cell suspension (from step 2) close
   together on the inverted lid of a 35 × 10–mm tissue culture plate.

5. Uncap PCR tube and place horizontally on a stereoscopic microscope. Immobilize by
   fixing to a piece of modeling clay or holding down with one finger. Adjust to clearly
   visualize the air-liquid interface near the upper tube wall.

6. Using a micropipet controller with mouthpiece, load the micropipet with medium. Transfer
   a small number of cells from the cell suspension to the first wash drop. Isolate one cell in
   the micropipet, then transfer it to the next wash drop. Finally pick the cell up in a minimal
   volume of medium for transfer to the reaction tube (step 7). Alternatively, isolate cells
   from a dilute suspension in a plate containing 2 ml medium and wash in another plate
   containing 2 ml medium before transferring to the reaction tube.

7. While observing the tube from the side and holding the micropipet holder in one hand,
   guide the micropipet containing the single cell into the reaction tube until the tip is near
   the liquid surface. While observing through the oculars, extend the tip into the lysis buffer
   and bring it into sharp focus near the upper wall of the tube. Expel fluid slowly under
   gentle pressure until the cell is seen leaving the micropipet as a projectile.

8. Repeat steps 5 to 7 as long as cells are easily drawn into the micropipet (usually five to ten cells if using drops, 30 to 40 if using plates).

9. Prepare negative controls (no cells) by transferring medium from the final wash drop or wash plate using same micropipet that was used for the cells.

10. When all cells are in reaction tubes, open tubes in a laminar-flow hood one at a time and add 50 µl PCR oil or liquid wax to completely cover reagents. Microcentrifuge a few seconds to collect wax or oil above lysis buffer. Store at −70°C until ready to proceed.

11. Incubate tubes 15 to 30 min in a heating block at 65°C, cool to room temperature, then add 5 µl neutralization buffer below the oil or wax interface. If necessary, store up to several months at −70°C.

BASIC PROTOCOL 2

FISH ANALYSIS OF SINGLE BLASTOMERES

FISH analysis of blastomeres for PGD is more difficult than FISH analysis of other cell types because of two unique constraints. First, only one or two cells are available for each specific analysis; therefore, hybridization efficiency (and thus blastomere preparation quality) is a critical factor in obtaining and interpreting results. Second, there is a narrow window of time in which the analysis must be performed (i.e., embryo biopsy on day 3 and microscopic analysis and embryo transfer on day 4).

Materials (see APPENDIX 1 for items with ✓)
    Blastomere in biopsy medium (see Support Protocol 2, step 16)
✓ Embryo culture and wash medium (see recipe for culture and micromanipulation media)
✓ Hypotonic/partial fixative solution
✓ 3:1 fixative solution
    LSI Hybridization Buffer (Vysis)
    Probes 1, 2, and 3 for FISH (Vysis)
    70%, 80%, 90%, and 100% ethanol
    Rubber cement
✓ 60% formamide wash solution, 45°C
✓ 2× SSC, pH 7.0
✓ 2× SSC, pH 7.0 containing 0.1% (v/v) NP-40 (Vysis)
    DAPI II (Vysis)

    Micropipet with 80-µm bore
    2-well Nunc dishes
    Clay Adams glass slides with etched rings (clean with ethanol prior to use)
    Coplin jars
    18-mm^2 glass coverslips (Fisher)
    Hybrite Hot Plate (Vysis) or 80°C water bath
    Moist chamber (Fig. 4.4.3)
    45°C water bath
    24 × 50–mm glass coverslips (Fisher)
    Fluorescence microscope

1. Once the blastomere has been biopsied, gently expel the blastomere into biopsy medium as far away from the embryo as possible. For embryos at developmental stages ≥8 cells, isolate a second blastomere in the same manner, if desired.

2. Release the embryo from the holding pipet and raise the micropipets out of the drop. Extract the embryo from the drop, being careful to avoid the biopsied blastomere(s), and

place the embryo into freshly prepared embryo culture and wash medium. Allow embryo to equilibrate with the culture medium 1 to 2 hr and return to incubator. Place the dish with the biopsied blastomere(s) on the warming plate until biopsy of all embryos is completed.

3. Using a micropipet with a bore size of 80 μm, place each blastomere in one well of a Nunc dish containing hypotonic/partial fixative solution and let sit 45 sec at room temperature.

4. Remove blastomere from Nunc dish and place on Clay Adams glass slide in one of the etched rings.

5. Remove excess fluid from the slide, but do not allow the blastomere to become dry. Gently add 2 to 3 drops of 3:1 fixative solution to blastomere on slide.

6. Locate blastomere under phase-contrast microscope, check for presence of intact nucleus, and note location of blastomere using stage coordinates or a slide locator. Identify slides using pencil.

7. Repeat steps 3 to 6 for each biopsied blastomere, placing each blastomere on a separate slide.

8. Determine formula for a master hybridization mix of sufficient volume to allow 10 μl of hybridization mix for each blastomere, according to the following composition for a hybridization mix containing three Vysis probes (also see manufacturer's instructions for appropriate probe volumes):

> 7 μl LSI Hybridization Buffer (Vysis)
> 1 μl probe 1
> 1 μl probe 2
> 1 μl probe 3 (or 1 μl H$_2$O if only 2 probes are required).

The combination of probes for chromosome rearrangements must be selected for each individual patient who is a carrier of chromosomal rearrangement. Vysis does however sell two probe mixes that can be used in PGD testing for aneuploidy

9. Warm the probes and hybridization buffer to room temperature by placing 5 min in a 37°C water bath. Microcentrifuge probes and buffer 1 to 3 sec.

10. Combine probes and buffer as determined in step 8 and mix well.

11. Microcentrifuge 1 to 3 sec.

This mix can be prepared a few days ahead of time and frozen until use.

12. Dehydrate slides with blastomeres sequentially in 70%, 80%, 90%, and 100% ethanol for 2 min each at room temperature in Coplin jars.

13. Place 10 μl appropriate hybridization mixture (steps 8 to 11) over etched ring containing blastomere. Cover with 18-mm^2 glass coverslip and seal carefully with rubber cement.

14. Denature blastomere DNA and probes at 80°C for 3 min.

15. Incubate overnight (shorter for some probes) at 37°C in moist chamber.

16. Remove rubber cement and coverslips. Wash slides once in 60% (or 50%; see manufacturer's instructions) formamide wash solution at 45°C for 10 min with agitation.

17. Transfer slides to 2× SSC. Let sit 10 min at room temperature.

18. Transfer slides to 2× SSC/0.1% NP-40. Let sit at room temperature for 5 min.

19. Counterstain with 10 μl DAPI II and cover with 24 × 50–mm glass coverslip. Examine slide with fluorescence microscope.

PROBE VALIDATION

For all PGD indications, routine chromosome analysis on peripheral blood lymphocytes (metaphase and interphase) from both members of the couple should be performed before commencing probe work-up in order to ensure that no chromosome rearrangement carriers go unidentified. Probes should also be validated on blastomeres if possible, or on other cell types (e.g., chorionic villus) if not.

Materials

Peripheral blood lymphocytes from each member of couple
Probes (Vysis) used in Basic Protocol 2
Chorionic villus cells (*UNIT 8.1*) from control specimen

1. Culture peripheral blood lymphocytes from both members of the couple and harvest metaphase cells (see *UNIT 4.1*).

2. Hybridize probes (together or separately) to five metaphase cells from each person's culture (*UNIT 4.4*).

3. Counterstain cells with DAPI (*UNIT 4.3*) to provide banded chromosomes, and examine slides with fluorescence microscope. Record results of probe localization.

4. Hybridize probes (together, not separately) to interphase cells from peripheral blood lymphocytes from each member of the couple (*UNIT 4.4*). Examine 200 to 250 cells to determine efficiency of probe hybridization in parental lymphocytes and record results:
Efficiency of hybridization = (no. cells with expected signal)/(total no. cells examined).

5. Hybridize probes (together, not separately) to interphase chorionic villus cells from a control specimen. Examine 200 to 250 interphase cells to determine efficiency of probe hybridization in this cell type.

> *Interphase chorionic villus cells may be obtained from a modified direct preparation (UNIT 8.1, Alternate Protocol). Follow steps 6 to 12a of that protocol beginning by placing 2 to 4 mg of chorionic villi into hypotonic solution (step 6). In step 8, a 20-min fixation should suffice. Slides are ready for hybridization once they are dry (end of step 12a).*

References: ESHRE PGD Consortium Steering Committee, 1999; Verlinski and Cieslak, 1993

Contributors: Patricia Minehart Miron, Samuel S. Chong, Robert E. Gore-Langton, and Mark R. Hughes

UNIT 9.10

Protein Truncation Test

NOTE: All reagents and equipment used for RNA preparation should be sterile, and all solutions should be treated with diethylpyrocarbonate (see recipe for DEPC-treated solutions in *APPENDIX 1*) to ensure they are RNase free. Gloves should be worn to prevent RNase contamination.

NOTE: Experiments involving PCR require extremely careful handling to prevent contamination.

NONRADIOACTIVE PROTEIN TRUNCATION TEST

NOTE: If DNA is to be used as the template, begin the experiment at step 9.

NOTE: It is important that the electroblotting be performed under cold conditions, with the gel and all solutions and equipment chilled.

Materials (*see* APPENDIX 1 *for items with* ✓)

   1 to 3 μg patient and control (i.e., nonmutant) RNA in ethanol (see Support Protocol) *or* DNA sample

✓ 3 M sodium acetate, pH 5.6 (pH adjusted with acetic acid)

   0.5 μg/μl random primer (e.g., Promega) or 10 pmol/μl gene-specific reverse primer

   Expand Reverse Transcriptase kit (Roche Diagnostics) containing 50 U/μl Expand reverse transcriptase and 5× buffer

   100 mM dithiothreitol (DTT; e.g., Life Technologies)

✓ 10 mM 4dNTP mix

   40 U/μl ribonuclease inhibitor (e.g., RNasin, Promega)

   20 pmol/μl gene-specific primers for PCR: forward, internal tailed forward, reverse, and internal reverse primers (Fig. 9.10.1)

   1.5 U/μl RNase H (e.g., Promega)

✓ 10× PCR buffer (e.g., SuperTaq buffer, HT Biotechnologies)

   5 U/μl *Taq* DNA polymerase (AmpliTaq, Perkin Elmer)

   Nonradioactive Protein Truncation Test kit (Roche Diagnostics) containing:

        Control DNA (200 ng/μl human genomic DNA)

        500 ng/μl 5′ control primer

        150 ng/μl 3′ control primer

        4× transcription mix

        T7 translation mix (including biotin-labeled lysine, if used)

        25 mM magnesium acetate solution

        25 mM sodium EDTA solution

        Sterile RNase-free water

        Combi-marker (biotin + color; prestained protein molecular weight marker can also be used)

        SDS sample buffer

✓ 30% (w/v) 37.5:1 acrylamide/bisacrylamide

✓ 1.5 M Tris·Cl, pH 6.6

✓ 10% SDS

   10% (w/v) ammonium persulfate (APS; freshly prepared)

   N,N,N′,N′-Tetramethylethylenediamine (TEMED)

✓ 1.0 M Tris·Cl, pH 6.8

✓ Running buffer

✓ Blotting buffer, ice cold

   PVDF Western blotting membrane (Roche Diagnostics) or equivalent

   Methanol

   Tris-buffered saline (TBS): 50 mM Tris·Cl, pH 7.5 (APPENDIX 1)/150 mM NaCl

   10× maleic acid stock solution: 1.0 M maleic acid (Fluka)/1.5 M NaCl, pH 7.5

   Chemiluminescence Blotting Substrate kit (Roche Diagnostics) containing:

        Luminescence substrate solution

        Starting solution B

        Blocking reagent

Streptavidin-POD conjugate (Roche Diagnostics)
Washing buffer: TBS containing 0.1% (v/v) Tween 20
Anti-HA high-affinity antibody (Roche Diagnostics)
Peroxidase-conjugated rabbit anti-rat immunoglobulins (Dako)

Pasteur pipets, sterile
Perkin Elmer 2400, 9600, or 9700 thermocycler
SDS-PAGE system (e.g., Mini Protean II, Bio-Rad), including electroblotting devices
   (Bio-Rad)
Plastic trays
Whatman 3MM filter paper, 7 × 10 cm
Plastic wrap (e.g., Saran wrap)
X-Omat AR film (Eastman Kodak) *or* luminescence detector (e.g., Lumi-Imager, Roche
   Diagnostics)

1. Take 1 to 3 μg ethanol-stored RNA and add 0.1 vol of 3 M sodium acetate, pH 5.6.
   Microcentrifuge 10 min at 12,000 × g, 4°C.

2. Remove the ethanol using a sterile Pasteur pipet. Air-dry the pellet, but do not
   over-dry.

3a. *For random priming:* Either add 2 μl of 0.5 μg/μl random primer and 8.5 μl water to
    the air-dried RNA pellet, or prepare a master mix of primer and water, and add 10.5 μl
    to the pellet. Mix and incubate 10 min at 65°C. Transfer the tube directly onto ice.

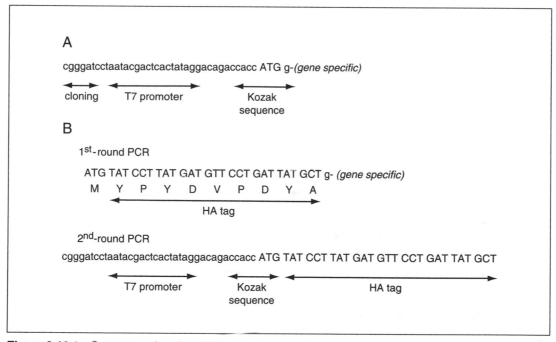

Figure 9.10.1   Sequence of a tailed PTT primer showing its individual elements. (**A**) Original PTT forward primer, containing a T7 RNA polymerase promoter sequence, a spacer, and a translation initiation (Kozak) sequence. Cloning is facilitated by an optional 5′ *Bam*HI site and an *Nco*I site covering the ATG codon. (**B**) PTT forward primer introducing an N-terminal tag in the translation product. In the first-round PCR, the HA tag (YPYDVPDYA) is introduced; in the second-round PCR this tag is used to add the other sequences required as in (A).

3b. *For gene-specific priming:* Add 2 µl of 10 pmol/µl gene-specific reverse primer and 8.5 µl H$_2$O to the air-dried RNA pellet. Mix and incubate 10 min at 65°C. Transfer the tube immediately onto ice.

4. Add the following to each RNA sample (20 µl final volume):

   4 µl 5× Expand Reverse Transcriptase buffer
   2 µl 0.1 M DTT
   2 µl 10 mM 4dNTP mix
   0.5 µl 40 U/µl RNasin
   1 µl 50 U/µl Expand Reverse Transcriptase.

   Incubate 10 min at 30°C and 60 min at 42°C.

5. *Optional:* Prepare RNA primers for second-strand synthesis by adding 4 U RNase H and incubating 20 min at 37°C. If necessary, store at −20°C.

6. Prepare 25 µl first-round PCR premix 1 containing:

   1 µl 10 mM 4dNTP mix
   2 to 4 µl cDNA (step 5)
   20 pmol gene-specific forward primer
   20 pmol gene-specific reverse primer
   H$_2$O to 25 µl.

7. Prepare 25 µl of premix 2 containing:

   5 µl 10× PCR buffer
   0.5 to 2 U *Taq* DNA polymerase
   H$_2$O to 25 µl.

8. Add 25 µl premix 2 to premix 1 and perform first-round PCR using the following parameters for a Perkin Elmer 2400, 9600, or 9700 thermocycler:

First cycle:	2 min	94°C
25 to 30 cycles:	15 sec	93°C
	30 sec	55° to 60°C
	1 min/kbp	72°C
Final cycle:	7 min	72°C

   Store PCR products at 4°C.

9. Prepare 25 µl second-round PCR premix 1 containing:

   5 µl 10× PCR buffer
   0.5 to 2 U *Taq* DNA polymerase
   H$_2$O to 25 µl.

10. Add to the first premix 25 µl of premix 2 containing:

    1 µl 10 mM 4dNTP mix
    20 to 30 pmol tailed internal forward primer
    20 to 30 pmol internal reverse primer
    1 µl 10×-diluted or undiluted first PCR reaction *or* 250 to 500 ng genomic DNA
    H$_2$O to 25 µl.

11. Perform second-round PCR using the amplification cycles described in step 8.

12. Analyze all PCR products by agarose gel electrophoresis to determine the amount and size of product(s) in each reaction. Note any abnormal sizes which indicate the presence of genetic rearrangements or mutations affecting RNA splicing, or genomic DNA contamination in the RT-PCR reaction.

13. Pipet the following into a 0.5-ml tube:

    1 µl PCR product (50 to 500 ng DNA)
    1.25 µl 4× T7 transcription mix
    2.75 µl H_2O.

    Incubate 15 min at 30°C in a heating block.

14. *Optional:* Check the transcription efficiency by analyzing 2.5 µl of the transcription reaction on a 1% agarose gel. Check for a clearly visible RNA band which runs below the PCR product (indicates success).

15. In a 0.5-ml tube, combine 2.5 µl transcription reaction (from step 13) and 10 µl translation mix (incorporates biotin-lysine). Incubate in a heating block 60 min at 30°C. Store samples at −20°C

16. Clean glass plates and spacers with water and ethanol, and assemble two gel cassettes on the gel casting stand (e.g., MiniPortean II gel system; see *APPENDIX 3F*).

17. Prepare a mix for two 10% gels:

    3.3 ml 30% (w/v) 37.5:1 acrylamide/bisacrylamide (adjust depending on product size)
    2.5 ml 1.5 M Tris·Cl, pH 8.8
    4 ml H_2O
    100 µl 10% SDS
    100 µl APS
    4 µl TEMED.

18. Pour each gel between the glass plates up to a level 2.5 cm from the upper edge of the small glass plate. Carefully overlay the gel with water to get a sharp meniscus. Leave 60 min for polymerization.

19. Prepare a mix for two stacking gels:

    670 µl 30% (v/v) 37.5:1 acrylamide/bisacrylamide
    500 µl 1 M Tris·Cl, pH 6.8
    2.7 ml H_2O
    40 µl 10% SDS
    40 µl 10% APS
    4 µl TEMED.

20. Remove the water from the top of each gel and carefully overlay the gel with the stacking mix. Carefully insert a comb without trapping air bubbles. Leave 15 min for polymerization.

21. Remove the comb and wash the wells with water. Place the gel in the electrophoresis tank and fill the chamber with running buffer.

22. Mix 2 µl of the translation reaction with 12 µl SDS sample buffer and denature the sample 4 min at 95°C in a heating block. Spin down briefly.

23. Load the gel with the samples, controls, and markers.

24. Prerun the gel 15 to 30 min at 70 V.

25. Raise the voltage to 120 V and run the gel 1 to 1.5 hr.

26. Remove the stacking gel and equilibrate the gel in ice-cold blotting buffer for at least 5 min.

27. Rinse the PVDF Western blotting membranes in methanol for 1 min and in blotting buffer for 5 min.

28. Take the gel out of the buffer, drain off excess fluid, and place a piece of 7×10–cm Whatman filter paper on the gel. Press carefully and remove air bubbles between gel and paper.

29. Pour blotting buffer in a plastic tray and place an opened blotting cassette in it. Soak the fiber pads with buffer. Prepare a blotting stack in the cassette, soaking the parts in blotting buffer and layering them in the following order (carefully remove air bubbles between each layer):

> Fiber pad
> Two pieces 7×10–cm Whatman 3MM filter paper
> Gel with paper side down
> Blotting membrane
> 1 piece Whatman 3MM filter paper
> Fiber pad.

30. Close the blotting cassette and insert it into the blotting holder. Put an ice tray in the buffer tank and fill the tank with blotting buffer.

31. Electroblot 1 hr at 120 V at a maximum of 250 mA.

32. Wash the membrane in TBS.

For chemiluminescence detection

33a. Block the membrane in 25 ml of $1\times$ maleic acid/2% blocking reagent for 40 min at room temperature or overnight at 4°C.

34a. Add 5 μl streptavidin-POD conjugate to the blocking solution and incubate the membrane 30 min at room temperature under gentle shaking.

35a. Prepare 5 ml detection solution by mixing 5 ml blotting substrate and 50 ml solution B. Store at room temperature in the dark until use.

36a. Wash the membrane four times, 10 min each, with 50 ml washing buffer.

37a. Pour the luminescence substrate solution in a small plastic tray and submerge the membrane for 1 min under gentle shaking.

38a. Wrap the membrane in Saran wrap, avoiding air bubbles.

39a. Expose to an X-ray film or analyze the blot using a luminescence detector. Examine the pattern of translation products.

For HA tag detection

33b. Block the membrane in 25 ml TBS/1% blocking reagent for 1 hr at room temperature or overnight at 4°C.

34b. Reduce the volume of blocking solution to 15 ml and add 0.2 μg/ml anti-HA antibody. Incubate 1 hr at room temperature under gentle shaking.

35b. Wash three times, 5 min each, with 50 ml washing buffer.

36b. Add 1.3 μg/ml anti-Ig peroxidase-conjugated antibody and incubate 30 min at room temperature

37b. Wash three times, 5 min each, with 50 ml washing buffer.

38b. Incubate the membrane 1 min in detection solution (see step 37a).

39b. Wrap the membrane in Saran wrap, avoiding air bubbles. Expose to an X-ray film or analyze the blot using a luminescence detector. Examine the pattern of translation products.

SUPPORT PROTOCOL

ISOLATION AND ANALYSIS OF RNA

CAUTION: Guanidinium thiocyanate is an irritant and phenol is toxic. Handle with care.

NOTE: Keep all RNA samples ice cold.

Materials (*see* APPENDIX 1 *for items with* ✓)

    Patient blood collected in a plastic EDTA-coated blood-collection tube (Greiner) *or* cultured cells that are 80% to 90% confluent in 75-cm^2 tissue culture flask

    Histopaque-1077 (Sigma)

  ✓ PBS, pH 7.8, sterile, 4°C

    Cycloheximide (e.g., Sigma; optional)

    RNAzolB (Campro Scientific) containing guanidinium thiocyanate and phenol

    Chloroform

    Isopropanol

    70% and 100% ethanol, 4°C

    $T_{10}E_{0.1}$ buffer, pH 8.0: 10 mM Tris·Cl/0.1 mM EDTA, pH 8.0

    1 M NaOH

    SeaKem LE agarose (FMC Bioproducts)

  ✓ 1× TBE buffer, sterile

  ✓ 10 mg/ml ethidium bromide

  ✓ 3 M sodium acetate (pH adjusted to 5.6 with acetic acid)

  ✓ Formamide loading buffer

    RNA size and quality standards

    12-ml white cap tubes

    Beckman GS-6R microcentrifuge and swing-out rotor

    Cell scraper, sterile

    CEP-swabs (Life Technologies)

    Spin-Ease extraction tubes (Life Technologies)

Prepare RNA from patient blood

1a. Collect 10 ml patient blood in an EDTA blood-collection tube.

2a. Fill two 12-ml tubes (white caps) with 5 ml Histopaque-1077. Slowly add 5 ml blood on top of the Histopaque in each tube. Centrifuge 20 min at 950 × *g* in a swinging-bucket rotor, room temperature, without brakes.

3a. Discard most of the first layer (serum and thrombocytes). Transfer the second (white) layer (lymphocytes) to a separate 10-ml tube and wash with 10 ml 4°C sterile PBS. Centrifuge 10 min at 550 × *g* in swinging-bucket rotor, room temperature.

4a. Remove the supernatant by carefully inverting the tube. Resuspend the small white (sometimes red) pellet in the remaining fluid by tapping the tube. Place the tube on ice. Proceed to step 5.

Prepare RNA from cultured cells

1b. *Optional:* Incubate cultured cells in the presence of 100 µg/ml cycloheximide 4 to 8 hr prior to harvesting.

> CAUTION: *Cycloheximide is very toxic; prepare the solution in a fume hood.*

2b. Remove the medium from cultured cells that are 80% to 90% confluent in a 75-cm^2 flask. Wash cells carefully with 4°C sterile PBS.

3b. Add 5 ml PBS. Put the flask on ice and collect the cells by scraping with a sterile cell scraper. Transfer the cells to a new 12-ml tube (white cap). Add 5 ml PBS to the original flask and continue scraping. Collect all cells.

4b. Centrifuge 10 min at 550 × *g* in swinging-bucket rotor, room temperature. Carefully remove the supernatant by inverting the tube. Place the tube on ice. Proceed to step 5.

Prepare RNA from buccal cells

1c. Carefully rinse the patient's mouth a few times with water to remove any contaminants (e.g., coffee or food leftovers).

2c. Collect buccal cells by carefully scraping both sides of the mouth using a sterile CEP-swab.

3c. Put the swab in a Spin-Ease extraction tube and microcentrifuge briefly. Transfer the tube with swab to a new 1.5-ml tube. Immediately add 300 µl RNAzolB to the swab and incubate 5 min.

4c. Microcentrifuge 1 min at 12,000 × *g*. Proceed to step 6.

5. Add 1.2 ml RNAzolB directly to the pellet (for cells isolated from 5 ml blood or cultured cells from one 75-cm^2 flask, use 1 to 1.3 ml RNAzolB). Lyse the cells by aspirating and ejecting them several times through the pipet tip. Transfer the lysate to a 1.5- to 2-ml microcentrifuge tube.

6. Add 0.1 ml chloroform per 1 ml lysate. Shake vigorously 15 sec. Incubate 5 min on ice.

7. Microcentrifuge the suspension 15 min at 12,000 × *g*, 4°C.

8. Carefully transfer the upper, clear, aqueous phase (0.6 to 0.7 ml) to a new microcentrifuge tube. Add 1 vol isopropanol. Incubate 15 min on ice. If necessary store at 4° or 20°C.

9. Microcentrifuge 15 min at 12,000 × *g*, 4°C.

10. Remove the supernatant and wash the RNA pellet with 200 µl of 70% ethanol.

11. Microcentrifuge 10 min at 12,000 × *g*, 4°C. Carefully remove all supernatant. Air dry the pellet, but do not over dry.

12. Dissolve the pellet in 50 µl $T_{10}E_{0.1}$ buffer. Add 120 µl of 100% ethanol and store at −80°C until further use.

13. Carefully clean an agarose gel electrophoresis tank, gel tray, and comb(s) to remove any residual RNase activity. Incubate the equipment 1 hr in 1 M NaOH. Wash extensively with sterile water.

14. Prepare 1.5% (w/v) SeaKem LE agarose in sterile TBE buffer and add 10 mg/ml ethidium bromide to a final concentration of 0.2 µg/ml.

15. Take an aliquot equal to 5% to 10% of the stored RNA sample. Add 3 M sodium acetate, pH 5.6, to a final concentration of 0.3 M. Microcentrifuge 10 min at 12,000 × *g*, 4°C.

16. Carefully discard the supernatant and dissolve the pellet in 5 μl formamide loading buffer. Load the sample on the gel. Also include an RNA standard to provide a marker for RNA size and quantity.

17. Run the gel. Analyze the RNA on a UV transilluminator. Determine RNA concentration by comparison with a standard. Verify that RNA quality is acceptable—i.e., 18S and 23S rRNA bands are clearly visible and no contaminating DNA (largest band near top of gel) is visible.

ALTERNATE PROTOCOL

RADIOACTIVE PROTEIN TRUNCATION TEST USING COUPLED TRANSCRIPTION/TRANSLATION

Coupled in vitro transcription/translation is a simplified (but more difficult to control) one-tube alternative to the uncoupled reaction. With this system, the authors have successfully synthesized products of up to 160 kDa (the product of a 4-kb sequence), although the amount of background products increased considerably in such translations.

Materials (see APPENDIX 1 for items with ✓)
> TnT T7-Coupled Rabbit Reticulocyte Lysate System (Promega) containing:
> > TnT T7 RNA polymerase
> > TnT rabbit reticulocyte lysate
> > TnT reaction buffer
> > Amino acid mix minus leucine, minus methionine, and minus cysteine
> > Luciferase-encoding control plasmid
> > Luciferase
> > Luciferase assay reagent
> > 5 mCi/ml [³H]leucine (160 Ci/mmol, Amersham)
> 1.5 U/μl ribonuclease inhibitor (e.g., RNasin, Promega)
> H₂O, sterile
> ✓ 2× SDS sample buffer
> Molecular weight markers: ¹⁴C-labeled (Amersham) or prestained (Bio-Rad) protein markers
> ✓ Staining solution
> ✓ Destaining solution
> Dimethylsulfoxide (DMSO)
> DMSO/PPO (1 M 2,5-diphenyloxazole in DMSO; reuse two to three times) or other autoradiography-enhancing reagent (e.g., Amersham's Amplify, NEN Life Science's EN³HANCE)

> Plastic box
> Whatman 3MM filter paper
> Gel-drying apparatus
> X-Omat AR film (Eastman Kodak)

1. Prepare cDNA from reverse-transcribed patient RNA and amplify by two rounds of PCR, or amplify DNA sample (see Basic Protocol, steps 1 to 12).

2. Remove the TnT T7-Coupled Rabbit Reticulocyte Lysate System reagents from −70°C storage. Place TnT T7 RNA polymerase directly on ice, and after thawing, store all other components on ice. Rapidly thaw the rabbit reticulocyte lysate by hand warming.

> *Do not leave T7 RNA polymerase on ice too long. Refreeze the lysate as soon as possible after preparing reaction mix.*

3. Prepare transcription/translation reaction in a 1.5-ml microcentrifuge tube; also include a control reaction using the luciferase-encoding control plasmid included in the kit (in place of the PCR product):

> 12.5 μl TnT rabbit reticulocyte lysate
> 1 μl TnT reaction buffer
> 0.5 μl TnT T7 RNA polymerase
> 0.5 μl amino acid mixture minus leucine
> 2 μl 5 mCi/ml [^{3}H]leucine
> 0.5 μl 40 U/μl RNasin
> 50 to 500 ng PCR product (from step 1; ≤8 μl)
> Sterile H_2O to 25 μl.

Mix and incubate 60 min at 30°C (30°C is optimal, but variations from 25° to 37°C do not greatly affect translation efficiency).

Good results are also obtained when the transcription/translation reaction is reduced to a 12-μl reaction volume with a ratio of PCR product to TnT mix of ≤1:2.

4. Add 1 vol of 2× SDS sample buffer to the transcription/translation products and mix. Store the reactions at −70°C.

5. Prepare SDS-PAGE gel and electrophorese products of transcription/translation reaction along with appropriate molecular weight markers (see Basic Protocol, steps 16 to 27).

6. Remove the gel, place it in a plastic box, cover with staining solution, and incubate 15 to 30 min with gentle agitation.

When ^{14}C-labeled or prestained protein markers are used, staining can be omitted.

7. Discard the staining solution. Rinse gel briefly with destaining solution, remove it, and repeat destaining until the protein markers are clearly visible.

8. Wash the gel 15 min with water.

9. Dehydrate the gel by covering with DMSO and incubating 10 min with gentle agitation. Repeat once.

CAUTION: DMSO is a hazardous organic chemical; therefore, steps 8 to 10 should be performed in a fume hood.

10. Incubate the gel twice, 10 min each, with DMSO/PPO.

11. Wash the gel, which will turn white, 10 to 15 min with water.

12. Place the gel on Whatman 3MM filter paper and dry ≥1 hr at 60° to 70°C in a gel-drying apparatus.

13. Autoradiograph the dried gel using Kodak X-Omat AR film (e.g., see CPMB APPENDIX 3A). Examine the pattern of translation products in the autoradiogram. Look for a 62-kDa protein that is also detectable using luciferase assay reagent if the TnT system control plasmid has been used.

Usually, products from 50 ng of a 1.5-kb PCR product in a 12-μl transcription/translation reaction can be easily detected after overnight exposure (≥8 hr).

References: den Dunnen and van Ommen, 1999; Roest et al., 1993

Contributors: Rolf Vossen and Johan T. den Dunnen

Genotyping of Apolipoprotein E (APOE): Comparative Evaluation of Different Protocols

Figure 9.11.1 presents various APOE alleles.

BASIC PROTOCOL

APOE GENOTYPING BY RFLP ANALYSIS

Materials (see APPENDIX 1 for items with ✓)

3 to 5 ng/μl genomic DNA (isolated from blood) in TE buffer, pH 7.4 (see APPENDIX 1)
APOE oligonucleotide primers:
> Forward: 5′-TAA GCT TGG CAC GGC TGT CCA AGG A-3′
> Reverse: 5′-ACA GAA TTC GCC CCG GCC TGG TAC ACT GCC-3′
✓ 100 mM 4dNTP mix (Roche; 25 mM each dNTP)
5 U/μl *Taq* DNA polymerase (Roche)
Dimethylsulfoxide (DMSO)
10× PCR amplification buffer containing 15 mM MgCl$_2$ (Roche)
10 U/μl *Hha*I restriction endonuclease (Promcga)
10× buffer C (Promega)
Low-melting agarose (e.g., Fisher)
NuSieve agarose (FMC Bioproducts)
✓ 10× TBE buffer
✓ 10 mg/ml ethidium bromide
✓ 6× gel loading buffer

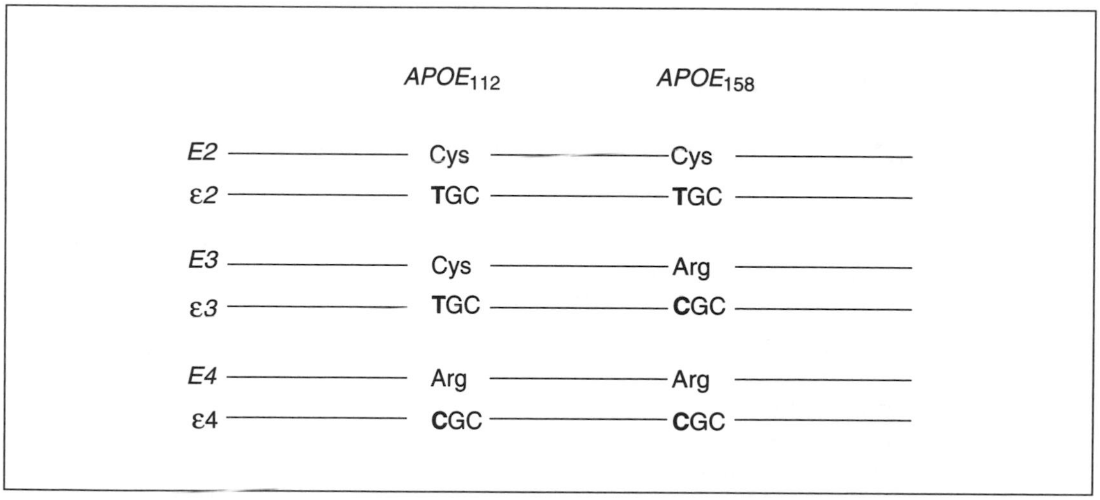

Figure 9.11.1 The various APOE alleles and genotypes are defined by the presence of two polymorphic loci at codons 112 and 158 of the APOE gene. The variable positions are shown with bold letters.

1. Set up amplification reactions in PCR tubes for a final volume of 25 μl using:

> 15 to 25 ng genomic DNA (add from 3 to 5 ng/μl genomic DNA preparation)
> 12.5 pmol APOE forward primer
> 12.5 pmol APOE reverse primer
> 6.25 nmol each dNTP (add from 100 mM 4dNTP mix)
> 1.25 U *Taq* DNA polymerase
> DMSO to 10% (v/v)
> $1\times$ PCR amplification buffer/1.5 mM $MgCl_2$ (add from $10\times$ PCR buffer/15 mM $MgCl_2$).

2. Perform PCR using the following amplification protocol:

Initial step:	10 min	94°C	(denaturation)
32 cycles:	30 sec	94°C	(denaturation)
	30 sec	56°C	(annealing)
	1 min	72°C	(extension)
Final step:	4 min	72°C	(extension)

3. Incubate the 25-μl PCR product 2 to 3 hr with 5 U of *Hha*I restriction endonuclease in $1\times$ buffer C at 37°C.

4. Prepare a 4% agarose gel solution by combining 1.5 g low-melting agarose and 4.5 g NuSieve agarose with 150 ml of $1\times$ TBE buffer. Mix thoroughly and heat in a microwave oven until the agarose has completely dissolved. Add 150 μl of 10 mg/ml ethidium bromide (0.15 mg), mix evenly with the liquid gel, and cast immediately in a gel tray of an appropriate size. Insert combs for 15 to 20 μl wells.

5. When the gel has polymerized, cover it completely with $1\times$ TBE buffer and remove the combs carefully. Load each well with a mixture of 10 μl cleaved PCR product (from step 3) and 2 μl of $6\times$ gel loading buffer.

6. Run the gel at 125 V for ~45 min and analyze immediately in a gel reader for different APOE genotypes, which are recognized by fragment lengths.

ALTERNATE PROTOCOL 1

APOE GENOTYPING BY REVERSE HYBRIDIZATION

Materials

> 3 to 5 ng/μl genomic DNA (isolated from blood) in TE buffer, pH 7.4 (see APPENDIX 1)
> INNO-LiPA ApoE kit (Innogenetics) including:
> > ApoE amplification buffer
> > ApoE primer mix
> > $MgCl_2$ solution
> > Glycerol solution
> > Membrane strips
> > Hybridization solution
> > Stringent wash solution
> > Denaturation solution
> > Conjugate, substrate
> > Conjugate diluent
> > Substrate solution
> > Substrate buffer

Rinse solution
Test troughs
Incubation trays
Taq DNA polymerase (Roche)

45°C water bath with shaking platform
Calibrated thermometer, high quality

1. Set up amplification reactions in PCR tubes for a final volume of 50 μl using:

   15 to 25 ng genomic DNA (add from 3 to 5 ng/μl genomic DNA preparation)
   10 μl amplification buffer
   10 μl ApoE primer mix
   10 μl MgCl$_2$ solution
   10 μl glycerol
   1 U *Taq* polymerase.

2. Perform PCR using the following amplification protocol:

Initial step:	5 min	95°C	(denaturation)
30 cycles:	30 sec	95°C	(denaturation)
	20 sec	60°C	(annealing)
	20 sec	72°C	(extension)
Final step:	10 min	72°C	(extension)

3. In a test trough, mix 10 μl PCR product with 10 μl denaturation solution and incubate 5 min at room temperature.

   Each of the following incubation, washing, and rinsing steps are carried out in a volume of 1 ml in a water bath on a shaking platform. The same test trough is used through the procedure and previous reagents must be removed before new ones are added.

4. Incubate the membrane strips in a mix of denatured DNA and the ready-to-use hybridization solution, for 30 min at 45°C (monitor with high-quality calibrated thermometer).

5. Wash the strips in wash solution at 45°C (monitor) once for 10 to 20 sec and then again for 10 min.

6. Rinse the strips twice in 1:5 (i.e., 20% rinse solution/80% water) rinse solution at room temperature, each time for 1 min.

7. Add 1 ml of 1:100 conjugate solution to each test trough and incubate the strips in this solution at room temperature for 30 min.

8. Rinse the strips twice in 1:5 rinse solution at room temperature, each time for 1 min.

9. Incubate the strips in the ready-to-use substrate buffer at room temperature for 1 min.

10. Add 1 ml of 1:100 substrate solution to each test trough and incubate the strips in this solution at room temperature for 30 min.

11. Rinse the strips twice in water at room temperature, each time for 3 min.

12. Read out the respective APOE genotypes from the membrane strip band pattern.

APOE GENOTYPING BY FLUORESCENCE POLARIZATION (FP)

Materials (*see* APPENDIX 1 *for items with* ✓)

> 3 to 5 ng/μl genomic DNA (isolated from blood) in TE buffer, pH 7.4 (see APPENDIX 1)
> APOE oligonucleotide primers for PCR:
> > Forward: 5′-TAA GCT TGG CAC GGC TGT CCA AGG A-3′
> > Reverse: 5′-ACA GAA TTC GCC CCG GCC TGG TAC ACT GCC-3′

✓ 40 mM 4dNTP mix (Roche; 10 mM each dNTP)
> 5 U/μl *Taq* DNA polymerase (Roche)
> Dimethylsulfoxide (DMSO)
> 10× PCR amplification buffer containing 15 mM $MgCl_2$ (Roche)
> Exo-SAP-IT (USB)
> Oligonucleotide primers for SNPs (elongation reaction):
> > 5′-GGC GCG GAC ATG GAG GAC GTG-3′ ($APOE_{112}$)
> > 5′-CGG CCT GGT ACA CTG CCA GGC-3′ ($APOE_{158}$)
> AcycloPrime-FP SNP Detection Kit (Perkin-Elmer Life Sciences) including:
> > 10× reaction buffer
> > AcycloTerminator mix
> > AcycloPol DNA polymerase

> 96-well PCR plates (MJ Research)
> Thermal cycler accommodating 96-well PCR plates
> 80°C water bath
> 96-well plate reader capable of fluorescence polarization measurements (e.g., Victor series from Perkin-Elmer Life Sciences)

1. Set up amplification reactions in wells of a 96-well PCR plate for a final volume of 30 μl using:

> 30 to 50 ng genomic DNA (add from 3 to 5 ng/μl genomic DNA preparation)
> 6 pmol APOE forward primer
> 6 pmol APOE reverse primer
> 0.5 nmol each dNTP (add from 40 mM 4dNTP mix)
> 1.25 U *Taq* DNA polymerase
> DMSO to 10% (v/v)
> 1× PCR amplification buffer/1.5 mM $MgCl_2$.

2. Perform PCR using the following amplification protocol:

Initial step:	10 min	94°C	(denaturation)
32 cycles:	30 sec	94°C	(denaturation)
	30 sec	56°C	(annealing)
	1 min	72°C	(extension)
Final step:	7 min	72°C	(extension)

3. Transfer 5 μl PCR product from each reaction to a well of a new PCR plate. Incubate with 2 μl of Exo-SAP-IT for 1 hr at 37°C.

4. Inactivate Exo-SAP-IT by incubating 15 min at 80°C.

5. Set up elongation reactions for each of the two APOE SNP sites in wells of a 96-well PCR plate for a final volume of 20 μl using:

> 7 μl of the 244-bp PCR-product/inactivated enzyme (step 4)
> 50 pmol SNP primer (APOE$_{112}$ or APOE$_{158}$)
> 2 μl 10× reaction buffer
> 1 μl AcycloTerminator mix
> 0.05 μl AcycloPol DNA polymerase.

6. Perform a primer elongation reaction using the following protocol:

Initial step:	2 min	95°C	(denaturation)
x cycles:	15 sec	95°C	(denaturation)
	30 sec	55°C	(annealing)
Final step:	2 min	15°C	(cooling)

where x is 30 cycles for APOE$_{112}$ or 60 cycles for APOE$_{158}$.

7. Centrifuge the samples at 150 × g, 4°C.

8. Analyze the samples on an FP plate reader.

ALTERNATE PROTOCOL 3

APOE GENOTYPING BY SNaPshot ANALYSIS

The amplification primers used in the SNaPshot procedure are different from those used in the Basic Protocol, since the protocols were developed in different laboratories.

Materials (see APPENDIX 1 for items with ✓)

> 3 to 5 ng/μl genomic DNA (isolated from blood) in TE buffer, pH 7.4 (see APPENDIX 1)
> APOE oligonucleotide primers:
> > Forward: 5′-CCA AGG AGC TGC AGG CGG CGC A-3′
> > Reverse: 5′-GCC CCG GCC TGG TAG ACT GCC A-3′
> ✓ 100 mM 4dNTP mix (Roche; 25 mM each dNTP)
> *Taq* DNA polymerase (Roche)
> Dimethylsulfoxide (DMSO)
> 10× PCR amplification buffer containing 15 mM MgCl$_2$ (Roche)
> QiaQuick DNA purification kit (Qiagen)
> Oligonucleotide primers for minisequencing:
> > 5′-CGG ACA TGG AGG ACG TG-3′ (APOE$_{112}$)
> > 5′-TTT TTT TTT TCC GAT GAC CTG CAG AAG-3′ (APOE$_{158}$)
> SNaPshot mix (containing fluorescent [F]ddNTPs; Applied Biosystems)
> Calf intestinal phosphatase (CIP; New England Biolabs)
> DNA size standard ladder (GeneScan-120 LIZ, Applied Biosystems)
> Hi-Di formamide (Applied Biosystems)
>
> 96- or 384-well PCR plates (MJ Research)
> 75°C water bath
> ABI PRISM 3100 Genetic Analyzer (Applied Biosystems)
> GeneMapper 2.0 or GeneScan software (Applied Biosystems)

1. Set up amplification reactions in 96- or 384-well PCR plates for a final volume of 25 µl using:

   30 to 40 ng genomic DNA (add from 3 to 5 ng/µl genomic DNA preparation)
   4 pmol forward APOE primer
   1.4 pmol reverse APOE primer
   5 nmol each dNTP (add from 100 mM 4dNTP mix)
   0.5 U *Taq* DNA polymerase
   DMSO to 10% (v/v)
   1× PCR amplification buffer/1.5 mM $MgCl_2$.

2. Perform PCR using the following amplification protocol:

Initial step	10 min	94°C	(denaturation)
35 cycles:	30 sec	94°C	(denaturation)
	30 sec	60°C	(annealing)
	30 sec	72°C	(extension)
Final step	10 min	72°C	(extension)

3. Remove excess dNTPs and primers with the QIAquick purification kit according to the kit instructions.

4. Set up minisequencing reactions in wells of 96- or 384-well PCR plates for a final volume of 10 µl using:

   1 µl of the 217-bp PCR product (step 3)
   2 pmol of each of the two minisequencing primers ($APOE_{112}$ and $APOE_{158}$)
   5 µl SNaPshot mix.

5. Perform PCR in 96- or 384-well plates using the following amplification protocol:

25 cycles:	10 sec	96°C	(denaturation)
	5 sec	50°C	(annealing)
	30 sec	60°C	(extension)

6. To minimize interference from the unincorporated [F]ddNTPs, incubate samples 1 hr at 37°C with 2.5 U calf intestinal phosphatase (CIP).

7. Inactivate CIP by incubating 15 min at 75°C.

8. Mix the SNaPshot product with a DNA size standard ladder and Hi-Di formamide and run on an ABI PRISM 3100 Genetic Analyzer.

9. Analyze the data with GeneMapper 2.0 or GeneScan software to identify APOE genotypes.

References: Eichner et al., 2002; Mayeux et al., 1998

Contributors: Martin Ingelsson, Youngah Shin, Michael C. Irizarry, Bradley T. Hyman, Lena Lilius, Charlotte Forsell, and Caroline Graff

CHAPTER 10

Cancer Genetics

R ecent advances in molecular genetics have demonstrated that cancer is a genetic disorder—sometimes inherited, but always associated with somatic genetic changes. Genetic studies have contributed enormously to our understanding of the pathogenesis of malignant diseases. Increasingly, they are finding their way into the armamentarium of the oncologist for diagnosis and management. The protocols in this chapter deal with methods for genetic analysis of cancer that have a place in routine clinical care.

UNIT 10.1 provides protocols for harvest and cytogenetic analysis of malignant hematological disease samples. Because of their relative accessibility to study, leukemias have the longest history of study for chromosome rearrangement. Identification of chromosome markers such as the Philadelphia chromosome can help in establishing a diagnosis and in following response to therapy. Many tumor-specific chromosome markers have been described in leukemias, and their recognition can be important in selecting appropriate means of therapy and in early detection of relapse. Cytogenetic analysis is also used in determining engraftment after bone marrow transplantation.

The study of chromosomes in solid tumors has been more difficult technically, and therefore less is known about tumor-specific rearrangements. Nevertheless, advances in cell culture and preparation of cells for chromosome analysis have enabled major progress in the examination of solid tumors. Techniques for metaphase harvest and cytogenetic analysis of cultures of solid tumors and lymphomas are provided in *UNIT 10.2*. These have proven very useful in distinguishing histologically similar tumors, such as Ewing sarcoma and neuroblastoma. It is likely that solid tumor analysis will increasingly play a role in clinical management as further technical advances are made.

The ability to recognize genetic changes at the molecular level promises to revolutionize the clinical evaluation of patients with cancer. Molecular studies obviate the need to culture cells and can detect minute numbers of malignant cells in small samples. The protocols in *UNIT 10.3* address the molecular analysis of genetic rearrangements in leukemias and lymphomas. Tumor-specific translocations juxtapose regions that are normally far apart in the genome. PCR assays have been devised to detect some of these translocations. Analysis of immunoglobulin or T cell receptor genes can provide evidence for clonality, and is useful in detecting tumor recurrence.

UNIT 10.4 describes protocols for molecular analysis of oncogene amplification. Amplification involves the replication of tens or hundreds of copies of a segment of the genome. It is one of many means of activation of proto-oncogenes and is often associated with enhanced degree of malignancy or tumor progression.

Current evidence suggests that malignancy is the consequence of an accumulation of genetic damage, representing a multi-step process. It can be expected that the study of these changes will increasingly assist in the diagnosis of cancer, and may ultimately provide clues to more specific modes of therapy.

Aside from inherited or acquired genetic changes associated with cancer, there are also epigenetic changes that may contribute to the malignant phenotype. *UNIT 10.5* provides protocols for the analysis of patterns of DNA methylation using PCR. This approach can be applied to the study of CpG methylation anywhere in the genome. It is finding applications both in the analysis of genetic changes in tumors and in the study of constitutional genetic disorders involving imprinted genes.

Contributor: Bruce R. Korf

UNIT 10.1

Metaphase Harvest and Cytogenetic Analysis of Malignant Hematological Specimens

Table 10.1.1 summarizes the major types of hematological malignancies, corresponding acceptable types of specimens, and appropriate cell culture methods.

NOTE: All reagents and equipment coming into contact with live cells must be sterile. All incubations are performed in a humidified 37°C, 5% CO_2 incubator unless otherwise specified.

BASIC PROTOCOL

PREPARATION OF CHROMOSOME SPREADS FROM BONE MARROW AND LEUKEMIC BLOOD SPECIMENS

Materials (*see* APPENDIX 1 *for items with* ✓)

✓ Complete culture medium A
✓ Complete culture medium B
  Bone marrow aspirate or leukemic blood, collected in a syringe or Vacutainer tube (Becton Dickinson) containing preservative-free heparin
  Bone marrow medium (e.g., Life Technologies, Sigma), optional
✓ 500 mg/ml ethidium bromide (Sigma) in HBSS (see recipe for HBSS)
  10 mg/ml Colcemid (Irving Scientific)
  0.075 M KCl solution, prepared fresh, 37°C

Table 10.1.1 Choice of Cell Culture Method[a]

Disease	Specimen	Type of culture	Protocol
ALL, AML, and MDS	Bone marrow or peripheral blood with >30% circulating blasts	24-hr, no stimulation	Basic Protocol
CML	Bone marrow or peripheral blood	24-hr, no stimulation	Basic Protocol
CLL	Bone marrow or peripheral blood	24-hr, no stimulation; 3-day PWM stimulation	Alternate Protocol 1
MM	Bone marrow or peripheral blood	24-hr, no stimulation; 5-day IL-4 stimulation	Alternate Protocol 2
Lymphoma	Lymph node	24-hr, no stimulation	Alternate Protocol 3
Lymphoma	Spleen	24-hr, no stimulation	Alternate Protocol 4

[a]Abbreviations: ALL, acute lymphocytic leukemia; AML, acute myeloid leukemia; CLL, chronic lymphoid leukemia; CML, chronic myeloid leukemia; IL-4, interleuken 4; MDS, myelodysplastic syndrome; MM, multiple myeloma; PWM, pokeweed mitogen.

3:1 (v/v) methanol/glacial acetic acid fixative, prepared fresh and tightly capped (to maintain its stability for several hours)

50-ml tissue culture flasks (Falcon)
15-ml culture tubes
Tabletop centrifuge (e.g., IEC HN-SII)
37°C, 5% CO_2 incubator
15-ml centrifuge tubes (Corning), sterile
85° to 90°C oven
Microwave, optional for STAT cases
Pasteur pipets
✓ Precleaned microscope slides (Gold Seal; Becton Dickinson)
Phase-contrast microscope

1. In a laminar flow hood, label two 50-ml tissue culture flasks with the following information: patient accession number, patient name, date, set-up time, and A or B to denote the complete culture medium.

2. Add 10 ml complete culture medium A to the first flask and 10 ml complete culture medium B to the second. If there is not enough specimen for two cultures (step 4), prepare only one.

3. Use gauze pad to remove cap on syringe or tube of a bone marrow aspirate or leukemic blood sample. Transfer sample to a sterile 15-ml culture tube. If sample was received in transport medium, lithium heparin, or a purple (EDTA) top tube, wash the sample twice in 10 ml bone marrow medium by centrifuging 10 min at 250 × g, room temperature.

4. Add 0.6 to 0.8 ml bone marrow or 1.0 ml leukemic blood to each culture flask and cap loosely and swirl to mix. If the white blood cell count is >100,000, however, add 0.1 to 0.2 ml bone marrow; if it is <1000, add 0.9 ml bone marrow.

5. Incubate cultures 15 to 24 hr at 37°C. Transfer to 15-ml centrifuge tubes. Add 100 ml of 500 mg/ml ethidium bromide in HBSS to each culture and swirl gently. Incubate 1 to 2 hr at 37°C. Add 100 ml of 10 mg/ml Colcemid to each culture and gently mix. Incubate 20 to 45 min at 37°C.

6. Centrifuge 10 min at 250 × g, room temperature. Remove and discard supernatant. Loosen each cell pellet by gently tapping on the tube or gently vortexing until the cells are resuspended and there is no sample lodged at the bottom of the tube.

7. Gently add, dropwise, 10 ml prewarmed 0.075 M KCl solution to each suspension, recap tubes, and gently invert to mix thoroughly. Incubate 15 to 20 min at 37°C.

8. Add dropwise 1 ml of 3:1 methanol/glacial acetic acid fixative to each culture. Recap tubes and gently invert. Centrifuge 10 min at 250 × g, room temperature. Remove and discard supernatants and loosen pellets. Gently add 10 ml of 3:1 methanol/glacial acetic acid fixative dropwise to the cells. Recap tubes and incubate ≥20 min at room temperature.

9. Centrifuge 10 min at 250 × g, room temperature. Remove and discard supernatants and loosen pellets. Gently add 10 ml of 3:1 methanol/glacial acetic acid fixative dropwise. Recap the tubes and gently invert. Repeat two additional times. Store fixed cells indefinitely at 4°C.

10a. *For STAT cases:* After overnight storage, prepare slides and age them in an 85° to 90°C oven for 1.5 to 2 hr or by microwaving the slides for 3.5 to 4.5 min. Allow slides to cool completely and proceed with Giemsa-trypsin banding (*UNIT 4.3*), decreasing trypsin times by ~10 sec to compensate for the freshness of the slides.

10b. *For all other cases:* Let tubes stand 15 min at room temperature. Centrifuge 10 min at 250 × *g*, room temperature. Remove and discard supernatant. Resuspend each pellet in ~0.5 to 3.0 ml fresh 3:1 methanol/glacial acetic acid fixative. Add enough fixative to obtain a suspension that is slightly cloudy in appearance.

11b. Draw a small sample of cell suspension into a Pasteur pipet. Do not suction cells past the stem of the Pasteur pipet as they will not be retrievable. From a height of ~12.7 cm (5 in.) to 61 cm (2 ft), drop four to five drops onto a precleaned microscope slide.

> *Laboratories vary in temperature and humidity; therefore, it is best to try several different methods to determine whether slides should be wet or dry and if they should be cold or at room temperature.*

12b. Check slide for number and quality of chromosome spreads using a phase-contrast microscope. If spreading is inadequate, make necessary adjustments as suggested in Table 10.1.2 (also see *UNIT 4.1*). Make four slides, two from each culture. Bake slides ~2 hr in an 85° to 90°C oven.

ALTERNATE PROTOCOL 1

PREPARATION OF CHROMOSOME SPREADS FROM CHRONIC LYMPHOCYTE LEUKEMIA BONE MARROW AND PERIPHERAL BLOOD SPECIMENS

Additional Materials (*also see Basic Protocol; see* APPENDIX 1 *for items with* ✓)

✓ Pokeweed mitogen (PWM): LECTIN-Phytolacca American Lectin (Sigma) reconstituted with 5 ml HBSS and stored at −20°C (see recipe for HBSS)

1. Prepare bone marrow aspirate or peripheral blood cultures as described (see Basic Protocol, steps 1 to 4), setting up a total of four cultures as follows:

> Two 24-hr cultures, no stimulation, in complete culture media A and B
> Two 3-day cultures, PWM stimulation, in complete culture media A and B.

2. Add 0.3 ml PWM per 10 ml complete culture medium (A or B) to the appropriate cultures. Mix by gently swirling contents of the flasks. Cap 3-day flasks tightly. Incubate unstimulated 24-hr cultures and 3-day PWM-stimulated cultures at 37°C.

3. Harvest cells (omitting ethidium bromide in the 3-day cultures), fix, and prepare chromosome spreads as described (see Basic Protocol, steps 5 to 12).

ALTERNATE PROTOCOL 2

PREPARATION OF CHROMOSOME SPREADS FROM MULTIPLE MYELOMA BONE MARROW AND PERIPHERAL BLOOD SPECIMENS

Additional Materials (*also see Basic Protocol*)

1 μg/500 μl interleukin 4 (IL-4; Sigma) in RPMI (Life Technologies; Irving Scientific), store in 100-μl aliquots at −20°C

1. Prepare bone marrow aspirate or peripheral blood cultures as described (see Basic Protocol, steps 1 to 4), setting up a total of three cultures as follows:

> Two 24-hr cultures, no stimulation, in complete culture media A and B
> One 5-day culture, IL-4 stimulation, in either complete culture medium A or B.

Table 10.1.2 Troubleshooting Guide for Chromosome Preparations

Problem	Possible cause	Solution
No or few metaphases	Poor growth	Check specimen type and diagnosis
	Inappropriate specimen (e.g., blood with <30% circulating blasts)	Repeat preparation with appropriate specimen
	Clotted specimen	More heparin and/or mix heparin with specimen thoroughly
	Aged specimen	Obtain specimen more quickly
	Hypocellular marrow	Obtain more specimen
	No mitogens added to stimulate mitosis	Repeat preparation and add appropriate mitogens
	Reagents not prepared fresh or properly	Check materials used
	Incubation temperature incorrect	Check incubator temperature
Scattered chromosomes	Cells broken during harvest procedures	Add hypotonic solution or fixative more slowly
	Cells broken during slidemaking	Drop from lower height; use cold, dry slides; increase drying time
	Hypotonic treatment time too long	Decrease hypotonic treatment time
Visible cytoplasm	Hypotonic treatment time not long enough	Increase hypotonic treatment time
	Drying problem during slidemaking	Fix in refrigerator overnight and repeatedly wash with fixative
Contracted chromosomes	Mitotic inhibitor treatment time too long or mitotic inhibitor too concentrated	Decrease mitotic inhibitor treatment time; reduce concentration of mitotic inhibitor
	No ethidium bromide treatment during harvest procedures	Add ethidium bromide
	Ethidium bromide treatment time not long enough	Increase ethidium bromide treatment time
	Majority of cells not in prometaphase stage	Decrease or increase incubation/culturing time
Chromatids separating	Ethidium bromide and/or Colcemid treatment time too long during harvest procedures	Decrease ethidium bromide and/or Colcemid treatment time
	Drying too fast	Decrease drying time; use cold, dry slides
Too many crossed chromosomes or chromosomes sticking together	Ethidium bromide treatment time during harvest procedure too long	Decrease ethidium bromide treatment time
	Ethidium bromide solution too concentrated	Decrease ethidium bromide solution concentration
	Hypotonic treatment time not long enough	Increase hypotonic treatment time
	Fixation problem during harvest procedure	Add fixative after hypotonic treatment drop by drop
	Insufficient fixation	Repeat fix washes
	Too many cells on slide	Drop fewer drops of cell suspension onto the slide; make cell suspension more dilute with fix
	Spreading problem during slidemaking	Drop onto almost dry, cold slides; drop from an increased height; blow onto slides during drying

10

2. Add 100 μl IL-4 in RPMI to the 5-day culture. Mix by gently swirling flasks. Cap 5-day flasks tightly. Incubate 24-hr and 5-day cultures at 37°C.

3. Harvest cells (omitting ethidium bromide in the 5-day culture), fix, and prepare chromosome spreads as described (see Basic Protocol, steps 5 to 12).

ALTERNATE PROTOCOL 3

PREPARATION OF CHROMOSOME SPREADS FROM LYMPH NODE SPECIMENS

Additional Materials (also see Basic Protocol; see APPENDIX 1 for items with ✓)

Lymph node specimen
Bone marrow medium (e.g., Life Technologies, Sigma)
✓ Collagenase solution

5-ml petri dish, 60 × 15–mm (Falcon)
Disposable scalpels

1. Write the following information on a 5-ml petri dish: accession number, patient name, set-up date and time.

2. Use a disposable scalpel or pipet to transfer a lymph node specimen (≤3 mm³) from its container to the labeled petri dish. Chop or mince specimen until no chunks of tissue remain. Add a small amount of bone marrow medium to the dish if the specimen begins to dry out.

3. Add the following to the dish:

> 4.0 ml bone marrow medium
> 1 ml collagenase solution
> 8 ml Colcemid.

> Swirl gently and incubate 15 to 24 hr at 37°C.

4. Transfer sample to a centrifuge tube and harvest cells, fix, and prepare chromosome spreads as described (see Basic Protocol, steps 6 to 12).

ALTERNATE PROTOCOL 4

PREPARATION OF CHROMOSOME SPREADS FROM SPLEEN

Additional Materials (also see Basic Protocol; see APPENDIX 1 for items with ✓)

Spleen specimen
✓ Collagenase solution, optional

5-ml petri dish, 60 × 15–mm (Falcon)
Disposable scalpels

1. Write the following information on a 5-ml petri dish: accession number and patient name. Use a disposable scalpel or pipet to transfer a spleen specimen (≤ 3 mm³) from its container to the labeled dish. Chop or mince specimen until no chunks of tissue remain. Add 0.5 ml collagenase solution if there are still small pieces.

2. Label two 50-ml tissue culture flasks with the following information: patient accession number, patient name, date, set-up time, and A or B to denote the medium. Add 10 ml complete culture medium A and B to the appropriate flasks.

3. Transfer half of the sample to each of the flasks and incubate 15 to 24 hr at 37°C. Transfer each sample to a centrifuge tube and harvest cells, fix, and prepare chromosome spreads as described (see Basic Protocol, steps 6 to 12).

Internet Resource

http://cgap.nci.nih.gov/Chromosomes/Mitelman

References: ISCN, 1991; Jaffe, 2001; Sandberg, 1990

Contributor: Paola Dal Cin

UNIT 10.2

Metaphase Harvest and Cytogenetic Analysis of Solid Tumor Cultures

NOTE: All reagents and equipment coming into contact with live cells must be sterile. All incubations are performed in a humidified 37°C, 5% CO_2 incubator unless otherwise specified.

BASIC PROTOCOL

CYTOGENETIC ANALYSIS OF METAPHASE CELLS FROM SOLID TUMOR, LYMPHOMA, OR EFFUSION SAMPLES

Materials

    10 μg/ml Colcemid (GIBCO/BRL)
    Actively growing tumor cell culture in 25-cm^2 tissue culture flask in 5 ml medium (see Support Protocols 1 to 3)
    1× trypsin/EDTA (GIBCO/BRL), for monolayer culture only
    0.067 M KCl
    Fixative: 3:1 (v/v) methanol/glacial acetic acid, prepared fresh

    15-ml centrifuge tubes
    Inverted microscope, for monolayer culture only
    Centrifuge: Fisher Marathon 21K, Becton Dickinson Primary Care Dynac II, or equivalent
    60°C or 70° to 75°C slide warmer

1. Add 10 μl of 10 μg/ml Colcemid to an actively growing tumor cell culture (final 0.02 μg/ml). Use 0.04 to 0.06 μg/ml for very slowly growing tumor populations. Incubate culture 15 to 18 hr.

2a. *For monolayer (attached) cultures:* When cells are ∼80% confluent, transfer medium to a 15-ml centrifuge tube. Add 2 ml of 1× trypsin/EDTA to flask, swirl briefly, and add to centrifuge tube. Add fresh 2 ml trypsin/EDTA to flask and incubate 5 min. Examine cells with an inverted microscope. Tap flask sharply to dislodge any attached cells. If many cells remain attached, incubate an additional 5 to 10 min. Add remaining cell mixture to the 15-ml tube.

2b. *For suspension cultures:* Within 1 day after culture establishment, transfer cells and medium to a 15-ml centrifuge tube.

3. Centrifuge 10 min at 300 × g, room temperature. Discard supernatant and break up cell pellet by flicking tube vigorously. Resuspend cells in 7 ml of 0.067 M KCl and incubate 2 min. Add seven drops of fixative to tube and invert several times to mix. Centrifuge 10 min at 300 × g and discard supernatant. Break up pellet and resuspend cells in 7 ml fixative. Incubate cells ≥1 hr (and up to 4 weeks) at −20°C. (Freezing is optional when metaphases are stained with quinacrine; UNIT 4.3.)

4. Centrifuge cells 10 min at 300 × g, discard supernatant, and resuspend in 7 ml fresh fixative. Centrifuge again and resuspend cells in several drops of fresh fixative. Prepare chromosome spreads (UNIT 4.1), adjusting temperature and humidity if necessary. If chromosomes are poorly spread and contained within a ball of cytoplasm, harvest additional cultures with a 10-min incubation in 0.067 M KCl. Incubate slides ≥24 hr (and <2 weeks) on a 60°C slide warmer or 1 hr on a 70° to 75°C slide warmer.

5. Stain slides using the GTG-banding method (UNIT 4.3). If the 70° to 75°C slide warmer is used, decrease trypsin incubation time by 50%.

6. Analyze ten or more metaphases from the neoplastic population and at least one diploid metaphase to establish the constitutional karyotype. If slides have few metaphase cells, carry out another overnight metaphase harvest using 2- or 3-fold more Colcemid (0.04 to 0.06 μg/ml).

SUPPORT PROTOCOL 1

DISAGGREGATION AND CULTURE OF SOLID TUMORS

Additional Materials (*also see Basic Protocol; see* APPENDIX 1 *for items with* ✓)

Tumor specimen (any solid carcinoma, sarcoma, germ cell, or neural tumor)
✓ Tumor transport solution, room temperature
✓ Complete RPMI/15% (v/v) FBS supplemented with 2.5 μg/ml amphotericin (store medium ≤1 month at 4°C), 37°C
✓ 10 mg/ml collagenase

Scalpel blades, sterile
25-cm² tissue culture flasks

1. Immediately place a surgically removed tumor specimen in tumor transport solution. Transport to the laboratory as rapidly as possible. If tumor specimens cannot be disaggregated immediately, hold ≤12 hr in transport solution at 4°C.

2. Mince specimens into fragments <1 mm³ with opposed scalpel blades. Add 5 to 10 ml prewarmed supplemented RPMI/15% FBS with 1 mg/ml collagenase. Incubate overnight.

3. Pipet tumor collagenase mixture up and down several times to disaggregate cell clusters and centrifuge 10 min at 300 × g. Discard supernatant and resuspend pellet in 5 ml supplemented medium without collagenase.

4. Transfer 2 ml to a 25-cm² tissue culture flask that is on its side and examine the flask with an inverted microscope. Adjust amount of cell suspension so that intact single cells and cell clusters cover ~25% to 50% of the flask surface. Distribute remaining material to additional flasks, establishing at least three cultures if possible. Adjust final volumes to 5 ml per flask with supplemented medium without collagenase. Incubate.

An enhanced RPMI/15% FBS medium that is superior in promoting short-term growth of ~25% of solid tumors contains 0.25% (v/v) aprotinin (Sigma), 1% (v/v) bovine pituitary extract (Collaborative Biomedical), and 0.5% (v/v) Mito + serum (Collaborative Biomedical). The enhanced medium should not be used during cell disaggregation (steps 2 and 3).

5. Examine cultures with an inverted microscope each day to observe which cell types are proliferating. Change medium when indicator color begins to turn yellow. If cultured cells appear to be fibroblasts, discard preparation. When the tumor cells are growing actively and cover ~80% of the flask surface (typically within 4 days), prepare for metaphase harvesting as described (see Basic Protocol).

SUPPORT PROTOCOL 2

CULTURE OF LYMPH NODE (LYMPHOMA) SPECIMENS FOR CYTOGENETIC ANALYSIS

This cytogenetic approach is identical for all lymphomas, including Hodgkin's disease and non-Hodgkin's lymphoma.

Additional Materials (also see Basic Protocol; see APPENDIX 1 for items with ✓)

    Lymph node specimen
✓ Tumor transport solution, ice cold
✓ Complete RPMI/15% (v/v) FBS supplemented with 2.5 μg/ml amphotericin (store medium ≤1 month at 4°C), 37°C
✓ 10 mg/ml collagenase
    Ficoll (Sigma)
✓ HBSS
    0.075 M KCl
    25-cm^2 tissue culture flasks *or* 60-mm petri dish

1. Place surgically removed lymph node specimen in ice-cold tumor transport solution and keep for 1 to 6 hr at 4°C.

2. Mince lymph node specimens into fragments <1 mm^3 with opposed scalpel blades. Add 5 ml prewarmed supplemented RPMI/15% FBS with 1 mg/ml collagenase to the minced tissue. Incubate 30 min.

3. Carefully layer tumor suspension over 5 ml Ficoll in a 15-ml centrifuge tube. Centrifuge 20 min at $700 \times g$. Remove Ficoll/collagenase interface (white, opaque layer of mononuclear cells), including 1 to 2 ml of adjacent Ficoll and collagenase, with a sterile pipet.

4. Add 10 ml HBSS to the mononuclear cells and centrifuge 10 min at $300 \times g$. Discard supernatant and resuspend mononuclear cell pellet in 5 ml supplemented RPMI/15% FBS without collagenase. Transfer to a 25-cm^2 flask or a 60-mm petri dish and add Colcemid to a final concentration of 0.01 μg/ml. Incubate overnight.

5. Proceed with metaphase cell harvesting of suspension cultures as described (see Basic Protocol, step 2b), but substitute 0.075 M KCl in the hypotonic treatment step.

SUPPORT PROTOCOL 3

PREPARATION OF EFFUSION (FLUID) SPECIMENS FOR CYTOGENETIC ANALYSIS

Additional Materials (also see Basic Protocol; see APPENDIX 1 for items with ✓)

    Malignant effusion sample
    10 μg/ml vinblastine (Lilly)
    10 mg/ml ethidium bromide solution, store at 4°C in dark or in foil-wrapped bottle

✓ Complete RPMI/15% FBS supplemented with 2.5 μg/ml amphotericin (store medium ≤1 month at 4°C), 37°C

0.85% (w/v) ammonium chloride, for bloody samples only

50-ml centrifuge tubes
25-cm² tissue culture flasks

1. Immediately transport malignant effusion sample to the laboratory before it has time to cool to room temperature. Place 30 ml effusion into a 50-ml centrifuge tube and add 150 μl of 10 μg/ml vinblastine and 30 μl of 10 mg/ml ethidium bromide. Incubate 1 hr and proceed with metaphase harvesting as described (see Basic Protocol, step 2b). Establish multiple harvests if possible.

2. Establish multiple (if possible) tissue cultures while the direct harvest is incubating by distributing remainder of malignant effusion to 50-ml tubes. Centrifuge 15 min at $300 \times g$ and discard supernatant.

3. If cell pellets are bloody, resuspend each pellet in 10 ml of 0.85% ammonium chloride and transfer to 15-ml centrifuge tube. Incubate 15 min at room temperature and centrifuge 10 min at $300 \times g$. Discard supernatant. Omit this step if pellets are not bloody.

4. Resuspend one cell pellet in 5 ml prewarmed supplemented RPMI/15% FBS and transfer to a 25-cm² flask. Assess cellularity using an inverted microscope. Plate remaining cell pellets in several more flasks, varying the number of cells in different flasks. Flasks should be 30% to 75% confluent once cells have attached. Incubate, changing medium when indicator turns yellow.

5. When tumor cells are growing actively and cover ~80% of the flask surface (generally after 2 to 5 days), prepare for metaphase harvesting of monolayer cultures as described (see Basic Protocol).

References: Harrison, 1992; Mandahl, 1992; Mitelman, 1991; Sandberg, 1990

Contributor: Jonathan Fletcher

Molecular Analysis of DNA Rearrangements in Leukemias and Non-Hodgkin's Lymphomas

STRATEGIC PLANNING

Because clonal antigen-receptor gene rearrangements are present in most lymphocytic neoplasms, this marker should be analyzed first in attempting to evaluate a lymphoid tissue biopsy specimen. Translocations may then be used as markers in secondary tests to subcategorize a tumor. There may be occasions when histologic and other data strongly predict the existence of a particular translocation and thus direct testing for the translocation is reasonable. Antigen-receptor gene rearrangements are not applicable to most cases of nonlymphocytic neoplasia, and in these cases chromosomal translocations constitute the only appropriate molecular marker for diagnosis.

Southern blot hybridization is the most reliable method for detecting rearrangements of immunoglobulin genes (heavy chain and κ and λ light chains) and the TCR-β and -γ chain genes.

As long as adequate test DNA is available, Basic Protocol 1 should be used to analyze all five genes as a stringent test for clonality because proliferations thought to be B or T cell in nature may contain clones of the other lineage. Tests for the various antigen-receptor genes can be run simultaneously, or if DNA is limited, Southern blots may be rehybridized sequentially for genes that are assayed using the same restriction enzymes.

PCR analysis for clonal rearrangements of some antigen-receptor genes (see Alternate Protocols 1 and 2) may also be used for rearrangements of immunoglobulin heavy chain genes and TCR-γ and -δ chain genes under certain circumstances: when there is limited DNA available, when the number of atypical cells within a specimen is small, and when a rapid and inexpensive screening test for clonal antigen-receptor gene rearrangements, particularly those for immunoglobulin heavy chain gene, is desired. However, because PCR does not detect all possible immunoglobulin heavy chain gene rearrangements, it cannot be considered the definitive method of analysis.

Southern blot hybridization may be used to detect chromosomal translocations only when breakpoints lie within regions of DNA no larger than ~25 kb. In translocations associated with tighter clustering of breakpoints, PCR is a faster and probably a less expensive method for detection. PCR is at least three orders of magnitude more sensitive than Southern blot hybridization. As sensitivity is of paramount importance in monitoring residual disease after therapy, PCR is the only option for detecting residual disease. For translocations associated with breakpoints that are highly scattered throughout the DNA, reverse transcription (RT)-PCR should be used (see Basic Protocol 2).

NOTE: PCR requires special precautions to prevent contamination. Familiarity with standard methods is essential.

NOTE: Experiments involving RNA require careful technique to prevent RNA degradation. Prepare all solutions in DEPC-treated water.

BASIC PROTOCOL 1

DETECTION OF CLONAL ANTIGEN-RECEPTOR GENE REARRANGEMENTS BY SOUTHERN BLOT HYBRIDIZATION

Southern blot hybridization can detect uniform rearrangements of immunoglobulin genes (heavy chain and κ and λ light chains) or T cell receptor genes (β and γ chains).

Materials *(see APPENDIX 1 for items with ✓)*

    10 μg DNA sample (*APPENDIX 3A*)
    Appropriate restriction endonucleases (Table 10.3.1) and buffers
    10 μg normal and rearranged control DNA (e.g., placenta and tumor DNA, respectively)
✓ 6× gel loading buffer
    Molecular size markers (e.g., *Hin*dIII digest of λ DNA)
    1.5 M NaCl/0.5 M NaOH
    0.5 M Tris·Cl/3 M NaCl, pH 7.0
✓ 3× and 2× SSC
✓ Prehybridization solution
    Probe DNA: 50 ng probe fragment (Table 10.3.1) or ~100 ng probe fragment with
        plasmid vector
    Random oligonucleotide primer labeling kit or equivalent reagents (*APPENDIX 3E*)
    20 μCi/μl [α-^{32}P]dCTP (6000 Ci/mmol)
✓ Hybridization solution

✓ High-salt wash solution
✓ Low-salt wash solution, 65°C

Recirculating pump for gel running buffer
Clean sponge
Whatman 3MM filter paper
UV cross-linker (e.g., UV Stratalinker; Stratagene)
Platform rocker
42° to 45°C incubator
G-50 Sephadex column (e.g., NuClean spin column; IBI or *APPENDIX 3E*)
100° and 65°C water baths
Lid locks for 1.5-ml centrifuge tubes
Plastic box with tight-fitting cover

1. Digest 10 μg DNA sample in a 40-μl reaction using appropriate restriction endonuclease (Table 10.3.1) and buffer at the appropriate temperature. For each enzyme, include control reactions with normal and rearranged DNA.

Table 10.3.1 Probes for Detecting Gene Rearrangements

Gene	Probe fragment (flanking restriction sites)	Appropriate restriction enzyme for digestion of tissue DNA	Size(s) of germline fragment(s) detected
Immunoglobulin heavy chain	J_H 3.5 kb (*Bgl*II)	*Bam*HI *Eco* RI *Hin*dIII	17 kb 16 kb 9.5 kb
Immunoglobulin light chain κ	C_κ 2.5 kb (*Eco*RI) J_κ 1.8 kb (*Sac*I)	*Bam*HI *Bam* HI *Sac*I	12 kb 12 kb 1.8 kb
Immunoglobulin light chain λ	C_λ 3 kb (*Bam*HI-*Hin*dIII)	*Eco*RI	8.0, 14, and 16 kb in most individuals; 13, 18, and 24 kb in individuals with Ig λ polymorphisms
TCR-α	—[b]	—[b]	—[b]
TCR-β	J_β1 0.55 kb (*Xba*I)	*Bgl*II *Hin*dIII	11 kb 5 kb
	J_β2 4.5 kb (*Eco*RI)	*Bgl*II	10 and 9.3 kb polymorphic fragments with about equal frequency
		*Hin*dIII	9 kb
TCR-γ	J_γ 2.2 kb (*Hin*dIII)	*Bgl*II *Bam*HI	12.6 and 9.5 kb 19 and 14 kb
TCR-δ[a]	J_δ 1 1.9 kb (*Bgl*II-*Eco*RI)	*Bgl*II *Eco*RV	4.9 kb 9.8 and 2.8 kb
	J_δ2 1.2 kb (*Eco*RI-*Sac*I)	*Bgl*II *Eco*RV	5.3 kb 9.8 kb
	J_δ3 2.3 kb (*Sac*I-*Eco*RV)	*Bgl*II	5.4 kb

[a] Southern blot analysis carried out only in cases of known T cell neoplasms that lack TCR-β or -γ chain gene clonal rearrangements or rare T cell neoplasms expressing TCR-γ δ.

[b] There are too many J_α segments to be analyzed routinely.

2. Pour a 300-ml 13 × 25–cm or a 350-ml 20 × 25–cm 0.8% (w/v) agarose gel in TAE buffer with 2 μg/ml ethidium bromide and use a comb with 8 × 1–mm slots (APPENDIX 3G).

3. Add 8 μl of 6× gel loading buffer to digest (1× final) and load into well. Include a lane with molecular size markers. Electrophorese 20 to 24 hr at 40 V, with a buffer recirculating pump, until high-molecular-weight DNA has migrated ≥3 cm from wells. Photograph gel with a ruler aligned with molecular weight markers. Leave gel on UV transilluminator 1 min to nick DNA. Cut off lower-left corner of gel for future orientation.

4. Incubate gel 30 min in 1.5 M NaCl/0.5 M NaOH with occasional shaking. Rinse gel in water. Incubate 30 min in 0.5 M Tris·Cl/3 M NaCl with occasional shaking. Rinse gel in water. Transfer DNA to a nylon membrane by overnight Southern blotting (APPENDIX 3G) using a soaked clean sponge half-submerged in buffer and topped with Whatman 3MM paper, rather than a solid support with wick.

5. Disassemble transfer set up. Label nylon membrane at the bottom with a pencil. If portions of the membrane will be hybridized with different probes, cut membrane with a clean, single-edged blade using gel wells as a guide. Rinse membrane in 2× SSC. Use a UV cross-linker to cross-link DNA to membrane. Store at room temperature between sheets of Whatman 3MM filter paper or plastic wrap until needed.

6. Wet membrane with 3× SSC solution and place in a heat-sealable bag. Double heat-seal three sides of the bag. Add 10 ml prehybridization solution. Remove bubbles by gently rolling a plastic pipet over the bag. Seal fourth side. Tape bag to a platform rocker in an incubator. Prehybridize 3 to 24 hr at 42° to 45°C.

7. Radiolabel probe DNA to a specific activity of 2×10^8 cpm/μg using random oligonucleotide primer labeling kit and 50 μCi [α-^{32}P]dCTP. Separate probe from unincorporated nucleotides by chromatography on a G-50 Sephadex column. Boil probe 10 min in tube with a lid lock. Cool on ice and microcentrifuge briefly at maximum speed.

8. Open hybridization bag and discard prehybridization solution. Add 10 ml hybridization solution and $10–40 \times 10^6$ cpm boiled radiolabeled probe per 100 cm^2 membrane. Remove air bubbles and seal bag. Tape bag to rocker and incubate overnight (≥8 hr) at 42°C.

9. Remove membrane from bag and place in 500 ml high-salt wash solution in a clean plastic box that has a tight-fitting cover. Shake 5 min at room temperature. Dispose radioactive hybridization and wash solutions appropriately and replace with 500 ml fresh high-salt wash solution. Shake 5 min at room temperature.

10. Drain solution and remove membrane. Add 1 liter 65°C low-salt wash solution to box and add membrane. Do not pour hot solution directly on membrane. Cover and incubate 20 to 30 min in a 65°C water bath with gentle shaking. When processing multiple filters, change positions of membranes after 15 min.

11. Pour off solution and repeat wash. Check that membrane radioactivity is near background with a Geiger counter. If autoradiogram has a high background, rewash membrane in low-salt wash solution 30 min at higher temperature (66° to 68°C). Check with Geiger counter and rewash again if needed.

12. Blot membrane dry on Whatman 3MM filter paper. Wrap in plastic wrap and set against X-ray film in a light-tight cassette with an intensifying screen. Expose film 3 to 4 days (or 8 to 12 hr for membranes with higher background reading; 7 days for cases with small tumor amounts) at −70°C. Develop film. If problems occur with hybridization, see Table 10.3.2.

Table 10.3.2 Troubleshooting Guide for Southern Blot Hybridization

Problem	Possible cause	Solution
Weak hybridization signal in specific lanes of the blot	Partially degraded specimen DNA	Check DNA in ethidium bromide–stained agarose gel and look for degradation. Either prepare new DNA sample or add more DNA to lane.
	Inadequately digested specimen DNA	Check DNA in ethidium bromide–stained agarose gel and look for incomplete digestion. Evaluate DNA purity and concentration by spectrophotometry readings. If impure, reextract and precipitate DNA; if too viscous, dilute. (1 μg/μl is a good working concentration.)
Weak hybridization signal in entire blot	Poor transfer of specimen DNA	Repeat electrophoresis and blot with care. UV-fix DNA to membrane.
	Low specific activity of probe	Strip and rehybridize membrane with fresh high-specific-activity probe
High background	Nonspecific hybridization	Rewash blot at higher stringency
		Strip and rehybridize blot with new purified probe free of unincorporated radionucleotide
		Redo blot with care not to fold or crush membrane

ALTERNATE PROTOCOL 1

DETECTION OF CLONAL IMMUNOGLOBULIN HEAVY CHAIN GENE REARRANGEMENTS BY PCR

Additional Materials (also see Basic Protocol 1; see APPENDIX 1 for items with ✓)

    1 to 2 ml bone marrow or 5 ml whole blood
✓ PBS
✓ Ficoll-Hypaque solution
    0.1 μg/μl normal and rearranged control DNA (e.g., polyclonal reactive lymphoid DNA and tumor DNA from a B cell line or clonal tissue)
✓ 10× PCR amplification buffer
✓ 25 mM MgCl$_2$
✓ 10 mM 4dNTP mix
    10 μM immunoglobulin heavy chain oligonucleotide outer and inner primers (V$_H$ and J$_H$)
    5 U/μl *Taq* DNA polymerase
    Mineral oil, if needed
    0.2 U/μl Klenow fragment of *E. coli* DNA polymerase I in 50 mM Tris·Cl (pH 8.0)/ 10 mM MgCl$_2$ (React 1 buffer; GIBCO/BRL)

    15-ml centrifuge tubes
    Centrifuge and rotor capable of producing 400 × *g*
    Thermal cycler (Perkin-Elmer Cetus 9600 or equivalent) with appropriate tubes

1. Place 1 to 2 ml bone marrow or 5 ml whole blood in a 15-ml centrifuge tube, add an equal volume of PBS, and mix well. Carefully layer Ficoll-Hypaque solution below sample/PBS

mixture (3 ml Ficoll-Hypaque/6 ml mixture) or layer sample mixture on top of Ficoll-Hypaque. Centrifuge 30 to 40 min at 400 × g, room temperature.

2. Remove upper layer of plasma and platelets, leaving mononuclear layer undisturbed. Use a fresh pipet to transfer mononuclear cell layer to a 15-ml centrifuge tube. Wash cells twice with PBS.

3. Isolate genomic DNA using standard methods (APPENDIX 3A). Resuspend DNA in sterile water to 0.1 μg/μl.

4. Prepare master mix by adding reagents in the following order (39.75 μl per reaction). Multiply volumes by the number of planned reactions (including normal, rearranged, and no-DNA controls) plus one to allow for pipetting loss. Use plugged pipet tips to reduce contamination of sample.

> 18.75 μl sterile H_2O
> 5 μl 10× PCR amplification buffer (1× final)
> 3 μl 25 mM $MgCl_2$ (1.5 mM final)
> 8 μl 10 mM 4dNTP mix (1.6 mM final)
> 2.5 μl 10 μM outer immunoglobulin heavy chain V_H primer (0.5 μM final)
> 2.5 μl 10 μM outer immunoglobulin heavy chain J_H primer (0.5 μM final).

Vortex master mix gently and irradiate 10 min in a UV cross-linker.

5. Add 0.25 μl of 5 U/μl *Taq* DNA polymerase per reaction to master mix. Vortex tube gently and aliquot 40 μl master mix into each reaction tube. Add 1 μg template DNA in 10 μl sterile water to sample and control tubes and 10 μl sterile water to no-DNA control tube. Add ~100 μl mineral oil overlay, if needed.

6. Place tubes in a thermal cycler and run amplification program as follows:

Initial step:	2 min	95°C	(denaturation)
29 cycles:	10 sec	95°C	(denaturation)
	30 sec	40°C	(annealing)
	30 sec	70°C	(extension)
1 cycle:	10 sec	95°C	(denaturation)
	30 sec	40°C	(annealing)
	7 min	75°C	(extension)
Final step:	indefinitely	4°C	(hold).

Cycling parameters are optimized for the Perkin-Elmer Cetus 9600 thermal cycler.

7. Remove 5 μl PCR product and electrophorese on a 2% (w/v) horizontal agarose gel in 0.5× TBE (APPENDIX 3G). Stain gel with 0.5 μg/ml ethidium bromide and visualize the 100-bp product on a UV transilluminator.

8. Prepare reaction tubes for second-round PCR as in steps 4 and 5, but use inner V_H and J_H primers. Dilute 1 μl first reaction product in 9 μl sterile water and add to reaction tube. Increase amount of first reaction product if gel shows nonspecific amplification or primer-dimers.

9. Place tubes in thermal cycler and run amplification program as above, except use a 58°C annealing temperature. Check reaction product; if present, proceed to step 10.

10. Combine 4 μl reaction product with 1 μl of 0.2 U/μl Klenow fragment. Incubate 1 hr at 37°C. Add 1 μl of 6× loading buffer to reaction mix and load onto a 12% (w/v) nondenaturing polyacrylamide gel (UNIT 7.2 & CPMB UNIT 2.7). Electrophorese 2800 V × hr (e.g., 200 V × 14 hr).

DETECTION OF CLONAL T CELL RECEPTOR-γ GENE REARRANGEMENTS BY PCR AND DENATURING GRADIENT GEL ELECTROPHORESIS

Additional Materials (*also see Basic Protocol 1; see* APPENDIX 1 *for items with* ✓)

✓ 10× PCR amplification buffer
✓ 25 mM MgCl$_2$
✓ 10 mM 4dNTP mix
 10 μM each outer and inner V$_{γ1}$ and J$_γ$ *or* outer V$_{γ9}$ and outer and inner J$_γ$ oligonucleotide primers (Fig. 10.3.1)
 5 U/μl *Taq* DNA polymerase
 0.1 μg/μl template DNA from Ficoll-Hypaque-prepared mononuclear cells (see Alternate Protocol 1, steps 1 to 3)
 0.1 μg/μl normal and rearranged control DNA (e.g., normal blood cell DNA and T cell clonal DNA from cell line or tumor)
 Mineral oil, if needed
✓ 3 M sodium acetate, pH 5.2
 100% (v/v) ethanol
✓ DGGE loading buffer, 60°C

Thermal cycler (Perkin-Elmer Cetus 9600 or equivalent) with appropriate tubes
0.5-ml microcentrifuge tubes
95° and 60°C water baths

1. Prepare PCR master mix by adding the following in order (39.75 μl per reaction). Multiply volumes by the number of planned reactions (including normal, rearranged, and no-DNA controls) plus one to allow for pipetting loss. Use plugged pipet tips to reduce contamination of sample.

> 21.25 μl sterile H$_2$O
> 5 μl 10× PCR amplification buffer (1× final)
> 3 μl 25 mM MgCl$_2$ (1.5 mM final)
> 8 μl 10 mM 4dNTP mix (1.6 mM final)
> 1.25 μl 10 μM outer V$_{γ1}$ primer (0.25 μM final)
> 1.25 μl 10 μM outer J$_γ$ primer (0.25 μM final).

Vortex master mix gently and irradiate 10 min in a UV cross-linker.

2. Add 0.25 μl of 5 U/μl *Taq* DNA polymerase per reaction to master mix. Vortex tube gently and aliquot 40 μl master mix into each reaction tube. Add 10 μl of 0.1 μg/μl template DNA to sample tubes and normal and rearranged control DNA to control tubes. Add 10 μl sterile water to no-DNA control tube. Add ~100 μl mineral oil overlay, if needed.

3. Place tubes in a thermal cycler and run amplification program:

Initial step:	1 min	95°C	(denaturation)
25 cycles:	15 sec	94°C	(denaturation)
	30 sec	55°C	(annealing)
	30 sec	70°C	(extension)
1 cycle:	7 min	75°C	(extension)
Final step:	indefinitely	4°C	(hold).

Cycling parameters are optimized for the Perkin-Elmer Cetus 9600 thermal cycler.

A

TCR $V_{\gamma 1-8}$ and J_γ primers

outer $V_{\gamma 1}$ primer	5′ – GAAGCTTCTAGCTTTCCTGTCTC – 3′
outer J_γ primer	5′ – CGTCGACAACAAGTGTTGTTCCAC – 3′
inner $V_{\gamma 1}$ primer	5′ – CTCGAGTGCGCTGCCTACAGAGAGG – 3′
inner J_γ primer	5′ – GGATCCACTGCCAAAGAGTTTCTT – 3′

B

TCR $V_{\gamma 9}$ and J_γ primers

outer $V_{\gamma 9}$ primer	5′ – GGAATTCCAAATTCTTGGTTTA – 3′
outer J_γ primer	use outer J_γ primer (as above)
inner $V_{\gamma 9}$ primer	use outer $V_{\gamma 9}$ primer
inner J_γ primer	use inner J_γ primer (as above)

Figure 10.3.1 Primer sequences for detecting TCR-γ chain gene rearrangements. (**A**) Sequences of outer and inner TCR-γ V_{γ}1-8 and J_γ primers. (**B**) Sequences of outer and inner V_{γ}9 and J_γ primers. A PCR assay using these primers is performed if no rearrangements are detected in the assay using $V_{\gamma 1}$ and J_γ primers.

4. Remove 5 μl PCR product and electrophorese on a 1.2% (w/v) horizontal agarose gel in 0.5× TBE (*APPENDIX 3G*). Stain gel with 0.5 μg/ml ethidium bromide and visualize the 500-bp product on a UV transilluminator.

5. Prepare master mix and reaction tubes for second-round PCR by adding in the following order (90 μl per reaction):

> 58.5 μl sterile H_2O
> 10 μl 10× PCR amplification buffer (1× final)
> 16 μl 10 mM 4dNTP mix (1.6 mM final)
> 2.5 μl 10 μM inner $V_{\gamma 1}$ primer (0.25 μM final)
> 2.5 μl 10 μM inner J_γ primer (0.25 μM final)
> 0.5 μl 5 U/μl *Taq* DNA polymerase.

Vortex tube gently and aliquot 90 μl master mix to each reaction tube.

6. Add 10 μl first-round PCR product to reaction tube. Place tubes in thermal cycler and run amplification program as in step 3, but use 15 cycles instead of 25. Use 10 μl reaction to check for 500-bp product.

7. Transfer remaining 90 μl to a 0.5-ml microcentrifuge tube and add 10 μl of 3 M sodium acetate and 250 μl of 100% ethanol. Incubate 30 min at $-20°C$. Microcentrifuge 30 min

at maximum speed, 4°C. Decant ethanol and air dry DNA pellet. Resuspend DNA in 5 µl water. Incubate 5 min at 95°C and 1 hr at 60°C to denature DNA.

8. Prepare a parallel denaturing gradient gel with a 30% to 60% (w/v) gradient of urea/formamide denaturing solution (*CPHG UNIT 7.5*). Add 5 µl of 60°C DGGE loading buffer to sample, microcentrifuge briefly at maximum speed, and immediately load on gel. Run gel 6 hr at 60°C, 150 V. If no clonal band is identified on DGGE, repeat entire protocol using the V primers.

BASIC PROTOCOL 2

DETECTION OF CHROMOSOMAL TRANSLOCATIONS BY REVERSE TRANSCRIPTASE–PCR

This protocol describes a reverse transcriptase (RT)-PCR assay for t(9;22)(q34;q11), found in chronic myelogenous leukemia and in some cases of acute lymphocytic leukemia. Figure 10.3.2 is a diagrammatic representation of the translocation and shows the positions and sequences of primers.

Materials (*see APPENDIX 1 for items with ✓*)

✓ DEPC-treated H$_2$O
  1 µg RNA pellet from bone marrow or blood samples (see Support Protocol)
✓ 5× reverse transcriptase buffer
✓ 2.5 mM 4dNTP mix (prepare in DEPC-treated H$_2$O)
✓ 0.1 M dithiothreitol (DTT)
  Ribonuclease inhibitor (recombinant RNasin; Promega)
  10 µM primer 1 (Fig. 10.3.2)
  200 U/µl reverse transcriptase (Superscript RNase H$^-$; GIBCO/BRL)
  10 µM paired first-round PCR primers (Fig. 10.3.2)
✓ 10× PCR amplification buffer
✓ 25 mM MgCl$_2$
  5 U/µl *Taq* DNA polymerase
  Mineral oil, if needed
  10 µM paired second-round PCR primers (Fig. 10.3.2)

  37°, 70° and 95°C water baths
  Thermal cycler (Perkin-Elmer Cetus 480 thermal cycler or equivalent) and appropriate tubes

1. Add 20 µl DEPC-treated water to tube containing 1 µg RNA pellet from bone marrow or blood samples. Incubate 10 min at 70°C. Cool on ice and microcentrifuge briefly at high speed. Store on ice.

2. Make reverse transcription mix as follows (38 µl per reaction):

     8 µl 5× reverse transcriptase buffer (1× final)
     15 µl DEPC-treated water
     8 µl 2.5 mM 4dNTP mix (0.5 mM final)
     4 µl 0.1 M DTT (10 mM final)
     1 µl ribonuclease inhibitor
     1 µl 10 µM primer 1 (0.25 µM final)
     1 µl 200 U/µl reverse transcriptase (5 U/µl final).

   Add 2 µl RNA and mix. Incubate 1 hr at 37°C. Heat tube 5 min at 95°C. Microcentrifuge briefly at high speed, room temperature. Store on ice.

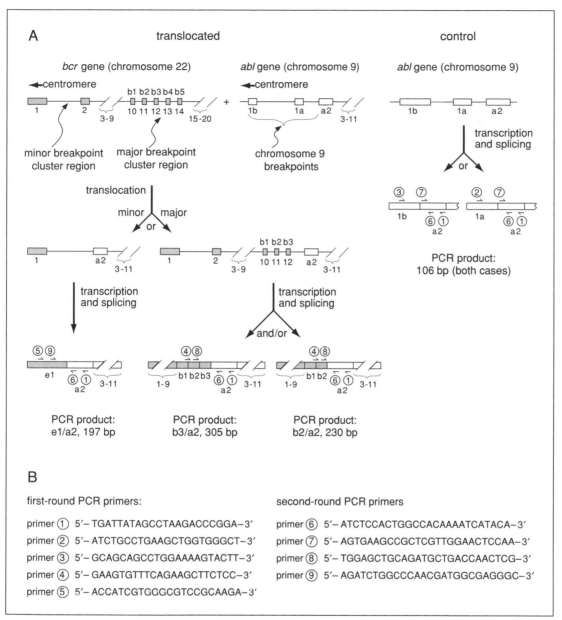

B

first-round PCR primers:

primer ① 5′– TGATTATAGCCTAAGACCCGGA–3′
primer ② 5′– ATCTGCCTGAAGCTGGTGGGCT–3′
primer ③ 5′– GCAGCAGCCTGGAAAAGTACTT–3′
primer ④ 5′– GAAGTGTTTCAGAAGCTTCTCC–3′
primer ⑤ 5′– ACCATCGTGGGCGTCCGCAAGA–3′

second-round PCR primers

primer ⑥ 5′– ATCTCCACTGGCCACAAAATCATACA–3′
primer ⑦ 5′– AGTGAAGCCGCTCGTTGGAACTCCAA–3′
primer ⑧ 5′– TGGAGCTGCAGATGCTGACCAACTCG–3′
primer ⑨ 5′– AGATCTGGCCCAACGATGGCGAGGGC–3′

Figure 10.3.2 Diagram of the rearrangements in translocation t(9;22). (**A**) Gene structure in translocated and control chromosomes. The *bcr* gene on chromosome 22 contains 20 exons. Exons 10 to 14, which comprise the major breakpoint region, are also called exons b1 to b5, respectively. The *abl* gene contains two alternative first exons, 1b and 1a. Whether or not the chromosome 9 breakpoint is 5′ to either of these exons, neither one has been detected in a fusion *bcr-abl* mRNA. In almost all reported cases, the *abl* portion of the fusion mRNA begins with the 5′ end of exon 2 (a2). The three possible *bcr-abl* PCR products, e1/a2, b3/a2 and b2/a2, are shown across the bottom of the diagram with their respective sizes. (**B**) Sequences of oligonucleotide primers for first- and second-round PCR for t(9;22). The bases given are identical to those in the human *abl* sequence (Shtivelman et al., 1985). First-round PCR antisense primer 1 is paired with sense primers 2 and 3 (control), 4 (major breakpoint), or 5 (minor breakpoint). Second-round PCR antisense primer 6 is paired with sense primers 7 (control), 8 (major breakpoint), or 9 (minor breakpoint). In this assay the *abl* gene is used for the control. Primers 2 and 3 are necessary because the *abl* gene has two alternative first exons. Primers 4 and 8 hybridize to exons in the major breakpoint region of the *bcr* gene. Primers 5 and 9 will hybridize to exon 1 of *bcr*, which is upstream of the minor breakpoint region in intron 1 of the *bcr* gene.

Table 10.3.3 First-Round PCR Master Mix for t(9;22) and t(14;18) Assays

Reaction component	t(9;22) assay[a]		t(14;18) assay[b]		
	Control (μl)	Fusion (μl)	Control (μl)	mbr fusion (μl)	mcr fusion (μl)
10 μM primer 1	2	2	2	2	—
10 μM primer 2	2	—	2	—	—
10 μM primer 3	2	—	—	2	2
10 μM primer 4	—	2	—	—	2
10 μM primer 5	—	2	—	—	—
10× PCR buffer	5	5	5	5	5
25 mM MgCl$_2$	3	3	3	4.5	4.5
Water	17.75	17.75	13.75	12.25	12.25
5 U/μl *Taq* polymerase	0.25	0.25	0.25	0.25	0.25
2.5 mM 4dNTP mix	—	—	4	4	4

[a] See Figure 10.3.2 for t(9;22) assay primer sequences.

[b] See Figure 10.3.3 for t(14;18) assay primer sequences.

3. For the appropriate translocation assay, prepare two master mix solutions (control and fusion assays; Table 10.3.3). Also prepare master mix solutions for no-RNA and appropriate positive and negative RNA controls. Add 18 μl reverse transcriptase reaction product to each master mix, overlay with ~100 μl mineral oil, if needed, place in a thermal cycler, and run amplification program:

Initial step:	4 min	94°C	(denaturation)
26 cycles:	1 min	94°C	(denaturation)
	1.5 min	61°C	(annealing and extension)
1 cycle:	10 min	72°C	(extension)
Final step:	indefinitely	4°C	(hold).

Cycling parameters are optimized for the Perkin-Elmer Cetus 480 thermal cycler.

4. Add 2 μl from the reaction to 198 μl water. Vortex and microcentrifuge briefly. Prepare two master mixes (control and fusion assays; Table 10.3.4). Add 10 μl diluted first-round PCR product to the appropriate master mix and run amplification program as in step 3, but use 30 cycles instead of 26.

5. Analyze 15 μl from each second-round PCR sample by electrophoresis on a 2.5% (w/v) agarose gel containing 0.5 μg/ml ethidium bromide (*APPENDIX 3G*). Major breakpoint translocations produce bands at 305 and/or 230 bp, and minor breakpoint translocations produce a band at 197 bp. A major and a minor breakpoint fusion mRNA, when present together, may also produce a band at 262 bp. The control produces a band at 106 bp.

SUPPORT PROTOCOL 1

RNA ISOLATION BY THE RAPID GUANIDINIUM METHOD

Materials

Mononuclear cell pellet isolated from 1 to 2 ml bone marrow or 5 ml whole blood (see Alternate Protocol 1, steps 1 and 2)

Table 10.3.4 Second-Round PCR Master Mix for t(9;22), and t(14;18) Assays

Reaction component	t(9;22) assay[a]		t(14;18) assay[b]		
	Control (μl)	Fusion (μl)	Control (μl)	mbr fusion (μl)	mcr fusion (μl)
10 μM primer 2	—	—	2	—	—
10 μM primer 5	—	—	2	2	—
10 μM primer 6	2	2	—	2	2
10 μM primer 7	2	—	—	—	2
10 μM primer 8	—	2	—	—	—
10 μM primer 9	—	2	—	—	—
10× PCR buffer	5	5	5	5	5
2.5 mM MgCl$_2$	3	3	3	4.5	4.5
2.5 mM 4dNTP mix	4	4	4	4	4
Water	23.75	21.75	23.75	22.25	22.25
5 U/μl *Taq* polymerase	0.25	0.25	0.25	0.25	0.25

[a]See Figure 10.3.2 for t(9;22) assay primer sequences.

[b]See Figure 10.3.3 for t(14;18) assay primer sequences.

✓ HBSS without Ca^{2+} or Mg^{2+}
✓ RNA isolation solution I
  2 M sodium acetate, pH 4
✓ Buffered phenol
  24:1 (v/v) chloroform/isoamyl alcohol
  100% and 70% (v/v) ethanol
✓ RNA isolation solution II
✓ DEPC-treated H$_2$O

1. Wash mononuclear cell pellet once with HBSS without Ca^{2+} or Mg^{2+}. Resuspend cells in HBSS and divide among three 1.5-ml microcentrifuge tubes. Microcentrifuge 2 min at $100 \times g$ to pellet cells. Discard supernatant. Use immediately or store ≤4 months at −70°C. Use one tube to prepare RNA and the other two to repeat assay.

2. Add 0.5 ml RNA isolation solution I to cell pellet. Disrupt cells by repeated aspiration. Add 0.05 ml of 2 M sodium acetate. Vortex and microcentrifuge briefly at maximum speed. Add 0.5 ml buffered phenol. Vortex and microcentrifuge. Add 0.1 ml of 24:1 chloroform/isoamyl alcohol. Vortex and microcentrifuge 15 min at maximum speed, 4°C.

3. Remove 0.4 ml of aqueous (upper) phase to a clean 1.5-ml tube that contains 0.8 ml of 100% ethanol, taking care not to disturb interface. Incubate >1 hr at −20°C. Microcentrifuge 15 min at maximum speed, 4°C. Discard supernatant.

4. Add 300 μl RNA isolation solution II to pellet and vortex to resuspend. Microcentrifuge briefly at maximum speed. Add 600 μl of 100% ethanol and mix. Incubate ≥1 hr at −20°C. Microcentrifuge 15 min at high speed, 4°C. Discard supernatant. Wash pellet with 0.5 ml of 70% ethanol. Discard ethanol. If pellet dislodges from bottom of tube, microcentrifuge 5 min at high speed, 4°C, before removing ethanol.

5. Cover open tubes with Parafilm and pierce several times with a fine needle. Dry pellet briefly under vacuum or dry on bench. Store dry RNA ≤2 months at −70°C or below, or dissolve in 20 μl DEPC-treated water if the assay will be carried out immediately.

DETECTION OF CHROMOSOMAL TRANSLOCATIONS IN DNA SAMPLES BY PCR

In t(14;18)(q32.3q21.3), found in ~90% of follicular lymphomas and ~30% of large cell lymphomas, the breakpoints in the *bcl*-2 gene on chromosome 18 are clustered in one of two regions of DNA (Fig. 10.3.3).

Additional Materials (*also see Basic Protocol 2*)

   0.1 μg/μl sample and control DNA from bone marrow or blood (see Alternate Protocol 1, steps 1 to 3)
   10 μM t(14;18)(q32.3q21.3) primers (Fig. 10.3.3)
   Internal oligonucleotide for hybridization probe (optional)
      mbr: 5′-GCCTGTTTCAACACAGACCC-3′
      mcr: 5′-GGACCTTCCTTGGTGTGTTG-3′

1. For each sample, prepare control and major breakpoint (mbr) master mix solutions according to Table 10.3.3. Use first-round PCR primers listed in Figure 10.3.3. Add 20 μl of 0.1 μg/μl sample and control DNA to appropriate tubes. Overlay with ~100 μl mineral oil, if needed, and place in a thermal cycler. Run amplification program:

Initial step:	4 min	94°C	(denaturation)
30 cycles:	1 min	94°C	(denaturation)
	1 min	55°C	(annealing)
	1 min	72°C	(extension)
1 cycle:	8 min	72°C	(extension)
Final step:	indefinitely	4°C	(hold).

   Cycling parameters are optimized for the Perkin-Elmer Cetus 480 thermal cycler.

2. Add 2 μl reaction mixture to 198 μl water. Vortex and microcentrifuge briefly. Make a master mix containing PCR components listed in Table 10.3.4. Use second-round nested primers shown in Figure 10.3.3. Add 10 μl diluted first-round PCR product to master mix and run amplification program as in step 1.

3. Analyze 15 μl from each second-round sample by electrophoresis on a 2.5% (w/v) agarose gel containing 0.5 μg/ml ethidium bromide (*APPENDIX 3G*). Major breakpoint translocation should produce bands of ~100 to 300 bp. The control should produce a band of 386 bp.

4. If the assay does not detect a translocation, repeat steps 1 to 3 using master mix volumes (Table 10.3.4) and primers (Fig. 10.3.3) specific for minor breakpoint (mcr) fusions and an annealing temperature of 58°C. Do not repeat the control (for DNA integrity). Minor breakpoint translocations should produce bands of ~200 to 1200 bp.

5. If there is a question as to whether a band represents a translocation, perform a Southern blot using an appropriate internal oligonucleotide as the hybridization probe (*APPENDIX 3G*).

References: Negrin and Blume, 1991; Sklar, 1992

Contributors: Janina Longtine, Edward Fox, Carol Reynolds, and Jeffrey Sklar

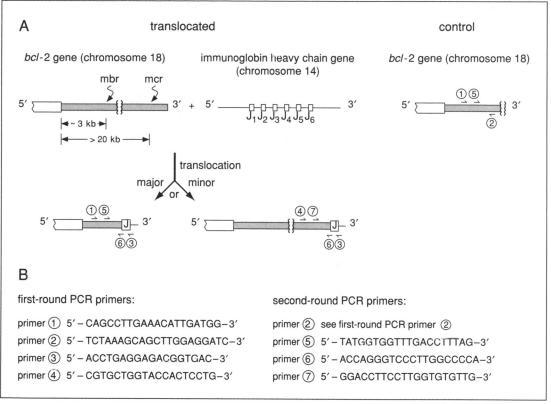

Figure 10.3.3 Diagram of the rearrangements in translocation t(14;18). (**A**) Gene structure in translocated and control chromosomes. The *bcl-2* gene, which breaks in either the major (mbr) or minor (mcr) breakpoint region, fuses with either a D (not shown) or J region in the immunoglobulin heavy chain gene. The *bcl-2* gene is also used as a control to ascertain that amplifiable DNA has been extracted. The position of each of the primers is shown. (**B**) Sequences of oligonucleotide primers for first- and second-round PCR for t(14;18). First-round PCR sense primer 1 (major breakpoint) is paired with antisense primers 2 (control) or 3 (J_H); antisense primer 3 is also paired with sense primer 4 (minor breakpoint). Second-round PCR sense primer 5 (major breakpoint) is paired with antisense primers 2 (control) or 6 (J_H); antisense primer 6 is also paired with sense primer 7 (minor breakpoint).

UNIT 10.4

Molecular Analysis of Gene Amplification in Tumors

NOTE: The quantification and analysis of any samples that have low-level amplification should be repeated to be certain that they are not false positives.

BASIC PROTOCOL

DETECTION OF GENE AMPLIFICATION BY SOUTHERN BLOT HYBRIDIZATION ANALYSIS

A troubleshooting guide for Southern blotting and hybridization is given in Table 10.4.1.

Table 10.4.1 Troubleshooting Guide for Southern Blotting and Hybridization Analysis

Problem	Possible cause	Solution
DNA preparation		
No DNA precipitate	Low DNA concentration	Incubate $\geq$2 hr at $-20°$C
DNA does not resuspend	Protein contamination	Homogenize new tumor
	Not enough TE buffer	Add more TE buffer; gently pipet
DNA concentration low	Necrotic or fibrotic tumor	Homogenize more tumor for DNA and/or use PCR method
DNA concentration erratic	Sample too concentrated	Dilute to 50–300 μg/ml
	Sample not homogeneous	Mix thoroughly on tube rocker
	Pipetting error	Check pipet calibration
No DNA pellet after digestion	Insufficient centrifugation	Microcentrifuge 15 min at maximum speed, 4°C
	Inadequate salt concentration	Use ⅟₁₀ vol 4 M NaCl
Electrophoresis		
DNA floats, does not spread	DNA not digested	Redigest sample
DNA spreads unevenly	Excess salt in pellet	Redigest sample
DNA dissipates out of well	Residual ethanol in pellet	Redigest sample
Degraded DNA sample	Necrotic tumor	Homogenize new tumor
	Contaminated reagents	Make new reagents
Incomplete digestion	Protein, salt, or OTC contamination	Redigest with pronase, reprecipitate, and wash thoroughly
	Inadequate enzyme concentration and/or activity	Use more (fresh) enzyme
Under or overloaded sample	DNA quantitation inaccurate	Requantitate DNA
Uneven DNA electrophoresis	Salt concentration wrong	Prepare fresh 10× TBE buffer; mix well
	Agarose not evenly distributed	Cast new gel
	Voltage too high or uneven	Check electrodes and power supply
Transfer		
Nitrocellulose membrane prewets unevenly or slowly	Membrane old or contaminated	Use new nitrocellulose membrane
Fragile nitrocellulose after baking	Gel not completely neutralized; membrane overbaked	Check pH of neutralizing solution; check oven temperature and function
Hybridization and autoradiogram analysis		
High background	Probe contains repetitive DNA	Use new probe or blocking DNA
	Probe is degraded	Check probe on gel
	Hybridization solution dried on filter	Wash at low salt, high temperature
Low signal	Insufficient probe concentration or size	Relabel or use different probe
	Excessive washing	Repeat transfer and wash at lower stringency
	Inadequate DNA transfer	Stain gel after transfer to check
	DNA degraded or undigested	Check DNA size and digestion on gel; make new DNA or use PCR
Aberrant bands	DNA digestion incomplete	Redigest with more enzyme
	DNA contains contaminant (e.g., plasmid)	Homogenize new tumor
	Polymorphism/rearrangement	Digest with other enzymes
Specks and splotches	Contaminated membrane	Use new membrane
	Glove powder	Wash gloves before handling
	Hybridization solution not equilibrated	Warm $\geq$1 hr at room temperature; mix
	Probe not evenly mixed	Mix hybridization solution thoroughly
	Bubbles in hybridization solution	Avoid bubbles, mix during hybridization

10

Materials *(see* APPENDIX 1 *for items with* ✓ *)*

    Tumor tissue (see Support Protocol)

    DNA Extraction Kit (Stratagene) or equivalent:

✓ DNA extraction solution: 50 mM Tris·Cl (pH 8.0)/20 mM EDTA (pH 8.0)/2% (w/v) SDS (see individual recipes)

    225 mg/ml pronase

    NaCl extraction solution: saturated NaCl solution

    10 mg/ml RNase

    100% and 70% ethanol, ice cold

✓ TE buffer, pH 8.0

✓ Calf thymus DNA standards: 10, 100, 250, and 500 μg/ml

✓ Capillary assay solution, prepared just before use

✓ 1× TNE buffer

    Single-copy control DNA

    Amplified control DNA

✓ 1 M spermidine

    10 U/μl restriction enzyme and appropriate 10× buffer

    1 mg/ml bovine serum albumin (BSA)

    4 M NaCl

✓ 6× gel loading buffer

    0.25 M HCl, for bands of interest >10 kb only

    Denaturing solution: 0.5 N NaOH/1 M NaCl, prepared fresh and equilibrated 30 min

✓ Neutralizing solution, prepared fresh

✓ 20× and 6× SSC

    Quik-Hyb solution (Stratagene)

✓ 10 mg/ml herring sperm DNA, sonicated and boiled

    Double-stranded DNA probe, labeled with [α-^{32}P]dCTP by random oligonucleotide priming to a specific activity $\geq 1 \times 10^9$ cpm/μg DNA and spin-column-purified (APPENDIX 3E)

✓ 20% SDS

    7.5-ml ground-glass tissue homogenizer or equivalent

    30- and 50-ml centrifuge tubes

    37° and 65°C water baths, with shaking

    Sorvall centrifuge and JS-5.2 rotor (or equivalent)

    Large-bore pipets

    Vacuum pump apparatus (creating $\geq$30 mmHg) or Speedvac evaporator

    2- or 5-ml O-ring screw-cap tubes

    Rocker platform

    Mini-fluorometer and 10-μl capillary assay tubes (Hoefer)

    0.5- and 1.5-ml microcentrifuge tubes

    10-μl capillary assay tube (Hoefer)

    Nitrocellulose membrane

    Blunt-ended forceps

    80°C vacuum oven or UV cross-linker (e.g., Stratalinker; Stratagene)

    Heat-sealable polyethylene bags

    Boiling water bath

    X-AR autoradiographic film

1. Cut one or more small pieces of tumor tissue into tiny bits with a scalpel. Homogenize pieces with a 7.5-ml ground-glass tissue homogenizer using several milliliters of DNA extraction solution. For red blood cells or bone marrow samples, freeze sample 1 to 2 hr at −80°C before homogenizing cell pellet. Homogenize liquid samples <3 ml directly.

2. Transfer homogenate to a 30-ml centrifuge tube and add DNA extraction solution to 14 ml total. Add 62.5 μl pronase (1 mg/ml final). Incubate overnight at 37°C, occasionally mixing the tube by inversion.

3. Chill tube ≥10 min on ice. Add 5 ml NaCl extraction solution and invert several times. Centrifuge tube 20 min at 2000 × g, 4°C. Use a large-bore pipet to transfer supernatant to a 50-ml centrifuge tube. Do not transfer any debris.

4. Add RNase to 20 μg/ml final. Vortex tube gently and incubate ≥15 min at 37°C. Chill tube on ice ≥10 min. Add 2.5 vol ice-cold 100% ethanol and invert tube several times. Look for a DNA precipitate to appear. To improve DNA recovery, incubate ≥2 hr at −20°C.

5. Centrifuge tube 20 min at 2000 × g, 4°C. Carefully pour off ethanol. Add ∼25 ml of 70% ethanol to DNA pellet and vortex gently. Centrifuge 10 min. Carefully pour off 70% ethanol and desiccate sample 15 min under a 30-mmHg vacuum or in a Speedvac evaporator at room temperature.

6. Add 200 to 2000 μl TE buffer (depending on size of DNA pellet) to tube to achieve a DNA concentration of 50 to 200 μg/ml (assume 100 mg sample yields 100 μg DNA). Transfer DNA to a 2- or 5-ml O-ring screw-cap tube and mix 15 to 30 min (or overnight) on a rocker platform to resuspend pellet. If pellet is not resuspended after overnight mixing, add more TE buffer. Store aqueous DNA indefinitely at 4°C, avoiding exposure to fluorescent light.

7. Incubate DNA sample and calf thymus DNA standards on a rocker platform ∼15 min at room temperature. During incubation, turn on a mini-fluorometer and warm up ≥15 min.

8. To appropriately labeled 0.5-ml microcentrifuge tubes (set up in duplicate) add 5 μl freshly prepared capillary assay solution and 2 μl of 1× TNE buffer. Then add 3 μl DNA sample or calf thymus DNA standards to the sample and standard tubes and an additional 3 μl of 1× TNE buffer to the blank capillary assay tube. Microcentrifuge tubes briefly. Mix and microcentrifuge again.

9. Transfer individual assay mixes from tubes to 10-μl capillary assay tubes. Calibrate mini-fluorometer with the blank tube to zero the machine and then with the 100 μg/ml calf thymus DNA standard to set the instrument to 100. Measure other standards and DNA sample (ideally between 50 and 200 μg/ml). For samples ≥300 μg/ml, dilute with TE buffer and reassay.

The differential PCR assay (see Alternate Protocol 2) should be used to analyze a sample that has a DNA concentration ≤1 μg/ml or that contains ≤5 μg total DNA.

10. Prepare the following sample and control digests in 1.5-ml microcentrifuge tubes:

> 5 μg DNA (sample, single-copy control, or amplified control)
> 1.2 μl 1 M spermidine (4 mM final)
> sterile H$_2$O to 230 μl.

11. Thoroughly thaw and resuspend 10× restriction enzyme buffer and 1 mg/ml BSA. If there is precipitate in the buffer, incubate briefly at 65°C. Add 30 μl of 10× buffer and 30 μl BSA to each sample and control tube. Microcentrifuge briefly at maximum speed. Vortex and microcentrifuge briefly again. Allow DNA to equilibrate in buffer mix for a few minutes. Add 10 μl of 10 U/μl restriction enzyme to each tube. Microcentrifuge briefly. Gently vortex and incubate as recommended by the manufacturer.

12. Microcentrifuge tubes briefly. Add 30 μl (1/10 vol) of 4 M NaCl and 825 μl (2.5 vol) ice-cold 100% ethanol to each tube. Invert tubes several times and vortex moderately. Incubate ∼30 min at −20°C.

13. Microcentrifuge tubes 15 min at maximum speed, 4°C. Remove and discard supernatant. Add ~1 ml of 70% ethanol. Microcentrifuge tubes 5 min. Carefully pour off 70% ethanol and drain the tubes. Desiccate DNA pellets ≥10 min under a 30-mmHg vacuum or in Speedvac at room temperature. Resuspend pellets in 10 μl TE buffer and 2 μl of 6× gel loading dye. Allow tubes to incubate several minutes at room temperature.

14. Prepare and pour a 0.6% to 2.0% (w/v) agarose gel (to separate DNA fragments of interest) in 1× TBE buffer (APPENDIX 3G). Load samples. Place single-copy and amplified DNA controls in the first and last lanes, respectively. Electrophorese gel at an appropriate voltage and time to provide clear separation of the bands of interest.

15. Stain gel in 0.5 μg/ml ethidium bromide ≤10 min with moderate shaking (APPENDIX 3G). Photograph gel on a UV transilluminator. If the DNA is intact and digestions are complete, proceed with transfer. For samples that are degraded, use differential PCR (see Alternate Protocol 2) to analyze the DNA.

16. To efficiently transfer bands >10 kb, depurinate gel 15 to 30 min in 0.25 M HCl. For all bands of interest, denature gel 30 min in denaturing solution with moderate shaking. Do not let dry areas form on top surface of gel. Neutralize gel 30 min in neutralizing solution with moderate shaking.

17. Transfer DNA to a nitrocellulose membrane in 6× SSC by Southern blotting for 4 hr to overnight (APPENDIX 3G). Handle membrane with blunt-ended forceps. Fix transferred DNA to the membrane by baking the membrane 2 hr at 80°C in a vacuum oven or by UV cross-linking according to manufacturer's instructions.

   Charged or uncharged nylon membrane could also be used with slight modifications of the procedure (APPENDIX 3G).

18. Prepare equilibrated Quik-Hyb solution containing 1 mg/ml herring sperm DNA. Place fixed membrane in a heat-sealable polyethylene bag with the hybridization solution, using ~140 μl Quik-Hyb solution/cm^2 membrane. Seal bag with a heat sealer. Prehybridize membrane ≥15 min at 65°C in a shaking water bath.

19. Prepare double-stranded [α-^{32}P]dCTP-labeled probe by boiling for 10 min (~1 × 10^6 dpm probe/ml Quik-Hyb). Open sealed bag and add boiled probe. Reseal bag and mix probe thoroughly in the solution. Hybridize 1 to 2 hr at 65°C in the shaking water bath.

20. Prepare three wash solutions of 2×, 0.5×, and 0.2× SSC in 0.2% SDS and heat to 55°C. Remove membrane from the bag and wash using three solutions in order, for 2 to 5 min each at 55°C. There should be few, if any, counts detected in areas of the filter that do not have DNA. An amplified gene control of 150 copies should register at least ~100 counts/sec by Geiger counter (e.g., mini-monitor, RPI); a single-copy control should register ~20 counts/sec.

21. Expose membrane to X-AR autoradiographic film with an intensifying screen for an appropriate period of time as indicated by the signal of the controls (usually overnight or 12 to 14 hr). Analyze film by densitometry.

ALTERNATE PROTOCOL 1

DETECTION OF GENE AMPLIFICATION BY SLOT BLOT HYBRIDIZATION ANALYSIS

The advantages of this procedure are that only 1 to 2 μg of undigested DNA is required and that film exposure takes only a few hours. However, this method lacks the sensitivity of Southern blots for detecting low levels of amplification, and the exact level of amplification may be underestimated.

Additional Materials (*also see Basic Protocol*)

    3.0 M NaOH, filter sterilized
    2.0 M ammonium acetate, pH 7.0, filter sterilized

    Dot or slot blotting manifold (e.g., Bio-Dot SF, Bio-Rad or Minifold 2, Schleicher & Schuell)
    Nitrocellulose *or* nylon membrane
    Whatman 3MM paper
    Heat lamp

1. Prepare tumor DNA as described (see Basic Protocol, steps 1 to 9).

2. Prepare serial two-fold dilutions of amplified control DNA using single-copy control DNA as the diluent. Do not dilute amplified DNA with TE buffer. Adjust concentrations so that equal volumes of control, standard, and sample DNA are applied to the blot.

3. Place tumor DNA sample, controls, and standards on a rocker platform and mix 10 min. In a 1.5-ml microcentrifuge tube, add 2 μg DNA to TE buffer to obtain 200 μl total volume. Include a reagent blank without DNA (TE buffer only).

4. Add 400 μl TE buffer and 60 μl of 3.0 M NaOH to each tube. Vortex tubes and microcentrifuge briefly. Incubate 1 hr at 65°C. Microcentrifuge tubes briefly and cool to room temperature (~15 min). Add 660 μl of 2.0 M ammonium acetate to each tube. Vortex tubes and microcentrifuge briefly.

5. Assemble a dot or slot blotting manifold according to manufacturer's instructions. Load 660 μl of each DNA sample per slot. Use a prewetted nitrocellulose membrane to obtain the cleanest signal for single-copy and low-level detection. Use a nylon membrane for multiple probing.

6. Apply house vacuum to manifold for ~1 min. Disassemble device and carefully peel membrane off the upper plate with blunt-ended filter forceps. Place membrane, DNA side up, on a piece of clean Whatman 3MM paper. Dry 5 min under a heat lamp.

7. Bake membrane 2 hr at 80°C in a vacuum oven. Prehybridize, hybridize, wash, expose membrane to autoradiographic film (for ~3 hr), and analyze film, as described (see Basic Protocol, steps 18 to 21).

ALTERNATE PROTOCOL 2

DETECTION OF GENE AMPLIFICATION BY DIFFERENTIAL PCR

For this protocol, a control DNA segment should come from a remote location on the same chromosome (i.e., the other chromosome arm). A control gene from another chromosome may lead to false positives and false negatives because of differences in chromosome numbers. This technique is not as sensitive as hybridization of Southern blots for detecting low levels of gene amplification.

NOTE: PCR requires special precautions to prevent contamination. Familiarity with standard methods is essential.

Additional Materials (also see Basic Protocol; see APPENDIX 1 for items with ✓)

✓ 10× *Taq* DNA polymerase buffer: 500 mM KCl/100 mM Tris·Cl, pH 8.3
  (store ≤18 months at −20°C; see individual recipes)
✓ 25 mM MgCl$_2$
✓ 1.25 mM 4dNTP mix
  20 μM target and control oligonucleotide primers, + and − strand sequences of each
  5 U/μl *Taq* DNA polymerase
  Mineral oil (not needed if thermal cycler has a heated lid)
  Low-molecular-weight DNA size markers (e.g., BioMarker Low; Bioventures)

  0.5-ml microcentrifuge tubes
  Thermal cycler

1. Prepare tumor DNA as described (see Basic Protocol, steps 1 to 9).

2. Prepare PCR reaction mixes by adding the following ingredients to a 0.5-ml microcentrifuge tube in the order listed (44.5 μl final):

     10 μl 10× *Taq* DNA polymerase buffer (1×)
     6 μl 25 mM MgCl$_2$ (1.5 mM)
     16 μl 1.25 mM 4dNTP mix, freshly diluted (200 μM)
     5 μl 20 μM target oligonucleotide primer, + strand sequence (1.0 μM)
     5 μl 20 μM target oligonucleotide primer, − strand sequence (1.0 μM)
     1 μl 20 μM control oligonucleotide primer, + strand sequence (0.2 μM)
     1 μl 20 μM control oligonucleotide primer, − strand sequence (0.2 μM)
     0.5 μl 5 U/μl *Taq* DNA polymerase (2.5 U).

   Place tube on ice.

   Optimum concentrations of primers and magnesium may vary, depending on the primers and template used (CPMB UNIT 15.1).

3. Prepare a water blank (no DNA) by adding 44.5 μl reaction mix to 55.5 μl sterile water in a 0.5-ml microcentrifuge tube. Gently mix. Overlay with 100 μl mineral oil. Place tube on ice.

4. Boil 10 to 100 ng tumor DNA 3 min and add to sterile water in a 0.5-ml microcentrifuge tube to a total volume of 55.5 μl. Add 44.5 μl reaction mix. Overlay with 100 μl mineral oil and place on ice. Include single-copy and amplified gene controls in each assay.

5. Determine optimal times and temperatures for denaturing, reannealing, and extension. Perform the PCR reaction in a thermal cycler using these conditions. Store PCR products indefinitely at 4°C until they are analyzed.

   Usual starting conditions are: 1.5 min at 93°C to denature, 1 min at 60°C to reanneal, and 1 min at 72°C to extend.

6. Prepare a 2% (w/v) agarose gel in 1× TBE buffer for products ≤1.0 kb (use a lower-percentage gel for larger products; APPENDIX 3G). Apply 3 μl of each amplified sample and control reaction DNA in loading buffer to the gel. Include aliquots from the water blank and low-molecular-weight DNA size standards. Electrophorese gel at ~84 V for ~1 hr (until the bromphenol blue reaches the end of gel). Stain gel in 0.5 μg/ml ethidium bromide solution, rinse in distilled water, and photograph on a UV transilluminator. The intensity of the internal control band will be decreased in proportion to the degree of amplification of the target gene.

OBTAINING AND PROCESSING TUMOR TISSUE

Materials

> Tumor tissue
> Liquid nitrogen or dry ice
> Cryotubes

1. If possible, obtain tumor tissue directly from the operating room. Remove staples or suture material prior to processing. If sample is already embedded in mounting medium for frozen sections (e.g., OTC), remove as much as possible prior to processing.

For solid tumor samples

2a. Select a representative sample of viable tumor (pink or red area), avoiding tumor capsule (tough, whitish area) and fibrosis or necrosis (brown or black areas). For metastatic tumor tissue, which can be used as long as a substantial portion consists of tumor, remove obvious surrounding normal tissue. Cut the sample into pieces ~5 mm^3 to facilitate subsequent storage and handling. If possible, also obtain and store a tissue section adjacent to the area taken for molecular analysis for follow-up comparisons.

3a. Freeze tissue pieces as quickly as possible, preferably by dropping directly into liquid nitrogen. Transfer to cryovials and store at −80°C. If liquid nitrogen is not available, freeze sample quickly by placing in a plastic tube or container on dry ice. Store at or below −80°C until ready for processing.

4a. If the tumor tissue is to be analyzed at another institution, place sample in a well-marked, sealed plastic tube or container, and ship on dry ice by overnight delivery.

For liquid tumor samples

2b. Confirm that sample contains a substantial amount of tumor tissue (>10% is necessary to detect 50- to 100-fold amplification). Freeze small-volume samples (<3 ml) quickly as above. If the sample is substantially >3 ml, process further to enrich for cellular elements.

3b. Ship small-volume samples overnight on dry ice as above. Ship large-volume liquid samples at room temperature by overnight carrier. The latter may be stored at 4°C for ≤2 days before processing for DNA extraction.

References: Kellems, 1993; Look et al., 1991

Contributors: Jonathon C. Wasson and Garrett M. Brodeur

UNIT 10.5

Methylation-Specific PCR

NOTE: PCR requires special precautions to prevent contamination. Familiarity with standard methods is essential.

BASIC PROTOCOL 1

DETERMINATION OF DNA METHYLATION PATTERNS BY METHYLATION-SPECIFIC PCR

Methylation-specific PCR (MSP) is a very specific, sensitive, rapid, and economical method to determine methylation patterns of CpG islands (Fig. 10.5.1). Typically, one to three CpG

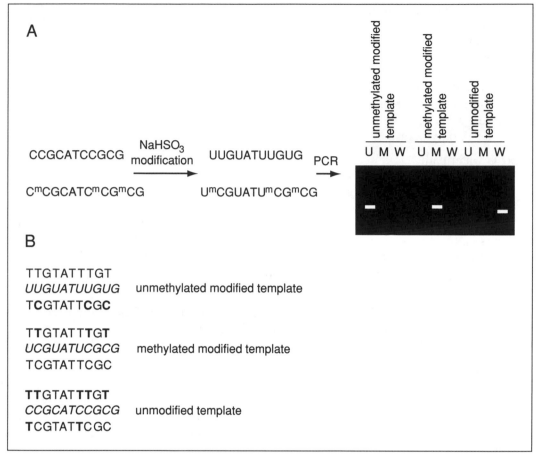

Figure 10.5.1 Methylation-specific PCR (MSP). (**A**) Single-stranded alleles of a gene, with methyl-cytosines denoted ^mC. Treatment of the single-stranded DNA with sodium bisulfite converts un-methylated cytosines to uracil, while methylated cytosines remain unchanged. This modified DNA is amplified with primers specific for the newly generated sequence, and the products analyzed by gel electrophoresis. (**B**) Potential template sequences following bisulfite modification. In each case the top line is unmethylated specific primer, the middle line (in italics) is the template, and the bottom line is the methylated specific primer, with mismatches shown in bold. A primer specific for the unmethy-lated, modified sequence matches only that sequence, and has numerous mismatches preventing annealing and elongation with either of the other templates. A primer specific for methylated (M), modified sequence also has mismatches with either unmodified (W, wild type) or unmethylated (U), modified template.

sites are included in each PCR primer near the 3′ end, and primer length should be adjusted to give nearly equal melting and annealing temperatures.

Materials (see APPENDIX 1 for items with ✓)

    Sample DNA
    Positive control DNA, previously determined to be methylated
    Negative control DNA, previously determined to be unmethylated
✓ 2 M and 3 M NaOH
    10 mM hydroquinone, prepared fresh
    3 M sodium bisulfite (pH 5.0; adjusted with 5 M NaOH), prepared fresh
    Mineral oil
    DNA Wizard cleanup kit (Promega) or equivalent
    80% (v/v) isopropanol

Water (distilled and autoclaved), 60° to 70°C
10 mg/ml glycogen
✓ 10 M ammonium acetate
100% and 70% (v/v) ethanol, ice cold
✓ 10× PCR amplification buffer with 15 mM MgCl$_2$
✓ 25 mM 4dNTP mix
300 ng/μl sense and antisense primers for the methylated and unmethylated gene of interest
Taq DNA polymerase

50°C water bath
Vacuum manifold
Thermal cycler with tube-controlled temperature monitoring and appropriate tubes

1. Dilute sample DNA (up to 1 μg) to 50 μl with water in a 1.5-ml microcentrifuge tube. If less DNA is used (including DNA microdissected from paraffin-embedded samples), expect to see less final product (i.e., very faint bands) or to use additional amplification cycles. Prepare separate dilutions of positive and negative control DNA, and process in parallel.

2. Add 5.5 μl of 2 M NaOH. Incubate 10 min at 37°C. Add 30 μl freshly prepared 10 mM hydroquinone and 520 μl freshly prepared 3 M sodium bisulfite. Mix well and layer with enough mineral oil to cover the surface of the aqueous phase (∼50 μl). Incubate 16 hr at 50°C.

3. Remove oil. Add 1 ml DNA Wizard cleanup reagent and add mixture to miniprep column included in kit. Apply vacuum, using a manifold. Wash with 2 ml of 80% isopropanol. Place column in a clean 1.5-ml tube and add 50 μl of 60° to 70°C water. Microcentrifuge tube and column 1 min. Add 5.5 μl of 3 M NaOH to each tube and incubate 5 min at room temperature.

4. Add 1 μl of 10 mg/ml glycogen and then add 17 μl of 10 M ammonium acetate and 3 vol ice-cold 100% ethanol. Precipitate DNA a few hours to overnight at −20°C, centrifuge 25 min, discard supernatant, wash with ice-cold 70% ethanol, and resuspend in 20 to 30 μl water. Treat single-stranded DNA like RNA (keep cold, minimize freeze-thaw cycles, and store at −70°C if possible).

5. Determine the number of samples to be analyzed, including a positive and no-DNA control, for both the unmethylated and methylated reactions. Prepare separate master mixes for the methylated and unmethylated PCR reactions (50 μl), each containing:

   5 μl 10× PCR amplification buffer
   2.5 μl 25 mM 4dNTP mix
   1 μl 300 ng/μl sense primer
   1 μl 300 ng/μl antisense primer
   28.5 μl water.

   Mix well.

6. Place 38-μl aliquots of each master mix into separate labeled PCR tubes. Add 2 μl bisulfite-modified DNA template (step 4) to each tube. Add ∼25 to 50 μl mineral oil to each tube, if needed, and place in a thermal cycler. Initiate PCR with a 5-min denaturation at 95°C.

7. For each sample, dilute 1.25 U *Taq* DNA polymerase in 10 μl sterile distilled water. Add to the 40 μl of mix through the oil layer and gently pipet up and down. Alternatively, use another form of hot-start PCR. Continue PCR amplification as follows:

35 cycles:	30 sec	95°C	(denaturation)
	30 sec	specific for primers	(annealing)
	30 sec	72°C	(elongation)
Final step:	4 min	72°C	(elongation).

Store at 4°C until analysis.

8. Prepare 6% to 8% (w/v) nondenaturing polyacrylamide gels with 1× TBE buffer (UNIT 7.2). Run vertical gels 1 to 2 hr at 10 V/cm, with reactions from each sample in adjacent lanes to allow for direct comparison between unmethylated and methylated alleles. Include positive and negative controls.

9. Stain gel with ethidium bromide and visualize and photograph under UV illumination (APPENDIX 3G). The range of products is typically 80 to 200 bp.

BASIC PROTOCOL 2

DETERMINATION OF METHYLATION OF CpG SITES WITHIN METHYLATION-SPECIFIC PCR PRODUCTS

Potentially suitable enzymes that can distinguish between methylated and unmethylated sequences are listed in Table 10.5.1. The PCR product must be checked to determine if these restriction sites are retained or generated by bisulfite treatment.

Materials (see APPENDIX 1 for items with ✓)

    Methylation-specific PCR (MSP) product (see Basic Protocol 1)
    Appropriate restriction endonuclease and buffer
    Bovine serum albumin (BSA), if needed
    10 mg/ml glycogen
    100% and 70% (v/v) ethanol, ice cold
✓ 1× formamide loading buffer

1. Place 10 µl MSP product in a 1.5-ml tube. Add 15 µl appropriate 10× restriction buffer, BSA if necessary, and water to 150 µl. Add 10 to 20 U appropriate restriction enzyme. Incubate reaction 4 to 6 hr under conditions recommended by the manufacturer.

2. Add 1 µl of 10 mg/ml glycogen (as a carrier) and 3 vol ice-cold 100% ethanol. Precipitate DNA several hours to overnight at −20°C. Microcentrifuge 25 min and discard supernatant. Wash with ice-cold 70% ethanol, microcentrifuge 5 min, discard supernatant, and dry pellet. Resuspend DNA in 10 to 12 µl of 1× formamide loading buffer.

Table 10.5.1 Enzymes for Restriction Analysis of Bisulfite-Modified PCR Products[a]

Endonuclease	Unmodified sequence	Methylated modified sequence	Unmethylated modified sequence
*Bst*UI	CGCG	CGCG*	TGTG
*Taq*I	TCGA or CCGA	TCGA*	TTGA
*Sna*BI	CACGCA	TACGTA*	TATGTA
*Hph*I	GGCGA	GGCGA	GGTGA*

[a]Underlined C sequences may alternatively be T in original sequence; asterisks denote sites recognized by the restriction endonuclease.

3. Analyze by nondenaturing polyacrylamide gel electrophoresis (*UNIT 7.2*). Generate a predicted restriction map of the PCR product to anticipate the sizes of the digest fragments and to select a percentage of polyacrylamide. Run undigested PCR product in the lane next to the digested product to compare restriction patterns. Stain gel with ethidium bromide and photograph under UV illumination (*APPENDIX 3G*). Analyze band pattern to identify sequence differences reflecting methylation.

References: Frommer et al., 1992; Herman et al., 1996; Kubota et al., 1997

Contributors: James G. Herman and Stephen B. Baylin

CHAPTER 11

Transcriptional Profiling

UNIT 11.1 describes methods for the preparation of mRNA for expression monitoring. This unit includes protocols for generating, purifying, and quantifying labeled cDNA probes for hybridization to oligonucleotide arrays. This unit also describes methods for processing and normalization of the raw gene expression data in preparation for clustering and further analysis.

UNIT 11.2 describes the use of cDNA microarrays for the profiling of human gene expression. A method is provided for producing cDNA microarrays, which uses a specialized printer to transfer amplified cDNAs onto a glass microscope slide. Preparation of labeled cDNA probes and their hybridization to the microarray are also described.

The expansion of DNA sequence and gene mapping data as well as their associated literature has been enormous during the last decade, and is likely to continue exponentially during the next few years with the flood of data generated by laboratories participating in the human genome project. With this overwhelming accumulation, the need to collect, organize, manipulate, and analyze the data becomes increasingly important. As a result, many different databases have been developed to store sequence and mapping information, and great progress has been made toward integrating these diverse information resources. Electronic access to the information is provided by a number of software systems designed not only to retrieve but also to manipulate and analyze the data. UNIT 11.3 provides an overview of the methods used to analyze the large quantities of expression data generated by microarray experiments.

Contributor: Nicholas C. Dracopoli

UNIT 11.1

Oligonucleotide Arrays for Expression Monitoring

NOTE: For all steps and recipes in this unit, use DEPC-treated water (APPENDIX 1) prepared from glass-distilled water.

BASIC PROTOCOL 1

AMPLIFICATION OF mRNA FOR EXPRESSION MONITORING AND HYBRIDIZATION TO OLIGONUCLEOTIDE ARRAY CHIPS

The overall method is described in Figure 11.1.1. A troubleshooting guide can be found at the end of this unit (Table 11.1.2).

Amplification is absolutely dependent on clean, intact starting RNA. From cultured cells, the authors use a protocol of lysis in guanidine followed by resin-based purification (such as the Qiagen RNeasy kit). The authors' results using one-step protocols (in which guanidine and phenol are combined) have been variable. Tissues are immediately snap-frozen in liquid

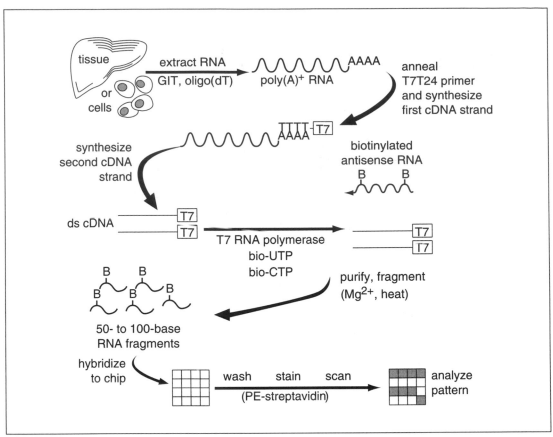

Figure 11.1.1 Chip analysis overview. Abbreviations: GIT, guanidine isothiocyanate; PE, phycoerythrin; SPRI, solid-phase reversible immobilization.

nitrogen and ground to a powder using a dry ice–embedded mortar and pestle. RNA is extracted in guanidine reagent in a Polytron mixer, followed by extraction with phenol (e.g., Promega RNAgents kit, Ambion Totally RNA kit).

This protocol will preferentially amplify the 3′ end of long mRNAs, because of the oligo(dT) primer. It is therefore best to use either custom or commercially available chips that display probes toward the 3′ end of the mRNA. In general, probe selection is restricted to the last 600 bases of coding sequence unless the 3′-untranslated region is >800 bases, in which case some untranslated sequence is also included.

Materials (*see* APPENDIX 1 *for items with* ✓)

SuperScript cDNA kit (Life Technologies), including:

     5× First Strand Buffer

     200 U/μl SuperScript II reverse transcriptase

     5× Second Strand Buffer

     10 mM dNTPs

     10 U/μl *E. coli* ligase

     2 U/μl *E. coli* RNase H

     10 U/μl *E. coli* DNA polymerase

     5 U/μl T4 DNA polymerase

T7T24 primer: 5′-GGCCAGTGAATTGTAATACGACTCACTATAGGGAGGCGG TTTTTTTTTTTTTTTTTTTTTTTTT-3′ (HPLC purification is recommended)

RNase inhibitor (Life Technologies or Ambion)
✓ DEPC-treated H_2O
Sample RNA: poly(A)$^+$ or total RNA
Amplification control transcript pool (see Support Protocol 1)
✓ 25:24:1 (v/v/v) phenol/chloroform/isoamyl alcohol (molecular biology grade)
✓ 7.5 M ammonium acetate
Absolute ethanol
70% (v/v) ethanol in DEPC-treated H_2O, prechilled to $-20°C$
10× transcription buffer (Ambion)
✓ 10× rNTP mix
✓ 100 mM dithiothreitol (DTT)
10 mM Bio-11-CTP and Bio-11-UTP (Enzo Diagnostics)
2500 U/µl T7 RNA polymerase (Epicentre)
RNeasy mini columns with RLT and RPE buffers and collection tubes (Qiagen)
✓ 5× fragmentation buffer
✓ 12× MES buffer
5 M NaCl
✓ 0.5 M EDTA
10% (v/v) Tween-20 (Pierce)
✓ 50 mg/ml bovine serum albumin (BSA)
10 mg/ml herring sperm DNA (Promega)
✓ 500 pM Bio948
20× hybridization control transcript pool (see Support Protocol 1)

Thermal cycler (e.g., Perkin-Elmer 9600 PCR machine with heated lid), for all
    temperature-controlled reactions
0.1- to 10-µl filtered micropipet tips (Continental)
Lyophilizer
Small, thin-walled PCR tubes
GeneChip (Affymetrix)
1- to 200-µl filtered gel-loading micropipet tips (Fisher)
40° and 50°C oven
Rotisserie-type rotator (Appropriate Technical Resources)

1. Set up a linked program on a thermal cycler as follows:

10 min	70°C
65 min	37°C (or 50°C for total RNA)
150 min	15.8°C
indefinitely	4°C (hold).

2. Prepare 10 µl first-strand reagent cocktail for sample RNA by combining the following, using 0.1- to 10-µl filtered micropipet tips:

4 µl 5× First Strand Buffer
200 pmol T7T24 primer
1 µl RNase inhibitor
1 µl 200 U/µl SuperScript II reverse transcriptase
DEPC-treated H_2O to 10 µl.

3. Combine each sample RNA with amplification control transcript pool. Use 5 µl amplification control pool per 1 µg sample poly(A)$^+$ RNA or 1 µl control pool per 10 to 20 µg total RNA. Lyophilize to reduce volume per tube to <10 µl, but do not dry completely. Adjust volume to 10 µl with DEPC-treated H_2O. Prewarm first-strand reagent cocktail in hand during this time.

4. Transfer each RNA to a small, thin-walled PCR tube and place in thermal cycler. Begin linked program. After the 70°C step has completed, allow 2 min for solution temperature to reach 37°C (or 50°C) and add prewarmed first-strand reagent cocktail (10 μl per tube). Hold at appropriate reaction temperature for remaining 60 min.

5. Prepare 130 μl second-strand reagent cocktail per tube (or prepare in advance and keep on ice ≤90 min):

> 91 μl DEPC-treated H_2O
> 30 μl 5× Second Strand Buffer
> 3 μl 10 mM dNTPs
> 1 μl 10 U/μl *E. coli* ligase
> 1 μl 2 U/μl *E. coli* RNase H
> 4 μl 10 U/μl *E. coli* DNA polymerase.

6. Add 130 μl second-strand reagent cocktail to tube (total 150 μl) after thermal cycler has shifted to 15.8°C. Mix by pipetting up and down. Incubate ≥2 hr at 15.8°C. Add 2 μl of 5 U/μl T4 DNA polymerase and incubate an additional 5 min. Remove samples and place on ice.

7. Add 150 μl of 25:24:1 phenol/chloroform/isoamyl alcohol, vortex, and microcentrifuge 5 min at 13,000 × g, room temperature. Carefully remove aqueous top phase to new tube, avoiding contamination with interface.

8. Add 70 μl of 7.5 M ammonium acetate and 0.5 ml absolute ethanol and mix. Microcentrifuge 20 min at 13,000 × g, room temperature. Remove supernatant. Wash pellet with 0.5 ml cold 70% ethanol. Vortex well. Microcentrifuge 10 min at 13,000 × g, remove supernatant, and allow pellet to air dry for several minutes. Resuspend cDNA pellet in 25 μl DEPC-treated H_2O and quantify cDNA (see Support Protocol 2).

9. For each cDNA, set up one reaction as follows (or prepare in advance except for cDNA, keep on ice ≤3 hr, and warm to room temperature before adding cDNA):

> 100 ng cDNA (not more)
> 6 μl 10× transcription buffer
> 6 μl 10× rNTP mix
> 3 μl 100 mM DTT
> 2.4 μl 10 mM Bio-11-UTP
> 2.4 μl 10 mM Bio-11-CTP
> 2 μl RNase inhibitor
> 2 μl 2500 U/μl T7 RNA polymerase
> DEPC-treated H_2O to 60 μl.

Incubate 8 hr to overnight at 37°C. Store ≤48 hr at −80°C, if needed, before purifying.

10. Bring volume of reaction to 100 μl with DEPC-treated H_2O, add 350 μl RLT buffer (without 2-mercaptoethanol), and mix. Add 250 μl absolute ethanol and mix. Apply to an RNeasy mini column placed in a collection tube. Microcentrifuge 15 sec at >8000 × g, room temperature.

11. Transfer column to a new collection tube. Add 500 μl RPE buffer. Microcentrifuge 15 sec at >8000 × g. Discard flowthrough and replace column on same collection tube. Add 500 μl RPE buffer and microcentrifuge 2 min at maximum speed.

12. Transfer column to a new collection tube. Add 50 μl DEPC-treated H_2O and microcentrifuge 1 min at >8000 × g. Repeat elution step, pooling second eluate with the first. Quantify yield of in vitro transcribed (IVT) RNA spectrophotometrically at 260 nm (*APPENDIX 3D*).

13. Bring volume of 10 μg RNA to 24 μl with DEPC-treated H_2O. Add 6 μl of 5× fragmentation buffer. Mix carefully and incubate 35 min at 95°C. Allow to cool to room temperature. Store fragmented RNA ≤1 year at −80°C. Thaw 5 min at 37°C before continuing.

14. Prepare 2× MES hybridization buffer stock:

   8.3 ml 12× MES buffer stock
   17.7 ml 5 M NaCl
   4.0 ml 0.5 M EDTA
   0.1 ml 10% Tween-20
   18.9 ml DEPC-treated H_2O
   1.0 ml 50 mg/ml BSA.

   Heat to 65°C for 10 min. Filter through 0.2-μm filter. Store at −20°C.

15. For each reaction, prepare 170 μl hybridization master mix as follows:

   100 μl 2× MES hybridization buffer stock
   1.7 μl 10 mg/ml herring sperm DNA
   20 μl 500 pM Bio948
   10 μl 20× hybridization control transcript pool
   38.3 μl DEPC-treated H_2O.

16. Add 170 μl hybridization master mix to 30 μl fragmented RNA. Heat to 99°C for 10 min and then move to 37°C for ≥5 min. Microcentrifuge 5 min at maximum speed.

17. Insert a filtered micropipet tip into the upper septum of a GeneChip to provide a vent. Fill the chip from the bottom septum with the 161.7 μl hybridization solution using a 1- to 200-μl filtered gel-loading micropipet tip. Remove vent and cover both septa with transparent tape. Use one GeneChip for each IVT RNA.

18. Incubate in a 40°C oven overnight (16 to 18 hr) on a rotisserie-type rotator at ~60 rpm. Transfer chip to a 50°C oven and continue rotating exactly 1 hr. Remove chip from oven and insert a filtered micropipet tip into the upper septum to vent chip.

19. Using a micropipettor with plunger fully depressed, insert a 200-μl gel-loading pipet tip into the lower septum. Holding chip vertically, slowly draw out all hybridization solution and store in a microcentrifuge tube at −20°C.

20. Fill chip with 1× MES hybridization buffer. Wash and stain chip according to manufacturer's protocols and then scan as soon as possible. If the chip cannot be washed right away, seal the two septa with transparent tape and store at 4°C for up to a few hours. If chip is stained but cannot be scanned right away, store for several hours at 4°C, wrapped in foil.

21. Prepare data and assess quantity (see Basic Protocol 2).

SUPPORT PROTOCOL 1

IN VITRO TRANSCRIPTION OF CONTROL GENES AND PREPARATION OF TRANSCRIPT POOLS

Additional Materials (also see Basic Protocol 1)
   Plasmids (Table 11.1.1; ATCC #87482 to #87490)
   25 mM 4rNTP mix: 25 mM each rGTP, rCTP, rATP, and UTP (Ultrapure; Amersham Biosciences) in DEPC-treated H_2O
   2500 U/μl T3 RNA polymerase (Enzo Diagnostics)

Table 11.1.1 Preparation of Plasmid Template Controls

Name[a]	ATCC #	Transcript size (kb)	Sense RNA		Antisense RNA	
			Linearize with	Polymerize with	Linearize with	Polymerize with
pGIBS-LYS[b]	87482	1.0	*Not*I	T3		
pGIBS-PHE[b]	87483	1.3	*Not*I	T3		
pGIBS-THR[b]	87484	2.0	*Not*I	T3		
pGIBS-TRP[b]	87485	2.5	*Not*I	T3		
pGIKS-BioB	87487	1.1			*Xho*I	T7
pGIKS-BioC	87488	0.8			*Xho*I	T7
pGIKS-BioD	87489	0.7			*Xho*I	T7
pGIKS-CRE	87490	1.0			*Xho*I	T7

[a] Abbreviations: BioB, BioC, and BioD are cloned fragments from the *E. coli bioB, bioC,* and *bioD* genes, respectively. LYS, PHE, THR, and TRP are fragments from the *Bacillus subtilis lysA, pheA, thrBC,* and *trpEDCF* genes, respectively. CRE is a fragment from the Cre recombinase derived from *E. coli* bacteriophage P1. pGIBS and pGIKS are derived from the Bluescript KS II vector (Stratagene).

[b] pGIBS-LYS, -PHE, -THR, and -TRP, contain a 40-nucleotide synthetic poly(A) tract at the $3'$ end of the respective genomic fragments derived from *B. subtilis*. Sense in vitro transcribed transcripts derived from *Not*I-linearized plasmid templates will contain the artificial poly(A) tail. Plasmids linearized with *Bam*HI prior to T3 transcription will generate sense transcripts without the synthetic poly(A) tract.

1. Prepare linearized plasmid templates (Table 11.1.1) and then purify (see Basic Protocol 1, steps 7 to 8). Resuspend pellet at $\sim$0.1 mg/ml in DEPC-treated H_2O and quantify DNA (see Support Protocol 2).

2. For labeled hybridization control transcripts, prepare four tubes containing the following (or prepare $\leq$3 hr in advance, hold on ice, and warm to room temperature before proceeding):

    6 µl 10× transcription buffer
    6 µl 10× rNTP mix
    3 µl 100 mM DTT
    2.4 µl 10 mM Bio-11-UTP
    2.4 µl 10 mM Bio-11-CTP
    2 µl RNase inhibitor
    2 µl 2500 U/µl T7 RNA polymerase
    DEPC-treated H_2O to 60 µl (including 100 ng DNA in step 4).

3. For unlabeled amplification transcripts, prepare four tubes as in step 2, except

    substitute 25 mM 4rNTP mix for 10× rNTP mix
    eliminate biotinylated nucleotides
    substitute T3 for T7 RNA polymerase.

4. Add 100 ng of each linearized plasmid DNA (not more) to a separate tube and incubate 8 hr to overnight at 37°C. Store in vitro transcribed (IVT) RNA $\leq$48 hr at −80°C, if desired. Purify products (see Alternate Protocol) and quantify by absorbance at 260 nm (*APPENDIX 3D*).

5. Dilute unlabeled amplification control transcripts to 200 nM stock solutions in DEPC-treated H_2O for long-term storage ($\leq$1 year at −80°C). Prepare the transcript pool by combining in DEPC-treated H_2O as follows (store $\leq$6 months at −80°C in aliquots of 50 to 100 µl):

                    10 pM LYS transcript
                    30 pM PHE transcript
                    90 pM THR transcript
                    180 pM TRP transcript.

6. Fragment biotin-labeled hybridization control transcripts (see Basic Protocol 1, step 13). Dilute to 20 nM stock solutions in 1× MES hybridization buffer with 0.1 mg/ml herring sperm DNA for long-term storage (≤1 year at −80°C). Prepare the 20× transcript pool by combining (in the same buffer with herring sperm DNA) as follows (store ≤6 months at −80°C):

                    30 pM fragmented BioB transcript
                    100 pM fragmented BioC transcript
                    500 pM fragmented BioD transcript
                    2 nM fragmented CRE transcript.

ALTERNATE PROTOCOL

SOLID-PHASE REVERSIBLE IMMOBILIZATION PURIFICATION OF cDNA AND IN VITRO TRANSCRIPTION PRODUCTS

This purification method for cDNA or IVT RNA is equivalent to that described in Basic Protocol 1 (steps 7 to 8 or steps 10 to 12) but can easily be adapted for automation.

Materials (see APPENDIX 1 for items with ✓)
        Carboxy-coated magnetic beads (PerSeptive BioSystems for cDNA purification; Bangs
            Laboratories for IVT purification)
    ✓ 0.5 M EDTA
        Sample to be purified: cDNA reaction (see Basic Protocol 1, step 6) or IVT RNA
            reaction (see Basic Protocol 1, step 9, or Support Protocol 1, step 4)
        2.5 M NaCl/20% (w/v) PEG 8000 (molecular biology grade; RNase free)
        70% (v/v) ethanol in DEPC-treated H_2O
        10 mM Tris acetate, pH 7.8 (RNase free)
        Magnetic stand (CPG)

To purify cDNA

1a. Place 10 μl PerSeptive carboxy-coated magnetic beads in a 1.5-ml microcentrifuge tube. Place tube on a magnetic stand and allow beads to separate to the side of the tube. Carefully remove supernatant with a micropipet.

2a. Add 10 μl of 0.5 M EDTA and resuspend beads by gentle vortexing or agitation. Replace tube on stand, wait for beads to separate, and remove supernatant. Repeat two more times. Resuspend beads in 10 μl of 0.5 M EDTA.

3a. Add 150 μl of 2.5 M NaCl/20% PEG 8000 and 10 μl resuspended beads to each 150-μl cDNA reaction and mix by gentle vortexing or agitation. Incubate 10 min at room temperature.

4a. Place tube on magnetic stand and allow beads to separate to the side of the tube (∼2 min for original separation, faster for washes). Draw off supernatant and then wash beads twice with 150 μl of 70% ethanol. Remove as much ethanol as possible and air dry 2 min.

5a. Add 25 μl of 10 mM Tris acetate, pH 7.8, and incubate 5 min at room temperature. Place tube on magnetic stand and save supernatant (i.e., eluted cDNA). Determine cDNA concentration by PicoGreen fluorescence (see Support Protocol 2).

To purify IVT RNA

1b. Place 20 μl Bangs Laboratories carboxy-coated magnetic beads in a 1.5-ml microcentrifuge tube. Place tube on a magnetic stand and allow beads to separate to the side of the tube. Carefully remove supernatant with a micropipet.

2b. Add 20 μl of 0.5 M EDTA and resuspend beads by gentle vortexing or agitation. Replace tube on stand, wait for beads to separate, and remove supernatant. Repeat two more times. Resuspend beads in 20 μl of 1.25 M NaCl/10% PEG.

3b. Add 60 μl of 2.5 M NaCl/20% PEG 8000 and 20 μl resuspended beads to each 60-μl IVT RNA reaction and mix by gentle vortexing or agitation. Incubate 10 min at room temperature.

4b. Place tube on magnetic stand and allow beads to separate to the side of the tube (∼2 min for original separation, faster for washes). Draw off supernatant and then wash beads twice with 150 μl of 70% ethanol. Remove as much ethanol as possible and air dry 3 min.

5b. Add 25 μl of 10 mM Tris acetate, pH 7.8, and incubate 5 min at room temperature. Place tube on magnetic stand and save supernatant (i.e., eluted RNA). Determine RNA concentration by absorbance at 260 nm (*APPENDIX 3D*).

SUPPORT PROTOCOL 2

QUANTITATION OF cDNA

Materials

    PicoGreen dsDNA Quantitation Kit (Molecular Probes):
        100 ng/μl standard DNA stock solution
        20× TE buffer
        PicoGreen reagent
    cDNA to be quantified (see Basic Protocol 1 or Alternate Protocol)

    Black-walled 96-well plate (Corning)
    Fluorimager (Molecular Dynamics, model FSI)

1. Prepare 1 ml of 2 μg/ml diluted standard DNA by diluting the stock solution in 1× TE buffer. Dilute PicoGreen reagent 1:200 (v/v) with 1× TE buffer. Prepare enough diluted reagent so that 100 μl can be placed in each well.

2. Pipet 100, 50, 20, 10, 5, 2, and 0 μl diluted standard DNA into seven wells in a black-walled 96-well plate. Bring each to 100 μl with 1× TE buffer (final 200, 100, 40, 20, 10, 4, and 0 ng per well). Pipet 2 μl cDNA to be quantified into additional wells, as needed.

3. Add 100 μl diluted PicoGreen reagent to each well and read fluorescence using a fluorimager.

4. Generate a volume report on each well according to manufacturer's instructions. Use volume number versus nanograms per well to generate a standard curve. Calculate cDNA concentration in test wells.

DATA REDUCTION, NORMALIZATION, AND QUALITY ASSESSMENT

1. The first data reduction process is handled by the Affymetrix GeneChip software. Using image pixel values, the software determines a fluorescence intensity value for each oligonucleotide probe and calculates a summary value for each transcript. The summary value is called "Average Difference" and is calculated by averaging the specific signal (the signal from a perfect match oligonucleotide minus the signal from a mismatch partner) across all probe pairs for a given transcript. Average Difference values are directly proportional to the frequency of a given transcript in the population of all transcripts. GeneChip software also provides a qualitative assessment, called "Absolute Decision," of whether a given transcript is "Present" or "Absent" in a sample.

2. Normalization of intensities from one array to the next is necessary to compare Average Differences for one gene over many experimental conditions. There are several methods to achieve this. The GeneChip software provides two methods: "normalization" is used to change the intensities of one array compared with another array of the same design (i.e., two chips displaying the same probes), whereas "scaling" modifies the Average Differences of experiments of any design to the same average intensity. In either case, the user can choose specific probe sets for this purpose or can use all probe sets on the array. A third method, outside the GeneChip software, is to generate a standard curve from the Average Difference values of the transcripts spiked into the hybridization at a known concentration (the hybridization control pool). This method allows each array to be normalized to the data generated within that array, independent of array design, and also allows for the conversion of Average Differences to absolute frequencies (mRNA molecules per million). For this conversion, negative Average Difference values are defaulted to zero. It is also advisable to develop a method for handling very low and zero Average Differences in a manner that reflects the lowest level of detectability, to preclude recording spurious frequency measurements. An estimate of detectability can be obtained from the Absent and Present calls of the known transcripts in the hybridization control pool. Frequency per million values below this estimate can then be flagged. GeneChip software flags values less than four times the noise value.

3. Quality assessment is done in three major areas:

   RNA quality. The commercially available GeneChip designs are based on publicly available data, from which the 3′-untranslated regions are not always available. If the probes are more than 1500 bases 5′ of the poly(A)$^+$ tail, false negative results can occur, even with high-quality RNA. If the RNA is even partially degraded, or if the reverse transcription reaction was inhibited by contaminants in the RNA, the risk of false negatives increases. RNA quality can be assessed using highly expressed genes, such as β-actin and glyceraldehyde-3-phosphate dehydrogenase, for which probes exist on the chips to monitor the presence of the 3′, middle, and 5′ regions of the transcripts. As the 5′-to-3′ ratio declines below 0.5, the possibility of false negative results should be seriously considered.

   Process quality. The availability of a series of unlabeled RNA transcripts that are spiked into the RNA sample of interest allows the investigator to monitor the success of the RNA amplification process. The members of the amplification control pool should have Average Difference values that correctly reflect their relative abundance in the pool. In addition, if normalization is applied in the GeneChip software, the probe sets used should have similar Average Difference values in the two arrays after normalization, and should not have an

11

Table 11.1.2 Troubleshooting Guide for Expression Monitoring

Problem	Possible cause	Action
Little or no cDNA yield	Beginning RNA quality was poor	Visualize size range of starting RNA on a gel. There should be a smear extending well above 5 kb, and not much <500 bases.
	RNA or reactants were RNase contaminated	Check RNA by incubating an aliquot at room temperature several hours and running on gel as above. Reactants can be incubated with test RNA and run on gel for visualization of degradation.
		Include RNase inhibitor in cDNA reaction.
	Enzyme inhibitors present in RNA preparation	Check signals for members of the amplification control transcript pool. If 5′ end signals are poor and there is no evidence for RNase contamination, reisolate sample RNA using a protocol with a more stringent purification method.
Little or no IVT yield	Too little cDNA template	Quantitate cDNA, use ≥50 ng in IVT reaction. If cDNA yield is low, refer to above section (little or no cDNA yield).
	Inhibitor of polymerase present	Residual phenol/chloroform or dNTPs can inhibit polymerase reaction. Reprecipitate cDNA, being careful to wash pellet with 70% ethanol.
	Wrong polymerase used	Be sure polymerase matches the incorporated promoter
	DTT inactive	Make fresh DTT stock
	Limiting nucleotide concentration	If making unlabeled RNA, be sure to substitute 25 mM 4rNTP solution for the 10 × rNTP mix used with biotin nucleotides.
Blank or low signal on GeneChip	Poor RNA quality	Even when cDNA and IVT yields seem adequate, poor RNA quality can be a cause of poor chip data. Look at housekeeping gene data for ratio of 5′ and 3′ signal. With oligo(dT)-primed cDNA, expect 5′ signal of β-actin and GAPDH[a] to be at least half of the 3′ signal.
	Stringency too high	Check incubator temperature and buffer solutions
	Poor staining of biotin with streptavidin conjugate, or bleaching of phycoerythrin	SA-PE[a] reagents can be variable. Try a different lot, or increase the amount of SA-PE used in the staining solution. Store SA-PE in the dark. Wrap chips in aluminum foil after staining to prevent bleaching before scanning.
	Scanning at wrong wavelength (even edge controls will be very light or absent)	Ensure that scanner is set at 560 nm and rescan

continued

Table 11.1.2 Troubleshooting Guide for Expression Monitoring, continued

Problem	Possible cause	Action
	Hybridization inhibitor in solution	Try using the same hybridization solution on a second chip of the same lot. If signal improves, it may be necessary to incorporate a prehybridization step.
	Scanner needs adjustment	Call for service; try another scanner if available
	Defective lot of GeneChips	Rehybridize saved hybridization solution to a lot of chips that has previously given strong signals. If signal improves, and hybridization inhibitor is not indicated, contact manufacturer.
A few very bright features per gene, the rest very quiet	Failure of fragmentation reaction	Check 5× fragmentation buffer and 94°C temperature control unit. If additional IVT RNA is available, try repeating the fragmentation reaction and check on a gel. Size should be ~20 to 50 bases.
High background values	Stringency too low	Try rewashing the chips at increased temperature and rescanning
	Contaminant	Recommend hybridization solutions be made with glass-distilled water. Some deionization systems can be problematic. If this is suspected, glass-distilled water is commercially available.
Sporadic pattern of round dots all over chip	SA-PE[a] aggregates	Manufacturer recommends the solution be vortexed and then microcentrifuged 2 min prior to taking aliquot
Round area in middle of chip has very low signal compared to rest of chip	Volume of hybridization solution is too low	Need a minimum of 180 μl. If hybridization solution cannot be remade, add 1× MES, 0.1 mg/ml herring sperm DNA, 0.5 mg/ml BSA to the saved hybridization solution to bring the volume to 200 μl and rehybridize to new chip.
	Rotator stopped during hybridization	Rehybridize saved solution to new chip, checking that rotator is functioning properly.

[a]GAPDH, glyceraldehyde 3-phosphate dehydrogenase; SA-PE, streptavidin-phycoerthyrin.

"Increase" or "Decrease" decision (as called by the GeneChip software) associated with them.

Array performance. Like poor RNA integrity, poor chip performance can lead to false negatives. The labeled hybridization control pool can provide an indication of the sensitivity of each chip, which can vary. The pool described in this protocol is widely used but has the drawback of poor representation below 5.0 pM. To robustly determine the limit of detectability, more control transcripts can be added. Alternatively, if normalization is done by the standard curve method, the frequency estimates of all genes on a single array can be paired with their Absolute Decision from GeneChip and an estimate of the limit of detectability can be obtained.

References: Golub et al., 1999; Lockhart et al., 1996; Schena et al., 1995

Contributors: Michael C. Byrne, Maryann Z. Whitley, and Maximillian T. Follettie

11

Profiling Human Gene Expression with cDNA Microarrays

NOTE: All centrifugations take place at room temperature (20° to 25°C) unless otherwise noted.

NOTE: All glassware that comes into contact with the array slides should be carefully cleaned and soaked in 1 N nitric acid before initial use and then shelved separately and used only for arrays.

NOTE: When working with RNA, use RNase-free water (e.g., DEPC-treated water, APPENDIX *1*) to make up all solutions, unless otherwise indicated.

BASIC PROTOCOL 1

cDNA AMPLIFICATION AND PRINTING

A cDNA microarray is made by applying amplified cDNA templates to a microscope slide using a slide printer. The variety of printers and pens for transferring PCR products from titer plates to slides precludes highly detailed descriptions of the process. The steps given here (step 31 to 33) provide a general description.

Materials (see APPENDIX 1 for items with ✓)

    Master plates: master set of clone-purified, sequence-verified human expressed sequence tags (ESTs; e.g., gf211 release, Research Genetics) in bacterial cells
✓ LB medium (e.g., Biofluids or APPENDIX *1*)
✓ 100 mg/ml carbenicillin
    70% and 100% (v/v) ethanol
    100% (v/v) denatured ethanol
    Super Broth (Biofluids)
    45% (w/v) glycerol (enzyme grade), sterilize by autoclaving and store at room temperature
    96-well alkaline lysis miniprep kit (Edge BioSystems)
        Lysis buffer, store at room temperature
        RNase solution
        Resuspension buffer
        Precipitation buffer
        Neutralization buffer
        Wide-bore pipet tips
        96-well receiving plates and filter plates
        Deep-well plates
✓ T low E buffer
    10× PCR buffer
    100 mM dATP
    100 mM dGTP
    100 mM dCTP
    100 mM dTTP
    1 mM AEK M13F PCR primer (5′-GTTGTAAAACGACGGCCAGTG-3′)
    1 mM AEK M13R PCR primer (5′-CACACAGGAAACAGCTATG-3′)
    5 U/μl *Taq* DNA polymerase (AmpliTaq; PE Biosystems)

✓ Ethanol/acetate solution
✓ 20× SSC
   Succinic anhydride (Sigma)
   1-Methyl-2-pyrrolidinone (Sigma)
✓ 1 M sodium borate, pH 8.0

   96-well round-bottom, deep-well, and V-bottom plastic cell culture plates (Corning)
   Centrifuge with horizontal microplate carrier with a depth capacity of 6.2 cm (e.g.,
      Sorvall Super T 21, Sorvall with ST-H750 microplate carrier rotor), room temperature
      and 4°C
   96-pin multi-blot replicator (V&P Scientific)
   1-gallon sealable storage bags (e.g., Glad Lock)
   Microporous tape sheets (e.g., AirPore Tape Sheets; Qiagen)
   Platform shaker with holders for deep-well plates, 37°C
   Sterile 96-well plate seals (e.g., Elkay Products)
   Thin-wall 96-well PCR plates and PCR plate sealer (e.g., CycleSeal plate sealer;
      Robbins Scientific)
   96-well thermal cycler (MJ Research)
   Microtiter plate washer (e.g., Immunowash Microplate washer; Bio-Rad)
   Heat sealable storage bags and heat sealer
   65°C incubator
   Robotic slide printer (e.g., GeneMachines, Genetic Microsystems, Genetix, Cartesian
      Technologies) and print pens (e.g., Majer Precision Engineering, TeleChem
      International)
   Diamond scribe for writing on slides
   Slide box, plastic with no paper or cork liners (e.g., PGC Scientifics)
   ~24 × 34 × 5–cm Pyrex baking dish
   30-slide stainless steel rack and small glass tank (Shandon/Lipshaw)
   Water bath, boiling
   1-liter glass tank

1. Seal low-density master plates and grow overnight at 37°C (most suppliers provide low-density cultures). Omit this step if cultures have already been grown to high density, as further growth will reduce viability.

2. Prepare replicate sets of standard 96-well round-bottom plates by labeling all plates and placing 100 µl LB medium containing 100 µg/ml carbenicillin in each well. (Check that the EST clones are in a vector conferring ampicillin resistance.) Mark plates for easy identification.

   To preserve the master plates, a set of working plates (steps 2 to 5) is used as the culture source.

3. Centrifuge master plates in a centrifuge with a horizontal microplate carrier for 2 min at 167 × g, to remove condensation and droplets from the seals.

4. Dip a 96-pin multi-blot replicator in a 100% ethanol bath, remove, and then flame the pins. Allow replicator to cool briefly. Dip replicator in the master plate and then into the corresponding LB daughter plate. Repeat as necessary for each plate to be inoculated.

5. Place inoculated LB daughter plates with lids into a 1-gallon sealable storage bag containing a moistened paper towel and grow overnight at 37°C.

6. Fill 96-well deep-well plates with 1 ml Super Broth containing 100 µg/ml carbenicillin per well. Using the replicator, inoculate deep-well plates directly from the freshly grown LB plates. Cover deep-well plates with microporous tape sheets and place the plastic lid over the sheet. Place plates on a platform shaker for 24 hr at 200 rpm, 37°C.

7. Add 50 µl of 45% sterile glycerol to each well of working plates and store frozen ($-80°C$).

8. Warm lysis buffer (from lysis miniprep kit) to 37°C until SDS dissolves. Add 1 ml RNase solution to 100 ml resuspension buffer and store at 4°C.

9. Add 350 µl of 100% denatured ethanol to each well of a 96-well receiving plate (from kit). Place filter plate on top and secure in place with tape. Handle with care as the wells will be very full.

10. Centrifuge bacterial cultures in the deep-well plates for 7 min at $1500 \times g$. Immediately invert briefly to remove supernatant and tap out excess medium on a clean paper towel. Do not delay or the pellets will loosen and may be dislodged.

11. Resuspend each pellet in 100 µl resuspension buffer with RNase. Vortex until entire pellet is thoroughly resuspended. Add 100 µl lysis buffer. Mix gently by rocking plate; avoid shearing the bacterial chromosomal DNA.

12. Add 100 µl precipitation buffer to each well. Mix briefly. Add 100 µl neutralization buffer to each well. Vortex. Transfer well contents to the prepared filter plate/receiving plate stack using wide-bore pipet tips provided in the kit.

13. Centrifuge stacked plates 12 min at $1500 \times g$. Remove stacked plates from centrifuge and discard filter plates. Decant ethanol and filtrate from receiver plate. Touch plate on clean paper towels to remove excess ethanol.

14. Add 500 µl of 70% ethanol to each well. Decant immediately. Touch on clean paper towels to remove excess ethanol. Place plates without lids in a clean drawer, cover with a clean paper towel, and allow to dry overnight.

15. Resuspend DNA pellet in 200 µl of T low E buffer. Seal top with sterile 96-well plate seals. Rehydrate for at least 2 days at 4°C before using. Store plasmid templates at $-20°C$.

16. For each 96-well plate to be amplified, prepare a PCR reaction master mix:

    1000 µl 10× PCR buffer
    20 µl 100 mM dATP
    20 µl 100 mM dGTP
    20 µl 100 mM dCTP
    20 µl 100 mM dTTP
    5 µl 1 mM AEK M13F primer
    5 µl 1 mM AEK M13R primer
    100 µl 5 U/µl *Taq* DNA polymerase
    8800 µl H_2O.

17. Label thin-wall 96-well PCR plates and mark the donor and recipient plates at the corner near the A1 well to facilitate correct orientation.

18. Pipet 100 µl PCR reaction master mix into each well. Gently tap plates to ensure that no air bubbles are trapped at the bottom of the wells. Add 1 µl purified EST plasmid template to each well. Watch that the pipet tips are all submerged in the PCR reaction mix when delivering the template.

19. Perform the following thermal cycling series:

1 cycle:	30 sec	96°C	(initial denaturation)
25 cycles:	30 sec	94°C	(denaturation)
	30 sec	55°C	(annealing)
	150 sec	72°C	(extension)
1 cycle:	5 min	72°C	(final extension).

After PCR, plates may be stored at 4°C while quality controls are performed.

20. Analyze 2 µl of each PCR product on a 2% agarose gel (see Support Protocol 1) with appropriate molecular weight markers. If amplified products from this template have been previously tested, analyze one row of wells from each amplified plate.

21. Analyze 1 µl of amplified products from one row of wells from each amplified plate by fluorometry (see Support Protocol 2).

22. To purify, fill a 96-well V-bottom plate with 200 µl per well ethanol/acetate solution. Transfer 100 µl PCR product into corresponding wells of V-bottom plate and mix by pipetting 75 µl four times in each well.

23. Incubate plates for ≤1 hr at −80°C or >1 hr to overnight at −20°C. Thaw completely.

24. Centrifuge 40 min at 2600 × g, 4°C. Aspirate supernatant from each well using a microtiter plate washer. Leave ~10 to 20 µl in the bottom of the well to avoid disturbing the pellet.

25. Wash pellets with 200 µl of 70% ethanol per well using the plate washer. Centrifuge plates 40 min at 2600 × g, 4°C. Aspirate supernatant from each well with the plate washer and allow plates to dry overnight in a closed drawer. Do not dry in a Speedvac evaporator.

26. Add 40 µl of 3× SSC per well. Seal plates with a foil sealer, taking care to achieve a tight seal over each well. Place plates in heat-sealable bags with paper towels moistened with 3× SSC and seal bags with a heat sealer. Place bags in a 65°C incubator for 2 hr and then turn off heat in the incubator and let plates cool gradually.

27. Analyze 1 µl purified, resuspended PCR product from one row of wells from each plate on a 2% agarose gel (see Support Protocol 1). Look for very intense bands, with no material failing to leave the loading well and no smearing. Store plates at −20°C.

28. Before transfering PCR products to slides, preclean print pens according to the manufacturer's specification.

29. Load printer slide deck with poly-L-lysine coated slides (see Support Protocol 3).

30. Thaw plates containing purified PCR products and centrifuge 2 min at 167 × g before opening. Transfer 5 to 10 µl PCR products to a plate that will serve as the source of solution for the printer.

    For quill-type pens, to produce consistent small spots, the volume of fluid in the print source should be sufficiently low that the pen is submerged to a depth of <1 mm.

31. Load pens with DNA solution and serially deposit solution on the first slide in the spotting pattern specified for the print. Use the repetitive test print to check the size and shape of the spotting pattern and its placement on the slide. Also verify that the pens are loading and spotting, and that a single loading will produce as many spots as are required to deliver material to every slide in the printer.

32. If one or more of the pens is not performing at the desired level, reclean or substitute another pen and test again. If all pens are performing, carry out the full print.

33. At the end of the print, remove slides from printer, label with the print identifier and the slide number by writing on the edge of the slide with a diamond scribe. Also etch a line that outlines the printed area of the slide onto the first slide to serve as a guide for the print area. Place slides in a dust-free slide box and age for 1 week.

34. Place slides, printed side face up, in an $\sim$24 $\times$ 34 $\times$ 5–cm Pyrex baking dish and cover with plastic wrap. Expose slides to 450 mJ of UV irradiation in a cross-linker. Transfer slides to a 30-slide stainless steel rack and place rack in a small glass tank.

35. Dissolve 6.0 g succinic anhydride in 325 ml of 1-methyl-2-pyrrolidinone in a glass beaker by stirring with a stir bar.

> CAUTION: *Nitrile gloves should be worn and work should be carried out in a chemical fume hood while handling 1-methyl-2-pyrrolidinone (a teratogen).*

36. Add 25 ml of 1 M sodium borate buffer to the beaker. Let solution mix for a few seconds, and then pour rapidly into glass tank with slides. Place glass tank on a platform shaker in a fume hood for 20 min, shaking at 70 to 90 cycles/min.

37. Submerge slides in a boiling water bath. Immediately turn off heating element and allow slides to stand in the water bath for 2 min. Transfer slides to a 1-liter glass tank filled with 100% ethanol and incubate 4 min.

38. Remove slides and centrifuge 3 min at 167 $\times$ g in a horizontal microtiter plate rotor. Transfer slides to a clean, dust-free slide box and let stand overnight before hybridizing.

BASIC PROTOCOL 2

RNA EXTRACTION AND LABELING

mRNAs from the cells of interest are labeled by reverse transcription with fluorescent nucleotides. Typically, the same results are produced with either poly(dT) or anchored poly(dT) primer when making cDNA. Some cell types may give more signal with the anchored poly(dT).

Materials *(see APPENDIX 1 for items with* ✓ *)*
    Cells harvested from tissue culture, cells in tissue culture, or whole frozen tissue
✓ PBS
    TRIzol reagent (Life Technologies)
    Chloroform
    100%, 75%, and 70% (v/v) ethanol
    RNeasy Maxi Kit (Qiagen)
        50-ml Maxi spin columns with collection tubes
        RW1 buffer
        RPE buffer
✓ DEPC-treated H$_2$O (provided with kit or APPENDIX 1)
✓ 3 M sodium acetate, pH 5.2
    Primer (select one): 2 μg/μl anchored oligo(dT) primer (5′-TTT TTT TTT TTT TTT TTT TTV N-3′; e.g., Genosys) or 1 μg/μl poly(dT)12-18 (Amersham Biosciences)
    200 U/μl Superscript II RNase H⁻ reverse transcriptase with 5× first strand buffer and 1 M DTT (Life Technologies)
✓ 10× low-T dNTP mix
    1 mM Cy5-dUTP or Cy3-dUTP, store −20°C (light sensitive)
    30 U/μl RNase inhibitor (e.g., RNasin; Promega)

11

✓ 0.5 M EDTA, pH 8.0
  1 N NaOH
✓ 1 M Tris·Cl, pH 7.5
✓ TE buffer, pH 7.5
  1 mg/ml human Cot-1 DNA (Life Technologies)

  Tissue homogenizer (e.g., Polytron PT1200; Brinkmann Instruments), for whole tissue
    only
  15-ml round-bottom polypropylene centrifuge tubes
  50-ml conical polypropylene centrifuge tubes
  Clinical centrifuge with horizontal rotor for 50-ml conical tubes
  Microcon YM-100 filter unit (Amicon), or equivalent
  0.2-ml thin-wall PCR tubes with caps
  Thermal cycler
  Fluorescence scanner (e.g., Storm system for gel analysis; Molecular Dynamics)

1a. *For cells harvested from tissue culture:* Wash cell pellet twice in PBS.

1b. *For cells in tissue culture:* Add 1 ml TRIzol per 2×10^7 cells and mix by shaking.

1c. *For whole frozen tissue:* Add 100 mg frozen tissue to 4 ml TRIzol and dissociate by
    homogenization with a tissue homogenizer.

2. Add 1/5 vol chloroform, shake 15 sec, and let stand for 3 min. Centrifuge 15 min at
   $12,000 \times g$, 4°C. Decant supernatant, place in a 15-ml polypropylene tube, and record the
   volume.

3. Add 1 vol of 70% ethanol to the supernatant dropwise while vortexing. To prevent RNA
   precipitation, be sure that each drop has been thoroughly mixed before adding the next.

4. Transfer supernatant from an extraction of 2×10^7 to 1×10^8 cells to an RNeasy Maxi
   column that has been placed in a 50-ml conical polypropylene centrifuge tube. Centrifuge
   5 min at $2880 \times g$ in a clinical centrifuge with a horizontal rotor. Collect eluate, pour it
   back onto the top of the column, and centrifuge again.

5. Discard eluate, add 15 ml RW1 buffer to column, and centrifuge. Repeat with two washes
   of 10 ml RPE buffer, centrifuging 10 min the last time. Discard eluate.

6. Put column in a fresh 50-ml centrifuge tube and add 1 ml DEPC-treated water to the
   column. Let stand 1 min and centrifuge 5 min at $2880 \times g$; do not discard eluate. Add
   another 1 ml DEPC-treated water to the column, let stand 1 min, and centrifuge 10 min
   into the same collection tube.

7. Dispense 400-μl aliquots of column eluate into 1.5-ml microcentrifuge tubes. Add 40 μl
   of 3 M sodium acetate, pH 5.2, and 1 ml of 100% ethanol to each tube, vortex, and let stand
   15 min at room temperature. Microcentrifuge 15 min at $12,000 \times g$, 4°C. Wash pellet two
   times in 75% ethanol and store in 75% ethanol indefinitely at −80°C.

8. Microcentrifuge 15 min at $12,000 \times g$, 4°C. Remove supernatant, air dry pellet, and then
   resuspend RNA at ∼1 mg/ml in DEPC-treated water. Determine concentration by reading
   A_{260} of 1 μl RNA in 100 μl of 50 mM NaOH.

9. Concentrate to >7 mg/ml by centrifuging at $500 \times g$ on a Microcon YM-100 filter unit,
   checking as necessary to determine the rate of concentration. Store at −80°C.

10. Anneal primer to RNA in a 0.2-ml thin-wall PCR tube. Use 2 μg/μl anchored primer or
    1 μg/μl poly(dT)12-18 primer (same volume used in both cases).

Component	For Cy5 labeling	For Cy3 labeling
total RNA (>7 mg/ml)	150 to 200 μg	50 to 80 μg
primer	1 μl	1 μl
DEPC-treated H2O	to 17 μl	to 17 μl.

Heat to 65°C for 10 min and cool on ice 2 min.

11. Prepare a master mix with the following components (23 μl per reaction). When handling Superscript polymerase, be very careful to suppress foaming to avoid denaturation.

> 8 μl 5× first strand buffer
> 4 μl 10× low-T dNTP mix
> 4 μl 1 mM Cy5- or Cy3-dUTP
> 4 μl 0.1 M DTT
> 1 μl 30 U/μl RNase inhibitor
> 2 μl 200 U/μl Superscript II.

12. Add 23 μl reaction mix to each sample, mix well by pipetting, and centrifuge briefly. Incubate 30 min at 42°C. Add 2 μl more Superscript II, mix well, and incubate 30 to 60 min at 42°C.

13. Add 5 μl of 0.5 M EDTA to stop the reaction. Add 10 μl of 1 N NaOH and incubate 60 min at 65°C to hydrolyze residual RNA. Cool to room temperature. Neutralize by adding 25 μl of 1 M Tris·Cl, pH 7.5.

> *The purity of the NaOH solution is crucial. Slight contamination or long storage in a glass vessel can produce a solution that will degrade the Cy5 dye molecule, turning the solution yellow. Some researchers achieve better results by reducing the time of hydrolysis to 30 min.*

14. Transfer labeled cDNA to a Microcon YM-100 cartridge and add 400 μl TE buffer and 20 μg human Cot-1 DNA. Pipet to mix and centrifuge 10 min at 500 × g. Add 200 μl TE buffer and concentrate in a Microcon YM-100 cartridge (~8 to 10 min at 500 × g) to about 20 to 30 μl. Be conservative with initial centrifugation times, as removing too much water will concentrate the cDNA onto the filter and inhibit its recovery. Alternatively, use a Speedvac evaporator, but do not evaporate to dryness.

15. Recover cDNA probe by inverting the concentrator over a clean collection tube and centrifuging 3 min at 500 × g.

> *In some cases, a gelatinous blue precipitate is recovered with Cy5-labeled cDNA. This signals the presence of contaminants. The more extreme the contamination, the greater the fraction of cDNA that will be captured in this gel. Even if heat solubilized, this material tends to produce uniform nonspecific binding to DNA targets.*

16. Run 2 to 3 μl probe on a 2% (w/v) agarose gel (6 cm wide × 8.5 cm long, with 2-mm-wide wells; APPENDIX 3G) in TAE buffer. For maximal sensitivity, use minimal dye in the loading buffer and do not add ethidium bromide to the gel or running buffer.

17. Scan gel on a fluorescence scanner (settings: red fluorescence, 200-μm resolution, 1000 V on PMT). Look for a dense smear of probe from 400 bp to >1000 bp, with little pile-up of low-molecular-weight transcripts. Weak labeling and significant levels of low-molecular-weight material indicate poor labeling. A fraction of the observed low-molecular-weight material is unincorporated fluor nucleotide.

HYBRIDIZATION AND DATA EXTRACTION

To estimate the abundance of particular messages from one RNA source relative to another reference RNA source, both labeled RNAs are hybridized to the same microarray. The cDNA immobilized on the slide is in sufficient excess that the fluorescence from each hybridized probe is proportional to their relative abundance in the cellular message pools.

The particulars of adjusting a microarray imaging device, identifying the signal in the image, and normalizing the values between the two channels vary considerably. This section (steps 9 and 10) is intended only to provide guidelines to the general considerations of imaging.

Materials *(see* APPENDIX 1 *for items with ✓)*

    Glass microarrays (see Basic Protocol 1)
    Cy3- and Cy5-labeled cDNAs (see Basic Protocol 2)
✓ DEPC-treated H_2O
    8 mg/ml poly(dA)40-60 (Amersham Biosciences)
✓ 4 mg/ml yeast tRNA
✓ 10 mg/ml human Cot-1 DNA
✓ 20× SSC
✓ 50× Denhardt solution
✓ 10% SDS
✓ 0.5× SSC/0.01% SDS wash buffer
✓ 0.06× SSC wash buffer

    0.2-ml thin-wall PCR tubes
    Thermal cycler
    24 × 50–mm glass coverslips
    Microarray hybridization chamber
    65°C water bath
    Clinical centrifuge with horizontal rotor for microtiter plates
    Microarray scanner
    Image analysis software

1. Determine the volume of hybridization solution required (critical). Generally, use 0.033 μl/mm^2 area of the coverslip used to cover the microarray (e.g., 40 μl for a 24 × 50–mm^2 coverslip).

2. For a 40-μl hybridization, pool the Cy3- and Cy5-labeled cDNAs into a single 0.2-ml thin-wall PCR tube and adjust volume to 30 μl by adding DEPC-treated water or removing water in a Speedvac evaporator; do not use high heat or heat lamps.

3. Combine the following components:

	For high sample blocking	For high array blocking
Cy5 + Cy3 probe	30 μl	28 μl
8 mg/ml poly(dA)	1 μl	2 μl
4 mg/ml yeast tRNA	1 μl	2 μl
10 mg/ml Cot-1 DNA	1 μl	0 μl
20× SSC	6 μl	6 μl
50× Denhardt solution	1 μl (optional)	2 μl.

When there is residual hybridization to control repeat DNA samples on the array, use the high sample blocking formulation. When there is diffuse background or a general haze on all of the array elements, use the high array blocking formulation. The authors generally try the high sample blocking formulation first.

4. Thoroughly mix components by pipetting, heat 2 min at 98°C in a thermal cycler, cool quickly to 25°C, and add 0.6 μl of 10% SDS. Centrifuge 5 min at 16,000 × g. Apply hybridization cocktail to a 24 × 50–mm glass coverslip and then touch to the inverted microarray, being careful to avoid bubbles. Practice this operation with buffer and plain slides before attempting actual samples.

5. Place slide in a leak-proof microarray hybridization chamber, add 5 μl of 3× SSC in the reservoir (if the chamber provides one) or at the scribed end of the slide, and seal the chamber. Submerge chamber in a 65°C water bath and hybridize slide 16 to 20 hr.

6. Remove hybridization chamber from the water bath, cool, and carefully dry off. Unseal chamber and remove the slide.

7. Place slide, with the coverslip still affixed, into a Coplin jar filled with 0.5× SSC/0.01% SDS wash buffer. Allow coverslip to fall from the slide and remove coverslip from the jar with forceps. Wash slide 2 to 5 min. Transfer slide to a fresh Coplin jar filled with 0.06× SSC. Wash slide 2 to 5 min.

   For more stringent washing, add a first wash that is 0.5× SSC/0.1% SDS or repeat the normal first wash twice.

8. Transfer slide to a slide rack and immediately centrifuge 3 min at low speed, 167 × g, in a clinical centrifuge equipped with a horizontal rotor for microtiter plates. Do not air dry.

9. Load slide in a microarray scanner and adjust the photomultiplier voltage and laser power so that the brightest signals produce slightly less than the maximum possible reading (i.e., 65,535 for a 16-bit scale) and the background values (between the arrayed spots) are consistently somewhat above zero. Collect data for the entire image.

10. Load the captured image into the image analysis software and examine images for overall quality of hybridization, noting the uniformity and level of background, the level and distribution of signals, the relative strengths of the signals in each channel, and the comparability of signals for most of the genes.

SUPPORT PROTOCOL 1

AGAROSE GEL ELECTROPHORESIS OF ESTs

Materials (see APPENDIX 1 for items with ✓)

   2% (w/v) agarose gel in 1× TAE buffer (APPENDIX 3G)
   ✓ 50× TAE buffer
   ✓ Loading buffer for ESTs
   PCR products in 96-well plate (see Basic Protocol 1, steps 20 and 27)
   ✓ 100-bp size standards

   Electrophoresis apparatus with capacity for four 50-well combs, (e.g., Owl Scientific)
   Disposable microtiter mixing trays (e.g., Becton Dickinson)
   Programmable, 12-channel pipettor with disposable tips (e.g., Matrix Technologies)

1. Cast a 2% agarose gel in 1× TAE buffer with four 50-well combs and submerge in an electrophoresis apparatus with sufficient 1× TAE buffer to just cover the surface of the gel (APPENDIX 3G).

2. Prepare a reservoir of loading buffer, using 12 wells of a disposable microtiter mixing tray.

3. Program a 12-channel pipettor to sequentially carry out the following steps:

    fill with 2 μl
    fill with 1 μl
    fill with 2 μl
    mix a volume of 5 μl five times
    expel 5 μl.

4. Load 2 μl PCR product from wells A1 to A12 of the 96-well plate. Load 1 μl of air. Load 2 μl loading buffer from the reservoir.

5. Place tips in clean wells of disposable microtiter mixing tray and allow pipettor to mix the sample and loading dye.

6. Place the pipettor in a 50-well row so that the tip containing the PCR product from well A1 is in the second well of the row, and the other tips are in every other succeeding well.

7. Repeat the process (changing tips each time), loading PCR plate row B starting in the third well, interleaved with row A, row C starting at well 26, and row D starting at well 27, interleaved with row C. Place 5 μl of 100-bp size standard in wells 1 and 50.

8. Repeat this process, loading samples from rows E, F, G, and H in the second, 50-well row of gel wells. Load samples from two 96-well PCR plates per gel, or load single-row samples from 16 PCR plates. To reduce diffusion and mixing, apply voltage to the gel for 1 min after loading each well strip.

9. Apply voltage to the gel and run until the bromphenol blue (faster band) has nearly migrated to the next set of wells. Take a digital photo of gel for future reference. Look for bands of fairly uniform brightness distributed in size between 600 to 2000 bp. Further computer analysis of such images can provide a list of band number and size.

SUPPORT PROTOCOL 2

FLUOROMETRIC DETERMINATION OF DNA CONCENTRATION

Materials (see APPENDIX 1 for items with ✓)
✓ Fluor buffer
  PCR products in 96-well plate (see Basic Protocol 1, step 21)
✓ TE buffer, pH 8
✓ 50, 100, 250, and 500 μg/ml dsDNA standards

  96-well plates for fluorescent detection (e.g., Dynex)
  12-channel multipipettor
  Fluorometer (e.g., PE Biosystems)
  Computer equipped with Microsoft Excel software, for fluorometer without automated analysis

1. Label a 96-well plate for fluorescence detection. With a 12-channel multipipettor, add 200 μl fluor buffer to each well. Add 1 μl PCR product from each well in a row of a PCR plate to a row of the fluorometry plate; use rows A to G of the fluorometry plate.

2. In the final row of the fluorometry plate, add 1 μl TE buffer to first and seventh wells, and add 1 μl of each of the series of dsDNA standards (50, 100, 250, and 500 μg/ml) to the subsequent wells.

3. Set a fluorometer for excitation at 346 nm and emission at 460 nm. Adjust as necessary to read the plate. If the fluorometer does not support automated analysis, export the data table to Microsoft Excel.

4. Test to see that the response for the standards is linear and reproducible from the range of 0 to 500 μg/ml of dsDNA. Subtract the average 0 μg/ml value from all other sample and control values. Calculate the concentration of dsDNA in the PCR reactions using the following equation:

$$[\text{dsDNA } (\mu\text{g/ml})] = [(\text{PCR sample value})/(\text{average 100 } \mu\text{g/ml value})] \times 100$$

SUPPORT PROTOCOL 3

COATING SLIDES WITH POLY-L-LYSINE

Materials (*see* APPENDIX 1 *for items with* ✓)
✓ Cleaning solution
✓ Poly-L-lysine solution

  Gold Seal microscope slides (Becton Dickinson)
  50-slide stainless steel racks and 50-slide glass tanks (Wheaton)
  25-slide plastic racks and 25-slide plastic boxes (Shandon Lipshaw)
  Plastic slide boxes with no paper or cork liners (e.g., PGC Scientific)

NOTE: Wear powder-free gloves when handling the slides. Change gloves frequently, as random contact with skin and surfaces transfers grease to the gloves. Clean all glassware and racks used for slide cleaning and coating with highly purified water only. Do not use detergent.

1. Place Gold Seal microscope slides into 50-slide racks and place racks in glass tanks with 500 ml cleaning solution. Place tanks on a platform shaker for 2 hr at 60 rpm.

2. Pour out cleaning solution and wash slides in water for 3 min. Repeat wash four times.

3. Transfer slides to 25-slide plastic racks and place in small plastic boxes. Submerge slides in 200 ml poly-L-lysine solution per box and shake for 1 hr at 60 rpm.

4. Rinse slides three times with water and submerge slides in water for 1 min.

5. Centrifuge slides 2 min at 400 × g and dry slide boxes used for coating.

6. Place slides back into slide box used for coating and let stand overnight before transferring to plastic slide boxes (without paper or cork liners) for storage. Let slides age 2 weeks at room temperature before printing on them.

References: DeRisi et al., 1997; Iyer et al., 1999; Lockhart et al., 1996; Schena et al., 1995

Contributors: Yuan Jiang, John Lueders, Arthur Glatfelter, Chris Gooden, and Michael Bittner

UNIT 11.3

Analysis of Expression Data: An Overview

Analysis of expression data generally involves two parts: normalizing the data and interpreting them. This unit covers mostly the latter, as it is less technology specific.

EXPERIMENTAL DESIGN

The primary consideration in designing an experiment should be what question or questions the experiment will attempt to answer. Typical goals of a microarray experiment are to find genes that are differentially expressed between two treatments or to study how expression levels in a set of genes change across a set of treatments. Because baseline gene expression levels are not usually known beforehand, most microarray experiments involve comparing gene expression levels between treatment samples and those representing control samples.

Once treatment and control conditions have been decided upon, the next step is specifying the units or samples to which the treatments will be applied and the rules by which these treatments are to be allocated. In the simplest case, mRNA samples are prepared for both treatment and control conditions, and then hybridized either to separate one-color chips or together on a two-color chip. More sophisticated designs are also possible, such as dye-swap and loop designs (Kerr and Churchill, 2001); these have the advantage of increased control over nonbiological sources of variation, such as labeling efficiencies, dye effects, hybridization differences, and other sources of error arising from the measurement process.

Because conventional statistical techniques depend on the existence of replicates, the number and structure of replicates needs to be determined during the design phase. Replication structure will typically be dictated by the goals of the experiment; the number of replicates to use can be estimated by deciding which statistical analyses will be used and performing a priori power calculations. The goal of these calculations is to estimate the number of replicates that will be needed to detect specified differences in gene expression levels. Early interaction and communication between biologists and statisticians is desirable and can result in higher-quality data that are more suitable for answering the questions of interest.

NORMALIZATION

Gene expression data come from a variety of sources and are measured in a variety of ways. The measurements are usually in arbitrary units, so normalization is necessary to compare values between different genes, samples, or experiments. The goal of normalization is usually to produce a dimensionless number that describes the relative concentration of each gene in each sample or experiment. The normalizations described here can be done either manually or automatically with specifically designed software.

One-Color Arrays

The most common type of transcriptional data comes from a one-sample, many-gene array (i.e., a one-color array). This typically associates one measurement per gene per array, indicating the relative abundance of mRNA for that sample. Some programs also produce related information such as a separate background signal (typically subtracted from the raw signal) and information relating to spot size and quality. This information can be used to evaluate the reliability of the data. Typically, these data come from software designed specifically for the scanner in question or from general software like ScanAlyze (Eisen Lab; *http://rana.lbl.gov/*).

Raw intensity measurements are usually in arbitrary units, which are not very useful in isolation. One approach for converting to more useful relative expression measurements is to determine and then divide out per-chip and per-gene effects. Gross exposure per chip can be calculated by taking the average or median of all the mRNA levels over a certain value (background) measured on a chip. It is an arbitrary number dependent upon processing (e.g., sample concentration or hybridization time) and is generally uninteresting biologically, although it can provide useful quality information. Generally, the higher the measured signals, the greater their reliability. Once this number has been determined, dividing raw values by the number results in measurements that have been adjusted for biases resulting from per-chip effects.

To account for per-gene effects, the normal concentration of mRNA for a given gene can be estimated in a similar way by looking at a large number of samples for a given gene. This is done by taking the median (or average) of the measurements for each gene, declaring that to be the normal concentration, and dividing all the measurements for a given gene by that value. In practice, this works quite well, especially with a larger number of samples or experiments. A dataset with a subset of replicates representing a single control condition can alternatively be keyed to represent normal concentration.

Other data transformations and normalizations can be applied where appropriate. For instance, negative intensity values can be set to zero or flagged as unreliable, negative control genes can be used as a measurement of background intensity, positive control (housekeeping) genes can be used as an estimate of per-chip effects, and various other procedures can be used to correct sources of error due to the measurement process. The result of dividing out per-chip and per-gene effects will be a number that measures the relative-to-normal level for each gene in each sample or experiment. It could indicate, for example, that a gene is expressed at 1.4 times its normal expression under a given treatment. These values can then be used in further data analyses.

Two-Color Arrays

In a two-color array there are two measurements for each gene, typically produced through hybridization of control and experimental samples, labeled with different colored dyes, to the same array. A ratio of the measured fluorescence levels for the signal and control provides a measurement of the relative expression of the gene in the sample of interest relative to the control. This dimensionless ratio can then be compared between chips and is directly comparable to the relative-to-normal concentration described above (One-Color Arrays). If the control signal is very low, then this ratio is likely meaningless as it will be dominated by noise; analysts should therefore disregard such readings.

There can still be similar extra information (e.g., background to subtract for both the signal and control, and quantitative or qualitative reliability indicators), which should be dealt with in the same way as in one-color experiments. Because of dye differences or different detection efficiencies of the scanners for different colors, it is often worthwhile to normalize with respect to the overall expression of each dye, in the same way as the per-chip normalization described above for the one-dye measurements. A normalization that uses Lowess curve fitting (Yang et al., 2002) also works well in this situation.

ANALYSIS

Identifying Differentially Expressed Genes

Once normalizations similar to those described above have been applied to gene expression data, the results are ratios that can be analyzed using standard statistical techniques. Although traditional techniques have focused on using fold change as a criterion for differential expression, the limitations of this approach have become apparent. For instance, the fold-change approach makes the assumption that a fixed fold change is equally significant for all genes in question. However, a two-fold change may be indicative of some true biological treatment effect in one gene, whereas in another, it may be within the naturally occurring range of variation of expression level for that gene or it may reflect the error inherent in the measurement process used. For these reasons, it is preferable to use traditional data-driven statistical techniques, where significant differential expression is based on the variation within replicate measurements. Most of these techniques depend upon the existence of replicate measurements to estimate variability and error. When replicates are not available, there are alternative methods that can be used to estimate error; these are described below.

When analyzing ratios resulting from microarray data, it is generally a good idea to first apply a log transform. This is because treatment effects on gene expression levels are generally believed to fit an additive model, with treatment effects being multiplicative. The log transform therefore places the data on a linear scale, and the resulting values are symmetric about zero. The most straightforward method of identifying differential expression is to apply a series of t tests on the log ratios, on a gene-by-gene basis. For two-color data, where the ratios contain expression values for both the control and treatment, a one-sample t test can be performed. When comparing across more than two treatments, for example, in a time-series experiment, this approach can be generalized by using analysis of variance (ANOVA) instead of simple t tests. For each gene, the means in all treatment conditions are compared simultaneously, and a single P value is generated. If the P value falls below the threshold, the gene being tested can be considered differentially expressed. More sophisticated ANOVA models that analyze the variance across an entire experiment at once have also been suggested (Kerr et al., 2000).

Microarray experiments typically result in expression information for thousands of genes. When performing univariate tests for every gene in an experiment, inflated experiment-wise error rates and false positives become issues that need to be addressed. In this situation, it is generally a good idea to apply some sort of multiple testing correction. A method for controlling the false discovery rate, such as Benjamini-Hochberg, represents a reasonable approach to controlling the yield of false positives and false negatives (Benjamini and Hochberg, 1995).

In cases where replicate measurements are not available, standard statistical formulas for variance and standard error cannot be used. It is still possible to estimate variances under these circumstances through the technique of pooling residuals from many different genes. It has been observed that gene expression variability is a function of the normal expression level. This quantity can be measured using control samples, or through normalization procedures similar to those detailed above. Because of this dependence, the pooling of error information is usually done locally. Another approach is to apply a variance-stabilizing transformation to the entire sample or experiment. Once this has been done, all genes can be assumed to have similar variance and thus all measurements in a given sample can be used to compute a common variance (Durbin et al., 2002). These techniques can also be used in cases where very few replicates are available, where they can lead to more reliable estimates of error. In all cases, true replicate measurements are always the best source of information about error and variance.

Clustering

Clustering is a generic name applied to the idea of grouping genes, usually based upon expression profiles. The general idea is that genes with similar expression profiles are likely to have a similar function or share other properties. In order to do this, the concept of similarity of expression profiles needs to be defined.

The objective is to define a function that produces a score of the similarity of the expression patterns of two genes. There are various ways to do this; the most common are distance formulas and various correlations. The simplest method for finding similar genes is to compare the expression pattern for a single gene against all the other genes in the experiment. This finds genes that have an expression profile similar to the gene of interest. Hopefully, the similar genes will somehow be related.

Often the goal is to find distinct groups of genes that have a certain, similar pattern. When one has no idea of what to look for in advance, all the genes can be divided up according to how similar they are to each other. There are many clustering algorithms; two of the most common are *k*-means and self-organizing maps. In both algorithms, the number of groups desired is roughly specified, and the genes are divided into approximately that number of hopefully distinct expression patterns. These algorithms are computationally intensive and are typically performed using software.

Another common method for clustering expression data is called hierarchical clustering or tree building. When a phylogenetic tree is constructed, organisms with similar properties are clustered together. A similar structure of genes can also be used to make a tree of genes, such that genes with similar expression patterns are grouped together. The more similar the expression patterns, the further down on the tree those genes will be joined. A similar tree can be made for experiments or samples, where experiments or samples that affect genes in similar ways can be clustered together. This technique has an advantage over the methods mentioned above in that the number of groups does not need to be specified in advance. Groups of genes can then be extracted as branches of the tree.

Classification and Class Prediction

Classification of tumors and other tissues is another potentially useful application of gene-expression data. These techniques can be used to find genes that are good predictors for cancers and other conditions, to verify tissue classifications obtained by other means, and for diagnostic purposes. The clustering techniques mentioned previously can be applied to samples instead of genes; when used in this context, they are examples of unsupervised learning (i.e., the identification of new or unknown classes using gene expression profiles). The term supervised learning refers to the classification of samples or tissues into known classes. In this setting, a set of tissue samples where the classification is previously known (e.g., cancerous tumors) is analyzed in a microarray experiment. The resulting gene expression data can then be used to classify or predict the class of new samples based on their gene expression levels.

There are a number of statistical techniques and algorithms that can be applied to perform class prediction. They include various types of discriminant analyses, nearest-neighbor techniques, classification trees, and others that fall under the general heading of machine learning (Dudoit et al., 2000). In all of these techniques, the basic steps followed are similar. First, predictor genes are chosen based on their expression profiles in the samples with known classification. These tend to be genes whose expression profiles are significantly different between the classes of interest, and thus are good for discriminating between those classes. Next, the expression profiles for these genes in the samples of unknown classification are examined. This information is then used to place the new or unknown samples into the appropriate classes. If the set of samples being classified have already been classified by alternative clinical methods, this can be used as a validation or verification of those methods. For samples not yet classified, this information is potentially valuable for diagnostic purposes.

Sequence Analysis

Genes with similar expression profiles may be regulated by common transcription factors. For organisms whose genomes are completely sequenced and mapped (e.g., *S. cerevisiae* and *C. elegans*), high-throughput computation now enables the search for candidate DNA-binding sites in upstream regions of genes clustered based on expression profile similarities (e.g., using Gene-Spring; *http://www.silicongenetics.com/cgi/SiG.cgi/Products/GeneSpring/index.smf*). Either a single putative sequence can be searched for in the upstream regions of a cluster of genes, or a whole-scale search can be employed to identify any short sequence (5 to 10 base pairs) occurring in the upstream regions of a gene cluster at unusually high frequencies (Wolfsberg et al., 1999; Gasch and Eisen, 2002).

Pathway and Ontology Analysis

Once genes are identified as strong candidates for differential expression across the conditions of interest, the results need to be interpreted in the context of known biology to extend molecular data to an understanding of higher-level biological effects. Comparing the list of gene profiles of interest against previously assembled lists of genes grouped by function, pathway of action, or cellular localization can provide useful insights. Facilitating the effort, the Gene Ontology (GO) Consortium has established standard hierarchical classifications for genes grouped by biological process, cellular localization, or molecular process with a fixed and controlled vocabulary for class names (Asburner et al., 2000). Furthermore, the group has embarked on gene curation efforts to assign genes to the defined classes. Investigators can now mine NCBI's LocusLink for gene classification information and effectively set up classifications based on GO annotations. Several online pathway databases have also come to the fore, specifically, the Kyoto Encyclopedia of Genes and Genomes (KEGG; Kanesha et al., 2002; *http://www.genome.ad.jp/kegg/*), Biocarta (*http://www.biocarta.com/genes/index.asp*), and GenMAPP (Dahlquist et al., 2002). Finally, statistics relating the expected probability of overlap to observed overlap between gene sets could also be used to examine the significance of potential relationships.

INFORMATICS AND DATABASES

When doing many experiments, especially in a large organization, a laboratory information management system (LIMS) database is useful to keep track of who has done what to which experiments, and other useful information. This database is often custom built to work with a particular laboratory's flow, but several commercial suppliers offer preconfigured LIMS databases. In a high-throughput work environment, a LIMS database can help by tracking a sample from its creation or isolation through to the data collected from a hybridized microarray. Often these data are useful for quality control (e.g., for tracking contaminated reagents). These management systems are often connected directly to array scanners so that the sample annotation, data collection, and subsequent data analysis are directed from a single platform.

Archiving of Data

In any database for microarray data, the actual results from microarray experiments should be stored in association with parameters that describe the experiments (i.e., the differences between the experiments). In addition, it is necessary to archive the results of statistical analyses (e.g., lists of genes with interesting behaviors across specific experimental parameters). There are two common techniques for this.

Text. The data may be stored as text files that have been produced by the image analysis software. This is an inexpensive, fast, and convenient method for individual scientists; however, it can make working in groups difficult and can increase the chances of losing historical data. Storing data in text files lacks the facility of a LIMS to provide a detailed description of the experiment associated with it. For the laboratory that is confined to using flatfile data storage, archiving can be improved by using a document management system like Pharmatrix Base4 (*http://www.opentext.com*) or a flatfile data repository like GeNet from Silicon Genetics (*http://www.silicongenetics.com/cgi/SiG.cgi/Products/GeNet/index.smf*).

SQL. Data may be stored in an Structured Query Language database, preferably associated with the LIMS for tracking production if it exists. A variety of analysis tools can then extract data from the database. This tends to be slow and expensive, but it makes backing up and archiving more reliable. The AADM database from Affymetrix is such a database, and it integrates with the Affymetrix LIMS. If parameters of the experiments are stored in this database, then they can be retrieved automatically by a number of data-analysis packages. Such databases provide little functionality, however, for storing the results of statistical analyses. A dedicated enterprise-level expression repository like GeNet can store both raw expression data and display the results of statistical analyses (e.g., hierarchical clustering dendrograms) in a single package.

Making Data Globally Accessible

FTP. Raw data files can be posted on an FTP server. This method is simple for the experimenter but hard for others to use, as it requires a detailed description of the data structure for the data to be useful.

Public databases. The NCBI (*http://www.ncbi.nlm.nih.gov/geo/*) and EBI (*http://www.ebi. ac.uk/arrayexpress/*) have created similar public databases. These solutions make the data available to anyone on the Internet and are thus reasonable for academics. In addition, GeNet provides users with an Internet-accessible repository that can be placed outside the firewall of one's own institution, so that guest users can access selected data via a web browser.

Contributors: Anoop Grewal, Peter Lambert, and Jordan Stockton

CHAPTER 12

Vectors for Gene Therapy

There are several critical factors in the development of strategies for gene transfer. Important considerations include the type and design of the vector, the method of delivery, the target organ system, and the biological end point. An increasing number of approaches have been developed and evaluated. Viral vectors (retrovirus, adenovirus, adeno-associated virus, and herpes virus) have been the predominant means for delivery of genes into cells. In response to growing concerns about the safety of viral vectors, considerable effort has been expended to optimize nonviral vectors (naked DNA, DNA-liposome complexes, and biolistics for particle-mediated delivery of DNA). These endeavors have succeeded in providing a safer alternative for a number of applications.

Several different systems are in use for somatic gene transfer. These include RNA viruses (retroviruses), DNA viruses (adenovirus, adeno-associated virus, and herpesvirus), and naked and complexed DNA. Each vector system has advantages and disadvantages that influence its use in a specific tissue or organ. Unfortunately, none of the vector systems is completely satisfactory, and no single vector is optimal for all gene therapy approaches. Studies are ongoing in virology, cell biology, and immunology to elucidate the behavior of vectors and the fate of DNA within somatic cells.

The importance of safety in the construction and handling of viral vectors must be stressed at the outset and we begin this chapter with a general consideration of biosafety issues in UNIT 12.1. This unit should help frame the issues for the reader considering use of gene transfer vectors. However, it is not meant to provide a complete consideration, and the interested reader is encouraged to make use of the referenced literature and Web-resources, as well as local institutional expertise. The reader is also reminded that the nature of the gene products expressed poses a potential safety concern. For example, the introduction of growth or angiogenic factors into vectors with the ability to infect human cells may pose significant biohazard risks. Additional safety issues are raised by the widespread use of permissive viral promoters capable of mediating transgene expression in many tissues other than the intended target. Consideration of the consequences of such ectopic expression and evaluation of systemic viral dissemination should be part of the safety evaluation. Although the institutional biosafety review committee must approve the construction and use of gene transfer vectors, the ultimate responsibility for safety resides with the investigator, who presumably best understands the biological systems involved.

UNIT 12.2 provides protocols for construction of adenoviral vectors. These vectors have become a mainstay of gene transfer and gene therapy studies, since adenoviruses efficiently infect many mammalian cells. The promises and pitfalls of these vectors are also discussed. UNIT 12.3 provides updated protocols for the construction of adeno-associated viral (AAV) vectors. Wild-type AAV is not a human pathogen, and AAV vectors can integrate into the infected cell genome to mediate long-term transgene expression in some systems. However, high-titer, helper-virus-independent production of AAV vectors has been problematic. Samulski and colleagues describe a plasmid-based system that provides an important step toward this goal.

Retroviruses, the first viral vectors used for human gene therapy studies, are discussed in UNIT 12.4. The biology of these vectors is probably the best understood of the vector systems, and they still fulfill important "niche" applications. Retroviral vectors have also been modified to create pseudotyped retroviral particles, which contain the envelope glycoprotein from the vesicular stomatitis virus. These pseudotyped retroviral vectors can be prepared at much higher titers than their wild-typed counterparts. Protocols for pseudotyped retroviral vectors are presented in UNIT 12.5. Protocols for high-titer lentiviral vectors are presented in UNIT 12.6. UNIT 12.7 provides protocols on the construction of replication-defective herpes simplex virus vectors and UNIT 12.8 provides protocols on gene delivery using helper virus-free HSV-1 amplicon.

Nonviral approaches to improving the efficiency of gene transfer include use of DNA-liposome complexes and particle-mediated DNA delivery. A protocol for liposome preparation is provided in UNIT 12.9.

Contributors: Anthony Rosenzweig and Elizabeth G. Nabel

UNIT 12.1

Biosafety in Handling Gene Transfer Vectors

RUNNING A BL2 LAB

It is easy to construct a BL2 lab. Any competent architect can read and follow the instructions in the CDC/NIH Biosafety manual or those provided by the NIH Office of Research Services. A standard BL2 lab has a sink, eyewash station, and drench shower; easily washed benches, floors, and ceilings; no nooks and crannies to trap noxious organisms; a fume hood; a biosafety cabinet (see below); access to an autoclave; means of discarding medical waste; and doors (marked with biohazard signs) that are closed when viable RG2 organisms are in use.

The harder part is setting up BL2 procedures and insuring they are followed. Four aspects of this are paramount: the training and behavior of the principal investigator (PI), staff training, use of personal protective equipment, and work procedures, particularly when using a biosafety cabinet.

The Principal Investigator

Responsibility for procedures and behavior in the laboratory is solely in the hands of the PI. A PI must visit the lab regularly, name someone to be responsible for laboratory safety on a day-to-day basis (a lab safety officer, e.g., lab manager), ensure the safety officer has had proper training in institutional safety and environmental policies, establish a training program for existing and new staff, make sure that staff do not work to exhaustion, restrict lab entry to those who know how to protect themselves from its risks, develop a systematic equipment maintenance schedule, and establish a system for renewing outdated or exhausted equipment. Experience shows that labs with a good lab manager tend to be safer and more productive.

Staff Training

As new staff begin working in a laboratory, a senior staff member must be assigned to teach them proper methods for working with biohazardous agents, how equipment they will use works, what to do in moments of crisis such as spills or exposure, and how to use protective equipment and clothing. Existing staff must be reminded of these matters on a regular basis, particularly when policies change.

Supplementary written materials must be made available to the staff; they should know where to find them and how to have any additional questions answered. In particular, a *biosafety manual* should be written to cover the procedures used in the laboratory and to summarize institutional policies related to work with biohazardous materials. The manual should include standard operating procedures, emergency procedures and phone numbers, a discussion of the hazards associated with the work in the laboratory, and a description of the protective equipment and garments available to the staff. A specimen biosafety guide from the Brigham and Women's Hospital in Boston can be found at *http://biosafety.bwh.harvard.edu/BWH_Biosafety_Manuals.htm*. Another style of manual, this one from the University of Georgia, is available on the Web at *http://www.ovpr.uga.edu/qau/qcg.html#top*.

Biosafety Cabinets

Most investigators think of a *laminar-flow tissue culture hood* as a means of protecting their cultures from contamination. They are wrong. Other, less elaborate equipment can be used to protect cultures. A laminar-flow tissue culture hood (more properly called a *biosafety cabinet*) is designed for two simultaneous functions: protecting the worker and protecting the work. Few people using biosafety cabinets have any idea of how they work. Users should know that the hood functions by using an air curtain at the front of the cabinet. The curtain prevents air inside the hood from getting out and air from outside the hood from getting in. Anything that interferes with the air curtain defeats the hood's ability to function properly. Passing a hand through the curtain causes a temporary disruption. Flames inside the hood produce a strong convection current that destroys the air curtain. Covering the front grilles eliminates the curtain.

Even a properly operating air curtain cannot prevent splashes and spills from escaping. That means the only protection from flying liquids is the cabinet's glass sash. Chair height should be adjusted so that the bottom of the sash is at about armpit level. That protects the worker's face while giving easy access to most of the work area.

Protective Equipment and Clothing

Gloves. Gloves protect an investigator only as long as they remain intact. Gloves deteriorate with use and time. Small pores enlarge with continuous expansion and contraction. Many liquids, including hand lotions, acids, bases, and some organic solvents degrade gloves, latex or artificial, with time. Frequent glove changes and hand washing should be part of the investigator's routine. Double gloving is clumsy and uncomfortable and is rarely necessary.

Unless the gloves are sterilized, they must be thought of as a source of contamination. In fact, nonsterilized latex gloves, made from tree sap, carry (and release) a wide variety of viable exotic spores and active nucleases and proteases.

Over the last decade the incidence of latex allergies has increased dramatically among laboratory and clinical workers. Widespread adoption of powder-free and reduced-latex-protein gloves has reduced the risk of this problem. However, a perverse fact of glove manufacturing is that the water-soluble allergens in the latex tend to congregate on the inside of the gloves and are available for absorption by sweaty hands—another reason for changing gloves and washing frequently.

Eye protection. If viable cultures are manipulated outside a biosafety cabinet, the investigator should wear some sort of face protection; a *face shield* or well-fitted *goggles* are appropriate. This is particularly important for procedures likely to generate aerosols. Eyes are a common entry portal for infectious agents.

Face (surgical) masks. Surgical masks are designed to protect the patient from the surgeon (who may have the sniffles), not the surgeon from a patient. Surgical masks stop big globs of mucus, but allow small aerosols to penetrate. For most laboratory work, they have no protective value except to prevent a contaminated finger from entering one's nose or mouth! If there is concern about aerosol generation during normal procedures, disposable *respirators*, often N95-class respirators, can be recommended and fitted by the Safety Office.

Sharps disposal. Sharp objects are the origin of most laboratory-acquired infections. If hollow-bore needles are to be used (for instance, in animal-based toxicity studies), use self-sheathing *Luer-lock* needles, do not remove them from a syringe before discarding, and throw out the entire apparatus (needle and syringe) in a heavy plastic "sharps" container. Scalpels, razor blades, and similar items should also be discarded in a sharps container. When full, the sealed container is considered medical waste and must be discarded accordingly.

Lab coats. Coats must be worn in a BL2 lab. They must be *removed when leaving*. Open lab coat sleeves can drag and spill bottles. The best (and most expensive) lab coats have elastic bands to close their ends. This both prevents bottle dragging and closes a route of exposure. A rubber band works, too. If the coat gets wet, stop working and change to a dry coat. Autoclave the coat if it becomes contaminated with culture.

Procedures

Decontamination. Autoclaving is the most reliable means of killing infectious organisms. However, simpler methods work well. Boiling is a fine way of killing organisms normally encountered in a lab working with human gene transfer vectors. Common laundry bleach (sodium hypochlorite) and phenolic lavatory disinfectants (check the label) are excellent decontaminating agents. They work well in vacuum traps.

Aerosols. Aerosols are suspensions of very small ($\sim$1-μm) invisible liquid droplets. As the lungs are an important entry point for infection, an unsuspecting investigator breathing in an invisible aerosol may become ill. There are several common sources. *Vortexing* a viral suspension yields a large aerosol. A tube that has recently been vortexed should be opened in a hood if it contains hazardous material. Flipping open a *snap-cap microcentrifuge tube* also generates an aerosol. Screw-cap microcentrifuge tubes do not suffer from this problem.

By far the largest aerosols one may encounter arise when a *centrifuge tube* breaks inside a swinging-bucket rotor. Released liquids are smashed into a very fine aerosol by the buckets. When the centrifuge chamber is opened, there is often no visible evidence of the aerosol. To prevent aerosol generation, one should buy clear plastic caps to cover each bucket. If a tube breaks, the capped bucket can be taken to a hood and decontaminated. Tubes may also break inside sealed fixed rotors (particularly in dirty fixed rotors). If biohazardous materials are involved, the entire rotor should be taken to a hood before opening.

Spills. Spills outside a biosafety cabinet usually generate a large aerosol. If the spill contains a vector or infectious agent, it is best to have everyone leave the lab until the aerosol settles (at least 30 min). Call for help. It may be necessary to don a disposable N95-class respirator before re-entering the lab. Heavy gloves, a disposable brush, disposable tongs, a dust pan, and absorbent towels will come in handy when picking up small spills. Make up a spill kit containing these items (and place in an easily accessible location) before it is needed! Larger spills may require a team effort. At every step, the cleanup team should avoid skin contact and work slowly to avoid generating new aerosols. Clean from the outside in.

Sleep. Getting proper rest is a necessity, not a luxury, for researchers. Stupid accidents happen to smart people when they are exhausted. Anyone who works 14 hr straight is increasing their risk and, perhaps, exposing others to a similar risk. It is crucial that others in the lab encourage a sleepy colleague to go home and get some rest.

Animal Biosafety Level 2

If animal testing is required, most requirements for a BL2-level animal facility are similar to laboratory requirements described above. One difference is a requirement that animal cages be decontaminated, preferably by autoclave, before they are washed. Bedding and other wastes must also be decontaminated before disposal. This can greatly increase the cost of animal husbandry. Many Institutional Biosafety Committees (IBCs) insist on BL2 containment and procedures during inoculation but switch to BL1 afterwards. NIH/CDC recommendations for animal husbandry when working with biohazardous agents can be found in Section 4 of Biosafety in Microbiological and Biomedical Laboratories, at *http://www.cdc.gov/od/ohs/biosfty/bmbl/section4.htm.*

SPECIFIC VECTORS

The NIH Office of Recombinant DNA Activities registers all human gene transfer protocols from NIH-funded institutions. The most up-to-date listing can be found at the NIH Office of Recombinant DNA Activities Web site ("Protocol List" at *http://www4.od.nih.gov/ oba/RAC/PROTOCOL.pdf).* Further discussion of vector features and associated laboratory and clinical safety issues is provided in *CPHG* Table 12.1.2.

INTERNET RESOURCES

http://www.bmbl.od.nih.gov/conyents.htm

http://www.fda.gov/cber/index.html

http://www4.od.nih.gov/oba

http://www4.od.nih.gov/oba/rac/guidelines_02/APPENDIX_B.htm

http://www4.od.nih.gov//oba/rac.guidelines_02/APPENDIX_M.htm

http://www.biosafety.bwh.harvard.edu

References: Fleming et al., 1995; Flint et al., 1999; National Research Council, 1989

Contributor: Andrew Braun

UNIT 12.2

Adenoviral Vectors

As outlined in Figure 12.2.1, the AdEasy system consists of three steps. In general, the production of adenoviruses is observed in 7 to 10 days after transfection. An alternative method which offers a simpler and more efficient approach to the production of recombinant adenovirus plasmids using the AdEasier cells is provided (Fig. 12.2.2).

The AdEasy system is freely available for academic researchers. For more detailed information, please visit the AdEasy Web site at *http://www.coloncancer.org/adeasy.htm.*

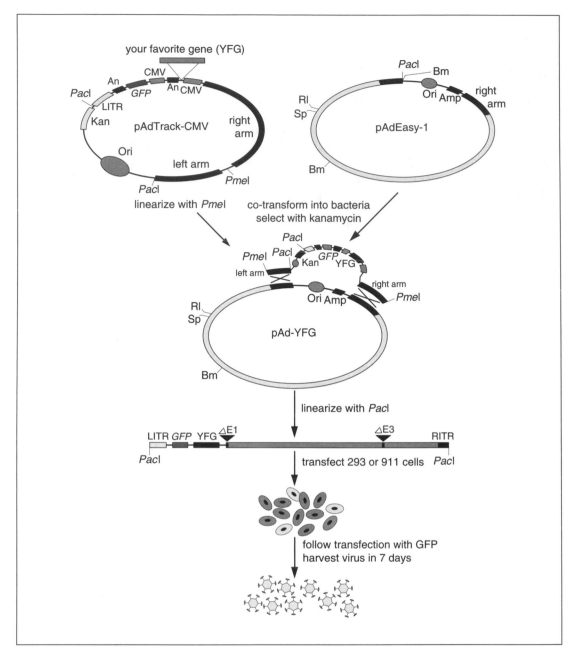

Figure 12.2.1 Schematic overview of the AdEasy technology. The gene of interest is first cloned into a shuttle vector (e.g., pAdTrack-CMV). The resultant plasmid is linearized by digesting with restriction endonuclease *Pme*I, and subsequently cotransformed into *E. coli*. BJ5183 cells with an adenoviral backbone plasmid (e.g., pAdEasy-1). Recombinants are selected for kanamycin resistance, and recombination confirmed by restriction endonuclease analyses. Finally, the linearized recombinant plasmid is transfected into adenovirus packaging cell lines (e.g., 293 cells). Recombinant adenoviruses are typically generated within 7 to 12 days. The "left arm" and "right arm" represent the regions mediating homologous recombination between the shuttle vector and the adenoviral backbone vector. Abbreviations: An, polyadenylation site; Bm, *Bam*HI, RI, *Eco*RI; LITR, left-hand inverted terminal repeat (ITR) and packaging signal; RITR, right-hand ITR; Sp, *Spe*I. Reproduced from He et al. (1998) with permission. Copyright (1998) National Academy of Sciences, U.S.A.

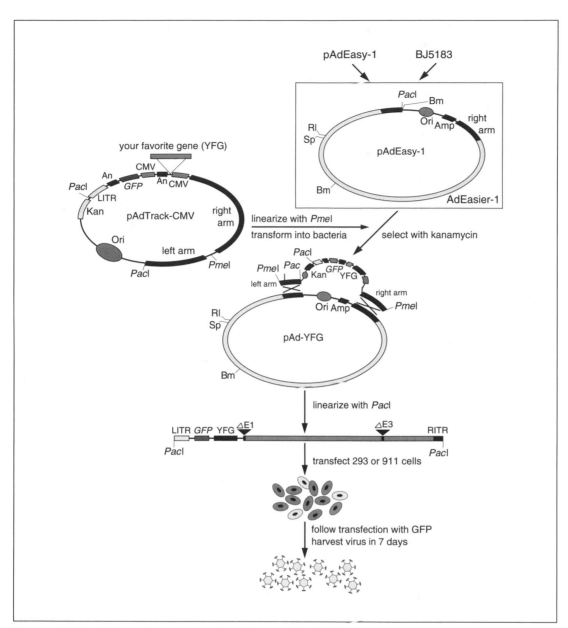

Figure 12.2.2 Schematic representation of the modified AdEasy technology. The revised methodology involves using the AdEasler cells that are BJ5183 derivatives containing pAdEasy backbone vectors (pAdEasy-1 shown) in order to increase the efficiency of generating recombinant adenovirus plasmids. The rest of methodology is virtually the same as the standard AdEasy system described in Figure 12.2.1. Adapted from He et al. (1998) with permission. Copyright (1998). National Academy of Sciences, U.S.A.

According to the National Institutes of Health biosafety guidelines based on risk assessment, manipulations on human adenoviruses should be performed in a laboratory operating at Biosafety Level 2 (BL2) as approved by the user's Institutional Biosafety Committee. The requirements include the use of laminar flow hoods, the establishment of proper procedures for decontamination and disposal of liquid and solid waste, and the disinfection of contaminated surfaces and equipment. Also see UNIT 12.1.

STRATEGIC PLANNING

Choice of Vectors

The AdEasy system provides four different shuttle vectors. As illustrated in Figure 12.2.3, the pShuttle vector is the basic shuttle with the maximal capacity for accommodating foreign genes and the flexibility of desired promoter for transgene expression.

The features and utility of different vectors are listed in Table 12.2.1 to help researchers choose an appropriate combination of vectors. Table 12.2.2 lists potential problems that can arise in using the AdEasy technology along with their possible causes and solutions.

Cloning Genes of Interest into Shuttle Vectors

If the gene of interest and the shuttle vector do not have correctly positioned restriction sites, it may be necessary to blunt-end one or both restriction sites with T4 DNA polymerase. In some cases, it may be more convenient to introduce new restriction sites at one or both ends by linker ligations or by PCR amplification. Introduction of restriction sites by PCR is quick and efficient, but the amplified genes must be verified by DNA sequencing.

If the pShuttle or pAdTrack vector is chosen, users have to provide a promoter and a polyadenylation signal for the transgene expression cassette. For all shuttle vectors, it is absolutely critical to include a consensus Kozak sequence in front of the coding sequences.

Because *Pme*I and *Pac*I sites are designed to linearize the final constructs for transformation and transfections, these sites should be avoided in the inserts. If they cannot be avoided, these vectors can still be used, but with more difficulty, by employing partial digestions, digestion with *Eco*RI and recA-assisted restriction endonuclease (RARE) cleavage, or site-directed mutagenesis to eliminate the restriction sites of interest.

If multiple gene expression cassettes are desired, it is critical to avoid cloning the same elements (e.g., CMV promoters) in head-to-head orientations. Deletion of the sequence between the two elements may occur if a homologous recombination event takes place. However, this type of unwanted recombination can be avoided by placing the repetitive elements in a head-to-tail orientation.

NOTE: All cell culture incubations are performed in a humidified 37°C, 5% CO_2 incubator unless otherwise specified.

NOTE: All solutions, reagents, and equipment coming into contact with cells must be sterile, and proper sterile and antiseptic techniques should be used accordingly. Biohazard wastes containing adenoviruses should be disinfected with chlorine bleach.

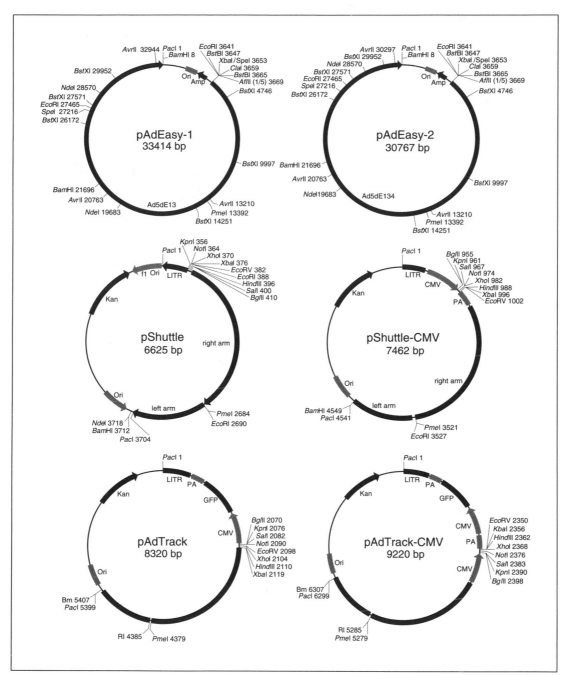

Figure 12.2.3 Shuttle vectors and adenoviral backbone plasmids used in the AdEasy technology. Abbreviations are defined in the legend to Figure 12.2.1. Reprinted from He et al. (1998) with permission. Copyright (1998) National Academy of Sciences, U.S.A.

Table 12.2.1 Selection of the AdEasy Vectors

Shuttle plasmid	Adenoviral backbone	Packaging cells	Maximum insert size	GFP tracer	Use
pAdTrack-CMV	pAdEasy-1	293 or 911	5.0 kb	Yes	Expression of gene of choice under CMV promoter with a built-in GFP tracer
pAdTrack-CMV	pAdEasy-2	911-E4	7.7 kb	Yes	Expression of gene of choice under CMV promoter with a built-in GFP tracer
pAdTrack	pAdEasy-1	293 or 911	5.9 kb	Yes	Expression of gene(s) of choice under your favorite promoter with a built-in GFP tracer
pAdTrack	pAdEasy-2	911-E4	8.6 kb	Yes	Expression of your favorite gene(s) under promoter of choice with a built-in GFP tracer
pShuttle-CMV	pAdEasy-1	293 or 911	6.6 kb	No	Expression of genes in the absence of GFP tracer
pShuttle-CMV	pAdEasy-2	911-E4	9.1 kb	No	Expression of genes in the absence of GFP tracer
pShuttle	pAdEasy-1	293 or 911	7.5 kb	No	Expression of large or multiple genes in the absence of GFP tracer
pShuttle	pAdEasy-2	911-E4	10.2 kb	No	Expression of large or multiple genes in the absence of GFP tracer

Table 12.2.2 Troubleshooting Guide for Using the AdEasy Technology

Problems	Possible cause(s)	Solution(s)
Low number or no colonies after cotransformation in BJ5183	Transformation conditions are not optimal	Follow the protocol provided in this unit
		Consult the manufacturer for specifications of the electroporator
	Incorrect antibiotic is used or antibiotic concentration is too high	Plate the transformation mix on LB plates containing 25 μg/ml kanamycin
	Wrong strain of bacteria is used, or BJ5183 cells are contaminated	Use BJ5183 for homologous recombination
		Grow BJ5183 cells in the presence of streptomycin to eliminate contaminating strains
	DNA preparations are not optimal	Purify the AdEasy backbone vectors by CsCl gradient
		Avoid gel-purifying the *Pme*I-digested shuttle vectors
		Purify shuttle plasmids using alkaline lysis miniprep procedure; avoid using commercial miniprep kits

continued

Problems	Possible cause(s)	Solution(s)
	Competence of BJ5183 is not sufficient	Keep BJ5183 cells concentrated
		Check the competence of BJ5183 cells
		Prepare high-quality competent BJ5183 by carefully following Support Protocol 3
		Avoid repeated freezing/thawing of the competent cell stock
		Obtain the competent BJ5183 commercially
		Introduce AdEasy plasmid into BJ5183 to make competent BJ5183/AdEasy cells (see text for precautions)
Too many colonies after cotransformation in BJ5183	Too much shuttle vector DNA is used for transformation	Reduce the quantity of shuttle plasmids used (usually 0.2–0.5 μg is sufficient)
	*Pme*I digestion is incomplete	Check digested products on agarose gel.
		Use less DNA.
		Make sure the enzyme is active.
	Incubation after transformation is too long	Minimize the length of incubation to no longer than 30 min at 37°C (in most cases, no incubation after electroporation is needed)
No virus plaques observed after transfection in 293 cells	Plasmid DNA preparation is not appropriate	Prepare DNA using CsCl gradient procedure
		Double check concentrations of the plasmids
	293 cell passages are too high	Use earlier passages or fresh stocks of 293 cells
	Recombinant plasmid is not linearized with *Pac*I	Digest the viral plasmid with *Pac*I
	Transfection efficiency is too low	Improve transfection efficiency by optimizing conditions or using different types of transfection reagents
	Did not wait long enough	Wait for longer time, e.g., ≤2 weeks (this is particularly important if transfection efficiency is low)
	There is a defect in the adenoviral backbone	Perform comprehensive restriction analysis of the recombinant plasmid along with control vectors
	Insert exceeds the packaging limit of adenovirus	Consult Table 12.4.1 for proper selection of the AdEasy vectors
No transgene expression detected	The integrity of transgene is not maintained	Make sure the transgene cassette is intact by restriction analysis or PCR
	Efficiency of transient transfection is not high enough	Improve transfection efficiency by optimizing conditions or using different types of transfection reagents
		Make sure detection system works properly by including positive controls
	Transgene is not efficiently expressed	Make sure to include a Kozak sequence in front of the coding sequence

BASIC PROTOCOL

GENERATION OF RECOMBINANT ADENOVIRAL VECTORS USING THE AdEasy METHOD

Viral production should be observed 7 to 10 days after transfection.

NOTE: Skip steps 6 to 10 if the AdEasier cells are used to generate recombinants as described (see Alternate Protocol).

Materials *(see APPENDIX 1 for items with ✓)*

Gene of interest
✓ LB medium with kanamycin
Restriction endonucleases (AdEasy specific, *Pac*I, and *Pme*I or *Eco*RI)
Shuttle vector DNA (Quantum Biotechnologies or Stratagene)
✓ 7.5 M ammonium acetate
seeDNA (Amersham Bioscience)
20 mg/ml glycogen (Roche Molecular)
✓ 25:24:1 (v/v/v) phenol/chloroform/isoamyl alcohol
70% and 100% ethanol
Electrocompetent BJ5183 cells (see Support Protocol 3, or Quantum Biotechnologies or Stratagene)
pAdEasy-1 supercoiled adenoviral backbone vector (Quantum Biotechnologies or Stratagene), CsCl purified
✓ LB/kanamycin plates
0.8% (w/v) agarose gel
Competent DH10B cells or other cells not prone to recombination
293 cells (E1-transformed human embryonic kidney cells)
LipofectAMINE reagent (Life Technologies)
Opti-MEM I medium (Life Technologies)
Dulbecco's modified Eagle medium (DMEM; Life Technologies)
Complete DMEM: DMEM with 10% FBS, 1% penicillin/streptomycin
HBSS or sterile PBS (Life Technologies)

15-ml conical tubes
37°C orbital shaker
2-mm electroporation cuvettes, ice cold
Bio-Rad Gene Pulser electroporator (or similar apparatus)
37°C bacteria incubator
25-cm^2 tissue culture flasks
37°C, 5% CO_2 incubator
Cell scrapers (rubber policeman)
50-ml conical centrifuge tubes
Dry ice/methanol bath

Clone the gene of interest directly into the chosen shuttle vector, taking into consideration the issues described above (see Strategic Planning). It is a good practice to confirm transgene expression in the shuttle vectors by transient transfection assays before starting adenovirus construction. Integrity of the transgenes in final recombinant adenoviral plasmids should also be analyzed by diagnostic restriction endonuclease digestions or PCR amplification.

1. Grow candidate gene-of-interest clones in 2 ml LB/kanamycin in a 15-ml conical tube and culture by shaking overnight in a 37°C orbital shaker.

2. Purify plasmid DNA using any standard alkaline lysis procedure. Confirm the presence and orientation of the transgene by restriction analysis and/or PCR amplification.

   All shuttle vectors and recombinant adenoviral plasmids confer resistance to kanamycin, except pAdEasy-1 and -2, which are ampicillin resistant.

3. Linearize the confirmed shuttle vector with *Pme*I or *Eco*RI restriction enzyme. To ensure a complete digestion, use a 100-μl restriction reaction (~0.1 to 0.5 μg DNA) with 5 μl enzyme. Ensure that the digestion is complete to minimize background, and verify on an agarose gel (APPENDIX 3G).

4. To the 100-μl DNA restriction solution, add 100 μl distilled, deionized water, 100 μl of 7.5 M ammonium acetate, and 2 μl seeDNA (or substitute with 2 μl of 20 mg/ml glycogen); extract with 300 μl of 25:24:1 phenol/chloroform/isoamyl alcohol, pH 8.0 (APPENDIX 3B).

5. Transfer the top layer of DNA solution to a clean tube. Precipitate with 600 μl of 100% ethanol by centrifuging 5 min at 16,000 × g, room temperature. Wash the pellet two times with 70% ethanol to eliminate residual salts. Resuspend DNA in 8 μl distilled, deionized water.

6. Prepare electrocompetent BJ5183 cells in 20 μl/tube aliquots kept at –80°C (see Support Protocol 3). When ready for transformation, thaw aliquots and keep competent cells on ice.

7. To 20 μl competent cells, add 8.0 μl *Pme*I-digested shuttle vector (from step 5) and 1.0 μl pAdEasy-1 supercoiled adenoviral backbone vector (stock 100 ng/μl). Limit the final volume to ≤30 μl.

8. Carefully transfer the bacteria/DNA mix to an ice-cold 2-mm electroporation cuvette, avoiding formation of bubbles, and keeping the cuvette on wet ice. Deliver the pulse at 2500 V, 200 Ω, and 25 μFD in a Bio-Rad Gene Pulser electroporator.

9. Resuspend transformation mix in 500 μl LB medium. Plate on 3 to 4 LB/kanamycin plates, and grow overnight (16 to 20 hr) at 37°C.

   Optionally, the transformed cells can be incubated 10 to 20 min at 37°C prior to plating.

10. Pick 10 to 20 of the smallest, well-isolated colonies, and grow each in 2 ml LB medium containing 25 μg/ml kanamycin for 10 to 15 hr in a 37°C orbital shaker.

    In order to obtain recombinant adenovirus plasmids more efficiently, users are encouraged to follow the alternative method listed at the end of this section (see Alternate Protocol).

11. Perform minipreps using any conventional alkaline lysis method. Check the size of supercoiled plasmids by running 1/5 of each miniprep on a 0.8% agarose gel (e.g., APPENDIX 3G). Potential recombinants run slower than the 12-kb band of a 1-kb ladder.

12. Perform *Pac*I restriction digestion on candidate clones. Correct recombinants usually yield a large fragment (~30 kb) and a smaller fragment of 3.0 or 4.5 kb.

13. Retransform 1 μl correct recombinant plasmids into DH10B (or other plasmid propagation strain not prone to recombination). Perform further restriction analysis on the clones to confirm their primary structure. Finally, purify plasmids by CsCl-banding or using commercial purification kits in preparation for transfection of 293 cells.

    If the background is high, consider one of the following modifications. (1) Do not incubate the bacteria mix after electroporation but directly plate them on LB/kanamycin plates; (2) try to reduce the amount of shuttle vector DNA used in the PmeI digestion; and (3) try to minimize the possibility of introducing nicks into the shuttle vector DNA (e.g., use the alkaline lysis procedure to prepare the shuttle plasmids).

Because of the higher frequency of recombination and rearrangement of plasmids in BJ5183 cells, one should not attempt to regrow the BJ5183 culture for the candidate recombinant clones. Instead, potential recombinant plasmids should be recovered from BJ5183 cells as early as possible (no later than 20 hr) and, once confirmed, should be retransformed into DH10B or other common strains used for plasmid propagation.

14. Plate 293 cells (or 911 cells) in one or two 25-cm^2 tissue culture flask(s) at 2×10^6 cells per flask, ~12 to 20 hr prior to transfection (confluency should be ~50% to 70% at the time of transfection).

15. On the day of transfection, digest recombinant adenoviral plasmids (~4 μg DNA per flask) with *Pac*I. To ensure complete digestion, carry out restriction reactions in 100-μl volumes. Precipitate digested plasmids with ethanol and resuspend in 20 μl sterile water.

16. Perform a standard LipofectAMINE reagent transfection according to manufacturer's manual. Mix 4 μg of *Pac*I-digested plasmid and 20 μl LipofectAMINE reagent for each 25-cm^2 tissue culture flask in 500 μl of Opti-MEM I medium, and incubate DNA/LipofectAMINE reagent mix for 15 to 30 min, room temperature.

17. While waiting for the incubation, remove growth medium from 25-cm^2 tissue culture flasks plated with 293 cells. Gently add 4 ml serum-free medium (e.g., plain DMEM or HBSS medium) to wash residual serum-containing medium. Remove DMEM and add 2.5 ml Opti-MEM I per 25-cm^2 tissue culture flask. Return for ~10 min to 37°C, 5% CO$_2$ incubator.

Special precautions are needed when washing the 293 cells because 293 cells are usually less adherent to the flasks.

18. Add DNA/LipofectAMINE mix dropwise to the 25-cm^2 tissue culture flasks, and return them to 37°C, 5% CO$_2$ humidified incubator.

19. Remove medium containing DNA/LipofectAMINE mix 4 to 6 hr later, and add 7 ml fresh complete DMEM.

Do not change the DNA/LipofectAMINE medium if a significant number of floating cells are observed. If a large number of floating cells are observed, add 6.0 ml complete DMEM to each flask and incubate 10 to 12 hr at 37°C. Remove the medium and add 7 ml fresh medium to each 25-cm^2 tissue culture flask.

20. If pAdTrack-based vectors are used, monitor transfection efficiency and virus production by GFP expression, visible with fluorescence microscopy. Maintain the transfected cells in the 37°C, 5% CO$_2$ incubator for 10 to 12 days. During this period, it is not necessary to change the medium.

In general, no obvious plaques or cytopathic effects (CPE) are observed by standard microscopy up to 2 weeks post-transfection. However, GFP plaques are usually observed under fluorescence microscopy starting 5 to 7 days after transfection. Whether "comet-like" plaques are observed largely depends on transfection efficiency. Lower transfection efficiency (10% to 30%) may produce comet-like plaques, whereas high transfection efficiency (>50%) may generate an intense "scattered stars" phenomenon.

21. Prepare viral lysates by scraping cells off flasks with a rubber policeman (do not use trypsin) at 10 to 12 days post-transfection and transfer to 50-ml conical tubes.

If the transfection efficiency is low (<30%), it is more desirable to harvest the cells at >15 days after transfection to ensure a reasonable initial viral titer. In this case, feed the cells with 2 ml fresh medium at ~10 days after transfection.

22. Centrifuge cells 10 min at $500 \times g$, 4°C, and resuspend pellet in 3.0 ml HBSS or sterile PBS.

23. Freeze cells in dry ice/methanol bath, and thaw in a 37°C water bath to release virus from cells. Vortex vigorously. Repeat freeze/thaw/vortex for three more cycles. Remove tubes from water bath as soon as they thaw to avoid warming virus supernatants, which can reduce titer.

24. Centrifuge samples briefly at 500 × *g*, 4°C, to pellet the cell debris. Store viral lysates at −20°C or −80°C if they are not immediately used for infection.

ALTERNATE PROTOCOL

GENERATE RECOMBINANT ADENOVIRUS PLASMIDS USING AdEasier CELLS

Stratagene is now selling competent AdEasier-1 cells (which they call BJ5183-AD-1). For users planning to make only a few adenoviral constructs, this is particularly convenient.

Additional Materials (also see Basic Protocol; see APPENDIX 1 *for items with ✓)*
  pAdEasy-2 plasmid (optional; Quantum Biotechnologies or Stratagene)
  ✓ LB agar plates containing 50 μg/ml ampicillin and 30 μg/ml streptomycin
  Restriction endonucleases (*Hin*dII or *Pst*I)
  ✓ LB medium without antibiotics
  ✓ LB medium containing 25 μg/ml kanamycin

1. Transform 50 ng pAdEasy-1 or pAdEasy-2 plasmid into electro-competent BJ5183 cells following the conditions described (see Basic Protocol, step 8).

2. Plate the transformation mix (usually 5% to 20%) on LB agar plates containing 50 μg/ul ampicillin and 30 μg/ml streptomycin. Incubate 15 to 20 hr at 37°C.

3. Pick 10 to 20 colonies and grow each in 2 ml LB medium containing ampicillin and streptomycin with continuous shaking at 37°C overnight.

4. Purify the plasmid DNA from each culture following any standard alkaline lysis procedure.

5. Use 20% to 30% miniprep DNA for restriction digestion (e.g., *Hin*dIII, *Pst*I) to confirm the integrity of the clones. Pick one confirmed clone (designated as AdEasier-1 or 2) for subsequent use.

6. Prepare electrocompetent AdEasier cells as described (see Support Protocol 3), except grow in LB medium containing ampicillin and streptomycin.

7. Store electrocompetent AdEasier cells in 20 μl/tube aliquots kept up to 6 months at −80°C. When ready for transformation, thaw aliquots and keep competent cells on ice.

8. Clone the gene into the shuttle vector and digest with *Pme*I as described (see Basic Protocol, steps 1 to 5).

9. To 20 μl electrocompetent AdEasier cells, add 8.0 μl *Pme*I-digested shuttle vector. Limit the final volume to ~30 μl.

10. Carefully transfer the bacteria/DNA mix to an ice-cold 2-mm electroporation cuvette, avoiding formation of bubbles and keeping the cuvette on wet ice. Deliver the pulse at 2500 V, 200 Ω, and 25 μFD in a Bio-Rad Gene Pulser electroporator.

11. Resuspend the transformation mix in 500 μl LB medium. Plate 10% to 20% of the transformation mix onto 1 to 2 LB/kanamycin plates, and grow overnight (16 to 20 hr) at 37°C.

12. Pick 10 to 20 of the smallest colonies, and grow each in 2 ml LB medium containing 25 µg/ml kanamycin for 10 to 15 hr in a 37°C orbital shaker. Purify, characterize, and regrow plasmid DNA as described (see Basic Protocol, steps 11 through 13).

Generate recombinant adenoviruses in the 293 packaging line by following the procedure described above (see Basic Protocol, steps 14 to 24).

SUPPORT PROTOCOL 1

PREPARATION AND PURIFICATION OF HIGH-TITER ADENOVIRUSES

In most cases, it takes two to four rounds of amplification to arrive at a large-scale preparation of high-titer viruses. However, the number of amplification rounds is largely dependent on the initial titers of the primary transfection lysates.

Additional Materials *(also see Basic Protocol; see APPENDIX 1 for items with ✓)*
Primary transfection viral supernatant (see Basic Protocol)
Cesium chloride (CsCl)
Mineral oil
Chlorine bleach
✓ 2× storage buffer
Blank solution: 1.35 g/ml CsCl mixed with equal volume 2× storage buffer
✓ TE buffer containing 0.1% SDS (see individual recipes)

75-cm² tissue culture flasks
Benchtop centrifuge
50-ml conical centrifuge tubes
Sorvall refrigerated centrifuge with HS-4 rotor
12-ml polyallomer tubes for SW 41 Ti rotor
Beckman ultracentrifuge (or equivalent) with SW 41 Ti rotor
Ring stand and clamp
3-ml syringe and 18-G needle

Round one: Amplify from primary transfection lysates

1. Plate 293 cells in 25-cm² tissue culture flasks at 80% to 90% confluency (~3 × 10⁶ cells/flask in 6 ml complete DMEM) at 12 to 15 hr prior to infection.

2. Infect 25-cm² tissue culture flasks of 293 cells by adding 30% to 50% of the primary transfection viral supernatants to each flask.

    The amount of the primary transfection lysates used in each infection is largely determined by their initial titers (usually in a range of 10^6 to 10^8 infectious particles/ml). The rest of the viral lysates should be kept at $-20°C$ or $-80°C$. A cytopathic effect (CPE) or genuine cell lysis should become evident at 2 to 4 days post-infection. Productive infections should be easily observed using the GFP expression incorporated in pAdTrack-based vectors.

3. Scrape and collect cells when 30% to 50% of the infected cells are detached, usually at 3 to 5 days post-infection.

4. Transfer the scraped cells to a 15-ml conical centrifuge tube, and centrifuge cells 10 min at ~500 × g, 4°C. Remove all but 5 ml medium and resuspend cells by vortexing.

5. Perform four cycles of freezing in a dry ice/methanol bath and thawing at 37°C to release the viruses from the cells. Perform the next round of amplification with cleared lysates or keep up to 1 year at −80°C.

12

Rounds 2 and 3: Perform intermediate-scale amplification

6. Plate 293 cells in 75-cm^2 tissue culture flasks at ~90% confluency 12 to 15 hr prior to infection (~5–7 × 10^6 cells/flask in 16 ml complete DMEM).

7. Add 2 to 4 ml viral lysate prepared in step 5 to one 75-cm^2 tissue culture flask of 293 cells. Return cells to 37°C, 5% CO$_2$ incubator.

8. Scrape and collect cells when 30% to 50% of the infected cells are detached, usually at 2 to 4 days post-infection.

9. Transfer the scraped cells to a 50-ml conical centrifuge tube, and centrifuge 10 min at ~500 × g, 4°C in a benchtop clinical centrifuge. Remove all but 10 ml medium and resuspend cells by vortexing. Perform four cycles of freezing in dry ice/methanol bath and thawing at 37°C to release the viruses from cells. Use cleared lysates for the next round (round 3) of amplification or keep at −80°C. Disinfect the virus-containing waste with chlorine bleach.

10. Using three to five 75-cm^2 flasks, repeat steps 6 to 9 for another round of amplification. When collecting the infected cells, resuspend in 25 ml sterile PBS or HBSS. Perform three to four cycles of freezing/thawing to release the viruses from cells. Use the cleared viral lysates for a final round of large-scale amplification or keep at −80°C.

> *Titers can be measured at any time, which is particularly easy with AdTrack-based vectors. Simply infect 293 cells with various dilutions of viral supernatant and count GFP-expressing cells 18 hr later. Without AdTrack, viruses can be plaque titered (see Support Protocol 2) or titered by limiting dilution using standard methods. After three rounds of amplification, viral titer should reach 10^9 to 10^{10} infectious particles (or plaque-forming units, pfu) per milliliter of lysate.*

Final round: Perform large-scale amplification and CsCl gradient purification

11. Plate 293 cells in 15 to 20 75-cm^2 tissue culture flasks to be 90% to 100% confluent at the time of infection (~1 × 10^7 cells/flask).

12. Infect cells with viral supernatant at a multiplicity of infection (MOI) of 10 pfu per cell.

13. When all infected cells have rounded up and about half of the cells are detached (usually at 3 to 4 days post-infection), harvest and combine infected cells from all flasks. Centrifuge 5 min at ~500 × g in a benchtop centrifuge and remove supernatant. Disinfect the virus-containing waste with chlorine bleach.

14. Resuspend the cell pellet in 8.0 ml sterile PBS. Perform four cycles of freezing in a dry ice/methanol bath and thawing at 37°C to release viruses from cells. Centrifuge viral lysate 5 min in a refrigerated centrifuge at 7000 × g, 4°C.

15. Weigh 4.4 g CsCl into a 50-ml conical tube, transfer 8.0 ml cleared virus supernatant to the tube (avoiding the pellet), and mix well by vortexing.

16. Transfer the CsCl solution (~10 ml, density of 1.35 g/ml) to a 12-ml polyallomer tube for an SW 41 Ti rotor. Overlay with 2 ml mineral oil to fill tube. Prepare a balance tube and ultracentrifuge for 18 to 24 hr at 176,000 × g, 10°C.

17. Remove tubes from ultracentrifuge and clamp onto a ring stand above a beaker of chlorine bleach. Note the position of the virus band, which appears as a narrow opaque white band ~1 to 2 cm below mineral oil interface. Collect virus fraction (~0.5 to 1.0 ml) with a 3-ml syringe and 18-G needle by puncturing the side of the tube under the band to extract it into syringe. Do not collect any bands above it.

18. Mix virus fraction with equal volume 2× storage buffer. Store virus stocks at −80°C. Check viral titer by GFP, plaque assays (see Support Protocol 2), or immunohistochemical staining, or simply read OD_{260}. To read OD, add 15 µl virus to 15 µl blank solution and 100 µl TE/0.1% SDS, vortex 30 sec, centrifuge 5 min, and measure OD_{260}.

One OD unit contains $\sim 10^{12}$ viral particles/ml (particles:infectious particles = $\sim 20:1$). It is best to keep the concentrated virus stock at $−80°C$ because the viral particles are generally more stable in high-salt conditions. For in vitro applications where the virus stock is highly diluted, the purified virus preparation can be directly used. However, because CsCl may interfere with or cause toxicity in some other applications, it is best to desalt the virus stocks immediately before use by using desalting columns or quick dialysis with agarose-tubes (see Support Protocol 5).

SUPPORT PROTOCOL 2

ADENOVIRUS PLAQUE ASSAY

Additional Materials *(also see Basic Protocol; see* APPENDIX 1 *for items with* ✓ *)*

    Adenovirus
    2.8% (w/v) Bacto agar (Becton Dickinson)
    2× Basal Medium Eagle (BME; Life Technologies)
    1 M HEPES
✓ 1.0 M $MgCl_2$
    Fetal bovine serum (FBS)
    100× penicillin/streptomycin solution (e.g., Life Technologies)
    100× neutral red stock (Life Technologies)

    6-well plates
    45°C water bath

1. Plate 293 cells in 6-well plates at 50% to 70% confluency (~ 2–5×10^5 cells/well in 5 ml complete DMEM).

2. Determine an appropriate range of 6 ten-fold dilutions based on the approximate adenovirus titer (typically, a range of 10^{-3} to 10^{-8} µl/well is chosen) in a 10- to 50-µl volume. Prepare enough of each dilution to run duplicate assays.

3. In each well of the 6-well plates, remove all but 2 ml complete DMEM. Add the serially diluted adenovirus to each well to infect for 6 to 16 hr. Set up duplicate wells for each dilution.

4. Prepare the overlay agar by autoclaving 100 ml of 2.8% Bacto agar and keep warm in a 45°C water bath.

5. Prepare a 100-ml overlay mix as follows (~ 25 ml overlay mix is needed for one 6-well plate):

    50.0 ml 2× Basal Medium Eagle (final 1×)
    2.0 ml 1.0 M HEPES (final 20.0 mM)
    1.25 ml 1.0 M $MgCl_2$ (final 12.5 mM)
    10.0 ml FBS (final 10% v/v)
    1.0 ml 100× penicillin/streptomycin solution (final 1×)
    Mix well and warm in a 37°C water bath.
    36.0 ml 2.8% Bacto-agar (final 1.0%).
    Mix well and swirl in a 37°C water bath.

6. Aspirate complete DMEM from wells, and overlay each well with 4 ml warmed overlay mixture by slowly adding the solution down the side of each well, taking care not to dislodge cells.

7. Allow agar to solidify 10 to 20 min at room temperature. To prevent the pre-drying of agar before plaque formation, add sterile PBS or HBSS in the space between wells, and wrap the plates with plastic wrap. Return plates to the 37°C incubator.

8. On day 5, overlay 1.5 ml agar overlay mix on top of existing agar in each well to feed cells and maintain monolayer integrity. After solidification, return plates to the incubator.

9. On day 9, prepare neutral red-containing agar overlay mix by adding 500 μl of 100× neutral red stock to 50 ml overlay mix. Overlay each well with 2 ml neutral red-containing agar mix.

10. Allow agar to solidify for 10 min at room temperature, and then return plates to the 37°C incubator.

11. After 12 to 20 hr, remove plates from incubator and hold up to light or place onto a light box, observing the monolayer from the bottom of the plate. For each well, count plaques, which will appear as clear pale orange areas amid a darker reddish-orange monolayer.

12. Determine plaque counts for each dilution by averaging the duplicate wells. This average will determine the titer of adenoviral stock as expressed in plaque forming units per milliliter (pfu/ml). For viruses encoding GFP, determine titer of infectious units (i.e., those resulting in expression of GFP) by counting GFP-expressing foci using fluorescence microscopy.

12

SUPPORT PROTOCOL 3

PREPARATION OF ELECTROCOMPETENT BJ5183 CELLS

Additional Materials (also see Basic Protocol; see APPENDIX 1 for items with ✓)
   ✓ LB medium containing 30 μg/ml streptomycin
   10% (v/v) sterile glycerol, ice cold
   10 ng/μl pAdEasy-1 plasmid DNA
   ✓ LB agar plates with 50 μg/ml ampicillin

   50-ml conical centrifuge tubes
   250-ml sterile centrifuge tubes (for IEC centrifuge)
   IEC centrifuge (or equivalent)
   1.5-ml microcentrifuge tubes, prechilled at −80°C

1. Use a fresh colony or frozen stock of BJ5183 cells to inoculate 10 ml LB medium containing 30 μg/ml streptomycin in a 50-ml conical tube. Grow cells overnight in a 37°C environmental shaker.

2. Dilute 1 ml of cells grown overnight into 1000 ml LB medium containing 30 μg/ml streptomycin. Shake vigorously for 4 to 6 hr with good aeration in 37°C environmental shaker, until A_{550} is ∼0.8.

3. Collect cells in four 250-ml sterile centrifuge tubes and incubate on ice 1 to 3 hr (the longer cells are incubated, the higher the competency). Centrifuge 10 min at 2600 × g, 4°C.

4. Remove supernatant. Wash the cell pellet by resuspending in 1000 ml sterile ice-cold 10% glycerol. Centrifuge cell suspension 20 min at 2500 × g, 4°C. Repeat one additional time.

5. Pour off most of the supernatant, then gently pipet off most of residual supernatant, leaving ~10 ml per 250-ml centrifuge tube. Combine cells and transfer cell suspension to a 50-ml centrifuge tube. Centrifuge for 10 min at 2500 × g, 4°C.

6. Remove most of the supernatant and add 40 ml ice-cold 10% glycerol. Resuspend cells and centrifuge 10 min at 2500 × g, 4°C.

7. Pipet out all but 2 ml of the supernatant and resuspend cell pellet. Pipet 20-μl aliquots per prechilled 1.5-ml microcentrifuge tube. Store the aliquots at −80°C.

8. To verify the competency of prepared BJ5183 cells, add 1.0 μl of 10 ng/μl pAdEasy-1 plasmid DNA to 20 μl BJ5183 competent cells.

9. Transfer cell/DNA mix to an ice-cold 2-mm cuvette. Perform electroporation with Bio-Rad gene pulser at 200 Ω/25 μF/2.5 kV.

10. In a 50-ml conical tube, add 1 ml LB medium to cells and shake for 1 hr at 37°C.

11. Make 100- and 1000-fold serial dilutions of the cells in LB medium. Plate 100 μl diluted cells on LB-agar plates with 50 μg/ml ampicillin. Incubate overnight in a 37°C incubator (titer should be >10^8 colonies/μg DNA).

SUPPORT PROTOCOL 4

PREPARATION OF ADENOVIRAL DNA

Materials (see APPENDIX 1 for items with ✓)
    Viral lysate or CsCl gradient purified virus stock
    10% SDS
    0.5 M EDTA
    20 mg/ml PCR grade proteinase K (Life Technologies)
✓ 7.5 M ammonium acetate
    seeDNA (Amersham Biosciences)
    PC-8 (Fisher) or 25:24:1 (v/v/v) phenol/chloroform/isoamyl alcohol (see APPENDIX 1)
    70% and 100% ethanol
    55°C water bath

1. To 100 μl viral lysate or 10 μl CsCl gradient purified virus stock, add 7 μl of 10% SDS, 3 μl of 0.5 M EDTA, and 20 μl of 20 mg/ml proteinase K. Mix well and incubate 3 hr in a 55°C water bath. Heat the mix 5 min at 95°C.

2. Bring the viral DNA solution to a total volume of 200 μl with deionized, distilled water, and then add 100 μl of 7.5 M ammonium acetate and 2 μl seeDNA.

3. Extract the mix two times with 300 μl PC-8. Transfer top phase to a new 1.5-ml microcentrifuge tube, avoiding the interface. Add 600 μl of 100% ethanol.

4. To precipitate viral DNA, microcentrifuge 10 min at maximum speed. Wash pellet two times with 70% ethanol. Dissolve viral DNA pellet in 50 μl water.

SUPPORT PROTOCOL 5

QUICK AGAROSE-TUBE DIALYSIS

Materials
    Agarose, molecular-biology grade
    2-ml microcentrifuge tubes
    200-μl filter tips

1. Prepare 1% agarose by melting the agarose in deionized, distilled water in a microwave oven at full power.

2. Make a dialysis-tube apparatus by pipetting 1 ml melted 1% agarose into a 2-ml micro-centrifuge tube. Stick a beveled 200-μl pipet tip to the very bottom of the tube.

3. After 1 hr at room temperature, remove the pipet tip. Add 50 μl ddH$_2$O to the hole to keep the gel wet. Store the tubes at 4°C.

4. To dialyze, remove the 50 μl water and add the virus stock solution (generally $\leq$25 μl) with a needle-nosed pipet tip.

5. After an appropriate time period (usually 10 to 20 min), remove the solution with a needle-nosed pipet tip and either use directly or add to a new agarose dialysis apparatus if further dialysis is desired.

References: Becker et al., 1994; He et al., 1998; Shenk, 1996

Contributor: Tong-Chuan He

UNIT 12.3

Production of Recombinant Adeno-Associated Viral Vectors

An overview of the rAAV production procedure is shown in Figure 12.3.1. This virus can then be used to infect tissue culture cells or can be infused into an intact animal (see, e.g., *UNITS 13.3 & 13.5*).

CAUTION: Virus work should be performed in a dedicated tissue culture hood and incubator separate from the hood and incubator used for the maintenance of laboratory cell lines. Proper disposal of virally contaminated materials should be performed.

NOTE: All equipment and reagents coming in contact with tissue culture cells (uninfected and infected) and the virus following CsCl gradients should be sterile.

NOTE: All tissue culture incubations are performed in a humidified 37°C, 5% CO$_2$ incubator unless otherwise specified.

BASIC PROTOCOL

PRODUCTION OF ADENOVIRUS-FREE rAAV BY TRANSIENT TRANSFECTION OF 293 CELLS

Materials (see APPENDIX 1 for items with ✓)

    pSub201 plasmid: used to clone the transgene between AAV termini (Fig. 12.3.2; ATCC #68065; see *http://www.med.unc.edu/genether/*)

    *Xba*I and *Hind*III restriction endonucleases, with appropriate buffers

    pXX2 plasmid: the AAV helper plasmid (UNC Vector Core Facility; see *http://www.med.unc.edu/genether/*)

    pXX6 plasmid: the adenoviral helper plasmid (UNC Vector Core Facility; see *http://www.med.unc.edu/genether/*)

    293 tissue culture cell line (ATCC #CRL 1573)

✓ Complete DMEM/10% FBS medium

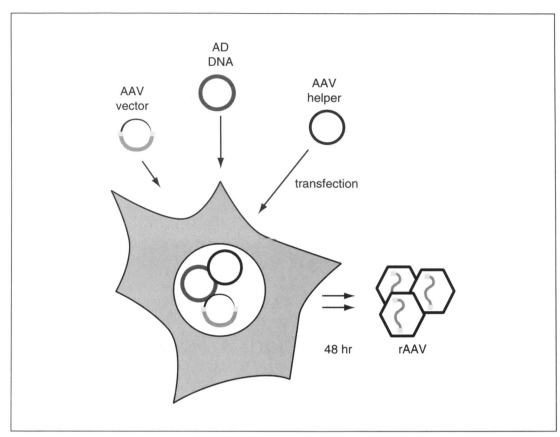

Figure 12.3.1 Generation of adenovirus-free recombinant adeno-associated virus. 293 cells (which supply the Ad *E1* gene) are transfected using three plasmids: the plasmid carrying the transgene (sub201-gene "*X*", AAV vector), the plasmid supplying the replication and capsid genes of AAV2 without terminal repeats (pXX2, AAV helper), and the plasmid supplying the adenovirus helper genes *E2*, *E4*, and *VA* RNA genes (pXX6, AD DNA), thereby generating Ad-free rAAV.

✓ Complete IMDM/10% FBS medium
   0.05% (w/v) porcine trypsin/0.02% (w/v) EDTA
✓ 2.5 M CaCl$_2$
✓ 2× HEPES-buffered saline (HeBS)
✓ Complete DMEM/2% FBS
   Dry ice/ethanol bath
   Ammonium sulfate [(NH$_4$)$_2$SO$_4$]
   OPTI-MEM I (Life Technologies)
✓ Saturated ammonium sulfate, pH 7.0, 4°C
✓ 1.37 g/ml and 1.5 g/ml density CsCl
   70% ethanol
✓ Phosphate-buffered saline (PBS)

   15-cm tissue culture plates
   50-ml disposable polystyrene and polypropylene centrifuge tubes
   Cell scrapers
   250-ml polypropylene centrifuge bottles
   Sorvall centrifuge with GS-3 and SS-34 rotors or equivalents
   Sonicator with a 3-mm diameter probe
   Tabletop centrifuge

50-ml high-speed polypropylene centrifuge tubes
Beckman ultracentrifuge with SW-41 rotor and 12.5-ml Beckman Ultra-Clear tubes
(or equivalent)
21-G needles
Pierce Slide-A-Lyzer dialysis cassettes (MWCO 10,000)

1. Digest the plasmid pSub201 (Fig. 12.3.2) with *Xba*I and *Hin*dIII restriction endonucleases
   (e.g., *CPMB UNIT 3.1*) to remove the *rep* and *cap* fragments, and gel purify the 4000-bp plasmid
   backbone containing the AAV ITRs (e.g., *CPMB UNIT 2.6*). Insert the desired transgene
   cassette into the *Xba*I sites to construct the rAAV vector plasmid (e.g., *CPMB UNIT 3.16*).
   Purify a large-scale plasmid preparation (at least 1 mg) of the rAAV vector and pXX2 and
   pXX6 plasmids by double CsCl gradient fractionation.

2. Seed 2×10^7 293 cells in 15-cm dishes to achieve 70% to 80% confluency the next morning.
   Maintain 293 cell line in complete DMEM/10% FBS medium and split a near confluent
   dish into three to four 15-cm dishes (20 dishes total for this protocol) by trypsinizing with
   0.05% trypsin/EDTA (also see *APPENDIX 3I*).

3. At a time point 12 to 16 hr following the split, replace the medium with 25 ml per 15-cm
   dish of complete IMDM/10% FBS. Incubate for 3 hr before the transfection of the cells.

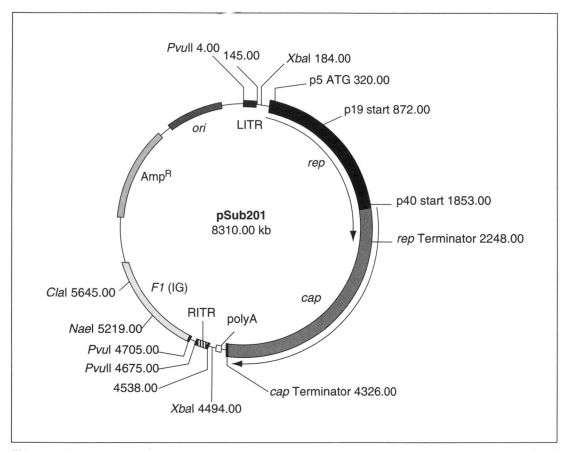

Figure 12.3.2 Plasmid Sub201. The plasmid contains the terminal repeats and the replication (*rep*)
and capsid (*cap*) genes of AAV Type 2. The *rep-cap* fragment can be replaced by the gene cassette of
interest, as only the terminal repeats are needed for packaging.

4. For transfection of four 15-cm dishes, combine the following in a disposable 50-ml polystyrene tube:

> 90 μg pXX6 helper plasmid (adenoviral helper genes)
> 30 μg rAAV vector plasmid
> 30 μg pXX2 helper plasmid (AAV helper genes)
> 0.4 ml 2.5 M $CaCl_2$
> Deionized distilled H_2O to 4 ml.

The total DNA is equal to 37.5 μg/plate and the ratio of AAV helper plasmid to Ad helper plasmid to rAAV vector plasmid is equal to a molar ratio of ~1:1:1.

5. Add 4 ml of 2× HeBS to the mixture from step 4, and pipet up and down with a 10-ml plastic pipet three times to mix. Incubate 1 to 5 min at room temperature. Observe a very fine precipitate under a microscope at 40× (if the precipitate is too coarse, reduce incubation time).

6. Vortex the suspension of DNA/$CaCl_2$ precipitate for 5 sec and then add 2 ml dropwise to the medium in each of four 15-cm plates of cells (from step 3). Swirl the plates to disperse evenly.

7. Repeat steps 4 to 6 to transfect groups of four dishes at a time until all 20 dishes have been transfected. Incubate cells 24 hr.

8. After the 24-hr incubation, aspirate the medium (two plates at a time so that the cells do not dry out) and add 25 ml of 37°C DMEM/2% FBS. Incubate the cells until 48 hr post-transfection.

9. Collect the cells and the supernatant from the tissue culture plates by scraping the cells with a cell scraper to collect cells and medium. Transfer suspension to 250-ml polypropylene centrifuge bottles.

10. Freeze and thaw the cell suspension (to liberate virus particles) three times by transferring the tubes between a dry ice/ethanol bath and a 37°C water bath (suspension can be stored up to 6 months at −20°C).

11. Centrifuge the bottles for 10 min at 3000 × g, 4°C, to pellet the cells. Decant the supernatant into fresh 250-ml polypropylene centrifuge tubes. Retain both the supernatant and the cell pellet, separately.

12. To precipitate the virus in the supernatant, add 78.25 g $(NH_4)_2SO_4$ per 250 ml of supernatant, mix thoroughly to dissolve the ammonium sulfate, and then continue precipitation on ice for 20 min. Centrifuge the tubes 10 min at 8300 × g, 4°C.

13. Slowly decant the supernatant into a container, separating it from the yellowish precipitate that has formed on the bottom and sides of the centrifuge vessel. Autoclave the supernatant (do not use bleach) and then dispose of it. Maintain the centrifuge bottle containing the precipitate on ice until the cell pellet from step 11 has been processed.

14. Resuspend the cell pellet from step 11 in 20 ml of OPTI-MEM I medium by pipetting up and down or vortexing, and transfer to a disposable 50-ml polypropylene centrifuge tube.

15. Using ear protection, sonicate the cell pellet (40 bursts, 50% duty, power level 2) at room temperature in a tissue culture hood dedicated for virus work. Observe a turbid cell lysate.

16. Centrifuge the tubes for 5 min at 3000 × g, room temperature, to pellet insoluble debris. Collect the clarified supernatant and transfer to a disposable 50-ml polypropylene tube.

17. Resuspend the cell pellet again in 20 ml of OPTI-MEM I and repeat steps 14 to 16.

18. Pool the supernatants from steps 16 and 17. Use these pooled supernatants to dissolve the ammonium sulfate precipitate in the centrifuge bottle (from step 13) by pipetting the solution down the side and the bottom of the bottle.

19. Measure the volume of the liquid from step 18 and add 1 vol of 4°C saturated ammonium sulfate, pH 7.0 per 3 vol of the virus-containing suspension. Mix well, and put on ice for 10 min.

20. Centrifuge for 10 min at 7700 × g, 4°C (a yellow precipitate should be seen at the bottom of the tubes). Transfer each supernatant to 50-ml high-speed polypropylene centrifuge tubes.

21. Using the volume of the lysate from step 18, add 2 vol of 4°C saturated $(NH_4)_2SO_4$ solution per 3 vol of the lysate and incubate the sample on ice for 20 min.

22. Centrifuge the tubes 20 min at 17,000 × g, 4°C, whereupon a large precipitate will be formed. Remove the supernatant and autoclave (do not add bleach) before discarding.

23. Dissolve pellets in 20 ml total (for 20 to 40 plates) of 1.37 g/ml CsCl solution. Resuspend the pellets as completely as possible so that, after mixing, no undissolved material can be seen.

24. Add 0.5 ml of 1.5 g/ml CsCl solution to each of two 12.5-ml Ultra-clear tubes. Overlay ~12 ml of virus sample upon 1.37 g/ml CsCl in each tube by holding the pipet to the side of the tube and slowly applying the sample. Move the pipet up along the side of the tube as the sample is dispensed.

25. Centrifuge the samples for 36 to 48 hr at 288,000 × g, 15°C. Decelerate with brake to 500 rpm, then turn brake off. Observe a possible diffuse AAV band in the middle of the tube.

26. Wipe the outside of the centrifuge tube with 70% ethanol and insert a 21-G needle ~1 cm from the bottom of the tube at a 90° angle and allow the gradient to drip into sterile microcentrifuge tubes. Collect 15 to 20 fractions of 10 drops each.

27. Assay 1 to 5 μl of each fraction by dot blot using an rAAV-specific probe to find the peak (see Support Protocol 1).

   The rAAV peak should band in the gradient with a density of 1.40 to 1.42 g/ml, and this can be checked with a refractometer (be sure to decontaminate the refractometer after use).

28. *Optional:* Add 1.37 g/ml CsCl solution to the pooled fractions to attain a final volume of 12 ml. Add 0.5 ml of 1.5 g/ml CsCl to one Ultra-Clear tube and overlay virus solution as in step 24, then reband, drip, and assay the gradient as in steps 25 to 27.

29. Dialyze the rAAV in MWCO 10,000 Slide-A-Lyzer dialysis cassettes against three 500-ml changes of sterile 1× PBS for at least 3 hr (or overnight) each at 4°C.

30. Divide the virus suspension into convenient aliquots (typically 100 to 200 μl) to avoid repeated freezing and thawing. Store virus >1 year at −20°C or −80°C.

ALTERNATE PROTOCOL

rAAV PURIFICATION USING HEPARIN SEPHAROSE COLUMN PURIFICATION

Additional Materials (also see Basic Protocol; see APPENDIX 1 for items with ✓)
   ✓ PBS-MK
   ✓ 15%, 25%, 40%, and 60% iodixanol

PBS-MK containing 1 M NaCl
0.5 M NaOH
20% (v/v) ethanol
Phenol red (Life Technologies)
Ethanol

Econopump peristaltic pump (Bio-Rad)
32.4-ml Optiseal tubes (Beckman)
50-μl borosilicate glass capillary pipets (Fisher)
Beckman ultracentrifuge with 70Ti rotor (or equivalent)
1-ml or 5-ml HiTrap heparin-Sepharose columns (Amersham Biosciences)
FPLC apparatus
Pierce Slide-A-Lyzer dialysis cassettes (MWCO 10,000)

1. Perform the transfection and carry out ammonium sulfate precipitation of the rAAV (see Basic Protocol, steps 1 to 22). Dissolve the ammonium sulfate pellet in PBS-MK to attain a total volume of 15 ml.

2. Set up the Econopump and insert 50-μl disposable borosilicate glass capillary pipets into the tubing. Calibrate the pump according to the instructions. Place 7.5 ml of the ammonium sulfate pellet dissolved in PBS-MK (step 1) in each of two Optiseal tubes.

3. Insert one glass capillary in each tube and, using the pump at a flow rate of 3 ml/min, underlay the virus solution with 6 ml of 15% iodixanol with 1 M NaCl without introducing air bubbles.

4. Using the technique in step 3, underlay the two solutions with 5 ml of 25% iodixanol, then with 5 ml of 40% iodixanol (which will later harbor the virus).

5. Underlay the gradient with 5 ml of 60% iodixanol. Carefully remove the glass capillary without disturbing the gradient. Slowly add PBS to the viral solution that forms the uppermost layer in the tube until the tube is filled to the top.

6. Insert a plug in each tube and centrifuge the tubes 1 hr at 500,000 × g, 15° to 25°C.

7. Carefully remove the tubes and, in a viral hood, unplug the sealing cap. Insert an 18-G needle attached to a 5-ml syringe just above the 60% iodixanol fraction. Extract the 40% iodixanol solution (clear band) leaving ~0.5 ml, so that none of the 25% iodixanol band is also extracted (Fig. 12.3.3). Store virus overnight at 4°C if necessary.

8. Perform a pump wash with buffer A (PBS-MK), then insert a heparin column in the FPLC.

   A virus prep from twenty 15-cm plates can be purified on a 1-ml heparin column; for larger preps use a 5-ml heparin column.

9. Equilibrate the column in five column volumes of buffer A (PBS-MK). For the injection, reduce the flow rate to 0.2 ml/min for a 1-ml column or 1 ml/min for a 5-ml column, then inject the sample. Collect fractions throughout the column run (wash, flowthrough, and elution).

10. Wash the column with 5 column volumes of buffer A (PBS-MK).

11. Elute the bound virus in a linear gradient over 5 column volumes from 0% to 100% buffer B (PBS-MK containing 1 M NaCl). Collect 0.5-ml fractions (for a 1-ml column) or 1-ml fractions (for a 5-ml column).

12. Wash the column with two column volumes of 100% buffer B (PBS-MK containing 1 M NaCl). Disconnect the heparin column and discard.

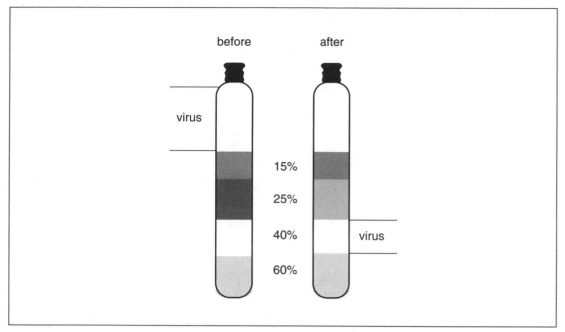

Figure 12.3.3 Iodixanol step gradient before and after centrifugation. The step gradient is generated by underlying the virus concentrate with 15%, 25%, 40% and 60% iodixanol solutions (see Alternate Protocol for instructions on forming the gradient). Percentages refer to the percent of iodixanol in each solution. After centrifugation the virus resides exclusively in the 40% iodixanol (clear) layer. Care should be taken when removing this layer not to remove any of the 25% layer as this contains cell debris.

13. Sterilize the FPLC by running 0.5 M NaOH through all the lines that came in contact with the virus. Fill all the lines with 20% ethanol for storage. Use a new heparin column for each purification to avoid cross-contamination between viral preps.

14. Assay fractions from throughout the purification for virus using a dot blot with an AAV-specific probe to find the viral peak (see Support Protocol 1).

15. Pool the fractions that have the highest rAAV titer and dialyze them in a MWCO 10,000 Slide-A-Lyzer dialysis cassette against two 1-liter changes of PBS for 2 hr each at 4°C.

16. Divide the virus suspension into 100- to 200-μl aliquots and store at −20°C or −80°C (viable for ≥1 year).

SUPPORT PROTOCOL 1

DETERMINATION OF rAAV TITERS BY THE DOT-BLOT ASSAY

A positive signal in this assay indicates that rAAV virions were produced, and quantitation yields a particle number in virions per milliliter.

Materials (*see* APPENDIX 1 *for items with* ✓)

Virus fractions or final virus preparation (e.g., see Basic Protocol or Alternate Protocol)
✓ DNase digestion buffer
✓ 0.1 M EDTA
✓ Proteinase solution
0.5 M NaOH
rAAV plasmid used to make recombinant virus (see Basic Protocol, step 1)

✓ TE buffer, pH 7.5
0.5 M NaOH containing 1 M NaCl
✓ 0.4 M Tris·Cl, pH 7.5
0.5 M NaCl containing 0.5 M Tris·Cl, pH 7.5
Radiolabeled probe to transgene (made using Boehringer Mannheim random-primed
DNA labeling kit according to manufacturer's instructions)

96-well plate
50°C water bath
Dot-blot apparatus
0.45-μm nylon membrane (Zeta Probe, Bio-Rad)

1. Prepare an experimental plan for the position of samples, controls, and DNA standards in duplicate in a 96-well format.

2. Place samples (usually 5 μl of CsCl- or column-purified fraction with 20 μl as the maximum if the virus is in CsCl) and controls in the appropriate wells of a 96-well plate. Add 50 μl DNase digestion buffer per well and incubate 1 hr at 37°C.

3. Stop the digestion by adding 10 μl of 0.1 M EDTA to each well. Mix well. Release virion DNA by adding 60 μl of proteinase solution. Incubate 30 min at 50°C.

4. Denature viral DNA by adding 120 μl of 0.5 M NaOH per well and incubate 10 min at room temperature.

5. Prepare linearized plasmid (from the plasmid used for transfection of this virus) at 0.1 μg/μl for DNA concentration standards. Make a five-fold dilution series in 100 μl TE buffer, pH 7.5, by adding 12.5 μl DNA to 112.5 μl TE buffer in position 1 and successively mixing 25 μl from each well to the next (625 ng, 125 ng, 25 ng, 5 ng, 1 ng, 200 pg, 40 pg).

6. Denature DNA standards by adding 100 μl of 0.5 M NaOH/1 M NaCl to each well.

7. Equilibrate nylon membrane in 0.4 M Tris·Cl, pH 7.5. Prepare a dot-blot manifold apparatus with a prewetted nylon membrane.

8. Add the denatured DNA samples without vacuum, and then apply the vacuum.

9. Disassemble the apparatus, remove membrane, and wash for 5 min with gentle shaking in 100 ml of 0.5 M NaCl/0.5 M Tris·Cl, pH 7.5.

10. Probe the filters with a radiolabeled probe specific for the rAAV sequences (e.g., APPENDIX 3G).

    The probe should be limited to the transgene cassette and should not include the plasmid backbone or ITR sequences.

11. Place the filters against film and expose (autoradiography). After developing the film, align the spots, excise the regions of the filter, and quantitate in a scintillation counter. Calculate how many molecules of the plasmid standards correspond to a given dilution of the rAAV stock.

    Remember to take into consideration that the plasmid standards are double-stranded, whereas the rAAV virions harbor only a single strand.

INFECTION OF CELLS IN VITRO WITH rAAV AND DETERMINATION OF TITER BY TRANSGENE EXPRESSION

Transducing titers can vary with cell type used, but with a standard cell line, transducing titers can be used to compare rAAV preps of the same transgene. This titer can be compared to the particle number to determine efficiency of a specific transgene in a particular cell type. Titers may also vary when comparing immortalized cell lines in culture to in vivo transduction.

NOTE: To assay transducing titer, cells can be coinfected with adenovirus (see *UNIT 12.2*), which acts as a helper virus and increases the transduction efficiency. The addition of adenovirus gives a better indication of the number of particles that are competent to transduce a cell by inducing an optimal environment for AAV infection. However, the adenovirus has a cytopathic effect because of its antigenicity and should not be used in vivo. Thus, a titer derived with use of adenovirus in vitro may not accurately reflect an in vivo competency. It is up to each investigator to establish a standard procedure for titering different rAAV preps.

Materials

Target cells
Tissue culture medium for target cells (see supplier's instructions for cells)
rAAV with appropriate transgene (see Basic Protocol or Alternate Protocol)
Tissue culture plates (multiwell plates recommended for assaying transducing titer)

1. Seed appropriate target cells into multiwell tissue culture plates with appropriate medium.

   The appropriate target cell type, number, and density will depend on the expression assay. Cells must not be treated with DNA synthesis inhibitors such as Ara-C at any time before infection. AAV is a single-stranded virus and requires second strand synthesis before gene expression is possible.

2. Infect cells by adding rAAV directly to the medium of the cells or mix rAAV with fresh medium immediately before adding it to the cells. For assaying transducing titer, infect cells with serial five-fold dilutions of the rAAV stock.

   Cells can also be infected at the time of plating. Incubation time depends on the assay. The appropriate amount of rAAV will depend upon the transgene, cells, and assay. Optionally, adenovirus type 5 can be added with rAAV at a multiplicity of infection (MOI) of 5 (see above note).

3. Carry out the specific assay for expression of the transgene (e.g., immunofluorescence, histological staining, drug resistance).

References: Samulski et al., 1999; Xiao et al., 1998

Contributors: Rebecca A. Haberman, Gabriele Kroner-Lux, and Richard Jude Samulski

UNIT 12.4

Production of Retroviral Vectors

An overview of virus production techniques is shown in Figure 12.4.1. The methods for generating virus from retroviral vector plasmids rely on the use of retrovirus packaging cells that synthesize all of the retroviral proteins but do not produce replication-competent virus. Packaging cell lines that are most useful for mammalian cell gene transfer are listed in Table 12.4.1.

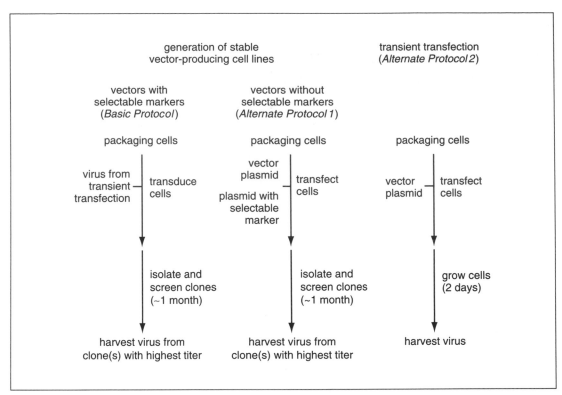

Figure 12.4.1 Production of virus from retrovirus vector plasmids. Select suitable packaging cell combinations for the transient transfection and transduction steps (Table 12.4.3).

NOTE: All solutions and equipment coming into contact with cells must be sterile, and proper sterile technique should be used accordingly.

NOTE: All culture incubations are performed in a humidified 37°C, 10% CO_2 incubator unless otherwise specified.

BASIC PROTOCOL

PRODUCTION OF STABLE CELL LINES TO GENERATE VECTORS WITH SELECTABLE MARKERS

Note that there are restrictions on the ability of virus from one packaging cell line to infect the same or other packaging cell lines, because of a phenomenon known as virus interference. Virus interference is caused by binding of envelope protein made by a packaging cell to a specific retrovirus receptor, resulting in a block to infection by other viruses that use the same receptor for entry. In addition, some retroviruses will not infect cells of the species used to make certain packaging cells (Table 12.4.2). A summary of virus pseudotypes that allow transduction of specific packaging cell types is given in Table 12.4.3, and this information should be considered in choosing packaging cell types for use in the following protocol.

A variety of retroviral vectors suitable for expression of inserted cDNAs are available from a number of laboratories. For most experimental purposes, a retroviral vector that also carries a selectable marker is desirable, although for human gene therapy purposes, vectors that carry

Table 12.4.1 Retrovirus Packaging Cell Lines

Pseudotype[a]	Name	Cell line origin	Maximum titer[b]	Drug resistance gene(s)[c]	Reference
Ecotropic	Ψ-2	Mouse	10^7	*gpt*	Mann et al. (1983)
	PsiCRE	Mouse	10^6	*hph, gpt*	Danos and Mulligan (1988)
	GP+E-86	Mouse	4×10^6	*gpt*	Markowitz et al. (1988a)
	PE501	Mouse	10^7	*tk*	Miller and Rosman (1989)
	ΩE	Mouse	? (high)	*gpt*	Morgenstern and Land (1990)
	ampli-GPE	Mouse	5×10^6	*neo*	Takahara et al. 1992
	BOSC 23	Human	10^7	*hph, gpt*	Pear et al. (1993)
Amphotropic	PA317	Mouse	4×10^7	*tk*	Miller and Buttimore (1986)
	ΨCRIP	Mouse	10^6	*hph, gpt*	Danos and Mulligan (1988)
	GP+*env*Am12	Mouse	10^6	*hph, gpt*	Markowitz et al. (1988b)
	DAN	Dog	4×10^4	*neo*	Dougherty et al. (1989)
	FLYA13	Human	10^7	*bsr, ble*	Cosset et al. (1995)
GALV	PG13	Mouse	3×10^6	*tk, dhfr**	Miller et al. (1991)
VSV	GP7C-tTA-G10	Mouse	10^6	*hph, pac*	Yang et al. (1995)
RD114	FLYRD18	Human	10^7	*bsr, ble*	Cosset et al. (1995)
10A1	PT67	Mouse	10^7	*tk, dhfr**	Miller and Chen (1996)

[a] Abbreviations: GALV, gibbon ape leukemia virus; RD114, RD114 endogenous cat retrovirus; VSV, vesicular stomatitis virus; 10A1, 10A1 murine leukemia virus.

[b] Highest reported titers (infectious u/ml). In some cases this value is from papers published after the initial report describing the cell line.

[c] Drug resistance gene(s) that are already present in the packaging cells due to their use for DNA transfer during cotransfection of defective helper virus constructs. Selection for vectors carrying these markers cannot be performed in these packaging cells. *ble*, a bacterial gene that confers resistance to bleomycin and phleomycin in mammalian cells; *bsr*, a bacterial gene that confers resistance to blasticidin S; *dhfr**, a mutant dihydrofolate reductase gene; *gpt*, xanthine-guanine phosphoribosyltransferase; *hph*, hygromycin phosphotransferase; *neo*, neomycin phosphotransferase; *pac*, puromycin N-acetyl phosphotransferase; *tk*, herpes simplex virus thymidine kinase.

only the therapeutic gene may be required to avoid possible immune reactions to selectable marker proteins.

Figure 12.4.2 depicts a set of retroviral vectors that contain selectable markers and unique cloning sites for insertion of cDNAs. The vectors are named according to the order of genetic elements in the vector: L, long terminal repeat (LTR); N, neomycin gene (*neo*); S, SV40 early

Table 12.4.2 Host Range of Retroviral Vectors

Vector pseudotype	Cells that can be transduced	
	Mouse	Human
Ecotropic	Yes	No
Amphotropic	Yes	Yes
GALV	No	Yes
VSV	Yes	Yes
RD114	No	Yes
10A1	Yes	Yes

Table 12.4.3 Packaging Cell Pseudotypes that Can Be Used for Infection of Other Packaging Cells

Target packaging cells		Virus pseudotypes that allow infection of the target packaging cells[a]
Pseudotype	Species	
Ecotropic	Mouse	Amphotropic, VSV, 10A1
Amphotropic	Mouse	Ecotropic, VSV, 10A1
	Human	GALV, VSV, RD114
	Dog	GALV, VSV, RD114
GALV	Mouse	Ecotropic, amphotropic, VSV, 10A1
VSV	Mouse	Ecotropic, amphotropic, 10A1
RD114	Human	Amphotropic, VSV, 10A1
10A1	Mouse	Ecotropic, VSV

[a]See Table 12.4.1.

promoter; C, human cytomegalovirus (CMV) immediate early promoter; HD, *hisD* gene; H, hygromycin-B-phosphotransferase (*hph*); and X, cloning site. With the exception of LN, the vectors contain two promoters, one of which drives expression of the selectable marker and the other expression of the inserted DNA. Transcription of inserted cDNAs is driven by strong viral promoters—either the retroviral LTR (LXSN, LXSHD, and LXSH), an immediate early promoter from human cytomegalovirus (LNCX, LHDCX), or the SV40 early promoter (LNSX). In general, the LTR and CMV promoters are very strong promoters in human cells, and the SV40 promoter is weaker. Vectors containing either of three different dominant selectable markers are shown. Selection for each of the markers is independent of the presence or absence of the other markers, allowing sequential use of vectors carrying different selectable markers to transfer multiple genes into cells.

Materials (*see* APPENDIX 1 *for items with* ✓)

Two retrovirus packaging cell lines: e.g., PE501 and PA317
✓ Complete Dulbecco's modified Eagle medium containing 4.5 g/liter glucose and 10% (v/v) FBS (DMEM/10% FBS)
Retrovirus vector plasmid containing selectable marker
4 mg/ml Polybrene (Sigma) in PBS (see APPENDIX 1 for PBS)
Drug appropriate for selectable marker: e.g., 0.75 mg/ml G418 (active compound), 4 mM histidinol, *or* 0.4 mg/ml hygromycin B
✓ Cell staining solution

6- and 10-cm tissue culture dishes
10-ml syringes
0.45-μm low-protein-binding cellulose acetate syringe filters
Cloning rings (*UNIT 3.1*)

1. On day 1, plate PE501 retrovirus packaging cells at 5×10^5 cells per 6-cm dish in complete DMEM/10% FBS. Incubate 1 day.

   Two different packaging cell lines are required; the virus produced by the first cell line must be capable of infecting the second cell line, i.e., they must be of different but compatible pseudotypes (see Table 12.4.3).

2. On day 2, replace the culture medium with 4 ml fresh tissue culture medium. Perform calcium phosphate-mediated transfection (see Support Protocol 1) using retroviral vector plasmid containing selectable marker. Incubate 1 day.

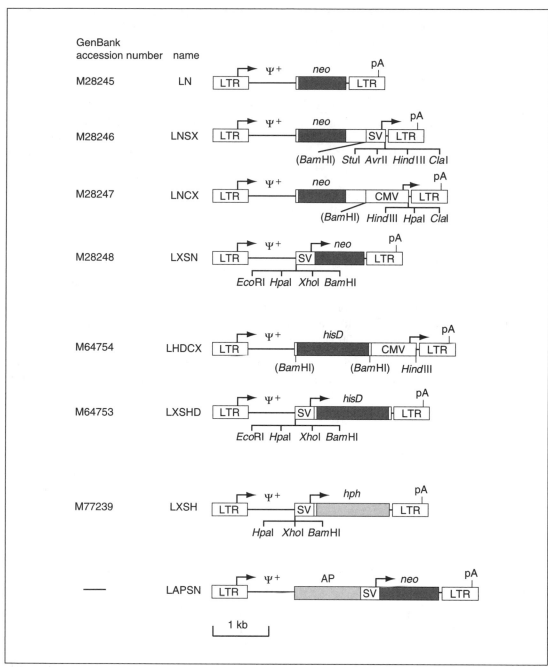

Figure 12.4.2 Retroviral vectors. Retroviral vectors that contain selectable drug markers—neomycin phosphotransferase (*neo*), histidinol dehydrogenase (*hisD*), or hygromycin phosphotransferase (*hph*)—are shown with their GenBank accession numbers for the complete vector sequences. Vector names are comprised of the abbreviations for the genetic elements they contain. The coding regions of these genes are shaded. Connecting lines indicate other viral sequences, and arrows indicate the cap sites of promotors and the direction of transcription. Restriction sites for cDNA insertion are indicated. Restriction sites in parentheses are discussed in the text. The vectors carrying *neo* and *hisD* have been described, LXSH was made from LXSN by replacement of the *neo* insert with *hph*, and LAPSN has been described. Abbreviations: C, cytomegalovirus (CMV) immediate early promotor; H, hygromycin-B-phosphotransferase (*hph*); HD, *hisD* gene; L, long terminal repeat (LTR); N, neomycin (*neo*) gene; pA, polyadenylation signals; S, SV40 early promoter; X, cloning site; ψ⁺, extended retroviral packaging signal.

3. On day 3, aspirate the medium from the transfected PE501 cells and add 4 ml fresh complete DMEM/10% FBS. Also for each dish of transfected PE501 cells, plate two dishes of PA317 cells at 10^5 cells per 6-cm dish in complete DMEM/10% FBS. Incubate 1 day.

4. On day 4, replace the medium on the PA317 cells with medium containing 4 μg/ml Polybrene (1000-fold dilution of stock solution).

5. Using 10-ml syringes, remove 3 ml virus-containing medium from each dish of transfected PE501 cells, and filter the medium through 0.45-μm low-protein-binding cellulose acetate syringe filters to remove live cells. Leave 1 ml medium on the dish to keep cells from drying out.

6. Infect one dish of PA317 cells with 1 ml filtered virus-containing medium from one dish of transfected PE501 cells, and another dish with 10 μl from the same PE501 dish. Incubate 5 days.

7. Trypsinize and plate each dish of transfected PE501 cells at a 1:20 dilution in 6-cm dishes containing medium supplemented with the drug appropriate for the selectable marker— e.g., 0.75 mg/ml G418, 4 mM histidinol, or 0.4 mg/ml hygromycin B. Incubate 5 days.

8. Stain the plate 5 min with 1.5 ml cell staining solution. Wash the stain off with water. Evaluate the dishes for colony formation as a measure of the efficiency of DNA transfection (transfection efficiency of ~1000 colonies/μg plasmid DNA is typical).

9. One day after infection (day 5), trypsinize the infected PA317 cells and plate the cells from each dish at dilutions of 9:10 and 1:10 into 10-cm dishes containing 10 ml medium supplemented with the appropriate selective drug (see step 7). Incubate 5 to 10 days.

10. After drug-resistant colonies have formed, isolate clones (~10 colonies) from the dishes that contain small numbers of colonies using cloning rings (*UNIT 3.1*). Expand the clones in medium containing the selective drug to provide enough cells for freezing and assaying.

11. When the clonal cell lines are sufficiently expanded, freeze one or two vials of each clone as a backup in case the culture is accidentally lost during screening of clones for the best vector producer.

12. Assay the titer of the vector by measuring transfer of the selectable marker encoded by the vector (see Support Protocol 2).

13. Identify those clones that transfer drug resistance, as detected by the growth of the target cells in selective medium, and analyze for expression of the cDNA inserted into the vector.

14. Prepare genomic DNA from the clones (e.g., *APPENDIX 3A*). Digest the DNA with a restriction endonuclease that cuts only in the viral LTRs and an endonuclease that cuts once in the vector and randomly in the surrounding genomic DNA.

15. Analyze the digested DNAs by Southern blot hybridization (*APPENDIX 3G*) using a cDNA sequence probe.

16. Screen the clones with a single vector provirus for production of helper virus (see Support Protocol 3) and discard any clones that produce helper virus.

17. Once a clonal cell line is identified that has the desired characteristics, thaw one of the backup vials (step 11), expand the cells, and freeze multiple vials for future use.

Most clones will produce the vector consistently for ≥2 months, and if vector titer declines, a new vial can be thawed for use.

18. To harvest vector from the cells, plate the cells in 10-cm tissue culture dishes in 10 ml medium and grow the cells until ~1 day post confluence. Feed the cells with 10 ml fresh medium, harvest vector-containing medium 12 to 24 hr later, and store at −70°C.

The half-life of vector at 37°C is ~4 hr, so harvests made at longer intervals are counterproductive in that defective virions are likely to accumulate in the medium. At least three 12-hr harvests can be made sequentially. Large amounts of the vector-containing medium can be frozen at −70°C indefinitely for future use. Vector stocks should be frozen and thawed quickly to preserve vector activity.

ALTERNATE PROTOCOL 1

PRODUCTION OF STABLE CELL LINES TO GENERATE VECTORS WITHOUT SELECTABLE MARKERS

In this procedure the pSV2neo plasmid is used for selection, but plasmids that express other selectable markers work equally well. PA317 packaging cells are used in this example, but other packaging cell lines can be substituted.

Materials *(see APPENDIX 1 for items with ✓)*

    Retrovirus packaging cells (e.g., PA317; see Table 12.4.1)
✓ Complete Dulbecco's modified Eagle medium containing 4.5 g/liter glucose and 10% (v/v) FBS (DMEM/10% FBS)
    Retrovirus packaging cells (e.g., PA317; see Table 12.4.1)
    Retrovirus vector plasmid DNA
    Plasmid DNA with selectable marker: e.g., pSV2neo
    0.75 mg/ml G418 (active drug) or other selective drug appropriate for the plasmid with selectable marker

6-cm tissue culture dishes
Cloning rings (UNIT 3.1), sterile

1. On day 1, plate PA317 retrovirus packaging cells at 5×10^5 cells per 6-cm tissue culture dish in 4 ml complete DMEM/10% FBS. Incubate 1 day.

2. On day 2, replace the culture medium with 4 ml fresh medium. Using calcium phosphate-mediated transfection (see Support Protocol 1), transfect the cells with 10 μg retrovirus vector plasmid DNA and either 0.1 or 0.2 μg pSV2neo plasmid DNA. Incubate 1 day.

3. On day 3, aspirate the medium from the transfected PA317 cells and add 4 ml fresh medium. Incubate 1 day.

4. On day 4, trypsinize the transfected cells and replate at dilutions of 1:2 to 1:10 in 10-cm dishes in medium containing 750 μg/ml G418. Incubate 5 to 10 days.

5. Using cloning rings (UNIT 3.1), isolate drug-resistant colonies from dishes containing small numbers of colonies. Expand the clones.

6. When the clones are sufficiently expanded, freeze one or two vials of each as a backup in case the cells are accidentally lost during screening of clones for the best vector producer.

7. If possible, test the clonal lines for production of the protein encoded by the cDNA included in the vector.

8. Prepare genomic DNA from the clones (e.g., APPENDIX 3A) and analyze the DNA by restriction endonuclease digestion and Southern blot hybridization (APPENDIX 3G).

9. Determine the vector titer.

10. Screen for helper virus using the marker rescue assay (see Support Protocol 3), expand the clones, and harvest the vector (see Basic Protocol, steps 16 to 18).

ALTERNATE PROTOCOL 2

PRODUCTION OF VECTOR BY TRANSIENT TRANSFECTION

Materials
    Retrovirus packaging cells (see Table 12.4.1) and appropriate tissue culture medium
    Retrovirus vector plasmid DNA
    6-cm tissue culture plates
    10-ml syringes
    0.45-μm low-protein-binding cellulose acetate syringe filters

1. On day 1, plate the retrovirus packaging cells at 5×10^5 per 6-cm dish in 4 ml of the appropriate tissue culture medium. Incubate 1 day.

2. On day 2, introduce the retrovirus vector plasmid DNA into the packaging cells by calcium phosphate–mediated transfection (see Support Protocol 1). Incubate 1 day.

3. On day 3, replace the medium with fresh culture medium. Incubate 1 day.

4. On day 4, after 12 to 24 hr, using a 10-ml syringe, harvest the culture medium and filter it through a 0.45-μm low-protein-binding cellulose acetate syringe filter to remove cells and debris. Use the virus-containing medium immediately to infect recipient cells or store it at $-70°C$ indefinitely.

SUPPORT PROTOCOL 1

CALCIUM PHOSPHATE–MEDIATED TRANSFECTION OF CULTURED CELLS

NOTE: All solutions should be sterilized by filtration through 0.22-μm sterile filters; if the plasmid DNA is not already sterile because it was ethanol precipitated during preparation, it should also be filtered.

Materials (see APPENDIX 1 for items with ✓)
    Retrovirus vector plasmid DNA
✓ 10 mM Tris·Cl, pH 7.5
    Retrovirus packaging cells (see Table 12.4.1)
✓ Complete Dulbecco's modified Eagle medium containing 4.5 g/liter glucose and 10% (v/v) FBS (DMEM/10% FBS)
✓ 2.0 M CaCl$_2$
    500 mM HEPES, pH 7.1
    2.0 M NaCl
✓ 150 mM sodium phosphate buffer, pH 7.0
    Sterile H$_2$O
    12 $\times$ 75–mm clear polystyrene tubes (Falcon)

1. Resuspend the purified retrovirus vector plasmid DNA in 10 mM Tris·Cl, pH 7.5, to a final concentration of ~1 μg/μl.

2. On the day before transfection, plate retrovirus packaging cells at 5×10^5 cells in 6-cm tissue culture dishes containing complete DMEM/10% FBS. Incubate 1 day.

3. Feed the retrovirus packaging cells with fresh medium.

4. For each plasmid, prepare and mix a DNA/$CaCl_2$ solution:

> 25 μl 2.0 M $CaCl_2$
> 10 μg plasmid DNA in 10 mM Tris·Cl, pH 7.5
> Sterile H_2O to 200 μl.

5. Prepare and mix fresh precipitation buffer:

> 100 μl 500 mM HEPES, pH 7.1
> 125 μl 2.0 M NaCl
> 10 μl 150 mM sodium phosphate buffer, pH 7.0
> Sterile H_2O to 1 ml.

6. Add the 200 μl DNA/$CaCl_2$ solution dropwise with constant agitation to 200 μl precipitation buffer in a clear 12 × 75–mm polystyrene tube. Incubate 30 min at room temperature. Observe a faint cloudiness, if the mixture remains clear or large clumps develop, prepare a new precipitation solution using fresh HEPES.

7. Check that the precipitate in the tube is a fine, not "clumpy," precipitate. Add the fine precipitate to a 6-cm dish of retroviral packaging cells and swirl the dish to distribute the precipitate. Incubate and proceed to select or screen the cells (see Basic Protocol or Alternate Protocol 2).

SUPPORT PROTOCOL 2

ASSAY TO TITER VECTORS CARRYING SELECTABLE MARKERS

It is more difficult to determine vector titer for vectors that do not carry marker genes: such titer determinations can involve Southern blot hybridization (APPENDIX 3G) of cells transduced with the vector to be assayed or another otherwise comparable vector that carries a selectable marker that allows direct determination of titer.

Materials (see APPENDIX 1 for items with ✓)

> Target cells susceptible to the vector pseudotype to be tested (e.g., NIH 3T3 or HeLa
>   cells) and appropriate tissue culture medium
> 4 mg/ml Polybrene (Sigma)
> Retroviral stock to be tested (e.g., from Basic Protocol or Alternate Protocol 1 or 2)
> Selective drug: 0.75 mg/ml G418, 0.4 mM hygromycin B, or 4 mM histidinol D
> ✓ Cell staining solution
> 6-cm tissue culture dishes

1. On day 1, plate target cells susceptible to vector pseudotype at 5×10^5 cells per 6-cm dish in the appropriate tissue culture medium. Incubate 1 day.

2. On day 2, change the medium on the cells to medium containing 4 μg/ml Polybrene and add varying amounts (0.01 μl to 100 μl) of retroviral stock to separate plates. Incubate 1 day.

3. On day 3, trypsinize and dilute the cells 1:20 into medium containing a selective drug at the appropriate concentration. Incubate 5 to 8 days.

4. After colonies have formed, stain the plate 5 min with 1.5 ml cell staining solution. Wash off stain with H_2O and count colonies. Calculate the virus titer.

> *Virus titer in colony-forming units per milliliter (CFU/ml) is calculated by dividing the number of colonies by the volume (in milliliters) of undiluted virus stock used for infection and multiplying by 20 to correct for the 1:20 cell dilution.*

MARKER RESCUE ASSAY FOR HELPER VIRUS

Although this assay is somewhat tedious and slow, it is reliable and very sensitive.

Materials

LAPSN vector (see Fig. 12.4.2)

Retrovirus packaging cells (Table 12.4.1) susceptible to LAPSN pseudotype and appropriate tissue culture medium

Naïve target cells that are susceptible to the vector pseudotype: e.g., NIH 3T3 or HeLa cells

Retrovirus stock to be tested (see Basic Protocol or Alternate Protocol 1 or 2), filtered through 0.45-μm low-protein-binding cellulose acetate syringe filter

4 mg/ml Polybrene (Sigma) in PBS (see APPENDIX 1 for PBS)

Positive control virus: amphotropic replication-competent "helper" virus

6-cm tissue culture dishes

10-ml syringes

0.45 μm low-protein-binding cellulose acetate syringe filters

1. Produce a virus stock containing the LAPSN vector (Fig. 12.4.2) which encodes alkaline phosphatase (AP) from retrovirus packaging cells that are transiently transfected with the virus (see Alternate Protocol 2) or from stable vector-producing cells (see Basic Protocol or Alternate Protocol 1).

2. Transduce naïve target cells with the virus. Passage the cells for 2 weeks to allow potential helper virus (which should not be present) to spread.

3. Assay the cells for vector production by infecting parental cells that do not contain the vector.

   Multiple aliquots of those cells that contain but do not release a retroviral vector (nonproducer cells) should be preserved in liquid nitrogen for use in future marker rescue assays. Nonproducer cells need to be generated only once.

4. On day 1, plate nonproducer cells containing the LAPSN vector at 5×10^5 cells per 6-cm tissue culture dish in the appropriate medium. Incubate 1 day.

5. On day 2, infect nonproducer cells by adding 1 ml filtered retrovirus stock to be tested, 3 ml tissue culture medium, and 4 μg/ml Polybrene. As a positive control, infect some dishes with a small amount of amphotropic replication-competent helper virus.

6. Beginning on day 3, incubate the cells 2 weeks to allow helper virus spread. Two to three times a week, trypsinize the cells and replate them at 1:10 to 1:40. Keep cells at a relatively high density.

7. On day 16, plate naïve target cells (same cells as the nonproducer cell line) susceptible to the vector pseudotype at 10^5 cells per 6-cm dish. Also, change the medium on the confluent dishes of nonproducer cells. Incubate 1 day.

8. On day 17, harvest medium from the nonproducer cells and filter through a 0.45-μm low-protein-binding syringe filter to remove cells and debris.

9. Infect naïve target cells using 1-ml samples of the medium, 3 ml tissue culture medium, and 4 μg/ml polybrene. Incubate 1 day.

10. On day 19, stain the infected target cells for alkaline phosphatase activity (see Support Protocol 4).

   A positive staining reaction indicates transfer of virus from the nonproducer vector-containing cells due to the presence of helper virus in the retrovirus stock.

SUPPORT PROTOCOL 4

STAINING CULTURED CELLS FOR ALKALINE PHOSPHATASE ACTIVITY

This protocol is written for cells cultured in 6-cm dishes, but the conditions can easily be adjusted for dishes of other sizes.

Materials (see APPENDIX 1 for items with ✓)

   Cell cultures transduced with alkaline phosphatase-encoding vector (e.g., LAPSN) in 6-cm tissue culture dishes
   0.25% (v/v) glutaraldehyde in PBS
   ✓ PBS
   ✓ Alkaline phosphatase staining buffer
   ✓ Alkaline phosphatase staining solution

1. Aspirate the medium from cell cultures transduced with an alkaline phosphatase–encoding vector. Fix cells with 3 ml of 0.25% glutaraldehyde for 5 to 10 min at room temperature. Use untransduced cells as a negative control and cells known to be producing alkaline phosphatase as a positive control.

2. Wash the cells two times with 2 to 3 ml PBS each wash. Add 2 ml PBS and heat 30 min at 65°C to inactivate cellular alkaline phosphatase activity.

3. After the dishes have cooled, aspirate the PBS and wash once with 2 ml alkaline phosphatase staining buffer to remove phosphates.

4. Add 1.5 ml alkaline phosphatase staining solution. Incubate ~4 hr (depending on cell type and background alkaline phosphatase activity) at room temperature.

5. Count alkaline phosphatase-positive foci of cells.

References: Cosset et al., 1995; Danos and Mulligan, 1988; Dougherty et al.,1989; Miller,1990

Contributor: A. Dusty Miller

UNIT 12.5

Production of Pseudotype-Retroviral Vectors

Retrovirus pseudotype is defined as the genome of one retrovirus encapsidated by the envelope protein of a second virus. The host range of the pseudotype is that of the virus donating the envelope protein. Depending on the envelope protein, therefore, pseudotype formation can either expand or limit the host range of the parental retrovirus.

To produce VSV-G pseudotyped retroviral vectors, packaging cell lines stably expressing MLV gag and pol and VSV-G proteins must be established. Because VSV-G is toxic to most mammalian cells tested, alternative procedures to produce VSV-G-pseudotyped retroviral vectors have been established.

NOTE: All reagents and equipment coming into contact with live cells must be sterile, and proper sterile technique should be used accordingly.

NOTE: All tissue culture incubations are performed in a humidified 37°C, 10% CO_2 incubator.

BASIC PROTOCOL

PSEUDOTYPE RETROVIRUS PRODUCTION BY TRANSIENT TRANSFECTION

In this protocol, the selectable marker is the *neo* gene that encodes *E. coli* neomycin phospho-transferase and the selective agent is G418 (see *UNIT 3.1* for details concerning this selectable marker). The Basic Protocol allows rapid production of small quantities of high-titer pseudo-typed retroviral vectors within 1 week.

Materials *(see APPENDIX 1 for items with ✓)*

Retroviral vector with 5′ U3 replaced by promoter of cytomegalovirus (CMV) immediate early (IE) gene
Cultures of 293GP and NIH3T3 cells
✓ Complete DMEM/10% FBS
Plasmid pCMV-G
✓ TE79/10 (Tris-buffered EDTA)
✓ 2 M $CaCl_2$
✓ 2× HeBS
✓ 0.1× HBSS
✓ 4 mg/ml polybrene
✓ 40 mg/ml G418
100% methanol
0.04% (w/v) Giemsa staining solution

Beckman polyallomer Quick-Seal centrifuge tubes
vTi50 rotor
High-speed ultracentrifuge
100-mm tissue culture plate
6-ml polypropylene tube, sterile
0.45-μm sterile nonpyrogenic syringe filter
14 × 89–mm Beckman Ultra-Clear centrifuge tubes, sterilized overnight with UV light in a laminar flow hood
SW41 rotor
Autoclavable wide-mouth centrifuge bottle (Nalgene)
Super-speed centrifuge (Sorvall)
GSA rotor

1. Clone (*UNIT 12.4*) the gene of interest into a modified retroviral vector as shown in Fig. 12.5.1, replacing the U3 region of the 5′ LTR with the CMV IE promotor.

2. Purify the vector containing the gene of interest by alkaline lysis followed by cesium chloride purification, using Beckman polyallomer Quick-Seal centrifuge tubes and an ultracentrifuge with vTi50 rotor.

3. Twenty-four hours prior to transfection, plate 293GP cells in 10 ml DMEM/10% FBS at a density of 1×10^6 per 100-mm tissue culture plate. Incubate.

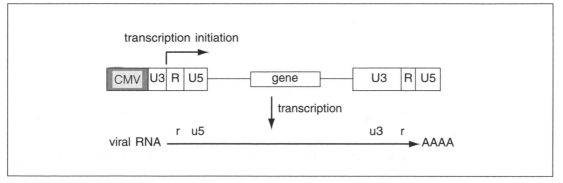

Figure 12.5.1 Schematic illustration of a retroviral vector containing the hybrid LTR (top) and its RNA transcript (bottom). Most of the U3 region of the 5′ LTR is replaced by the CMV IE promoter. Because the R and U5 regions in the CMV-LTR hybrid remain intact, the transcription initiation site in the hybrid CMV-LTR is identical to that in the wild-type LTR. The arrow above the hybrid LTR indicates the transcription initiation site and the direction of transcription.

4. On the day of transfection, mix in a sterile 6-ml polypropylene tube 20 μg cesium chloride-purified vector DNA and 20 μg pCMV-G containing the VSV-G gene under the control of the CMV IE promoter. Adjust the volume to 437 μl with TE79/10.

5. Add 63 μl of 2 M CaCl$_2$ and mix well, then add 500 μl of 2× HeBS with constant agitation.

6. Allow calcium phosphate–DNA precipitate to form for 30 min at room temperature. Add the precipitate to tissue culture plates in which 293GP cells are at least 50% confluent (70% to 80% is ideal; cell density is critical for virus production). Incubate 8 hr.

7. Replace the culture medium with 6 ml fresh DMEM/10% FBS and continue incubation.

8. Between 24 and 96 hr after transfection, collect the culture supernatant, filter it through a sterile 0.45-μm syringe filter, and store at −70°C.

   Massive cell death and detachment from tissue culture plates becomes apparent at 48 hr after transfection, as a result of VSV-G overexpression. High-titer virus can be harvested 24 hr after transfection, however, before cell death becomes apparent. Culture supernatant can continue to be collected until <30% of the transfected cells remains attached to the plate. To further increase virus yield, culture supernatant can be collected up to three times per day at 4-hr intervals. It is important, however, that the poorly attached 293GP cells not be disturbed during virus collection and medium replacement. Detachment of the transfected cells will result in massive cell loss.

For small-scale (<100 ml) concentration of pseudotyped retroviral vectors

9a. Thaw the frozen virus at 37°C and transfer to 14 × 89–mm Beckman Ultra-Clear centrifuge tubes that have been UV-sterilized overnight in a laminar flow hood. Centrifuge 90 min at 50,000 × g, 4°C.

10a. Discard the supernatant culture medium in a laminar flow hood. Resuspend the virus pellet at 4°C in DMEM/10% FBS or 0.1× HBSS to 0.5% to 1% of the original volume. Disperse virus aggregates occasionally by gentle pipetting. Incubate overnight.

11a. Divide virus into aliquots (<50% of infectious virus routinely recovered) and store at −70°C. To further increase virus titers, perform a second ultracentrifugation step (titers as high as 10^9 cfu/ml have been demonstrated).

For large-scale concentration of the pseudotyped retroviral vectors

9b. Thaw the frozen virus at 37°C and transfer to sterile 250-ml wide-mouth centrifuge bottles. Sediment virus by centrifuging 15 hr at 13,000 × g (9000 rpm in a GSA rotor), 4°C.

10b. Resuspend virus pellet at 4°C with occasional gentle pipetting. Incubate 4 hr.

11b. Divide virus into aliquots (~20% to 40% of the input infectious virus is routinely recovered) and store at −70°C.

12. Twenty-four hours prior to transfection, plate murine NIH3T3 cells in DMEM/10% FBS at a density of 2×10^5 per 100-mm tissue culture plate.

13. Infect cells with 0.1, 1, and 10 μl of the harvested virus in the presence of 4 μg/ml polybrene. Incubate overnight.

14. Replace the culture medium with fresh DMEM/10% FBS supplemented with 800 μg/ml G418. Repeat every 3 days.

15. Two weeks after infection, fix G418-resistant colonies with 100% methanol for 10 min, stain with 0.04% Giemsa staining solution for 10 min, and count. Multiply the number of G418-resistant colonies by the corresponding dilution factor to determine the titer of the undiluted virus stock, which is expressed as cfu/ml.

ALTERNATE PROTOCOL

PSEUDOTYPE RETROVIRUS PRODUCTION FROM STABLE PRODUCER CELLS

This protocol is useful for large-scale production of pseudotyped retroviral vectors.

Additional Materials *(also see Basic Protocol; see* APPENDIX 1 *for items with* ✓ *)*
　　Culture of the amphotrophic packaging cell line PA317
　　Retroviral vector with the *neo* gene containing the gene of interest
✓ 80 mg/ml hygromycin
✓ 1 mg/ml puromycin
✓ 1 mg/ml tetracycline
✓ 200 μM β-estradiol
　　6- and 24-well tissue culture plates

1. Twenty-four hours prior to transfection, plate out PA317 cells in 10 ml DMEM/10% FBS at a density of 1×10^6 per 100-mm tissue culture plate. Incubate.

2. Transfect the cells with 20 μg DNA of the retroviral construct containing the gene of interest and the *neo* selectable marker (see Basic Protocol, steps 4 to 7).

3. Forty-eight hours after transfection, harvest virus, filter it through a 0.45-μm filter, and store at −70°C.

   Depending on the retroviral construct, the titer generated from transient transfection of PA317 cells ranges from 10^2 to 10^3 cfu/ml.

4. Twenty-four hours prior to retrovirus infection, plate out 293GP/G-21 cells at a density of 2×10^5 per 100-mm tissue culture plate in 10 ml DMEM/10% FBS supplemented with 800 mg/ml hygromycin, 1 mg/ml puromycin, and 2 mg/ml tetracycline.

5. Use the amphotropic virus harvested in step 3 to infect 293GP/G-21 cells in the medium described in step 4 plus 4 mg/ml polybrene. Twenty-four hours after infection, place the cells in the medium described in step 4 supplemented with 800 mg/ml G418.

6. Replace the culture medium with fresh medium described in step 4 supplemented with 800 mg/ml G418 every 3 days.

7. After 2-week culture for selection, pick G418-resistant colonies with an automatic pipettor and transfer into 24-well tissue culture plates. Subsequently transfer the clones from 24-well plates into 6-well plates, and then into 100-mm tissue culture plates, as needed.

8. To identify high-titer producer clones, plate the cells at a density of 2×10^6 per 100-mm plate in DMEM/10% FBS supplemented with 800 mg/ml hygromycin, 1 mg/ml puromycin, 800 mg/ml G418, and 2 mg/ml tetracycline 24 hr before induction of virus production.

9. On the day of induction, replace the culture medium with 10 ml DMEM/10% FBS. To remove residual tetracycline, incubate the cells at least 30 min at 37°C and replace medium again with fresh DMEM/10% FBS. Repeat this procedure three times.

10. To induce virus production, add β-estradiol to 10 ml DMEM/10% FBS at a concentration of 2 mM.

11. Seventy-two hours after the addition of β-estradiol, harvest the virus in the culture supernatant and filter through a 0.45-mm syringe filter. Determine the virus titer in NIH3T3 cells (see Basic Protocol, steps 12 to 15).

12. To mass-produce virus once the high-titer producer clone has been isolated, plate, wash, and induce the producer cells as described in steps 8 to 10. Starting 48 hr after β-estradiol induction, collect virus in the culture supernatant three times a day for up to 2 weeks and subject the pooled culture medium to ultracentrifugation to further concentrate the virus (see Basic Protocol, steps 9a to 11a or 9b to 11b).

The induced cells continue to produce high-titer virus (10^5 to 10^6 cfu/ml) for up to 2 weeks.

References: Chen et al., 1996b; Miyanohara et al.,1995; Yee et al., 1994

Contributor: Jiing-Kuan Yee

12

UNIT 12.6

Production of High-Titer Lentiviral Vectors

In contrast to oncoretroviruses, lentiviruses such as the human immunodeficiency virus (HIV) are a subfamily of retroviruses that can infect both growth-arrested and dividing cells. Accordingly, lentiviral vectors can efficiently transduce nonreplicating cells both in tissue culture and in vivo.

When producing vector stocks, it is mandatory to avoid the emergence of replication-competent recombinants. In the retroviral genome, a single RNA molecule that also contains critical *cis*-acting elements carries all the coding sequences. Biosafety of a vector production system is therefore best achieved by distributing the sequences encoding its various components over as many independent units as possible, to maximize the number of recombination events that would be required to recreate a replication-competent virus.

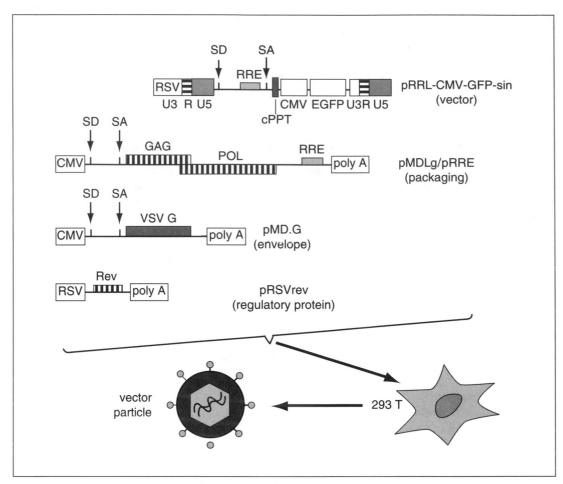

Figure 12.6.1 Production of HIV vectors: Schematic description of the protocol for HIV vector production. Vectors are produced by transfecting 293T cells with four plasmids, the main features of which are shown. The vector plasmid (top) is characterized by the presence of a chimeric RSV/HIV 5′ LTR and a partially deleted 3′ LTR. The Gag and Pol proteins of HIV-1 and the G protein of VSV are encoded by two independent plasmids (second and third from top). The fourth plasmid (bottom) encodes the Rev post-transcriptional regulator, essential for Gag/Pol expression and vector production. Abbreviations: CMV, immediate early promoter from the human cytomegalovirus; cPPT, central polypurine tract; PolyA, polyadenylation signal; RRE, Rev responsive element; RSV, U3 region (promoter) from the Rous Sarcoma Virus; SD and SA, splice donor and acceptor sites, respectively; sin, self-inactivating.

The plasmids employed in this unit are depicted in Figure 12.6.1. The pMDLg/pRRE construct encodes the HIV-1 Gag and Gag/Pol proteins. The pMD.G plasmid provides the VSV G envelope protein. The pRSVrev plasmid encodes the HIV-1 Rev protein. Finally, pRRL-CMV-GFP-sin encodes the vector RNA molecules that are packaged into the particles. On this plasmid, the CMV promoter and the sequence encoding the green fluorescent protein (GFP) can be replaced by other transcriptional cassettes and genes of interest. Importantly, these four plasmids are designed with a quasi-absence of sequence overlap, to minimize the chance of recombination. More detailed information on the sequence of these plasmids is available on the Web site *http://www.tronolab.unige.ch/x_home2.htm*. This Web site can also be used to request all the plasmids mentioned in this unit.

CAUTION: VSV G–pseudotyped lentiviral vectors have a broad tropism, both in vitro and in vivo; biosafety precautions need to take into account the nature of the transgene. A BL2 laboratory is required. Procedures using lentiviral vectors must be reviewed and approved by the local biosafety committee of the institution where they are conducted. See *UNIT 12.1* for an overview of biosafety issues associated with gene therapy vectors.

NOTE: All solutions and equipment coming into contact with living cells must be sterile, and aseptic technique should be used accordingly.

BASIC PROTOCOL

PRODUCTION OF HIGH-TITER HIV-1-BASED VECTOR STOCKS BY TRANSIENT TRANSFECTION OF 293T CELLS

Materials (see APPENDIX 1 for items with ✓)

    293T cells (information on laboratories providing 293T cells is available at *http://www.tronolab.unige.ch*)

    Dulbecco's modified Eagle medium/10% FBS (DMEM-10)

    pMD.G (encoding the VSV G envelope protein in TE)

    pRRL-CMV-GFP-sin (vector in TE)

    pMDLgag/polRRE (encoding the HIV-1 Gag and Pol proteins in TE)

    pRSVrev (encoding the HIV-1 Rev protein in TE)

✓ TE buffer, pH 8.0

✓ 0.5 M $CaCl_2$

✓ 2× HeBS

    70% ethanol in a spray bottle

✓ PBS, pH 7.4

    14% (v/v) bleach

    10-cm tissue culture dishes

    37°C humidified incubators, 10% and 5% CO_2

    0.5- and 1.5-ml microcentrifuge tubes, sterile, disposable

    15- and 50-ml conical centrifuge tubes, sterile

    125-ml filter bottles, 0.45-μm pore size

    75-cm^2 tissue culture flasks

    30-ml (25 × 89–mm) and 5-ml (13 × 51–mm) disposable conical polyallomer ultracentrifuge tubes (Beckman)

    Ultracentrifuge with SW28 and SW55 rotors (Beckman)

1. Maintain 293T cells in Dulbecco's modified Eagle medium/10% FBS (DMEM-10) medium, in 10-cm tissue culture dishes. Culture cells in a 37°C humidified incubator with a 10% CO_2 atmosphere, and split 1:4 to 1:6, three times per week (e.g., every Monday, Wednesday and Friday).

2. Two days before the transfection, prepare 10 dishes of 2×10^6 293T cells. Incubate overnight in a 37°C humidified incubator with a 10% CO_2 atmosphere. On the following day, prepare 20 dishes of 293T cells by splitting the 10 dishes 1:2. Seed cells in 10 ml of DMEM-10. Incubate overnight in a 37°C humidified incubator with a 10% (or 5%) CO_2 atmosphere.

3. Adjust the DNA concentration (*APPENDIX 3D*) of all plasmids (i.e., pMD.G, pRRL-CMV-GFP-sin, pMDLg/pRRE, and pRSVrev) to 1 mg/ml in TE buffer, pH 8.0.

4. Mix 100 μl of pMD.G, 400 μl of pRRL-CMV-GFP-sin, 240 μl of pMDLg/pRRE, and 60 μl of pRSVrev in a sterile 1.5-ml microcentrifuge tube.

These amounts are for 20 plates. They correspond to 5 μg pMD.G, 20 μg pRLL-CMV-GFP-sin, 12 μg pMDLg/pRRE, and 3 μg pRSVrev per plate.

5. Start transfection late in the afternoon, add 210 μl of sterile distilled water to each of 20 disposable 1.5-ml microcentrifuge tubes. Add 40 μl of the DNA mixture to each 1.5-ml tube and vortex.

6. Add 250 μl of 0.5 M $CaCl_2$ to each 1.5-ml tube prepared in step 5 and vortex.

7. Add 500 μl of 2× HeBS to twenty 15-ml sterile conical tubes, then slowly transfer, dropwise, the 500 μl of DNA/$CaCl_2$ mixture from each tube prepared in step 6, while vigorously vortexing.

8. Leave the precipitates (1 ml final volume per tube) at room temperature for 30 min.

9. Add the 1 ml of precipitate from step 8 dropwise to each of the 20 culture dishes prepared in step 2. Mix by gentle swirling until the medium has recovered a uniformly red color.

10. Place the dishes overnight in a 37°C humidified incubator with a 5% CO_2 atmosphere. Early the next morning, aspirate the medium and gently add 10 ml of fresh DMEM-10, prewarmed to 37°C. Incubate for 28 hr.

11. Transfer the culture medium from each plate to four 50-ml centrifuge tubes. Close the tubes, and spray them with 70% ethanol before taking them out of the hood. Centrifuge the tubes for 2 min at $500 \times g$, 4°C, to pellet detached cells.

12. Connect a 125-ml filter bottle to a vacuum line. Filter 100 ml of culture medium through the 0.45-μm pore size membrane. Use a second filter bottle to clear the remaining 100 ml of culture medium. Pool the filtered medium in a 75-cm² tissue culture flask.

13. *Optional:* Save a 1-ml aliquot of the filtered vector stock in a 1.5-ml microcentrifuge tube for titration (see Support Protocols 1 and 2). Keep this aliquot at 4°C until used.

Titration is optional, however it is very important if the final titer is $<10^8$ TU/ml. In this case, it is important to determine which step of the procedure was not optimized. Titer of the sample taken at step 13 indicates whether vector production was successful. A problem would be indicated by a titer of $<10^6$ TU/ml or a recovery of <50% after centrifugation. Comparing the vector particle numbers at steps 13 and 18 allows calculation of the recovery achieved after the first centrifugation. To calculate particle recovery, divide the particle number after the centrifugation by the particle number before the centrifugation. Particle number in a suspension is calculated by multiplying the titer (TU/ml) by the volume (ml) of the suspension. See Table 12.6.1.

VSV G–pseudotyped lentivectors have a halflife of 24 hr at 37°C. Aliquots of vector stocks should be stored at 4°C for up to 24 hr, or frozen at −80°C for longer storage.

14. Transfer the filtered culture medium into six 30-ml disposable conical polyallomer ultracentrifuge tubes. Place the tubes in an SW28 rotor. Close the ultracentrifuge buckets before taking them out of the hood. Ultracentrifuge for 90 min at $72,100 \times g$, 16°C.

Table 12.6.1 Recovery of Vector Particles Throughout a Typical Purification Procedure

Sample	Protocol step #	Titer (TU/ml)	Volume (ml)	Particle number	Recovery (%)
Crude stock	13	10^6	180	1.8×10^8	N.A.
Intermediate stock	18	2×10^7	5	10^8	55
Final stock	21	2×10^9	0.2	0.4×10^8	40

15. Open the ultracentrifuge buckets in the hood. Use forceps to take the tubes carefully out of the ultracentrifuge buckets. Invert the tubes to transfer the supernatants to a 75-cm^2 tissue culture flask. Save a small aliquot of the supernatant for titration (<5% of the pre-centrifugation viral particles should be present in the supernatant). To the remaining supernatant, add $^1/_6$ volume of 14% bleach, mix, wait 1 hr, and discard. Keep the tubes inverted and wipe the walls of the tubes with paper towels to eliminate as much as possible of the supernatant. Hold paper towels with forceps. Do not wipe the conical part of the tubes. Put the empty tubes on ice. Add 600 μl of PBS to each of the resulting vector pellets.

16. Resuspend pellets by pipetting up and down 20 times with a 1000-μl pipet tip, avoiding foaming. Leave on ice 30 min.

17. Pipet up and down again 20 times. Pool the resuspended vector particles from the six tubes into one 5-ml polyallomer centrifuge tube. Add sufficient PBS to fill the tube.

   A single round of centrifugation may yield titers high enough for in vitro experiments. In this case, the vector particles are pooled, divided into aliquots in 0.5-ml microcentrifuge tubes, and stored at −80°C.

18. Dilute a 5-μl aliquot of the vector suspension into 195 μl DMEM-10. Keep this aliquot at 4°C until used for titration (see Support Protocols 1 and 2).

19. Ultracentrifuge for 90 min at 76,000 × *g*, 16°C. Open the centrifuge buckets in the hood. Use forceps to take the tube out of the bucket. Invert the tube on a 50-ml centrifuge tube to discard the supernatant. Save a small aliquot of the supernatant for titration (<5% of the pre-centrifugation viral particles should be present in the supernatant). To the remaining supernatant, add $^1/_6$ volume of 14% bleach, mix, wait 1 hr, and discard. Wipe the wall of the tube (see step 15). Add 210 μl of PBS to the vector particles at the bottom of the tube.

20. Resuspend the vector particles with a 1000-μl pipet tip by pipetting up and down 20 times. Incubate on ice for 2 hr. Complete resuspension by repeating pipetting.

21. Dilute 5 μl of the concentrated vector stock in 495 μl of DMEM-10. Mix 100 μl of this first dilution with 400 μl of fresh DMEM-10. Keep these 1:100 and 1:500 diluted stocks at 4°C until used for titration (see Support Protocol 1 and 2). Calculate the % recovery using the aliquots saved at steps 13, 18, and 21 (Table 12.6.1 shows some typical results of a viral purification process).

22. Divide the concentrated stock (step 21) in 10 aliquots of 20 μl. Store the aliquots at −80°C.

SUPPORT PROTOCOL 1

TITRATION OF LENTIVIRUS GFP VECTOR STOCKS

The titration procedure for lentiviral vectors resembles those used for other types of vectors (see *UNITS 12.7 & 12.8* for herpes vectors and *UNIT 12.3* for AAV vectors). Several parameters can affect the apparent titer of a vector stock, such as the cell line used as the target cells, the volume of medium to which vector particles are added, the duration of the infection, and the presence of polycations favoring the binding of vector particles to the surface of target cells. The procedure can be used with any cell line. However some cell lines are poorly transduced by lentivectors and the expected yield noted here for HeLa cells is not valid for all cell lines. The best alternative to HeLa cells are 293T cells because the cell line is in culture in the laboratory anyway for the vector production. However, even though 293T cells are transduced as efficiently as HeLa cells, they have the disadvantage of poor adherence characteristics. While this is not a problem for the GFP detection protocol (Support Protocol 1), it is a point to keep in mind when using the *LacZ* staining procedure (Support Protocol 2). In either case, if using 293T cells, all reagents must be added by very gentle pipetting.

Additional Materials *(also see Basic Protocol)*

HeLa cells (ATCC #CCL-2)

Vector (see Basic Protocol)

0.25% trypsin/0.53 mM EDTA (without dye; Life Technologies; commercially available 10× stock diluted in PBS)

6-well tissue culture plates

Fluorescence-activated cell sorter (FACS; Becton Dickinson) and appropriate tubes

1. The day before titration, accurately seed 0.5×10^5 (to produce 1×10^5 cells per well on the following day) HeLa cells in 2 ml of DMEM-10 into all wells of a 6-well culture plate, ensuring a uniform spread of cells on the bottom of the wells. Prepare one plate for each vector stock to be titrated. Incubate overnight at 37°C, 10% CO_2.

2. To five wells, add aliquots of the vector to be titrated: use 50 μl and 25 μl of the undiluted stock, and 100 μl, 50 μl, and 25 μl of a 1:50 diluted stock (corresponding to 2.0, 1.0, and 0.5 μl of undiluted vector). Do not infect the cells in the last well; these are controls. Incubate 2 days.

3. Before the fluorescence-activated cell sorter (FACS) analysis, remove the culture medium, wash once with 2 ml PBS, and add 500 μl of 0.25% colorless trypsin/0.53 mM EDTA. Incubate 5 min at 37°C to detach the cells. Pipet up and down with a 1000-μl pipet tip to disrupt clumps. Transfer cells to a FACS tube containing 500 μl PBS.

   If desired, an aliquot of the cells can be kept in culture. If the FACS analysis is not done within 1 hr, cells can be fixed in 4% (w/v) paraformaldehyde solution (see recipe) for 30 min and kept for at least 1 week at 4°C.

4. Determine the percentage of GFP-positive cells by FACS analysis.

5. Calculate the titer in transducing units (TU)/ml, according to the formula:

$$\frac{(1 \times 10^5 \text{ seeded cells} \times \% \text{ GFP-positive cells}) \times 1000}{\mu l \text{ of vector}}$$

   For accurate titer calculations, the number of GFP-positive cells in 2 wells infected with 2 consecutive dilutions must be close to the expected 1:2 ratio. This linearity is observed when <15% of the target cells are transduced.

SUPPORT PROTOCOL 2

TITRATION OF LENTIVIRUS *LacZ* VECTOR STOCKS

LacZ-containing vectors (such as some of those listed at *http://www.tronolab.unige.ch*) are best titrated using the histochemical detection of β-galactosidase. Standard 96-well plates are convenient because wells are entirely visible in a single microscopic field at 40× magnification. Since cells double every 24 hr, most transduction events appear as clusters of 2, 3, or 4 blue cells. Single blue cells are rare but present. The number of blue cells must be low enough to facilitate the identification of clusters corresponding to single transduction events. Transduction events are counted in wells containing between 10 and 100 blue foci.

Additional Materials *(also see Basic Protocol; see APPENDIX 1 for items with ✓)*

HeLa cells (ATCC #CCL-2)

Vector (see Basic Protocol)

✓ 4% (w/v) paraformaldehyde solution

✓ Xgal staining solution

96-well tissue culture plates
37°C humidified incubator with 10% CO_2 atmosphere
Multichannel pipettor with appropriate tips
Inverted microscope

1. On the day before titration, prepare a 96-well plate (1 plate is sufficient to titrate four vector stocks) by seeding 5000 HeLa cells in 200 μl DMEM-10 medium in each of the 96 wells, ensuring a uniform spread of cells on the bottom of the wells. Incubate overnight in a 37°C, 10% CO_2 incubator. Perform each titration in duplicate.

2. Add 2 μl of vector diluted in 200 μl of DMEM-10 to the wells of the second column (400 μl total in wells), leaving the wells of the first column uninfected as a control.

3. Prepare a serial dilution by transferring 200 μl of DMEM-10 medium from the wells in one column to the wells of the following column with a multichannel pipettor. Start with the second column, and continue until all but the first column wells contain vector, mixing the medium by pipetting up and down several times at each passage. Incubate for 2 days at 37°C.

4. To detect *lacZ*-positive cells, remove the medium, wash once with 400 μl PBS, fix cells with 250 μl 4% (w/v) paraformaldehyde solution for exactly 5 min, wash two times with PBS, and add 250 μl of Xgal staining solution. Incubate at 37°C and check plates regularly.

5. Replace the Xgal staining solution with 250 μl PBS. Store plates at 4°C for at least 1 week if necessary.

6. Count transduction events in two wells containing between 10 and 100 blue foci on an inverted microscope. Divide the number of transduction events by the dilution factor and multiply by 1000 to calculate the titer in TU/ml.

References: Cisterni et al., 2000; Deglon et al., 2000

Contributors: Romain Zufferey and Didier Trono

12

UNIT 12.7

Construction of Replication-Defective Herpes Simplex Virus Vectors

The authors have developed methods to delete HSV-1 IE gene functions and to subsequently introduce foreign genes into the HSV-1 genome using homologous recombination. Because some of the HSV-1 IE genes—infected cell proteins 4 (*ICP4*) and 27 (*ICP27*)—are essential for growth of the virus in cell culture, complementing cell lines expressing these gene products in *trans* are required for isolation and propagation of replication-defective virus mutants.

For related information on gene delivery using helper virus–free HSV-1 amplicon vectors, see *UNIT 12.8*.

CAUTION: Radioactive, biological, and chemical substances require special handling. In addition, all work with viruses should be performed in a biosafety hood and incubator separate from those used for normal tissue culture.

NOTE: All solutions and equipment coming into contact with cells must be sterile, and proper sterile technique should be used accordingly.

NOTE: All culture incubations are performed in a humidified 37°C, 5% CO_2 incubator unless otherwise specified.

NOTE: When isolating and handling RNA, wear gloves, keep solutions cold and samples on ice when possible, and use new sterile or DEPC-treated microcentrifuge tubes and pipettor tips.

BASIC PROTOCOL 1

CONSTRUCTION OF HSV-1 IE GENE-COMPLEMENTING CELL LINES

Because the herpes simplex virus (HSV) immediate early (IE) genes contribute to the toxicity of the virus, it is necessary to sequentially delete these cytotoxic genes from the viral vector genome. To minimize the frequency of homologous recombination between the deletion virus and the HSV-1 gene that has been inserted into the complementing cell line, it is imperative that the plasmid(s) used in this protocol to produce the complementing cell line be constructed in such a manner as to contain the complementing IE gene devoid of sequences homologous with the deletion virus. Nonhomologous recombination may still occur between the deletion virus and the complementing cell line, although the frequency of this occurrence is extremely low ($<10^{-12}$).

Many of the viral IE gene products are cytotoxic, so it is desirable to engineer the complementing cell line to express the complementing gene(s) only when the cell is infected.

In engineering the complementing cell line, separate plasmids encoding either the complementing HSV-1 IE gene(s) or a selectable marker (neomycin phosphotransferase II or puromycin *N*-acetyltransferase) can be used. Alternatively, the HSV-1 IE gene(s) can be cloned into the plasmid encoding the selectable marker, increasing the chance of obtaining resistant clones that possess and express the HSV-1 IE gene-complementing functions. When generating a line that complements two essential HSV-1 IE genes, the drug resistance gene should be cloned between the two HSV-1 genes in order to increase the chance of isolating drug-resistant clones that express both HSV-1 gene products.

Materials (see APPENDIX 1 for items with ✓)

    Vero cells (African green monkey kidney cells; ATCC #CCC81) or other appropriate cell line

    Trypsin/EDTA: 0.05% (w/v) trypsin/0.3 mM EDTA (Life Technologies)

✓ Complete Eagle modified essential medium with and without 10% (v/v) FBS (complete MEM/10% FBS and complete MEM)

    Gel-purified plasmid DNA fragment

✓ 2× HEPES-buffered saline (HeBS), pH 7.05

✓ 2 M CaCl$_2$

    20% (v/v) glycerol in 2× HeBS

    1 mg/ml neomycin sulfate (G418, Life Technologies) *or* 10 μg/ml puromycin (Clontech), in complete MEM/10% FBS

    Freezing medium: complete MEM/10% FBS containing 10% (v/v) dimethylsulfoxide (DMSO) or glycerol

✓ PBS, pH 7.5

✓ DNA extraction buffer

✓ 10× PCR amplification buffer with 15 mM MgCl$_2$

✓ 10 mM 4dNTP mix

    100 ng/μl *ICP27* primer pair:

        5′ primer: 5′-GCC GCC GCG ACG ACC TGG AAT-3′

        3′ primer: 5′-TGT GGG GCG CTG GTT GAG GAT-3′

10 U/μl *Taq* DNA polymerase
DNA probe specific for *ICP27* sequences
HSV *ICP27* mutant
HSV IE gene mutant
✓ 1% (w/v) methylcellulose overlay
✓ 1% (w/v) crystal violet staining solution

30-, 60-, and 100-mm tissue culture dishes
25-cm^2 tissue culture flasks
Beckman GPR refrigerated tabletop centrifuge and GH-3.7 swinging-bucket rotor
96-well round-bottom tissue culture plates
25-cm^2 tissue culture flasks
Cell scrapers

1. Three days before transfection, split Vero cells 1:10 using trypsin/EDTA. Incubate 2 days.

2. On the day before transfection, split the cells ~1:2 and plate into complete MEM/10% FBS in three to five 60-mm tissue culture dishes at a cell density of 1×10^6 cells per dish. Incubate 1 day.

3. For each dish, prepare transfection mixture by adding 2 to 10 μg gel-purified plasmid DNA to 500 μl of 2× HeBS. Mix and incubate 20 min on ice. Perform the transfection in triplicate.

    The plasmid must contain the entire ICP27 coding sequence under the control of an HSV IE gene promoter. Plasmids may be obtained from a herpes virus laboratory or constructed from cloned HSV libraries using the HSV-1 sequence in GenBank. Plasmids containing drug resistance markers can be obtained commercially (e.g., from Clontech, Stratagene, or USB).

4. Add 34 μl of 2 M CaCl$_2$ dropwise to the transfection mixture. Mix gently and incubate 20 min at room temperature. Aspirate the medium from the dish and rinse once with 1 ml of 2× HeBS. Be careful to not allow the cells to dry out.

5. Pipet transfection mixture up and down to break up large clumps of precipitate. Carefully add transfection mixture to cell monolayer one dish at a time. Incubate dishes 10 min at 37°C.

6. Add 2.5 ml complete MEM/10% FBS to each dish and incubate 6 hr. Aspirate medium carefully (without causing the monolayer to lift) and wash dish once with 1 ml of 2× HeBS.

7. Slowly and carefully add 2 ml of 20% glycerol to each dish. Incubate *exactly* 3 min at room temperature.

8. Carefully aspirate all of the glycerol solution and wash the monolayer three times with 2 ml complete MEM/10% FBS. Carefully add 3 ml complete MEM/10% FBS. Incubate 24 hr.

9. After 24 hr, aspirate the medium and add 3 ml complete MEM/10% FBS. Incubate 48 hr.

10. After 48 hr, remove cells from the dish using 1.5 ml trypsin/EDTA. Pellet the cells by centrifuging 5 min at ~1000 × *g*, 4°C.

For neomycin-resistant clones

11a. Resuspend the pellet in 10 ml complete MEM/10% FBS containing 1 mg/ml neomycin. Plate 4×10^5 cells in each of ten 100-mm tissue culture dishes.

12a. Incubate until drug-resistant clones appear, replacing the medium every 3 days with fresh complete MEM/10% FBS containing 1 mg/ml neomycin.

The selection process begins in the second to third week. The appearance of the clones is rather obvious and they tend to grow quickly.

IMPORTANT NOTE: *Do not remove cells from neomycin-containing medium.*

For puromycin-resistant clones

11b. Resuspend pellet in 3 ml complete MEM/10% FBS. Plate cells in 100-mm tissue culture dish. Incubate 2 to 3 days (two to three rounds of division) until cultures are nearly confluent.

12b. Replace the medium with 3 ml complete MEM/10% FBS containing 10 µg/ml puromycin. Incubate until drug-resistant clones appear, replacing the medium every 3 days with fresh complete MEM/10% FBS containing 10 µg/ml puromycin.

Selection begins in the first 24 hr following drug addition. Few cells will survive the drug treatment, although those that do survive will replicate and form foci.

To eliminate false positives, keep the concentration of puromycin at 10 µg/ml, even though the resistant cells grow poorly in the presence of the drug. Puromycin-resistant clones do not grow as well as neomycin-resistant clones do in the presence of drug.

13. After foci have appeared, gently aspirate the medium. Add 20 µl trypsin/EDTA dropwise on top of each isolated focus, keeping each focus well separated.

14. Using a 200-µl micropipettor tip, scrape each individual focus of cells from the plate. Transfer the cells from each focus to a separate 15-ml tube that contains 10 ml complete MEM/10% FBS with the appropriate drug. Plate 100 µl in each well of a 96-well round-bottom tissue culture plate. Incubate to allow the clones to replicate.

15. Expand each drug-resistant clone (i.e., the cells from each well of the 96-well plates that exhibits growth in the presence of the drug) in 25-cm^2 tissue culture flasks in complete MEM/10% FBS containing the appropriate drug. Once 25-cm^2 flasks are seeded with puromycin-resistant clones, reduce the puromycin concentration to 5 to 7 µg/ml to enable the clones to grow faster.

16. Freeze clones away as stocks in freezing medium that contains drug at the appropriate concentration. Store at $-80°C$ in a liquid nitrogen dewar.

17. Add 5×10^5 cells from a drug-resistant clone to a 30-mm tissue culture dish. Incubate overnight.

18. Aspirate the medium and wash once with 1 ml PBS. Scrape the cells off the plate and transfer them to a 1.5-ml microcentrifuge tube. Microcentrifuge 3 min at maximum speed, 25°C, to pellet the cells.

19. Aspirate the supernatant. Add 200 µl DNA extraction buffer to lyse the cells. Incubate 1 to 12 hr (until there is no pellet or particular matter present) at 37°C. Boil the lysate 15 min.

20. Prepare PCR reaction mix (100 µl per reaction):

> 10 µl 10× PCR amplification buffer with 15 mM MgCl$_2$
> 4 µl 10 mM dNTP mix
> 2 µl 100 ng/µl *ICP27* primer pair (200 ng)
> 2 µl 10 U/µl *Taq* DNA polymerase
> 10 µl lysate
> 72 µl sterile H$_2$O.

21. Using the following amplification program, amplify each lysate:

$$
\begin{array}{llll}
30 \text{ cycles:} & 1 \text{ min} & 95°C & \text{(denaturation)} \\
& 1 \text{ min} & 60°C & \text{(annealing)} \\
& 10 \text{ min} & 60°C & \text{(extension).}
\end{array}
$$

22. Analyze PCR reaction products by agarose gel electrophoresis and Southern blot hybridization (*APPENDIX 3G*) using a DNA probe specific for *ICP27* sequences.

    Amplified ICP27 product obtained using the ICP27 gene primers specified in the Materials list is 219 bp long.

23. Plate 1.5×10^6 cells from drug-resistant *ICP27*-positive clones into 5 ml complete MEM/10% FBS in 60-mm tissue culture dishes. Set up one to two plates for each mutant tested. Also set up one to two plates of Vero cells as a negative (noncomplementing) control. Incubate 24 hr.

24. Infect culture with 1×10^3 plaque-forming units (pfu) of a known HSV *ICP27* mutant virus in 500 μl complete MEM without serum. Incubate 60 to 90 min, rocking the plate every 15 min to distribute the inoculum.

25. After adsorption, aspirate the virus inoculum and add 3 ml of 1.0% methylcellulose overlay. Incubate until plaques are visible. Once plaques are clearly visible, aspirate the methylcellulose. Stain the plate with 1 ml of 1% crystal violet staining solution 5 min at room temperature.

26. Aspirate the staining solution. Rinse the plate gently with tap water to remove excess stain. Air dry. Identify complementing cell lines.

    If the neomycin-or puromycin-resistant clones are capable of complementing the desired IE gene product (ICP4, ICP27, or ICP0), then plaques should appear on these plates. For clones that complement two HSV-1 IE gene functions, single-gene deletion mutants for both IE genes should be able to grow on the complementing line. Cell-line isolates capable of complementing the desired IE gene mutant can be further analyzed to determine the clone that best complements the deletion mutant.

27. Add 1.5×10^6 cells from drug-resistant *ICP27*-positive complementing clones in 3 ml complete MEM/10% FBS to 60-mm tissue culture dish. Incubate 2 to 4 hr.

28. Infect cultures with 5×10^6 and 1×10^7 pfu (MOI = 5 and 10) of known HSV IE gene mutant in 500 μl complete MEM without serum. Incubate 60 to 90 min to allow virus to adsorb to cells, rocking the plate every 15 min to distribute the inoculum evenly.

29. Aspirate the virus inoculum and add 3 ml complete MEM/10% FBS. Incubate 18 to 24 hr.

30. Harvest the culture by scraping the cells off the surface of the plate with a cell scraper. Transfer to a 15-ml tube.

31. Prepare midistocks for titering (see Support Protocol 1, steps 4 to 9). Titer virus stocks for three dilutions (see Support Protocol 2).

    Calculate the number of particles produced per cell to determine the burst size. Each drug-resistant clone that complements single IE gene mutants will display a different burst size value.

32. Select the clone in which the virus replicates the fastest with the greatest burst size for future use. Freeze stocks of cells at $-80°C$ in a liquid nitrogen dewar.

BASIC PROTOCOL 2

CONSTRUCTION OF REPLICATION-DEFECTIVE VECTORS

An example of the construction of a replication-defective genomic HSV-1 vector deleted for an essential IE gene by homologous recombination is found in Figure 12.7.1.

Materials (see APPENDIX 1 for items with ✓)

    IE gene–complementing cell line (see Basic Protocol 1)

    Trypsin/EDTA: 0.05% (w/v) trypsin/5.3 mM EDTA solution (Life Technologies)

    ✓ Complete Eagle modified essential medium with and without 10% (v/v) FBS (complete MEM/10% FBS and complete MEM)

    Plasmid with reporter cassette (e.g., HCMV IEp-*lacZ*-BGH pA; see Fig. 12.7.1) inserted in place of deleted HSV-1 genes

    Restriction endonuclease with 8-bp recognition sequence, and appropriate buffer

    Viral DNA (see Support Protocol 3)

    ✓ 2× HEPES-buffered saline (HeBS), pH 7.05

    ✓ 2 M $CaCl_2$

    20% (v/v) glycerol in 2× HeBS

    ✓ 0.1% Xgal staining solution

    ✓ PBS, pH 7.5

    ✓ DNA extraction buffer

    ✓ 10× PCR amplification buffer with 15 mM $MgCl_2$

    ✓ 10 mM 4dNTP mix

    100 ng/µl *ICP27* primer pair:

        5′ primer: 5′-GCC GCC GCG ACG ACC TGG AAT-3′

        3′ primer: 5-TGT GGG GCG CTG GTT GAG GAT-3′

    100 ng/µl glycoprotein B (*gB*) primer pair:

        5′ primer: 5′-ATT CTC CTC CGA CGC CAT ATC CAC CAC CTT-3′

        3′ primer: 5′-AGA AAG CCC CCA TTG GCC AGG TAG T-3′

    10 U/µl *Taq* DNA polymerase

    Sterile H_2O

    DNA probes specific for *ICP27* or *gB* sequences

    30- and 60-mm tissue culture dishes

    Wide-bore pipet tips (Bio-Rad)

    Cell scrapers

    Beckman GPR refrigerated tabletop centrifuge and GH-3.7 swinging-bucket rotor

    Sonicator with cup horn (VirTis)

    Rocker platform (e.g., Nutator, Becton Dickinson Primary Care Diagnostics)

    Multichannel pipettor and reagent reservoirs (Costar)

    96-well flat-bottom tissue culture plate

1. Three days prior to transfection, split cultures of IE gene–complementing cell line 1:10 using trypsin/EDTA. Grow the cells in complete MEM/10% FBS. Incubate 2 days.

2. One day prior to transfection, split the cells again using trypsin/EDTA. Plate 1×10^6 cells into 3 ml complete MEM/10% FBS in each of four (or eight if duplicates are desired) 60-mm tissue culture dishes. Incubate 1 day.

3. Linearize the plasmid with reporter cassette by digesting with a restriction endonuclease that cuts between the flanking HSV-1 sequence and any *E. coli* DNA present within the bacterial vector.

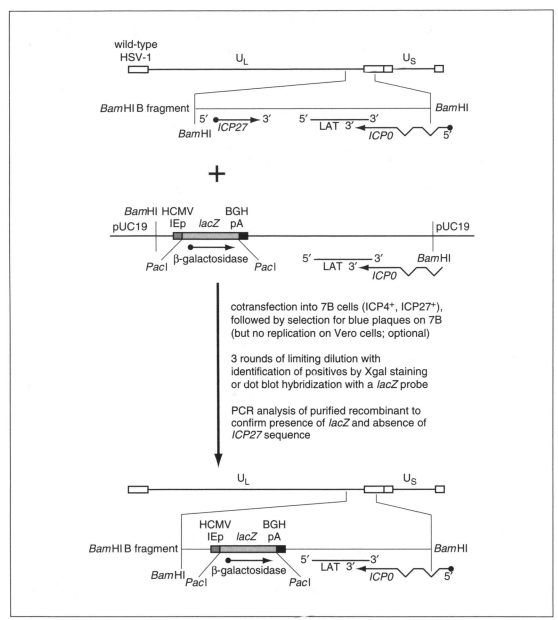

Figure 12.7.1 Example of the construction of a replication-defective genomic HSV-1 vector deleted for an essential IE gene by homologous recombination (see Basic Protocol 2). Purified wild-type HSV-1 DNA is cotransfected into the 7B *ICP4/ICP27*-complementing cell line along with a plasmid containing the *Bam*HI B fragment of the HSV-1 genome with the *ICP27* IE gene sequences in the plasmid replaced with the HCMV IEp-*lacZ*-BGHpA expression cassette. The plasmid has been designed with *Pac*I restriction endonuclease recognition sites flanking the *lacZ* reporter gene, although the virus wild-type genome is devoid of *Pac*I sites. Homologous recombination between the HSV-1 flanking sequences in the plasmid and those resident within the wild-type viral genome results in a recombinant deleted for *ICP27* that grows on the 7B complementing line but not on Vero cells. Because the recombinant also contains the HCMV IEp-*lacZ*-BGH pA expression cassette, it can be readily isolated and purified using limiting dilution with Xgal staining (blue plaques versus clear plaques for wild-type parental virus). The absence of ICP27 sequence and the presence of the expression cassette can be confirmed by PCR. Abbreviations: BGH pA, bovine growth hormone polyadenylation and cleavage sequence; HCMV IEp, human cytomegalovirus immediate-early promoter; *ICP*, infected cell protein; LAT, latency-associated transcript; UL, unique long segment of HSV-1 genome; US, unique short segment of HSV-1 genome.

4. Using wide-bore pipet tips, prepare a transfection mixture for each dish by adding 1 to 5 μg (concentration that yields ≥200 plaques by transfection) viral DNA to an amount of linearized plasmid DNA equal to 10 and 50 genome equivalents of viral DNA.

One genome equivalent of viral DNA is 152 kb.

5. Add 600 μl of 2× HeBS to each tube and mix and incubate 20 min on ice. Then, add 41 μl of 2 M $CaCl_2$ dropwise, mixing gently, and incubate 20 min at room temperature.

6. Aspirate the medium from the dish one dish at a time. Wash once with 1 ml of 2× HeBS. Aspirate the wash, but do not dry out before transfection mixture is added.

7. Pipet transfection mixture up and down to break up large clumps of precipitate. Carefully add transfection mixture to cell monolayer one dish at a time. Incubate 40 min.

8. Carefully add 4 ml complete MEM/10% FBS to each dish without disturbing transfection mixture. Incubate 4 hr.

9. After 4 hr, aspirate the medium/transfection mixture. Wash once with 1 ml of 2× HeBS without causing the monolayer to lift off of plate. Aspirate the 2× HeBS wash. Slowly and carefully add 2 ml of 20% glycerol solution. Incubate dish *exactly* 4 min at room temperature.

10. Carefully remove all of the glycerol solution by aspiration. Wash the monolayer three times with 2 ml complete MEM/10% FBS. Be sure monolayer remains intact.

11. Carefully add 4 ml complete MEM/10% FBS. Incubate, inspecting the cultures two times daily under a microscope for the development of cytopathic effect (CPE), which indicates the presence of infectious foci.

CPE is detected as rounded, highly vacuolated cells. It usually takes 3 to 5 days to observe CPE, depending on the virus and cell type used to propagate the recombinant.

12. After plaques open (dying cells are lysed in the center of the plaque), remove the medium and store the medium temporarily at 4°C.

13. Scrape the cells in 1 ml complete MEM/10% FBS and transfer them to a 15-ml tube. Centrifuge 5 min at 2000 × g, 4°C. Resuspend pellet in 250 μl MEM/10% FBS.

14. Freeze (−80°C) and thaw (37°C) the pellet three times to lyse the cells and isolate the virus from the pellet. Sonicate the lysate for 5 sec using a cup-horn sonicator. Centrifuge 5 min at 2000 × g, 4°C.

15. Combine this supernatant with the medium removed in step 12. Store at −80°C for use as a recombinant stock.

16. Titer the recombinant stock (see Support Protocol 2).

17. Add 30 pfu recombinant virus to 3 ml of 1×10^6 IE gene–complementing cells in complete MEM/10% FBS in a 15-ml conical polypropylene tube. Wrap the cap of the tube with Parafilm and incubate the tube 1 hr on a rocker platform at 37°C.

18. Following adsorption, add 7 ml complete MEM/10% FBS and transfer to a multichannel pipet reservoir. Using a multichannel pipettor, add 100 μl to each well of a 96-well flat-bottom tissue culture plate. Incubate 2 to 5 days and check two times daily for the appearance of plaques; do not let the infection proceed too far.

19. Identify wells containing only single plaques. Transfer the medium from single-plaque wells to a new 96-well tissue culture plate and store at −80°C (virus stock).

20. Using a multichannel pipettor, overlay each well of the scored plate with 100 μl Xgal staining solution. Incubate the plates overnight at 37°C (in some cases, blue plaques can be observed within 1 to 2 hr).

21. Subject positive isolates to two additional rounds of limiting dilution analysis (steps 16 to 20) to ensure purity of the stock.

22. Following the final purification, use the virus stock to produce a midistock (see Support Protocol 1).

23. Plate 5×10^5 cells from a drug-resistant IE gene–complementing (e.g., *ICP27*-positive) clone into a 30-mm tissue culture dish. Incubate overnight (dishes should be subconfluent to confluent the following day).

24. Aspirate medium and add 1×10^7 pfu virus from the midistock preparation (from step 22; MOI = 10) in 500 μl complete MEM without serum. Incubate 60 to 90 min, rocking the dish every 15 min to distribute the inoculum evenly.

25. Following adsorption, aspirate the inoculum and add 2 ml complete MEM/10% FBS. Incubate 12 to 16 hr.

26. Aspirate the medium and wash once with 1 ml PBS, pH 7.5. Scrape the cells off the dish and transfer them to a 1.5-ml microcentrifuge tube. Microcentrifuge 3 min at maximum speed, room temperature, to pellet the cells.

27. Aspirate the supernatant. Add 200 μl DNA extraction buffer and incubate 1 to 2 hr at 37°C to lyse the cells. Boil the lysate 15 min.

28. Prepare PCR reaction mixture (90 μl per reaction):

    10 μl 10× PCR amplification buffer with 15 mM $MgCl_2$
    4 μl 10 mM 4dNTP mix
    2 μl 100 ng/μl *ICP27* or *gB* primer pair
    2 μl 10 U/μl *Taq* DNA polymerase
    72 μl sterile H_2O.

29. Using the following amplification program, amplify 10 μl of each boiled lysate in a final volume of 100 μl in PCR assay buffer containing *ICP27* or *gB* primer pair:

30 cycles:	1 min	95°C	(denaturation)
	1 min	60°C	(annealing)
	10 min	71°C	(extension).

30. Using a probe specific for *ICP27* or *gB*, analyze PCR reaction product by Southern blot hybridization (APPENDIX 3G).

    Amplified ICP27 and gB products obtained using the primers described in the Materials list are 219 and 191 bp long respectively.

BASIC PROTOCOL 3

INSERTION OF FOREIGN GENE SEQUENCE INTO A REPLICATION-DEFECTIVE GENOMIC HSV VECTOR

An example of the construction of a replication-defective genomic HSV-1 vector containing foreign gene sequenes by homologous recombination is found in Figure 12.7.2.

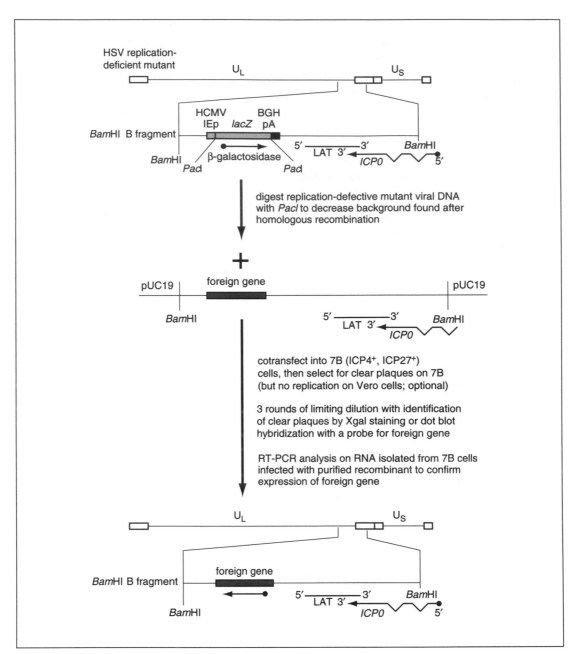

Figure 12.7.2 Example of the construction of a replication-defective genomic HSV-1 vector containing foreign gene sequences by homologous recombination (see Basic Protocol 3). Foreign gene cassettes can be introduced into the *ICP27*− replication-defective mutant by homologous recombination following digestion of the mutant viral DNA with the *Pac*I restriction endonuclease. *Pac*I digestion reduces the background levels of parental virus and provides free ends, thereby increasing the frequency of recombination. The *Bam*HI B plasmid construct in which the foreign gene of interest replaces the *ICP27* gene sequences is cotransfected along with the *ICP27*− replication-defective mutant virus DNA into *ICP4/ICP27*-complementing 7B cells. Following homologous recombination, clear plaque isolates are purified from blue plaque parental virus by limiting dilution. RT-PCR analysis is employed to verify the expression of foreign gene. Abbreviations: BGH pA, bovine growth hormone polyadenylation and cleavage sequence; HCMV IEp, human cytomegalovirus immediate early promotor; *ICP*, infected cell protein; LAT, latency-associated transcript; UL, unique long segment of HSV-1 genome; US, unique short segment of HSV-1 genome.

The recombination frequency obtained using this approach is at least ten-fold greater than that of standard marker transfer.

Dot-blot or Southern blot hybridization (APPENDIX 3G) can be used to confirm the presence of the gene expression cassette within the genome of the replication-defective genomic HSV-1 vector. Alternatively, PCR can be employed to detect the presence of foreign gene sequences. Reverse transcription (RT)-PCR is used to verify expression of the foreign gene product from the replication-defective virus vector or expression can be monitored by Northern blot (APPENDIX 3H), immunoblot (CPMB UNIT 10.8) or ELISA (CPMB UNIT 11.3), depending on the gene product.

Materials (see APPENDIX 1 for items with ✓)

IE gene–complementing cells (see Basic Protocol 1)
Replication-defective HSV vector (see Basic Protocol 2) digested with *PacI* under the
    conditions recommended by the manufacturer
Plasmid containing foreign gene cassette, linearized using the appropriate restriction
    endonuclease
Subconfluent monolayer cultures of IE gene–complementing cell line (see Basic
    Protocol 1) in 24-well tissue culture plates
✓ Complete Eagle modified essential medium with and without 10% (v/v) FBS (complete
    MEM/10% FBS and complete MEM)
✓ Tris-buffered saline (TBS), pH 7.5
✓ DNA extraction buffer
✓ 25:24:1 (v/v/v) phenol/chloroform/isoamyl alcohol
Chloroform
Isopropanol, $-20°C$
70% ethanol
Total RNA isolation reagent (TRI reagent, Molecular Research Center)
1-Bromo-3-chloropropane (BCP, Aldrich)
RNase-free DNase I (Worthington)
10 U/μl reverse transcriptase and buffer (e.g., SuperScript II and 5× SuperScript II
    buffers, Life Technologies)
✓ 1.0 mM DTT
100 ng/μl *lacZ* primer pair:

> 5′ primer: 5′-ACC CCT TCA TTG ACC TCA AC-3′
> 3′ primer: 5′-ATT GGG GGT AGG AAC ACG-3′

100 ng/μl *GAPDH* primer pair:

> 5′ primer: 5′-TTG CTG ATT CGA GGG GTT AAC CGT CAC GAG-3′
> 3′ primer: 5′-ACC AGA TGA TCA CAC TGC GGT GAT TAC GAT-3′

100 ng/μl foreign gene primer pair
✓ 10 mM 4dNTP mix
Placental RNase inhibitor: 50 to 100 U/μl RNasin (Promega)
Sterile H$_2$0
✓ 10× PCR amplification buffer with 15 nM CaCl$_2$
DNA probe for foreign gene sequences
10 U/μl *Taq* DNA polymerase

Rocker platform (e.g., Nutator, Becton Dickinson Primary Care Diagnostics)
30-mm tissue culture dishes
96-well flat-bottom and 24-well tissue culture plates
1.5-ml microcentrifuge tubes, sterile
Pellet pestles (Kontes Glass)

1. Cotransfect cultures of IE gene–complementing cells with replication-defective HSV digested with *Pac*I and the plasmid containing the foreign gene cassette linearized using the appropriate restriction endonuclease. Isolate single-plaque wells in a 96-well flat-bottom tissue culture plate (see Basic Protocol 2, steps 1 to 18).

2. Transfer 80 μl medium from each single-plaque well into a new 96-well plate. Store the new plate at −80°C (virus stock).

3. For each single-plaque well, mix the remaining 20 μl of medium (containing virus) with 100 μl complete MEM/10% FBS. Aspirate the medium from one well of a subconfluent monolayer culture of IE gene–complementing cells in a 24-well tissue culture plate and add the 120 μl of virus-containing medium. Incubate 60 to 90 min, rocking the plate every 15 min to distribute the inoculum evenly.

4. Following adsorption, aspirate the inoculum and add 1 ml fresh complete MEM/10% FBS. Incubate 2 to 3 days. Examine the cultures for cytopathic effect (CPE) as evidenced by cell rounding (see Basic Protocol 2, step 11).

5. Scrape each well with a micropipettor tip to remove the monolayer. Transfer the suspension to a 1.5-ml microcentrifuge tube. Microcentrifuge 10 min at maximum speed, room temperature, to pellet the cells and virus.

6. Carefully aspirate the supernatant and wash the pellet once with 500 μl TBS, pH 7.5. Microcentrifuge 5 min at maximum speed to pellet cells and virus. Aspirate the supernatant.

7. Resuspend the pellet in 200 μl DNA extraction buffer. Incubate overnight at 37°C on a rocker platform.

8. Extract the lysate once with 200 μl of 25:24:1 phenol/chloroform/isoamyl alcohol. Vortex vigorously to mix the two phases.

9. Microcentrifuge 5 min at maximum speed, room temperature, to separate the two phases. Remove the lower (organic) phase and retain the upper aqueous phase and interface.

10. Extract the aqueous phase once with 200 μl chloroform. Vortex to mix the two phases. Microcentrifuge 2 min at maximum speed, room temperature, to separate the two phases. Transfer the upper (aqueous) phase to a new tube.

   The DNA is extremely viscous and will enter the pipet as a visible slurry.

11. Add 500 μl −20°C isopropanol to precipitate viral DNA. Microcentrifuge 15 min at maximum speed, room temperature. Carefully aspirate supernatant.

12. Wash the pellet once with 500 μl of 70% ethanol, vortexing briefly. Microcentrifuge 5 min. Aspirate supernatant and air dry pellet.

13. Resuspend pellet in 50 μl water. Analyze the viral DNA by dot-blot or Southern blot hybridization (*APPENDIX 3G*) using DNA probes specific for the inserted foreign gene sequences.

14. Plate 1×10^5 cells from an IE gene–complementing cell line in 30-mm tissue culture dish. Incubate overnight (plates should be subconfluent to confluent after incubation).

15. Aspirate the medium and add 1×10^7 pfu of recombinant virus containing foreign gene cassette (from step 2; MOI = 10; see Support Protocol 2 for titering virus stock) in 500 μl complete MEM without serum. Incubate 60 to 90 min, rocking the dish every 15 min to distribute the inoculum evenly.

16. Following adsorption, aspirate the inoculum and add 2 ml complete MEM/10% FBS. Incubate 12 to 16 hr.

12

17. Lyse the cells by adding 0.8 ml TRI reagent to each dish. Scrape the dish and transfer cell lysate to a sterile 1.5-ml microcentrifuge tube. Homogenize the lysate using a pellet pestle and incubate 5 min at room temperature.

18. Add 0.1 ml BCP to lysate and vortex 15 sec. Incubate 2 to 3 min at room temperature.

19. Microcentrifuge 15 min at maximum speed, room temperature. Transfer upper (aqueous; RNA) phase to a new 1.5-ml microcentrifuge tube.

20. Precipitate the RNA by adding 0.5 ml $-20°C$ isopropanol and incubating 10 min at room temperature. Microcentrifuge 10 min at maximum speed, room temperature.

21. Wash the RNA pellet with 70% ethanol. Microcentrifuge 5 min at maximum speed, room temperature. Aspirate supernatant and air dry pellet.

22. Dissolve extracted RNA in 50 μl water. Add 1000 U RNase-free DNase I to destroy the DNA template. Incubate 60 min at 37°C.

23. Incubate 10 min at 75°C to inactivate the enzyme. Prepare RT-PCR assay buffer (20 μl per reaction):

>   4 μl 5× SuperScript II buffer
>   2 μl 1.0 mM DTT
>   1 μl 100 ng/μl primer for foreign gene, *lacZ*, or *GADPH* positive
>   cellular control
>   1 μl 10 mM 4dNTP mix
>   1 μl 50 to 100 U/μl RNasin
>   1 μl 10 U/μl SuperScript II reverse transcriptase
>   5 μl RNA sample
>   5 μl sterile H$_2$O
>   Incubate 30 min at 37°C.

24. Prepare PCR assay buffer (80 μl per reaction):

>   10 μl 10× PCR amplification buffer with 15 mM MgCl$_2$
>   4 μl 10 mM 4dNTP mix
>   1 μl 100 ng/μl foreign gene, *lacZ*, or *GAPDH* primer pair
>   1 μl 10 U/μl *Taq* DNA polymerase
>   64 μl sterile H$_2$O.

25. Bring the volume of the RT-PCR reaction to 100 μl with PCR assay buffer. Amplify using the following program:

30 cycles:	1 min	95°C	(denaturation)
	1 min	60°C	(annealing)
	10 min	71°C	(extension).

26. Analyze RT-PCR reaction products by Southern blot hybridization (APPENDIX 3G) using a DNA probe specific for the foreign gene of interest, *lacZ* (negative control), or the cellular gene control *GAPDH*.

> *Amplified GAPDH and lacZ products obtained using the primers described in the Materials list are 614- and 324-bp long respectively. The lacZ fragment should not be detected if the foreign gene has been inserted into the rector.*

PREPARATION OF HERPES SIMPLEX VIRUS STOCK

Materials (*see* APPENDIX 1 *for items with* ✓)

> Permissive cells appropriate for growth of virus
> Virus stock of known titer (see Support Protocol 2)
> ✓ Complete Eagle modified essential medium with and without 10% (v/v) FBS (complete
>   MEM/10% FBS and complete MEM)
> 1 M HEPES (*N*-2-hydroxyethylpiperazine-*N′*-2-ethanesulfonic acid)
>
> 25-cm^2 tissue culture flasks
> 15- and 50-ml conical polypropylene centrifuge tubes
> Beckman GPR refrigerated tabletop centrifuge and GH-3.7 swinging-bucket rotor
> Cup-horn sonicator (VirTis)
> 50-ml polypropylene Oak Ridge tubes
> Beckman J2-21M preparative centrifuge and JA-20 rotor (preparative)
> 850-cm^2 tissue culture roller bottles
> Roller bottle apparatus
> Cell scrapers

1. Plate 1.0×10^6 permissive cells in a 25-cm^2 tissue culture flask. Incubate 1 to 2 days (until cells are actively dividing).

2. On the day of infection, aspirate the medium from the cell monolayer. Add 3×10^4 pfu of virus (MOI = 0.01) in 1 ml complete MEM without serum. Incubate 60 to 90 min to allow virus to adsorb to cells, rocking the flask every 15 min to distribute the inoculum evenly.

3. Aspirate the inoculum and add 10 ml complete MEM/10% FBS. Incubate until all the cells exhibit CPE—i.e., have rounded up and are starting to come off the surface of the flask; examine cells two times daily.

4. Dislodge the cells from the plastic surface either by tapping or scraping the flask. Transfer the cell suspension to a 15-ml conical polypropylene centrifuge tube. Centrifuge 5 min at $1500 \times g$, 4°C.

5. Decant the supernatant into a new tube and store temporarily at 4°C. Resuspend the pellet in complete MEM/10% FBS to a final volume of 1 ml.

6. Freeze (-80°C) and thaw (37°C) the infected cell suspension three times. Sonicate the suspension 5 sec using a cup-horn sonicator and centrifuge 5 min at $2000 \times g$, 4°C, to pellet debris.

7. Combine supernatant with the supernatant stored at 4°C (step 5) and transfer both to a 50-ml polypropylene Oak Ridge tube. Centrifuge 30 min at $48,000 \times g$, 4°C, to pellet the virus.

8. Carefully aspirate the supernatant and resuspend the pellet in complete MEM/10% FBS to a final volume of 1 ml. Titer the virus (see Support Protocol 2).

9. Divide the virus preparation into aliquots in 2-ml cryotubes and store at -80°C for use in preparing a maxistock.

10. Plate 2.0×10^7 permissive cells in each of two 850-cm^2 roller bottles. Incubate 1 to 2 days.

11. On the day of infection, decant medium from the bottles. Add 1×10^6 pfu virus (MOI = 0.01) in 20 ml complete MEM without serum to each bottle. Allow the virus to adsorb 60 to 90 min at 37°C on a roller bottle apparatus.

12. After adsorption, decant the inoculum and add 100 to 125 ml complete MEM/10% FBS per bottle. Incubate at 37°C until all the cells have rounded up and are starting to detach from the surface. Observe cells at least two times daily to obtain the exact harvest time.

13. Remove the cells from the surface of the roller bottle using a cell scraper. Transfer the suspension to 50-ml conical polypropylene centrifuge tubes. Centrifuge 5 min at $1500 \times g$, 4°C, to pellet infected cells.

14. Decant the supernatants into new tubes and store temporarily at 4°C. Resuspend each pellet in complete MEM/10% FBS to give a final volume of 2 ml.

15. Freeze (-80°C) and thaw (37°C) the infected cell suspension three times without cracking the tubes. Sonicate the suspension 5 sec in a cup-horn sonicator and centrifuge 5 min at $2000 \times g$, 4°C, to pellet cell debris.

16. Combine supernatant with the supernatant stored at 4°C (step 14). Transfer to 50-ml polypropylene Oak Ridge tubes on ice. Centrifuge 30 min at $48,000 \times g$, 4°C, to pellet virus.

17. Carefully aspirate the supernatant. Resuspend the pellets in complete MEM/10% FBS to a final volume of 1 to 2 ml. Store 100-µl aliquots in 2-ml cryotubes at -80°C.

   It is important to maintain stocks at a low passage in order to reduce the chance of rescuing wild-type virus during the propagation of viruses carrying deletions of essential gene(s).

18. Titer the virus stock (see Support Protocol 2).

19. *Optional*: To purify the virus further, resuspend the pellet (step 17) in PBS and purify using a sucrose, T-10 dextran, or Nycodenz gradient (e.g., CPMB UNIT 16.11). Add glycerol to 10% (v/v) before storage.

   For sucrose gradient purification, use a linear 30% to 65% gradient and centrifuge 16 hr at $71,000 \times g$, 4°C.

SUPPORT PROTOCOL 2

PLAQUE ASSAY TO TITER VIRUS

The assay is performed in duplicate or triplicate.

Materials *(see APPENDIX 1 for items with ✓)*
   Permissive cells appropriate for growth of virus
   Virus stock to be titrated
✓ Complete Eagle modified essential medium with and without 10% (v/v) FBS (complete MEM/10% FBS and complete MEM)
✓ 1% (w/v) methylcellulose overlay
✓ 1% (w/v) crystal violet solution
   12-well tissue culture plates

1. Plate 1.0×10^5 permissive cells per well in each well of a 12-well tissue culture plate. Incubate overnight until the cells form a nearly confluent monolayer.

2. The next day, prepare a series of ten-fold dilutions (10^{-5} to 10^{-10}) of the virus stock in 1 ml complete MEM without serum. Add 100 µl of each dilution to duplicate wells of

permissive cells. Allow the virus to adsorb 60 to 90 min at 37°C, rocking the plates every 15 min to distribute the inoculum evenly.

3. After adsorption, aspirate the virus inoculum. Add 1 ml of 1.0% methylcellulose overlay. Incubate 3 to 5 days until well-defined plaques appear.

4. When plaques are clearly visible, aspirate the methylcellulose. Stain 5 min with 1 ml of 1% crystal violet at room temperature.

5. Aspirate the staining solution and rinse the plates gently with tap water to remove excess stain. Air dry.

6. Count the number of plaques per well. Determine the average number of plaques for each dilution and multiply by a factor of 10 to get the number of plaque-forming units per milliliter (pfu/ml) for each dilution. Multiply this number by 10 to the power of the dilution to get the titer in pfu per milliliter for the virus stock.

SUPPORT PROTOCOL 3

ISOLATION OF VIRAL DNA

Infectivity of viral DNA is measured as a ratio of the number of plaques produced per microgram of viral DNA following transfection of the viral DNA into permissive cells. Methods for the purification of viral DNA must be optimized to yield ~100 to 1000 plaques per microgram of purified viral DNA.

Materials (see APPENDIX 1 for items with ✓)
    Monolayers of permissive cells, in 150-cm^2 tissue culture flasks and 60-mm tissue culture dishes
    Stock of HSV-1 virus of known titer (see Support Protocol 1)
✓ Complete Eagle modified essential medium with and without 10% (v/v) FBS (complete MEM/10% FBS and complete MEM)
✓ Tris-buffered saline (TBS), pH 7.5
✓ DNA extraction buffer
✓ 25:24:1 (v/v/v) phenol/chloroform/isoamyl alcohol
    Chloroform
    Isopropanol, −20°C
    70% ethanol
    Labeled total HSV-1 viral DNA
✓ 1% (w/v) methylcellulose overlay
✓ 1% (w/v) crystal violet staining solution

    150-cm^2 tissue culture flask
    Cell scrapers (optional)
    15-ml conical polypropylene tubes
    Beckman GPR refrigerated tabletop centrifuge and GH-3.7 swinging-bucket rotor or equivalent
    Rocker platform (e.g., Nutator, Becton Dickinson Primary Care Diagnostics)
    Heat-sealed Pasteur pipet
    Wide-bore pipet tips (Bio-Rad)

1. Aspirate medium from a subconfluent to confluent monolayer of permissive cells in a 150-cm^2 tissue culture flask (1.7×10^7 cells).

Cells should be split 1:2 at both 3 and 1 days prior to infection to ensure that cells are actively replicating.

The choice of cells employed will depend on whether it is necessary to complement an essential IE gene deletion mutant, a multiple gene mutant, or a wild-type virus.

2. Infect cells with 5.0×10^7 pfu of virus (MOI = 3) in 5 ml complete MEM without serum. Incubate 60 to 90 min at 37°C, rocking the flask every 15 min to distribute the inoculum evenly.

3. After adsorption, aspirate the inoculum and add 20 ml complete MEM/10% FBS. Incubate ~18 to 24 hr. Periodically examine the flask for the development of cytopathic effect (CPE), as evidenced by the presence of rounded cells that are still adherent to the flask.

4. Dislodge the cells by sharply tapping the flask or using a scraper. Transfer the cell suspension to a 15-ml conical polypropylene centrifuge tube. Centrifuge 5 to 10 min at $2000 \times g$, 4°C.

5. Remove supernatant and resuspend cell pellet in 5 ml TBS, pH 7.5.

6. Combine the 5-ml cell suspensions from two tubes into one tube. Pellet the cells by centrifuging 5 to 10 min at 1500 to $2000 \times g$, 4°C.

7. Add 5 ml DNA extraction buffer and invert the tube several times to resuspend the cells. Incubate on a rocker platform overnight at 37°C.

   IMPORTANT NOTE: *Do not vortex the cell suspension to avoid shearing viral DNA.*

8. Add 5 ml of 25:24:1 phenol:chloroform:isoamyl alcohol and gently mix the contents of the tube by inversion, do not vortex.

9. Centrifuge 5 min at $2000 \times g$, room temperature. Remove and discard the lower (organic) phase, but leave interface.

10. Extract the aqueous phase with 5 ml chloroform and mix gently by inversion. Centrifuge 5 min at $2000 \times g$, room temperature. Transfer as much of the aqueous phase as possible to a new 15-ml tube.

    The viral DNA is present at the interface. It is extremely viscous and will enter the pipet as a visible slurry.

11. Add 10 ml −20°C isopropanol and mix well to precipitate. Spool the DNA on a heat-sealed Pasteur pipet. Remove the spooled DNA and transfer the pipet to a new 15-ml tube. Break off the pipet so that only the tip where the DNA is spooled remains in the tube. Let the DNA air dry overnight, then resuspend it in 1.0 ml water using a wide-bore pipet tip.

12. Determine the DNA concentration spectrophotometrically at 260 nm and 280 nm in a UV spectrophotometer (see APPENDIX 3D; clean preparations will have an $A_{260/280}$ ratio >1.7).

13. For Southern blot hybridization, digest 1-μg quantities of the viral DNA preparation with several diagnostic restriction endonucleases and separate the fragments by agarose gel electrophoresis (APPENDIX 3G). Blot and hybridize using the conditions described in APPENDIX 3G, except use total HSV-1 viral DNA as a probe, to determine if the viral DNA is intact.

14. Transfect cultures of permissive cells in 60-mm tissue culture dishes with 1 μg viral DNA from step 11 (see Basic Protocol 2, steps 1 to 10, *except* omit plasmid cotransfection). Carefully add 4 ml complete MEM/10% FBS. Incubate 24 hr.

15. On the day following transfection, remove the medium and add 3 ml of 1% methylcellulose. Incubate and inspect periodically for the appearance of well-defined plaques.

12

16. Once plaques have formed, aspirate the overlay and add 2 ml of 1% crystal violet staining solution. Incubate 5 min at room temperature. Aspirate the staining solution, rinse with tap water, and air dry the plates.

17. Count the number of plaques per plate and determine the number of infectious particles per microgram DNA.

   Good preparations will yield 100 to 1000 plaques per microgram of DNA.

References: Alvira et al., 1999; DeLuca et al., 1985; Krisky et al., 1998; Rasty et al., 1995; Samaniego et al., 1998

Contributors: William F. Goins, Peggy Marconi, David Krisky, Darren Wolfe, Joseph C. Glorioso, Ramesh Ramakrishnan, and David J. Fink

UNIT 12.8

Gene Delivery Using Helper Virus–Free HSV-1 Amplicon Vectors

Sets of cosmids that overlap and represent the entire HSV-1 genome can form, via homologous recombination, circular replication-competent viral genomes, which give rise to infectious virus progeny. However, if the DNA cleavage/packaging (*pac*) signals are deleted, reconstituted virus genomes are not packageable, but still provide all the helper functions required for the packaging of cotransfected amplicon DNA. The resulting stocks of packaged amplicon vectors are essentially free of contaminating helper virus (Fig. 12.8.1).

CAUTION: HSV-1 is a human pathogen. Follow biosafety level 2 practices when working with HSV-1 or vectors that are based on HSV-1. Wear safety glasses and gloves at all times.

NOTE: All solutions and equipment coming into contact with cells must be sterile, and proper sterile technique should be used accordingly.

NOTE: All cell culture incubations are performed in a humidified 37°C, 5% CO_2 incubator unless otherwise stated.

BASIC PROTOCOL

PREPARATION OF HELPER VIRUS–FREE AMPLICON STOCKS

The use of an HSV-1 amplicon vector that expresses a reporter gene is highly recommended when establishing this protocol in the laboratory. The gene for green fluorescent protein (GFP) is an ideal reporter.

Materials (see APPENDIX 1 for items with ✓)

   2-2 cells (Smith et al., 1992)
   Dulbecco's modified Eagle medium (Life Technologies) with 10% and 6% fetal bovine
      serum (DMEM/10% FBS and DMEM/6% FBS)
   G418 (Geneticin; Life Technologies)
   0.25% trypsin/0.02% EDTA (Life Technologies)
   Opti-MEM I reduced-serum medium (Life Technologies)
   *Pac*I-digested cosmid DNA of set C6Δa48Δa (see Support Protocol 1)
   HSV-1 amplicon DNA (maxiprep DNA isolated from *E. coli*)

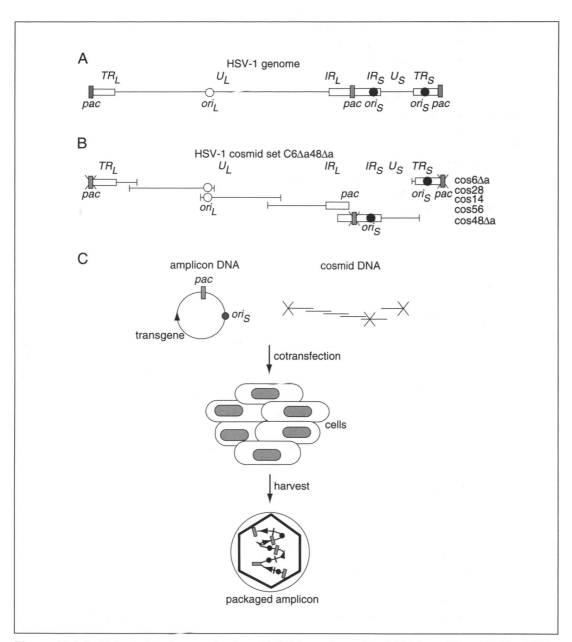

Figure 12.8.1 Helper virus-free packaging of HSV-1 amplicons into HSV-1 particles. (**A**) The HSV-1 genome (152 kbp) is composed of unique long (U_L) and unique short (U_S) segments (horizontal lines), which are flanked by inverted repeats (open rectangles): IR_S, internal repeat of the short segment; TR_S, terminal repeat of the short segment; IR_L, internal repeat of the long segment; TR_L, terminal repeat of the long segment. The origins of DNA replication, ori_S (solid circle) and ori_L (open circle), and the DNA cleavage/packaging signals, *pac* (solid rectangles), are also shown. (**B**) Schematic diagram of HSV-1 cosmid set C6Δa48Δa with deleted *pac* signals (X) (Fraefel et al., 1996; Cunningham and Davison, 1993), which includes cos6Δa, cos14, cos28, cos48Δa, and cos56. (**C**) For the helper virus-free packaging of amplicons into HSV-1 particles, cells that are susceptible for HSV-1 replication are cotransfected with amplicon DNA and DNA from cosmid set C6Δa48Δa. In the absence of the *pac* signals, this cosmid set cannot generate a packageable HSV-1 genome, but can provide all the functions required for the replication and packaging of the cotransfected amplicon DNA, which contains a functional *pac* signal. The resulting vector stocks are essentially free of helper virus.

LipofectAMINE reagent (Life Technologies)
10%, 30% and 60% (w/v) sucrose in PBS
✓ Phosphate-buffered saline (PBS)

75-cm² tissue culture flasks
Humidified 37°C, 5% CO_2 incubator
60-mm-diameter tissue culture dishes
15-ml conical centrifuge tubes
Dry ice/ethanol bath
Probe sonicator
0.45-μm syringe-tip filters (use Sarstedt polyethersulfone membrane filters)
20-ml disposable syringes
30-ml centrifuge tubes (Beckman Ultra-Clear 25 × 89–mm and 14 × 95–mm)
Sorvall SS-34 rotor
Fiber-optic illuminator
Ultracentrifuge with Beckman SW28 and SW40 rotors

1. Maintain 2-2 cells in DMEM/10% FBS containing 500 μg/ml G418. Propagate the culture two times a week by splitting ~1/5 in fresh medium (20 ml) into a new 75-cm² tissue culture flask.

2. On the day before transfection, remove culture medium, wash each plate twice with PBS, add a thin layer of trypsin/EDTA, and incubate 10 min at 37°C to allow cells to detach from plate. Count cells using a hemacytometer (see APPENDIX 3I) and plate 1.2×10^6 cells per 60-mm-diameter tissue culture dish in 3 ml DMEM/10% FBS.

3. For each 60-mm dish to be transfected, place 100 μl Opti-MEM I reduced-serum medium into each of two 15-ml conical tubes. To one tube, add 2 μg of the PacI-digested cosmid DNA mixture (0.4 μg of each of the five clones; see Support Protocol 1) and 0.6 μg amplicon DNA. To the other tube, add 12 μl LipofectAMINE.

4. Combine the contents of the two tubes. Mix well (without vortexing) and incubate 45 min at room temperature.

5. Wash the cultures prepared the day before (step 2) once with 2 ml Opti-MEM I. Add 1.1 ml Opti-MEM I to the tube from step 4 containing the DNA-LipofectAMINE transfection mixture (1.3 ml total volume). Aspirate all medium from the culture (cells should be confluent), add the transfection mixture, and incubate 5.5 hr.

6. Aspirate the transfection mixture and wash the cells three times with 2 ml Opti-MEM I. After aspirating the last wash, add 3.5 ml DMEM/6% FBS and incubate 2 to 3 days.

   Two days after transfection, at least 50% of the cells should show cytopathic effects (the cells should round up but remain stuck to the plate). The use of an amplicon vector that expresses the gene for GFP can be an invaluable tool to monitor transfection efficiency and spread of the vector during the course of the packaging experiment.

7. Scrape cells into the medium using a rubber policeman. Transfer the suspension to a 15-ml conical centrifuge tube and perform three freeze-thaw cycles using a dry ice/ethanol bath and a 37°C water bath.

8. Place the tube containing the cells into a beaker of ice water. Submerge the tip of the sonicator probe ~0.5 cm into the cell suspension and sonicate 20 sec with 20% output energy.

9. Remove cell debris by centrifugation for 10 min at $1400 \times g$, 4°C and filter the supernatant through a 0.45-μm syringe-tip filter attached to a 20-ml disposable syringe into a new 15-ml conical tube. Remove a sample for titration (see Support Protocol 2), then divide

the remaining stock into 1-ml aliquots, freeze them in a dry ice/ethanol bath, and store at −80°C (at least 6 months). Alternatively, concentrate (steps 10a and 11a) or purify and concentrate (steps 10b to 13b) the stock before storage.

Concentrate by centrifugation

10a. Transfer the vector solution from step 9 to a 30-ml centrifuge tube and spin 2 hr at 20,000 × *g*, 4°C.

11a. Resuspend the pellet in a small volume (e.g., 300 μl) of 10% sucrose. Remove a sample of the stock for titration (see Support Protocol 2), then divide into aliquots (e.g., 30 μl) and freeze in a dry ice/ethanol bath. Store at −80°C (for at least 6 months).

Purify and concentrate using a discontinuous sucrose gradient

10b. Prepare a sucrose gradient in a Beckman Ultra-Clear 25 × 89–mm centrifuge tube by layering the following solutions in the tube:

> 7 ml 60% sucrose
> 7 ml 30% sucrose
> 3 ml 10% sucrose (top layer).

11b. Carefully add the vector stock from step 9 (up to 20 ml) on top of the gradient and centrifuge 2 hr at 100,000 × *g*, 4°C.

12b. Aspirate the 10% and 30% sucrose layers from the top and collect the virus band at the interface between the 30% and 60% layers. Transfer to a Beckman Ultra-Clear 14 × 95–mm centrifuge tube, add ~15 ml PBS, and pellet virus particles for 1 hr at 100,000 × *g*, 4°C.

> *The interface between the 30% and 60% sucrose layers appears as a cloudy band when viewed with a fiber-optic illuminator.*

13b. Resuspend the pellet in a small volume (e.g., 300 μl) of 10% sucrose. Divide into aliquots (e.g., 30 μl) and freeze in a dry ice/ethanol bath. Store at −80°C (stable for at least 6 months). Before freezing, retain a sample of the stock for titration (see Support Protocol 2).

SUPPORT PROTOCOL 1

PREPARATION OF HSV-1 COSMID DNA FOR TRANSFECTION

The cosmid clones are stored at −80°C in SOB medium supplemented with 7% dimethyl sulfoxide (DMSO).

Materials *(see* APPENDIX 1 *for items with* ✓ *)*

> *E. coli* clones of HSV-1 cosmid set C6Δa48Δa, which includes cos6Δa, cos14, cos28, cos48Δa, and cos56 (see Fig. 12.8.1)
> ✓ SOB medium containing 50 μg/ml ampicillin (SOB/amp)
> Dimethylsulfoxide (DMSO)
> Plasmid Maxi Kit (Qiagen), which includes Qiagen-tip 500 columns and buffers P1, P2, P3, QBT, QC, and QF (prewarm buffer QF to 65°C)
> Isopropanol
> 70% (v/v) ethanol
> ✓ TE buffer, pH 7.5
> Restriction endonucleases *Dra*I, *Kpn*I, and *Pac*I
> High-molecular-weight DNA standard (Life Technologies)
> 1-kb DNA ladder (Life Technologies)
> Electrophoresis-grade agarose

✓ TAE electrophoresis buffer
1 mg/ml ethidium bromide in H_2O
✓ 25:24:1 (v/v) phenol/chloroform/isoamyl alcohol
24:1 (v/v) chloroform/isoamyl alcohol
100% ethanol
✓ 3 M sodium acetate, pH 5.5

17 × 100–mm graduated snap-cap tubes (e.g., Falcon 2059), sterile
Sorvall GSA and SS-34 rotors
65° and 37°C water baths
250-ml polypropylene centrifuge tubes
30-ml centrifuge tubes

1. For each of the five clones of HSV-1 cosmid set C6Δa48Δa, prepare a 17 × 100–mm sterile snap-cap tube containing 5 ml sterile SOB/amp medium. Inoculate each with a loop of frozen long-term culture of the appropriate cosmid clone. Incubate 8 hr at 37°C in a shaker.

2. Transfer 1 ml of each culture to an individual 2-liter flask containing 300 ml sterile SOB/amp, and incubate another 12 to 16 hr at 37°C, with shaking.

    It is often convenient to grow the 5-ml preculture during the day and the larger culture overnight. Do not exceed the stated incubation times, as this may increase the risk of introducing modifications into the cosmids. For the same reason, avoid colony-purifying the clones.

3. Place 1-ml aliquots of bacterial culture in cryogenic storage vials and add 70 μl DMSO to each. Mix well and freeze at −80°C for long-term storage (up to several years).

4. Pellet the remaining bacterial culture in 250-ml polypropylene centrifuge tubes by centrifugation for 10 min at 4000 × g, 4°C.

5. Decant medium and invert each tube on a paper towel for 1 to 2 min to drain all liquid. Resuspend the pellets in 15 ml buffer P1 and add 15 ml buffer P2. Mix by inverting each tube four to six times, and incubate 5 min at room temperature.

6. Add 15 ml buffer P3 and mix immediately by inverting the tubes six times. Incubate the tubes for 20 min on ice. Invert the tube once more and centrifuge 30 min at 16,000 × g, 4°C.

7. Meanwhile, equilibrate five Qiagen-tip 500 columns with 10 ml buffer QBT, and allow the columns to empty by gravity flow. Label each column with the name of the clone and drape with a small piece of a Kimwipe tissue.

8. Carefully filter the supernatants from step 6 through the tissue into the Qiagen-tip 500 columns, and allow the liquid to enter the resin by gravity flow.

9. Wash each column twice with 30 ml buffer QC, and then elute DNA with 15 ml prewarmed (65°C) buffer QF into a 30-ml centrifuge tube.

10. Precipitate the DNA with 10.5 ml (0.7 vol) isopropanol, and immediately centrifuge 30 min at 20,000 × g, 4°C.

11. Carefully remove the supernatants without disturbing the pellets from step 10 and mark the locations of the pellets on the outside of the tubes. Wash the pellets with chilled 70% ethanol and, if necessary, re-pellet at the same settings as in step 10.

12. Aspirate the supernatants completely, but avoid drying the pellets. Resuspend in 200 μl TE buffer (pH 7.5), transfer the DNA solutions to microcentrifuge tubes, and store at 4°C. After characterization of the DNA (concentration and restriction enzyme analysis; steps 13 and 14), store aliquots of 10 to 50 μg at −20°C.

13. Determine the absorbance of the DNA solutions from step 12 at 260 nm (A_{260}) and 280 nm (A_{280}) using a UV spectrophotometer.

   A value of 1.0 for A_{260} is equivalent to 50 μg/ml of double-stranded DNA. Additionally, the ratio between A_{260} and A_{280} provides information about DNA purity. Typically, pure DNA preparations have a A_{260}/A_{280} value of 1.8; do not use DNA preparations with a ratio below this value.

14. In parallel reactions, digest 2 μg of each cosmid DNA from step 12 with 10 U each of *Dra*I and *Kpn*I restriction endonuclease for 2 hr at 37°C. Separate the fragments overnight by electrophoresis on a 0.4% agarose gel at 40 V in TAE electrophoresis buffer, using high-molecular-weight DNA and 1-kb DNA ladder as size standards. Stain with ethidium bromide and compare restriction fragment patterns with those shown in Figure 12.8.2 (treat gel with care; 0.4% gels are very delicate).

15. Pool 10 μg of each of the five cosmid DNAs in a microcentrifuge tube and digest with 50 U of *Pac*I restriction endonuclease in a total volume of 100 μl for ≥3 hr at 37°C, using high-molecular-weight DNA and 1-kb DNA ladder as size standards. Confirm completion of the digest by electrophoresis of 1- to 2-μl aliquots of the reaction mixture on a 0.4% agarose gel (see Fig. 12.8.2).

   The HSV-1 inserts in the cosmids are flanked by unique Pac I restriction sites. Digestion with Pac I liberates the HSV-1 insert (35 to 40 kb) from the cosmid backbone (∼7 kb), which facilitates homologous recombination between the clones following transfection.

16. Extract DNA in the reaction mixtures, first with 100 μl (1 vol) of 25:24:1 (v/v/v) phenol/chloroform/isoamyl alcohol, and then with 100 μl (1 vol) of 24:1 (v/v) chloroform/isoamyl alcohol. Precipitate DNA by adding 250 μl (2.5 vol) 100% ethanol and 10 μl (0.1 vol) 3 M sodium acetate (pH 5.5), then incubate overnight at −20°C. Do not vortex, but gently mix by tapping with fingers to avoid damaging large DNA fragments.

   IMPORTANT NOTE: *Because the DNA will be transfected into mammalian cells (see Basic Protocol), perform the following manipulations under sterile conditions.*

17. Microcentrifuge the tubes 10 min at 20,000 × g, room temperature, carefully dispose of the supernatant, and wash the pellet once with 70% ethanol. Decant the ethanol, allow the pellets to dry for 1 min, and resuspend (with minimal pipetting) in 100 μl TE buffer (pH 7.5).

18. Measure DNA concentration as described in step 13, and store 10-μg aliquots at −20°C until required.

SUPPORT PROTOCOL 2

TITRATION OF AMPLICON STOCKS

NOTE: The titers expressed as transducing units per milliliter (t.u./ml) are relative and do not necessarily reflect numbers of infectious vector particles per milliliter. Factors influencing relative transduction efficiencies include: (1) the cells used for titration, (2) the promoter regulating the expression of the transgene, (3) the transgene, and (4) the sensitivity of the detection method.

Materials *(see APPENDIX 1 for items with ✓)*

   Vero (clone 76; ECACC #85020205), BHK (clone 21; ECACC #85011433), or 293 (ATCC #1573) cells
   DMEM (e.g., Life Technologies) supplemented with 10% and 2% FBS (DMEM/10% FBS and DMEM/2% FBS)

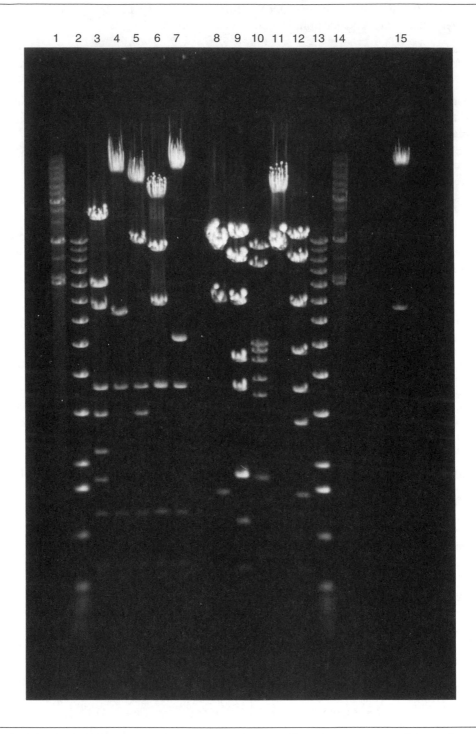

Figure 12.8.2 Analytical agarose gel showing restriction endonuclease patterns of clones from cosmid set C6△a48△a. Cosmids were either individually digested with *Dra*I (lanes 3-7) and *Kpn*I (lanes 8-12), or pooled and digested with *Pac*I (lane 15). The reaction mixtures were loaded on a 0.4% agarose gel, and the fragments were separated overnight at 40 V in TAE electrophoresis buffer and stained with ethidium bromide. Lanes 3 and 8, cos6△a; lanes 4 and 9, cos14; lanes 5 and 10, cos28; lanes 6 and 11, cos48△a; lanes 7 and 12, cos56; lanes 1 and 14, high-molecular-weight DNA standard (Life Technologies); lanes 2 and 13, 1-kb DNA ladder (Life Technologies).

✓ PBS

Samples collected from vector stocks (see Basic Protocol, steps 9, 11a, or 13b)

✓ 4% (w/v) paraformaldehyde solution, pH 7.0

✓ X-gal staining solution, GST solution, *or* appropriate primary and secondary antibodies

24-well tissue culture plates

Humidified 37°C, 5% CO_2 incubator

Inverted fluorescence microscope

Inverted light microscope

1. Plate cells (e.g., Vero 76, BHK 21, or 293 cells) at a density of 1.0×10^5 per well of a 24-well tissue culture plate in 0.5 ml DMEM/10% FBS. Incubate overnight.

2. Aspirate the medium and wash each well once with PBS. Remove PBS and add 0.1-µl, 1-µl, or 5-µl samples collected from vector stocks, diluted to 250 µl each in DMEM/2% FBS.

3. Incubate 1 to 2 days and remove the inoculum. Fix cells for 20 min at room temperature with 250 µl of 4% paraformaldehyde, pH 7.0. Wash the fixed cells three times with PBS, then proceed (depending on the transgene) with a detection protocol such as green fluorescence (step 4a), X-gal staining (step 4b), or immunocytochemical staining (steps 4c to 6c).

Detect cells expressing the gene for green fluorescent protein (GFP)

4a. Examine the culture from step 3 (before or after fixation) using an inverted fluorescence microscope. Count green fluorescent cells and determine the vector titer in t.u./ml by multiplying the number of transgene-positive cells by the dilution factor.

Detect cells expressing the E. coli lacZ gene

4b. Add 250 µl X-gal staining solution per well of the 24-well tissue culture plate from step 3, and incubate 4 to 12 hr (depending on the cell type and the promoter regulating expression of the transgene) at 37°C.

5b. Stop the staining reaction by washing the cells three times with PBS. Count blue cells using an inverted light microscope, and determine the vector titer in t.u./ml by multiplying the number of transgene-positive cells by the dilution factor.

Detect transgene-expressing cells by immunocytochemical staining

4c. Add 250 µl GST solution per well of the 24-well tissue culture plate from step 3 (to block nonspecific binding sites and to permeabilize cell membranes) and let stand 30 min at room temperature. Replace the blocking solution with the primary antibody (diluted in GST) and incubate overnight at 4°C.

   The optimal dilution of the primary antibody is determined empirically but is typically 1/100 to 1/10,000.

5c. Wash the cells three times with PBS, leaving the solution in the well for 10 min each time. Add secondary antibody (diluted in GST) and incubate at least 4 hr at room temperature.

6c. Wash the cells twice with PBS and develop according to the appropriate visualization protocol. Count transgene-positive cells using an inverted light microscope and determine the vector titer as t.u./ml by multiplying the number of the transgene-positive cells by the dilution factor.

   Conjugated secondary antibodies are commercially available. Information concerning the optimal dilution as well as the visualization protocol (which depends on the conjugation) is usually provided by the antibody supplier.

References: Cunningham and Davison, 1993; Pechan et al., 1996; Spaete and Frenkel, 1982

Contributor: Cornel Fraefel

UNIT 12.9

Liposome Vectors for In Vivo Gene Delivery

NOTE: A variety of cationic liposome formulations are commercially available for lipofection of cells in tissue culture. Optimal formulations are highly cell line–dependent and typically do not correlate with optimal formulations used for gene transfer of corresponding tissues in vivo. Thus, for in vitro transfection the reader is advised to test a variety of the commercially available liposomes following the manufacturer's instructions.

BASIC PROTOCOL

LIPOSOME PREPARATION BY THIN-FILM HYDRATION FOLLOWED BY EXTRUSION

Thin-film hydration followed by extrusion is one of many methods that can be used to prepare liposomes. This method was selected because it provides a quick and easy protocol for the preparation of a concentrated and homogeneous suspension of small unilamellar liposomes suitable for laboratory-scale animal experiments.

A cationic liposome preparation with a 3:2 molar ratio of DC-Chol/DOPE represents the optimal formulation to be used with the lipoplex for direct intratumor injection. The optimal liposome preparation to be used with the LPD lipopolyplex is composed of a 1:1 molar ratio of DOTAP/cholesterol. Both cationic liposome stock solutions are stable for several months (6 to 12) when stored at 4°C, and therefore can be prepared in advance of lipoplex and lipopolyplex production.

Materials (see APPENDIX 1 for items with ✓)

   Chloroform
✓ 2.0 mg/ml DC-Chol stock solution
   20.0 mg/ml DOPE (1,2-dioleoyl-3-phosphatidylethanolamine) in chloroform (Avanti Polar Lipids)
   20.0 mg/ml DOTAP (1,2-dioleoyl-3-trimethylammonium propane) in chloroform (Avanti Polar Lipids)
✓ 20.0 mg/ml cholesterol stock solution
✓ 5.2% (w/v) dextrose

   730.0-ml Corex glass centrifuge tube
   N_2 gas tank
   Vacuum desiccator
   Bath sonicator
   Water bath, 65°C
   LiposoFast extruder (Avestin)
   1.0-, 0.4-, and 0.1-μm polycarbonate membrane filters

1. Rinse a 30.0-ml Corex tube three times with chloroform.

2. Mix lipids for the appropriate liposome preparation in the Corex tube.

a. *For DC-Chol/DOPE liposomes (3:2 molar ratio):* Add 1.07 ml DC-Chol stock solution and 0.1 ml DOPE stock solution, and mix by swirling or vortexing.

b. *For DOTAP/cholesterol liposomes (1:1 molar ratio):* Add 1.0 ml DOTAP stock solution and 0.55 ml cholesterol stock solution, and mix by swirling or vortexing.

3. Evaporate chloroform to form a thin lipid film on the glass by blowing N_2 gas down the side of the tube while rotating it by hand.

4. Dry film to completion by vacuum desiccation for 2 to 3 hr. To prevent loss of lipid film under vacuum, cover tube with aluminum foil and poke small holes in foil with 26-G 1/2 needle.

5. Add 2.0 ml of 5.2% dextrose to film.

6. Suspend lipids in solution by vortexing several times until nearly all of the lipid film is in suspension.

   Lipids may remain on the sides of the tube, appearing as a solid white precipitate. Several quick (10- to 15-sec) bursts in a bath sonicator will help to suspend the lipids in solution.

7. Incubate suspension either 2 to 3 hr at room temperature or overnight at 4°C to allow complete hydration of lipids. Cover with parafilm. Heat lipid suspension 5 to 10 min in 65°C water bath.

8. To disperse lipid aggregates, sonicate lipid suspension in bath sonicator until all aggregates disappear, then return to water bath.

9. Place two 1.0-μm polycarbonate filters in extruder. Heat extruder to 65°C for 5 min.

10. Extrude the lipid dispersion by passing the suspension through the extruder ten times. Return extruate to water bath.

    Following extrusion the lipid suspension should be transformed from an opaque, cloudy solution to a transparent, cloudy suspension.

11. Repeat steps 9 and 10 using 0.4- and 0.1-μm polycarbonate filters sequentially to obtain small unilamellar liposomes with a mean diameter of 100 to 200 nm. Store liposomes at 4°C (up to 12 months).

    The liposomes are ready for mixing with DNA for gene delivery (CPHG UNIT 12.8).

References: Felgner et al., 1987; Gao and Huang, 1991, 1995

Contributors: Mark Whitmore, Song Li, and Leaf Huang

12

CHAPTER 13

Delivery Systems for Gene Therapy

Gene therapy is an approach to the treatment of human diseases based upon the transfer of genetic material into somatic cells of an individual. Gene transfer can be achieved directly in vivo by administration of gene-bearing viral or nonviral vectors into blood or tissues, or indirectly ex vivo through the introduction of genetic material into cells manipulated in the laboratory followed by delivery of the gene-containing cells back to the individual. In this regard, gene therapy is a set of approaches to the delivery of recombinant DNA to somatic cells, and as such is a natural progression in the application of recombinant DNA technology to human medicine. Viewed broadly, gene therapy is simply an extension of conventional medical therapies in which genetic material rather than protein is the therapeutic agent. By altering the genetic material within a cell, gene therapy may correct underlying disease pathophysiology. In principle, gene transfer should be applicable to many tissues and disease processes. It offers the potential to cure inherited disorders and/or to be used as an adjuvant to conventional therapies for many diseases for which current therapeutic approaches are ineffective or for which the prospects for effective therapies are low. However, gene therapy is still in its infancy, and its promise has not yet been fulfilled.

Two critical steps comprise gene therapy: delivery of the gene to appropriate cells, and gene maintenance and expression. Chapter 12 discusses vectors for somatic gene transfer, including naked or complexed DNA, RNA viruses (retroviruses), and DNA viruses (adenovirus, adeno-associated virus, and herpes virus). Each vector system has advantages and disadvantages that influence their selection for delivery to specific tissues. Unfortunately, none of the available vector systems is satisfactory for delivery to all tissues, and hence the specific characteristics of each vector must be considered for each particular clinical application.

Expression of transferred genes is essential to successful gene therapy. Although much may be known about the DNA sequences that direct high-level gene expression in tissue culture cells or transgenic mice, the same principles may not apply to the expression of recombinant genes in animal models of human disease. These difficulties may reflect undefined cellular mechanisms that repress virally introduced genes, a selective disadvantage of cells expressing transferred genes, a lack of appropriate positive regulatory sequences in the constructs, or other unknown factors. These and other issues must be considered when evaluating delivery and expression of transferred genes in specific tissues.

Although specific hypotheses can be tested in animal models, it is important to recognize that phenotypic differences are likely to exist between animal models of disease and human patients and that the principles of disease pathogenesis may vary between species. Nonetheless, animal models provide an important link in the development of gene therapy approaches, especially for the elucidation of disease pathophysiology and for the design of therapeutic approaches in preclinical settings.

UNIT 13.1 describes approaches for gene delivery to normal and diseased blood vessels: catheter-mediated gene delivery is discussed in several animal models of human vascular disease. Gene delivery to skeletal muscle has proven to be an important vehicle for vaccination and generation of recombinant proteins that are secreted in the circulation for treatment of systemic disorders; these protocols are discussed in UNIT 13.2. Protocols for ex vivo and in vivo gene delivery to the brain are described in UNIT 13.3. These include specialized methods to address the complexity of the central nervous system tissue and the anatomical location of different regions of the brain.

A cancer gene therapy technique, discussed in *UNIT 13.4*, is the transfer of genes for cytokines or other immunomodulatory products to cancer cells in vivo or ex vivo in an attempt to stimulate immune recognition of gene-modified and other cancer cells. Hematopoietic stem cells are an important vehicle for gene delivery to treat malignancies, systemic disorders, and possibly AIDS.

Gene delivery to the airway is described in *UNIT 13.5*, which presents the preparation and transduction of commonly used airway models including primary and polarized airway epithelial cells, human bronchial xenografts, and rodent lung.

Protocols for optimized in vivo hepatic gene transfer are described in *UNIT 13.6*. Hepatic gene transfer has proven extremely effective in animal models and is currently being evaluated in clinical trials for a variety of metabolic disorders. In rodents, a single tail vein injection of an adenoviral vector can transduce most hepatocytes in vivo. This provides a convenient model for assessing vector design as well as for evaluating the effects of specific transgenes in genetic mouse models of human disease.

Contributors: Anthony Rosenzweig and Elizabeth G. Nabel

UNIT 13.1

Gene Transfer to Arteries

STRATEGIC PLANNING

Using the protocols in this unit, vectors (viral or nonviral) or stably transduced endothelial or smooth muscle cells can be directly introduced by surgical methods into normal or injured arteries. Consideration of a suitable model for arterial gene delivery should include choice of animal species, arterial site, and vector (including regulatory sequences), choice of surgical method, and whether to use an injured or uninjured artery. The details provided here are for pig, rabbit, and mouse arteries, but they can be applied to other species. All methods described in this unit may be used for delivery of viral and nonviral vectors; the first protocol can also be used for delivery of stably transduced endothelial or smooth muscle cells.

Optimization of gene transfer is necessary for each vector and animal model. Parameters to be optimized include concentration and titer of vector, instillation time and pressure, choice of delivery catheter, adequacy of side-branch ligation, and degree of vessel injury. These features should be considered and optimized with reporter-gene vectors prior to evaluation of other biologically active constructs. A broad range of amounts of DNA and viral vectors (2- to 3-log-fold) should be evaluated during optimization. Generally, DNA doses in the 1 to 1000 μg/ml range and adenoviral vector doses in the $\sim 10^9$ to 10^{10} pfu/ml range should be considered as starting points. Reporter genes suitable for arterial gene transfer include human placental (heat-stable) alkaline phosphatase (hpAP), *E. coli* β-galacatosidase (*lacZ*), chloramphenicol acetyl transferase (CAT), or luciferase. Care should be taken to avoid background signal in arterial tissue with hpAP or *lacZ*. To assess for optimal gene expression, sacrifice of animals at 2 to 5 days would be appropriate for either viral or nonviral vectors. Gene expression might be assessed by histochemistry (for hpAP and *lacZ*), or by immunohistochemistry or biochemical methods (for CAT and luciferase).

On the day of gene delivery, final preparation of vectors or cells should be performed ~ 10 to 15 min prior to instillation into an artery—including fixing of DNA-liposome complexes, dilution of adenoviral vectors, or removal of transfected cells from tissue culture plates.

NOTE: All protocols involving live animals should meet NIH and research institute or university guidelines and should be approved by the institutional animal care and use committee (IUCAC).

NOTE: All surgical instruments must be sterile and all procedures must be performed in a sterile manner in a sterile surgical facility unless otherwise specified.

BASIC PROTOCOL 1

SURGICAL GENE DELIVERY TO PORCINE ARTERIES

The porcine iliofemoral artery is useful because it is surgically accessible and is similar in size and structure to human arteries. To transfect both left and right arteries in the same animal, the time of exposure to anesthesia can be reduced by preparing the second iliac artery and inserting the second catheter during the 20-min vector instillation in the first artery.

Materials *(see APPENDIX 1 for items with ✓)*
>   Juvenile domestic pig (3 months old; 15 to 25 kg) of either sex
>   Aspirin
>   Telazol
>   Xylazine
>   1% isoflurane
> ✓ PBS, sterile
>   Viral or nonviral vector encoding recombinant gene solution (Chapter 12) *or* transduced endothelial or smooth muscle cells in suspension (0.8 ml total volume for each artery)
>
>   Endotracheal tube (suitable for use with pig) and apparatus for delivering 1% isoflurane anesthesia
>   Standard surgical instruments
>   Balfour retractor
>   2-0 silk ties
>   Double balloon catheters (USCI Bard; one catheter for each artery)
>   Pressure transducers and hemodynamic monitor
>   1-0 vicryl sutures
>   Skin staples (e.g., Autoclip 9 mm; Becton Dickinson)

1. Give a juvenile domestic pig 10 mg aspirin per kg body weight orally 2 days prior to surgery and continue three times weekly for the duration of the study.

2. Anesthetize the pig by intramuscular injection of 6.0 mg Telazol and 2.2 mg xylazine per kg body weight. Intubate with an endotracheal tube and continue anesthesia by administering 1% isoflurane through the tube throughout the surgical procedure.

3. Prep and drape the pig and then make a midline abdominal incision extending caudally to the pubis through the skin and fascia and divide the abdominal musculature in the midline. Avoid the midline urethra in male pigs. Open the peritoneal cavity and retract the intestines cranially using a Balfour retractor.

4. Using a combination of blunt and sharp dissection, isolate each iliac and femoral artery from its cranial extent caudally to beyond the bifurcation of the femoral artery. Divide and remove overlying lymph nodes (more common in males) and adherent tissue. Dissect on the vessel surface in a direction parallel to the artery to avoid cutting side branches.

13

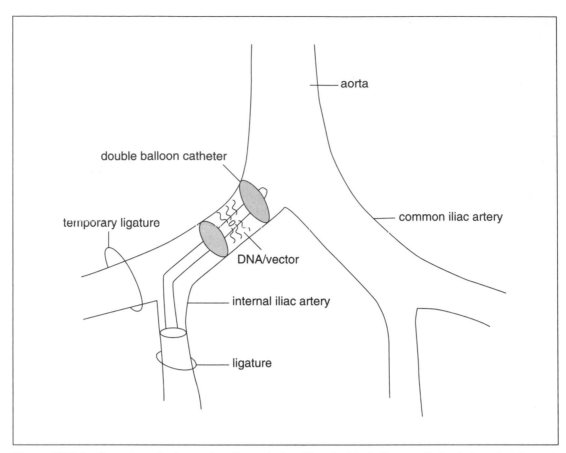

Figure 13.1.1 Gene transfer in porcine iliac arteries. The double balloon catheter is inserted through an arteriotomy in the internal iliac artery and advanced into the common iliac artery. Balloon injury with the distal balloon is followed by further advancement of the catheter so that the balloons straddle the injured area. Instillation of the gene vector follows.

5. With careful dissection, ligate all side branches throughout the course of each vessel using 2-0 silk ties. Failure to ligate all side branches will significantly decrease the chances of maintaining a steady infusion pressure during gene delivery.

6. Ligate the internal iliac artery at its most caudal point with 2-0 silk ties. Make an arteriotomy of the internal iliac artery just proximal to the ligature using fine scissors. Gently dilate the vessel by expanding the scissors. Insert a double balloon catheter (Fig. 13.1.1).

Steps 7 to 9 are for inducing arterial injury. To transfer genes to normal vessels, skip to step 10.

7. Using visual guidance, advance the catheter so as to position the distal balloon in the site to be injured. This maneuver requires straightening of the vessel with gentle traction.

8. Fill the tubing to a pressure transducer with sterile PBS. Zero the strain gauge on the pressure transducer and attach the connections to the distal balloon lumen. Inflate the distal balloon with 1 cc of air for 1 to 5 min (for mild to severe injury). The pressure gauge should read ~500 mmHg.

9. Deflate and advance the catheter such that the central portion of the catheter between the two balloons straddles the injured area of the artery.

10. Inflate the distal balloon with 1 cc of air. Flush the artery segment with 3 to 5 ml sterile PBS to remove blood contents.

11. Fill the dead space of the catheter (0.8 ml) with viral or nonviral vector solution or suspension containing transduced cells.

12. Inflate the proximal balloon with 1 cc of air.

13. Instill solution to fill the injured area of the artery between the two balloons (0.1 to 0.3 ml). Measure the instillation pressure and adjust to ~150 mmHg (mean arterial pressure of the pig is 90 mmHg) by infusing additional vector. Instill vector for 20 min, monitoring pressure during infusion.

14. Deflate proximal balloon first and then distal balloon. Remove the catheter and ligate the internal iliac artery with 2-0 silk suture. Check for restoration of arterial blood flow.

15. Close the peritoneum and the muscle with 1-0 vicryl continuous sutures. Close the fascial layer with 1-0 vicryl continuous sutures. Staple skin closed.

BASIC PROTOCOL 2

SURGICAL GENE DELIVERY TO STENTED PORCINE ILIAC ARTERIES

Materials

    3.5-mm NC Ranger angioplasty-balloon catheter mounted with stent (7-cell, 16-mm NIR stent; Boston Scientific)

1. Surgically expose and isolate the iliofemoral artery as described in Basic Protocol 1, steps 1 to 6.

2. Insert a 3.5-mm NC Ranger angioplasty-balloon catheter mounted with stent, pass it in a retrograde approach into each femoral artery (Fig. 13.1.2), and expand for 30 sec at 8 atm pressure.

3. Introduce the double balloon catheter into the stented vessel. Advance the catheter such that its central portion between the two balloons straddles the stent-deployed segment. The balloon/artery ratio should be 1.1:1.0. Do not ligate treated iliac artery during the procedure.

4. Inflate proximal balloon with 1 cc of air. Flush the arterial segment between the two balloons with 0.1 to 0.3 ml sterile PBS to remove blood contents. Inflate distal balloon, adjust pressure to 150 mmHg, and infuse 500 ml viral or nonviral vector solution slowly over 20 min.

5. Deflate the proximal balloon first and then the distal balloon and remove catheters. Check for restoration of blood flow in the treated vessel.

6. Close the peritoneum, the muscle, and the fascial layer and staple the skin as in Basic Protocol 1, step 15.

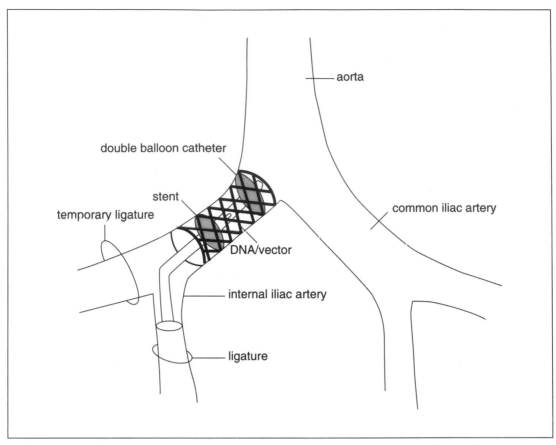

Figure 13.1.2 Gene transfer in porcine stented iliac arteries. The angioplasty-balloon catheter mounted with stent is inserted through an arteriotomy in the internal iliac artery and advanced into the common artery. Following stent deployment, gene vector is instilled by the double balloon catheter.

BASIC PROTOCOL 3

SURGICAL GENE DELIVERY TO ATHEROSCLEROTIC RABBIT ILIAC ARTERIES

In rabbits, gene delivery is performed in one of the iliac arteries to minimize distal limb ischemia.

Materials *(see* APPENDIX 1 *for items with* ✓ *)*

New Zealand white rabbits (3 to 4 kg) of either sex
Aspirin
Ketamine
Xylazine
1% isoflurane
0.5% cholesterol/2.3% peanut oil rabbit chow (Purina Test Diets)
10 U/ml heparin
✓ PBS, sterile
Viral or nonviral vector encoding recombinant gene solution (0.7 ml total volume;
    Chapter 12)

Endotracheal tube (suitable for use with rabbit) and apparatus for delivering 1% isoflurane anesthesia
Standard surgical instruments
2-0 and 4-0 silk ties
3F Fogarty balloon catheters (Baxter)
2-0 PDS suture
Rabbit collar
Balfour retractor
Surgical scalpel blade, no. 11
Human angioplasty balloon catheters (2- to 3-mm-sized balloons)
Pressure transducer and hemodynamic monitor
5-0 prolene sutures
2-0 vicryl sutures

1. Give a New Zealand white rabbit 10 mg aspirin per kg body weight orally 2 days prior to surgery and continue three times weekly for the duration of the study.

2. Sedate rabbit by intramuscular injection of 35 mg ketamine and 5 mg xylazine per kg body weight. Intubate with an endotracheal tube and maintain anesthesia by administering 1% isoflurane through the tube throughout the surgical procedure.

3. Surgically expose the distal femoral artery above the knee. Ligate the distal portion with a 2-0 silk tie. Loop a tie proximally. Create an arteriotomy proximal to the ligature and advance a deflated 3F Fogarty balloon catheter 10 to 15 cm into the aorta.

4. Inflate the balloon and withdraw the catheter slowly, feeling resistance on the inflated balloon. When the catheter reaches the arteriotomy site, deflate the balloon and repeat this process two additional times. After final withdrawal of the catheter, tie off the proximal tie.

5. Close muscle and fascia layers and skin with 2-0 PDS interrupted sutures. To prevent the rabbit from gnawing at the sutures, attach a collar to the rabbit's neck until sutures resorb.

6. On the day after surgery, begin feeding a 0.5% cholesterol/2.3% peanut oil chow and continue feeding for 3 weeks. Measure serum cholesterol and triglyceride levels before and after feeding regimen. Expected levels before and after are 54 ± 4 and 1107 ± 44 mg/dl for serum cholesterol and 36 ± 1 and 218 ± 15 mg/dl for serum triglyceride (Simari et al., 1996).

7. Sedate, intubate, and anesthetize the rabbit as in step 2. Prep and drape the rabbit and make a midline abdominal incision extending caudally to the pubis through the skin and fascia. Divide the abdominal musculature in the midline. Open the peritoneal cavity and retract the intestines cranially using a Balfour retractor.

8. Using a combination of blunt and sharp dissection, isolate the distal aorta and the injured atherosclerotic iliac artery. Divide and remove overlying lymph nodes (more common in males) and adherent tissue. Dissect on the vessel surface in a direction parallel to the artery to avoid cutting side branches.

9. With careful dissection, ligate all side branches throughout the course of the injured vessel using 4-0 silk ties. Failure to ligate all side branches will significantly decrease the chance of maintaining a steady infusion pressure during gene delivery.

10. Administer 20 U heparin per kg body weight by intravenous injection. Loop ties around the distal aorta and around each iliac artery at the bifurcation and distal to the previously injured atherosclerotic site. Temporarily secure the tie on the aorta and the noninjured artery.

13

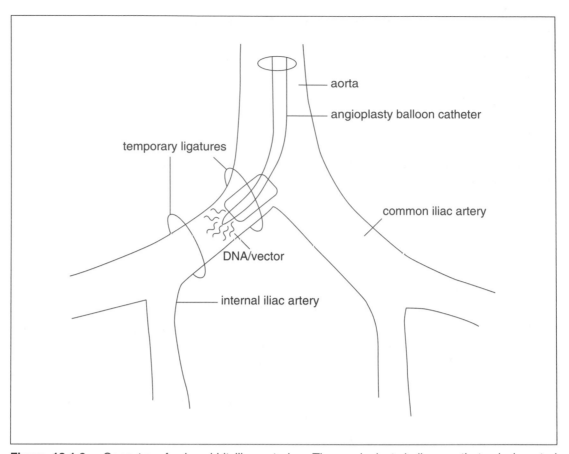

Figure 13.1.3   Gene transfer in rabbit iliac arteries. The angioplasty-balloon catheter is inserted through an arteriotomy in the aorta and advanced into the iliac artery. Following balloon injury the catheter is withdrawn to the position shown here and the gene vector is instilled into the isolated arterial segment.

11. Make an arteriotomy in the distal aorta with the tip of a no. 11 scalpel blade. Insert an appropriately sized human angioplasty balloon catheter into the aorta and advance it to the atherosclerotic site (Fig. 13.1.3). Use a larger balloon to achieve a greater degree of injury.

12. Inflate angioplasty balloon to nominal pressure (as determined by manufacturer) for 60 sec. Deflate and repeat two additional times. Vary the inflation time and pressure to achieve greater degrees of injury.

13. Withdraw balloon to a position that allows the distal catheter tip to enter the injured area and be situated within the area of the proximal iliac tie to allow for infusion of gene vector.

14. Inflate balloon catheter to a minimal pressure (1 to 2 atm). Secure catheter with the overlying temporary ligature. Flush artery segment with 3 to 5 ml sterile PBS to remove blood contents. Fill dead space of catheter with 0.7 ml of a viral or nonviral vector solution. Temporarily secure the tie distal to site of injury.

15. Instill solution to fill dead space between the ligatures (0.1 to 0.3 ml). Measure the instillation pressure using a pressure transducer. Adjust to ~150 mmHg by infusing additional vector solution. Instill vector for 20 min, monitoring pressure during infusion.

16. Release distal and then proximal ligatures. Check for restoration of arterial blood flow following release. If flow is not present, instill a heparin flush into the noninjured iliac artery and aorta and then pass the Fogarty catheter retrograde through the aorta to disrupt and remove thrombus formed during the procedure.

17. Remove balloon catheter and ligate and repair the aorta with 5-0 prolene suture. Close the peritoneum and muscle with 2-0 vicryl continuous sutures. Close the fascial layer with 2-0 vicryl interrupted sutures. Close the skin with 2-0 PDS interrupted sutures.

BASIC PROTOCOL 4

SURGICAL GENE DELIVERY TO NORMAL MURINE CAROTID ARTERIES

The anatomic locaton of the carotid artery allows temporary cessation of flow for gene delivery while collateral flow from the contralateral artery prevents ischemia. Flow can be resumed in an antegrade fashion either through the transduced artery or into the transduced artery but not through it.

Materials

Adult C57Bl/6 mice (25 to 35 g)
Ketamine
Xylazine
Normal saline
Lactated Ringers solution (Baxter Healthcare)
Viral or nonviral vector encoding recombinant gene solution (10 μl per artery; Chapter 12)
M-199 medium (Invitrogen Life Technologies), for resumption of flow through the common carotid artery only

Enrofloxacin (Bayer)
25- and 33-G needles
Surgical microdissecting microscope
Surgical scalpel blade, no. 15
Agricola microdissecting retractor
Portable cautery unit
26-mm Straight Schwartz Micro-Serrefines microvascular clamps (Roboz Surgical Instruments)
6-0 silk sutures
Microinjection syringe
7-0 prolene sutures, for resumption of flow through the common carotid artery only
Microdissecting scissors, for resumption of flow through the common carotid artery only
Silicone catheters (tubing: 0.12-in. i.d., 0.025-in. o.d.; Helix Medical), for resumption of flow through the common carotid artery only
8-0 vicryl sutures
Nexabond topical skin closure system
Warmed mouse cage

1. Anesthetize an adult C57Bl/6 mouse by intramuscular injection of 80 mg ketamine and 12 mg xylazine per kg body weight using a 25-G needle. Prep and drape the mouse. Place mouse supine under a surgical microdissecting microscope and ensure adequate visualization.

13

2. Make an anterior midline ventral neck incision from the cricoid to the manubrium with a no. 15 surgical scalpel blade. Incise the strap muscles in the midline and retract laterally with an Agricola microdissecting retractor. Through this incision, isolate the common carotid vessels with gentle blunt dissection, protecting the vagus and recurrent laryngeal nerves.

3. Electrocoagulate small side branches, if present, with a portable cautery unit ≥ 1 mm away from the carotid artery. Moisten tissues periodically with normal saline. Failure to ligate all side branches will result in leakage of the vector solution and may decrease transgene expression.

For flow into the common carotid artery

4a. Clamp the common carotid artery ~7 to 8 mm proximal to the bifurcation with a 26-mm Straight Schwartz Micro-Serrefines microvascular clamp. Apply another microvascular clamp to both the internal and external carotid arteries ~2 to 3 mm distal to the bifurcation of the common carotid artery.

5a. Loop a 6-0 silk suture around the common carotid artery 1 mm proximal to the bifurcation with the tie ready but not tightened.

6a. Make a puncture in the carotid vessel 1 mm proximal to the bifurcation using a 33-G needle. Expel blood from the clamped vessel through the puncture site by gently milking the vessel with a cotton swab dampened with Lactated Ringers solution towards the puncture site.

7a. Reinsert a 33-G needle attached to a microinjection syringe containing 10 μl of a viral or nonviral vector solution into the puncture site in the common carotid artery. Inject solution gently and very slowly into the vessel.

8a. Tighten the 6-0 silk suture placed previously and simultaneously remove the needle. (The suture and puncture should be at the same place so that tightening of the suture seals off the puncture site.) Leave vessel mildly distended with vector solution for 20 min.

9a. Remove both clamps to restore blood flow into the ligated distal carotid artery. Proceed to step 10.

For flow through the common carotid artery

4b. Clamp the common carotid artery ~7 to 8 mm proximal to the bifurcation with a 26-mm Straight Schwartz Micro-Serrefines microvascular clamp. Occlude the internal carotid artery just distal to the carotid bifurcation with a 7-0 prolene suture. Occlude the external carotid artery 2 to 3 mm distal to the bifurcation with a 7-0 prolene suture.

5b. Using microdissecting scissors, make an arteriotomy in the proximal external carotid. Introduce a silicone catheter through the arteriotomy in the proximal external carotid and advance the catheter to 1 mm proximal to the bifurcation. Flush blood from this section by injecting 20 to 30 μl M-199 medium, and then remove catheter.

6b. Introduce a second catheter attached to a microinjection syringe filled to the tip with a viral or nonviral vector solution through the arteriotomy into the common carotid artery just distal to the bifurcation. Fix catheter tip with a tie placed just proximal to the arteriotomy (this isolates a segment of the common carotid artery 3 to 4 mm in length). Infuse 4 to 5 μl vector solution into the common carotid artery. Leave artery distended for 20 min.

7b. Withdraw vector solution into the syringe and then withdraw the catheter. Briefly release the clamp on the common carotid artery to flush residual vector solution through the arteriotomy.

13

8b. Advance the ligature proximally that had been used to secure the catheter and tie off the external carotid artery at its origin.

9b. Restore blood flow by removing the ligature on the internal carotid artery and the clamp on the common carotid artery.

10. Close the subcutaneous fascial layer with 8-0 vicryl continuous sutures. Close the skin with Nexabond topical skin closure system or with 6-0 vicryl sutures.

11. Administer enrofloxacin (2.5 mg/kg) and Lactated Ringers solution (1 ml) subcutaneously postoperatively. Allow mouse to recover in a warmed cage.

BASIC PROTOCOL 5

SURGICAL GENE DELIVERY INTO INJURED MURINE FEMORAL ARTERIES

Reporter gene expression can be observed 3 days after gene delivery, whereas therapeutic gene delivery can be analyzed up to 4 weeks postinjury.

Material

Heparin
8-week-old C57Bl/6 mice (25 to 30 g) of either sex
Xylazine
Ketamine
Viral or nonviral vector encoding recombinant gene solution (Chapter 12)

37°C warm plate
Dissecting microscope (e.g., X6 to X25 stereomicroscope WILD M650; Wild Heerbrugg)
Standard surgical instruments
Guide wire (0.010-in. o.d; ACS Hi-Torque Intermediate)
Fine plastic tubing (0.010-in. i.d., 0.012-in. o.d.; VWR)
6-0 silk tie (Ethicon)
Auto-injection device (e.g., Model 55-1111; Harvard Apparatus)
Pressure transducer and hemodynamic monitor
Skin staples (e.g., Autoclip 9 mm, Becton Dickinson)

1. Administer 100 U heparin per kg body weight intravenously via tail vein to an 8-week-old C57Bl/6 mouse. Anesthetize mouse with 5 mg xylazine and 80 mg ketamine per kg body weight. Place anesthetized mouse on its back on a 37°C warm plate set up under a dissecting microscope and secure its extremities using tape.

2. Expose left femoral and saphenous arteries via a longitudinal incision on the anterolateral surface of the leg. Introduce a flexible guide wire into the saphenous artery and then insert it through up to the femoral artery ~1.5 cm (Fig. 13.1.4).

3. Pass the wire along the vessel three times under rotation and remove it. Insert a piece of fine plastic tubing containing viral or nonviral vector solution into the injured artery. Temporarily secure the proximal femoral artery with a looped 6-0 silk tie.

4. Flush the injured segment with 10 U/ml heparin and temporarily tie off the distal saphenous artery with 6-0 silk tie.

5. Instill ~0.3 ml vector solution through the tubing using an auto-injection device for 15 min and use a pressure transducer to adjust the pressure to 100 to 150 mmHg during infusion.

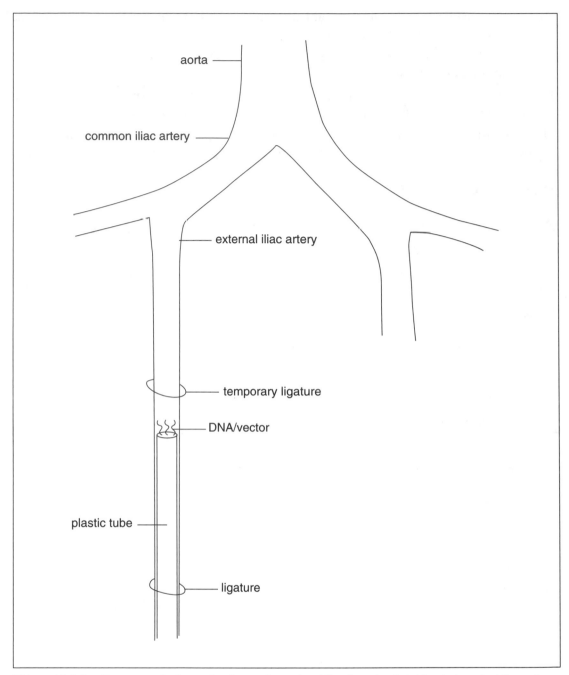

Figure 13.1.4 Gene transfer in murine femoral arteries. The fine plastic tubing is inserted through an arteriotomy in the saphenous artery and advanced into the femoral artery. Following flexible guide wire injury, the gene vector is instilled into the isolated arterial segment.

6. Release the proximal ligature and restore blood flow. Remove tubing and tie off distal artery. Check for restoration of arterial blood flow. Staple skin closed.

7. Administer 100 U heparin per kg body weight intravenously via tail vein 1 day after surgery.

References: Akyürek et al., 2000; Chang et al., 1995

Contributors: Levent M. Akyürek, Ripudamanjit Singh, Hong San, Elizabeth G. Nabel, and Robert D. Simari

UNIT 13.2

Gene Delivery to Muscle

Muscles are suitable targets for delivery of retrovirus-transduced myoblasts and DNA vaccines; both methods are described here.

NOTE: All protocols involving live animals should meet NIH and research institute or university guidelines and should be approved by the institutional animal care and use committee (IUCAC).

NOTE: All solutions and equipment coming into contact with cells must be sterile, and proper sterile technique should be used accordingly. All culture incubations are performed in a humidified 37°C, 5% CO_2 incubator unless otherwise specified.

BASIC PROTOCOL 1

ISOLATION AND GROWTH OF MOUSE PRIMARY MYOBLASTS (WITHOUT CELL SORTING)

Primary myoblasts can be isolated from mice of any age, but isolation from neonatal mice gives a greater yield of myogenic cells.

Materials (*see* APPENDIX 1 *for items with* ✓)

Neonatal mice, preferably 1 to 3 days old
70% (v/v) ethanol in squirt bottle
✓ PBS, sterile
✓ Collagenase/dispase/$CaCl_2$ solution
✓ F-10–based primary myoblast growth medium
✓ F-10/DMEM–based primary myoblast growth medium
✓ Differentiation medium (optional)

Instruments for decapitation or apparatus for CO_2 inhalation
Sharp curved surgical scissors, sterile
Stereo dissecting microscope
Fine forceps (two pairs), sterile
Tissue culture plates
Sterile razor blades
80-μm nylon mesh (e.g., Nitex; Tetko) and sterile funnel, optional
Tabletop centrifuge (e.g., IEC HN-SII)
✓ 35-, 60-, 100-, and 150-mm collagen-coated tissue culture plates
Inverted microscope with phase optics

1. Sacrifice one to five neonatal mice by decapitation or CO_2 inhalation.

2. Rinse limbs with 70% ethanol and remove them with sterile sharp curved surgical scissors. Working under a stereo dissecting microscope, dissect muscle away from skin and bone with sterile fine forceps. As successive limbs are being processed, store the muscle tissue in a tissue culture plate in a drop of sterile PBS, maintaining sterility in the accumulated tissue.

3. Add enough PBS to keep tissue moist and then mince tissue to a slurry in the culture plate using razor blades. Add ~2 ml collagenase/dispase/CaCl$_2$ solution per gram tissue (~0.5 ml for one to five pups) and continue mincing for several minutes. Carry out this and all subsequent steps in a sterile tissue culture hood.

4. Transfer minced tissue to a sterile tube and incubate at 37°C until the mixture is a fine slurry (usually ~20 min). Gently triturate with a plastic pipet several times during the incubation to break up clumps.

5. *Optional:* Filter slurry through a piece of 80-μm nylon mesh in a sterile funnel to remove large pieces of tissue. This is not necessary, and more cells may emerge from these large pieces over the next 2 days in culture.

6. Centrifuge cells 5 min at 350 × g, room temperature. Remove supernatant and resuspend pellet in 2 to 4 ml (depending on the amount of tissue being processed) F-10-based primary myoblast growth medium. Plate in a 35- or 60-mm collagen-coated culture plate.

7. Incubate, replacing medium every 2 days with fresh F-10-based primary myoblast growth medium and observing cells with an inverted microscope. Do not leave cells in same culture plate for >5 days, regardless of cell density. Do not grow at less than ~10% confluence, but do not allow overcrowding either, as cells may start to differentiate or die. Split at ≤1:5 dilution.

8. When cells are ready to be split, remove from plate by aspirating the growth medium, rinsing with room temperature PBS so that a small amount of PBS is left in the plate, then sharply hitting the vertical edge of the plate in a sideways fashion against the edge of a table top to dislodge the cells. It may be necessary to incubate in PBS for several minutes before the cells can be knocked off the plastic surface. Do not use trypsin or EDTA.

9. Preplate cells by transferring to a fresh collagen-coated plate and incubate 15 min. Swirl the liquid in the plate, then tilt the plate toward the pipet and jostle the dish from side to side to keep myoblasts in suspension while slowly removing cells. Transfer resuspended cells to a new collagen-coated plate.

10. Repeat steps 8 and 9 as needed during the initial week of culture expansion or until all fibroblasts (the very flat cells) have been eliminated from the culture, leaving behind only myoblasts (compact cells that are much smaller in diameter). Continue culturing using F-10/DMEM-based primary myoblast growth medium instead of F-10-based medium. Change this medium every 3 days instead of every 2 days, but avoid overgrowth of cells. Visually monitor cultures to ensure that fibroblasts are not evident.

    Primary myoblasts can be frozen for storage using standard cell culture protocols (APPENDIX 3I).

    After ~1 week of growth, the myoblasts usually go through a crisis period during which a significant fraction of the population dies. The remaining myoblasts soon repopulate.

11. Replace medium in an ~50% to 70% confluent test plate of myoblasts with differentiation medium. Resume incubation, changing medium daily. Examine with inverted microscope for the appearance of large multinucleated myotubes, which should be obvious within several days to 1 week. After ~1 week in differentiation medium, myotubes can sometimes be observed to twitch randomly as the contractile machinery assembles.

PURIFICATION OF PRIMARY MYOBLASTS USING CELL SORTING

Myoblasts can also be isolated from mixed cultures using cell-sorting techniques. This is the method of choice for isolation of human primary myoblasts and can be used for mouse primary myoblasts. For human myoblasts, mixed cell populations from prior enzymatic digestions are stained with the 5.1H11 antibody, which recognizes human neural cell adhesion molecules. This antibody is available as a hybridoma from the Developmental Studies Hybridoma Bank. Cells are incubated with undiluted hybridoma supernatant for 20 min, followed by 20 min in 7 μg/ml biotinylated anti-mouse IgG (Vector Laboratories) and 20 min in 10 μg/ml Texas Red–avidin (Molecular Probes), all at room temperature. Cells are then subjected to fluorescence-activated cell sorting (FACS; see Support Protocol) to isolate the stained myoblasts (Webster et al., 1988). Both FACS and magnetic isolation have been used to isolate mouse primary myoblasts using the CA5.5 antibody against mouse α7-integrin (Blanco-Bose et al., 2001). This antibody is available as ascites fluid from Sierra Biosource and may require further purification using a protein G column to give reproducible results.

BASIC PROTOCOL 2

INFECTION OF PRIMARY MYOBLASTS WITH RETROVIRUS

Transduction of the cells with more than one type of virus (e.g., a virus with a marker gene and a virus with a gene encoding a therapeutic protein) can be achieved by subjecting cells to sequential rounds of infection with each virus. A single cell can be infected by more than one virus, but the infection efficiency goes down with each successive exposure. Therefore, if one virus carries a selectable gene and the other virus does not, the cells should be infected in multiple rounds with the nonselectable virus until >99% of cells have been infected, and then they should be infected once with the selectable virus. Selection then follows.

Materials (see APPENDIX 1 for items with ✓)

    Primary myoblasts in culture (see Basic Protocol 1)
✓ F-10/DMEM-based primary myoblast growth medium
✓ 6-well or 60-mm collagen-coated tissue culture plates
    Retroviral supernatant of choice (UNIT 12.4), fresh or frozen at $-80°C$
    800 μg/ml Polybrene (hexadimethrine bromide)
    Tabletop centrifuge equipped with microplate carrier (e.g., Beckman GPR with GH-3.7 rotor), 32°C

1. Grow primary myoblasts at 37°C, 5% CO_2 in F-10/DMEM-based primary myoblast growth medium in either 6-well or 60-mm collagen-coated tissue culture plates until cells are ~10% confluent (see Basic Protocol 1). Cells should be in log phase throughout the infection. Be certain that cells have room to grow for 2 to 3 days but are not plated so sparsely that they do not grow well.

2. Supplement undiluted retroviral supernatant with 800 μg/ml Polybrene to final concentration of 8 μg/ml. Remove growth medium from cells to be infected, replace with supplemented viral supernatant. Return cells to incubator for ~15 min to equilibrate CO_2, and then carefully wrap plate in Parafilm.

3. Centrifuge plate 30 min at $1100 \times g$, 32°C. Remove viral supernatant, replace with F-10/DMEM-based primary growth medium, and return plate to incubator.

13

4. Repeat steps 2 and 3 with new viral supernatant every 8 hr, as many times as desired.

   Infection efficiency starts at ~60% to 90% after one infection and typically reaches ~100% after four infections.

5. *Optional:* If a virus carrying the *lacZ* reporter gene was used, prepare a subpopulation of the cells with the highest *lacZ* expression by sorting with FACS (see Support Protocol).

BASIC PROTOCOL 3

IMPLANTATION OF MYOBLASTS INTO SKELETAL MUSCLE

The steps described here include the authors' conditions for anesthetizing mice, but any standard method of mouse anesthesia is appropriate. Metofane is no longer produced in the United States but is available from Medical Developments, Australia. Isoflurane is another option, as are injectable anesthetics like pentobarbital or Avertin.

Immunosuppression is best avoided, but may necessary when implanting allogeneic cells or cells that produce a foreign protein. To implant syngeneic cells of unknown gender in the absence of immunosuppression, use only male recipient mice.

Materials *(see* APPENDIX 1 *for items with* ✓ *)*

   Retrovirus-infected myoblasts to be injected (see Basic Protocol 2), sorted to isolate high expression (optional; see Support Protocol)
✓ F-10/DMEM-based primary myoblast growth medium
✓ PBS, sterile
   0.5% (w/v) BSA in PBS, sterile
   Metofane (methoxyflurane)
   Mouse, immunosuppressed if needed
   70% (v/v) ethanol in squirt bottle

   Tabletop centrifuge (e.g., IEC HN-SII)
   Gauze or cheesecloth, optional
   Glass chamber with grid or divider *or* commercially available inhalation chamber, optional
   Electric clippers or sharp curved surgical scissors
   Ear punch, optional
   50-μl Hamilton syringe with metal needle adapter removed
   Mouse cage with bedding or paper towels

1. Passage retrovirus-infected myoblasts (see Basic Protocol 2) within 48 hr before implantation and give cells fresh medium 12 to 24 hr in advance of implantation.

2. Trypsinize retrovirus-infected cells (APPENDIX 3I) and resuspend in 10 ml F-10/DMEM-based primary myoblast growth medium. Count cells in a hemacytometer (APPENDIX 3I).

3. Centrifuge cells 4 min at $350 \times g$, room temperature. Remove supernatant and resuspend in 50 ml PBS. Centrifuge again. Resuspend in 1 ml PBS and transfer to a 1.5-ml microcentrifuge tube. Microcentrifuge 30 sec at 4000 rpm. Remove supernatant and resuspend pellet at 10^8 cells/ml in 0.5% BSA in PBS, breaking up as many cell clumps as possible with a pipet tip. If many mice are to be injected, trypsinize cells in batches and do not hold >1 hr on ice before injection.

4a. *To use an inhalation cone*: Make a Metofane inhalation cone in a fume hood by pressing a crumpled lint-free tissue into the bottom of a 50-ml conical screw-cap polypropylene tube and adding a few drops of Metofane onto the tissue. Hold a mouse by its tail for several seconds and then introduce the opening of the tube in front of the mouse. The mouse will ultimately climb inside to avoid being suspended. Hold it there until mouse loses consciousness. Keep mouse with its head in tube during following procedures

If this tube is kept closed, it can be used for several experiments without replacing the Metofane.

CAUTION: *This procedure should be done in a well-ventilated area. If Metofane vapors are detected, proper ventilation hood conditions should be observed.*

4b. *To use an inhalation chamber*: Prepare chamber in a fume hood by soaking a piece of gauze or cheesecloth in Metofane and placing it in a glass chamber (not plastic) under a grid or so that it is physically separated from mice that are put inside. (Metofane is a skin irritant.) Place a mouse in the inhalation chamber and close lid. It will take ~1 min for the mouse to lose consciousness. Transfer mouse to inhalation cone (step 4a) and keep its head in tube during following procedures.

5. Shave hair off of the region of the leg to be injected. Use a commercially available beard trimmer, which is very quick and effective and does not nick the skin. Alternatively, use a sharp pair of curved scissors, holding the leg taut and moving the scissors, with scissors curving upwards, against the direction of hair growth with each cut. If necessary, use an ear punch to permanently identify the mice while they are unconscious.

6. Place mouse on a piece of clean disposable bench paper. Sterilize hind legs by swabbing with a lint-free tissue soaked in 70% ethanol.

7. Gently pull the cell suspension (step 3) up and down with a micropipettor or equivalent to break up clumps. Load a 50-μl Hamilton syringe with cells as follows to avoid air bubbles: use the micropipettor to fill the plastic reservoir of the needle (~75 μl), insert the syringe into the needle barrel while holding the needle over the tube of cells (on ice) to catch the overflow, and then very gently draw the desired amount of cell suspension into the syringe.

8a. *To inject into the lateral gastrocnemius:* Turn the mouse on one side with its head away from you. Hold the foot between your thumb and ring finger, placing your index and middle fingers above and below the thigh to support the leg. Note the position of the lateral gastrocnemius, which runs between the knee and the ankle on the dorsal and lateral side of the leg. Holding the needle at a 45° angle to the skin, slowly push the needle through the skin and 1 to 2 mm deep into the muscle posterior to the shin. Gradually inject 5 μl into the muscle, wait a few seconds, then slowly withdraw the needle.

If it is desirable to achieve the maximum level of myoblast incorporation throughout the leg, for the lower leg muscles, 50 μl total of a 10^5 cells/μl suspension works well. The cell suspension can be delivered in ten 5-μl injections per leg, although the suspension may leak out through the needle wounds. Alternatively, 50 μl cell suspension can be delivered in as many small injections throughout the leg as possible, each ~1 μl in volume. This results in myoblast fusion into a greater number of host fibers, although it takes more time and there is more potential damage to the muscle.

8b. *To inject into the tibialis anterior:* Turn the mouse on its back and gently pull the leg down towards you by the foot. Note the position of the tibialis anterior, which runs between the knee and the ankle on the lateral side of the tibia. Find the kneecap and inject the cells roughly one-third of the way down from the knee to the ankle.

9. Transfer mouse to a cage with bedding or paper towels. Allow animal to recover (typically 2 to 5 min when Metofane is used). Monitor expression of recombinant gene and or marker using an appropriate technique.

13

DIRECT INJECTION OF PLASMID DNA INTO MUSCLE

The protocol for the direct injection of plasmid DNA into skeletal muscle is very similar to that for the injection of myoblasts. The procedure works well with a plasmid concentration of 1 mg/ml in PBS and an injection volume of 5 to 10 μl. Cleanliness of the DNA is important. If plasmid has been purified on adsorbent columns—e.g., Qiagen—it must be treated to remove bacterial endotoxin, which can cause inflammation at the injection site. Qiagen sells a kit that removes endotoxin as the plasmid is being purified. To achieve optimal uptake of the plasmid, the muscle can be damaged at the injection site with a small piece of dry ice ~1 week before the injection, although this entails two rounds of incision and suturing. Muscle toxins such as cardiotoxin or bupivacaine are also sometimes used to damage the muscle before plasmid injection. However, most investigators simply inject DNA solution with no prior manipulation. Exposure of injected muscle to a transient high voltage can increase the efficiency of DNA uptake by several orders of magnitude (e.g., Aihara and Miyazaki, 1998; Draghia-Akli et al., 1999; Mir et al., 1999).

SUPPORT PROTOCOL

ISOLATION OF *lacZ*-LABELED CELLS BY FLUORESCENCE-ACTIVATED CELL SORTING (FACS)

For details on FACS, see Coligan et al. (2004).

Materials (see APPENDIX 1 for items with ✓)

   200 mM fluorescein di-β-D-galactopyranoside (FDG; 100×; Molecular Probes) in 1:1
      (v/v) DMSO/water (store at −20°C in the dark)
   Cells to be sorted: *lacZ*-expressing cells (see Basic Protocol 2) and control cells not
      expressing *lacZ*
✓ FACS staining buffer, room temperature to 37°C
✓ FACS stop buffer, ice cold
✓ F-10- or F-10/DMEM-based primary myoblast growth medium
✓ Collagen-coated tissue culture plates

   Tabletop centrifuge (e.g., IEC HN-SII)
   Fluorescence-activated cell sorter (e.g., Becton Dickinson FACStar)

NOTE: Carry out preparations in parallel for *lacZ*-expressing cells and control cells not expressing *lacZ*.

1. Dilute 1 μl of 200 mM FDG in 99 μl water and prewarm in a 37°C water bath.

2. Trypsinize 1×10^4 to 5×10^6 cells to be sorted (*APPENDIX 3I*), centrifuge 4 min at $350 \times g$, room temperature, and then resuspend in 100 μl FACS staining buffer. Warm cells 10 min in a 37°C water bath and then add 100 μl prewarmed FDG. Incubate 1 min at 37°C.

3. Add 1.8 ml ice-cold FACS stop buffer and place on ice. Keep cells cold and sort as quickly as possible.

4. Set a fluorescence-activated cell sorter to detect forward scatter, propidium iodide, and fluorescein. Sort the cells, discarding propidium iodide–positive (i.e., dead) cells and any fluorescein-positive cells below the desired threshold. If the population does not form

two discernible peaks, use the control cell sample to indicate background fluorescence levels.

5. Have the sorter deposit desired cells (generally the 5% to 10% of cells that show highest expression) in a tube of F-10- or F-10/DMEM-based primary myoblast growth medium. The volume of medium into which the cells are deposited depends on the expected number of sorted cells; try to collect them at reasonably high density.

6. Plate sorted cells on collagen-coated plates and return to cell incubator.

BASIC PROTOCOL 4

DNA VACCINE ADMINISTRATION BY INTRAMUSCULAR INJECTION OF THE QUADRICPEPS MUSCLES IN THE MOUSE

Materials (*see* APPENDIX 1 *for items with* ✓)

Mouse (C57BL/6 and BALB/c strains are commonly used)
✓ Anesthetic solution
✓ DNA vaccine solution, diluted as appropriate ($\leq$2 mg/ml; Fig. 13.2.1)
70% (v/v) ethanol
0.3- and 1-ml syringes
26-G and 28-G, 0.5-in. needles
Electric clippers
Cotton balls or gauze pads

1. Anesthetize a mouse by intraperitoneal injection of 0.3 ml anesthetic solution per 18 to 20 g of total body weight, using a 1-ml syringe and a 26-G needle. This will sedate the mouse for 15 to 20 min.

2. Load a 0.3-ml syringe with a 28-G needle with sufficient DNA vaccine solution to perform the immunizations (generally 50 μl/leg or 100 μl/mouse).

3. Position mouse on its back with its head raised (to provide room to grip the skin on the back of the neck) and shave off the fur on the skin covering the quadriceps muscle on the upper thigh with electric clippers.

4. With one hand, hold mouse by the back of the neck tightly enough to allow little room for movement. With the other hand, hold the leg near the foot and ankle area and bend it at the knee, thereby positioning the quadriceps muscle for easy access. Wet the area of the quadriceps muscle lightly with a cotton ball dampened with 70% ethanol. If possible, have one person position the mouse while another performs the injection.

5. Position the quadriceps between the thumb and index finger and insert needle of syringe into the muscle at an ~45° angle to a depth of ~2 to 3 mm, with syringe pointed toward the knee. If the needle strikes the femoral bone, it has gone too far. Slowly dispense DNA vaccine solution and then withdraw needle from muscle. A swelling in the muscle should be seen or felt if the injection was done correctly.

6. Repeat injection procedure for other quadriceps.

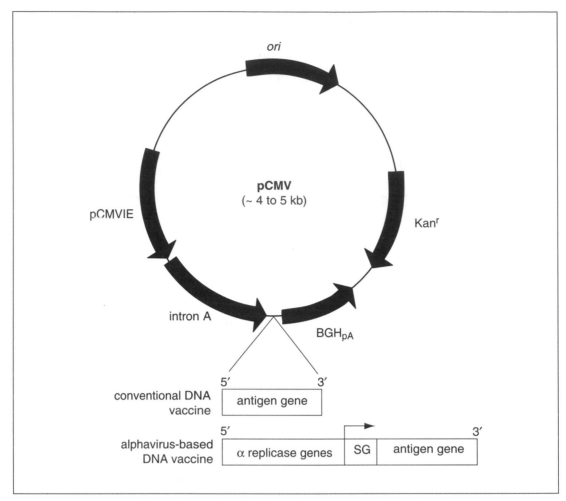

Figure 13.2.1 A typical DNA vaccine vector contains several important elements, including the promoter (human cytomegalovirus major intermediate/early promoter or pCMVIE), transcription terminator (polyadenylation signal from bovine growth hormone or BGH$_{pA}$), bacterial origin of replication (*ori*), antibiotic resistance gene (Kan^r), and antigen gene. Conventional DNA vaccines encode simply the antigen gene, whereas alphavirus-based DNA vaccines encode an alphavirus RNA replicon with a subgenomic (SG) promoter driving expression of the antigen.

ALTERNATE PROTOCOL 3

DNA VACCINE ADMINISTRATION BY INTRAMUSCULAR INJECTION OF THE ANTERIOR TIBIALIS IN THE MOUSE

Vaccination into the anterior tibialis is performed as in Basic Protocol 4, except that the area from the knee to the ankle on the outside of the leg is prepared and injected. After loading the syringe (50 µl/leg for both legs), the needle is covered with a piece of 0.38-mm-i.d. polyethylene tubing so that only the bevel of the needle is exposed. The DNA solution is injected with moderate pressure and steadiness. This method tends to be more reproducible because the depth of inoculation is controlled, ensuring delivery to the center of the muscle.

References: Aihara and Miyazaki, 1998; Amara et al., 2001; Barr and Leiden, 1991; Shiver et al., 2002; Ulmer et al., 1993; Wolff et al., 1990

Contributors: Matthew L. Springer, Thomas A. Rando, and Helen M. Blau (myoblast injection); Jeffrey Ulmer and Rino Rappuoli (DNA vaccination)

UNIT 13.3

Ex Vivo and In Vivo Gene Delivery to the Brain

An overview of the approach involved in ex vivo gene therapy is shown in Figure 13.3.1; that for in vivo gene therapy is shown in Figure 13.3.2. Established cell lines and primary cells can be used, although the former tend to form tumors upon transplantation, limiting their use.

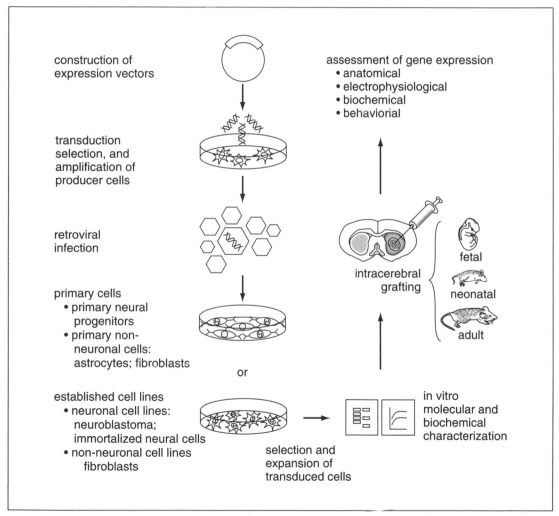

Figure 13.3.1 Ex vivo gene transfer strategy. Retroviral vectors constructed with desired transgenes are transduced into packaging cell lines. Stably transfected cells are selected and expanded, and recombinant virus is collected from the packaging cells and used to infect primary or established cell lines. After selection for the stable transfectants, cells are expanded, characterized for the expression of the transgene, and implanted into the brains of fetal, neonatal, or adult animals. The continued expression of the foreign gene in vivo can be assessed by a combination of methods listed in Table 13.3.1.

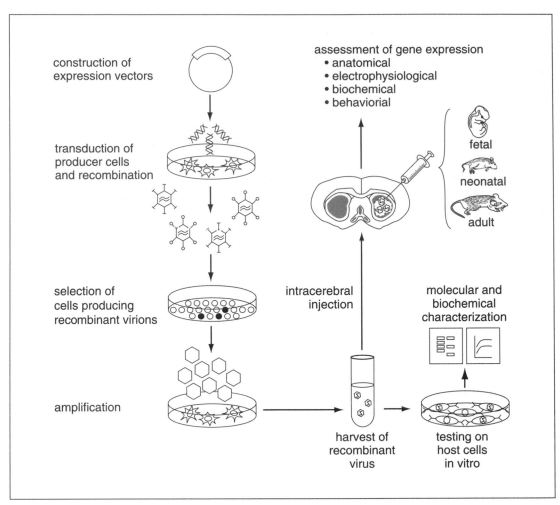

Figure 13.3.2 In vivo gene transfer strategy. Vectors are constructed with desired transgenes and used to transduce the producer cells. Stably transfected cells are selected and expanded, and recombinant virus is collected from the producer cells and directly injected into specific areas of the brain. A number of viral vectors can be used for this purpose. The continued expression of the foreign gene in vivo can be assessed by a combination of methods listed in Table 13.3.1.

Because stereotaxic coordinates will vary from one strain of rats to another, the coordinates for a particular strain should be determined in a pilot experiment: inject a small amount of dye (e.g., DiI, Molecular BioProbes) in the rat brain at particular coordinates, sacrifice the animal, and then determine the location of the injection.

The correlation between the continued production of transgenes and their efficacy in ameliorating a functional deficit in an animal model should be examined by behavioral testing, biochemical analysis of the transgene products, immunohistochemistry, in situ hybridization, or electron microscopy at appropriate time points after surgery (Table 13.3.1).

NOTE: All protocols involving live animals should meet NIH and research institute or university guidelines and should be approved by the institutional animal care and use committee (IUCAC).

Table 13.3.1 Assessment of Gene Expression by Cells Grafted into the Brain

Level	Analysis	Method
Anatomical	Survival of grafted cells	Thionine staining Immunohistochemistry
	Graft/host interaction	Electron microscopy
	Postsynaptic alteration	Autoradiography
Electrophysiological	Graft/host interaction	Stimulation-evoked activity
	Spontaneous activity of graft and host cells	In vivo recording
Biochemical	Expression of gene products	Microdialysis (for detection of secreted transgene)
		Biochemical analysis of grafted cells after removal from brain
Behavioral	Integrated/conditioned activity	Water maze Paw reaching
	Spontaneous/unconditioned activity	Hyper/hypoactivity Sensory neglect
	Drug-induced activity	Rotational asymmetry

NOTE: When injecting cells, there is a tendency for the injection needle to clog after prolonged use. Regularly rinse needle with ethanol followed by saline solution and check that cell suspension is freely ejected from needle.

NOTE: All solutions and equipment coming into contact with cells must be sterile and proper sterile technique should be used accordingly (*APPENDIX 3I*). All culture incubations are performed in a humidified 37°C, 5% CO_2 incubator (using prewarmed media) unless otherwise specified.

BASIC PROTOCOL 1

IMPLANTATION OF GENETICALLY MODIFIED CELLS INTO THE ADULT RAT BRAIN

Materials (see *APPENDIX 1* for items with ✓)

Culture of genetically modified cells (see Support Protocol)
Adult host rat
✓ Anesthetic solution
Betadine solution (10% povidone iodine; J.A. Webster)
Opthalmic ointment (Mycitracin; Upjohn)
Antibiotic powder (Neo-Predef; Upjohn)

Rat brain atlas (Paxinos and Watson, 1986)
Ear punch (Fisher) or ear tags (Harvard Bioscience)
Stereotaxic frame with electrode manipulator and Hamilton syringe clamp (Small Animal Sterotaxic, model no. 900, and 45° tip ear bars, model no. 955; David Kopf Instruments)

Surgical instruments:
     Scalpel with rounded no. 10 blades
     Bulldog clamps
     Swabs
     Medium forceps
     7-mm wound clips and wound clip applicator
Dental drill with size 2 carbide burr
26-G needles
5-μl Hamilton syringe and needles ($\leq$26-G)
Cage for postoperative recovery

1. Start cultures of genetically modified cells for grafting so that cells are 80% to 90% confluent at time of harvest. At time of surgery, prepare cell suspension (see Support Protocol) while anesthetizing and prepping animal (step 3). Resuspend suspension once every 30 to 60 min during surgical operation.

2. Referring to a rat brain atlas, determine the coordinates of the injection site(s). Assume stereotaxic frame of reference measured from the bregma for the anterior-posterior and medial-lateral coordinates, and from the dura for the vertical coordinates.

3. Anesthetize an adult host rat with intramuscular injection of 0.25 ml anesthetic solution per 160 g body weight and mark animal with an ear punch or ear tags. Shave skull, sterilize skin with Betadine solution, and apply ophthalmic ointment in eyes to protect from drying.

4. Place anesthetized rat in a stereotaxic frame. To maintain rodent brain in the correct position, insert ear bars in auditory duct, carefully lay top of upper incisors on incisor bar, and fix by tightening thumb screws on the calculated graduations. Pull mouth bar forward to exert correct tension on the incisors.

5. Using scalpel with rounded no. 10 blade, make a single incision in the skin covering the skull from a point on the midline between the eyes to a point between the ears. Separate skin flaps with bulldog clamps and then clean and dry skull surface with swabs to remove blood and connective tissue.

6. Identify the bregma. Taking coordinates from stereotaxic brain atlas and pilot experiment, drill a 1.0-mm-wide hole over the desired point into the cranium using a dental drill until dura is visible. Cut dura with point of a 26-G needle.

7. Redissociate cell suspension by gently tapping tube. Draw up required volume (1 to 3 μl; 75,000 to 300,000 cells per injection site) of cell suspension into a 5-μl Hamilton syringe mounted on an electrode manipulator on the stereotaxic frame, avoiding air bubbles.

8. Lower syringe so as to insert needle to the specific vertical distance below the dura. Slowly inject solution at $\leq$1 μl/min. When injection is completed, raise needle 1 mm, leave in place 2 min, and then gently withdraw over a 1-min period.

9. If there is more than one injection site, drill a 1.0-mm hole at the new coordinates and repeat injection procedure. Avoid injecting >10 μl at each session per rat.

10. Remove rat from stereotaxic frame, clean skull, and sprinkle open wound with antibiotic powder. Put skin flaps together, close skin incision with 7-mm wound clips, and transfer animal to a recovery cage.

13

IMPLANTATION OF GENETICALLY MODIFIED CELLS INTO THE NEONATAL RAT BRAIN

Take the following steps to reduce the risk of postoperative maternal neglect and cannibalism. Acclimatize pregnant rats to animal housing facility where work will be done several days before delivery. Do not use hyperactive mothers. Wear gloves while handling pups and use pups that have had their first postpartum feeding. Place paper towels in the cage and then remove several pups at a time. Return pups to cage as soon as possible after surgery, but only after removing all superficial blood. Keep cages away from noise and traffic and minimize cage cleaning during the first postoperative week.

Materials

Culture of genetically modified cells (see Support Protocol)
Newborn rats (P0 to P2 pups)
100% (v/v) ethanol
Antibiotic powder (Neo-Predef; Upjohn)
Surgical glue (Vetbond Tissue Adhesive no. 1469; 3M or J.A. Webster)

Stereotaxic frame with electrode manipulator and Hamilton syringe clamp (Small Animal Sterotaxic, model no. 900, and 45° tip ear bars, model no. 955; David Kopf Instruments)
Operating microscope
Surgical instruments:
> Scalpel with rounded no. 10 and sharp no. 11 blades and handles
> Microdissecting forceps
> Swabs
Neonatal rat brain atlas (Paxinos et al., 1994; Altman and Bayer, 1995)
5-μl Hamilton syringe and 26-G needles *or* glass micropipet
37°C heating pad or heat lamp

1. Start cultures of genetically modified cells for grafting so that cells are 80% to 90% confluent at time of harvest. At time of surgery, prepare cell suspension (see Support Protocol) while anesthetizing and prepping animal (step 2). Resuspend suspension once every 30 to 60 min during surgical operation.

2. Put a newborn rat pup on wet ice for 5 min to anesthetize it, and then sterilize skin around head with 100% ethanol. Place anesthetized pup in a stereotaxic frame (see Basic Protocol 1, step 4) under an operating microscope.

3. Using a scalpel with a rounded no. 10 blade, make a single incision in the skin covering the skull from a point on the midline between the eyes to a point between the ears. Retract skin with microdissecting forceps and then clean and dry skull surface with swabs to remove blood and connective tissue.

4. Identify the bregma. Taking coordinates from a neonatal rat brain atlas, identify and mark the injection site on the skull and make a 1-mm cut on the cartilaginous skull using a scalpel with a no. 11 blade.

5. Redissociate cell suspension by gently tapping tube. Draw up required volume (1 to 2 μl) into a Hamilton syringe mounted on an electrode manipulator in the stereotaxic frame, avoiding air bubbles.

13

6. Lower syringe so as to insert needle to the specific vertical distance. Inject required volume slowly over a 1-min period. When injection is completed, hold needle in place ~1 to 2 min and then raise needle slowly over a 1- to 2-min period.

7. If there is more than one injection site, repeat steps 4 to 6 at new sites.

8. Clean surgical area and sprinkle with antibiotic powder, put skin flaps together, and close skin incision with surgical glue. Revive pup on a 37°C heating pad or under a heat lamp; when active, return to its mother for care.

BASIC PROTOCOL 3

IMPLANTATION OF GENETICALLY MODIFIED CELLS INTO THE FETAL RAT BRAIN

Materials (see APPENDIX 1 for items with ✓)

    Culture of genetically modified cells (see Support Protocol)
    Timed pregnant rat (E14 and up)
✓ Anesthetic solution
    75% (v/v) ethanol
    Ophthalmic ointment (Mycitracin; Upjohn)
    Antibiotic powder (Neo-Predef; Upjohn)

    37°C hot plate
    Sterile towels
    Surgical instruments:
        Sharp scissors
        Tissue forceps
        Needle holder
        5-0 chromic catgut
        Swabs
    Optic fiber (Donlan Jenner Industries)
    Prenatal rat brain atlas (Altman and Bayer, 1995)
    5-μl Hamilton syringe and 30-G needle *or* glass micropipet
    Cage for postoperative recovery

1. Start cultures of genetically modified cells for grafting so that cells are 80% to 90% confluent at time of harvest. At time of surgery, prepare cell suspension (see Support Protocol) while anesthetizing and prepping animal (step 2). Resuspend suspension once every 30 to 60 min during surgical operation.

2. Anesthetize a timed pregnant rat with intramuscular injection of 0.25 ml anesthetic solution per 160 g body weight. Shave the abdomen and sterilize with 75% ethanol. Apply ophthalmic ointment to eyes to protect from drying. Place anesthetized rat on a 37°C hot plate covered with sterile towels.

3. Make a midline laparotomy with scissors and remove uterine horn from abdominal cavity. Transilluminate uterine sac with an optic fiber to orient anatomical structures of embryo and identify telencephalic vesicles and calvarian sutures, referring to a prenatal rat brain atlas.

4. Redissociate cell suspension by gently tapping tube. Draw up required volume (1 to 3 μl) of cell suspension into a 5-μl Hamilton syringe or glass micropipet, avoiding air bubbles.

13

5. Manually inject required volume slowly into the brain parenchyma or ventricle of embryo while a second experimenter holds embryo in place. When injection is completed, hold needle in place ~1 to 2 min and then withdraw slowly over a 1- to 2-min period.

6. If there is more than one injection site, repeat steps 4 and 5 at new sites. Then repeat injection procedure for each embryo.

7. Place uterine horn back into the abdomen. Close peritoneum and muscle layers using 5-0 chromic catgut, cover the closed muscle layer with antibiotic powder, and repair skin incision with 5-0 chromic catgut. Transfer mother to a recovery cage and allow her to recuperate and give birth.

BASIC PROTOCOL 4

IMPLANTATION OF COLLAGEN-EMBEDDED GENETICALLY MODIFIED FIBROBLASTS INTO A CAVITY IN THE ADULT RAT BRAIN

Collagen plugs are used for experiments that specifically require an aspirative lesion. The lesion cavity interrupts pathways between brain regions and the collagen plug containing cells expressing the transgene of interest is implanted in the cavity to create a physical bridge between the two regions.

Materials (see APPENDIX 1 for items with ✓)
    Culture of genetically modified fibroblasts (see Support Protocol)
✓ PBS
✓ DMEM/10% FBS medium, 37°C
    0.1 M NaOH
✓ 0.3% collagen
    Adult host rat
    Antibiotic powder (Neo-Predef; Upjohn)

    1.5-ml conical screw-cap tubes (Nalgene)
    37°C, 10% CO_2 incubator
    Dental drill with size 2 carbide burr
    27-G needle
    Modified Pasteur pipet attached to vacuum source (see step 6)
    Surgical instruments:
        Medium and fine forceps
        7-mm wound clips and wound clip applicator
    Hydrated gel foam (Upjohn)
    Cage for postoperative recovery

1. Wash a culture of genetically modified fibroblasts with PBS and trypsinize (APPENDIX 3I). Resuspend cells in 37°C DMEM/10% FBS medium. Count cells (see Support Protocol) and adjust cell density such that 162 μl contains ~1.2×10^5 cells. Divide into 162-μl aliquots in 1.5-ml conical screw-cap tubes.

2. For each aliquot, add 6.9 μl of 0.1 M NaOH and mix by pipetting solution gently up and down. If color does not change to purple, add more 0.1 M NaOH.

3. Immediately add 81 μl of 0.3% collagen and mix by pipetting up and down. Cap tube and twist briskly between palms to separate collagen plug from the sides of the tube. Loosen cap and incubate overnight in a 37°C, 10% CO_2 incubator. If transgene product is secreted from the cells, use the supernatant to quantify the amounts made.

4. Anesthetize and prepare an adult host rat for surgery (see Basic Protocol 1, steps 2 to 5).

5. Identify the bregma. Taking coordinates from the rat brain atlas, drill a 1.5-mm-wide hole over the desired point into the cranium using a dental drill. Cut dura with the point of a 27-G needle.

6. Make an aspirative lesion cavity in the desired area of the brain by gently sucking out the tissue with a modified Pasteur pipet attached to a vacuum source.

   The pipet is modified by bending the tip and making a hole at the top of the pipet. When this is attached to a vacuum source, the flow of liquid can be controlled by opening and closing the hole with the fingertip. This design allows for better control of the aspiration of the tissue during lesioning.

7. With fine forceps, remove the 2- to 3-mm^3 collagen plug containing fibroblasts from the tube, implant it into the wound, and then push plug down in place with hydrated gel foam. The plug should fill the cavity; if necessary, the plug can be trimmed to the size of the cavity, but the number of cells implanted will be unknown.

8. Remove rat from stereotaxic frame, clean skull, and cover wound with antibiotic powder. Put skin flaps together, close skin incision with wound clips, and transfer the animal to a recovery cage.

BASIC PROTOCOL 5

DIRECT INJECTION OF RECOMBINANT VIRAL VECTORS INTO THE RAT BRAIN

Materials (*see* APPENDIX 1 *for items with* ✓)

Adult rat
Recombinant virus suspension (concentrated to high titer; see Chapter 12) in PBS
   (APPENDIX 1)
Antibiotic powder (Neo-Predef; Upjohn)

Rat brain atlas (Paxinos and Watson, 1986)
Surgical instruments:
   Medium forceps
   7-mm wound clips and wound clip applicator (or sutures)
5-µl Hamilton syringe with beveled 26-G needle
Cage for postoperative recovery

1. Anesthetize adult rat, prepare for surgery, and drill hole in the skull (see Basic Protocol 1, steps 2 to 6). Injected virus can diffuse 3 to 4 mm in the brain.

2. Slowly inject 2 to 4 µl of recombinant virus suspension into the target region over a period of ~4 min with a 5-µl Hamilton syringe. When injection is completed, raise needle 1 mm, leave needle in place an additional 2 min to minimize cell diffusion up the needle track, and then gently withdraw over a 1-min period.

   Virus titers used in the authors' laboratory are 10^9 to 10^{11} virus particles/ml for adenovirus, 10^{11} virus particles/ml for adeno-associated virus, and 10^9 virus particles/ml for HIV or MoMuLV.

3. If there is more than one injection site, repeat steps 5 and 6 at new sites. Avoid injecting >10 µl at each session per rat.

4. Remove rat from stereotaxic frame, clean skull, and sprinkle open wound with antibiotic powder. Put skin flaps together, close skin incision with wound clips or sutures, and transfer animal to a recovery cage.

13

PREPARATION OF GENETICALLY MODIFIED CELLS FOR GRAFTING

Materials (see APPENDIX 1 for items with ✓)

Recombinant fibroblasts or neural progenitor cells transfected with retroviral vector
(*UNIT 12.4*)

✓ DMEM/10% FBS (for fibroblasts) or DMEM/F12/N2/FGF (for neural progenitor cells)
✓ PBS
ATV trypsin (Irvine Scientific)
✓ D-PBS
FGF-2 (Collaborative Research or R&D Systems), for neural progenitor cells only
75-cm^2 tissue culture flasks or 10-cm tissue culture plates
IEC Clinical centrifuge (Fisher) or equivalent
Coulter counter (Coulter), for fibroblasts only

For fibroblasts

1a. Grow recombinant fibroblasts in DMEM/10% FBS in 75-cm^2 flasks or 10-cm plates until cultures are 80% to 90% confluent (*APPENDIX 3I*). Do not use quiescent cells for grafting as these survive poorly in the brain.

2a. Wash cells twice, each time using 10 ml PBS. Add 1 to 2 ml ATV trypsin to the flask and let sit at room temperature for 2 to 3 min. Slap gently on the side of the flask to dislodge cells.

3a. Resuspend cells in 8 to 9 ml DMEM/10% FBS, transfer to a 15-ml centrifuge tube, and centrifuge 3 min at 1000 × *g*, room temperature. Remove supernatant. Resuspend cells in 1 ml D-PBS by pipetting up and down with a 1-ml pipet. Add 4 ml D-PBS and wash cells twice, centrifuging 3 min at 1000 × *g*. Before last centrifugation, count cells in a Coulter counter. Resuspend cells in 0.4 ml D-PBS, transfer to a 0.5-ml microcentrifuge tube, and microcentrifuge 1 min at 2000 × *g*.

4a. Remove supernatant and resuspend fibroblasts in D-PBS at a concentration of 50,000 to 100,000 cells/μl. Keep cells on ice or at room temperature. Graft cells within 3 hr.

For neural progenitor cells

1b. Grow cells in DMEM/F12/N2/FGF medium in 75-cm^2 flasks until cultures are 80% to 90% confluent (*APPENDIX 3I*).

2b. Remove culture medium, add 1 to 2 ml ATV trypsin, and dislodge cells by gentle slapping of the flask.

3b. Resuspend cells in 8 to 9 ml DMEM/F12/N2/FGF, transfer to a 15-ml centrifuge tube, and centrifuge 3 min at 1000 × *g*, room temperature. Remove supernatant. Resuspend cells in 1 ml D-PBS by pipetting up and down with a medium- to small-bore Pasteur pipet. Add 4 ml D-PBS and wash cells twice, centrifuging 3 min at 1000 × *g*. Before last centrifugation, count cells in a hemacytometer (*APPENDIX 3I*). Resuspend cells in 0.4 ml D-PBS, transfer to a 0.5-ml microcentrifuge tube, and microcentrifuge 1 min at 2000 × *g*.

4b. Remove supernatant and resuspend cells in D-PBS containing 20 ng/ml FGF-2 at a concentration of 50,000 cells/μl. Keep cells on ice or at room temperature. Graft cells within 3 hr.

References: Akli et al., 1993; Fisher et al., 1991; Kawaja et al., 1992

Contributors: Jaana Suhonen, Jasodhara Ray, Ulrike Blömer, and Fred H. Gage

Human Hematopoietic Cell Culture, Transduction, and Analyses

NOTE: All solutions and equipment coming into contact with live cells must be sterile, and proper sterile technique should be used accordingly. All culture incubations are performed in a humidified 37°C, 5% (and not higher) CO_2 incubator unless otherwise specified.

NOTE: Marrow cells are sensitive to even minor contaminants, so all lots of reagents should be tested for toxicity (see Support Protocol 5). Pretested media and individual cell culture components are also commercially available from Stem Cell Technologies.

BASIC PROTOCOL

RETROVIRAL-MEDIATED TRANSDUCTION OF CD34$^+$ CELLS WITH STROMAL SUPPORT

Retroviral-mediated transduction is described in Figure 13.4.1. The stromal support layer for transduction can be from the same donor (autologous) or a second donor (allogeneic). When transducing cells from a patient with a genetic disorder, it is imperative to use autologous stroma to avoid contamination with normal copies of the gene of interest.

Materials *(see* APPENDIX 1 *for items with* ✓ *)*

    Human bone marrow, umbilical cord blood, or mobilized peripheral blood treated with anticoagulant
✓ HBSS
✓ PBS (4° and 37°C) and 10× PBS
    Percoll (Sigma)
    Ficoll-Hypaque (Amersham Biosciences)
✓ BBMM
    Collagen-treated 75-cm^2 flasks with 0.2-µm filter-vented caps, for bone marrow only
    Anti-CD34 antibody (anti-HPCA-1 or HPCA-2, Becton Dickinson Immunocytometry; or purified anti-CD34, Ancell), filter sterilized
✓ 5% (v/v) FBS (heat-inactivated)/PBS, 4°C (see individual recipes)
    Goat anti-mouse IgG magnetic beads (e.g, Dynabeads; Dynal)
✓ Transduction medium, room temperature to 37°C
    Retroviral vector supernatant (preferably >5 × 10^6 infectious particles/ml; see Support Protocol 2)
    1 mg/ml protamine sulfate (Lyphomed)
✓ LBMM
✓ Methylcellulose medium
✓ 50 mg/ml G418 sulfate (geneticin), screened (see Support Protocol 5)
    100 mM HEPES buffer solution (Life Technologies)

    Cesium radiation source (e.g., Gammacell 1000, Nordion International)
    15- and 50-ml polypropylene centrifuge tubes
    Beckman GS-6R centrifuge and GH 3.7 rotor (or equivalent) with adaptors for 15- and 50-ml tubes, room temperature and 4°C
    Rotating platform
    25- and 75-cm^2 tissue culture flasks with 0.2-µm filter-vented caps (Costar)
    Magnetic device for immunoselection (e.g., MPC-1; Dynal)

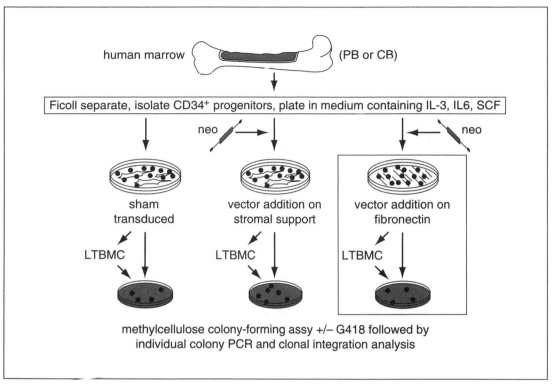

Figure 13.4.1 Transduction of hematopoietic cells. A sample of human hematopoietic cells is obtained from bone marrow, cord blood (CB), or mobilized peripheral blood (PB). Red blood cells are removed by Ficoll or Ficoll/Percoll density gradient centrifugation. CD34+ progenitors are enriched by immuno-magnetic separation and plated in medium containing the cytokines interleukin 3 (IL-3), interleukin 6 (IL-6), and stem cell factor (SCF). A support layer of irradiated allogeneic human stromal cells is used to colocalize retroviral vector particles and target cells during the transduction. For the Alternate Proto-col, shown in the box, plates coated with the carboxy-terminal domain of plasma fibronectin are used in place of stromal support. Sterile-filtered supernatant from retroviral vector–producing fibroblasts is added at 24-hr intervals for 3 days. The example shown uses the LN retroviral vector (see Fig. 12.4.2), which carries the *neo* gene and encodes resistance to the selective agent geneticin (G418). Following transduction, cells are plated in a methylcellulose-based colony-forming unit (cfu) assay with and without G418. The remaining cells are maintained in long-term bone marrow culture (LTBMC). Samples of the transduced cells are removed from the LTBMC during weekly feedings and can be used for PCR or Southern, northern, and western blot analysis. The cfu can be replated from LTBMC following 5 weeks of culture. Colonies are individually plucked from the methylcellulose medium following 14 days of growth and are then subjected to PCR for the *neo* gene and inverse PCR for clonal integration analysis.

1-ml syringe
35 × 10–mm gridded tissue culture dishes (Nunc)
35-mm and 10-cm tissue culture dishes

1. *Day 0:* Prepare primary human stromal cells, irradiate them with a cesium radiation source, and plate them (see Support Protocol 4). Always include a sham-transduced flask, which will receive medium alone (no viral supernatant).

2. *Day 1:* Obtain a bone marrow, cord blood, or mobilized peripheral blood cell sample. Dilute bone marrow or blood with an equal volume of HBSS or PBS.

3. Add 90 ml Percoll to 10 ml of $10\times$ PBS to make "100% Percoll." Add 100 ml of 100% Percoll to 39 ml HBSS to generate 72% Percoll (scale up or down as necessary).

4. Carefully layer 15 ml of 72% Percoll in the bottom of a 50-ml polypropylene centrifuge tube, followed by 5 ml Ficoll-Hypaque and 30 ml diluted cell suspension. Do not mix layers.

5. Centrifuge 15 min at $730 \times g$, room temperature. Remove two-thirds of the yellow serum layer from the top and discard. Remove the buffy coat layer (opaque mononuclear cell fraction), taking ≤ 10 ml, and transfer to a new 50-ml tube.

6. Add 40 ml HBSS or PBS to the mononuclear cells and mix. Centrifuge 5 min at $400 \times g$, room temperature. Discard supernatant and tap bottom of the tube to dislodge cells from tight cell pellet. For cells obtained from bone marrow samples, proceed with step 7; for cells obtained from cord blood or mobilized peripheral blood sample, proceed to step 9.

7. Resuspend bone marrow cell pellet in 1 ml BBMM. Count cells and assess viability using trypan blue exclusion (*APPENDIX 3I*). Dilute with BBMM to 1×10^6 cells/ml. Remove any remaining clumps.

8. Plate 15 ml cell suspension in a collagen-treated 75-cm^2 tissue culture flask with 0.2-μm filter vented cap. Put flask into an incubator and allow stromal cells to adhere for 2 hr. Remove nonadherent hematopoietic cells and place in a 50-ml centrifuge tube. Flush flask several times with 5 ml PBS and add to the 50-ml tube.

   Stromal cells are removed because they are CD34$^+$ and will invalidate transduction results. The flask and adhering cells can be used to initiate stromal monolayers (see Support Protocol 4, step 3).

9. Centrifuge stromal cell–depleted mononuclear cells 5 min at $400 \times g$, 4°C. Remove supernatant and then tap the bottom of the tube immediately to dislodge cells from the tight pellet and avoid clumping.

10. Wash mononuclear cells in 25 ml cold PBS, centrifuge 5 min at $400 \times g$, 4°C, and discard supernatant. Resuspend pellet in 1 ml cold PBS and count an aliquot by trypan blue exclusion. Remove any remaining clumps. Keep cells at 4°C until sorting is finished.

11. Add 1 ml cold PBS per 1×10^7 cells and transfer to a 15-ml centrifuge tube. Add sterile anti-CD34 antibody at 5 μg/ml cell suspension (or as determined from preliminary titration experiment). Incubate 30 min on a slowly rotating platform at 4°C.

    If small numbers of cells are used ($<1 \times 10^7$ cells total), perform antibody binding and Dynabead binding in 0.5 ml PBS in a 15-ml tube standing upright in the cold room. Tap tube to mix gently at 10-min intervals during binding. Wash steps and concentrations will remain the same.

12. Centrifuge cells 5 min at $400 \times g$, 4°C, and remove supernatant. Wash twice by resuspending in 12 ml cold 5% FBS/PBS, centrifuging, and removing the supernatant.

13. Resuspend cell pellet in cold PBS at 1×10^7 cells/ml. Add 20 μl goat anti-mouse IgG magnetic beads/ml cell suspension. Incubate 30 to 60 min on a slowly rotating platform, 4°C.

14. Add cold PBS to a final volume of 7 ml. Place tube in a magnetic device for immunoseparation. After 2 min, carefully remove PBS and transfer to a tube labeled CD34$^-$ cells. Remove tube from magnetic holder and wash CD34$^+$ cells and beads with 7 ml cold PBS. Put tube back into magnetic holder for 2 min and then discard PBS fraction.

Magnetic beads can be removed with chymopapain (see Support Protocol 3). If possible, however, leave beads on cells. Beads will drop off after 1 to 2 days of culture, and do not hamper transduction, growth of cfu, or engraftment. Chymopapain treatment lowers transduction and cleaves certain cell-surface determinants.

15. Immediately resuspend CD34$^+$ cells (typically 1% of bone marrow) and beads in 2 ml transduction medium and count cells. For each 25-cm^2 tissue culture flask of irradiated human bone marrow stromal cells (step 1), dilute 1×10^4 to 1×10^7 CD34$^+$ progenitors to 5 ml with transduction medium (1×10^6 cells/ml maximum).

16. Remove medium from irradiated stromal cells and immediately add 5 ml CD34$^+$ cell suspension. Do not let stromal layer dry out. Return flasks to incubator. Ensure that medium is warm, pH is neutral, and that the CO_2 in the incubator is 4.5% to 5.0% (not higher).

17. Thaw retroviral vector supernatant as quickly as possible. Remove flask from incubator and add 5 ml retroviral supernatant and 40 μl of 1 mg/ml protamine sulfate (40 μg/ml final). Return flask to incubator as quickly as possible and incubate 24 hr.

18. Carefully collect supernatant from transduction flask and then add 5 ml transduction medium prewarmed to room temperature (minimum) or 37°C (maximum). Return flask to incubator.

19. Centrifuge nonadherent cells in the removed supernatant 5 min at $400 \times g$, 4°C. Discard supernatant. Resuspend pellet by tapping tube with finger, so cells and beads do not clump.

20. Add 5 ml retroviral supernatant, which has just been thawed and warmed to 37°C, to the cells. Resuspend cells in the supernatant and return them to the transduction flask. Add 40 μl protamine sulfate. Return flask to incubator immediately and incubate 24 hr.

21. Repeat steps 18 to 20 once more.

22. Incubate an additional 24 hr, until cells are ready to be harvested from the transduction flask for colony-forming unit (cfu) plating assay to determine the efficiency of transduction.

    A cfu assay using G418 selection is described below, or another selective agent can be used (e.g., taxol or colchicine for the mdr gene). For vectors that lack selectable markers, single-colony PCR can be used (see Support Protocols 7 and 8).

23. Flush transduced cells from the stromal layer by drawing the medium in the flask into a 10-ml pipet and then letting the medium cascade over the adherent layer. Do not spray the stream of medium directly at stroma, or it may lift up in chunks. Transfer nonadherent cells (with medium) to a new flask. Refeed adherent layer with LBMM.

    The adherent layer will become a vigorously growing long-term bone marrow culture (LTBMC) over the course of the following week. To plate cfu from the adherent layer, if desired, see Support Protocol 6.

24. Incubate nonadherent cells ≥2 hr to deplete dissociated stromal cells by adherence. If there are many large stromal cells on the bottom of the flask, move nonadherent cells to a new flask and repeat. If there are few adherent cells, proceed to next step.

25. Collect nonadherent cells. Centrifuge 5 min at $400 \times g$, room temperature. Remove supernatant and resuspend pellet in 1 ml BBMM. Count cells and assess viability by trypan blue exclusion. Cell count is very important to the outcome of the plating, so count in quadruplicate. Calculate the volume of cells to plate: (cells required in each plate/cells per ml) $\times 1000 = $ μl cell suspension to add to each plate. If necessary, condense or dilute so that the volume is between 10 and 80 μl (critical).

13

For $CD34^+$ cells from adult bone marrow or mobilized peripheral blood and transduced ex vivo for 72 hr, plate 500 and 1000 cells per dish. For cord blood progenitors, marrow cells from infants, or freshly isolated human $CD34^+$ cells from any source (for screening or baseline cfu levels), plate 250 and 500 cells per dish.

When assaying unseparated mononuclear cell preparations, 5×10^4 cells per plate are used, and when assaying $CD34^+$ cells, 500 and 1000 cells are used (to yield 50 to 100 colonies per plate).

26. Using sterile technique, draw exactly 1 ml methylcellulose medium into a 1-ml syringe. Dispense into a 35×10–mm gridded tissue culture dish. Prepare eight gridded tissue culture dishes (four dishes at each of two cell concentrations) for each flask (transduced and sham).

27. To four dishes, add 18 μl of 50 mg/ml G418 stock. To the other four dishes, add 18 μl of 100 mM HEPES buffer solution.

28. Add the calculated volume of hematopoietic cell suspension to the appropriate dish. Mix extremely well by letting the medium pool on one side of the plate, shaking gently, and then rolling it around.

29. Place each pair of dishes into a 10-cm tissue culture dish with an extra open 35-mm dish containing 2 ml sterile water. Incubate plates 14 to 21 days in a fully humidified 37°C incubator with 4.5% to 5% CO_2.

30. Plate remainder of cells from each sample onto a second layer of irradiated stroma in 10 ml LBMM.

    Cells left over from cfu plating can be maintained in LTBMC (in addition to the plate from step 23) to allow subsequent testing of vector expression.

31. Maintain LTBMC by removing mature cells that are released into the medium and refeeding adherent layer, which contains the most primitive hematopoietic cells. During the first 2 weeks of culture, feed every 5 days by removing 75% of spent medium and nonadherent cells and replacing it with an equal volume of fresh LBMM.

    The proliferating progenitor cells adhere to or are underneath the stromal layer. During the maximal expansion period (2 to 4 weeks of culture), it is acceptable to remove 100% of the medium, gently flush the adherent layer to dislodge loosely attached cells, and then refeed. This allows harvest of maximal numbers of mature cells and does not deplete the culture of the most primitive adherent cells, which quickly replenish the flask with nonadherent progeny.

    Nonadherent cells can be harvested from both flasks to begin assays for DNA, RNA, or protein, or for immunohistochemistry within 2 weeks (see Support Protocol 9).

32. At the end of the incubation period, count and classify cfu from plates made in step 29 using the following criteria:

    Burst-Forming Unit–Erythrocyte (BFU-E): At least 3 large clusters of erythroid cells, some of which are red in color; cells in clusters are not perfectly round and seem compressed against one another; colonies may contain a few round white cells, which are erythroid progenitors, between erythroid clusters.

    Colony-Forming Unit–Granulocyte/Macrophage (CFU-GM): Colony containing ≥ 50 cells; all are white with no erythroid development; some cells may be up to three times the size of others and have clear circles within them—these are mature macrophages with lysosomes.

    Colony-Forming Unit–Granulocyte/Erythrocyte/Macrophage (CFU-GEM): Contains all types of cells described above; may also contain megakaryocytes (CFU-GEMM), which are extremely large cells with multilobed nuclei.

33. If G418 is used to quantify the extent of transfer of the *neo* gene, calculate the percentage of resistant colonies from each set of plates that had the same concentration of cells plated from a given sample, by comparing plates with and without G418 addition: (number of colonies in plates with G418/number of colonies in plates without G418) $\times$ 100 = % G418^r cfu.

ALTERNATE PROTOCOL

RETROVIRAL-MEDIATED TRANSDUCTION ON FIBRONECTIN-COATED DISHES

Additional Materials (also see Basic Protocol; see APPENDIX 1 *for items with* ✓ *)*

Recombinant fibronectin C-terminal fragment (retronectin; Bio-Whittaker) or C-terminal fragment of plasma fibronectin containing CS-1, RGD, and heparin-binding domains, lyophilized

✓ 2% (w/v) BSA: deionized 10% BSA diluted 1:5 in PBS (see individual recipes)

Cell dissociation buffer (Life Technologies)

6-well tissue culture plate or 35-mm tissue culture dishes

1. Prepare CD34$^+$ cells for transduction (see Basic Protocol, steps 2 to 15), except adjust cell density to 1.5×10^5 cells/ml.

2. Resuspend lyophilized recombinant fibronectin C-terminal fragment in PBS at 50 μg/ml. Freeze and store at $-20°C$ (avoid repeated freeze-thaw cycles) or store ≤ 1 month at 4°C without a significant loss of activity.

3. Add 2 ml fibronectin suspension to the well of a 6-well plate or a 35-mm tissue culture dish. Incubate 1 hr at room temperature. Remove fibronectin from plate and block with 2 ml of 2% BSA. Incubate 30 min at room temperature. For convenience, plates can be kept in fibronectin up to 12 hr or in BSA up to 4 hr.

4. Remove BSA and rinse plate very gently with PBS by flushing from the edge. Immediately add cells to be transduced and the retroviral supernatant. Do not let plates dry out. For each sample to be transduced, add 2 ml transduction medium containing the CD34$^+$ progenitors and 2 ml retroviral vector supernatant to one well of the 6-well plate. Incubate 24 hr.

5. Carefully remove medium and centrifuge 5 min at $400 \times g$, room temperature. Remove supernatant and tap the tube to keep cell pellet from clumping. Return cells to the well in 2 ml fresh 37°C transduction medium and 2 ml viral supernatant. Incubate 24 hr.

6. Dislodge progenitors from the fibronectin layer by treatment with cell dissociation buffer, following manufacturer's instructions. Wash, count, and plate cells in methylcellulose medium and screen for colony-forming units (see Basic Protocol, steps 25 to 33).

SUPPORT PROTOCOL 1

MAINTENANCE OF VECTOR-PRODUCING FIBROBLASTS

Grow vector-producing fibroblasts (VPF) in 75-cm^2 tissue culture flasks in 15 ml D10HG (*APPENDIX 1*). Keep cells subconfluent by passaging them using trypsin when they reach 80% confluency. Freeze cells from early passages to bank as seed stocks. If VPF are grown for weeks at a time in vitro, they must be reselected to ensure that they have retained the vector and packaging plasmids. Information about selection methods appropriate for each cell line can be obtained from ATCC.

Supernatant collected from VPF must have a titer of $\geq 5 \times 10^6$ infectious U/ml, as measured on 3T3 cells. If the titer is lower, higher-titer clones should be rederived from the initial pool of packaging cells (*UNIT 12.4*).

SUPPORT PROTOCOL 2

COLLECTION OF CELL-FREE SUPERNATANT FOR TRANSDUCTION

Materials (see APPENDIX 1 for items with ✓)
  Culture of vector-producing fibroblasts (VPF; see Support Protocol 1)
✓ D10HG medium, 32°C
  Trypsin/versene solution (Bio-Whittaker)

  75-cm^2 tissue culture flasks with 0.2-μm filter-vented caps (Costar)
  32°C humidified, 4.5% to 5.0% CO_2 incubator
  0.45-μm Uniflo-25 syringe filter with calcium acetate membrane (Schleicher & Schuell)
  10-ml polypropylene tubes

1. Trypsinize a culture of VPF (*APPENDIX 3I*). Plate new cultures at a concentration of 10^6 cells/75-cm^2 tissue culture flask in 15 ml prewarmed D10HG. Lay flask on its side and incubate 1 to 2 days at 37°C until cells are almost confluent.

2. Replace medium in each flask with prewarmed D10HG. Transfer cultures to a 32°C incubator for production of supernatant and incubate 48 hr.

3. Collect supernatant. Discard confluent monolayers of VPF. Filter supernatant through a 0.45-μm Uniflo-25 syringe filter with a calcium acetate membrane. Store supernatant in 10-ml polypropylene tubes at −70°C in 5-ml aliquots until needed for transduction.

SUPPORT PROTOCOL 3

ENZYMATIC REMOVAL OF MAGNETIC BEADS FROM IMMUNOMAGNETICALLY SELECTED CELLS

The CD33 and CD14 epitopes are cleaved by chymopapain, so subsequent FACS analysis based on those epitopes is not possible. However, the CD2, CD3, CD34 (HPCA-2 and 8G12), and CD45 epitopes remain.

The percentage of purity obtained in the CD34 isolation can be obtained by using this technique on a small fraction of the sorted cells, and then staining the cells with HPCA-2 antibody (Becton Dickinson Immunocytometry) and analyzing by FACS.

NOTE: Cells that will be used for transduction should not be treated with chymopapain to remove magnetic beads.

Materials
  Immunomagnetically selected CD34$^+$ cells (see Basic Protocol)
  RPMI medium containing 1% (w/v) human serum albumin (RPMI/1% HSA)
  2500 U/ml chymopapain (Chymodactin, Baxter)
  Beckman GS-6R centrifuge and GH 3.7 rotor (or equivalent)

1. Resuspend immunomagnetically selected CD34$^+$ cells in RPMI/1% HSA. Centrifuge 5 min at $400 \times g$, room temperature. Repeat twice more.

2. Resuspend cells in 1.8 ml RPMI/1% HSA and add 200 μl of 2500 U/ml chymopapain (500 U total). Incubate 15 min at room temperature. Wash cells twice as in step 1.

3. Resuspend cells in a medium appropriate for the method of analysis.

SUPPORT PROTOCOL 4

ESTABLISHING PRIMARY HUMAN MARROW STROMAL MONOLAYERS FROM HARVESTED BONE MARROW

Materials (see APPENDIX 1 for items with ✓)

Screens used to filter harvested bone marrow or bone marrow aspirate
✓ DOM
✓ PBS
Trypsin/versene solution (Bio-Whittaker)
✓ HBSS
✓ D10HG medium
✓ BBMM

25- and 75-cm² tissue culture flasks with 0.2-μm filter-vented caps (Costar)
Cesium radiation source (e.g., Gammacell 1000, Nordion International)

1. Collect bony spicules from screen used to filter harvested bone marrow and let spicules settle by gravity sedimentation.

   If screens are unavailable, isolate mononuclear fractions by Ficoll-Hypaque/Percoll gradient centrifugation (see Basic Protocol, steps 2 to 8) and use these starting with step 3 below. Freeze nonadherent hematopoietic cells (see Support Protocol 10).

2. Plate spicules in 75-cm² tissue culture flasks with 15 ml DOM. Incubate at 37°C.

3. Remove 100% of nonadherent cells 2 to 12 hr after plating and discard. Add 15 ml DOM to adherent layer and return flask to incubator.

4. When the stromal cells reach 80% confluency, remove medium from flask and discard. Rinse flask with 15 ml PBS and discard. Add 2 ml trypsin/versene solution and tip flask back and forth gently to completely coat adherent layer. Remove excess trypsin, leaving ~500 μl in the flask. Incubate 10 to 15 min at 37°C, turning the flask every 3 to 4 min to coat all surfaces.

5. Resuspend cells from each flask in 45 ml DOM. Transfer 15 ml to each of three new 75-cm² tissue culture flasks. Discard original flask. Incubate stromal cells until they reach 80% confluency and passage once again.

6. Repeat steps 4 and 5 once more to generate passage 3 stroma. If monolayer is not a smooth, homogeneous stromal cell population, repeat passage up to three more times.

7. Wash cells thoroughly with HBSS to remove horse serum, trypsinize cells from one 80% confluent flask of passage 3 to 6 stromal cells, and resuspend cells in 10 ml D10HG medium.

8. Irradiate cells in suspension with 2000 Rad (20 Gy). Plate irradiated stromal cells from each flask into six 25-cm² filter vent cap flasks (each will contain ~2 × 10⁵ cells) in BBMM. Use stromal cells for transduction (see Basic Protocol, step 1) or as a feeder layer for long-term bone marrow cultures (see Basic Protocol, step 30). Stromal layers can be used for up to 2 weeks postirradiation.

SUPPORT PROTOCOL 5

SCREENING MEDIA AND INDIVIDUAL COMPONENTS

Many of the reagents used in human marrow culture can display nonspecific toxicity for hematopoietic cells, although they may be acceptable for growing other cell types. Therefore, even tissue culture–grade ingredients should be screened. Screenings are done for the methylcellulose stocks, FBS, BSA, and G418.

Materials (*see* APPENDIX 1 *for items with* ✓)

 Different lots or preparations of component to be tested
✓ Methylcellulose medium, prepared without component to be tested
 CD34$^+$ progenitor cells (see Basic Protocol, steps 10 to 15)
 35 × 10–mm gridded tissue culture dishes (Nunc)

1. Add an appropriate amount of component to be tested from different lots or preparations into an appropriate volume of methylcellulose medium to achieve the desired final concentration. Do not include G418 or other selection component unless cells transduced with the appropriate vector are used for screening.

2. Mix sample well and then dispense 1-ml aliquots (in duplicate or quadruplicate) into 35 × 10–mm gridded tissue culture dishes. Add 500 to 1000 CD34$^+$ progenitor cells to each plate and mix well. Incubate 14 days.

3. Count colony-forming units (cfu; see Basic Protocol, step 32). When screening samples of G418, compare the growth of hematopoietic progenitors that have been transduced by a vector carrying the *neo* gene with that of nontransduced or sham-transduced cells.

4. Select the lot or preparation of component that supports growth of the largest number of mixed-lineage colonies. Select the lot of G418 that permits maximal cfu formation from the transduced cell population but provides complete killing of the nontransduced cells.

SUPPORT PROTOCOL 6

PLATING COLONY-FORMING CELLS FROM CULTURES

It is important not to contaminate the colony-forming assay with stromal cells. Even though irradiated, the stromal cells may be able to divide slowly in the methylcellulose medium.

Materials (*see* APPENDIX 1 *for items with* ✓)

 Long-term bone marrow culture (LTBMC), 5 weeks after plating (see Basic Protocol, step 31) *or* posttransduction cultures (see Basic Protocol, step 23)
 Trypsin/versene solution (Bio-Whittaker)
✓ LBMM
✓ Methylcellulose medium

 Magnetic device for immunoselection (e.g., MPC-1; Dynal)
 25-cm^2 tissue culture flask with 0.2-μm filter-vented cap (Costar)
 Beckman GS-6R centrifuge and GH 3.7 rotor (or equivalent)
 35 × 10-mm gridded tissue culture dishes (Nunc)

1. Trypsinize stromal layer from a LTBMC 5 weeks after plating using trypsin/versene solution. Vigorously pipet collected cells up and down to break up the layers and clumps. If Dynabeads are present inside macrophages (phagocytosed), remove with magnetic device and discard.

2. Plate cells in fresh LBMM in a clean 25-cm^2 tissue culture flask. Incubate 1 to 2 hr. Gently rinse adherent layer with the medium (do not spray medium at the adherent cells) to collect the nonadherent cells. Transfer medium to a clean 25-cm^2 tissue culture flask and incubate 1 to 2 hr. Discard first flask.

3. Collect nonadherent cells by gentle flushing. Transfer medium to a 15-ml centrifuge tube and centrifuge 5 min at 400 × g, room temperature. Resuspend pellet in 1 ml LBMM and count cells (APPENDIX 3I). Plate 25,000 and 50,000 cells per 35 ×10–mm gridded tissue culture dish in 1 ml methylcellulose medium with appropriate supplements (see Basic Protocol, steps 25 to 28).

> *If the cell number is too low to plate the specified numbers of cells, the stromal layer may have been too confluent (so culture nutrients were depleted), feeding may have been too infrequent (when deprived of nutrients, the default pathway for progenitors is differentiation to macrophages), or IL-3 may have been present throughout the 5 weeks of culture, driving early proliferation and terminal differentiation of progenitor cells. It is best to add IL-3 only during transduction.*

SUPPORT PROTOCOL 7

PREPARING WHOLE-CELL LYSATES OF INDIVIDUAL CFU COLONIES FOR PCR

Although the whole-cell lysis procedure is quicker and more convenient, PCR from extracted DNA (see Support Protocol 8) gives more reproducible results. In the whole-cell lysate, ionic strength will vary depending on cell number in the original colony, and this can affect the outcome of the PCR, which is sensitive to cation concentration.

To verify that each cfu has arisen from a hematopoietic progenitor bearing a unique proviral integrant, use the inverse PCR technique described in Support Protocol 8.

Materials (see APPENDIX 1 for items with ✓)

    Cultures of cfu in methylcellulose medium (see Basic Protocol, step 32)
✓ PBS
✓ RBC lysis buffer, for cfu containing developing erythroblasts only
✓ Whole-cell lysis buffer
    10 mg/ml proteinase K (Sigma) in H$_2$O

    Inverted phase-contrast microscope
    20- to 200-μl plugged pipet tips, sterile
    Boiling water bath

1. Place a culture dish of cfu in methylcellulose medium on the stage of an inverted phase-contrast microscope. Using a sterile 20- to 200-μl plugged pipet tip, isolate single colonies containing ≥200 cells from methylcellulose medium in a volume of 20 to 40 μl. Use a clean pipet tip for each colony. Keep a record of those colonies that contain erythrocytes (they will need a different buffer in step 4).

> *If the colonies have been grown from a single cfu deposited by an automated cell deposition unit (ACDU) of a FACSVantage (Becton Dickinson Immunocytometry) into a 96-well plate in methylcellulose medium, the cells may be flushed out with PBS. This method provides more convincing isolation of single colonies without the possibility of cross-contamination by neighboring cfu.*

> *If individual cells are cultured in 96-well plates for isolation of individual clones from methylcellulose medium, fill outer rows of wells with water to provide extra humidity.*

2. Once the colony is in the pipet tip, flush it very well with 1 ml PBS into a 1.5-ml microcentrifuge tube. Incubate 1 hr at room temperature to dissolve methylcellulose.

13

3. Microcentrifuge 5 min at 10,000 rpm, room temperature. Carefully remove supernatant. Microcentrifuge again briefly. A tiny triangular pellet should be visible. Carefully remove liquid; it is acceptable to leave ≤10 μl PBS on the pellet to avoid cell loss. Proceed with lysis or store pellet at −20°C.

4. If the colony contained developing erythroblasts (BFU-E or CFU-MIX; see Basic Protocol, step 32), add 200 μl RBC lysis buffer to pellet and mix cells with a pipet tip. Microcentrifuge 2 min at 3000 rpm. If pellet is still red, repeat lysis procedure. When pellet is white, wash once with PBS.

5. Resuspend colony pellets in 20 μl whole-cell lysis buffer for standard PCR. Add 2 μl proteinase K. Incubate tube 1 hr at 56°C. If the colonies are large (e.g., ≥2000 cells), add an additional 2 μl proteinase K after 30 min and continue digestion.

6. Boil lysates 10 min. Store whole-cell lysates at −20°C until they are used for PCR (generally 1 to 2 μl of lysate per reaction).

SUPPORT PROTOCOL 8

ANALYSIS OF CLONAL INTEGRATION IN INDIVIDUAL COLONIES

Use of the inverse polymerase chain reaction (PCR) to determine clonal integration pattern is shown in Figure 13.4.2 (also see Nolta et al., 1996).

Materials (see APPENDIX 1 for items with ✓)

Individual well-isolated colonies (see Support Protocol 7)
✓ Proteinase K digestion buffer
  10 mg/ml proteinase K (Sigma) in H_2O
✓ 25:24:1 (v/v/v) phenol/chloroform/isoamyl alcohol
  20 mg/ml glycogen (Boehringer Mannheim)
✓ 10 M ammonium acetate
  100% and 70% (v/v) ethanol
✓ TE buffer, pH 7.5
  0.1 M spermidine (Sigma)
  React 2 buffer (Life Technologies)
  5 U/μl TaqI restriction enzyme (Life Technologies)
  5 U/μl T4 DNA ligase and 5× buffer (Life Technologies)
  50 pmol/μl PCR primers (Operon):

> INVa: 5′-AGGAACTGCTTACCAACA-3′
> INVb: 5′-CTGTTCCTTGGGGAGGGT-3′
> INVc: 5′-TCCTGACCTTGATCTGA-3′
> INVd: 5′-CTGAGTGATTGACTACC-3′

✓ 10× PCR amplification buffer (also from Perkin-Elmer)
✓ 25 mM $MgCl_2$
✓ 2 mM 4dNTP mix
  5 U/μl AmpliTaq DNA polymerase (Perkin-Elmer)
  SeaKem LE agarose (FMC Bioproducts)
  NuSieve agarose (FMC Bioproducts)
  1×10^7 cpm [^{32}P]ATP end-labeled oligonucleotide probe (APPENDIX 3E) specific for LTR
    sequences: 5′-GGCAAGCTAGCTTAAGT-3′
  Circumvent Thermal Cycle Sequencing Kit (New England Biolabs)

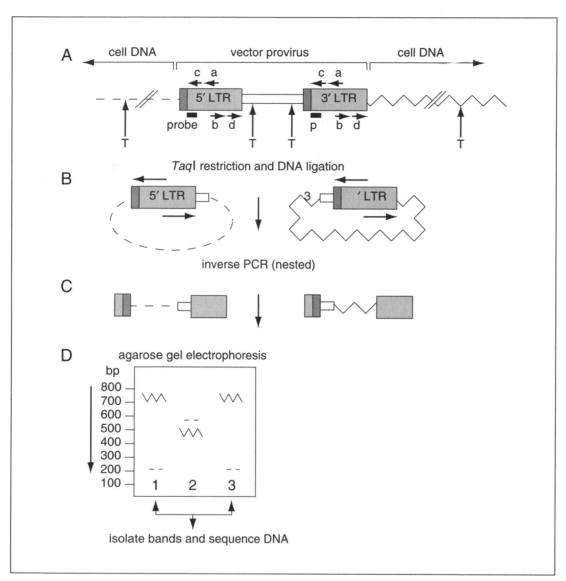

Figure 13.4.2 Inverse PCR to determine clonal integration pattern. Genomic DNA is isolated from individual colonies grown in methylcellulose medium. (**A**) The DNA is digested with *Taq*I, which cuts (T) once in the vector internal to each long terminal repeat (LTR) and again in the human cellular DNA flanking the integrated vector at the nearest *Taq*I restriction enzyme recognition sequence. Two fragments of unique length are obtained per proviral integrant, each containing one intact LTR and a human DNA segment of variable size. (**B**) The cut DNA segments are self-ligated. (**C**) Two rounds of nested PCR are performed, initially amplifying outward from the ends of the LTR into the flanking cellular DNA using primers INVa (a) and INVb (b) for the first round, and INVc (c) and INVd (d) for the second round. (**D**) The PCR products are separated by electrophoresis in an agarose gel, and bands with similar electrophoretic mobility are isolated and sequenced to confirm identity.

12° to 15°, 56°, and 65°C water baths (or thermal cycler programmed accordingly)
Thermal cycler and appropriate PCR tubes
Nylon membrane (e.g., Biotrace; Pall Biodyne)
Scalpel, sterile
Spin-X columns (Costar)

1. Collect individual, well-isolated cfu. Pellet cells and lyse red blood cells (see Support Protocol 7, steps 1 to 4). Lyse individual pellets (100 to 1000 cells) in 200 μl proteinase K digestion buffer containing 10 μg proteinase K. Incubate 2 hr at 56°C.

2. Extract lysate with 200 μl of 25:24:1 phenol/chloroform/isoamyl alcohol. Add 2 μg glycogen, 18 μl of 10 M ammonium acetate, and 500 μl of 100% ethanol. Incubate ≥2 hr at −20°C or 10 min in a dry ice/ethanol bath.

3. Microcentrifuge 5 min at 10,000 rpm. Remove supernatant and rinse pellet in 70% ethanol. Dry pellet on benchtop 1 to 2 hr (do not use a vacuum) and resuspend in 25 μl TE buffer.

4. Prepare *Taq*I digestion mixture:

   3 μl 0.1 M spermidine
   10 μl React 2 buffer
   2 μl 5 U/μl *Taq*I restriction enzyme
   75 μl H$_2$O.

   Add 10 μl resuspended pellet and incubate 2 hr at 65°C. After the first hour of incubation, add an additional 2 μl *Taq*I.

5. To 8 μl digested DNA add 2 μl of 5× T4 DNA ligase buffer. Heat 14 min at 56°C. Let stand 15 min to cool. Add 1 μl of 5 U/μl T4 DNA ligase and incubate 2 hr at 12° to 15°C.

6. Mix equal volumes of 50 pmol/μl INVa and INVb primers (INVa + INVb mix). Mix 2 vol of 10× PCR amplification buffer with 3 vol of 25 mM MgCl$_2$ (PCR buffer master mix).

7. Prepare first-round PCR reaction mix (for multiple reactions, multiply by $n + 1$):

   5 μl PCR buffer master mix
   5 μl 2 mM 4dNTP mix (0.2 mM each dNTP final)
   0.5 μl 5 U/μl AmpliTaq DNA polymerase
   1 μl INVa + INVb mix.

8. Add 11.5 μl first-round PCR reaction mix to a PCR tube. Add 36.5 μl water and 2 μl circularized DNA (step 5) to tube and overlay with 50 μl mineral oil. Amplify in a thermal cycler using the following program:

First cycle:	5 min	95°C	(denaturation)
	2 min	50°C	(annealing)
	4 min	72°C	(extension)
29 cycles:	1 min	95°C	(denaturation)
	2 min	50°C	(annealing)
	4 min	72°C	(extension).

9. Mix equal volumes of 50 pmol/μl INVc and INVd (INVc + INVd mix). Perform second-round nested PCR using the same reaction mixture except use 2 μl first-round PCR product as template and 1 μl INVc + INVd mix as primer.

10. Electrophorese resulting PCR products (15 μl per lane) on a 2% (w/v) gel (1% SeaKem LE agarose and 1% NuSieve agarose), transfer gel bands to a nylon membrane, and hybridize (*APPENDIX 3G*) with a [^{32}P]ATP end-labeled oligonucleotide probe specific for LTR sequences.

11. To verify the identity of clonal patterns (bands having same apparent molecular weight), reamplify first-round PCR product in five tubes using second-round reaction mix. Run 15 μl product in each of 10 lanes on a 1% (w/v) agarose gel. Excise corresponding inverse PCR product bands, chop finely with a sterile scalpel, and recover amplified DNA from agarose by centrifugation through a Spin-X column twice for 5 min each.

12. Precipitate DNA and resuspend in 8 μl TE buffer. Sequence DNA using a Circumvent Thermal Cycle Sequencing Kit and 1 μl inverse PCR primer INVc (Nolta et al., 1996).

SUPPORT PROTOCOL 9

HARVESTING CELLS FOR ANALYSIS OF LONG-TERM BONE MARROW CULTURES

Cell pellets are collected from long-term bone marrow cultures (LTBMC) during the weekly medium changes ($\sim 1 \times 10^6$ cells/sample) and are washed once in 10 ml PBS prior to extraction of nucleic acids or protein preparation for further analysis.

To prepare RNA, pellets are resuspended in 200 μl PBS and kept on ice until cells are lysed in 4 M guanidine isothiocyanate solution (*UNIT 10.3*). It is important to resuspend cells into a slurry prior to lysis and not to leave them as a packed pellet. Cells must be lysed simultaneously to prevent RNA digestion by RNase that has not yet been inactivated by the guanidine isothiocyanate. RNA lysates are then stored at $-70°C$ until extracted. RNA extraction , gel electrophoresis, and nucleic acid transfer are done by standard methods (see *UNIT 10.3* and *APPENDICES 3G & 3H*).

To prepare DNA for Southern blot hybridization (*APPENDIX 3G*), nonadherent cells from cultures are centrifuged, washed once in PBS, and stored as a pellet at $-20°C$ for up to several years; these pellets will yield DNA adequate for Southern blotting or PCR.

To prepare proteins from nonadherent cells for protein analysis, pellets are washed four times in 10 ml PBS. After the final centrifugation, all PBS is removed from the tube, and the tube is subjected to a second brief spin to bring down any remaining drops of fluid from the side of the 15-ml centrifuge tube. All PBS is removed with a micropipet tip, and cells are stored as a dry pellet at $-70°C$. Even 100 μl of PBS on the pellet can substantially dilute protein lysis buffer, which will be added for enzyme assays. If cells will be processed within 1 hr, they can be kept on ice until assayed.

SUPPORT PROTOCOL 10

FREEZING HEMATOPOIETIC CELLS

Materials (see *APPENDIX 1* for items with ✓)

    Bone marrow, cord blood, or peripheral blood cells
    Autologous serum or heat-inactivated human AB serum
✓ Freezing medium, 4°C

    2-ml cryovials (Nalge)
    Rate-controlled cell freezer
    Liquid nitrogen tanks for cell storage

1. Resuspend bone marrow, cord blood, or peripheral blood cells at 4×10^7 cells/ml in undiluted autologous serum. Freeze CD34-enriched cells at 1×10^7 cells/ml.

2. Put 1 ml cells into a 2-ml cryovial and put cryovial in the refrigerator or on ice until a rate-controlled cell freezer has been cooled to 6°C.

3. Add 1 ml cold freezing medium to cells and mix. Put vial in the rate-controlled cell freezer and freeze at −1°C/min. When freezing cycle is completed (at −90°C), store vial in liquid nitrogen.

If a rate-controlled cell freezer is not available, cells can be frozen (with slightly lower viability) by placing vials in the vapor phase of a liquid nitrogen tank overnight, then transferring them to the liquid phase.

SUPPORT PROTOCOL 11

THAWING HEMATOPOIETIC CELLS

Materials *(see APPENDIX 1 for items with ✓)*

  1 mg/ml DNase (Sigma)
✓ Thawing medium
  Cryopreserved hematopoietic cells in 2-ml cryovials (see Support Protocol 10)
✓ PBS

  15-ml centrifuge tubes, sterile
  37°C water bath

1. Add 100 μl of 1 mg/ml DNase to 10 ml thawing medium in a sterile 15-ml centrifuge tube.

2. Rapidly thaw one 2-ml tube of cryopreserved hematopoietic cells and dilute directly into the thawing medium without centrifuging. Incubate 1 hr in a 37°C water bath.

3. Wash cells twice with PBS before processing further.

Alternatively, plate cells directly from the cryovial into BBMM medium to let clumps dissociate. This method is more effective for very large clumps of cells.

References: Hanenberg et al., 1996; Moore et al., 1992; Nolta et al., 1996

Contributors: Jan A. Nolta and Donald B. Kohn

UNIT 13.5

Gene Delivery to the Airway

STRATEGIC PLANNING

Choice of Airway Model System

The choice of an airway model system is dependent on the level of differentiation required to address the hypotheses at hand. Several model systems have been used that offer flexibility for genetic modification using recombinant vector systems. These model systems include: (1) proliferating cultures of primary airway epithelial cells, (2) polarized airway epithelial monolayers, (3) human bronchial xenografts, and (4) intact lung in animal models. As human airways are most suitable for addressing aspects of human genetic diseases and gene therapy, this unit has emphasized strategies for generating human airway models. Several factors, listed

Table 13.5.1 Choice of Airway Model System

Airway model system	Advantages	Disadvantages
Proliferating cultures of human airway cells (Basic Protocols 1 and 2)	Ease of use; rate of gene transfer is typically higher than in differentiated models of the airway	Cells are undifferentiated
Polarized human airway epithelial monolayers (Basic Protocols 3 and 4)	Good differentiation; experimentally very flexible for functional measurements (e.g., bioelectric properties)	Requires freshly isolated airway cells; differentiated monolayers may not contain all cell types found in vivo
Human bronchial xenografts (Basic Protocols 5 and 6)	Excellent differentiation which appears to contain the native distribution of all cell types in the human airway; partially reconstitutes submucosal glands; partial immune response present	Technically difficult; expensive; immune incompetent with respect to T cells
Intact lung of animal models (Basic Protocol 7)	Can assess aspects of intact lung function and gene delivery	Not human; airway cell types may differ from that of human

in Table 13.5.1, may influence the choice of model systems for research, including the extent of epithelial differentiation and whether it is necessary to study human airway cells in the context of genetic diseases.

Choice of Vector System

The choice of vector system for genetically modifying airway model systems is an important consideration in the overall design of experiments aimed at addressing disease pathophysiology, airway biology, and gene therapy. Several vector systems have been successfully used in airway models including recombinant adenovirus (UNIT 12.2), recombinant retrovirus (UNIT 12.4), recombinant adeno-associated virus (UNIT 12.3), and cationic liposome/DNA complexes. Proliferating cultures of primary airway epithelial cells are more efficiently transduced with all of these vectors as compared to more differentiated models. The relative efficiency of each of these vector systems for use in the different model systems is outlined in Table 13.5.2.,

Table 13.5.2 Choice of Vector System[a]

Airway model system	Efficiency of gene transfer
Proliferating cultures (Basic Protocols 1 and 2)	Adenovirus > retrovirus > adeno-associated virus
Polarized monolayers (Basic Protocols 3 and 4)	Adenovirus > retrovirus > adeno-associated virus
Bronchial xenografts (ex vivo gene delivery; Basic Protocols 5 and 6)	Retrovirus > adeno-associated virus > adenovirus
Bronchial xenografts (in vivo gene delivery; Basic Protocols 5 and 6)	Adenovirus > retrovirus > adeno-associated virus
Lung of animal models (Basic Protocol 7)	Adenovirus > adeno-associated virus > retrovirus

[a]Other considerations influencing vector choice, such as the potential for integration, stability of transgene expression, and the potential for inflammatory responses, are described in CPHG Chapter 12.

although there are some exceptions. For example, recombinant adenoviral vectors may have a higher level of toxicity resulting from cryptic expression of viral genes. Hence, in some instances other less-efficient gene transfer methods may be experimentally more preferable, depending on considerations of vector biology. The level of cellular proliferation, which may affect the extent of transduction with certain viral vectors, should also be a consideration. Furthermore, the extent of gene transfer with some of the existing vector systems is currently only limited by achievable titers. For example, new technologies for concentrating retroviral stocks have allowed for higher transduction efficiencies than were previously attainable. In this unit, the most commonly used viral vectors for gene transfer to the airway are described.

Choosing a Reporter Gene

An important question when evaluating gene delivery to the airway is the efficiency of various vector systems for expressing encoded transgenes. Several reporter genes have been used historically—e.g., chloramphenicol acetyltransferase (CAT), firefly luciferase, human growth factor, β-glucuronidase, green fluorescent protein (GFP), β-galactosidase (β-gal), and alkaline phosphatase (AP). For additional discussion of reporter genes, see CPMB Chapter 9. As histologic evaluation of transgene expression is more effective in assessing gene transfer at the cellular level, the last three reporter genes mentioned above are the most commonly used. GFP provides an additional advantage of allowing transgene expression to be assessed in viable cells. This is often very useful in evaluating the kinetics of gene delivery and transgene expression in a noninvasive manner. However, the application of this approach is most often limited to primary culture models. Another important factor in choosing an appropriate reporter gene is the level of endogenous enzymatic activity. Although experimental approaches have been designed to reduce the extent of background staining, both β-gal and AP have endogenous activities in airway epithelia. Several approaches to minimizing complications of background staining include defined, short reaction times, inhibition and/or inactivation of endogenous enzymatic activities, and the stringent use of negative control vectors.

NOTE: All protocols involving live animals should meet NIH and research institute or university guidelines and should be approved by the institutional animal care and use committee (IUCAC).

NOTE: All solutions and equipment coming into contact with living cells must be sterile, and aseptic technique should be used accordingly. All culture incubations are performed in a humidified 37°C, 5% CO_2 incubator unless otherwise specified.

BASIC PROTOCOL 1

ISOLATION OF HUMAN PRIMARY AIRWAY EPITHELIAL CELLS

Materials (see APPENDIX 1 for items with ✓)

Human lung (keep on ice), from lung-transplant donor specimen (preferred) or postmortem lung (harvested <6 hr following death)
✓ Medium A, 4°C
Protease type XIV (e.g., Sigma), 4°C
✓ FBS
Ham's F-12 medium (Life Technologies) with and without 10% FBS, 4° and 37°C
✓ Medium B, 37°C
✓ Medium C
0.1% trypsin/EDTA (Life Technologies)
✓ Trypsin inhibitor buffer

Cryopreservation medium: medium C containing 10% (v/v) dimethyl sulfoxide and 10% (v/v) FBS, 4°C

Dissecting equipment including forceps, scalpel, and hemostat
100- and 150-mm tissue culture dishes (uncoated plastic)
15- and 50-ml conical centrifuge tubes
Platform rocker
Tabletop centrifuge
3-cm^2 piece of 500-μm nichrome or copper wire mesh, sterile
2-ml cryovials (e.g., Nunc)

1. Dissect bronchial airways from a human lung in a tissue culture hood on ice and place airways in medium A at 4°C. Dissection can easily be performed down to the fifth-order bronchus.

2. Place dissected airways into a 150-mm tissue culture dish with excess medium A. Remove excess adventitia from outside the airways and cut bronchial segments into 1- to 2-cm rings. Cut bronchial rings longitudinally to expose airway surface. Remove excess mucous secretions.

3. Place specimens in a 50-ml conical tube with 35 ml medium A (add enough tissue to bring total volume to 45 ml) and incubate 30 min with rocking at 4°C. Decant medium, replace with fresh medium A, and again rock 30 min at 4°C. Repeat for a total of six washes. If tissue is severely infected at time of harvest, it may be necessary to increase the number of washing steps for a total of 5 to 6 hr.

4. Place samples in a fresh 50-ml conical tube containing 35 ml medium A supplemented with 0.1% (w/v) protease type XIV. Incubate 36 hr at 4°C. Add FBS to a final concentration of 10% (make sure tube is <75% full) and shake gently for 30 sec.

5. Allow 1 min for tissue to settle to the bottom of the tube. Pipet medium (containing liberated airway cells) into a fresh tube on ice and place remaining tissue into a 100-mm tissue culture plate.

6. Using blunt side of a scalpel, scrape surface of airways, rinse with 4°C Ham's F-12 medium/10% FBS, and combine washed cells in a 50-ml conical tube. Save a piece of airway tissue at −80°C for genotyping, if necessary.

7. Combine all supernatants containing epithelial cells and centrifuge 5 min at ∼300 × g, 4°C. Decant medium and resuspend cells in a total of 10 ml of Ham's F-12 medium/10% FBS. Filter through a sterile 500-μm wire mesh into a 15-ml conical tube.

8. Wash cells two more times by centrifugation in the same medium. Prior to the last washing, remove a 10-μl aliquot of cell suspension and quantitate the total yield of viable isolated cells by trypan blue exclusion (*APPENDIX 3I*; typical yield from one lobe, ∼2–4 × 10^7 cells). Do not count red blood cells.

9. Resuspend final pellet in 10 ml of 37°C medium B and plate at ∼1–2 × 10^6 cells per 100-mm tissue culture dish in 10 ml medium B per dish. Incubate 24 hr. If cystic fibrosis (CF) cells are used, double antibiotic concentrations for first 24 hr.

10. Aspirate cells that have not adhered to plate and wash adherent cells twice in 37°C Ham's F-12 medium. Feed cells with fresh medium B and incubate 48 hr (72 hr for heavily contaminated cells).

Amphotericin B is highly toxic to cells and time of exposure should be kept to a minimum unless problems are encountered with fungal or bacterial contamination. Acquiring tissue directly from the transplantation room (rather than the pathology lab) will reduce fungal contamination.

11. Feed cells with medium C (at this point clones of expanding epithelial cells should be visible). Feed every 3 days with fresh medium C. Typically, cells are ready for cryopreservation, passaging, or transplantation into xenograft models (see Basic Protocol 5) 5 days after plating (~80% confluency). Do not allow cells to become >80% confluent or they will begin to differentiate.

12. *Optional:* Remove medium and incubate cells 1 to 3 min with 5 ml of 0.1% trypsin/EDTA at 37°C, while closely monitoring detachment of cells (*APPENDIX 3I*). Add 5 ml trypsin inhibitor buffer. Harvest cells immediately by gentle tapping of the plate. Centrifuge cells 5 min at 300 × *g*, 4°C, and wash twice in medium C. Plate cells at a 1:5 dilution and propagate in medium C as described in step 11.

    Typically, cells can be expanded one time without loss of ability to differentiate in a xenograft model.

13. Harvest cells by treating with trypsin/EDTA and gently tapping the plate. Centrifuge 5 min at 300 × *g*, 4°C, and resuspend pellets in 1 ml of 4°C cryopreservation medium for each 100-mm plate of cells. Divide into 1-ml aliquots in 2-ml cryogenic vials. Freeze overnight at −80°C and then transfer to liquid nitrogen (also see *APPENDIX 3I*).

14. To use cryopreserved cells, quick-thaw at 37°C and immediately place in 37°C growth medium. It is not necessary to remove DMSO.

BASIC PROTOCOL 2

TRANSDUCTION OF PRIMARY AIRWAY EPITHELIAL CELLS

Materials (*see APPENDIX 1 for items with* ✓)

    8 mg/ml polybrene (Sigma) in H$_2$O (store in aliquots at −20°C)
    100-mm plate of primary airway epithelial cells (see Basic Protocol 1), freshly isolated or passaged once
    Ham's F-12 medium (Life Technologies)
✓ Medium C

    0.45-μm filter syringe filter, for retroviral infection only
    12,000- to 14,000-Da MWCO dialysis membrane (Life Technologies), for rAAV infection only

To transduce with retrovirus

1a. Harvest retroviral supernatants from confluent monolayers of clonal ψ Crip producer cells or by other classical techniques including transient transfection (*UNIT 12.4*). For clonal producer cell lines, place 5 ml DMEM/10% calf serum growth medium on a 100-mm confluent monolayer 16 to 18 hr prior to harvesting.

2a. Pass retroviral supernatant through a 0.45-μm filter to remove cellular debris. Supplement with 8 mg/ml polybrene to final concentration of 2 μg/ml.

3a. Remove medium from a 100-mm plate of freshly isolated primary airway epithelial cells on the second day after seeding (~10% confluency) and add 10 ml retroviral supernatant. Incubate 2 hr. Wash cells twice with Ham's F-12 medium and then add hormonally defined medium C and resume incubation. Infect cells up to three times on sequential days.

    After infection, cells can be used to generate genetically modified airway epithelium in bronchial xenografts (see Basic Protocol 6). Typically, retroviral titers of 1 × 10^6 cfu/ml are capable of transducing primary airway cells at an efficiency of 10% to 30% after three serial infections.

To transduce with rAAV

1b. Prepare and purify rAAV (*UNIT 12.3*). Dialyze against 500 ml Ham's F-12 medium at 4°C overnight using a 12,000- to 14,000-Da MWCO dialysis membrane. Mix 5×10^9 DNA particles of rAAV (typically 5 to 100 μl) with 1.5 ml medium C at room temperature.

2b. Remove growth medium from a 100-mm plate of primary airway epithelial cells (freshly isolated or passaged once) containing ~5×10^5 cells. Add 1.5 ml of rAAV/medium C mixture and incubate 1 hr with rotation every 5 to 10 min. Add 5 ml medium C and continue incubation for an additional 24 hr.

> *A single infection typically gives an efficiency of 25% transgene-expressing cells. Infections can be carried out on sequential days to improve the level of gene transfer.*

BASIC PROTOCOL 3

GENERATION OF POLARIZED AIRWAY EPITHELIAL MONOLAYERS

Freshly isolated uncultured airway cells provide the highest level of differentiation. Cells that have been passaged once can also yield good results, depending on the quality of airway tissue used to generate primary cells.

Materials (*see APPENDIX 1 for items with* ✓)

    Glacial acetic acid
    Collagen, Type VI, acid-soluble, from human placenta (Sigma), screened for highest potency and lowest cell toxicity
✓ PBS
✓ 5% serum airway medium
✓ Ussing chamber culture medium

    0.2-μm filter (Millipore Millex GS or equivalent Nalgene filter)
    12-mm Millicell-HA culture plate inserts (Millipore)
    100-mm tissue culture dishes
    Tabletop centrifuge
    24-well tissue culture plates

1. Add 50 μl glacial acetic acid to 25 ml distilled water, add 12.5 mg Type VI human placenta collagen, and dissolve by gently stirring 30 min at 37°C. Dilute 1:10 with distilled water and pass through a 0.2-μm filter (50 μg/ml final; store ≤1 month at 4°C).

2. Coat a 12-mm Millicell-HA culture plate insert by placing 500 μl diluted collagen on the surface and incubating ≥18 hr (but <1 week) at room temperature.

3. Just before use, remove all liquid collagen and air dry 3 to 5 min. Rinse once with PBS and twice with 5% serum airway medium.

4. Generate primary airway epithelial cells (see Basic Protocol 1, steps 1 to 8) *or* use passage-1 cells (see Basic Protocol 1, steps 1 to 12; not as effective) and proceed to step 6.

5. Resuspend cell pellet in 5% serum airway medium by gentle pipetting. Plate in 100-mm tissue culture dishes and incubate 1 to 3 hr to allow fibroblasts to attach. Collect supernatant, containing nonattached airway primary epithelial cells. Count viable cells by trypan blue exclusion (*APPENDIX 3I*) and centrifuge 5 min at $300 \times g$.

6. Place 500 μl of 5% serum airway medium in each well of a 24-well tissue culture plate and place collagen-coated Millicell inserts in the wells. Seed 3×10^5 airway primary cells (5×10^5 cells/cm^2) in each insert in a total volume of 100 to 500 μl of 5% serum airway medium (Fig. 13.5.1). Incubate 24 hr.

13

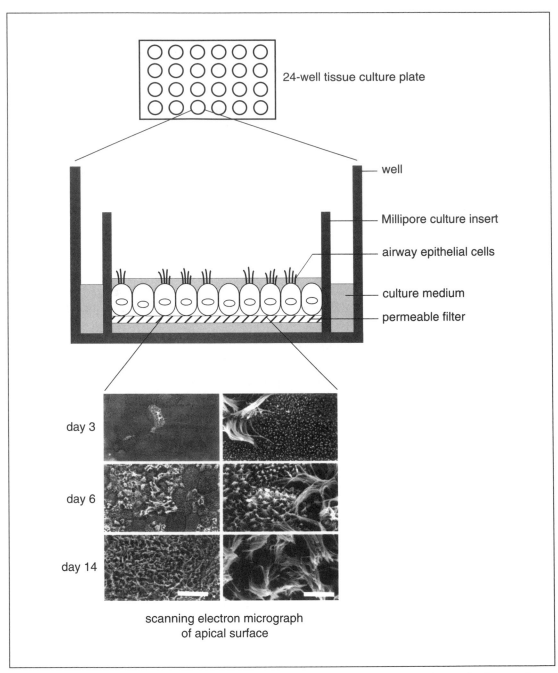

24-well tissue culture plate

well

Millipore culture insert

airway epithelial cells

culture medium

permeable filter

day 3

day 6

day 14

scanning electron micrograph
of apical surface

Figure 13.5.1 In vitro cultures of polarized airway epithelial cells. The monolayer of primary airway epithelial cells is grown on collagen-coated Millipore cell culture inserts (pore size 0.4 μm, total growth area 0.6 cm^2) placed in wells of a 24-well tissue culture plate. The support filter is made from mixed esters of cellulose nitrate and acetate and subsequently coated with liquid collagen. During culturing, the basolateral side is exposed to Ussing chamber culture medium, while the apical surface is exposed to the air. Scanning electron micrographs kindly provided by Zabner et al. (1996) with permission from American Society for Microbiology. Bar: right, 5 μm; left, 37.5 μm.

7. Remove medium on the apical (mucosal) side and replace 5% serum airway medium on the basal (basolateral) side with Ussing chamber culture medium. Continue to incubate cells with this air-liquid interface. Maintain CO_2 at 7.5% during the first week of culture to increase the efficiency of monolayer formation with higher cross-membrane resistance. Well-differentiated monolayers of airway epithelia are obtained after 2 weeks in culture. Representative cultures demonstrate a transepithelial resistance of >1000 ohm/cm^2.

BASIC PROTOCOL 4

GENE TRANSFER TO POLARIZED AIRWAY EPITHELIA

Some modifications that may improve infection efficiency are described in *CPHG UNIT 13.9*.

Materials (see APPENDIX 1 for items with ✓)

✓ HBSS, for recombinant retrovirus only
  25% and 40% (w/v) sucrose in HBSS, for recombinant retrovirus only
✓ Lactose storage buffer, for recombinant retrovirus only
  8 mg/ml polybrene in H_2O (store in aliquots at $-20°C$), for recombinant retrovirus only
✓ Ussing chamber culture medium
  Polarized airway epithelial cells grown on Millicell-HA culture plate inserts (see Basic Protocol 3)
  Ham's F-12 medium (Life Technologies), for recombinant adeno-associated virus (rAAV) only
✓ PBS, for adenovirus only

  0.45-μm filter (Millipore), for recombinant retrovirus only
  Sorvall RC-26 Plus centrifuge with SS-34 rotor (or equivalent), 4°C, and centrifuge tubes accommodating 40 ml, for recombinant retrovirus only
  Beckman ultracentrifuge with SW 41 rotor (or equivalent), 4°C, and SW 41 centrifuge tubes, for recombinant retrovirus only
  Filtron 100K concentrator (PALL), for recombinant retrovirus only
  58°C water bath, for rAAV only
  12,000- to 14,000-Da MWCO dialysis membrane (Life Technologies), for rAAV only

To transduce with recombinant retrovirus

1a. Generate retroviral supernatants to be used for gene-transfer studies (*UNIT 12.4*). For higher-titer preparations, clarify retrovirus-containing culture medium by filtering through a 0.45-μm filter. Titer should be $>10^8$ cfu/ml.

2a. Centrifuge retroviral supernatant 16 hr at $7000 \times g$ in a Sorvall RC-26 Plus centrifuge with an SS-34 rotor (40 ml per tube) at 4°C. Remove supernatant and resuspend viral pellet in 0.5 ml HBSS.

3a. Generate a 25% to 40% sucrose gradient by placing 4 ml of 40% sucrose in HBSS in bottom of an SW 41 centrifuge tube and layering an equal volume of 25% sucrose on top by slow pipetting. Layer viral suspension on top of gradient and centrifuge 1.5 hr at $265,000 \times g$, 4°C.

4a. Collect virus from interface and exchange buffer to lactose storage buffer by diafiltration through a Filtron 100K concentrator according to manufacturer's instructions. Titer viral stock (*UNIT 12.4*) and store in aliquots at $-80°C$.

5a. Add 8 mg/ml polybrene to a final concentration of 8 μg/ml. Mix retroviral stock with an equal volume of Ussing chamber culture medium and apply 100 μl to the apical surface

13

of polarized airway epithelial cells. Alternatively, invert filter and apply 100 μl virus to the basal side of the epithelium. Incubate 4 hr.

Typically, retroviral transduction from the apical side of a polarized culture gives little gene transfer, whereas infection from the basolateral side is ~1% to 5% efficient.

6a. Remove virus-containing medium and return epithelial monolayer to normal culturing conditions (air on the apical surface and Ussing chamber culture medium on the basolateral surface).

7a. Analyze cultures functionally and/or examine transgene expression.

To transduce with rAAV

1b. Purify rAAV through three rounds of CsCl density ultracentrifugation (UNIT 12.3). Heat 60 min at 58°C to inactivate contaminating helper adenovirus. Dialyze against 500 ml Ham's F-12 medium at 4°C overnight using a 12,000- to 14,000-Da MWCO dialysis membrane.

2b. Mix 5 to 20 μl rAAV ($\sim 1 \times 10^{10}$ particles) with 100 μl Ussing chamber culture medium. Apply 100 μl directly to apical surface of polarized airway epithelial cells. Alternatively, invert filter and apply 100 μl for basolateral infection. Incubate 24 hr.

Basolateral infection provides ~100-fold higher efficiency (~1% transgene-expressing cells by 30 to 40 days after infection) than apical infection.

3b. Remove virus-containing medium and return epithelial monolayer to normal culturing conditions (air on the apical surface and Ussing chamber culture medium in the basolateral compartment).

4b. Analyze cultures functionally and/or examine transgene expression (begins at 4 days and peaks at 40 days after infection).

To transduce with adenovirus

1c. Generate recombinant adenovirus (UNIT 12.2). Dilute adenoviral stock in PBS to 5×10^7 pfu/100 μl (multiplicity of infection [MOI] ~50 pfu/cell). Apply 100 μl onto apical side of polarized airway epithelial cells. Incubate >12 hr.

2c. Remove virus-containing medium and wash apical surface twice with Ussing chamber culture medium. Return chamber inserts to normal culturing conditions (air on the apical surface and Ussing chamber culture medium on the basolateral compartment).

3c. Analyze cultures functionally and/or examine transgene expression (peaks by 3 days after transfection).

Adenovirus infection in poorly differentiated epithelia (within 3 days of seeding) will result in ~50% to 60% transgene-expressing cell.

BASIC PROTOCOL 5

GENERATION OF HUMAN BRONCHIAL XENOGRAFTS

Human proximal airway xenografts can be generated from multiple tissue sources, including nasal, tracheal, and bronchial epithelium. The most widely used bronchial xenografts are generated from primary airway epithelial cultures, which are seeded into denuded rat tracheas (Fig. 13.5.2).

Materials *(see APPENDIX 1 for items with ✓)*

Fisher 344 rats, 200 to 250 g, male (Harlan Bioproducts for Science or Charles River Labs)

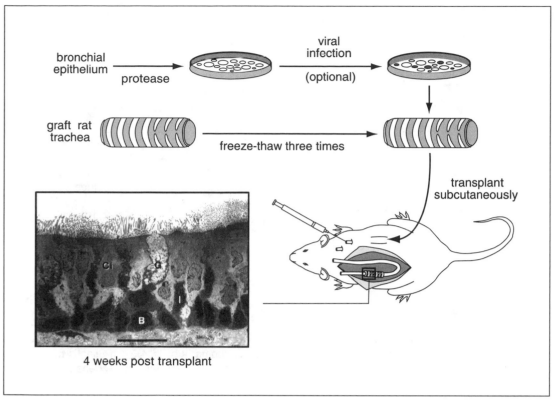

Figure 13.5.2 Human bronchial xenograft model. Subcutaneous implants of xenograft human airways are generated from primary bronchial human airway cells transplanted onto denuded rat tracheas. Primary cells can be genetically manipulated ex vivo with vectors such as recombinant retrovirus and adeno-associated virus (see Basic Protocol 2). Alternatively, xenografts can be generated from genetically unaltered primary cells and gene-transfer studies performed in vivo following reconstitution of grafts. Typically, a fully differentiated mucociliary epithelium is obtained by 4 weeks posttransplantation. B, basal cell; Ci, ciliated cell; I, intermediate cell; G, goblet cell. Bar = 20 μm.

70% (v/v) ethanol
MEM (Life Technologies), 4°C
Primary bronchial airway epithelial cells (see Basic Protocol 1) at ~80% confluency
✓ Medium C, 4°C
nu/nu athymic mice, 20 to 25 g, male (Harlan Bioproducts for Science)
Ketamine
Xylazine
✓ PBS
Povidone-iodine
Ham's F-12 medium (Life Technologies), room temperature

Silastic tubing (0.030-in. i.d. × 0.065-in. o.d.; Dow Corning)
Teflon tubing (0.031-in. i.d. × 0.063-in. o.d.; Thomas)
Adapter (0.8-mm barb-to-barb connector; Bio-Rad)
0.035-in.-diameter Chromel A steel wire (Hoskins Mfg.)
100-mm tissue culture plates
Gas sterilization pouch (M.D. Industries) and gas sterilization apparatus
2-ml screw-cap tubes (Sarstedt)
2-0 braided silk suture (e.g., Ethicon)

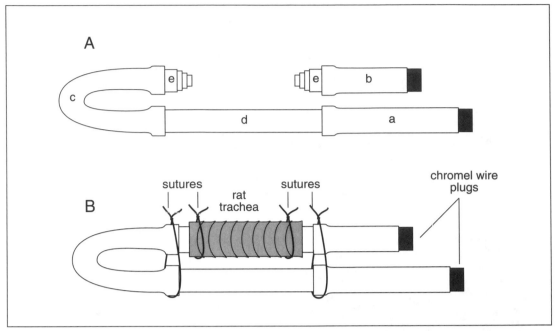

Figure 13.5.3 Xenograft cassettes are generated from a series of defined types of tubing as defined in (**A**). Tube a, 2.5-cm Silastic tubing; tube b, 1.9-cm Silastic tubing; tube c, 4.5-cm Silastic tubing; tube d, 3.2-cm Teflon tubing; tube e, adapter. (**B**) Tubing is connected to denuded rat tracheas by a series of suture ligations.

Hemostats
Small airtight transfer chamber to equilibrate xenograft cassettes in 5% CO_2
Water bottle
Sterile surgical drapes for mouse surgery
Small forceps (two per mouse) and sharp scissors (one per mouse), sterilized in self-sealable autoclave sterilization pouch (M.D. Industries)
Disposable skin stapler 35R (American Cyanamid)
Sterile cages for mice
0.75-in., 21-G Surflo winged infusion set (Terumo Medical)

1. Cut tubing and adapters for xenograft cassettes to length and assemble as outlined in Figure 13.5.3A. Place cassette in a 100-mm tissue culture plate. Seal in a sterilization pouch and gas sterilize.

2. Euthanize male Fisher 344 rats by CO_2 asphyxiation and pin to a Styrofoam bed.

3. While working in a laminar flow hood (or under good clean conditions at an open bench), clean neck and chest of euthanized rats with excess 70% ethanol and excise trachea from the pharynx to the carina (bifurcation of bronchi at the end of the trachea). Do not nick arteries. Immediately place each excised trachea in a separate 2-ml screw-cap tube and keep on ice until all tracheas have been harvested.

4. Denude tracheas of all viable epithelium by three rounds of freezing at −80°C and thawing at room temperature. Clean tracheas of excessive fat and cut to size (typically at the first and thirteenth tracheal ring). Rinse each tracheal lumen with 10 ml of cold MEM. Pair tracheas for length and combine pairs in the same tube. Store at −80°C.

All procedures that follow are performed in a laminar flow hood under sterile conditions.

5. Using 2-0 braided silk sutures, ligate rat tracheas to the adapter (e) attached to tubing b as shown in Figure 13.5.3B. Securely tie sutures with triple knots and loop them around the tubing and trachea a total of three times.

6. Inject $1–2 \times 10^6$ primary bronchial airway epithelial cells in 20 µl cold medium C under sterile conditions into the open end of the rat trachea using a 20-µl micropipettor. Insert pipet tip as deeply as possible into trachea and slowly withdraw it as cells are injected into trachea.

7. With care not to allow cells to leak out, ligate remaining open end of rat trachea to adapter (e) attached to tubing c as shown in Figure 13.5.3B. Secure length of the rat trachea by stretching it to physiologic length and clamping tubing b and tubing a with a hemostat. Tie remaining two sutures as shown in Figure 13.5.3B to secure adapters to tubing d.

8. Place xenograft cassettes with seeded cells into a 100-mm tissue culture plate with 1 to 2 ml medium C. Incubate plate 1 to 2 hr in a humidified 37°C, 5% CO_2 incubator to equilibrate pH. For transport, use a small airtight sterile container equilibrated in the incubator with the dishes of xenografts to maintain CO_2.

9. Anesthetize male *nu/nu* athymic mouse by intraperitoneal injection of 100 mg/kg ketamine and 20 mg/kg xylazine in PBS. Place mouse on a warm water bottle covered with a sterile drape and remove sterile surgical instruments from autoclave pouches.

10. Clean sites of surgical incisions with povidone-iodine followed by alcohol. Make four incisions as shown in Figure 13.5.4B. Make very small incisions on the neck of the mouse (0.16 cm) with just enough width to pass the tubing. Make two incisions on the flanks of

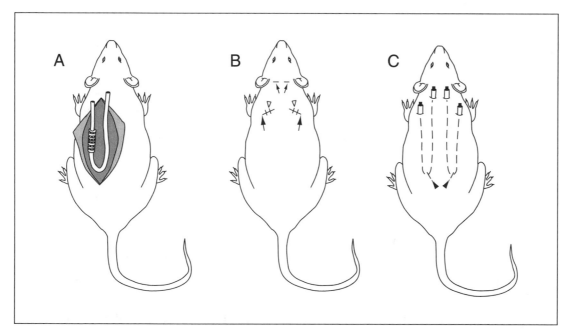

Figure 13.5.4 Transplantation of bronchial xenografts in *nu/nu* mice. (**A**) Subcutaneous transplantation of the xenograft cassette. (**B**) Four incisions are made as marked by arrows and the xenograft cassette guided subcutaneously using forceps, so that one port exits through the back of the neck and the other port through the main incision. Surgical staples are used to close incisions as marked by open arrowheads. (**C**)The resultant xenografted mouse after 1 week posttransplantation, when staples are removed. Staples marked by closed arrowheads in panel C are used to maintain the position of the cassette and prevent subcutaneous migration (it may be necessary to leave these staples in for 2 to 3 weeks).

the mouse, ~1 cm long. Separate skin from muscle by blunt dissection and place xenograft cassette subcutaneously, tunneling the distal end of tubing a out of the incision behind the neck using forceps to guide the xenograft tubing (Fig. 13.5.4C).

11. Use two or three staples to close each of the largest incisions. Use an additional staple to anchor each xenograft to the skin at the loop of tubing c. Take care not to puncture the xenograft tubing while inserting this staple (Fig. 13.9.4C, solid arrowheads).

12. Transfer mouse to a sterile cage. Keep mouse warm and monitor until it is awake. House mice separately as they may chew each other's tubing and maintain in pathogen-free housing. Autoclave all caging, food, and water before cage changing.

13. For the first 3 weeks after transplantation, irrigate xenografts weekly with 1 ml Ham's F-12 medium using a Surflo winged infusion set with a 0.75-in., 21-G needle, to remove excess mucous secretions, using the following technique. If needed, anesthetize mouse with a 25% to 50% dose of ketamine/xylazine.

    a. Attach a 21-G needle to a 1-ml syringe (with plunger removed) and fill with room temperature Ham's F-12 medium. Insert needle into the tubing from one end of the xenograft.

    b. Insert a winged 21-G infusion needle attached to a second syringe (with plunger attached and pushed all the way in) into opposite end of the xenograft.

    c. Apply negative pressure by withdrawing the plunger of the second syringe to aspirate liquid through lumen of xenograft. Finally, aspirate air into lumen of xenograft.

14. After 3 weeks, irrigate xenografts twice per week.

BASIC PROTOCOL 6

GENE TRANSFER TO HUMAN BRONCHIAL XENOGRAFTS

Materials

    Primary proliferating human airway cells (for ex vivo transduction with retrovirus; see Basic Protocol 1) or mouse with fully differentiated xenograft (for in vivo transduction with adenovirus; see Basic Protocol 5)
    Retroviral supernatant (*UNIT 12.4*) or purified recombinant adenovirus (*UNIT 12.2*)
    Ham's F-12 medium (Life Technologies)

For ex vivo transduction with retrovirus

1a. Transduce primary proliferating human airway cells with recombinant retrovirus (see Basic Protocol 2). Infect cells three times for best transduction.

2a. Trypsinize primary airway epithelial cells (typically 5 days after seeding; see Basic Protocol 1, step 12) and seed into xenograft cassettes (see Basic Protocol 5).

3a. Transplant xenograft cassettes into *nu/nu* mice and maintain xenografts (see Basic Protocol 5).

4a. After differentiation (typically 4 weeks), either use grafts for functional measurements or harvest them for histologic assessment of transgene expression (see Support Protocol and *CPHG UNIT 13.9*). Typically, the efficiency of transduction in primary cells is maintained in the final reconstituted xenografts.

13

For in vivo transduction with adenovirus

1b. Generate recombinant adenovirus and purify through two rounds of CsCl density centrifugation followed by desalting with by Sephadex G-50 filtration (not by dialysis) into Ham's F-12 medium (*UNIT 12.2*).

2b. Immediately instill 100 to 150 μl purified virus (5×10^{10} pfu for ~10% infection efficiency) into the lumen of one end of the fully differentiated xenograft (4 weeks after transplantation) and close ends of the xenograft by insertion of Chromel wire plugs (see Basic Protocol 5). For >95% infection efficiency, carry out three 16-hr infections every other day.

3b. Infect xenografts for 16 hr and then irrigate with 1 ml Ham's F-12 medium followed by air (see Basic Protocol 5, step 13).

4b. Typically, if functional measurements are to be performed on xenograft airways, allow xenografts to recover from acute toxicity of viral infection for 48 to 72 hr.

BASIC PROTOCOL 7

IN VIVO GENE DELIVERY TO THE LUNG

Materials *(see APPENDIX 1 for items with ✓)*

Cotton rats (80 to 120 g; Virion Systems)
Isoflurane
Purified recombinant adenovirus (*UNIT 12.2*)
✓ PBS
Pentobarbital
100% optimal cutting temperature (OCT) medium (Baxter) *and* 50% (v/v) OCT medium in PBS
Pulverized dry ice/isopentane slurry

Gauze pads
50-ml conical centrifuge tubes
Isopropyl alcohol swabs
Scalpel, forceps (2), and small sharp scissors
1-ml syringe with 30-G needle
Disposable skin stapler 35R (American Cyanamid)
18-G AngioCath (Becton Dickinson) and 3-ml syringe
Plastic embedding blocks (Baxter)
Cryostat

1. Anesthetize an 80- to 120-g cotton rat with isoflurane by placing an isoflurane-soaked gauze pad into the bottom of a 50-ml conical tube and positioning the tube over the nose of the rat. Judge the depth of anesthesia by noting the respiration rate and control anesthesia by manipulating the distance of the conical tube from the nose of the animal.

2. Clean neck with an isopropyl alcohol swab and make a 1-cm midline incision with a sharp scalpel. Expose trachea by blunt dissection and make a small (3-mm) incision through the muscle surrounding the trachea. Instill $1–5 \times 10^{10}$ infectious particles in a total volume of 100 to 150 μl PBS through a 1-ml syringe with a 30-G needle, directly through the tracheal wall without an incision, followed by 300 μl of air.

3. Close incision with two staples and return animal to cage.

For mice, reduce the total dose and volume of virus according to total kg weight of the animal. Typically a 25-g mouse tolerates a dose of 2×10^9 pfu recombinant virus instilled in a total volume of 25 to 35 μl.

The volumes stated in this protocol will give gene transfer to both airways and alveolar regions. To enhance gene transfer to the airways and limit alveolar gene transfer, smaller inoculum volumes can be used with the same dose of virus (50 μl total for rats; 15 μl for mice).

4. Euthanize animals by an overdose (200 mg/kg) of pentobarbital, administered intraperitoneally.

5. Open the thorax and expose the proximal end of the trachea. Attach an 18-G AngioCath to a 3-ml syringe filled with a 37°C 1:1 mixture of OCT embedding medium and PBS. Insert the end of the AngioCath tubing into the proximal end of the trachea and ligate with a triple-knotted suture.

6. Apply gentle pressure to syringe plunger until lungs are visibly inflated within the thorax (typically 2 ml OCT/PBS). Do not overinflate. Remove heart-lung cassette and quickly cool in ice-cold PBS for 10 min.

7. Remove heart-lung cassette from the ice-cold PBS, place on an ice-cooled surface, and trim away non-lung tissue.

8. Cut lung into large (~1-cm) wedges and place in plastic embedding blocks filled with 100% OCT embedding medium, so that the cut face is in contact with the bottom of the block.

9. Freeze blocks in a pulverized dry ice/isopentane slurry. Manipulate tissue samples during freezing to retain placement on the bottom of the block.

10. Cut 6-μm frozen sections using a cryostat and process for multiple histologic analyses such as immunocytochemistry, histochemistry, and in situ hybridization.

SUPPORT PROTOCOL 1

HARVESTING OF HUMAN BRONCHIAL XENOGRAFTS FOR MORPHOLOGIC ANALYSIS TO EVALUATE TRANSGENE EXPRESSION

Materials (see APPENDIX 1 for items with ✓)

    Mice harboring xenografts (see Basic Protocols 5 and 6)
✓ PBS
    Optimal cutting temperature (OCT) medium (Baxter)
    Pulverized dry ice/isopentane slurry
    Dissecting equipment: small sharp scissors, forceps, and razor blades
    Plastic embedding block (Baxter)
    Cryostat

1. Euthanize mice harboring xenografts by CO_2 asphyxiation, carefully excise xenograft cassette, and place on a flat surface lined with Parafilm. Using two razor blades, carefully slice the xenograft trachea into 2-mm rings by opposing stokes of the cutting edge of the blades. Discard ends of xenograft tissue where tubing is attached.

2. Rinse rings in PBS and blot on lint-free tissues to remove excess mucus from the lumen. Do not clamp epithelial surface with forceps.

    The tissue can now either be directly histochemically stained or fresh-frozen in OCT embedding medium as described in the following steps.

3. Slowly lower xenograft rings, with cut face down, into an OCT-filled plastic embedding block. Take care to allow OCT to enter the lumen of the xenograft ring, or frozen sections will be difficult to cut.

4. Push rings to bottom of block with blunt forceps. Do not enter the lumen of the xenograft while positioning sample. Up to six rings can be embedded into one block.

5. Place tissue blocks in a pulverized dry ice/isopentane slurry and store at −80°C until sectioning.

6. Section tissue blocks into 6-μm frozen sections on a cryostat. Avoid repeated freeze-thawing of freshly cut sections before processing for histochemical staining. Once sections are placed on slides they should be kept within the cryostat (−18°C) until staining and not be air dried. Alternatively, slides can be stored at −80°C for several weeks without substantial loss of β-galactosidase activity.

References: Engelhardt and Wilson, 1997; Engelhardt et al., 1992

Contributors: Dongsheng Duan, Yulong Zhang, and John F. Engelhardt

UNIT 13.6

Gene Delivery to the Liver

With a single intravenous injection of recombinant adenovirus (*UNIT 12.2*), nearly 100% of hepatocytes can be transduced. The most commonly used serotype of adenovirus, human type 5, is capable of infecting hepatocytes from multiple species, including mouse, rat, rabbit, and nonhuman and human primates. Quantities given here are for human type 5 adenovirus. Table 13.6.1 presents a troubleshooting guide for gene delivery to the liver.

NOTE: All protocols involving live animals should meet NIH and research institute or university guidelines and should be approved by the institutional animal care and use committee (IUCAC).

13

Table 13.6.1 Troubleshooting Guide for Gene Delivery to the Liver

Problem	Possible cause	Solution
No transgene expression	Virus is not viable	Use virus that has not been freeze-thawed multiple times. Use new preparation of virus. Store on ice for shorter time before injection.
Difficulty with injections		Practice, especially for tail-vein injections in mice. Alternatively, try a different site of injection, such as the jugular vein in mice.
Low levels of transgene expression	Virus is less active than expected	Use virus that has not been freeze-thawed multiple times. Use new preparation of virus. Store on ice for shorter time before injection.
	Promoter is not very active in the liver	Try a different promoter, such as CMV, or a liver-specific promoter/enhancer combination (RSV promoter, e.g., works much less efficiently in the liver than the CMV promoter).

BASIC PROTOCOL 1

RECOMBINANT ADENOVIRUS DELIVERY TO MOUSE LIVER BY TAIL VEIN INJECTION

Tail vein injections are generally preferred in mice, as they require neither surgery nor anesthesia. Alternative sites of injection include the jugular vein.

Materials (*see* APPENDIX 1 *for items with* ✓)

Adult mice, at least 6 weeks of age, 20 to 25 g
70% (v/v) ethanol in squirt bottle
✓ Recombinant adenovirus suspension: 1×10^{12} particles/ml in sterile PBS

Mouse cage with wire lid
Heat lamp
Mouse restrainer
Cotton gauze pads
1-ml syringe with 27-G (1/2-in. or ∼1.25-cm) needle

1. Place up to six mice in a cage and shine a heat lamp ∼6 to 10 in. (15 to 25 cm) above cage through the wire lid to warm them up. Watch mice; in ∼5 min, they will be huddled together in a corner of the cage, and will be ready to inject. Do not continue to actively heat the mice at this point.

2. Place one mouse into a restrainer. Apply 70% ethanol to the tail and wipe off excess with gauze pad.

3. Draw at least 0.2 ml adenovirus suspension into a 1-ml syringe with a 27-G needle and carefully remove air from the needle.

4. Hold the end of the tail so that it is extended and twist it slightly to one side to see the vein. Insert needle at a shallow angle (as the tail vein is relatively close to the surface) and inject 0.1 ml diluted recombinant adenovirus into the tail vein. The vein should transiently shift to a clear color. If any local swelling is observed, stop the injection, as the virus is being injected into the tail tissue, not the vein.

5. Withdraw needle and immediately apply pressure with gloved finger until bleeding stops. Return mouse to original cage.

BASIC PROTOCOL 2

RECOMBINANT ADENOVIRUS DELIVERY TO RABBIT LIVER BY PERIPHERAL EAR VEIN INJECTION

Materials (*see* APPENDIX 1 *for items with* ✓)

Adult rabbits, ∼2 kg
70% (v/v) ethanol, optional
✓ Recombinant adenovirus suspension: $0.75–1.0 \times 10^{13}$ particles in ∼3 ml sterile PBS

Restraining cage for rabbits
Gauze pads
5-ml syringe and 21-G butterfly needle
Petroleum jelly

1. Place an adult rabbit in a restraining cage. Dilate vessels in the ear by applying gauze pads dipped in warm water. Application of 70% ethanol may also be helpful. Note that the vein runs along the outer edge of the ear, and the artery runs along the center.

2. Attach a 21-G butterfly needle to a 5-ml syringe and draw ~3 ml adenovirus suspension through the needle, leaving no air in the butterfly. Some air may be contained in the syringe to allow injection of full amount of solution.

3. Insert butterfly needle into the ear vein and slide in far enough so that little effort is required to keep the needle in place. Begin injecting adenovirus at a rate of ~1 ml/min. Leave a small amount of solution in the butterfly needle, taking care not to inject any air.

4. Remove needle and immediately apply pressure to the ear using a gauze pad until bleeding ceases.

5. Gently rinse ear with cool water. Apply a small amount of petroleum jelly to the ear. Observe rabbit for next 10 to 15 min. If it seems lethargic, touch or pet the rabbit until it is alert and sitting up or moving about the cage.

SUPPORT PROTOCOL

HARVESTING LIVER TISSUE FOR TRANSGENE EXPRESSION ANALYSIS

If the transgene encodes a protein whose product is secreted into the circulation, expression can be confirmed by bleeding the animal and checking for protein in the plasma by either immunoblotting (e.g., *CPMB UNIT 10.8*) or enzymatic assay (e.g., *CPMB UNIT 11.2*). For intracellular or membrane proteins, liver tissue should be harvested and transgene expression analyzed by immunohistochemistry or immunofluorescence (e.g., *CPMB* Chapter 14), enzymatic assay (usually fresh-frozen sections or paraffin sections), or immunoblotting. Multiple time points may be taken to monitor the level and duration of transgene expression.

Materials

Mouse or rabbit injected with recombinant vector (see Basic Protocol 1 or 2)
70% (v/v) ethanol
Buffered formalin
OCT (optimal cutting temperature) compound
2-Methylbutane

Dissecting board
Forceps and scalpel
Sterile plastic petri dishes
Razor blades
Plastic molds
Cryostat
Dry ice
Cryotubes
Liquid nitrogen

1. Euthanize a mouse or rabbit ~3 to 5 days after adenovirus injection. Place animal on its back and use tape to fasten legs to a dissecting board, keeping legs outstretched.

2. Apply 70% ethanol to the belly of the animal to wet down the hair. Use forceps to pick up the skin and cut down the midline, dissecting skin away from abdominal wall. Using forceps to lift the abdominal wall, cut through the midline, taking care to avoid internal organs. Make additional lateral cuts and fold back skin and abdominal wall.

3. Dissect out liver carefully, avoiding holding liver tissue itself with forceps. Place in a sterile plastic petri dish. Cut pieces of liver tissue using two clean razor blades; hold one along the line to be cut, and then drag the second razor blade adjacent to the first.

4. Preserve liver tissue samples in one or more of the following ways.

   a. *For paraffin sections.* Cut small pieces (a few mm^2) and place in buffered formalin. Fix overnight or longer before embedding.

   b. *For fresh-frozen sections.* Cut pieces to be placed in a plastic mold (for mouse liver, 22 × 22–mm molds can hold one large or two smaller pieces of liver) and cover with OCT compound. Place a shallow container in the fume hood and fill to a depth of ∼1 in. (∼2.5 cm) with 2-methylbutane and dry ice pellets or ground dry ice. When cold, place plastic mold in the container so that the level of cold 2-methylbutane is approximately level with the OCT. When fully frozen, store sample at −70°C. Warm to −20°C before cutting on a cryostat.

   c. *For RNA purification (northern blots) or protein preparations (immunoblots).* Cut pieces of liver tissue from several different lobes, place in labeled cryotubes, and drop into liquid nitrogen. Store at −70°C.

References: Ferry and Heard, 1998; Hitt et al., 1997

Contributor: Karen Kozarsky

13

APPENDIX 1

Reagents and Solutions

This appendix includes recipes for all reagents and solutions used in *Short Protocols in Human Genetics*. Solutions are listed alphabetically, with the unit(s) in which the solutions are used listed in parentheses. These unit references are critical because solutions from different units may have the same name but different recipes. No unit numbers are indicated for commonly used solutions. Some solutions (e.g., PBS) may have one commonly used formulation followed by one or more that are used in specific units. Recipes for some solutions include reagents for which separate recipes are included in this appendix; these are indicated by "see recipe" or a ✓. General information about molarities and specific gravities of acids and bases can be found in Table A.1.1. When preparing solutions, use deionized, distilled water and (for most applications) reagents of the highest grade available. Sterilization is recommended for most applications and is generally accomplished by autoclaving or filtration.

CAUTION: Follow standard laboratory safety guidelines and heed manufacturers' precautions when working with hazardous chemicals.

Acetone/formaldehyde fixative (UNIT 4.7)

Prepare 0.007% to 0.03% (v/v) formaldehyde in acetone (10 to 40 μl of 37% formaldehyde in 50 ml acetone). Optimize the formaldehyde concentration, using higher concentrations to improve nuclear and chromosome morphology, and lower concentrations to improve immunostaining. A good initial concentration is 0.015%.

Acid precipitation solution (APPENDIX 3E)
1 M HCl
0.1 M sodium pyrophosphate
Store indefinitely at room temperature

Acid Tyrode's solution (UNIT 9.9)
800 mg NaCl (137 mM final)
20 mg KCl (2.7 mM final)
400 mg polyvinylpyrrolidone (PVP; average mol. wt. ~40,000; 0.1 μM final)
20 mg CaCl$_2$·2H$_2$O (1.36 mM final)
10 mg MgCl$_2$·6H$_2$O (0.49 mM final)
5 mg NaH$_2$PO$_4$·H$_2$O (or 4 mg NaH$_2$PO$_4$; 0.36 mM final)
100 mg glucose (5.5 mM final)
Milli-Q water to 100 ml
Adjust to pH 2.3 with dilute HCl (acceptable working range pH 2.2 to 2.4)

Table A.1.1 Molarities and Specific Gravities of Concentrated Acids and Bases

Acid/base	Molecular weight	% by weight	Molarity (approx.)	1 M solution (ml/liter)	Specific gravity
Acids					
Acetic acid (glacial)	60.05	99.6	17.4	57.5	1.05
Formic acid	46.03	90	23.6	42.4	1.205
		98	25.9	38.5	1.22
Hydrochloric acid	36.46	36	11.6	85.9	1.18
Nitric acid	63.01	70	15.7	63.7	1.42
Perchloric acid	100.46	60	9.2	108.8	1.54
		72	12.2	82.1	1.70
Phosphoric acid	98.00	85	14.7	67.8	1.70
Sulfuric acid	98.07	98	18.3	54.5	1.835
Bases					
Ammonium hydroxide	35.0	28	14.8	67.6	0.90
Potassium hydroxide	56.11	45	11.6	82.2	1.447
		50	13.4	74.6	1.51
Sodium hydroxide	40.0	50	19.1	52.4	1.53

Filter sterilize through 0.22-μm filter, discarding small volume of initial filtrate
Store up to 1 month in aliquots at 4°C

New lots should be tested to ensure effectiveness in zona drilling (i.e., the controlled dissolving of a hole in the zona pellucida).

Acrylamide/bisacrylamide mix, 37.5:1 (30%)
(*UNIT 9.10*)
30% (w/v) acrylamide
0.8% (w/v) bisacrylamide
H_2O to 500 ml
Filter
Store in the dark at 4°C

CAUTION: *Acrylamide is a neurotoxin; always wear gloves when handling acrylamide, preparing solutions, or pouring gels. Polyacrylamide is not toxic.*

Acrylamide stock solution, 50% (w/v) (*UNIT 7.2*)
49 g acrylamide
1 g bisacrylamide
H_2O to 100 ml
Store up to 30 days at 4°C

CAUTION: *Acrylamide is a neurotoxin; always wear gloves when handling acrylamide, preparing solutions, or pouring gels. Polyacrylamide is not toxic.*

AHC medium and plates (*UNIT 5.1*)
1.7 g nitrogen base without amino acids and without ammonium sulfate (Difco)
5 g ammonium sulfate
10 g casein hydrolysate-acid (salt- and vitamin-free; U.S. Biochemical)
20 mg adenine hemisulfate (Sigma)
Dissolve in 950 ml H_2O
Adjust pH to 5.8 with 5 M HCl (~150 μl)
Add 18 g agar (if preparing plates)
Autoclave 20 min
Add 50 ml sterile 40% (w/v) dextrose
For liquid medium, store at 4°C
For plates, pour 80 to 100 ml medium per 150-mm plate and store at 4°C

Alkaline buffer (*UNIT 4.5*)
1 mM sodium borate adjusted to pH 10 to 11 with NaOH
0.8% (w/v) KCl
Mix in 1:1 ratio just before use

KCl concentration and pH of working solution can be adjusted to optimal values as determined empirically.

Alkaline phosphatase buffer, pH 9.5 (*UNIT 4.4*)
✓ 0.1 M Tris·Cl, pH 9.5
0.1 M NaCl
50 mM $MgCl_2$ (add immediately before use)
Store ≤1 year (without $MgCl_2$) at room temperature

Alkaline phosphatase–conjugated $(CTG)_{10}$ oligonucleotide (*UNIT 9.4*)
Conjugate CTG_{10} oligonucleotide to alkaline phosphatase using a commercial kit (e.g., E-Link Plus oligonucleotide labeling kit, Cambridge Research). Monitor efficiency of the oligonucleotide conjugation using the spectrophotometer-based enzyme assay described by the manufacturer. Aliquot conjugated oligonucleotide into single-use tubes (e.g., 20 μl/tube) to avoid repeated freeze/thaw cycles. Store ≤1 year at −20°C.

Alkaline phosphatase staining buffer (*UNIT 12.4*)
100 mM NaCl
✓ 50 mM $MgCl_2$
✓ 100 mM Tris·Cl, pH 8.5
Store indefinitely in a foil-wrapped glass container at room temperature

Alkaline phosphatase staining solution (*UNIT 12.4*)
Prepare a solution containing 1× NBT solution (see recipe) and 1× BCIP solution (see recipe) in alkaline phosphatase staining buffer (see recipe). Prepare just before use.

Ammonium acetate, 10 M
Dissolve 385.4 g ammonium acetate in 150 ml H_2O
Add H_2O to 500 ml
Filter sterilize

Ammonium sulfate, saturated, pH 7.0 (*UNIT 12.3*)
Add 450 g $(NH_4)_2SO_4$ to 500 ml water. Heat on a stir plate until completely dissolved. Filter through Whatman paper while still warm and allow to cool (upon cooling, crystals will form which should not be removed). Adjust pH to pH 7.0 with ammonium hydroxide. Store up to 1 year at 4°C.

Amplification buffer, 10× (*UNITS 1.3 & 9.9*)
✓ 100 mM Tris·Cl, pH 8.3
500 mM KCl (exclude for potassium-free buffer)
25 mM $MgCl_2$
0.1% (w/v) gelatin
Autoclave and store in 1.5-ml aliquots (to minimize contamination) at room temperature for up to 3 months.

Buffer without any added potassium is used when this ion is provided by the KOH and KCl present in the sperm lysis and neutralization solutions.

Anesthetic solution (*UNIT 13.2*)
Mix 2 ml Ketaset (100 mg/ml ketamine·HCl; Fort Dodge Animal Health) and 1 ml 20 mg/ml xylazine (Rompun; Bayer Corp. Agricultural Division) with 27 ml sterile PBS for vaccination (see recipe). Store up to 1 week at 4°C.

Anesthetic solution (*UNIT 13.3*)
7.5 ml 10 mg/ml ketamine (Fort Dodge Laboratories)
0.75 ml 100 mg/ml acepromazine maleate (Fermenta Animal Health)
1.9 ml 20 mg/ml xylazine (Rompun; Bayer Corp. Agricultural Division)
0.9% NaCl to 20 ml
Prepare fresh before injecting

Antifade mounting media (*UNITS 4.4, 4.7, 4.8 & 8.6*)

DABCO mounting medium: Dissolve 0.233 g 1,4-diazobicyclo-[2.2.2]octane (Sigma; 0.21 M final) in 800 μl water. Add 200 μl 1 M Tris·Cl, pH 8.0 (see recipe; 0.02 M final) and 9 ml glycerol (90% final). Mix. Store 100-μl aliquots wrapped in foil at −20°C. Thaw and use once.

Phenylenediamine dihydrochloride: Dissolve 50 mg *p*-phenylenediamine dihydrochloride in 5 ml PBS (9 mM final). Adjust to pH 8 with 0.5 M carbonate/bicarbonate buffer, pH 9.0 (see recipe). Add to 45 ml glycerol (90% final), mix, and filter through 0.22-μm filter. Store in small aliquots in the dark at −20°C. Thaw and use at once.

Vectashield: Purchase from Vector Laboratories; follow manufacturer's instructions.

APAAP substrate solution (*UNIT 4.7*)

Naphthol AS-MX phosphate solution:
0.0020 g Naphthol AS-MX phosphate
　(acid-free; 0.03 M final)
200 μl *N,N*-dimethylformamide
Store in glass tubes up to 2 months at 4°C

Levamisole hydrochloride solution:
0.0024 g levamisole hydrochloride (Sigma; 9.9
　mM final)
✓ 1.0 ml Tris/acetic acid substrate buffer, pH 8.2
Prepare fresh

Fast red solution:
0.0100 g fast red
✓ 1.0 ml Tris/acetic acid substrate buffer, pH 8.2
Prepare fresh

Working solution:
✓ 780 μl Tris/acetic acid substrate buffer, pH 8.2
20 μl Naphthol AS-MX phosphate solution
100 μl levamisole hydrochloride solution
100 μl fast red solution

Mix in the order listed. Filter through a 0.45-μm filter directly onto slides. Use at 20°C immediately upon mixing.

Bacto Trypsin solution (*UNIT 4.3*)

Reconstitute Bacto trypsin powder (Difco) in 10 ml water in the Bactro trypsin vial. Divide into 0.4-ml aliquots and store ≤6 months frozen at −20°C. To use, mix 0.3 ml stock solution with 70 ml Sorensen phosphate buffer, pH 6.8 (see recipe), in a Coplin jar. To increase trypsin concentration, increase stock to as much as 1.5 ml. Prepare fresh daily.

BBMM (basal bone marrow medium) (*UNIT 13.4*)
✓ 375 ml 1× IMDM
✓ 100 ml FBS, heat inactivated 1 hr at 56°C
✓ 25 ml deionized 10% (w/v) BSA
5 ml 200 mM L-glutamine
2.5 ml 10,000 U/ml penicillin/10,000 μg/ml
　streptomycin (Life Technologies)
✓ 500 μl 0.1 M 2-ME
✓ 500 μl 1 mM hydrocortisone
Store up to 1 month at 4°C

The lots of FBS and BSA must be screened (see Support Protocol 5).

BCIP solution, 100× (*UNIT 12.4*)

Dissolve 5-bromo-4-chloro-3-indoyl phosphate (BCIP) in water at 10 mg/ml. Store in a foil-wrapped glass container (i.e., in the dark) at 20°C (stable for months).

BEGM (bronchial epithelial growth medium) (*UNIT 13.5*)

Add the following components of the BEGM SingleQuot kit (Clonetics) to 500 ml BEBM (bronchial epithelial basal medium, also from Clonetics):
2 ml of 13 mg/ml bovine pituitary extract (BPE; 0.052 mg/ml final)
0.5 ml of 0.5 mg/ml hydrocortisone (0.0005 mg/ml final)
0.5 ml of 0.5 μg/ml human recombinant epidermal growth factor (hEGF; 0.0005 μg/ml final)
0.5 ml of 0.5 mg/ml epinephrine (0.0005 mg/ml final)
0.5 ml of 10 mg/ml transferrin (0.01 mg/ml final)
0.5 ml of 5 mg/ml insulin (0.005 mg/ml final)
0.5 ml of 0.1 μg/ml retinoic acid (0.0001 μg/ml final)
0.5 ml of 6.5 μg/ml triiodothyronine (0.0065 μg/ml final)

Some variability is seen in the quality of BPE from Clonetics. It may be necessary to prescreen BPE batches for those that provide optimal growth performance.

Bio948, 500 pM (*UNIT 11.1*)
✓ DEPC-treated water containing:
500 pM biotinylated control oligonucleotide
Bio948
(5′-GTCAAGATGCTACCGTTCAG-3′)
✓ 1× MES buffer
0.1 mg/ml herring sperm DNA (Promega)
0.5 mg/ml BSA (Life Technologies)
Store up to 1 year at −20°C

Bio-Gel P-60 (*UNIT 4.5*)

Slowly add 2 g Bio-Gel P-60 (Bio-Rad) to 300 ml H₂O. Suspend gel beads thoroughly and autoclave. Store ≤6 months at room temperature.

Biotin detection solution (*UNIT 4.4*)

Dilute fluorescein-avidin DCS *or* rhodamine-avidin D (Vector Laboratories) to 2 μg/ml in 4× SSC (see recipe)/1% (w/v) BSA (fraction V). Prepare fresh daily.

Biotin/digoxigenin detection solution (*UNIT 4.4*)

Dilute fluorescein-avidin DCS (Vector Laboratories) and rhodamine-conjugated Fab fragment of sheep anti-digoxigenin (Boehringer Mannheim) to 2 μg/ml each in 4× SSC (see recipe)/1% (w/v) BSA (fraction V). Prepare fresh daily.

Biotinylated alkaline phosphatase solution *(UNIT 4.4)*
  ✓ PBS containing:
  1% (w/v) BSA
  0.1% (v/v) Tween 20
  2.5 µg/ml biotinylated alkaline phosphatase
  Prepare fresh

Biotinylated anti-avidin antibody *(UNIT 8.6)*

Reconstitute 0.5 mg biotinylated anti–avidin D (Vector Laboratories) in 1 ml sterile water and store in 110-µl aliquots at −20°C. Before use, dilute 100 µl of 500 µg/ml stock in 9.9 ml PNM buffer (see recipe). Prepare fresh (final 5 µg/ml).

Biotinylated horseradish peroxidase solution *(UNIT 4.4)*
  ✓ PBS containing:
  1% (w/v) BSA
  0.1% (v/v) Tween 20
  3 µg/ml biotinylated horseradish peroxidase
  Prepare fresh

Blocking solution *(UNIT 2.1)*

Prepare 1% (w/v) solution of blocking reagent for nucleic acid hybridization (Boehringer Mannheim) in detection buffer (see recipe) by gently heating and stirring ~15 min. Make fresh immediately before use.

Blotting buffer *(UNIT 9.10)*
  25 mM Tris, ultrapure (ICN Biochemicals)
  190 mM glycine
  20% (v/v) methanol
  Prepare 1 liter fresh and chill at least 1 hr in ice water before use

Bovine serum albumin, see BSA

BrdU stock solution, 2 mM *(UNIT 8.4)*

Dissolve 6.142 mg of 5-bromo-2′-deoxyuridine (e.g., Aldrich or Sigma) in 10 ml H$_2$O. Filter sterilize with a 0.22-µm filter. Store 1-ml aliquots 2 months in sterile vials wrapped in aluminum foil at −20°C. The powder is stable at least 2 years stored in a desiccator in the dark at −20°C.

BSA (bovine serum albumin), 10% (w/v)

Dissolve 10 g BSA fraction V (e.g., Sigma) in 100 ml H$_2$O and filter sterilize using a low-protein-binding 0.22-µm filter. Store indefinitely at 4°C.

BSA is available in varying forms that differ in fraction of origin, preparation, purity, pH, and cost. Use a form that is appropriate for the application.

BSA, 10% (w/v), deionized *(UNIT 13.4)*

Pour 100 g Sigma fraction V BSA powder carefully onto the surface of 440 ml sterile water and let stand 24 to 36 hr at 4°C, without stirring, until BSA is totally dissolved. Deionize by adding 15 g resin beads (AG 501-X8 [D]; Bio-Rad). Stir slowly in cold room until all beads have changed color (usually in a few hours). Add another 15 g resin beads and stir in cold room

overnight. Repeat addition of resin beads if there are no blue beads remaining the next morning. Pour BSA through a funnel lined with gauze into a well-rinsed (no soap residue) graduated cylinder and record volume. Add an equal volume of 2× IMDM (see recipe). Filter solution through Whatman paper fitted into a funnel. Check pH to be sure that it is close to 7.0; if not, add 7.5% (w/v) sodium bicarbonate solution (tissue culture grade; Sigma) to raise pH. Sterilize by filtration through 0.8-µm and then 0.2-µm disposable filter units (Nalge). Store 25-ml aliquots in plastic tubes up to 1 year at −20°C.

Buffered phenol, see Phenol

CaCl$_2$, 0.5 M *(UNIT 12.6)*
  36.7 g CaCl$_2$·2H$_2$O
  H$_2$O to 500 ml
  Filter sterilize through a 0.45-µm nitrocellulose filter
  Store up to 2 months at 4°C in 50-ml aliquots

CaCl$_2$, 2 M *(UNITS 12.4, 12.5, 12.7)*
  29.41 g CaCl$_2$·2H$_2$O
  H$_2$O to 100 ml
  Filter sterilize using a 0.2-µm nitrocellulose filter
  Store up to 1 year at 4°C

CaCl$_2$, 2.5 M *(UNIT 12.3)*
  183.7 g CaCl$_2$·2H$_2$O (Sigma; tissue culture grade)
  H$_2$O to 500 ml
  Filter sterilize through a 0.45-µm nitrocellulose filter (Nalgene)
  Store at −20°C in 10-ml aliquots (can be frozen and thawed repeatedly)

Calf thymus DNA standard solutions *(APPENDIX 3D)*

Kits containing calf thymus DNA standard for fluorometry are available (Fluorometry Reference Standard Kits; Hoefer). Premeasured, CsCl-gradient-purified DNA of defined GC content, for use in absorption and fluorometric spectroscopy, is available from Sigma (e.g., calf thymus DNA, 42% GC; *Clostridium perfringens* DNA, 26.5% GC).

Calf thymus DNA standards *(UNIT 10.4)*

Use 1 mg/ml calf thymus DNA stock solution to prepare the following standards. Store tightly capped ≤6 months at 4°C.

For 10 µg/ml: Mix 10 µl stock solution with 100 µl 10× TNE buffer (see recipe) and 890 µl sterile water.

For 100 µg/ml: Mix 10 µl stock solution with 10 µl 10× TNE buffer and 80 µl sterile water.

For 250 µg/ml: Mix 25 µl stock solution with 10 µl 10× TNE buffer and 65 µl sterile water.

For 500 µg/ml: Mix 50 µl stock solution with 10 µl 10× TNE buffer and 40 µl sterile water.

Capillary assay solution *(UNIT 10.4)*
  ✓ 100 µl 10× TNE buffer
  20 µl 1 mg/ml Hoechst 33258 dye stock solution

880 μl H₂O

880 μl H_2O

Prepare immediately before use (while fluorometer is warming up)

Carbenicillin, 100 mg/ml (UNIT 11.2)
1 g carbenicillin (Life Technologies)
10 ml sterile water
Sterile filter with a 0.2-μm filter
Store up to 2 months at $-20°C$

Carbonate/bicarbonate buffer, pH 9.0, 0.5 M (UNIT 4.4)
0.42 g $NaHCO_3$
10 ml H_2O
Adjust pH to 9 with NaOH
Store ≤1 year at room temperature

Cell lysis buffer (UNIT 5.1)
✓ 1 liter 0.5 M EDTA, pH 8.0
10 g sodium N-lauroylsarcosine (Sarkosyl; 1% final)
25 μg/ml proteinase K (add just before use)
Store indefinitely at $25°C$

Cell lysis solution (UNIT 1.3)
9 parts 222 mM fresh autoclaved KOH
✓ 1 part 500 mM DTT
Prepare just before use

Cell shocking solution (UNIT 9.6)
8.0 g/liter NH_4Cl (149 mM)
1.0 g/liter EDTA (2.6 mM)
0.1 g/liter KH_2PO_4 (0.7 mM)
Adjust pH to 7.0 with NaOH
Autoclave to sterilize
Store ≤1 year at room temperature (if sterility is maintained)

Cell staining solution (UNIT 12.4)
1 g/liter Coomassie brilliant blue G
40% (v/v) ethanol
10% (v/v) acetic acid
50% (v/v) H_2O
Store indefinitely at room temperature

Chang medium, 20% (UNIT 8.2)
80 ml α-MEM (Irvine Scientific or Life Technologies)
20 ml FBS (heat-inactivated 1 hr at $56°C$; Irvine Scientific or Whittaker Bioproducts)
2.5 ml Chang A supplement (Irvine Scientific)
22 ml Chang B medium (Irvine Scientific)
1 ml 200 mM L-glutamine (Life Technologies)
1 ml penicillin/streptomycin stock (Life Technologies; 5000 U/ml stock concentration each)

Remove 20 ml from a bottle containing 100 ml α-MEM and then add the other ingredients to the remaining 80 ml medium. Make fresh each week under sterile conditions and store at $4°C$. Check the sterility of the newly made medium by placing 3 ml into a sterile tube and incubating overnight in a humidified $37°C$, 5% CO_2 incubator. Medium should still be clear (no turbidity) after 24 hr.

Testing components: Test each new lot of Chang medium for its ability to support growth by cul-turing half of a sample from a routine amniocentesis (i.e., 14 to 22 gestational weeks) in 100% Chang medium (see recipe). Test new batches of penicillin/streptomycin, L-glutamine, FBS, and α-MEM separately in 20% Chang medium. Use only half of an amniotic fluid sample to test any new reagent.

Growth of the sample in medium containing a reagent from a new lot should not be any different from growth of the same sample in medium containing the reagent from a previous lot. Growth in 100% Chang medium should be 2 to 3 days faster than growth in 20% Chang medium. Chromosome breakage is greatly increased in cultures grown in 100% Chang medium.

Chang medium, 100% (UNITS 8.2 & 8.7)
One bottle Chang B medium (Irvine Scientific)
One bottle Chang A supplement (Irvine Scientific)
1 ml penicillin/streptomycin (Life Technologies; 5000 U/ml each)
1 ml 200 mM L-glutamine (Life Technologies)

Chelex 100 slurry (APPENDIX 3C)
Prepare a 1:1 (w/v) mixture of Chelex 100 (50 to 100 mesh, Bio-Rad) in deionized, distilled water and autoclave. Store up to 3 months at room temperature. Shake the slurry to produce a uniform suspension before dispensing.

Cholesterol stock solution, 20.0 mg/ml (51.7 mM) (UNIT 12.9)
Rinse a glass test tube three times with chloroform and air dry. Weigh 100.0 mg cholesterol into the tube. Dissolve in ~4.0 ml chloroform, then dilute to 5.0 ml with more chloroform. Store up to 3 months at $-20°C$ in a sealed, light-protected glass container.

Church's hybridization buffer (UNIT 5.2)
500 ml 1.0 M sodium phosphate buffer, pH 7.2 (500 mM final)
✓ 350 ml 20% (w/v) SDS (7% final)
✓ 100 ml 10% (w/v) BSA (1% final)
✓ 2 ml 0.5 M EDTA, pH 7.6 (1 mM final)
48 ml H_2O

For 1.0 M sodium phosphate buffer: Add 268.08 g $NaH_2PO_4 \cdot 7H_2O$ to 800 ml water and warm, if necessary, to dissolve solid. Adjust pH to 7.2 with concentrated phosphoric acid (~4 ml). Adjust final volume to 1 liter. Store indefinitely at room temperature.

For hybridization buffer: Add SDS, BSA, EDTA, and water. Store up to 3 months at room temperature.

Cleaning solution (UNIT 11.2)
Dissolve 100 g NaOH in 400 ml water. Add 600 ml of 100% ethanol and stir until the solution clears. If the solution does not clear, add water until it does. Store up to 24 hr at room temperature.

Cloning cylinders (*UNIT 3.1*)

Glass or metal cloning cylinders can be purchased (e.g., Bellco Glass) and should be thoroughly rinsed in distilled water. They should then be placed in a glass petri plate and sterilized by autoclaving or stored in 95% ethanol and sterilized by flaming immediately prior to use. After use, the cylinders may be cleaned with 10 N KOH (to remove vacuum grease) and reused. Store indefinitely at room temperature.

Alternatively, disposable cloning cylinders can be made by cutting a 200-μl pipettor tip 1 cm from the wide end using a scalpel, and sterilizing the resulting 1-cm cylinder by autoclaving.

Cold nucleotide mix, 10× (*UNIT 2.1*)

✓ 2 mM each dGTP, dATP, and dTTP (see dNTPs)
✓ 25 μM dCTP (see dNTPs)
   Store up to 6 months at −80°C

Collagen, 0.3% (*UNIT 13.3*)

Dissolve 10 mg rat tail collagen Type I (Sigma) in 3.3 ml 0.1% acetic acid (15 μl glacial acetic acid in 14.985 ml autoclaved double-distilled water; filter through 0.22-μm filter). Vortex solution three to five times during the day and let shake at room temperature overnight to dissolve collagen completely. Prepare fresh.

Freshness of collagen is important for the quality of the plug; do not keep collagen >2 weeks.

Collagen-coated tissue culture plates (*UNIT 13.2*)

Add 1 ml concentrated acetic acid to 179 ml water and sterilize using a 0.2-μm filter. Add 20 ml of 0.1% sterile calf skin collagen (Sigma) in 0.1 N acetic acid. Place in plastic tissue culture plates and incubate overnight at room temperature (not in a cell incubator). Remove collagen solution from plates (solution can be reused; store at 4°C). Rinse plates thoroughly with sterile water and allow to dry. Store at room temperature (stable at least 6 months.)

Collagenase, 10 mg/ml (*UNIT 10.2*)

Dissolve lyophilized collagenase in sterile unsupplemented RPMI 1640 to give a stock solution of 10 mg/ml. Add 0.05% (w/v) DNase type I (Sigma) and pass through a 0.45-μm filter. Store in 0.5- to 1-ml aliquots ≤6 months at −20°C. Do not refreeze after thawing.

Type B collagenase from Boehringer Mannheim and Life Technologies collagenase work equally well.

Collagenase/dispase/CaCl₂ solution (*UNIT 13.2*)

   1.5 U/ml collagenase D (Roche Molecular
      Biochemicals)
   2.4 U/ml dispase, grade II (Roche Molecular
      Biochemicals)
   2.5 mM CaCl₂
   Prepare immediately before use from 2× collagenase and dispase stocks (stored in aliquots at −20°C)

Collagenase solution (*UNIT 8.1*)

Dissolve collagenase type IV powder (Worthington) in Eagle's MEM to give a concentration of 100 U/ml (based on the specific activity provided by the supplier). Filter sterilize, aliquot to 5-ml bottles, and store ≤2 months at −20°C.

Collagenase solution (*UNIT 8.3*)

Dissolve collagenase CLSI (Worthington) or collagenase type II or IV in serum-free complete culture medium or HBSS (see recipe) at 3 mg/ml. Filter sterilize and store in 1-ml aliquots at −5°C.

Collagenase solution (*UNIT 10.1*)

Resuspend 100 mg DNase (Sigma) in 10 ml HBSS (see recipe). Divide into 5-ml aliquots and store at −20°C. Add one aliquot to 95 ml of RPMI 1640. Add an entire 1-g vial of BMR type B collagenase (Roche Diagnostics). Cover and shake to dissolve completely. Rinse collagenase vial to dissolve any residual powder. Filter solution using a 0.45-μm filter (Fisher; three or four filters will be necessary). Store up to 6 months at −20°C in 1-ml aliquots and thaw as needed.

Colorimetric substrate (*UNIT 2.2*)

Dissolve one 5-mg tablet of phosphatase substrate (Sigma; store up to 6 months at −20°C) in 5 ml diethanolamine buffer (see recipe). Prepare fresh before use.

Complete culture medium A (*UNIT 10.1*)

   To a sterile 100-ml bottle add:
   100 ml RPMI 1640 with HEPES (Life
      Technologies)
✓ 20 ml heat-inactivated FBS
   6.2 ml Origen GCT-CM (Igen International,
      Fisher)
   1 ml glutamine (Life Technologies; dissolve
      completely)
   1 ml penicillin/streptomycin/neomycin (PSN)
      antibiotic mixture (Life Technologies)
   Store up to 6 months at −20°C and thaw as
      needed

Complete culture medium B (*UNIT 10.1*)

   Prepare as for complete culture medium A (above), but omit Origen GCT-CM.

Complete DMEM

Dulbecco's modified Eagle medium, high-glucose formulation (e.g., Life Technologies), containing:

✓ FBS (heat-inactivated) at desired concentration
   (optional)
   1% (v/v) nonessential amino acids
✓ 2 mM L-glutamine
   100 U/ml penicillin
   100 μg/ml streptomycin sulfate
   Filter sterilize and store ≤1 month at 4°C

Complete IMDM/10% FBS (*UNIT 12.3*)

Iscove's modified Dulbecco's medium (Life Technologies) containing:

✓ 10% (v/v) FBS
100 U/ml penicillin
100 μg/ml streptomycin sulfate
Store up to 2 months at 4°C

Complete MEM *(UNIT 12.7)*

Eagle modified essential medium (MEM) containing:

1× nonessential amino acids
100 U/ml penicillin G
100 μg/ml streptomycin sulfate
2 mM glutamine
10% (v/v) FBS (for serum-containing medium only)
Filter sterilize through 0.2-μm filters (Nalgene)
Store 1 to 2 months at 4°C

All ingredients listed are available from Life Technologies.

Complete MEM without serum is used for virus adsorption and for preparing methylcellulose overlay (see recipe).

Complete RPMI

RPMI 1640 medium (e.g., Life Technologies) containing:

✓ FBS (heat-inactivated) at desired concentration (optional)
✓ 2 mM L-glutamine
100 U/ml penicillin
100 μg/ml streptomycin sulfate
Filter sterilize and store ≤1 month at 4°C

Complete RPMI *(UNIT 4.2)*

RPMI 1640 medium (Life Technologies) with 300 mg/liter L-glutamine and 25 mM HEPES
1% (v/v) 10 mg/ml gentamicin solution (Sigma; final 100 μg/ml)
Store up to expiration date on bottle at 4°C

Control primers *(UNIT 9.7)*

In general, use control primers A and B. Use C and D to amplify an ARMS product that is >300 bp. Use E and F when the ARMS product has a high GC content (>65%) and is >300 bp.
Control primer A
5′-CCCACCTTCCCCTCTCTCCAGGCAAAT GGG-3′
Control primer B
5′-GGGCCTCAGTCCCAACATGGCTAAGAG GTG-3′

These primers amplify a fragment of the ATT gene and generate a 360-bp PCR product.

Control primer C
5′-CCAAGCCCAACCTTAAGAAGAAAATT GGAG-3′
Control primer D
5′-CCAAACCCACGGTACGCATGGGAACA CTGC-3′

These primers amplify a fragment of the APC gene and generate a 813-bp PCR product.

Control primer E
5′-CTCACCTGCGTCAGGAGAGCACACAC TTGC-3′

Control primer F
5′-CATCGAGACCTCGGCCAAGACCCGGC AGG-3′

These primers amplify a fragment of the RAS gene and generate a 825-bp PCR product.

Crystal violet staining solution, 1% (w/v) *(UNIT 12.7)*
1 g crystal violet (Sigma)
100 ml 50:50 (v/v) methanol/H$_2$O
Store 1 to 2 years at room temperature

CsCl$_2$ gradient solutions *(UNIT 12.3)*
50 g CsCl (for 1.37 g/ml) or 60 g CsCl (for 1.5 g/ml)
✓ PBS to 100 ml

Check density of each solution by weighing 1 ml. Filter sterilize. Store up to 1 year at room temperature.

CSPD, 0.2 mM *(UNIT 2.1)*

Prepare 25 mM stock solution of disodium 3-(4-methoxyspiro{1,2-dioxetane-3,2′-(5′-chloro)tricyclo[3.3.1.1^{3,7}]decan}-4-yl)phenyl phosphate (CSPD; Tropix) and store at 4° or −20°C. Immediately before use, dilute 80 μl in 10 ml substrate buffer (see recipe).

Culture and micromanipulation media *(UNIT 9.9)*

Embryo culture and wash medium: Prepare bicarbonate-buffered human tubal fluid (HTF) medium (Irvine Scientific), or other suitable bicarbonate-buffered IVF culture medium, supplemented with one of the following proteins sources: 10% synthetic serum substitute (Irvine Scientific), 10% of a 5% solution of human serum albumin (Irvine Scientific), or 10% maternal serum. Store up to 1 week at 4°C once protein supplement has been added. Unopened medium may be used up to 50 days after the date of manufacture if stored at 4°C.

Biopsy medium: Prepare HEPES-buffered HTF medium or other suitable HEPES-buffered medium supplemented with a protein source as described above. Store up to 1 week at 4°C.

Blastomere wash medium: Prepare HEPES-buffered HTF medium or other suitable HEPES-buffered medium without protein supplement (or with a purified protein source as described above, provided that the protein is free of genomic DNA and does not interfere with DNA amplification or analysis). Store up to 1 week at 4°C.

Filter sterilize all media through a 0.22-μm filter, discarding a small volume of the initial filtrate.

Zona drilling is slower but more precise in HEPES-buffered than bicarbonate-buffered medium. HEPES buffering also provides the convenience of air equilibration and better pH stability during biopsy. Protein sources are chosen for their ability to support advanced embryo development and for absence of DNA when performing molecular analysis.

Use Milli-Q purified water (meeting College of American Pathologists Type I reagent-grade water standards) or other embryo-tested ultrapure water in all reagents.

Cycle sequencing reaction buffer, 10× *(UNIT 7.5)*
- ✓ 100 mM Tris·Cl, pH 9.0
- 500 mM KCl
- 40 mM MgCl$_2$
- 20 µM *each* dATP, dCTP, dGTP (*or* 7-deaza-dGTP), and dTTP
- Store in aliquots at −20°C

Cyclosporin A, 0.2 mg/ml *(APPENDIX 3J)*

Add 2.0 ml of 50 mg/ml cyclosporin A stock solution (Sandimmune IV; Sandoz) to 500 ml of 37°C complete DMEM (see recipe). Store in a cool dark place in 10-ml aliquots at −70°C.

D10HG medium *(UNIT 13.4)*
- 450 ml Dulbecco's modified Eagle medium (DMEM) with high glucose
- ✓ 50 ml FBS, heat inactivated 1 hr at 56°C
- 5 ml 200 mM L-glutamine
- 2.5 ml 10,000 U/ml penicillin/10,000 µg/ml streptomycin (Life Technologies)
- Store up to 1 month at 4°C

DAPI in antifade mounting medium *(UNIT 4.8)*

Dissolve 1 mg 4′,6-diamidino-2-phenylindole (DAPI) in 10 ml water. Add a few drops of methanol before adding water to help dissolve DAPI. Aliquot into tubes wrapped in aluminum foil and store at −20°C (stable 1 year or more). Before use, add 1 to 2 µl DAPI stock solution to 1 ml phenylenediamine antifade mounting medium (see recipe for antifade mounting media). Protect from light at −20°C. Replace as solution darkens.

DAPI staining solution *(UNIT 4.3)*

Dissolve 0.2 µg/ml 4′,6-diamidino-2-phenylindole (DAPI; Sigma; 5.7×10^{-7} M final) in McIlvaine buffer, pH 7.0 (see recipe). Store in a foil-covered Coplin jar ≥1 week at 4°C.

DAPI staining solution *(UNIT 4.4)*

Dissolve 1 mg 4′,6-diamidino-2-phenylindole (DAPI) in 10 ml water. Add a few drops of methanol before adding water to help dissolve DAPI. Aliquot into aluminum foil–wrapped tubes and store a year or more at −20°C. Before use, dilute 1/1000 in PBS (see recipe). Store in aluminum foil–wrapped tubes several weeks at 4°C.

DAPI staining solution *(UNIT 8.6)*

Mix 1.5 mg 4′,6-diamidino-2-phenylindole (DAPI) in 0.5 ml H$_2$O and store indefinitely at −20°C. Before use, mix 10 µl DAPI stock in 15 ml phenylenediamine dihydrochloride antifade mounting medium (final 2 µg/ml DAPI; see recipe for antifade mounting media). Store in aliquots in 1-ml vials up to 6 months at −20°C.

DC-Chol stock solution, 2.0 mg/ml (3.7 mM) *(UNIT 12.9)*

Rinse a glass test tube three times with chloroform and air dry. Weigh 10.0 mg (dimethylaminoethane)carbamoyl-cholesterol powder (DC-chol) into the tube. Dissolve in ~4.0 ml chloroform, then bring to 5.0 ml with more chloroform. Store up to 3 months at −20°C in a sealed, light-protected glass container.

DC-cholesterol powder can be purchased from Sigma or prepared according to Gao and Huang (1991).

DdeI digestion buffer, 5× *(UNIT 9.9)*
- Prepare using HPLC-grade H$_2$O
- 500 mM NaCl
- ✓ 250 mM Tris·Cl, pH 7.9
- 50 mM MgCl$_2$
- 5 mM DTT
- Store up to 1 year at −20°C

Deionized formamide, see Formamide

Denaturing acrylamide gel solution *(APPENDIX 3F)*
- 25.2 g urea (ultrapure)
- 6.0 ml 38% acrylamide/2% bisacrylamide
- ✓ 6.0 ml 10× TBE
- 27 ml H$_2$O (total volume 60 ml)
- Filter solution through Whatman no. 1 filter paper
- Store 2 to 4 weeks at 4°C

Quantities are for a single 4% acrylamide sequencing gel. For 6% or 8% gels, increase acrylamide/bisacrylamide to 9.0 and 12.0 ml, respectively, and decrease water to keep total volume constant.

Solutions of acrylamide deteriorate quickly, especially when exposed to light or left at room temperature.

Denaturing buffer *(UNIT 9.3)*
- 0.4 M NaOH
- 3.0 M NaCl
- Store up to 1 year at room temperature

Denaturation solution *(UNIT 4.8)*

Combine 70 ml formamide (Ultrapure or deionized; see recipe) and 10 ml of 20× SSC, pH 7.0 (see recipe). If necessary, adjust pH to 7.0 with HCl, then bring final volume to 100 ml with water. Store at 4°C. Prepare fresh solution weekly, or more frequently with heavy use (>50 slides/week).

Denaturation solution *(UNIT 8.6)*
- 28 ml formamide (deionized and molecular biology certified; 70% v/v final)
- ✓ 4 ml 20× SSC
- 8 ml H$_2$O
- Adjust pH to 7.0 with HCl
- Store up to 1 week at 4°C

Denaturing solution *(UNIT 9.2)*
- 1 N NaOH
- 2.0 M NaCl

✓ 0.025 EDTA
0.0001% (w/v) bromphenol blue (add just before use)
Prepare fresh

Denaturing solution *(APPENDIX 3G)*
20 ml 10 M NaOH
120 ml 5 M NaCl
H_2O to 1 liter
Prepare fresh

Denhardt's solution, 100×
10 g Ficoll 400
10 g polyvinylpyrrolidone
10 g BSA (fraction V)
H_2O to 500 ml
Filter sterilize
Store at $-20°C$ in 25-ml aliquots

DEPC-treated H_2O *(UNITS 7.1 & 11.1)*

Add 0.2 ml diethylpyrocarbonate to 100 ml sterile H_2O and shake vigorously. Autoclave to inactivate residual DEPC. Prepare fresh before use.

CAUTION: *DEPC is a suspected carcinogen and should be handled carefully.*

DEPC-treated solutions *(UNIT 9.10)*

Add diethylpyrocarbonate (DEPC, Sigma) to 0.05% to 0.1% (v/v). Let stand overnight, then autoclave to inactivate residual DEPC. Store at room temperature.

CAUTION: *DEPC is a suspected carcinogen and should be handled carefully.*

Destaining solution *(UNIT 9.10)*
10% (v/v) methanol
10% (v/v) acetic acid
80% (v/v) H_2O
Store at room temperature

Detection buffer *(UNIT 2.1)*
✓ 100 mM Tris·Cl, pH 7.4
150 mM NaCl
0.3% (v/v) Tween 20

Make from sterile stock solutions and store indefinitely at room temperature. Inspect solution for bacterial contamination before use.

Detection solution *(UNIT 4.8)*
5 μg/ml fluorescein-avidin DCS (Vector Laboratories)
1 μg/ml rhodamine-conjugated Fab fragment of sheep anti-digoxigenin (Boehringer Mannheim)
✓ 4× SSC, pH 7.0
1% (w/v) BSA
Prepare immediately before use

Developing solution *(UNIT 2.1)*
120 g sodium carbonate (0.28 M final)
2 ml 37% formaldehyde (0.02% final)
H_2O to 4 liters
Prepare fresh

Dextran sulfate, 60% *(UNIT 10.3)*
24 g dextran sulfate
26.6 ml H_2O

Heat 4 to 5 hr in 55°C water bath to dissolve
Store ≤6 months at 4°C

Dextrose, 5.2% *(UNIT 12.9)*

Dissolve 2.6 g dextrose in ∼40 ml sterile water, then dilute to 50.0 ml with water (can also be prepared at 5× concentration). Sterile filter through a 0.2-μm membrane. Store in 5.0-ml aliquots indefinitely at $-20°C$.

DGGE loading buffer *(UNIT 10.3)*
20% (w/v) sucrose
✓ 10 mM Tris·Cl, pH 7.8
✓ 1 mM EDTA, pH 8.0
0.5% (w/v) bromphenol blue

Dialysis buffer *(APPENDIX 3A)*
Stock solution:
✓ 360 ml 1 M Tris·Cl, pH 7.5
✓ 72 ml 0.5 M EDTA, pH 8.0
72 ml 5 M NaCl (0.72 M final)
Filter sterilize
Store at room temperature

Working solution: At time of dialysis, dilute 49 ml stock in 3500 ml water. Store at 4°C.

Dialysis buffer I *(UNIT 5.3)*
✓ 20 mM Tris·Cl, pH 7.5
1 mM Na_2EDTA
500 μM spermine
Store up to 1 month at 4°C

Dialysis buffer II *(UNIT 5.3)*
✓ 20 mM Tris·Cl, pH 7.6
✓ 1 mM EDTA, pH 8.0
100 μM spermine
Prepare fresh from stock solutions

Diethanolamine buffer *(UNIT 2.2)*
100 mg $MgCl_2·6H_2O$
97 ml diethanolamine (Sigma)
800 ml H_2O
Adjust to pH 9.8 with 1 M HCl
Dilute to 1 liter with H_2O
Store in a dark bottle up to 6 months at room temperature

Diethylpyrocarbonate, see DEPC

Differentiation medium *(UNIT 13.2)*
95 ml Dulbecco's modified Eagle medium (DMEM; Life Technologies)
5 ml horse serum (HyClone; 5% final)
1 ml 100× penicillin/streptomycin solution (Life Technologies; 100 U/ml penicillin and 100 μg/ml streptomycin final)
Store at 4°C; stable at least 1 month

Digoxigenin detection solution *(UNIT 4.4)*

Dilute 2 μg/ml fluorescein- *or* rhodamine-conjugated Fab fragment of sheep anti-digoxigenin (Boehringer Mannhein) in 4× SSC (see recipe)/ 1% (w/v) BSA (fraction V). Prepare fresh daily.

Disodium phosphate buffer, pH 7.0 *(UNIT 4.3)*
0.2 g KCl (2 mM final)
8.0 g NaCl (0.14 M final)

0.2 g KH$_2$PO$_4$ (1.4 mM final)
1.16 g Na$_2$HPO$_4$ (8 mM final)
1 liter H$_2$O
Adjust pH to 7.0 with monobasic or dibasic
   phosphate solution if needed
Store ≤6 months at room temperature

Distamycin A staining solution (UNIT 4.3)

Dissolve 0.2 mg/ml distamycin A (Sigma; 3.5 ×
10^{-4} M final) in McIlvaine buffer, pH 7.0 (see
recipe). Store protected from light in small
aliquots ≤6 months at −20°C.

Dithiothreitol, see DTT

DMEM/10% FBS medium (UNIT 13.3)

*Dulbecco's modified essential medium (Irvine
Scientific) containing:*

2 mM L-glutamine (Irvine Scientific)
2.5 g/ml Fungizone (Irvine Scientific)
50 μg/ml gentamycin (Irvine Scientific)
10% fetal bovine serum (FBS; Irvine Scientific)
Store ≤2 months at 4°C

DMEM/F12/N2/FGF medium (UNIT 13.3)

*1:1 mixture of DMEM and F12 media (Irvine
Scientific) containing:*

2.5 mM L-glutamine (Irvine Scientific)
3.1 g/liter D-glucose (Irvine Scientific)
5 μg/ml insulin (Sigma)
100 μg/ml transferrin (human iron-free; Sigma)
20 nM progesterone (Sigma)
100 μM putrescine dihydrochloride (Sigma)
30 nM sodium selenite
Store ≤2 months at 4°C

DNA extraction buffer (UNIT 12.7)

✓ 50 mM Tris·Cl, pH 8.0
✓ 2 mM EDTA
0.5% (w/v) Tween 20 (Sigma)
Store up to 1 year at room temperature
Just before use, add 400 μg/ml proteinase K

*SDS can be substituted for Tween 20 in this
buffer.*

DNA vaccine solution (UNIT 13.2)

Prepare concentrated DNA stock solutions at 2
mg/ml in sterile saline (140 mM NaCl) or PBS
for vaccination (see recipe); store stocks indefinitely at −70°C. Dilute to the desired final concentration in the same buffer prior to injections.
For mice, ≤200 μg DNA may be injected on a
single occasion.

*For purification and preliminary testing of DNA
vaccine, see discussion in CPHG UNIT 13.2.*

DNase I, 1 mg/ml (UNIT 4.4)

Dissolve 1 mg DNase I (Worthington) in 1 ml
0.15 M NaCl/90% (v/v) glycerol. Store a year or
more at −20°C in 20-μl aliquots. Immediately
prior to use, dilute 1/1000 (1 μg/ml) or 1/250
(4 μg/ml) in cold sterile water. Refreeze stock
solution, but discard unused working solution.

DNase I solution (UNIT 4.5)
   Stock solution:
✓ 20 mM Tris·Cl, pH 7.5
1 mM MgCl$_2$
50 mM NaCl
1 mg DNase I (Boehringer Mannheim)
Mix by inverting tube (do not vortex)
Store overnight at 4°C to allow reagents to
   dissolve completely
Add 1 vol glycerol and mix thoroughly for
   long-term storage (≥1 year) at −20°C

*Periodic recalibration of activity may be neces-
sary.*

Working solution: Dilute 1:5000 with DNase
diluent (see recipe) immediately before use.

*Amount of dilution may vary with activity of
DNase stock.*

DNase digestion buffer (UNIT 12.3)
✓ 10 mM Tris·Cl, pH 7.5
✓ 10 mM MgCl$_2$
2 mM CaCl$_2$
50 U/ml DNase I
Prepare fresh

DNase diluent (UNIT 4.5)
✓ 7.5 ml 1 M Tris·Cl, pH 7.6 to 7.8 (500 mM final)
0.75 ml 1 M MgCl$_2$ (50 mM final)
6.75 ml H$_2$O
Store in 1-ml aliquots at −20°C (stable at least
   1 year)

dNTP mix, 1.25 mM (UNIT 9.5)

Purchase 25 mM dATP, dCTP, dGTP, and dTTP
from Perkin-Elmer. Microcentrifuge each for 30
sec at maximum speed. Combine 50 μl of each
dNTP and add 800 μl double-distilled water (total
1 ml at 1.25 mM each dNTP). Divide into 100-μl
aliquots.

dNTPs: dATP, dTTP, dCTP, and dGTP

Concentrated stocks: Purchase deoxyribonucle-
oside triphosphates (dNTPs) from a commercial
supplier (Amersham Pharmacia Biotech is recommended) either as ready-made 100 mM solutions (preferred for shipping and storage) or
in lyophilized form. If lyophilized, dissolve in
deionized H$_2$O to an expected concentration of
30 mM, then adjust to pH 7.0 with 1 M NaOH.
Determine the actual concentration of each dNTP
by UV spectrophotometry at 260 nm using the
following extinction coefficients: A, 15,200; C,
7,050; G, 12,010; T, 8,400.

4dNTP mix: Prepare mixed dNTP solutions containing equimolar amounts of all four DNA precursors. For example, 2 mM 4dNTP mix contains
2 mM *each* dATP, dCTP, dGTP, and dTTP.

3dNTP mix: Prepare as for 4dNTP mixes, but
omit one particular dNTP as required, e.g., for
a labeling reaction. For example, 2 mM 3dNTP
mix (minus dATP) contains 2 mM *each* dCTP,
dGTP, and dTTP.

Store all dNTPs and dNTP mixes as aliquots at
−20°C (stable for ≤1 year).

DOM (Dexter's original medium) (UNIT 13.4)
✓ 350 ml 1× IMDM
75 ml horse serum (HS), heat inactivated 1 hr at
56°C
✓ 75 ml FBS, heat inactivated 1 hr at 56°C
5 ml 200 mM L-glutamine
2.5 ml 10,000 U/ml penicillin/10,000 μg/ml
streptomycin (Life Technologies)
✓ 500 μl 0.1 M 2-ME
✓ 500 μl 1 mM hydrocortisone
Store up to 1 month at 4°C

*Individual lots of serum must be screened (see
Support Protocol 5).*

D-PBS (Dulbecco's PBS) (UNIT 13.3)
✓ PBS
0.1 g/liter $CaCl_2$
0.1 g/liter $MgCl_2 \cdot 6H_2O$
1.0 g/liter D-glucose
0.036 g/liter sodium pyruvate
Store ≤2 months at 4°C

dsDNA standards (UNIT 11.2)
For 500 μg/ml, use the 0.5 mg/ml dsDNA from
Life Technologies. For 50, 100, and 250 μg/ml,
dilute as appropriate in TE buffer (see recipe). It
is good practice to check both the integrity (i.e.,
on an agarose gel; APPENDIX 3G) and the concentration (i.e., absorbance; APPENDIX 3D) of the
standards before use.

DTT (dithiothreitol), 1 M
15.45 g DTT
100 ml H_2O
Filter sterilize; do not autoclave
Store at −20°C

E. coli DNA polymerase I buffer, 10× (APPENDIX 3E)
✓ 500 mM Tris·Cl, pH 7.5
100 mM $MgCl_2$
10 mM DTT
0.5 mg/ml BSA or gelatin
Store ≤1 month at 4°C

EcoRI endonuclease/methylase buffer, 10× (UNIT 5.4)
100 μl 32 mM S-adenosylmethionine (New
England Biolabs; final 0.8 mM)
80 μl 1 M $MgCl_2$ (final 20 mM)
800 μl 5 M NaCl (final 1 M)
✓ 2 ml 1 M Tris·Cl, pH 7.5 (final 0.5 M)
40 μl 1 M dithiothreitol (DTT; final 10 mM)
980 μl sterile H_2O (total volume 4 ml)
Store up to 1 year at −80°C in small aliquots
(0.21 ml)

EDTA, 0.5 M, pH 8.0
186.1 g $Na_2EDTA \cdot 2H_2O$
(ethylenediaminetetraacetic acid)
700 ml H_2O
✓ Stir while adding 50 ml 10 M NaOH
Add H_2O to 1 liter

*Begin titrating before the sample is completely
dissolved. EDTA, even in the disodium salt form,
is difficult to dissolve at this concentration unless
the pH is increased to between 7 and 8. Heating
the solution may also help to dissolve EDTA.*

EDTA buffer with 0.025% (w/v) colchicine (UNIT 4.2)
8.0 g NaCl
0.2 g KH_2PO_4
0.2 g KCl
1.15 g Na_2HPO_4
0.2 g EDTA, disodium salt

Dissolve in 1 liter deionized H_2O and autoclave.
Add 5 μl of 0.5% (w/v) colchicine solution (see
recipe) to 100 ml buffer just before use and warm
to 37°C.

β-Estradiol, 200 mM (UNIT 12.5)
24.8 mg (active) β-estradiol (water soluble;
Sigma)
H_2O to 20 ml
Filter sterilize using a 0.2-μm nitrocellulose
filter
Store up to 1 year at 4°C

Ethanol/acetate solution (UNIT 11.2)
3 M sodium acetate stock: Prepare 408.24 g/liter
(3 M) sodium acetate trihydrate. Prepare 3 M
acetic acid by diluting 172.4 ml glacial acetic acid
in water to 1 liter. Titrate the pH of 3 M sodium
acetate solution to 6.0 with the 3 M acetic acid solution. Filter sterilize using a 0.2-μm filter. Store
up to 6 months at room temperature.

Working solution: Dilute 50 ml 3 M sodium acetate stock with 950 ml 100% ethanol. Store up
to 2 weeks at room temperature.

Ethidium bromide assay solution (APPENDIX 3D)
✓ 10 ml 10× TNE buffer
89.5 ml H_2O
Filter through a 0.45-μm filter
✓ Add 0.5 ml 1 mg/ml ethidium bromide

Ethidium bromide solution
Prepare a 10 mg/ml stock solution H_2O (e.g.,
0.2 g in 20 ml). Mix well and store at 4°C in the
dark or in a foil-wrapped bottle. Do not sterilize.
Before use, dilute to 0.5 μg/ml or other desired
concentration in electrophoresis buffer (e.g.,
1× TAE or TBE; see recipes) or in H_2O.

*Ethidium bromide working solution is used to
stain agarose gels to permit visualization of nucleic acids under UV light. Gels should be placed
in a glass dish containing sufficient working solution to cover them and shaken gently or allowed
to stand for 10 to 30 min. If necessary, gels can be
destained by shaking in electrophoresis buffer or
H_2O for an equal length of time to reduce background fluorescence. Alternatively, a gel can be
run directly in ethidium bromide at the appropriate working concentration.*

CAUTION: *Ethidium bromide is a toxic and
powerful mutagen. Gloves should be worn when
working with solution or gel and a mask should*

A1

be worn when weighing out solid. Keep separate waste containers for disposal of contaminated solid and liquid material.

F-10-based primary myoblast growth medium
(UNIT 13.2)

> 400 ml 1× Ham F-10 nutrient mixture (Life Technologies)
> 100 ml fetal bovine serum (HyClone; 20% final)
> 50 μl 25 μg/ml basic fibroblast growth factor, human (bFGF; Promega) in 0.5% (w/v) BSA/1× PBS (see recipes); freeze aliquots only once at −20°C (2.5 ng bFGF/ml final)
> 5 ml 100× penicillin/streptomycin (Life Technologies; 100 U/ml penicillin and 100 μg/ml streptomycin final)
> Store at 4°C; stable at least 1 month

F-10/DMEM-based primary myoblast growth medium *(UNIT 13.2)*

> Prepare F-10-based primary myoblast growth medium (see recipe) but use 200 ml 1× Ham F-10 nutrient mixture and 200 ml Dulbecco's Modified Eagle Medium (DMEM; Life Technologies).

FACS staining buffer *(UNIT 13.2)*

> 4% (v/v) fetal bovine serum (HyClone)
> 10 mM HEPES
> ✓ PBS (1× final)
> Store indefinitely at 4°C

FACS stop buffer *(UNIT 13.2)*

> 1.8 μl 1 mg/ml propidium iodide (1 μg/ml final)
> 36 μl 50 mM phenylethyl-β-D-thiogalactoside (PETG; 1 mM final; optional)
> ✓ 1.8 ml FACS staining buffer, ice-cold
> Prepare fresh

> *PETG inhibits endogenous β-galactosidase activity of the cells, but this is often not necessary.*

FBS (fetal bovine serum)

> Purchase FBS (shipped on dry ice) and store frozen until needed (≤1 year at −20°C). After thawing, store 3 to 4 weeks at 4°C. If FBS will not be used within this time, aseptically divide into aliquots and freeze. Keep one aliquot for use at 4°C and discard after 3 to 4 weeks. Avoid thawing and refreezing.

> Heat-inactivated FBS can be purchased commercially or made in the laboratory by heating serum 30 to 60 min in a 56°C water bath.

Fibroblast culture medium *(UNIT 4.5)*

> Prepare complete α-MEM (Life Technologies) with and without 10% FBS. Store up to several weeks at 4°C.

> *If desired, appropriate selective agents may be included to select for somatic cell hybrids (UNIT 3.1).*

Ficoll-Hypaque solution, 1.077 g/liter *(UNIT 10.3)*

> Dissolve 64 g Ficoll (mol. wt. 400,000; Sigma) and 0.7 g NaCl (12 mM final) in 600 ml water using a magnetic stirrer at low speed. Add 99 g

sodium diatrizoate (0.16 M final; Sigma). When all components are in solution, add water to 1 liter. Filter sterilize with a 0.22-μm filter. Store ≤6 months at 4° to 25°C, protected from direct light.

Ficoll-Hypaque solution may also be purchased (Ficoll-Paque; Amersham Biosciences).

Ficoll loading buffer, 10× *(UNIT 9.3)*

> 25% (v/v) Ficoll in H$_2$O (Pharmacia Biotech)
> 0.25% (w/v) bromphenol blue
> ✓ 10 mM EDTA
> Store ≤6 months at −20°C

Fixative solution, 3:1 *(UNIT 9.9)*

> 3 ml 100% methanol
> 1 ml glacial acetic acid
> Make fresh each day

Fluor buffer *(UNIT 11.2)*

> 25 μl Hoechst 33258 solution (from FluoReporter Blue dsDNA Quantitation Kit; Molecular Probes)
> 10 ml TNE buffer (from FluoReporter Blue dsDNA Quantitation Kit)
> 10 ml water

> *Hoechst 33258 solution contains the dye at an unspecified concentration in 1:4 DMSO/H$_2$O. TNE buffer is 10 mM Tris·Cl (pH 7.4)/2 M NaCl/1 mM EDTA.*

Fluorescein-labeled avidin *(UNIT 8.6)*

> Dissolve 1 mg fluorescein-avidin DCS (Vector Laboratories) in 1 ml H$_2$O. Store indefinitely at −20°C. Before use, dilute 50 μl stock in 9.95 ml PNM buffer (see recipe; final 5 μg/ml). Store indefinitely at −20°C.

Fluorodeoxyuridine (FUdR), 10 μM *(UNIT 8.1)*

> Add 1.2 mg FUdR to 5 ml H$_2$O. Dilute 1/100 with distilled H$_2$O and filter sterilize. Prepare 1-ml aliquots in 1.5-ml microcentrifuge tubes and store ≤6 months at −20°C.

> CAUTION: *FUdR is hazardous.*

Formaldehyde-buffered acetone fixative *(UNIT 4.7)*
Buffer:

> 0.2 g Na$_2$HPO$_4$·2H$_2$O (4.8 mM)
> 1 g KH$_2$PO$_4$ (24.5 mM)
> 300 ml distilled H$_2$O
> Store ≤6 months at 4°C

> *Working solution:*

> 15 ml buffer
> 22.5 ml acetone (7.75 M final)
> 12.5 ml 37% formaldehyde (9.3% v/v final)
> Store up to 1 week at 4°C in a Coplin jar

Formaldehyde loading buffer *(APPENDIX 3H)*

> ✓ 1 mM EDTA, pH 8.0
> 0.25% (w/v) bromphenol blue
> 0.25% (w/v) xylene cyanol
> 50% (v/v) glycerol
> Store up to 3 months at room temperature

Formamide, deionized

Stir 500 ml formamide (purity is critical) with 50 g of 20- to 50-mesh ion-exchange resin (Bio-Rad) for 30 min at room temperature. Filter twice through Whatman no. 1 filter paper. Store in 50-ml aliquots at −20°C. Thaw and use once.

CAUTION: *Formamide is hazardous.*

Formamide loading buffer, 2×

✓ Deionized formamide
0.05% (w/v) bromphenol blue
0.05% (w/v) xylene cyanol FF
✓ 20 mM EDTA
Do not sterilize
Store at −20°C

Formamide prehybridization/hybridization solution (APPENDIX 3H)

✓ 5× SSC
✓ 5× Denhardt's solution
50% (w/v) formamide
✓ 1% (w/v) SDS
✓ 100 μg/ml salmon sperm DNA

Combine first four ingredients. Just before use, denature sonicated salmon sperm DNA (see recipe) by boiling 10 min and then add to FPH solution.

Commercial formamide is usually satisfactory for use. If the liquid has a yellow color, deionize as follows: add 5 g of mixed-bed ion-exchange resin [e.g., Bio-Rad AG 501-X8 or 501-X8(D) resins] per 100 ml formamide, stir at room temperature for 1 hr, and filter through Whatman no. 1 paper.

CAUTION: *Formamide is a teratogen. Handle with care.*

Formamide wash solution, 50% and 65% (v/v) (UNIT 8.6)

20 or 26 ml formamide (deionized and molecular biology certified)
✓ 4 ml 20× SSC
H_2O to 40 ml total
Adjust pH to 7.0 with HCl
Store up to 1 week at 4°C

Formamide wash solution, 60% (v/v) (UNIT 9.9)

30 ml formamide (deionized and molecular biology certified, American Bioanalytical)
✓ 5 ml 20× SSC
15 ml H_2O
Adjust pH to 7.0 to 7.1 with 6 N HCl
Make fresh each day

Fragmentation buffer, 5× (UNIT 11.1)

Dissolve 6.06 g Tris base (Sigma; molecular biology grade) in 175 ml DEPC-treated water (see recipe). Adjust pH to 8.1 with glacial acetic acid. Add 12.3 g potassium acetate (from an unopened or dedicated bottle) and 8.04 g magnesium acetate (Sigma; molecular biology grade) and adjust volume to 250 ml (final pH ~8.4). Filter sterilize with a 0.2-μm filter. Store ≤6 months at −20°C.

Freezing medium (APPENDIX 3J)

✓ 45 ml complete RPMI
✓ 45 ml FBS, heat-inactivated
Filter sterilize through 0.2-μm filter
Add 10 ml DMSO under sterile conditions
Store indefinitely at −20°C

Freezing medium (UNIT 13.4)

✓ 80 ml 1× IMDM
20 ml DMSO
Sterile filter
Add 10,000 U/ml heparin (20 U/ml final)
Store up to 1 year at 4°C
Add 10 μg/ml sterile-filtered DNase immediately prior to use

G418, 40 mg/ml (UNIT 12.5)

4 g (active) G418 (Geneticin; Life Technologies)
0.1 M HEPES, pH 7.5, to 100 ml
Filter sterilize using a 0.2-μm nitrocellulose filter
Store up to 6 months at 4°C

G418 sulfate, 50 mg/ml (UNIT 13.4)

Prepare a 100 mM buffer stock by diluting 1 M HEPES (Sigma) 10-fold in water. Use the purity (milligram active protein/gram weight as listed in the accompanying data sheet) of the purchased geneticin (e.g., Life Technologies) to calculate the total milligrams of active protein. Dilute geneticin in 100 mM HEPES to 50 mg active protein/ml. Be sure that the electrode on the pH meter has been soaked in water and rinsed thoroughly to remove potentially toxic contaminants and then adjust pH to 7.30 using ~1 drop of 10 N NaOH/ml (pH will increase steadily). Store small aliquots up to 3 to 5 years at −20°C. Avoid repeated freeze-thaw cycles. Store thawed aliquots up to 2 months at 4°C.

Gel loading buffer, 6×

0.25% (w/v) bromphenol blue
0.25% (w/v) xylene cyanol FF
40% (w/v) sucrose *or* 15% (w/v) Ficoll 400 *or* 30% (v/v) glycerol
Store at 4°C (room temperature if Ficoll is used)

This buffer does not need to be sterilized. Sucrose, Ficoll 400, and glycerol are essentially interchangeable in this recipe. Other concentrations (e.g., 10×) can be prepared for convenience.

Giemsa stain (UNIT 3.1)

Mix 2 g Giemsa stain (Gurrs improved R66; Bio/Medical Specialties) with 32 ml glycerol prewarmed to 60°C. Add an additional 100 ml prewarmed glycerol and stir 1 hr at room temperature. Wrap container in foil and incubate overnight at 60°C. Add 132 ml absolute ethanol and stir an additional 2 to 3 hr at room temperature. Filter through a 0.45-mm filter (Nalgene type CN). Age solution ≥2 weeks prior to use. Store several months at room temperature.

Giemsa stain, 5% (v/v) (UNIT 4.7)

Dilute Giemsa stain (Merck) 1:20 (v/v) in Sorensen buffer (see recipe). Store up to 3 days at 4°C.

L-Glutamine, 0.2 M (100×)

Thaw frozen L-glutamine (Life Technologies), aliquot aseptically into usable portions, then re-freeze. For convenience, L-glutamine can be stored in 1-ml aliquots if 100-ml bottles of medium are used, and in 5-ml aliquots if 500-ml bottles are used. Store ≤1 year at −20°C.

Growth medium (APPENDIX 3J)

✓ 500 ml complete RPMI
✓ 100 ml FBS, heat-inactivated
5 ml 100× glutamine solution (Life Technologies)
5 ml 100× antibiotic/fungizone solution
5 ml 100× HEPES (Life Technologies)
Filter sterilize through a 0.2-μm filter
Store ≤1 week at 4°C

Test medium sterility by pipetting 3 ml into a 60 × 15–mm petri dish and another 0.5 ml into liquid LB medium (see recipe). Incubate both ≥24 hr at 37°C and check for contamination.

GST solution (UNIT 12.8)

✓ PBS containing:
2% (v/v) goat serum
0.2% (v/v) Triton X-100
Store up to 1 month at 4°C

GTE solution (UNIT 5.1)

9.1 ml sterile H_2O
250 μl 2 M glucose
✓ 400 μl 0.5 M EDTA, pH 8.0
✓ 250 μl 1 M Tris·Cl, pH 8.0
Store indefinitely at 4°C

HBSS (Hanks balanced salt solution)

0.40 g KCl (5.4 mM)
0.09 g Na_2HPO_4·$7H_2O$ (0.3 mM)
0.06 g KH_2PO_4 (0.4 mM)
0.35 g $NaHCO_3$ (4.2 mM)
0.14 g $CaCl_2$ (1.3 mM)
0.10 g $MgCl_2$·$6H_2O$ (0.5 mM)
0.10 g $MgSO_2$·$7H_2O$ (0.6 mM)
8.0 g NaCl (137 mM)
1.0 g D-glucose (5.6 mM)
0.01 g phenol red (0.01%; optional)
Add H_2O to l liter and adjust to pH 7.4
Filter sterilize and store at 4°C

HBSS can also be purchased from a number of commercial suppliers. HBSS may be made or purchased with or without $CaCl_2$ and $MgCl_2$. These components are optional and usually have no effect on an experiment; in a few cases, however, their presence may be detrimental. Consult individual protocols to see if the presence or absence of these components is recommended.

HBSS, 0.1× (UNIT 12.5)

✓ 1 part 1× HBSS (made without glucose or phenol red)

✓ 9 parts PBS, pH 7.5 (adjust pH with HCl)
Filter sterilize using a 0.2-μm nitrocellulose filter
Store up to 1 year at 4°C

HCl, 1 M

Mix in the following order:
913.8 ml H_2O
86.2 ml concentrated HCl (Table A.1.1)
Do not sterilize

HeBS (HEPES-buffered saline), 2× (UNITS 12.3 & 12.6)

16.4 g NaCl (0.28 M final)
11.9 g HEPES (0.05 M final)
0.21 g Na_2HPO_4 (1.5 mM final)
800 ml H_2O
Titrate to pH 7.05 with 5 N NaOH
Add H_2O to 1 liter
Filter sterilize through a 0.45-μm nitrocellulose filter
Test for transfection efficiency
Store at −20°C in 50-ml aliquots

The pH is extremely important for efficient transfection. The optimal pH range is 7.05 to 7.12.

There can be wide variability in transfection efficiency between batches of 2× HeBS. Efficiency should be checked with each new batch. 2× HeBS can be tested rapidly by mixing 0.5 ml of 2× HeBS with 0.5 ml of 250 mM $CaCl_2$, and vortexing. A fine precipitate should develop that is readily visible in the microscope. Transfection efficiency must still be confirmed, but if the solution does not form a precipitate in this test, there is something wrong.

HeBS, 2× (UNIT 12.5)

2.382 g HEPES
3.28 g NaCl
42.4 mg Na_2HPO_4
42.5 H_2O to 200 ml
Adjust pH to 7.12 with 5 M NaOH
Filter sterilize using a 0.2-μm nitrocellulose filter
Store up to 6 months at 4°C

HeBS, 2×, pH 7.05 (UNIT 12.7)

20 mM HEPES
135 mM NaCl
5 mM KCl
5.5 mM dextrose
0.7 mM Na_2HPO_4
Adjust pH to 7.05
Divide into aliquots and store 6 to 12 months at −20°C

Accurate pH adjustment is critical.

Herring sperm DNA, 10 mg/ml (UNITS 2.4 & 10.4)

Add 1 g desiccated herring sperm DNA to 100 ml sterile water and stir overnight. Sonicate at 100 W for 5- to 10-sec intervals until DNA solution is no longer viscous. Boil 10 min, aliquot, and store indefinitely at –20°C.

Herring sperm DNA, 2 mg/ml *(APPENDIX 3G)*

Add 200 mg high-molecular-weight herring sperm DNA to 50 ml water and stir overnight to resuspend. Shear DNA by sonicating it or by drawing it repeatedly through an 18-G needle into a 60-ml syringe and then through a 23-G needle. Boil DNA 10 min and dilute to a final volume of 100 ml. Store in 10-ml aliquots ≤1 year at −20°C.

High-salt wash solution *(UNIT 10.3)*
✓ 10 ml 10% (w/v) SDS (0.1% final)
✓ 100 ml 20× SSC (2× final)
  H₂O to 1 liter
  Use at room temperature

Hoechst 33258 assay solution *(APPENDIX 3D)*

Dissolve Hoechst 33258 in water at 1 mg/ml. Store stock ≥6 months at 4°C. Before use, add 10 μl of 1 mg/ml stock to 100 ml of 1× TNE buffer (see recipe) that has been filtered through a 0.45-μm filter.

Hoechst 33258 staining solutions A and B *(UNIT 4.3)*
Stock solution: Prepare 50 μg/ml Hoechst 33258 (Calbiochem or Sigma) in water. Store wrapped in foil ≤3 months at 4°C.

Staining solution A: Dilute with PBS (see recipe) to 0.5 μg/ml. Store wrapped in foil ≤1 month at 4°C.

Staining solution B: Dilute with PBS (see recipe) to 25 μg/ml. Store wrapped in foil ≤1 month at 4°C.

Hoechst 33258 stock solution, 250 μg/ml *(UNIT 8.4)*

Add 25 mg Hoechst 33258 powder (2-[2-{4-hydroxyphenyl}-6-benzimidazolyl]-6-[1-methyl-4-piperazyl]-benzimidazole) to 100 ml H₂O in a glass bottle. Wrap in foil and store ≤1 month at 4°C.

HPB (high-phosphate buffer), 5× *(UNIT 5.2)*
  2.5 M NaCl
  0.5 M Na₂HPO₄
  0.026 M EDTA
  Adjust pH to 7.0
  Autoclave
  Store ≤6 months at room temperature

Human C₀t-1 DNA, 10 mg/ml *(UNIT 11.2)*

Add 925 μl of 100% ethanol and 75 μl of 3 M sodium acetate, pH 5.2 (see recipe), to 500 μl of 1 μg/μl human C₀t-1 DNA (Life Technologies). Centrifuge at 14,000 × g. Aspirate supernatant and allow pellet to air dry for 5 min. Resuspend pellet in 50 μl DEPC-treated water (see recipe). Store up to 6 months at −20°C.

Hybridization buffer *(UNIT 9.3)*
✓ 3× SSPE
  7.0% (w/v) SDS
  0.5% (w/v) Carnation nonfat dry milk
  0.5 mg/ml sonicated, denatured salmon sperm DNA
✓ 5 mM EDTA

6% (w/v) polyethylene glycol 8000 (Baker)
Store at −20°C
Prewarm and filter (0.45-μm) just before use

Hybridization cocktail *(UNIT 4.5)*
  6.8 ml formamide
  2.5 ml 50% (w/v) dextran sulfate (autoclave and store at 4°C; stable >1 year)
✓ 0.63 ml 20× SSC
  Mix thoroughly and store up to 3 months at 4°C

The final concentrations for hybridization cocktail as diluted in the protocol are 55% formamide, 10% dextran sulfate, and 1× SSC.

Hybridization cocktail, pH 7.0 *(APPENDIX 3G)*
  500 ml formamide
✓ 10 ml 100× Denhardt's solution
✓ 50 ml 10% (w/v) SDS
  150 ml 50% (w/v) dextran sulfate (500,000 molecular weight)
✓ 250 ml 20× SSC
✓ 40 ml 2 mg/ml herring sperm DNA
  Store at room temperature
  Check pH with pH paper before using

Hybridization solution *(UNIT 5.2)*
✓ 0.5% SDS
✓ 0.5× HPB
✓ 5.5× SSC
  Make fresh daily

Hybridization solution *(UNIT 8.6)*
  97 μl formamide (deionized and molecular biology certified)
✓ 15 μl 20× SSC
  36.5 μl H₂O
✓ 1.5 μl 10 mg/ml sonicated salmon sperm DNA
  Store up to 1 month at 4°C

Hybridization solution *(UNIT 10.3)*

Combine ingredients of prehybridization solution (see recipe) but omit water. Add 1.1 ml 60% dextran sulfate (see recipe). Prepare fresh and keep at room temperature.

Hybridization wash buffer *(UNIT 4.8)*

Combine 50 ml formamide (Ultrapure or deionized; see recipe) and 10 ml of 20× SSC, pH 7.0 (see recipe). Adjust pH to 7.0 if necessary with HCl, then bring final volume to 100 ml with water. Store at 4°C. Prepare fresh solution weekly, or more frequently with heavy use (>50 slides/week).

Hydrion buffer, pH 6.8 *(UNIT 4.3)*

Dissolve one Hydrion capsule, pH 6.8 (Metric-pak, Micro Essential Laboratory), in 100 ml water. For working solution, dilute 5 ml stock in 95 ml water. Store both solutions ≤6 months at room temperature.

Hydrocortisone, 1 mM *(UNIT 13.4)*

Add 36.3 mg hydrocortisone (tissue culture grade, MW 362.5; Sigma) to 400 μl of 100% ethanol and warm 5 min at 37°C. Bring total volume to 10 ml with sterile water and mix as well as possible. Dilute 1 ml into 9 ml sterile water to generate a 1 mM working solution. Store up to 1 month at 4°C.

Hydrocortisone does not actually dissolve in ethanol; it forms a slurry. However, it should be completely dissolved in the 1 mM working solution.

Hygromycin, 80 mg/ml (*UNIT 12.5*)
10^7 U hygromycin (Calbiochem)
H_2O to 350 ml
Filter sterilize using a 0.2-μm nitrocellulose filter
Store up to 6 months at 4°C

Hypotonic solution (*UNIT 8.6*)
1 part 0.75 M (5.6 mg/ml) KCl
1 part 0.8% (w/v) trisodium citrate dihydrate
1 part H_2O
Prepare fresh before each use

Hypotonic/partial fixative solution (*UNIT 9.9*)

Dissolve 1.2 g trisodium citrate dihydrate (Sigma) in 80 ml H_2O, then bring final volume to 100 ml with H_2O (final 1.2% [w/v]). Store up to 1 week at room temperature. To use, dilute 3 parts with 1 part of 3:1 fixative solution (see recipe). Make fresh each day.

IMDM (Iscove's modified Dulbecco's medium), 2× (*UNIT 13.4*)

Add contents of a 1-liter envelope of powdered IMDM (Life Technologies) to a well-rinsed medium bottle containing 450 ml water. Rinse envelope with the medium and add to bottle. Add 40.3 ml 7.5% (w/v) sodium bicarbonate solution (tissue culture grade; Sigma) and then bring total volume to 500 ml. Filter sterilize and store up to 1 month at 4°C. If a precipitate starts to form (usually after 1 month of storage), prepare new medium.

Avoid using bottles that have been washed; soap residue may be toxic to hematopoietic cells.

Immunoperoxidase substrate solution (*UNIT 4.7*)
10 mg 3-amino-9-ethylcarbazole (Sigma; 0.9 mM final)
3 ml *N,N*-dimethylformamide (0.72 M final)
50 ml 0.02 M sodium acetate buffer (0.02 M final)
50 μl H_2O_2, added just before use (8.3 mM final)

Iodixanol gradient solutions (*UNIT 12.3*)
15% iodixanol:
✓ 50 ml 10× PBS
✓ 0.5 ml 1 M $MgCl_2$
0.5 ml 2.5 M KCl
100 ml 5 M NaCl
125 ml Optiprep (iodixanol; Nycomed)

0.75 ml 0.5% phenol red stock
H_2O to 500 ml
Filter sterilize

25% iodixanol:
✓ 50 ml 10× PBS
✓ 0.5 ml 1 M $MgCl_2$
0.5 ml 2.5 M KCl
200 ml Optiprep (iodixanol; Nycomed)
1.0 ml 0.5% phenol red stock
H_2O to 500 ml
Filter sterilize

40% iodixanol:
✓ 50 ml 10× PBS
✓ 0.5 ml 1 M $MgCl_2$
0.5 ml 2.5 M KCl
333 ml Optiprep (iodoxinol; Nycomed)
H_2O to 500 ml
Filter sterilize

Phenol red is omitted to clearly distinguish this solution (which will contain virus) from the others.

60% iodixanol:

500 ml Optiprep (iodixanol; Nycomed)
0.25 ml 0.5% phenol red stock
✓ 0.5 ml 1 M $MgCl_2$
0.5 ml 2.5 M KCl

Since Optiprep is 60% iodixanol, this recipe actually contains 54% iodixanol.

Store all iodixanol solutions up to 6 months at 4°C.

Lactose storage buffer (*UNIT 13.5*)
✓ 25 mM Tris·Cl, pH 7.4
60 mM NaCl
1 mg/ml arginine
50 mg/ml lactose
Filter sterilize using 0.2-μm filter
Store up to 6 months at −20°C

N-Lauroylsarcosine, 10% (*UNITS 5.2 & 5.4*)

Dissolve 100 g *N*-lauroylsarcosine (Sarkosyl) in 700 ml water (may require heating). Adjust volume to 1 liter with water. Store up to 6 months at room temperature.

CAUTION: *Take care not to inhale particles of dry N-lauroylsarcosine.*

LB medium and plates
10 g tryptone
5 g yeast extract
5 g NaCl
950 ml H_2O
1 ml 1 M NaOH
H_2O to 1 liter

For liquid medium: Autoclave 25 min and then cool to ≤50°C. Add antibiotics, color-detection reagents, and nutritional supplements as needed (Table A.1.2). If not for immediate use, store

Table A.1.2 Additives for Bacterial Media

Additive[a]	Stock conc. (mg/ml)	Final conc. (μg/ml)
Antibiotics		
Ampicillin	4	50
Carbenicillin (in 50% ethanol)	4	50
Chloramphenicol (in methanol)	10	20
Gentamycin	10	15
Kanamycin	10	30
Streptomycin	50	30
Tetracycline(in 70% ethanol)[b]	12	12
Color-detection reagents		
Isopropyl-1-thio-β-D-galactoside (IPTG)	23.8 (100 mM)	23.8 (100 μM)
5-Bromo-4-chloro-3-indolyl-β-D-galactoside (Xgal) dissolved in *N,N*-dimethylformamide	20	20

[a]All additives should be dissolved in sterile distilled H_2O unless otherwise indicated. All additives should be stored at 4°C, except tetracycline and carbenicillin, which should be stored at −20°C.

[b]Light-sensitive; store stock solutions and plates in the dark.

(without antibiotics) ≤1 year at room temperature.

The original recipe for LB medium does not contain NaOH. There are many different recipes for LB that differ only in the amount of NaOH added. Even though the pH is adjusted to ∼7 with NaOH, the medium is not very highly buffered; thus, as a culture growing in LB medium nears saturation, the medium pH drops.

For plates: Before autoclaving, add 15 g agar or agarose to above ingredients and heat with stirring to dissolve agar. Autoclave, cool to ∼50°C, and add antibiotics, color-detection reagents, and nutritional supplements as needed. Pour 30 to 40 ml into 100-mm sterile disposable petri dishes and allow to solidify. Store at 4°C, wrapped in the original bags.

The medium will stay liquid indefinitely at 50°C but will rapidly solidify if its temperature falls much below 45°C.

For most applications, it is advisable to dry the plates before storage by leaving them out at room temperature for 2 or 3 days or by leaving them with the lids off for 30 min in a 37°C incubator or laminar flow hood.

LBMM (long-term bone marrow medium) *(UNIT 13.4)*
✓ 500 ml BBMM
   50 μl 50 U/μl recombinant human interleukin 6 (hIL-6; Biosource International), sterile

100 μl 50 ng/μl recombinant human granulocyte/macrophage colony-stimulating factor (hGM-CSF; BioSource International), sterile
50 μl 50 ng/μl recombinant human stem cell factor (hSCF; Biosource International), sterile
20 μl 50 ng/μl recombinant human interleukin 3 (hIL-3; Biosource International), sterile (optional)
Store up to 2 weeks at 4°C

Recombinant human IL-3 (10 ng/ml) is added when maximal proliferation of the cultured cells is required—for example, if many assays must be done from the collected nonadherent cells during the following week. However, the authors have noticed that the continued presence of IL-3 in the cultures tends to decrease the lifespan of the long-term culture-initiating cells in vitro.

LCR mix *(UNIT 2.2)*
   20 μl 1 M KCl
✓ 20 μl 10× thermostable ligase buffer (2× final)
   20 μl 10 mM NAD^+ (Sigma, store up to 6 months at –20°C)
   40 μl 0.1% (v/v) Triton X-100
   Prepare fresh

Lithium acetate/Tris/EDTA solution *(UNIT 5.3)*
   3 ml 1 M lithium acetate (filter sterilize and store at room temperature)

✓ 3 ml 100 mM Tris·Cl, pH 7.5
✓ 3 ml 10 mM EDTA, pH 8.0
21 ml H₂O
Prepare fresh

Loading buffer, 1× (UNIT 9.3)
80% (v/v) formamide
0.1% (w/v) bromphenol blue
0.1% (w/v) xylene cyanol FF
10 mM NaOH
✓ 1 mM EDTA
Store ≤6 months at −20°C

CAUTION: *Formamide is hazardous; use appropriate precautions.*

Loading buffer (UNIT 9.7)
✓ 70% (v/v) 1× TBE buffer
30% (v/v) glycerol
0.1% (w/v) bromphenol blue

Loading buffer for ESTs (UNIT 11.2)
4.0 ml glycerol (enzyme grade)
✓ 0.9 ml DEPC-treated H₂O
0.1 ml 0.25% (w/v) xylene cyanol FF/0.25% (w/v) bromphenol blue
Store up to 1 month at room temperature

Low-salt wash solution (UNIT 10.3)
✓ 20 ml 10% (w/v) SDS (0.1% final)
✓ 10 ml 20× SSC (0.1× final)
H₂O to 2 liters
Heat in two 1-liter flasks in 65°C water bath

Low-T dNTP mix, 10× (UNIT 11.2)
25 μl 100 mM dGTP (0.5 mM in 1×)
25 μl 100 mM dATP (0.5 mM in 1×)
25 μl 100 mM dCTP (0.5 mM in 1×)
10 μl 100 mM dTTP (0.2 mM in 1×)
✓ 415 μl DEPC-treated H₂O
Store up to 2 months at −20°C

LPS solution, 750 μg/ml (UNIT 4.2)
Dissolve 25 mg lipopolysaccharide (LPS; *E. coli* serotype 0111:B4; Sigma) in 33.3 ml complete RPMI (see recipe). Divide into 3- to 4-ml aliquots and store up to 6 weeks at −20°C. For a final concentration of 50 μg/ml, use 0.1 ml LPS in a 1.5-ml culture.

Lymphocyte culture medium (UNIT 4.5)
✓ Complete RPMI
15% FBS
2% (v/v) phytohemagglutinin (Life Technologies)
5 U/ml heparin
Store several weeks at 4°C

Lysis buffer (UNIT 4.5)
✓ 2.5 ml 20% (w/v) SDS
✓ 10 ml 0.5 M EDTA
✓ 20 ml 1 M Tris·Cl, pH 7.4
67.5 ml H₂O
Store at room temperature (stable at least 1 year)

m-AMSA, 5 mg/ml (UNIT 4.5)
Dissolve *N*-[4-(9-acridinylamino)-3-methoxy-phenyl]methanesulfonamide (m-AMSA; Drug

Synthesis Branch, National Cancer Institute) at 10 mg/ml in DMSO and dilute with 1 vol water. Filter sterilize and store at 4°C.

MAC hypotonic solution (UNIT 4.7)
Stock solution:

0.14702 g CaCl₂·2 H₂O (1 mM final)
0.3728 g KCl (5 mM final)
0.1626 g MgCl₂·6 H₂O (0.8 mM final)
0.5844 g NaCl (10 mM final)
3.4230 g sucrose (10 mM final)
3.74 ml glycerol (50 mM final)

Dissolve each component separately in a small volume of sterile water, then combine in a volumetric flask and bring total volume to 1 liter. Adjust pH to 7.0. Filter through a 1.2-μm filter and store up to several months at 4°C.

Working solution:

0.8 parts stock solution
✓ 1.2 parts complete RPMI/5% FBS
Prepare fresh (can be stored up to 1 week at 4°C under sterile conditions)

This solution contains physiological concentrations of ions, sucrose, and glycerol.

Master hybridization mix (UNIT 4.4)
✓ 1 ml 20 × SSC
0.5 ml 20 mg/ml nuclease-free BSA
1.5 ml sterile H₂O
2 ml 50% (w/v) dextran sulfate (Pharmacia Biotech, mol. wt. 500,000, autoclaved)
Store at 4°C and use up to 6 weeks

Master hybridization solution (UNIT 4.8)
✓ 5 ml formamide (Ultrapure or deionized)
✓ 1.0 ml 20× SSC, pH 7.0
1 g dextran sulfate (Sigma)
Heat 2 to 3 hr at 70°C, vortexing periodically to dissolve
Check pH with pH paper; adjust to 7.0 if necessary
Bring volume to 10 ml with H₂O
Store in 1-ml aliquots ≤1 year at −20°C

MboI buffer, 10× and 1× (UNIT 5.4)
For 10× buffer: Prepare 100 mM Tris·Cl, pH 8.0 (see recipe), containing 1 M NaCl and store up to 1 year at room temperature.

For 1× buffer: Dilute 1 part 10× buffer in 9 parts water and add 10 mM 2-mercaptoethanol. Prepare immediately before use.

McIlvaine buffer, pH 5.6, 7.0, and 7.5 (UNIT 4.3)
For pH 5.6: 43 ml of 0.1 M citric acid
For pH 7.0: 18.1 ml of 0.1 M citric acid
For pH 7.5: 8 ml of 0.1 M citric acid

For each pH, bring the above volumes to 100 ml total with 0.2 M Na₂HPO₄. Check pH and adjust with either citric acid or Na₂HPO₄, if necessary. Store ≤6 months at room temperature.

2-ME (2-mercaptoethanol), 0.1 M (*UNIT 13.4*)

In a fume hood, dilute 0.7 ml of ~14.3 M 2-mercaptoethanol (Sigma) to 100 ml with water. Filter sterilize and store 2-ml aliquots ≤3 years at −20°C. Store thawed aliquot at 4°C (stable at least 2 months).

Medium A (*UNIT 13.5*)

Modified Eagle medium (MEM; Life Technologies) containing:

50 U/ml penicillin
50 μg/ml streptomycin
80 μg/ml tobramycin (Eli Lilly)
100 μg/ml ceftazidime (Eli Lilly)
100 μg/ml Primaxin (Merck)
5.0 μg/ml amphotericin B
10 μg/ml DNase I (Type II-S; from bovine pancreas; Sigma)
0.5 mg/ml dithiothreitol
Prepare fresh and keep at 4°C

Medium B (*UNIT 13.5*)

✓ BEGM supplemented with:
50 U/ml penicillin
50 μg/ml streptomycin
80 μg/ml tobramycin
100 μg/ml ceftazidime
100 μg/ml Primaxin
5.0 μg/ml amphotericin B
Store up to 2 weeks at 4°C

Medium C (*UNIT 13.5*)

✓ BEGM supplemented with:
50 U/ml penicillin
50 μg/ml streptomycin
40 μg/ml tobramycin
50 μg/ml ceftazidime
50 μg/ml Primaxin
Store up to 2 weeks at 4°C

MES buffer, 12× (*UNITS 2.4 & 11.1*)

70.4 g 2-(*N*-morpholino)ethanesulfonic acid (MES) free acid monohydrate (Sigma)
193.3 g MES sodium salt (Sigma)
✓ 800 ml DEPC-treated H_2O

Mix and adjust volume to 1000 ml. The pH should be between 6.5 and 6.7. Filter through a 0.20-μM filter; do not autoclave. Store at 2° to 8°C. Discard solution if yellow.

Methotrexate, 10 μM (*UNIT 4.1*)

Dissolve 0.5 mg methotrexate (amethopterin; Sigma) in 100 ml HBSS (see recipe). Sterilize by filtration through a 0.22-μm filter. Store 5-ml aliquots in red-top Vacutainers (untreated tubes) or any suitable sterile untreated tube for ≤1 year at −20°C. Store aliquot in use ≤1 month at 4°C.

CAUTION: *Methotrexate is hazardous.*

Methylcellulose medium (*UNIT 13.4*)

✓ 40 ml 2× IMDM
✓ 30 ml FBS, heat inactivated 1 hr at 56°C
✓ 10 ml deionized 10% BSA
2 mM L-glutamine

100 U/ml penicillin
10 μg/ml streptomycin
✓ 100 μl 0.1 M 2-ME
✓ 100 μl 1 mM hydrocortisone
Filter sterilize
✓ Add 35 ml 3.9% (w/v) methylcellulose stock solution
Mix well by shaking
Add 30 μl 50 U/μl recombinant human interleukin 6 (hIL-6; Biosource International)
Add 20 μl 50 ng/μl recombinant human interleukin 3 (hIL-3; Biosource International)
Add 100 μl 50 ng/μl recombinant human stem cell factor (hSCF; Biosource International)
Add 100 μl 50 ng/μl recombinant human granulocyte/macrophage colony-stimulating factor (hGM-CSF; Biosource International)
Add 1 ml 2 U/μl recombinant human erythropoietin (Biosource International)
Keep at 4°C and let bubbles rise and disappear before using
Store up to 2 weeks at 4°C

For plating fresh noncultured cells, add erythropoietin on days 4 to 7. When plating cells from a 72-hr ex-vivo transduction, erythropoietin can be added directly.

Methylcellulose overlay, 1% (w/v) (*UNIT 12.7*)

Add 25 g methylcellulose (Aldrich) to 100 ml PBS, pH 7.5 (see recipe), in a 500-ml sterile bottle containing a stir bar. Autoclave 45 min on liquid cycle. After the solution cools, add 350 ml complete MEM (see recipe) without serum. Mix well and stir overnight at 4°C. Once the methylcellulose has dissolved, add 50 ml FBS (see recipe). Store 1 to 3 months at 4°C.

Methylcellulose stock solution, 3.9% (*UNIT 13.4*)

Weigh 11.8 g prescreened (see Support Protocol 5) methylcellulose powder (4000 centipoise; Sigma, Fisher) and break up any clumps. Add to 150 ml boiling water in a well-rinsed medium bottle and mix well. Allow suspension to cool until it can be touched without burning the hand. Then add 150 ml ice-cold water (no ice crystals) all at once. Mix as well as possible. Let stand 1 hr at room temperature to let bubbles rise and then store overnight at 4°C. Warm to room temperature and then autoclave to sterilize. If fluid level is significantly below starting volume, adjust volume with sterile water and mix very well by stirring with a sterile 25-ml pipet.

A medium bottle that has not been washed should be used because soap residue may be toxic to the cells.

The extremely viscous solution can be slowly removed from the bottle with a 25-ml pipet and bulb pipettor. Much of the methylcellulose will cling to the inside of the pipet, so when it is dispensed, extra time should be allowed for the pipet to empty completely.

MgCl₂, 1 M
20.3 g MgCl₂·6H₂O
H₂O to 100 ml

MgCl₂ is extremely hygroscopic. Do not store opened bottles for long periods of time.

MOPS running buffer, 10× (APPENDIX 3H)
0.4 M 3-(N-morpholino)-propanesulfonic acid (MOPS), pH 7.0
✓ 0.5 M sodium acetate
✓ 0.01 M EDTA
Store up to 3 months at 4°C in the dark; discard if it turns yellow

NaOH, 10 M
Dissolve 400 g NaOH in 450 ml H2O
Add H₂O to 1 liter
Do not sterilize

NaOH/SDS solution (UNIT 5.1)
4.65 ml sterile water
100 µl 10 N NaOH
250 µl 20% (w/v) SDS
Prepare fresh

NBT/BCIP substrate solution (UNIT 4.4)
Add 220 µl of a 75 mg/ml nitroblue tetrazolium (NBT) solution in dimethylformamide to 50 ml alkaline phosphatase buffer, pH 9.5 (see recipe), and mix gently (do not vortex). Add 170 µl of a 50 mg/ml 5-bromo-4-chloro-3-indolyl phosphate (BCIP) solution in dimethylformamide and mix gently again (final 330 µg/ml NBT and 170 µg/ml BCIP). Prepare fresh each time.

NBT solution, 100× (UNIT 12.4)
50 mg/ml nitroblue tetrazolium
70% (v/v) dimethylsulfoxide
✓ 100 mM Tris·Cl, pH 8.5
Store in a foil-wrapped glass container (i.e., in the dark) at −20°C (stable for months)

Neutralization buffer (UNITS 1.3 & 9.9)
✓ 900 mM Tris·Cl, pH 8.3
300 mM KCl
200 mM HCl
Autoclave and store at 4°C up to several months

Neutralizing buffer, 1 M phosphate, pH 6.8 (UNIT 9.3)
79.25 g Na₂HPO₄
60.25g NaH₂PO₄·H₂O
Bring to 1 liter with H₂O
Store up to 1 year at room temperature

Neutralizing solution (UNIT 10.4)
60 g Tris base (0.5 M final)
88 g NaCl (1.5 M final)
Adjust pH to 7.0 with ~36 ml concentrated (10.5 N) HCl
Add H₂O to 1 liter
Prepare fresh for each transfer

Neutralizing solution (APPENDIX 3G)
✓ 250 ml 2 M Tris·Cl, pH 7.6

300 ml 5 M NaCl
H₂O to 1 liter
Prepare fresh

Nick translation buffer (UNIT 4.5)
✓ 500 µl 1 M Tris·Cl, pH 7.6 to 7.8
50 µl 1 M MgCl₂
✓ 10 µl 25 mM 4dNTP mix
440 µl H₂O
Store in 100-µl aliquots at −20°C (stable at least 1 year)

Nick translation buffer, 10× (UNIT 4.4)
2 µl 100 mM dATP (0.2 mM final)
2 µl 100 mM dCTP (0.2 mM final)
2 µl 100 mM dGTP (0.2 mM final)
✓ 500 µl 1 M Tris·Cl, pH 7.8 (0.5 M final)
50 µl 1 M MgCl₂ (0.05 M final)
7 µl 2-mercaptocthanol (0.1 M final)
2 µl 50 µg/ml BSA (0.1 µg/ml final)
435 µl sterile distilled H₂O
Store 50-µl aliquots ≤1 year at −20°C

NKM buffer (APPENDIX 3A)
28 ml 5 M NaCl
30 ml 1 M KCl
1.5 ml 1 M MgCl₂
Add H₂O to 1 liter
Filter sterilize
Store in 500-ml sterile bottles at 4°C

Nuclear lysis buffer (UNIT 9.6)
Per liter:

50 ml 2 M Tris·Cl, pH 7.5
80 ml 5 M NaCl
✓ 4 ml 500 mM EDTA
Adjust pH to 8.0 with NaOH
Autoclave to sterilize
Store ≤1 year at room temperature (if sterility is maintained)

Nucleotide mix, 10× (UNIT 4.8)
5 µl 10 mM dATP (200 µM final)
5 µl 10 mM dCTP (200 µM final)
5 µl 10 mM dGTP (200 µM final)
✓ 125 µl 1 M Tris·Cl, pH 7.2 (500 µM final)
✓ 12.5 µl 1 M MgCl₂ (200 µM final)
1.7 µl 14.7 M 2-mercaptoethanol (100 mM final)
2.5 µl 10 mg/ml bovine serum albumin (100 µg/ml final)
93.3 µl H₂O (final volume 250 µl)
Store in aliquots ≤1 year at −20°C

OLA ligation mix (UNIT 2.2)
15 µl 1 M KCl
✓ 120 µl 10× thermostable ligase buffer
120 µl 10 mM NAD (Sigma)
345 µl 0.1% (v/v) Triton X-100
Prepare fresh

Oligo wash buffer (UNIT 2.1)
✓ 1% (w/v) SDS
✓ 70 mM sodium phosphate buffer, pH 7.2

Make from sterile stock solutions and store indefinitely at room temperature. Inspect solution for bacterial contamination before use.

Oligonucleotide primers for LCR (UNIT 2.2)

Six oligonucleotide primers must be synthesized to genotype a biallelic nucleotide variation. The primers should be designed to generate a single base 3′ overhang at the junction (see Fig. 2.2.2). All primers have a T_m between 60° and 65°C. Label the joining primers with ^{32}P at their 5′ ends using T4 polynucleotide kinase (see APPENDIX 3E); after incubating 45 min at 37°C, add cold ATP to 1 mM to complete phosphorylation of the primers. Store primers up to 2 years at −20°C.

Overgo labeling buffer (UNIT 5.2)

Prepare solutions A, B and C, then mix in a 2:5:3 ratio. Store in 0.5-ml aliquots ≤3 months at −20°C.

Solution A:

✓ 1 ml 1.25 M Tris·Cl, pH 8.0 with 0.125 M MgCl$_2$ (see recipes)
18 μl 2-mercaptoethanol
5 μl 0.1 M dGTP
5 μl 0.1 M dTTP
Store ≤1 year at −80°C

Add all dNTPs except for the one(s) that have the α-^{32}P label. The above solution is for labeling with both [α-^{32}P]dATP and [α-^{32}P]dCTP.

Solution B:

2 M HEPES, pH 6.6
Store ≤1 year at room temperature

Solution C:

✓ 3 mM Tris·Cl, pH 8.0
✓ 0.2 mM EDTA
Store ≤1 year at room temperature

Paraformaldehyde solution, 4% (w/v) (UNITS 12.6 & 12.8)

Add 20 g paraformaldehyde to 300 ml H$_2$O and heat to 55° to 60°C. Slowly add 1 M NaOH dropwise over ~10 min until the solution becomes clear, then cool the solution to room temperature. Use pH paper to check that the pH is 7.0 to 7.5 (add more NaOH if necessary). Add 100 ml of 0.5 M sodium phosphate buffer, pH 7.0 (see recipe), and then add water to a final volume of 500 ml (final 0.1 M phosphate, pH 7.0 to 7.5). Store up to 1 week at 4°C.

CAUTION: *Paraformaldehyde is toxic, and the preparation process involving heating results in considerable vaporization, which increases the hazard. It is essential to use appropriate safety procedures, such as working in a fume hood.*

PBS (phosphate-buffered saline), pH ~7.3

8.0 g NaCl (137 mM)
0.2 g KCl (2.7 mM)
2.16 g Na$_2$HPO$_4$·7H$_2$O (8.0 mM)
0.2 g KH$_2$PO$_4$ (1.5 mM)
H$_2$O to 1 liter

Filter sterilize and store at 4°C

PBS can be prepared as a 10× stock. Prepackaged PBS (pH 7.4), which is reconstituted by adding water, is commercially available from Sigma. This is very convenient if used in large quantities.

PBS, pH 7.8 (UNIT 9.10)

144 mM NaCl
10 mM KH$_2$PO$_4$
Adjust pH to 7.8 with K$_2$HPO$_4$
Store at room temperature

PBS, pH 7.5 (UNIT 12.7)

135 mM NaCl
2.5 mM KCl
1.5 mM KH$_2$PO$_4$
8.0 mM Na$_2$HPO$_4$, pH 7.5
Store 1 to 2 years at room temperature

PBS for vaccination (UNIT 13.2)

✓ 5 mM sodium phosphate buffer, pH 7.2
140 mM NaCl
Sterilize with a 0.2-μm filter
Store up to 1 month at 4°C

PBS-MK (UNIT 12.3)

✓ 50 ml 10× PBS
✓ 0.5 ml 1 M MgCl$_2$
0.5 ml 2.5 M KCl
H$_2$O to 500 ml
Filter sterilize
Store up to 2 years at 4°C

PCR amplification buffer, 10×

500 mM KCl
✓ 100 mM Tris·Cl, pH 8.3
15 mM MgCl$_2$
0.1% (w/v) gelatin
Store in aliquots at −20°C
Sterilize by autoclaving or prepare from sterile water and stock solutions

MgCl$_2$ is used at 15 mM in many PCR assays. The optimal concentration depends upon the sequence and primer of interest and may have to be determined experimentally.

PCR amplification buffer, 10× (UNIT 10.3)

✓ 100 mM Tris·Cl, pH 8.3
500 mM KCl
Store ≤1 year at −20°C

PCR buffer with nonionic detergents (UNIT 3.1)

50 mM KCl
✓ 10 mM Tris·Cl, pH 8.3
2.5 mM MgCl$_2$
0.45% (v/v) Nonidet P-40 (NP-40)
0.45% (v/v) Tween 20
100 μg/ml proteinase K (add just before use)
Stable many months at room temperature without proteinase K

PEG 1000 solution, 50% (w/v) (UNIT 3.1)

Heat solid Koch light polyethylene glycol (PEG) 1000 (Research Products International) to 65°C until just melted, then weigh out amount needed

A1

by pipetting liquefied PEG into a tared glass vial. Sterilize by autoclaving and mix 5 g with 5 ml serum-free medium. Typically, PEG will not significantly change pH of medium; if it does, readjust to original value using sterile 0.1 N NaOH. Store solid PEG at room temperature. Store solution no more than a few days at 4°C.

PEG solution, 40% (UNIT 5.3)

40% (w/v) PEG (polyethylene glycol) 4000
✓ 10 mM Tris·Cl, pH 7.5
100 mM lithium acetate
Autoclave and store several months at room temperature

Pepsin solution (UNIT 8.5)

Prepare 200 ml of 0.9% (w/v) NaCl. Take 20 ml and adjust to pH 1.5 with concentrated HCl, then add 1 g pepsin and stir until dissolved. Add enough reserved 0.9% NaCl to bring volume to 200 ml. Store 5-ml single-use aliquots in plastic tubes at −20°C. Do not refreeze or reuse.

PET solution (APPENDIX 3C)

✓ 50 mM Tris·Cl, pH 8.5 at 25°C
✓ 1 mM EDTA
0.5% Tween 20
Autoclave
Store up to 6 months at room temperature
Before use add 1/100 vol 20 mg/ml proteinase K

The proteinase K stock solution is prepared in deionized, distilled, autoclaved water and can be stored in small aliquots up to 2 months at −20°C.

PHA solution (UNIT 4.2)

Dissolve 2 mg phytohemagglutinin (PHA; purified for best results; Abbott Laboratories) in 17.8 ml complete RPMI (see recipe). Divide into 3- to 4-ml aliquots and store up to 6 weeks at −20°C. For a final concentration of 7.5 μg/ml, use 0.1 ml PHA solution in a 1.5-ml culture.

For optimum results, perform a dose-response curve (3 to 9 μg/ml in duplicate) with each new lot of PHA to assay the mitotic index and strain response. Use blood pooled from approximately six mice for each strain, with C57BL/6 as a control.

The authors have found that a concentration of 7.5 μg/ml gives the best results regardless of the mitogenic units marked on the vial.

Phenol, buffered

Thaw distilled phenol (stored frozen at −20°C) in a 65°C water bath. Add an equal volume of 1 M Tris·Cl, pH 7.5 (see recipe), mix thoroughly, and allow to stand overnight. The next day, remove the top layer of the Tris/phenol mixture with a pipet. Add water to provide a top layer. Label and date the bottle. Store at 4°C.

Degradation of phenol can be detected by a change in color from clear to a light yellow, progressing to a light pink.

Phenol/chloroform/isoamyl alcohol, 25:24:1 (v/v/v)

✓ 25 vol buffered phenol (bottom phase)
24 vol chloroform
1 vol isoamyl alcohol
Store in brown glass bottle or clear glass bottle wrapped in aluminum foil ≤2 months at 4°C

In some cases, 1:1 (v/v) phenol/chloroform may be used.

Phenylmethylsulfonyl fluoride, see PMSF

Phosphate buffer, pH 8.0, 0.5 M (UNIT 4.5)

473.5 ml 0.5 M Na$_2$HPO$_4$
26.5 ml 0.5 M NaH$_2$PO$_4$
Autoclave and store at room temperature (stable at least 1 year)

Phosphate-buffered saline, see PBS

PIB (pronuclear injection buffer) (UNIT 5.3)

Add 10 ml of 1 M Tris·Cl, pH 7.5 (see recipe) and 0.2 ml of 0.5 M EDTA (see recipe) to 80 ml deionized water and bring the volume to 100 ml with deionized water. Filter the solution through a 0.2-μm endotoxin-free sterile cellulose acetate filter (e.g., ValuPrep sterile syringe filter, VWR Scientific). Store 1-ml aliquots up to 6 months at room temperature. Do not reuse aliquots.

PMSF solution, 100 mM (UNIT 5.4)

Prepare a 100 mM stock of phenylmethylsulfonyl fluoride in isopropanol and store up to 1 year at −20°C. Mix well by vortexing and dilute as appropriate immediately before use.

PMSF/TE wash buffer (UNIT 5.1)

Mix 20 mg PMSF in 1 ml of 100% isopropanol and warm 10 to 15 min in a 65°C water bath to dissolve. Store at 4°C (stable for many months). Before use, prepare fresh PMSF/TE wash buffer adding 200 μl PMSF stock to 100 ml TE buffer (see recipe).

PN buffer (UNIT 4.5)

✓ 100 ml 0.5 M phosphate buffer, pH 8.0
2.5 ml Nonidet P-40 (NP-40)
397.5 ml H$_2$O
Store at room temperature (stable at least 1 year)

PN buffer (UNIT 4.8)

Solution A: 0.1 M Na$_2$HPO$_4$·7H$_2$O/0.1% Nonidet P-40

Solution B: 0.1 M NaH$_2$PO$_4$·H$_2$O/0.1% Nonidet P-40

Take a desired amount of solution A and add solution B slowly to bring pH to 8.0. This should require ~1/5 solution B. Store up to 1 month at room temperature.

PN buffer (UNIT 8.6)

Solution A: Dissolve 6.9 g NaH_2PO_4 (anhydrous) in 100 ml H_2O (0.5 M final).

Solution B: Dissolve 71 g Na_2HPO_4 in 1000 ml H_2O (0.5 M final).

Titrate solution B to pH 8.0 with 30 to 40 ml solution A. Add Nonidet P-40 to a final of 0.5% (e.g., 5.2 ml NP-40 to ~1040 ml of solution).

PNM buffer (UNIT 8.6)

✓ 500 ml PN buffer
25 g nonfat dry milk (5% w/v final)
0.1 g sodium azide (3 mM final)

Let mixture stand 72 hr at room temperature, then transfer supernatant to 50-ml conical tubes. Store up to 1 week at 4°C.

Polybrene, 4 mg/ml (UNIT 12.5)

0.4 g hexadimethrine bromide (polybrene; Sigma)
H_2O to 100 ml
Filter sterilize using a 0.2-μm nitrocellulose filter
Store up to 1 year at 4°C

Poly-L-lysine-coated slides (UNIT 4.7)

Prepare an 0.05% (w/v) solution of poly-L-lysine (mol. wt. 70,000 to 150,000; Sigma) in water. Add one drop of this solution to an ethanol-cleaned slide. Place a second ethanol-cleaned slide on top of the first and spread the drop to coat both slides. Air dry 10 min. Store up to 6 months at −20°C wrapped in foil.

Poly-L-lysine solution (UNIT 11.2)

35 ml poly-L-lysine (0.1%, w/v; Sigma)
✓ 35 ml PBS
280 ml H_2O
Store up to 24 hr at room temperature

Potassium acetate buffer, 0.1 M

Solution A: 11.55 ml glacial acetic acid per liter (0.2 M)

Solution B: 19.6 g potassium acetate ($KC_2H_3O_2$) per liter (0.2 M)

Referring to Table A.1.3 for desired pH, mix the indicated volumes of solutions A and B, then dilute with H_2O to 100 ml. Filter sterilize if necessary. Store up to 3 months at room temperature.

For additional details, see sodium acetate buffer.

Potassium acetate solution (UNIT 5.1)

29.5 g potassium acetate
Sterile water to 88.5 ml
11.5 ml glacial acetic acid

Potassium phosphate buffer, 0.1 M

Solution A: 27.2 g KH_2PO_4 per liter (0.2 M)

Solution B: 34.8 g K_2HPO_4 per liter (0.2 M)

Referring to Table A.1.4 for desired pH, mix the indicated volumes of solutions A and B, then dilute with H_2O to 200 ml. Filter sterilize if necessary. Store up to 3 months at room temperature.

For additonal details, see sodium phosphate buffer.

Preavidin block solution (UNIT 4.5)

✓ 3 ml 20× SSC, pH 7.0
600 μl 25% (w/v) BSA
3 ml 25% (w/v) nonfat dry milk
8.4 ml H_2O
Mix, then centrifuge 5 min at 1800 × g to clarify
Filter sterilize through a 0.45-μm filter
Store ≤6 months in 1-ml aliquots at 4°C

Store 25% BSA at −20°C in 1- to 25-ml aliquots (stable at least 1 year).

Precleaned microscope slides (UNIT 10.1)

Place microscope slides (Gold Seal; Becton Dickinson) in a container of 7× cleaning solution (ICN Biomedicals) diluted to 1× with warm water. Scrub each slide with a test-tube brush. Rinse slides individually with warm water. Place all slides in a clean beaker and rinse twice with warm water, then twice with room temperature distilled water. Fill beaker containing slides with distilled water. Store at ~4°C. Rerinse all slides that are not used the same day. After 2 days, rewash all unused slides.

Prehybridization/hybridization buffer (UNIT 9.4)

✓ 5 ml 20× SSC
✓ 1 ml 10% (w/v) SDS
✓ 1 ml 100× Denhardt's solution
13 ml H_2O
Store several months at room temperature

Prehybridization/hybridization buffer (UNIT 9.8)

Dissolve 33.5 g sodium phosphate (dibasic heptahydrate, pH 7; Sigma) and 3.75 ml phosphoric acid (Fisher) in 250 ml water. Add 2 ml of 0.5 M EDTA (1 mM final) and 50 ml of 5 M NaCl (250 mM final). Bring volume to 700 ml with water. Heat solution to 65°C and, while stirring, add 70 g SDS (7% final) followed by 100 g polyethylene glycol 8000 (Sigma; 10% final). Continue to stir and allow solution to cool to room temperature. Add 100 ml of 10% (w/v) BSA (Sigma; 1% final). Bring to 1 liter with water. Divide into aliquots and store ≤6 months at −20°C.

Prehybridization solution (UNIT 2.1)

0.25 M NaCl
✓ 130 mM sodium phosphate buffer, pH 7.2
✓ 5% (w/v) SDS
10% (w/v) PEG (polyethylene glycol) 8000

Make from sterile stock solutions and store indefinitely at room temperature. Inspect solution for bacterial contamination before use.

Prehybridization solution (UNIT 5.2)

✓ 5× Denhardt's solution
✓ 0.5% SDS
✓ 0.5× HPB

✓ 5.5× SSC
Make fresh daily

Prehybridization solution (UNIT 10.3)
✓ 5 ml deionized formamide (50% final)
✓ 1.5 ml 20× SSC (3× final)
✓ 1 ml 50× Denhardt's solution (5× final)
0.2 ml 50 mM sodium pyrophosphate (1 mM final)
✓ 1 ml 10% (w/v) SDS (1% final)
0.2 5 mg/ml salmon sperm DNA (100 μg/ml final)
1.1 ml H₂O

Boil salmon sperm DNA 5 min before adding to solution. Make up solution just prior to use and keep at room temperature.

Primers (UNIT 9.5)

Labeled forward primer: Met-F: 5′-TCCAG AATCTGTTCCAGAGCGTGC-3′ labeled with LI-COR IRDye 700 (available from LI-COR).

Unlabeled reverse primer: Met-R1: 5′-GCTGT GAAGGTTGCTGTTCCTCAT-3′ (available from Life Technologies).

Pronuclear injection buffer, see PIB

Propidium iodide staining solution (UNIT 4.4)

Dissolve 1 mg propidium iodide in 10 ml H₂O (0.15 mM final). Aliquot into aluminum foil–wrapped tubes and store ≤1 year at −20°C. To use, dilute 1/1000 in PBS (see recipe). Store in aluminum foil–wrapped tube ≤6 months at 4°C.

Proteinase K, 2 mg/ml (APPENDIX 3A)
49 ml H₂O
0.5 ml 1 M CaCl₂ (10 mM final)
✓ 0.5 ml 1 M Tris·Cl, pH 7.5
100 mg proteinase K
Store in aliquots at −20°C

Proteinase K digestion buffer (UNIT 13.4)
✓ 10 mM Tris·Cl, pH 7.40
0.15 M NaCl
✓ 0.01 M EDTA, pH 8.0
✓ 0.1% (w/v) SDS
Store up to 1 year at room temperature

Proteinase K lysis solution (UNIT 5.4)
10 ml filtered 10% N-lauroylsarcosine (sodium salt; Sigma)
✓ 40 ml 0.5 M EDTA, pH 8.0
100 mg proteinase K (Boehringer Mannheim)
Prepare fresh and use immediately after adding proteinase K

Proteinase K solution (UNIT 4.8)
Prepare 1 mg/ml proteinase K (e.g., Sigma) in H₂O. Store in aliquots at −20°C. Before use, dilute 3 μl in 50 ml of 20 mM Tris·Cl, pH 7.5 (see recipe) with 2 mM CaCl.

Proteinase solution (UNIT 12.3)
1 M NaCl
1% (w/v) N-lauroylsarcosine (Sarkosyl)

100 g/ml proteinase K
Make fresh each time

Puromycin, 1 mg/ml (UNIT 12.5)
0.1 g puromycin (Sigma)
H₂O to 100 ml
Filter sterilize using a 0.2-μm nitrocellulose filter
Store up to 6 months at 4°C

Quinacrine dihydrochloride, 0.5% (w/v) (UNIT 4.7)

Dissolve 1 g quinacrine dihydrochloride (Sigma) in 50 ml water (42 mM final). Store in foil-wrapped Coplin jar up to 4 weeks at 4°C.

CAUTION: *Quinacrine dihydrochloride is hazardous; use appropriate precautions.*

Quinacrine staining solution (UNITS 4.3)

Dissolve 0.5 g quinacrine dihydrochloride (Sigma) or 2.5 mg quinacrine mustard (Sigma) in 50 ml water. Store in a foil-wrapped Coplin jar ~1 month (depending on frequency of use) at 4°C. Do not use if fluorescence pales or bacterial growth is detected.

Quinacrine mustard produces brighter fluorescence, but cannot be washed from chromosome preparations as quinacrine dihydrochloride can. Quinacrine mustard is also more expensive and a more potent carcinogen.

Rabbit anti-mouse Ig antiserum solution, 1:25 (v/v) dilution (UNIT 4.7)
1 part rabbit anti-mouse immunoglobulin (Dakopatts)
1 part FBS filtered with a 1.2-μm membrane
✓ 23 parts TBS/FBS
Make enough for two rounds of incubation of all slides

RBC lysis buffer (UNIT 13.4)
0.32 M sucrose
✓ 10 mM Tris·Cl, pH 7.5
5 mM MgCl₂
1% (v/v) Triton X-100
Store up to 6 months at room temperature

RBC lysis solution, 10× (UNIT 5.4)
9.54 g NH₄Cl (1.78 M final)
0.237 g NH₄HCO₃ (0.03 M final)
Dissolve in 100 ml
Sterilize using 0.2-μm filter
Store up to 1 month at 4°C in a tightly closed container
Dilute to 1× concentration with sterile water before use

Recombinant adenovirus suspension (UNIT 13.6)

Purify recombinant adenovirus by CsCl gradients followed by removal of CsCl by column purification or by dialysis (UNIT 12.2). Virus that is in PBS may be used immediately after purification. If virus has been stored (e.g., in PBS/10% glycerol at −70°C), thaw virus on ice, dilute to appropriate concentration with sterile PBS, and keep on ice until use. Use diluted virus as soon as possible (within 1 to 2 hr).

Restriction buffer, 5× *(UNIT 9.3)*
✓ 0.5 M Tris·Cl, pH 8.0
0.5 M NaCl
0.1 M MgCl₂
Store up to 1 year at −20°C

Resuspension buffer *(APPENDIX 3A)*
20 ml 5 M NaCl
✓ 10 ml 1 M Tris·Cl, pH 7.5
1.5 ml 1 M MgCl₂
Add H₂O to 1 liter
Filter sterilize
Store in 500-ml sterile bottles at 4°C

Resuspension solution *(UNIT 5.4)*
✓ 15 mM Tris·Cl, pH 8.0
✓ 10 mM EDTA
100 mg/ml DNase-free RNase A
Sterilize using 0.2-μm filter
Store up to 6 months at 4°C

Reverse transcriptase buffer, 5× *(UNIT 10.3)*
✓ 0.25 M Tris·Cl, pH 8.3
0.20 M KCl
✓ 0.03 M MgCl₂
0.5 mg/ml BSA
Store ≤6 months at 4°C

RNA isolation solution I *(UNIT 10.3)*
25 g guanidinium thiocyanate (4 M final)
29.3 ml H₂O
1.76 ml 0.75 M sodium citrate, pH 7.0 (25 mM final)
2.64 ml 10% (w/v) *N*-lauroylsarcosine (0.5% final)
Dissolve at 65°C
Store at room temperature (stable ≥3 months)

CAUTION: *Do not inhale guanidinium thiocyanate and wear gloves when handling it. Do not expose guanidinium thiocyanate to acid conditions, because this may produce cyanide gas.*

RNA isolation solution II *(UNIT 10.3)*
✓ 10 ml RNA isolation solution I
72 μl 2-mercaptoethanol (0.1 M final)
Make fresh before each extraction

rNTP mix, 10× *(UNIT 11.1)*
✓ DEPC-treated water containing:
30 mM rGTP (Ultrapure; Amersham Biosciences)
15 mM rATP (Ultrapure; Amersham Biosciences)
12 mM rCTP (Ultrapure; Amersham Biosciences)
12 mM UTP (Ultrapure; Amersham Biosciences)
Store ≤3 months at −20°C in small aliquots (e.g., 50 to 100 μl)

Running buffer *(UNIT 9.10)*
25 mM Tris base
200 mM glycine

0.1% (w/v) SDS
Store at room temperature

Salmon sperm DNA, sonicated, 10 mg/ml
Dissolve 10 mg salmon sperm DNA (Worthington) in 1 ml sterile water in a polycarbonate tube. Sonicate five times, 30 sec each time, at maximum power, chilling tube on ice between bursts. Check molecular size of DNA by gel electrophoresis *(APPENDIX 3G)*; it should be 200 to 400 bp. Store in 50-μl aliquots ≤1 year at −20°C.

SCE buffer *(UNIT 5.1)*
1 M sorbitol
0.1 M sodium citrate
✓ 0.06 M EDTA, pH 7.0
Filter sterilize
Store indefinitely at room temperature

SCE solution *(UNIT 5.3)*
1 M sorbitol
100 mM sodium citrate, pH 5.8
10 mM Na₂EDTA
30 mM 2-mercaptoethanol
Store at room temperature (stable 1 month)

SD dropout plates and media *(UNIT 5.3)*
Prepare 10× YNB medium:

1.7 g yeast nitrogen base without amino acids and ammonium sulfate (YNB −AA/AS; Difco)
5 g ammonium sulfate
100 ml deionized H₂O
Autoclave and store several months at room temperature

Prepare SD dropout plates:

Add 250 ml H₂O to a 500-ml flask
Add 10 g Bacto-agar (Difco)
Autoclave, then while still hot (50° to 60°C), add:
50 ml 10× YNB medium
50 ml 20% (w/v) glucose
Add nutrient supplements *less* indicated dropout nutrients:
15 mg L-adenine hemisulfate (Ade; final 30 μg/ml)
10 mg L-histidine-HCl (His; final 20 μg/ml)
20 mg L-leucine (Leu; final 40 μg/ml)
20 mg L-lysine-HCl (Lys; final 40 μg/ml)
30 mg L-tryptophan (Trp; final 60 μg/ml)
5 mg uracil (Ura; final 10 μg/ml)
Add sterile H₂O to 500 ml
Mix well and pour into sterile 100-mm² petri plates
Allow to harden overnight
Store up to several months at 4°C

For example, for SD −Ura −Trp plates, use all nutrients except uracil and tryptophan. For liquid medium, simply omit agar from recipe.

SDS, 20% (w/v)
Dissolve 20 g SDS (sodium dodecyl sulfate or sodium lauryl sulfate) in H₂O to 100 ml total with stirring. Iit may be necessary to heat the solution

slightly to fully dissolve the powder. Filter sterilize using a 0.45-μm filter.

SDS sample buffer, 2× (UNIT 9.10)
✓ 100 mM Tris·Cl, pH 6.8
4% (w/v) SDS
0.1% (w/v) bromphenol blue
20% (v/v) glycerol
Store at room temperature
Add 200 mM dithiothreitol or 8% (v/v) 2-mercaptoethanol immediately before use

Sephadex G-50 (UNIT 4.5)
Slowly add 2 g Sephadex G-50 (Pharmacia Biotech) to 300 ml H_2O. Suspend gel beads thoroughly and autoclave. Store up to 6 months (as long as sterility is maintained) at room temperature.

Sequenase reaction buffer, 5× (UNIT 2.1)
✓ 200 mM Tris·Cl, pH 7.5
100 mM $MgCl_2$
250 mM NaCl
Store 4 to 8 weeks at 4°C

Serum airway medium, 5% (UNIT 13.5)
279 ml DMEM (high-glucose formulation; Life Technologies)
279 ml Ham's F-12 nutrient mixture (Life Technologies)
6 ml 100× MEM nonessential amino acids (Life Technologies)
✓ 30 ml FBS
0.72 ml 0.12 U/ml insulin (Regular Iletin I; Eli Lilly)
100 U/ml penicillin
100 μg/ml streptomycin
Store up to 2 weeks at 4°C

The insulin (Regular Iletin I from Eli Lilly) is packaged in 10-ml vials at 100 U/ml.

Silanized glass microscope slides (UNIT 8.5)
Prepare a 2% (v/v) silanization solution by mixing 2 ml of 3-aminopropyltriethoxysilane (Sigma) and 98 ml acetone. Dip slides into silanization solution for 20 sec. Wash in two changes of acetone, 15 sec each, then air dry. Store slides indefinitely at room temperature.

Silver nitrate staining solution (UNIT 2.1)
Dissolve 1 g silver nitrate (store desiccated; handle with caution) in 1 liter sterile, deionized, double-distilled water (0.006 M final). Store in a dark bottle at room temperature.

Size standards, 100 bp (UNIT 11.2)
50 μl 1 mg/ml DNA ladder (Life Technologies)
✓ 5 μl 1 M Tris·Cl, pH 8.0
✓ 5 μl 0.5 M EDTA, pH 8.0
✓ 440 μl loading buffer for ESTs
Store up to 1 month at 4°C

Slide-making supplies (UNIT 8.1)
Bent-pipet rake: Heat the tip of a Pasteur pipet in a flame and bend it to form a 90° angle ∼2.5 cm (1 in.) from the tip.

Slide-making apparatus: Plans for this apparatus are available from the author (L.J.).

SOB/amp medium, per liter (UNIT 12.8)
20 g tryptone
5 g yeast extract
0.5 g NaCl
H_2O to 950 ml
10 ml 250 mM KCl
Adjust pH to 7.0 with 5 N NaOH
Autoclave and store up to 6 months at room temperature
Just before use, add 5 ml sterile 2 M $MgCl_2$ and 1 ml of 50 mg/ml ampicillin

SOC medium
35 g yeast extract
20 g tryptone
0.5 g NaCl
2.5 ml 1 M KCl
H_2O to 960 ml
Adjust pH to 7.0 with 5 M NaOH (∼0.2 ml)
Autoclave 25 min
Cool to 50°C, then add:
10 ml 1 M $MgCl_2$, sterile
10 ml 1 M $MgSO_4$, sterile
20 ml 1 M glucose, filter-sterilized
Store up to 1 year at 4°C

Sodium acetate, 3 M, pH 5.2
408 g sodium acetate trihydrate ($NaC_2H_3O_2·3H_2O$)
800 ml H_2O
Adjust pH to 5.2 with 3 M acetic acid
Add H_2O to 1 liter
Filter sterilize

Sodium acetate buffer, 0.1 M
Solution A: 11.55 ml glacial acetic acid per liter (0.2 M)

Solution B: 27.2 g sodium acetate trihydrate ($NaC_2H_3O_2·3H_2O$) per liter (0.2 M)

Referring to Table A.1.3 for desired pH, mix the indicated volumes of solutions A and B, then dilute with H_2O to 100 ml. Filter sterilize if necessary. Store up to 3 months at room temperature.

To prepare buffers with pH values between those listed in Table A.1.3, prepare the closest higher pH, then titrate with solution A.

Buffer may also be made as a 5- or 10-fold concentrate, if desired. Because acetate buffers show concentration-dependent pH changes, the pH of the concentrate should be checked by diluting an aliquot.

Sodium bicarbonate/sodium carbonate buffer, pH 9.0, 1 M (UNIT 2.2)
7.98 g $NaHCO_3$
0.55 g Na_2CO_3

Table A.1.3 Preparation of 0.1 M Sodium and Potassium Acetate Buffers[a]

Desired pH	Solution A (ml)	Solution B (ml)
3.6	46.3	3.7
3.8	44.0	6.0
4.0	41.0	9.0
4.2	36.8	13.2
4.4	30.5	19.5
4.6	25.5	24.5
4.8	20.0	30.0
5.0	14.8	35.2
5.2	10.5	39.5
5.4	8.8	41.2
5.6	4.8	45.2

[a] Adapted by permission from CRC, 1975.

Table A.1.4 Preparation of 0.1 M Sodium and Potassium Phosphate Buffers[a]

Desired pH	Solution A (ml)	Solution B (ml)	Desired pH	Solution A (ml)	Solution B (ml)
5.7	93.5	6.5	6.9	45.0	55.0
5.8	92.0	8.0	7.0	39.0	61.0
5.9	90.0	10.0	7.1	33.0	67.0
6.0	87.7	12.3	7.2	28.0	72.0
6.1	85.0	15.0	7.3	23.0	77.0
6.2	81.5	18.5	7.4	19.0	81.0
6.3	77.5	22.5	7.5	16.0	84.0
6.4	73.5	26.5	7.6	13.0	87.0
6.5	68.5	31.5	7.7	10.5	90.5
6.6	62.5	37.5	7.8	8.5	91.5
6.7	56.5	43.5	7.9	7.0	93.0
6.8	51.0	49.0	8.0	5.3	94.7

[a] Adapted by permission from CRC, 1975.

H_2O to 100 ml
Store up to 1 year at room temperature

Sodium borate, 1 M, pH 8.0 *(UNIT 11.2)*
  61.83 g boric acid
  ✓ 900 ml DEPC-treated water
  Adjust pH to 8.0 with 1 N NaOH
  Bring to 1 liter
  Sterilize with a 0.2-μm filter
  Store up to 6 months at room temperature

Sodium heparin solution, 500 USP U/ml *(UNIT 4.2)*
  Combine 0.5 ml of 10,000 USP U/ml sterile sodium heparin (J.A. Webster) with 9.5 ml complete RPMI (see recipe) under aseptic conditions. Store up to 1 week at 4°C.

Sodium phosphate buffer, 0.1 M
  Solution A: 27.6 g $NaH_2PO_4 \cdot H_2O$ per liter (0.2 M)

  Solution B: 53.65 g $Na_2HPO_4 \cdot 7H_2O$ per liter (0.2 M)

  Referring to Table A.1.4 for desired pH, mix the indicated volumes of solutions A and B, then dilute with H_2O to 200 ml. Filter sterilize, if necessary. Store up to 3 months at room temperature.

To prepare buffers with pH values between those listed in Table A.1.4, prepare the closest higher pH, then titrate with solution A.

Buffer may also be made as a 5- or 10-fold concentrate, if desired. Because phosphate buffers show concentration-dependent pH changes, the pH of the concentrate should be checked by diluting an aliquot.

Sonicated genomic DNA *(UNIT 4.5)*

Dissolve 10 mg genomic DNA specific for the species of the target or probe DNA in 10 ml water and place in a −20°C freezer until nearly frozen. Sonicate 30 sec at maximum frequency and power setting. To check single-strand size, melt 1 μg for 3 min at 100°C, cool on ice, run on a 2% agarose gel with a 100- to 1000-bp molecular size marker, and stain with ethidium bromide. Marker DNA does not have to be denatured. If fragment size is too large, place DNA in −20°C freezer until nearly frozen again and repeat sonication, using longer sonication times. Store at 4°C ($\leq$3 months) or −20°C (stable $\geq$ 1 year).

Sonicated salmon sperm DNA, 10 mg/ml *(UNIT 4.4)*

Dissolve 10 mg salmon sperm DNA (Worthington) in 1 ml sterile water in a polycarbonate tube. Sonicate five times, 30 sec each, at maximum power. Chill tube on ice between bursts. Check molecular size of DNA by gel electrophoresis (should be 200 to 400 bp). Store in 50-μl aliquots $\leq$1 year at −20°C.

Sorensen buffer *(UNIT 4.7)*

66.7 mM KH_2PO_4 (9.078 g/liter)
66.6 mM $Na_2HPO_4 \cdot 2\ H_2O$ (11.870 g/liter; Merck)
Adjust the pH to 6.8
Store up to 4 months at 4°C

Sorensen phosphate buffer, pH 6.8 *(UNIT 4.3)*

Solution A: 9.08 g/liter (66 mM) KH_2PO_4 (pH ∼4.5)

Solution B: 16.89 g/liter (62 mM) $Na_2HPO_4 \cdot 7H_2O$ (pH ∼9.0)

Mix 510 ml solution A and 490 ml solution B for pH 6.8 buffer. If desired, adjust pH from 5.0 to 8.2 by adjusting volumes of solutions A and B. Store all solutions for $\leq$6 months at room temperature.

Spermidine, 1 M *(UNIT 10.4)*

Add 34 ml sterile water to 5 g spermidine free base [N-(3-aminopropyl)-1,4-butanediamine]. Aliquot and store indefinitely at −20°C.

SSC, 20×

175 g NaCl (3 M)
88 g trisodium citrate dihydrate ($Na_3C_6H_5O_7 \cdot 2H_2O$; 0.3 M)
H2O to 800 ml
Adjust pH to 7.0 with 1 M HCl
Add H_2O to 1 liter

SSPE, 20×

800 ml H_2O
175 g NaCl (3 M)
7.4 g EDTA (20 mM)
24.0 g NaH_2PO_4 (0.20 mM)
Adjust pH with 10 N NaOH (∼6.5 ml for pH 7.4)
Adjust volume to 1 liter with H_2O

SSPE-T, 10×, 6×, 1× *(UNIT 2.4)*

900 mM NaCl
60 mM NaH_2PO_4
✓ 6 mM EDTA, pH 7.4
0.005% Triton X-100
Dilute with water to appropriate concentration
Store up to 1 month at room temperature

Staining solution *(UNIT 2.4)*

Prepare 2.2 μg/ml streptavidin R-phycoerythrin (Molecular Probes) and 0.5 mg/ml acetylated BSA in 6× SSPE-T (see recipe). Prepare fresh.

Staining solution *(UNIT 9.10)*

50% (v/v) methanol
0.05% (w/v) Coomassie brilliant blue R
10% (v/v) acetic acid
40% (v/v) H_2O
Store at room temperature

Standard enzyme diluent *(APPENDIX 3E)*

✓ 200 mM Tris·Cl, pH 7.5
500 μg/ml BSA (Pentax Fraction V)
10 mM 2-mercaptoethanol
Store $\leq$1 month at 4°C

Stop solution *(UNIT 7.5)*

90% (v/v) formamide
0.1% (w/v) bromphenol blue
0.1% (w/v) xylene cyanol

Storage buffer, 2× *(UNIT 12.2)*

✓ 10 mM Tris·Cl, pH 8.0
100 mM NaCl
0.1% (w/v) bovine serum albumin (BSA)
50% (v/v) glycerol
Filter to sterilize
Store up to 1 year at 4°C

Streptavidin solution *(UNIT 4.4)*

✓ PBS containing:
1% (w/v) BSA
0.1% (v/v) Tween 20
3 μg/ml streptavidin
Prepare fresh

Streptavidin-coated microtiter plates *(UNIT 2.2)*

Add 50 μl of 25 μg/ml streptavidin in 1× PBS, pH 7.0 (see recipe), to each well of a flat-bottom 96-well microtiter plate. Wrap plate with plastic wrap and store for a minimum of 12 hr and a maximum of 6 months at 4°C.

Stripping solution *(APPENDIX 3G)*

✓ 5 ml 20× SSC
✓ 10 ml 10% (w/v) SDS
H_2O to 1 liter
Prepare fresh

Stripping solution (APPENDIX 3H)
- ✓ 1% (w/v) SDS
- ✓ 0.1× SSC
- ✓ 40 mM Tris·Cl, pH 7.5 to 7.8

Store up to 1 year at room temperature
Where formamide stripping is desired, mix above solution 1:1 with formamide just before use.

Substrate buffer (UNIT 2.1)
- ✓ 100 mM Tris·Cl, pH 9.5

100 mM NaCl
50 mM MgCl$_2$

Make from sterile stock solutions and store indefinitely at room temperature. Inspect solution for bacterial contamination before use.

Sucrose lysis solution (APPENDIX 3A)

109.4 g sucrose
700 ml H$_2$O
- ✓ 10 ml 1 M Tris·Cl, pH 7.5

5 ml 1 M MgCl$_2$
10 ml Triton X-100
Add H$_2$O to 1 liter
Filter sterilize
Store in sterile 500-ml bottles at 4°C

T low E buffer (UNIT 11.2)
- ✓ 10 ml 1 M Tris·Cl, pH 8.0
- ✓ 0.2 ml 0.5 M EDTA, pH 8.0
- ✓ 900 ml DEPC-treated H$_2$O

Autoclave and store up to 6 months at room temperature

T4 DNA ligase buffer, 5× (UNIT 5.4)
- ✓ 2 M Tris·Cl, pH 7.6

50 mM MgCl$_2$
5 mM ATP
5 mM dithiothreitol (DTT)
25% (w/v) polyethylene glycol 8000 (PEG 8000)
Store up to 1 year at −20°C

T4 polynucleotide kinase buffer, 10× (UNIT 2.1)
- ✓ 500 mM Tris·Cl, pH 7.5
- ✓ 1 mM EDTA

500 mM MgCl$_2$
50 mM dithiothreitol
1 mM spermidine
Store ≤1 month at 4°C or ≤6 months at −20°C

T4 polynucleotide kinase buffer, 10× (UNIT 9.2)
- ✓ 700 mM Tris·Cl, pH 7.6

100 mM MgCl$_2$
50 mM spermidine
Store frozen at −20°C

T4 polynucleotide kinase buffer, 10× (APPENDIX 3E)
- ✓ 500 mM Tris·Cl, pH 7.5

500 mM MgCl$_2$
50 mM dithiothreitol
0.5 mg/ml BSA or gelatin
Store ≤1 month at 4°C or ≤6 months at −20°C

T4 polynucleotide kinase reaction buffer, 10× (UNIT 7.5)
- ✓ 500 mM Tris·Cl, pH 7.5

80 mM MgCl$_2$
20 mM dithiothreitol
Store in aliquots at −20°C

TAE (Tris/acetate/EDTA) electrophoresis buffer, 50×

242 g Tris base
57.1 ml glacial acetic acid (Table A.1.1)
37.2 g Na$_2$EDTA·2H$_2$O (2 mM)
H$_2$O to 1 liter

This solution does not normally need to be sterilized. The Tris base and acetic acid correspond to 40 mM Tris·acetate.

TAE electrophoresis buffer, 10× (UNIT 12.8)

24.2 Tris base
5.71 ml glacial acetic acid
3.72 g Na$_2$EDTA·2H$_2$O
H$_2$O to 1 liter
Store indefinitely at room temperature

TBE (Tris/borate/EDTA) electrophoresis buffer, 10×

108 g Tris base (890 mM)
55 g boric acid (890 mM)
- ✓ 40 ml 0.5 M EDTA, pH 8.0 (20 mM)

H$_2$O to 1 liter

10× and 5× TBE tend to precipitate over time. If convenient, dilute to 2× or 1× immediately, or stir continuously. This solution does not normally need to be sterilized.

TBS (Tris-buffered saline), pH 7.5 (UNIT 12.7)
- ✓ 50 mM Tris·Cl, pH 7.5
- ✓ 1 mM EDTA

150 mM NaCl
Store 1 to 2 years at room temperature

TBS/FBS (UNIT 4.7)

10 ml 0.5 M Tris base
90 ml 0.15 M NaCl (0.135 M final)
Adjust pH to 7.6
Store up to 6 months at 4°C
Before use, add 2.5% (v/v) FBS filtered with a 1.2-μm filter

TE buffer
- ✓ 10 ml 1 M Tris·Cl, pH 7.4, 7.5, or 8.0 (or other desired pH; 10 mM final)
- ✓ 2 ml 0.5 M EDTA, pH 8.0 (1 mM final)

H$_2$O to 1 liter

TE79/10 (UNIT 12.5)
- ✓ 1 mM Tris·Cl, pH 7.9
- ✓ 0.1 mM EDTA

Filter sterilize using a 0.2-μm nitrocellulose filter (Nalgene)
Store up to 1 year at 4°C

TEN (Tris/EDTA/NaCl) solution, 10× (APPENDIX 3A)
- ✓ 9 ml 1 M Tris·Cl, pH 7.5
- ✓ 24 ml 0.5 M EDTA, pH 8.0

9 ml 5 M NaCl
Filter sterilize
Store at room temperature

Terminal transferase reaction buffer, pH 6.6, 5×
(UNIT 2.1)
    1 M potassium cacodylate (handle with caution)
✓ 125 mM Tris·Cl, pH 6.6
    1.25 mg/ml BSA
    Store indefinitely at −20°C

Terrific broth (TB)
    900 ml H_2O
    12 g tryptone
    24 g yeast extract
    4 ml glycerol
    Autoclave 20 min and allow broth to cool to
        <60°C
    Add 100 ml sterile potassium phosphate solution
    Store ≤1 year at 4°C

Potassium phosphate solution is made by dissolving 2.31 g (0.17 M) KH_2PO_4 and 12.54 g (0.72 M) K_2HPO_4 in H_2O, then autoclaving 20 min to sterilize.

Tetracycline, 1 mg/ml *(UNIT 12.5)*
    0.1 g tetracycline (Sigma)
    50% (v/v) ethanol to 100 ml
    Store up to 6 months at −20°C

Thawing medium *(UNIT 13.4)*
    RPMI medium containing:
✓ 30% (v/v) FBS, heat inactivated 1 hr at 56°C
    20 U/ml heparin
    100 U/ml penicillin
    10 μg/ml streptomycin
    2 mM L-glutamine
    Store up to 1 month at 4°C

Thermostable ligase buffer, 10× *(UNIT 2.2)*
✓ 200 mM Tris·Cl, pH 8.0
    100 mM $MgCl_2$
    10 mM dithiothreitol (DTT)
    Store up to 1 year at −20°C

Thymidine, 1 mM *(UNIT 4.1)*
    Dissolve 25 mg thymidine (Sigma) in 100 ml HBSS (see recipe). Sterilize by passing through a 0.22-μm filter. Aliquot 5 ml into red-top Vacutainers (sterile untreated containers) and store ≤1 year at −20°C. Store aliquot in use ≤1 month at 4°C.

Thymidine, 1 mM *(UNITS 4.3 & 8.1)*
    Dissolve 2.4 mg thymidine in 10 ml H_2O. Filter sterilize and prepare 1-ml aliquots in 1.5-ml microcentrifuge tubes. Store ≤6 months at −20°C.

TMAC solution *(UNIT 9.2)*
    Dissolve 657.6 g tetramethylammonium chloride (mol. wt. = 109.6) in 1 liter H_2O. Filter solution through Whatman no. 1 filter paper and determine the precise concentration by measuring the refractive index (n) of a three-fold diluted solution. The molarity (M) of the diluted solution = $53.6 \times (n − 1.331)$. Store at room temperature in brown bottles.

    CAUTION: *TMAC is hazardous; use appropriate precautions.*

TMAC hybridization solution *(UNIT 9.2)*
✓ 6 ml TMAC solution (~3 M final)
✓ 20 μl 0.5 M EDTA (1 mM final)
    1 ml 0.1 M Na_3PO_4, pH 6.8 (10 mM final)
    100 μl 10% (w/v) SDS (0.1% final)
✓ 1 ml 50× Denhardt's solution (5× final)
    40 μl 10 mg/ml yeast RNA (40 μg/ml final)
    1.84 ml H_2O

TMAC wash solution *(UNIT 9.2)*
✓ 600 ml TMAC solution (~3 M final)
✓ 2 ml 0.5 M EDTA (1 mM final)
    100 ml 0.1 M Na_3PO_4, pH 6.8 (10 mM final)
    10 ml 10% (w/v) SDS (0.1% final)
    288 ml H_2O

TNE buffer, 10× *(UNIT 10.4)*
    12.1 g Tris base (100 mM final)
    3.7 g Na_2EDTA (10 mM final)
    58.4 g NaCl (1 M final)
    Add H_2O to 1 liter
    Adjust pH to 7.4
    Filter sterilize with a 0.2-μm filter
    Store ≤6 months at 4°C

TNE buffer, 10× *(APPENDIX 3D)*
    100 mM Tris base
✓ 10 mM EDTA
    2.0 M NaCl
    Adjust pH to 7.4 with concentrated HCl
    As needed, dilute with H_2O to desired concentration

Transduction medium *(UNIT 13.4)*
✓ 100 ml BBMM
    20 μl 50 ng/μl interleukin 3 (IL-3; Biosource International; 10 ng/ml final)
    66 μl of 50 U/μl interleukin 6 (IL-6; Biosource International; 50 ng/ml final)
    100 μl of 50 ng/μl stem cell factor (SCF; Biosource International; 50 ng/ml final)

Transfer buffer *(UNIT 9.3)*
    3 M NaCl
    8 mM NaOH
    Store up to 6 months at room temperature

Transformation medium *(APPENDIX 3J)*
    Per sample:
    5 ml complete Iscove's modified Dulbecco's medium (IMDM; Sigma) with 15% FBS (see recipe)
    1 ml Epstein-Barr virus (EBV) stock *(APPENDIX 3J)*
✓ 1 ml 0.2 mg/ml cyclosporin A
    Prepare fresh for each use

Cell pellet size may vary between samples, requiring more or less transformation medium. Adjust the volume of transformation medium proportionally.

Tris-buffered saline, see TBS

Tris·Cl, 1 M
    Dissolve 121 g Tris base [tris(hydroxymethyl) aminomethane] in 800 ml H_2O
    Adjust to desired pH with concentrated HCl

A1

Mix and add H$_2$O to 1 liter
Store up to 6 months at 4°C or room temperature

Always use high-quality Tris. Lower-quality Tris can be recognized by its yellow appearance when dissolved.

Approximately 70 ml HCl is needed for a pH 7.4 solution, or ~42 ml for a pH 8.0 solution.

IMPORTANT NOTE: *The pH of Tris buffers changes significantly with temperature, decreasing 0.028 pH units per 1°C. Tris-buffered solutions should be adjusted to the desired pH at the temperature at which they will be used. Because the pK$_a$ of Tris is 8.08, Tris should not be used as a buffer below pH ~7.2 or above pH ~9.0.*

Tris/acetic acid substrate buffer, 0.1 M, pH 8.2
(*UNIT 4.7*)
12.11 g Tris base (0.1 M final)
2.89 ml glacial acetic acid (0.05 M final)
1 liter H$_2$O
Adjust pH to 8.2
Store up to 2 to 3 months at 4°C
Use at 20°C

Tris/NaCl wash buffer (*UNIT 2.2*)
✓ 100 mM Tris·Cl, pH 7.5
150 mM NaCl
0.05% (v/v) Tween 20
Store up to 6 months at room temperature

Trypsin/EDTA solution
✓ Sterile HBSS (see recipe) or 0.9% (w/v) NaCl
0.25% (w/v) trypsin
0.2% (w/v) EDTA
Store ≤1 year at −20°C

Specific applications may require different concentrations of trypsin; the appropriate units should be consulted for details.

Trypsin/EDTA solution is available in various concentrations including 10×, 1×, and 0.25% (w/v). It is received frozen from the manufacturer and can be thawed and aseptically divided into smaller volumes. Preparing trypsin/EDTA from powdered stocks may reduce its cost; however, most laboratories prefer commercially prepared solutions for convenience.

Trypsin/EDTA/HBSS solution (*UNIT 8.1*)
Add 10 ml of 0.065% (v/v) trypsin/EDTA solution (see recipe) to 70 ml Ca^{2+}- and Mg^{2+}-free HBSS (see recipe). Store 20-ml aliquots at 4°C for 10 to 14 days.

Trypsin inhibitor buffer (*UNIT 13.5*)
Prepare 1 mg/ml trypsin inhibitor (Type I-S, from soybean; Sigma) in Ham's F-12 medium (Life Technologies). Filter through a 0.2-μm filter and store in aliquots up to 6 months at −20°C.

Trypsin solution (*UNIT 4.3*)
Dissolve 1.25 g powdered trypsin (Sigma) in 200 ml water and stir gently 6 to 8 hr to dissolve. Divide into 2-ml aliquots and store frozen at −20°C. To use, combine 2 ml stock with 48 ml disodium

phosphate buffer (see recipe) in a Coplin jar. Prepare fresh daily.

Trypsin solution, 0.0125% (w/v) (*UNIT 4.2*)
Combine 1.0 ml of 2.5% (w/v) trypsin (Sigma) with 99 ml deionized water (0.025%). Dilute 1:1 (v/v) with deionized water (0.0125%). Store up to 2 months at 0°C.

Trypsin/Giemsa solution (*UNIT 4.2*)
1.0 ml Gurr's Improved Giemsa R66 (Bio/medical Specialties)
45 ml Gurr's phosphate buffer, pH 6.8 (Bio/medical Specialities)
✓ 4 drops (from a Pasteur pipet) of 0.0125% (w/v) trypsin solution
Prepare fresh

Tumor transport solution (*UNIT 10.2*)
✓ 100 ml HBSS
5 μg/ml amphotericin
1% (v/v) penicillin/streptomycin solution (Life Technologies)
Divide into 5- to 10-ml aliquots in 50-ml centrifuge tubes
Store ≤2 months at room temperature

The 50-ml centrifuge tubes make it easy to drop in tumor specimens upon collection.

Ussing chamber culture medium (*UNIT 13.5*)
49% (v/v) DMEM
49% (v/v) Ham's F-12 nutrient mixture (Life Technologies)
2% (v/v) Ultraser G (Sepracor)
Store up to 2 weeks at 4°C

Wash buffer, 0.06× SSC (*UNIT 11.2*)
✓ 3 ml 20× SSC
✓ 997 ml DEPC-treated water
Filter sterilize on a 0.5-μm filter device
Store up to 2 months at room temperature

Wash buffer, 0.5× SSC/0.01% (v/v) SDS (*UNIT 11.2*)
✓ 25 ml 20× SSC
✓ 974 ml DEPC-treated water
Filter sterilize on a 0.5-μm filter device
✓ Add 1 ml 10% SDS and mix well
Store up to 2 months at room temperature

Well-to-read acrylamide gel, 6% (w/v) (*UNIT 2.3*)
Referring to *APPENDIX 3F* for gel preparation, for each 24-cm gel, add 50 g urea and 15 ml of 40% acrylamide solution to 28 ml water. Stir using low heat until all ingredients are dissolved. Deionize the solution by adding 0.66 g Amberlite resin and stirring for 5 min. Filter 10 ml of 10× TBE electrophoresis buffer (see recipe) through a vacuum filter, then filter the deionized urea/acrylamide solution into the buffer and degas the solution. Assemble the glass plates. Add 250 μl ammonium persulfate and 28 μl TEMED per gel to start polymerization. Pour gel. Allow 2 hr for complete polymerization of the gel.

Whole-cell lysis buffer *(UNIT 13.4)*

    50 mM KCl
✓ 10 mM Tris·Cl, pH 8.3
    1.5 mM $MgCl_2$
    0.1 mg/ml gelatin
    0.45% (v/v) Igepal CA-630 (Sigma)
    0.45% (w/v) Tween 20 (polyethoxysorbitan)
    Store 1-ml aliquots up to several years at $-20°C$

Wright stain *(UNIT 4.3)*

Dissolve 0.3 g Wright stain (Sigma or Fisher) in 100 ml of 100% methanol. Stir at room temperature ≥30 min or overnight. Let sit 1 to 2 hr. Filter through Whatman no. 1 filter paper into dark bottle. Store ~1 month at room temperature. To use, dilute 1:4 in 5% (v/v) Hydrion buffer working solution (see recipe). Use immediately while still warm.

Xgal staining solution *(UNITS 12.6 & 12.8)*

✓ PBS, pH 7.4, containing:
    20 mM $K_3Fe(CN)_6$
    20 mM $K_4Fe(CN)_6·3H_2O$
✓ 2 mM $MgCl_2$
    Filter sterilize and store up to 1 year at 4°C

Before use, equilibrate solution to 37°C and add 20 μl/ml of 50 mg/ml of 5-bromo-4-chloro-3-indolyl-β-D-galactopyranoside (Xgal) in dimethylsulfoxide. Store Xgal solution in 1-ml aliquots up to several years at $-20°C$ in the dark.

Xgal staining solution, 0.1% (w/v) *(UNIT 12.7)*

Dissolve 2 mg 5-bromo-4-chloro-3-indoyl-β-D-galactopyranoside (Xgal, Boehringer Mannheim) in 250 μl dimethyl formamide (DMF, Sigma) or dimethylsulfoxide (DMSO, Fisher). Mix 1.6 ml water with 150 μl of 1 M Tris·Cl, pH 8.0 (see recipe; 0.075 M final) containing 10.5 mg potassium ferrocyanide [$K_4Fe(CN)_6·3H_2O$; 12 mM final] and 8.8 mg potassium ferricyanide [$K_3Fe(CN)_6$; 13 mM final]. Mix the two solutions together to use. Prepare fresh.

Yeast tRNA, 4 mg/ml *(UNIT 11.2)*

Resuspend yeast tRNA at 10 mg/ml (based on supplier's quantitation) in DEPC-treated water (see recipe) in a 1.5-ml polypropylene conical centrifuge tube. Add 0.5 vol buffered phenol (see recipe) and vortex. Add 0.5 vol chloroform and vortex again. Centrifuge 5 min at 10,000 × g. Transfer aqueous layer to a new 1.5-ml tube and extract twice with 1 vol chloroform. Transfer aqueous layer to a new 1.5-ml tube. Add 0.1 vol of 3 M sodium acetate, pH 5.2 (see recipe). Add 2 vol 100% (v/v) ethanol. Centrifuge 5 min. Aspirate supernatant and add 1 vol of 70% (v/v) ethanol. Centrifuge 5 min, aspirate supernatant, and allow pellet to dry. Resuspend in DEPC-treated water at the original volume. Determine the RNA concentration by spectrometry *(APPENDIX 3D)*. Dilute to 4 mg/ml and store frozen at $-20°C$.

YPD medium and plates *(UNITS 5.1 & 5.3)*

Place 10 g yeast extract and 20 g peptone in a 2-liter flask. Add 20 g agar if preparing plates. Add 900 ml water and autoclave 20 min. Add 100 ml 20% (w/v) dextrose that has been filter sterilized or autoclaved separately. For liquid medium, distribute into sterile bottles (50 to 100 ml) and store at room temperature. For plates, pour 80 to 100 ml medium per 150-mm plate and store at 4°C.

APPENDIX 2

Useful Information and Data

APPENDIX 2A

Overview of Human Repetitive DNA Sequences

This section contains brief descriptions of the most abundant classes of repetitive DNA in the human genome. Figure A.2A.1 shows the chromosomal distribution of these classes of repeats for human chromosome 16. Refer to *CPHG APPENDIX 1B* for references on this topic.

Telomere repeat. The tandemly repeating unit TTAGGG is located at the very ends of the linear DNA molecules that make up human and other vertebrate chromosomes. The telomere repeat $(TTAGGG)_n$ extends for 5000 to 12,000 bp; its structure contains guanine-guanine base pairs and is different from that of normal DNA. It is replicated in an unusual fashion by the enzyme telomerase, which prevents chromosome shortening during replication.

Subtelomeric repeats. These are classes of repetitive sequences interspersed in the last 500,000 bases of nonrepetitive DNA adjacent to the telomere. Some sequences are chromosome-specific; others seem to be present near the ends of many or all human chromosomes. Subtelomeric DNA also contains other classes of interspersed repetitive DNA that are not subtelomere-specific.

Microsatellite repeats. This class includes a variety of simple di-, tri-, tetra-, and pentanucleotide tandem repeats (also known as simple sequence repeats, SSRs) that are dispersed in the euchromatic arms of most chromosomes. The dinucleotide repeat $(GT)_n$ or $(CA)_n$ is the most common dispersed repeat, occurring on average every 30,000 bases in the human genome, for a total copy number of 100,000. GT repeats range in size from $\sim$20 to 60 bp and appear in most eukaryotic genomes. The repeat length, n, of microsatellites is highly variable among individuals in the population; in consequence, microsatellite repeats provide an abundant class of simple sequence length polymorphisms (SSLPs) or genetic markers that can be typed using the polymerase chain reaction (*UNIT 2.1*).

Minisatellite repeats. This is a class of dispersed tandem repeats, also known as variable number tandem repeats or VNTRs, in which the repeating unit is 30 to 35 bp in length and variable in sequence except that it contains a conserved core sequence of 10 to 15 bp. The tandem repeat length for minisatellites is highly variable, ranging in size from 200 to several thousand base pairs. Minisatellites occur at $\sim$1000 sites in the genome, with a tendency to be more concentrated toward the telomeric ends of chromosomes. The core sequences of many minisatellites are highly conserved, and probes from these core sequences have been used to detect polymorphisms in many species of higher eukaryotes. Hypervariable minisatellite repeats have been shown to be useful for DNA-based individual identification or DNA fingerprinting.

Alu repeats. These are the most abundant interspersed repeat in the human genome. The *Alu* sequence has an average length of 282 bp and occurs on average once every 3300 bp in the human genome, for a total copy number of 500,000 to 1,000,000. Because of its short length, *Alu* is classified as a short interspersed repeated sequence (SINE). Because of their high frequency in the human genome and their lack of conservation in orders preceding

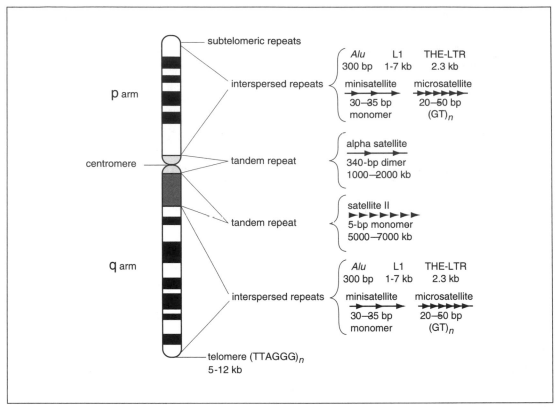

Figure A.2A.1 Chromosomal organization of the major classes of repetitive DNA. Distribution of major classes of repetitive DNA along human chromosome 16. The length of the segment between arrowheads correlates with the size of the repeat unit.

the primates, *Alu* repeats are a useful target for interspersed repetitive sequence PCR (IRS-PCR). IRS-PCR permits the specific amplification of human sequences from a background of nonhuman DNA—e.g., from a human/rodent somatic cell hybrid (*UNIT 3.1*) or a human sequence-bearing YAC clone in a yeast host.

L1 repeats. This long interspersed repeated sequence (LINE) is, at full length, ~7000 bp long. L1s (also called *Kpn*I repeats) have a common sequence at the 3′ end but are variably shortened at the 5′ end, and thus range in size up to 7000 bases. There are ~3500 full-length L1s and 100,000 truncated copies in the human genome. Full-length L1s (3.5 percent of the total) are a divergent group of class II retrotransposons ("jumping genes") that can move around the genome and are thought to be remnants of retroviruses. Class II retrotransposons do not contain long terminal repeats, have at least one open-reading frame, and contain a poly(A) tail. A full-length, functional L1 has been discovered; it encodes a functional reverse transcriptase—an enzyme essential to the process by which the L1s are copied and reinserted into the genome. Primers designed from the L1 repeat have also been shown to be useful for IRS-PCR.

THE-LTR repeats. This is a superfamily of interspersed repetitive elements that number 40,000 to 100,000 in mammalian genomes. This superfamily includes some MER elements (described below) that have been shown to be derived from parts of an ancestral THE1 element. THE-LTR repeats are retrotransposon-like elements with a 1600-bp internal sequence flanked by long terminal repeats (LTRs; 350 to 600 bp). The internal sequence contains a long open reading frame and is often truncated or excised.

MER repeats. These are families of middle-reiteration-frequency repeats (200 to 10,000 copies per haploid genome). More than 30 MER families have been identified to date. The repeat families are of unknown origin and variable size, and presumably represent a "fossil record" of evolutionary processes that shaped the development of the human genome.

Alpha satellite DNA. These are a family of related repeats that occur as long tandem arrays at the centromeric region of all human chromosomes. The repeating unit is a 340-bp dimer consisting of 169- and 171-bp monomers that are 73% similar. Alpha satellite DNA (also known as alphoid DNA) occurs on both sides of the centromeric constriction and extends over a region of ~1000 to 5000 bp. Alpha satellite DNA in other primates is related to that in humans. Subfamilies of alpha satellite DNA that are specific for single or a few chromosomes have been identified for many human chromosomes. Chromosome-specific centromere markers developed from these subfamilies are useful cytogenetic markers for analysis of chromosomal abnormalities using fluorescence in situ hybridization (FISH).

Satellite I, II, and III repeats. These are three classical human satellite DNAs that can be isolated from the bulk of genomic DNA by centrifugation in buoyant-density gradients because their densities differ from the densities of other DNA sequences. Satellite I is rich in A and T nucleotides and is composed of alternating arrays of 17- and 25-bp repeating units. Satellites II and III are both derived from the simple five-base repeating unit ATTCC. Satellite II is more highly diverged from the basic repeating unit than satellite III. Satellites I, II, and III occur as long tandem arrays in the heterochromatic regions of chromosomes 1, 9, 16, 17, and Y and the satellite regions of the short (p) arms of chromosomes 13, 14, 15, 21, and 22. Chromosome-specific probes have also been developed from these classical satellite repeats.

C_0t_1 **DNA.** This fraction of repetitive DNA is separable from other genomic DNA because of its faster reannealing (renaturation) kinetics. C_0t_1 DNA contains sequences that have copy numbers of 10,000 or greater. C_0t_1 is available commercially and is useful for preannealing to DNA probes to block highly repetitive sequences from hybridizing to target DNA.

Contributor: Norman A. Doggett

APPENDIX 2B

ISCN Standard Idiograms

Chromosome banding is used mainly to identify both normal and rearranged chromosomes, to define chromosome breakpoints, and to describe the specific location of DNA sequences on chromosomes (see *UNITS 4.3 & 4.4*). A nomenclature has been developed to standardize the identification of chromosomes and the naming of chromosome bands. The system currently in use is *An International System for Human Cytogenetic Nomenclature* (1995), which includes a chromosome band nomenclature, as well as standard idiograms, which are "diagrammatic representations of a karyotype, which may be based on measurements of the chromosomes" (ISCN, 1995). The idiograms presented here, with the permission of S. Karger, D. Adler, and *Cytogenetics and Cell Genetics*, are drawings of G-banded chromosomes with band numbers indicated. Heterochromatic regions, which contain classes of repetitive DNA and can show individual differences in size, are indicated by patterned areas. These include the centromeric regions of all chromosomes, the large blocks of heterochromatin on chromosomes 1, 9, 16, and distal Yq, and the short arms of the acrocentric chromosomes 13, 14, 15, 21, and 22.

The idiograms presented in Figures A.2B.1 to A.2B.24 represent the standard ISCN (1995) resolutions at the 400-, 550-, and 850-band-per-haploid-genome levels (left to right). Band numbers (lowest at the centromere to highest at the telomere) are indicated to the left of each band. The line at the centromere (indentation) separates the p from the q arm. The 850-band-level idiograms also indicate differences in staining intensities among the G-bands. These idiograms are available in digitized format at *http://www.pathology.washington.edu/cytopages* to facilitate the manipulation of these diagrams (using Adobe Illustrator or other appropriate programs) to create representations of rearranged chromosomes. An example of such manipulation is shown in Figure A.2B.25.

CHROMOSOME BAND NOMENCLATURE

ISCN 1995 also includes the following detailed instructions for naming individual chromosome bands; this material is reprinted by permission of S. Karger (numbers correspond to ISCN 1995 section designations and are included for reference).

2.3.1 Identification and Definition of Chromosome Landmarks, Bands, and Regions

Each chromosome in the human somatic cell complement is considered to consist of a continuous series of bands, with no unbanded areas. As defined earlier, a **band** is a part of a chromosome clearly distinguishable from adjacent parts by virtue of its lighter or darker staining intensity. The bands are allocated to various regions along the chromosome arms, and the regions are delimited by specific **landmarks**. These are defined as consistent and distinct morphologic features important in identifying chromosomes. Landmarks include the ends of the chromosome arms, the centromere, and certain bands. The bands and the regions are numbered from the centromere outward. A **region** is defined as any area of a chromosome lying between two adjacent landmarks.

2.3.2 Designation of Regions and Bands

Regions and bands are numbered consecutively from the centromere outward along each chromosome arm. The symbols **p** and **q** are used to designate, respectively, the short and long arms of each chromosome. The centromere (cen) itself is designated 10, the part facing the short arm is p10, and the part facing the long arm is q10. Thus, the two regions adjacent to the centromere are labeled as 1 in each arm; the next, more distal regions as 2, and so on. A band used as a landmark is considered as belonging entirely to the region distal to the landmark and is accorded the band number of 1 in that region.

In designating a particular band, four items are required: (1) the chromosome number, (2) the arm symbol, (3) the region number, and (4) the band number within that region. These items are given in order without spacing or punctuation. For example, 1p33 indicates chromosome 1, short arm, region 3, band 3.

Subdivision of an Existing Landmark or Band

Whenever an existing band is subdivided, a decimal point is placed after the original band designation and is followed by the number assigned to each **sub-band**. The sub-bands are numbered sequentially from the centromere outward. For example, if the original band 1p31 is subdivided into three equal or unequal sub-bands, the sub-bands are labeled 1p31.1, 1p31.2, and 1p31.3, sub-band 1p31.1 being proximal and 1p31.3 distal to the centromere. If a sub-band is subdivided, additional digits, but no further punctuation, are used; e.g., sub-band 1p31.1 might be further subdivided into 1p31.11, 1p31.12, etc. Although in principle a band can be subdivided into any number of new bands at any one stage, a band is usually subdivided into three sub-bands.

Contributors: Rhona R. Schreck, Christine M. Distèche, and David Adler

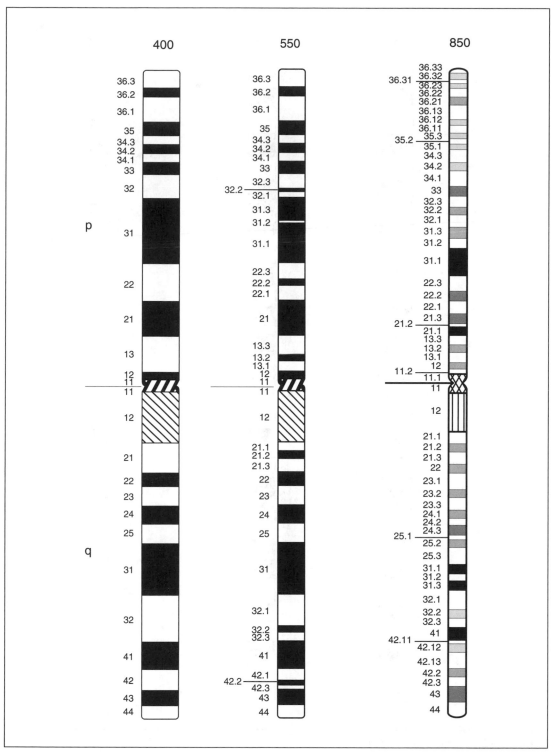

Figure A.2B.1 Chromosome 1 idiogram. For this and all subsequent figures, the 400- and 550-level idiograms are based on ISCN (1981) nomenclature; the location and width of bands are not based on any measurements. The 850-band idiogram replaces the ISCN (1981) version. The band numbers are the same, but relative widths of euchromatic bands are based on measurements and the staining intensities reflect GTG bands (Francke, 1981).

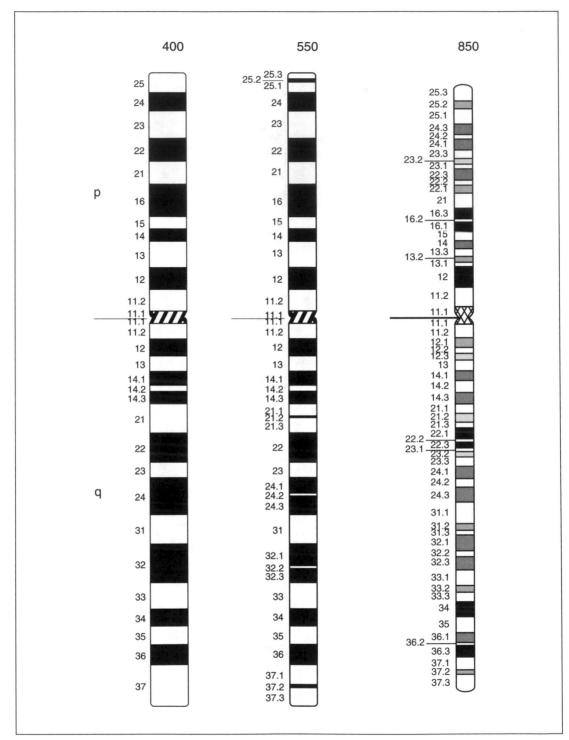

Figure A.2B.2 Chromosome 2 idiogram.

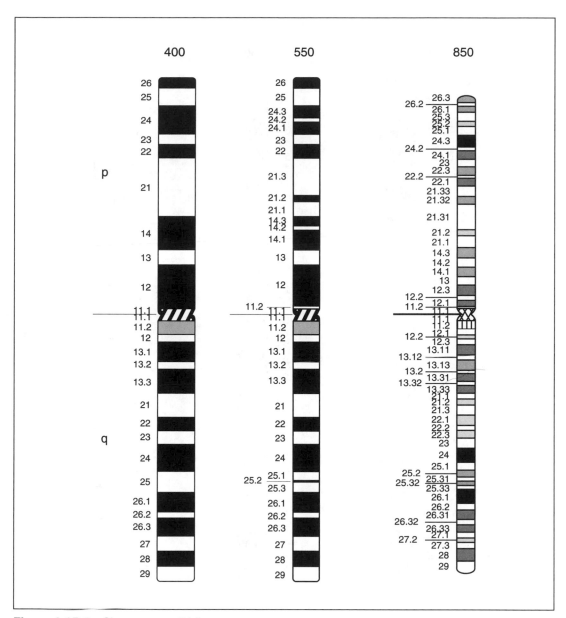

Figure A.2B.3 Chromosome 3 idiogram.

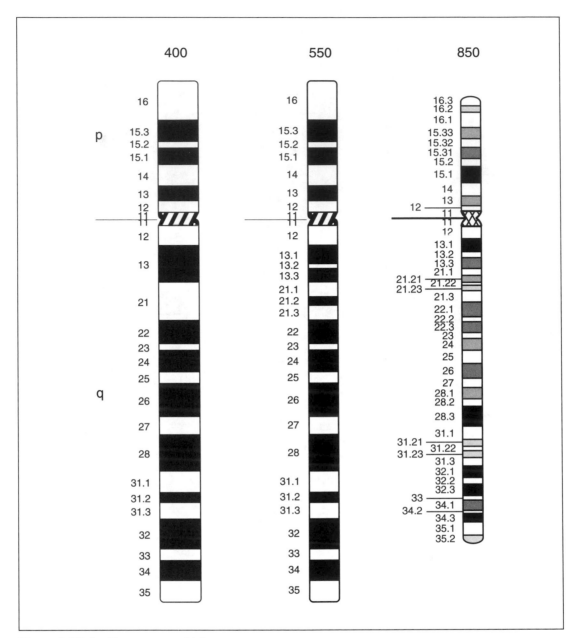

Figure A.2B.4 Chromosome 4 idiogram.

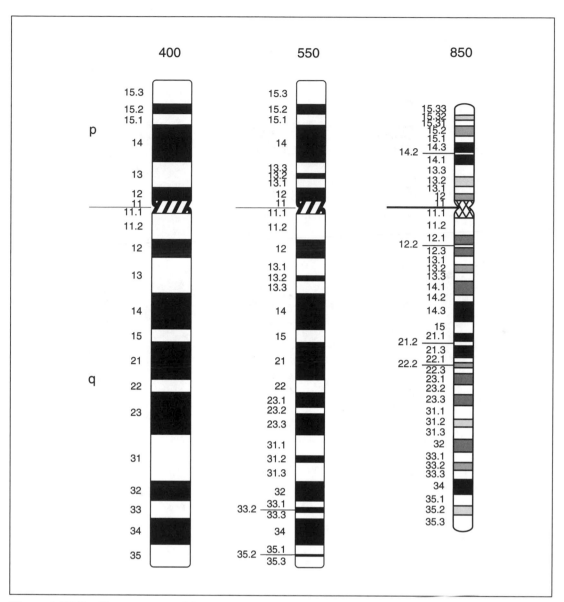

Figure A.2B.5 Chromosome 5 idiogram.

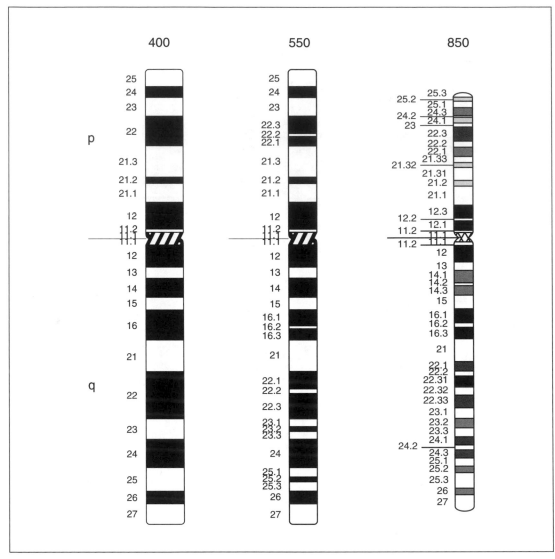

Figure A.2B.6   Chromosome 6 idiogram.

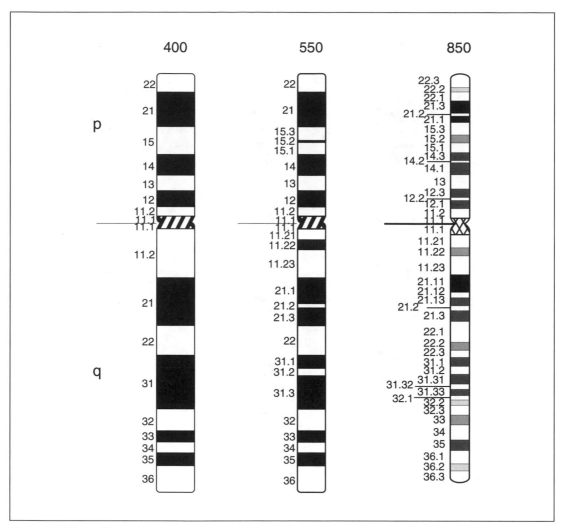

Figure A.2B.7 Chromosome 7 idiogram.

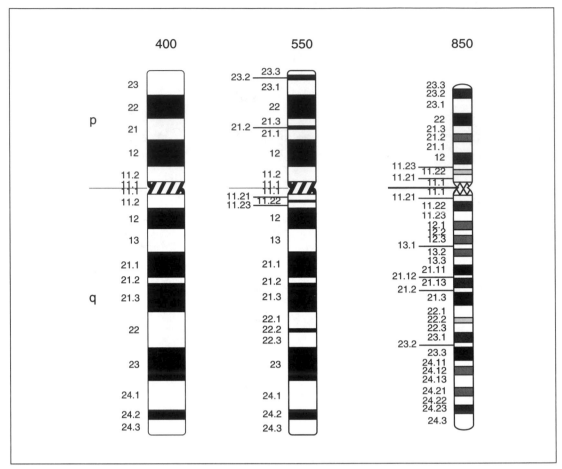

Figure A.2B.8 Chromosome 8 idiogram.

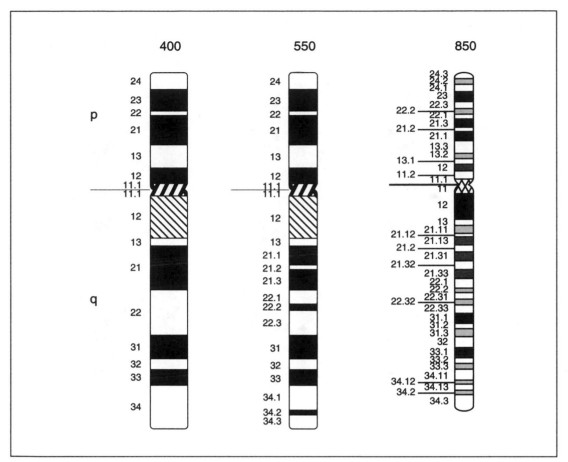

Figure A.2B.9 Chromosome 9 idiogram.

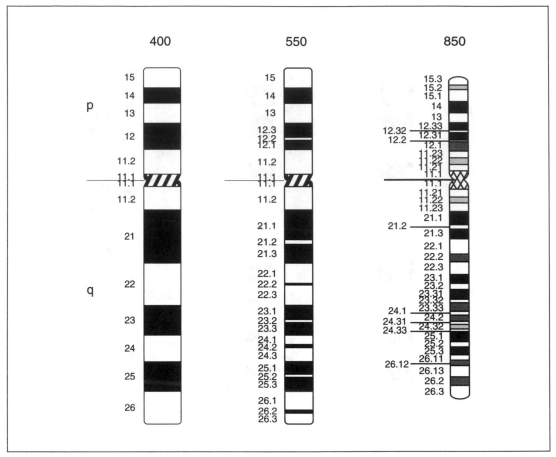

Figure A.2B.10 Chromosome 10 idiogram.

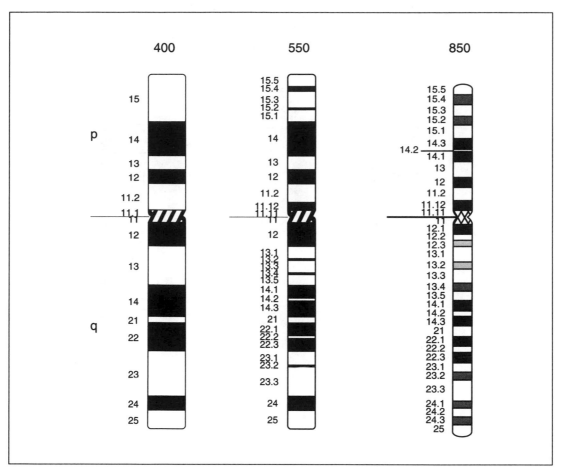

Figure A.2B.11 Chromosome 11 idiogram.

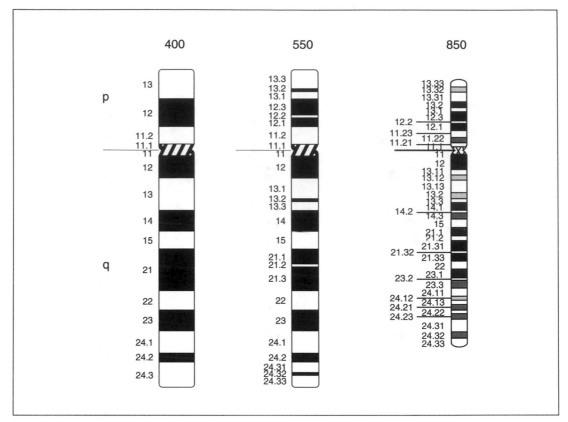

Figure A.2B.12   Chromosome 12 idiogram.

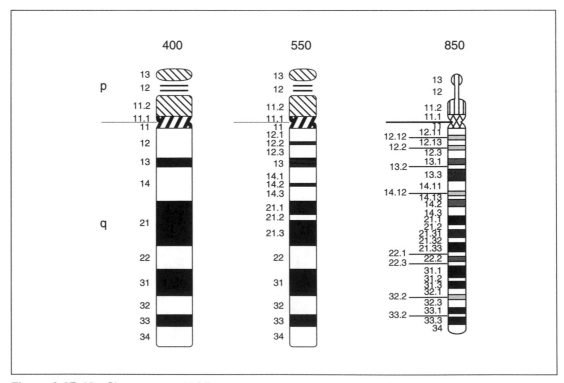

Figure A.2B.13   Chromosome 13 idiogram.

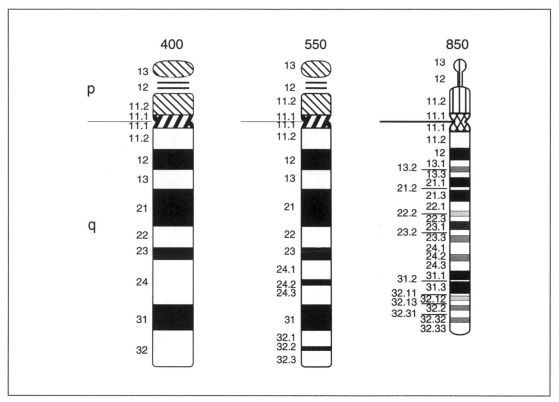

Figure A.2B.14 Chromosome 14 idiogram.

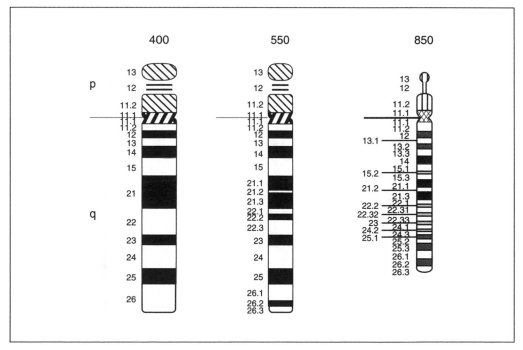

Figure A.2B.15 Chromosome 15 idiogram.

A2

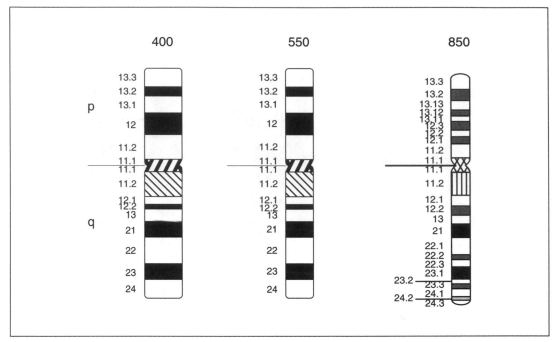

Figure A.2B.16   Chromosome 16 idiogram.

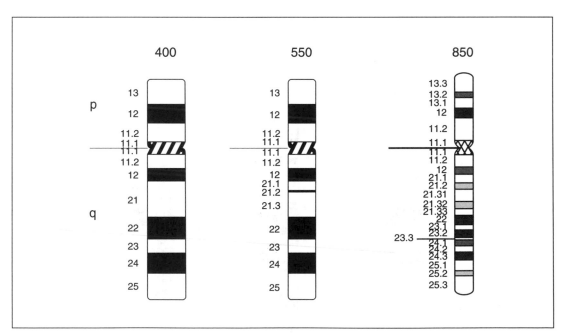

Figure A.2B.17   Chromosome 17 idiogram.

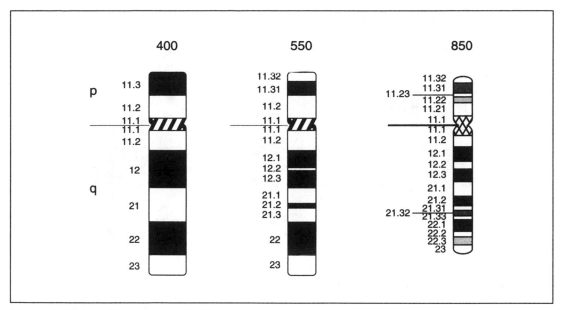

Figure A.2B.18 Chromosome 18 idiogram.

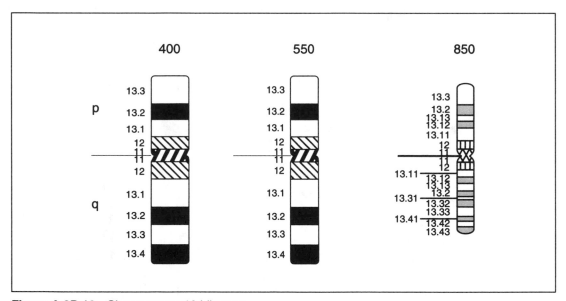

Figure A.2B.19 Chromosome 19 idiogram.

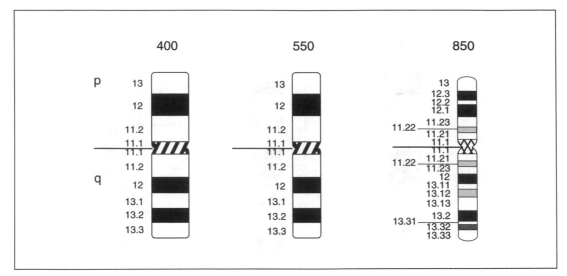

Figure A.2B.20   Chromosome 20 idiogram.

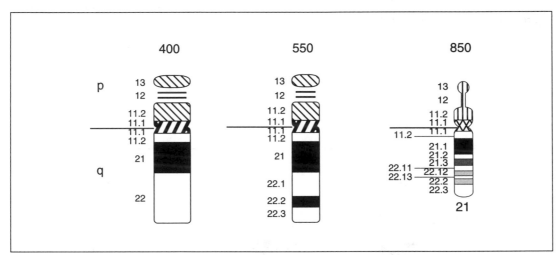

Figure A.2B.21   Chromosome 21 idiogram.

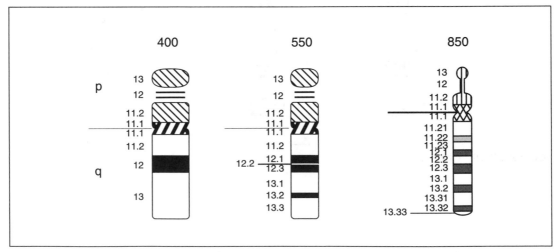

Figure A.2B.22   Chromosome 22 idiogram.

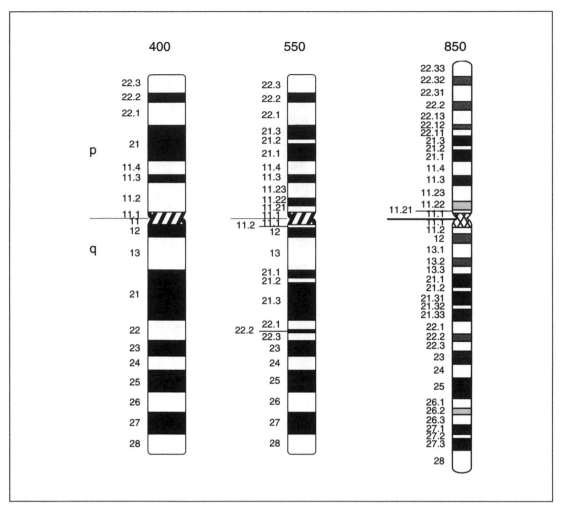

Figure A.2B.23 X chromosome idiogram.

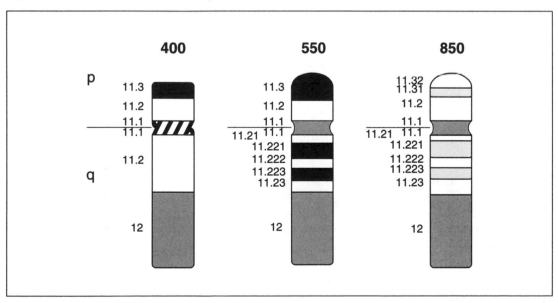

Figure A.2B.24 Y chromosome idiogram. As suggested by observations of Magenis and Barton (1987), the number of bands on the euchromatic portion of the long arm has been expanded.

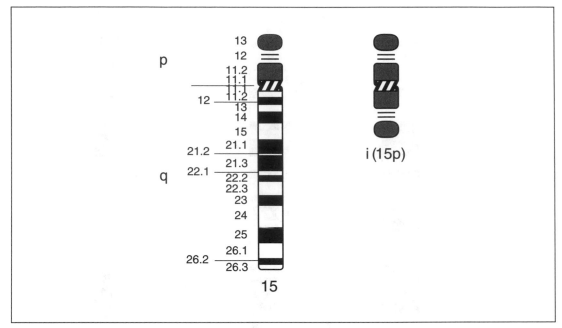

p

q

13
12
11.2
11.1
11.1
11.2
13
14
15
21.1
21.3
22.2
22.3
23
24
25
26.1
26.3

12
12
21.2
22.1
26.2

i (15p)

15

Figure A.2B.25 Isochromosome derived from the short arm of chromosome 15. Idiograms of a normal copy of chromosome 15 (left) and an isochromosome for the short arm of chromosome 15 [i(15p)] (right).

Genetic Linkage Reference Maps: Access to Internet-Based Resources

CHLC

The primary URL for the The Cooperative Human Linkage Center (CHLC) home page is *http://www.chlc.org*. This site provides descriptions of CHLC activities and resources, several different types of genetic maps, information on markers (e.g., primer sequence, allele size, allele sizes for several CEPH individuals, heterozygosity, and repeat type), lists of publications, and links to other useful sites.

Généthon

The primary URL for Genethon is *http://www.genethon.fr/php/index.php*. This site provides descriptions of Genethon activities and access to genetic, physical, and transcript maps. The information on the genetic maps includes the final Généthon map. Detailed information on each marker (primer sequence, allele sizes, allele sizes for CEPH 1347-02, and heterozygosity) is also available, along with links to other sites.

Marshfield

The primary URL for Marshfield Medical Research Foundation (MMRF), Center for Medical Genetics, is *http://www.marshfieldclinic.org/research/genetics/*. This site provides information on the activities of the Center for Medical Genetics and access to maps and markers generated and used by the Center, including detailed information on the genotyped markers, the maps, the multiplex screening sets, and the laboratory methods used. Individual genotypes for the CEPH families genotyped at the MMRF are also available.

Utah

The primary URL for the Eccles Institute of Human Genetics is *http://www.genetics.utah.edu/*. This site provides information about the Eccles Institute, along with the maps generated using markers developed by the Utah Marker Development Group (UMDG) at the Institute. Marker data include primer, sequence, and number of alleles. CEPH allele sizes are not available.

LDB

The primary URL for the Location DataBase is *http://cedar.genetics.soton.ac.uk/public_html/ldb.html*. This site provides summary maps of all the chromosomes. These maps are developed and updated using a statistical algorithm that attempts to integrate data from many different maps and sources, including genetic maps, radiation hybrid maps, cytogenetic maps, and physical maps. Links to the primary information are provided when possible.

CEPH

The primary URL for CEPH (Centre D'Etude du Polymorphisme Humain) is *http://www.cephb.fr*. This site provides information about CEPH and its activities and access to the CEPH genotype database. This database includes RFLP and VNTR data, along with extensive microsatellite marker data. The database provides information on markers, including allele sizes and frequencies and heterozygosity (when known). It can also provide the actual genotypes for individuals in the CEPH pedigrees—extremely useful information for promoting consistency of allele calling between gels, projects, and laboratories. Note that it does not provide primer sequence data or genetic maps.

Contributor: Jonathan L. Haines

Radioisotope Data

Table A.2D.1 Physical Characteristics of Commonly Used Radionuclides[a]

Nuclide	Half-life	Emission	Energy, max (MeV)	Range of emission, max	Approx. specific activity at 100% enrichment (Ci/mg)	Atom resulting from decay	Target organ
^{3}H	12.43 years	β	0.0186	0.42 cm (air)	9.6	$^3_2$He	Whole body
^{14}C	5370 years	β	0.156	21.8 cm (air)	4.4 mCi/mg	$^{14}_7$N	Bone, fat
^{32}P[b]	14.3 days	β	1.71	610 cm (air)	285	$^{33}_{16}$S	Bone
				0.8 cm (water)			
				0.76 cm (Plexiglas)			
^{33}P[b]	25.4 days	β	0.249	49 cm	156	$^{33}_{16}$S	Bone
^{35}S	87.4 days	β	0.167	24.4 cm (air)	43	$^{35}_{17}$Cl	Testes
^{125}I[c]	60 days	γ	0.27–0.035	0.2 mm (lead)	14.2	$^{125}_{52}$Te	Thyroid
^{131}I[c]	8.04 days	β	0.606	165 cm (air)	123	$^{130}_{54}$Xe	Thyroid
		γ	0.364	2.4 cm (lead)			

[a]Table compiled based on information in Lederer et al. (**??**) and Shleien (**??**).

[b]Recommended shielding is Plexiglas; half-value layer measurement is 1 cm.

[c]Recommended shielding is lead; half-value layer measurement is 0.02 mm.

Table A.2D.2 Shielding Radioactive Emission[a] of β Emitters

Energy (MeV)	Mass (mg)/cm² to reduce intensity by 50%	Thickness (mm) to reduce intensity by 50%			
		Water	Glass	Lead	Plexiglas
0.1	1.3	0.013	0.005	0.0011	0.0125
1.0	48	0.48	0.192	0.042	0.38
2.0	130	1.3	0.52	0.115	1.1
5.0	400	4.0	1.6	0.35	4.2

[a]From Dawson et al. (1986). Reprinted with permission.

Centrifuges and Rotors

Centrifugation runs described in this book usually specify a relative centrifugal force (RCF; measured in × g), corresponding to a speed (in rpm) for a particular centrifuge and rotor model. As available equipment will vary from laboratory to laboratory, the investigator must be able to adapt these specifications to other centrifuges and rotors.

The relationship between RCF and speed (rpm) is determined by the following equation:

$$RCF = 1.12r \, (rpm/1000)^2$$

where r is the rotating radius between the particle being centrifuged and the axis of rotation. In most cases, an accurate conversion from speed to relative centrifugal force (or vice versa) can be obtained using the maximum value of r—or r_{max}—equal to the distance between the axis of rotation and the bottom of the centrifuge tube as it sits in the well or bucket of the rotor.

Table A.2E.1 provides r_{max} values for commonly used rotors manufactured by Du Pont (Sorvall), Beckman, Fisher, and IEC. There are situations (e.g., where an adapter is used to fit a smaller tube into a larger rotor well) where r_{max} will not accurately represent the effective rotating radius. In such cases, the manual for the rotor should be consulted to obtain the appropriate value of r.

As an alternative to use of the above equation, the nomograms in Figures A.2E.1 (for centrifuge runs <21,000 rpm) and A.2E.2 (for faster spins) make it possible to determine the RCF where speed and r_{max} are known, or the speed where RCF and r_{max} are known. This is done by aligning a ruler across the two known values and reading the unknown value at the point where the ruler crosses the remaining column.

Table A.2E.1 Maximum Rotating Radii for Common Rotors, Grouped by Centrifuge Model

Rotor Model[a]	r_{max} (mm)	Rotor Model[a]	r_{max} (mm)
For Sorvall centrifuge models GLC-1, GLC-2, GLC-2B, GLC-3, GLC-4, RT-6000B, T-6000, T-6000B		SV-288	90
		TZ-28	95
A/S400	140	**For Sorvall ultracentrifuges**	
H-1000B	186	T-865	91
HL-4 with 50-ml bucket	180	T-865.1	87.1
HL-4 with 100-ml bucket	204	T-875	87.1
HL-4 with Omni-Carrier	163	T-880	84.7
M and A-384 (inner row)	91	T-1270	82
M and A-384 (outer row)	121	TFT-80.2	65.5
SP/X and A-500 (inner row)	82	TFT-80.4	60.1
SP/X and A-500 (outer row)	123	**For Beckman GP series centrifuges**	
		GA-10	123
For Sorvall centrifuge models RC-3, RC-3B, RC-3C		GA-24	123
		GA-24 with adapter for 10-ml tubes	108
H-2000B	261	GH-3.7 (buckets)	204
H-4000 and HG-4L	230	GH-3.7 (microplate carrier)	168
H-6000A	260	GH-3.8 (buckets)	204
HL-8 with Omni-Carrier	221	GH-3.8 (microplate carrier)	168
HL-8 with 50-ml bucket	238	**For Beckman TJ-6 series centrifuges**	
HL-8 with 100-ml bucket	247	TA-10	123
HL-2 and HL-2B	166	TA-24	108
LA/S400	140	TA-24 with adapter for 10-ml tubes	123
For Sorvall centrifuge models RC-2, RC-2B, RC-5, RC-5B, RC-5C		TH-4 (stainless steel buckets)	186
		TH-4 (100-ml tube holders)	201
GSA	145	TH-4 (microplate carrier)	165
GS-3	151	**For Beckman AccuSpin**	
HB-4	147	AA-10	123
HS-4 with 250-ml bucket	172	AA-24	108
SA-600	129	AA-24 with adapter for 10-ml tubes	123
SE-12	93	AH-4	163
SH-80	101	**For Beckman J6 series centrifuges**	
SM-24 (inner row)	91	JR-3.2	206
SM-24 (outer row)	110	JS-2.9	265
SS-34	107	JS-3.0	254
SV-80	101		

continued

Table A.2E.1 Maximum Rotating Radii for Common Rotors, Grouped by Centrifuge Model continued

Rotor Model[a]	r_{max}(mm)	Rotor Model[a]	r_{max}(mm)
JS-4.0	226	Type 50 Ti	80.8
JS-4.2	254	Type 50.2 Ti	107.9
JS-4.2SM	248	Type 50.3 Ti	79.5
JS-5.2	226	Type 50.4 Ti (inner row)	96.4
Microplate carrier (6-bucket rotors)	214	Type 50.4 Ti (outer row)	111.4
Microplate carrier (4-bucket rotors)	192	Type 55.2 Ti	100.3
		Type 60 Ti	89.9
For Beckman J2-21 series centrifuges		Type 65	77.7
JA-10	158	Type 70 Ti	91.9
JA-14	137	Type 70.1 Ti	82.0
JA-17	123	Type 75 Ti	79.7
JA-18	132	Type 80 Ti	84.0
JA-18.1 (25° angle)	112	VAC 50	86.4
JA-18.1 (45° angle)	116	VC 53	78.8
JA-20	108	VTi 50	86.6
JA-20.1	115	VTi 65	85.4
JA-21	102	VTi 65.2	87.9
JCF-Z	89	VTi 80	71.1
JCF-Z with small pellet core	81		
JE-6B	125	*For Beckman Airfuge ultracentrifuge*	
JS-7.5	165	A-95	17.6
JS-13	142	A-100/18	14.6
JS-13.1	140	A-100/30	16.5
JV-20	93	A-110	14.7
		ACR-90 (2.4-ml liner)	11.8
For Beckman series L7 and L8 ultracentrifuges		ACR-90 (3.5-ml liner)	13.4
SW 25.1	129.2	Batch rotor	14.6
SW 28	161.0	EM-90	13.0
SW 28.1	171.3		
SW 30	123.0	*For Beckman TL-100 series ultracentrifuges*	
SW 30.1	123.0	TLA-100	38.9
SW 40 Ti	158.8	TLA 100.1	38.9
SW 41 Ti	153.1	TLA-100.2	38.9
SW 50.1	107.3	TLA-100.3	48.3
SW 55 Ti	108.5	TLA-45	55.1
SW 60 Ti	120.3	TLS-55	76.4
SW 65 Ti	89.0	TLV-100	35.7
Type 15	142.1		
Type 19	133.4	*Miscellaneous centrifuges and rotors[b]*	
Type 21	121.5	Clay Adams Dynac	—[c]
Type 25	100.4	Fisher Centrific	113
Type 30	104.8	Fisher Marathon 21K with 4-place rotor	160
Type 30.2	94.2	IEC Clinical centrifuge with 4-place swinging-bucket rotor	155
Type 35	104.0		
Type 40	80.8	IEC general-purpose centrifuge models HN, HN-SII, and Centra-4	—[c]
Type 40.3	79.5		
Type 42.1	98.6		
Type 42.2 Ti	104		
Type 45 Ti	103		
Type 50	70.1		

[a] Sorvall centrifuges and rotors are a product of Du Pont Company Medical Products, Beckman centrifuges are a product of Beckman Instruments, IEC centrifuges are a product of International Equipment Co., Clay Adams Dynac centrifuges are a product of Becton Dickinson Labware, and Fisher centrifuges are a product of Fisher Scientific. For ordering information see *APPENDIX 4*.

[b] These instruments are often loosely referred to as "clinical," "tabletop," or "low-speed" centrifuges.

[c] These instruments accept a wide range of trunnion-ring rotors with variable rotating radii, as well as fixed-angle and swinging-bucket rotors that in turn accept a variety of adapters making it possible to spin different numbers tubes of various sizes. For instance, the commonly used IEC 958 trunnion-ring rotor may be adjusted to radii ranging from 137 to 181 mm, depending on the trunnion-ring chosen. It is therefore necessary to consult the manual for the specific system being used to obtain an accurate speed to RCF conversion.

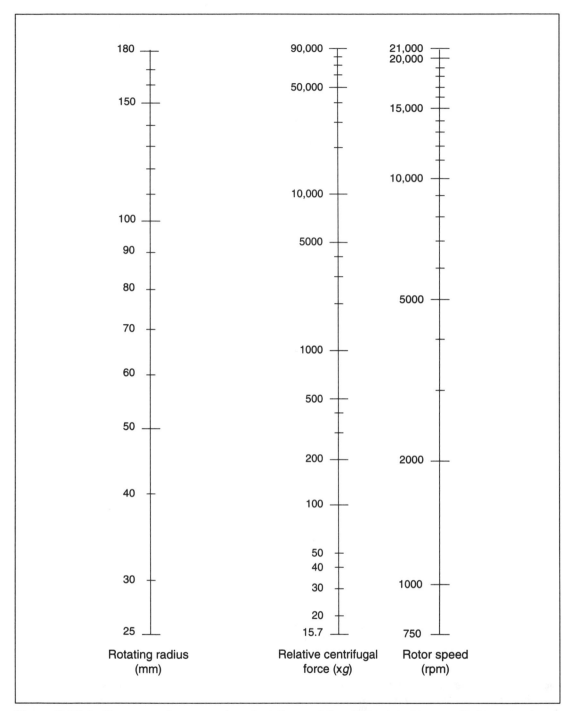

Figure A.2E.1 Nomogram for conversion of relative centrifugal force to rotor speed in low-speed centrifuge runs. To determine an unknown value in a given column, align ruler through known values in the other two columns. The desired value is found at the intersection of the ruler with the column of interest. For faster centrifugations, use Figure A.2E.2. A more precise conversion can be obtained using the equation at the beginning of this appendix. See Table A.2E.1 for rotating radii of commonly used rotors.

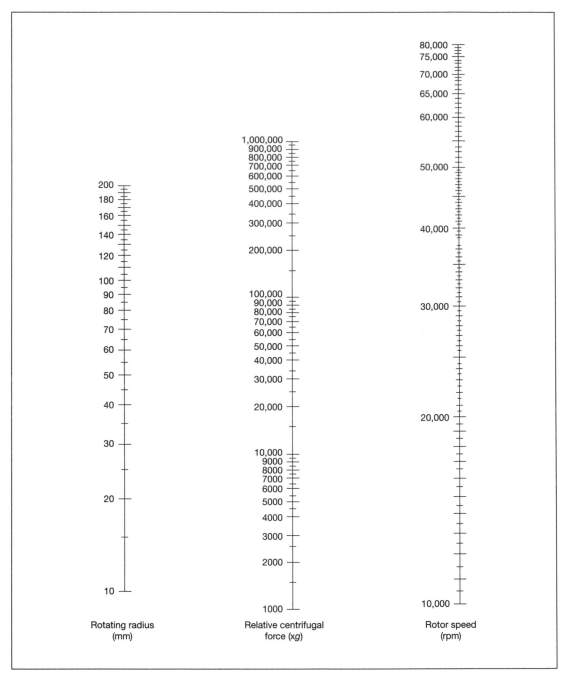

Figure A.2E.2 Nomogram for conversion of relative centrifugal force to rotor speed in high-speed centrifuge runs. For slower centrifugations and instructions for using the nonogram, use Figure A.2E.1. A more precise conversion can be obtained using the equation at the beginning of this appendix. See Table A.2E.1 for rotating radii of commonly used rotors.

Commonly Used Techniques

APPENDIX 3A

Isolation of Genomic DNA from Mammalian Cells

BASIC PROTOCOL 1

DNA ISOLATION FROM WHOLE BLOOD

Materials (*see* APPENDIX 1 *for items with* ✓)
  10 ml fresh or thawed frozen whole blood
✓ NKM buffer, 4°C
✓ Resuspension buffer
✓ 10× TEN solution
✓ 2 mg/ml proteinase K
✓ 10% (w/v) SDS
✓ Buffered phenol
✓ 25:24:1 (v/v) phenol/chloroform/isoamyl alcohol
  24:1 (v/v) chloroform/isoamyl alcohol
✓ TE buffer, pH 8.0
✓ Dialysis buffer

  Refrigerated centrifuge
  Orbital shaker
  Dialysis tubing (MWCO 5000)
  1.5-ml storage vials

1. Bring 1 to 15 ml fresh or thawed whole blood (scale proportionally for smaller quantities) to 40 ml with 4°C NKM buffer. Vortex, then centrifuge 30 min at 3500 × g, 4°C. Remove all but ~10 ml supernatant. Resuspend pellet and vortex (and pipet up and down with a Pasteur pipet if necessary) to break up any clumps.

2. Add resuspension buffer to 40 ml, balance tubes, and centrifuge 30 min at 3500 × g, 4°C. Remove all but ~4 ml supernatant. Add:

     0.5 ml 10× TEN solution
     0.25 ml 2 mg/ml proteinase K
     0.5 ml 10% SDS.

   Mix gently and incubate overnight at 37°C to allow digestion.

3. Extract DNA twice with an equal volume of buffered phenol, once with an equal volume of 25:24:1 phenol/chloroform/isoamyl alcohol, and twice with an equal volume of 24:1 chloroform/isoamyl alcohol. For each extraction: mix gently on an orbital shaker 30 min at room temperature, centrifuge 15 min at 1500 × g, room temperature, and remove and

discard organic phase (usually the bottom layer; phases may invert in presence of high salt).

4. Transfer DNA solution into dialysis tubing. Dialyze four times, 12 hr each time, with dialysis buffer. Save 0.3 ml of the last change of dialysis buffer. Transfer dialyzed DNA solution to 1.5-ml microcentrifuge tubes.

5. Quantitate DNA by diluting 0.1 ml DNA solution in 0.2 ml TE buffer and reading the A_{260} and A_{280} of the diluted DNA (A_{260}/A_{280} ratio ≥ 1.7 is usually suitable; APPENDIX 3D) as well as the last change of dialysis buffer and unused dialysis buffer (which should give the same absorbance readings if dialysis is complete).

6. Dilute DNA appropriately with TE buffer and divide into aliquots in 1.5-ml storage vials that can be tightly sealed. Add 40 µl chloroform to each aliquot to kill any contaminating microorganisms. Label tubes, seal tightly, and store up to several years at 4°C. If DNA samples are to be used for PCR only, store at −70°C until use.

Avoid freezing if DNA is to be used for Southern analysis to prevent DNA degradation during freeze-thaw cycles.

ALTERNATE PROTOCOL 1

DNA ISOLATION FROM CELL PELLETS

Additional Materials *(also see Basic Protocol 1; see APPENDIX 1 for items with* ✓ *)*
   Cultured cells (APPENDIX 3I & 3J)
   ✓ Phosphate-buffered saline (PBS), sterile
   ✓ Sucrose lysis solution
   50-ml centrifuge tubes

1a. *If using a fresh cell culture:* Collect ~50 ml suspension culture containing 5×10^7 cells, or trypsinize and then collect an equivalent number of adherent cells, in a 50-ml centrifuge tube. Centrifuge 10 min at $1500 \times g$, 4°C. Remove supernatant. Resuspend pellet in 1.0 ml sterile PBS and transfer into a 1.5-ml microcentrifuge tube. Centrifuge 10 min at 1500 $\times$ g. Remove supernatant and freeze pellet in dry ice/ethanol bath or at −70°C. Thaw before proceeding to step 2.

1b. *If using previously frozen cell aliquot:* Thaw cells and centrifuge 10 min at 1500 $\times$ g. Remove supernatant.

2. Resuspend pellet in 40 ml sucrose lysis solution. Vortex to mix. Centrifuge 30 min at 3500 $\times$ g, 4°C. Proceed with DNA extraction and dialysis (see Basic Protocol 1, steps 2 to 5).

ALTERNATE PROTOCOL 2

RECOVERY OF GENOMIC DNA BY HIGH-SALT PRECIPITATION

It is possible to dispense with the lengthy dialysis used in Basic Protocol 1 and to recover DNA directly using a high-salt/ethanol precipitation. DNA fragments tend to be smaller than those isolated using dialysis, but they are acceptable for most purposes.

Additional Materials *(also see Basic Protocol 1)*
   5 M NaCl
   100% and 70% ethanol

1. Isolate and phenol extract DNA (see Basic Protocol 1, steps 1 to 3). Add 5 M NaCl to give a final concentration of 0.4 M. Add 2 vol of 100% ethanol, mix by gentle swirling, and centrifuge 30 min at 3500 × g, 4°C.

   If the ethanol is added slowly and DNA is sufficiently concentrated, the latter will immediately salt out upon addition of ethanol and may be visualized as a clear viscous material at the ethanol interface. DNA may be recovered directly by spooling, then lyophilized.

2. Remove supernatant, add 5 ml of 70% ethanol, and centrifuge 30 min at 3500 × g. Remove supernatant and dry sample under a stream of nitrogen gas or in a lyophilizer. Resuspend DNA in 1 ml TE buffer and let stand overnight at room temperature to allow DNA to dissolve.

3. Add 0.1 ml DNA to 0.2 ml TE buffer. Determine DNA concentration by reading absorption at 260 and 280 nm (APPENDIX 3D).

BASIC PROTOCOL 2

ISOLATION OF DNA FROM BUCCAL SWABS

Materials (see APPENDIX 1 for items with ✓)
   50 mM NaOH
✓ 1 M Tris·Cl, pH 6.5
   70% ethanol

   CYTO-PAK with Cyto-soft brush (CP-5B; Medical Packaging) *or* Caliber
      Cotton-Tipped Applicator swabs (size 6 in. L; Allegiance Healthcare)
   95°C heating block

1. Have the subject lightly rinse out his or her mouth with water twice. Remove a brush or swab from its sleeve, avoiding contamination as much as possible, and roll it against the inside of the subject's cheek for ~30 sec. Place the brush back into the sleeve. Store up to 1 week at 4°C, but extract DNA as soon as possible.

   The yield of DNA from buccal epithelial cells can be highly variable because of variations in both the subject and the technique of the collector. It is wise when possible to obtain two brushes or two to four swabs. Care should be taken not to roll or brush too hard, as repeated hard strokes can cause abrasion, soreness, or even bleeding.

2. Pipet 600 μl of 50 mM NaOH into a 1.5-ml microcentrifuge tube. Cut off the sample-containing end of the collection brush or swab using scissors and place it in the microcentrifuge tube using sterile forceps (two swab tips may be placed in one tube). Sterilize scissors both before and after use by washing in 70% ethanol.

3. Vigorously vortex the tube and place in a heating block at 95°C for 5 min to complete cell lysis. Remove the brush or swabs from the microcentrifuge tube using sterile forceps, leaving behind the residual solution containing the DNA. Neutralize the solution by adding 60 μl of 1 M Tris·Cl, pH 6.5, and vortex the tube again.

   DNA is now ready for quantitation, or it can be used directly for PCR. Buccal DNA samples to be used for PCR may be stored at either 4° or −20°C.

   Since the DNA is not highly purified, exact quantitation is difficult and usually not necessary. Typically 5 to 10 μl of the DNA solution is sufficient for a successful PCR reaction. Estimates of yield can be made using the DNA Dipstick kit (Invitrogen). DNA yields are highly variable, but 4 to 8 μg/brush is typical.

DNA isolated from buccal cells should be used as soon as possible. While long-term storage (>1 month) and use of buccal DNA is possible, the quality of results can decline, sometimes dramatically, and long-term storage is not recommended.

Reference: Richards et al., 1994

Contributors: John R. Gilbert and Jeffrey M. Vance

APPENDIX 3B

Extraction and Precipitation of DNA

BASIC PROTOCOL 1

PHENOL EXTRACTION

Materials *(see* APPENDIX 1 *for items with* ✓ *)*

DNA sample to be extracted (monovalent cation concentration ≤0.5 M)
✓ 1:1 (v/v) phenol/chloroform *or* 25:24:1 (v/v/v) phenol/chloroform/isoamyl alcohol
24:1 (v/v) chloroform/isoamyl alcohol
✓ TE buffer, pH 8.0

1. Estimate volume of DNA sample, add 1 vol of 1:1 phenol/chloroform or 25:24:1 phenol/chloroform/isoamyl alcohol, and mix to form an emulsion.

   Size of extracted DNA fragments is affected by mixing technique because vigorous mixing can shear DNA. In general, vortexing is used for fragments ≤10 kb, hand mixing for fragments of 10 to 30 kb, and a gentle rotating platform for fragments >30 kb.

   Samples <50 μl should be diluted to a larger volume (e.g., ≥100 μl) with TE buffer because small volumes are hard to extract. Samples ≤400 μl can be extracted in a single microcentrifuge tube. Larger samples can be extracted in polypropylene tubes with tight-fitting caps.

2. Microcentrifuge 15 sec at maximum speed, room temperature.

   The centrifugation should be long enough to obtain clear separation of the organic and aqueous phases. Protein will form a visible white precipitate at the interface. Larger volumes should be centrifuged 5 min at 1200 × g, room temperature.

   If the salt concentration in the original solution is <0.5 M or the sucrose concentration is <10% (w/v), the DNA-containing aqueous phase should be uppermost. Otherwise, the organic and aqueous phases may invert. It is wise to save the organic phase until ethanol precipitation verifies the presence of DNA.

3. Transfer aqueous phase to a fresh microcentrifuge tube with a pipettor (preferably with a large bore, cut pipet tip).

4. *Optional (to maximize yield):* Back-extract the remaining organic phase and interface by adding an equal volume of TE buffer, pH 8.0. Mix well and microcentrifuge as in step 2. Combine the aqueous phase with that collected in step 3.

5. Re-extract pooled aqueous phases with an equal volume of 1:1 phenol/chloroform chloroform or 25:24:1 phenol/chloroform/isoamyl alcohol. Mix, microcentrifuge, and collect aqueous phase as in steps 2 and 3. Repeat extraction of the aqueous phases until there is no visible protein at the interface.

6. Extract pooled aqueous phases once with an equal volume of 24:1 chloroform/isoamyl alcohol to remove residual phenol. Mix and microcentrifuge as in step 2. Recover and ethanol precipitate (Basic Protocol 2) the aqueous phase.

> *If high-molecular-weight DNA is required, residual organic solvent should be removed by dialyzing the aqueous phase against large volumes of low-ionic-strength buffer at 4°C. Alternatively, the aqueous phase can be extracted twice with 1 vol each time of diethyl ether that has been equilibrated by vortexing with water or TE buffer, and the ether removed by evaporation.*

BASIC PROTOCOL 2

ETHANOL PRECIPITATION OF DNA

Materials (see APPENDIX 1 *for items with* ✓)

  DNA sample to be precipitated
  Appropriate monovalent cation solution (Table A.3B.1)
  100% and 70% ethanol, ice cold
✓ TE buffer, pH 8.0

1. Estimate DNA sample volume and add monovalent cation solution to give the desired cation concentration (Table A.3B.1).

> *If the monovalent cation concentration of sample DNA is >0.5 M, it should be diluted with TE buffer to give a final concentration of 0.3 M. If sample DNA contains >10 mM $MgCl_2$, it should be diluted with TE buffer to ≤10 mM $MgCl_2$ (final) before ethanol precipitation. DNA samples containing >1 mM phosphate or >10 mM EDTA should not be ethanol precipitated; phosphate ions or EDTA can be removed by conventional or spin-column chromatography (APPENDIX 3E).*

> *Samples ≤400 μl can be precipitated in a 1.5-ml microcentrifuge tube; larger volumes should be precipitated in a silanized thick-walled Corning glass centrifuge tube.*

2. Add 2 to 2.5 vol (calculated after addition of monovalent cation solution) of ice-cold 100% ethanol. Mix well. Incubate 15 to 30 min in crushed dry ice, at least 30 min in a −20°C

Table A.3B.1 Monovalent Cations Used for Precipitation of DNA

Monovalent cation solution	Stock solution (M)	Final conc. (M)	Comments
Ammonium acetate	10.0	2.0–2.5	Use to prepare DNA for PCR; to precipitate oligonucleotides >50 bases; to reduce coprecipitation of dNTPs and small oligonucleotides. NH_4^+ inhibits T4 polynucleotide kinase.
Potassium acetate, pH 5.5	3.0	0.3	Use to isolate plasmids by isopropanol precipitation. Inhibits restriction enzymes.
Sodium acetate, pH 5.2	3.0	0.3	Use to precipitate oligonucleotides <50 bases and to prepare DNA for most applications. Can interfere with PCR.
Sodium chloride	5.0	0.2-0.5	Use to precipitate samples containing SDS. Difficult to remove due to low solubility in 70% ethanol. Na^+ inhibits T4 DNA ligase.

freezer or at least 15 min in a $-70°C$ freezer. If desired, store ethanol solution of DNA indefinitely at $-20°$ or $-70°C$.

To improve recovery from solutions with low concentrations of DNA (i.e., <10 μg/ml), the following modifications can be made: (1) MgCl$_2$ can be added to ≤10 mM (final); (2) the amount of ethanol can be increased to 3 vol and the incubation carried out for 30 min in dry ice; or (3) 10 μg tRNA from E. coli, yeast, or bovine liver can be added to the solution and the precipitation carried out as in step 2. Carrier tRNA should not be used if the DNA will be phosphorylated using T4 polynucleotide kinase. If the DNA fragments are <100 bases or <0.1 μg/ml, the incubation time should be increased to ~1 hr.

Alternatively, an equal volume of isopropanol can be used for precipitation. However, isopropanol is less volatile than ethanol and therefore harder to remove. Moreover, it is difficult to lyophilize isopropanol-precipitated DNA. Isopropanol coprecipitates salts and sucrose, so isopropanol precipitation should generally be followed by ethanol precipitation.

3. Microcentrifuge 10 min at maximum speed, 4°C.

   A pellet of ≥10 μg DNA should be visible. The pellet can be located by orienting the hinge of the microcentrifuge tube up during centrifugation.

   For preparations of small DNA fragments or solutions with low concentrations of DNA, microcentrifuge for a longer period. When processing solutions in Corning glass tubes, centrifuge 15 min at 8000 × g in a fixed-angle rotor, at 4°C.

4. Use a Pasteur pipet to carefully draw off supernatant from side of tube opposite the DNA pellet. Add 1 ml of ice-cold 70% ethanol, mix, and microcentrifuge 5 min at maximum speed, 4°C. Repeat once. Save supernatants until quantitative recovery of DNA is verified.

   The DNA pellet will not adhere tightly to the wall of the tube (≤50% may be on the wall), so aspirate supernatant carefully and rinse walls with a small volume of ethanol.

   Use ice-cold 95% (v/v) ethanol to wash the DNA pellet if fragments are <200 bases.

5. Dry pellet in a Speedvac evaporator or in a desiccator under vacuum, or by air drying.

   Air-dried high-molecular-weight DNA is easier to redissolve.

6. Dissolve DNA pellet at desired concentration in water if it is going to be used for enzyme reactions requiring specific buffers. Dissolve in TE buffer, pH 8.0, if it is going to be stored indefinitely.

   Small quantities of DNA at low concentrations (e.g., ~25 μg at <1 mg/ml) should dissolve rapidly. Higher concentrations may need to be vortexed and heated 5 min at 65°C to dissolve the DNA. Genomic DNA may require several days of gentle mixing to dissolve.

   DNA may not dissolve easily in solutions containing MgCl$_2$ or >0.1 M NaCl. It is easier to dissolve DNA in low-ionic-strength buffer and adjust the concentration of MgCl$_2$ or NaCl after DNA is in solution.

   DNA can be stored in TE buffer, pH 8.0, for short periods at 4°C, or up to several years in appropriately sized aliquots at $-20°$ or $-70°C$. Avoid repeated freezing and thawing.

Reference: Wallace, 1987

Preparation of DNA from Fixed, Paraffin-Embedded Tissue

DNA purified from fixed, paraffin-embedded tissue (PET) is not generally suitable for Southern blot analysis. However, the polymerase chain reaction (PCR) can be used to amplify short regions of this DNA.

BASIC PROTOCOL

DNA ISOLATION USING MIXED-BED CHELATING RESIN

Materials (see APPENDIX 1 for items with ✓)

Paraffin-embedded tissue sections, ~5 to 10 μm thick
✓ PET solution
✓ Chelex 100 slurry
65°C and boiling water baths

1. Place 2 to 10 tissue sections, each 5- to 10-μm thick, in a sterile 1.5-ml microcentrifuge tube. Add 200 μl PET solution to the sample and microcentrifuge briefly at room temperature. Incubate samples at 65°C until the paraffin has melted (~5 min). Vortex gently and briefly, then microcentrifuge briefly at room temperature. Return tubes to 65°C overnight.

   If this digestion does not produce amplifiable DNA, add one or more aliquots of proteinase K at 2-hr intervals to improve digestion.

2. Add 100 μl Chelex 100 slurry. Place the tubes in a boiling water bath for 10 min. Microcentrifuge tubes 5 min at maximum speed, room temperature; the DNA solution is below a thin layer of paraffin and above the resin. Break through the hardened paraffin layer using a micropipet tip. Remove the solution by aspiration to a clean microcentrifuge tube, taking care not to remove any of the Chelex beads. If any Chelex is removed with the DNA solution, repeat the centrifugation and aspiration, because the Chelex may inhibit the subsequent PCR reaction. Store DNA up to 1 month at 4 °C or for longer periods at −20°C or less.

3. *Optional:* Quantitate the DNA spectrophotometrically or by a dye-binding method (APPENDIX 3D).

ALTERNATE PROTOCOL

DNA ISOLATION FROM NONOPTIMALLY FIXED TISSUE

This protocol should be used if the Basic Protocol fails or when a specimen has been fixed in a manner that is known to limit PCR amplification of the extracted DNA for PCR.

Additional Materials (also see Basic Protocol; see APPENDIX 1 for items with ✓)

Xylenes
100% and 70% ethanol
✓ 25:24:1 (v/v/v) phenol/chloroform/isoamyl alcohol
24:1 (v/v) chloroform/isoamyl alcohol

✓ 3 M sodium acetate, pH 5.2 (filter sterilize with 0.45-μm filter and store up to 6 months at room temperature)

✓ TE buffer, pH 8.0

1. Place 2 to 10 tissue sections in a sterile 1.5-ml microcentrifuge tube (see Basic Protocol, step 1). If four or more sections are being extracted, heat 1 min at 65°C to melt paraffin. Add 1 ml xylenes and vortex. Microcentrifuge 10 min at maximum speed, room temperature. Remove the supernatant by aspiration or decantation. If the pellet is large, repeat the xylene treatment and centrifugation.

2. Add 1 ml of 100% ethanol. Vortex briefly. Microcentrifuge 10 min at maximum speed, room temperature. Remove the supernatant by aspiration or decantation. Add 1 ml of 70% ethanol. Vortex briefly. Microcentrifuge 10 min at maximum speed, room temperature. Remove supernatant by aspiration or decantation. Dry under vacuum or allow to air dry on benchtop.

3. Add 200 μl PET solution. Incubate 5 min at 65°C. Vortex gently to mix, then microcentrifuge briefly at room temperature. Incubate overnight at 65°C.

4. Add 100 μl Chelex 100 slurry. Place sample in boiling water bath for 10 min. Microcentrifuge 5 min at maximum speed, room temperature. Remove the solution by aspiration to a clean microcentrifuge tube, taking care not to remove any of the Chelex beads.

5. Extract the DNA solution with 200 μl of 25:24:1 phenol/chloroform/isoamyl alcohol (*APPENDIX 3B*). Place the aqueous phase in a clean microcentrifuge tube and repeat extraction. Extract the aqueous phase with 24:1 chloroform/isoamyl alcohol. Add 1/10 vol of 3 M sodium acetate, pH 5.2. Add 2 vol of 100% ethanol (*APPENDIX 3B*). Allow to stand 15 min at room temperature. Microcentrifuge 15 min at maximum speed, room temperature. Remove supernatant by decantation or aspiration.

6. Wash pellet with 70% ethanol and repeat microcentrifugation. Remove the 70% ethanol, then dry pellet briefly under vacuum or allow it to air dry on bench. Resuspend pellet in 50 μl TE buffer or water (or other volume depending on size of pellet). Quantitate DNA spectrophotometrically or by a dye-binding method (*APPENDIX 3D*).

Reference: Mies, 1994

Contributor: Edward A. Fox

APPENDIX 3D

Quantitation of DNA and RNA with Absorption and Fluorescence Spectroscopy

These three procedures cover a range from 5 to 10 ng/ml DNA to 50 μg/ml DNA (see Table A.3D.1).

BASIC PROTOCOL

DETECTION OF NUCLEIC ACIDS USING ABSORPTION SPECTROSCOPY

A_{260} measurements are quantitative for relatively pure nucleic acid preparations in microgram quantities. Absorbance readings cannot discriminate between DNA and RNA; however, the

Table A.3D.1 Properties of Absorbance and Ethidium Bromide Fluorescence Spectrophotometric Assays for DNA and RNA

Property	Absorbance (A_{260})	Fluorescence	
		H33258	EtBr
Sensitivity (μg/ml)			
DNA	1–50	0.01–15	0.1–10
RNA	1–40	n.a.	0.2-10
Ratio of signal			
DNA/RNA	0.8	400	2.2

ratio of A at 260 and 280 nm can be used as an indicator of nucleic acid purity. Proteins, for example, have a peak absorption at 280 nm that will reduce the A_{260}/A_{280} ratio. Absorbance at 325 nm indicates particulates in the solution or dirty cuvettes; contaminants containing peptide bonds or aromatic moieties such as protein and phenol absorb at 230 nm.

Materials (*see* APPENDIX 1 *for items with* ✓)

✓ 1× TNE buffer
DNA sample to be quantitated
✓ Calf thymus DNA standard solutions

Matched quartz semi-micro spectrophotometer cuvettes (1-cm pathlength)
Single- or dual-beam spectrophotometer (ultraviolet to visible)

1. Pipet 1.0 ml of 1× TNE buffer into a quartz cuvette. Place the cuvette in a single- or dual-beam spectrophotometer, read at 325 nm (note contribution of the blank relative to distilled water if necessary), and zero the instrument. Use this blank solution as the reference in double-beam instruments. For single-beam spectrophotometers, remove blank cuvette and insert cuvette containing DNA sample or standard suspended in the same solution as the blank. Take reading. Repeat this process at 280, 260, and 230 nm.

2. To determine the concentration (C) of DNA present, use the A_{260} reading in conjunction with one of the following equations:

Single-stranded DNA:
$$C(\text{pmol/μl}) = \frac{A_{260}}{10 \times S}$$

$$C(\text{μg/ml}) = \frac{A_{260}}{0.027}$$

Double-stranded DNA:
$$C(\text{pmol/μl}) = \frac{A_{260}}{13.2 \times S}$$

$$C(\text{μg/ml}) = \frac{A_{260}}{0.020}$$

Single-stranded RNA:
$$C(\text{μg/ml}) = \frac{A_{260}}{0.025}$$

Oligonucleotide:
$$C(\text{pmol/μl}) = A_{260} \times \frac{100}{1.5N_A + 0.71N_C + 1.20N_G + 0.84N_T}$$

Table A.3D.2 Molar Extinction Coefficients of DNA Bases[a]

Base	$\varepsilon_{260\,nm}^{IM}$
Adenine	15,200
Cytosine	7,050
Guanosine	12,010
Thymine	8,400

[a]Measured at 260 nm; see Wallace and Miyada, 1987.

where S represents the size of the DNA in kilobases and N is the number or residues of base A, G, C, or T.

For double- or single-stranded DNA and single-stranded RNA, these equations assume a 1-cm-pathlength spectrophotometer cuvette and neutral pH. For oligonucleotides, concentrations are calculated in the more convenient units of pmol/μl. The base composition of the oligonucleotide has significant effects on absorbance, because the total absorbance is the sum of the individual contributions of each base (Table A.3D.2).

3. Use the A_{260}/A_{280} ratio and readings at A_{230} and A_{325} to estimate the purity of the nucleic acid sample.

Ratios of 1.8 to 1.9 and 1.9 to 2.0 indicate highly purified preparations of DNA and RNA, respectively. Contaminants that absorb at 280 nm (e.g., protein) will lower this ratio.

Absorbance at 230 nm reflects contamination of the sample by phenol or urea, whereas absorbance at 325 nm suggests contamination by particulates and dirty cuvettes. Light scatter at 325 nm can be magnified 5-fold at 260 nm.

Typical values at the four wavelengths for a highly purified preparation are shown in Table A.3D.3.

ALTERNATE PROTOCOL 1

DNA DETECTION USING THE DNA-BINDING FLUOROCHROME HOECHST 33258

Specific for nanogram amounts of DNA, the Hoechst 33258 fluorochrome has little affinity for RNA and works equally well with either whole-cell homogenates or purified preparations

Table A.3D.3 Spectrophotometric Measurements of Purified DNA[a]

Wavelength (nm)	Absorbance	A_{260}/A_{280}	Conc. (μg/ml)
325	0.01	—	—
280	0.28	—	—
260	0.56	2.0	28
230	0.30	—	—

[a]Typical absorbance readings of highly purified calf thymus DNA suspended in 1× TNE buffer. The concentration of DNA was nominally 25 μg/ml.

of DNA. The fluorochrome is, however, sensitive to changes in DNA composition, with preferential binding to AT-rich regions.

Additional Materials *(also see Basic Protocol; see* APPENDIX 1 *for items with* ✓ *)*

✓ Hoechst 33258 assay solution

Dedicated filter fluorometer (Hoefer TKO100) *or* scanning fluorescence spectrophotometer (Shimadzu model RF-5000 or Perkin-Elmer model LS-5B or LS-3B) capable of an excitation wavelength of 365 nm and emission wavelength at 460 nm

Fluorometric square glass cuvettes *or* disposable acrylic cuvettes (Sarstedt)

Teflon stir rod

1. Prepare the scanning fluorescence spectrophotometer by setting the excitation wavelength to 365 nm and the emission wavelength to 460 nm (not necessary with dedicated filter fluorometer).

2. Pipet 2.0 ml Hoechst 33258 assay solution into cuvette and place in sample chamber. Take a reading without DNA and use as background. If the fluorometer has a concentration readout mode or is capable of creating a standard curve, set instrument to read 0 with the blank solution. Otherwise note the readings in relative fluorescence units. Be sure to take a blank reading for each cuvette used, as slight variations can cause changes in the background reading.

3. With the cuvette still in the sample chamber, add 2 μl DNA standard to the blank Hoechst 33258 assay solution. Mix in the cuvette with a Teflon stir rod or by capping and inverting the cuvette. Read emission in relative fluorescence units or set the concentration readout equal to the final DNA concentration. Repeat measurements with remaining DNA standards using fresh assay solution (take background zero reading and zero instrument if needed).

   If necessary, the DNA standards should be quantitated by A_{260} measurement (Basic Protocol) before being used here.

   Small-bore tips designed for loading sequencing gels minimize errors of pipetting small volumes. Pre-rinse tips with sample and make sure no liquid remains outside the tip after drawing up the sample.

   Read samples in duplicate or triplicate, with a blank reading taken each time. Unusual or unstable blank readings indicate a dirty cuvette or particulate material in the solution, respectively.

4. Repeat step 3 with unknown samples.

   A dye concentration of 0.1μg/ml is adequate for final DNA concentrations up to ~500 ng/ml. Increasing the working dye concentration to 1 μg/ml Hoechst 33258 will extend the assay's range to 15 μg/ml DNA, but will limit sensitivity at low concentrations (5 to 10 ng/ml). Sample volumes of ≤10 μl can be added to the 2.0-ml aliquot of Hoechst 33258 assay solution.

ALTERNATE PROTOCOL 2

DNA AND RNA DETECTION WITH ETHIDIUM BROMIDE FLUORESCENCE

In contrast to the fluorochrome Hoechst 33258, ethidium bromide is relatively unaffected by differences in the base composition of DNA. Ethidium bromide is not as sensitive as Hoechst 33258 and, although capable of detecting nanogram levels of DNA, will also bind to RNA. In preparations of DNA with minimal RNA contamination or with DNA samples having an unusually high guanine and cytosine (GC) content where the Hoechst 33258 signal can be quite low, ethidium bromide offers a relatively sensitive alternative to the more popular Hoechst 33258 DNA assay. A fluorometer capable of an excitation wavelength of 302 or 546 nm and an emission wavelength of 590 nm is required for this assay.

Additional Materials *(also see Basic Protocol; see* APPENDIX 1 *for items with* ✓ *)*
   ✓ Ethidium bromide assay solution

1. Pipet 2.0 ml ethidium bromide assay solution into cuvette of appropriate composition and place in sample chamber. Set excitation wavelength to 302 nm (use quartz cuvette) or 546 nm (use glass cuvette) and emission wavelength to 590 nm. Take an emission reading without DNA and use as background. If the instrument has a concentration readout mode or is capable of creating a standard curve, set instrument to read 0 with the blank solution. Otherwise note the readings in relative fluorescence units.

2. Read and calibrate samples as described in step 3 of Alternate Protocol 1. Read emissions of the unknown samples as in step 4 of Alternate Protocol 1.

   A dye concentration of 5 μg/ml in the ethidium bromide assay solution is appropriate for final DNA concentrations up to 1000 ng/ml; 10 μg/ml ethidium bromide in the ethidium bromide assay solution will extend the assay's range to 10 μg/ml DNA, but is only used for DNA concentrations >1 μg/ml. Sample volumes of up to 10 μl can be added to the 2.0-ml aliquot of ethidium bromide assay solution.

Reference: Labarca and Paigen,1980

Contributor: Sean R. Gallagher

APPENDIX 3E

Enzymatic Labeling of DNA

BASIC PROTOCOL 1

UNIFORM LABELING OF DNA BY NICK TRANSLATION

Nick translation is frequently used for the preparation of sequence-specific probes for screening libraries, for genomic DNA blots, and for in situ hybridization.

Materials *(see* APPENDIX 1 *for items with* ✓ *)*
   ✓ 0.5 mM 3dNTP mix (minus dATP)
   ✓ 5 to 15 U *E. coli* DNA polymerase I and 10× buffer (see APPENDIX 1 for buffer)
   100 μCi [α-^{32}P]dATP (3000 Ci/mmol)

✓ 1 mg/ml DNase I stock solution, diluted 1/10,000 in standard enzyme diluent (see
   APPENDIX 1) just before use
   DNA to be labeled
✓ 0.5 M EDTA
   10 mg/ml tRNA
✓ TE buffer, pH 8.0

1. Prepare the following reaction mix (25 µl total) on ice:

   2.5 µl 0.5 mM 3dNTP mix
   2.5 µl 10× *E. coli* DNA polymerase I buffer
   10 µl (100 µCi) [α-^{32}P]dATP
   1 µl diluted DNase I
   1 µl (5 to 15 U) *E. coli* DNA polymerase I
   H$_2$O to 25 µl.

   Add 0.25 µg DNA to 8 µl reaction mix. Immediately incubate 15 to 45 min at 12° to
   14°C.

   *The amount of DNase I is critical for optimal nick translation and it is recommended that a
   DNase calibration curve be established for each batch of enzyme. For reproducible results, it is
   best to prepare and store a 1 mg/ml DNase I solution.*

   *For DNA purified by electrophoresis in a low gelling/melting temperature agarose gel, add 8 µl
   of molten gel slice (at 37°C). Although the resulting mixture will solidify, the nick translation
   reaction will take place (sometimes at a reduced efficiency).*

2. Stop reaction with 1 µl of 0.5 M EDTA, 3 µl of 10 mg/ml tRNA, and 100 µl TE buffer,
   pH 8.0. Phenol extract (*APPENDIX 3B*) and transfer aqueous phase to a fresh tube. If DNA
   was added as a molten gel slice, remelt the stopped reaction mixture by incubating 10 min
   at 70°C.

3. Separate the labeled DNA from the unincorporated radioactive precursors by spin-column
   chromatography (Support Protocol 1). If desired, remove a 1-µl aliquot before loading
   sample on spin column and determine ^{32}P incorporation by acid precipitation (Support
   Protocol 2; specific activity should be 10^8 cpm/µg).

ALTERNATE PROTOCOL

LABELING DNA BY RANDOM OLIGONUCLEOTIDE-PRIMED SYNTHESIS

This method is an alternative to nick translation for producing uniformly radioactive DNA
of high specific activity. It employs the Klenow fragment (the C-terminal 70% of *E. coli*
polymerase II).

Additional Materials (also see Basic Protocol 1; see APPENDIX 1 for items with ✓)
✓ 10× *E. coli* polymerase I buffer
   3 to 8 U/µl Klenow fragment
   Random hexanucleotides

1. Digest DNA with an appropriate restriction endonuclease. Purify by gel electrophoresis (*APPENDIX 3G*) or ethanol precipitation (*APPENDIX 3B*). Resuspend in TE buffer,
   pH 8.0.

2. Prepare the following reaction mix on ice:

    2.5 μl 0.5 mM 3dNTP mix
    2.5 μl 10× *E. coli* polymerase I buffer
    5 μl (50 μCi) [α-^{32}P]dATP
    1 μl (3 to 8 U) Klenow fragment.

3. Combine 30 to 100 ng DNA with 1 to 5 μg random hexanucleotides (14 μl total volume). Boil 2 to 3 min and place on ice.

4. Add 11 μl of the reaction mix from step 2 to the denatured DNA and immediately incubate 2 to 4 hr at room temperature. Stop reaction by adding:

    1 μl 0.5 M EDTA
    3 μl 10 mg/ml tRNA
    100 μl TE buffer, pH 8.0.

Phenol extract and transfer aqueous phase to a fresh tube. If DNA was added as a molten gel slice, remelt the stopped reaction mixture by incubation at 70°C for 10 min.

5. Separate labeled DNA from unincorporated radioactive precursors by spin-column chromatography (Support Protocol 1). If desired, remove a 1-μl aliquot before loading sample on spin column and determine ^{32}P incorporation by acid precipitation (Support Protocol 2; specific activity should be 10^8 cpm/μg).

BASIC PROTOCOL

5′-END-LABELING OLIGONUCLEOTIDES USING T4 POLYNUCLEOTIDE KINASE

Materials (*see APPENDIX 1 for items with* ✓)

  ✓ 50 mM Tris·Cl, pH 7.5
    10 mM MgCl$_2$
    5 mM DTT
    1 to 50 pmol dephosphorylated DNA, 5′ ends
    50 pmol (150 μCi) [γ-^{32}P]ATP (specific activity >3000 Ci/mmol)
    50 μg/ml BSA
  ✓ 20 U T4 polynucleotide kinase and 10× buffer (see *APPENDIX 1* for buffer)
  ✓ 0.5 M EDTA
    75°C water bath or heating block

1. Set up the following in a 30-μl reaction:

    50 mM Tris·Cl, pH 7.5
    10 mM MgCl$_2$
    5 mM DTT
    1 to 50 pmol dephosphorylated DNA, 5′ ends
    50 pmol (150 μCi) [γ-^{32}P]ATP
    50 μg/ml BSA
    H$_2$O to 30 μl.

Add 20 U T4 polynucleotide kinase and incubate 1 hr at 37°C.

2. Stop the reaction by adding 1 μl of 0.5 M EDTA or by heating to 75°C for 10 min. Phenol/chloroform extract the sample (*APPENDIX 3B*). Separate labeled DNA from unincorporated radioactive precursors by spin-column chromatography (Support Protocol 1). Alternatively, add this reaction mixture to a PCR reaction (*UNIT 2.1*) to produce labeled amplification products without purification.

A3

SPIN-COLUMN PROCEDURE FOR SEPARATING RADIOACTIVELY LABELED DNA FROM UNINCORPORATED DNTP PRECURSORS

Materials *(see* APPENDIX 1 *for items with* ✓ *)*

    Resin: e.g., Sephadex G-50 (Pharmacia) or Bio-Gel P-60 (Bio-Rad)
✓ TE buffer, pH 8.0
    Radioactive sample: reaction mixture containing radioactive precursors (Basic Protocol 1 or 2 or Alternate Protocol)

    500-ml screw-cap bottle
    IEC clinical centrifuge
    5-ml disposable syringe
    Silanized glass wool
    50-ml polypropylene tube

1. Add 30 g resin to 300 ml TE buffer in a 500-ml screw-cap bottle and heat several hours at 65°C (or overnight at room temperature). Cool to room temperature, decant excess TE buffer, and replace with 1/2 vol fresh buffer (relative to volume of resin). Store at room temperature or 4°C indefinitely.

2. Plug bottom of a 5-ml disposable syringe with clean, silanized glass wool. Fill syringe with an even suspension of the swelled resin, and place in a 50-ml polypropylene tube. Centrifuge 2 to 3 min at 250 × g to pack the column (adjust time and speed so that resin is packed neither too tightly nor too loosely).

3. Dilute radioactive sample with TE buffer, pH 8.0, to 100 μl and load in the center of the column. Place syringe in a new 50-ml polypropylene tube, and centrifuge 5 min at 700 to 1100 × g. Save liquid at bottom of tube containing the labeled DNA. Dispose of all radioactive waste properly.

MEASURING RADIOACTIVITY IN DNA AND RNA BY TRICHLOROACETIC ACID (TCA) PRECIPITATION

Materials *(see* APPENDIX 1 *for items with* ✓ *)*

    Radioactive sample: reaction mixture containing radioactive precursors (Basic Protocol 1 or 2 or Alternate Protocol)
✓ 500 μg/ml sonicated salmon sperm DNA in TE buffer, pH 8.0 (see individual recipes)
✓ 10% (w/v) trichloroacetic acid (TCA) *or* acid precipitation solution (see APPENDIX 1), ice-cold
    100% ethanol
    Scintillation fluid (toluene-based)

    Glass microfiber filters (2.4-cm diameter, Whatman GF/A)
    Filtration device

1. Add a known volume (typically 1 μl) of a reaction mixture containing radioactive precursors to a disposable glass tube containing 100 μl of 500 μg/ml sonicated salmon sperm DNA in TE buffer, pH 8.0. Spot 10 μl of the mixture onto a glass microfiber filter and set aside for use in step 3. To the remaining 90 μl, add 1 ml of ice-cold 10% TCA or acid precipitation solution and incubate 5 to 10 min on ice.

2. Collect precipitate by filtering solution through a second glass microfiber filter. Rinse tube with 3 ml of 10% TCA or acid precipitation solution and pour through filter. Wash filter four more times with 3 ml solution, followed by 3 ml ethanol.

3. Dry both filters under a heat lamp (unnecessary with scintillation fluids that can accommodate aqueous samples), and place them in separate vials containing 3 ml of a toluene-based scintillation fluid. Measure radioactivity in a liquid scintillation counter. Determine incorporation of radioactivity into nucleic acid from ratio of cpm on second filter (which measures radioactivity in nucleic acid) to cpm on first filter (which measures total radioactivity in sample).

Contributors: Stanley Tabor and Kevin Struhl

APPENDIX 3F

Denaturing Polyacrylamide Gel Electrophoresis

BASIC PROTOCOL

Polyacrylamide gels that contain a high concentration of urea as a denaturant are capable of resolving short (<500 nucleotides) single-stranded fragments of DNA or RNA that differ in length by as little as one nucleotide.

Materials (see APPENDIX 1 for items with ✓)

    70% ethanol *or* isopropanol in squirt bottle
    5% (v/v) dimethyldichlorosilane (Sigma) in $CHCl_3$
✓ Denaturing acrylamide gel solution
    TEMED
    10% (w/v) ammonium persulfate (make fresh weekly and store at 4°C)
✓ 1× TBE buffer, pH 8.3 to 8.9
    Samples for electrophoresis containing formamide and marker dyes (e.g., *UNITS 2.1, 2.2,* or *9.3*)

    30 × 40–cm front and back gel plates
    0.2- to 0.4-mm uniform-thickness spacers
    Large book-binder clamps
    60-ml syringe
    0.2- to 0.4-mm shark's-tooth *or* preformed-well combs
    Sequencing gel electrophoresis apparatus
    Pasteur pipet *or* Beral thin stem (Beral Enterprises)
    Power supply with leads
    95°C heating block or water bath
    46 × 57–cm gel blotting paper (e.g., Whatman 3MM)
    Kodak XAR-5 X-ray film

1. Meticulously wash front and back 30 × 40–cm gel plates with soap and water. Rinse well with deionized water and dry. Wet plates with 70% ethanol or isopropanol in a squirt bottle and wipe dry with Kimwipe or other lint-free paper towel. Apply a film of 5% dimethyldichlorosilane in $CHCl_3$ to one side of each plate by wetting a Kimwipe with the solution and wiping carefully. After the film dries, wipe plate with 70% ethanol or isopropanol and dry with a Kimwipe. Check plates for dust and other particulates.

2. Assemble gel plates according to manufacturer's instructions, with the silanized surfaces facing inward. Use 0.2- to 0.4-mm uniform-thickness spacers and large book-binder clamps, making certain side and bottom spacers fit tightly together.

3. Prepare 60 ml of desired denaturing acrylamide gel solution in a 100-ml beaker (optionally, heat solution to ≤55°C to speed dissolution of urea and cool to room temperature before adding TEMED). Thoroughly mix 60 μl TEMED, then 0.6 ml of 10% ammonium persulfate, into acrylamide solution immediately before pouring gel.

4. Pour gel immediately. Gently pull acrylamide solution into a 60-ml syringe, avoiding bubbles. With short plate on top, raise upper edge of gel sandwich to 45° angle from the benchtop and slowly expel acrylamide between plates along one side. Adjust angle of plates so gel solution flows slowly down one side.

5. When solution reaches top of short plate, lower gel sandwich so that top edge is ∼5 cm above benchtop. Place an empty disposable pipet-tip rack or stopper underneath the sandwich to maintain the low angle. Insert flat side of a 0.2- to 0.4-mm shark's-tooth comb into the solution 2 to 3 mm below top of short plate, being very careful to avoid bubbles. Alternatively, insert teeth of preformed-well comb into gel solution. Use book-binder clamps to pinch combs between plates so that no solidified gel forms between combs and plates. Layer extra acrylamide gel solution onto comb to ensure full coverage. Rinse syringe with water to remove acrylamide.

6. When gel polymerizes, remove bottom spacer or tape at bottom of gel sandwich. Remove extraneous polyacrylamide from around combs with razor blade. Clean spilled urea and acrylamide solution from outer plate surfaces with water. Remove shark's-tooth comb gently from gel sandwich without stretching or tearing top of gel. Clean comb with water so it will be ready to be reinserted in step 8. If preformed-well comb was used, take care to prevent tearing of polyacrylamide wells; this comb will not be reinserted.

7. Fill bottom reservoir of gel apparatus with 1× TBE buffer so that gel plates will be submerged 2 to 3 cm in buffer. Place gel sandwich in electrophoresis apparatus and clamp plates to support. Sweep out any air bubbles at bottom of gel by squirting buffer between plates using syringe with a bent 20-G needle.

8. Pour 1× TBE buffer into top reservoir to ∼3 cm above top of gel. Rinse top of gel with 1× TBE buffer using a Pasteur pipet or Beral thin stem. Reinsert teeth of cleaned shark's-tooth comb into gel sandwich with points just barely sticking into gel. Using a Pasteur pipet or Beral thin stem, rinse wells thoroughly with 1× TBE buffer to remove stray fragments of polyacrylamide. Omit this step if a preformed-well comb is used.

9. Preheat gel ∼30 min by setting power supply to 45 V/cm, 1700 V, 70 W constant power. Rinse wells with 1× TBE buffer just prior to loading gels, to remove urea that has leached into them. Heat samples 2 min at 95°C in covered microcentrifuge tubes, then place on ice. Load 2 to 3 μl sample per well. Rinse sequencing pipet tip twice in lower reservoir after dispensing from each reaction tube.

10. Run gels at 45 to 70 W constant power. Maintain a gel temperature of ∼65°C. Observe migration of marker dyes (Table A.3F.1) to determine length of electrophoresis.

Temperatures >65°C can result in cracked plates or smeared bands; too low a temperature can lead to incomplete denaturation. To ensure even conduction of the heat generated during electrophoresis, an aluminum plate (0.4-cm thick, 34 × 22–cm) can be clamped onto the front glass plate with the same book-binder clamps used to hold the gel sandwich to the apparatus. The aluminum plate must be positioned so that it does not touch any buffer during electrophoresis.

11. Fill dry-ice traps attached to gel dryer (if required) and preheat dryer to 80°C.

Table A.3F.1 Migration of Marker Dyes Relative to Oligonucleotide Length (in Bases)

Polyacrylamide	Bromphenol blue	Xylene cyanol
5%	35 b	130 b
6%	26 b	106 b
8%	19 b	75 b
10%	12 b	55 b

12. After electrophoresis is complete, drain buffer from upper and lower reservoirs of apparatus and discard liquid as radioactive waste. Remove gel sandwich from apparatus and place under cold running tap water until surfaces of both glass plates are cool. Lay sandwich flat on paper towels with short plate up. Remove excess liquid and remaining clamps or tape. Remove one side spacer and insert long metal spatula between glass plates where spacer had been. Pry plates apart by gently rocking spatula (gel should stick to bottom plate; if it sticks to the top plate, flip sandwich over). Slowly lift top plate from the side with inserted spatula, gradually increasing the angle until the top plate is completely separated from gel.

13. Once plates are separated, remove second side spacer and any extraneous bits of polyacrylamide around gel. Hold two pieces of dry 46 × 57–cm blotting paper together as one piece. Beginning at one end of gel and working slowly towards the other, lay paper on top of gel. Take care to prevent air bubbles from forming between paper and gel. Peel blotting paper up; gel should come off plate with it. Gradually curl paper and gel away from plate as it is being pulled away.

14. Place paper and gel on preheated gel dryer. Cover with plastic wrap. Remove any bubbles between plastic wrap and gel by gently rubbing covered surface of gel from middle toward edges with a Kimwipe. Dry gel thoroughly 20 min to 1 hr at 80°C (when gel is completely dry, the plastic will easily peel off without sticking).

15. Remove plastic wrap and place dried gel in X-ray cassette with Kodak XAR-5 film in direct contact with gel. Autoradiograph at room temperature. After sufficient exposure time (usually overnight), remove X-ray film and process.

Contributors: Lisa M. Albright and Barton E. Slatko

APPENDIX 3G

Analysis of DNA by Southern Blot Hybridization

BASIC PROTOCOL

Materials *(see* APPENDIX 1 *for items with* ✓ *)*

Agarose, electrophoresis grade (low/gelling melting temperature agarose is not recommended for blotting)

✓ Electrophoresis buffer: TAE buffer *or* TBE buffer (see individual recipes)

✓ 10 mg/ml ethidium bromide

DNA molecular size markers: e.g., λ phage cut with *Hin*dIII, yielding fragments of 23.13, 9.42, 6.56, 4.36, 2.32, 2.03, 0.56, and 0.13 kb (store at 4°C)

Sample DNA

✓ 10× gel loading buffer

✓ Denaturing solution
✓ Neutralizing solution
✓ 20× and 2× SSC
✓ Hybridization cocktail, 42°C
✓ 2 mg/ml herring sperm DNA
  10^8 to 10^9 cpm/μg labeled single-stranded probe DNA (*APPENDIX 3E*)
✓ 2× SSC/0.1% (w/v) SDS (see individual recipes)
✓ 0.2× SSC/0.1% (w/v) SDS (see individual recipes), room temperature and 65°C
  Fluorescent or radioactive ink
✓ Stripping solution, 90°C

Gel casting platform
Gel comb
Horizontal gel electrophoresis apparatus
DC power supply
UV transilluminator
UV-blocking goggles or face shield
Polaroid MP4 camera with orange filter (Kodak Wratten no. 23A), UV-blocking filter
  (Kodak Wratten no. 2B), and film cassettes
65°, 90°, and 100°C water baths
Glass or plastic box or dish, preferably ≥3-liter volume
Two rectangular plastic or glass plates, one 20 cm wide and long enough to bridge the
  sides of the box or dish (Fig. A.3G.1) and one 20 × 20 cm
Whatman 3MM filter paper, 20-cm-wide roll and sheets
Nylon or nitrocellulose membrane, 20 × 20 cm
Vacuum oven, 80°C
Glass hybridization tube or plastic hybridization bag
42°C hybridization oven with platform shaker or rotator

NOTE: If glass hybridization tubes and hybridization oven are used, they should be obtained from the same manufacturer (e.g., the Techne Hybridization Oven, VWR Scientific or Fisher).

1. Set up gel casting platform and gel comb in the desired configuration. If gel casting platform does not have walls on all four sides (i.e., it has open ends), make walls for missing sides with masking tape.

   A ~20 × 20–cm gel with a thickness of ~0.5 to 1.0 cm is suitable for Southern blotting of genomic DNA; it requires 250 to 300 ml agarose. The teeth of the gel comb should be thick enough and deep enough to leave slots of 70-μl volume, but there should be 1 to 2 mm between the teeth and the floor of gel casting platform so wells will have gel bottoms.

2. Weigh an appropriate quantity of electrophoresis-grade agarose for the gel. Pour agarose into a 500-ml flask and add electrophoresis buffer (TAE or TBE) to the appropriate final volume. Swirl flask to mix. Heat flask ~2 to 3 min in a microwave oven at full power to melt agarose, swirling intermittently to mix thoroughly, but do not allow to boil.

   For fragments of 15 to 25 kb, a 0.5% (w/v) gel is preferable; for fragments 2.5 to 15 kb, a 1% (w/v) gel is appropriate; and for fragments of 0.5 to 2.5 kb, a 2% (w/v) gel is appropriate.

3. Allow solution to become cool to the touch (~55°C). Add 25 to 30 μl of 10 mg/ml ethidium bromide (final concentration of 1 μg/ml) and swirl to mix. Pour 55°C agarose into gel mold to cast gel. Be certain gel comb is positioned appropriately before gel begins to harden.

4. When gel has hardened completely (~45 min), carefully remove gel comb without damaging wells and remove ends from casting platform. Leave gel supported on the floor of the casting platform.

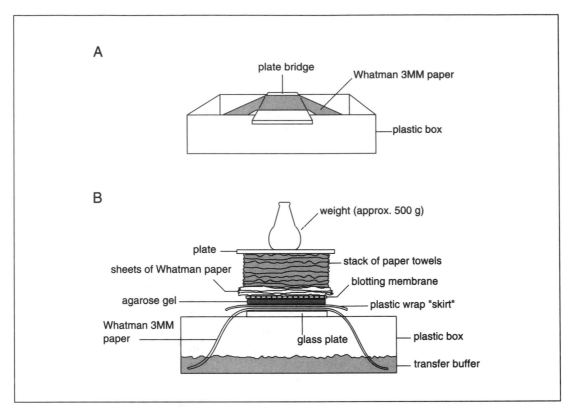

Figure A.3G.1 Upward capillary Southern transfer. **(A)** The plastic box used for transfer is bridged with a glass plate. A sheet of Whatman 3MM filter paper is soaked with 20× SSC and spread across the bridge with both ends hanging down into buffer. **(B)** The completed transfer stack. The box contains 20× SSC, which will be drawn by capillary action up the wetted Whatman 3MM filter paper, through gel and membrane, and into upper sheets of Whatman 3MM filter paper and paper towels. The weight may be a flask filled with water. It should weigh ~500 g.

The ends of the gel must be open so they make electrical contact with the electrophoresis buffer when placed in the electrophoresis tank. However, sides of the casting platform need not be removed; the floor and sides of the casting box can provide useful support for the gel.

Do not touch gel with bare hands because it contains ethidium bromide, which is mutagenic. Also, skin oils should not be allowed to contaminate the gel.

5. Place casting platform containing set gel in the electrophoresis tank. Add the same electrophoresis buffer (TAE or TBE) used to cast the gel, so as to completely cover the gel to a depth of ~1 cm.

   During long electrophoresis runs, some buffer evaporation may occur. If gel becomes exposed above buffer with the current running, the gel will melt.

   Gels for electrophoresis of plasmid or phage DNA should not be run in the same apparatus as gels for hybridization because even small amounts of residual plasmid or phage DNA, should they remain in the box, may potentially contaminate the gel and be picked up by the new membrane. Always wash electrophoresis tank with soap and water (not alcohol) between uses.

   Be certain that the electrophoresis tank is in a stable location where it will not be bumped or knocked, and where it will not present an electric shock hazard.

6. Dilute 1 μg DNA molecular size marker (e.g., λ phage DNA cut with *Hind*III) in 20 to 30 μl water and add 10× loading buffer to 1× final concentration. Load DNA samples

(maximum 10 to 15 μg DNA) containing 1× loading buffer onto gel and load DNA molecular size markers into the last well of the gel.

7. Attach electrodes on gel box to DC power supply. Be sure to attach the negative lead to the end of the box nearest the gel wells because DNA will move toward the positive electrode. Turn on power supply and set voltage at 30 to 40 V.

The gel box should be covered with the lid provided by the manufacturer. This reduces the risk of electric shock to other investigators and may also reduce evaporation of buffer. The gel should be observed briefly after power is applied to ensure that gel does not overheat due to insufficient buffer volume and that samples are running in the right direction.

For a 20-cm genomic Southern gel, 40 V is the largest potential that should be applied. Higher voltages, though they will increase the migration rate of DNA and therefore shorten running time, will also decrease resolution of different fragment sizes. If expected fragments are similar in size, a lower voltage may be more appropriate.

8. When bromphenol blue dye in loading buffer has advanced to the bottom of the gel, turn off power supply and remove gel, supported by the casting platform or a glass plate, from the tank.

9. Wearing UV-blocking goggles or face shield, photograph gel (minimizing time of personnel exposure to UV light). Place gel face-up on the clean surface of a UV transilluminator, directly over the light. Keep surface of the transilluminator clean with soap and water so contents of previous gels do not contaminate the new gel. Mount a Polaroid camera (e.g., MP4) on a stand pointing downward at the face of the gel. Use an orange filter and a UV-blocking filter to produce a good black-and-white photograph. Place a clear plastic ruler next to the gel to aid in measuring distances in the picture.

10. Place gel with its support into a box or dish. Add 1.5 liters denaturing solution and incubate 30 min on platform shaker at room temperature. Repeat once with fresh denaturing solution and twice with neutralizing solution.

Some investigators treat the gel with 0.25 M HCl for 30 min before denaturing. Acid hydrolysis results in cleavage of DNA fragments into smaller pieces without altering their location in the gel. Smaller fragments transfer to the blotting membrane more efficiently; this technique may improve results when larger DNA fragments are being analyzed.

11. Remove gel with its support and rinse box with water. Add 20× SSC to 5 to 8 cm deep and place the Plexiglas or glass rectangle across the box, bridging it from side to side. Cut a 60-cm piece of Whatman 3MM filter paper from the roll and soak it in 20× SSC. Place filter paper across bridge so that both ends are submerged in buffer (Fig. A.3G.1A).

12. Wet a 10-ml pipet in 20× SSC and roll it firmly across filter paper where it crosses the bridge to force out any air bubbles between paper and bridge (at all subsequent steps, be careful to exclude air bubbles). Place agarose gel face down onto the filter paper where it crosses the bridge (Fig. A.3G.1B). Remove air bubbles from under the gel.

13. Cut a piece of nylon or nitrocellulose membrane to the same size as the gel and place it in a clean glass dish. Cover with water and soak 5 min.

Nylon membranes are more durable and can be rehybridized several times. Each membrane may have specific instructions for use that apply to that product only. Handle the membrane as little as possible and then only with gloved hands or, preferably, clean blunt-ended forceps.

14. Discard water and soak membrane 10 min in 20× SSC. Place wet membrane on gel. Remove air bubbles between membrane and gel. Cut off one corner of the gel and the corresponding corner of the membrane to keep track of orientation (remember gel is inverted). Soak five squares of Whatman 3MM filter paper in 20× SSC, place them on top of the membrane one at a time, and roll out air bubbles.

15. Place a "skirt" of plastic wrap around edges of gel. Make a stack of paper towels cut or folded to the shape of the membrane. Pile towels ≥10 cm high on top of filter paper. Lay a glass or Plexiglas plate on top of paper towels and place an ∼500-g weight on the plate (Fig. A.3G.1B). Incubate overnight to allow adequate time for DNA transfer.

    If any of the paper components of the stack above the gel touch the filter paper below the gel, buffer will be able to bypass the gel entirely (referred to as a "short circuit"). Plastic wrap around edges of the gel prevents short circuiting.

    Small DNA fragments will transfer more quickly than large ones; 15- to 20-kb fragments require overnight transfer. Check completeness of transfer after the stack has been taken apart by restaining gel with ethidium bromide and examining it under UV light.

16. Remove weight, glass plate, paper towels, and filter paper. Mark positions of wells on membrane with a soft lead pencil and remove membrane from gel. Rinse membrane in 2× SSC and air dry 15 to 30 min on paper towels. When membrane is completely dry, bind DNA to membrane by baking nitrocellulose membrane 2 hr under vacuum at 80°C. Wrap nylon membrane in UV-transparent plastic wrap and expose to UV light ∼2 min, with even and equal exposure of all areas of the membrane. Hybridize membrane immediately or store dry between sheets of filter paper up to several months at room temperature.

17. Wet membrane in 2× SSC and place the membrane in a glass hybridization tube. Add 10 ml 42°C hybridization cocktail and prehybridize 1 hr on a rotator in 42°C hybridization oven.

    An alternative to the hybridization tube is a heat-sealable plastic bag large enough to hold the membrane spread out flat. For hybridization in a bag, use ∼40 ml hybridization cocktail. When sealing the membrane in the bag, and subsequently when adding the probe, it is extremely important to exclude all air bubbles.

18. Add 2 ml of 2 mg/ml herring sperm DNA and ∼10^7 cpm labeled DNA probe/100-cm^2 membrane to a glass tube with screw cap. Boil mixture 10 min to denature, and place on ice. Add probe to hybridization container and seal. Incubate overnight at 42°C with continuous rotation (for tubes) or agitation (for plastic bags) to ensure that all parts of membrane are exposed to probe.

19. Discard hybridization cocktail (and all subsequent wash solutions) as radioactive waste. Add 40 ml of 2× SSC/0.1% SDS, and incubate with rotation or agitation 15 min at room temperature. Repeat wash once with fresh 2× SSC/0.1% SDS. Do not allow membrane to dry.

    If a hybridization tube is used, wash steps can be conducted in the tube. If a bag is used, the membrane should be removed from the bag with forceps and washed in a flat-bottom glass dish (such as a baking dish) with enough of each wash solution to completely cover the membrane (usually 500 to 1000 ml).

20. Replace wash solution with 0.2× SSC/0.1% SDS and incubate 15 min at room temperature. In a 65°C oven or a baking dish supported in a 65°C water bath, incubate 15 min in 0.2× SSC/0.1% SDS, 65°C. Repeat wash once with fresh 65°C 0.2× SSC/0.1% SDS.

    Hybridization of probe to nonidentical sequences on the membrane is reduced by lowering the salt concentration and increasing the temperature of wash solutions. Experimentation is often necessary in order to find optimal wash conditions for a given probe. For some probes, it may be difficult to find conditions that effectively remove nonspecific binding without also removing useful signal. The most common adjustment is to vary the temperature of the final wash.

21. Discard last wash solution. Allow membrane to drain, but do not allow it to dry. Use forceps to place membrane, DNA-side-up, on a piece of discarded X-ray film cut to 20 × 20 cm. Wrap membrane and film in plastic wrap, leaving no holes or openings. Mark one corner of the wrapped membrane with fluorescent or radioactive ink, to leave a mark on the developed film indicating the orientation of the sheet of film relative to the membrane. In a darkroom, place membrane face-up in an empty film cassette and cover with a sheet of X-ray film (e.g., Kodak X-Omat R). Place an intensifying screen on top of film, close cassette, and expose 24 to 48 hr at −70°C.

22. Open cassette in a darkroom and develop film (preferably through an automatic developer). When a suitable exposure has been achieved, remove membrane from cassette and mark locations of gel wells (from the marks on the membrane) on the film with a marking pen.

23. Remove membrane from plastic wrap and film backing (do not allow to dry) and place it in 90°C stripping solution. Incubate 15 min at 90°C. To be certain that all probe has been removed, expose to X-ray film again; if signal persists, repeat stripping procedure or use harsher treatment (e.g., NaOH). Reuse membrane immediately or dry and store between sheets of filter paper at room temperature.

Reference: Southern, 1975

Contributor: John Jarcho

APPENDIX 3H

Analysis of RNA by Northern Blot Hybridization

NOTE: To inhibit RNase activity, all solutions for northern blotting should be prepared using sterile deionized water that has been treated with DEPC (*APPENDIX 1*).

BASIC PROTOCOL

NORTHERN HYBRIDIZATION OF RNA FRACTIONATED BY AGAROSE-FORMALDEHYDE GEL ELECTROPHORESIS

Materials *(see APPENDIX 1 for items with ✓)*
- ✓ 10× MOPS running buffer
  12.3 M (37%) formaldehyde, pH >4.0
  RNA sample: total cellular RNA or poly(A)$^+$ RNA
  Formamide
- ✓ Formaldehyde loading buffer
  Staining solutions (select one):
    - ✓ 0.5 M ammonium acetate with and without 0.5 μg/ml ethidium bromide
    - ✓ 1.1 M formaldehyde/10 mM sodium phosphate (pH 7.0; *APPENDIX 1* for sodium phosphate) with and without 10 μg/ml acridine orange
  0.05 M NaOH/1.5 M NaCl
- ✓ 0.5 M Tris·Cl (pH 7.4)/1.5 M NaCl (see *APPENDIX 1* for Tris · Cl)
- ✓ 20× SSC
- ✓ 0.03% (w/v) methylene blue in 0.3 M sodium acetate, pH 5.2 (optional; see *APPENDIX 1* for sodium acetate)

DNA suitable for use as probe template
✓ Formamide prehybridization/hybridization (FPH) solution
✓ 2× SSC, 0.2× SSC, and 0.1× SSC in 0.1% (w/v) SDS
✓ Stripping solution

✓ RNase-free glass dishes (see *APPENDIX 1* for DEPC treatment)
UV transilluminator, calibrated
Nylon or nitrocellulose membrane
Hybridization tubes or heat-sealable plastic bags for hybridization
65°, 80°, and 100°C water baths
Plastic dish to accommodate membrane
X-ray film and cassettes for autoradiography
3MM Whatman paper
Vacuum oven
UV-transparent plastic wrap (e.g., Saran Wrap or other polyvinylidene wrap)
Radiation monitor

1. Prepare a 1% agarose gel by dissolving 1.0 g agarose in 72 ml water. Cool to 60°C and add 10 ml of 10× MOPS running buffer and 18 ml of 12.3 M formaldehyde. Allow gel to set. Place solidified gel in tank and add 1× MOPS running buffer to a depth of ∼1 mm over the gel.

   Scale recipe as needed. The gel should be 2 to 6 mm thick with wells large enough to hold 60 μl sample. There should be enough lanes for duplicate samples.

   CAUTION: Formaldehyde is toxic through skin contact and inhalation of vapors. All operations involving formaldehyde should be carried out in a fume hood.

2. Adjust volume of each RNA sample (0.5 to 10 μg for each duplicate lane) to 11 μl with water, then add:

   5 μl 10× MOPS running buffer
   9 μl 12.3 M formaldehyde
   25 μl formamide.

   Mix by vortexing, microcentrifuge briefly (5 to 10 sec), and incubate 15 min at 55°C.

3. Add 10 μl formaldehyde loading buffer, vortex, microcentrifuge briefly, and load onto gel so that one set of all samples is followed by a duplicate set. Run gel at 5 V/cm until the bromphenol blue dye has migrated one-half to two-thirds the length of the gel (∼3 hr).

   Molecular-weight markers are not usually needed on total RNA gels, as rRNA molecules are stained as sharp bands and can be used as internal markers. If poly(A)+ RNA is being fractionated, commercial RNAs (e.g., 0.24- to 9.5-kb RNA ladder; Life Technologies) can be used.

4a. *To stain with ethidium bromide:* Remove gel, cut off one duplicate set of lanes to be stained, and wash twice, 20 min each, in an RNase-free glass dish containing 0.5 M ammonium acetate. Replace solution with 0.5 μg/ml ethidium bromide in 0.5 M ammonium acetate and allow to stain 40 min. If necessary, destain up to 1 hr in 0.5 M ammonium acetate.

4b. *To stain with acridine orange:* Remove gel, cut off one duplicate set of lanes to be stained, and stain 2 min in 1.1 M formaldehyde/10 mM sodium phosphate containing 10 μg/ml acridine orange. If necessary, destain 20 min in the same buffer without acridine orange.

5. Visualize RNA on a UV transilluminator and photograph with a ruler laid alongside the gel so that band positions can later be identified on the membrane.

6. Place unstained portion of gel in an RNase-free glass dish and rinse with several changes of deionized water. Treat as follows:

   30 min in 10 gel vol 0.05 M NaOH/1.5 M NaCl (partial hydrolysis)
   20 min in 10 gel vol 0.5 M Tris·Cl (pH 7.4)/1.5 M NaCl (neutralization)
   45 min in 10 gel vol 20× SSC.

7. Set up transfer stack as for Southern blotting (*APPENDIX 3G*) and transfer overnight.

8. Recover membrane and flattened gel together. Mark in pencil the position of the wells on the membrane, and ensure that the up-down and back-front orientations are recognizable (e.g., by cutting one corner of membrane). Rinse membrane in 2× SSC and then allow to dry on a sheet of Whatman 3MM paper.

9. Place membrane between two sheets of Whatman 3MM filter paper and cross-link by baking 2 hr at 80°C under vacuum. Alternatively, for nylon only, wrap dry membrane in UV-transparent plastic wrap, place RNA-side-down on a UV transilluminator (254 nm), and irradiate for the appropriate length of time (*APPENDIX 3G*).

10. If desired, check transfer efficiency by staining the gel in ethidium bromide or acridine orange (step 4) *or* (for nylon only) by staining the membrane 45 sec in 0.03% (w/v) methylene blue in 0.3 M sodium acetate, pH 5.2, and destaining 2 min in water.

11. Prepare RNA or DNA probe (*APPENDIX 3E*), ideally 100 to 1000 bp in length, labeled to a specific activity of $>10^8$ dpm/μg. Remove unincorporated nucleotides (*APPENDIX 3E*).

12. Wet blot in 6× SSC, place RNA-side-up in a hybridization tube or heat-sealable plastic bag, and add ~1 ml FPH solution per 10 cm^2 membrane. Incubate with rotation 3 hr at 42°C (for DNA probe) or 60°C (for RNA probe). For nylon, reduce time to 15 min.

13. If the probe is double-stranded, denature by heating 10 min in a 100°C water bath or incubator. Transfer to ice.

14. Add probe to the tube or bag at 10 ng probe/ml if the specific activity is 10^8 dpm/μg or 2 ng/ml if the specific activity is 10^9 dpm/μg. Incubate overnight.

15. Pour off hybridization solution. If hybridization was performed in a bag, transfer membrane to a plastic dish for washes. Wash as follows:

    1 vol 2× SSC/0.1% SDS, twice for 5 min each
    1 vol 0.2× SSC/0.1% SDS, twice for 5 min each (low-stringency wash)
    prewarmed (42°C) 0.2× SSC/0.1% SDS, twice for 15 min each
       (moderate-stringency wash; optional)
    prewarmed (68°C) 0.1× SSC/0.1% SDS, twice for 15 min each (high-stringency
       wash; optional).

16. Rinse in 2× SSC at room temperature, blot excess liquid, cover in UV-transparent plastic wrap, and perform autoradiography.

17. Place membrane in a hybridization bag or open container with stripping solution without formamide (enough to cover membrane). Place bag in 80°C water for 5 min. Pour out solution and repeat wash three to four times.

18. Monitor radioactivity. If further stripping is needed, place membrane in a bag with fresh stripping solution and place bag in boiling water for 5 min. Pour out solution and repeat three to four times.

19. Monitor radioactivity. If further stripping is needed, place membrane in a bag with fresh stripping solution containing formamide. Place bag in 65°C water for 5 min. Pour out solution, and repeat wash three times with formamide and one time without formamide.

20. Place membrane on filter paper to remove excess solution. Wrap membrane in plastic wrap and perform autoradiography or chemiluminescent detection to verify probe removal.

References: Alwine et al., 1977; Thomas, 1980

Contributors: Terry Brown and Karol Mackey

APPENDIX 3I

Techniques for Mammalian Cell Tissue Culture

STERILE TECHNIQUE

All cells handled in the lab are potentially infectious and should be handled with caution. Protective apparel such as gloves, lab coats or aprons, and eyewear should be worn when appropriate. Care should be taken when handling sharp objects such as needles, scissors, scalpel blades, and glass that could puncture the skin. Sterile disposable plastic supplies may be used to avoid the risk of broken or splintered glass.

Frequently, specimens received in the laboratory are not sterile and cultures prepared from these specimens may become contaminated with bacteria, fungus, or yeast. Solid tissues such as skin biopsies and products of conception (*UNIT 8.3*) can be soaked in antibiotic/antimycotic solutions prior to establishing the culture in an attempt to prevent culture contamination. In addition, antibiotics (penicillin, streptomycin, kanamycin, or gentamycin) and fungicides (amphotericin B or mycostatin) may be added to tissue culture medium to combat potential contaminants (see Table A.3I.1). An antibiotic/antimycotic solution or lyophilized powder that contains penicillin, streptomycin, and amphotericin B is available from Sigma and other suppliers. The solution can be used to wash specimens prior to culture and can be added to medium used for tissue culture.

All materials that come into direct contact with cultures must be sterile. Sterile disposable dishes, flasks, pipets, etc. can be obtained directly from manufacturers. Reusable glassware must be washed, rinsed thoroughly, then sterilized by autoclaving or by dry heat before reusing. With dry heat, glassware should be heated 90 min to 2 hr at 160°C to ensure sterility. Materials that may be damaged by very high temperatures can be autoclaved 20 min at 120°C and 15 psi. All media, reagents, and other solutions that come into contact with the cultures must also be sterile; media may be obtained as a sterile liquid from the manufacturer, autoclaved if not heat-sensitive, or filter sterilized. Supplements can be added to media prior to filtration, or they can be added aseptically after filtration. Filters with 0.20- to 0.22-μm pore size should be used to remove small gram-negative bacteria from culture media and solutions.

Table A.3I.1 Working Concentrations of Antibiotics and Fungicides for Mammalian Cell Culture

Additive	Final concentration
Penicillin	50–100 U/ml
Streptomycin sulfate	50–100 μg/ml
Kanamycin	100 μg/ml
Gentamycin	50 μg/ml
Mycostatin	20 μg/ml
Amphotericin B	0.25 μg/ml

Contamination can occur at any step in handling cultured cells. Care should be taken when pipetting media or other solutions for tissue culture. The necks of bottles and flasks, as well as the tips of the pipets, should be flamed before the pipet is introduced into the bottle. If the pipet tip comes into contact with the benchtop or any other nonsterile surface, it should be discarded and a fresh pipet obtained. Forceps and scissors used in tissue culture can be rapidly sterilized by dipping in 70% alcohol and flaming.

Although tissue culture work can be done on an open bench if aseptic methods are strictly enforced, biological safety cabinets are recommended to protect the cultures as well as the laboratory worker. In a laminar flow hood, the flow of air protects the work area from dust and contamination and acts as a barrier between the work surface and the worker. Many different styles of safety hoods are available and the laboratory should consider the types of samples being processed and the types of potential pathogenic exposure in making a selection. Manufacturer recommendations should be followed regarding routine maintenance checks on air flow and filters. For day-to-day use, the cabinet should be turned on for at least 5 min prior to beginning work. All work surfaces both inside and outside of the hood should be kept clean and disinfected daily and after each use.

Some safety cabinets are equipped with ultraviolet (UV) lights for decontamination of work surfaces. The current recommendation, however, is that work surfaces be wiped down with ethanol instead of relying on UV lamps, although some labs use the lamps in addition to ethanol wipes to decontaminate work areas. A special metering device is available to measure the output of UV lamps, and the lamps should be replaced when they fall below the minimum requirements for protection.

Cultures should be checked routinely for contamination. Indicators in the tissue culture medium change color when contamination is present. Medium that contains phenol red changes to yellow because of increased acidity. Cloudiness and turbidity are also observed in contaminated cultures. Once contamination is confirmed with a microscope, infected cultures are generally discarded. Keeping contaminated cultures increases the risk of contaminating other cultures. Sometimes a contaminated cell line can be salvaged by treating it with various combinations of antibiotics and antimycotics in an attempt to eradicate the infection. However, such treatment may adversely affect cell growth and is often unsuccessful in ridding cultures of contamination.

CULTURE MEDIUM PREPARATION

Powdered medium must be reconstituted with tissue culture grade water according to manufacturer's directions. Distilled or deionized water is not of sufficiently high quality for medium preparation; double- or triple-distilled water or commercially available tissue culture water should be used. The medium should be filter-sterilized and transferred to sterile bottles. Prepared medium can generally be stored ≤1 month in a 4°C refrigerator. Laboratories using large volumes of medium may choose to prepare their own medium from standard recipes. Basic media such as Eagle minimal essential medium (MEM), Dulbeccos modified Eagle medium (DMEM; *APPENDIX 1*), RPMI 1640 (*APPENDIX 1*), Chang medium (*APPENDIX 1*), and Ham F10 nutrient mixture (e.g., Life Technologies) are composed of amino acids, glucose, salts, vitamins, and other nutrients. A basic medium is supplemented by addition of L-glutamine, antibiotics (typically penicillin and streptomycin sulfate), and usually serum to formulate a "complete medium." Where serum is added, the amount is indicated as a percentage of fetal bovine serum (FBS) or other serum. Some media are also supplemented with antimycotics, nonessential amino acids, and various growth factors. Supplements should be added to medium prior to sterilization or filtration, or added aseptically just before use.

The optimum pH for most mammalian cell cultures is 7.2 to 7.4. Adjust pH of the medium as necessary after all supplements are added. Buffers such as bicarbonate and HEPES are routinely used in tissue culture medium to prevent fluctuations in pH that might adversely affect cell growth. HEPES is especially useful in solutions used for procedures that do not take place in a controlled CO_2 environment.

Most cultured cells will tolerate a wide range of osmotic pressure and an osmolarity between 260 and 320 mOsm/kg is acceptable for most cells. The osmolarity of human plasma is ~290 mOsm/kg, and this is probably the optimum for human cells in culture as well (Freshney, 1993).

Fetal bovine serum (FBS; sometimes known as fetal calf serum, FCS) is the most frequently used serum supplement. Calf serum, horse serum, and human serum are also used; some cell lines are maintained in serum-free medium (Freshney, 1993). Complete medium is supplemented with 5% to 30% (v/v) serum, depending on the requirements of the particular cell type being cultured. Serum that has been heat-inactivated (30 min to 1 hr at 56°C) is generally preferred, because heat treatment inactivates complement and is thought to reduce the number of contaminants. Serum is obtained frozen, then is thawed, divided into smaller portions, and refrozen until needed. There is considerable lot-to-lot variation in FBS. Most suppliers will provide a sample of a specific lot and reserve a supply of that lot while the serum is tested for its suitability. The suitability of a serum lot depends upon the use. Frequently the ability of serum to promote cell growth equivalent to a laboratory standard is used to evaluate a serum lot. Once an acceptable lot is identified, enough of that lot should be purchased to meet the culture needs of the laboratory for an extended period of time.

Commercially prepared media containing L-glutamine are available, although many laboratories choose to obtain medium without L-glutamine, and then add it to a final concentration of 2 mM just before use. Breakdown of L-glutamine is temperature- and pH-dependent; to prevent degradation, 100× L-glutamine (*APPENDIX 1*) should be stored frozen in aliquots until needed.

In addition to practicing good aseptic technique, most laboratories add antimicrobial agents to medium to further reduce the risk of contamination. A combination of penicillin and streptomycin is the most commonly used antibiotic additive; kanamycin and gentamycin are used alone. Mycostatin and amphotericin B are the most commonly used fungicides. Table A.3I.1 lists the final concentrations for the most commonly used antibiotics and antimycotics. Combining antibiotics in tissue culture media can be tricky, as some antibiotics are not compatible and one may inhibit the action of another. Furthermore, combined antibiotics may be more cytotoxic at lower concentrations than the individual antibiotics. In addition, prolonged use of antibiotics may cause cell lines to develop antibiotic resistance. For this reason, some laboratories add antibiotics and/or fungicides to medium when initially establishing a culture but eliminate them from medium used in later subcultures.

All tissue culture medium, whether commercially prepared or prepared within the laboratory, should be tested for sterility prior to use. A small aliquot from each lot of medium is incubated 48 hr at 37°C and monitored for evidence of contamination such as turbidity (infected medium will be cloudy) and color change (if phenol red is the indicator, infected medium will turn yellow). Any contaminated medium should be discarded.

BASIC PROTOCOL

TRYPSINIZING AND SUBCULTURING CELLS FROM A MONOLAYER

Materials (see *APPENDIX 1 for items with* ✓)

    Primary cultures of cells
✓ HBSS *without* Ca^{2+} and Mg^{2+}, 37°C

✓ 0.25% (w/v) trypsin/EDTA solution, 37°C
✓ Complete DMEM or RPMI with 10% to 15% (v/v) fetal bovine serum, 37°C

Sterile Pasteur pipets
37°C warming tray *or* incubator
Tissue culture plasticware or glassware including pipets and 25-cm^2 flasks or 60-mm petri plates, sterile

NOTE: All incubations are performed in a humidified 37°C, 5% CO_2 incubator unless otherwise specified.

1. Remove all medium from primary culture with a sterile Pasteur pipet. Wash adhering cell monolayer once or twice with a small volume of 37°C HBSS without Ca^{2+} and Mg^{2+} to remove any residual FBS that may inhibit the action of trypsin. If this is the first medium change, rather than discarding medium that is removed from primary culture, put it into a fresh dish or flask; the medium contains unattached cells that may attach and grow, thereby providing a back-up culture.

2. Add enough 37°C trypsin/EDTA solution to culture to cover adhering cell layer. Place plate on a 37°C warming tray 1 to 2 min. Tap bottom of plate on the countertop to dislodge cells. Check culture with an inverted microscope to be sure that cells are rounded up and detached from the surface. If cells are not sufficiently detached, return plate to warming tray for an additional minute or two.

3. Add 2 ml 37°C complete medium. Draw cell suspension into a Pasteur pipet and rinse cell layer two or three times to dissociate cells and to dislodge any remaining adherent cells. As soon as cells are detached, add serum or medium containing serum to inhibit further trypsin activity that might damage cells. If cultures are to be split 1/3 or 1/4 rather than 1/2, add sufficient medium such that 1 ml of cell suspension can be transferred into each fresh culture vessel.

4. Add an equal volume of cell suspension to fresh plates or flasks that have been labeled with the patient name, lab number, date of subculture, and passage number.

   Alternatively, cells can be counted using a hemacytometer or Coulter counter and diluted to the desired density so a specific number of cells can be added to each culture vessel. A final concentration of $\sim 5 \times 10^4$ cells/ml is appropriate for most subcultures.

   For primary cultures and early subcultures, 60-mm petri plates or 25-cm^2 flasks are generally used; larger petri plates or flasks (e.g., 150-mm plates or 75-cm^2 flasks) may be used for later subcultures.

5. Add 4 ml (9 ml if using 75-cm^2 flasks) of fresh medium to each new culture. Incubate in a humidified 37°C, 5% CO_2 incubator. If necessary, feed subconfluent cultures after 3 or 4 days by removing old medium and adding fresh 37°C medium. Passage secondary culture when it becomes confluent by repeating steps 1 to 4, and continue to passage as necessary.

SUPPORT PROTOCOL 1

FREEZING HUMAN CELLS GROWN IN MONOLAYER CULTURES

Materials (*see* APPENDIX 1 *for items with* ✓)
Log-phase monolayer culture of cells in petri plate
✓ Complete DMEM or RPMI

Freezing medium: complete medium with 10% to 20% (v/v) FBS and 5% to 10% (v/v) dimethyl sulfoxide (DMSO), 4°C

Liquid nitrogen

Benchtop clinical centrifuge (e.g., Fisher Centrific or Clay Adams Dynac) with 45°C fixed-angle or swinging-bucket rotor

1. Trypsinize cells (preferably in log phase; Basic Protocol, steps 1 to 3) from up to three plates from the same culture of the same patient. Transfer cell suspension to a sterile centrifuge tube and add 2 ml complete medium with serum. Centrifuge 5 min at 300 to $350 \times g$, room temperature.

2. Remove supernatant and add 1 ml (~3 ml for cells from a nearly confluent 25-cm^2 flask) of 4°C freezing medium. Resuspend pellet. Add 4 ml of 4°C freezing medium, mix cells thoroughly, and place on wet ice. Count cells using a hemacytometer (Support Protocol 3). Dilute with more freezing medium as necessary to get a final cell concentration of 10^6 or 10^7 cells/ml.

3. Pipet 1-ml aliquots of cell suspension into labeled 2-ml cryovials. Tighten caps on vials. Place vials 1 hr to overnight in a −70°C freezer, then transfer to liquid nitrogen storage freezer.

ALTERNATE PROTOCOL

FREEZING CELLS GROWN IN SUSPENSION CULTURE

Freezing cells from suspension culture is similar in principle to freezing cells from monolayer. The major difference is that suspension cultures need not be trypsinized. For materials, see Support Protocol 1.

1. Transfer cell suspension to a centrifuge tube and spin 10 min at 300 to $350 \times g$, room temperature. Remove supernatant and resuspend pellet in 4°C freezing medium at a density of 10^6 to 10^7 cells/ml.

   Some laboratories freeze lymphoblastoid lines at the higher cell density because they plan to recover them in a larger volume of medium and because there may be a greater loss of cell viability upon recovery as compared to other types of cells (e.g., fibroblasts).

2. Transfer 1-ml aliquots of cell suspension into cryovials and freeze as for monolayer cultures.

SUPPORT PROTOCOL 2

THAWING AND RECOVERING HUMAN CELLS

When cryopreserved cells are needed for study, they should be thawed rapidly and plated at high density to optimize recovery.

CAUTION: Protective clothing, particularly insulated gloves and goggles, should be worn when removing frozen vials or ampules from the liquid nitrogen freezer. The room containing the liquid nitrogen freezer should be well ventilated. Care should be taken not to spill liquid nitrogen on the skin.

Materials

Vial of frozen cells (see Support Protocol 1 or Alternate Protocol)

70% (v/v) ethanol

✓ Complete medium/20% FBS, 37°C

NOTE: All incubations are performed in a humidified 37°C, 5% CO_2 incubator unless otherwise specified.

1. Remove vial from liquid nitrogen freezer and immediately place it into a 37°C water bath. Agitate vial continuously until medium is thawed (usually <60 sec).

    Cells should be thawed as quickly as possible to prevent formation of ice crystals that can cause cell lysis. Try to avoid getting water around the cap of the vial.

2. Wipe top of vial with 70% ethanol, or submerge in 70% ethanol and then air dry, before opening.

3. Transfer thawed cell suspension into a sterile centrifuge tube containing 2 ml warm complete medium/20% FBS. Centrifuge 10 min at 150 to 200 × g, room temperature. Discard supernatant.

4. Gently resuspend cell pellet in small amount (~1 ml) of complete medium/20% FBS and transfer to properly labeled culture plate containing the appropriate amount (generally 5 to 20 ml) of medium.

    Cultures are reestablished at a higher cell density than that used for original cultures because there is some cell death associated with freezing.

5. Check cultures after ~24 hr to ensure that cells have attached to the plate. Change medium after 5 to 7 days or when pH indicator (e.g., phenol red) in medium changes color. Keep cultures in medium with 20% FBS until cell line is reestablished.

    If recovery rate is extremely low, only a subpopulation of the original culture may be growing; be extra careful of this when working with cell lines known to be mosaic.

SUPPORT PROTOCOL 3

DETERMINING CELL NUMBER AND VIABILITY WITH A HEMACYTOMETER AND TRYPAN BLUE STAINING

A hemacytometer is a thick glass slide with a central area designed as a counting chamber; the one described here is the Improved Neubauer from Baxter Scientific (Fig. A.3I.1). The central portion of the slide is the counting platform, which is bordered by a 1-mm groove. The central platform is divided into two counting chambers by a transverse groove. Each counting chamber consists of a silver footplate on which is etched a 3 × 3–mm grid. This grid is divided into nine secondary squares, each 1 × 1 mm. The four corner squares and the central square are used for determining the cell count. The corner squares are further divided into 16 tertiary squares and the central square into 25 tertiary squares to aid in cell counting. Accompanying the hemacytometer slide is a thick, even-surfaced coverslip. Ordinary coverslips may have uneven surfaces which can introduce errors in cell counting; therefore, it is imperative that the coverslip provided with the hemacytometer be used in determining cell number. Cell suspension is applied to a defined area and counted so cell density can be calculated.

Materials

70% (v/v) ethanol
Cell suspension
✓ 0.4% (w/v) trypan blue *or* 0.4% (w/v) nigrosin prepared in HBSS (see *APPENDIX 1* for HBSS)

Hemacytometer with coverslip (Improved Neubauer, Baxter Scientific)
Hand-held counter

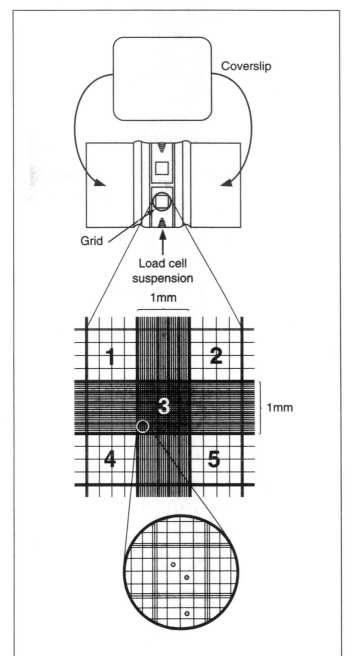

Figure A.3I.1 Hemacytometer slide (Improved Neubauer) and coverslip. The coverslip is applied to the slide and cell suspension is added to counting chamber using a Pasteur pipet. Each counting chamber has a 3 × 3-mm grid (enlarged). The four corner squares (1, 2, 4, and 5) and the central square (3) are counted on each side of the hemacytometer (numbers added).

1. Clean surface of hemacytometer slide and coverslip with 70% alcohol and allow to dry. Wet edge of coverslip slightly with tap water and press over grooves on hemacytometer so that it rests evenly over the silver counting area.

2. For cells grown in monolayer cultures, detach cells from surface of dish using trypsin as described in the Basic Protocol. Dilute cells as needed to obtain a uniform suspension (maximum 20 to 50 cells per 1-mm square of the hemacytometer is recommmended). Disperse any clumps.

3. Use a sterile Pasteur pipet to transfer cell suspension to edge of hemacytometer counting chamber. Hold tip of pipet under the coverslip and dispense one drop of suspension; allow it to be drawn under the coverslip by capillary action.

The hemacytometer should be considered nonsterile. If cell suspension is to be used for cultures, do not reuse the pipet and do not return any excess cell suspension in the pipet to the original suspension.

4. Fill second counting chamber. Allow cells to settle for a few minutes before beginning to count. Blot off excess liquid. View slide on microscope with 100× magnification (i.e., 10× ocular with a 10× objective). Position slide to view the large central area of the grid (section 3 in Fig. A.3I.1); this area is bordered by a set of three parallel lines.

The central area of the grid should almost fill the microscope field. Subdivisions within the large central area are also bordered by three parallel lines and each subdivision is divided into sixteen smaller squares by single lines. Cells within this area should be evenly distributed without clumping. If cells are not evenly distributed, wash and reload hemacytometer.

5. Use a hand-held counter to count cells in each of the corner and central squares (Fig. A.3I.1, squares numbered 1 to 5). Repeat counts for other counting chamber.

Five squares are counted from each of the two counting chambers for a total of ten squares counted. Count cells touching the middle line of the triple line on the top and left of the squares. Do not count cells touching the middle line of the triple lines on the bottom or right side of the square.

6. Determine cells per ml by the following calculations:

cells/ml = average count per square × dilution factor × 10^4

total cells = cells/ml × total original volume of cell suspension from which sample was taken.

10^4 is the volume correction factor for the hemacytometer: each square is 1×1 mm and the depth is 0.1 mm.

7. Determine number of viable cells by adding 0.5 ml of 0.4% trypan blue (or 0.4% nigrosin), 0.3 ml HBSS, and 0.1 ml cell suspension to a small tube. Mix thoroughly and let stand 5 min before loading hemacytometer. Count total number of cells and total number of viable (unstained) cells. Calculate percent viable cells as follows:

$$\% \text{ viable cells} = \frac{\text{number of unstained cells}}{\text{total number of cells}} \times 100$$

8. Decontaminate coverslip and hemacytometer by rinsing with 70% ethanol and then deionized water. Air dry and store for future use.

SUPPORT PROTOCOL 4

PREPARING CELLS FOR TRANSPORT

Both monolayer and suspension cultures can easily be shipped in 25-cm² tissue culture flasks. Cells are grown to near confluency in a monolayer or to desired density in suspension. Medium is removed from monolayer cultures and the flask is filled with fresh medium. Fresh medium is added to suspension cultures to fill the flask. *It is essential that the flasks be completely filled with medium to protect cells from drying if flasks are inverted during transport.* The cap is tightened and taped securely in place. The flask is sealed in a leak-proof plastic bag or other leak-proof container designed to prevent spillage in the event that the flask should become damaged. The primary container is then placed in a secondary insulated container to protect it from extreme temperatures during transport. A biohazard label is affixed to the outside of the package. Generally, cultures are transported by same-day or overnight courier.

Cells can also be shipped frozen. The vial containing frozen cells is removed from the liquid nitrogen freezer and placed immediately on dry ice in an insulated container to prevent thawing during transport.

Reference: Lee, 1991

Contributor: Mary C. Phelan

APPENDIX 3J

Establishment of Permanent Cell Lines by Epstein-Barr Virus Transformation

NOTE: All reagents and equipment coming into contact with live cells must be sterile. All incubations are performed in a humidified 37°C, 5% CO_2 incubator unless otherwise specified.

BASIC PROTOCOL

EPSTEIN-BARR VIRUS TRANSFORMATION OF CULTURED LYMPHOCYTES

This protocol describes EBV transformation and culture of human B lymphocytes for the establishment of permanent cell lines. Such cell lines are useful as a renewable source of DNA in family genotype studies.

Materials (see APPENDIX 1 for items with ✓)

  Whole blood collected in 8.5-ml yellow-top Vacutainer tubes with ACD solution A (Becton Dickinson Labware)
✓ HBSS without Ca^{2+} or Mg^{2+}
✓ Transformation medium, 37°C
✓ Growth medium, 37°C

  Leuco-prep tubes (Becton Dickinson Labware)
  15-ml conical centrifuge tube, sterile
  25- and 75-cm² tissue culture flasks

1. Collect two 8.5-ml Vacutainers of blood from each patient. Use blood immediately or store ≤3 days at room temperature. Do not store at 4°C.

2. Transfer whole blood into Leuco-prep tube (use one 16 × 125–mm tube for each 8.5-ml Vacutainer). Centrifuge 30 min at 1000 × *g*, room temperature; blood will form the following layers from top to bottom—serum, white blood cells (WBCs), gel layer, and red blood cells (RBCs).

3. Transfer WBCs to a 15-ml conical centrifuge tube. Bring volume to 10 ml with HBSS without Ca^{2+} or Mg^{2+}. Centrifuge WBC suspension 10 min at 150 × *g*, room temperature. After transfer is complete, dispose of Leuco-prep tubes as biohazard.

4. Discard supernatant and resuspend pellet in 5 ml HBSS without Ca^{2+} or Mg^{2+}. Centrifuge 10 min at 150 × *g*. Discard supernatant and resuspend cell pellet in 6 to 8 ml transformation

medium. Aliquot equal volumes into two 25-cm² tissue culture flasks labeled A and B. Incubate cultures 7 to 10 days in an upright position before feeding.

Formation of cell clumps indicates healthy growth. A change in the color of the medium is due to a change in pH and indicates that the cells need to be fed.

5. After ~1 week, remove ~1.5 ml of transformation medium from surface of culture and replace it with an equal amount of growth medium. Incubate for another week. Add 0.2 ml fresh growth medium once a week. Incubate cultures until medium clears and turns yellow, and small clumps of cells become visible (after 2 weeks, ~60% to 70% of cultures should be ready).

6. Add 0.5 to 1.0 ml fresh growth medium and incubate. Inspect cultures every 2 to 3 days. When medium turns yellow and clumps become larger and more numerous, add 3 ml fresh growth medium. At the next feeding add 5 ml fresh growth medium.

7. When the volume of medium totals ~12 to 15 ml and the cells are ready to be fed, transfer culture into 75-cm² flask containing 10 ml fresh growth medium. Pipet 2 ml of medium back into the 25-cm² flask and mark it with a "R" (for reserve). Add 3 ml growth medium into the reserve flask. Incubate both flasks. Continue to incubate reserve flask until stock is frozen away; it is not usually necessary to feed this flask.

8. When cultures need to be fed, add 15 ml growth medium to 75-cm² flask. Continue to feed culture every 2 to 3 days until volume of medium in 75-cm² flask totals 50 ml. Freeze culture (Support Protocol 1).

Flasks and tubes should be disposed of as biohazards.

To prepare a cell pellet for DNA isolation (APPENDIX 3A), grow cultures until volume of medium is 60 to 70 ml. Add 10 ml growth medium and transfer 50 ml cell suspension to a centrifuge tube.

Transformation of B cells with EBV is a relatively inefficient process, with ~1 in 100 cells being transformed.

It is recommended that ≥5 vials of immortalized cells, each containing 1 to 2 × 10⁷ cells, be stored for future culture or DNA isolation (Support Protocol 1).

SUPPORT PROTOCOL 1

PREPARATION OF EPSTEIN-BARR VIRUS STOCK

Materials *(see APPENDIX 1 for items with ✓)*

    EBV-infected B95-8 cells (ATCC)
✓ Complete Iscove's modified Dulbecco's medium (IMDM; e.g., Sigma) with 10% (v/v) fetal bovine serum (see APPENDIX 1 for FBS), 37°C

    25- and 75-cm² tissue culture flasks
    50-ml centrifuge tubes
    95% (v/v) dry ice/ethanol bath *or* liquid nitrogen
    0.45-μm filter

1. Thaw EBV-infected B95-8 cells and set up cultures in 25-cm² flasks with 3 ml complete IMDM/10% FBS and incubate in horizontal position.

2. Feed culture by adding 3 ml fresh IMDM/10% FBS when medium becomes yellow and incubate until the desired volume is reached. Incubate an additional 1 to 2 weeks, without

feeding, until medium turns a bright lemon color. To change culture to a larger flask, or divide into several flasks, agitate the flask to detach attached cells and decant. Feed as indicated above.

3. To harvest the culture, transfer cell suspension into a 50-ml centrifuge tube. Freeze the suspension in a 95% dry ice/ethanol bath until solid and thaw it in a 37°C water bath three times to burst cells and release virus.

4. Centrifuge suspension 10 min at $300 \times g$, 4°C. Filter supernatant through a 0.45-μm filter. Aliquot to desired volume and store in liquid nitrogen or −80°C freezer and thaw just before use.

   Aliquots 1-, 5-, and 50-ml in volume may be frozen and an appropriate-sized vial thawed depending on the number of cell lines to be transformed.

   EBV may be titered by the method of Miller and Lipman (1973). The titer of frozen stocks should be stable for a minimum of 1 year. EBV stocks derived from B95-8 supernatants typically contain $\sim 10^3$ transforming units per ml in a good preparation. Because in practice EBV titers may be variable between preparations, each preparation should be tested for transforming ability. Once frozen at −80°C or below, a preparation should be stable ≥1 year.

SUPPORT PROTOCOL 2

FREEZING AND RECULTURING OF EPSTEIN-BARR VIRUS-TRANSFORMED LYMPHOCYTES

Materials *(see APPENDIX 1 for items with* ✓ *)*

 Cultures of EBV-transformed lymphocytes (Basic Protocol)
✓ Freezing medium
✓ Complete RPMI, serum-free

 50-ml centrifuge tubes
 15-ml centrifuge tubes
 1.8-ml cryovials (USA/Scientific Plastics)

1. Transfer culture of EBV-transformed lymphocytes into 50-ml centrifuge tube. Transfer 0.5 ml of culture into a 15-ml centrifuge tube and determine cell viability by trypan blue exclusion (*APPENDIX 3I*), which should be ≥70%.

2. Centrifuge culture suspension 15 min at $300 \times g$. Remove supernatant and resuspend pellet in 5 ml freezing medium. Aliquot 1 ml cells into each of 5 cryovials labeled with sample number and date.

   Freezing medium contains 10% DMSO as a cryopreservative, which is necessary if cells are to be recultured. If cells are to be used only for DNA isolation, DMSO should be omitted. If cells frozen with DMSO are to be used directly for DNA isolation, DMSO should be removed by washing cells with an isotonic solution prior to lysis.

   It is recommended that ≥5 vials of immortalized cells, each containing 1 to 2×10^7 cells, be stored for future culture or DNA isolation (see Support Protocol 1).

3. Store cryovial overnight at −70°C. Transfer cryovial to liquid nitrogen storage tank and record the location.

4. To reculture frozen cells, remove vials from freezer and place on dry ice. Transfer tubes to 37°C water bath and thaw as quickly as possible to minimize formation of ice crystals. Dilute contents of each vial with 6 to 7 ml serum-free complete RPMI. Centrifuge at 300

× *g* just long enough to loosely pellet cells. Pour off supernatant and resuspend pellet in 3 ml growth medium. Continue culturing as in steps 6 to 8 of the Basic Protocol.

Contributor: John Gilbert

APPENDIX 3K

Karyotyping

BASIC PROTOCOL

The chromosome banding procedures described in Chapter 4 permit identification of individual chromosomes, each characterized by its unique banding pattern. This identification process results in establishment of the karyotype, which is "a systematized array of the chromosomes of a single cell ... that ... can typify the chromosomes of an individual or even a species" (ISCN, 1995). A standard nomenclature has been developed to assign identification numbers to each human chromosome on the basis of size, shape, and banding pattern (ISCN, 1995).

To establish the chromosome constitution of a nonmosaic individual or homogeneous cell line, one or preferably two karyotypes should be prepared. For mosaic or chimeric individuals, or acquired chromosome changes in tumors and heterogeneous cell lines, at least one karyotype should be prepared for each different chromosome complement identified. Additional analysis will depend upon initial findings and the purpose of the analysis. For routine clinical analysis, it is usual to completely analyze three cells through the microscope or from images in addition to those karyotyped. In addition, a minimum of 15 to 20 cells should be counted to establish cell number and to identify the presence of other cell lines. For cancer cases or transformed lines, more cells may need to be analyzed, but for confirmation of previously identified abnormalities a smaller analysis may suffice.

Chromosome recognition and identification is a skill that requires some experience. A novice should begin with normal cells and ask a trained cytogeneticist to confirm karyotype correctness. Some basic guidelines are presented below.

1. Chromosomes in somatic cells usually come in pairs of identical size and band pattern, with the exception of sex chromosomes in the heterogametic sex.

2. Chromosomes are grouped by shape as follows:

    metacentric—centromere in center of chromosome, two arms of equal length
    submetacentric—centromere displaced slightly from center, one arm (q) longer
    acrocentric—centromere displaced near one end, one arm (p) very short
    telocentric—centromere at one end (note: no human chromosome is telocentric).

3. Human chromosomes are grouped by size as follows:

    Nos. 1 to 3—large metacentric chromosomes (group A)
    Nos. 4 to 5—large submetacentric chromosomes (group B)
    Nos. 6 to 12 and X—medium submetacentric chromosomes (group C)
    Nos. 13 to 15—medium acrocentric chromosomes (group D)
    Nos. 16 to 18—small submetacentric chromosomes (group E)
    Nos. 19 and 20—small metacentric chromosomes (group F)
    Nos. 21, 22, and Y—small acrocentric chromosomes (group G).

A3

4. Chromosomes are oriented with p arm (short or "petit" arm) on top (in metacentric chromosomes this arm is chosen by convention).

5. Within groups, refer to chromosome idiograms (*APPENDIX 2B*) for landmarks.

6. Sex chromosomes (X and Y) can be assigned at end of relevant group or placed separately as a pair.

7. Some minor cell-to-cell variability can be expected, which is due to differential contraction (condensation) of the chromosomes.

8. Differences between homologs may be due to polymorphisms. In humans, these occur primarily in the centromeric region of chromosomes, in the heterochromatic region of chromosomes 1, 9, and 16—located adjacent to the centromere; in the short arm of the acrocentric chromosomes (13, 14, 15, 21, and 22); and in the heterochromatin on the long arm of the Y chromosome.

The basic steps in establishing a karyotype are as follows:

1. Count the number of chromosomes in a photographic print of banded metaphase cell (*UNIT 4.3*) to establish chromosome number.

2. Cut images of individual chromosomes out of metaphase print.

    Multiple prints may be needed if chromosomes overlap each other.

3. Group chromosomes from print by size and shape.

4. Identify and pair homologous chromosomes by comparing band patterns.

5. Arrange and mount pairs of homologs in order (largest chromosomes have lowest number) by comparison with chromosome idiograms (see *APPENDIX 2B*).

    Mounting is usually done by using glue or double-stick tape to affix the chromosomes on a form indicating the chromosome number under each pair of homologs arranged in groups.

6. Locate regions of interest (breakpoints or hybridization sites) and name according to system of nomenclature symbols given in ISCN, 1995 (*APPENDIX 2B*).

7. Describe chromosome complement using standard nomenclature (ISCN, 1995).

Reference: ISCN, 1995

Contributors: Rhona R. Schreck and Christine Distèche

Selected Suppliers of Reagents and Equipment

L isted below are addresses and phone numbers of commercial suppliers who have been recommended for particular items used in our manuals because: (1) the particular brand has actually been found to be of superior quality, or (2) the item is difficult to find in the marketplace. Consequently, this compilation may not include some important vendors of biological supplies. For comprehensive listings, see *Linscott's Directory of Immunological and Biological Reagents* (Santa Rosa, CA), *The Biotechnology Directory* (Stockton Press, New York), the annual Buyers' Guide supplement to the journal *Bio/Technology*, as well as various sites on the Internet.

A4

A.C. Daniels
72-80 Akeman Street
Tring, Hertfordshire, HP23 6AJ, UK
(44) 1442 826881
FAX: (44) 1442 826880

A.D. Instruments
5111 Nations Crossing Road #8
Suite 2
Charlotte, NC 28217
(704) 522-8415   FAX: (704) 527-5005
http://www.us.endress.com

A.J. Buck
11407 Cronhill Drive
Owings Mill, MD 21117
(800) 638-8673   FAX: (410) 581-1809
(410) 581-1800
http://www.ajbuck.com

A.M. Systems
131 Business Park Loop
P.O. Box 850
Carlsborg, WA 98324
(800) 426-1306   FAX: (360) 683-3525
(360) 683-8300
http://www.a-msystems.com

Aaron Medical Industries
7100 30th Avenue North
St. Petersburg, FL 33710
(727) 384-2323   FAX: (727) 347-9144
http://www.aaronmed.com

Abbott Laboratories
100 Abbott Park Road
Abbott Park, IL 60064
(800) 323-9100   FAX: (847) 938-7424
http://www.abbott.com

ABCO Dealers
55 Church Street Central Plaza
Lowell, MA 01852
(800) 462-3326   (978) 459-6101
http://www.lomedco.com/abco.htm

Aber Instruments
5 Science Park
Aberystwyth, Wales SY23 3AH, UK
(44) 1970 636300
FAX: (44) 1970 615455
http://www.aber-instruments.co.uk

ABI Biotechnologies
See Perkin-Elmer

ABI Biotechnology
See Apotex

Access Technologies
Subsidiary of Norfolk Medical
7350 N. Ridgeway
Skokie, IL 60076
(877) 674-7131   FAX: (847) 674-7066
(847) 674-7131
http://www.norfolkaccess.com

Accurate Chemical and Scientific
300 Shames Drive
Westbury, NY 11590
(800) 645-6264   FAX: (516) 997-4948
(516) 333-2221
http://www.accuratechemical.com

AccuScan Instruments
5090 Trabue Road
Columbus, OH 43228
(800) 822-1344   FAX: (614) 878-3560
(614) 878-6644
http://www.accuscan-usa.com

AccuStandard
125 Market Street
New Haven, CT 06513
(800) 442-5290   FAX: (877) 786-5287
http://www.accustandard.com

Ace Glass
1430 NW Boulevard
Vineland, NJ 08360
(800) 223-4524   FAX: (800) 543-6752
(609) 692-3333

ACO Pacific
2604 Read Avenue
Belmont, CA 94002
(650) 595-8588   FAX: (650) 591-2891
http://www.acopacific.com

Acros Organic
See Fisher Scientific

Action Scientific
P.O. Box 1369
Carolina Beach, NC 28428
(910) 458-0401   FAX: (910) 458-0407

AD Instruments
1949 Landings Drive
Mountain View, CA 94043
(888) 965-6040   FAX: (650) 965-9293
(650) 965-9292
http://www.adinstruments.com

Adaptive Biosystems
15 Ribocon Way
Progress Park
Luton, Bedsfordshire LU4 9UR, UK
(44)1 582-597676
FAX: (44)1 582-581495
http://www.adaptive.co.uk

Adobe Systems
1585 Charleston Road
P.O. Box 7900
Mountain View, CA 94039
(800) 833-6687   FAX: (415) 961-3769
(415) 961-4400
http://www.adobe.com

Advanced Bioscience Resources
1516 Oak Street, Suite 303
Alameda, CA 94501
(510) 865-5872   FAX: (510) 865-4090

Advanced Biotechnologies
9108 Guilford Road
Columbia, MD 21046
(800) 426-0764   FAX: (301) 497-9773
(301) 470-3220
http://www.abionline.com

Advanced ChemTech
5609 Fern Valley Road
Louisville, KY 40228
(502) 969-0000
http://www.peptide.com

Advanced Machining and Tooling
9850 Businesspark Avenue
San Diego, CA 92131
(858) 530-0751   FAX: (858) 530-0611
http://www.amtmfg.com

Advanced Magnetics
See PerSeptive Biosystems

Advanced Process Supply
See Naz-Dar-KC Chicago

Advanced Separation Technologies
37 Leslie Court
P.O. Box 297
Whippany, NJ 07981
(973) 428-9080   FAX: (973) 428-0152
http://www.astecusa.com

Advanced Targeting Systems
11175-A Flintkote Avenue
San Diego, CA 92121
(877) 889-2288   FAX: (858) 642-1989
(858) 642-1988
http://www.ATSbio.com

Advent Research Materials
Eynsham, Oxford OX29 4JA, UK
(44) 1865-884440
FAX: (44) 1865-84460
http://www.advent-rm.com

Advet
Industrivagen 24
S-972 54 Lulea, Sweden
(46) 0920-211887
FAX: (46) 0920-13773

Aesculap
1000 Gateway Boulevard
South San Francisco, CA 94080
(800) 282-9000
http://www.aesculap.com

Affinity Chromatography
307 Huntingdon Road
Girton, Cambridge CB3 OJX, UK
(44) 1223 277192
FAX: (44) 1223 277502
http://www.affinity-chrom.com

Affinity Sensors
See Labsystems Affinity Sensors

Affymetrix
3380 Central Expressway
Santa Clara, CA 95051
(408) 731-5000   FAX: (408) 481-0422
(800) 362-2447
http://www.affymetrix.com

Agar Scientific
66a Cambridge Road
Stansted CM24 8DA, UK
(44) 1279-813-519
FAX: (44) 1279-815-106
http://www.agarscientific.com

A/G Technology
101 Hampton Avenue
Needham, MA 02494
(800) AGT-2535   FAX: (781) 449-5786
(781) 449-5774
http://www.agtech.com

Agen Biomedical Limited
11 Durbell Street
P.O. Box 391
Acacia Ridge 4110
Brisbane, Australia
61-7-3370-6300   FAX: 61-7-3370-6370
http://www.agen.com

Agilent Technologies
395 Page Mill Road
P.O. Box 10395
Palo Alto, CA 94306
(650) 752-5000
http://www.agilent.com/chem

Agouron Pharmaceuticals
10350 N. Torrey Pines Road
La Jolla, CA 92037
(858) 622-3000   FAX: (858) 622-3298
http://www.agouron.com

Agracetus
8520 University Green
Middleton, WI 53562
(608) 836-7300   FAX: (608) 836-9710
http://www.monsanto.com

AIDS Research and Reference
Reagent Program
U.S. Department of Health and
Human Services
625 Lofstrand Lane
Rockville, MD 20850
(301) 340-0245   FAX: (301) 340-9245
http://www.aidsreagent.org

AIN Plastics
249 East Sanford Boulevard
P.O. Box 151
Mt. Vernon, NY 10550
(914) 668-6800   FAX: (914) 668-8820
http://www.tincna.com

Air Products and Chemicals
7201 Hamilton Boulevard
Allentown, PA 18195
(800) 345-3148   FAX: (610) 481-4381
(610) 481-6799
http://www.airproducts.com

ALA Scientific Instruments
1100 Shames Drive
Westbury, NY 11590
(516) 997-5780   FAX: (516) 997-0528
http://www.alascience.com

Aladin Enterprises
1255 23rd Avenue
San Francisco, CA 94122
(415) 468-0433   FAX: (415) 468-5607

Aladdin Systems
165 Westridge Drive
Watsonville, CA 95076
(831) 761-6200   FAX: (831) 761-6206
http://www.aladdinsys.com

Alcide
8561 154th Avenue NE
Redmond, WA 98052
(800) 543-2133   FAX: (425) 861-0173
(425) 882-2555
http://www.alcide.com

Aldevron
3233 15th Street, South
Fargo, ND 58104
(877) Pure-DNA   FAX: (701) 280-1642
(701) 297-9256
http://www.aldevron.com

Aldrich Chemical
P.O. Box 2060
Milwaukee, WI 53201
(800) 558-9160   FAX: (800) 962-9591
(414) 273-3850   FAX: (414) 273-4979
http://www.aldrich.sial.com

Alexis Biochemicals
6181 Cornerstone Court East, Suite 103
San Diego, CA 92121
(800) 900-0065   FAX: (858) 658-9224
(858) 658-0065
http://www.alexis-corp.com

Alfa Aesar
30 Bond Street
Ward Hill, MA 10835
(800) 343-0660   FAX: (800) 322-4757
(978) 521-6300   FAX: (978) 521-6350
http://www.alfa.com

Alfa Laval
Avenue de Ble 5 - Bazellaan 5
BE-1140 Brussels, Belgium
32(2) 728 3811
FAX: 32(2) 728 3917 or 32(2) 728 3985
http://www.alfalaval.com

Alice King Chatham Medical Arts
11915-17 Inglewood Avenue
Hawthorne, CA 90250
(310) 970-1834   FAX: (310) 970-0121
(310) 970-1063

Allegiance Healthcare
800-964-5227
http://www.allegiance.net

Allelix Biopharmaceuticals
6850 Gorway Drive
Mississauga, Ontario
L4V 1V7 Canada
(905) 677-0831   FAX: (905) 677-9595
http://www.allelix.com

Allentown Caging Equipment
Route 526, P.O. Box 698
Allentown, NJ 08501
(800) 762-CAGE   FAX: (609) 259-0449
(609) 259-7951
http://www.acecaging.com

Alltech Associates
Applied Science Labs
2051 Waukegan Road
P.O. Box 23
Deerfield, IL 60015
(800) 255-8324   FAX: (847) 948-1078
(847) 948-8600
http://www.alltechweb.com

Alomone Labs
HaMarpeh 5
P.O. Box 4287
Jerusalem 91042, Israel
972-2-587-2202   FAX: 972-2-587-1101
US: (800) 791-3904
FAX: (800) 791-3912
http://www.alomone.com

Alpha Innotech
14743 Catalina Street
San Leandro, CA 94577
(800) 795-5556   FAX: (510) 483-3227
(510) 483-9620
http://www.alphainnotech.com

Altec Plastics
116 B Street
Boston, MA 02127
(800) 477-8196   FAX: (617) 269-8484
(617) 269-1400

Alza
1900 Charleston Road
P.O. Box 7210
Mountain View, CA 94043
(800) 692-2990   FAX: (650) 564-7070
(650) 564-5000
http://www.alza.com

Alzet
c/o Durect Corporation
P.O. Box 530
10240 Bubb Road
Cupertino, CA 95015
(800) 692-2990   (408) 367-4036
FAX: (408) 865-1406
http://www.alzet.com

Amac
160B Larrabee Road
Westbrook, ME 04092
(800) 458-5060   FAX: (207) 854-0116
(207) 854-0426

Amaresco
30175 Solon Industrial Parkway
Solon, Ohio 44139
(800) 366-1313   FAX: (440) 349-1182
(440) 349-1313

Ambion
2130 Woodward Street, Suite 200
Austin, TX 78744
(800) 888-8804   FAX: (512) 651-0190
(512) 651-0200
http://www.ambion.com

American Association of
Blood Banks
College of American Pathologists
325 Waukegan Road
Northfield, IL 60093
(800) 323-4040   FAX: (847) 8166
(847) 832-7000
http://www.cap.org

American Bio-Technologies
See Intracel Corporation

American Bioanalytical
15 Erie Drive
Natick, MA 01760
(800) 443-0600   FAX: (508) 655-2754
(508) 655-4336
http://www.americanbio.com

American Cyanamid
P.O. Box 400
Princeton, NJ 08543
(609) 799-0400   FAX: (609) 275-3502
http://www.cyanamid.com

American HistoLabs
7605-F Airpark Road
Gaithersburg, MD 20879
(301) 330-1200   FAX: (301) 330-6059

American International Chemical
17 Strathmore Road
Natick, MA 01760
(800) 238-0001   (508) 655-5805
http://www.aicma.com

American Laboratory Supply
See American Bioanalytical

American Medical Systems
10700 Bren Road West
Minnetonka, MN 55343
(800) 328-3881   FAX: (612) 930-6654
(612) 933-4666
http://www.visitams.com

American Qualex
920-A Calle Negocio
San Clemente, CA 92673
(949) 492-8298   FAX: (949) 492-6790
http://www.americanqualex.com

American Radiolabeled Chemicals
11624 Bowling Green
St. Louis, MO 63146
(800) 331-6661   FAX: (800) 999-9925
(314) 991-4545   FAX: (314) 991-4692
http://www.arc-inc.com

American Scientific Products
See VWR Scientific Products

American Society for
Histocompatibility and
Immunogenetics
P.O. Box 15804
Lenexa, KS 66285
(913) 541-0009   FAX: (913) 541-0156
http://www.swmed.edu/home_pages/
ASHI/ashi.htm

**American Type Culture Collection
(ATCC)**
10801 University Boulevard
Manassas, VA 20110
(800) 638-6597   FAX: (703) 365-2750
(703) 365-2700
http://www.atcc.org

Amersham
See Amersham Pharmacia Biotech

Amersham International
Amersham Place
Little Chalfont, Buckinghamshire
HP7 9NA, UK
(44) 1494-544100
FAX: (44) 1494-544350
http://www.apbiotech.com

Amersham Medi-Physics
Also see Nycomed Amersham
3350 North Ridge Avenue
Arlington Heights, IL 60004
(800) 292-8514   FAX: (800) 807-2382
http://www.nycomed-amersham.com

Amersham Pharmacia Biotech
800 Centennial Avenue
P.O. Box 1327
Piscataway, NJ 08855
(800) 526-3593   FAX: (877) 295-8102
(732) 457-8000
http://www.apbiotech.com

A4

Amgen
1 Amgen Center Drive
Thousand Oaks, CA 91320
(800) 926-4369   FAX: (805) 498-9377
(805) 447-5725
http://www.amgen.com

Amicon
Scientific Systems Division
72 Cherry Hill Drive
Beverly, MA 01915
(800) 426-4266   FAX: (978) 777-6204
(978) 777-3622
http://www.amicon.com

Amika
8980F Route 108
Oakland Center
Columbia, MD 21045
(800) 547-6766   FAX: (410) 997-7104
(410) 997-0100
http://www.amika.com

Amoco Performance Products
See BPAmoco

AMPI
See Pacer Scientific

Amrad
576 Swan Street
Richmond, Victoria 3121, Australia
613-9208-4000
FAX: 613-9208-4350
http://www.amrad.com.au

Amresco
30175 Solon Industrial Parkway
Solon, OH 44139
(800) 829-2805   FAX: (440) 349-1182
(440) 349-1199

Anachemia Chemicals
3 Lincoln Boulevard
Rouses Point, NY 12979
(800) 323-1414   FAX: (518) 462-1952
(518) 462-1066
http://www.anachemia.com

Ana-Gen Technologies
4015 Fabian Way
Palo Alto, CA 94303
(800) 654-4671   FAX: (650) 494-3893
(650) 494-3894
http://www.ana-gen.com

Analox Instruments USA
P.O. Box 208
Lunenburg, MA 01462
(978) 582-9368   FAX: (978) 582-9588
http://www.analox.com

Analytical Biological Services
Cornell Business Park 701-4
Wilmington, DE 19801
(800) 391-2391   FAX: (302) 654-8046
(302) 654-4492
http://www.ABSbioreagents.com

Analytical Genetics Testing Center
7808 Cherry Creek S. Drive, Suite 201
Denver, CO 80231
(800) 204-4721   FAX: (303) 750-2171
(303) 750-2023
http://www.geneticid.com

AnaSpec
2149 O'Toole Avenue, Suite F
San Jose, CA 95131
(800) 452-5530   FAX: (408) 452-5059
(408) 452-5055
http://www.anaspec.com

Ancare
2647 Grand Avenue
P.O. Box 814
Bellmore, NY 11710
(800) 645-6379   FAX: (516) 781-4937
(516) 781-0755
http://www.ancare.com

Ancell
243 Third Street North
P.O. Box 87
Bayport, MN 55033
(800) 374-9523   FAX: (651) 439-1940
(651) 439-0835
http://www.ancell.com

Anderson Instruments
500 Technology Court
Smyrna, GA 30082
(800) 241-6898   FAX: (770) 319-5306
(770) 319-9999
http://www.graseby.com

Andreas Hettich
Gartenstrasse 100
Postfach 260
D-78732 Tuttlingen, Germany
(49) 7461 705 0
FAX: (49) 7461 705-122
http://www.hettich-centrifugen.de

Anesthetic Vaporizer Services
10185 Main Street
Clarence, NY 14031
(719) 759-8490
http://www.avapor.com

Animal Identification and
Marking Systems (AIMS)
13 Winchester Avenue
Budd Lake, NJ 07828
(908) 684-9105   FAX: (908) 684-9106
http://www.animalid.com

Annovis
34 Mount Pleasant Drive
Aston, PA 19014
(800) EASY-DNA   FAX: (610)
361-8255
(610) 361-9224
http://www.annovis.com

Apotex
150 Signet Drive
Weston, Ontario
M9L 1T9 Canada
(416) 749-9300   FAX: (416) 749-2646
http://www.apotex.com

Apple Scientific
11711 Chillicothe Road, Unit 2
P.O. Box 778
Chesterland, OH 44026
(440) 729-3056   FAX: (440) 729-0928
http://www.applesci.com

Applied Biosystems
See PE Biosystems

Applied Imaging
2380 Walsh Avenue, Bldg. B
Santa Clara, CA 95051
(800) 634-3622   FAX: (408) 562-0264
(408) 562-0250
http://www.aicorp.com

Applied Photophysics
203-205 Kingston Road
Leatherhead, Surrey, KT22 7PB
UK
(44) 1372-386537

Applied Precision
1040 12th Avenue Northwest
Issaquah, Washington 98027
(425) 557-1000
FAX: (425) 557-1055
http://www.api.com/index.html

Appligene Oncor
Parc d'Innovation
Rue Geiler de Kaysersberg, BP 72
67402 Illkirch Cedex, France
(33) 88 67 22 67
FAX: (33) 88 67 19 45
http://www.oncor.com/prod-app.htm

Applikon
1165 Chess Drive, Suite G
Foster City, CA 94404
(650) 578-1396   FAX: (650) 578-8836
http://www.applikon.com

Appropriate Technical Resources
9157 Whiskey Bottom Road
Laurel, MD 20723
(800) 827-5931   FAX: (410) 792-2837
http://www.atrbiotech.com

APV Gaulin
100 S. CP Avenue
Lake Mills, WI 53551
(888) 278-4321   FAX: (888) 278-5329
http://www.apv.com

Aqualon
See Hercules Aqualon

A4

Aquarium Systems
8141 Tyler Boulevard
Mentor, OH 44060
(800) 822-1100   FAX: (440) 255-8994
(440) 255-1997
http://www.aquariumsystems.com

Aquebogue Machine and Repair Shop
Box 2055
Main Road
Aquebogue, NY 11931
(631) 722-3635   FAX: (631) 722-3106

Archer Daniels Midland
4666 Faries Parkway
Decatur, IL 62525
(217) 424-5200
http://www.admworld.com

Archimica Florida
P.O. Box 1466
Gainesville, FL 32602
(800) 331-6313   FAX: (352) 371-6246
(352) 376-8246
http://www.archimica.com

Arcor Electronics
1845 Oak Street #15
Northfield, IL 60093
(847) 501-4848

Arcturus Engineering
400 Logue Avenue
Mountain View, CA 94043
(888) 446 7911   FAX: (650) 962 3039
(650) 962 3020
http://www.arctur.com

Ardais Corporation
One Ledgemont Center
128 Spring Street
Lexington, MA 02421
(781) 274-6420   (781) 274-6421
http://www.ardais.com

Argonaut Technologies
887 Industrial Road, Suite G
San Carlos, CA 94070
(650) 998-1350   FAX: (650) 598-1359
http://www.argotech.com

Ariad Pharmaceuticals
26 Landsdowne Street
Cambridge, MA 02139
(617) 494-0400   FAX: (617) 494-8144
http://www.ariad.com

Armour Pharmaceuticals
See Rhone-Poulenc Rorer

Aronex Pharmaceuticals
8707 Technology Forest Place
The Woodlands, TX 77381
(281) 367-1666   FAX: (281) 367-1676
http://www.aronex.com

Artisan Industries
73 Pond Street
Waltham, MA 02254
(617) 893-6800
http://www.artisanind.com

ASI Instruments
12900 Ten Mile Road
Warren, MI 48089
(800) 531-1105   FAX: (810) 756-9737
(810) 756-1222
http://www.asi-instruments.com

Aspen Research Laboratories
1700 Buerkle Road
White Bear Lake, MN 55140
(651) 264-6000   FAX: (651) 264-6270
http://www.aspenresearch.com

Associates of Cape Cod
704 Main Street
Falmouth, MA 02540
(800) LAL-TEST   FAX: (508) 540-8680
(508) 540-3444
http://www.acciusa.com

Astra Pharmaceuticals
See AstraZeneca

AstraZeneca
1800 Concord Pike
Wilmington, DE 19850
(302) 886-3000   FAX: (302) 886-2972
http://www.astrazeneca.com

AT Biochem
30 Spring Mill Drive
Malvern, PA 19355
(610) 889-9300   FAX: (610) 889-9304

ATC Diagnostics
See Vysis

ATCC
See American Type Culture Collection

Athens Research and Technology
P.O. Box 5494
Athens, GA 30604
(706) 546-0207   FAX: (706) 546-7395

Atlanta Biologicals
1425-400 Oakbrook Drive
Norcross, GA 30093
(800) 780-7788 or (770) 446-1404
FAX: (800) 780-7374 or (770) 446-1404
http://www.atlantabio.com

Atomergic Chemical
71 Carolyn Boulevard
Farmingdale, NY 11735
(631) 694-9000   FAX: (631) 694-9177
http://www.atomergic.com

Atomic Energy of Canada
2251 Speakman Drive
Mississauga, Ontario
L5K 1B2 Canada
(905) 823-9040   FAX: (905) 823-1290
http://www.aecl.ca

ATR
P.O. Box 460
Laurel, MD 20725
(800) 827-5931   FAX: (410) 792-2837
(301) 470-2799
http://www.atrbiotech.com

Aurora Biosciences
11010 Torreyana Road
San Diego, CA 92121
(858) 404-6600   FAX: (858) 404-6714
http://www.aurorabio.com

Automatic Switch Company
A Division of Emerson Electric
50 Hanover Road
Florham Park, NJ 07932
(800) 937-2726   FAX: (973) 966-2628
(973) 966-2000
http://www.asco.com

Avanti Polar Lipids
700 Industrial Park Drive
Alabaster, AL 35007
(800) 227-0651   FAX: (800) 229-1004
(205) 663-2494   FAX: (205) 663-0756
http://www.avantilipids.com

Aventis
BP 67917
67917 Strasbourg Cedex 9, France
33 (0) 388 99 11 00
FAX: 33 (0) 388 99 11 01
http://www.aventis.com

Aventis Pasteur
1 Discovery Drive
Swiftwater, PA 18370
(800) 822-2463   FAX: (570) 839-0955
(570) 839-7187
http://www.aventispasteur.com/usa

Avery Dennison
150 North Orange Grove Boulevard
Pasadena, CA 91103
(800) 462-8379   FAX: (626) 792-7312
(626) 304-2000
http://www.averydennison.com

Avestin
2450 Don Reid Drive
Ottawa, Ontario
K1H 1E1 Canada
(888) AVESTIN   FAX: (613) 736-8086
(613) 736-0019
http://www.avestin.com

AVIV Instruments
750 Vassar Avenue
Lakewood, NJ 08701
(732) 367-1663   FAX: (732) 370-0032
http://www.avivinst.com

Axon Instruments
1101 Chess Drive
Foster City, CA 94404
(650) 571-9400   FAX: (650) 571-9500
http://www.axon.com

Azon
720 Azon Road
Johnson City, NY 13790
(800) 847-9374   FAX: (800) 635-6042
(607) 797-2368
http://www.azon.com

A4

BAbCO
1223 South 47th Street
Richmond, CA 94804
(800) 92-BABCO   FAX: (510) 412-8940
(510) 412-8930
http://www.babco.com

Bacharach
625 Alpha Drive
Pittsburgh, PA 15238
(800) 736-4666   FAX: (412) 963-2091
(412) 963-2000
http://www.bacharach-inc.com

Bachem Bioscience
3700 Horizon Drive
King of Prussia, PA 19406
(800) 634-3183   FAX: (610) 239-0800
(610) 239-0300
http://www.bachem.com

Bachem California
3132 Kashiwa Street
P.O. Box 3426
Torrance, CA 90510
(800) 422-2436   FAX: (310) 530-1571
(310) 539-4171
http://www.bachem.com

Baekon
18866 Allendale Avenue
Saratoga, CA 95070
(408) 972-8779   FAX: (408) 741-0944

Baker Chemical
See J.T. Baker

Bangs Laboratories
9025 Technology Drive
Fishers, IN 46038
(317) 570-7020   FAX: (317) 570-7034
http://www.bangslabs.com

Bard Parker
See Becton Dickinson

Barnstead/Thermolyne
P.O. Box 797
2555 Kerper Boulevard
Dubuque, IA 52004
(800) 446-6060   FAX: (319) 589-0516
http://www.barnstead.com

Barrskogen
4612 Laverock Place N
Washington, DC 20007
(800) 237-9192   FAX: (301) 464-7347

BAS
See Bioanalytical Systems

BASF
Specialty Products
3000 Continental Drive North
Mt. Olive, NJ 07828
(800) 669-2273   FAX: (973) 426-2610
http://www.basf.com

Baum, W.A.
620 Oak Street
Copiague, NY 11726
(631) 226-3940   FAX: (631) 226-3969
http://www.wabaum.com

Bausch & Lomb
One Bausch & Lomb Place
Rochester, NY 14604
(800) 344-8815   FAX: (716) 338-6007
(716) 338-6000
http://www.bausch.com

Baxter
Fenwal Division
1627 Lake Cook Road
Deerfield, IL 60015
(800) 766-1077   FAX: (800) 395-3291
(847) 940-6599   FAX: (847) 940-5766
http://www.powerfulmedicine.com

Baxter Healthcare
One Baxter Parkway
Deerfield, IL 60015
(800) 777-2298   FAX: (847) 948-3948
(847) 948-2000
http://www.baxter.com

Baxter Scientific Products
See VWR Scientific

Bayer
Agricultural Division
Animal Health Products
12707 Shawnee Mission Pkwy.
Shawnee Mission, KS 66201
(800) 255-6517   FAX: (913) 268-2803
(913) 268-2000
http://www.bayerus.com

Bayer
Diagnostics Division (Order Services)
P.O. Box 2009
Mishiwaka, IN 46546
(800) 248-2637   FAX: (800) 863-6882
(219) 256-3390
http://www.bayer.com

Bayer Diagnostics
511 Benedict Avenue
Tarrytown, NY 10591
(800) 255-3232   FAX: (914) 524-2132
(914) 631-8000
http://www.bayerdiag.com

Bayer Plc
Diagnostics Division
Bayer House, Strawberry Hill
Newbury, Berkshire RG14 1JA, UK
(44) 1635-563000
FAX: (44) 1635-563393
http://www.bayer.co.uk

BD Immunocytometry Systems
2350 Qume Drive
San Jose, CA 95131
(800) 223-8226   FAX: (408) 954-BDIS
http://www.bdfacs.com

BD Labware
Two Oak Park
Bedford, MA 01730
(800) 343-2035   FAX: (800) 743-6200
http://www.bd.com/labware

BD PharMingen
10975 Torreyana Road
San Diego, CA 92121
(800) 848-6227   FAX: (858) 812-8888
(858) 812-8800
http://www.pharmingen.com

BD Transduction Laboratories
133 Venture Court
Lexington, KY 40511
(800) 227-4063   FAX: (606) 259-1413
(606) 259-1550
http://www.translab.com

BDH Chemicals
Broom Road
Poole, Dorset BH12 4NN, UK
(44) 1202-745520
FAX: (44) 1202- 2413720

BDH Chemicals
See Hoefer Scientific Instruments

BDIS
See BD Immunocytometry Systems

Beckman Coulter
4300 North Harbor Boulevard
Fullerton, CA 92834
(800) 233-4685   FAX: (800) 643-4366
(714) 871-4848
http://www.beckman-coulter.com

Beckman Instruments
Spinco Division/Bioproducts Operation
1050 Page Mill Road
Palo Alto, CA 94304
(800) 742-2345   FAX: (415) 859-1550
(415) 857-1150
http://www.beckman-coulter.com

Becton Dickinson Immunocytometry & Cellular Imaging
2350 Qume Drive
San Jose, CA 95131
(800) 223-8226   FAX: (408) 954-2007
(408) 432-9475
http://www.bdfacs.com

Becton Dickinson Labware
1 Becton Drive
Franklin Lakes, NJ 07417
(888) 237-2762   FAX: (800) 847-2220
(201) 847-4222
http://www.bdfacs.com

Becton Dickinson Labware
2 Bridgewater Lane
Lincoln Park, NJ 07035
(800) 235-5953   FAX: (800) 847-2220
(201) 847-4222
http://www.bdfacs.com

Becton Dickinson Primary
Care Diagnostics
7 Loveton Circle
Sparks, MD 21152
(800) 675-0908   FAX: (410) 316-4723
(410) 316-4000
http://www.bdfacs.com

Behringwerke Diagnostika
Hoechster Strasse 70
P-65835 Liederback, Germany
(49) 69-30511   FAX: (49) 69-303-834

Bellco Glass
340 Edrudo Road
Vineland, NJ 08360
(800) 257-7043   FAX: (856) 691-3247
(856) 691-1075
http://www.bellcoglass.com

Bender Biosystems
See Serva

Beral Enterprises
See Garren Scientific

Berkeley Antibody
See BAbCO

Bernsco Surgical Supply
25 Plant Avenue
Hauppague, NY 11788
(800) TIEMANN   FAX: (516) 273-6199
(516) 273-0005
http://www.bernsco.com

Beta Medical and Scientific
(Datesand Ltd.)
2 Ferndale Road
Sale, Manchester M33 3GP, UK
(44) 1612 317676
FAX: (44) 1612 313656

**Bethesda Research Laboratories
(BRL)**
See Life Technologies

Biacore
200 Centennial Avenue, Suite 100
Piscataway, NJ 08854
(800) 242-2599   FAX: (732) 885-5669
(732) 885-5618
http://www.biacore.com

Bilaney Consultants
St. Julian's
Sevenoaks, Kent TN15 0RX, UK
(44) 1732 450002
FAX: (44) 1732 450003
http://www.bilaney.com

Binding Site
5889 Oberlin Drive, Suite 101
San Diego, CA 92121
(800) 633-4484   FAX: (619) 453-9189
(619) 453-9177
http://www.bindingsite.co.uk

BIO 101
See Qbiogene

Bio Image
See Genomic Solutions

Bioanalytical Systems
2701 Kent Avenue
West Lafayette, IN 47906
(800) 845-4246   FAX: (765) 497-1102
(765) 463-4527
http://www.bioanalytical.com

Biocell
2001 University Drive
Rancho Dominguez, CA 90220
(800) 222-8382   FAX: (310) 637-3927
(310) 537-3300
http://www.biocell.com

Biocoat
See BD Labware

BioComp Instruments
650 Churchill Road
Fredericton, New Brunswick
E3B 1P6 Canada
(800) 561-4221   FAX: (506) 453-3583
(506) 453-4812
http://131.202.97.21

BioDesign
P.O. Box 1050
Carmel, NY 10512
(914) 454-6610   FAX: (914) 454-6077
http://www.biodesignofny.com

BioDiscovery
4640 Admiralty Way, Suite 710
Marina Del Rey, CA 90292
(310) 306-9310   FAX: (310) 306-9109
http://www.biodiscovery.com

Bioengineering AG
Sagenrainstrasse 7
CH8636 Wald, Switzerland
(41) 55-256-8-111
FAX: (41) 55-256-8-256

Biofluids
Division of Biosource International
1114 Taft Street
Rockville, MD 20850
(800) 972-5200   FAX: (301) 424-3619
(301) 424-4140
http://www.biosource.com

BioFX Laboratories
9633 Liberty Road, Suite S
Randallstown, MD 21133
(800) 445-6447   FAX: (410) 498-6008
(410) 496-6006
http://www.biofx.com

BioGenex Laboratories
4600 Norris Canyon Road
San Ramon, CA 94583
(800) 421-4149   FAX: (925) 275-0580
(925) 275-0550
http://www.biogenex.com

Bioline
2470 Wrondel Way
Reno, NV 89502
(888) 257-5155   FAX: (775) 828-7676
(775) 828-0202
http://www.bioline.com

Bio-Logic Research & Development
1, rue de I-Europe
A.Z. de Font-Ratel
38640 CLAIX, France
(33) 76-98-68-31
FAX: (33) 76-98-69-09

Biological Detection Systems
See Cellomics or Amersham

Biomeda
1166 Triton Drive, Suite E
P.O. Box 8045
Foster City, CA 94404
(800) 341-8787   FAX: (650) 341-2299
(650) 341-8787
http://www.biomeda.com

BioMedic Data Systems
1 Silas Road
Seaford, DE 19973
(800) 526-2637   FAX: (302) 628-4110
(302) 628-4100
http://www.bmds.com

Biomedical Engineering
P.O. Box 980694
Virginia Commonwealth University
Richmond, VA 23298
(804) 828-9829   FAX: (804) 828-1008

Biomedical Research Instruments
12264 Wilkins Avenue
Rockville, MD 20852
(800) 327-9498
(301) 881-7911
http://www.biomedinstr.com

Bio/medical Specialties
P.O. Box 1687
Santa Monica, CA 90406
(800) 269-1158   FAX: (800) 269-1158
(323) 938-7515

BioMerieux
100 Rodolphe Street
Durham, North Carolina 27712
(919) 620-2000
http://www.biomerieux.com

BioMetallics
P.O. Box 2251
Princeton, NJ 08543
(800) 999-1961   FAX: (609) 275-9485
(609) 275-0133
http://www.microplate.com

A4

Biomol Research Laboratories
5100 Campus Drive
Plymouth Meeting, PA 19462
(800) 942-0430   FAX: (610) 941-9252
(610) 941-0430
http://www.biomol.com

Bionique Testing Labs
Fay Brook Drive
RR 1, Box 196
Saranac Lake, NY 12983
(518) 891-2356   FAX: (518) 891-5753
http://www.bionique.com

Biopac Systems
42 Aero Camino
Santa Barbara, CA 93117
(805) 685-0066   FAX: (805) 685-0067
http://www.biopac.com

Bioproducts for Science
See Harlan Bioproducts for Science

Bioptechs
3560 Beck Road
Butler, PA 16002
(877) 548-3235   FAX: (724) 282-0745
(724) 282-7145
http://www.bioptechs.com

BIOQUANT-R&M Biometrics
5611 Ohio Avenue
Nashville, TN 37209
(800) 221-0549   (615) 350-7866
FAX: (615) 350-7282
http://www.bioquant.com

Bio-Rad Laboratories
2000 Alfred Nobel Drive
Hercules, CA 94547
(800) 424-6723   FAX: (800) 879-2289
(510) 741-1000   FAX: (510) 741-5800
http://www.bio-rad.com

Bio-Rad Laboratories
Maylands Avenue
Hemel Hempstead, Herts HP2 7TD, UK
http://www.bio-rad.com

Bioreclamation
492 Richmond Road
East Meadow, NY 11554
(516) 483-1196   FAX: (516) 483-4683
http://www.bioreclamation.com

BioRobotics
3-4 Bennell Court
Comberton, Cambridge CB3 7DS, UK
(44) 1223-264345
FAX: (44) 1223-263933
http://www.biorobotics.co.uk

BIOS Laboratories
See Genaissance Pharmaceuticals

Biosearch Technologies
81 Digital Drive
Novato, CA 94949
(800) GENOME1   FAX: (415) 883-8488
(415) 883-8400
http://www.biosearchtech.com

BioSepra
111 Locke Drive
Marlborough, MA 01752
(800) 752-5277   FAX: (508) 357-7595
(508) 357-7500
http://www.biosepra.com

Bio-Serv
18th Street, Suite 1
Frenchtown, NJ 08825
(908) 996-2155   FAX: (908) 996-4123
http://www.bio-serv.com

BioSignal
1744 William Street, Suite 600
Montreal, Quebec
H3J 1R4 Canada
(800) 293-4501   FAX: (514) 937-0777
(514) 937-1010
http://www.biosignal.com

Biosoft
P.O. Box 10938
Ferguson, MO 63135
(314) 524-8029   FAX: (314) 524-8129
http://www.biosoft.com

Biosource International
820 Flynn Road
Camarillo, CA 93012
(800) 242-0607   FAX: (805) 987-3385
(805) 987-0086
http://www.biosource.com

BioSpec Products
P.O. Box 788
Bartlesville, OK 74005
(800) 617-3363   FAX: (918) 336-3363
(918) 336-3363
http://www.biospec.com

Biosure
See Riese Enterprises

Biosym Technologies
See Molecular Simulations

Biosys
21 quai du Clos des Roses
602000 Compiegne, France
(33) 03 4486 2275
FAX: (33) 03 4484 2297

Bio-Tech Research Laboratories
NIAID Repository
Rockville, MD 20850
http://www.niaid.nih.gov/ncn/repos.htm

Biotech Instruments
Biotech House
75A High Street
Kimpton, Hertfordshire SG4 8PU, UK
(44) 1438 832555
FAX: (44) 1438 833040
http://www.biotinst.demon.co.uk

Biotech International
11 Durbell Street
Acacia Ridge, Queensland 4110
Australia
61-7-3370-6396
FAX: 61-7-3370-6370
http://www.avianbiotech.com

Biotech Source
Inland Farm Drive
South Windham, ME 04062
(207) 892-3266   FAX: (207) 892-6774

Bio-Tek Instruments
Highland Industrial Park
P.O. Box 998
Winooski, VT 05404
(800) 451-5172   FAX: (802) 655-7941
(802) 655-4040
http://www.biotek.com

Biotecx Laboratories
6023 South Loop East
Houston, TX 77033
(800) 535-6286   FAX: (713) 643-3143
(713) 643-0606
http://www.biotecx.com

BioTherm
3260 Wilson Boulevard
Arlington, VA 22201
(703) 522-1705   FAX: (703) 522-2606

Bioventures
P.O. Box 2561
848 Scott Street
Murfreesboro, TN 37133
(800) 235-8938   FAX: (615) 896-4837
http://www.bioventures.com

BioWhittaker
8830 Biggs Ford Road
P.O. Box 127
Walkersville, MD 21793
(800) 638-8174   FAX: (301) 845-8338
(301) 898-7025
http://www.biowhittaker.com

Biozyme Laboratories
9939 Hibert Street, Suite 101
San Diego, CA 92131
(800) 423-8199   FAX: (858) 549-0138
(858) 549-4484
http://www.biozyme.com

Bird Products
1100 Bird Center Drive
Palm Springs, CA 92262
(800) 328-4139   FAX: (760) 778-7274
(760) 778-7200
http://www.birdprod.com/bird

B & K Universal
2403 Yale Way
Fremont, CA 94538
(800) USA-MICE   FAX: (510) 490-3036

A4

BLS Ltd.
Zselyi Aladar u. 31
1165 Budapest, Hungary
(36) 1-407-2602   FAX: (36) 1-407-2896
http://www.bls-ltd.com

Blue Sky Research
3047 Orchard Parkway
San Jose, CA 95134
(408) 474-0988   FAX: (408) 474-0989
http://www.blueskyresearch.com

Blumenthal Industries
7 West 36th Street, 13th floor
New York, NY 10018
(212) 719-1251   FAX: (212) 594-8828

BOC Edwards
One Edwards Park
301 Ballardvale Street
Wilmington, MA 01887
(800) 848-9800   FAX: (978) 658-7969
(978) 658-5410
http://www.bocedwards.com

Boehringer Ingelheim
900 Ridgebury Road
P.O. Box 368
Ridgefield, CT 06877
(800) 243-0127   FAX: (203) 798-6234
(203) 798-9988
http://www.boehringer-ingelheim.com

Boehringer Mannheim
Biochemicals Division
See Roche Diagnostics

Boekel Scientific
855 Pennsylvania Boulevard
Feasterville, PA 19053
(800) 336-6929   FAX: (215) 396-8264
(215) 396-8200
http://www.boekelsci.com

Bohdan Automation
1500 McCormack Boulevard
Mundelein, IL 60060
(708) 680-3939   FAX: (708) 680-1199

BPAmoco
4500 McGinnis Ferry Road
Alpharetta, GA 30005
(800) 328-4537   FAX: (770) 772-8213
(770) 772-8200
http://www.bpamoco.com

Brain Research Laboratories
Waban P.O. Box 88
Newton, MA 02468
(888) BRL-5544   FAX: (617) 965-6220
(617) 965-5544
http://www.brainresearchlab.com

Braintree Scientific
P.O. Box 850929
Braintree, MA 02185
(781) 843-1644   FAX: (781) 982-3160
http://www.braintreesci.com

Brandel
8561 Atlas Drive
Gaithersburg, MD 20877
(800) 948-6506   FAX: (301) 869-5570
(301) 948-6506
http://www.brandel.com

Branson Ultrasonics
41 Eagle Road
Danbury, CT 06813
(203) 796-0400   FAX: (203) 796-9838
http://www.plasticsnet.com/branson

B. Braun Biotech
999 Postal Road
Allentown, PA 18103
(800) 258-9000   FAX: (610) 266-9319
(610) 266-6262
http://www.bbraunbiotech.com

B. Braun Biotech International
Schwarzenberg Weg 73-79
P.O. Box 1120
D-34209 Melsungen, Germany
(49) 5661-71-3400
FAX: (49) 5661-71-3702
http://www.bbraunbiotech.com

B. Braun-McGaw
2525 McGaw Avenue
Irvine, CA 92614
(800) BBRAUN-2 (800) 624-2963
http://www.bbraunusa.com

B. Braun Medical
Thorncliffe Park
Sheffield S35 2PW, UK
(44) 114-225-9000
FAX: (44) 114-225-9111
http://www.bbmuk.demon.co.uk

Brenntag
P.O. Box 13788
Reading, PA 19612-3788
(610) 926-4151   FAX: (610) 926-4160
http://www.brenntagnortheast.com

Bresatec
See GeneWorks

Bright/Hacker Instruments
17 Sherwood Lane
Fairfield, NJ 07004
(973) 226-8450   FAX: (973) 808-8281
http://www.hackerinstruments.com

Brinkmann Instruments
Subsidiary of Sybron
1 Cantiague Road
P.O. Box 1019
Westbury, NY 11590
(800) 645-3050   FAX: (516) 334-7521
(516) 334-7500
http://www.brinkmann.com

Bristol-Meyers Squibb
P.O. Box 4500
Princeton, NJ 08543
(800) 631-5244   FAX: (800) 523-2965
http://www.bms.com

Broadley James
19 Thomas
Irvine, CA 92618
(800) 288-2833   FAX: (949) 829-5560
(949) 829-5555
http://www.broadleyjames.com

Brookhaven Instruments
750 Blue Point Road
Holtsville, NY 11742
(631) 758-3200   FAX: (631) 758-3255
http://www.bic.com

Brownlee Labs
See Applied Biosystems
Distributed by Pacer Scientific

Bruel & Kjaer
Division of Spectris Technologies
2815 Colonnades Court
Norcross, GA 30071
(800) 332-2040   FAX: (770) 847-8440
(770) 209-6907
http://www.bkhome.com

Bruker Analytical X-Ray Systems
5465 East Cheryl Parkway
Madison, WI 53711
(800) 234-XRAY   FAX: (608) 276-3006
(608) 276-3000
http://www.bruker-axs.com

Bruker Instruments
19 Fortune Drive
Billerica, MA 01821
(978) 667-9580   FAX: (978) 667-0985
http://www.bruker.com

BTX
Division of Genetronics
11199 Sorrento Valley Road
San Diego, CA 92121
(800) 289-2465   FAX: (858) 597-9594
(858) 597-6006
http://www.genetronics.com/btx

Buchler Instruments
See Baxter Scientific Products

Buckshire
2025 Ridge Road
Perkasie, PA 18944
(215) 257-0116

Burdick and Jackson
Division of Baxter Scientific Products
1953 S. Harvey Street
Muskegon, MI 49442
(800) 368-0050   FAX: (231) 728-8226
(231) 726-3171
http://www.bandj.com/mainframe.htm

Burleigh Instruments
P.O. Box E
Fishers, NY 14453
(716) 924-9355   FAX: (716) 924-9072
http://www.burleigh.com

A4

Burns Veterinary Supply
1900 Diplomat Drive
Farmer's Branch, TX 75234
(800) 92-BURNS   FAX: (972) 243-6841
http://www.burnsvet.com

Burroughs Wellcome
See Glaxo Wellcome

The Butler Company
5600 Blazer Parkway
Dublin, OH 43017
(800) 551-3861   FAX: (614) 761-9096
(614) 761-9095
http://www.wabutler.com

Butterworth Laboratories
54-56 Waldegrave Road
Teddington, Middlesex
TW11 8LG, UK
(44)(0)20-8977-0750
FAX: (44)(0)28-8943-2624
http://www.butterworth-labs.co.uk

Buxco Electronics
95 West Wood Road #2
Sharon, CT 06069
(860) 364-5558   FAX: (860) 364-5116
http://www.buxco.com

C/D/N Isotopes
88 Leacock Street
Pointe-Claire, Quebec
H9R 1H1 Canada
(800) 697-6254   FAX: (514) 697-6148

C.M.A./Microdialysis AB
73 Princeton Street
North Chelmsford, MA 01863
(800) 440-4980   FAX: (978) 251-1950
(978) 251-1940
http://www.microdialysis.com

Calbiochem-Novabiochem
P.O. Box 12087-2087
La Jolla, CA 92039
(800) 854-3417   FAX: (800) 776-0999
(858) 450-9600
http://www.calbiochem.com

California Fine Wire
338 South Fourth Street
Grover Beach, CA 93433
(805) 489-5144   FAX: (805) 489-5352
http://www.calfinewire.com

Calorimetry Sciences
155 West 2050 North
Spanish Fork, UT 84660
(801) 794-2600   FAX: (801) 794-2700
http://www.calscorp.com

Caltag Laboratories
1849 Bayshore Highway, Suite 200
Burlingame, CA 94010
(800) 874-4007   FAX: (650) 652-9030
(650) 652-0468
http://www.caltag.com

Cambrex Corporation
1 Meadowlands Plaza
East Rutherford, NJ 07073
(201) 804-3000   FAX: (201) 804-9852
http://www.cambrex.com

Cambridge Electronic Design
Science Park, Milton Road
Cambridge CB4 0FE, UK
44 (0) 1223-420-186
FAX: 44 (0) 1223-420-488
http://www.ced.co.uk

Cambridge Isotope Laboratories
50 Frontage Road
Andover, MA 01810
(800) 322-1174   FAX: (978) 749-2768
(978) 749-8000
http://www.isotope.com

Cambridge Research Biochemicals
See Zeneca/CRB

Cambridge Technology
109 Smith Place
Cambridge, MA 02138
(617) 441-0600   FAX: (617) 497-8800
http://www.camtech.com

Camlab
Nuffield Road
Cambridge CB4 1TH, UK
(44) 122-3424222
FAX: (44) 122-3420856
http://www.camlab.co.uk/home.htm

Campden Instruments
Park Road
Sileby Loughborough
Leicestershire LE12 7TU, UK
(44) 1509-814790
FAX: (44) 1509-816097
http://www.campden-inst.com/home.htm

Cappel Laboratories
See Organon Teknika Cappel

Carl Roth GmgH & Company
Schoemperlenstrasse 1-5
76185 Karlsrube
Germany
(49) 72-156-06164
FAX: (49) 72-156-06264
http://www.carl-roth.de

Carl Zeiss
One Zeiss Drive
Thornwood, NY 10594
(800) 233-2343   FAX: (914) 681-7446
(914) 747-1800
http://www.zeiss.com

Carlo Erba Reagenti
Via Winckelmann 1
20148 Milano
Lombardia, Italy
(39) 0-29-5231
FAX: (39) 0-29-5235-904
http://www.carloerbareagenti.com

Carolina Biological Supply
2700 York Road
Burlington, NC 27215
(800) 334-5551   FAX: (336) 584-76869
(336) 584-0381
http://www.carolina.com

Carolina Fluid Components
9309 Stockport Place
Charlotte, NC 28273
(704) 588-6101   FAX: (704) 588-6115
http://www.cfcsite.com

Cartesian Technologies
17851 Skypark Circle, Suite C
Irvine, CA 92614
(800) 935-8007
http://cartesiantech.com

Cayman Chemical
1180 East Ellsworth Road
Ann Arbor, MI 48108
(800) 364-9897   FAX: (734) 971-3640
(734) 971-3335
http://www.caymanchem.com

CB Sciences
One Washington Street, Suite 404
Dover, NH 03820
(800) 234-1757   FAX: (603) 742-2455
http://www.cbsci.com

CBS Scientific
P.O. Box 856
Del Mar, CA 92014
(800) 243-4959   FAX: (858) 755-0733
(858) 755-4959
http://www.cbssci.com

CCR (Coriell Cell Repository)
See Coriell Institute for
Medical Research

CE Instruments
Grand Avenue Parkway
Austin, TX 78728
(800) 876-6711   FAX: (512) 251-1597
http://www.ceinstruments.com

Cedarlane Laboratories
5516 8th Line, R.R. #2
Hornby, Ontario
L0P 1E0 Canada
(905) 878-8891   FAX: (905) 878-7800
http://www.cedarlanelabs.com

CEL Associates
P.O. Box 721854
Houston, TX 77272
(800) 537-9339   FAX: (281) 933-0922
(281) 933-9339
http://www.cel-1.com

Cel-Line Associates
See Erie Scientific

A4

Celite World Minerals
130 Castilian Drive
Santa Barbara, CA 93117
(805) 562-0200   FAX: (805) 562-0299
http://www.worldminerals.com/celite

Cell Genesys
342 Lakeside Drive
Foster City, CA 94404
(650) 425-4400   FAX: (650) 425-4457
http://www.cellgenesys.com

Cell Signaling Technology
166B Cummings Center
Beverly, MA 01915
(877) 616-CELL   FAX: (978) 867-2388
(978) 867-2488
http://www.cellsignal.com

Cell Systems
12815 NE 124th Street, Suite A
Kirkland, WA 98034
(800) 697-1211   FAX: (425) 820-6762
(425) 823-1010

Cellmark Diagnostics
20271 Goldenrod Lane
Germantown, MD 20876
(800) 872-5227   FAX: (301) 428-4877
(301) 428-4980
http://www.cellmark-labs.com

Cellomics
635 William Pitt Way
Pittsburgh, PA 15238
(888) 826-3857   FAX: (412) 826-3850
(412) 826-3600
http://www.cellomics.com

Celltech
216 Bath Road
Slough, Berkshire SL1 4EN, UK
(44) 1753 534655
FAX: (44) 1753 536632
http://www.celltech.co.uk

Cellular Products
872 Main Street
Buffalo, NY 14202
(800) CPI-KITS   FAX: (716) 882-0959
(716) 882-0920
http://www.zeptometrix.com

CEM
P.O. Box 200
Matthews, NC 28106
(800) 726-3331

Centers for Disease Control
1600 Clifton Road NE
Atlanta, GA 30333
(800) 311-3435   FAX: (888) 232-3228
(404) 639-3311
http://www.cdc.gov

CERJ
Centre d'Elevage Roger Janvier
53940 Le Genest Saint Isle
France

Cetus
See Chiron

Chance Propper
Warly, West Midlands B66 1NZ, UK
(44)(0)121-553-5551
FAX: (44)(0)121-525-0139

Charles River Laboratories
251 Ballardvale Street
Wilmington, MA 01887
(800) 522-7287   FAX: (978) 658-7132
(978) 658-6000
http://www.criver.com

Charm Sciences
36 Franklin Street
Malden, MA 02148
(800) 343-2170   FAX: (781) 322-3141
(781) 322-1523
http://www.charm.com

Chase-Walton Elastomers
29 Apsley Street
Hudson, MA 01749
(800) 448-6289   FAX: (978) 562-5178
(978) 568-0202
http://www.chase-walton.com

ChemGenes
Ashland Technology Center
200 Homer Avenue
Ashland, MA 01721
(800) 762-9323   FAX: (508) 881-3443
(508) 881-5200
http://www.chemgenes.com

Chemglass
3861 North Mill Road
Vineland, NJ 08360
(800) 843-1794   FAX: (856) 696-9102
(800) 696-0014
http://www.chemglass.com

Chemicon International
28835 Single Oak Drive
Temecula, CA 92590
(800) 437-7500   FAX: (909) 676-9209
(909) 676-8080
http://www.chemicon.com

Chem-Impex International
935 Dillon Drive
Wood Dale, IL 60191
(800) 869-9290   FAX: (630) 766-2218
(630) 766-2112
http://www.chemimpex.com

Chem Service
P.O. Box 599
West Chester, PA 19381-0599
(610) 692-3026   FAX: (610) 692-8729
http://www.chemservice.com

Chemsyn Laboratories
13605 West 96th Terrace
Lenexa, KS 66215
(913) 541-0525   FAX: (913) 888-3582
http://www.tech.epcorp.com/ChemSyn/
chemsyn.htm

Chemunex USA
1 Deer Park Drive, Suite H-2
Monmouth Junction, NJ 08852
(800) 411-6734
http://www.chemunex.com

Cherwell Scientific Publishing
The Magdalen Centre
Oxford Science Park
Oxford OX44GA, UK
(44)(1) 865-784-800
FAX: (44)(1) 865-784-801
http://www.cherwell.com

ChiRex Cauldron
383 Phoenixville Pike
Malvern, PA 19355
(610) 727-2215   FAX: (610) 727-5762
http://www.chirex.com

Chiron Diagnostics
See Bayer Diagnostics

Chiron Mimotopes Peptide Systems
See Multiple Peptide Systems

Chiron
4560 Horton Street
Emeryville, CA 94608
(800) 244-7668   FAX: (510) 655-9910
(510) 655-8730
http://www.chiron.com

Chrom Tech
P.O. Box 24248
Apple Valley, MN 55124
(800) 822-5242   FAX: (952) 431-6345
http://www.chromtech.com

Chroma Technology
72 Cotton Mill Hill, Unit A-9
Brattleboro, VT 05301
(800) 824-7662   FAX: (802) 257-9400
(802) 257-1800
http://www.chroma.com

Chromatographie
ZAC de Moulin No. 2
91160 Saulx les Chartreux
France
(33) 01-64-54-8969
FAX: (33) 01-69-0988091
http://www.chromatographie.com

Chromogenix
Taljegardsgatan 3
431-53 Mlndal, Sweden
(46) 31-706-20-70
FAX: (46) 31-706-20-80
http://www.chromogenix.com

Chrompack USA
c/o Varian USA
2700 Mitchell Drive
Walnut Creek, CA 94598
(800) 526-3687   FAX: (925) 945-2102
(925) 939-2400
http://www.chrompack.com

A4

Chugai Biopharmaceuticals
6275 Nancy Ridge Drive
San Diego, CA 92121
(858) 535-5900   FAX: (858) 546-5973
http://www.chugaibio.com

Ciba-Corning Diagnostics
See Bayer Diagnostics

Ciba-Geigy
See Ciba Specialty Chemicals or
Novartis Biotechnology

Ciba Specialty Chemicals
540 White Plains Road
Tarrytown, NY 10591
(800) 431-1900   FAX: (914) 785-2183
(914) 785-2000
http://www.cibasc.com

Ciba Vision
Division of Novartis AG
11460 Johns Creek Parkway
Duluth, GA 30097
(770) 476-3937
http://www.cvworld.com

Cidex
Advanced Sterilization Products
33 Technology Drive
Irvine, CA 92618
(800) 595-0200 (949) 581-5799
http://www.cidex.com/ASPnew.htm

Cinna Scientific
Subsidiary of Molecular Research Center
5645 Montgomery Road
Cincinnati, OH 45212
(800) 462-9868   FAX: (513) 841-0080
(513) 841-0900
http://www.mrcgene.com

Cistron Biotechnology
10 Bloomfield Avenue
Pine Brook, NJ 07058
(800) 642-0167   FAX: (973) 575-4854
(973) 575-1700
http://www.cistronbio.com

Clark Electromedical Instruments
See Harvard Apparatus

Clay Adam
See Becton Dickinson Primary Care
Diagnostics

**CLB (Central Laboratory of
the Netherlands)**
Blood Transfusion Service
P.O. Box 9190
1006 AD Amsterdam, The Netherlands
(31) 20-512-9222
FAX: (31) 20-512-3332

Cleveland Scientific
P.O. Box 300
Bath, OH 44210
(800) 952-7315   FAX: (330) 666-2240
http://www.clevelandscientific.com

Clonetics
Division of BioWhittaker
http://www.clonetics.com
Also see BioWhittaker

Clontech Laboratories
1020 East Meadow Circle
Palo Alto, CA 94303
(800) 662-2566   FAX: (800) 424-1350
(650) 424-8222   FAX: (650) 424-1088
http://www.clontech.com

Closure Medical Corporation
5250 Greens Dairy Road
Raleigh, NC 27616
(919) 876-7800   FAX: (919) 790-1041
http://www.closuremed.com

CMA Microdialysis AB
73 Princeton Street
North Chelmsford, MA 01863
(800) 440-4980   FAX: (978) 251-1950
(978) 251 1940
http://www.microdialysis.com

Cocalico Biologicals
449 Stevens Road
P.O. Box 265
Reamstown, PA 17567
(717) 336-1990   FAX: (717) 336-1993

Coherent Laser
5100 Patrick Henry Drive
Santa Clara, CA 95056
(800) 227-1955   FAX: (408) 764-4800
(408) 764-4000
http://www.cohr.com

Cohu
P.O. Box 85623
San Diego, CA 92186
(858) 277-6700   FAX: (858) 277-0221
http://www.COHU.com/cctv

Cole-Parmer Instrument
625 East Bunker Court
Vernon Hills, IL 60061
(800) 323-4340   FAX: (847) 247-2929
(847) 549-7600
http://www.coleparmer.com

**Collaborative Biomedical Products
and Collaborative Research**
See Becton Dickinson Labware

Collagen Aesthetics
1850 Embarcadero Road
Palo Alto, CA 94303
(650) 856-0200   FAX: (650) 856-0533
http://www.collagen.com

Collagen Corporation
See Collagen Aesthetics

College of American Pathologists
325 Waukegan Road
Northfield, IL 60093
(800) 323-4040   FAX: (847) 832-8000
(847) 446-8800
http://www.cap.org/index.cfm

Colonial Medical Supply
504 Wells Road
Franconia, NH 03580
(603) 823-9911   FAX: (603) 823-8799
http://www.colmedsupply.com

Colorado Serum
4950 York Street
Denver, CO 80216
(800) 525-2065   FAX: (303) 295-1923
http://www.colorado-serum.com

Columbia Diagnostics
8001 Research Way
Springfield, VA 22153
(800) 336-3081   FAX: (703) 569-2353
(703) 569-7511
http://www.columbiadiagnostics.com

Columbus Instruments
950 North Hague Avenue
Columbus, OH 43204
(800) 669-5011   FAX: (614) 276-0529
(614) 276-0861
http://www.columbusinstruments.com

Compu Cyte Corp.
12 Emily Street
Cambridge, MA 02139
(800) 840-1303   FAX: (617) 577-4501
(617) 492-1300
http://www.compucyte.com

Compugen
25 Leek Crescent
Richmond Hill, Ontario
L4B 4B3 Canada
800-387-5045   FAX: (905) 707-2020
(905) 707-2000
http://www.compugen.com/locations.htm

Computer Associates International
One Computer Associates Plaza
Islandia, NY 11749
(631) 342-6000   FAX: (631) 342-6800
http://www.cai.com

Connaught Laboratories
See Aventis Pasteur

Connectix
2955 Campus Drive, Suite 100
San Mateo, CA 94403
(800) 950-5880   FAX: (650) 571-0850
(650) 571-5100
http://www.connectix.com

Contech
99 Hartford Avenue
Providence, RI 02909
(401) 351-4890   FAX: (401) 421-5072
http://www.iol.ie/~burke/contech.html

Continental Laboratory Products
5648 Copley Drive
San Diego, CA 92111
(800) 456-7741   FAX: (858) 279-5465
(858) 279-5000
http://www.conlab.com

ConvaTec
Professional Services
P.O. Box 5254
Princeton, NJ 08543
(800) 422-8811
http://www.convatec.com

Cooper Instruments & Systems
P.O. Box 3048
Warrenton, VA 20188
(800) 344-3921   FAX: (540) 347-4755
(540) 349-4746
http://www.cooperinstruments.com

Cooperative Human Tissue Network
(866) 462-2486
http://www.chtn.ims.nci.nih.gov

Cora Styles Needles 'N Blocks
56 Milton Street
Arlington, MA 02474
(781) 648-6289   FAX: (781) 641-7917

Coriell Cell Repository (CCR)
See Coriell Institute for
Medical Research

Coriell Institute for Medical Research
Human Genetic Mutant Repository
401 Haddon Avenue
Camden, NJ 08103
(856) 966-7377   FAX: (856) 964-0254
http://arginine.umdnj.edu

Corion
8 East Forge Parkway
Franklin, MA 02038
(508) 528-4411   FAX: (508) 520-7583
(800) 598-6783
http://www.corion.com

Corning and Corning Science Products
P.O. Box 5000
Corning, NY 14831
(800) 222-7740   FAX: (607) 974-0345
(607) 974-9000
http://www.corning.com

Costar
See Corning

Coulbourn Instruments
7462 Penn Drive
Allentown, PA 18106
(800) 424-3771   FAX: (610) 391-1333
(610) 395-3771
http://www.coulbourninst.com

Coulter Cytometry
See Beckman Coulter

Covance Research Products
465 Swampbridge Road
Denver, PA 17517
(800) 345-4114   FAX: (717) 336-5344
(717) 336-4921
http://www.covance.com

Coy Laboratory Products
14500 Coy Drive
Grass Lake, MI 49240
(734) 475-2200   FAX: (734) 475-1846
http://www.coylab.com

CPG
3 Borinski Road
Lincoln Park, NJ 07035
(800) 362-2740   FAX: (973) 305-0884
(973) 305-8181
http://www.cpg-biotech.com

CPL Scientific
43 Kingfisher Court
Hambridge Road
Newbury RG14 5SJ, UK
(44) 1635-574902
FAX: (44) 1635-529322
http://www.cplscientific.co.uk

CraMar Technologies
8670 Wolff Court, #160
Westminster, CO 80030
(800) 4-TOMTEC
http://www.cramar.com

Crescent Chemical
1324 Motor Parkway
Hauppauge, NY 11788
(800) 877-3225   FAX: (631) 348-0913
(631) 348-0333
http://www.creschem.com

Crist Instrument
P.O. Box 128
10200 Moxley Road
Damascus, MD 20872
(301) 253-2184   FAX: (301) 253-0069
http://www.cristinstrument.com

Cruachem
See Annovis
http://www.cruachem.com

CS Bio
1300 Industrial Road
San Carlos, CA 94070
(800) 627-2461   FAX: (415) 802-0944
(415) 802-0880
http://www.csbio.com

CS-Chromatographie Service
Am Parir 27
D-52379 Langerwehe, Germany
(49) 2423-40493-0
FAX: (49) 2423-40493-49
http://www.cs-chromatographie.de

Cuno
400 Research Parkway
Meriden, CT 06450
(800) 231-2259   FAX: (203) 238-8716
(203) 237-5541
http://www.cuno.com

Curtin Matheson Scientific
9999 Veterans Memorial Drive
Houston, TX 77038
(800) 392-3353   FAX: (713) 878-3598
(713) 878-3500

CWE
124 Sibley Avenue
Ardmore, PA 19003
(610) 642-7719   FAX: (610) 642-1532
http://www.cwe-inc.com

Cybex Computer Products
4991 Corporate Drive
Huntsville, AL 35805
(800) 932-9239   FAX: (800) 462-9239
http://www.cybex.com

Cygnus Technology
P.O. Box 219
Delaware Water Gap, PA 18327
(570) 424-5701   FAX: (570) 424-5630
http://www.cygnustech.com

Cymbus Biotechnology
Eagle Class, Chandler's Ford
Hampshire SO53 4NF, UK
(44) 1-703-267-676
FAX: (44) 1-703-267-677
http://www.biotech@cymbus.com

Cytogen
600 College Road East
Princeton, NJ 08540
(609) 987-8200   FAX: (609) 987-6450
http://www.cytogen.com

Cytogen Research and Development
89 Bellevue Hill Road
Boston, MA 02132
(617) 325-7774   FAX: (617) 327-2405

CytRx
154 Technology Parkway
Norcross, GA 30092
(800) 345-2987   FAX: (770) 368-0622
(770) 368-9500
http://www.cytrx.com

Dade Behring
Corporate Headquarters
1717 Deerfield Road
Deerfield, IL 60015
(847) 267-5300   FAX: (847) 267-1066
http://www.dadebehring.com

Dagan
2855 Park Avenue
Minneapolis, MN 55407
(612) 827-5959   FAX: (612) 827-6535
http://www.dagan.com

Dako
6392 Via Real
Carpinteria, CA 93013
(800) 235-5763   FAX: (805) 566-6688
(805) 566-6655
http://www.dakousa.com

A4

Dako A/S
42 Produktionsvej
P.O. Box 1359
DK-2600 Glostrup, Denmark
(45) 4492-0044   FAX: (45) 4284-1822

Dakopatts
See Dako A/S

Dalton Chemical Laboratoris
349 Wildcat Road
Toronto, Ontario
M3J 253 Canada
(416) 661-2102   FAX: (416) 661-2108
(800) 567-5060 (in Canada only)
http://www.dalton.com

Damon, IEC
See Thermoquest

Dan Kar Scientific
150 West Street
Wilmington, MA 01887
(800) 942-5542   FAX: (978) 658-0380
(978) 988-9696
http://www.dan-kar.com

DataCell
Falcon Business Park
40 Ivanhoe Road
Finchampstead, Berkshire
RG40 4QQ, UK
(44) 1189 324324
FAX: (44) 1189 324325
http://www.datacell.co.uk
In the US:
(408) 446-3575   FAX: (408) 446-3589
http://www.datacell.com

DataWave Technologies
380 Main Street, Suite 209
Longmont, CO 80501
(800) 736-9283   FAX: (303) 776-8531
(303) 776-8214

Datex-Ohmeda
3030 Ohmeda Drive
Madison, WI 53718
(800) 345-2700   FAX: (608) 222-9147
(608) 221-1551
http://www.us.datex-ohmeda.com

DATU
82 State Street
Geneva, NY 14456
(315) 787-2240   FAX: (315) 787-2397
http://www.nysaes.cornell.edu/datu

David Kopf Instruments
7324 Elmo Street
P.O. Box 636
Tujunga, CA 91043
(818) 352-3274   FAX: (818) 352-3139

Decagon Devices
P.O. Box 835
950 NE Nelson Court
Pullman, WA 99163
(800) 755-2751   FAX: (509) 332-5158
(509) 332-2756
http://www.decagon.com

Decon Labs
890 Country Line Road
Bryn Mawr, PA 19010
(800) 332-6647   FAX: (610) 964-0650
(610) 520-0610
http://www.deconlabs.com

Decon Laboratories
Conway Street
Hove, Sussex BN3 3LY, UK
(44) 1273 739241
FAX: (44) 1273 722088

Degussa
Precious Metals Division
3900 South Clinton Avenue
South Plainfield, NJ 07080
(800) DEGUSSA   FAX: (908) 756-7176
(908) 561-1100
http://www.degussa-huls.com

Deneba Software
1150 NW 72nd Avenue
Miami, FL 33126
(305) 596-5644   FAX: (305) 273-9069
http://www.deneba.com

Deseret Medical
524 West 3615 South
Salt Lake City, UT 84115
(801) 270-8440   FAX: (801) 293-9000

Devcon Plexus
30 Endicott Street
Danvers, MA 01923
(800) 626-7226   FAX: (978) 774-0516
(978) 777-1100
http://www.devcon.com

Developmental Studies Hybridoma Bank
University of Iowa
436 Biology Building
Iowa City, IA 52242
(319) 335-3826   FAX: (319) 335-2077
http://www.uiowa.edu/~dshbwww

DeVilbiss
Division of Sunrise Medical Respiratory
100 DeVilbiss Drive
P.O. Box 635
Somerset, PA 15501
(800) 338-1988   FAX: (814) 443-7572
(814) 443-4881
http://www.sunrisemedical.com

Dharmacon Research
1376 Miners Drive #101
Lafayette, CO 80026
(303) 604-9499   FAX: (303) 604-9680
http://www.dharmacom.com

DiaCheM
Triangle Biomedical
Gardiners Place
West Gillibrands, Lancashire
WN8 9SP, UK
(44) 1695-555581
FAX: (44) 1695-555518
http://www.diachem.co.uk

Diagen
Max-Volmer Strasse 4
D-40724 Hilden, Germany
(49) 2103-892-230
FAX: (49) 2103-892-222

Diagnostic Concepts
6104 Madison Court
Morton Grove, IL 60053
(847) 604-0957

Diagnostic Developments
See DiaCheM

Diagnostic Instruments
6540 Burroughs
Sterling Heights, MI 48314
(810) 731-6000   FAX: (810) 731-6469
http://www.diaginc.com

Diamedix
2140 North Miami Avenue
Miami, FL 33127
(800) 327-4565   FAX: (305) 324-2395
(305) 324-2300

DiaSorin
1990 Industrial Boulevard
Stillwater, MN 55082
(800) 328-1482   FAX: (651) 779-7847
(651) 439-9719
http://www.diasorin.com

Diatome US
321 Morris Road
Fort Washington, PA 19034
(800) 523-5874   FAX: (215) 646-8931
(215) 646-1478
http://www.emsdiasum.com

Difco Laboratories
See Becton Dickinson

Digene
1201 Clopper Road
Gaithersburg, MD 20878
(301) 944-7000 (800) 344-3631
FAX: (301) 944-7121
http://www.digene.com

Digi-Key
701 Brooks Avenue South
Thief River Falls, MN 56701
(800) 344-4539   FAX: (218) 681-3380
(218) 681-6674
http://www.digi-key.com

A4

Digitimer
37 Hydeway
Welwyn Garden City, Hertfordshire
AL7 3BE, UK
(44) 1707-328347
FAX: (44) 1707-373153
http://www.digitimer.com

Dimco-Gray
8200 South Suburban Road
Dayton, OH 45458
(800) 876-8353   FAX: (937) 433-0520
(937) 433-7600
http://www.dimco-gray.com

Dionex
1228 Titan Way
P.O. Box 3603
Sunnyvale, CA 94088
(408) 737-0700   FAX: (408) 730-9403
http://dionex2.promptu.com

Display Systems Biotech
1260 Liberty Way, Suite B
Vista, CA 92083
(800) 697-1111   FAX: (760) 599-9930
(760) 599-0598
http://www.displaysystems.com

Diversified Biotech
1208 VFW Parkway
Boston, MA 02132
(617) 965-8557   FAX: (617) 323-5641
(800) 796-9199
http://www.divbio.com

DNA ProScan
P.O. Box 121585
Nashville, TN 37212
(800) 841-4362   FAX: (615) 292-1436
(615) 298-3524
http://www.dnapro.com

DNAStar
1228 South Park Street
Madison, WI 53715
(608) 258-7420   FAX: (608) 258-7439
http://www.dnastar.com

DNAVIEW
Attn: Charles Brenner
http://www.wco.com
~cbrenner/dnaview.htm

Doall NYC
36-06 48th Avenue
Long Island City, NY 11101
(718) 392-4595   FAX: (718) 392-6115
http://www.doall.com

Dojindo Molecular Technologies
211 Perry Street Parkway, Suite 5
Gaitherbusburg, MD 20877
(877) 987-2667
http://www.dojindo.com

Dolla Eastern
See Doall NYC

Dolan Jenner Industries
678 Andover Street
Lawrence, MA 08143
(978) 681-8000 (978) 682-2500
http://www.dolan-jenner.com

Dow Chemical
Customer Service Center
2040 Willard H. Dow Center
Midland, MI 48674
(800) 232-2436   FAX: (517) 832-1190
(409) 238-9321
http://www.dow.com

Dow Corning
Northern Europe
Meriden Business Park
Copse Drive
Allesley, Coventry CV5 9RG, UK
(44) 1676 528 000
FAX: (44) 1676 528 001

Dow Corning
P.O. Box 994
Midland, MI 48686
(517) 496-4000
http://www.dowcorning.com

Dow Corning (Lubricants)
2200 West Salzburg Road
Auburn, MI 48611
(800) 248-2481   FAX: (517) 496-6974
(517) 496-6000

Dremel
4915 21st Street
Racine, WI 53406
(414) 554-1390
http://www.dremel.com

Drummond Scientific
500 Parkway
P.O. Box 700
Broomall, PA 19008
(800) 523-7480   FAX: (610) 353-6204
(610) 353-0200
http://www.drummondsci.com

Duchefa Biochemie BV
P.O. Box 2281
2002 CG Haarlem, The Netherlands
31-0-23-5319093
FAX: 31-0-23-5318027
http://www.duchefa.com

Duke Scientific
2463 Faber Place
Palo Alto, CA 94303
(800) 334-3883   FAX: (650) 424-1158
(650) 424-1177
http://www.dukescientific.com

Duke University Marine Laboratory
135 Duke Marine Lab Road
Beaufort, NC 28516-9721
(252) 504-7503   FAX: (252) 504-7648
http://www.env.duke.edu/marinelab

DuPont Biotechnology Systems
See NEN Life Science Products

DuPont Medical Products
See NEN Life Science Products

DuPont Merck Pharmaceuticals
331 Treble Cove Road
Billerica, MA 01862
(800) 225-1572   FAX: (508) 436-7501
http://www.dupontmerck.com

DuPont NEN Products
See NEN Life Science Products

Dynal
5 Delaware Drive
Lake Success, NY 11042
(800) 638-9416   FAX: (516) 326-3298
(516) 326-3270
http://www.dynal.net

Dynal AS
Ullernchausen 52,
0379 Oslo, Norway
47-22-06-10-00   FAX: 47-22-50-70-15
http://www.dynal.no

Dynalab
P.O. Box 112
Rochester, NY 14692
(800) 828-6595   FAX: (716) 334-9496
(716) 334-2060
http://www.dynalab.com

Dynarex
1 International Boulevard
Brewster, NY 10509
(888) DYNAREX   FAX: (914) 279-9601
(914) 279-9600
http://www.dynarex.com

Dynatech
See Dynex Technologies

Dynex Technologies
14340 Sullyfield Circle
Chantilly, VA 22021
(800) 336-4543   FAX: (703) 631-7816
(703) 631-7800
http://www.dynextechnologies.com

Dyno Mill
See Willy A. Bachofen

E.S.A.
22 Alpha Road
Chelmsford, MA 01824
(508) 250-7000   FAX: (508) 250-7090

E.W. Wright
760 Durham Road
Guilford, CT 06437
(203) 453-6410   FAX: (203) 458-6901
http://www.ewwright.com

E-Y Laboratories
107 N. Amphlett Boulevard
San Mateo, CA 94401
(800) 821-0044   FAX: (650) 342-2648
(650) 342-3296
http://www.eylabs.com

A4

Eastman Kodak
1001 Lee Road
Rochester, NY 14650
(800) 225-5352   FAX: (800) 879-4979
(716) 722-5780   FAX: (716) 477-8040
http://www.kodak.com

ECACC
See European Collection of Animal
Cell Cultures

EC Apparatus
See Savant/EC Apparatus

Ecogen, SRL
Gensura Laboratories
Ptge. Dos de Maig
9(08041) Barcelona, Spain
(34) 3-450-2601   FAX: (34) 3-456-0607
http://www.ecogen.com

Ecolab
370 North Wabasha Street
St. Paul, MN 55102
(800) 35-CLEAN   FAX: (651) 225-3098
(651) 352-5326
http://www.ecolab.com

ECO PHYSICS
3915 Research Park Drive, Suite A-3
Ann Arbor, MI 48108
(734) 998-1600   FAX: (734) 998-1180
http://www.ecophysics.com

Edge Biosystems
19208 Orbit Drive
Gaithersburg, MD 20879-4149
(800) 326-2685   FAX: (301) 990-0881
(301) 990-2685
http://www.edgebio.com

Edmund Scientific
101 E. Gloucester Pike
Barrington, NJ 08007
(800) 728-6999   FAX: (856) 573-6263
(856) 573-6250
http://www.edsci.com

EG&G
See Perkin-Elmer

Ekagen
969 C Industry Road
San Carlos, CA 94070
(650) 592-4500   FAX: (650) 592-4500

Elcatech
P.O. Box 10935
Winston-Salem, NC 27108
(336) 544-8613   FAX: (336) 777-3623
(910) 777-3624
http://www.elcatech.com

Electron Microscopy Sciences
321 Morris Road
Fort Washington, PA 19034
(800) 523-5874   FAX: (215) 646-8931
(215) 646-1566
http://www.emsdiasum.com

Electron Tubes
100 Forge Way, Unit F
Rockaway, NJ 07866
(800) 521-8382   FAX: (973) 586-9771
(973) 586-9594
http://www.electrontubes.com

Elicay Laboratory Products, (UK) Ltd.
4 Manborough Mews
Crockford Lane
Basingstoke, Hampshire
RG 248NA, England
(256) 811-118   FAX: (256) 811-116
http://www.elkay-uk.co.uk

Eli Lilly
Lilly Corporate Center
Indianapolis, IN 46285
(800) 545-5979   FAX: (317) 276-2095
(317) 276-2000
http://www.lilly.com

ELISA Technologies
See Neogen

Elkins-Sinn
See Wyeth-Ayerst

EMBI
See European Bioinformatics Institute

EM Science
480 Democrat Road
Gibbstown, NJ 08027
(800) 222-0342   FAX: (856) 423-4389
(856) 423-6300
http://www.emscience.com

EM Separations Technology
See R & S Technology

Endogen
30 Commerce Way
Woburn, MA 01801
(800) 487-4885   FAX: (617) 439-0355
(781) 937-0890
http://www.endogen.com

ENGEL-Loter
HSGM Heatcutting Equipment
& Machines
1865 E. Main Street, No. 5
Duncan, SC 29334
(888) 854-HSGM   FAX: (864) 486-8383
(864) 486-8300
http://www.engelgmbh.com

Enzo Diagnostics
60 Executive Boulevard
Farmingdale, NY 11735
(800) 221-7705   FAX: (516) 694-7501
(516) 694-7070
http://www.enzo.com

Enzogenetics
4197 NW Douglas Avenue
Corvallis, OR 97330
(541) 757-0288

The Enzyme Center
See Charm Sciences

Enzyme Systems Products
486 Lindbergh Avenue
Livermore, CA 94550
(888) 449-2664   FAX: (925) 449-1866
(925) 449-2664
http://www.enzymesys.com

Epicentre Technologies
1402 Emil Street
Madison, WI 53713
(800) 284-8474   FAX: (608) 258-3088
(608) 258-3080
http://www.epicentre.com

Erie Scientific
20 Post Road
Portsmouth, NH 03801
(888) ERIE-SCI   FAX: (603) 431-8996
(603) 431-8410
http://www.eriesci.com

ES Industries
701 South Route 73
West Berlin, NJ 08091
(800) 356-6140   FAX: (856) 753-8484
(856) 753-8400
http://www.esind.com

ESA
22 Alpha Road
Chelmsford, MA 01824
(800) 959-5095   FAX: (978) 250-7090
(978) 250-7000
http://www.esainc.com

Ethicon
Route 22, P.O. Box 151
Somerville, NJ 08876
(908) 218-0707
http://www.ethiconinc.com

Ethicon Endo-Surgery
4545 Creek Road
Cincinnati, OH 45242
(800) 766-9534   FAX: (513) 786-7080

Eurogentec
Parc Scientifique du Sart Tilman
4102 Seraing, Belgium
32-4-240-76-76   FAX: 32-4-264-07-88
http://www.eurogentec.com

European Bioinformatics Institute
Wellcome Trust Genomes Campus
Hinxton, Cambridge CB10 1SD, UK
(44) 1223-49444
FAX: (44) 1223-494468

European Collection of Animal
Cell Cultures (ECACC)
Centre for Applied Microbiology
& Research
Salisbury, Wiltshire SP4 0JG, UK
(44) 1980-612 512
FAX: (44) 1980-611 315
http://www.camr.org.uk

A4

Evergreen Scientific
2254 E. 49th Street
P.O. Box 58248
Los Angeles, CA 90058
(800) 421-6261   FAX: (323) 581-2503
(323) 583-1331
http://www.evergreensci.com

Exalpha Biologicals
20 Hampden Street
Boston, MA 02205
(800) 395-1137   FAX: (617) 969-3872
(617) 558-3625
http://www.exalpha.com

Exciton
P.O. Box 31126
Dayton, OH 45437
(937) 252-2989   FAX: (937) 258-3937
http://www.exciton.com

Extrasynthese
ZI Lyon Nord
SA-BP62
69730 Genay, France
(33) 78-98-20-34
FAX: (33) 78-98-19-45

Factor II
1972 Forest Avenue
P.O. Box 1339
Lakeside, AZ 85929
(800) 332-8688   FAX: (520) 537-8066
(520) 537-8387
http://www.factor2.com

Falcon
See Becton Dickinson Labware

Febit AG
Kafertaler Strasse 190
D-68167 Mannheim
Germany
(49) 621-3804-0
FAX: (49) 621-3804-400
http://www.febit.com

Fenwal
See Baxter Healthcare

Filemaker
5201 Patrick Henry Drive
Santa Clara, CA 95054
(408) 987-7000 (800) 325-2747

Fine Science Tools
202-277 Mountain Highway
North Vancouver, British Columbia
V7J 3P2 Canada
(800) 665-5355   FAX: (800) 665 4544
(604) 980-2481   FAX: (604) 987-3299

Fine Science Tools
373-G Vintage Park Drive
Foster City, CA 94404
(800) 521-2109   FAX: (800) 523-2109
(650) 349-1636   FAX: (630) 349-3729

Fine Science Tools
Fahrtgasse 7-13
D-69117 Heidelberg, Germany
(49) 6221 905050
FAX: (49) 6221 600001
http://www.finescience.com

Finn Aqua
AMSCO Finn Aqua Oy
Teollisuustiez, FIN-04300
Tuusula, Finland
358 025851   FAX: 358 0276019

Finnigan
355 River Oaks Parkway
San Jose, CA 95134
(408) 433-4800   FAX: (408) 433-4821
http://www.finnigan.com

Dr. L. Fischer
Lutherstrasse 25A
D-69120 Heidelberg
Germany
(49) 6221-16-0368
http://home.eplus-online.de/
electroporation

Fisher Chemical Company
Fisher Scientific Limited
112 Colonnade Road
Nepean, Ontario K2E 7L6 Canada
(800) 234-7437   FAX: (800) 463-2996
http://www.fisherscientific.com

Fisher Scientific
2000 Park Lane
Pittsburgh, PA 15275
(800) 766-7000   FAX: (800) 926-1166
(412) 562-8300
http://www3.fishersci.com

W.F. Fisher & Son
220 Evans Way, Suite #1
Somerville, NJ 08876
(908) 707-4050   FAX: (908) 707-4099

Fitzco
5600 Pioneer Creek Drive
Maple Plain, MN 55359
(800) 367-8760   FAX: (612) 479-2880
(612) 479-3489
http://www.fitzco.com

5 Prime → 3 Prime
See 2000 Eppendorf-5 Prime
http://www.5prime.com

Flambeau
15981 Valplast Road
Middlefield, Ohio 44062
(800) 232-3474   FAX: (440) 632-1581
(440) 632-1631
http://www.flambeau.com

Fleisch (Rusch)
2450 Meadowbrook Parkway
Duluth, GA 30096
(770) 623-0816   FAX: (770) 623-1829
http://ruschinc.com

Flow Cytometry Standards
P.O. Box 194344
San Juan, PR 00919
(800) 227-8143   FAX: (787) 758-3267
(787) 753-9341
http://www.fcstd.com

Flow Labs
See ICN Biomedicals

Flow-Tech Supply
P.O. Box 1388
Orange, TX 77631
(409) 882-0306   FAX: (409) 882-0254
http://www.flow-tech.com

Fluid Marketing
See Fluid Metering

Fluid Metering
5 Aerial Way, Suite 500
Sayosett, NY 11791
(516) 922-6050   FAX: (516) 624-8261
http://www.fmipump.com

Fluorochrome
1801 Williams, Suite 300
Denver, CO 80264
(303) 394-1000   FAX: (303) 321-1119

Fluka Chemical
See Sigma-Aldrich

FMC BioPolymer
1735 Market Street
Philadelphia, PA 19103
(215) 299-6000   FAX: (215) 299-5809
http://www.fmc.com

FMC BioProducts
191 Thomaston Street
Rockland, ME 04841
(800) 521-0390   FAX: (800) 362-1133
(207) 594-3400   FAX: (207) 594-3426
http://www.bioproducts.com

Forma Scientific
Milcreek Road
P.O. Box 649
Marietta, OH 45750
(800) 848-3080   FAX: (740) 372-6770
(740) 373-4765
http://www.forma.com

Fort Dodge Animal Health
800 5th Street NW
Fort Dodge, IA 50501
(800) 685-5656   FAX: (515) 955-9193
(515) 955-4600
http://www.ahp.com

Fotodyne
950 Walnut Ridge Drive
Hartland, WI 53029
(800) 362-3686   FAX: (800) 362-3642
(262) 369-7000   FAX: (262) 369-7013
http://www.fotodyne.com

A4

Fresenius HemoCare
6675 185th Avenue NE, Suite 100
Redwood, WA 98052
(800) 909-3872
(425) 497-1197
http://www.freseniusht.com

Fresenius Hemotechnology
See Fresenius HemoCare

Fuji Medical Systems
419 West Avenue
P.O. Box 120035
Stamford, CT 06902
(800) 431-1850   FAX: (203) 353-0926
(203) 324-2000
http://www.fujimed.com

Fujisawa USA
Parkway Center North
Deerfield, IL 60015-2548
(847) 317-1088   FAX: (847) 317-7298

Ernest F. Fullam
900 Albany Shaker Road
Latham, NY 12110
(800) 833-4024   FAX: (518) 785-8647
(518) 785-5533
http://www.fullam.com

Gallard-Schlesinger Industries
777 Zechendorf Boulevard
Garden City, NY 11530
(516) 229-4000   FAX: (516) 229-4015
http://www.gallard-schlessinger.com

Gambro
Box 7373
SE 103 91 Stockholm, Sweden
(46) 8 613 65 00
FAX: (46) 8 611 37 31
In the US: **COBE Laboratories**
225 Union Boulevard
Lakewood, CO 80215
(303) 232-6800   FAX: (303) 231-4915
http://www.gambro.com

Garner Glass
177 Indian Hill Boulevard
Claremont, CA 91711
(909) 624-5071   FAX: (909) 625-0173
http://www.garnerglass.com

Garon Plastics
16 Byre Avenue
Somerton Park, South Australia 5044
(08) 8294-5126   FAX: (08) 8376-1487
http://www.apache.airnet.com.au/
~garon

Garren Scientific
9400 Lurline Avenue, Unit E
Chatsworth, CA 91311
(800) 342-3725   FAX: (818) 882-3229
(818) 882-6544
http://www.garren-scientific.com

GATC Biotech AG
Jakob-Stadler-Platz 7
D-78467 Constance, Germany
(49) 07531-8160-0
FAX: (49) 07531-8160-81
http://www.gatc-biotech.com

Gaussian
Carnegie Office Park
Building 6, Suite 230
Carnegie, PA 15106
(412) 279-6700   FAX: (412) 279-2118
http://www.gaussian.com

G.C. Electronics/A.R.C. Electronics
431 Second Street
Henderson, KY 42420
(270) 827-8981   FAX: (270) 827-8256
http://www.arcelectronics.com

GDB (Genome Data Base, Curation)
2024 East Monument Street, Suite 1200
Baltimore, MD 21205
(410) 955-9705   FAX: (410) 614-0434
http://www.gdb.org

GDB (Genome Data Base, Home)
Hospital for Sick Children
555 University Avenue
Toronto, Ontario
M5G 1X8 Canada
(416) 813-8744   FAX: (416) 813-8755
http://www.gdb.org

Gelman Sciences
See Pall-Gelman

Gemini BioProducts
5115-M Douglas Fir Road
Calabasas, CA 90403
(818) 591-3530   FAX: (818) 591-7084

Gen Trak
5100 Campus Drive
Plymouth Meeting, PA 19462
(800) 221-7407   FAX: (215) 941-9498
(215) 825-5115
http://www.informagen.com

Genaissance Pharmaceuticals
5 Science Park
New Haven, CT 06511
(800) 678-9487   FAX: (203) 562-9377
(203) 773-1450
http://www.genaissance.com

GENAXIS Biotechnology
Parc Technologique
10 Avenue Ampere
Montigny le Bretoneux
78180 France
(33) 01-30-14-00-20
FAX: (33) 01-30-14-00-15
http://www.genaxis.com

GenBank
National Center for Biotechnology
Information
National Library of Medicine/NIH
Building 38A, Room 8N805
8600 Rockville Pike
Bethesda, MD 20894
(301) 496-2475   FAX: (301) 480-9241
http://www.ncbi.nlm.nih.gov

Gene Codes
640 Avis Drive
Ann Arbor, MI 48108
(800) 497-4939   FAX: (734) 930-0145
(734) 769-7249
http://www.genecodes.com

Genemachines
935 Washington Street
San Carlos, CA 94070
(650) 508-1634   FAX: (650) 508-1644
(877) 855-4363
http://www.genemachines.com

Genentech
1 DNA Way
South San Francisco, CA 94080
(800) 551-2231   FAX: (650) 225-1600
(650) 225-1000
http://www.gene.com

General Scanning/GSI Luminomics
500 Arsenal Street
Watertown, MA 02172
(617) 924-1010   FAX: (617) 924-7327
http://www.genescan.com

General Valve
Division of Parker Hannifin Pneutronics
19 Gloria Lane
Fairfield, NJ 07004
(800) GVC-VALV
FAX: (800) GVC-1-FAX
http://www.pneutronics.com

Genespan
19310 North Creek Parkway, Suite 100
Bothell, WA 98011
(800) 231-2215   FAX: (425) 482-3005
(425) 482-3003
http://www.genespan.com

Gene Therapy Systems
10190 Telesis Court
San Diego, CA 92122
(858) 457-1919   FAX: (858) 623-9494
http://www.genetherapysystems.com

Genethon Human Genome
Research Center
1 bis rue de l'Internationale
91000 Evry, France
(33) 169-472828
FAX: (33) 607-78698
http://www.genethon.fr

A4

Genetic Microsystems
34 Commerce Way
Wobum, MA 01801
(781) 932-9333　FAX: (781) 932-9433
http://www.genticmicro.com

Genetic Mutant Repository
See Coriell Institute for
Medical Research

Genetic Research Instrumentation
Gene House
Queenborough Lane
Rayne, Braintree, Essex CM7 8TF, UK
(44) 1376 332900
FAX: (44) 1376 344724
http://www.gri.co.uk

Genetics Computer Group
575 Science Drive
Madison, WI 53711
(608) 231-5200　FAX: (608) 231-5202
http://www.gcg.com

**Genetics Institute/American
Home Products**
87 Cambridge Park Drive
Cambridge, MA 02140
(617) 876-1170　FAX: (617) 876-0388
http://www.genetics.com

Genetix
63-69 Somerford Road
Christchurch, Dorset BH23 3QA, UK
(44) (0) 1202 483900
FAX: (44)(0) 1202 480289
In the US: (877) 436 3849
US FAX: (888) 522 7499
http://www.genetix.co.uk

Gene Tools
One Summerton Way
Philomath, OR 97370
(541) 9292-7840　FAX: (541)
9292-7841
http://www.gene-tools.com

Geneva Bioinformatics (GeneBio) S.A.
25 Avenue de Champel
CH—1206 Geneva, Switzerland
(41) 22-702-9900
FAX: (41) 22-702-9999
http://www.genebio.com

GeneWorks
P.O. Box 11, Rundle Mall
Adelaide, South Australia 5000, Australia
1800 882 555　FAX: (08) 8234 2699
(08) 8234 2644
http://www.geneworks.com

Genome Systems (INCYTE)
4633 World Parkway Circle
St. Louis, MO 63134
(800) 430-0030　FAX: (314) 427-3324
(314) 427-3222
http://www.genomesystems.com

Genomic Solutions
4355 Varsity Drive, Suite E
Ann Arbor, MI 48108
(877) GENOMIC　FAX: (734) 975-4808
(734) 975-4800
http://www.genomicsolutions.com

Genomyx
See Beckman Coulter

Genosys Biotechnologies
1442 Lake Front Circle, Suite 185
The Woodlands, TX 77380
(281) 363-3693　FAX: (281) 363-2212
http://www.genosys.com

Genotech
92 Weldon Parkway
St. Louis, MO 63043
(800) 628-7730　FAX: (314) 991-1504
(314) 991-6034

GENSET
876 Prospect Street, Suite 206
La Jolla, CA 92037
(800) 551-5291　FAX: (619) 551-2041
(619) 515-3061
http://www.genset.fr

Gensia Laboratories Ltd.
19 Hughes
Irvine, CA 92718
(714) 455-4700　FAX: (714) 855-8210

Genta
99 Hayden Avenue, Suite 200
Lexington, MA 02421
(781) 860-5150　FAX: (781) 860-5137
http://www.genta.com

GENTEST
6 Henshaw Street
Woburn, MA 01801
(800) 334-5229　FAX: (888) 242-2226
(781) 935-5115　FAX: (781) 932-6855
http://www.gentest.com

Gentra Systems
15200 25th Avenue N., Suite 104
Minneapolis, MN 55447
(800) 866-3039　FAX: (612) 476-5850
(612) 476-5858
http://www.gentra.com

Genzyme
1 Kendall Square
Cambridge, MA 02139
(617) 252-7500　FAX: (617) 252-7600
http://www.genzyme.com
See also R&D Systems

Genzyme Genetics
One Mountain Road
Framingham, MA 01701
(800) 255-7357　FAX: (508) 872-9080
(508) 872-8400
http://www.genzyme.com

George Tiemann & Co.
25 Plant Avenue
Hauppauge, NY 11788
(516) 273-0005　FAX: (516) 273-6199

GIBCO/BRL
A Division of Life Technologies
1 Kendall Square
Grand Island, NY 14072
(800) 874-4226　FAX: (800) 352-1968
(716) 774-6700
http://www.lifetech.com

Gilmont Instruments
A Division of Barnant Company
28N092 Commercial Avenue
Barrington, IL 60010
(800) 637-3739　FAX: (708) 381-7053
http://barnant.com

Gilson
3000 West Beltline Highway
P.O. Box 620027
Middletown, WI 53562
(800) 445-7661
(608) 836-1551
http://www.gilson.com

Glas-Col Apparatus
P.O. Box 2128
Terre Haute, IN 47802
(800) Glas-Col　FAX: (812) 234-6975
(812) 235-6167
http://www.glascol.com

Glaxo Wellcome
Five Moore Drive
Research Triangle Park, NC 27709
(800) SGL-AXO5　FAX: (919) 248-2386
(919) 248-2100
http://www.glaxowellcome.com

Glen Mills
395 Allwood Road
Clifton, NJ 07012
(973) 777-0777　FAX: (973) 777-0070
http://www.glenmills.com

Glen Research
22825 Davis Drive
Sterling, VA 20166
(800) 327-4536　FAX: (800) 934-2490
(703) 437-6191　FAX: (703) 435-9774
http://www.glenresearch.com

Glo Germ
P.O. Box 189
Moab, UT 84532
(800) 842-6622　FAX: (435) 259-5930
http://www.glogerm.com

Glyco
11 Pimentel Court
Novato, CA 94949
(800) 722-2597　FAX: (415) 382-3511
(415) 884-6799
http://www.glyco.com

A4

Gould Instrument Systems
8333 Rockside Road
Valley View, OH 44125
(216) 328-7000   FAX: (216) 328-7400
http://www.gould13.com

Gralab Instruments
See Dimco-Gray

GraphPad Software
5755 Oberlin Drive #110
San Diego, CA 92121
(800) 388-4723   FAX: (558) 457-8141
(558) 457-3909
http://www.graphpad.com

Graseby Anderson
See Andersen Instruments
http://www.graseby.com

Grass Instrument
A Division of Astro-Med
600 East Greenwich Avenue
W. Warwick, RI 02893
(800) 225-5167   FAX: (877) 472-7749
http://www.grassinstruments.com

Greenacre and Misac Instruments
Misac Systems
27 Port Wood Road
Ware, Hertfordshire SF12 9NJ, UK
(44) 1920 463017
FAX: (44) 1920 465136

Greer Labs
639 Nuway Circle
Lenois, NC 28645
(704) 754-5237
http://greerlabs.com

Greiner
Maybachestrasse 2
Postfach 1162
D-7443 Frickenhausen, Germany
(49) 0 91 31/80 79 0
FAX: (49) 0 91 31/80 79 30
http://www.erlangen.com/greiner

GSI Lumonics
130 Lombard Street
Oxnard, CA 93030
(805) 485-5559   FAX: (805) 485-3310
http://www.gsilumonics.com

GTE Internetworking
150 Cambridge Park Drive
Cambridge, MA 02140
(800) 472-4565   FAX: (508) 694-4861
http://www.bbn.com

GW Instruments
35 Medford Street
Somerville, MA 02143
(617) 625-4096   FAX: (617) 625-1322
http://www.gwinst.com

H & H Woodworking
1002 Garfield Street
Denver, CO 80206
(303) 394-3764

Hacker Instruments
17 Sherwood Lane
P.O. Box 10033
Fairfield , NJ 07004
800-442-2537   FAX: (973) 808-8281
(973) 226-8450
http://www.hackerinstruments.com

Haemenetics
400 Wood Road
Braintree, MA 02184
(800) 225-5297   FAX: (781) 848-7921
(781) 848-7100
http://www.haemenetics.com

Halocarbon Products
P.O. Box 661
River Edge, NJ 07661
(201) 242-8899   FAX: (201) 262-0019
http://halocarbon.com

Hamamatsu Photonic Systems
A Division of Hamamatsu
360 Foothill Road
P.O. Box 6910
Bridgewater, NJ 08807
(908) 231-1116   FAX: (908) 231-0852
http://www.photonicsonline.com

Hamilton Company
4970 Energy Way
P.O. Box 10030
Reno, NV 89520
(800) 648-5950   FAX: (775) 856-7259
(775) 858-3000
http://www.hamiltoncompany.com

Hamilton Thorne Biosciences
100 Cummings Center, Suite 102C
Beverly, MA 01915
http://www.hamiltonthorne.com

Hampton Research
27631 El Lazo Road
Laguna Niguel, CA 92677
(800) 452-3899   FAX: (949) 425-1611
(949) 425-6321
http://www.hamptonresearch.com

Harlan Bioproducts for Science
P.O. Box 29176
Indianapolis, IN 46229
(317) 894-7521   FAX: (317) 894-1840
http://www.hbps.com

Harlan Sera-Lab
Hillcrest, Dodgeford Lane
Belton, Loughborough
Leicester LE12 9TE, UK
(44) 1530 222123
FAX: (44) 1530 224970
http://www.harlan.com

Harlan Teklad
P.O. Box 44220
Madison, WI 53744
(608) 277-2070   FAX: (608) 277-2066
http://www.harlan.com

Harrick Scientific Corporation
88 Broadway
Ossining, NY 10562
(914) 762-0020   FAX: (914) 762-0914
http://www.harricksci.com

Harrison Research
840 Moana Court
Palo Alto, CA 94306
(650) 949-1565   FAX: (650) 948-0493

Harvard Apparatus
84 October Hill Road
Holliston, MA 01746
(800) 272-2775   FAX: (508) 429-5732
(508) 893-8999
http://harvardapparatus.com

Harvard Bioscience
See Harvard Apparatus

Haselton Biologics
See JRH Biosciences

Hazelton Research Products
See Covance Research Products

Health Products
See Pierce Chemical

Heat Systems-Ultrasonics
1938 New Highway
Farmingdale, NY 11735
(800) 645-9846   FAX: (516) 694-9412
(516) 694-9555

Heidenhain Corp
333 East State Parkway
Schaumberg, IL 60173
(847) 490-1191   FAX: (847) 490-3931
http://www.heidenhain.com

HEKA Instruments
33 Valley Rd.
Southboro, MA 01960
(866) 742-0606 FAX: (508) 481-8945
http://www.heka.com

Hellma Cells
11831 Queens Boulevard
Forest Hills, NY 11375
(718) 544-9166   FAX: (718) 263-6910
http://www.helmaUSA.com

Hellma
Postfach 1163
D-79371 Mullheim/Baden, Germany
(49) 7631-1820
FAX: (49) 7631-13546
http://www.hellma-worldwide.de

Henry Schein
135 Duryea Road, Mail Room 150
Melville, NY 11747
(800) 472-4346   FAX: (516) 843-5652
http://www.henryschein.com

Heraeus Kulzer
4315 South Lafayette Boulevard
South Bend, IN 46614
(800) 343-5336
(219) 291-0661
http://www.kulzer.com

Heraeus Sepatech
See Kendro Laboratory Products

Hercules Aqualon
Aqualon Division
Hercules Research Center, Bldg. 8145
500 Hercules Road
Wilmington, DE 19899
(800) 345-0447   FAX: (302) 995-4787
http://www.herc.com/aqualon/pharma

Heto-Holten A/S
Gydevang 17-19
DK-3450 Allerod, Denmark
(45) 48-16-62-00
FAX: (45) 48-16-62-97
Distributed by ATR

Hettich-Zentrifugen
See Andreas Hettich

Hewlett-Packard
3000 Hanover Street
Mailstop 20B3
Palo Alto, CA 94304
(650) 857-1501   FAX: (650) 857-5518
http://www.hp.com

HGS Hinimoto Plastics
1-10-24 Meguro-Honcho
Megurouko
Tokyo 152, Japan
3-3714-7226   FAX: 3-3714-4657

Hitachi Scientific Instruments
Nissei Sangyo America
8100 N. First Street
San Elsa, CA 95314
(800) 548-9001   FAX: (408) 432-0704
(408) 432-0520
http://www.hii.hitachi.com

Hi-Tech Scientific
Brunel Road
Salisbury, Wiltshire, SP2 7PU
UK
(44) 1722-432320
(800) 344-0724 (US only)
http://www.hi-techsci.co.uk

Hoechst AG
See Aventis Pharmaceutical

Hoefer Scientific Instruments
Division of Amersham-Pharmacia
Biotech
800 Centennial Avenue
Piscataway, NJ 08855
(800) 227-4750   FAX: (877) 295-8102
http://www.apbiotech.com

Hoffman-LaRoche
340 Kingsland Street
Nutley, NJ 07110
(800) 526-0189   FAX: (973) 235-9605
(973) 235-5000
http://www.rocheUSA.com

Holborn Surgical and Medical
Instruments
Westwood Industrial Estate
Ramsgate Road
Margate, Kent CT9 4JZ UK
(44) 1843 296666
FAX: (44) 1843 295446

Honeywell
101 Columbia Road
Morristown, NJ 07962
(973) 455-2000   FAX: (973) 455-4807
http://www.honeywell.com

Honeywell Specialty Films
P.O. Box 1039
101 Columbia Road
Morristown, NJ 07962
(800) 934-5679   FAX: (973) 455-6045
http://www.honeywell-specialtyfilms.com

Hood Thermo-Pad Canada
Comp. 20, Site 61A, RR2
Summerland, British Columbia
V0H 1Z0 Canada
(800) 665-9555   FAX: (250) 494-5003
(250) 494-5002
http://www.thermopad.com

Horiba Instruments
17671 Armstrong Avenue
Irvine, CA 92714
(949) 250-4811   FAX: (949) 250-0924
http://www.horiba.com

Hoskins Manufacturing
10776 Hall Road
P.O. Box 218
Hamburg, MI 48139
(810) 231-1900   FAX: (810) 231-4311
http://www.hoskinsmfgco.com

Hosokawa Micron Powder Systems
10 Chatham Road
Summit, NJ 07901
(800) 526-4491   FAX: (908) 273-7432
(908) 273-6360
http://www.hosokawamicron.com

HT Biotechnology
Unit 4
61 Ditton Walk
Cambridge CB5 8QD, UK
(44) 1223-412583

Hugo Sachs Electronik
Postfach 138
7806 March-Hugstetten, Germany
D-79229(49) 7665-92000
FAX: (49) 7665-920090

Human Biologics International
7150 East Camelback Road, Suite 245
Scottsdale, AZ 85251
(480) 990-2005   FAX: (480)-990-2155
http://www.humanbiological.com

Human Genetic Mutant Cell
Repository
See Coriell Institute for
Medical Research

HVS Image
P.O. Box 100
Hampton, Middlesex TW12 2YD, UK
FAX: (44) 208 783 1223
In the US: (800) 225-9261
FAX: (888) 483-8033
http://www.hvsimage.com

Hybaid
111-113 Waldegrave Road
Teddington, Middlesex TW11 8LL, UK
(44) 0 1784 42500
FAX: (44) 0 1784 248085
http://www.hybaid.co.uk

Hybaid Instruments
8 East Forge Parkway
Franklin, MA 02028
(888)4-HYBAID   FAX: (508) 541-3041
(508) 541-6918
http://www.hybaid.com

Hybridon
155 Fortune Boulevard
Milford, MA 01757
(508) 482-7500   FAX: (508) 482-7510
http://www.hybridon.com

HyClone Laboratories
1725 South HyClone Road
Logan, UT 84321
(800) HYCLONE   FAX: (800) 533-9450
(801) 753-4584   FAX: (801) 750-0809
http://www.hyclone.com

Hyseq
670 Almanor Avenue
Sunnyvale, CA 94086
(408) 524-8100   FAX: (408) 524-8141
http://www.hyseq.com

IBA GmbH
1508 South Grand Blvd.
St. Louis, MO 63104
(877) 422-4624   FAX: (888) 531-6813
http://www.iba-go.com

IBF Biotechnics
See Sepracor

IBI (International Biotechnologies)
See Eastman Kodak
For technical service (800) 243-2555
(203) 786-5600

ICN Biochemicals
See ICN Biomedicals

A4

ICN Biomedicals
3300 Hyland Avenue
Costa Mesa, CA 92626
(800) 854-0530   FAX: (800) 334-6999
(714) 545-0100   FAX: (714) 641-7275
http://www.icnbiomed.com

ICN Flow and Pharmaceuticals
See ICN Biomedicals

ICN Immunobiochemicals
See ICN Biomedicals

ICN Radiochemicals
See ICN Biomedicals

ICONIX
100 King Street West, Suite 3825
Toronto, Ontario
M5X 1E3 Canada
(416) 410-2411   FAX: (416) 368-3089
http://www.iconix.com

ICRT (Imperial Cancer Research Technology)
Sardinia House
Sardinia Street
London WC2A 3NL, UK
(44) 1712-421136
FAX: (44) 1718-314991

Idea Scientific Company
P.O. Box 13210
Minneapolis, MN 55414
(800) 433-2535   FAX: (612) 331-4217
http://www.ideascientific.com

IEC
See International Equipment Co.

IITC
23924 Victory Boulevard
Woodland Hills, CA 91367
(888) 414-4482 (818) 710-1556
FAX: (818) 992-5185
http://www.iitcinc.com

IKA Works
2635 N. Chase Parkway, SE
Wilmington, NC 28405
(910) 452-7059   FAX: (910) 452-7693
http://www.ika.net

Ikegami Electronics
37 Brook Avenue
Maywood, NJ 07607
(201) 368-9171   FAX: (201) 569-1626

Ikemoto Scientific Technology
25-11 Hongo
3-chome, Bunkyo-ku
Tokyo 101-0025, Japan
(81) 3-3811-4181
FAX: (81) 3-3811-1960

Imagenetics
See ATC Diagnostics

Imaging Research
c/o Brock University
500 Glenridge Avenue
St. Catharines, Ontario
L2S 3A1 Canada
(905) 688-2040   FAX: (905) 685-5861
http://www.imaging.brocku.ca

Imclone Systems
180 Varick Street
New York, NY 10014
(212) 645-1405   FAX: (212) 645-2054
http://www.imclone.com

IMCO Corporation LTD., AB
P.O. Box 21195
SE-100 31
Stockholm, Sweden
46-8-33-53-09   FAX: 46-8-728-47-76
http://www.imcocorp.se

Imgenex Corporation
11175 Flintkote Avenue
Suite E
San Diego, CA 92121
(888) 723-4363   FAX: (858) 642-0937
(858) 642.0978
http://www.imgenex.com

IMICO
Calle Vivero, No. 5-4a Planta
E-28040, Madrid, Spain
(34) 1-535-3960   FAX: (34) 1-535-2780

Immunex
51 University Street
Seattle, WA 98101
(206) 587-0430   FAX: (206) 587-0606
http://www.immunex.com

Immunochemistry Technologies
9401 James Avenue, South
Suite 155
Bloomington, MN 55431
(800) 829-3194   FAX: (952) 888-8988
(952) 888-8788
http://www.immunochemistry.com

Immunocorp
1582 W. Deere Avenue
Suite C
Irvine, CA 92606
(800) 446-3063
http://www.immunocorp.com

Immunotech
130, av. Delattre de Tassigny
B.P. 177
13276 Marseilles Cedex 9
France
(33) 491-17-27-00
FAX: (33) 491-41-43-58
http://www.immunotech.fr

Imperial Chemical Industries
Imperial Chemical House
Millbank, London SW1P 3JF, UK
(44) 171-834-4444
FAX: (44)171-834-2042
http://www.ici.com

Inceltech
See New Brunswick Scientific

Incstar
See DiaSorin

Incyte
6519 Dumbarton Circle
Fremont, CA 94555
(510) 739-2100   FAX: (510) 739-2200
http://www.incyte.com

Incyte Pharmaceuticals
3160 Porter Drive
Palo Alto, CA 94304
(877) 746-2983   FAX: (650) 855-0572
(650) 855-0555
http://www.incyte.com

Individual Monitoring Systems
6310 Harford Road
Baltimore, MD 21214

Indo Fine Chemical
P.O. Box 473
Somerville, NJ 08876
(888) 463-6346   FAX: (908) 359-1179
(908) 359-6778
http://www.indofinechemical.com

Industrial Acoustics
1160 Commerce Avenue
Bronx, NY 10462
(718) 931-8000   FAX: (718) 863-1138
http://www.industrialacoustics.com

Inex Pharmaceuticals
100-8900 Glenlyon Parkway
Glenlyon Business Park
Burnaby, British Columbia
V5J 5J8 Canada
(604) 419-3200   FAX: (604) 419-3201
http://www.inexpharm.com

Ingold, Mettler, Toledo
261 Ballardvale Street
Wilmington, MA 01887
(800) 352-8763   FAX: (978) 658-0020
(978) 658-7615
http://www.mt.com

Innogenetics N.V.
Technologie Park 6
B-9052 Zwijnaarde
Belgium
(32) 9-329-1329   FAX: (32) 9-245-7623
http://www.innogenetics.com

Innovative Medical Services
1725 Gillespie Way
El Cajon, CA 92020
(619) 596-8600   FAX: (619) 596-8700
http://www.imspure.com

A4

Innovative Research
3025 Harbor Lane N, Suite 300
Plymouth, MN 55447
(612) 519-0105   FAX: (612) 519-0239
http://www.inres.com

Innovative Research of America
2 N. Tamiami Trail, Suite 404
Sarasota, FL 34236
(800) 421-8171   FAX: (800) 643-4345
(941) 365-1406   FAX: (941) 365-1703
http://www.innovrsrch.com

Inotech Biosystems
15713 Crabbs Branch Way, #110
Rockville, MD 20855
(800) 635-4070   FAX: (301) 670-2859
(301) 670-2850
http://www.inotechintl.com

INOVISION
22699 Old Canal Road
Yorba Linda, CA 92887
(714) 998-9600   FAX: (714) 998-9666
http://www.inovision.com

Instech Laboratories
5209 Militia Hill Road
Plymouth Meeting, PA 19462
(800) 443-4227   FAX: (610) 941-0134
(610) 941-0132
http://www.instechlabs.com

Instron
100 Royall Street
Canton, MA 02021
(800) 564-8378   FAX: (781) 575-5725
(781) 575-5000
http://www.instron.com

Instrumentarium
P.O. Box 300
00031 Instrumentarium
Helsinki, Finland
(10) 394-5566
http://www.instrumentarium.fi

Instruments SA
Division Jobin Yvon
16-18 Rue du Canal
91165 Longjumeau, Cedex, France
(33)1 6454-1300
FAX: (33)1 6909-9319
http://www.isainc.com

Instrutech
20 Vanderventer Avenue, Suite 101E
Port Washington, NY 11050
(516) 883-1300   FAX: (516) 883-1558
http://www.instrutech.com

Integrated DNA Technologies
1710 Commercial Park
Coralville, IA 52241
(800) 328-2661   FAX: (319) 626-8444
http://www.idtdna.com

Integrated Genetics
See Genzyme Genetics

Integrated Scientific Imaging Systems
3463 State Street, Suite 431
Santa Barbara, CA 93105
(805) 692-2390   FAX: (805) 692-2391
http://www.imagingsystems.com

Integrated Separation Systems (ISS)
See OWL Separation Systems

IntelliGenetics
See Oxford Molecular Group

Interactiva BioTechnologie
Sedanstrasse 10
D-89077 Ulm, Germany
(49) 731-93579-290
FAX: (49) 731-93579-291
http://www.interactiva.de

Interchim
213 J.F. Kennedy Avenue
B.P. 1140
Montlucon
03103 France
(33) 04-70-03-83-55
FAX: (33) 04-70-03-93-60

Interfocus
14/15 Spring Rise
Falcover Road
Haverhill, Suffolk CB9 7XU, UK
(44) 1440 703460
FAX: (44) 1440 704397
http://www.interfocus.ltd.uk

Intergen
2 Manhattanville Road
Purchase, NY 10577
(800) 431-4505   FAX: (800) 468-7436
(914) 694-1700   FAX: (914) 694-1429
http://www.intergenco.com

Intermountain Scientific
420 N. Keys Drive
Kaysville, UT 84037
(800) 999-2901   FAX: (800) 574-7892
(801) 547-5047   FAX: (801) 547-5051
http://www.bioexpress.com

International Biotechnologies (IBI)
See Eastman Kodak

International Equipment Co. (IEC)
See Thermoquest

International Institute for the
Advancement of Medicine
1232 Mid-Valley Drive
Jessup, PA 18434
(800) 486-IIAM   FAX: (570) 343-6993
(570) 496-3400
http://www.iiam.org

International Light
17 Graf Road
Newburyport, MA 01950
(978) 465-5923   FAX: (978) 462-0759

International Market Supply (I.M.S.)
Dane Mill
Broadhurst Lane
Congleton, Cheshire CW12 1LA, UK
(44) 1260 275469
FAX: (44) 1260 276007

International Marketing Services
See International Marketing Ventures

International Marketing Ventures
6301 Ivy Lane, Suite 408
Greenbelt, MD 20770
(800) 373-0096   FAX: (301) 345-0631
(301) 345-2866
http://www.imvlimited.com

International Products
201 Connecticut Drive
Burlington, NJ 08016
(609) 386-8770   FAX: (609) 386-8438
http://www.mkt@ipcol.com

Intracel Corporation
Bartels Division
2005 Sammamish Road, Suite 107
Issaquah, WA 98027
(800) 542-2281   FAX: (425) 557-1894
(425) 392-2992
http://www.intracel.com

Invitrogen
1600 Faraday Avenue
Carlsbad, CA 92008
(800) 955-6288   FAX: (760) 603-7201
(760) 603-7200
http://www.invitrogen.com

In Vivo Metric
P.O. Box 249
Healdsburg, CA 95448
(707) 433-4819   FAX: (707) 433-2407

IRORI
9640 Towne Center Drive
San Diego, CA 92121
(858) 546-1300   FAX: (858) 546-3083
http://www.irori.com

Irvine Scientific
2511 Daimler Street
Santa Ana, CA 92705
(800) 577-6097   FAX: (949) 261-6522
(949) 261-7800
http://www.irvinesci.com

ISC BioExpress
420 North Kays Drive
Kaysville, UT 84037
(800) 999-2901   FAX: (800) 574-7892
(801) 547-5047
http://www.bioexpress.com

ISCO
P.O. Box 5347
4700 Superior
Lincoln, NE 68505
(800) 228-4373   FAX: (402) 464-0318
(402) 464-0231
http://www.isco.com

A4

Isis Pharmaceuticals
Carlsbad Research Center
2292 Faraday Avenue
Carlsbad, CA 92008
(760) 931-9200
http://www.isip.com

Isolabs
See Wallac

ISS
See Integrated Separation Systems

J & W Scientific
See Agilent Technologies

J.A. Webster
86 Leominster Road
Sterling , MA 01564
(800) 225-7911   FAX: (978) 422-8959
http://www.jawebster.com

J.T. Baker
See Mallinckrodt Baker
222 Red School Lane
Phillipsburg, NJ 08865
(800) JTBAKER   FAX: (908) 859-6974
http://www.jtbaker.com

Jackson ImmunoResearch
Laboratories
P.O. Box 9
872 W. Baltimore Pike
West Grove, PA 19390
(800) 367-5296   FAX: (610) 869-0171
(610) 869-4024
http://www.jacksonimmuno.com

The Jackson Laboratory
600 Maine Street
Bar Harbor, ME 04059
(800) 422-6423   FAX: (207) 288-5079
(207) 288-6000
http://www.jax.org

Jaece Industries
908 Niagara Falls Boulevard
North Tonawanda, NY 14120
(716) 694-2811   FAX: (716) 694-2811
http://www.jaece.com

Jandel Scientific
See SPSS

Janke & Kunkel
See Ika Works

Janssen Life Sciences Products
See Amersham

Janssen Pharmaceutica
1125 Trenton-Harbourton Road
Titusville, NJ 09560
(609) 730-2577   FAX: (609) 730-2116
http://us.janssen.com

Jasco
8649 Commerce Drive
Easton, MD 21601
(800) 333-5272   FAX: (410) 822-7526
(410) 822-1220
http://www.jascoinc.com

Jena Bioscience
Loebstedter Str. 78
07749 Jena, Germany
(49) 3641-464920
FAX: (49) 3641-464991
http://www.jenabioscience.com

Jencons Scientific
800 Bursca Drive, Suite 801
Bridgeville, PA 15017
(800) 846-9959   FAX: (412) 257-8809
(412) 257-8861
http://www.jencons.co.uk

JEOL Instruments
11 Dearborn Road
Peabody, MA 01960
(978) 535-5900   FAX: (978) 536-2205
http://www.jeol.com/index.html

Jewett
750 Grant Street
Buffalo, NY 14213
(800) 879-7767   FAX: (716) 881-6092
(716) 881-0030
http://www.JewettInc.com

John's Scientific
See VWR Scientific

John Weiss and Sons
95 Alston Drive
Bradwell Abbey
Milton Keynes, Buckinghamshire
MK1 4HF UK
(44) 1908-318017
FAX: (44) 1908-318708

Johnson & Johnson Medical
2500 Arbrook Boulevard East
Arlington, TX 76004
(800) 423-4018
http://www.jnjmedical.com

Johnston Matthey Chemicals
Orchard Road
Royston, Hertfordshire SG8 5HE, UK
(44) 1763-253000
FAX: (44) 1763-253466
http://www.chemicals.matthey.com

Jolley Consulting and Research
683 E. Center Street, Unit H
Grayslake, IL 60030
(847) 548-2330   FAX: (847) 548-2984
http://www.jolley.com

Jordan Scientific
See Shelton Scientific

Jorgensen Laboratories
1450 N. Van Buren Avenue
Loveland, CO 80538
(800) 525-5614   FAX: (970) 663-5042
(970) 669-2500
http://www.jorvet.com

JRH Biosciences and JR Scientific
13804 W. 107th Street
Lenexa, KS 66215
(800) 231-3735   FAX: (913) 469-5584
(913) 469-5580

Jule Bio Technologies
25 Science Park, #14, Suite 695
New Haven, CT 06511
(800) 648-1772   FAX: (203) 786-5489
(203) 786-5490
http://hometown.aol.com/precastgel/
index.htm

K.R. Anderson
2800 Bowers Avenue
Santa Clara, CA 95051
(800) 538-8712   FAX: (408) 727-2959
(408) 727-2800
http://www.kranderson.com

Kabi Pharmacia Diagnostics
See Pharmacia Diagnostics

Kanthal H.P. Reid
1 Commerce Boulevard
P.O. Box 352440
Palm Coast, FL 32135
(904) 445-2000   FAX: (904) 446-2244
http://www.kanthal.com

Kapak
5305 Parkdale Drive
St. Louis Park, MN 55416
(800) KAPAK-57   FAX: (612) 541-0735
(612) 541-0730
http://www.kapak.com

Karl Hecht
Stettener Str. 22-24
D-97647 Sondheim
Rhon, Germany
(49) 9779-8080   FAX: (49) 9779-80888

Karl Storz
Koningin-Elisabeth Str. 60
D-14059 Berlin, Germany
(49) 30-30 69 09-0
FAX: (49) 30-30 19 452
http://www.karlstorz.de

KaVo EWL
P.O. Box 1320
D-88293 Leutkirch im Allgau, Germany
(49) 7561-86-0   FAX: (49) 7561-86-371
http://www.kavo.com/english/
startseite.htm

Keithley Instruments
28775 Aurora Road
Cleveland, OH 44139
(800) 552-1115   FAX: (440) 248-6168
(440) 248-0400
http://www.keithley.com

Kemin
2100 Maury Street, Box 70
Des Moines, IA 50301
(515) 266-2111   FAX: (515) 266-8354
http://www.kemin.com

Kemo
3 Brook Court, Blakeney Road
Beckenham, Kent BR3 1HG, UK
(44) 0181 658 3838
FAX: (44) 0181 658 4084
http://www.kemo.com

Kendall
15 Hampshire Street
Mansfield, MA 02048
(800) 962-9888   FAX: (800) 724-1324
http://www.kendallhq.com

Kendro Laboratory Products
31 Pecks Lane
Newtown, CT 06470
(800) 522-SPIN   FAX: (203) 270-2166
(203) 270-2080
http://www.kendro.com

Kendro Laboratory Products
P.O. Box 1220
Am Kalkberg
D-3360 Osterod, Germany
(55) 22-316-213
FAX: (55) 22-316-202
http://www.heraeus-instruments.de

Kent Laboratories
23404 NE 8th Street
Redmond, WA 98053
(425) 868-6200   FAX: (425) 868-6335
http://www.kentlabs.com

Kent Scientific
457 Bantam Road, #16
Litchfield, CT 06759
(888) 572-8887   FAX: (860) 567-4201
(860) 567-5496
http://www.kentscientific.com

Keuffel & Esser
See Azon

Keystone Scientific
Penn Eagle Industrial Park
320 Rolling Ridge Drive
Bellefonte, PA 16823
(800) 437-2999   FAX: (814) 353-2305
(814) 353-2300 Ext 1
http://www.keystonescientific.com

Kimble/Kontes Biotechnology
1022 Spruce Street
P.O. Box 729
Vineland, NJ 08360
(888) 546-2531   FAX: (856) 794-9762
(856) 692-3600
http://www.kimble-kontes.com

Kinematica AG
Luzernerstrasse 147a
CH-6014 Littau-Luzern, Switzerland
(41) 41 2501257   FAX: (41) 41
2501460
http://www.kinematica.ch

Kin-Tek
504 Laurel Street
LaMarque, TX 77568
(800) 326-3627
FAX: (409) 938-3710
http://www.kin-tek.com

Kipp & Zonen
125 Wilbur Place
Bohemia, NY 11716
(800) 645-2065   FAX: (516) 589-2068
(516) 589-2885
http://www.kippzonen.thomasregister.
com/olc/kippzonen

Kirkegaard & Perry Laboratories
2 Cessna Court
Gaithersburg, MD 20879
(800) 638-3167   FAX: (301) 948-0169
(301) 948-7755
http://www.kpl.com

Kodak
See Eastman Kodak

Kontes Glass
See Kimble/Kontes Biotechnology

Kontron Instruments AG
Postfach CH-8010
Zurich, Switzerland
41-1-733-5733   FAX: 41-1-733-5734

David Kopf Instruments
P.O. Box 636
Tujunga, CA 91043
(818) 352-3274   FAX: (818) 352-3139

Kraft Apparatus
See Glas-Col Apparatus

Kramer Scientific Corporation
711 Executive Boulevard
Valley Cottage, NY 10989
(845) 267-5050   FAX: (845) 267-5550

Kulite Semiconductor Products
1 Willow Tree Road
Leonia, NJ 07605
(201) 461-0900   FAX: (201) 461-0990
http://www.kulite.com

Lab-Line Instruments
15th & Bloomingdale Avenues
Melrose Park, IL 60160
(800) LAB-LINE   FAX: (708) 450-5830
FAX: (800) 450-4LAB
http://www.labline.com

Lab Products
742 Sussex Avenue
P.O. Box 639
Seaford, DE 19973
(800) 526-0469   FAX: (302) 628-4309
(302) 628-4300
http://www.labproductsinc.com

LabRepco
101 Witmer Road, Suite 700
Horsham, PA 19044
(800) 521-0754   FAX: (215) 442-9202
http://www.labrepco.com

Lab Safety Supply
P.O. Box 1368
Janesville, WI 53547
(800) 356-0783   FAX: (800) 543-9910
(608) 754-7160   FAX: (608) 754-1806
http://www.labsafety.com

Lab-Tek Products
See Nalge Nunc International

Labconco
8811 Prospect Avenue
Kansas City, MO 64132
(800) 821-5525   FAX: (816) 363-0130
(816) 333-8811
http://www.labconco.com

Labindustries
See Barnstead/Thermolyne

Labnet International
P.O. Box 841
Woodbridge, NJ 07095
(888) LAB-NET1   FAX: (732) 417-1750
(732) 417-0700
http://www.nationallabnet.com

LABO-MODERNE
37 rue Dombasle
Paris
75015 France
(33) 01-45-32-62-54
FAX: (33) 01-45-32-01-09
http://www.labomoderne.com/fr

Laboratory of Immunoregulation
National Institute of Allergy and
Infectious Diseases/NIH
9000 Rockville Pike
Building 10, Room 11B13
Bethesda, MD 20892
(301) 496-1124

Laboratory Supplies
29 Jefry Lane
Hicksville, NY 11801
(516) 681-7711

Labscan Limited
Stillorgan Industrial Park
Stillorgan
Dublin, Ireland
(353) 1-295-2684
FAX: (353) 1-295-2685
http://www.labscan.ie

Labsystems
See Thermo Labsystems

Labsystems Affinity Sensors
Saxon Way, Bar Hill
Cambridge CB3 8SL, UK
44 (0) 1954 789976
FAX: 44 (0) 1954 789417
http://www.affinity-sensors.com

A4

Labtronics
546 Governors Road
Guelph, Ontario
N1K 1E3 Canada
(519) 763-4930   FAX: (519) 836-4431
http://www.labtronics.com

Labtronix Manufacturing
3200 Investment Boulevard
Hayward, CA 94545
(510) 786-3200   FAX: (510) 786-3268
http://www.labtronix.com

Lafayette Instrument
3700 Sagamore Parkway North
P.O. Box 5729
Lafayette, IN 47903
(800) 428-7545   FAX: (765) 423-4111
(765) 423-1505
http://www.lafayetteinstrument.com

Lambert Instruments
Turfweg 4
9313 TH Leutingewolde
The Netherlands
(31) 50-5018461   FAX: (31)
50-5010034
http://www.lambert-instruments.com

Lampire Biological Laboratories
P.O. Box 270
Pipersville, PA 18947
(215) 795-2538   FAX: (215) 795-0237
http://www.lampire.com

Lancaster Synthesis
P.O. Box 1000
Windham, NH 03087
(800) 238-2324   FAX: (603) 889-3326
(603) 889-3306
http://www.lancastersynthesis-us.com

Lancer
140 State Road 419
Winter Springs, FL 32708
(800) 332-1855   FAX: (407) 327-1229
(407) 327-8488
http://www.lancer.com

LaVision GmbH
Gerhard-Gerdes-Str. 3
D-37079
Goettingen, Germany
(49) 551-50549-0
FAX: (49) 551-50549-11
http://www.lavision.de

Lawshe
See Advanced Process Supply

Laxotan
20, rue Leon Blum
26000 Valence, France
(33) 4-75-41-91-91
FAX: (33) 4-75-41-91-99
http://www.latoxan.com

LC Laboratories
165 New Boston Street
Woburn, MA 01801
(781) 937-0777   FAX: (781) 938-5420
http://www.lclaboratories.com

LC Packings
80 Carolina Street
San Francisco, CA 94103
(415) 552-1855   FAX: (415) 552-1859
http://www.lcpackings.com

LC Services
See LC Laboratories

LECO
3000 Lakeview Avenue
St. Joseph, MI 49085
(800) 292-6141   FAX: (616) 982-8977
(616) 985-5496
http://www.leco.com

Lederle Laboratories
See Wyeth-Ayerst

Lee Biomolecular Research
Laboratories
11211 Sorrento Valley Road, Suite M
San Diego, CA 92121
(858) 452-7700

The Lee Company
2 Pettipaug Road
P.O. Box 424
Westbrook, CT 06498
(800) LEE-PLUG   FAX: (860) 399-7058
(860) 399-6281
http://www.theleeco.com

Lee Laboratories
1475 Athens Highway
Grayson, GA 30017
(800) 732-9150   FAX: (770) 979-9570
(770) 972-4450
http://www.leelabs.com

Leica
111 Deer Lake Road
Deerfield, IL 60015
(800) 248-0123   FAX: (847) 405-0147
(847) 405-0123
http://www.leica.com

Leica Microsystems
Imneuenheimer Feld 518
D-69120
Heidelberg, Germany
(49) 6221-41480
FAX: (49) 6221-414833
http://www.leica-microsystems.com

Leinco Technologies
359 Consort Drive
St. Louis, MO 63011
(314) 230-9477   FAX: (314) 527-5545
http://www.leinco.com

Leitz U.S.A.
See Leica

LenderKing Metal Products
8370 Jumpers Hole Road
Millersville, MD 21108
(410) 544-8795   FAX: (410) 544-5069
http://www.lenderking.com

Letica Scientific Instruments
Panlab s.i., c/Loreto 50
08029 Barcelona, Spain
(34) 93-419-0709
FAX: (34) 93-419-7145
http://www.panlab-sl.com

Leybold-Heraeus Trivac DZA
5700 Mellon Road
Export, PA 15632
(412) 327-5700

LI-COR
Biotechnology Division
4308 Progressive Avenue
Lincoln, NE 68504
(800) 645-4267   FAX: (402) 467-0819
(402) 467-0700
http://www.licor.com

Life Science Laboratories
See Adaptive Biosystems

Life Science Resources
Two Corporate Center Drive
Melville, NY 11747
(800) 747-9530   FAX: (516) 844-5114
(516) 844-5085
http://www.astrocam.com

Life Sciences
2900 72nd Street North
St. Petersburg, FL 33710
(800) 237-4323   FAX: (727) 347-2957
(727) 345-9371
http://www.lifesci.com

Life Technologies
9800 Medical Center Drive
P.O. Box 6482
Rockville, MD 20849
(800) 828-6686   FAX: (800) 331-2286
http://www.lifetech.com

Lifecodes
550 West Avenue
Stamford, CT 06902
(800) 543-3263   FAX: (203) 328-9599
(203) 328-9500
http://www.lifecodes.com

Lightnin
135 Mt. Read Boulevard
Rochester, NY 14611
(888) MIX-BEST   FAX: (716) 527-1742
(716) 436-5550
http://www.lightnin-mixers.com

Linear Drives
Luckyn Lane, Pipps Hill
Basildon, Essex SS14 3BW, UK
(44) 1268-287070
FAX: (44) 1268-293344
http://www.lineardrives.com

A4

Linscott's Directory
4877 Grange Road
Santa Rosa, CA 95404
(707) 544-9555 FAX: (415) 389-6025
http://www.linscottsdirectory.co.uk

Linton Instrumentation
Unit 11, Forge Business Center
Upper Rose Lane
Palgrave, Diss, Norfolk IP22 1AP, UK
(44) 1-379-651-344
FAX: (44) 1-379-650-970
http://www.lintoninst.co.uk

List Biological Laboratories
501-B Vandell Way
Campbell, CA 95008
(800) 726-3213 FAX: (408) 866-6364
(408) 866-6363
http://www.listlabs.com

LKB Instruments
See Amersham Pharmacia Biotech

Lloyd Laboratories
604 West Thomas Avenue
Shenandoah, IA 51601
(800) 831-0004 FAX: (712) 246-5245
(712) 246-4000
http://www.lloydinc.com

Loctite
1001 Trout Brook Crossing
Rocky Hill, CT 06067
(860) 571-5100 FAX: (860)571-5465
http://www.loctite.com

Lofstrand Labs
7961 Cessna Avenue
Gaithersburg, MD 20879
(800) 541-0362 FAX: (301) 948-9214
(301) 330-0111
http://www.lofstrand.com

Lomir Biochemical
99 East Main Street
Malone, NY 12953
(877) 425-3604 FAX: (518) 483-8195
(518) 483-7697
http://www.lomir.com

LSL Biolafitte
10 rue de Temara
7810C St.-Germain-en-Laye, France
(33) 1-3061-5260
FAX: (33) 1-3061-5234

Ludl Electronic Products
171 Brady Avenue
Hawthorne, NY 10532
(888) 769-6111 FAX: (914) 769-4759
(914) 769-6111
http://www.ludl.com

Lumigen
24485 W. Ten Mile Road
Southfield, MI 48034
(248) 351-5600 FAX: (248) 351-0518
http://www.lumigen.com

Luminex
12212 Technology Boulevard
Austin, TX 78727
(888) 219-8020 FAX: (512) 258-4173
(512) 219-8020
http://www.luminexcorp.com

LYNX Therapeutics
25861 Industrial Boulevard
Hayward, CA 94545
(510) 670-9300 FAX: (510) 670-9302
http://www.lynxgen.com

Lyphomed
3 Parkway North
Deerfield, IL 60015
(847) 317-8100 FAX: (847) 317-8600

M.E.D. Associates
See Med Associates

Macherey-Nagel
6 South Third Street, #402
Easton, PA 18042
(610) 559-9848 FAX: (610) 559-9878
http://www.macherey-nagel.com

Macherey-Nagel
Valencienner Strasse 11
P.O. Box 101352
D-52313 Dueren, Germany
(49) 2421-969141
FAX: (49) 2421-969199
http://www.macherey-nagel.ch

Mac-Mod Analytical
127 Commons Court
Chadds Ford, PA 19317
800-441-7508 FAX: (610) 358-5993
(610) 358-9696
http://www.mac-mod.com

Mallinckrodt Baker
222 Red School Lane
Phillipsburg, NJ 08865
(800) 582-2537 FAX: (908) 859-6974
(908) 859-2151
http://www.mallbaker.com

Mallinckrodt Chemicals
16305 Swingley Ridge Drive
Chesterfield, MD 63017
(314) 530-2172 FAX: (314) 530-2563
http://www.mallchem.com

Malven Instruments
Enigma Business Park
Grovewood Road
Malven, Worchestershire
WR 141 XZ, United Kingdom

Marinus
1500 Pier C Street
Long Beach, CA 90813
(562) 435-6522 FAX: (562) 495-3120

Markson Science
c/o Whatman Labs Sales
P.O. Box 1359
Hillsboro, OR 97123
(800) 942-8626 FAX: (503) 640-9716
(503) 648-0762

Marsh Biomedical Products
565 Blossom Road
Rochester, NY 14610
(800) 445-2812 FAX: (716) 654-4810
(716) 654-4800
http://www.biomar.com

Marshall Farms USA
5800 Lake Bluff Road
North Rose, NY 14516
(315) 587-2295
e-mail: info@marfarms.com

Martek
6480 Dobbin Road
Columbia, MD 21045
(410) 740-0081 FAX: (410) 740-2985
http://www.martekbio.com

Martin Supply
Distributor of Gerber Scientific
2740 Loch Raven Road
Baltimore, MD 21218
(800) 282-5440 FAX: (410) 366-0134
(410) 366-1696

Mast Immunosystems
630 Clyde Court
Mountain View, CA 94043
(800) 233-MAST FAX: (650) 969-2745
(650) 961-5501
http://www.mastallergy.com

Matheson Gas Products
P.O. Box 624
959 Route 46 East
Parsippany, NJ 07054
(800) 416-2505 FAX: (973) 257-9393
(973) 257-1100
http://www.mathesongas.com

Mathsoft
1700 Westlake Avenue N., Suite 500
Seattle, WA 98109
(800) 569-0123 FAX: (206) 283-8691
(206) 283-8802
http://www.mathsoft.com

Matreya
500 Tressler Street
Pleasant Gap, PA 16823
(814) 359-5060 FAX: (814) 359-5062
http://www.matreya.com

Matrigel
See Becton Dickinson Labware

A4

Matrix Technologies
22 Friars Drive
Hudson, NH 03051
(800) 345-0206   FAX: (603) 595-0106
(603) 595-0505
http://www.matrixtechcorp.com

MatTek Corp.
200 Homer Avenue
Ashland, Massachusetts 01721
(508) 881-6771   FAX: (508) 879-1532
http://www.mattek.com

Maxim Medical
89 Oxford Road
Oxford OX2 9PD
United Kingdom
44 (0)1865-865943
FAX: 44 (0)1865-865291
http://www.maximmed.com

Mayo Clinic
Section on Engineering
Project #ALA-1, 1982
200 1st Street SW
Rochester, MN 55905
(507) 284-2511   FAX: (507) 284-5988

McGaw
See B. Braun-McGaw

McMaster-Carr
600 County Line Road
Elmhurst, IL 60126
(630) 833-0300   FAX: (630) 834-9427
http://www.mcmaster.com

McNeil Pharmaceutical
See Ortho McNeil Pharmaceutical

MCNC
3021 Cornwallis Road
P.O. Box 12889
Research Triangle Park, NC 27709
(919) 248-1800   FAX: (919) 248-1455
http://www.mcnc.org

MD Industries
5 Revere Drive, Suite 415
Northbrook, IL 60062
(800) 421-8370   FAX: (847) 498-2627
(708) 339-6000
http://www.mdindustries.com

MDS Nordion
447 March Road
P.O. Box 13500
Kanata, Ontario
K2K 1X8 Canada
(800) 465-3666   FAX: (613) 592-6937
(613) 592-2790
http://www.mds.nordion.com

MDS Sciex
71 Four Valley Drive
Concord, Ontario
Canada L4K 4V8
(905) 660-9005   FAX: (905) 660-2600
http://www.sciex.com

Mead Johnson
See Bristol-Meyers Squibb

Med Associates
P.O. Box 319
St. Albans, VT 05478
(802) 527-2343   FAX: (802) 527-5095
http://www.med-associates.com

Medecell
239 Liverpool Road
London N1 1LX, UK
(44) 20-7607-2295
FAX: (44) 20-7700-4156
http://www.medicell.co.uk

Media Cybernetics
8484 Georgia Avenue, Suite 200
Silver Spring, MD 20910
(301) 495-3305   FAX: (301) 495-5964
http://www.mediacy.com

Mediatech
13884 Park Center Road
Herndon, VA 20171
(800) cellgro
(703) 471-5955
http://www.cellgro.com

Medical Systems
See Harvard Apparatus

Medifor
647 Washington Street
Port Townsend, WA 98368
(800) 366-3710   FAX: (360) 385-4402
(360) 385-0722
http://www.medifor.com

MedImmune
35 W. Watkins Mill Road
Gaithersburg, MD 20878
(301) 417-0770   FAX: (301) 527-4207
http://www.medimmune.com

MedProbe AS
P.O. Box 2640
St. Hanshaugen
N-0131 Oslo, Norway
(47) 222 00137   FAX: (47) 222 00189
http://www.medprobe.com

Megazyme
Bray Business Park
Bray, County Wicklow
Ireland
(353) 1-286-1220
FAX: (353) 1-286-1264
http://www.megazyme.com

Melles Griot
4601 Nautilus Court South
Boulder, CO 80301
(800) 326-4363   FAX: (303) 581-0960
(303) 581-0337
http://www.mellesgriot.com

Menzel-Glaser
Postfach 3157
D-38021 Braunschweig, Germany
(49) 531 590080
FAX: (49) 531 509799

E. Merck
Frankfurterstrasse 250
D-64293 Darmstadt 1, Germany
(49) 6151-720

Merck
See EM Science

Merck & Company
Merck National Service Center
P.O. Box 4
West Point, PA 19486
(800) NSC-MERCK
(215) 652-5000
http://www.merck.com

Merck Research Laboratories
See Merck & Company

Merck Sharpe Human Health Division
300 Franklin Square Drive
Somerset, NJ 08873
(800) 637-2579   FAX: (732) 805-3960
(732) 805-0300

Merial Limited
115 Transtech Drive
Athens, GA 30601
(800) MERIAL-1   FAX: (706) 548-0608
(706) 548-9292
http://www.merial.com

Meridian Instruments
P.O. Box 1204
Kent, WA 98035
(253) 854-9914   FAX: (253) 854-9902
http://www.minstrument.com

Meta Systems Group
32 Hammond Road
Belmont, MA 02178
(617) 489-9950   FAX: (617) 489-9952

Metachem Technologies
3547 Voyager Street, Bldg. 102
Torrance, CA 90503
(310) 793-2300   FAX: (310) 793-2304
http://www.metachem.com

Metallhantering
Box 47172
100-74 Stockholm, Sweden
(46) 8-726-9696

MethylGene
7220 Frederick-Banting, Suite 200
Montreal, Quebec
H4S 2A1 Canada
http://www.methylgene.com

Metro Scientific
475 Main Street, Suite 2A
Farmingdale, NY 11735
(800) 788-6247   FAX: (516) 293-8549
(516) 293-9656

Metrowerks
980 Metric Boulevard
Austin, TX 78758
(800) 377-5416
(512) 997-4700
http://www.metrowerks.com

Mettler Instruments
Mettler-Toledo
1900 Polaris Parkway
Columbus, OH 43240
(800) METTLER   FAX: (614) 438-4900
http://www.mt.com

Miami Serpentarium Labs
34879 Washington Loop Road
Punta Gorda, FL 33982
(800) 248-5050   FAX: (813) 639-1811
(813) 639-8888
http://www.miamiserpentarium.com

Michrom BioResources
1945 Industrial Drive
Auburn, CA 95603
(530) 888-6498   FAX: (530) 888-8295
http://www.michrom.com

Mickle Laboratory Engineering
Gomshall, Surrey, UK
(44) 1483-202178

Micra Scientific
A division of Eichrom Industries
8205 S. Cass Ave, Suite 111
Darien, IL 60561
(800) 283-4752   FAX: (630) 963-1928
(630) 963-0320
http://www.micrasci.com

MicroBrightField
74 Hegman Avenue
Colchester, VT 05446
(802) 655-9360   FAX: (802) 655-5245
http://www.microbrightfield.com

Micro Essential Laboratory
4224 Avenue H
Brooklyn, NY 11210
(718) 338-3618   FAX: (718) 692-4491

Micro Filtration Systems
7-3-Chome, Honcho
Nihonbashi, Tokyo, Japan
(81) 3-270-3141

Micro-Metrics
P.O. Box 13804
Atlanta, GA 30324
(770) 986-6015   FAX: (770) 986-9510
http://www.micro-metrics.com

Micro-Tech Scientific
140 South Wolfe Road
Sunnyvale, CA 94086
(408) 730-8324   FAX: (408) 730-3566
http://www.microlc.com

Microbix Biosystems
341 Bering Avenue
Toronto, Ontario
M8Z 3A8 Canada
1-800-794-6694   FAX: 416-234-1626
1-416-234-1624
http://www.microbix.com

MicroCal
22 Industrial Drive East
Northampton, MA 01060
(800) 633-3115   FAX: (413) 586-0149
(413) 586-7720
http://www.microcalorimetry.com

Microfluidics
30 Ossipee Road
P.O. Box 9101
Newton, MA 02164
(800) 370-5452   FAX: (617) 965-1213
(617) 969-5452
http://www.microfluidicscorp.com

Microgon
See Spectrum Laboratories

Microlase Optical Systems
West of Scotland Science Park
Kelvin Campus, Maryhill Road
Glasgow G20 0SP, UK
(44) 141-948-1000
FAX: (44) 141-946-6311
http://www.microlase.co.uk

Micron Instruments
4509 Runway Street
Simi Valley, CA 93063
(800) 638-3770   FAX: (805) 522-4982
(805) 552-4676
http://www.microninstruments.com

Micron Separations
See MSI

Micro Photonics
4949 Liberty Lane, Suite 170
P.O. Box 3129
Allentown, PA 18106
(610) 366-7103   FAX: (610) 366-7105
http://www.microphotonics.com

MicroTech
1420 Conchester Highway
Boothwyn, PA 19061
(610) 459-3514

Midland Certified Reagent Company
3112-A West Cuthbert Avenue
Midland, TX 79701
(800) 247-8766   FAX: (800) 359-5789
(915) 694-7950   FAX: (915) 694-2387
http://www.mcrc.com

Midwest Scientific
280 Vance Road
Valley Park, MO 63088
(800) 227-9997   FAX: (636) 225-9998
(636) 225-9997
http://www.midsci.com

Miles
See Bayer

Miles Laboratories
See Serological

Miles Scientific
See Nunc

Millar Instruments
P.O. Box 230227
6001-A Gulf Freeway
Houston, TX 77023
(713) 923-9171   FAX: (713) 923-7757
http://www.millarinstruments.com

MilliGen/Biosearch
See Millipore

Millipore
80 Ashbury Road
P.O. Box 9125
Bedford, MA 01730
(800) 645-5476   FAX: (781) 533-3110
(781) 533-6000
http://www.millipore.com

Miltenyi Biotec
251 Auburn Ravine Road, Suite 208
Auburn, CA 95603
(800) 367-6227   FAX: (530) 888-8925
(530) 888-8871
http://www.miltenyibiotec.com

Miltex
6 Ohio Drive
Lake Success, NY 11042
(800) 645-8000   FAX: (516) 775-7185
(516) 349-0001

Milton Roy
See Spectronic Instruments

Mini-Instruments
15 Burnham Business Park
Springfield Road
Burnham-on-Crouch, Essex CM0 8TE,
UK
(44) 1621-783282
FAX: (44) 1621-783132
http://www.mini-instruments.co.uk

Mini Mitter
P.O. Box 3386
Sunriver, OR 97707
(800) 685-2999   FAX: (541) 593-5604
(541) 593-8639
http://www.minimitter.com

Mirus Corporation
505 S. Rosa Road
Suite 104
Madison, WI 53719
(608) 441-2852 FAX: (608) 441-2849
http://www.genetransfer.com

Misonix
1938 New Highway
Farmingdale, NY 11735
(800) 645-9846 FAX: (516) 694-9412
http://www.misonix.com

Mitutoyo (MTI)
See Dolla Eastern

MJ Research
Waltham, MA 02451
(800) PELTIER FAX: (617) 923-8080
(617) 923-8000
http://www.mjr.com

Modular Instruments
228 West Gay Street
Westchester, PA 19380
(610) 738-1420 FAX: (610) 738-1421
http://www.mi2.com

Molecular Biology Insights
8685 US Highway 24
Cascade, CO 80809-1333
(800) 747-4362 FAX: (719) 684-7989
(719) 684-7988
http://www.oligo.net

Molecular Biosystems
10030 Barnes Canyon Road
San Diego, CA 92121
(858) 452-0681 FAX: (858) 452-6187
http://www.mobi.com

Molecular Devices
1312 Crossman Avenue
Sunnyvale, CA 94089
(800) 635-5577 FAX: (408) 747-3602
(408) 747-1700
http://www.moldev.com

Molecular Designs
1400 Catalina Street
San Leandro, CA 94577
(510) 895-1313 FAX: (510) 614-3608

Molecular Dynamics
928 East Arques Avenue
Sunnyvale, CA 94086
(800) 333-5703 FAX: (408) 773-1493
(408) 773-1222
http://www.apbiotech.com

Molecular Probes
4849 Pitchford Avenue
Eugene, OR 97402
(800) 438-2209 FAX: (800) 438-0228
(541) 465-8300 FAX: (541) 344-6504
http://www.probes.com

Molecular Research Center
5645 Montgomery Road
Cincinnati, OH 45212
(800) 462-9868 FAX: (513) 841-0080
(513) 841-0900
http://www.mrcgene.com

Molecular Simulations
9685 Scranton Road
San Diego, CA 92121
(800) 756-4674 FAX: (858) 458-0136
(858) 458-9990
http://www.msi.com

**Monoject Disposable Syringes
& Needles/Syrvet**
16200 Walnut Street
Waukee, IA 50263
(800) 727-5203 FAX: (515) 987-5553
(515) 987-5554
http://www.syrvet.com

Monsanto Chemical
800 North Lindbergh Boulevard
St. Louis, MO 63167
(314) 694-1000 FAX: (314) 694-7625
http://www.monsanto.com

Moravek Biochemicals
577 Mercury Lane
Brea, CA 92821
(800) 447-0100 FAX: (714) 990-1824
(714) 990-2018
http://www.moravek.com

Moss
P.O. Box 189
Pasadena, MD 21122
(800) 932-6677 FAX: (410) 768-3971
(410) 768-3442
http://www.mosssubstrates.com

Motion Analysis
3617 Westwind Boulevard
Santa Rosa, CA 95403
(707) 579-6500 FAX: (707) 526-0629
http://www.motionanalysis.com

Mott
Farmington Industrial Park
84 Spring Lane
Farmington, CT 06032
(860) 747-6333 FAX: (860) 747-6739
http://www.mottcorp.com

MSI (Micron Separations)
See Osmonics

Multi Channel Systems
Markwiesenstrasse 55
72770 Reutlingen, Germany
(49) 7121-503010
FAX: (49) 7121-503011
http://www.multichannelsystems.com

Multiple Peptide Systems
3550 General Atomics Court
San Diego, CA 92121
(800) 338-4965 FAX: (800) 654-5592
(858) 455-3710 FAX: (858) 455-3713
http://www.mps-sd.com

Murex Diagnostics
3075 Northwoods Circle
Norcross, GA 30071
(707) 662-0660 FAX: (770) 447-4989

MWG-Biotech
Anzinger Str. 7
D-85560 Ebersberg, Germany
(49) 8092-82890 FAX: (49)
8092-21084
http://www.mwg·biotech.com

Myriad Industries
3454 E Street
San Diego, CA 92102
(800) 999-6777 FAX: (619) 232-4819
(619) 232-6700
http://www.myriadindustries.com

Nacalai Tesque
Nijo Karasuma, Nakagyo-ku
Kyoto 604, Japan
81-75-251-1723
FAX: 81-75-251-1762
http://www.nacalai.co.jp

Nalge Nunc International
Subsidiary of Sybron International
75 Panorama Creek Drive
P.O. Box 20365
Rochester, NY 14602
(800) 625-4327 FAX: (716) 586-8987
(716) 264-9346
http://www.nalgenunc.com

Nanogen
10398 Pacific Center Court
San Diego, CA 92121
(858) 410-4600 FAX: (858) 410-4848
http://www.nanogen.com

Nanoprobes
95 Horse Block Road
Yaphank, NY 11980
(877) 447-6266 FAX: (631) 205-9493
(631) 205-9490
http://www.nanoprobes.com

Narishige USA
1710 Hempstead Turnpike
East Meadow, NY 11554
(800) 445-7914 FAX: (516) 794-0066
(516) 794-8000
http://www.narishige.co.jp

Nasco-Fort Atkinson
P.O. Box 901
901 Janesville Ave.
Fort Atkinson, WI 53538-0901
(800) 558-9595 FAX: (920) 563-8296
http://www.enasco.com

A4

National Bag Company
2233 Old Mill Road
Hudson, OH 44236
(800) 247-6000   FAX: (330) 425-9800
(330) 425-2600
http://www.nationalbag.com

National Band and Tag
Department X 35, Box 72430
Newport, KY 41032
(606) 261-2035   FAX: (800) 261-8247
https://www.nationalband.com

National Biosciences
See Molecular Biology Insights

National Diagnostics
305 Patton Drive
Atlanta, GA 30336
(800) 526-3867   FAX: (404) 699-2077
(404) 699-2121
http://www.nationaldiagnostics.com

National Disease Research Exchange
1880 John F. Kennedy Blvd., 11th Fl.
Philadelphia, PA 19103
(800) 222-6374
http://www.ndri.com

**National Institute of Standards
and Technology**
100 Bureau Drive
Gaithersburg, MD 20899
(301) 975-NIST   FAX: (301) 926-1630
http://www.nist.gov

National Instruments
11500 North Mopac Expressway
Austin, TX 78759
(512) 794-0100   FAX: (512) 683-8411
http://www.ni.com

National Labnet
See Labnet International

National Scientific Instruments
975 Progress Circle
Lawrenceville, GA 300243
(800) 332-3331   FAX: (404) 339-7173
http://www.nationalscientific.com

National Scientific Supply
1111 Francisco Bouldvard East
San Rafael, CA 94901
(800) 525-1779   FAX: (415) 459-2954
(415) 459-6070
http://www.nat-sci.com

Naz-Dar-KC Chicago
Nazdar
1087 N. North Branch Street
Chicago, IL 60622
(800) 736-7636   FAX: (312) 943-8215
(312) 943-8338
http://www.nazdar.com

NB Labs
1918 Avenue A
Denison, TX 75021
(903) 465-2694   FAX: (903) 463-5905
http://www.nblabslarry.com

NEB
See New England Biolabs

NEN Life Science Products
549 Albany Street
Boston, MA 02118
(800) 551-2121   FAX: (617) 451-8185
(617) 350-9075
http://www.nen.com

NEN Research Products, Dupont (UK)
Diagnostics and Biotechnology Systems
Wedgewood Way
Stevenage, Hertfordshire SG1 4QN, UK
44-1438-734831
44-1438-734000
FAX: 44-1438-734836
http://www.dupont.com

Neogen
628 Winchester Road
Lexington, KY 40505
(800) 477-8201   FAX: (606) 255-5532
(606) 254-1221
http://www.neogen.com

Neosystems
380, 11012 Macleod Trail South
Calgary, Alberta
T2J 6A5 Canada
(403) 225-9022   FAX: (403) 225-9025
http://www.neosystems.com

Neuralynx
2434 North Pantano Road
Tucson, AZ 85715
(520) 722-8144   FAX: (520) 722-8163
http://www.neuralynx.com

Neuro Probe
16008 Industrial Drive
Gaithersburg, MD 20877
(301) 417-0014   FAX: (301) 977-5711
http://www.neuroprobe.com

Neurocrine Biosciences
10555 Science Center Drive
San Diego, CA 92121
(619) 658-7600   FAX: (619) 658-7602
http://www.neurocrine.com

Nevtek
HCR03, Box 99
Burnsville, VA 24487
(540) 925-2322   FAX: (540) 925-2323
http://www.nevtek.com

New Brunswick Scientific
44 Talmadge Road
Edison, NJ 08818
(800) 631-5417   FAX: (732) 287-4222
(732) 287-1200
http://www.nbsc.com

New England Biolabs (NEB)
32 Tozer Road
Beverly, MA 01915
(800) 632-5227   FAX: (800) 632-7440
http://www.neb.com

New England Nuclear (NEN)
See NEN Life Science Products

New MBR
Gubelstrasse 48
CH8050 Zurich, Switzerland
(41) 1-313-0703

Newark Electronics
4801 N. Ravenswood Avenue
Chicago, IL 60640
(800) 4-NEWARK   FAX: (773)
907-5339
(773) 784-5100
http://www.newark.com

Newell Rubbermaid
29 E. Stephenson Street
Freeport, IL 61032
(815) 235-4171   FAX: (815) 233-8060
http://www.newellco.com

Newport Biosystems
1860 Trainor Street
Red Bluff, CA 96080
(530) 529-2448   FAX: (530) 529-2648

Newport
1791 Deere Avenue
Irvine, CA 92606
(800) 222-6440   FAX: (949) 253-1800
(949) 253-1462
http://www.newport.com

Nexin Research B.V.
P.O. Box 16
4740 AA Hoeven, The Netherlands
(31) 165-503172
FAX: (31) 165-502291

NIAID
See Bio-Tech Research Laboratories

Nichiryo
230 Route 206
Building 2-2C
Flanders, NJ 07836
(877) 548-6667   FAX: (973) 927-0099
(973) 927-4001
http://www.nichiryo.com

Nichols Institute Diagnostics
33051 Calle Aviador
San Juan Capistrano, CA 92675
(800) 286-4NID   FAX: (949) 240-5273
(949) 728-4610
http://www.nicholsdiag.com

Nichols Scientific Instruments
3334 Brown Station Road
Columbia, MO 65202
(573) 474-5522   FAX: (603) 215-7274
http://home.beseen.com
technology/nsi_technology

A4

Nicolet Biomedical Instruments
5225 Verona Road, Building 2
Madison, WI 53711
(800) 356-0007   FAX: (608) 441-2002
(608) 273-5000
http://nicoletbiomedical.com

N.I.G.M.S. (National Institute of General Medical Sciences)
See Coriell Institute for
Medical Research

Nikon
Science and Technologies Group
1300 Walt Whitman Road
Melville, NY 11747
(516) 547-8500   FAX: (516) 547-4045
http://www.nikonusa.com

Nippon Gene
1-29, Ton-ya-machi
Toyama 930, Japan
(81) 764-51-6548
FAX: (81) 764-51-6547

Noldus Information Technology
751 Miller Drive
Suite E-5
Leesburg, VA 20175
(800) 355-9541   FAX: (703) 771-0441
(703) 771-0440
http://www.noldus.com

Nonlinear Dynamics
See NovoDynamics

Nordion International
See MDS Nordion

North American Biologicals (NABI)
16500 NW 15th Avenue
Miami, FL 33169
(800) 327-7106 (305) 625-5305
http://www.nabi.com

North American Reiss
See Reiss

Northwestern Bottle
24 Walpole Park South
Walpole, MA 02081
(508) 668-8600   FAX: (508) 668-7790

NOVA Biomedical
Nova Biomedical 200
Prospect Street Waltham, MA 02454
(800) 822-0911   FAX: (781) 894-5915
http://www.novabiomedical.com

Novagen
601 Science Drive
Madison, WI 53711
(800) 526-7319   FAX: (608) 238-1388
(608) 238-6110
http://www.novagen.com

Novartis
59 Route 10
East Hanover, NJ 07936
(800)526-0175   FAX: (973) 781-6356
http://www.novartis.com

Novartis Biotechnology
3054 Cornwallis Road
Research Triangle Park, NC 27709
(888) 462-7288   FAX: (919) 541-8585
http://www.novartis.com

Nova Sina AG
Subsidiary of Airflow Lufttechnik GmbH
Kleine Heeg 21
52259 Rheinbach, Germany
(49) 02226 920-0
FAX: (49) 02226 9205-11

Novex/Invitrogen
1600 Faraday
Carlsbad, CA 92008
(800) 955-6288   FAX: (760) 603-7201
http://www.novex.com

Novo Nordisk Biochem
77 Perry Chapel Church Road
Franklington, NC 27525
(800) 879-6686   FAX: (919) 494-3450
(919) 494-3000
http://www.novo.dk

Novo Nordisk BioLabs
See Novo Nordisk Biochem

Novocastra Labs
Balliol Business Park West
Benton Lane
Newcastle-upon-Tyne
Tyne and Wear NE12 8EW, UK
(44) 191-215-0567
FAX: (44) 191-215-1152
http://www.novocastra.co.uk

NovoDynamics
123 North Ashley Street
Suite 210
Ann Arbor, MI 48104
(734) 205-9100   FAX: (734) 205-9101
http://www.novodynamics.com

Novus Biologicals
P.O. Box 802
Littleton, CO 80160
(888) 506-6887   FAX: (303) 730-1966
http://www.novus-biologicals.com/
main.html

NPI Electronic
Hauptstrasse 96
D-71732 Tamm, Germany
(49) 7141-601534
FAX: (49) 7141-601266
http://www.npielectronic.com

NSG Precision Cells
195G Central Avenue
Farmingdale, NY 11735
(516) 249-7474   FAX: (516) 249-8575
http://www.nsgpci.com

Nu Chek Prep
109 West Main
P.O. Box 295
Elysian, MN 56028
(800) 521-7728   FAX: (507) 267-4790
(507) 267-4689

Nuclepore
See Costar

Numonics
101 Commerce Drive
Montgomeryville, PA 18936
(800) 523-6716   FAX: (215) 361-0167
(215) 362-2766
http://www.interactivewhiteboards.com

NYCOMED AS Pharma
c/o Accurate Chemical & Scientific
300 Shames Drive
Westbury, NY 11590
(800) 645-6524   FAX: (516) 997-4948
(516) 333-2221
http://www.accuratechemical.com

Nycomed Amersham
Health Care Division
101 Carnegie Center
Princeton, NJ 08540
(800) 832-4633   FAX: (800) 807-2382
(609) 514-6000
http://www.nycomed-amersham.com

Nyegaard
Herserudsvagen 5254
S-122 06 Lidingo, Sweden
(46) 8-765-2930

Ohmeda Catheter Products
See Datex-Ohmeda

Ohwa Tsusbo
Hiby Dai Building
1-2-2 Uchi Saiwai-cho
Chiyoda-ku
Tokyo 100, Japan
03-3591-7348   FAX: 03-3501-9001

Oligos Etc.
9775 S.W. Commerce Circle, C-6
Wilsonville, OR 97070
(800) 888-2358   FAX: (503)
6822D1635
(503) 6822D1814
http://www.oligoetc.com

Olis Instruments
130 Conway Drive
Bogart, GA 30622
(706) 353-6547 (800) 852-3504
http://www.olisweb.com

Olympus America
2 Corporate Center Drive
Melville, NY 11747
(800) 645-8160   FAX: (516) 844-5959
(516) 844-5000
http://www.olympusamerica.com

Omega Engineering
One Omega Drive
P.O. Box 4047
Stamford, CT 06907
(800) 848-4286   FAX: (203) 359-7700
(203) 359-1660
http://www.omega.com

Omega Optical
3 Grove Street
P.O. Box 573
Brattleboro, VT 05302
(802) 254-2690   FAX: (802) 254-3937
http://www.omegafilters.com

Omnetics Connector Corporation
7260 Commerce Circle
East Minneapolis, MN 55432
(800) 343-0025 (763) 572-0656
Fax: (763) 572-3925
http://www.omnetics.com/main.htm

Omni International
6530 Commerce Court
Warrenton, VA 20187
(800) 776-4431   FAX: (540) 347-5352
(540) 347-5331
http://www.omni-inc.com

Omnion
2010 Energy Drive
P.O. Box 879
East Troy, WI 53120
(262) 642-7200   FAX: (262) 642-7760
http://www.omnion.com

Omnitech Electronics
See AccuScan Instruments

Oncogene Research Products
P.O. Box Box 12087
La Jolla, CA 92039-2087
(800) 662-2616   FAX: (800) 766-0999
http://www.apoptosis.com

Oncogene Science
See OSI Pharmaceuticals

Oncor
See Intergen

Online Instruments
130 Conway Drive, Suites A & B
Bogart, GA 30622
(800) 852-3504 (706) 353-1972
(706) 353-6547
http://www.olisweb.com

Operon Technologies
1000 Atlantic Avenue
Alameda, CA 94501
(800) 688-2248   FAX: (510) 865-5225
(510) 865-8644
http://www.operon.com

Optiscan
P.O. Box 1066
Mount Waverly MDC, Victoria
Australia 3149
61-3-9538 3333   FAX: 61-3-9562 7742
http://www.optiscan.com.au

Optomax
9 Ash Street
P.O. Box 840
Hollis, NH 03049
(603) 465-3385   FAX: (603) 465-2291

Opto-Line Associates
265 Ballardvale Street
Wilmington, MA 01887
(978) 658-7255   FAX: (978) 658-7299
http://www.optoline.com

Orbigen
6827 Nancy Ridge Drive
San Diego, CA 92121
(866) 672-4436 (858) 362-2030
(858) 362-2026
http://www.orbigen.com

Oread BioSaftey
1501 Wakarusa Drive
Lawrence, KS 66047
(800) 447-6501   FAX: (785) 749-1882
(785) 749-0034
http://www.oread.com

Organomation Associates
266 River Road West
Berlin, MA 01503
(888) 978-7300   FAX: (978)838-2786
(978) 838-7300
http://www.organomation.com

Organon
375 Mount Pleasant Avenue
West Orange, NJ 07052
(800) 241-8812   FAX: (973) 325-4589
(973) 325-4500
http://www.organon.com

Organon Teknika (Canada)
30 North Wind Place
Scarborough, Ontario
M1S 3R5 Canada
(416) 754-4344   FAX: (416) 754-4488
http://www.organonteknika.com

Organon Teknika Cappel
100 Akzo Avenue
Durham, NC 27712
(800) 682-2666   FAX: (800) 432-9682
(919) 620-2000   FAX: (919) 620-2107
http://www.organonteknika.com

Oriel Corporation of America
150 Long Beach Boulevard
Stratford, CT 06615
(203) 377-8282   FAX: (203) 378-2457
http://www.oriel.com

OriGene Technologies
6 Taft Court, Suite 300
Rockville, MD 20850
(888) 267-4436   FAX: (301) 340-9254
(301) 340-3188
http://www.origene.com

OriginLab
One Roundhouse Plaza
Northhampton, MA 01060
(800) 969-7720   FAX: (413) 585-0126
http://www.originlab.com

Orion Research
500 Cummings Center
Beverly, MA 01915
(800) 225-1480   FAX: (978) 232-6015
(978) 232-6000
http://www.orionres.com

Ortho Diagnostic Systems
Subsidiary of Johnson & Johnson
1001 U.S. Highway 202
P.O. Box 350
Raritan, NJ 08869
(800) 322-6374   FAX: (908) 218-8582
(908) 218-1300

Ortho McNeil Pharmaceutical
Welsh & McKean Road
Spring House, PA 19477
(800) 682-6532
(215) 628-5000
http://www.orthomcneil.com

Oryza
200 Turnpike Road, Unit 5
Chelmsford, MA 01824
(978) 256-8183   FAX: (978) 256-7434
http://www.oryzalabs.com

OSI Pharmaceuticals
106 Charles Lindbergh Boulevard
Uniondale, NY 11553
(800) 662-2616   FAX: (516) 222-0114
(516) 222-0023
http://www.osip.com

Osmonics
135 Flanders Road
P.O. Box 1046
Westborough, MA 01581
(800) 444-8212   FAX: (508) 366-5840
(508) 366-8212
http://www.osmolabstore.com

Oster Professional Products
150 Cadillac Lane
McMinnville, TN 37110
(931) 668-4121   FAX: (931) 668-4125
http://www.sunbeam.com

A4

Out Patient Services
1260 Holm Road
Petaluma, CA 94954
(800) 648-1666   FAX: (707) 762-7198
(707) 763-1581

OWL Scientific Plastics
See OWL Separation Systems

OWL Separation Systems
55 Heritage Avenue
Portsmouth, NH 03801
(800) 242-5560   FAX: (603) 559-9258
(603) 559-9297
http://www.owlsci.com

Oxford Biochemical Research
P.O. Box 522
Oxford, MI 48371
(800) 692-4633   FAX: (248) 852-4466
http://www.oxfordbiomed.com

Oxford GlycoSystems
See Glyco

Oxford Instruments
Old Station Way
Eynsham
Witney, Oxfordshire OX8 1TL, UK
(44) 1865-881437
FAX: (44) 1865-881944
http://www.oxinst.com

Oxford Labware
See Kendall

Oxford Molecular Group
Oxford Science Park
The Medawar Centre
Oxford OX4 4GA, UK
(44) 1865-784600
FAX: (44) 1865-784601
http://www.oxmol.co.uk

Oxford Molecular Group
2105 South Bascom Avenue, Suite 200
Campbell, CA 95008
(800) 876-9994   FAX: (408) 879-6302
(408) 879-6300
http://www.oxmol.com

OXIS International
6040 North Cutter Circle
Suite 317
Portland, OR 97217
(800) 547-3686   FAX: (503) 283-4058
(503) 283-3911
http://www.oxis.com

Oxoid
800 Proctor Avenue
Ogdensburg, NY 13669
(800) 567-8378   FAX: (613) 226-3728
http://www.oxoid.ca

Oxoid
Wade Road
Basingstoke, Hampshire RG24 8PW, UK
(44) 1256-841144
FAX: (4) 1256-814626
http://www.oxoid.ca

Oxyrase
P.O. Box 1345
Mansfield, OH 44901
(419) 589-8800   FAX: (419) 589-9919
http://www.oxyrase.com

Ozyme
10 Avenue Ampere
Montigny de Bretoneux
78180 France
(33) 13-46-02-424
FAX: (33) 13-46-09-212
http://www.ozyme.fr

PAA Laboratories
2570 Route 724
P.O. Box 435
Parker Ford, PA 19457
(610) 495-9400   FAX: (610) 495-9410
http://www.paa-labs.com

Pacer Scientific
5649 Valley Oak Drive
Los Angeles, CA 90068
(323) 462-0636   FAX: (323) 462-1430
http://www.pacersci.com

Pacific Bio-Marine Labs
P.O. Box 1348
Venice, CA 90294
(310) 677-1056   FAX: (310) 677-1207

Packard Instrument
800 Research Parkway
Meriden, CT 06450
(800) 323-1891   FAX: (203) 639-2172
(203) 238-2351
http://www.packardinst.com

Padgett Instrument
1730 Walnut Street
Kansas City, MO 64108
(816) 842-1029

Pall Filtron
50 Bearfoot Road
Northborough, MA 01532
(800) FILTRON   FAX: (508) 393-1874
(508) 393-1800

Pall-Gelman
25 Harbor Park Drive
Port Washington, NY 11050
(800) 289-6255   FAX: (516) 484-2651
(516) 484-3600
http://www.pall.com

PanVera
545 Science Drive
Madison, WI 53711
(800) 791-1400   FAX: (608) 233-3007
(608) 233-9450
http://www.panvera.com

Parke-Davis
See Warner-Lambert

Parr Instrument
211 53rd Street
Moline, IL 61265
(800) 872-7720   FAX: (309) 762-9453
(309) 762-7716
http://www.parrinst.com

Partec
Otto Hahn Strasse 32
D-48161 Munster, Germany
(49) 2534-8008-0
FAX: (49) 2535-8008-90

PCR
See Archimica Florida

PE Biosystems
850 Lincoln Centre Drive
Foster City, CA 94404
(800) 345-5224 FAX: (650) 638-5884
(650) 638-5800
http://www.pebio.com

Pel-Freez Biologicals
219 N. Arkansas
P.O. Box 68
Rogers, AR 72757
(800) 643-3426   FAX: (501) 636-3562
(501) 636-4361
http://www.pelfreez-bio.com

Pel-Freez Clinical Systems
Subsidiary of Pel-Freez Biologicals
9099 N. Deerbrook Trail
Brown Deer, WI 53223
(800) 558-4511   FAX: (414) 357-4518
(414) 357-4500
http://www.pelfreez-bio.com

Peninsula Laboratories
601 Taylor Way
San Carlos, CA 94070
(800) 650-4442   FAX: (650) 595-4071
(650) 592-5392
http://www.penlabs.com

Pentex
24562 Mando Drive
Laguna Niguel, CA 92677
(800) 382-4667   FAX: (714) 643-2363
http://www.pentex.com

PeproTech
5 Crescent Avenue
P.O. Box 275
Rocky Hill, NJ 08553
(800) 436-9910   FAX: (609) 497-0321
(609) 497-0253
http://www.peprotech.com

A4

Peptide Institute
4-1-2 Ina, Minoh-shi
Osaka 562-8686, Japan
81-727-29-4121   FAX: 81-727-29-4124
http://www.peptide.co.jp

Peptide Laboratory
4175 Lakeside Drive
Richmond, CA 94806
(800) 858-7322   FAX: (510) 262-9127
(510) 262-0800
http://www.peptidelab.com

Peptides International
11621 Electron Drive
Louisville, KY 40299
(800) 777-4779   FAX: (502) 267-1329
(502) 266-8787
http://www.pepnet.com

Perceptive Science Instruments
2525 South Shore Boulevard, Suite 100
League City, TX 77573
(281) 334-3027   FAX: (281) 538-2222
http://www.persci.com

Perimed
4873 Princeton Drive
North Royalton, OH 44133
(440) 877-0537   FAX: (440) 877-0534
http://www.perimed.se

Perkin-Elmer
761 Main Avenue
Norwalk, CT 06859
(800) 762-4002   FAX: (203) 762-6000
(203) 762-1000
http://www.perkin-elmer.com
See also PE Biosystems

PerSeptive Bioresearch Products
See PerSeptive BioSystems

PerSeptive BioSystems
500 Old Connecticut Path
Framingham, MA 01701
(800) 899-5858   FAX: (508) 383-7885
(508) 383-7700
http://www.pbio.com

PerSeptive Diagnostic
See PE Biosystems
(800) 343-1346

Pettersson Elektronik AB
Tallbacksvagen 51
S-756 45 Uppsala, Sweden
(46) 1830-3880   FAX: (46) 1830-3840
http://www.bahnhof.se/~pettersson

Pfanstiehl Laboratories, Inc.
1219 Glen Rock Avenue
Waukegan, IL 60085
(800) 383-0126   FAX: (847) 623-9173
http://www.pfanstiehl.com

PGC Scientifics
7311 Governors Way
Frederick, MD 21704
(800) 424-3300   FAX: (800) 662-1112
(301) 620-7777   FAX: (301) 620-7497
http://www.pgcscientifics.com

Pharmacia Biotech
See Amersham Pharmacia Biotech

Pharmacia Diagnostics
See Wallac

Pharmacia LKB Biotech
See Amersham Pharmacia Biotech

Pharmacia LKB Biotechnology
See Amersham Pharmacia Biotech

Pharmacia LKB Nuclear
See Wallac

Pharmaderm Veterinary Products
60 Baylis Road
Melville, NY 11747
(800) 432-6673
http://www.pharmaderm.com

Pharmed (Norton)
Norton Performance Plastics
See Saint-Gobain Performance Plastics

PharMingen
See BD PharMingen

Phenomex
2320 W. 205th Street
Torrance, CA 90501
(310) 212-0555   FAX: (310) 328-7768
http://www.phenomex.com

PHLS Centre for Applied
Microbiology and Research
See European Collection of Animal
Cell Cultures (ECACC)

Phoenix Flow Systems
11575 Sorrento Valley Road, Suite 208
San Diego, CA 92121
(800) 886-3569   FAX: (619) 259-5268
(619) 453-5095
http://www.phnxflow.com

Phoenix Pharmaceutical
4261 Easton Road, P.O. Box 6457
St. Joseph, MO 64506
(800) 759-3644   FAX: (816) 364-4969
(816) 364-5777
http://www.phoenixpharmaceutical.com

Photometrics
See Roper Scientific

Photon Technology International
1 Deerpark Drive, Suite F
Monmouth Junction, NJ 08852
(732) 329-0910   FAX: (732) 329-9069
http://www.pti-nj.com

Physik Instrumente
Polytec PI
23 Midstate Drive, Suite 212
Auburn, MA 01501
(508) 832-3456   FAX: (508) 832-0506
http://www.polytecpi.com

Physitemp Instruments
154 Huron Avenue
Clifton, NJ 07013
(800) 452-8510   FAX: (973) 779-5954
(973) 779-5577
http://www.physitemp.com

Pico Technology
The Mill House, Cambridge Street
St. Neots, Cambridgeshire
PE19 1QB, UK
(44) 1480-396-395
FAX: (44) 1480-396-296
http://www.picotech.com

Pierce Chemical
P.O. Box 117
3747 Meridian Road
Rockford, IL 61105
(800) 874-3723   FAX: (800) 842-5007
FAX: (815) 968-7316
http://www.piercenet.com

Pierce & Warriner
44, Upper Northgate Street
Chester, Cheshire CH1 4EF, UK
(44) 1244 382 525
FAX: (44) 1244 373 212
http://www.piercenet.com

Pilling Weck Surgical
420 Delaware Drive
Fort Washington, PA 19034
(800) 523-2579   FAX: (800) 332-2308
http://www.pilling-weck.com

PixelVision
A division of Cybex Computer Products
14964 NW Greenbrier Parkway
Beaverton, OR 97006
(503) 629-3210   FAX: (503) 629-3211
http://www.pixelvision.com

P.J. Noyes
P.O. Box 381
89 Bridge Street
Lancaster, NH 03584
(800) 522-2469   FAX: (603) 788-3873
(603) 788-4952
http://www.pjnoyes.com

Plas-Labs
917 E. Chilson Street
Lansing, MI 48906
(800) 866-7527   FAX: (517) 372-2857
(517) 372-7177
http://www.plas-labs.com

A4

Plastics One
6591 Merriman Road, Southwest
P.O. Box 12004
Roanoke, VA 24018
(540) 772-7950   FAX: (540) 989-7519
http://www.plastics1.com

Platt Electric Supply
2757 6th Avenue South
Seattle, WA 98134
(206) 624-4083   FAX: (206) 343-6342
http://www.platt.com

Plexon
6500 Greenville Avenue
Suite 730
Dallas,TX 75206
(214) 369-4957   FAX: (214) 369-1775
http://www.plexoninc.com

Polaroid
784 Memorial Drive
Cambridge, MA 01239
(800) 225-1618   FAX: (800) 832-9003
(781) 386-2000
http://www.polaroid.com

Polyfiltronics
136 Weymouth St.
Rockland, MA 02370
(800) 434-7659   FAX: (781) 878-0822
(781) 878-1133
http://www.polyfiltronics.com

Polylabo Paul Block
Parc Tertiare de la Meinau
10, rue de la Durance
B.P. 36
67023 Strasbourg Cedex 1
Strasbourg, France
33-3-8865-8020
FAX: 33-3-8865-8039

PolyLC
9151 Rumsey Road, Suite 180
Columbia, MD 21045
(410) 992-5400   FAX: (410) 730-8340

Polymer Laboratories
Amherst Research Park
160 Old Farm Road
Amherst, MA 01002
(800) 767-3963   FAX: (413) 253-2476
http://www.polymerlabs.com

Polymicro Technologies
18019 North 25th Avenue
Phoenix, AZ 85023
(602) 375-4100   FAX: (602) 375-4110
http://www.polymicro.com

Polyphenols AS
Hanabryggene Technology Centre
Hanaveien 4-6
4327 Sandnes, Norway
(47) 51-62-0990
FAX: (47) 51-62-51-82
http://www.polyphenols.com

Polysciences
400 Valley Road
Warrington, PA 18976
(800) 523-2575   FAX: (800) 343-3291
http://www.polysciences.com

Polyscientific
70 Cleveland Avenue
Bayshore, NY 11706
(516) 586-0400   FAX: (516) 254-0618

Polytech Products
285 Washington Street
Somerville, MA 02143
(617) 666-5064   FAX: (617) 625-0975

Polytron
8585 Grovemont Circle
Gaithersburg, MD 20877
(301) 208-6597   FAX: (301) 208-8691
http://www.polytron.com

Popper and Sons
300 Denton Avenue
P.O. Box 128
New Hyde Park, NY 11040
(888) 717-7677   FAX: (800) 557-6773
(516) 248-0300   FAX: (516) 747-1188
http://www.popperandsons.com

Porphyrin Products
P.O. Box 31
Logan, UT 84323
(435) 753-1901   FAX: (435) 753-6731
http://www.porphyrin.com

Portex
See SIMS Portex Limited

Powderject Vaccines
585 Science Drive
Madison, WI 53711
(608) 231-3150   FAX: (608) 231-6990
http://www.powderject.com

Praxair
810 Jorie Boulevard
Oak Brook, IL 60521
(800) 621-7100
http://www.praxair.com

Precision Dynamics
13880 Del Sur Street
San Fernando, CA 91340
(800) 847-0670   FAX: (818) 899-4-45
http://www.pdcorp.com

Precision Scientific Laboratory
Equipment
Division of Jouan
170 Marcel Drive
Winchester, VA 22602
(800) 621-8820   FAX: (540) 869-0130
(540) 869-9892
http://www.precisionsci.com

Primary Care Diagnostics
See Becton Dickinson Primary
Care Diagnostics

Primate Products
1755 East Bayshore Road, Suite 28A
Redwood City, CA 94063
(650) 368-0663   FAX: (650) 368-0665
http://www.primateproducts.com

5 Prime → 3 Prime
See 2000 Eppendorf-5 Prime
http://www.5prime.com

Princeton Applied Research
PerkinElmer Instr.: Electrochemistry
801 S. Illinois
Oak Ridge, TN 37830
(800) 366-2741   FAX: (423) 425-1334
(423) 481-2442
http://www.eggpar.com

Princeton Instruments
A division of Roper Scientific
3660 Quakerbridge Road
Trenton, NJ 08619
(609) 587-9797   FAX: (609) 587-1970
http://www.prinst.com

Princeton Separations
P.O. Box 300
Aldephia, NJ 07710
(800) 223-0902   FAX: (732) 431-3768
(732) 431-3338

Prior Scientific
80 Reservoir Park Drive
Rockland, MA 02370
(781) 878-8442   FAX: (781) 878-8736
http://www.prior.com

PRO Scientific
P.O. Box 448
Monroe, CT 06468
(203) 452-9431   FAX: (203) 452-9753
http://www.proscientific.com

**Professional Compounding Centers
of America**
9901 South Wilcrest Drive
Houston, TX 77099
(800) 331-2498   FAX: (281) 933-6227
(281) 933-6948
http://www.pccarx.com

Progen Biotechnik
Maass-Str. 30
69123 Heidelberg, Germany
(49) 6221-8278-0
FAX: (49) 6221-8278-23
http://www.progen.de

Prolabo
A division of Merck Eurolab
54 rue Roger Salengro
94126 Fontenay Sous Bois Cedex
France
33-1-4514-8500
FAX: 33-1-4514-8616
http://www.prolabo.fr

Proligo
2995 Wilderness Place
Boulder, CO 80301
(888) 80-OLIGO   FAX: (303) 801-1134
http://www.proligo.com

Promega
2800 Woods Hollow Road
Madison, WI 53711
(800) 356-9526   FAX: (800) 356-1970
(608) 274-4330   FAX: (608) 277-2516
http://www.promega.com

Protein Databases (PDI)
405 Oakwood Road
Huntington Station, NY 11746
(800) 777-6834   FAX: (516) 673-4502
(516) 673-3939

Protein Polymer Technologies
10655 Sorrento Valley Road
San Diego, CA 92121
(619) 558-6064   FAX: (619) 558-6477
http://www.ppti.com

Protein Solutions
391 G Chipeta Way
Salt Lake City, UT 84108
(801) 583-9301   FAX: (801) 583-4463
http://www.proteinsolutions.com

Prozyme
1933 Davis Street, Suite 207
San Leandro, CA 94577
(800) 457-9444   FAX: (510) 638-6919
(510) 638-6900
http://www.prozyme.com

PSI
See Perceptive Science Instruments

Pulmetrics Group
82 Beacon Street
Chestnut Hill, MA 02167
(617) 353-3833   FAX: (617) 353-6766

Purdue Frederick
100 Connecticut Avenue
Norwalk, CT 06850
(800) 633-4741   FAX: (203) 838-1576
(203) 853-0123
http://www.pharma.com

Purina Mills
LabDiet
P. O. Box 66812
St. Louis, MO 63166
(800) 227-8941   FAX: (314) 768-4894
http://www.purina-mills.com

Qbiogene
2251 Rutherford Road
Carlsbad, CA 92008
(800) 424-6101   FAX: (760) 918-9313
http://www.qbiogene.com

Qiagen
28159 Avenue Stanford
Valencia, CA 91355
(800) 426-8157   FAX: (800) 718-2056
http://www.qiagen.com

Quality Biological
7581 Lindbergh Drive
Gaithersburg, MD 20879
(800) 443-9331   FAX: (301) 840-5450
(301) 840-9331
http://www.qualitybiological.com

Quantitative Technologies
P.O. Box 470
Salem Industrial Park, Bldg. 5
Whitehouse, NJ 08888
(908) 534-4445   FAX: 534-1054
http://www.qtionline.com

Quantum Appligene
Parc d'Innovation
Rue Geller de Kayserberg
67402 Illkirch, Cedex, France
(33) 3-8867-5425
FAX: (33) 3-8867-1945
http://www.quantum-appligene.com

Quantum Biotechnologies
See Qbiogene

Quantum Soft
Postfach 6613
CH-8023
Zurich, Switzerland
FAX: 41-1-481-69-51
profit@quansoft.com

Questcor Pharmaceuticals
26118 Research Road
Hayward, CA 94545
(510) 732-5551   FAX: (510) 732-7741
http://www.questcor.com

Quidel
10165 McKellar Court
San Diego, CA 92121
(800) 874-1517   FAX: (858) 546-8955
(858) 552-1100
http://www.quidel.com

R-Biopharm
7950 Old US 27 South
Marshall, MI 49068
(616) 789-3033   FAX: (616) 789-3070
http://www.r-biopharm.com

R. C. Electronics
6464 Hollister Avenue
Santa Barbara, CA 93117
(805) 685-7770   FAX: (805) 685-5853
http://www.rcelectronics.com

R & D Systems
614 McKinley Place NE
Minneapolis, MN 55413
(800) 343-7475   FAX: (612) 379-6580
(612) 379-2956
http://www.rndsystems.com

R & S Technology
350 Columbia Street
Peacedale, RI 02880
(401) 789-5660   FAX: (401) 792-3890
http://www.septech.com

RACAL Health and Safety
See 3M
7305 Executive Way
Frederick, MD 21704
(800) 692-9500   FAX: (301) 695-8200

Radiometer America
811 Sharon Drive
Westlake, OH 44145
(800) 736-0600   FAX: (440) 871-2633
(440) 871-8900
http://www.rameusa.com

Radiometer A/S
The Chemical Reference Laboratory
kandevej 21
DK-2700 Brnshj, Denmark
45-3827-3827   FAX: 45-3827-2727

Radionics
22 Terry Avenue
Burlington, MA 01803
(781) 272-1233   FAX: (781) 272-2428
http://www.radionics.com

Radnoti Glass Technology
227 W. Maple Avenue
Monrovia, CA 91016
(800) 428-l4l6   FAX: (626) 303-2998
(626) 357-8827
http://www.radnoti.com

Rainin Instrument
Rainin Road
P.O. Box 4026
Woburn, MA 01888
(800)-4-RAININ   FAX: (781) 938-1152
(781) 935-3050
http://www.rainin.com

Rank Brothers
56 High Street
Bottisham, Cambridge
CB5 9DA UK
(44) 1223 811369
FAX: (44) 1223 811441
http://www.rankbrothers.com

Rapp Polymere
Ernst-Simon Strasse 9
D 72072 Tubingen, Germany
(49) 7071-763157
FAX: (49) 7071-763158
http://www.rapp-polymere.com

Raven Biological Laboratories
8607 Park Drive
P.O. Box 27261
Omaha, NE 68127
(800) 728-5702   FAX: (402) 593-0995
(402) 593-0781
http://www.ravenlabs.com

A4

Razel Scientific Instruments
100 Research Drive
Stamford, CT 06906
(203) 324-9914   FAX: (203) 324-5568

RBI
See Research Biochemicals

Reagents International
See Biotech Source

Receptor Biology
10000 Virginia Manor Road, Suite 360
Beltsville, MD 20705
(888) 707-4200   FAX: (301) 210-6266
(301) 210-4700
http://www.receptorbiology.com

Regis Technologies
8210 N. Austin Avenue
Morton Grove, IL 60053
(800) 323-8144   FAX: (847) 967-1214
(847) 967-6000
http://www.registech.com

Reichert Ophthalmic Instruments
P.O. Box 123
Buffalo, NY 14240
(716) 686-4500   FAX: (716) 686-4545
http://www.reichert.com

Reiss
1 Polymer Place
P.O. Box 60
Blackstone, VA 23824
(800) 356-2829   FAX: (804) 292-1757
(804) 292-1600
http://www.reissmfg.com

Remel
12076 Santa Fe Trail Drive
P.O. Box 14428
Shawnee Mission, KS 66215
(800) 255-6730   FAX: (800) 621-8251
(913) 888-0939   FAX: (913) 888-5884
http://www.remelinc.com

Reming Bioinstruments
6680 County Route 17
Redfield, NY 13437
(315) 387-3414   FAX: (315) 387-3415

RepliGen
117 Fourth Avenue
Needham, MA 02494
(800) 622-2259   FAX: (781) 453-0048
(781) 449-9560
http://www.repligen.com

Research Biochemicals
1 Strathmore Road
Natick, MA 01760
(800) 736-3690   FAX: (800) 736-2480
(508) 651-8151   FAX: (508) 655-1359
http://www.resbio.com

Research Corporation Technologies
101 N. Wilmot Road, Suite 600
Tucson, AZ 85711
(520) 748-4400   FAX: (520) 748-0025
http://www.rctech.com

Research Diagnostics
Pleasant Hill Road
Flanders, NJ 07836
(800) 631-9384   FAX: (973) 584-0210
(973) 584-7093
http://www.researchd.com

Research Diets
121 Jersey Avenue
New Brunswick, NJ 08901
(877) 486-2486   FAX: (732) 247-2340
(732) 247-2390
http://www.researchdiets.com

Research Genetics
2130 South Memorial Parkway
Huntsville, AL 35801
(800) 533-4363   FAX: (256) 536-9016
(256) 533-4363
http://www.resgen.com

Research Instruments
Kernick Road Pernryn
Cornwall TR10 9DQ, UK
(44) 1326-372-753
FAX: (44) 1326-378-783
http://www.research-instruments.com

Research Organics
4353 E. 49th Street
Cleveland, OH 44125
(800) 321-0570   FAX: (216) 883-1576
(216) 883-8025
http://www.resorg.com

Research Plus
P.O. Box 324
Bayonne, NJ 07002
(800) 341-2296   FAX: (201) 823-9590
(201) 823-3592
http://www.researchplus.com

Research Products International
410 N. Business Center Drive
Mount Prospect, IL 60056
(800) 323-9814   FAX: (847) 635-1177
(847) 635-7330
http://www.rpicorp.com

Research Triangle Institute
P.O. Box 12194
Research Triangle Park, NC 27709
(919) 541-6000   FAX: (919) 541-6515
http://www.rti.org

Restek
110 Benner Circle
Bellefonte, PA 16823
(800) 356-1688   FAX: (814) 353-1309
(814) 353-1300
http://www.restekcorp.com

Rheodyne
P.O. Box 1909
Rohnert Park, CA 94927
(707) 588-2000   FAX: (707) 588-2020
http://www.rheodyne.com

Rhone Merieux
See Merial Limited

Rhone-Poulenc
2 T W Alexander Drive
P.O. Box 12014
Research Triangle Park, NC 08512
(919) 549-2000   FAX: (919) 549-2839
http://www.Rhone-Poulenc.com
Also see Aventis

Rhone-Poulenc Rorer
500 Arcola Road
Collegeville, PA 19426
(800) 727-6737   FAX: (610) 454-8940
(610) 454-8975
http://www.rp-rorer.com

Rhone-Poulenc Rorer
Centre de Recherche de Vitry-Alfortville
13 Quai Jules Guesde, BP14 94403
Vitry Sur Seine, Cedex, France
(33) 145-73-85-11
FAX: (33) 145-73-81-29
http://www.rp-rorer.com

Ribi ImmunoChem Research
563 Old Corvallis Road
Hamilton, MT 59840
(800) 548-7424   FAX: (406) 363-6129
(406) 363-3131
http://www.ribi.com

RiboGene
See Questcor Pharmaceuticals

Ricca Chemical
448 West Fork Drive
Arlington, TX 76012
(888) GO-RICCA   FAX: (800)
RICCA-93
(817) 461-5601
http://www.riccachemical.com

Richard-Allan Scientific
225 Parsons Street
Kalamazoo, MI 49007
(800) 522-7270   FAX: (616) 345-3577
(616) 344-2400
http://www.rallansci.com

Richelieu Biotechnologies
11 177 Hamon
Montral, Quebec
H3M 3E4 Canada
(802) 863-2567   FAX: (802) 862-2909
http://www.richelieubio.com

Richter Enterprises
20 Lake Shore Drive
Wayland, MA 01778
(508) 655-7632   FAX: (508) 652-7264
http://www.richter-enterprises.com

Riese Enterprises
BioSure Division
12301 G Loma Rica Drive
Grass Valley, CA 95945
(800) 345-2267 FAX: (916) 273-5097
(916) 273-5095
http://www.biosure.com

Robbins Scientific
1250 Elko Drive
Sunnyvale, CA 94086
(800) 752-8585 FAX: (408) 734-0300
(408) 734-8500
http://www.robsci.com

Roboz Surgical Instruments
9210 Corporate Boulevard, Suite 220
Rockville, MD 20850
(800) 424-2984 FAX: (301) 590-1290
(301) 590-0055

Roche Diagnostics
9115 Hague Road
P.O. Box 50457
Indianapolis, IN 46256
(800) 262-1640 FAX: (317) 845-7120
(317) 845-2000
http://www.roche.com

Roche Molecular Systems
See Roche Diagnostics

Rocklabs
P.O. Box 18-142
Auckland 6, New Zealand
(64) 9-634-7696
FAX: (64) 9-634-7696
http://www.rocklabs.com

Rockland
P.O. Box 316
Gilbertsville, PA 19525
(800) 656-ROCK FAX: (610) 367-7825
(610) 369-1008
http://www.rockland-inc.com

Rohm
Chemische Fabrik
Kirschenallee
D-64293 Darmstadt, Germany
(49) 6151-1801 FAX: (49) 6151-1802
http://www.roehm.com

Roper Scientific
3440 East Brittania Drive, Suite 100
Tucson, AZ 85706
(520) 889-9933 FAX: (520) 573-1944
http://www.roperscientific.com

Rosetta Inpharmatics
12040 115th Avenue NE
Kirkland, WA 98034
(425) 820-8900 FAX: (425) 820-5757
http://www.rii.com

ROTH-SOCHIEL
3 rue de la Chapelle
Lauterbourg
67630 France
(33) 03-88-94-82-42
FAX: (33) 03-88-54-63-93

Rotronic Instrument
160 E. Main Street
Huntington, NY 11743
(631) 427-3898 FAX: (631) 427-3902
http://www.rotronic-usa.com

Roundy's
23000 Roundy Drive
Pewaukee, WI 53072
(262) 953-7999 FAX: (262) 953-7989
http://www.roundys.com

RS Components
Birchington Road
Weldon Industrial Estate
Corby, Northants NN17 9RS, UK
(44) 1536 201234
FAX: (44) 1536 405678
http://www.rs-components.com

Rubbermaid
See Newell Rubbermaid

SA Instrumentation
1437 Tzena Way
Encinitas, CA 92024
(858) 453-1776 FAX: (800) 266-1776
http://www.sainst.com

Safe Cells
See Bionique Testing Labs

Sage Instruments
240 Airport Boulevard
Freedom, CA 95076
831-761-1000 FAX: 831-761-1008
http://www.sageinst.com

Sage Laboratories
11 Huron Drive
Natick, MA 01760
(508) 653-0844 FAX: 508-653-5671
http://www.sagelabs.com

Saint-Gobain Performance Plastics
P.O. Box 3660
Akron, OH 44309
(330) 798-9240 FAX: (330) 798-6968
http://www.nortonplastics.com

San Diego Instruments
7758 Arjons Drive
San Diego, CA 92126
(858) 530-2600 FAX: (858) 530-2646
http://www.sd-inst.com

Sandown Scientific
Beards Lodge
25 Oldfield Road
Hampden, Middlesex TW12 2AJ, UK
(44) 2089 793300
FAX: (44) 2089 793311
http://www.sandownsci.com

Sandoz Pharmaceuticals
See Novartis

Sanofi Recherche
Centre de Montpellier
371 Rue du Professor Blayac
34184 Montpellier, Cedex 04
France
(33) 67-10-67-10
FAX: (33) 67-10-67-67

Sanofi Winthrop Pharmaceuticals
90 Park Avenue
New York, NY 10016
(800) 223-5511 FAX: (800) 933-3243
(212) 551-4000
http://www.sanofi-synthelabo.com/us

Santa Cruz Biotechnology
2161 Delaware Avenue
Santa Cruz, CA 95060
(800) 457-3801 FAX: (831) 457-3801
(831) 457-3800
http://www.scbt.com

Sarasep
(800) 605-0267 FAX: (408) 432-3231
(408) 432-3230
http://www.transgenomic.com

Sarstedt
P.O. Box 468
Newton, NC 28658
(800) 257-5101 FAX: (828) 465-4003
(828) 465-4000
http://www.sarstedt.com

Sartorius
131 Heartsland Boulevard
Edgewood, NY 11717
(800) 368-7178 FAX: (516) 254-4253
http://www.sartorius.com

SAS Institute
Pacific Telesis Center
One Montgomery Street
San Francisco, CA 94104
(415) 421-2227 FAX: (415) 421-1213
http://www.sas.com

Savant/EC Apparatus
A ThermoQuest company
100 Colin Drive
Holbrook, NY 11741
(800) 634-8886 FAX: (516) 244-0606
(516) 244-2929
http://www.savec.com

Savillex
6133 Baker Road
Minnetonka, MN 55345
(612) 935-5427

A4

Scanalytics
Division of CSP
8550 Lee Highway, Suite 400
Fairfax, VA 22031
(800) 325-3110   FAX: (703) 208-1960
(703) 208-2230
http://www.scanalytics.com

Schering Laboratories
See Schering-Plough

Schering-Plough
1 Giralda Farms
Madison, NJ 07940
(800) 222-7579   FAX: (973) 822-7048
(973) 822-7000
http://www.schering-plough.com

Schleicher & Schuell
10 Optical Avenue
Keene, NH 03431
(800) 245-4024   FAX: (603) 357-3627
(603) 352-3810
http://www.s-und-s.de/english-index.html

Science Technology Centre
1250 Herzberg Laboratories
Carleton University
1125 Colonel Bay Drive
Ottawa, Ontario
K1S 5B6 Canada
(613) 520-4442   FAX: (613) 520-4445
http://www.carleton.ca/universities/stc

Scientific Instruments
200 Saw Mill River Road
Hawthorne, NY 10532
(800) 431-1956   FAX: (914) 769-5473
(914) 769-5700
http://www.scientificinstruments.com

Scientific Solutions
9323 Hamilton
Mentor, OH 44060
(440) 357-1400   FAX: (440) 357-1416
http://www.labmaster.com

Scion
82 Worman's Mill Court, Suite H
Frederick, MD 21701
(301) 695-7870   FAX: (301) 695-0035
http://www.scioncorp.com

Scott Specialty Gases
6141 Easton Road
P.O. Box 310
Plumsteadville, PA 18949
(800) 21-SCOTT   FAX: (215) 766-2476
(215) 766-8861
http://www.scottgas.com

Scripps Clinic and Research
Foundation
Instrumentation and Design Lab
10666 N. Torrey Pines Road
La Jolla, CA 92037
(800) 992-9962   FAX: (858) 554-8986
(858) 455-9100
http://www.scrippsclinic.com

SDI Sensor Devices
407 Pilot Court, 400A
Waukesha, WI 53188
(414) 524-1000   FAX: (414) 524-1009

Sefar America
111 Calumet Street
Depew, NY 14043
(716) 683-4050   FAX: (716) 683-4053
http://www.sefaramerica.com

Seikagaku America
Division of Associates of Cape Cod
704 Main Street
Falmouth, MA 02540
(800) 237-4512   FAX: (508) 540-8680
(508) 540-3444
http://www.seikagaku.com

Sellas Medizinische Gerate
Hagener Str. 393
Gevelsberg-Vogelsang, 58285
Germany
(49) 23-326-1225

Sensor Medics
22705 Savi Ranch Parkway
Yorba Linda, CA 92887
(800) 231-2466   FAX: (714) 283-8439
(714) 283-2228
http://www.sensormedics.com

Sensor Systems LLC
2800 Anvil Street, North
Saint Petersburg, FL 33710
(800) 688-2181   FAX: (727) 347-3881
(727) 347-2181
http://www.vsensors.com

SenSym/Foxboro ICT
1804 McCarthy Boulevard
Milpitas, CA 95035
(800) 392-9934   FAX: (408) 954-9458
(408) 954-6700
http://www.sensym.com

Separations Group
See Vydac

Sepracor
111 Locke Drive
Marlboro, MA 01752
(877)-SEPRACOR (508) 357-7300
http://www.sepracor.com

Sera-Lab
See Harlan Sera-Lab

Sermeter
925 Seton Court, #7
Wheeling, IL 60090
(847) 537-4747

Serological
195 W. Birch Street
Kankakee, IL 60901
(800) 227-9412   FAX: (815) 937-8285
(815) 937-8270

Seromed Biochrom
Leonorenstrasse 2-6
D-12247 Berlin, Germany
(49) 030-779-9060

Serotec
22 Bankside
Station Approach
Kidlington, Oxford OX5 1JE, UK
(44) 1865-852722
FAX: (44) 1865-373899
In the US: (800) 265-7376
http://www.serotec.co.uk

Serva Biochemicals
Distributed by Crescent Chemical

S.F. Medical Pharmlast
See Chase-Walton Elastomers

SGE
2007 Kramer Lane
Austin, TX 78758
(800) 945-6154   FAX: (512) 836-9159
(512) 837-7190
http://www.sge.com

Shandon/Lipshaw
171 Industry Drive
Pittsburgh, PA 15275
(800) 245-6212   FAX: (412) 788-1138
(412) 788-1133
http://www.shandon.com

Sharpoint
P.O. Box 2212
Taichung, Taiwan
Republic of China
(886) 4-3206320
FAX: (886) 4-3289879
http://www.sharpoint.com.tw

Shelton Scientific
230 Longhill Crossroads
Shelton, CT 06484
(800) 222-2092   FAX: (203) 929-2175
(203) 929-8999
http://www.sheltonscientific.com

Sherwood-Davis & Geck
See Kendall

Sherwood Medical
See Kendall

A4

Shimadzu Scientific Instruments
7102 Riverwood Drive
Columbia, MD 21046
(800) 477-1227   FAX: (410) 381-1222
(410) 381-1227
http://www.ssi.shimadzu.com

Sialomed
See Amika

Siemens Analytical X-Ray Systems
See Bruker Analytical X-Ray Systems

Sievers Instruments
Subsidiary of Ionics
6060 Spine Road
Boulder, CO 80301
(800) 255-6964   FAX: (303) 444-6272
(303) 444-2009
http://www.sieversinst.com

SIFCO
970 East 46th Street
Cleveland, OH 44103
(216) 881-8600   FAX: (216) 432-6281
http://www.sifco.com

Sigma-Aldrich
3050 Spruce Street
St. Louis, MO 63103
(800) 358-5287   FAX: (800) 962-9591
(800) 325-3101   FAX: (800) 325-5052
http://www.sigma-aldrich.com

Sigma-Aldrich Canada
2149 Winston Park Drive
Oakville, Ontario
L6H 6J8 Canada
(800) 5652D1400   FAX: (800)
2652D3858
http://www.sigma-aldrich.com

Silenus/Amrad
34 Wadhurst Drive
Boronia, Victoria 3155 Australia
(613)9887-3909   FAX: (613)9887-3912
http://www.amrad.com.au

Silicon Genetics
2601 Spring Street
Redwood City, CA 94063
(866) SIG SOFT   FAX: (650) 365 1735
(650) 367 9600
http://www.sigenetics.com

SIMS Deltec
1265 Grey Fox Road
St. Paul, Minnesota 55112
(800) 426-2448   FAX: (615) 628-7459
http://www.deltec.com

SIMS Portex
10 Bowman Drive
Keene, NH 03431
(800) 258-5361   FAX: (603) 352-3703
(603) 352-3812
http://www.simsmed.com

SIMS Portex Limited
Hythe, Kent CT21 6JL, UK
(44)1303-260551
FAX: (44)1303-266761
http://www.portex.com

Siris Laboratories
See Biosearch Technologies

Skatron Instruments
See Molecular Devices

SLM Instruments
See Spectronic Instruments

SLM-AMINCO Instruments
See Spectronic Instruments

Small Parts
13980 NW 58th Court
P.O. Box 4650
Miami Lakes, FL 33014
(800) 220-4242   FAX: (800) 423-9009
(305) 558-1038   FAX: (305) 558-0509
http://www.smallparts.com

Smith & Nephew
11775 Starkey Road
P.O. Box 1970
Largo, FL 33779
(800) 876-1261
http://www.smith-nephew.com

SmithKline Beecham
1 Franklin Plaza, #1800
Philadelphia, PA 19102
(215) 751-4000   FAX: (215) 751-4992
http://www.sb.com

Solid Phase Sciences
See Biosearch Technologies

SOMA Scientific Instruments
5319 University Drive, PMB #366
Irvine, CA 92612
(949) 854-0220   FAX: (949) 854-0223
http://somascientific.com

Somatix Therapy
See Cell Genesys

Sonics & Materials
53 Church Hill Road
Newtown, CT 06470
(800) 745-1105   FAX: (203) 270-4610
(203) 270-4600
http://www.sonicsandmaterials.com

Sonosep Biotech
See Triton Environmental Consultants

Sorvall
See Kendro Laboratory Products

Southern Biotechnology Associates
P.O. Box 26221
Birmingham, AL 35260
(800) 722-2255   FAX: (205) 945-8768
(205) 945-1774
http://SouthernBiotech.com

SPAFAS
190 Route 165
Preston, CT 06365
(800) SPAFAS-1   FAX: (860) 889-1991
(860) 889-1389
http://www.spafas.com

Specialty Media
Division of Cell & Molecular Technologies
580 Marshall Street
Phillipsburg, NJ 08865
(800) 543-6029   FAX: (908) 387-1670
(908) 454-7774
http://www.specialtymedia.com

Spectra Physics
See Thermo Separation Products

Spectramed
See BOC Edwards

SpectraSource Instruments
31324 Via Colinas, Suite 114
Westlake Village, CA 91362
(818) 707-2655   FAX: (818) 707-9035
http://www.spectrasource.com

Spectronic Instruments
820 Linden Avenue
Rochester, NY 14625
(800) 654-9955   FAX: (716) 248-4014
(716) 248-4000
http://www.spectronic.com

Spectrum Medical Industries
See Spectrum Laboratories

Spectrum Laboratories
18617 Broadwick Street
Rancho Dominguez, CA 90220
(800) 634-3300   FAX: (800) 445-7330
(310) 885-4601   FAX: (310) 885-4666
http://www.spectrumlabs.com

Spherotech
1840 Industrial Drive, Suite 270
Libertyville, IL 60048
(800) 368-0822   FAX: (847) 680-8927
(847) 680-8922
http://www.spherotech.com

SPSS
233 S. Wacker Drive, 11th floor
Chicago, IL 60606
(800) 521-1337   FAX: (800) 841-0064
http://www.spss.com

SS White Burs
1145 Towbin Avenue
Lakewood, NJ 08701
(732) 905-1100   FAX: (732) 905-0987
http://www.sswhiteburs.com

Stag Instruments
16 Monument Industrial Park
Chalgrove, Oxon OX44 7RW, UK
(44) 1865-891116
FAX: (44) 1865-890562

A4

Standard Reference Materials Program
National Institute of Standards and Technology
Building 202, Room 204
Gaithersburg, MD 20899
(301) 975-6776   FAX: (301) 948-3730

Starna Cells
P.O. Box 1919
Atascandero, CA 93423
(805) 466-8855   FAX: (805) 461-1575
(800) 228-4482
http://www.starnacells.com

Starplex Scientific
50 Steinway
Etobieoke, Ontario
M9W 6Y3 Canada
(800) 665-0954   FAX: (416) 674-6067
(416) 674-7474
http://www.starplexscientific.com

State Laboratory Institute of Massachusetts
305 South Street
Jamaica Plain, MA 02130
(617) 522-3700   FAX: (617) 522-8735
http://www.state.ma.us/dph

Stedim Labs
1910 Mark Court, Suite 110
Concord, CA 94520
(800) 914-6644   FAX: (925) 689-6988
(925) 689-6650
http://www.stedim.com

Steinel America
9051 Lyndale Avenue
Bloomington, MN 55420
(800) 852 4343   FAX: (952) 888-5132
http://www.steinelamerica.com

Stem Cell Technologies
777 West Broadway, Suite 808
Vancouver, British Columbia
V5Z 4J7 Canada
(800) 667-0322   FAX: (800) 567-2899
(604) 877-0713   FAX: (604) 877-0704
http://www.stemcell.com

Stephens Scientific
107 Riverdale Road
Riverdale, NJ 07457
(800) 831-8099   FAX: (201) 831-8009
(201) 831-9800

Steraloids
P.O. Box 689
Newport, RI 02840
(401) 848-5422   FAX: (401) 848-5638
http://www.steraloids.com

Steris Corporation
5960 Heisley Road
Mentor, Ohio 44060
(800) 548-4873
440-354-2600
http://www.steris.com

Sterling Medical
2091 Springdale Road, Ste. 2
Cherry Hill, NJ 08003
(800) 229-0900   FAX: (800) 229-7854
http://www.sterlingmedical.com

Sterling Winthrop
90 Park Avenue
New York, NY 10016
(212) 907-2000   FAX: (212) 907-3626

Sternberger Monoclonals
10 Burwood Court
Lutherville, MD 21093
(410) 821-8505   FAX: (410) 821-8506
http://www.sternbergermonoclonals.com

Stoelting
502 Highway 67
Kiel, WI 53042
(920) 894-2293   FAX: (920) 894-7029
http://www.stoelting.com

Stovall Lifescience
206-G South Westgate Drive
Greensboro, NC 27407
(800) 852-0102   FAX: (336) 852-3507
http://www.slscience.com

Stratagene
11011 N. Torrey Pines Road
La Jolla, CA 92037
(800) 424-5444   FAX: (888) 267-4010
(858) 535-5400
http://www.stratagene.com

Strategic Applications
530A N. Milwaukee Avenue
Libertyville, IL 60048
(847) 680-9385   FAX: (847) 680-9837

Strem Chemicals
7 Mulliken Way
Newburyport, MA 01950
(800) 647-8736   FAX: (800) 517-8736
(978) 462-3191   FAX: (978) 465-3104
http://www.strem.com

StressGen Biotechnologies Biochemicals Division
120-4243 Glanford Avenue
Victoria, British Columbia
V8Z 4B9 Canada
(800) 661-4978   FAX: (250) 744-2877
(250) 744-2811
http://www.stressgen.com

Structure Probe/SPI Supplies (Epon-Araldite)
P.O. Box 656
West Chester, PA 19381
(800) 242-4774   FAX: (610) 436-5755
http://www.2spi.com

Sud-Chemie Performance Packaging
101 Christine Drive
Belen, NM 87002
(800) 989-3374   FAX: (505) 864-9296
http://www.uniteddesiccants.com

Sumitomo Chemical
Sumitomo Building
5-33, Kitahama 4-chome
Chuo-ku, Osaka 541-8550, Japan
(81) 6-6220-3891
FAX: (81)-6-6220-3345
http://www.sumitomo-chem.co.jp

Sun Box
19217 Orbit Drive
Gaithersburg, MD 20879
(800) 548-3968   FAX: (301) 977-2281
(301) 869-5980
http://www.sunboxco.com

Sunbrokers
See Sun International

Sun International
3700 Highway 421 North
Wilmington, NC 28401
(800) LAB-VIAL   FAX: (800) 231-7861
http://www.autosamplervial.com

Sunox
1111 Franklin Boulevard, Unit 6
Cambridge, Ontario
N1R 8B5 Canada
(519) 624-4413   FAX: (519) 624-8378
http://www.sunox.ca

Supelco
See Sigma-Aldrich

SuperArray
P.O. Box 34494
Bethesda, MD 20827
(888) 503-3187   FAX: (301) 765-9859
(301) 765-9888
http://www.superarray.com

Surface Measurement Systems
3 Warple Mews, Warple Way
London W3 ORF, UK
(44) 20-8749-4900
FAX: (44) 20-8749-6749
http://www.smsuk.co.uk/index.htm

SurgiVet
N7 W22025 Johnson Road, Suite A
Waukesha, WI 53186
(262) 513-8500 (888) 745-6562
FAX: (262) 513-9069
http://www.surgivet.com

Sutter Instruments
51 Digital Drive
Novato, CA 94949
(415) 883-0128   FAX: (415) 883-0572
http://www.sutter.com

Swiss Precision Instruments
1555 Mittel Boulevard, Suite F
Wooddale, IL 60191
(800) 221-0198   FAX: (800) 842-5164

Synaptosoft
3098 Anderson Place
Decatur, GA 30033
(770) 939-4366   FAX: 770-939-9478
http://www.synaptosoft.com

SynChrom
See Micra Scientific

Synergy Software
2457 Perkiomen Avenue
Reading, PA 19606
(800) 876-8376   FAX: (610) 370-0548
(610) 779-0522
http://www.synergy.com

Synteni
See Incyte

Synthetics Industry
Lumite Division
2100A Atlantic Highway
Gainesville, GA 30501
(404) 532-9756   FAX: (404) 531-1347

Systat
See SPSS

Systems Planning and Analysis (SPA)
2000 N. Beauregard Street
Suite 400
Alexandria, VA 22311
(703) 931-3500
http://www.spa-inc.net

3M Bioapplications
3M Center
Building 270-15-01
St. Paul, MN 55144
(800) 257-7459   FAX: (651) 737-5645
(651) 736-4946

T Cell Diagnostics and T Cell Sciences
38 Sidney Street
Cambridge, MA 02139
(617) 621-1400

TAAB Laboratory Equipment
3 Minerva House
Calleva Park
Aldermaston, Berkshire RG7 8NA, UK
(44) 118 9817775
FAX: (44) 118 9817881

Taconic
273 Hover Avenue
Germantown, NY 12526
(800) TAC-ONIC   FAX: (518) 537-7287
(518) 537-6208
http://www.taconic.com

Tago
See Biosource International

TaKaRa Biochemical
719 Alliston Way
Berkeley, CA 94710
(800) 544-9899   FAX: (510) 649-8933
(510) 649-9895
http://www.takara.co.jp/english

Takara Shuzo
Biomedical Group Division
Seta 3-4-1
Otsu Shiga 520-21, Japan
(81) 75-241-5100
FAX: (81) 77-543-9254
http://www.Takara.co.jp/english

Takeda Chemical Products
101 Takeda Drive
Wilmington, NC 28401
(800) 825-3328   FAX: (800) 825-0333
(910) 762-8666   FAX: (910) 762-6846
http://takeda-usa.com

TAO Biomedical
73 Manassas Court
Laurel Springs, NJ 08021
(609) 782-8622   FAX: (609) 782-8622

Tecan US
P.O. Box 13953
Research Triangle Park, NC 27709
(800) 33-TECAN   FAX: (919) 361-5201
(919) 361-5208
http://www.tecan-us.com

Techne
University Park Plaza
743 Alexander Road
Princeton, NJ 08540
(800) 225-9243   FAX: (609) 987-8177
(609) 452-9275
http://www.techneusa.com

Technical Manufacturing
15 Centennial Drive
Peabody, MA 01960
(978) 532-6330   FAX: (978) 531-8682
http://www.techmfg.com

Technical Products International
5918 Evergreen
St. Louis, MO 63134
(800) 729-4451   FAX: (314) 522-6360
(314) 522-8671
http://www.vibratome.com

Technicon
See Organon Teknika Cappel

Techno-Aide
P.O. Box 90763
Nashville, TN 37209
(800) 251-2629   FAX: (800) 554-6275
(615) 350-7030
http://www.techno-aid.com

Ted Pella
4595 Mountain Lakes Boulevard
P.O. Box 492477
Redding, CA 96049
(800) 237-3526   FAX: (530) 243-3761
(530) 243-2200
http://www.tedpella.com

Tekmar-Dohrmann
P.O. Box 429576
Cincinnati, OH 45242
(800) 543-4461   FAX: (800) 841-5262
(513) 247-7000   FAX: (513) 247-7050

Tektronix
142000 S.W. Karl Braun Drive
Beaverton, OR 97077
(800) 621-1966   FAX: (503) 627-7995
(503) 627-7999
http://www.tek.com

Tel-Test
P.O. Box 1421
Friendswood, TX 77546
(800) 631-0600   FAX: (281)482-1070
(281)482-2672
http://www.isotex-diag.com

TeleChem International
524 East Weddell Drive, Suite 3
Sunnyvale, CA 94089
(408) 744-1331   FAX: (408) 744-1711
http://www.gst.net/~telechem

Terrachem
Mallaustrasse 57
D-68219 Mannheim, Germany
0621-876797-0   FAX: 0621-876797-19
http://www.terrachem.de

Terumo Medical
2101 Cottontail Lane
Somerset, NJ 08873
(800) 283-7866   FAX: (732) 302-3083
(732) 302-4900
http://www.terumomedical.com

Tetko
333 South Highland Manor
Briarcliff, NY 10510
(800) 289-8385   FAX: (914) 941-1017
(914) 941-7767
http://www.tetko.com

TetraLink
4240 Ridge Lea Road
Suite 29
Amherst, NY 14226
(800) 747-5170   FAX: (800) 747-5171
http://www.tetra-link.com

A4

TEVA Pharmaceuticals USA
1090 Horsham Road
P.O. Box 1090
North Wales, PA 19454
(215) 591-3000   FAX: (215) 721-9669
http://www.tevapharmusa.com

Texas Fluorescence Labs
9503 Capitol View Drive
Austin, TX 78747
(512) 280-5223   FAX: (512) 280-4997
http://www.teflabs.com

The Nest Group
45 Valley Road
Southborough, MA 01772
(800) 347-6378   FAX: (508) 485-5736
(508) 481-6223
http://world.std.com/~nestgrp

ThermoCare
P.O. Box 6069
Incline Village, NV 89450
(800) 262-4020
(775) 831-1201

Thermo Labsystems
8 East Forge Parkway
Franklin, MA 02038
(800) 522-7763   FAX: (508) 520-2229
(508) 520-0009
http://www.finnpipette.com

Thermometric
Spjutvagen 5A
S-175 61 Jarfalla, Sweden
(46) 8-564-72-200

Thermoquest
IEC Division
300 Second Avenue
Needham Heights, MA 02194
(800) 843-1113   FAX: (781) 444-6743
(781) 449-0800
http://www.thermoquest.com

Thermo Separation Products
Thermoquest
355 River Oaks Parkway
San Jose, CA 95134
(800) 538-7067   FAX: (408) 526-9810
(408) 526-1100
http://www.thermoquest.com

Thermo Shandon
171 Industry Drive
Pittsburgh, PA 15275
(800) 547-7429   FAX: (412) 899-4045
http://www.thermoshandon.com

Thermo Spectronic
820 Linden Avenue
Rochester, NY 14625
(585) 248-4000   FAX: (585) 248-4200
http://www.thermo.com

Thomas Scientific
99 High Hill Road at I-295
Swedesboro, NJ 08085
(800) 345-2100   FAX: (800) 345-5232
(856) 467-2000   FAX: (856) 467-3087
http://www.wheatonsci.com/html/nt/
Thomas.html

Thomson Instrument
354 Tyler Road
Clearbrook, VA 22624
(800) 842-4752   FAX: (540) 667-6878
(800) 541-4792   FAX: (760) 757-9367
http://www.hplc.com

Thorn EMI
See Electron Tubes

Thorlabs
435 Route 206
Newton, NJ 07860
(973) 579-7227   FAX: (973) 383-8406
http://www.thorlabs.com

Tiemann
See Bernsco Surgical Supply

TILL Photonics GmbH
Lochhamer Schlag 19
D-82166 Grafelfing
Germany
(49) 89-895-662-0
FAX: (49) 89-895-662-101
http://www.till-photonics.com/

Timberline Instruments
1880 South Flatiron Court, H-2
P.O. Box 20356
Boulder, CO 80308
(800) 777-5996   FAX: (303) 440-8786
(303) 440-8779
http://www.timberlineinstruments.com

TissueInformatics
711 Bingham Street, Suite 202
Pittsburgh, PA 15203
(418) 488-1100   FAX: (418) 488-6172
http://www.tissueinformatics.com

Tissue-Tek
A Division of Sakura Finetek USA
1750 West 214th Street
Torrance, CA 90501
(800) 725-8723   FAX: (310) 972-7888
(310) 972-7800
http://www.sakuraus.com

Tocris Cookson
114 Holloway Road, Suite 200
Ballwin, MO 63011
(800) 421-3701   FAX: (800) 483-1993
(636) 207-7651   FAX: (636) 207-7683
http://www.tocris.com

Tocris Cookson
Northpoint, Fourth Way
Avonmouth, Bristol BS11 8TA, UK
(44) 117-982-6551
FAX: (44) 117-982-6552
http://www.tocris.com

Tomtec
See CraMar Technologies

TopoGen
P.O. Box 20607
Columbus, OH 43220
(800) TOPOGEN
FAX: (800) ADD-TOPO
(614) 451-5810   FAX: (614) 451-5811
http://www.topogen.com

Toray Industries, Japan
Toray Building 2-1
Nihonbash-Muromach
2-Chome, Chuo-Ku
Tokyo, Japan 103-8666
(03) 3245-5115   FAX: (03) 3245-5555
http://www.toray.co.jp

Toray Industries, U.S.A.
600 Third Avenue
New York, NY 10016
(212) 697-8150   FAX: (212) 972-4279
http://www.toray.com

Toronto Research Chemicals
2 Brisbane Road
North York, Ontario
M3J 2J8 Canada
(416) 665-9696   FAX: (416) 665-4439
http://www.trc-canada.com

TosoHaas
156 Keystone Drive
Montgomeryville, PA 18036
(800) 366-4875   FAX: (215) 283-5035
(215) 283-5000
http://www.tosohaas.com

Towhill
647 Summer Street
Boston, MA 02210
(617) 542-6636   FAX: (617) 464-0804

Toxin Technology
7165 Curtiss Avenue
Sarasota, FL 34231
(941) 925-2032   FAX: (9413) 925-2130
http://www.toxintechnology.com

Toyo Soda
See TosoHaas

Trace Analytical
3517-A Edison Way
Menlo Park, CA 94025
(650) 364-6895   FAX: (650) 364-6897
http://www.traceanalytical.com

Transduction Laboratories
See BD Transduction Laboratories

A4

Transgenomic
2032 Concourse Drive
San Jose, CA 95131
(408) 432-3230   FAX: (408) 432-3231
http://www.transgenomic.com

Transonic Systems
34 Dutch Mill Road
Ithaca, NY 14850
(800) 353-3569   FAX: (607) 257-7256
http://www.transonic.com

Travenol Lab
See Baxter Healthcare

Tree Star Software
20 Winding Way
San Carlos, CA 94070
800-366-6045
http://www.treestar.com

Trevigen
8405 Helgerman Court
Gaithersburg, MD 20877
(800) TREVIGEN   FAX: (301)
216-2801
(301) 216-2800
http://www.trevigen.com

Trilink Biotechnologies
6310 Nancy Ridge Drive
San Diego, CA 92121
(800) 863-6801   FAX: (858) 546-0020
http://www.trilink.biotech.com

Tripos Associates
1699 South Hanley Road, Suite 303
St. Louis, MO 63144
(800) 323-2960   FAX: (314) 647-9241
(314) 647-1099
http://www.tripos.com

Triton Environmental Consultants
120-13511 Commerce Parkway
Richmond, British Columbia
V6V 2L1 Canada
(604) 279-2093   FAX: (604) 279-2047
http://www.triton-env.com

Tropix
47 Wiggins Avenue
Bedford, MA 01730
(800) 542-2369   FAX: (617) 275-8581
(617) 271-0045
http://www.tropix.com

TSI Center for Diagnostic Products
See Intergen

2000 Eppendorf-5 Prime
5603 Arapahoe Avenue
Boulder, CO 80303
(800) 533-5703   FAX: (303) 440-0835
(303) 440-3705

Tyler Research
10328 73rd Avenue
Edmonton, Alberta
T6E 6N5 Canada
(403) 448-1249   FAX: (403) 433-0479

UBI
See Upstate Biotechnology

Ugo Basile Biological Research Apparatus
Via G. Borghi 43
21025 Comerio, Varese, Italy
(39) 332 744 574
FAX: (39) 332 745 488
http://www.ugobasile.com

UltraPIX
See Life Science Resources

Ultrasonic Power
239 East Stephenson Street
Freeport, IL 61032
(815) 235-6020   FAX: (815) 232-2150
http://www.upcorp.com

Ultrasound Advice
23 Aberdeen Road
London N52UG, UK
(44) 020-7359-1718
FAX: (44) 020-7359-3650
http://www.ultrasoundadvice.co.uk

UNELKO
14641 N. 74th Street
Scottsdale, AZ 85260
(480) 991-7272   FAX: (480)483-7674
http://www.unelko.com

Unifab Corp.
5260 Lovers Lane
Kalamazoo, MI 49002
(800) 648-9569   FAX: (616) 382-2825
(616) 382-2803

Union Carbide
10235 West Little York Road, Suite 300
Houston, TX 77040
(800) 568-4000   FAX: (713) 849-7021
(713) 849-7000
http://www.unioncarbide.com

United Desiccants
See Sud-Chemie Performance Packaging

United States Biochemical
See USB

United States Biological (US Biological)
P.O. Box 261
Swampscott, MA 01907
(800) 520-3011   FAX: (781) 639-1768
http://www.usbio.net

Universal Imaging
502 Brandywine Parkway
West Chester, PA 19380
(610) 344-9410   FAX: (610) 344-6515
http://www.image1.com

Upchurch Scientific
619 West Oak Street
P.O. Box 1529
Oak Harbor, WA 98277
(800) 426-0191   FAX: (800) 359-3460
(360) 679-2528   FAX: (360) 679-3830
http://www.upchurch.com

Upjohn
Pharmacia & Upjohn
http://www.pnu.com

Upstate Biotechnology (UBI)
1100 Winter Street, Suite 2300
Waltham, MA 02451
(800) 233-3991   FAX: (781) 890-7738
(781) 890-8845
http://www.upstatebiotech.com

USA/Scientific
346 SW 57th Avenue
P.O. Box 3565
Ocala, FL 34478
(800) LAB-TIPS   FAX: (352) 351-2057
(3524) 237-6288
http://www.usascientific.com

USB
26111 Miles Road
P.O. Box 22400
Cleveland, OH 44122
(800) 321-9322   FAX: (800) 535-0898
FAX: (216) 464-5075
http://www.usbweb.com

USCI Bard
Bard Interventional Products
129 Concord Road
Billerica, MA 01821
(800) 225-1332   FAX: (978) 262-4805
http://www.bardinterventional.com

UVP (Ultraviolet Products)
2066 W. 11th Street
Upland, CA 91786
(800) 452-6788   FAX: (909) 946-3597
(909) 946-3197
http://www.uvp.com

V & P Scientific
9823 Pacific Heights Boulevard, Suite T
San Diego, CA 92121
(800) 455-0644   FAX: (858) 455-0703
(858) 455-0643
http://www.vp-scientific.com

V-A Optical Labs
60 Red Hill Ave.
San Anselmo, CA 94960
(415) 459-1919   FAX: (415) 459-7216
http://www.vaoptical.com/

A4

Valco Instruments
P.O. Box 55603
Houston, TX 77255
(800) FOR-VICI   FAX: (713) 688-8106
(713) 688-9345
http://www.vici.com

Valpey Fisher
75 South Street
Hopkin, MA 01748
(508) 435-6831   FAX: (508) 435-5289
http://www.valpeyfisher.com

Value Plastics
3325 Timberline Road
Fort Collins, CO 80525
(800) 404-LUER   FAX: (970) 223-0953
(970) 223-8306
http://www.valueplastics.com

Vangard International
P.O. Box 308
3535 Rt. 66, Bldg. #4
Neptune, NJ 07754
(800) 922-0784   FAX: (732) 922-0557
(732) 922-4900
http://www.vangard1.com

Varian Analytical Instruments
2700 Mitchell Drive
Walnut Creek, CA 94598
(800) 926-3000   FAX: (925) 945-2102
(925) 939-2400
http://www.varianinc.com

Varian Associates
3050 Hansen Way
Palo Alto, CA 94304
(800) 544-4636   FAX: (650) 424-5358
(650) 493-4000
http://www.varian.com

Vector Core Laboratory/
National Gene Vector Labs
University of Michigan
3560 E MSRB II
1150 West Medical Center Drive
Ann Arbor, MI 48109
(734) 936-5843   FAX: (734) 764-3596

Vector Laboratories
30 Ingold Road
Burlingame, CA 94010
(800) 227-6666   FAX: (650) 697-0339
(650) 697-3600
http://www.vectorlabs.com

Vedco
2121 S.E. Bush Road
St. Joseph, MO 64504
(888) 708-3326   FAX: (816) 238-1837
(816) 238-8840
http://database.vedco.com

Ventana Medical Systems
3865 North Business Center Drive
Tucson, AZ 85705
(800) 227-2155   FAX: (520) 887-2558
(520) 887-2155
http://www.ventanamed.com

Verity Software House
P.O. Box 247
45A Augusta Road
Topsham, ME 04086
(207) 729-6767   FAX: (207) 729-5443
http://www.vsh.com

Vernitron
See Sensor Systems LLC

Vertex Pharmaceuticals
130 Waverly Street
Cambridge, MA 02139
(617) 577-6000   FAX: (617) 577-6680
http://www.vpharm.com

Vetamac
Route 7, Box 208
Frankfort, IN 46041
(317) 379-3621

Vet Drug
Unit 8
Lakeside Industrial Estate
Colnbrook, Slough SL3 0ED, UK

Vetus Animal Health
See Burns Veterinary Supply

Viamed
15 Station Road
Cross Hills, Keighley
W. Yorkshire BD20 7DT, UK
(44) 1-535-634-542
FAX: (44) 1-535-635-582
http://www.viamed.co.uk

Vical
9373 Town Center Drive, Suite 100
San Diego, CA 92121
(858) 646-1100   FAX: (858) 646-1150
http://www.vical.com

Victor Medical
2349 North Watney Way, Suite D
Fairfield, CA 94533
(800) 888-8908   FAX: (707) 425-6459
(707) 425-0294

Virion Systems
9610 Medical Center Drive, Suite 100
Rockville, MD 20850
(301) 309-1844   FAX: (301) 309-0471
http://www.radix.net/~virion

VirTis Company
815 Route 208
Gardiner, NY 12525
(800) 765-6198   FAX: (914) 255-5338
(914) 255-5000
http://www.virtis.com

Visible Genetics
700 Bay Street, Suite 1000
Toronto, Ontario
M5G 1Z6 Canada
(888) 463-6844 (416) 813-3272
http://www.visgen.com

Vitrocom
8 Morris Avenue
Mountain Lakes, NJ 07046
(973) 402-1443   FAX: (973) 402-1445

VTI
7650 W. 26th Avenue
Hialeah, FL 33106
(305) 828-4700   FAX: (305) 828-0299
http://www.vticorp.com

VWR Scientific Products
200 Center Square Road
Bridgeport, NJ 08014
(800) 932-5000   FAX: (609) 467-5499
(609) 467-2600
http://www.vwrsp.com

Vydac
17434 Mojave Street
P.O. Box 867
Hesperia, CA 92345
(800) 247-0924   FAX: (760) 244-1984
(760) 244-6107
http://www.vydac.com

Vysis
3100 Woodcreek Drive
Downers Grove, IL 60515
(800) 553-7042   FAX: (630) 271-7138
(630) 271-7000
http://www.vysis.com

W&H Dentalwerk Burmoos
P.O. Box 1
A-5111 Burmoos, Austria
(43) 6274-6236-0
FAX: (43) 6274-6236-55
http://www.wnhdent.com

Wako BioProducts
See Wako Chemicals USA

Wako Chemicals USA
1600 Bellwood Road
Richmond, VA 23237
(800) 992-9256   FAX: (804) 271-7791
(804) 271-7677
http://www.wakousa.com

Wako Pure Chemicals
1-2, Doshomachi 3-chome
Chuo-ku, Osaka 540-8605, Japan
81-6-6203-3741   FAX: 81-6-6222-1203
http://www.wako-chem.co.jp/egaiyo/
index.htm

Wallac
See Perkin-Elmer

A4

Wallac
A Division of Perkin-Elmer
3985 Eastern Road
Norton, OH 44203
(800) 321-9632   FAX: (330) 825-8520
(330) 825-4525
http://www.wallac.com

Waring Products
283 Main Street
New Hartford, CT 06057
(800) 348-7195   FAX: (860) 738-9203
(860) 379-0731
http://www.waringproducts.com

Warner Instrument
1141 Dixwell Avenue
Hamden, CT 06514
(800) 599-4203   FAX: (203) 776-1278
(203) 776-0664
http://www.warnerinstrument.com

Warner-Lambert
Parke-Davis
201 Tabor Road
Morris Plains, NJ 07950
(973) 540-2000   FAX: (973) 540-3761
http://www.warner-lambert.com

Washington University Machine Shop
615 South Taylor
St. Louis, MO 63310
(314) 362-6186   FAX: (314) 362-6184

Waters Chromatography
34 Maple Street
Milford, MA 01757
(800) 252-HPLC   FAX: (508) 478-1990
(508) 478-2000
http://www.waters.com

Watlow
12001 Lackland Road
St. Louis, MO 63146
(314) 426-7431   FAX: (314) 447-8770
http://www.watlow.com

Watson-Marlow
220 Ballardvale Street
Wilmington, MA 01887
(978) 658-6168   FAX: (978) 988 0828
http://www.watson-marlow.co.uk

Waukesha Fluid Handling
611 Sugar Creek Road
Delavan, WI 53115
(800) 252-5200   FAX: (800) 252-5012
(414) 728-1900   FAX: (414) 728-4608
http://www.waukesha-cb.com

WaveMetrics
P.O. Box 2088
Lake Oswego, OR 97035
(503) 620-3001   FAX: (503) 620-6754
http://www.wavemetrics.com

Weather Measure
P.O. Box 41257
Sacramento, CA 95641
(916) 481-7565

Weber Scientific
2732 Kuser Road
Hamilton, NJ 08691
(800) FAT-TEST   FAX: (609) 584-8388
(609) 584-7677
http://www.weberscientific.com

Weck, Edward & Company
1 Weck Drive
Research Triangle Park, NC 27709
(919) 544-8000

Wellcome Diagnostics
See Burroughs Wellcome

Wellington Laboratories
398 Laird Road
Guelph, Ontario
N1G 3X7 Canada
(800) 578-6985   FAX: (519) 822-2849
http://www.well-labs.com

Wesbart Engineering
Daux Road
Billingshurst, West Sussex
RH14 9EZ, UK
(44) 1-403-782738
FAX: (44) 1-403-784180
http://www.wesbart.co.uk

Whatman
9 Bridewell Place
Clifton, NJ 07014
(800) 631-7290   FAX: (973) 773-3991
(973) 773-5800
http://www.whatman.com

Wheaton Science Products
1501 North 10th Street
Millville, NJ 08332
(800) 225-1437   FAX: (800) 368-3108
(856) 825-1100   FAX: (856) 825-1368
http://www.algroupwheaton.com

Whittaker Bioproducts
See BioWhittaker

Wild Heerbrugg
Juerg Dedual Gaebrisstrasse 8 CH
9056 Gais, Switzerland
(41) 71-793-2723
FAX: (41) 71-726-5957
http://www.homepage.swissonline.net/
dedual/wild`heerbrugg

Willy A. Bachofen AG Maschinenfabrik
Utengasse 15/17
CH4005 Basel, Switzerland
(41) 61-681-5151
FAX: (41) 61-681-5058
http://www.wab.ch

Winthrop
See Sterling Winthrop

Wolfram Research
100 Trade Center Drive
Champaign, IL 61820
(800) 965-3726   FAX: (217) 398-0747
(217) 398-0700
http://www.wolfram.com

World Health Organization
Microbiology and Immunology Support
20 Avenue Appia
1211 Geneva 27, Switzerland
(41-22) 791-2602
FAX: (41-22) 791-0746
http://www.who.org

World Precision Instruments
175 Sarasota Center Boulevard
International Trade Center
Sarasota, FL 34240
(941) 371-1003   FAX: (941) 377-5428
http://www.wpiinc.com

Worthington Biochemical
Halls Mill Road
Freehold, NJ 07728
(800) 445-9603   FAX: (800) 368-3108
(732) 462-3838   FAX: (732) 308-4453
http://www.worthington-biochem.com

WPI
See World Precision Instruments

Wyeth-Ayerst
2 Esterbrook Lane
Cherry Hill, NJ 08003
(800) 568-9938   FAX: (858) 424-8747
(858) 424-3700

Wyeth-Ayerst Laboratories
P.O. Box 1773
Paoli, PA 19301
(800) 666-7248   FAX: (610) 889-9669
(610) 644-8000
http://www.ahp.com

Xenotech
3800 Cambridge Street
Kansas City, KS 66103
(913) 588-7930   FAX: (913) 588-7572
http://www.xenotechllc.com

Xeragon
19300 Germantown Road
Germantown, MD 20874
(240) 686-7860   FAX: (240)686-7861
http://www.xeragon.com

Xillix Technologies
300-13775 Commerce Parkway
Richmond, British Columbia
V6V 2V4 Canada
(800) 665-2236   FAX: (604) 278-3356
(604) 278-5000
http://www.xillix.com

A4

Xomed Surgical Products
6743 Southpoint Drive N
Jacksonville, FL 32216
(800) 874-5797   FAX: (800) 678-3995
(904) 296-9600   FAX: (904) 296-9666
http://www.xomed.com

Yakult Honsha
1-19, Higashi-Shinbashi 1-chome
Minato-ku Tokyo 105-8660, Japan
81-3-3574-8960

Yamasa Shoyu
23-8 Nihonbashi Kakigaracho
1-chome, Chuoku
Tokyo, 103 Japan
(81) 3-479 22 0095
FAX: (81) 3-479 22 3435

Yeast Genetic Stock Center
See ATCC

Yellow Spring Instruments
See YSI

YMC
YMC Karasuma-Gojo Building
284 Daigo-Cho, Karasuma Nisihiirr
Gojo-dori Shimogyo-ku
Kyoto, 600-8106, Japan
(81) 75-342-4567
FAX: (81) 75-342-4568
http://www.ymc.co.jp

YSI
1725-1700 Brannum Lane
Yellow Springs, OH 45387
(800) 765-9744   FAX: (937) 767-9353
(937) 767-7241
http://www.ysi.com

Zeneca/CRB
See AstraZeneca
(800) 327-0125   FAX: (800) 321-4745

Zivic-Miller Laboratories
178 Toll Gate Road
Zelienople, PA 16063
(800) 422-LABS   FAX: (724) 452-4506
(800) MBM-RATS
FAX: (724) 452-5200
http://zivicmiller.com

Zymark
Zymark Center
Hopkinton, MA 01748
(508) 435-9500   FAX: (508) 435-3439
http://www.zymark.com

Zymed Laboratories
458 Carlton Court
South San Francisco, CA 94080
(800) 874-4494   FAX: (650) 871-4499
(650) 871-4494
http://www.zymed.com

Zymo Research
625 W. Katella Avenue, Suite 30
Orange, CA 92867
(888) 882-9682   FAX: (714) 288-9643
(714) 288-9682
http://www.zymor.com

Zynaxis Cell Science
See ChiRex Cauldron

A4

References

AABB. 1997. Parentage Testing Standards Committee Annual Report Summary. American Association of Blood Banks, Arlington, Va.

Abbott, C. and Povey, S. 1991. Development of human chromosome-specific PCR primers for the characterization of somatic cell hybrids. *Genomics* 9:73–77.

Abbs, S., Yau, S.C., Clark, S., Mathew, C.G., and Bobrow, M. 1991. A convenient multiplex PCR system for the detection of dystrophin gene deletions: A comparative analysis with cDNA hybridization shows mistypings by both methods. *J. Med. Genet.* 28:304–311.

Abecasis, G.R., Cookson, W.O.C., and Cardon, L.R. 2000. Pedigree tests of transmission disequilibrium. *Eur. J. Hum. Genet.* 8:545–551.

ACMG (American College of Medical Genetics). 1993. Prenatal interphase fluorescence in situ hybridization (FISH) policy statement. *Am. J. Hum. Genet.* 53:526–627.

Aihara, H. and Miyazaki, J. 1998. Gene transfer into muscle by electroporation in vivo. *Nat. Biotechnol* 16:867–870.

Akeson, E.C. and Davisson, M.T. 2000. Centromeric heterochromatin variants. In Genteic Variants and Strains of the Laboratory Mouse, Vol. 2 (M.F. Lyon, S. Rastan, and S.D.M. Brown, eds.) pp 1506–1509. Oxford University Press, Oxford.

Akli, S., Caillaud, C., Vigne, E., Stratford-Perricaudet, L.D., Ponaru, L., Perricaudet, M., Kahn, A., and Peschanski, M.R. 1993. Transfer of a foreign gene into the brain using adenovirus vectors. *Nat. Genet.* 3:224–228.

Akyurek, L.M., Yang, Z.Y., Aoki, K., San, H., Nabel, G.J., Parmacek, M.S., and Nabel, E.G. 2000. SM22alpha promoter targets gene expression to vascular smooth muscle cells in vitro and in vivo. *Mol. Med* 6:983–991.

Albertsen, H.M., Abderrahim, H., Cann, H.M., Dausset, J., Le Paslier, D., and Cohen, D. 1990. Construction and characterization of a yeast artificial chromosome library containing seven haploid equivalents. *Proc. Natl. Acad. Sci. U.S.A.* 87:4256–4260.

Allen, R.W., Wallhermfechtel, M., and Miller, W.V. 1990. The application of restriction fragment length polymorphism mapping to parentage testing. *Transfusion* 30:552–564.

Allen, R., Zoghbi, H., Moseley, A., Rosenblatt, H., and Belmont, J. 1992. Methylation of *Hpa*II and *Hha*I sites near the polymorphic CA repeat in the human androgen-receptor gene correlates with X chromsome inactivation. *Am. J. Hum. Genet.* 51:1229–1239.

Almasy, L. and Blangero, J. 1998. Multipoint quantitative-trait linkage analysis in general pedigrees. *Am. J. Hum. Genet.* 62:1198–1211.

Altman, J. and Bayer, S. 1995. Atlas of Prenatal Rat Brain Development. CRC Press, Boca Raton, Fla.

Altschul, S., Gish, W., Miller, W., Myers, E., and Lipman, D. 1990. Basic local alignment search tool. *J. Mol. Biol.* 215:403–410.

Alvira, M.R., Cohen, J.B., Goins, W.F., and Glorioso, J.C. 1999. Genetic studies exposing the splicing events involved in HSV-1 latency associated transcript (LAT) production during lytic and latent infection. *J. Virol.* 73:3866–3876.

Alwine, J.C., Kemp, D.J., and Stark, G.R. 1977. Method for detection of specific RNAs in agarose gels by transfer to diazobenzyloxymethyl-paper and hybridization with DNA probes. *Proc. Natl. Acad. Sci. U.S.A.* 74:5350–5354.

Amara, R.R., Villinger, F., Altman, J.D., Lydy, S.L., O'Neil, S.P., Staprans, S.I., Montefiori, D.C., Xu, Y., Herndon, J.G., Wyatt, L.S., Candido, M.A., Kozyr, N.L., Earl, P.L., Smith, J.M., Ma, H.L., Grimm, B.D., Hulsey, M.L., Miller, J., McClure, H.M., McNicholl, J.M., Moss, B., and Robinson, H.L. 2001. Control of a mucosal challenge and prevention of AIDS by a multiprotein DNA/MVA vaccine. *Science* 292:69–74.

Antoch, M.P., Song, E.J., Chang, A.M., Vitaterna, M.H., Zhao, Y., Wilsbacher, L.D., Sangoram, A.M., King, D.P., Pinto, L.H., and Takahashi, J.S. 1997. Functional identification of the mouse circadian Clock gene by transgenic BAC rescue. *Cell* 89:655–667.

Antonarakis, S.E. and the Nomenclature Working Group. 1998. Recommendations for a nomenclature system for human gene mutations. *Hum. Mutat.* 11:1–3.

Arnheim, N. and Shibata, D. 1997. DNA mismatch repair in mammals: Role in disease and meiosis. *Curr. Opin. Genet. Dev.* 7:364–370.

Ashburner, M., Ball, C.A., Blake, J.A., Botstein, D., Butler, H., Cherry, J.M., Davis, A.P., Dolinski, K., Dwight, S.S., Eppig, J.T., Harris, M.A., Hill, D.P., Issel-Tarver, L., Kasarskis, A., Lewis, S., Matese, J.C., Richardson, J.E., Ringwald, M., Rubin, G.M., and Sherlock, G. 2000. Gene Ontology: Tool for the unification of biology. The Gene Ontology Consortium. *Nat. Genet* 25:25–29.

Athwal, R.S., Smarsh, M., Searle, B.M., and Deo, S.S. 1985. Integration of a dominant selectable marker into human chromosomes and transfer of marked chromosomes to mouse cells by microcell fusion. *Somatic Cell Mol. Genet.* 11:177–187.

Auerbach, A.D. 1993. Fanconi anemia diagnosis and the diepoxybutane (DEB) test. *Exp. Hematol.* 21:731–733.

Auerbach, A.D., Rogatko, A., and Schroeder-Kurth, T.M. 1989. International Fanconi Anemia Registry: Relation of clinical symptoms to diepoxybutane sensitivity. *Blood* 73:391–396.

Ausubel, F.M., Brent, R., Kingston, R.E., Moore, D.D., Seidman, J.G., Smith, J.A., and Struhl, K. 2004. The Polymerase Chain Reaction. In Current Protocols in Molecular Biology. Chapter 15. John Wiley & Sons, New York.

Barch, M.J., (ed.) 1997. The ACT Cytogenetics Laboratory Manual, 3rd ed. Lippincott-Raven, Philadelphia.

Barr, E. and Leiden, J.M. 1991. Systemic delivery of recombinant proteins by genetically modified myoblasts. *Science* 254:1507–1509.

Baxevanis, A.D., Boguski, M.S., and Ouellette, B.F.F. 1997. Computational analysis of DNA and protein sequences. In Genome Analysis: A Laboratory Manual (B. Birren, E.D. Green, S. Kapholz, R.M. Myers, and J. Roskams, eds.) pp. 533–586. Cold Spring Harbor Laboratory Press, Cold Spring Harbor, N.Y.

Becker, T.C., Noel, R.J., Coats, W.S., Gomez-Foix, A.M., Alam, T., Gerard, R.D., and Newgard, C.B. 1994. Use of recombinant adenovirus for metabolic engineering of mammalian cells. *Methods Cell Biol.* 43:161–189.

Beggs, A.H. and Kunkel, L.M. 1990. A polymorphic CACA repeat in the 3′ untranslated region of dystrophin. *Nucl. Acids Res.* 18:1931.

Beggs, A.H., Koenig, M., Boyce, F.M., and Kunkel, L.M. 1990. Detection of 98% of DMD/BMD gene deletions by polymerase chain reaction. *Hum. Genet.* 86:45–48.

Beggs, A.H., Hoffman, E.P., Snyder, J.R., Arahata, K., Specht, L., Shapiro, F., Angelini, C., Sugita, H., and Kunkel, L.M. 1991. Exploring the molecular basis for variability among patients with Becker muscular dystrophy: Dystrophin gene and protein studies. *Am. J. Hum. Genet.* 49:54–67.

Benjamini, Y. and Hochberg, Y. 1995. Controlling the false discovery rate: A practical and powerful approach to multiple testing. *J. R. Statist. Soc. B* 57:2889–3000.

Bentley, D.R., Todd, C., Collins, J., Holland, J., Dunham, I., Hassock, S., Bankier, A., and Giannelli, F. 1992. The development and application of automated gridding for efficient screening of yeast and bacterial ordered libraries. *Genomics* 12:534–541.

Beroud, C., Collod-Beroud, G., Boileau, C., Soussi, T., and Junien, C. 2000. UMD (Universal Mutation Database): A generic software to build and analyze locus-specific databases. *Hum. Mutat.* 15:86–94.

Birren, B. and Lai, E. 1993. Pulsed field gel electrophoresis: A practical guide. Academic Press, San Diego.

Blackwelder, W.C. and Elston, R.C. 1985. A comparison of sib pair linkage tests for disease susceptibility loci. *Genet. Epidemiol.* 2:5–97.

Blanco-Bose, W.E., Yao, C.C., Kramer, R.H., and Blau, H.M. 2001. Purification of mouse primary myoblasts based on alpha 7 integrin expression. *Exp. Cell Res.* 265:212–220.

Blonden, L.A., den Dunnen, J.T., van Paassen, H.M., Wapenaar, M.C., Grootscholten, P.M., Ginjaar, H.B., Bakker, E., Pearson, P.L., and van Ommen, G.J. 1989. High resolution deletion breakpoint mapping in the DMD gene by whole cosmid hybridization. *Nucl. Acids Res.* 17:5611–5621.

Boguski, M.S., Lowe, T.M., and Tolstoshev, C.M. 1993. dbEST: Database for "Expressed Sequence Tags". *Nat. Genet.* 4:332–333.

Botstein, D., Falco, S.C., Stewart, S.E., Brennan, M., Schere, S., Stinchcomb, D.T., Struhl, K., and Davis, R.W. 1979. Sterile host yeasts (SHY): A eukaryotic system of biological containment for recombinant DNA experiments. *Gene* 8:17–24.

Boyce, F.M., Beggs, A.H., Feener, C., and Kunkel, L.M. 1991. Dystrophin is transcribed in brain from a distant upstream promoter. *Proc. Natl. Acad. Sci. U.S.A.* 88:1276–1280.

Brenner, C.H. 1997. Symbolic kinship program. *Genetics* 145:535–542.

Brenner, C.H. and Morris, J. 1989. Paternity index calculations in single locus hypervariable probes: Validation and other studies. *In* Proceedings for the International Symposium on Human Identification Data Acquisitions and Statistical Analysis, pp. 21–53. Promega Corporation, Madison, WI.

Brambati, B., Oldrini, A., Ferrazzi, J., and Lanzani, A. 1987. Chorionic villus sampling: An analysis of the obstetric experience of 1000 cases. *Prenatal Diagn.* 7:157–169.

Brown, A. and McKie, M. 2000. MuStaR and other software for locus-specific mutation databases. *Hum. Mutat.* 15:76–85.

Burge, C. and Karlin, S. 1997. Prediction of complete gene structures in human genomic DNA. *J. Mol. Biol* 268:78–94.

Burke, D., Carle, G., and Olson, M.V. 1987. Cloning of large segments of DNA into yeast by means of artificial chromosome vectors. *Science* 236:806–812.

Burset, M. and Guigo, R. 1996. Evaluation of gene structure prediction programs. *Genomics* 34:353–367.

Cabin, D.E., Hawkins, A., Griffin, C., and Reeves, R.H. 1995. YAC transgenic mice in the study of the genetic basis of Down syndrome. *Progr. Clin. Biol. Res.* 393:213–226.

Cayatte, A.J., Du, Y., Oliver-Krasinski, J., Lavielle, G., Verbeuren, T.J., and Cohen, R.A. 2000. The thromboxane receptor antagonist S18886 but not aspirin inhibits atherogenesis in apo E-deficient mice. *Arterioscler. Thromb. Vasc. Biol* 20:1724–1728.

Chamberlain, J.S., Gibbs, R.A., Ranier, J.E., Nguyen, P.N., and Caskey, C.T. 1988. Deletion screening of the Duchenne muscular dystrophy lo-cus via multiplex DNA amplification. *Nucl. Acids Res.* 16:11141–11156.

Chamberlain, J.S., Gibbs, R.A., Ranier, J.E., and Caskey, C.T. 1990. Multiplex PCR for the diagnosis of Duchenne muscular dystrophy. *In* PCR Protocols: A Guide to Methods and Applications (M.A. Innis, D.H. Gelfand, J.J. Sninsky, and T.J. White, eds.) pp. 272–281. Academic Press, San Diego.

Chang, M.W., Barr, E., Seltzer, J., Jiang, Y.-Q., Nabel, G.J., Nabel, E.G., Parmacek, M.S., and Leiden, J.M. 1995. Cytostatic gene therapy for vascular proliferative disorders with a constitutively active form of the retinoblastoma gene product. *Science* 267:518–522.

Chen, J. N., Haffter, P., Odenthal, J., Vogelsang, E., Brand, M., van Eeden, F.J., Furutani-Seiki, M., Granato, M., Hammerschmidt, M., Heisenberg, C.P., Jiang, Y.J., Kane, D.A., Kelsh, R.N., Mullins, M.C., and Nusslein-Volhard, C. 1996a. Mutations affecting the cardiovascular system and other internal organs in zebrafish. *Development* 123:293–302.

Chen, S.T., Iida, A., Guo, L., Friedmann, T., and Yee, J.K. 1996b. Generation of packaging cell lines for VSV-G pseudotyped retroviral vectors using a modified tetracycline-inducible system. *Proc. Natl. Acad. Sci. U.S.A.* 93:10057–10062.

Chen, X., Levine, L., and Kwok, P.-Y. 1999. Fluorescence polarization in homogeneous nucleic acid analysis. *Genome Res.* 9:492–498.

Cheng, S., Higuchi, R., and Stoneking, M. 1994. Complete mitochondrial genome amplification. *Nat. Genet.* 7:350–351.

Cisterni, C., Henderson, C.E., Aebischer, P., Pettmann, B., and Deglon, N. 2000. Efficient gene transfer and expression of biologically active glial cell line-derived neurotrophic factor in rat motoneurons transduced with lentiviral vectors. *J. Neurochem.* 74:1820–1828.

Claustres, M., Horaitis, O., Vanevski, M., and Cotton, R.G.H. 2002. Time for a unified system of mutation description and reporting: A review of locus specific mutation databases. *Genome Res.* 12:680–688.

Claverie, J.M. 1997a. Computational methods for the identification of genes in vertebrate genomic sequences. *Hum. Mol. Genet.* 6:1735–1744.

Claverie, J.M. 1997b. Exon detection by similarity searches. *Methods Mol. Biol.* 68:283–313.

Claverie, J.M. 1998. Computational methods for exon detection. *Mol. Biotechnol.* 10:27–48.

Coffey, A.J., Roberts, R.G., Green, E.D., Cole, C.G., Butler, R., Anand, R., Giannelli, F., and Bentley, D.R. 1992. Construction of a 2.6-Mb contig in yeast artificial chromosomes spanning the human dystrophin gene using an STS-based approach. *Genomics* 12:474–484.

Coligan, J.E., Kruisbeek, A.M., Margulies, D.H., Shevach, E.M., and Strober, W. (eds.) 2004. Current Protocols in Immunology. John Wiley & Sons, Hoboken, N.J.

Cosset, F.-L., Takeuchi, Y., Battini, J.L., Weiss, R.A., and Collins, M.K. 1995. High-titer packaging cells producing recombinant retroviruses resistant to human serum. *J. Virol.* 69:7430–7436.

Cotton, R.G. and Scriver, C.R. 1998. Proof of "disease causing" mutation. *Hum. Mutat.* 12:1–3.

Covone, A.E., Caroli, F., and Romeo, G. 1992. Screening Duchenne and Becker muscular dystrophy patients for deletions in 30 exons of the dystrophin gene by three-multiplex PCR. *Am. J. Hum. Genet.* 51:675–677.

Cowell, J.K. 1984. A photographic representation of the variability in the G-banded structure of the chromosomes in the mouse karyotype. A guide to the identification of the individual chromosomes. *Chromosoma (Berl)* 89:294–320.

CRC (Chemical Rubber Company). 1975. CRC Handbook of Biochemistry and Molecular Biology, Physical and Chemical Data, 3rd ed., Vol. 1. CRC Press, Boca Raton, Fla.

Cremer, T., Lichter, P., Borden, J., Ward, D.C., and Manuelidis, L. 1988. Detection of chromosome aberrations in metaphase and interphase tumor cells by in situ hybridization using chromosome-specific library probes. *Hum. Genet.* 80:235–246.

Cunningham, C. and Davison, A.J. 1993. A cosmid-based system for constructing mutants of herpes simplex virus type 1. *Virology* 197:116–124.

Dahlquist, K.D., Salomonis, N., Vranizan, K., Lawlor, S.C., and Conklin, B.R. 2002. GenMAPP, a new tool for viewing and analyzing microarray data on biological pathways. *Nat. Genet.* 31:19–20.

Danos, O. and Mulligan, R.C. 1988. Safe and efficient generation of recombinant retroviruses with amphotropic and ecotropic host ranges. *Proc Natl. Acad. Sci. U.S.A.* 85:6460–6464.

Davis, S. and Weeks, D.E. 1997. Comparison of nonparametric statistics for detection of linkage analysis in nuclear families: Single marker evaluation. *Am. J. Hum. Genet.* 61:1431–1444.

Dawson, M.C., Elliott, D.C., Elliott, W.H., and Jones, K.M., (eds.). 1986. Data for Biochemical Research. Alden Press, London.

Dean, M., White, M.B., Amos, J., Gerrard, B., Stewart, C., Khaw, K.-T., and Leppert, M. 1990. Multiple mutations in highly conserved residues are found in mildly affected cystic fibrosis patients. *Cell* 61:863–870.

Deglon, N., Tseng, J.L., Bensadoun, J.C., Zurn, A.D., Arsenijevic, Y., Pereira de Almeida, L., Zufferey, R., Trono, D., and Aebischer, P. 2000. Self-inactivating lentiviral vectors with enhanced transgene expression as potential gene transfer system in Parkinson's disease. *Hum. Gene Ther.* 11:179–190.

DeLuca, N.A., McCarthy, A.M., and Schaffer, P.A. 1985. Isolation and characterization of deletion mutants of herpes simplex virus type 1 in the gene encoding immediate-early regulatory protein ICP4. *J. Virol.* 56:558–570.

den Dunnen, J.T. and Antonarakis, S.E. 2000. Mutation nomenclature extensions and suggestions to describe complex mutations: A discussion. *Hum. Mutat.* 15:7–12.

den Dunnen, J.T. and van Ommen, G.J.B. 1999. The Protein Truncation Test: A review. *Hum. Mutat.* 14:95–102.

DeRisi, J.L., Iyer, V.R., and Brown, P.O. 1997. Exploring the metabolic and genetic control of gene expression on a genomic scale. *Science* 278:680–686.

DeVries, D.D., Went, L.N., Bruyn, G.W., Scholte, H.R., Hofstra, R.M.W., Bolhuis, P.A., and van Oost, B.A. 1996. Genetic and biochemical impairment of mitochondrial complex I activity in a family with Lebec hereditary optic neuropathy and hereditary spastic dystonia. *Am. J. Hum. Genet.* 58:703–711.

Dib, C., Faure, S., Fizames, C., Samson, D., Drouot, N., Vignal, A., Millasseau, P., Marc, S., Hazan, J., Seboun, E., Lathrop, M., Gyapay, G., Morissette, J., and Weissenbach, J. 1996. A comprehensive genetic map of the human genome based on 5,264 microsatellites. *Nature* 380:152–154.

Dougherty, J.P., Wisniewski, R., Yang, S., Rhode, B.W., and Temin, H.M. 1989. New retrovirus helper cells with almost no nucleotide sequence homology to retrovirus vectors. *J. Virol.* 63:3209–3212.

Draghia-Akli, R., Fiorotto, M.L., Hill, L.A., Malone, P.B., Deaver, D.R., and Schwartz, R.J. 1999. Myogenic expression of an injectable protease-resistant growth hormone-releasing hormone augments long-term growth in pigs. *Nat. Biotechnol.* 17:1179–1183.

Du Manoir, S., Schroeck, E., Bentz, M., Speicher, M.R., Joos, S., Ried, T., Lichter, P., and Cremer, T. 1995. Quantitative analysis of comparative genomic hybridization. *Cytometry* 19.1:27–41.

Dubois, B.L. and Naylor, S.L. 1993. Characterization of NIGMS human/rodent somatic cell hybrid mapping panel 2 by PCR. *Genomics* 16:315–319.

Dudoit, S., Fridlyand, J., and Speed, T. 2000. Comparison of Discrimination Methods for the Classification of Tumors Using Gene Expression Data. Tech. Rep. 576, Dept. of Statistics, University of California, Berkeley, Calif.

Durbin, B.P., Hardin, J.S., Hawkins, D.M., and Rocke, D.M. 2002. A variance-stabilizing transformation for gene expression microarray data. *Bioinformatics* 18:S105–S110.

Eichinger, D.J. and Boeke, J.D. 1988. The DNA intermediate in yeast Ty1 element transposition copurifies with virus-like particles: Cell-free Ty1 transposition. *Cell* 54:955–966.

Eichner, J.E., Terence Dunn, S., Perveen, G., Thompson, D.M., Stewart, K.E., and Stroehla, B.C. 2002. Apolipoprotein E polymorphism and cardiovascular disease: A HuGE review. *Am. J. Epidemiol.* 487–495.

Elston, R.C., Guo, X., and Williams, L.V. 1996. Two-stage global search designs for linkage analysis using pairs of affected relatives. *Genet. Epidemiol.* 13:535–558.

Endean, D. 1989. RFLP analysis and paternity testing: Observations and caveats. *In* Proceedings for the International Symposium on Human Identification Data Acquisition and Statistical Analysis, pp. 55–75. Promega Corporation, Madison, Wis.

Endean, D. 1995. Apparent false exclusion observed in various DNA-VNTR systems reported by AABB-accredited parentage testing laboratories. *In* American Association of Blood Banks, Accreditation Requirements Manual, 2nd ed., pp. 66. American Association of Blood Banks, Arlington, Va.

Engelhardt, J.F. and Wilson, J.M. 1997. Explant models of the airway. *In* The Lung (R.G. Crystal, J.B. West, P.J. Barnes, and E.R. Weibel, eds.) pp. 345–352. Raven Press, New York.

Engelhardt, J.F., Yankaskas, J.R., and Wilson, J.M. 1992. In vivo retroviral gene transfer into human bronchial epithelia of xenografts. *J. Clin. Invest.* 90:2598–2607.

ESHRE PGD Consortium Steering Committee. 1999. ESHRE Preimplantation Genetic Diagnosis (PGD) Consortium: Preliminary assessment of data from January 1997-September 1998. *Europ. Soc. Hum. Reprod. Embryol.* 14:3138–3148.

Evans, E.P. 1996. Standard idiogram. *In* Genetic Variants and Strains of the Laboratory Mouse, Vol. 2 (M.F. Lyon, S. Rastan, and S.D.M. Brown, eds.) pp. 1446–1448. Oxford University Press, Oxford.

Fabrizi, G.M., Tiranti, V., Mariotti, C. Guazzi, G.C., Malandrini, A., DiDonato, S., and Zeviani, M. 1995. Sequence analysis of mitochondrial DNA in a new inherited encephalomyopathy. *J. Neurol.* 242:490–496.

Fan, J.-B., Chen, X., Halushka, M.K., Berno, A., Huang, X., Ryder, T., Lipshutz, R.J., Lockhart, D.J., and Chakravarti, A. 2000. Parallel genotyping of human SNPs using generic high-density oligonucleotide tag arrays. *Genome Res.* 10:853–860.

Felgner, P.L., Gadek, T.R., Holm, M., Roman, R., Chan, H.W., Wenz, M., Northrop, J.P., Ringold, G.M., and Danielsen, M. 1987. Lipofection: A highly efficient, lipid-mediated DNA-transfection procedure. *Proc. Natl. Acad. Sci. U.S.A.* 84:7413–7417.

Ferrie, R.M., Schwarz, M.J., Robertson, N.H., Vaudin, S., Super, M., Malone, G., and Little, S. 1992. Development, multiplexing and application of ARMS tests for common mutations in the CFTR gene. *Am. J. Hum. Genet.* 51:251–262.

Ferry, N. and Heard, J.M. 1998. Liver-directed gene transfer vectors. *Hum. Gene Ther.* 9:1974–1981.

Fisher, L.J., Jinnah, H.A., Kale, L.C., Higgins, G.A., and Gage, F.H. 1991. Survival and function of intrastriatally grafted primary fibroblasts genetically modified to produce L-dopa. *Neuron* 6:371–380.

Fleming, D.O., Richardson, J.H., Tulis, J.I., and Vesley, D. (eds.) 1995. Laboratory Safety: Principles and Practices, 2nd ed. ASM Press (American Society for Microbiology), Washington, D.C.

Fletcher, J.A., Pinkus, G.S., Weidner, N., and Morton, C.C. 1991. Lineage-restricted clonality in biphasic solid tumors. *Am. J. Path.* 138:1199–1207.

Flint, S.J., Enquist, L.W., Krug, R.M., Racaniello, V.R., and Skalka, A.M. 1999. Virology, Molecular Biology, Pathogenesis and Control. ASM Press (American Society for Microbiology), Washington, D.C.

Fraefel, C., Song, S., Lim, F., Lang, P., Yu, L., Wang, Y., Wild, P., and Geller, A.I. 1996. Helper virus-free transfer of herpes simplex virus type 1 plasmid vectors into neural cells. *J. Virol.* 70:7190–7197.

Francke, U. 1981. High resolution ideograms of trypsin-Giemsa banded human chromosomes. *Cytogenet. Cell Genet.* 31:24–32.

Freshney, R.I. 1993. Culture of Animal Cells. A Manual of Basic Techniques, 3rd ed. Wiley-Liss, New York.

Frommer, M., McDonald, L.E., Millar, D.S., Collis, C.M., Watt, F., Grigg, G.W., Molloy, P.L., and Paul, C.L. 1992. A genomic sequencing protocol that yields a positive display of 5-methylcytosine residues in individual DNA strands. *Proc. Natl. Acad. Sci. U.S.A* 89:1827–1831.

Fu, K., Hartlen, R., Johns, T., Genge, A., Karpati, G., and Shoubridge, E.A. 1996. A novel heteroplasmic tRNA$^{leu(CUN)}$ mtDNA point mutation in a sporadic patient with mitochondrial encephalomyopathy segregates rapidly in skeletal muscle and suggests an approach to therapy. *Hum. Mol. Genet.* 5:1835–1840.

Gao, X. and Huang, L. 1991. A novel cationic liposome reagent for efficient transfection of mammalian cells. *Biochem. Biophys. Res. Commun.* 179:280–285.

Gao, X. and Huang, L. 1995. Cationic liposome-mediated gene transfer. *Gene Ther.* 2:710–722.

Garber, R.A. and Morris, J.W. 1983. General equations for the average power of exclusion for genetic systems of n codominant alleles in one-parent and no-parent cases of disputed parentage. *In* Inclusion Probabilities in Parentage Testing (R.H. Walker, ed.). American Association of Blood Banks, Arlington, Va.

Gasch, A.P. and Eisen, M. 2002. Exploring the conditional coregulation of yeast through fuzzy *k*-means clustering. *Genome Biol.* 3:RESEARCH0059.

Gauderman, W.J. 2002. Sample size requirements for association studies of gene-gene interaction. *Am. J. Epidemiol.* 155:478–484.

Gemmill, R. 1990. Pulsed field gel electrophoresis. *Adv. Electrophor.* 4:1–48.

Gillin, F.D., Roufa, D.J., Beaudet, A.L., and Caskey, C.T. 1972. 8-Azaguanine resistance in mammalian cells. I. Hypoxanthine-guanine phosphoribosyltransferase. *Genetics* 72:239–252.

Gjertson, D.W. and Endean, D. 1996. Promega Statistics Workshop, Scottsdale, Ariz.

Golub, T.R., Slonim, D.K. Tamayo, P., Huard, C., Gaasenbeek, M., Mesirov, J.P., Coller, H., Loh, M.L., Downing, J.R., Caligiuri, M.A., et al. 1999. Molecular classification of cancer: Class discovery and class prediction by gene expression monitoring. *Science* 286:531–537.

Goradia, T.M. and Lange, K. 1990. Multilocus ordering strategies based on sperm typing. *Ann. Hum. Genet.* 54:49–77.

Gritz, L., and Davies J. 1983. Plasmid-encoded hygromycin B resistance: The sequence of hygromycin B phosphotransferase gene and its expression in *Escherichia coli* and *Saccharomyces cerevisiae*. *Gene* 25:179–188.

Guigo, R. 1997. Computational gene identification. *J. Mol. Med* 75:389–393.

Haines, J.L. and Pericak-Vance, M.A. 1998. Approaches to Gene Mapping in Complex Human Diseases. John Wiley & Sons, New York.

Hammans, S.R. Sweeney, M.G. Brockington, M., Morgan, H.J.A., and Harding, A.E. 1991. Mitochondrial encephalopathies: Molecular genetic diagnosis from blood samples. *Lancet* 337:1311–1313.

Hanenberg, H., Xiao, X.L., Dilloo, D., Hashino, K., Kato, I., and Williams, D.A. 1996. Colocalization of retrovirus and target cells on specific fibronectin fragments increases genetic transduction of mammalian cells. *Nat. Med.* 2:876–882.

Harrison, C.J. 1992. The lymphomas and chronic lymphoproliferative disorders. *In* Human Cytogenetics: A Practical Approach (D.E. Rooney and B.H. Czepulkowski, eds.) pp. 97–120. IRL Press, Oxford.

Haseman, J.K. and Elston, R.C. 1972. The investigation of linkage between a quantitative trait and a marker locus. *Behav. Genet.* 2:3–19.

Hastbacka, J., De La Chapelle, A., Kaitila, I., Sistonen, P., Weaver, A., and Lander, E. 1992. Linkage disequilibrium mapping in isolated founder populations: Diastrophic dysplasia in Finland. *Nat. Genet.* 2:204–211.

He, T.C., Zhou, S., da Costa, L.T., Yu, J., Kinzler, K.W., and Vogelstein, B. 1998. A simplified system for generating recombinant adenoviruses. *Proc. Natl. Acad. Sci. U.S.A.* 95:2509–2514.

Heiniger, H.J., Taylor, B.A., Hard, E.J., and Meier, H. 1975. Heritability of the phytohemagglutinin responsiveness of lymphocytes and its relationship to leukemogenesis. *Cancer Res.* 35:825–831.

Heng, H.H.Q. and Tsui, L.-C. 1994. Free chromatin mapping by FISH. *In* Methods of Molecular Biology: In Situ Hybridization Protocols (K.H.A. Choo, ed.) pp. 109–122. Humana Press, Clifton, N.J.

Henke, J., Fimmers, R., Baur, M.P., and Henke, L. 1993. DNA minisatellite mutations: Recent investigations concerning distribution and impact on parentage testing. *Int. J. Leg. Med.* 105:217–222.

Hentemann, M., Reiss, J., Wagner, M., and Cooper, D.N. 1990. Rapid detection of deletions in the Duchenne muscular dystrophy gene by PCR amplification of deletion-prone exon sequences. *Hum. Genet.* 84:228–232.

Herman, J.G., Graff, J.R., Myohanen, S., Nelkin, B.D., and Baylin, S.B. 1996. Methylation-specific PCR: A novel PCR assay for methylation status of CpG islands. *Proc. Natl. Acad. Sci. U.S.A.* 93:9821–9826.

Hirschhorn, J.N., Lohmueller, K., Byrne, E., and Hirschhorn, K. 2002. A comprehensive review of genetic association studies. *Genet. Med.* 4:45–61.

Hitt, M.M., Addison, C.L., and Graham, F.L. 1997. Human adenovirus vectors for gene transfer into mammalian cells. *Adv. Pharmacol.* 40:137–206.

Hogan, B., Beddington, R., Costaatini, F., and Lacy, E. 1994. Manipulating the Mouse Embryo, 2nd ed. Cold Spring Harbor Laboratory Press, Cold Spring Harbor, N.Y.

Houwen, R.H.J., Baharloo, S., Blankenship, K., Raeymaekers, P., Juyn, J., Sandkuijl, A., and Freimer, N.B. 1994. Genome screening by searching for shared segments: Mapping a gene for benign recurrent intrahepatic cholestasis. *Nat. Genet.* 8:380–386.

Hsu, T.M. and Kwok, P.-Y. 2003. Homogeneous primer extension assay with fluorescence polarization detection. *Methods Mol. Biol.* 212:177–187.

Hsu, L.Y.F., Kaffe, S., Jenkins, E.C., Alonso, L., Benn, P.A., David, K., Hirschorn, K., Lieber, E., Shanske, A., Shapiro, L.R., Schutta, E., and Warburton, D. 1992. Proposed guidelines for diagnosis of chromosome mosaicism in amniocytes based on data derived from chromosome mosaicism and pseudomosaicism studies. *Prenatal Diag.* 12:555–573.

Hsu, T.M., Chen, X., Duan, S., Miller, R., and Kwok, P.-Y. 2001. A universal SNP genotyping assay with fluorescence polarization detection. *BioTechniques* 31:560–570.

Huxley, C., Hagino, Y., Schlessinger, D., and Olson, M.V. 1991. The human HPRT gene on a yeast artificial chromosome is functional when transferred to mouse cells by cell fusion. *Genomics* 9:742–750.

Ioannou, P.A., Amemiya, C.T., Garnes, J., Kroisel, P.M., Shizuya, H., Chen, C., Batzer, M.A., and de Jong, P.J. 1994. A new bacteriophage P1-derived vector for the propagation of large human DNA fragments. *Nat. Genet.* 6:84–89.

ISCN. 1981. An International System for Human Cytogenetic Nomenclature—High-Resolution Banding (D.G. Harnden, J.E. Lindsten, K. Buckton, and H.P. Klinger, eds.). S. Karger, Basel, Switzerland.

ISCN. 1991. Guidelines for cancer cytogenetics, Supplement to An International System for Human Cytogenetic Nomenclature (F. Mitelman, ed.). S. Karger, Basel.

ISCN. 1995. An International System for Human Cytogenetic Nomenclature (F. Mitelman, ed.). S. Karger, Basel, Switzerland, and Farmington, Conn.

Iyer, V.R., Eisen, M.B., Ross, D.T., Schuler, G., Moore, T., Lee, J.C.F. Trent, J.M., Staudt, L.M., Hudson, J. Jr., Boguski, M.S., Lashkari, D., Shalon, D., Botstein, D., and Brown, P.O. 1999. The transcriptional program in the response of human fibroblasts to serum. *Science* 283:83–87.

Jaffe, E.S., Harris, N.L., Stein, H., and Vardiman, J.W. 2001. World Health Organization Classification of Tumors. Pathology and Genetics of Tumours of Haematopoietic and Lymphoid Tissues. IARC Press, Washington, D.C.

Kallioniemi, O.P., Kallioniemi, A., Sudar, D., Rutovitz, D., Gray, J., Waldman, F., and Pinkel, D. 1993. Comparative genomic hybridization: A rapid new method for detecting and mapping DNA amplification in tumors. *Sem. Cancer Biol.* 4:41–46.

Kallioniemi, O.P., Kallioniemi, A., Piper, J., Isola, J., Waldman, F.M., Gray, J.W., and Pinkel, D. 1994. Optimizing comparative genomic hybridization for analysis of DNA sequence copy number changes in solid tumors. *Genes Chrom. Cancer* 10:231–243.

Kanesha, M. et al. 2002. The KEGG databases at GenomeNet. *Nucl. Acids Res.* 30:42–46.

Kans, J.A. and Ouellette, B.F.F. 1998. Submitting DNA sequences to the databases. *In* Bioinformatics: A Practical Guide to the Analysis of Genes and Proteins (A.D. Baxevanis and B.F.F. Ouellette, eds.) pp. 319–353. John Wiley & Sons, New York.

Kao, F.-T., Jones, C., and Puck, T.T. 1976. Genetics of somatic mammalian cells: Genetic, immunologic, and biochemical analysis with Chinese hamster cell hybrids containing selected human chromosomes. *Proc. Natl. Acad. Sci. U.S.A.* 73:193–197.

Karlin, S. and Altschul, S.F. 1990. Methods for assessing the statistical significance of molecular sequence features by using general scoring schemes. *Proc. Natl. Acad. Sci. U.S.A* 87:2264–2268.

Karlin, S. and Altschul, S.F. 1993. Applications and statistics for multiple high-scoring segments in molecular sequences. *Proc. Natl. Acad. Sci. U.S.A* 90:5873–5877.

Kawaja, M.D., Fisher, L.J., Schinstine, M., Jinnah, H.A., Ray, J., Chen, L.S., and Gage, F.H. 1992. Grafting genetically modified cells within the rat central nervous system: Methodological considerations. *In* Neural

Transplantation: A Practical Approach (S.B. Dunnett and A. Bjorklund, eds.) pp. 20–55. Oxford University Press, New York.

Keightly, J.A., Hoffbuh, K.C., Burton, M.D., Salas, V.M., Johnston, W.S.W., Penn, A.M.W., Buist, N.R.M., and Kennaway, N.G. 1996. A microdeletion in cytochrome *c* oxidase (COX) subunit III associated with COX deficiency and recurrent myoglobulinuria. *Nat. Genet.* 12:410–415.

Kellems, R.E., (ed.) 1993. Gene Amplification in Mammalian Cells. Marcel Dekker, New York.

Kent, W.J. 2002. BLAT—The BLAST-like alignment tool. *Genome Res.* 12:656–664.

Kerr, M.K. and Churchill, G.A. 2001. Statistical design and the analysis of gene expression microarrays. *Genet. Res.* 77:123–128.

Kerr, M.K., Martin, M., and Churchill, G.A. 2000. Analysis of variance for gene expression microarray data. *J. Comput. Biol.* 7:819–837.

Khoury, M.J., Beaty, T.H. and Cohen, B.H. 1993. Fundamentals of Genetic Epidemiology, Chapters 7 and 8. Oxford University Press, New York.

Kilimann, M.K., Pizzuti, A., Grompe, M., and Caskey, C.T. 1992. Point mutations and polymorphisms in the human dystrophin gene identified in genomic DNA sequences amplified by multiplex PCR. *Hum. Genet.* 89:253–258.

Klinger, K., Landes, G., Shook, D., Harvey, R., Lopez, L., Locke, P., Lerner, T., Osathanondh, R., Leverone, B., Houseal, T., Pavelka, K., and Dackowski, W. 1992. Rapid detection of chromosome aneuploidies in uncultured amniocytes by using fluorescence in situ hybridization (FISH). *Am. J. Hum. Genet.* 51:55–65.

Knuutila, S., Nylund, S.J., Wessman, M., and Larramendy, M.L. 1994a. Analysis of genotype and phenotype on the same interphase or mitotic cell. A manual of MAC (Morphology Antibody Chromosomes) methodology. *Cancer Genet. Cytogenet.* 72:1–15.

Koenig, M., Beggs, A.H., Moyer, M., Scherpf, S., Heindrich, K., Bettecken, T., Meng, G., Muller, C.R., Lindlof, M., Kaariainen, H., de la Chapelle, A., Kiuru, A., Savontaus, M.-L., Gilgenkrantz, H., Recan, D., Chelly, J., Kaplan, J.-C., Covone, A.E., Archidiacono, N., Romeo, G., Liechti-Gallati, S., Schneider, V., Braga, S., Moser, H., Darras, B.T., Murphy, P., Francke, U., Chen, J.D.,

Morgan, G., Denton, M., Greenberg, C.R., van Ommen, G.J.B., and Kunkel, L.M. 1989. The molecular basis for Duchenne versus Becker muscular dystrophy: Correlation of severity with type of deletion. *Am. J. Hum. Genet.* 45:498–506.

Kogan, S.C., Doherty, M., and Gitschies, J. 1987. An improved method for prenatal diagnosis of genetic diseases by analysis of amplified DNA sequences. *New Engl. J. Med.* 317:985–990.

Krisky, D.M., Marconi, P.C., Oligino, T.J., Rouse, R.J.D., Fink, D.J., Cohen, J.B., Watkins, S.C., and Glorioso, J.C. 1998. Development of herpes simplex virus replication-defective multigene vectors for combination gene therapy applications. *Gene Ther.* 5:1517–1530.

Kruglyak, L. and Lander, E.S. 1995. Complete multipoint sib-pair analysis of qualitative and quantitative traits. *Am. J. Hum. Genet.* 57:439–454.

Kubota, T., Das, S., Christian, S.L., Baylin, S.B., Herman, J.G., and Ledbetter, D.H. 1997. Methylation-specific PCR simplifies imprinting analysis. *Nat. Genet* 16:16–17.

Kunkel, L.M., Snyder, J.R., Beggs, A.H., Boyce, F.M., and Feener, C.A. 1991. Searching for dystrophin gene deletions in patients with atypical presentations. *In* Etiology of Human Diseases at the DNA Level (J. Lindsten and U. Petterson, eds.) pp. 51–60. Raven Press, New York.

Kurdi-Haidar, B., Levine, F., Roemer, K., LaPorte, P., and Friedmann, T. 1993. Provirus-anchored long range mapping of mammalian genomes. *Genomics* 15:305–310.

Kwok, P.-Y. 2002. SNP genotyping with fluorescence polarization detection. *Hum. Mutat.* 19:315–323.

Labarca, C. and Paigen, K. 1980. A simple, rapid, and sensitive DNA assay procedure. *Anal. Biochem.* 102:344–352.

Landegren, U., Nilsson, M., and Kwok, P.Y. 1998. Reading bits of genetics information: Methods for single-nucleotide polymorphism analysis. *Genome Res.* 8:769–776.

Lander, E.S. and Botstein, D. 1987. Homozygosity mapping: A way to map human recessive traits with the DNA of inbred children. *Science* 236:1567–1570.

Larsen, L.A., Christiansen, M., Vuust, J., and Andersen, P.S. 1999. High-throughput single-strand con-

formation polymorphism analysis by automated capillary electrophoresis: Robust multiplex analysis and pattern-based identification of allelic variants. *Hum. Mutat.* 13:318–327.

Latt, S.A. 1976. Optical studies of metaphase chromosome organization. *Annu. Rev. Biophys. Bioeng.* 5:1–37.

Latt, S.A., Shreck, R.R., D'Andrea, A., Kaiser, T.N., Schlesinger, F., Lester, S., and Sakai, K. 1984. Detection, significance, and mechanism of sister chromatid formation: Past experiments, current concepts, future challenges. *In* Sister Chromatid Exchanges (R.R. Tice and A. Hollaender, eds.) pp. 11–40. Plenum, New York.

Lederer, C.M., Hollander, J.M., and Perlman, I., (eds.). 1967. Table of Radioisotopes, 6th ed. John Wiley & Sons, New York.

Lee, E.C. 1991. Cytogenetic analysis of continuous cell lines. *In* The ACT Cytogenetics Laboratory Manual, 2nd ed. (M.J. Barch, ed.) pp. 107–148. Raven Press, New York.

Lee, J.L., Warburton, D., and Robertson, E.J. 1990. Cytogenetic methods for the mouse: Preparation of chromosomes, karyotyping, and in situ hybridization. *Anal. Biochem.* 189:1–17.

Lenk, U., Hanke, R., and Speer, A. 1994. Carrier detection in DMD families with point mutations using PCR-SSCP and direct sequencing. *Neuromusc. Dis.* 4:411–418.

Lichter, P., Cremer, T., Borden, J., Manuelidis, L., and Ward, D.C. 1988. Delineation of individual human chromosomes in metaphase and interphase cells by in situ suppression hybridisation using recombinant DNA libraries. *Hum. Genet.* 80:224–234.

Lichter, P., Tang, C.C., Call, K., Hermanson, G., Evans, G., Housman, D., and Ward, D.C. 1990. High-resolution mapping of human chromosome 11 by in situ hybridization with cosmid clones. *Science* 247:64–69.

Lindblad-Toh, K., Winchester, E., Daly, M.J., Wang, D.G., Hirschhorn, J.N., Laviolette, J.P., Ardlie, K., Reich, D.E., Robinson, E., Sklar, P., Shah, N., Thomas, D., Fan, J.B., Gingeras, T., Warrington, J., Patil, N., Hudson, T.J., and Lander, E.S. 2000. Large-scale discovery and genotyping of single-nucleotide polymorphisms in the mouse. *Nat. Genet.* 24:381–386.

Litt, M. and Luty, J.A. 1989. A hypervariable microsatellite revealed by in vitro amplification of a dinucleotide repeat within the cardiac muscle actin gene. *Am. J. Hum. Genet.* 44:397–401.

Lockhart, D.J., Dong, H., Byrne, M.C., Follettie, M.T., Gallo, M.V., Chee, M.S., Mittmann, M., Wang, C., Kobayashi, M., Horton, H., and Brown, E.L. 1996. Expression monitoring by hybridization to high-density oligonucleotide arrays. *Nat. Biotechnol.* 14:1675–1680.

Look, A.T., Hayes, F.A., Shuster, J.J., Douglass, E.C., Castleberry, R.P., Bowman, L.C., Smith, E.I., and Brodeur, G.M. 1991. Clinical relevance of tumor cell ploidy and N-myc gene amplification in childhood neuroblastoma: A Pediatric Oncology Group study. *J. Clin. Oncol.* 9:581–591.

Maddalena, A., Richards, C.S., McGinniss, M.J., Brothman, A., Desnick, R.J., Grier, R.E., Hirsch, B., Jacky, P., McDowell, G.A., Popovich, B., Watson, M., and Wolff, D.J. 2001. Technical standards and guidelines for fragile X: The first of a series of disease-specific supplements to the standards and guidelines for clinical genetics laboratories of the American College of Medical Genetics. *Genet. in Med.* 3:200–205.

Madej, T., Gibrat, J.-F., and Bryant, S. 1995. Threading a database of protein cores. *Proteins* 23:356–369.

Magenis, E. and Barton, S.J. 1987. Delineation of human prometaphase paracentromeric regions using sequential GTG- and C-banding. *Cytogenet. Cell Genet.* 45:132–140.

Mahadevan, M., Tsilfidis, C., Sabourin, L., Shutler, G., Amemiya, C., Jansen, G., Neville, C., Narang, M., Barcelo, J., O'Hoy, K., Leblond, S., Earle-MacDonald, J., De Jong, P.J., Wieringa, B., and Korneluk, R. 1992. Myotonic dystrophy mutation: An unstable CTG repeat in the 3' untranslated region of the gene. *Science* 255:1253–1255.

Mandahl, N. 1992. Methods in solid tumor cytogenetics. *In* Human Cytogenetics: A Practical Approach (D.E. Rooney and B.H. Czepulkowski, eds.) pp. 155–188. IRL Press, Oxford.

Mann, R., Mulligan, R.C., and Baltimore, D. 1983. Construction of a retrovirus packaging mutant and its use to produce helper-free defective retrovirus. *Cell* 33:153–159.

Markowitz, D., Goff, S., and Bank, A. 1988a. A safe packaging line for gene transfer: Separating viral genes

on two different plasmids. *J. Virol.* 62:1120–1124.

Markowitz, D., Goff, S., and Bank, A. 1988b. Construction and use of a safe and efficient amphotropic packaging cell line. *Virology* 167:400–406.

Martin, E.R., Monks, S.A., Warren, L.L., and Kaplan, N.L. 2000. A test for linkage and association in general pedigrees: The pedigree disequilibrium test (PDT). *Am. J. Hum. Genet.* 67:146–154.

Mayeux, R., Saunders, A., Shea, S., Mirra, S., Evans, D., Roses, A., Hyman, B., Crain, B., Tang, M.-X., and Phelps, C. 1998. Utility of the apolipoprotein E genotype in the diagnosis of Alzheimer's disease. *New Engl. J. Med.* 338:506–511.

McPherson, J.D., Marra, M., Hillier, L., et al. 2001. A physical map of the human genome. *Nature* 409:934–941.

Meng, G., Kress, W., Scherpf, S., Bettecken, T., Feichtinger, W., Schempp, W., Schmid, M., and Muller, C.R. 1991. A comparison of the dystrophin gene structure in primates and lower vertebrates. *In* Muscular Dystrophy Research: From Molecular Diagnosis Toward Therapy (C. Angelini, G.A. Danieli, and D. Fontanari, eds.) pp. 23–30. Excerpta Medica, New York.

Mies, C. 1994. Molecular biological analysis of paraffin-embedded tissues. *Hum. Pathol.* 25:555–560.

Miller, A.D. 1990. Retrovirus packaging cells. *Hum. Gene Ther.* 1:5–14.

Miller, A.D. and Buttimore, C. 1986. Redesign of retrovirus packaging cell lines to avoid recombination leading to helper virus production. *Mol. Cell. Biol.* 6:2895–2902.

Miller, A.D. and Chen, F.C. 1996. Retrovirus packaging cells based on 10A1 murine leukemia virus for production of vectors that use multiple receptors for cell entry. *J. Virol.* 70:5564–5571.

Miller, G. and Lipman, M. 1973. Release of infectious Epstein-Barr virus by transformed marmoset leucocytes. *Proc. Natl. Acad. Sci. U.S.A.* 70:190–194.

Miller, A.D. and Rosman, G.J. 1989. Improved retroviral vectors for gene transfer and expression. *Biotechniques* 7:980–990.

Miller, A.D., Garcia, J.V., von Suhr, N., Lynch, C.M., Wilson, C., and Eiden, M.V. 1991. Construction and properties of retrovirus packaging cells based on gibbon ape leukemia virus. *J. Virol.* 65:2220–2224.

Mir, L.M., Bureau, M.F., Gehl, J., Rangara, R., Rouy, D., Caillaud, J.M., Delaere, P., Branelle, D., Schwartz, B., and Scherman, D. 1999. High-efficiency gene transfer into skeletal muscle mediated by electric pulses. *Proc. Natl. Acad. Sci. U.S.A.* 96:4262–4267.

Mitelman, F. 1991. Catalog of Chromosome Aberrations in Cancer. Wiley-Liss, New York.

Miyanohara, A., Yee, J.K., Bouic, K., LaPorte, P., and Friedmann, T. 1995. Efficient in vivo transduction of the neonatal mouse liver with pseudotyped retroviral vectors. *Gene Ther.* 2:138–142.

Moore, K.A., Deisseroth, A.B., Reading, C.L., Williams, D.E., and Belmont, J.W. 1992. Stromal support enhances cell-free retroviral vector transduction of human bone marrow long-term culture-initiating cells. *Blood* 79:1393–1399.

Morgenstern, J.P. and Land, H. 1990. Advanced mammalian gene transfer: High titer retroviral vectors with multiple drug selection markers and a complementary helper-free packaging cell line. *Nucl. Acids Res.* 18:3587–3596.

Morris, J. 1993. Application of genetic testing to disputed paternity. *Clin. Chem.* 39:716–717.

Morris, J.W., Sanda, A.I., and Glassberg, J. 1989. Biostatistical evaluation from continuous allele frequency distribution deoxyribonucleic acid (DNA) probes in reference to disputed paternity and identity. *J. Forensic Sci.* 34:1311–1317.

Murray, V. 1989. Improved double-stranded DNA sequencing using the linear polymerase chain reaction. *Nucl. Acids Res.* 17:8889.

National Research Council [NRC (U.S.); Committee on Hazardous Biological Substances in the Laboratory]. 1989. Biosafety in the Laboratory: Prudent Practices for the Handling and Disposal of Infectious Materials. National Academy Press, Washington, D.C.

Negrin, R.S. and Blume, K.G. 1991. The use of polymerase chain reaction for the detection of minimal residual malignant disease. *Blood* 78:255–258.

Nelson, D.L., Ledbetter, S.A., Corbo, L., Victoria, M.F., Ramirez-Solis, R., Webster, T.D., Ledbetter, D.H., and Caskey, C.T. 1989. *Alu* polymerase chain reaction: A method for rapid isolation of human-specific sequences from complex DNA sources. *Proc. Natl. Acad. Sci. U.S.A.* 86:6686–6690.

Nelson, M.R., Kardia, S.L.R., Ferrell, R.E., and Sing, C.F. 2001. A combinatorial partitioning method (CPM) to identify multi-locus genotypic partitions that predict quantitative trait variations. *Genome. Res.* 11:458–470.

Newton, C.R., Graham, A., Heptinstall, L.E., Powell, S.J., Summers, C., Kalsheker, N., Smith, J., and Markham, A.F. 1989. Analysis of any point mutation in DNA: The amplification refractory mutation system (ARMS). *Nucl. Acid Res.* 17:2503 2516.

Nickerson, D.A., Kaiser, R., Lappin, S., Stewart, J., Hood, L., and Landegren, U. 1990. Automated DNA diagnostics using an ELISA-based oligonucleotide ligation assay. *Proc. Natl. Acad. Sci. U.S.A.* 87:8923–8927.

Nickerson, D.A., Tobe, V.O., and Taylor, S.L. 1997. PolyPhred: Automating the detection and genotyping of single nucleotide substitutions using fluorescence-based resequencing. *Nucl. Acids Res.* 25:2745–2751.

Nolta, J.A., Dao, M.A., Wells, S., Smogorzewska, E.M., and Kohn, D.B. 1996. Transduction of pluripotent human hematopoietic stem cells demonstrated by clonal analysis after engraftment in immune-deficient mice. *Proc. Natl. Acad. Sci. U.S.A* 93:2414–2419.

Osoegawa, K., Woon, P.-Y., Zhao, B., Frengen, E., Tateno, M., Catanese, J.J., and de Jong, P.J. 1998. An improved approach for construction of bacterial artificial chromosome libraries. *Genomics* 52:1–8.

Ott, J. 1999. Analysis of Human Genetic Linkage, 3rd ed. Johns Hopkins University Press, Baltimore.

Parentage Testing Accreditation Requirements Manual, 2nd ed. 1995. American Association of Blood Banks, Bethesda, Md.

Parra, I. and Windle, B. 1993. High resolution visual mapping of stretched DNA by fluorescent hybridization. *Nat. Genet.* 5:17–21.

Pavan, W.J., Hieter, P., and Reeves, R.H. 1990. Modification and transfer into an embryonal carcinoma cell line of a 360-kilobase human-derived yeast artificial chromosome. *Mol. Cell. Biol.* 10:4163–4169.

Paxinos, G. and Watson, C. 1986. The Rat Brain in Stereotaxic Coordinates. Academic Press, San Diego.

Paxinos, G., Ashwell, K.W.S., and Tork, I. 1994. Atlas of the Developing Rat Nervous System. Academic Press, San Diego.

Pear, W.S., Nolan, G.P., Scott, M.L., and Baltimore, D. 1993. Production of high-titer helper-free retroviruses by transient transfection. *Proc. Natl. Acad. Sci. U.S.A.* 90:8392–8396.

Pechan, P.A., Fotaki, M., Thompson, R.L., Dunn, R.J., Chase, M., Chiocca, E.A., and Breakefield, X.O. 1996. A novel "piggyback" packaging system for herpes simplex virus amplicon vectors. *Hum. Gene Ther.* 7:2003–2013.

Perry, P. and Wolff, S. 1974. New Giemsa method for the differential staining of sister chromatids. *Nature* 257:156–158.

Pinkel, D., Landegent, J., Collins, C., Fuscoe, J., Segraves, R., Lucas, J., and Gray, J.W. 1988. Fluorescence in situ hybridization with human chromosome-specific libraries: Detection of trisomy 21 and translocations of chromosome 4. *Proc. Natl. Acad. Sci. U.S.A.* 85:9138–9142.

Priest, J.H. 1991. Prenatal chromosome diagnosis and cell culture. *In* The ACT Cytogenetics Laboratory Manual, 2nd ed. (M.J. Barch, ed.) pp. 149–204. Raven Press, New York.

Prior, T.W., Bartolo, C., Pearl, D.K., Papp, A.C., Snyder, P.J., Sedra, M.S., Burghes, A.H.M., and Mendell, J.R. 1995. Spectrum of small mutations in the dystrophin coding region. *Am. J. Hum. Genet.* 57:22–33.

Pritchard, J.K., Stephens, M., Rosenberg, N.A., and Donnelly, P. 2000. Association mapping in structured populations. *Am. J. Hum. Genet.* 67:170–181.

Rabinowitz, D. and Laird, N. 2000. A unified approach to adjusting association tests for population admixture with arbitrary pedigree structure and arbitrary missing marker information. *Hum. Hered.* 50:211–223.

Ranade, K., Chang, M.-S., Ting, C.-T., Pei, D., Hsiao, C.-F., Olivier, M., Pesich, R., Hebert, J., Chen, Y.-I., Dzau, V.J., Curb, D., Olshen, R., Risch, N., Cox, D.R., and Botstein, D. 2001. High-throughput genotyping with single nucleotide polymorphisms. *Genome Res.* 11:1262–1268.

Rasty, S., Goins, W.F., and Glorioso, J.C. 1995. Site-specific integration of multigenic shuttle plasmids into the herpes simplex virus type 1 genome using a cell-free Cre-*lox* recombination system. *Methods Mol.Genet.* 7:114–130.

Richards, B., Skoletsky, J., Shuber, A., Balfour, R., Stern, R., Dorkin, H., Parad, R., Witt, D., and Klinger, K. 1994. Multiplex PCR amplification from the *CFTR* gene using DNA prepared from buccal brushes/swabs. *Hum. Mol. Genet.* 2:159–163.

Risch, N. 1990. Linkage strategies for genetically complex traits. I. Multilocus models. *Am. J. Hum. Genet.* 16:222–228.

Ritchie, M.D., Hahn, L.W., Roodi, N., Bailey, L.R., Dupont, W.D., Parl, F.F., and Moore, J.H. 2001. Multifactor-dimensionality reduction reveals high-order interactions among estrogen-metabolism genes in sporadic breast cancer. *Am. J. Hum. Genet.* 69:138–147.

Roberts, R.G., Montandon, A.J., Bobrow, M., and Bently, D.R. 1989. Detection of novel genetic markers by mismatch analysis. *Nucl. Acids Res.* 17:5961–5971.

Roberts, R.G., Bobrow, M., and Bentley, D.R. 1992. Point mutations in the dystrophin gene. *Proc. Natl. Acad. Sci. U.S.A.* 89:2331–2335.

Roest, P.A.M., Roberts, R.G., Sugino, S., van Ommen, G.J.B., and den Dunnen, J.T. 1993. Protein truncation test (PTT) for rapid detection of translation-terminating mutations. *Hum. Mol. Genet.* 2:1719–1721.

Rogic, S., Mackworth, A., and Ouellette, B.F.F. 2001. Evaluation of gene-finding programs on mammalian sequences. *Genome Res* 11:817–832.

Rooney, D.E. and Czepalkowski, B.H. Prenatal diagnosis and tissue culture. 1992. *In* Human Cytogenetics: A Practical Approach, 2nd ed., Vol. 1 (D.E. Rooney and B.H. Czepalkowski, eds.) pp. 55–89. IRL Press at Oxford University Press, Oxford.

Rousseau, F., Heitz, D., Biancalana, V., Blumenfeld, S., Kretz, C., Bou, J., Tommerup, N., Van Der Hagen, C., DeLozier-Blanchet, C., Croquette, M.-F., Gilgenkrantz, S., Jalbert, P., Voelckel, M.-A., Oberl, I., and Mandel, J.-L. 1991. Direct diagnosis by DNA analysis of the fragile X syndrome of mental retardation. *New Engl. J. Med.* 325:1673–1681.

Sahar, E. and Latt, S.A. 1980. Energy transfer and binding competition between dyes used to enhance staining differentiation in metaphase chromosomes. *Chromosoma* 79:1–28.

Sakuraba, H., Ishii, K., Shimmoto, M., Yamada, H., and Suzuki, Y. 1991. A screening for dystrophin gene deletions in Japanese patients with

Duchenne/Becker muscular dystrophy by the multiplex polymerase chain reaction. *Brain Dev.* 13:339–342.

Samaniego, L.A., Neiderhiser, L., and DeLuca, N.A. 1998. Persistence and expression of the herpes simplex virus genome in the absence of immediate-early proteins. *J. Virol.* 72:3307–3320.

Sambrook, J., Fritsch, E.F., and Maniatis, T. 1989. Molecular Cloning. A Laboratory Manual, 2nd ed. Cold Spring Harbor Laboratory Press, Cold Spring Harbor, N.Y.

Samulski, R.J., Sally, M., and Muzyczka, N. 1999. Adeno-associated viral vectors. *In* The Development of Human Gene Therapy (T. Friedman, ed.) pp. 36:131–172. Cold Spring Harbor Laboratory Press, Cold Spring Harbor, N.Y.

Sandberg, A.A. 1990. The Chromosomes in Human Cancer and Leukemia, 2nd ed. Elsevier, New York.

Santorelli, F.M., Shanske, S., Macaya, A., DeVivo, D.C., and DiMauro, S. 1993. The mutation at nt 8993 of mitochondrial DNA is a common cause of Leigh's syndrome. *Ann. Neurol.* 34:827–834.

Santorelli, F.M., Suk-Chun, M., Vasquez-Acevedo, M., Gonzalez-Astiazarin, A., Ridaura-Sanz, C., Gonzalez-Halphen, D., and DiMauro, S. 1995. A novel mitochondrial DNA point mutation associated with mitochondrial encephaolcardiomyopathy. *Biochem. Biophys. Res. Commun.* 216:835–840.

Santorelli, F.M., Schlessel, J.S., Slonim, A.E., and DiMauro, S. 1996. Novel mutation in the mitochondrial DNA tRNA glycine gene associated with sudden unexpected death. *Ped. Neurol.* 15:145–149.

Sawyer J.R., Moore, M.M., and Hozier, J.C. 1987. High resolution G-banded chromosomes of the mouse. *Chromosoma (Berl)* 95:350–358.

Schena, M., Shalon, D., Davis, R.W., and Brown, P.O. 1995. Quantitative monitoring of gene expression patterns with a complementary DNA microarray. *Science* 270:467–470.

Schofield, D.E. and Fletcher, J.A. 1992. Trisomy 12 in pediatric granulosa-stromal cell tumors: Demonstration by a modified method of fluorescence in situ hybridization on paraffin-embedded material. *Am. J. Path.* 141:1265–1269.

Schröck, E., du Manoir, S., Veldman, T., Schoell, B., Wienberg, J., Ferguson-Smith, M.A., Ning, Y., Ledbetter, D.H., Bar-Am, I., Soenksen, D., Garini, Y., and Ried, T. 1996. Multicolor spectral karyotyping of human chromosomes. *Science* 273:494–496.

Schwartz, D.C. and Cantor, C.R. 1984. Separation of yeast chromosome-sized DNAs by pulsed field gradient gel electrophoresis. *Cell* 37:67–75.

Schweitzer, D. 1981. Counterstain enhanced chromosome banding. *Hum. Genet.* 57:1–14.

Scriver, C.R., Nowacki, P.M., and Lehvaslaiho, H. 1999. Guidelines and recommendations for content, structure, and deployment of mutation databases. *Hum. Mutat.* 13:344–350.

Scriver, C.R., Nowacki, P.M., and Lehvaslaiho, H. 2000. Guidelines and recommendations for content, structure, and deployment of mutation databases. II. Journey in progress *Hum. Mutat.* 15:13–15.

Shenk, T. 1996. Adenoviridae: The viruses and their replication. *In* Fields Virology (B.N. Fields, D.M. Kulpe, P.M. Howley, R.M. Chanok, J.L. Melnick, T.P. Monath, B. Roizman, and S.E. Straus, eds.) pp. 2111–2148. Lippincott, Philadelphia.

Shiver, J.W., Fu, T.M., Chen, L., Casimiro, D.R., Davies, M.E., Evans, R.K., Zhang, Z.Q., Simon, A.J., Trigona, W.L., Dubey, S.A., Huang, L., Harris, V.A., Long, R.S., Liang, X., Handt, L., Schleif, W.A., Zhu, L., Freed, D.C., Persaud, N.V., Guan, L., Punt, K.S., Tang, A., Chen, M., Wilson, K.A., Collins, K.B., Heidecker, G.J., Fernandez, V.R., Perry, H.C., Joyce, J.G., Grimm, K.M., Cook, J.C., Keller, P.M., Kresock, D.S., Mach, H., Troutman, R.D., Isopi, L.A., Williams, D.M., Xu, Z., Bohannon, K.E., Volkin, D.B., Montefiori, D.C., Miura, A., Krivulka, G.R., Lifton, M.A., Kuroda, M.J., Schmitz, J.E., Letvin, N.L., Caulfield, M.J., Bett, A.J., Youil, R., Kaslow, D.C., and Emini, E.A. 2002. Replication-incompetent adenoviral vaccine vector elicits effective anti-immunodeficiency-virus immunity. *Nature* 415:331–335.

Shizuya, H., Birren, B., Kim, U.-J., Mancino, V., Stepak, T., Tachiiri, Y., and Simon, M. 1992. Cloning and stable maintenance of 300-kilobase-pair fragments of human DNA in *Escherichia coli* using an f-factor-based vector. *Proc. Natl. Acad. Sci. U.S.A.* 89:8794–8797.

Shleien, B., (ed.) 1987. Radiation Safety Manual for Users of Radioisotopes in Research and Academic Institutions. Nucleon Lectern Associates, Olney, Md.

Shoffner, J.M. and Wallace, D.C. 1995. Oxidative phosphorylation diseases. *In* The Metabolic and Molecular Bases of Inherited Disease (C.R. Scriver, A.L. Beaudet, W.S. Sly, and D. Valle, eds.) pp. 1535–1610. McGraw-Hill, New York.

Shtivelman, E., Lifshitz, B., Gale, R.P., and Canaani, E. 1985. Fused transcripts of *abl* and *bcr* genes in chronic myelogenous leukaemia. *Nature* 315:550–554.

Shuber, A.P., Skoletsky, J., Stern, R., and Handelin, B.L. 1992. Efficient 12-mutation testing in the CFTR gene: A general model for complex mutation analysis. *Hum. Molec. Genet.* 2:159–163.

Shuber, A.P., Michalowsky, L.A., Nass, G.S., Skoletsky, J., Hire, L.M., Kotsopoulos, S.K., Phipps, M.F., Barberio, D.M., and Klinger, K.W. 1997. High throughput parallel analysis of hundreds of patient samples for more than 100 mutations in multiple disease genes. *Hum. Mol. Genet.* 6:337–47.

Sikorski, R.S. and Hieter, P. 1989. A system of shuttle vectors and yeast host strains designed for efficient manipulation of DNA in *S. cerevisiae*. *Genetics* 122:19–27.

Silvestri, G., Servidel, S., Ranas, M., Ricci, E., Spinazzola, A., Paris, E., and Tonall, P. 1996. A novel mitochondrial DNA point mutation in the tRNA^{Ile} gene is associated with progressive external ophthalmoplegia. *Biochem. Biophys. Res. Commun.* 220:622–627.

Simari, R.D., San, H., Rekhter, M., Ohno, T., Gordon, D., Nabel, G.J., and Nabel, E.G. 1996. Regulation of cellular proliferation and intimal formation following balloon injury in atherosclerotic rabbit arteries. *J. Clin. Invest* 98:225–235.

Sklar, J. 1992. Antigen receptor genes: Structure, function, and techniques for analysis of their rearrangements. *In* Neoplastic Hematopathology (D.M. Knowles, ed.) pp. 215–244. Williams & Wilkens, Baltimore.

Smith, D.R. 1990. Genomic long-range restriction mapping. *Methods* 1:195–203.

Smith, I.L., Hardwicke, M.A., and Sandri-Goldin, R.M. 1992. Evidence that the herpes simplex virus immediate-early protein ICP27 acts post-transcriptionally during infection to regulate gene expression. *Virology* 186:74–86.

Snyder, E.E. and Stormo, G.D. 1993. Identification of coding regions in genomic DNA sequences: An application of dynamic programming and neural networks. *Nucl. Acids Res* 21:607–613.

Snyder, E.E. and Stormo, G.D. 1997. Identifying genes in genomic DNA sequences. *In* DNA and Protein Sequence Analysis (M.J. Bishop and C.J. Rawlings, eds.) pp. 209–224. Oxford University Press, New York.

Southern, E.M. 1975. Detection of specific sequences among DNA fragments separated by gel electrophoresis. *J. Mol. Biol.* 98:503–517.

Spaete, R.R. and Frenkel, N. 1982. The herpes simplex virus amplicon: A new eucaryotic defective-virus cloning-amplifying vector. *Cell* 30:295–304.

Speel, E.J.M., Schutte, B., Wiegant, J., Ramaekers, F.C., and Hopman, A.H.N. 1992. A novel fluorescence detection method for in situ hybridization, based on the alkaline phosphatase-fast red reaction. *J. Histochem. Chytochem.* 40:1299–1308.

Speer, A., Rosenthal, A., Billwitz, H., Hanke, R., Forrest, S.M., Love, D., Davies, K.E., and Coutelle, C. 1989. DNA amplification of a further exon of Duchenne muscular dystrophy locus increase possibilities for deletion screening. *Nucl. Acids Res.* 17:4892.

Speicher, M.R., Ballard, S.G., and Ward, D.C. 1996. Karyotyping human chromosomes by combinatorial multi-fluor FISH. *Nat. Genet.* 12:368–375.

Sternberg, N. 1990. Bacteriophage P1 cloning system for the isolation, amplification, and recovery of DNA fragments as large as 100 kilobase pairs. *Proc. Natl. Acad. Sci. U.S.A.* 87:103–107.

Sumner, A.T. 1990. Chromosome Banding. Unwin Hyman, London.

Sumner, A.T., and Evans, H.J. 1973. Mechanism involved in the banding of chromosomes with quinicrine and Giemsa. II. The interaction of the dyes with the chromosomal components. *Exp. Cell Res.* 81:223–236.

Takahara, Y., Hamada, K., and Housman, D.E. 1992. A new retrovirus packaging cell for gene transfer constructed from amplified long terminal repeat-free chimeric proviral genes. *J. Virol.* 66:3725–3732.

Taylor, R.W., Chinney, P.F., Haldane, F., Morris, A.A.M., Bindoff, L.A., Wilson, J., and Turnbull, D.M. 1996. MELAS associated with a mutation in the valine transfer RNA gene of mitochondrial DNA. *Ann. Neuorol.* 40:459–462.

Terwilliger, J.D. and Ott, J. 1994. Handbook of Human Genetic Linkage. Johns Hopkins University Press, Baltimore.

Theune, S., Fung, J., Todd, S., Sakaguchi, A.Y., and Naylor, S.L. 1991. PCR primers for human chromosomes: Reagents for the rapid analysis of somatic cell hybrids. *Genomics* 9:511–516.

Thomas, P.S. 1980. Hybridization of denatured RNA and small DNA fragments transferred to nitrocellulose. *Proc. Natl. Acad. Sci. U.S.A.* 77:5201–5205.

Traver, M. 1996. Promega Statistics Workshop, Scottsdale, Ariz.

Ulmer, J.B., Donnelly, J.J., Parker, S.E., Rhodes, G.H., Felgner, P.L., Dwarki, V.J., Gromkowski, S.H., Deck, R.R., De Witt, C.M., Friedman, A., Hawe, L., Leander, K.R., Martinez, D., Perry, H.C., Shiver, J.W., Montgomery, D.L., and Liu, M.A. 1993. Heterologous protection against influenza by injection of DNA encoding a viral protein. *Science* 259:1745–1749.

Verlinsky, Y. and Cieslak, J. 1993. Embryological and technical aspects of preimplantation genetic diagnosis. *In* Preimplantation Diagnosis of Genetic Diseases. A New Technique in Assisted Reproduction (Y. Verlinsky and A.M. Kuliev, eds.) pp. 49–67. Wiley-Liss, New York.

Vollrath, D. and Davis, R.W. 1987. Resolution of DNA molecules greater than 5 megabases by contour-clamped homogeneous electric fields. *Nucl. Acids Res.* 15:7865–7876.

Wallace, D.M. 1987. Precipitation of nucleic acids. *Methods Enzymol.* 152:41–48.

Wallace, R.B., and Miyada C.G. 1987. Oligonucleotide probes for the screening of recombinant DNA libraries. *Methods Enzymol.* 152:432–442.

Wallace, D.C. Lott, M.T. Brown, M.D., Huoponen, K., and Torroni, A. 1995. Report of the committee on human mitochondrial DNA. *In* Human Gene Mapping 1994, a Compendium (A.J. Cuticchia, ed.) pp. 910–954. Johns Hopkins University Press, Baltimore.

Wang, D.G., Fan, J.B., Siao, C.J., Berno, A., Young, P., Sapolsky, R., Ghandour, G., Perkins, N., Winchester, E., Spencer, J., Kruglyak, L., Stein, L., Hsie, L., Topaloglou, T., Hubell, E., Robinson, E., Mittman, M., Morris, M.S., Shen, N., Kilburn, D., Rioux, J., Nusbaum, C., Rozen, S., Hudson, T.J., Lipshutz, R., Chee, M., and Lander, E.S. 1998. Large-scale identification, mapping and genotyping of single-nucleotide polymorphisms in the human genome. *Science* 280:1077–1082.

Warburton, D., Gersen, S., Yu, M.-T., Jackson, C., Handelin, B., and Housman, D. 1990. Monochromosomal rodent-human hybrids from microcell fusion of human lymphoblastoid cells containing an inserted dominant selectable marker. *Genomics* 6:358–366.

Walz, T., Geisel, J., Bodis, M., Knapp, J.P., and Herrmann, W. 2000. Fluorescence-based single-strand conformation polymorphism analysis of mutations by capillary electrophoresis. *Electrophoresis* 21:375–379.

Weber, J.L. and May, P.E. 1989. Abundant class of human DNA polymorphisms which can be typed using the polymerase chain reaction. *Am. J. Hum. Genet.* 44:388–396.

Webster, C., Pavlath, G.K., Parks, D.R., Walsh, F.S., and Blau, H.M. 1988. Isolation of human myoblasts with the fluorescence-activated cell sorter. *Exp. Cell Res* 174:252–65.

Weissenbach, J., Gyapay, G., Dib, C., Vignal, A., Morissette, J., Millasseau, P., Vaysseix, G., and Lathrop, M. 1992. A second-generation linkage map of the human genome. *Nature* 359:794–801.

Welsh, P.S., Metzger, D.A., and Higuchi, R. 1991. Chelex 100 as a medium for simple extraction of DNA. *BioTechniqes* 10:506–513.

Wiedmann, M., Wilson, W.J., Czajka, J., Luo, J., Barany, F., and Batt, C.A. 1994. Ligase chain reaction (LCR)—Overview and applications. *PCR Methods Appl.* 3:S51–S64.

Wolff, J.A., Malone, R.W., Williams, P., Chong, W., Acsadi, G., Jani, A., and Felgner, P.L. 1990. Direct gene transfer into mouse muscle in vivo. *Science* 247:1465–1468.

Wolfsberg, T.G., Gabrielian, A.E., Campbell, M.J., Cho, R.J., Spouge, J.L., and Landsman, D. 1999. Candidate regulatory sequence elements for cell cycle-dependent transcription in *Saccharomyces cerevisiae*. *Genome Res* 9:775–792.

Xiao, X., Li, J., and Samulski, R.J. 1998. Production of high-titer recombinant adeno-associated virus vectors in the absence of helper adenovirus. *J. Virol.* 72:2224–2232.

Yang, Y., Vanin, E.F., Whitt, M.A., Fornerod, M., Zwart, R., Schneiderman, R.D., Grosveld, G., and Nienhuis, A.W. 1995. Inducible, high-level production of infectious murine leukemia retroviral vector particles pseudotyped with vesicular stomatitis virus G envelope protein. *Hum. Gene Ther.* 6:1203–1213.

Yang, Y.H., Dudoit, S., Luu, P., Lin, D.M., Peng, V., Ngai, J., and Speed, T.P. 2002. Normalization for cDNA microarray data: A robust composite method addressing single and multiple slide systematic variation. *Nucl. Acids Res* 4:e15.

Yee, J.K., Miyanohara, A., LaPorte, P., Bouic, K., Burns, J.C., and Friedmann, T. 1994. A general method for the generation of high-titer, pantropic retroviral vectors: Highly efficient infection of primary hepatocytes. *Proc. Natl. Acad. Sci. U. S. A.* 91:9564–9568.

Zabner, J., Zeiher, B.G., Friedman, E., and Welsh, M.J. 1996. Adenovirus-mediated gene transfer to ciliated airway epithelia requires prolonged incubation time. *J. Virol.* 70:6994–7003.

Zhang, M.Q. 1997. Identification of protein coding regions in the human genome by quadratic discriminant analysis. *Proc. Natl. Acad. Sci. U.S.A* 94:565–568.

Zhang, J. and Madden, T.L. 1997. PowerBLAST: A new network BLAST application for interactive or automated sequence analysis and annotation. *Genome Res.* 7:649–656.

Zhang, Z., Schaffer, A.A., Miller, W., Madden, T.L., Lipman, D.J., Koonin, E.V., and Altschul, S.F. 1998a. Protein sequence similarity searches using patterns as seeds. *Nucl. Acids Res.* 26:3986–3990.

INDEX

Page numbers in this book are hyphenated: the number before the hyphen refers to the chapter and the number after the hyphen refers to the page within the chapter (e.g., 12-3 is page 3 of Chapter 12). A range of pages is indicated by an arrow connecting the page numbers (e.g., 12-3→12-5 refers to pages 3 through 5 of Chapter 12).

A

ABI PRISM 7700, 2-33, 2-35→2-36
ABI prism 3100 genetic analyzer, SSCP mutations, 7-12→7-13
ABI 3700 DNA analyzer, heterozygote mutation detection, 7-16→7-19
ABI 310 genetic analyzer, SSCP mutations, 7-11→7-12
Absorption spectroscopy of nucleic acids, A3-8→A3-10
Accession number, sequence similarity searches, 6-37
Acrocentric chromosomes, A3-37
ACT (analysis of complex traits) 1-56 (fig.)
Acute lymphocytic leukemia (ALL), choice of culture method, 10-2 (table)
Acute myeloid leukemia (AML)
    characterized, 8-31
    choice of culture method, 10-2 (table)
Additives, bacterial media, A1-17 (table)
AdEasy system
    adenovirus plaque assay, 12-18→12-19
    choice of vectors, 12-8, 12-10 (table)
    cloning genes of interest into shuttle vectors, 12-8, 12-9 (fig.)
    generation of recombinant adenoviral vectors, 12-12→12-16
    overview, 12-6→12-7 (figs.)
    preparation of
        adenoviral DNA, 12-20
        electrocompetent BJ5183 cells, 12-19→12-20
        high-titer adenoviruses, 12-16→12-18
        quick agarose-tube dialysis, 12-20→12-21
    troubleshooting guide, 12-10→12-11 (table)
Adenine phosophoribosyltransferase (APRT), as selectable marker, 3-3
Adeno-associated virus, recombinant (rAAV), production of, 12-21→12-25
Adenoviral vectors, see AdEasy system
Adenovirus, recombinant, delivery to the liver
    mouse tail vein injection, 13-60
    rabbit ear vein injection, 13-60→13-61
    troubleshooting guide, 13-59 (table)
Adoption studies, linkage analysis, 1-4
Aerosols, biosafety procedures, 12-4
Affected family-based controls (AFBAC), in tests of disease association, 1-80→1-81
Affected pedigree member (APM), linkage analysis, 1-56 (fig.)
Affected relative pair, linkage studies, 1-12
Affected sib pairs (ASP)
    defined, 1-56 (fig.)
    linkage studies
        characterized, 1-12
        simple testing, 1-60→1-61
        using SIBPAL, 1-61→1-67
        without computer analysis, 1-55, 1-60→1-61
Affymetrix GeneChip, for expression monitoring, 11-9
Agarose, films, in single-sperm typing, 1-20→1-21
Agarose blocks, high-molecular-weight DNA embedded in, 5-4→5-5
    preparation from
        animal tissue cells, 5-39→5-40
        lymphocytes, 5-38→5-39

Agarose-formaldehyde gel electrophoresis, A3-23→A3-26
Agarose gel electrophoresis, of ESTs, 11-20→11-21
Agarose plates, rapid estimation of DNA concentration on, 5-28→5-29
Age-of-onset (AO), two-point linkage analysis, 1-49→1-54
Airway, gene delivery to
    airway model system choice, 13-44→13-45
    bronchial xenografts, human
        generation of, 13-52→13-56
        harvesting, 13-58→13-59
    human primary airway epithelial cells isolation of, 13-46→13-48
        transduction of, 13-48→13-49
    in vivo delivery to the lung, 13-57→13-58
    polarized airway epithelial monolayers, generation of, 13-49→13-51
    reporter gene choice, 13-46
    vector system choice, 13-45→13-46
Alkaline lysis
    modified miniprep for DNA recovery from BAC/PAC clones, 5-44→5-45
    preparation of BAC/PAC vector for cloning, 5-36→5-37
Alkaline phosphatase
    enzymatic detection for FISH, 4-29 (table), 4-30→4-31
    retroviral vectors, staining with, 12-39
Alkaline phosphatase anti-alkaline phosphatase (APAAP), sequential MAC analysis
    characterized, 4-48, 4-49 (table), 4-50→4-52
    schematic illustration of, 4-52 (fig.)
ALL, see Acute lymphocytic leukemia
Allele frequencies, multipoint linkage analysis, 1-40 (fig.)
Allele-specific microarrays for SNP genotyping, 2-25→2-30
Allele-specific oligonucleotides (ASOs)
    mutation analysis
        screening PCR-amplified DNA with, 9-12→9-14
        troubleshooting guide, 9-15 (table)
    probes, paternity testing, 9-31
Allelic association
    case-control studies, 1-78→1-79
    family-based studies, 1-77, 1-83
    haplotype relative risk (HRR), 1-80
    transmission disequilibrium test (TDT), 1-80→1-81
Alpha satellite DNA repeats, A2-3
Alpha satellite probes for interphase FISH, 8-29→8-30
Alternate hypothesis, linkage analysis 1-59→1-60
Alu
    databases, for nucleotide sequence, 6-25, 6-31
    gene, 5-21 (table), 5-23
    repeats, A2-1→A2-2
Aminoglycoside phosphotransferase (*neo*, G418, APH), as selectable marker, 3-3
AML, see Acute myeloid leukemia
Ammonium acetate, to precipitate DNA, A3-5 (table)
Amniocentesis, 8-36→8-37

Amniocytes
    cultures of, 8-34 (table)
    preparation for interphase FISH
        amniotic fluid cells, 8-29→8-31
        attached to a surface, 8-28
        uncultured, 8-26→8-28
Amniotic fluid samples
    high-resolution chromosome banding, using ethidium bromide, 8-10
    interphase FISH, 8-29→8-31
    passaging cells from
        coverslip monolayer culture, 8-10→8-11
        flasks, 8-13→8-14
    preparation, culture, harvest and analysis
        flask method, 8-11→8-13
        in situ coverslip method, 8-5→8-10
Amplicons, HSV-1
    cosmid DNA preparation for transfection, 12-69→12-70
    helper-free stocks, 12-66, 12-68→12-69
    titration, 12-71→12-74
Amplification
    cDNA microarrays, 11-12→11-16
    high-titer adenoviruses, 12-16→12-17
    mRNA, hybridization to oligonucleotide arrays for expression monitoring, 11-1→11-5
    of target sequences
        genomic DNA, 7-5→7-6
        modified second-round of cDNA, 7-5
        RNA from lymphocytes, 7-2→7-5
Amplification-refractory mutation system (ARMS), point mutations
    analysis of
        multiple mutations by multiplex ARMS, 9-43→9-44
        single mutations, 9-40→9-43
    characterized, 9-38→9-40
    primer additional mismatch selection, 9-40 (table)
    rapid DNA extraction from mouth wash and blood samples, 9-44
    troubleshooting guide, 9-45 (table)
Amplified fragment length polymorphisms (AmpFLPS), paternity testing, 9-31→9-32
AmpliTaq Gold, 2-28
Aneuploidy
    detection by FISH, 8-22→8-26
    in sections from paraffin-embedded tissue, 8-22→8-26
Animal safety procedures, 12-5
Animal tissue, large-insert cloning, DNA preparation from, 5-39→5-40
Antibiotics
    added to medium, A1-17 (table)
    in tissue culture, A3-26 (table)
Antigen-receptor gene rearrangements, molecular analysis by Southern blot hybridization, 10-11→10-14
APAAP, see Alkaline phosphatase anti-alkaline phosphatase
APM, see Affected pedigree method
Apolipoprotein E (APOE), genotyping by
    fluorescence polarization (FP), 9-86→9-87
    reverse hybridization, 9-84→9-85
    RFLP analysis, 9-83→9-84
    SNaPshot analysis, 9-87→9-88
Arab Genetic Disease database, 7-27

Arrayed libraries, screening by hybridization, 5-8→5-9
Arteries, gene transfer to
  murine carotid, normal, 13-9→13-11
  murine femoral, injured, 13-11→13-13
  porcine iliac, stented, 13-5, 13-6 (fig.)
  porcine iliofemoral, 13-3→13-5
  rabbit iliac, artherosclerotic, 13-6→13-9
  strategic planning, 13-2→13-3
Ascertainment bias, linkage analysis, 1-7, 1-13→1-15
Ascite effusions, preparing chromosome slides of, 4-4
Association, population, 1-77
Automated fluorescent genotyping, see Fluorescent genotyping
Autosomal dominant diseases
  heterozygote mutation detection, 7-13→7-19
  linkage analysis
    multipoint, 1-40→1-49
    nuclear family segregation, 1-16 (fig.)
    three-generation family segregation, 1-15 (fig.)
    two-point, 1-31→1-39, 1-49→1-53
Avidin, detection of biotinylated probes in ISH, 4-37, 4-39

B

Bacterial artificial chromosome (BAC/PAC)
  clone identification, 6-59
  cloning strategies, 5-1
  defined, 6-38→6-39
  libraries, construction of, 5-32→5-45
    linearization and gel purification of, 5-30→5-31
    into mammalian cells and mouse embryos, 5-29→5-30
    library construction, 5-32→5-36
    modified alkaline lysis miniprep for recovery from BAC/PAC, 5-44→5-45
    partial digestion, size fractionation of genomic DNA, 5-40→5-44
    preparation
      of BAC/PAC clones using pCYPAC2, pPAC4, or pBACe3.6 vector, 5-32→5-36
      of cloning vectors, 5-36→5-38
      of high-molecular-weight DNA, in agarose blocks, 5-4→5-5
      for restriction digestion and CHEF gel analysis, 5-7→5-8
    screening libraries by hybridization
      arraying colonies and DNA at low and high density, 5-8→5-14
      colony blot preparation, 5-14→5-15, 5-17→5-19
      colony hybridization, 5-15→5-16
Banding, see Chromosome binding
Basic Local Alignment Search Tool (BLAST)
  cDNA analysis, 6-61
  genomic DNA sequence analysis, 6-33→6-34, 6-38
  heterozygous mutation detection, 7-15
  sequence analysis, database submissions, 6-20
  for sequence similarity searches
    accessing programs and documentation, 6-22→6-23
    basic vs. advanced searches, 6-23
    BLASTN, 6-23 (table), 6-31→6-32
    BLASTP, 6-23 (table), 6-26→6-30, 6-33
    BLASTX, 6-23 (table), 6-30, 6-33→6-34
    using FASTA format, 6-26, 6-34
    filtering strategies, 6-26→6-27, 6-32→6-32
    mRNA sequences, analysis of, 6-33
    NCBI databases, 6-24→6-26
    parameters, 6-36→6-37
    PSI-BLAST, 6-32→6-33
    query sequence, formatting, 6-26

results, 6-27, 6-33
  sequence alignment algorithms, 6-34→6-36
  sequence identifier syntax, 6-37→6-38
  TBLASTN, 6-23 (table), 6-26, 6-31, 6-33
  TBLASTX, 6-23 (table)
Becker muscular dystrophy (BMD)
  identifying gene deletions in, 9-1→9-11
  see also Dystrophin gene deletions
Behavioral disorders, linkage analysis, 1-6
Bilineal pedigrees, in linkage analysis, 1-17
Biopsy
  embryo, preimplantation genetic diagnosis, 9-64→9-68
  samples, preparation of, for culture, 8-17
Biosafety cabinets, 12-3
Biotin
  labeling of DNA probes by nick translation, 4-31→4-32
  primers for ligation assays, 2-16→2-17
  probes for DIRVISH, 4-40→4-42
Biotin-labeled probes for interphase FISH, 8-30→8-31
BLAST, see Basic Local Alignment Search Tool
BLASTN, genomic DNA analysis
  characterized, 6-33→6-34
  sequence alignment algorithms, 6-35→6-36
  sequence similarity searches, 6-23 (table), 6-26, 6-31→6-32
Blastomere isolation, in preimplantation genetic diagnosis, 9-68→9-69, 9-71→9-72
BLASTP
  mRNA sequence analysis, 6-33
  sequence alignment algorithms, 6-35
  sequence similarity searches, 6-23 (table), 6-27→6-30, 6-32
BLASTX
  sequence alignment algorithms, 6-35
  genomic DNA analysis, 6-34
  sequence similarity searches, 6-23 (table), 6-30
BLAT search, 6-38, 6-59→6-60
Blood
  collection
    from orbital sinus of plexus of mouse, 4-8 (fig.)
    tail vein method, 4-10
  culture and metaphase harvest from, 4-1→4-14
  DNA extraction in ARMS test, 9-44
  fetal, DEB-treated cultures, 8-37
  leukemic, chromosome preparation, 10-2→10-4, 10-6→10-7
  peripheral, culture and harvest of, 4-1→4-14
Bloom's syndrome (BS), sister-chromatid exchange in, 8-22
BLOSUM matrices, for BLAST sequence alignment algorithms, 6-29, 6-35
  similarity searches, 6-34
Blotting and hybridization, BAC/PAC colony preparation, 5-14→5-15
BL2 laboratory
  animal safety, 12-5
  biosafety cabinets, 12-3
  principal investigator (PI), 12-2
  procedures, 12-4→12-5
  protective clothing and equipment, 12-3→12-4
  staff training, 12-2→12-3
BMD, see Becker muscular dystrophy
Bone marrow
  chromosome spread preparation, 10-2→10-4, 10-6→10-7
  culture, long-term, 13-31 (fig.), 13-33, 13-43

mitotic chromosome preparations from, 4-12
  See also Hematopoietic cells
Brain, gene delivery to
  direct injection of recombinant viral vectors into rat brain, 13-28
  gene expression assessment, 13-23 (table)
  genetically modified cells for grafting, preparation of, 13-29
  implantation of collagen-embedded genetically modified fibroblasts into cavity into adult rat brain, 13-27→13-28
  implantation of genetically modified cells into rat brain
    adult, 13-23→13-24
    fetal, 13-26→13-27
    neonatal, 13-25→13-26
  transfer strategies, 13-21 (fig.), 13-22 (fig.)
BrdU, see Bromodeoxyuridine
Bright-field microscopy, giemsa banding (G-banding), 4-17
Bromodeoxyuridine (BrdU)
  analysis of sister-chromatid exchanges, 8-18 (fig.), 8-19, 8-21
  chromosome replication banding techniques
    B-pulse banding, 4-18→4-19
    T-pulse banding, 4-19→4-20
  visualization methods, 4-20→4-22
  free chromatin, preparation of, 4-35→4-36
Bromo-4-chloro-3-indolyl phosphate (BCIP), as substrate for ISH, 4-29 (table)
Bromophenol blue
  detection of biotinylated probes in ISH, 4-41
  migration in PAGE, A3-18 (table)
Buccal swabs, DNA isolation from, A3-3→A3-4
Burst-forming unit-erythrocyte (BFU-E), 13-34

C

Calcium phosphate-mediated transfection, 12-36→12-37
Cancer
  cytogenetic analysis
    of hematological specimens, 10-2→10-7
    of solid tumor cultures, 10-7→10-10
  methylation-specific PCR, 10-30→10-34
  molecular analysis of
    DNA rearrangements in leukemias and non-Hodgkin's lymphoma, 10-10→10-23
    gene amplification in tumors, 10-23→10-30
Candidate genes
  identification in genomic DNA
    gene identification, 6-2→6-13
    human genome, accessing, 6-38→6-61
    searching NCBI databases using Entrez, 6-61→6-74
    sequence databases, 6-14→6-21
    sequence similarity searches, 6-22→6-38
  mutation detection by
    amplification of sequences, 7-2→7-6
    cycle sequencing, 7-19→7-22
    heterozygote detection using automated fluorescence-based sequencing, 7-13→7-19
    human mutation databases, 7-22→7-29
    overview, 7-1→7-2
    single-strand conformation polymorphism analysis, 7-6→7-13
Capillary electrophoresis, paternity testing, 9-32

Capillary electrophoresis single-strand
conformation polymorphism analysis
(CE-SSCP)
automated using ABI 310 genetic
analyzer, 7-11→7-12
automated using ABI prism 3100 genetic
analyzer, 7-12→7-13
benefits of, 7-8
sample preparation, 7-9→7-11
Carotid arteries, murine model of gene
transfer to, 13-9→13-11
Case-control association studies, disease
association
design of, 1-77→1-78
in population-based samples, 1-78→1-79
CD34⁺ cells, retroviral-mediated transduction,
13-30→13-37
cDNA
amplification by PCR, mutation detection,
7-5
cloning, 6-60
microarray analysis, gene expression in
transcriptional profiling,
11-12→11-13
quantitation of, 11-8
in retroviral vector construction,
12-30→12-32, 12-33 (fig.)
solid-phase reversible immobilization
(SPRI) purification of, 11-7→11-8
Celera, Genome Browser, 6-40
Cell cultures
involving DEB, 8-33→8-34
synchronization, in cytogenetic analysis,
4-3 (fig.)
Cell number determination, A3-31→A3-33
Cell pellets, DNA isolation from, A3-2
Cell sorting
FACS to isolate lacZ-labeled cells,
13-18→13-19
myoblast purification with, 13-15
Cells
LacZ-labeled, isolation by FACS,
13-18→13-19
myoblasts, in gene delivery to muscle,
13-13→13-21
Centre d'Etude du Polymorphisme Humain
(CEPH), genetic maps, A2-23
Centrifuges, rotor speed conversions,
A2-24→A2-28 (tables)
CEPH, see Centre d'Etude du
Polymorphisme Huamain family anal-
ysis
Cesium chloride, in adenovirus purification,
12-17→12-18
CFTR, see Cystic fibrosis transmembrane
conductance regulation
CFU, see Colony-forming unit
CGG repeats, in fragile X syndrome assess-
ment, 9-16→9-18
CHEF, see Contour-clamped homogeneous
electric field electrophoresis
CHLC, see Cooperative Human Linkage
Center
Chorionic villus cells, cultures of, 8-34 (table),
8-36→8-37
Chorionic villus samples (CVS)
direct preparation, 8-17
preparing metaphase spreads, culture
methods, 8-2→8-5
Chromosomal aneuploidy, see Aneuploidy
Chromosomal phase, in linkage analysis,
1-15→1-16
Chromosomal translocations, detection by
PCR, 10-18→10-22
Chromosome banding
applications, 4-15 (table)
C-banding, 4-56→4-57
with distamycin-DAPI staining, 4-22
G-banding (giemsa), 4-14, 4-16→4-17,
4-56→4-57

high-resolution, 8-10
with Hoechst-distamycin staining,
4-22→4-23
with in situ hybridization, 4-62→4-64
ISCN idiograms, A2-5→A2-22
nomenclature, A2-4
Q-banding (quinacrine), 4-15→4-16
R-banding (RHG), 4-15 (table)
replication banding
B-pulse, 4-18→4-19
T-pulse, 4-19→4-20
visualization, 4-20→4-22
Chromosome breakage, frequency in peripheral blood lymphocytes and cultured
fetal cells, 8-34 (table)
Chromosome paints/painting
M-FISH system, 4-43→4-45
probes for, 4-42→4-43
Rx-FISH, 4-46→4-47
SKY system, 4-43→4-45
Chromosome preparations
from bone marrow, 4-12, 10-2→10-4,
10-6→10-7
from fetal liver, 4-10→4-11
from leukemic blood specimens,
10-2→10-4, 10-6→10-7
from lymph node specimens, 10-6
peripheral blood, 4-2→4-8
for PFGE, from S. cerevisiae, 5-5→5-7
solid adult tissues or tissues, 4-13
from spleen, 10-6→10-7
troubleshooting guide, 10-5 (table)
Chromosomes, karyotyping, A3-37→A3-38
Chronic lymphocytic leukemia (CLL)
choice of culture method, 10-2 (table)
preparation of chromosome spreads, 10-4
Chronic myeloid leukemia (CML), choice of
culture method, 10-2 (table)
CILINK, 1-36, 1-37 (table)
Clean room, pre-PCR, 2-24→2-25
Cleavage, sites for restriction endonucleases,
5-3 (table)
Clinical molecular genetics
ARMS analysis of point mutations,
9-38→9-45
characterized, 9-1→9-2
fragile X syndrome, 9-16→9-20
genotyping of apolipoprotein E (APOE),
9-83→9-88
multiple point mutations using ASOs),
9-12→9-16
multiplex PCR, identifying dystrophin
gene deletions, 9-2→9-11
myotonic dystrophy, trinucleotide repeats,
9-21→9-26
oxidative phosphorylation diseases for
detection of mitochondrial DNA
mutations, 9-46→9-58
paternity, analysis of, 9-28→9-38
preimplantation genetic diagnosis, single-cell DNA and FISH analysis
for, 9-59→9-73
protein truncation test, 9-73→9-82
CLL, see Chronic lymphocytic leukemia
CLODSCORE, 1-36, 1-37 (table)
Cloning
cylinders, for colony isolation, 3-6→3-7
of somatic cell hybrid populations, 3-7
vectors, see Vectors
Clustering, in transcriptional profiling,
11-25→11-26
CMAP, 1-36, 1-37 (table)
CML, see Chronic myeloid leukemia
Coding sequences, identification in sequenced DNA, 6-18→6-19
Collagen-embedded cells, implantation in
brain, 13-27→13-28
Colony-forming unit (CFU) assay
cell preparation for, 13-39→13-40
plating for, 13-38→13-39

Colony-forming unit-granulocyte/erythro-
cyte/macrophage (CFU-GEM), 13-34
Colony-forming unit-granulocyte/macrophage
(CFU-GM), 13-34
Colony hybridization, screening large-insert
libraries
arrayed libraries, 5-8→5-14
preparing BAC/PAC colony blots
basic protocol, 5-14→5-15
with overgo oligonucleotide probes,
5-17→5-18
screening protocol, 5-15→5-16
troubleshooting guide, 5-16 (table)
Colony isolation, using cloning cylinders,
3-6→3-7
Combinatorial partitioning method (CPM),
1-97
Commonly used techniques
denaturing polyacrylamide gel electrophoresis, A3-16→A3-18
enzymatic labeling of DNA, A3-12→A3-16
establishment of permanent cell lines by
Epstein-Barr virus transformation,
A3-34→A3-37
extraction of DNA, A3-4→A3-5
isolation of genomic DNA from mammalian cells, A3-1→A3-4
karyotyping, A3-37→A3-38
mammalian cell tissue culture techniques,
A3-26→A3-34
northern blot hybridization, A3-23→A3-26
precipitation of DNA, A3-5→A3-6
preparation of DNA from fixed, paraffin-embedded tissue, A3-7→A3-8
quantitation of DNA and RNA,
A3-8→A3-12
Southern blot hybridization,
A3-18→A3-23
Comparative genomic hybridization (CGH)
using directly labeled DNA, 4-64→4-66
using in directly labeled DNA, 4-67
microscopy, imaging, and image analysis,
4-69→4-70
preparation of
genomic DNA, 4-68
labeled DNA probes, 4-68→4-69
metaphase chromosomes, 4-67→4-68
Computer programs and Web sites
Affymetrix GeneChip, 11-9
BLASTP, 6-23 (table), 6-27→6-30,
6-32→6-33, 6-35
BLASTX, 6-23 (table), 6-30, 6-34→6-35
CILINK, 1-36, 1-37 (table)
CLODSCORE, 1-36, 1-37 (table)
CMAP, 1-36, 1-37 (table)
Entrez, 6-14→6-15, 6-61→6-76,
6-69→6-74
FGENEH/FGENES, 6-5, 6-6 (fig.), 6-13
GeneID, 6-10
GeneMachine, 6-13
GeneParser, 6-10, 6-12
Genescan 672 Collection, 2-22→2-23
GENSCAN, 6-6→6-9, 6-12→6-13
GRAIL, 6-3→6-4, 6-12→6-13
heterozygous mutation detection, 7-15
HMMgene 6-11, 6-13
HOMOG, 1-31
human mutation databases, 7-26→7-29
ILINK, 1-36, 1-37 (table)
LCP, 1-29, 1-43
LINKLODS, 1-31, 1-40, 1-43
LINKMAP, 1-31, 1-36, 1-37 (table), 1-40,
1-43
LODSCORE, 1-36, 1-37 (table)
MacVector, 2-1
MAKEPED, 1-29, 1-31→1-36
MAPFUN, 1-40, 1-46→1-47
MDR, 1-94
MENDEL, 1-25→1-26
MLINK, 1-29, 1-31, 1-37, 1-43, 1-46

multifactor dimensionality reduction (MDR), 1-94
MZEF, 6-5, 6-12→6-13
OLIGO, 2-1
Power, 1-92→1-93
PowerBLAST, 6-13
PREPLINK, 1-29, 1-52, 1-54→1-55
PRIMER, 2-1
PrimerExpress, 2-33
PROCRUSTES, 6-9→6-10, 6-12
Sequin, 6-13, 6-17, 6-19
SIBPAL, 1-59, 1-61→1-67
TBLASTN, 6-23 (table), 6-32, 6-34, 6-35
TBLASTX, 6-23 (table), 6-34
THREELOC, 1-24→1-26
TWOLOC, 1-24→1-26
UNKNOWN, 1-31
XGRAIL, 6-4, 6-13
See also, Databases
Concanavalin A (Con-A), chromosome preparation for karyotyping, 4-7
Conception, products of
preparation of, for culture, 8-16→8-17
skin, preparation of, 8-17
tissue biopsies, 8-17
Contigs
as ISH probes, 4-48
location changes, 6-60→6-61
NCBI database, 6-61
Contour clamped homogeneous electric field electrophoresis (CHEF) gel analysis
genomic DNA, partial digestion and size fractionation, 5-41
of restriction-digested BAC DNA, 5-7→5-8
Cooperative Human Linkage Center (CHLC), genetic maps, A2-22
Cosmids
HSV-1 DNA, preparation for transfection, 12-69→12-71
as ISH probes, 4-48, 4-62
Coupled transcription/translation, radioactive protein truncation test, 9-81→9-82
CpG, methylation-specific PCR (MSP)
characteristics of, 10-30
determination of methylation in PCR products, 10-33→10-34
$C_o t_1$ DNA, A2-3
Cytogenetic analysis
chromosome preparation from
bone marrow, 10-2→10-4, 10-6→10-7
leukemic blood specimens, 10-2→10-4, 10-6→10-7
lymph node specimens, 10-6
spleen, 10-6→10-7
of hematological specimens, 10-2→10-7
of solid tumor cultures, 10-7→10-10
CsCl, see Cesium chloride
Cubby searches, 6-66→6-69
Culture
of hematopoietic cells
for colony-forming assay, 13-38→13-40
in long-term bone marrow culture, 13-31 (fig.), 13-33, 13-43
screening media, 13-38
stromal layer, 13-30→13-35, 13-37
of primary airway epithelial cells, 13-46→13-49
of samples for metaphase analysis
amniotic fluid samples, 8-5→8-14
biopsy samples, 8-14→8-17
chorionic villus samples (CVS), 8-2→8-5
products of conception, 8-14→8-17
Cultured cells
fetal, chromosome breakage, 8-34 (table)
retroviral vector production, 12-36→12-37
Cycle sequencing, mutation detection
basic protocol, 7-19, 7-21→7-22
troubleshooting, 7-20 (table)
Cystic fibrosis

clinical molecular genetics, 9-1
segregation analysis, 1-5
Cystic fibrosis transmembrane regulator (CFTR), preimplantation genetic diagnosis, 9-62 (table)
Cytogenetic analysis
amniocyte preparation for interphase FISH, 8-26→8-31
amniotic fluid samples, preparation and culture, 8-5→8-14
chromosomal aneuploidy using paraffin-embedded tissue, determination of, 8-22→8-26
chromosome banding techniques, 4-1, 4-14→4-23
chromosome preparation from
bone marrow, 4-12
fetal liver, 4-10→4-11
peripheral blood cells, 4-2→4-8
solid adult tissues or tumors, 4-13
comparative genomic hybridization, 4-1, 4-64→4-70
diagnosis of Fanconi anemia by diepoxybutane analysis, 8-31→8-37
FISH analysis
high resolution, 4-1, 4-33→4-42
multicolor approaches for simultaneous human genome analysis, 4-42→4-48
in situ hybridization, to metaphase chromosomes and interphase nuclei, 4-24→4-33
metaphase spreads from chorionic villus samples (CVS), 8-2→8-5
morphology antibody chromosome (MAC) technique, phenotype and genotype determination, 4-1, 4-48→4-64
overview of, 4-1
products of conception for chromosome analysis, preparation and culture of, 8-14→8-17
sister-chromatid exchanges, 8-18→8-22
solid tissue for chromosome analysis, preparation and culture of, 8-14→8-17
Cytogenetics, somatic cell hybrids, 3-7→3-8
Cytomegalovirus (CMV) promoter, in retroviral vectors, 12-32, 12-33 (fig.), 12-40→12-41
Cytopathic effect, replication-defective HSV vector construction, 12-56
Cytospin, MAC analysis
applications, 4-49 (table), 4-56
slide preparation, 4-57→4-58

D

Databases
GenMAPP, 11-27
human mutation, 7-22→7-29
Kyoto Encyclopedia of Genes and Genomes (KEGG), 11-27
laboratory information management system (LIMS), 11-27→11-28
NCBI, LocusLink, 11-27
SQL, 11-27→11-28
for transcriptional profiling, 11-27→11-28
See also, Computer programs and databases
dbSNP, 7-26→7-27
DDBJ, see DNA Database of Japan
DEB, see Diepoxybutane
Decontamination, biosafety procedures, 12-4
Deletions
characterized, 9-1
dystrophin, 9-2→9-11
Delivery systems, for gene therapy
to airway, 13-44→13-59
to arteries, 13-2→13-13

to brain, ex vivo and in vivo, 13-21→13-29
human hematopoietic cell culture, transduction, and analyses, 13-30→13-44
to liver, 13-59→13-62
to muscle, 13-13→13-21
Denaturing gradient gel electrophoresis (DGGE), for detecting clonal T cell receptor-γ gene rearrangements, 10-16→10-18
Destaining slides of chromosome preparations, 4-23
Diabetes mellitus, insulin-dependent, linkage testing, 1-60
Dialysis, quick agarose-tube, 12-20→12-21
Diaminobenzidine tetrahydrochloride (DAB), as substrate for ISH, 4-29 (table)
Diepoxybutane (DEB), diagnosis of Fanconi anemia
overview, 8-31→8-32
test for postnatal diagnosis, 8-32→8-35
test for prenatal diagnosis, 8-36→8-37
test using fibroblast cultures, 8-35→8-36
working with and disposal of, 8-33→8-34
Differentially expressed genes, analysis of, 11-24→11-25
Differential PCR, for detection of gene amplification, 10-28→10-29
Digestion of DNA with restriction enzymes, for library construction, 5-7→5-8
Digoxigenin
labeled probes
for DIRVISH, 4-40→4-42
for interphase FISH, 8-30→8-31
for ISH, 4-28→4-29
labeling of DNA probes by nick translation, 4-31→4-32
Direct visualization hybridization (DIRVISH)
high resolution mapping, to stretched DNA, 4-37→4-40
preparation of stretched DNA for mapping, 4-36→4-37
DIRVISH, see Direct visualization hybridization
Disaggregation of solid tumors for culture, 10-8→10-9
Discrete traits, association studies, 1-92→1-96
Disease association studies
family-based, 1-76→1-83
gene-gene interactions
discrete traits, 1-92→1-96
quantitative traits, 1-96→1-97
Disease genes, genetic associations, multiallelic tests of, 1-85→1-86
Diseases, recurrence risk, 1-4→1-5
Disruption of tissues for metaphase analysis
enzymatic, 8-15→8-16
mechanical, 8-14→8-15
Distamycin, staining of chromosomes
distamycin-DAPI staining, 4-22
destaining, 4-23
Hoechst 33258-distamycin staining, 4-22→4-23
Dizygotic (DZ) twins
in linkage analysis, 1-4
paternity testing, 9-36
DMD, see Duchenne muscular dystrophy
DNA
adenoviral, 12-20
BAC, 5-1, 5-7→5-8, 5-30→5-31, 5-44→5-45
comparative genomic hybridization (CGH)
directly labeled, 4-64→4-66
indirectly labeled, 4-67
cosmid, 12-69→12-71
enzymatic labeling
nick translation, A3-12→A3-13
random-primed synthesis, A3-13→A3-14

with T4 polynucleotide kinase, A3-14
isolation
 from buccal swabs, A3-3 →A3-4
 from cell pellets, A3-2
 from mouth wash or blood, 9-44
 partial digestion and size fractionation, 5-43
 from whole blood, A3-1 →A3-2
methylation
 determination of CpG sites within methylation-specific PCR, 10-33→10-34
 pattern determination by methylation-specific PCR, 10-30→10-33
PAC clones, 5-44→5-45
YAC, 5-1, 5-21→5-24, 5-26
PCR-amplified, multiple point mutation analysis using ASOs, 9-12→9-15
phenol extraction, A3-4→A3-5
plasmid, in gene delivery by direct injection, 13-18
preparation, yeast
 ethidium bromide/agarose plates, 5-28→5-29
 gel-purified, 5-27→5-28
 from spheroplasts, 5-19→5-23
quantitation, A3-9→A3-12
separation, see Electrophoresis
Southern blot hybridization, A3-18→A3-23
vaccine administration by intramuscular injection
 anterior tibialis, 13-20→13-21
 quadriceps, 13-19→13-20
 vaccine vector, 13-20 (fig.)
viral, isolation of, 12-64→12-65
whole-genome amplification, single-sperm typing, 1-23→1-24
DNA Data base of Japan (DDBJ), 6-15→6-17, 6-25
Donor cells, for hybrid cell line construction, 3-5→3-6
Dot blots
 determination of rAAV titers by, 12-27→12-28
 multiple point mutation analysis, 9-14
Down syndrome, 1-1
Digoxigenin
 labeled primers for ligation assays, 2-17
 labeling of DNA probes, using terminal transferase, 2-11→2-12
Drosophila melanogaster genome, for nucleotide sequence, 6-25
Duchenne muscular dystrophy (DMD)
 identification, 9-1→9-11
 preimplantation genetic diagnosis, 9-59→9-63
 see also Dystrophin gene deletions
Dystrophin gene deletions
 correspondence to clinical phenotypes, 9-4 (fig.)
 detection by multiplex PCR, 9-2→9-7
 for diagnostic multiplex mixes, 9-7→9-11
 distinguishing between BMD and DMD, 9-4 (fig.), 9-6 (fig.)
 form for recording results, 9-6 (fig.)

E

Eccles Institute of Human Genetics, genetic maps, A-23
Effusion, specimens for cytogenetic analysis, 10-9→10-10
Electrocompetent cells, preparation of, 12-19→12-20
Electronic PCR (e-PCR), 6-38
Electrophoresis
 agarose-formaldehyde gel electrophoresis, A3-23→A3-26
 denaturing PAGE, A3-16→A3-17
 DNA migration in PAGE, A3-18 (table)
 paternity testing, 9-32→9-33

of pooled PCR products for fluorescent genotyping, 2-21→2-24
EMBL, see European Molecular Biology Laboratory
Embryo
 biopsy, preimplantation genetic diagnosis, 9-64→9-68
 introduction of YAC DNA into, 5-19→5-23
End-labeling, of primers for PCR genotyping, 2-2→2-3
Ensembl, human genome data
 annotations, 6-47 (table)
 customized display, 6-52
 display formats, 6-53
 downloading sequence, 6-53
 FTP format, 6-53
 gene models, 6-53
 information retrieval, 6-57
 queries, 6-50→6-52, 6-57→6-60
 views, 6-51 (table)
 web site services, 6-48 (table)
Entrez
 Cubby searches, 6-66→6-69
 intro to, 6-14→6-16
 integrated database info retrieval description, 6-14→6-15
 querying
  basic protocol, 6-61→6-66
  Boolean search statements, 6-63, 6-64 (table), 6-65 (fig)
  combining queries, 6-69→6-72
  individual database records, 6-63
  related material, 6-65
  structure examinations, 6-72→6-74
  types of, 6-61→6-66
Enzymatic disruption of tissues for metaphase analysis, 8-15→8-16
Enzyme-substrate combinations, detection of ISH probes, 4-29 (table)
Epithelial cells, primary airway
 isolation of, 13-46→13-48
 transduction of, 13-48→13-49
Epstein-Barr virus (EBV)
 preparation of virus stock, A3-35 → A3-36
 transformation of cultured lymphocytes, A3-34→A3-35
Escherichia coli databases, for nucleotide sequences, 6-25
Ethanol precipitation, A3-5→A3-6
Ethidium bromide
 free chromatin preparation, 4-35→4-36
 high-resolution chromosome banding, 8-10
 northern hybridization, A3-24
 to quantify DNA and RNA, A3-9 (table), A3-12
 rapid estimation of DNA concentration on, 5-28→5-29
 for Southern blotting, A3-18→A3-22
European Molecular Biology Laboratory (EMBL), 6-14, 6-16→6-17, 6-25
Ex vivo techniques
 gene delivery, to the brain, 13-21→13-29
 transduction with retrovirus, human bronchial xenografts, 13-56→13-57
ExoFish, 6-38
Expressed-sequence tags (ESTs)
 database for nucleotide sequences, 6-25, 6-31, 6-33
 genomic DNA analysis, 6-33→6-34
 PCR products, agarose gel electrophoresis, 11-20→11-21
 sequence database data submission, 6-14, 6-18, 6-21
Expression
 accessibility of data, 11-28
 analysis
  classification and class prediction, 11-26
  clustering, 11-25→11-27

identifying differentially expressed genes, 11-24→11-25
 pathway/ontology analysis, 11-27
 sequence analysis, 11-26→11-27
 archiving of data, 11-27→11-28
 cDNA microarrays, transcription profiling
  cDNA amplification and printing, 11-12→11-16
  hybridization and data extraction, 11-19→11-20
  RNA extraction and labeling, 11-16→11-18
 databases, 11-27→11-28
 experimental design, 11-23
 monitoring
  oligonucleotide arrays for, see Oligonucleotide arrays for expression monitoring
  troubleshooting guide for, 11-10→11-11 (table)
 normalization, 11-23→11-24
 pattern analysis
  one-color array experiments, 11-23→11-24
  two-color array experiments, 11-24
Extension mix for SBE genotyping assays, 2-31 (table)
Extraction, of RNA, 11-16→11-18

F

FACS, see Fluorescence-activated cell sorting
Family-based studies
 case-control compared with, 1-77→1-80
 of disease associations
  alleles, 1-76, 1-83
  association, 1-77, 1-80, 1-83
  haplotype relative risk, 1-80
  linkage, 1-77, 1-80→1-83
  linkage disequilibrium, 1-77
  marker locus, 1-76
Fanconi anemia (FA), diagnosis of
 criteria, 8-31
 diepoxybutane test
  chromosome breakage, Giemsa staining, 8-34→8-35
  using fibroblast cultures, 8-35→8-36
  postnatal diagnosis, 8-32→8-34
  prenatal diagnosis, 8-36→8-37
Fast blue BN, Fast green BN, as substrates for ISH, 4-29 (table)
FASTA format, for BLAST sequence searches, 6-26, 6-34
Femoral arteries, gene transfer to injured, 13-11→13-13
Fetal liver, mitotic chromosome preparations from, 4-10→4-11
FGENEH/FGENES, in gene identification, 6-5, 6-6 (fig.), 6-13
Fibroblasts
 collagen-embedded, implanted into adult rat brain, 13-27→13-28
 cultures of, 8-34 (table), 8-35→8-36
 high-resolution FISH analysis, 4-33→4-34
 replication banding, 4-19→4-20
 tissue biopsies, 8-17
 vector-producing, 13-35→13-36
Fibronectin, in transduction of hematopoietic cells, 13-35
File transfer protocol (FTP)
 applications, 11-28
 human genome data, 6-50, 6-53, 6-56
Filter hybridization, screening large-insert libraries, 5-8→5-10
Filtering
 to screen genomic libraries, 6-26
 of sequence databases for BLAST searches, 6-26→6-27
Fingerprinting, defined, 6-38
First-degree relatives, linkage studies, 1-5, 1-14
FISH, see Fluorescence in situ hybridization

Flask method for amniotic fluid cultures, 8-11→8-13
Fluorescence, quantitation of DNA
  ethidium bromide, A3-9 (table), A3-12
  Hoechst 33258, A3-10→A3-11
Fluorescence-activated cell sorting (FACS)
  LacZ-labeled cells, isolation, 13-18→13-19
  primary myoblasts, purification of, 13-15
  single-sperm typing, 1-21→1-23
Fluorescence-based sequencing, mutation detection
  characterized, 7-13
  identification from sequence traces, 7-15→7-16
  troubleshooting guide, 7-14 (table)
Fluorescence in situ hybridization (FISH)
  defined, 6-39
  high-resolution, preparation of
    biotin and digoxigenin labeling of DIRVISH probes, 4-40→4-42
    free chromatin, 4-33→4-36
    stretched cellular DNA for DIRVISH mapping, 4-36→4-40
  interphase FISH
    of amniocytes, 8-26→8-28
    of amniotic fluid cells, 8-29→8-31
  to metaphase chromosomes, 4-24→4-28
  multicolor, for simultaneous human genome analysis, 4-42→4-48
  ordering sequences in interphase nuclei, 4-32→4-33
  preimplantation genetic diagnosis
    probe validation, 9-73
    single blastomeres, 9-71→9-72
  preparation of nuclear suspensions, 8-22→8-25
Fluorescence microscopy
  distamycin-DAPI staining, 4-22
  quinacrine banding (Q-banding), 4-16
Fluorescence polarization (FP), APOE genotyping by, 9-86→9-87
Fluorescent genotyping
  PCR amplification
    gel electrophoresis of products, 2-21→2-24
    pooling labeled products, 2-19→2-21
    of SSLPs for, 2-17→2-19
Fluorescent polarization detection, primer extension high-throughput genotyping
  characterized, 2-36
  primer extension assay with 4-plex PCR, 2-39→2-41
  primer extension assay with single-plex PCR, 2-37→2-39
  strategic planning, 2-37
Fluorometry, DNA concentration determination in gene expression with cDNA microarrays, 11-21→11-21
Fragile X syndrome
  CGG repeats
    detection by Southern blot hybridization, 9-19
    fragile X region, 9-16
  PCR amplification, 9-16→9-18
  restriction enzyme map, 9-20 (fig.)
  Southern blot hybridization, detection of amplification and methylation, 9-19→9-20
Free chromatin preparation, high-resolution FISH
  with alkaline buffer, 4-33→4-34
  from lymphocytes by drug treatment, 4-35→4-36
  mapping, 4-36
  optimization of, 4-34→4-35
Freezing
  of EBV-transformed lymphocytes, A3-36→A3-37
  hematopoietic cells, 13-43→13-44
  of mammalian cells grown in cultures, A3-29→A3-30

FTP, see File transfer protocol
Fungicides, in tissue culture, A3-26 (table)
FVIII gene, in preimplantation genetic diagnosis, 9-62 (tables), 9-63

G

G1 and G2 cells, in interphase chromosome mapping with FISH, 4-32→4-33
G-11 staining of metaphase chromosomes, in hybrid cell lines, 3-7→3-8
Gel casting plate for PFGE, 5-28 (fig.)
Gel electrophoresis, see Electrophoresis
GenBank, NCBI sequence database
  BankIt, 6-17→6-18
  contig location annotation, 6-60
  submitting data to, 6-16→6-18, 6-20→6-21
Gene amplification, detection in tumors by
  differential PCR, 10-28→10-29
  slot blot hybridization, 10-27→10-28
  Southern blot hybridization, 10-23→10-27
Gene delivery
  to airway, 13-44→13-59
    airway model system choice, 13-44→13-45
    generation of human bronchial xenografts, 13-52→13-56
    generation of polarized airway epithelial monolayers, 13-49→13-51
    harvesting bronchial xenografts for morphologic analysis to evaluate transgene expression, 13-58→13-59
    in vivo delivery to the lung, 13-57→13-58
    isolation of human primary airway epithelial cells, 13-46→13-48
    reporter gene choice, 13-46
    transduction of primary airway epithelial cells, 13-48→13-49
    vector system choice, 13-45→13-46
  using helper virus-free HSV-1 amplicon vectors
    preparation of amplicon stocks, 12-66→12-69
    preparation for HSV-1 cosmid DNA for transfection, 12-69→12-71
  in vivo, liposome vectors for, 12-74→12-75
  LacZ-labeled cells, isolation by FACS, 13-18→13-19
  to liver, 13-59→13-62
  myoblast implantation into skeletal muscle, 13-16→13-17
  plasmid DNA, direct injection into muscle, 13-18
  See also Gene transfer
Gene expression, see Expression
Gene-gene interactions
  detection methods in association studies of discrete traits
    logistic regression, 1-92→1-93
    multifactor dimensionality reduction (MDR), 1-93→1-96
  detection methods in association of quantitative traits
    combinatorial partitioning method (CPM), 1-97
    linear regression, 1-96→1-97
  overview of, 1-90→1-92
GENEHUNTER/GENEHUNTER PLUS, 1-55, 1-57, 1-59, 1-67→1-71
GeneID, in gene identification, 6-10
Gene identification
  effectiveness of methods, 6-11→6-12
  methods of, 6-2→6-11
  strategies and considerations of methods, 6-12→6-14
GeneMachine, in gene identification, 6-13
GeneParser, in gene identification, 6-10, 6-12
Gene Ontology (GO), 11-27

Generalized transmission disequilibrium test statistic (GTDT), 1-84→1-85, 1-89→1-90
Gene rearrangements, in cancer, 10-11→10-18
Gene Recognition and Analysis Internet Link (GRAIL)
  for GC content, 6-12→6-13
  versions and features of, 6-3→6-4
Genescan 672 Collection software, 2-22→2-23
GeneSweep, multiplex analysis of SSLPs, 2-9
Gene therapy, vectors, characterized, 12-1→12-2
Gene transfer
  to arteries, 13-2→13-13
  to brain, ex vivo and in vivo
    into adult, 13-23→13-24
    cell preparation for grafting, 13-29
    of collagen-embedded fibroblasts into adult, 13-27→13-28
    direct injection of vectors, 13-28
    into fetus, 13-26→13-27
    into neonate, 13-25→13-26
    strategy, 13-21→13-22
  into hematopoietic cells, 13-30→13-44
  to human bronchial xenografts, 13-56→13-57
  to polarized airway epithelia, 13-51→13-52
Généthon, genetic maps, A2-22
Genetic maps and databases, Internet resources, A2-22→A2-23. See also Maps and mapping
GenMAPP, 11-27
Genome survey sequences (GSSs)
  database for nucleotide sequences, 6-25, 6-33
  genomic DNA analysis, 6-33→6-34
  sequence database data submission, 6-14, 6-18, 6-21
Genomic DNA
  amplification of sequences from affected individuals, 7-5→7-7,
  candidate gene identification
    human genome, accessing, 6-38→6-61
    methods of, 6-1→6-14
    searching NCBI databases using Entrez, 6-61→6-74
    sequence databases, 6-14→6-21
    sequence similarity searches, 6-22→6-38
  high-molecular-weight, mammalian, 5-4→5-5
  inverse PCR, clonal integration analysis, 13-41 (fig.)
  isolation from mammalian cells, A3-1→A3-4
  partial digestion, 5-40→5-44
  paternity testing, 9-33→9-34
  PFGE for long-range restriction mapping, 5-2→5-4
  preparation
    using agarose gels, 5-4→5-7
    for comparative genomic hybridization, 4-68
    from whole blood in paternity testing, 9-33→9-34
  recovery by high-salt precipitation, A3-2→A3-3
  retroviral vector construction, 12-34→12-35
  size fractionation, 5-40→5-44
  trinucleotide repeat analysis in myotonic dystrophy assessment, 9-24→9-26
  X chromosome inactivation assay, 9-26→9-27
Genotype, defined, 1-1
Genotyping

fluorescent, automated, 2-17→2-24
high-throughput
    with primer extension fluorescent polar-
        ization detection, 2-36→2-41
    using TaqMan assay, 2-33→2-41
by ligation assays, 2-12→2-17
MAC analysis, 4-50 (table), 4-56→4-57,
    4-62→4-64
PCR methods of, 2-1→2-12
SNP, using DNA microarrays, 2-24→2-33
GENSCAN, in gene identification, 6-6→6-9,
    6-12→6-13
Gi, 6-39
Gibbon chromosome-specific paints, for
    Rx-FISH, 4-46→4-47
Giemsa banding (G-banding)
    aging slides
        with heat, 4-17
        with hydrogen peroxide, 4-18
    applications, 4-15 (table)
    karotyping, mitotic chromosome prepara-
        tions for, 4-14
    MAC analysis, 4-56→4-57, 4-62→4-64
    methods, 4-16→4-17
Giemsa stain
    for chromosome breakage, 8-34→8-35
    for sister-chromatid exchanges, 8-18
        (fig.), 8-20, 8-21 (table)
Goodness-of-fit test, ASP linkage studies,
    1-60
Grafting, gene delivery to brain of adult rat,
    13-23→13-24, 13-29
GRAIL, see Gene Recognition and Analysis
    Internet Link
Grandparents, in paternity testing, 9-36
Green fluorescent protein (GFP) vector
    lentivirus stocks, titration of,
        12-47→12-48
GTG banding, see Giemsa banding

H

Haplotype
    linkage analysis, 1-16
    transmission, single-sperm cell analysis,
        1-26
Haplotype relative risk (HRR), disease as so-
    ciations, 1-80
Hardy-Weinberg equilibrium, 1-91
Haseman-Elston regression (H-E), linkage
    analysis, 1-56 (fig.), 1-62, 1-67
Helper virus, marker rescue assay for,
    12-38→12-39
Helper virus-free amplicon stocks
    packaging system, 12-67 (fig.)
    preparation of, 12-66, 12-68→12-69
Hemacytometer, A3-31→A3-33
Hematopoietic cells, transduction of
    analysis of clonal integration in individual
        colonies, 13-40→13-43
    CD34+ cells with stromal support,
        13-30→13-37
    collection of cell-free supernatant for,
        13-36
    on fibronectin-coated dishes, 13-35
    freezing cells, 13-43→13-44
    in long-term bone marrow culture, har-
        vesting cells for, 13-43
    plate colony-forming cells from cultures,
        13-38→13-39
    preparing whole-cell lysates of individual
        cfu colonies for PCR,
        13-39→13-43
    primary human marrow stromal
        monolayers from harvested bone
        marrow, 13-37
    screening media, 13-38
    thawing cells, 13-44
    vector-producing fibroblast culture,
        13-35→13-36

Hemophilia, preimplantation genetic diagno-
    sis, 9-59
Heritability, 1-5
Herpes simplex virus (HSV)
    amplicons, HSV-1
        cosmid DNA preparation for
            transfection, 12-69→12-70
        titration, 12-71→12-74
    DNA isolation, 12-64→12-65
    helper virus-free packaging system, 12-67
        (fig.)
    IE gene-complementing cell lines,
        12-50→12-53
    replication-defective, construction of
        basic protocol, 12-54→12-57
        HSV-1 IE gene-complementing cell
            lines, 12-50→12-53
        stock preparation, 12-62→12-63
        titer, plaque assay, 12-63→12-64
Heterozygosity, linkage analysis, 1-16
Heterozygote mutation detection
    using automated fluorescence-based se-
        quencing
        characterized, 7-13
        identification from sequence traces,
            7-15→7-16
        troubleshooting guide, 7-14 (table)
        large-volume sequencing in ABI 3700
            DNA analyzer, 7-16→7-19
HEXA gene, in preimplantation genetic diag-
    nosis, 9-62→9-63
HGVS, 7-26→7-28
High-molecular-weight DNA
    from animal tissue, 5-39→5-40
    from lymphocytes, 5-38→5-39
    preparation of agarose blocks
        mammalian genomic DNA embedded
            in, 5-4→5-5
        standards for using S. cerevisiae chro-
            mosomes and YACs, 5-5→5-7
High-resolution chromosome banding, 8-10
High-throughput genome sequences (HTGS)
    database, for nucleotide sequence, 6-31
    genomic DNA analysis, 6-33→6-34
    sequencing keywords, 6-40 (table)
    submitting data, 6-21, 6-25
High-throughput genotyping
    with primer extension fluorescent polar-
        ization detection
        characterized, 2-36
        primer extension assay with 4-plex
            PCR, 2-39→2-41
        primer extension assay with single-plex
            PCR, 2-37→2-39
        strategic planning, 2-37
    using TaqMan assay, 2-33→2-41
High-titer adenoviruses, 12-16→12-18
hisD, see Histidinol dehydrogenase
Histidinol dehydrogenase (hisD)
    retroviral vector construction, 12-32,
        12-33 (fig.)
    as selectable marker, 3-3
HLA
    disease association, 1-78
    linkage testing, 1-60
HMM gene, in gene identification, 6-11, 6-13
Hoechst-distamycin staining of chromo-
    somes, 4-15 (table), 4-22→4-23
Hoechst 33258
    in analysis of sister-chromatid exchanges,
        8-18 (fig.), 8-20, 8-21 (table)
    detection of DNA, A3-10→A3-11
HOMOG, 1-31
Homologous recombination in yeast, intro-
    duction of mammalian selectable
    marker by, 5-24→5-27
Homolog search, human genome,
    6-58→6-59
Homozygosity mapping, DNA pooling in,
    1-71→1-76
Horseradish peroxidase (HRP)
    enzymatic detection for FISH, 4-29→4-30

immunophenotyping using, 4-53→4-55
HPH, see Hygromycin phosphotransferase
HUGO, see Human Genome Organisation
Human gene mutation database (HGMD),
    7-25→7-27
Human genome, accessing
    contigs, 6-60
    data
        accessing, 6-46→6-56
        annotation, 6-43→6-46
        assembling the sequence, 6-41→6-43
        input, 6-40→6-41
    information retrieval, 6-56→6-60
    searching NCBI databases using Entrez,
        6-66→6-72
        combining queries, 6-69→6-72
        Cubby, 6-66→6-69
        examining structures, 6-72→6-74
        queries, 6-61→6-66
    terminology, 6-38→6-39
    web sites, 6-39 (table)
Human genome, multicolor FISH analysis,
    4-42→4-48
Human genome variation database
    (HGVbase), 7-26
Human Genome Organisation (HUGO),
    Mutation Database Initiative (MDI),
    7-23, 7-24 (fig.), 7-25, 7-27
Human immunodeficiency virus (HIV), vec-
    tors, HIV-based, lentiviral,
    12-43→12-47
Human mutation databases
    accessing, 7-25
    data submission, 7-23, 7-25
    ethnic, 7-27
    locus-specific databases (LSDBs), 7-22
        (table), 7-23, 7-27→7-29
    national, 7-27
    overview of, 7-22 (table), 7-26→7-27
    types of, 7-22→7-23
Human repeats, filtering sequence databases
    for, 6-33
Human repetitive DNA sequences,
    A2-1→A2-3
Human SNP, mapping assay, using
    allele-specific microarrays, 2-26, 2-29
HuSNP, see Human SNP
Hybridization
    allele-specific, for SNP genotyping,
        2-27→2-29
    comparative genomic, see Comparative
        genomic hybridization
    conditions, stripping probes and
        rehybridizing, 9-15→9-16
    generic microarrays, for SNP genotyping,
        2-32
    for large-insert library screening
        arraying colonies and DNA at low and
            high density, 5-8→5-14
        by colony hybridization, 5-15→5-19
        preparing BAC/PAC colony blots for,
            5-14→5-15
    mRNA in expression monitoring,
        11-1→11-5
    in myotonic dystrophy assessment,
        9-21→9-26
    reverse, APOE genotyping, 9-84→9-85
Hygromycin-B-phosphotransferase (HPH)
    retroviral vector construction, 12-32,
        12-33 (fig.)
    as selectable marker, 3-3
Hypoxanthine-guanine
    phosphoribosyltransferase (HPRT),
    as selectable marker, 3-2

I

Identifier syntax for sequence databases,
    6-37→6-38
Identity-by-descent (IBD), in linkage analysis,
    1-55→1-61, 1-67

Iden ti ty-by-state (IBS), in link age anal y sis, 1-55, 1-57→1-59
Iliac ar ter ies, gene trans fer to
  por cine model, 13-5, 13-6 (fig.)
  rab bit atherosclerotic model, 13-6→13-9
ILINK, 1-36, 1-37 (ta ble)
Iliofemoral ar ter ies, por cine model of gene trans fer to, 13-3→13-5
Im age anal y sis, com par a tive genomic hy brid-iza tion (CGH), 4-69→4-70
Im aging, com par a tive genomic hy brid i za tion (CGH), 4-69→4-70
Immunofluorescence, MAC phenotyping, 4-49 (ta ble)
Im mu no glob u lin genes, PCR to de tect gene re ar rangements, 10-12 (table), 10-14→10-15
Immunoperoxidase, MAC anal y sis phenotyping, 4-49 (ta ble)
  sche matic il lus tra tion, 4-54 (fig.)
Immunophenotyping, MAC anal y sis
  APAAP, 4-48, 4-49 (ta ble), 4-50→4-52
  with fluo res cence de tec tion, 4-55→4-56
  with HRPO de tec tion, 4-53→4-55
In fec tion, in vi tro, of rAAV, 12-29
Informatics
  data col lec tion, link age stud ies, 1-11→1-12
  in transcriptional pro fil ing, 11-27→11-28
In for ma tion con tent of ped i grees, in link age anal y sis, 1-15→1-17
In situ coverslips, amniotic fluid cul tures, 8-5→8-10
In situ cul tures, for MAC anal y sis, prep a ra tion, 4-59→4-60
In situ hy brid iza tion (ISH) of chro mo some
  preparations
  amplification
    of biotinylated sig nals, 4-28
    from digoxigenin-la beled probes, 4-28→4-29
  bi o tin and digoxigenen la bel ing of probes by nick trans la tion, 4-31→4-32
  char ac ter ized, 3-7
  en zy matic de tec tion of nonisotopically la-beled probes with
    al ka line phosphatase, 4-29 (ta ble), 4-30→4-31
    horse rad ish peroxidase, 4-29→4-30
  FISH, to metaphase chro mo somes, 4-24→4-28
  for MAC anal y sis, with APAAP immunostaining, 4-50→4-53
  or der ing se quences in interphase nu clei by FISH, 4-32→4-33
In su lin-de pend ent di a be tes mellitus (IDDM), see Di a be tes mellitus
In ter nal la bel ing, in PCR geno typ ing, 2-4
Internet re sources
  ge netic maps and da ta bases, A2-22→A2-23
  heterozygote mu ta tion de tec tion, hu man gene mu ta tion de tec tion, 7-23, 7-25→7-29
Interphase cytogenetics, FISH us ing amniocytes
  at tached to a sur face, 8-28
  un cul tured, 8-26→8-28
Interphase nu clei, or der ing se quences by FISH, 4-32→4-33
In tra mus cu lar vac cine in jec tion of the
  an te rior tibialis, 13-20→13-21
  quadriceps, 13-19→13-20
Introns, da ta base sub mis sions, 6-19
In verse PCR, for clonal in te gra tion anal y sis, 13-40→13-43
In vi tro tran scrip tion (IVT)
  con trol genes, oligonucleotide ar rays for expression moni tor ing, 11-5→11-7
  prod ucts, solid-phase re vers i ble im mo bi li za tion (SPRI) of, 11-7→11-8

In vivo tech niques, for gene de liv ery
  to the brain, 13-21→13-29
  to the lung, 13-57→13-58
ISCN
  idiograms, A2-3→A2-22
  rules for des ig nat ing, karyotyping, A3-37→A3-38
IVT, see In vi tro tran scrip tion

K

Karyotyping
  mouse, 4-7→4-14
  pro ce dures, A3-37→A3-38
  spec tral (SKY), 4-43→4-45
Kinase re ac tions, to la bel oligonucleotides, A3-14
Kurtosis, in sta tis tics, 1-59
Kyoto En cy clo pe dia of Genes and Genomes (KEGG), 11-27

L

L1 re peats, A2-2
Labeling
  digoxigenin, of probes us ing ter mi nal transferase, 2-11→2-12
  nick trans la tion, A3-12→A3-13
  ran dom-primed syn the sis, A3-13→A3-14
  RNA ex trac tion, 11-16→11-18
    with T4 polynucleotide kinase, A3-14
  procedures
    end-la bel ing prim ers for PCR, 2-2→2-3
Laboratory
  BL2, 12-2→12-5
  setup for PCR, 2-25
Lab o ra tory in for ma tion man age ment sys tem (LIMS), 11-27
LacZ
  la beled cells, iso la tion by FACS, 13-18→13-19
  vec tor lentivirus stocks, ti tra tion of, 12-47→12-49
Large-in sert clon ing and anal y sis
  con struc tion of BAC/PAC librar ies, 5-32→5-45
  in tro duc tion into mam ma lian cells and em bryos, 5-19→5-31
  over view, 5-1→5-2
  pulsed-field gel elec tro pho re sis for long-range re stric tion map ping, 5-2→5-8
  screen ing large-in sert librar ies, by hy brid iza tion, 5-8→5-19
LCP, 1-29, 1-43
Lentiviral vec tors, high-ti ter
  HIV-1-based, pro duc tion of, 12-43→12-47
  ti tra tion of stocks
    GFP vec tor, 12-47→12-48
    LacZ, 12-47→12-49
Leukemias, DNA re ar range ments in, 10-10→10-23
Li braries, large-in sert
  BAC/PAC clones
    prep a ra tion us ing pCYPAC2, pPAC4, or PBACe3.6 vec tor, 5-32→5-36
    prep a ra tion of vec tor for clon ing, 5-36→5-38
    con struc tion, in BACs and PACs, 5-32→5-38
    screen ing by hy brid iza tion
      ar ray ing col o nies and DNA at low and high den sity, 5-8→5-14
      pre par ing BAC and PAC col ony blots, 5-14→5-15
      screen ing BAC, PAC, and P1 librar ies by col ony hy brid iza tion, 5-15→5-17
Ligase chain re ac tion (LCR), 2-14→2-16
Ligation as says
  LCR, 2-14→2-16

OLA, 2-12→2-214
  prep a ra tion of mod i fied oligonucleotides for, 2-16→2-17
Lin ear re gres sion anal y sis, gene-gene in ter ac tions, 1-96→1-97
Linkage anal y sis
  adop tion study, 1-4
  age-dependent penetrance, 1-49→1-53
  data col lec tion for
    informatics, 1-11→1-12
    stra te gic ap proach to, 1-6→1-11
  heritability, 1-5
  link age maps, homozygosity us ing pooled DNA, 1-71→1-76
  us ing LINKAGE pro grams
    al go rithm for, 1-30 (fig.)
    ex am ples, 1-31→1-55
    li a bil ity classes, 1-28
    loci, types of, 1-28
    lod score, 1-13, 1-27→1-28
    loops, 1-28→1-29
    map func tion, 1-29
    pro grams, 1-29→1-31
  mea sures of, multiallelic tests, 1-84→1-85
  model-free, see Model-free link age analysis
  multipoint, of autosomal dom i nant dis ease by as sess ment of ad di tional mark ers to establish dis ease gene lo ca tion, 1-40→1-49
  ped i gree se lec tion, 1-12→1-14
  re cur rence risk, 1-4→1-5
  seg re ga tion anal y sis, 1-5→1-6
  sim u la tion stud ies, 1-17→1-18, 1-67
  sin gle-sperm typing
    multiplex am pli fi ca tion from single sperm cells, 1-23→1-24
    PCR am pli fi ca tion of ge netic mark ers from sin gle sperm cells, 1-19→1-23
  sperm typ ing
    data anal y sis, 1-24→1-26
    sin gle, 1-19→1-24
  study de sign
    as cer tain ment bias, 1-7, 1-13→1-15
    flow chart, 1-13 (fig.)
    in for ma tion con tent of ped i grees, 1-15→1-17
    linkage equi lib rium, 1-17→1-18
    sam ple, 1-11 (fig.)
    sim u la tion stud ies, sam ple size and power de ter mi na tion, 1-18
    stra te gic ap proach, 1-12→1-14
  twin anal y sis, 1-4
  two-point
    autosomal dom i nant dis ease with un linked marker, 1-31→1-40
    X-linked re ces sive dis ease, 1-54→1-55
LINKLODS, 1-31, 1-40, 1-43
LINKMAP, 1-31, 1-36, 1-37 (ta ble), 1-40, 1-43
Lipofection, of mam ma lian cell, with YAC DNA, 5-23→5-24
Lipopolysaccharide (LPS), chro mo some prep a ra tion for karyotyping, 4-7
Liposome vec tors, in vivo gene de liv ery, 12-74→12-75
Liver, gene de liv ery to
  har vest ing for transgene ex pres sion anal y sis, 13-61→13-62
  to mouse liver by tail vein in jec tion, 13-60
  to rab bit liver by pe riph eral ear vein in jec tion, 13-60→13-61
  trou ble shoot ing guide, 13-59 (ta ble)
Lo ca tion Da ta Base, ge netic maps, A2-23
LocusLink, 11-27
Locus-specific mu ta tion da ta bases (LSDBs), 7-22 (ta ble), 7-23, 7-27→7-29
Lod score, 1-13, 1-27→1-28
LODSCORE, 1-36, 1-37 (ta ble)
Lo gis tic re gres sion anal y sis, gene-gene interactions, 1-92→1-93

Long-interspersed repeated sequence, type of IRS, A2-2
Long-range restriction mapping
basic protocol, 5-2→5-4
preparation of BAC DNA, restriction digestion, and CHEF gel analysis, 5-7→5-8
preparation of high-molecular-weight in agarose blocks
mammalian genomic DNA, 5-4→5-5
standards using S. Cerevisiase chromosomes and YACs, 5-5→5-7
YAC purification, 5-27→5-28
Long tandem repeats (LTRs), paternity testing, 9-31
Low-complexity regions, filtering sequence databases for, 6-23, 6-32
Lung, in vivo gene delivery, 13-57→13-58
Lymph nodes
chromosome preparation from, 10-6
specimens for cytogenetic analysis, 10-9
Lymphoblastoid cells, isolation in preimplantation genetic diagnosis, 9-69→9-71
Lymphocytes
chromosome-breakage analysis, 8-34 (table)
free chromatin preparation, high-resolution FISH, 4-35→4-36
isolation, in preimplantation genetic diagnosis, 9-69→9-71
large-insert cloning, DNA preparation from, 5-38→5-40
replication banding, 4-18→4-19
RNA amplification from, 7-2→7-5
transformation with EBV, A3-34→A3-36
Lymphoma, detection of chromosomal translocations by PCR, 10-22
Lysis
modified alkaline, 5-44→5-45
single-sperm typing, 1-21→1-23

M

MacVector, 2-1
Magnetic beads
in CD34⁺ isolation, 13-36→13-37
enzymatic removal, immunomagnetically selected cells, 13-36→13-37
MAKEPRED, 1-29, 1-31→1-36
Mammalian cell tissue culture
cell number determination, A3-31→A3-33
freezing, A3-29→A3-30
media preparation, A3-27→A3-28, See also APPENDIX 1
sterile technique, A3-26→A3-27
thawing and recovering, A3-30→A3-31
transportation, A3-33→A3-34
trypsinizing and subculturing, A3-28→A3-29
Mammalian DNA, preparation for PFGE, 5-4→5-5
MAPFUN, 1-40, 1-46→1-47
Map function, 1-29
Maps and mapping
data collection, clinical and epidemiological, for linkage studies, 1-3→1-12
direct visualization hybridization, 4-36→4-37
high-resolution
by DIRVISH, 4-37→4-40
FISH, 4-36
Mendelian traits
allelic association, 1-3
heterozygosity (HET), 1-1
simple vs. complex, 1-1, 1-3
phenotype vs. genotype, 1-1
polymorphisms, 1-3
positional cloning, 1-2 (fig.)
restriction, long-range, 5-2→5-8
single-sperm typing, 1-24→1-25

Markers
alleles, multiple, 1-83→1-85
analysis, for somatic cell hybrid analysis, 3-8→3-9
haplotypes, association tests, 1-86
loci for disease association, 1-76
retroviral vectors
construction process, 12-30→12-35
rescue assay, 12-38→12-39
single-sperm typing, 1-19→1-23
for somatic cell hybrids, 3-2→3-3
YAC inserts, tagging, 5-20→5-21
Marshfield Medical Research Foundation, genetic maps, A2-22→A2-23
Maximum-likelihood
segregation analysis method, 1-5→1-6
single-sperm typing, 1-24→1-25
maxTDT statistic, 1-85→1-86, 1-89
MDS, see Myelodisplastic syndrome
Mean, in statistics, 1-59
Means test, ASP linkage studies, 1-60, 1-66 (fig.)
Mechanical disruption of tissues for metaphase analysis, 8-14→8-15
Media
bacterial, additives for, A1-17 (table)
mammalian cells, A3-27→A3-28, See also APPENDIX 1
Medication history, sample optical scanning form, 1-10 (fig.)
MEDLINE, 6-14→6-15, 6-17, 6-63
MENDEL, 1-25→1-26
Mendelian inheritance, characterized, 1-1
MER repeats, A2-2→A2-3
Metabolic disorders, linkage analysis, 1-6
Metacentric chromosomes, A3-37
Metaphase analysis, of solid tissue, 8-14→8-16
Metaphase chromosomes
analysis of sister-chromatid exchanges, 8-18→8-22
preparation for
comparative genomic hybridization, 4-67→4-68
karyotyping, 4-7→4-13
Metaphase spreads
analysis of
malignant hematological specimens, 10-2→10-7
sister-chromatid exchanges, 8-17→8-22
solid tumor cultures, 10-7→10-10-8
preparation from
amniotic fluid cultures, 8-5→8-13
chorionic villus samples (CVS), 8-2→8-5
products of conception, 8-14→8-17
tissue biopsy, 8-17
Methylation-specific PCR (MSP), determination of
DNA methylation patterns, 10-30→10-33
methylation of CpG sites within PCR products, 10-33→10-34
M-FISH, characteristics of, 4-43→4-45
Microarrays
cDNA, profiling human gene expression with
agarose gel electrophoresis of ESTs, 11-20→11-21
cDNA amplification and printing, 11-12→11-16
fluorometric determination of DNA concentration, 11-21→11-22
hybridization and data extraction, 11-19→11-20
RNA extraction and labeling, 11-16→11-18
SNP genotype using
allele-specific, 2-25→2-30
generic, 2-30→2-33
Microsatellite repeats, A2-1
Microscopy, comparative genomic hybridization (CGH), 4-69→4-70

Minisatellite repeats, A2-1
Mitochondrial DNA (mtDNA), mutations
detection using mismatch oligonucleotide, 9-53 (fig.)
polymorphisms encountered with EcoRV and BamHI, 9-50 (table)
probes for detecting, 9-51→9-52
rearrangements by Southern blot hybridization, 9-46→9-50
restriction analysis of PCR products, 9-52→9-58
summary of, 9-47→9-48 (table)
testing parameters, 9-54→9-57
Mito database, for nucleotide sequence, 6-25
Mitotic chromosomes
chromosome preparation from
bone marrow, 4-12, 10-2→10-4, 10-6→10-7
fetal liver, 4-10→4-11
leukemic blood specimens, 10-2→10-4, 10-6→10-7
lymph node specimens, 10-6
peripheral blood, 4-7→4-8
solid adult tissues or tumors, 4-13
spleen, 10-6→10-7
collecting blood by tail vein method, 4-10
setting up timed mouse matings, 4-12
MLINK, 1-29, 1-31, 1-37, 1-43, 1-46
MM, see Multiple myeloma
Model-free linkage analysis
characterized, 1-55
examples of
affected sib pairs (ASP), using SIBPAL, 1-61→1-67
affected sib pairs (ASP), without computer analysis, 1-55, 1-60→1-61
identity by state (IBS) and identity by descent (IBD), 1-55→1-61, 1-67
GENEHUNTER and GENEHUNTER PLUS, 1-55, 1-57, 1-59, 1-67→1-71
SimIBD, 1-55
decision tree, 1-56 (fig.)
hypothesis testing, 1-59→1-1-60
vs. model-dependent, vs. nonparametric tests, 1-57
statistical terms: mean, variance skewness, and kurtosis, 1-59
Model-free tests
for genetic linkage
Haseman-Elston (H-E), 1-56 (fig.)
identity by descent (IBD), 1-55→1-61, 1-67
identity by state (IBS), 1-55→1-59
vs. model dependent vs. nonparametric tests, 1-57
Models, for the human airway system, 13-44→13-45
Molar extinction coefficients of DNA bases, A3-10 (table)
Monolayer cells
freezing, A3-29→A3-30
polarized airway epithelial, 13-49→13-51
primary human marrow stromal, 13-37
in somatic cell hybrid construction, 3-4→3-5
thawing and recovery, A3-30→A3-31
trypsinizing and subculturing, 3-28→A3-29
Monolayer cultures, cytogenetic analysis of metaphase cells from solid tumor, lymphoma, or effusion samples, 10-7→10-8
Monovalent cations, used to precipitate DNA, A3-5 (table)
Monozygous (MZ) twins, in linkage analysis, 1-4
Morphology antibody chromosome (MAC), analysis of phenotype and genotype on the same cell
combinations of preparative and analytical techniques, 4-49 (table)

genotyping techniques
   chromosome banding, G- or C-,
      4-56→4-57
   in situ hybridization, of previously
      GTG-banded chromosomes,
      4-62→4-64
   listing of, 4-50 (table)
   overview, 4-48
   phenotyping techniques
      fluorescence detection, 4-55→4-56
      HRP-based detection, 4-53→4-55
      listing of, 4-49 (table)
      sequential using APAAP immunostaining
        and in situ hybridization, 4-48,
        4-50→4-53
   specimen preparation
      blood and bone marrow smears,
        4-61→4-62
      cytospin slides, 4-57→4-58
      in situ cultures, 4-59→4-60
      tissue sections, 4-60→4-61
Mouse
   adenovirus delivery via tail vein injection,
      13-60
   cytogenetic analysis
      chromosome preparation, 4-1→4-14
      gene transfer models, 13-9→13-21
   embryos, introduction of BAC/PAC into,
      5-29→5-30
   human bronchial xenografts to,
      13-55→13-56, 13-58→13-59
Mouthwash, DNA extraction in ARMS test,
   9-44
mRNA
   amplification and labeling for hybridization
      to oligonucleotide array chips,
      11-1→11-5
   database submissions, 6-19
   expression profiling, RNA extraction and
      labeling, 11-16→11-18
   sequence analysis, BLAST search re-
      sults, 6-33
M13 sequence ladder, preparation of,
   2-10→2-11
Multifactor dimensionality reduction (MDR),
   1-93→1-96
Multiple myeloma (MM)
   choice of culture method, 10-2 (table)
   preparation of chromosome spreads,
      10-4, 10-6
Multiple point mutations, simultaneous detection using ASOs, 9-12→9-16
Multiplex analysis of SSLPs, using
   nonradioactive techniques, 2-6→2-10
Multiplex ARMS test, 9-43→9-44
Multiplex PCR, see Polymerase chain reaction
Multipoint linkage analysis, 1-40→1-49
Murine studies, see Mouse
Muscle
   gene transfer in, 13-13→13-21
   infection of primary myoblasts with retro-
      virus, 13-15→13-16
   intramuscular injection of vaccine
      anterior tibialis, 13-20→13-21
      quadriceps, 13-19→13-20
   isolation
      and growth of mouse primary
        myoblasts, 13-13→13-15
      of LacZ-labeled cells by fluores-
        cence-activated cell sorting
        (FACS), 13-18→13-19
   skeletal
      myoblast implantation into,
        13-16→13-17
      plasmid DNA injection into, 13-18
MuStaR, 7-29
Mutation analysis, multiple point mutations,
   simultaneous detection using ASOs,
   9-12→9-15
Mutations
   Paternity Index, 9-34→9-37, 9-38 (table)

See also Deletions; Point mutations; Rearrangements
MUTbase, 7-29
Myelodisplastic syndrome (MDS), choice of
   culture method, 10-2 (table)
Myoblasts
   implantation into skeletal muscle,
      13-16→13-17
   infection with retrovirus, 13-15→13-16
   isolation and growth without cells sorting,
      13-13→13-15
   purification with cell sorting, 13-15
Myotonic dystrophy (DM), trinucleotide repeats
   classification system, 9-24 (table)
   CTG trinucleotides, 9-22→9-23
   hybridization analysis of
      CTG-PCR products, 9-21→9-23
      genomic DNA, 9-24→9-26
   radioactive detection, 9-24
   rapid transfer of PCR product using a
      vacuum blotter, 9-23
MZEF, in gene identification, 6-5, 6-12→6-13

N

Naphtol-AS-MX-phosphate, as substrate for
   ISH, 4-29 (table)
National Center for Biological Information
   (NCBI)
   annotations, 6-47 (table)
   genomic DNA, 6-61→6-74
   Human Genome Resources, 6-53, 6-54
      (table)
   human sequence data
      customized display, 6-55→6-56
      downloading sequence, 6-56
      FTP format, 6-56
      gene model evidence, 6-56
      information retrieval, 6-57
      overview of, 6-53
      queries, 6-53→6-55, 6-57→6-60
   LocusLink, 11-27
   Molecular Modeling Database (MMDB),
      6-16
   sequencing databases
      for use in BLAST searches,
        6-22→6-27, 6-29, 6-37
      contig sequences, 6-60→6-61
      generally, 6-14
   Sequin, 6-13, 6-17, 6-19
   web site services, 6-48 (table)
National Human Genome Research Institute
   (NHGRI), 7-27
NCBI, see National Center for Biotechnology
   Information
neo gene, 5-21 (table), 5-25 (fig.), 5-27,
   12-31, 12-33 (fig.), 12-51→12-53
Nick translation, labeling
   ISH probes, 4-31→4-32
   standard procedures, A3-12→A3-13
Nitro-blue tetrazolium (NBT), as substrate for
   ISH, 4-29 (table)
Non-Hodgkin lymphoma, molecular analysis
   of DNA rearrangements in
   detection of clonal antigen-receptor gene
      rearrangements by Southern blot
      hybridization, 10-11→10-14
   overview of, 10-10→10-11
Nonparametric analysis of linkage data, 1-12
Normalization, for expression analysis,
   11-23→11-24
Northern blot hybridization, A3-23→A3-26
Nuclear families, in linkage studies, 1-14,
   1-16→1-17, 1-90
Nucleotide, sequence databases, 6-24→6-25
Null hypothesis, linkage analysis, 1-59→1-60

O

OLIGO, 2-1
Oligogenic disease, 1-1

Oligonucleotide arrays for expression monitoring
   amplification of mRNA for expression
      monitoring and hybridization to
      oligonucleotide array chips
      basic protocol, 11-1→11-5
   data reduction, normalization, and quality
      assessment, 11-9, 11-11
   in vitro transcription of control genes and
      preparation of transcript pools,
      11-5→11-7
   quantitation of cDNA, 11-8
   solid-phase reversible immobilization puri-
      fication of cDNA and in vitro
      transcription products, 11-7→11-8
   troubleshooting guide, 11-10→11-11 (ta-
      ble)
Oligonucleotide ligation assay (OLA),
   2-12→2-14
Oligonucleotides
   allele-specific (ASO), 9-12→9-16, 9-31
   for hybridization, parparation with
      overgo oligonucleotide probes,
      5-17→5-19   labeling
      nick translation, A3-12→A3-13
      random-primed synthesis,
        A3-13→A3-14
      with T4 polynucleotide kinase, A3-14
   for ligation assays, 2-16→2-17
   overgo, as hybridization probes,
      5-17→5-19
OMIM, see Online Mendelian Inheritance in
   Man
Oncogenes, in cancer, 10-1
One-color array, for transcriptional profiling,
   11-23→11-24
Online Mendelian Inheritance in Man (OMIM),
   6-15, 7-25→7-27
Ontology analysis of expression, 11-27
Open reading frames (ORFs), database sub-
   missions, 6-19→6-20
Optical scanning, sample form, 1-10 (fig.)
Ouabain, as selectable marker, 3-2
Oxidative phosphorylation (OXPHOS) dis-
   eases, detection of mtDNA mutations
   probe preparation using long-range PCR,
      9-51→9-52
   screening for mitochondrial DNA
      point mutations by restriction analysis
        of PCR products, 9-52→9-58
      rearrangements by Southern blot
        hybridization, 9-46→9-50

P

PAC, see Bacterial artificial chromosome
Packaging, retroviral cell lines, 12-31 (table)
PAGE, see Polyacrylamide gel
   electrophoresis
PAM matrices, for BLAST sequence similarity
   searches, 6-34
Paraffin-embedded tissue, DNA preparation
   from, A3-7→A3-8
Paraffin-embedded tissues
   nuclear suspensions for FISH, 8-22→8-25
   sections, 8-25→8-26
Parametric analysis of linkage data, 1-12
Parent-offspring relationship, 1-5
Partial digestion of genomic DNA, 5-40→5-44
Passaging cells, amniotic fluid cultures,
   8-10→8-11, 8-13→8-14
Paternity index (PI), in paternity testing,
   9-34→9-37, 9-38 (table)
Paternity testing
   analysis by polymorphic loci by PCR,
      9-30→9-33
   fatherless cases, 9-36
   interpretation, statistical evaluation, and
      reporting of DNA profiles,
      9-34→9-38
   motherless cases, 9-36

prepa ra tion of genomic DNA from whole
blood, 9-33→9-34
prior prob ability, 9-38
related alleged father, 9-37
single exclusions, 9-37→9-38
twin typing, 9-36→9-37
VNTR analy sis, by RFLP tech nol ogy,
9-28→9-30
Path way analy sis of ex pression, 11-27
Pa tient his to ries, sam ple forms, 1-8→1-10
(fig.)
Pax genes, 6-58
pBACe3.6 vec tor, 5-32→5-35
PCR, *see* Polymer ase chain re ac tion
pCYPAC2 vec tor, 5-32→5-35
pdb databases, for nu cle o tide se quences,
6-25
Pedigree
analy sis, for link age stud ies
autosomal dom i nant dis ease,
1-5→1-17, 1-33 (fig.), 1-41
(fig.), 1-51 (fig.), 1-53 (figs.)
us ing GENEHUNTER PLUS, 1-68 (fig.)
iden tity-by-de scent (IBD), 1-58 (fig.)
iden tity-by-state (IBS), 1-58 (fig.)
us ing SIBPAL, 1-63 (fig.)
TDT ap plications, 1-90
X-linked re ces sive dis ease, 1-54→1-55
extension, 1-17
se lec tion and in for ma tion con tent, for linkage stud ies, 1-12→1-14, 1-17
Ped i gree dis equi lib rium test (PDT), 1-90
Penetrance
age-de pendent, 1-49→1-53
gene-gene in ter ac tion, 1-91→1-92, 1-94
(table)
re duced, 1-6, 1-32 (ta ble)
PEP, *see* Primer-ex ten sion preamplification
Pep tide se quence da ta bases, for BLASTP
and BLASTX searches, 6-24
Peripheral blood
for chro mo some prep a ra tions from
chronic lym pho cyte leu ke mia, 1-4
mul ti ple myeloma bone mar row, 1-4,
1-6
lympho cytes, chromosome breakage,
8-34 (ta ble)
Perma nent cell lines, es tab lished by EBV
transformation, A3-34→A3-37
Phase-con trast mi cros copy
chro mo some prep a ra tion from bone
mar row, 4-12
chro mo some spreading, 4-5
sin gle-sperm typ ing, 1-21
sis ter-chromatid ex changes, 8-20→8-21
Phe nol ex trac tion of DNA, A3-4→A3-5
Phe no type, de fined, 1-1
Phenotyping, MAC analy sis, 4-49 (ta ble),
4-53→4-56
Phytohemagglutinin (PHA)
chro mo some prep a ra tion for karyotyping,
4-7→4-8
replication banding, 4-18
Plaque as say, adenovirus, 12-18→12-19
Plasmid DNA, di rect in jec tion into mus cle for
gene de liv ery, 13-18
Pleural ef fu sions, preparing chromosome
slides of, 4-4
Point mu tations
analy sis by ARMS test, 9-38→9-45
characterized, 9-38→9-39
mtDNA, screen ing by re stric tion analy sis
of PCR prod ucts, 9-52→9-58
multiple, si mul ta neous de tec tion us ing
ASOs, 9-12→9-15
Poke weed mitogen
choice of cul ture method, 10-2 (ta ble)
chro mo some prep a ra tion for karyotyping,
4-7
Polyacrylamide gel elec tro pho re sis (PAGE)
dena tur ing PAGE, A3-16→A3-17

DNA mi gra tion in PAGE, A3-18 (ta ble)
Poly-L-lysine, coat ing slides with, 11-22
Poly mer ase chain re ac tion (PCR)
amplification
of RNA, 7-2→7-5
SSLPs, 2-2→2-2-4, 2-17→2-19
in X chro mo some in ac ti va tion,
9-27→9-28
APOE geno typ ing by
fluorescence polarization, 9-86→9-87
reverse hy brid ization, 9-85
RFLP analy sis, 9-84
SNaP shot analy sis, 9-88
ARMS test for sin gle point mu ta tions,
9-41→9-43
cDNA am pli fi ca tion, 11-12→11-15
char ac ter iza tion of cell hy brids, 3-8,
3-10→3-13 (ta ble)
chro mo so mal translocations, de tec tion
by, 10-22
de gen er ate oligonucleotide primed
(DOP), 4-42
for de tect ing mutations
CE-SSCP, 7-10→7-13
heterozygote mu ta tion de tec tion,
7-17→7-18
SSLPs, 7-6→7-8
dif fer ential, 10-28→10-29
DNA prep a ra tion from par af fin-em bed ded
tis sue, A3-7→A3-8
for fragile X syn drome as sess ment,
9-16→9-18
gene re ar range ment de tec tion,
10-16→10-18
in geno typing
characterized, 2-1
4-plex, 2-39→2-41
sin gle-plex, 2-37→2-39
hematopoietic cells for, 13-39→13-40
inverse, clonal in te gra tion analy sis,
13-40→13-43
labeling, in X chro mo some in ac ti va tion,
9-27→9-28
methylation-spe cific PCR prod ucts,
10-33→10-34
mtDNA screen ing by re stric tion analy sis,
9-52→9-58
mul ti ple point mu ta tions analy sis,
9-14→9-15
multiplex
ARMS test, 9-38→9-45
for dystrophin gene de le tions,
9-2→9-11
sin gle-sperm typ ing, 1-23
SSLP amplification, 2-6→2-10, 2-31
for myo tonic dys tro phy as sess ment,
9-21→9-22, 9-24→9-26
pa ter nity test ing, 9-30→9-33
preimplantation ge netic di ag no sis (PGD),
spe cific sin gle-copy gene loci from
single diploid cells, 9-59→9-63
rep li ca tion-de fec tive HSV vec tor con struc-
tion, 12-52→12-53, 12-57, 12-61
in sin gle-sperm typ ing, 1-19→1-21
SSLP genotyping
us ing end-la beled prim ers, 2-2→2-3
us ing in ter nal la bel ing, 2-4
us ing nonradiative mul ti plex analy sis,
2-6→2-10
us ing nonradioactive sil ver stain ing,
2-4→2-6
pooled DNA, 1-74→1-76
ther mal cy cling pa ram e ters, 2-3 (ta ble)
ther mal cy cling for
cDNA am pli fi ca tion, 11-15
SSLPs, am pli fi ca tion of, 2-3 (ta ble)
Poly mor phic loci analy sis in pa ter nity test ing,
by PCR, 9-30→9-33
Polymorphism, characterized, 1-3
Pooled DNA
am pli fi ca tion and analy sis of, 1-74→1-76
tech niques, over view of, 1-71→1-73

Population-based con tin gency statistic
(PBCS), 1-78→1-80
Pop u la tion struc ture, ef fect on analy sis in
disease as so cia tions, 1-79→1-80
Por cine mod els of gene trans fer to ar ter ies
iliac, stented, 13-5, 13-6 (fig.)
iliofemoral, 13-3→13-5
Po si tion-spe cific it er ated BLAST, *see*
PSI-BLAST
Po si tional clon ing, flow chart for, 1-2 (fig.)
Po tas sium ac e tate, to pre cip i tate DNA, A3-5
(table)
PowerBLAST, in gene iden ti fi ca tion, 6-13
Power de ter mi na tion, in link age analy sis,
1-18
Power pro gram, 1-92→1-93
pPAC4 vec tor, 5-32→5-35
Pre cip i ta tion of DNA
ethanol, A3-5→A3-6
high-salt, A3-2→A3-3
monovalent cat ions, A3-5 (ta ble)
Preimplantation ge netic di ag no sis (PGD)
em bryo bi opsy, 9-64→9-68
FISH analy sis, 9-71→9-72
probe val i da tion, 9-73
sin gle blastomeres
FISH analy sis, 9-71→9-73
iso la tion from af fected em bryos,
9-68→9-69
sin gle dip loid cells
analy sis of spe cific gene loci, 9-59,
9-61→9-63
whole-genome amplification by
primer-extension
preamplification (PEP), 9-64
sin gle lym pho cytes/lymphoblastoid cells,
9-69→9-71
Prenatal studies and di ag no ses
amniocentesis, 8-36→8-37
cho ri onic villus sam ples (CVS), 8-2→8-5,
8-17
of chro mo somal aneuploidy, 8-22→8-26
clin i cal guide lines for interphase FISH,
8-26→8-31
interphase analy sis of amniocytes, 8-28
prep a ra tions for interphase FISH
amniocytes at tached to a sur face, 8-28
un cultured amniocytes, 8-26→8-28
PREPLINK, 1-29, 1-52, 1-54→1-55
PRIMER, 2-1
Primer-ex ten sion preamplification (PEP),
sin gle-diploid cells
analy sis of spe cific gene loci, 9-59,
9-61→9-63
whole-ge nome am pli fi ca tion by, 9-64
Primer-ex ten sion preamplification (PEP),
sin gle-sperm typ ing, 1-23→1-24
Primers
am pli fi ca tion, dystrophin gene exons,
9-9→9-11 (ta ble)
ARMS, 9-38→9-39, 9-40 (ta ble),
9-42→9-43
biotinylated, for li ga tion as says,
2-16→2-17
di ag nos tic mul ti plex mixes, 9-7→9-9
high-through put genotyping, 2-37→2-41
for li ga tion as says, 2-16→2-17
mul ti plex ARMS test, 9-43
for mul ti plex PCR re ac tions, 9-5
for PCR
cy cle se quenc ing, 7-21
heterozygote mu ta tion de tec tion,
7-17→7-18
RNA am pli fi ca tion, 7-2→7-3
YAC DNA de tec tion , 5-21
for T cell re cep tor-γ-chain gene re arrange ments, 10-17 (fig.)
for t(9;22), 10-19 (fig.)
Print ing, cDNA microarrays, 11-12→11-16
Probes
for DIRVISH, 4-40→4-42

gene rearrangement, detection of, 10-12 (table)
hybridized, detection of, 4-24 (fig.)
interphase FISH, 8-29→8-31
for ISH, 4-28→4-32
labeled DNA, for comparative genomic hybridization, 4-68→4-69
labeling methods, A3-12→A3-16
for northern blot hybridization, A3-24→A3-25
overgo oligonucleotide, 5-17→5-19
for PCR, TasqMan genotyping, 2-34→2-35
preimplantation genetic diagnosis, 9-73
for Southern blot hybridization, A3-19→A3-23
stripping, mutation analysis, 9-15→9-16
for YAC DNA detection, 5-21 (table)
PROCRUSTES, in gene identification, 6-9→6-10, 6-12
Proportions test, ASP linkage studies, 1-60
Protective clothing and equipment, 12-3→12-4
Protein truncation test (PTT)
isolation and analysis of RNA, 9-79→9-81
nonradioactive, 9-74→9-79
radioactive using coupled transcription/translation, 9-81→9-82
PSI-BLAST
sequence alignment algorithms, 6-35
sequence similarity searches, 6-32→6-33
PubMed database, 6-15→6-16, 6-29, 6-65→6-66 (figs.), 7-26
Pulsed-field gel electrophoresis (PFGE), genomic mapping, 5-2→5-8
Purification
of adenovirus-free rAAV using heparin Sepharose column purification, 12-25→12-27
high-titer adenoviruses, 12-16→12-18
solid-phase reversible immobilization (SPRI), 11-7→11-8
of YAC DNA
by lipofection, 5-23→5-24
PFGE, 5-27→5-28
Puromycin, 12-52→12-53
PWM, see Poke weed mitogen

Q

Quality control, mutation detection databases, 7-25
Quantitation
cDNA, solid-phase reversible immobilization (SPRI) purification of, 11-8
DNA and RNA, A3-8→A3-12
Quantitative traits, association studies, 1-96→1-97
Quinacrine banding (Q-banding) of chromosomes, 4-15→4-16

R

Rabbit
adenovirus delivery to, 13-60→13-61
models of gene transfer to atherosclerotic iliac arteries, 13-6→13-9
Radiation hybrid (RH) map, 6-39
Radioactivity, measuring, by TCA precipitation, A3-15→A3-16
Radioisotopes
physical characteristics, A2-24 (table)
shielding requirements, A2-24 (table)
Random hexamer priming, A3-13→A3-14
Random oligonucleotide-primed synthesis, see Random hexamer priming
Rapid transfer, of PCR product using vacuum blotter, 9-23
Rats, gene delivery to brain
genetically modified cells for grafting, 13-29

implantation of genetically modified fibroblasts into rat brain
adult, 13-23→13-24, 13-27→13-28
fetal, 13-26→13-27
neonatal, 13-25→13-26
Reagents and solutions, A1-1→A1-32
Recessive disease, X-linked, linkage analysis, 1-54→1-55
Recombinant adeno-associated viral (rAAV) vectors
adenovirus-free, production of, 12-21→12-25
in vitro infection, 12-29
packaging, 12-23 (fig.)
titer determination, 12-27→12-28
Recombinant adenoviral vectors, generation of using
AdEasier cells, 12-15→12-16
AdEasy method, 12-12→12-15
Recombination, homologous, modifying YACs, 5-24→5-27
Region, chromosome, defined, A2-4
Rehybridization, mutation analysis, 9-15→9-16
Repetitive DNA, human, A2-1→A2-3
Replica plating, YAC clones, 5-9→5-11
Replication banding of chromosomes
B-pulse for
fibroblasts, 4-19
lymphocytes, 4-18
BrdU, visualization by
fluorescent dye, 4-20
heat treatment and Giemsa staining, 4-21→4-22
light treatment and Giemsa staining, 4-21
T-pulse for
fibroblasts, 4-19→4-20
lymphocytes, 4-19
Reporter genes, for a model airway system, choosing, 13-46
Reporter molecules, for ISH, 4-25 (fig.)
Restriction enzymes
gene rearrangement detection, 10-12 (table)
cleavage sites and reaction temperatures for, 5-3 (table)
PCR product analysis, 10-33→10-34
with rare cleavage specificities, 5-3 (table)
Restriction fragment length polymorphism (RFLP)
APOE genotyping, 9-84→9-85
paternity testing, 9-28→9-30
Restriction mapping
fragile X region, 9-20 (fig.)
long-range
basic protocol, 5-2→5-4
preparation of BAC DNA, restriction digestion, and CHEF gel analysis, 5-7→5-8
preparation of high-molecular-weight DNA in agarose blocks
mammalian genomic DNA, 5-4→5-5
standards using S. cerevisiae chromosomes and YACs, 5-5→5-7
Retroviral vectors
amphotropic, 12-31 (table), 12-32 (table)
applications, 12-29
calcium phosphate-mediated transfection, 12-36→12-37
ecotropic, 12-31 (table), 12-32 (table)
generation of
without selectable markers, 12-35→12-36
with selectable markers, 12-30→12-35
host range of, 12-31 (table)
marker rescue assay, 12-38→12-39
packaging cell lines, 12-31 (table)
production by transient transfection, 12-36
pseudotypes
applications, 12-32 (table), 21-39→12-40

from stable producer cells, 12-42→12-43
by transient transfection, 12-40→12-42
staining for alkaline phosphatase activity, 12-39
titer vectors carrying selectable markers, 12-37
virus production from plasmids, 12-30 (fig.)
Retrovirus
high-titer lentiviral vectors, 12-43→12-49
myoblast infection for gene delivery to muscle, 13-15→13-16
packaging cell lines, 12-31 (table)
transduction to
human bronchial xenografts, 13-56→13-57
polarized airway epithelia, 13-51→13-52
primary airway epithelial cells with, 13-48→13-49
Reverse hybridization, APOE genotyping, 9-84→9-85
Reverse transcriptase PCR (RT-PCR)
chromosomal translocation detection, 10-18→10-20
replication-defective HSV vector construction, 12-61→12-62
Reverse transcription, RNA extraction and labeling, 11-16
RFLP, see Restriction fragment length polymorphism
RNA
amplification, from lymphocytes, 7-2→7-5
fractionation, A3-23→A3-26
isolation
and analysis for PTT, 9-79→9-81
by rapid guanidium method, 10-20→10-22
northern blot hybridization, A3-23→A3-26
quantitation, A3-8→A3-12
Robotic devices, replica plating, 5-11→5-14
Rotors
conversion tables and nomograms, A2-27→A2-28
maximum radii, A2-25→A2-26
Rx-FISH, characteristics of, 4-46→4-47

S

Saccharomyces cerevisiae, high-molecular-weight DNA, preparation for PFGE, 5-5→5-7
Safety issues
BL2 laboratory, 12-2→12-5
spills, 12-4
SAGE, see Statistical Analysis for Genetic Epidemiology
Samples
in linkage analysis, 1-4, 1-18
single-sperm typing, 1-24
Satellite DNA repeats, A2-3
Screening, retroviral-mediated transduction of hematopoietic cells, 13-38
Segregation distortion, single-sperm cell analysis, 1-26
Selectable markers
in retroviral vectors
construction of, 12-30→12-35
titer, 12-37
for somatic cell hybrids, 3-2→3-3
YACs modified by homologous recombination, 5-24→5-27
Sequence alignment algorithms, for BLAST searches, 6-34→6-36
Sequence analysis
identifier syntax for, 6-37→6-38
strategies, for BLAST sequence similarity search
BLAST parameters, 6-36→6-37
filtering for human repeats, 6-33
filtering for low-complexity regions, 6-32

for genomic DNA sequences, 6-33→6-34
for mRNA sequences, 6-33
short sequences, 6-34
in transcriptional profiling, 11-26→11-27
Sequence databases
intro to Entrez, 6-14→6-16
submitting data
to existing Genbank entry, updates/corrections, 6-20→6-21
expressed-sequence tags (ESTs), 6-18, 6-21
general considerations, 6-16→6-17
genome survey sequences (GSSs), 6-18, 6-21
high-throughput genome sequences (HTGSs), 6-21
instructions and tips for preparing sequence submissions, 6-18→6-20
methods of, 6-17→18
sequence-tagged sites (STSs), 6-18, 6-21
Sequence identifier syntax for BLAST searches, 6-37→6-38
Sequence searching algorithms, defined, 6-39
Sequence similarity searching using BLAST programs, 6-22→6-38
Sequence-tagged sites (STSs)
database for nucleotide sequences, 6-25, 6-33
defined, 6-40
genomic DNA analysis, 6-33→6-34
sequence database data submissions, 6-14, 6-18, 6-21
Sequin, see National Center for Biotechnology Information, Sequin
Short sequences, BLAST searches, 6-34
Short tandem repeat polymorphic markers (STRPS), see Simple sequence length polymorphism
Short-tandem repeats (STRs), paternity testing, 9-32
Shuttle vectors, 12-8, 12-9 (fig.)
Sibling transmission/disequilibrium test (S-TDT)
characterized, 1-86→1-89
combining data with transmission/disequilbrium tests, 1-88→1-89
pedigrees, 1-90
validity, as test of disease associations, 1-89
SIBPAL, 1-59, 1-61→1-67
Silver staining, for SSLP genotyping by PCR, 2-4→2-6
Simple sequence length polymorphism (SSLP)
characterized, 1-73
genotyping, by PCR
using allele-specific microarrays, 2-25→2-30
basic protocol, 2-17→2-19
using generic microarrays and SBE, 2-30→2-33
strategic planning, 2-24→2-25
TaqMan assay, 2-33→2-36
pooled DNA samples
amplification and analysis of, 1-74→1-76
homozygosity mapping of, 1-72 (fig.)
Simulation studies in linkage analysis, 1-17→1-18, 1-67
Single-base extension, for SNP genotype, 2-30→2-33
Single nucleotide polymorphisms (SNPs)
databases, 7-22→7-23, 7-26→7-27
defined, 6-39
disease association, 1-76, 6-58
gene-gene interactions, 1-91→1-92
genotyping

using allele-specific microarrays, 2-25→2-30
using generic microarrays and single-base extension, 2-30→2-33
using TaqMan assay, 2-33→2-36
Single-sperm typing
applications, 1-19
data analysis, 1-24→1-26
PCR amplification of genetic markers
basic protocol, 1-19→1-20
isolation of lysed single sperm cells by FACS, 1-21→1-23
isolation of single sperm cells from agarose films, 1-20→1-21
multiplex amplification from single sperm cells, 1-23
whole-genome DNA amplification from single cells by primer-extension preamplification (PEP), 1-23→1-24
segregation distortion based on haplotype transmission, 1-26
Single-strand conformation polymorphism (SSCP)
mutation detection, basic protocol, 7-6→7-8
mutation detection using capillary electrophoresis
using ABI 310 genetic analyzer, 7-11→7-12
using ABI prism 3100 genetic analyzer, 7-12→7-13
benefits of, 7-8
sample preparation, 7-9→7-11
Sister-chromatid exchanges (SCE)
in mammalian metaphase chromosomes, 8-18→8-22
troubleshooting guide, 8-21 (table)
Skeletal muscle, see Muscle
Slides
aging
with heat, 4-17
with hydrogen peroxide, 4-18
coating with poly-L-lysine, 11-22
preparation for chromosomes protocol, 4-4→4-6
Slot blot hybridization, to detect gene amplification, 10-27→10-28
SNaP shot analysis, APOE genotyping, 9-87→9-88
Software, see Computer programs and Web sites
SOLAR (sequential oligogenic linkage analysis routines), linkage analysis program, 1-56(fig.), 1-57
Solid-phase reversible immobilization (SPRI) purification of cDNA and in vitro transcription products, 11-7→11-8
Solid tissues
adult, mitotic chromosome preparations from, 4-13
disruption and culture for metaphase chromosome analysis
enzymatic, 8-15→8-16
mechanical, 8-14→8-15
Somatic cell hybrids
cell lines used in fusion, 3-4 (table)
characterization of hybrids
G-11 staining of metaphase chromosomes, 3-7→3-8
in situ hybridization, 3-7
marker analysis, 3-8→3-9
PCR primers, 3-8, 3-10→3-13 (table)
chromosome slide preparation, 4-4
colony isolation using cloning cylinders, 3-6→3-7
construction by whole-cell fusion
of monolayer cells, 3-4→3-5
of suspension donor cells, 3-5→3-6
selectable markers for mammalian cells
adenine phosphoribosyltransferase (APRT), 3-3

aminoglycoside phosphotransferase (neo, G418, APH), 3-3
histidinol dehydrogenase (his), 3-3
hygromycin-B-phosphotransferase (HPH), 3-3
hypoxanthine-guanine phosphoribosyltransferase (HPRT), 3-2
oubain, 3-2
thymidine kinase (TK), 3-2
xanthine-guanine phosphoribosyltransferase (XGPRT, gpt), 3-3
subcloning hybrid cell populations, 3-7
Southern blot hybridization
of BAC DNA, 5-30
detection of
gene amplification, 10-23→10-27
gene rearrangements, basic protocol, 10-11→10-13
fragile X syndrome, 9-19→9-20
procedures, A3-18→A3-23
retroviral vector production, 12-34
troubleshooting guide, 10-14 (table)
viral DNA isolation, 12-65
Spectral karyotyping (SKY), characteristics of, 4-43→4-45
Spectrophotometric measurements of purified DNA, A3-10 (table)
S phase of cell cycle, accumulation of cells, 4-3 (fig.)
Spheroplast, fusion of mammalian and yeast cells, 5-19→5-23
Spin-column, to remove unincorporated precursors, A3-15
Spleen, chromosome preparation from, 10-6→10-7
SSAHA, 6-39
Staining
chromosome banding techniques, 4-14→4-23
G-11, of metaphase chromosomes in hybrid cell lines, 3-7→3-8
using silver for SSLP genotyping, 2-4→2-6
single-sperm typing, 1-22→1-23
Statistical Analysis for Genetic Epidemiology (SAGE)
Family Structure Program (FSP), 1-62, 1-64
linkage analysis, 1-57, 1-62
Statistical Analysis System (SAS), linkage analysis, 1-60
Sterile technique, A3-26→A3-27
Streptavidin, enzymatic detection in ISH, 4-30→4-31
Stripping probes, mutation analysis, 9-15→9-16
Stromal cells, in transduction of hematopoietic cells, 13-30→13-35, 13-37
Sub-band, chromosome, defined, A2-4
Subculturing from a monolayer, A3-28→A3-29
Submetacentric chromosomes, A3-37
Subtelomeric repeats, A2-1
Suspension cells, freezing cells in, A3-30
SV40 promoter, 12-31→12-32

T

TAG-SBE genotyping assay, for SNP genotype, 2-30 (fig.)
Tail vein blood collection, 4-10
Taq DNA polymerase, mitochondrial point mutations, 9-51, 9-53, 9-58
TaqMan assay, in high-throughput genotyping, 2-33→2-36
TBLASTN
mRNA sequence analysis, 6-34
sequence alignment algorithms, 6-35

sequence similarity searches, 6-23 (table), 6-31
TBLASTX
   genomic DNA analysis, 6-34
   sequence similarity searches, 6-23 (table)
TCA, see Trichloroacetic acid
T cell receptor genes, PCR, for detecting gene rearrangements, 10-11, 10-16→10-18
TCR genes, see T cell receptor genes
TDT, see Transmission disequilibrium test
Telocentric chromosomes, A3-37
Telomere repeat, A2-1
Temperature, for restriction endonucleases with rare cleavage specificities, 5-3 (table)
Terminal transferase, for labeling probes with digoxigenin, 2-11→2-12
T4 polynucleotide kinase
   to end-label primers, A3-14
   for PCR, 9-24
Thawing
   hematopoietic cells, 13-44
   and recovering human cells, A3-30→A3-31
THE-LTR repeats, A2-2
Thermal cycling, see Polymerase chain reaction
Thin-film hydration, for liposome preparation, 12-74→12-75
THREELOC, 1-24→1-26
Thymidine kinase (TK), as selectable marker, 3-2
Tissue(s), preparation of samples, 8-17
Tissue sections
   for MAC analysis, 4-60→4-61
   preparation for FISH, 8-22→8-26
Titer
   adeno-associated virus
      by dot-blot assay, 12-27→12-28
      by in vitro infection of cells, 12-29
   retroviral vectors, 12-37→12-38
Transcriptional profiling
   computer tools for, 11-9
   expression data analysis, 11-22→11-28
   data analysis
      classification and class prediction, 11-26
      clustering, 11-25→11-27
      identifying differentially expressed genes, 11-24→11-25
      pathway/ontology analysis, 11-27
      sequence analysis, 11-26→11-27
   data characterization/normalization
      one-color array experiments, 11-23→11-24
      two-color array experiments, 11-24
   data reduction, 11-9
   human gene expression with cDNA microarrays, 11-12→11-22
   oligonucleotide arrays for expression monitoring, 11-1→11-11
Transduction
   adeno-associated virus, 12-29
   primary airway epithelial cells, 13-48→13-49
   retroviral-mediated hematopoietic cells
      analysis of clonal integration in individual colonies, 13-40→13-43
      CD34+ cells with stromal support, 13-30→13-35
      collection of cell-free supernatant for, 13-36
      enzymatic removal of magnetic beads from selected cells, 13-36→13-37
      establishing primary human marrow stromal monolayers, 13-37
      on fibronectin-coated dishes, 13-35
      freezing cells, 13-43→13-44
      in long-term bone marrow culture, harvesting cells for, 13-43

plate colony-forming cells from cultures, 13-38→13-39
preparing whole-cell lysates of individual cfu colonies for PCR, 13-39→13-40
screening media, 13-38
thawing cells, 13-44
vector-producing fibroblast culture, 13-35→13-36
Transfection
   high-titer adenoviruses, 12-16
   preparation of HSV-1 cosmid DNA for, 12-69→12-71
   retroviral vector production, calcium phosphate-mediated, 12-36→12-37
   transient
      adenovirus-free rAAV production, 12-21→12-25
      pseudotype retroviral vector production, 12-40→12-42
      retroviral vector production, 12-36
Transgene expression
   analysis
      harvesting liver tissue for, 13-61→13-62
      of human bronchial xenografts, 13-58→13-59
   in vitro infection of cells with rAAV and determination of titer by, 12-29
Translocation, chromosomal, detection by PCR, 10-18→10-22
Transmission disequilibrium test (TDT)
   characterized, 1-80→1-81
   comparison with population relative risk test, 1-81→1-82
   for disease associations, 1-80→1-81
   generalization of, 1-83→1-86, 1-89→1-90
   incomplete genotype data, 1-82→1-83
   mode of inheritance, 1-82
   parental genotypes, 1-83
   pedigrees, 1-90
   segregation distortion, 1-82
Transportation of cells, A3-33→A3-34
Trichloroacetic acid (TCA) precipitation, to measure incorporation of radioactivity, A3-15→A3-16
Trinucleotide repeats, analysis in myotonic dystrophy (DM), 9-23→9-26
TRP gene, 5-21 (table), 5-25 (fig.), 5-27
Trypan blue, for determining cell numbers, A3-31→A3-33
Trypsinization, of cell monolayers, A3-28→A3-29
Tumors
   classification of, 11-26
   cytogenetic analysis, of metaphase cells
      basic protocol, 10-7→10-8
      disaggregation and culture of solid tumors, 10-8→10-9
      effusion specimens, 10-9→10-10
      lymph node specimens, 10-9
   DNA rearrangements in leukemias and non-Hodgkin lymphomas, 10-10→10-23
   gene rearrangements, 10-11→10-18
   mitotic chromosome preparations from, 4-13
Turner syndrome, 1-1
Twin studies, linkage analysis, 1-4
Twin typing, paternity testing, 9-36→9-37
Two-color array, for transcriptional profiling, 11-24
TWOLOC, 1-24→1-26
Two-point linkage analysis, 1-31→1-39, 1-49→1-53

U

UCSC Genome Browser
   annotations
      features of, 6-43→6-44, 6-47 (table)
      gene models, 6-44, 6-50
   applications, 6-46→6-47

BLAT, 6-59, 6-60
customized display, 6-49→6-50
downloading sequence, 6-50
FTP format, 6-50
information retrieval, 6-56→6-57
queries, 6-46→6-49, 6-57→6-60
sequence assembly, 6-41→6-43
web site, 6-39, 6-48 (table)
Unilineal pedigrees, in linkage analysis, 1-17
Universal mutation database (UMD), 7-29
UNKNOWN, 1-31
URA3 gene, 5-21 (table), 5-27

V

Vaccine, intramuscular injection of the anterior tibialis, 13-20→13-21
   quadriceps, 13-19→13-20
Validation, probes, in preimplantation genetic diagnosis, 9-73
Variable number of tandom repeat (VNTR), paternity testing, in RFLP technology, 9-28→9-30
Variance, in statistics, 1-59
Vector Black Blue, and Red, as substrate for ISH, 4-29 (table)
Vector database, for nucleotide sequence, 6-25
Vector-producing fibroblasts, 13-35→13-36
Vectors
   adeno-associated, 12-21→12-29
   adeno, 12-5, 12-12→12-21
   for direct injection into rat brain, 13-28
   for DNA vaccine, 13-20 (fig.)
   herpes simplex virus, 12-49→12-66
   HSV-1 amplicon, 12-66→12-74
   lentiviral, 12-43→12-49
   liposome, 12-74→12-75
   for a model airway system, choosing, 13-45→13-46
   pBACe3.6, 5-32→6-35
   pCYPAC2, 5-32→6-35
   pPAC4, 5-32→5-35
   pseudo-type retroviral, 12-39→12-43
   retroviral, 12-29→12-39
   transfer, biosafety in handling, 12-2→12-5
Vesicular stomatitis virus (VSV), pseudotype retroviral vectors, 12-39→12-40, 12-44
Viral vectors
   adeno-associated, 12-21→12-29
   adenoviral, 12-5→12-21
   herpes simplex, replication defective construction of, 12-49→12-57
   foreign gene sequence insertion into, 12-57→12-66
   high-titer lentiviral, 12-43→12-49
   pseudotype retroviral, 12-39→12-43
   retroviral, 12-29→12-39
Viruses
   Epstein-Barr, A3-34→A3-36
   herpes simplex, 12-49→12-66
   high-titer lentiviral, 12-43→12-49
   vesicular stomatitis (VSV), 12-39→12-40, 12-44
   viral DNA, isolation of, 12-64→12-65

VNTR, see Variable number of tandem repeat
VSV, see Vesicular stomatitis virus

W

WebBLAST, 6-15→6-16
Web sites, see Computer programs and Web sites
Whole-cell fusion for somatic cell hybrids
   of monolayer cells, 3-4→3-5
   of suspension donor cells, 3-5→3-6
Whole-genome amplification
   single diploid cells by PEP, 9-64
   single-sperm typing, 1-23→1-24

Whole genome shot gun (WGS), 6-38, 6-40
Wilson disease, sample patient history form,
    1-8→1-9 (fig.)
Wright staining, chromosome banding, 4-15
    (table)

X

Xanthine-guanine phosphoribosyltransferase
    (XGPRT, gpt), as selectable marker,
    3-3
X chromosome, inactivation
    assay, 9-26→9-27

PCR amplification and labeling of DNA
    templates, 9-27→9-28
Xenografts, human bronchial, 13-52→13-59
XGRAIL, in gene identification, 6-4, 6-13
X-linked disorders, see Heterozygote muta-
    tion detection
Xylene cyanol, migration in PAGE, A3-18
    (table)

Y

Yeast artificial chromosome (YAC)

high-molecular-weight DNA, preparation
    for PFGE, 5-5→5-7
libraries, arrayed, 5-8→5-9
probes for FISH, 4-48
purification by PFGE, 5-27→5-28
rapid estimation on ethidium bro-
    mide/agarose plates, 5-28→5-29
replica plating, 5-9→5-14
Yeast databases, for nucleotide sequences,
    6-25

Z

ZFX/ZFY gene, in preimplantation genetic di-
    agnosis, 9-62 (tables), 9-63